Intermediate Algebra

Third Edition

Elayn Martin-Gay

University of New Orleans

PEARSON

Prentice
Hall

Upper Saddle River, New Jersey 07458

Library of Congress Cataloging-in-Publication Data

Martin-Gay, K. Elayn
  Intermediate algebra / Elayn Martin-Gay.—3rd ed.
    p. cm.
  Includes index.
  ISBN 0-13-186830-6 (alk. paper) — ISBN 0-13-186845-4 (alk. paper)
  1. Algebra—Problems, exercises, etc.   I. Title.

QA152.3.M36 2007
512.9—dc22   2005057731

Executive Editor: *Paul Murphy*
Editor in Chief: *Christine Hoag*
Project Manager: *Mary Beckwith*
Production Management: *Elm Street Publishing Services, Inc.*
Senior Managing Editor: *Linda Mihatov Behrens*
Executive Managing Editor: *Kathleen Schiaparelli*
Media Project Manager, Developmental Math: *Audra J. Walsh*
Media Production Editor: *Jenelle Woodrup*
Assistant Managing Editor, Science and Math Print Supplements: *Karen Bosch*
Manufacturing Buyer: *Alan Fischer*
Manufacturing Manager: *Alexis Heydt-Long*
Director of Marketing: *Patrice Jones*
Senior Marketing Manager: *Kate Valentine*
Marketing Assistant: *Jennifer de Leeuwerk*
Development Editor: *Laura Wheel*
Editor in Chief, Development: *Carol Trueheart*
Editorial Assistant: *Abigail Rethore*
Art Director: *Maureen Eide*
Interior/Cover Designer: *Suzanne Behnke*
Art Editor: *Thomas Benfatti*
Creative Director: *Juan R. López*
Director of Creative Services: *Paul Belfanti*
Cover Photo: *Stockdisc/Getty Images*
Manager, Cover Visual Research & Permissions: *Karen Sanatar*
Director, Image Resource Center: *Melinda Reo*
Manager, Rights and Permissions: *Zina Arabia*
Manager, Visual Research: *Beth Brenzel*
Image Permission Coordinator: *Craig Jones*
Photo Researcher: *Rachel Lucas*
Composition: *Interactive Composition Corporation*
Art Studio: *Scientific Illustrators and Laserwords*

© 2007, 2003, 1999 by Prentice-Hall, Inc.
Pearson Prentice Hall
Pearson Education, Inc.
Upper Saddle River, New Jersey 07458

Photo Credits appear on page I9, which constitutes a continuation of the copyright page.

Printed in the United States of America
10 9 8 7 6 5 4 3

ISBN: (paperback) 0-13-186830-6; (case bound) 0-13-186829-2

Pearson Education LTD., *London*
Pearson Education Australia PTY., Limited, *Sydney*
Pearson Education Singapore, Pte. Ltd.
Pearson Education North Asia Ltd., *Hong Kong*
Pearson Education Canada, Ltd., *Canada*
Pearson Educacíon de Mexico, S.A. de C.V.
Pearson Education—Japan, *Tokyo*
Pearson Education Malaysia, Pte. Ltd.

To my mother, Barbara M. Miller,
and her husband, Leo Miller, and to the memory of
my father, Robert J. Martin

Contents

1   Real Numbers and Algebraic Expressions   1

2   Equations, Inequalities, and Problem Solving   69

3 Graphs and Functions   153

4 Systems of Equations and Inequalities   257

5 Polynomials and Polynomial Functions   318

6 Rational Expressions 402

7 Rational Exponents, Radicals, and Complex Numbers 477

8 Quadratic Equations and Functions 552

9 Exponential and Logarithmic Functions 634

10 Conic Sections 711

Tools to Help Students Succeed

Your textbook includes a number of features designed to help you succeed in this math course—as well as the next math course you take. These features include:

Feature	Benefit	Page
Well-crafted Exercise Sets: We learn math by doing math	The exercise sets in your text offer an ample number of exercises carefully ordered so you can master basic mathematical skills and concepts while developing all-important problem solving skills. Exercise sets include Mixed Practice exercises to help you master multiple key concepts, as well as Mental Math, Writing, Applications, Concept Check, Concept Extension, and Review exercises.	179–182
Solutions-to-Selected Exercises: Built-in solutions at the back of the text	If you need to review problems you find difficult, this built-in solutions manual at the back of the text provides the step-by-step solutions to every other odd-numbered exercise in the exercise sets.	A43
Study Skills Builders: Maximize your chances for success	Study Skills Builders reinforce the material in *Section 1.1—Tips for Success in Mathematics.* Study Skills Builders are a great resource for study ideas and self-assessment to maximize your opportunity for success in this course. Take your new study skills with you to help you succeed in your next math course.	190, 242
The Bigger Picture: Succeed in this math course and the next one you take	The Bigger Picture focuses on the key concept of this course—solving equations and inequalities—and asks you to keep an ongoing outline so you can recognize and solve different types of equations and inequalities. A strong foundation in solving equations and inequalities will help you succeed in this algebra course, as well as the next math course you take.	137
Examples: Step-by-step instruction for you	Examples in the text provide you with clear, concise step-by-step instructions to help you learn. Annotations in the examples provide additional instruction.	177
Helpful Hints: Help where you'll need it most	Helpful Hints provide tips and advice at exact locations where students need it most. Strategically placed where you might have the most difficulty, Helpful Hints will help you work through common trouble spots.	175
Practice Problems: Immediate reinforcement	Practice Problems offer immediate reinforcement after every example. Try each Practice Problem after studying the corresponding example to make sure you have a good working knowledge of the concept.	177
Integrated Review: Mid-chapter progress check	To ensure that you understand the key concepts covered in the first sections of the chapter, work the exercises in the Integrated Review before you continue with the rest of the chapter.	200
Vocabulary Check: Key terms and vocabulary	Make sure you understand key terms and vocabulary in each chapter with the Vocabulary Check.	237
Chapter Highlights: Study smart	Chapter Highlights outline the key concepts of the chapter along with examples to help you focus your studying efforts as you prepare for your test.	237
Chapter Test: Take a practice test	In preparation for your classroom test, take this practice test to make sure you understand the key topics in the chapter. Be sure to use the **Chapter Test Prep Video CD** included with this text to see the author present a fully worked-out solution to each exercise in the Chapter Test.	250

Martin-Gay's CD VIDEO RESOURCES Help Students Succeed

Martin-Gay's Chapter Test Prep Video CD (available with this text)

- Provides students with help during their most "teachable moment"—while they are studying for a test.

- Text author Elayn Martin-Gay presents step-by-step solutions to the exact exercises found in each Chapter Test in the book.

- Easy video navigation allows students to instantly access the worked-out solutions to the exercises they want to review.

- A close-captioned option for the hearing impaired is provided.

Martin-Gay's CD Lecture Series (with Tips for Success in Mathematics)

- Text author Elayn Martin-Gay presents the key concepts from every section of the text in 10–15 minute mini-lectures.

- Students can easily review a section or a specific topic before a homework assignment, quiz, or test.

- Includes fully worked-out solutions to exercises marked with a CD Video icon (⊙) in each section.

- Includes *Section 1.1, Tips for Success in Mathematics.*

- A close-captioned option for the hearing impaired is provided.

- Ask your bookstore for information about Martin-Gay's *Intermediate Algebra,* Third Edition, CD Lecture Series, or visit www.prenhall.com.

Additional Resources to Help You Succeed

Student Study Pack
A single, easy-to-use package—available bundled with your textbook or by itself—for purchase through your bookstore. This package contains the following resources to help you succeed:

Student Solutions Manual
- Contains worked-out solutions to odd-numbered exercises from each section exercise set, Practice Problems, Mental Math exercises, and all exercises found in the Chapter Review and Chapter Tests.

Prentice Hall Math Tutor Center
- Staffed by qualified math instructors who provide students with tutoring on examples and odd-numbered exercises from the textbook. Tutoring is available via telephone, fax, email, or the Internet.

Martin-Gay's CD Lecture Series
- Text author Elayn Martin-Gay presents the key concepts from every section of the text with 10–15 minute mini-lectures. Students can easily review a section or a specific topic before a homework assignment, quiz, or test.
- Includes fully worked-out solutions to exercises marked with a CD Video icon ( ) in each section. Also includes *Section 1.1, Tips for Success in Mathematics.*
- A close-captioned option for the hearing impaired is provided.

Online Homework and Tutorial Resources

MyMathLab *MyMathLab*

MyMathLab is a series of text specific, easily customizable, online courses for Prentice Hall textbooks in mathematics and statistics. MyMathLab is powered by Course Compass™—Pearson Education's online teaching and learning environment—and by MathXL®—our online homework, tutorial, and assessment system. MyMathLab gives instructors the tools they need to deliver all or a portion of their course online, whether students are in a lab setting or working from home. MyMathLab provides a rich and flexible set of course materials, featuring free-response exercises that are algorithmically generated for unlimited practice and mastery. Students can also use online tools, such as video lectures, animations, and a multimedia textbook, to independently improve their understanding and performance. MyMathLab is available to qualified adopters. For more information, visit our Web site at www.mymathlab.com or contact your Prentice Hall sales representative. (MyMathLab must be set up and assigned by your instructor.)

MathXL® www.mathxl.com *Math XL*

MathXL is a powerful online homework, tutorial, and assessment system that accompanies the text. With MathXL, instructors can create, edit, and assign online homework and tests using algorithmically generated exercises correlated to your textbook. All student work is tracked in MathXL's online gradebook. Students can take chapter tests in MathXL and receive personalized study plans based on their test results. The study plan diagnoses weaknesses and links students directly to tutorial exercises for the objectives they need to study and retest. Students can also access supplemental animations and video clips directly from selected exercises. MathXL is available to qualified adopters. For more information, visit our Web site at www.mathxl.com, or contact your Prentice Hall sales representative for a product demonstration. (MathXL must be set up and assigned by your instructor.)

Preface

Intermediate Algebra, **Third Edition** was written to provide a solid foundation in algebra as well as to develop problem solving skills. Specific care was taken to make sure students have the most up-to-date relevant text preparation for their next mathematics course or for nonmathematical courses that require an understanding of algebraic fundamentals. I have tried to achieve this by writing a user-friendly text that is keyed to objectives and contains many worked-out examples. As suggested by AMATYC and the NCTM Standards (plus Addenda), real-life and real-data applications, data interpretation, conceptual understanding, problem solving, writing, cooperative learning, appropriate use of technology, mental mathematics, number sense, estimation, critical thinking, and geometric concepts are emphasized and integrated throughout the book.

The many factors that contributed to the success of the previous editions have been retained. In preparing the Third Edition, I considered comments and suggestions of colleagues, students, and many users of the prior edition throughout the country.

What's New in the Third Edition?

Enhanced Exercise Sets

- **NEW!** Three forms of mixed sections of exercises have been added to the Third Edition.
  - **Mixed Practice** exercises combining objectives within a section
  - **Mixed Practice** exercises combining previous sections
  - **Mixed Review** exercises included at the end of the Chapter Review

  These exercises require students to determine the problem type and strategy needed in order to solve it. In doing so, students need to think about key concepts to proceed with a correct method of solving—just as they would need to do on a test.

- **NEW! Concept Check exercises** have been added to the section exercise sets. These exercises are related to the Concept Check(s) found within the section. They help students measure their understanding of key concepts by focusing on common trouble areas. These exercises may ask students to identify a common error, and/or provide an explanation.

- **NEW! Concept Extensions** (formerly Combining Concepts) have been revised. These exercises extend the concepts and require students to combine several skills or concepts to solve the exercises in this section.

Increased Emphasis on Study Skills and Student Success

- **NEW! Study Skills Builders** (formerly Study Skill Reminders) Found at the end of many exercise sets, Study Skills Builders allow instructors to assign exercises that will help students improve their study skills and take responsibility for their part of the learning process. Study Skills Builders reinforce the material found in Section 1.1, "Tips for Success in Mathematics," and serve as an excellent tool for self-assessment.

- **NEW! The Bigger Picture** is a recurring feature beginning in Section 2.6, that focuses on the key concepts of the course—solving equations and inequalities. It helps students develop an outline to recognize and solve different types of equations and inequalities. By working the exercises and developing this outline throughout the text, students can begin to transition from thinking "section by section" to thinking about how the mathematics in this course is part of the "bigger picture" of mathematics in general. A completed outline is provided in Appendix C so students have a model for their work.

- **NEW! Chapter Test Prep Video CD** provides students with help during their most "teachable moment"—while they are studying for a test. Included with every copy of the student edition of the text, this video CD provides fully worked-out solutions by the author to every exercise from each Chapter Test in the text. The easy video navigation allows students to instantly access the solutions to the exercises they want to review. The problems are solved by the author in the same manner as in the text.

- **NEW! Chapter Test files in TestGen** provide algorithms specific to each exercise from each Chapter Test in the text. Allows for easy replication of Chapter Tests with consistent, algorithmically generated problem types for additional assignments or assessment purposes.

Content Changes in the Third Edition

- Increased coverage on functions throughout the text. For example, see Section 3.6, Exercises 79 through 86, 92, and 93.

- New Section 3.7, Finding Domain and Range from Graphs and Graphing Piecewise-Defined Functions and Section 3.8, Shifting and Reflecting Graphs of Functions have been added to the third edition for those wishing to include coverage of these topics in their course.

- The distance formula and midpoint formula have been moved earlier in the text and are now covered in Section 7.3.

- The chapter on logarithms (now Chapter 9) precedes the chapter on Conic Sections (now Chapter 10).

- More challenging exercises have been included to ensure a student's understanding of a concept and functions. For example, for new function exercises see Section 5.1, Exercises 89 through 94; Section 5.2, Exercises 93 through 96; Section 7.1, Exercises 91 through 94. See Section 6.1, Exercises 79 through 82 for new conceptual problems.

- All exercise sets have been reviewed and updated to ensure that even- and odd-numbered exercises are paired.

- New Appendix E, Stretching and Compressing Graphs of Absolute Value Functions.

Key Pedagogical Features

The following key features have been retained and/or updated for the Third Edition of the text:

Problem Solving Process This is formally introduced in Chapter 2 with a four-step process that is integrated throughout the text. The four steps are **Understand, Translate, Solve,** and **Interpret.** The repeated use of these steps in a variety of examples shows their wide applicability. Reinforcing the steps can increase students' comfort level and confidence in tackling problems.

Exercise Sets Revised and Updated The exercise sets have been carefully examined and extensively revised. Special focus was placed on making sure that even- and odd-numbered exercises are paired.

Examples Detailed step-by-step examples were added, deleted, replaced, or updated as needed. Many of these reflect real life. Additional instructional support is provided in the annotated examples.

Practice Problems Throughout the text, each worked-out example has a parallel Practice Problem. These invite students to be actively involved in the learning process. Students should try each Practice Problem after finishing the corresponding example. Learning by doing will help students grasp ideas before moving on to other concepts. Answers to the Practice Problems are provided at the bottom of each page.

Helpful Hints Helpful Hints contain practical advice on applying mathematical concepts. Strategically placed where students are most likely to need immediate

Exercise Icons These icons facilitate the assignment of specialized exercises and let students know what resources can support them.

CD Video icon: exercise worked on Martin-Gay's CD Lecture Series.

Triangle icon: identifies exercises involving geometric concepts.

Pencil icon: indicates a written response is needed.

Calculator icons: optional exercises intended to be solved using a scientific or graphing calculator.

Group Activities Found at the end of each chapter, these activities are for individual or group completion, and are usually hands-on or data-based activities that extend the concepts found in the chapter allowing students to make decisions and interpretations and to think and write about algebra.

Optional: Calculator Exploration Boxes and Calculator Exercises The optional Calculator Explorations provide key strokes and exercises at appropriate points to provide an opportunity for students to become familiar with these tools. Section exercises that are best completed by using a calculator are identified by ▦ or ▦ for ease of the assignment.

A Word about Textbook Design and Student Success

The design of developmental mathematics textbooks has become increasingly important. As students and instructors have told Prentice Hall in focus groups and market research surveys, these textbooks cannot look "cluttered" or "busy." A "busy" design can distract a student from what is most important in the text. It can also heighten math anxiety.

As a result of the conversations and meetings we have had with students and instructors, we concluded the design of this text should be understated and focused on the most important pedagogical elements. Students and instructors helped us to identify the primary elements that are central to student success. These primary elements include:

- Exercise Sets

- Examples and Practice Problems

- Helpful Hints

- Rules, Property, and Definition boxes

As you will notice in this text, these primary features are the most prominent elements in the design. We have made every attempt to make sure these elements are the features the eye is drawn to. The remaining features, the secondary elements in the design, blend into the "fabric" or "grain" of the overall design. These secondary elements complement the primary elements without becoming distractions.

Prentice Hall's thanks goes to all of the students and instructors (as noted by the author in Acknowledgments) who helped us develop the design of this text. At every step in the design process, their feedback proved valuable in helping us to make the right decisions. Thanks to your input, we're confident the design of this text will be both practical and engaging as it serves its educational and learning purposes.

Sincerely,

Paul Murphy

Executive Editor
Developmental Mathematics
Prentice Hall

reinforcement, Helpful Hints help students avoid common trouble areas and mistakes.

Concept Checks   This feature allows students to gauge their grasp of an idea as it is being presented in the text. Concept Checks stress conceptual understanding at the point-of-use and help suppress misconceived notions before they start. Answers appear at the bottom of the page. Exercises related to Concept Checks are now included in the exercise sets.

Selected Solutions   Solutions to every-other odd exercise are included in the back of the text. This built-in solutions manual allows students to check their work.

Integrated Reviews   A unique, mid-chapter exercise set that helps students assimilate new skills and concepts that they have learned separately over several sections. These reviews provide yet another opportunity for students to work with "mixed" exercises as they master the topics.

Vocabulary Check   Provides an opportunity for students to become more familiar with the use of mathematical terms as they strengthen their verbal skills. These appear at the end of each chapter before the Chapter Highlights.

Chapter Highlights   Found at the end of every chapter, these contain key definitions and concepts with examples to help students understand and retain what they have learned and help them organize their notes and study for tests.

Chapter Review   The end of every chapter contains a comprehensive review of topics introduced in the chapter. The Chapter Review offers exercises keyed to every section in the chapter, as well as Mixed Review **(NEW!)** exercises that are not keyed to sections.

Chapter Test and Chapter Test Prep Video CD   The Chapter Test is structured to include those problems that involve common student errors. The **Chapter Test Prep Video CD** gives students instant author access to a step-by-step video solution of each exercise in the Chapter Test.

Cumulative Review   Follows every chapter in the text (except Chapter 1). Each odd-numbered exercise contained in the Cumulative Review is an earlier worked example in the text that is referenced in the back of the book along with the answer. The even exercises are new.

Mental Math   Found at the beginning of an exercise set, these mental warm-ups reinforce concepts found in the accompanying section and increase student's confidence before they tackle an exercise set.

Writing Exercises   These exercises occur in almost every exercise set and require students to provide a written response to explain concepts or justify their thinking.

Applications   Real-world and real-data applications have been thoroughly updated and many new applications are included. These exercises occur in almost every exercise set and show the relevance of mathematics and help students gradually, and continuously develop their problem solving skills.

Review Exercises (formerly Review and Preview exercises)   These exercises occur in each exercise set (except in Chapter 1) and are keyed to earlier sections. They review concepts learned earlier in the text that will be needed in the next section or chapter.

Exercise Set Resource Icons at the opening of each exercise set remind students of the resources available for extra practice and support:

CD/Video for Review

MyMathLab

Math XL
MathXL®

PH Math/Tutor Center

Student Solutions Manual

See Student Resource descriptions pages xviii–xix for details on the individual resources available.

Instructor and Student Resources

The following resources are available to help instructors and students use this text more effectively.

Instructor Resources

Annotated Instructor's Edition (0-13-186845-4)

- Answers to all exercises printed on the same text page
- Teaching Tips throughout the text placed at key points
- Includes Vocabulary Check at the beginning of relevant sections
- General tips and suggestions for classroom or group activities

Instructor Solutions Manual (0-13-196301-5)

- Solutions to the even- and odd-numbered exercises
- Solutions to every Mental Math exercise
- Solutions to every Practice Problem
- Solutions to every exercise in the Integrated Reviews, Chapter Reviews, Chapter Tests, and Cumulative Reviews

Instructor's Resource Manual with Tests (0-13-173075-7)

- **NEW!** Includes Mini-Lectures for every section from the text
- Group Activities
- Free Response Test Forms, Multiple Choice Test Forms, Cumulative Tests, and Additional Exercises
- Answers to all items

TestGen (0-13-220826-1)

- Enables instructors to build, edit, print, and administer tests
- Features a computerized bank of questions developed to cover all text objectives
- Available on dual-platform Windows/Macintosh CD-Rom

Instructor Adjunct Resource Kit (0-13-188758-0)

The Martin-Gay Instructor/Adjunct Resource Kit (IARK) contains tools and resources to help adjuncts and instructors succeed in the classroom. The IARK includes:

- Instructor-to-Instructor CD Videos that offer tips, suggestions, and strategies for engaging students and presenting key topics
- PDF files of the Instructor Solutions Manual and the Instructor's Resource Manual
- TestGen

MyMathLab Instructor Version (0-13-147898-2)
MyMathLab www.mymathlab.com

MyMathLab is a series of text specific, easily customizable, online courses for Prentice Hall textbooks in mathematics and statistics. MyMathLab is powered by Course Compass™—Pearson Education's online teaching and learning environment—and by MathXL®—our online homework, tutorial, and assessment system. MyMathLab gives instructors the tools they need to deliver all or a portion of their course online, whether students are in a lab setting or working from home. MyMathLab provides a rich and flexible set of course materials, featuring free-response exercises that are algorithmically generated for unlimited practice and mastery. Students can also use online tools, such as video lectures, animations, and a multimedia textbook, to independently improve their understanding and performance. Instructors can use

MyMathLab's homework and test managers to select and assign online exercises correlated directly to the text, and they can import TestGen tests into MyMathLab for added flexibility. MyMathLab's online gradebook—designed specifically for mathematics and statistics—automatically tracks students' homework and test results and gives the instructor control over how to calculate final grades. Instructors can also add offline (paper-and-pencil) grades to the gradebook. MyMathLab is available to qualified adopters. For more information, visit our website at www.mymathlab.com or contact your Prentice Hall sales representative.

MathXL Instructor Version (0-13-147895-8)
MathXL® www.mathxl.com

MathXL is a powerful online homework, tutorial, and assessment system that accompanies the text. With MathXL, instructors can create, edit, and assign online homework and tests using algorithmically generated exercises correlated to your textbook. All student work is tracked in MathXL's online gradebook. Students can take chapter tests in MathXL and receive personalized study plans based on their test results. The study plan diagnoses weaknesses and links students directly to tutorial exercises for the objectives they need to study and retest. Students can also access supplemental animations and video clips directly from selected exercises. MathXL is available to qualified adopters. For more information, visit our Web site at www.mathxl.com, or contact your Prentice Hall sales representative for a product demonstration.

Interact Math® Tutorial Web site www.interactmath.com

Get practice and tutorial help online! This interactive tutorial Web site provides algorithmically generated practice exercises that correlate directly to the exercises in your textbook. You can retry an exercise as many times as you like with new values each time for unlimited practice and mastery. Every exercise is accompanied by an interactive guided solution that gives you helpful feedback if you enter an incorrect answer, and you can also view a worked-out sample problem that steps you through an exercise similar to the one you're working on.

Student Resources

Student Solutions Manual (0-13-219595-X)

- Solutions to the odd-numbered section exercises
- Solutions to the Practice Problems
- Solutions to every Mental Math exercise
- Solutions to every exercise found in the Chapter Reviews and Chapter Tests

Martin-Gay's CD Lecture Series (0-13-188759-9)

- Perfect for review of a section or a specific topic, these mini-lectures by Elayn Martin-Gay cover the key concepts from each section of the text in approximately 10–15 minutes.
- Includes fully worked-out solutions to exercises in each section marked with a 💿
- Includes coverage of Section 1.1, "Tips for Success Mathematics"
- Closed-captioned for the hearing impaired

Prentice Hall Math Tutor Center (0-13-064604-0)

- Staffed by qualified math instructors who provide students with tutoring on examples and odd-numbered exercises from the textbook
- Tutoring is available via telephone, fax, e-mail, or the Internet
- Whiteboard technology allows tutors and students to see problems worked while they "talk" in real time over the Internet during tutoring sessions

Intermediate Algebra, Third Edition *Student Study Pack (0-13-174314-7)*

The Student Study Pack includes:

- Martin-Gay's CD Lecture Series
- Student Solutions Manual
- Prentice Hall Math Tutor Center access code

Chapter Test Prep Video CD—Standalone (0-13-173276-5)

- Includes fully worked-out solutions to every problem from each Chapter Test in the text.

MathXL Tutorials on CD—Standalone (0-13-173318-4)

- Provides algorithmically generated practice exercises that correlate to exercises at the end of sections.
- Every exercise is accompanied by an example and a guided solution, selected exercises include a video clip.
- The software recognizes student errors and provides feedback. It can also generate printed summaries of student's progress.

Interact Math® Tutorial Web Site www.interactmath.com

Get practice and tutorial help online! This interactive tutorial Web site provides algorithmically generated practice exercises that correlate directly to the exercises in your textbook. You can retry an exercise as many times as you like with new values each time for unlimited practice and mastery. Every exercise is accompanied by an interactive guided solution that gives you helpful feedback if you enter an incorrect answer, and you can also view a worked-out sample problem that steps you through an exercise similar to the one you're working on.

Acknowledgments

There are many people who helped me develop this text, and I will attempt to thank some of them here. Cindy Trimble and Carrie Green were *invaluable* for contributing to the overall accuracy of the text. Chris Callac, Laura Wheel, and Lori Mancuso were *invaluable* for their many suggestions and contributions during the development and writing of this Third Edition. Ingrid Benson provided guidance throughout the production process.

A special thanks to my editor, Paul Murphy, for all of his assistance, support, and contributions to this project. A very special thank you goes to my project manager, Mary Beckwith, for being there 24/7/365, as my students say. Last, my thanks to the staff at Prentice Hall for all their support: Linda Behrens, Alan Fischer, Patty Burns, Tom Benfatti, Paul Belfanti, Maureen Eide, Suzanne Behnke, Kate Valentine, Patrice Jones, Chris Hoag, Paul Corey, and Tim Bozik.

I would like to thank the following reviewers for their input and suggestions:

Rosalie Abraham, *Florida Community College—Jacksonville*

Ana Bacica, *Brazesport College*

Nelson Collins, *Joliet Junior College*

Nancy Desilet, *Carroll Community College*

Elizabeth Eagle, *University of North Carolina—Charlotte*

Dorothy French, *Community College of Philadelphia*

Sharda Gudehithla, *Wilbur Wright College*

Pauline Hall, *Iowa State University*

Debra R. Hill, *University of North Carolina—Charlotte*

Glenn Jablonski, *Triton College*

Sue Kellicut, *Seminole Community College*

Jean McArthur, *Joliet Junior College*

Mary T. McMahon, *North Central College*

Owen Mertens, *Missouri State University*

Jeri Rogers, *Seminole Community College*

William Stammerman, *Des Moines Area Community College*

Patrick Stevens, *Joliet Junior College*

Arnavaz Taraporevala, *New York City College of Technology*

I would also like to thank the following dedicated group of instructors who participated in our focus groups, Martin-Gay Summits, and our design review for this edition of the text. Their feedback and insights have helped to strengthen this edition of the text. These instructors include:

Cedric Atkins, *Mott Community College*

Laurel Berry, *Bryant & Stratton*

Bob Brown, *Community College of Baltimore County–Essex*

Lisa Brown, *Community College of Baltimore County–Essex*

Gail Burkett, *Palm Beach Community College*

Cheryl Cantwell, *Seminole Community College*

Jackie Cohen, *Augusta State College*

Janice Ervin, *Central Piedmont Community College*

Pauline Hall, *Iowa State College*

Sonya Johnson, *Central Piedmont Community College*

Irene Jones, *Fullerton College*

Nancy Lange, *Inver Hills Community College*

Jean McArthur, *Joliet Junior College*

Marica Molle, *Metropolitan Community College*

Linda Padilla, *Joliet Junior College*

Carole Shapero, *Oakton Community College*

Jennifer Strehler, *Oakton Community College*

Tanomo Taguchi, *Fullerton College*

Leigh Ann Wheeler, *Greenville Technical Community College*

Valerie Wright, *Central Piedmont Community College*

A special thank you to those students who participated in our design review: Katherine Browne, Mike Bulfin, Nancy Canipe, Ashley Carpenter, Jeff Chojnachi, Roxanne Davis, Mike Dieter, Amy Dombrowski, Kay Herring, Todd Jaycox, Kaleena Levan, Matt Montgomery, Tony Plese, Abigail Polkinghorn, Harley Price, Eli Robinson, Avery Rosen, Robyn Schott, Cynthia Thomas, and Sherry Ward.

Additional Acknowledgments

As usual, I would like to thank my husband, Clayton, for his constant encouragement. I would also like to thank my children, Eric and Bryan, for providing most of the cooking and humor in our household. I would also like to thank my extended family for their help and wonderful sense of humor. Their contributions are too numerous to list. They are Rod and Karen Pasch; Peter, Michael, Christopher, Matthew, and Jessica Callac; Stuart and Earline Martin; Josh, Mandy, Bailey, Ethan, and Avery Barnes; Mark, Sabrina, and Madison Martin; Leo and Barbara Miller; and Jewett Gay.

Elayn Martin-Gay

About the Author

Elayn Martin-Gay has taught mathematics at the University of New Orleans for more than 25 years. Her numerous teaching awards include the local University Alumni Association's Award for Excellence in Teaching, and Outstanding Developmental Educator at University of New Orleans, presented by the Louisiana Association of Developmental Educators.

Prior to writing textbooks, Elayn Martin-Gay developed an acclaimed series of lecture videos to support developmental mathematics students in their quest for success. These highly successful videos originally served as the foundation material for her texts. Today, the videos are specific to each book in the Martin-Gay series. The author has also created Chapter Test Prep Videos to help students during their most "teachable moment"—as they prepare for a test, along with Instructor-to-Instructor videos that provide teaching tips, hints, and suggestions for each developmental mathematics course, including basic mathematics, prealgebra, beginning algebra, and intermediate algebra.

Elayn is the author of 10 published textbooks as well as multimedia interactive mathematics, all specializing in developmental mathematics courses. She has participated as an author across the broadest range of educational materials: textbooks, videos, tutorial software, and Interactive Math courseware. All of these components are designed to work together. This offers an opportunity of various combinations for an integrated teaching and learning package offering great consistency for the student.

Applications Index

Intermediate Algebra

1

Real Numbers and Algebraic Expressions

In arithmetic, we add, subtract, multiply, divide, raise to powers, and take roots of numbers. In algebra, we add, subtract, multiply, divide, raise to powers, and take roots of variables. Letters, such as *x*, that represent numbers are called variables. Understanding algebraic expressions made up of combinations of variables and numbers depends on your understanding of arithmetic expressions. This chapter reviews the arithmetic operations on real numbers and the corresponding algebraic expressions. After this review, we will be prepared to explore how widely useful these algebraic expressions are for problem solving in diverse situations.

Hoover Dam is evidence of America's ability to construct monumental projects in the midst of difficult conditions. In 1931, during the height of the Depression, masses of American workers came to the Black Canyon on the Arizona–Nevada border to tame the Colorado River. It took less than five years, in a harsh and barren land, to build the most amazing dam of its time. The Hoover Dam is a curved gravity dam. Lake Mead pushes against the dam, creating compressive forces that travel along the great curved wall. The canyon walls push back, counteracting these forces. This action squeezes the concrete in the arch together, making the dam very rigid. This way, Lake Mead is securely held in check.

Hoover Dam is a National Historic Landmark and has been rated by the American Society of Civil Engineers as one of America's Seven Modern Civil Engineering Wonders.

In Exercise 107 on page 50, we will use scientific notation to write the very large number of cubic feet of water capacity of Lake Mead. (*Source:* U.S. Bureau of Reclamation, PBS: "Building Big")

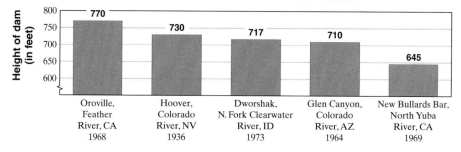

Tallest Dams in the United States

Height of dam (in feet)

Oroville, Feather River, CA 1968	Hoover, Colorado River, NV 1936	Dworshak, N. Fork Clearwater River, ID 1973	Glen Canyon, Colorado River, AZ 1964	New Bullards Bar, North Yuba River, CA 1969
770	730	717	710	645

1.1 TIPS FOR SUCCESS IN MATHEMATICS

Before reading this section, remember that your instructor is your best source of information. Please see your instructor for any additional help or information.

Objective Ⓐ Getting Ready for This Course

Now that you have decided to take this course, remember that a *positive attitude* will make all the difference in the world. Your belief that you can succeed is just as important as your commitment to this course. Make sure that you are ready for this course by having the time and positive attitude that it takes to succeed.

Next, make sure that you have scheduled your math course at a time that will give you the best chance for success. For example, if you are also working, you may want to check with your employer to make sure that your work hours will not conflict with your course schedule.

On the day of your first class period, double-check your schedule and allow yourself extra time to arrive on time in case of traffic problems or difficulty locating your classroom. Make sure that you bring at least your textbook, paper, and a writing instrument. Are you required to have a lab manual, graph paper, calculator, or some other supply besides this text? If so, also bring this material with you.

Objective Ⓑ General Tips for Success

Below are some general tips that will increase your chance for success in a mathematics class. Many of these tips will also help you in other courses you may be taking.

Exchange names and phone numbers or e-mail addresses with at least one other person in class. This contact person can be a great help if you miss an assignment or want to discuss math concepts or exercises that you find difficult.

Choose to attend all class periods. If possible, sit near the front of the classroom. This way, you will see and hear the presentation better. It may also be easier for you to participate in classroom activities.

Do your homework. You've probably heard the phrase "practice makes perfect" in relation to music and sports. It also applies to mathematics. You will find that the more time you spend solving mathematics exercises, the easier the process becomes. Be sure to schedule enough time to complete your assignments before the next class period.

Check your work. Review the steps you made while working a problem. Learn to check your answers in the original problems. You may also compare your answers with the answers to selected exercises section in the back of the book. If you have made a mistake, try to figure out what went wrong. Then correct your mistake. If you can't find what went wrong, don't erase your work or throw it away. Bring your work to your instructor, a tutor in a math lab, or a classmate. It is easier for someone to find where you had trouble if they look at your original work.

Learn from your mistakes. Everyone, even your instructor, makes mistakes. Use your errors to learn and to become a better math student. The key is finding and understanding your errors. Was your mistake a careless one, or did you make it because you can't read your own math writing? If so, try to work more slowly or write more neatly and make a conscious effort to carefully check your work. Did you make a mistake because you don't understand a concept? If so, take the time to review the concept or ask questions to better understand it.

Know how to get help if you need it. It's all right to ask for help. In fact, it's a good idea to ask for help whenever there is something that you don't understand. Make sure you know when your instructor has office hours and how to find his or her office. Find out whether math tutoring services are available on your campus. Check

on the hours, location, and requirements of the tutoring service. Know whether software is available and how to access this resource.

Organize your class materials, including homework assignments, graded quizzes and tests, and notes from your class or lab. All of these items will make valuable references throughout your course and when studying for upcoming tests and the final exam. Make sure that you can locate these materials when you need them.

Read your textbook before class. Reading a mathematics textbook is unlike reading a novel or a newspaper. Your pace will be much slower. It is helpful to have paper and a pencil with you when you read. Try to work out examples on your own as you encounter them in your text. You should also write down any questions that you want to ask in class. When you read a mathematics textbook, sometimes some of the information in a section will be unclear. But after you hear a lecture or watch a videotape on that section, you will understand it much more easily than if you had not read your text beforehand.

Don't be afraid to ask questions. You are not the only person in class with questions. Other students are normally grateful that someone has spoken up.

Hand in assignments on time. This way you can be sure that you will not lose points for being late. Show every step of a problem and be neat and organized. Also be sure that you understand which problems are assigned for homework. If allowed, you can always double-check the assignment with another student in your class.

Objective C **Using This Text**

There are many helpful resources that are available to you in this text. It is important that you become familiar with and use these resources. They should increase your chances for success in this course.

- *Practice Problems.* Each example in every section has a parallel Practice Problem. As you read a section, try each Practice Problem after you've finished the corresponding example. This "learn-by-doing" approach will help you grasp ideas before you move on to other concepts.

- *Chapter Test Prep Video CD.* The book contains a CD. This CD contains all of the Chapter Test exercises worked out by the author. This supplement is very helpful before a classroom chapter test.

- *Lecture Video CDs.* Exercises marked with a ⊙ are fully worked out by the author on video CDs. Check with your instructor for the availability of these video CDs.

- *Symbols at the beginning of an exercise set.* If you need help with a particular section, the symbols listed at the beginning of each exercise set will remind you of the numerous supplements available.

- *Objectives.* The main section of exercises in each exercise set is referenced by an objective, such as **A** or **B**, and also an example(s). There is also often a section of exercises entitled "Mixed Practice," which is referenced by two or more objectives or sections. These are mixed exercises written to prepare you for your next exam. Use all of this referencing if you have trouble completing an assignment from the exercise set.

- *Icons (Symbols.)* Make sure that you understand the meaning of the icons that are beside many exercises. ⊙ tells you that the corresponding exercise may be viewed on the video segment that corresponds to that section. ⟍ tells you that this exercise is a writing exercise in which you should answer in complete sentences. △ tells you that the exercise involves geometry.

- *Integrated Reviews.* Found in the middle of each chapter, these reviews offer you a chance to practice—in one place—the many concepts that you have learned separately over several sections.

- *End of Chapter Opportunities.* There are many opportunities at the end of each chapter to help you understand the concepts of the chapter.

  Chapter Highlights contain chapter summaries and examples.

  Chapter Reviews contain review problems. The first part is organized section by section and the second part contains a set of mixed exercises.

  Chapter Tests are sample tests to help you prepare for an exam. The Chapter Test Prep Video CD, found in this text, contains all the Chapter Test exercises worked by the author.

  Cumulative Reviews are reviews consisting of material from the beginning of the book to the end of that particular chapter.

- *Study Skill Builders.* This feature is found at the end of many exercise sets. In order to increase your chance of success in this course, please read and answer the questions in these Study Skill Builders.

- *The Bigger Picture.* This feature contains the directions for building an outline to be used throughout the course. The purpose of this outline is to help you make the transition from thinking "section by section" to thinking about how the mathematics in this course is part of a bigger picture.

See the Preface at the beginning of this text for a more thorough explanation of the features of this text.

Objective D Getting Help

If you have trouble completing assignments or understanding the mathematics, get help as soon as you need it! This tip is presented as an objective on its own because it is so important. In mathematics, usually the material presented in one section builds on your understanding of the previous section. This means that if you don't understand the concepts covered during a class period, there is a good chance that you will not understand the concepts covered during the next class period. If this happens to you, get help as soon as you can.

Where can you get help? Many suggestions have been made in this section on where to get help, and now it is up to you to do it. Try your instructor, a tutoring center, or a math lab, or you may want to form a study group with fellow classmates. If you do decide to see your instructor or go to a tutoring center, make sure that you have a neat notebook and are ready with your questions.

Objective E Preparing for and Taking an Exam

Make sure that you allow yourself plenty of time to prepare for a test. If you think that you are a little "math anxious," it may be that you are not preparing for a test in a way that will ensure success. The way that you prepare for a test in mathematics is important. To prepare for a test:

1. Review your previous homework assignments.
2. Review any notes from class and section-level quizzes you have taken. (If this is a final exam, also review chapter tests you have taken.)
3. Review concepts and definitions by reading the Highlights at the end of each chapter.
4. Practice working out exercises by completing the Chapter Review found at the end of each chapter. (If this is a final exam, go through a Cumulative Review. There is one found at the end of each chapter except Chapter 1. Choose the review found at the end of the latest chapter that you have covered in your course.) *Don't stop here!*
5. It is important that you place yourself in conditions similar to test conditions to find out how you will perform. In other words, as soon as you feel that you know the material, get a few blank sheets of paper and take a sample test. There is a Chapter Test available at the end of each chapter, or you can work selected

problems from the Chapter Review. Your instructor may also provide you with a review sheet. During this sample test, do not use your notes or your textbook. Then check your sample test. If you are not satisfied with the results, study the areas that you are weak in and try again.

6. On the day of the test, allow yourself plenty of time to arrive at where you will be taking your exam.

When taking your test:

1. Read the directions on the test carefully.
2. Read each problem carefully as you take the test. Make sure that you answer the question asked.
3. Watch your time and pace yourself so that you can attempt each problem on your test.
4. If you have time, check your work and answers.
5. Do not turn your test in early. If you have extra time, spend it double-checking your work.

Objective 🄵 Managing Your Time

As a college student, you know the demands that classes, homework, work, and family place on your time. Some days you probably wonder how you'll ever get everything done. One key to managing your time is developing a schedule. Here are some hints for making a schedule:

1. Make a list of all of your weekly commitments for the term. Include classes, work, regular meetings, extracurricular activities, etc. You may also find it helpful to list such things as laundry, regular workouts, grocery shopping, etc.
2. Next, estimate the time needed for each item on the list. Also make a note of how often you will need to do each item. Don't forget to include time estimates for the reading, studying, and homework you do outside of your classes. You may want to ask your instructor for help estimating the time needed.
3. In the exercise set that follows, you are asked to block out a typical week on the schedule grid given. Start with items with fixed time slots like classes and work.
4. Next, include the items on your list with flexible time slots. Think carefully about how best to schedule items such as study time.
5. Don't fill up every time slot on the schedule. Remember that you need to allow time for eating, sleeping, and relaxing! You should also allow a little extra time in case some items take longer than planned.
6. If you find that your weekly schedule is too full for you to handle, you may need to make some changes in your workload, classload, or in other areas of your life. You may want to talk to your advisor, manager or supervisor at work, or someone in your college's academic counseling center for help with such decisions.

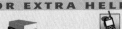
1. What is your instructor's name?

2. What are your instructor's office location and office hours?

3. What is the best way to contact your instructor?

4. Do you have the name and contact information of at least one other student in class?

5. Will your instructor allow you to use a calculator in this class?

6. Is tutorial software available to you? If so, what type and where?

7. Is there a tutoring service available on campus? If so, what are its hours? What services are available?

8. Have you attempted this course before? If so, write down ways that you might improve your chances of success during this second attempt.

9. List some steps that you can take if you begin having trouble understanding the material or completing an assignment.

10. How many hours of studying does your instructor advise for each hour of instruction?

11. What does the ✎ in this text mean?

12. What does the ⚙ in this text mean?

13. What does the △ in this text mean?

14. Search the minor columns in your text. What are Practice Problems?

15. When might be the best time to work a Practice Problem?

16. Where are the answers to Practice Problems?

17. What answers are contained in this text and where are they?

18. What solutions are contained in this text and where are they?

19. What and where are Integrated Reviews?

20. What video CD is contained in this book, where is it, and what material is on it?

21. Chapter Highlights are found at the end of each chapter. Find the Chapter 1 Highlights and explain how you might use it and how it might be helpful.

22. Chapter Reviews are found at the end of each chapter. Find the Chapter 1 Review and explain how you might use it and how it might be useful.

23. Chapter Tests are found at the end of each chapter. Find the Chapter 1 Test and explain how you might use it and how it might be helpful when preparing for an exam on Chapter 1. Include how the Chapter Test Prep Video in the book may help.

24. Read or reread objective **F** and fill out the schedule grid below.

	Monday	Tuesday	Wednesday	Thursday	Friday	Saturday	Sunday
7:00 a.m.							
8:00 a.m.							
9:00 a.m.							
10:00 a.m.							
11:00 a.m.							
12:00 p.m.							
1:00 p.m.							
2:00 p.m.							
3:00 p.m.							
4:00 p.m.							
5:00 p.m.							
6:00 p.m.							
7:00 p.m.							
8:00 p.m.							
9:00 p.m.							

1.2 ALGEBRAIC EXPRESSIONS AND SETS OF NUMBERS

Objectives

A Identify Natural Numbers, Whole Numbers, Integers, Rational, and Irrational Real Numbers.

B Write Phrases as Algebraic Expressions.

Recall that letters that represent numbers are called **variables.** An **algebraic expression** is formed by numbers and variables connected by the operations of addition, subtraction, multiplication, division, raising to powers, or taking roots. For example,

$$2x, \qquad \frac{x+5}{6}, \qquad \sqrt{y} - 1.6, \qquad \text{and} \qquad z^3$$

are algebraic expressions or, more simply, expressions. (Recall that the expression $2x$ means $2 \cdot x$.)

Algebraic expressions occur often during problem solving. For example, the B747-400 aircraft costs $8443 per hour to operate. The algebraic expression $8443t$ gives the total cost to operate the aircraft for t hours. (*Source: The World Almanac, 2005*)

To find the cost to operate the aircraft for 5.2 hours, for example, we replace the variable t with 5.2 and perform the indicated operation. This process is called **evaluating an expression**, and the result is called the **value** of the expression for the given replacement value.

In our example, when $t = 5.2$ hours,

$$8443t = 8443(5.2) = 43{,}903.6$$

Thus, it costs $43,903.60 to operate the B747-400 aircraft for 5.2 hours.

When evaluating an expression to solve a problem, we often need to think about the kind of number that is appropriate for the solution. For example, if we are asked to determine the maximum number of parking spaces for a parking lot to be constructed, an answer of $98\frac{1}{10}$ is not appropriate because $\frac{1}{10}$ of a parking space is not realistic.

We practice writing algebraic expressions in Objective **B** and evaluating algebraic expressions in Section 1.5.

Objective A Identifying Common Sets of Numbers

Let's review some common sets of numbers and their graphs on a **number line.** To construct a number line, we draw a line and label a point 0 with which we associate the number 0. This point is called the **origin.** If we choose a point to the right of 0 and label it 1, the distance from 0 to 1 is called the **unit distance** and can be used to locate more points. The **positive numbers** lie to the right of the origin, and the **negative numbers** lie to the left of the origin. The number 0 is neither positive nor negative.

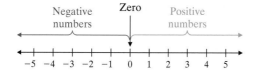

✔**Concept Check** Use the definitions of positive numbers, negative numbers, and zero to describe the meaning of *nonnegative numbers.*

A number is **graphed** on a number line by shading the point on the number line that corresponds to the number. Some common sets of numbers and their graphs are shown next.

Identifying Numbers

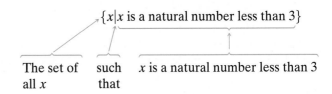

Natural numbers: $\{1, 2, 3, \dots\}$

Whole numbers: $\{0, 1, 2, 3, \dots\}$

Integers: $\{\dots, -3, -2, -1, 0, 1, 2, 3, \dots\}$

Each listing of three dots, $\dots$, is called an **ellipsis** and means to continue in the same pattern.

A **set** is a collection of objects. The objects of a set are called its **members or elements.** When the elements of a set are listed, such as those displayed in the box, the set is written in **roster** form. A set can also be written in **set builder notation,** which describes the members of the set but does not list them. The following set is written in set builder notation.

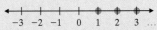

$$\{x \,|\, x \text{ is a natural number less than 3}\}$$

The set of all x such that x is a natural number less than 3

This same set written in roster form is $\{1, 2\}$.

A set that contains *no* elements is called the **empty set** or **null set,** symbolized by $\{\ \}$ or $\varnothing$.

$$\{x \,|\, x \text{ is a month with 32 days}\} \text{ is } \varnothing \text{ or } \{\ \}$$

because no month has 32 days. The set has no elements.

EXAMPLES Write each set in roster form.

1. $\{x \,|\, x \text{ is a whole number between 1 and 6}\}$

$\{2, 3, 4, 5\}$

2. $\{x \,|\, x \text{ is a natural number greater than 100}\}$

$\{101, 102, 103, \dots\}$

▢ **Work Practice Problems 1–2**

The symbol $\in$ is used to denote that an element is in a particular set. The symbol $\in$ is read as "is an element of." For example, the true statement "3 is an element of $\{1, 2, 3, 4, 5\}$" can be written in symbols as

$$3 \in \{1, 2, 3, 4, 5\}$$

The symbol $\notin$ is read as "is not an element of." In symbols, we write the true statement "p is not an element of $\{a, 5, g, j, q\}$" as

$$p \notin \{a, 5, g, j, q\}$$

Copyright 2007 Pearson Education, Inc.

EXAMPLES Determine whether each statement is true or false.

3. $3 \in \{x \mid x \text{ is a natural number}\}$ True, since 3 is a natural number and therefore an element of the set.

4. $7 \notin \{1, 2, 3\}$ True, since 7 is not an element of the set $\{1, 2, 3\}$.

■ **Work Practice Problems 3–4**

PRACTICE PROBLEMS 3–4

Determine whether each statement is true or false.

3. $0 \in \{x \mid x \text{ is a natural number}\}$
4. $9 \notin \{4, 6, 8, 10\}$

We can use set builder notation to describe three other common sets of numbers.

Identifying Numbers

Real numbers: $\{x \mid x \text{ corresponds to a point on the number line}\}$

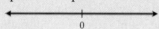

Rational numbers: $\left\{\dfrac{a}{b} \;\middle|\; a \text{ and } b \text{ are integers and } b \neq 0\right\}$

Irrational numbers: $\{x \mid x \text{ is a real number and } x \text{ is not a rational number}\}$

Notice that every integer is also a rational number since each integer can be written as the quotient of itself and 1:

$$3 = \frac{3}{1}, \qquad 0 = \frac{0}{1}, \qquad -8 = \frac{-8}{1}$$

Not every rational number, however, is an integer. The rational number $\frac{2}{3}$, for example, is not an integer. Some square roots are rational numbers and some are irrational numbers. For example, $\sqrt{2}$, $\sqrt{3}$, and $\sqrt{7}$ are irrational numbers but $\sqrt{25}$ is a rational number because $\sqrt{25} = 5 = \frac{5}{1}$. The number π is an irrational number, To help you make the distinction between rational and irrational numbers, here are a few examples of each.

Rational Numbers		Irrational Numbers
Number	**Equivalent Quotient of Integers,** $\dfrac{a}{b}$	
$-\dfrac{2}{3}$	$\dfrac{-2}{3}$ or $\dfrac{2}{-3}$	$\sqrt{5}$
$\sqrt{36}$	$\dfrac{6}{1}$	$\dfrac{\sqrt{6}}{7}$
5	$\dfrac{5}{1}$	$-\sqrt{3}$
0	$\dfrac{0}{1}$	π
1.2	$\dfrac{12}{10}$	$\dfrac{2}{\sqrt{3}}$
$3\dfrac{7}{8}$	$\dfrac{31}{8}$	

Some rational and irrational numbers are graphed below.

Irrational Numbers

$-\sqrt{5}$ $\sqrt{2}$ π

$-3 \quad -2 \quad -1 \quad 0 \quad 1 \quad 2 \quad 3 \quad 4$

Rational Numbers $-1.5 \qquad \dfrac{1}{2} \quad \dfrac{5}{4}$

Answers

3. false, **4.** true

Every rational number can be written as a decimal that either repeats or terminates. For example,

$$\frac{1}{2} = 0.5 \qquad\qquad \frac{5}{4} = 1.25$$

$$\frac{2}{3} = 0.6666666\ldots = 0.\overline{6} \qquad \frac{1}{11} = 0.090909\ldots = 0.\overline{09}$$

An irrational number written as a decimal neither terminates nor repeats. When we perform calculations with irrational numbers, we often use decimal approximations that have been rounded. For example, consider the following irrational numbers along with a four-decimal-place approximation of each:

$$\pi \approx 3.1416 \qquad \sqrt{2} \approx 1.4142$$

Earlier we mentioned that every integer is also a rational number. In other words, all the elements of the set of integers are also elements of the set of rational numbers. When this happens, we say that the set of integers, set I, is a **subset** of the set of rational numbers, set Q. In symbols,

$$\underbrace{I \subseteq Q}_{\text{is a subset of}}$$

The natural numbers, whole numbers, integers, rational numbers, and irrational numbers are each a subset of the set of real numbers. The relationships among these sets of numbers are shown in the following diagram.

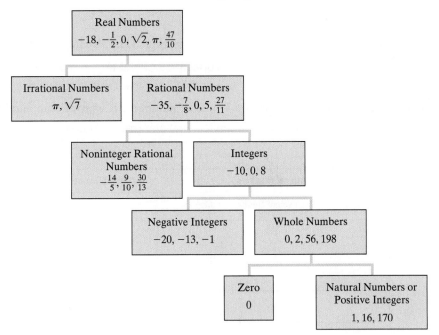

PRACTICE PROBLEMS 5–7

Determine whether each statement is true or false.

5. 0 is a real number.

6. Every integer is a rational number.

7. $\sqrt{3}$ is a rational number.

EXAMPLES Determine whether each statement is true or false.

5. 3 is a real number.

True. Every whole number is a real number.

6. Every rational number is an integer.

False. The number $\frac{2}{3}$, for example, is a rational number, but it is not an integer.

7. $\frac{1}{5}$ is an irrational number.

False. The number $\frac{1}{5}$ is a rational number since it is in the form $\frac{a}{b}$ with a and b integers and $b \neq 0$.

■ **Work Practice Problems 5–7**

Answers

5. true, **6.** true, **7.** false

Objective   **Writing Phrases as Algebraic Expressions**

Often, solving problems involves translating a phrase into an algebraic expression. The following is a list of key words and phrases and their translations.

Selected Key Words/Phrases and Their Translations			
Addition	**Subtraction**	**Multiplication**	**Division**
sum	difference of	product	quotient
plus	minus	times	divide
added to	subtracted from	multiply	into
more than	less than	twice	ratio
increased by	decreased by	of	
total	less		

EXAMPLES   Write each phrase as an algebraic expression. Use the variable x to represent each unknown number.

8. Eight times a number $8 \cdot x$ or $8x$

9. Three more than eight times a number $8x + 3$

10. The quotient of a number and -7 $x \div -7$ or $\dfrac{x}{-7}$

11. One and six-tenths subtracted from twice a number $2x - 1.6$ or $2x - 1\dfrac{6}{10}$

12. Six less than a number $x - 6$

13. Twice the sum of four and a number $2(4 + x)$

▌ **Work Practice Problems 8–13**

PRACTICE PROBLEMS 8–13

Write each phrase as an algebraic expression. Use the variable x to represent each unknown number.

8. Seventeen times a number

9. Five more than six times a number

10. The quotient of six and a number

11. One-fourth subtracted from three times a number

12. Eleven less than a number

13. Three times the difference of a number and ten

Answers
8. $17 \cdot x$ or $17x$,   9. $6x + 5$,
10. $\dfrac{6}{x}$ or $6 \div x$,   11. $3x - \dfrac{1}{4}$,
12. $x - 11$,   13. $3(x - 10)$

1.2 EXERCISE SET

FOR EXTRA HELP

 Student Solutions Manual

 PH Math/Tutor Center

 CD/Video for Review

 Math XL MathXL®

 MyMathLab MyMathLab

Objective A *Write each set in roster form. See Examples 1 and 2.*

1. $\{x \mid x$ is a natural number less than 6$\}$

2. $\{x \mid x$ is a natural number greater than 6$\}$

3. $\{x \mid x$ is a natural number between 10 and 17$\}$

4. $\{x \mid x$ is an odd natural number$\}$

5. $\{x \mid x$ is a whole number that is not a natural number$\}$

6. $\{x \mid x$ is a natural number less than 1$\}$

7. $\{x \mid x$ is an even whole number less than 9$\}$

8. $\{x \mid x$ is an odd whole number less than 9$\}$

List the elements of the set $\left\{3, 0, \sqrt{7}, \sqrt{36}, \dfrac{2}{5}, -134\right\}$ *that are also elements of the given set. See Examples 3 and 4.*

9. Whole numbers

10. Integers

11. Natural numbers

12. Rational numbers

13. Irrational numbers

14. Real numbers

Place $\in$ *or* $\notin$ *in the space provided to make each statement true. See Examples 3 through 7.*

15. -11 $\quad \{x \mid x$ is an integer$\}$

16. -6 $\quad \{2, 4, 6, \ldots\}$

17. 0 $\quad \{x \mid x$ is a positive integer$\}$

18. 12 $\quad \{1, 2, 3, \ldots\}$

19. 12 $\quad \{1, 3, 5, \ldots\}$

20. $\dfrac{1}{2}$ $\quad \{x \mid x$ is an irrational number$\}$

Determine whether each statement is true or false. See Examples 5 through 7.

21. Every whole number is a real number.

22. Every irrational number is a real number.

23. Some real numbers are irrational numbers.

24. Some real numbers are whole numbers.

25. Every whole number is a natural number.

26. Every irrational number is a rational number.

27. $\dfrac{1}{2}$ is a real number.

28. $-\dfrac{4}{5}$ is a real number.

Mixed Practice *Determine whether each statement is true or false. See Examples 3 through 7.*

29. $0 \in \{x \mid x$ is an integer$\}$

30. $0 \in \{x \mid x$ is a whole number$\}$

31. $\sqrt{7} \notin \{x \mid x$ is an irrational number$\}$

32. $-\dfrac{7}{11} \notin \{x \mid x$ is a rational number$\}$

12

33. $\{x \mid x$ is a day of the week starting with the letter $B\}$ is $\varnothing$.

34. $\{x \mid x$ is a rational number and an irrational number$\}$ is $\varnothing$.

35. Some real numbers are integers.

36. Every integer is a real number.

37. Some rational numbers are irrational numbers.

38. Every real number is also an integer.

Objective **B** *Write each phrase as an algebraic expression. Use the variable x to represent each unknown number. See Examples 8 through 13.*

39. Twice a number

40. Six times a number

41. Ten less than a number

42. A number minus seven

43. The sum of a number and two

44. The difference of twenty-five and a number

45. A number divided by eleven

46. The quotient of a number and thirteen

47. Four subtracted from a number

48. Seventeen subtracted from a number

49. A number plus two and three-tenths

50. Fifteen and seven-tenths plus a number

51. A number less than one and one-third

52. Two and three-fourths less than a number

53. Nine times a number

54. Nine minus a number

55. Nine added to a number

56. Nine divided by a number

57. Five more than twice a number

58. One more than six times a number

59. Twelve minus three times a number

60. Four subtracted from three times a number

61. One plus twice a number

62. Three less than twice a number

63. Ten subtracted from five times a number

64. Four minus three times a number

65. The quotient of five and the difference of four and a number

66. The quotient of four and the sum of a number and one

67. Twice the sum of a number and three

68. Eight times the difference of a number and nine

For Exercises 69 through 72, fill in the blank so that each is a true statement. There are many possible correct answers.

69. _____ $\in \{1, 3, 5, 7\}$

70. _____ $\notin \{1, 3, 5, 7\}$

71. _____ $\subseteq \{1, 3, 5, 7\}$

72. _____ $\not\subseteq \{1, 3, 5, 7\}$

73. Name a whole number that is not a natural number.

74. Name a rational number that does not simplify to an integer.

Concept Extensions

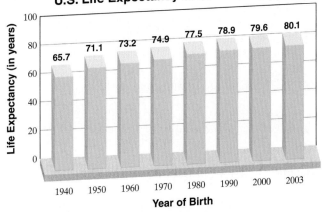

U.S. Life Expectancy at Birth for Females

Life Expectancy (in years): 65.7, 71.1, 73.2, 74.9, 77.5, 78.9, 79.6, 80.1

Year of Birth: 1940, 1950, 1960, 1970, 1980, 1990, 2000, 2003

Source: Social Security Administration and World Almanac, 2005

75. The bar graph to the left shows the U.S. life expectancy at birth for females born in the years shown. Use the graph to calculate the *increase* in life expectancy over each ten-year period shown.

Year	Increase in Life Expectancy (in years) from 10 Years Earlier
1950	
1960	
1970	
1980	
1990	
2000	

76. In your own words, explain why every natural number is also a rational number but not every rational number is a natural number.

77. In your own words, explain why every irrational number is a real number but not every real number is an irrational number.

 STUDY SKILLS BUILDER

Learning New Terms?

Many of the terms used in this text may be new to you. It will be helpful to make a list of new mathematical terms and symbols as you encounter them and to review them frequently. Placing these new terms (including page references) on 3 × 5 index cards might help you later when you're preparing for a quiz.

Answer the following.

1. Name one way you might place a word and its definition on a 3 × 5 card.

2. How do new terms stand out in this text so that they can be found?

1.3 EQUATIONS, INEQUALITIES, AND PROPERTIES OF REAL NUMBERS

Objectives

A Write Sentences as Equations.

B Use Inequality Symbols.

C Find the Opposite, or Additive Inverse, and the Reciprocal, or Multiplicative Inverse, of a Number.

D Identify and Use the Commutative, Associative, and Distributive Properties.

Objective **A** Writing Sentences as Equations

When writing sentences as equations, we use the symbol = to translate the phrase **"is equal to."** All of the following key words and phrases also mean equality.

Equality

equals	is/was	represents	is the same as
gives	yields	amounts to	is equal to

EXAMPLES Write each sentence as an equation.

1. The sum of x and 5 is 20

$$x + 5 = 20$$

2. The difference of 8 and x is the same as the product of 2 and x.

$$8 - x = 2 \cdot x$$

3. The quotient of z and 9 amounts to 9 plus z.

$$z \div 9 = 9 + z$$

or

$$\frac{z}{9} = 9 + z$$

Work Practice Problems 1–3

PRACTICE PROBLEMS 1–3

Write each sentence as an equation.

1. The difference of x and 7 is 45.

2. The product of 5 and x amounts to the sum of x and 14.

3. The quotient of y and 23 is the same as 20 subtracted from y.

Objective **B** Using Inequality Symbols

If we want to write in symbols that two numbers are not equal, we can use the symbol ≠, which means **"is not equal to."** For example,

$$3 \neq 2$$

Graphing two numbers on a number line gives us a way to compare two numbers. For any two real numbers graphed on a number line, the number to the left is less than the number to the right. This means that the number to the right is greater than the number to the left.

The symbol < means **"is less than."** Since -4 is to the left of -1 on the number line, we write $-4 < -1$. The symbol > means **"is greater than."** Since -1 is to the right of -4 on the number line, we write $-1 > -4$.

$$-4 < -1 \text{ or } -1 > -4$$

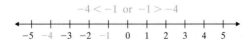

Notice that since $-4 < -1$, then we also know that $-1 > -4$. This is true for any two numbers, say, a and b.

If $a < b$, then also $b > a$. For example, since $-4 < -1$, then $-1 > -4$.

Answers

1. $x - 7 = 45$, **2.** $5x = x + 14$,

3. $\dfrac{y}{23} = y - 20$

15

PRACTICE PROBLEMS 4–11

Insert $<$, $>$, or $=$ between each pair of numbers to form a true statement.

4. 7 -7
5. -1 11
6. -10 -12
7. -3.25 -3.025
8. 7.206   7.2060
9. 18.6 -14.2
10. $\dfrac{4}{7}$ $\dfrac{5}{7}$
11. $\dfrac{3}{8}$ $\dfrac{1}{3}$

EXAMPLES   Insert $<$, $>$, or $=$ between each pair of numbers to form a true statement.

4. -5 5 -5 is to the left of 5 on a number line, so $-5 < 5$.
5. 3 -7 3 is to the right of -7, so $3 > -7$.
6. -16 -6 -16 is to the left of -6, so $-16 < -6$.
7. -2.5 -2.1 -2.5 is to the left of -2.1, so $-2.5 < -2.1$.
8. 6.36   6.360   The true statement is $6.36 = 6.360$.
9. 4.3 -5.2 4.3 is to the right of -5.2, so $4.3 > -5.2$.
10. $\dfrac{5}{8}$ $\dfrac{3}{8}$ The denominators are the same, so $\dfrac{5}{8} > \dfrac{3}{8}$ since $5 > 3$.
11. $\dfrac{2}{3}$ $\dfrac{3}{4}$ By dividing, we see that $\dfrac{3}{4} = 0.75$ and $\dfrac{2}{3} = 0.666\ldots$.

                 Thus $\dfrac{2}{3} < \dfrac{3}{4}$ since $0.666\ldots < 0.75$.

▣ **Work Practice Problems 4–11**

Helpful Hint

When inserting the $>$ or $<$ symbol, think of the symbols as arrowheads that "point" toward the smaller number when the statement is true.

In addition to $<$ and $>$, there are the inequality symbols $\leq$ and $\geq$. The symbol $\leq$ means **"is less than or equal to,"** and the symbol $\geq$ means **"is greater than or equal to."**

PRACTICE PROBLEMS 12–15

Determine whether each statement is true or false.

12. $-11 \leq 16$
13. $-7 \leq -7$
14. $-7 \geq -7$
15. $-25 \leq -30$

EXAMPLES   Determine whether each statement is true or false.

12. $-9 \leq 7$ True, since $-9 < 7$ is true.
13. $-5 \leq -5$ True, since $-5 = -5$ is true.
14. $-5 \geq -5$ True, since $-5 = -5$ is true.
15. $-24 \geq -20$ False, since neither $-24 > -20$ nor $-24 = -20$ is true.

▣ **Work Practice Problems 12–15**

Objective ⓒ Finding Opposites and Reciprocals

Of all the real numbers, two of them stand out as extraordinary: 0 and 1. Zero is the only real number that can be added to *any* real number and result in the same real number. Also, 1 is the only real number that can be multiplied by *any* real number and result in the same real number. This is why 0 is called the **additive identity** and 1 is called the **multiplicative identity.**

Identity Properties

For any real number a,

       Identity Property of 0: $a + 0 = 0 + a = a$

Also,

       Identity Property of 1: $a \cdot 1 = 1 \cdot a = a$

Answers

4. $>$,   5. $<$,   6. $>$,   7. $<$,   8. $=$,
9. $>$,   10. $<$,   11. $>$,   12. true,
13. true,   14. true,   15. false

We use the identity property of 1 when we say that x, for example, means $1 \cdot x$ or $1x$. We also use this property when we write equivalent expressions. For example,

$$\underbrace{\frac{2}{3} = \frac{2}{3} \cdot 1}_{\text{identity property of 1}} = \frac{2}{3} \cdot \frac{5}{5} = \frac{10}{15} \quad \frac{5}{5} \text{ is another name for 1.}$$

Two numbers whose sum is the additive identity 0 are called **opposites** or **additive inverses** of each other. Each real number has a unique opposite.

Opposites or Additive Inverses

If a is a real number, then the unique **opposite,** or **additive inverse,** of a is written as $-a$ and the following is true:

$$a + (-a) = 0 \qquad (\text{Also, } -a + a = 0)$$

On a number line, we picture a real number and its opposite as being the same distance from 0 but on opposite sides of 0.

The opposite of 6 is -6.

The opposite of $\dfrac{2}{3}$ is $-\dfrac{2}{3}$.

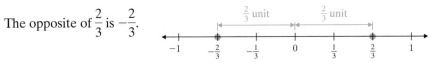

The opposite of -4 is 4.

We stated that the opposite or additive inverse of a number a is $-a$. This means that the opposite of -4 is $-(-4)$. But we stated above that the opposite of -4 is 4. This means that $-(-4) = 4$, and in general, we have the following property.

Double Negative Property

For every real number a, we have $-(-a) = a$.

EXAMPLES Find the opposite, or additive inverse, of each number.

16. 8   The opposite of 8 is -8.

17. $-\dfrac{1}{5}$ The opposite of $-\dfrac{1}{5}$ is $-\left(-\dfrac{1}{5}\right)$ or $\dfrac{1}{5}$.

18. 0   The opposite of 0 is -0, or 0.

19. -3.5 The opposite of -3.5 is $-(-3.5)$ or 3.5.

◼ **Work Practice Problems 16–19**

Two numbers whose product is 1 are called **reciprocals** or **multiplicative inverses** of each other. Just as each real number has a unique opposite, each nonzero real number also has a unique reciprocal.

PRACTICE PROBLEMS 16–19

Find the opposite, or additive inverse, of each number.

16. 7   **17.** $\dfrac{2}{13}$

18. $-\dfrac{5}{7}$ **19.** -4.7

Answers

16. -7,   **17.** $-\dfrac{2}{13}$,   **18.** $\dfrac{5}{7}$,   **19.** 4.7

Reciprocals or Multiplicative Inverses

If a is a nonzero real number, then its **reciprocal,** or **multiplicative inverse,** is $\dfrac{1}{a}$ and the following is true:

$$a \cdot \frac{1}{a} = 1 \qquad \left(\text{Also, } \frac{1}{a} \cdot a = 1 \right)$$

PRACTICE PROBLEMS 20–22

Find the reciprocal, or multiplicative inverse, of each number.

20. 13

21. −5

22. $\dfrac{2}{3}$

EXAMPLES Find the reciprocal, or multiplicative inverse, of each number.

20. 11   The reciprocal of 11 is $\dfrac{1}{11}$.

21. −9   The reciprocal of −9 is $-\dfrac{1}{9}$.

22. $\dfrac{7}{4}$ The reciprocal of $\dfrac{7}{4}$ is $\dfrac{4}{7} \left(\text{since } \dfrac{7}{4} \cdot \dfrac{4}{7} = 1 \right)$.

▢ **Work Practice Problems 20–22**

Helpful Hint

The number 0 has no reciprocal. Why? There is no number that when multiplied by 0 gives a product of 1.

✔ **Concept Check**   Can a number's additive inverse and multiplicative inverse ever be the same? Explain.

Objective D Using the Commutative, Associative, and Distributive Properties

In addition to these special real numbers, all real numbers have certain properties that allow us to write equivalent expressions—that is, expressions that have the same value. These properties will be especially useful in Chapter 2 when we solve equations.

The **commutative properties** state that the order in which two real numbers are added or multiplied does not affect their sum or product.

Commutative Properties

For any real numbers a and b,

       Addition: $a + b = b + a$

       Multiplication: $a \cdot b = b \cdot a$

For example,

$$7 + 11 = 18 \quad \text{and} \quad 11 + 7 = 18 \quad \text{Addition}$$
$$7 \cdot 11 = 77 \quad \text{and} \quad 11 \cdot 7 = 77 \quad \text{Multiplication}$$

The **associative properties** state that regrouping numbers that are added or multiplied does not affect their sum or product.

Associative Properties

For real numbers $a, b,$ and c,

       Addition: $(a + b) + c = a + (b + c)$

       Multiplication: $(a \cdot b) \cdot c = a \cdot (b \cdot c)$

Answers

20. $\dfrac{1}{13}$,   **21.** $-\dfrac{1}{5}$,   **22.** $\dfrac{3}{2}$

✔ **Concept Check Answer**

no; answers may vary

For example,

$$(2 + 3) + 7 = 5 + 7 = 12 \quad \text{Addition}$$
$$2 + (3 + 7) = 2 + 10 = 12$$
$$(2 \cdot 3) \cdot 7 = 6 \cdot 7 = 42 \quad \text{Multiplication}$$
$$2 \cdot (3 \cdot 7) = 2 \cdot 21 = 42$$

EXAMPLE 23 Use the commutative property of addition to write an expression equivalent to $7x + 5$.

Solution: $7x + 5 = 5 + 7x$

▣ **Work Practice Problem 23**

EXAMPLE 24 Use the associative property of multiplication to write an expression equivalent to $4 \cdot (9y)$. Then simplify this equivalent expression.

Solution: $4 \cdot (9y) = (4 \cdot 9)y = 36y$

▣ **Work Practice Problem 24**

The **distributive property** states that multiplication distributes over addition. In Section 1.4, we learn that subtraction is defined in terms of addition. Because of this, we can also say that multiplication distributes over subtraction.

Distributive Properties

For real numbers a, b, and c,

$$a(b + c) = ab + ac$$

Also,

$$a(b - c) = ab - ac$$

For example,

$$3(6 + 2) = 3(8) = 24$$
$$3(6 + 2) = 3(6) + 3(2) = 18 + 6 = 24$$

EXAMPLES Use the distributive property to multiply.

25. $3(2x - y) = 3 \cdot 2x - 3 \cdot y = 6x - 3y$

26. $4(2y + 5) = 4 \cdot 2y + 4 \cdot 5 = 8y + 20$

27. $0.7x(y - 2) = 0.7x \cdot y - 0.7x \cdot 2 = 0.7xy - 1.4x$

▣ **Work Practice Problems 25–27**

✔ **Concept Check** Is the statement below true? Why or why not?

$$6(2a)(3b) = 6(2a) \cdot 6(3b)$$

PRACTICE PROBLEM 23

Use the commutative property of addition to write an expression equivalent to $9 + 4x$.

PRACTICE PROBLEM 24

Use the associative property of multiplication to write an expression equivalent to $5 \cdot (6x)$. Then simplify this equivalent expression.

Helpful Hint

The distributive property also applies to sums or differences of more than two terms.
For example,

$$a(b + c + d) = ab + ac + ad$$

PRACTICE PROBLEMS 25–27

Use the distributive property to multiply.
25. $7(4x - y)$
26. $8(3 + x)$
27. $5x(y - 4)$

Answers
23. $4x + 9$, **24.** $(5 \cdot 6)x = 30x$,
25. $28x - 7y$, **26.** $24 + 8x$,
27. $5xy - 20x$

✔ **Concept Check Answer**
no; $6(2a)(3b) = 6(6ab) = 36ab$

Objective Ⓐ *Write each sentence as an equation. See Examples 1 through 3.*

1. The sum of 10 and x is -12.

2. The difference of y and 3 amounts to 12.

3. Twice x plus 5 is the same as -14.

4. Three more than the product of 4 and c is 7.

5. The quotient of n and 5 is 4 times n.

6. The quotient of 8 and y is 3 more than y.

7. The difference of z and one-half is the same as the product of z and one-half.

8. Five added to one-fourth q is the same as 4 more than q.

Objective Ⓑ *Insert $<$, $>$, or $=$ between each pair of numbers to form a true statement. See Examples 4 through 11.*

9. $0 \quad -2$

10. $-5 \quad 0$

11. $-16 \quad -17$

12. $-14 \quad -24$

13. $\dfrac{12}{3} \quad \dfrac{12}{2}$

14. $\dfrac{20}{5} \quad \dfrac{20}{4}$

15. $7.4 \quad 7.40$

16. $\dfrac{12}{4} \quad \dfrac{15}{5}$

17. $8.6 \quad -3.5$

18. $-4.7 \quad 3.8$

19. $\dfrac{7}{11} \quad \dfrac{9}{11}$

20. $\dfrac{9}{20} \quad \dfrac{3}{20}$

21. $\dfrac{1}{2} \quad \dfrac{5}{8}$

22. $\dfrac{3}{4} \quad \dfrac{7}{8}$

23. $-7.9 \quad -7.09$

24. $-13.07 \quad -13.7$

Determine whether each statement is true or false. See Examples 12 through 15.

25. $-6 \leq 0$

26. $0 \leq -4$

27. $-3 \geq -3$

28. $-8 \leq -8$

29. $-14 \geq -1$

30. $-14 \leq -1$

31. $-3 \leq -3$

32. $-8 \geq -8$

Objective Ⓒ *Write the opposite (or additive inverse) of each number. See Examples 16 through 19.*

33. 9

34. 15

35. -6.2

36. -7.8

37. $\dfrac{4}{7}$

38. $\dfrac{9}{5}$

39. $-\dfrac{5}{11}$

40. $-\dfrac{14}{3}$

41. 0

42. 10.3

Write the reciprocal (or multiplicative inverse) of each number if one exists. See Examples 20 through 22.

43. 5

44. 9

45. -8

46. -4

47. $-\dfrac{1}{4}$

48. $\dfrac{1}{9}$

49. 0

50. $\dfrac{0}{6}$

51. $\dfrac{7}{8}$

52. $-\dfrac{23}{5}$

Mixed Practice *Fill in the chart. See Examples 16 through 22.*

	Number	Opposite	Reciprocal
53.	25		
54.	7		
55.		10	
56.			$-\dfrac{1}{6}$
57.	$-\dfrac{1}{7}$		
58.	$\dfrac{1}{11}$		
59.	0		
60.	1		
61.			$\dfrac{19}{16}$
62.		$\dfrac{36}{13}$	

Objective D *Use a commutative property to write an equivalent expression. See Example 23.*

63. $7x + y$

64. $3a + 2b$

65. $z \cdot w$

66. $r \cdot s$

67. $\dfrac{1}{3} \cdot \dfrac{x}{5}$

68. $\dfrac{x}{2} \cdot \dfrac{9}{10}$

Use an associative property to write an equivalent expression. See Example 24.

69. $5 \cdot (7x)$

70. $3 \cdot (10z)$

71. $(x + 1.2) + y$

72. $5q + (2r + s)$

73. $(14z) \cdot y$

74. $(9.2x) \cdot y$

Use the distributive property to multiply. See Examples 25 through 27.

75. $3(x + 5)$

76. $7(y + 2)$

77. $4(z - 6)$

78. $2(7 - y)$

79. $8(2a + b)$

80. $9(c + 7d)$

81. $6x(y - 4)$

82. $11y(z - 2)$

83. $0.4(2x + 5y)$

84. $0.5(3a - 4b)$

85. $\dfrac{1}{2}(4x - 9y)$

86. $\dfrac{1}{3}(4x + 9y)$

87. $2(6x + 5y + 2z)$

88. $5(3a + b + 9c)$

Complete each statement to illustrate the given property.

89. $3x + 6 = $ _____ Commutative property of addition

90. $8 + 0 = $ _____ Additive identity property

91. $\dfrac{2}{3} + \left(-\dfrac{2}{3}\right) = $ _____ Additive inverse property

92. $4(x + 3) = $ _____ Distributive property

93. $7 \cdot 1 = $ _____ Multiplicative identity property

94. $0 + 5.4 = $ _____ Additive identity property

95. $10(2y) = $ _____ Associative property of multiplication

96. $9y + (x + 3z) = $ _____ Associative property of addition

Concept Extensions

In each statement, a property of real numbers has been incorrectly applied. Correct the right-hand side of each statement. See the second Concept Check in this section.

97. $3(x + 4) = 3x + 4$

98. $5(7y) = (5 \cdot 7)(5 \cdot y)$

99. $4 + 8y = 4y + 8$

100. Name the only real number that has no reciprocal, and explain why this is so.

101. Name the only real number that is its own opposite, and explain why this is so.

102. Is subtraction commutative? Explain why or why not.

103. Is division commutative? Explain why or why not.

104. Evaluate $12 - (5 - 3)$ and $(12 - 5) - 3$. Use these two expressions and discuss whether subtraction is associative.

105. Evaluate $24 \div (6 \div 3)$ and $(24 \div 6) \div 3$. Use these two expressions and discuss whether division is associative.

106. To demonstrate the distributive property geometrically, represent the area of the larger rectangle in two ways: First as length a times width $b + c$, and second as the sum of the areas of the smaller rectangles.

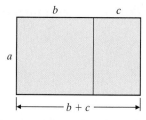

  STUDY SKILLS BUILDER

Are You Familiar with Your Textbook Supplements?

There are many student supplements available for additional study. Below, I have listed some of these. See the preface of this text or your instructor for further information.

- *Chapter Test Prep Video CD.* This material is found in your textbook and is fully explained. The CD contains video clip solutions to the Chapter Test exercises in this text and are excellent help when studying for chapter tests.

- *Lecture Video CDs.* These video segments are keyed to each section of the text. The material is presented by me, Elayn Martin-Gay, and I have placed a video icon by each exercise in the text that I have worked on the video.

- *The Student Solutions Manual.* This contains worked out solutions to odd-numbered exercises as well as every exercise in the Integrated Reviews, Chapter Reviews, Chapter Tests, and Cumulative Reviews.

- *Prentice Hall Tutor Center.* Mathematics questions may be phoned, faxed, or emailed to this center.

- *MyMathLab, MathXL, and Interact Math.* These are computer and Internet tutorials. This supplement may already be available to you somewhere on campus, for example at your local learning resource lab. Take a moment and find the name and location of any such lab on campus.

As usual, your instructor is your best source of information.

Let's see how you are doing with textbook supplements:

1. Name one way the Chapter Test Prep Video can help you prepare for a chapter test.

2. List any textbook supplements that you have found useful.

3. Have you located and visited a learning resource lab located on your campus?

4. List the textbook supplements that are currently housed in your campus' learning resource lab.

1.4 OPERATIONS ON REAL NUMBERS

Objective A Finding the Absolute Value of a Number

In Section 1.3, we used the number line to compare two real numbers. The number line can also be used to visualize distance, which leads to the concept of absolute value. The **absolute value** of a number is the distance between the number and 0 on a number line. The symbol for absolute value is $|\ |$. For example, since -4 and 4 are both 4 units from 0 on the number line, each has an absolute value of 4.

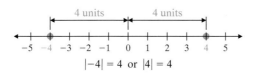

$$|-4| = 4 \text{ or } |4| = 4$$

An equivalent definition of the absolute value of a real number a is given next.

Absolute Value

The absolute value of a, written as $|a|$, is

$$|a| = \begin{cases} a \text{ if } a \text{ is 0 or a positive number} \\ -a \text{ if } a \text{ is a negative number} \end{cases}$$

↑
the opposite of

EXAMPLES Find each absolute value.

1. $|3| = 3$
2. $|0| = 0$
3. $\left|-\dfrac{1}{7}\right| = -\left(-\dfrac{1}{7}\right) = \dfrac{1}{7}$

   ↑
   the opposite of
4. $-|2.7| = -2.7$
5. $-|-8| = -8$ Since $|-8|$ is 8, we have $-|-8| = -8$.

 Work Practice Problems 1–5

> **Helpful Hint**
>
> Since distance is always positive or zero, the absolute value of a number is always positive or zero.

✔ **Concept Check** Explain how you know that $|14| = -14$ is a false statement.

Objective B Adding and Subtracting Real Numbers

When solving problems, we often need to add real numbers. For example, if the New Orleans Saints lose 5 yards in one play, then lose another 7 yards in the next play, their total loss may be described by $-5 + (-7)$.

The addition of two real numbers may be summarized by the following.

Adding Real Numbers

1. To add two numbers with the *same* sign, add their absolute values and attach their common sign.

2. To add two numbers with *different* signs, subtract the smaller absolute value from the larger absolute value and attach the sign of the number with the larger absolute value.

For example, to add $-5 + (-7)$, we first add their absolute values.

$$|-5| = 5, \quad |-7| = 7, \quad \text{and} \quad 5 + 7 = 12$$

Next, we attach their common negative sign.

$$-5 + (-7) = -12$$

(This represents a total loss of 12 yards for the New Orleans Saints.)

To find $(-4) + 3$, we first subtract their absolute values. (Subtract smaller absolute value from larger absolute value.)

$$|-4| = 4, \quad |3| = 3, \quad \text{and} \quad 4 - 3 = 1$$

Next, we attach the sign of the number with the larger absolute value.

$$(-4) + 3 = -1$$

PRACTICE PROBLEMS 6–11

Add.

6. $-7 + (-10)$

7. $8 + (-12)$

8. $-14 + 20$

9. $-4.6 + (-1.9)$

10. $-\dfrac{2}{3} + \dfrac{1}{6}$

11. $-\dfrac{1}{7} + \dfrac{1}{2}$

EXAMPLES Add.

6. $-3 + (-11) = -(3 + 11) = -14$ Add their absolute values, or $3 + 11 = 14$. Then attach the common negative sign.

7. $3 + (-7) = -4$ Subtract their absolute values, or $7 - 3 = 4$. Since -7 has the larger absolute value, the answer is -4.

8. $-10 + 15 = 5$

9. $-8.3 + (-1.9) = -10.2$

10. $-\dfrac{1}{4} + \dfrac{1}{2} = -\dfrac{1}{4} + \dfrac{1}{2} \cdot \dfrac{2}{2} = -\dfrac{1}{4} + \dfrac{2}{4} = \dfrac{1}{4}$

11. $-\dfrac{2}{3} + \dfrac{3}{7} = -\dfrac{2}{3} \cdot \dfrac{7}{7} + \dfrac{3}{7} \cdot \dfrac{3}{3} = -\dfrac{14}{21} + \dfrac{9}{21} = -\dfrac{5}{21}$

▨ **Work Practice Problems 6–11**

Subtraction of two real numbers may be defined in terms of addition.

Subtracting Real Numbers

If a and b are real numbers, then the difference of a and b, written $a - b$, is defined by

$$a - b = a + (-b)$$

In other words, to subtract a second real number from a first, we add the first number and the opposite of the second number.

Answers

6. -17, **7.** -4, **8.** 6, **9.** -6.5,

10. $-\dfrac{1}{2}$, **11.** $\dfrac{5}{14}$

EXAMPLES Subtract.

Add the opposite.

12. $2 - 8 = 2 + (-8) = -6$

Add the opposite.

13. $-8 - (-1) = -8 + (1) = -7$

14. $10.7 - (-9.8) = 10.7 + 9.8 = 20.5$

15. $-\dfrac{2}{3} - \dfrac{1}{4} = -\dfrac{2}{3} + \left(-\dfrac{1}{4}\right) = -\dfrac{2}{3} \cdot \dfrac{4}{4} + \left(-\dfrac{1}{4} \cdot \dfrac{3}{3}\right) = -\dfrac{8}{12} + \left(-\dfrac{3}{12}\right) = -\dfrac{11}{12}$

▣ **Work Practice Problems 12–15**

To add or subtract three or more real numbers, we add or subtract from left to right.

EXAMPLES Simplify each expression.

16. $11 + 2 - 7 = 13 - 7 = 13 + (-7) = 6$

17. $-5 - 4 + 2 = -5 + (-4) + 2 = -9 + 2 = -7$

▣ **Work Practice Problems 16–17**

Objective Ⓒ Multiplying and Dividing Real Numbers

To discover sign patterns when you multiply real numbers, recall that multiplication by a positive integer is the same as repeated addition. For example,

$$3(2) = 2 + 2 + 2 = 6$$
$$3(-2) = (-2) + (-2) + (-2) = -6$$

Notice here that $3(-2) = -6$. This illustrates that the product of two numbers with different signs is negative. We summarize sign patterns for multiplying any two real numbers as follows.

Multiplying Two Real Numbers

1. The product of two numbers with the *same* sign is positive.

2. The product of two numbers with *different* signs is negative.

Also recall that the product of zero and any real number a is zero.

Product Property of 0

$$0 \cdot a = 0 \qquad \text{Also,} \qquad a \cdot 0 = 0$$

EXAMPLES Multiply.

18. $-8(-1) = 8$

19. $2\left(-\dfrac{1}{6}\right) = \dfrac{2}{1} \cdot \left(-\dfrac{1}{6}\right) = -\dfrac{2}{6} = -\dfrac{1}{3}$

20. $-1.2(0.3) = -0.36$

21. $7(-6) = -42$

22. $-\dfrac{1}{3}\left(-\dfrac{1}{2}\right) = \dfrac{1}{6}$

23. $(-4.6)(-2.5) = 11.5$

24. $0(-6) = 0$

▣ **Work Practice Problems 18–24**

PRACTICE PROBLEMS 12–15

Subtract.

12. $7 - 14$

13. $-10 - (-2)$

14. $13.3 - (-8.9)$

15. $-\dfrac{1}{3} - \dfrac{1}{2}$

PRACTICE PROBLEMS 16–17

Simplify each expression.

16. $18 + 3 - 4$

17. $-3 - 11 + 7$

PRACTICE PROBLEMS 18–24

Multiply.

18. $-4(-2)$

19. $5\left(-\dfrac{1}{10}\right)$

20. $-3.2(0.1)$

21. $8(-6)$

22. $-\dfrac{2}{5}\left(-\dfrac{1}{3}\right)$

23. $(-1.3)(-1.5)$

24. $0(-10)$

Answers

12. -7, **13.** -8, **14.** 22.2, **15.** $-\dfrac{5}{6}$,

16. 17, **17.** -7, **18.** 8, **19.** $-\dfrac{1}{2}$,

20. -0.32, **21.** -48, **22.** $\dfrac{2}{15}$,

23. 1.95, **24.** 0

Recall that $\dfrac{8}{4} = 2$ because $2 \cdot 4 = 8$. Likewise, $\dfrac{8}{-4} = -2$ because $(-2)(-4) = 8$.

Also, $\dfrac{-8}{4} = -2$ because $(-2)4 = -8$, and $\dfrac{-8}{-4} = 2$ because $2(-4) = -8$. From these examples, we can see that the sign patterns for division are the same as for multiplication.

Dividing Two Real Numbers

1. The quotient of two numbers with the *same* sign is positive.

2. The quotient of two numbers with *different* signs is negative.

Notice from the previous reasoning that we cannot divide by 0. Why? If $\dfrac{5}{0}$ did exist, it would equal a number such that the number times 0 would equal 5. There is no such number, so we cannot define division by 0. We say, for example, that $\dfrac{5}{0}$ is **undefined.**

PRACTICE PROBLEMS 25–30

Divide.

25. $\dfrac{45}{-9}$

26. $\dfrac{-16}{-4}$

27. $\dfrac{25}{-5}$

28. $\dfrac{-3}{0}$

29. $\dfrac{0}{-3}$

30. $\dfrac{-1}{-4}$

EXAMPLES Divide.

25. $\dfrac{20}{-4} = -5$

26. $\dfrac{-9}{-3} = 3$

27. $\dfrac{-40}{10} = -4$

28. $\dfrac{-8}{0}$ is undefined.

29. $\dfrac{0}{-8} = 0$

30. $\dfrac{-10}{-80} = \dfrac{1}{8}$

◼ **Work Practice Problems 25–30**

With sign rules for division, we can understand why the positioning of the negative sign in a fraction does not change the value of the fraction. For example,

$$\dfrac{-12}{3} = -4, \qquad \dfrac{12}{-3} = -4, \qquad \text{and} \qquad -\dfrac{12}{3} = -4$$

Since all these fractions equal -4, we can say that

$$\dfrac{-12}{3} = \dfrac{12}{-3} = -\dfrac{12}{3}$$

In general, the following holds true:

If a and b are real numbers and $b \neq 0$, then

$$\dfrac{a}{-b} = \dfrac{-a}{b} = -\dfrac{a}{b}$$

Also recall that division by a nonzero real number b is the same as multiplication by its reciprocal $\dfrac{1}{b}$. In other words,

$$a \div b = a \cdot \dfrac{1}{b}$$

Answers

25. -5, **26.** 4, **27.** -5,
28. undefined, **29.** 0, **30.** $\dfrac{1}{4}$

EXAMPLES Divide.

31. $-\dfrac{1}{10} \div \left(-\dfrac{2}{5}\right) = -\dfrac{1}{10} \cdot \left(-\dfrac{5}{2}\right) = \dfrac{5}{20} = \dfrac{1}{4}$

32. $-\dfrac{1}{4} \div \dfrac{3}{7} = -\dfrac{1}{4} \cdot \dfrac{7}{3} = -\dfrac{7}{12}$

▣ **Work Practice Problems 31–32**

Objective D Simplifying Expressions Containing Exponents

Recall that when two numbers are multiplied, each number is called a **factor** of the product. For example, in $3 \cdot 5 = 15$, the 3 and 5 are factors.

A natural number *exponent* is a shorthand notation for repeated multiplication of the same factor. This repeated factor is called the **base,** and the number of times it is used as a factor is indicated by the **exponent.** For example,

$$\overset{\text{exponent}}{4^{3}} = 4 \cdot 4 \cdot 4 = 64$$

base — 4 is a factor 3 times.

Exponents

If a is a real number and n is a natural number, then the **nth power of a,** or **a raised to the nth power,** written as a^{n}, is the product of n factors, each of which is a.

$$\overset{\text{exponent}}{a^{n}} = \underbrace{a \cdot a \cdot a \cdot a \cdot \,\cdots\, \cdot a}_{a \text{ is a factor } n \text{ times.}}$$

base —

It is not necessary to write an exponent of 1. For example, 3 is assumed to be 3^{1}.

EXAMPLES Find the value of each expression.

33. $3^{2} = 3 \cdot 3 = 9$

34. $-5^{2} = -(5 \cdot 5) = -25$

35. $-5^{3} = -(5 \cdot 5 \cdot 5) = -125$

36. $\left(\dfrac{1}{2}\right)^{4} = \left(\dfrac{1}{2}\right)\left(\dfrac{1}{2}\right)\left(\dfrac{1}{2}\right)\left(\dfrac{1}{2}\right) = \dfrac{1}{16}$

37. $(-5)^{2} = (-5)(-5) = 25$

38. $(-5)^{3} = (-5)(-5)(-5) = -125$

▣ **Work Practice Problems 33–38**

✔**Concept Check** When $(-8.2)^{7}$ is evaluated, will the value be positive or negative? How can you tell without making any calculations?

Helpful Hint

Be very careful when finding the value of expressions such as -5^{2} and $(-5)^{2}$.

$$-5^{2} = -(5 \cdot 5) = -25 \qquad \text{and} \qquad (-5)^{2} = (-5)(-5) = 25$$

Without parentheses, the base to square is 5, not -5.

PRACTICE PROBLEMS 31–32

Divide.

31. $-\dfrac{3}{4} \div \left(-\dfrac{3}{8}\right)$

32. $-\dfrac{1}{11} \div \dfrac{2}{7}$

PRACTICE PROBLEMS 33–38

Simplify each expression.

33. 4^{2}

34. -3^{2}

35. -3^{3}

36. $\left(\dfrac{1}{3}\right)^{4}$

37. $(-3)^{2}$

38. $(-3)^{3}$

Answers

31. 2, **32.** $-\dfrac{7}{22}$, **33.** 16, **34.** -9,

35. -27, **36.** $\dfrac{1}{81}$, **37.** 9, **38.** -27

✔ **Concept Check Answer**

negative; the exponent is an odd number

Objective **E** Finding the Root of a Number

The opposite of squaring a number is taking the **square root** of a number. For example, since the square of 4, or 4^2, is 16, we say that a square root of 16 is 4. The notation $\sqrt{a}$ is used to denote the **positive,** or **principal square root** of a nonnegative number a. We then have in symbols that

$$\sqrt{16} = 4$$

PRACTICE PROBLEMS 39–41

Find each root.

39. $\sqrt{36}$

40. $\sqrt{4}$

41. $\sqrt{\dfrac{1}{49}}$

EXAMPLES Find each root.

39. $\sqrt{9} = 3$, since 3 is positive and $3^2 = 9$.

40. $\sqrt{25} = 5$, since $5^2 = 25$.

41. $\sqrt{\dfrac{1}{4}} = \dfrac{1}{2}$, since $\left(\dfrac{1}{2}\right)^2 = \dfrac{1}{4}$.

■ **Work Practice Problems 39–41**

We can find roots other than square roots. Since 2 cubed, written as 2^3, is 8, we say that the **cube root** of 8 is 2. This is written as

$$\sqrt[3]{8} = 2$$

Also, since $3^4 = 81$ and 3 is positive,

$$\sqrt[4]{81} = 3$$

PRACTICE PROBLEMS 42–44

Find each root.

42. $\sqrt[3]{64}$

43. $\sqrt[4]{1}$

44. $\sqrt[5]{243}$

EXAMPLES Find each root.

42. $\sqrt[3]{27} = 3$, since $3^3 = 27$.

43. $\sqrt[5]{1} = 1$, since $1^5 = 1$.

44. $\sqrt[4]{16} = 2$, since 2 is positive and $2^4 = 16$.

■ **Work Practice Problems 42–44**

Of course, as mentioned in Section 1.2, not all roots simplify to rational numbers. We study radicals further in Chapter 7.

Answers

39. 6, **40.** 2, **41.** $\dfrac{1}{7}$, **42.** 4, **43.** 1,

44. 3

1.4 EXERCISE SET

Student Solutions Manual   PH Math/Tutor Center   CD/Video for Review   Math XL MathXL®   MyMathLab MyMathLab

Objective **A** *Find each absolute value. See Examples 1 through 5.*

1. $|2|$

2. $|8|$

3. $|-4|$

4. $|-6|$

5. $|0|$

6. $|-1|$

7. $-|3|$

8. $-|11|$

9. $-\left|-\dfrac{2}{9}\right|$

10. $-\left|-\dfrac{5}{13}\right|$

Objective **B** *Add or subtract as indicated. See Examples 6 through 17.*

11. $-3 + 8$

12. $12 + (-7)$

 13. $-14 + (-10)$

14. $-5 + (-9)$

15. $-4.3 - 6.7$

16. $-8.2 - (-6.6)$

 17. $13 - 17$

18. $15 - (-1)$

19. $\dfrac{11}{15} - \left(-\dfrac{3}{5}\right)$

20. $\dfrac{7}{10} - \dfrac{4}{5}$

21. $19 - 10 - 11$

22. $-13 - 4 + 9$

23. $-14 - 7$

24. $-6 - 31$

25. $-\dfrac{4}{5} - \left(-\dfrac{3}{10}\right)$

26. $-\dfrac{5}{2} - \left(-\dfrac{2}{3}\right)$

27. Subtract 14 from 8.

28. Subtract 9 from -3.

Objective **C** *Multiply or divide as indicated. See Examples 18 through 32.*

 29. $-5 \cdot 12$

30. $-3 \cdot 8$

31. $-8(-10)$

32. $-4(-11)$

33. $-17 \cdot 0$

34. $-5 \cdot 0$

35. $0(-1)$

36. $0(-34)$

37. $\dfrac{-9}{3}$

38. $\dfrac{-20}{5}$

39. $\dfrac{-12}{-4}$

40. $\dfrac{-36}{-6}$

41. $3\left(-\dfrac{1}{18}\right)$

42. $5\left(-\dfrac{1}{50}\right)$

43. $(-0.7)(-0.8)$

44. $(-0.9)(-0.5)$

45. $9.1 \div (-1.3)$

46. $22.5 \div (-2.5)$

47. $-4(-2)(-1)$

48. $-5(-3)(-2)$

Objective **D** *Find the value of each expression. See Examples 33 through 38.*

49. -7^2

50. $(-7)^2$

51. $(-6)^2$

52. -6^2

53. $(-2)^3$

54. -2^3

55. $\left(-\dfrac{1}{3}\right)^3$

56. $\left(-\dfrac{1}{2}\right)^4$

Objective **E** *Find each root. See Examples 39 through 44.*

57. $\sqrt{49}$

58. $\sqrt{81}$

59. $\sqrt{64}$

60. $\sqrt{100}$

61. $\sqrt{\dfrac{1}{9}}$

62. $\sqrt{\dfrac{1}{25}}$

63. $\sqrt[3]{64}$

64. $\sqrt[5]{32}$

65. $\sqrt{\dfrac{4}{25}}$

66. $\sqrt{\dfrac{4}{81}}$

Objectives **A** **B** **C** **D** **E** **Mixed Practice** *Perform the indicated operations. See Examples 1 through 44.*

67. $-4 + 7$

68. $-9 + 15$

69. $-9 + (-3)$

70. $-17 + (-2)$

71. $6(-3)$

72. $5(-4)$

73. $-9 \cdot 8$

74. $-6 \cdot 7$

75. $(-11)^2$

76. -11^2

77. $\dfrac{16}{-2}$

78. $\dfrac{35}{-7}$

79. $-18 \div 6$

80. $-42 \div 6$

81. $-\dfrac{2}{7} \cdot \left(-\dfrac{1}{6}\right)$

82. $\dfrac{5}{9} \cdot \left(-\dfrac{3}{5}\right)$

83. $-2(-3.6)$

84. $-5(-4.2)$

85. $-4 - (-19)$

86. $-5 - (-17)$

87. $6.3 - 18.5$

88. $15.9 - 21.7$

89. $\dfrac{0}{-5}$

90. $\dfrac{0}{-11}$

91. $\dfrac{-18}{0}$

92. $\dfrac{-22}{0}$

93. $\sqrt{\dfrac{1}{16}}$

94. $\sqrt{\dfrac{1}{49}}$

95. $\dfrac{-6}{7} \div 2$

96. $\dfrac{-9}{13} \div (-3)$

97. $\dfrac{-5.2}{-1.3}$

98. $\dfrac{-4.2}{-1.4}$

99. $-25 \div (-5)$

100. $-88 \div (-11)$

101. $\sqrt[4]{81}$

102. $\sqrt[3]{1}$

103. $\sqrt[3]{8}$

104. $\sqrt[3]{125}$

105. $-\dfrac{1}{6} \div \dfrac{9}{10}$

106. $\dfrac{4}{7} \div \left(-\dfrac{1}{8}\right)$

107. $-\dfrac{2}{3} \cdot \left(\dfrac{6}{4}\right)$

108. $\dfrac{5}{6} \cdot \left(\dfrac{-12}{15}\right)$

109. $\dfrac{3}{5} \div \left(-\dfrac{2}{5}\right)$

110. $\dfrac{2}{7} \div \left(-\dfrac{1}{14}\right)$

111. $16 - 8 - 9$

112. $-14 - 3 + 6$

113. $-5 + (-7) - 10$

114. $-8 + (-10) - 6$

115. $-7(-1)(5)$

116. $-6(2)(-3)$

117. $-6(-5)(0)$

118. $4(-3)(0)$

119. -4^4

120. $(-4)^4$

Concept Extensions

121. Explain why -3^2 and $(-3)^2$ simplify to different numbers.

122. Explain why -3^3 and $(-3)^3$ simplify to the same number.

Answer each statement true or false.

123. $-7 - 1 < -7(-1)$

124. $\dfrac{-100}{-5} > -100 + (-5)$

Each circle below represents a whole, or 1. Determine the unknown fractional part of each circle.

125.

126.

127. Most of Mauna Kea, a volcano on Hawaii, lies below sea level. If this volcano begins at 5998 meters below sea level and then rises 10,203 meters, find the height of the volcano above sea level.

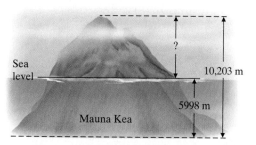

128. The highest point on land on Earth is the top of Mt. Everest, in the Himalayas, at an elevation of 29,028 feet above sea level. The lowest point on land is the Dead Sea, between Israel and Jordan, at 1312 feet below sea level. Find the difference in elevations. (*Source:* National Geographic Society)

A fair game is one in which each team or player has the same chance of winning. Suppose that a game consists of three players taking turns spinning a spinner. If the spinner lands on yellow, player 1 gets a point. If the spinner lands on red, player 2 gets a point, and if the spinner lands on blue, player 3 gets a point. After 12 spins, the player with the most points wins. Use this information to answer Exercises 129 through 133.

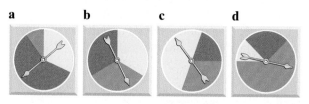

129. Which spinner would lead to a fair game?

130. If you are player 2 and want to win the game, which spinner would you choose?

131. If you are player 1 and want to lose the game, which spinner would you choose?

132. Is it possible for the game to end in a three-way tie? If so, list the possible ending scores.

133. Is it possible for the game to end in a two-way tie? If so, list the possible ending scores.

Use a calculator to approximate each square root. Round to four decimal places.

134. $\sqrt{10}$

135. $\sqrt{273}$

136. $\sqrt{7.9}$

137. $\sqrt{19.6}$

Investment firms often advertise their gains and losses in the form of bar graphs such as the one that follows. This graph shows investment risk over time for common stocks by showing average annual compound returns for 1 year, 5 years, 10 years, 15 years, and 20 years. For example, after 1 year, the annual compound return in percent for an investor is anywhere from a gain of 56% to a loss of 64%. Use this graph to answer Exercises 138 through 142.

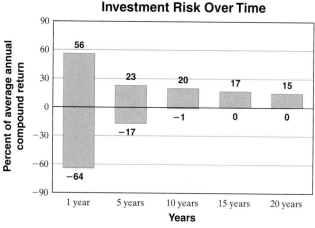

Source: Yahoo finance; mutual funds

138. A person investing in common stocks may expect at most an average annual gain of what percent after 15 years?

139. A person investing in common stocks may expect to lose at most an average per year of what percent after 5 years?

140. Find the difference in percent of the highest average annual return and the lowest average annual return after 20 years.

141. Find the difference in percent of the highest average annual return and the lowest average annual return after 5 years.

142. Do you think that the type of investment shown in the figure is recommended for short-term investments or long-term investments? Explain your answer.

Real Numbers

Write each set by listing its elements.

1. $\{x \,|\, x$ is a natural number less than 4$\}$

2. $\{x \,|\, x$ is an odd whole number less than 6$\}$

3. $\{x \,|\, x$ is an even natural number greater than 7$\}$

4. $\{x \,|\, x$ is a whole number between 10 and 15$\}$

Write each phrase as an algebraic expression. Let x represent the unknown number.

5. Twice the difference of a number and three

6. The quotient of six and the sum of a number and ten

Insert $<$, $>$, or $=$ between each pair of numbers to form a true statement.

7. $-4 \qquad -6$

8. $8.6 \qquad 8.600$

9. $\dfrac{9}{10} \qquad \dfrac{11}{10}$

10. $-6.1 \qquad -6.01$

Write each sentence as an equation.

11. The product of 5 and x is the same as 20.

12. The sum of a and 12 amounts to 14.

13. The quotient of y and 10 is the same as the product of y and 10.

14. The sum of x and 1 equals the difference of x and 1.

Perform each indicated operation.

15. $-4 + 7$

16. $-11 + 20$

17. $-4(7)$

18. $-11(20)$

19. $-8 - (-13)$

20. $-12 - 16$

21. $\dfrac{-20}{-4}$

22. $\dfrac{-18}{6}$

23. -5^2

24. $(-5)^2$

25. $-6 - 1 + 20$

26. $18 - 4 - 19$

27. $\dfrac{0}{-3}$

28. $\dfrac{5}{0}$

29. $-4(3)(2)$

30. $-5(-1)(6)$

31. $-\dfrac{1}{2} \cdot \dfrac{6}{7}$ **32.** $\dfrac{4}{5} \cdot \left(-\dfrac{1}{8}\right)$ **33.** $\dfrac{3}{10} - \dfrac{4}{5}$ **34.** $-\dfrac{2}{3} - \dfrac{1}{4}$

35. $\dfrac{1.6}{-0.2}$ **36.** $\dfrac{-4.8}{16}$ **37.** $6.7 - (-1.3)$ **38.** $-4.6 + 9$

39. $\dfrac{1}{2} + \left(-\dfrac{7}{8}\right)$ **40.** $\dfrac{1}{2} \div \left(-\dfrac{7}{8}\right)$ **41.** $\sqrt{49}$ **42.** $\sqrt[3]{27}$

43. -2^2 **44.** $(-2)^3$ **45.** $\sqrt{\dfrac{1}{81}}$ **46.** $\sqrt{\dfrac{1}{100}}$

Fill in the chart.

	Number	Opposite	Reciprocal
47.	-6		
48.	4		
49.			$\dfrac{7}{5}$
50.		$\dfrac{7}{30}$	

Use the distribution property to multiply.

51. $9(m + 5)$ **52.** $11(7 + r)$

53. $3(2y - 3x)$ **54.** $8(4m - 7n)$

55. $0.2(3a + 7)$ **56.** $0.6(2n + 5)$

57. $\dfrac{1}{5}(10x - 19y + 20)$ **58.** $\dfrac{1}{2}(10x - 19y + 20)$

31. _____

32. _____

33. _____

34. _____

35. _____

36. _____

37. _____

38. _____

39. _____

40. _____

41. _____

42. _____

43. _____

44. _____

45. _____

46. _____

51. _____

52. _____

53. _____

54. _____

55. _____

56. _____

57. _____

58. _____

1.5 ORDER OF OPERATIONS AND ALGEBRAIC EXPRESSIONS

In this section, we review order of operations and algebraic expressions.

Objective **A** Using the Order of Operations

The expression $3 + 2 \cdot 30$ represents the total number of compact disks (CDs) shown.

Expressions containing more than one operation are written to follow a particular agreed-upon **order of operations.** For example, when we write $3 + 2 \cdot 30$, we mean to multiply first, and then add.

Order of Operations

Simplify expressions using the following order.

1. If grouping symbols such as parentheses are present, simplify expressions within those first, starting with the innermost set.
2. Evaluate exponential expressions, roots, and absolute values.
3. Perform multiplications or divisions in order from left to right.
4. Perform additions or subtractions in order from left to right.

Helpful Hint

Fraction bars, radical signs, and absolute value bars can sometimes be used as grouping symbols. For example,

	Fraction Bar	Radical Sign	Absolute Value Bars
Grouping Symbol	$\dfrac{-1-7}{6-11}$	$\sqrt{15+1}$	$\lvert -7.2 - \sqrt{4} \rvert$
Not Grouping Symbol	$-\dfrac{8}{9}$	$\sqrt{9}$	$\lvert -3.2 \rvert$

PRACTICE PROBLEM 1

Simplify: $15 - 2 \cdot 5$

EXAMPLE 1 Simplify: $3 + 2 \cdot 30$

Solution: First we multiply; then we add.

$$3 + 2 \cdot 30 = 3 + 60 = 63$$

■ **Work Practice Problem 1**

PRACTICE PROBLEM 2

Simplify: $2 + 5(2 - 6)^2$

EXAMPLE 2 Simplify: $1 + 2(1 - 4)^2$

Solution: Remember order of operations so that you are *not* tempted to add 1 and 2 first. Unless there are grouping symbols, addition is last in order of operations.

$$
\begin{aligned}
1 + 2(1 - 4)^2 &= 1 + 2(-3)^2 && \text{Simplify inside parentheses first.}\\
&= 1 + 2(9) && \text{Write } (-3)^2 \text{ as 9.}\\
&= 1 + 18 && \text{Multiply.}\\
&= 19 && \text{Add.}
\end{aligned}
$$

■ **Work Practice Problem 2**

EXAMPLE 3 Simplify: $\dfrac{|-2|^3 + 1}{-7 - \sqrt{4}}$

Solution: Here, the fraction bar serves as a grouping symbol. We simplify the numerator and the denominator separately. Then we divide.

$$\frac{|-2|^3 + 1}{-7 - \sqrt{4}} = \frac{2^3 + 1}{-7 - 2} \quad \text{Write } |-2| \text{ as 2 and } \sqrt{4} \text{ as 2.}$$

$$= \frac{8 + 1}{-9} \quad \text{Write } 2^3 \text{ as 8.}$$

$$= \frac{9}{-9} = -1 \quad \text{Simplify the numerator; then divide.}$$

Work Practice Problem 3

Besides parentheses, other symbols used for grouping expressions are brackets [ ], braces { }, radical signs, and absolute value bars. Brackets [ ] and braces { } are commonly used when we group expressions that already contain parentheses.

EXAMPLE 4 Simplify: $3 - [6(4 - 6) + 2(5 - 9)]$

Solution:

$$3 - [6(4 - 6) + 2(5 - 9)] = 3 - [6(-2) + 2(-4)] \quad \begin{array}{l}\text{Simplify within the}\\\text{innermost sets of}\\\text{parentheses.}\end{array}$$

$$= 3 - [-12 + (-8)]$$

$$= 3 - [-20]$$

$$= 23$$

Work Practice Problem 4

EXAMPLE 5 Simplify: $\dfrac{-5\sqrt{30 - 5} + (-2)^2}{4^2 + |7 - 10|}$

Solution: Here, the fraction bar, radical sign, and absolute value bars serve as grouping symbols. Thus, we simplify within the radical sign and absolute value bars first, remembering to calculate above and below the fraction bar separately.

$$\frac{-5\sqrt{30 - 5} + (-2)^2}{4^2 + |7 - 10|} = \frac{-5\sqrt{25} + (-2)^2}{4^2 + |-3|} = \frac{-5 \cdot 5 + 4}{16 + 3} = \frac{-25 + 4}{16 + 3}$$

$$= \frac{-21}{19} \text{ or } -\frac{21}{19}$$

Work Practice Problem 5

✔**Concept Check** True or false? If two different people use the order of operations to simplify a numerical expression and neither makes a calculation error, it is not possible that they each obtain a different result. Explain.

Objective B Evaluating Algebraic Expressions

Recall from Section 1.2 that an algebraic expression is formed by numbers and variables connected by the operations of addition, subtraction, multiplication, division, raising to powers, or taking roots. Also, if numbers are substituted for the variables in an algebraic expression and the operations performed, the result is called the **value** of the expression for the given replacement values. This entire process is called **evaluating an expression.**

PRACTICE PROBLEM 3

Simplify: $\dfrac{|-3|^2 + 5}{\sqrt{9} - 10}$

PRACTICE PROBLEM 4

Simplify:

$7 - [2(1 - 3) + 5(10 - 12)]$

Helpful Hint

When grouping symbols occur within grouping symbols, remember to perform operations on the innermost set first.

PRACTICE PROBLEM 5

Simplify: $\dfrac{|9 - 16| + 5^2}{-3\sqrt{8 + 1} + (-4)^2}$

Answers

3. -2, **4.** 21, **5.** $\dfrac{32}{7}$

✔ **Concept Check Answer**

true; answers may vary

PRACTICE PROBLEM 6

Evaluate each expression when $x = 5$ and $y = -4$.

a. $2x - 6y$

b. $-3y^2$

c. $\dfrac{\sqrt{x}}{y} + \dfrac{y}{x}$

EXAMPLE 6 Evaluate each expression when $x = 4$ and $y = -3$.

a. $3x - 7y$ **b.** $-2y^2$ **c.** $\dfrac{\sqrt{x}}{y} - \dfrac{y}{x}$

Solution: For each expression, replace x with 4 and y with -3.

a. $3x - 7y = 3 \cdot 4 - 7(-3)$ Let $x = 4$ and $y = -3$.

      $= 12 - (-21)$ Multiply.

      $= 12 + 21$ Write as an addition.

      $= 33$ Add.

b. $-2y^2 = \underbrace{-2(-3)^2}$ Let $y = -3$.

> **Helpful Hint**
>
> In $-2(-3)^2$, the exponent 2 goes with the base of -3 only.

      $= -2(9)$ Write $(-3)^2$ as 9.

      $= -18$ Multiply.

c. $\dfrac{\sqrt{x}}{y} - \dfrac{y}{x} = \dfrac{\sqrt{4}}{-3} - \dfrac{-3}{4}$

      $= -\dfrac{2}{3} + \dfrac{3}{4}$ Write $\sqrt{4}$ as 2.

      $= -\dfrac{2}{3} \cdot \dfrac{4}{4} + \dfrac{3}{4} \cdot \dfrac{3}{3}$ The LCD is 12.

      $= -\dfrac{8}{12} + \dfrac{9}{12}$ Write each fraction with a denominator of 12.

      $= \dfrac{1}{12}$ Add.

Work Practice Problem 6

PRACTICE PROBLEM 7

Use the algebraic expression given in Example 7 to complete the following table.

Degrees Fahrenheit	x	-13	0	41
Degrees Celsius	$\dfrac{5(x-32)}{9}$			

EXAMPLE 7 Converting Degrees Fahrenheit to Degrees Celsius

The algebraic expression $\dfrac{5(x - 32)}{9}$ represents the equivalent temperature in degrees Celsius when x is degrees Fahrenheit. Complete the following table by evaluating this expression at the given values of x.

Degrees Fahrenheit	x	-4	10	32
Degrees Celsius	$\dfrac{5(x-32)}{9}$			

Solution: To complete the table, we evaluate $\dfrac{5(x - 32)}{9}$ at each given replacement value.

When $x = -4$,

$$\frac{5(x - 32)}{9} = \frac{5(-4 - 32)}{9} = \frac{5(-36)}{9} = -20$$

When $x = 10$,

$$\frac{5(x - 32)}{9} = \frac{5(10 - 32)}{9} = \frac{5(-22)}{9} = \frac{-110}{9} \text{ or } -12\frac{2}{9}$$

When $x = 32$,

$$\frac{5(x - 32)}{9} = \frac{5(32 - 32)}{9} = \frac{5 \cdot 0}{9} = 0$$

Answers

6. a. 34,   **b.** -48 **c.** $-\dfrac{5\sqrt{5} + 16}{20}$,

7. $-25; -\dfrac{160}{9}$ or $-17\dfrac{7}{9}; 5$

The completed table is as follows:

Degrees Fahrenheit	x	-4	10	32
Degrees Celsius	$\dfrac{5(x-32)}{9}$	-20	$\dfrac{-110}{9}$ or $-12\dfrac{2}{9}$	0

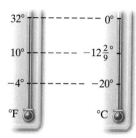

Thus, $-4°$F is equivalent to $-20°$C, $10°$F is equivalent to $-\dfrac{110°}{9}$ C or $-12\dfrac{2°}{9}$ C, and $32°$F is equivalent to $0°$C.

■ Work Practice Problem 7

Objective C Simplifying Algebraic Expressions by Combining Like Terms

Often, an algebraic expression may be **simplified** by removing grouping symbols and combining any like terms. The **terms** of an expression are the addends of the expression. For example, in the expression $3x^2 + 4x$, the terms are $3x^2$ and $4x$.

Expression	**Terms**
$-2x + y$	$-2x, \quad y$
$3x^2 - \dfrac{y}{5} + 7$	$3x^2, \quad -\dfrac{y}{5}, \quad 7$

Helpful Hint

The expression $3x^2 - \dfrac{y}{5} + 7$ can be written as $3x^2 + \left(-\dfrac{y}{5}\right) + 7$ so that the addends (terms) are

$$3x^2 \qquad -\dfrac{y}{5} \qquad 7.$$

Terms with the same variable(s) raised to the same power(s) are called **like terms.** We can add or subtract like terms by using the distributive property. This process is called **combining like terms.**

EXAMPLES Simplify by combining like terms.

8. $3x - 5x + 4 = (3 - 5)x + 4$ Use the distributive property.
$$= -2x + 4$$

9. $y + 3y = 1y + 3y = (1 + 3)y$
$$= 4y$$

10. $7x + 9x - 6 - 10 = (7 + 9)x + (-6 - 10)$
$$= 16x - 16$$

■ Work Practice Problems 8–10

PRACTICE PROBLEMS 8–10

Simplify by combining like terms.

8. $9x - 15x + 7$

9. $8y + y$

10. $4x + 12x - 9 - 10$

Answers

8. $-6x + 7$, **9.** $9y$, **10.** $16x - 19$

The associative and commutative properties may sometimes be needed to re-arrange and group like terms when we simplify expressions.

PRACTICE PROBLEMS 11–12

Simplify.

11. $-4x + 7 - 5x - 8$

12. $5y - 6y + 2 - 11 + y$

EXAMPLES Simplify.

11. $-7x + 5 + 3x - 2 = -7x + 3x + 5 - 2$ Use the commutative property.
$= (-7 + 3)x + (5 - 2)$ Use the distributive property.
$= -4x + 3$ Simplify.

12. $3y - 2y - 5 - 7 + y = 3y - 2y + y - 5 - 7$ Use the commutative property.
$= (3 - 2 + 1)y + (-5 - 7)$ Use the distributive property.
$= 2y - 12$ Simplify.

◻ **Work Practice Problems 11–12**

PRACTICE PROBLEMS 13–16

Simplify by using the distributive property to multiply and then combining like terms.

13. $-3(y + 1) + 4$

14. $8x + 2 - 4(x - 9)$

15. $(3.2x - 4.1) - (-x + 7.6)$

16. $\frac{1}{5}(15m - 40n)$

$-\frac{1}{4}(8m - 4n + 1) + \frac{1}{5}$

EXAMPLES Simplify by using the distributive property to multiply and then combining like terms.

13. $-2(x + 3) + 7 = -2(x) + (-2)(3) + 7 = -2x - 6 + 7 = -2x + 1$

14. $7x + 3 - 5(x - 4) = 7x + 3 - 5x + 20$ Use the distributive property.
$= 2x + 23$ Combine like terms.

15. $(2.1x - 5.6) - (-x - 5.3) = (2.1x - 5.6) - 1(-x - 5.3)$
$= 2.1x - 5.6 + 1x + 5.3$ Use the distributive property.
$= 3.1x - 0.3$ Combine like terms.

16. $\frac{1}{2}(4a - 6b) - \frac{1}{3}(9a + 12b - 1) + \frac{1}{4}$

$= 2a - 3b - 3a - 4b + \frac{1}{3} + \frac{1}{4}$ Use the distributive property.

$= -a - 7b + \frac{7}{12}$ Combine like terms.

◻ **Work Practice Problems 13–16**

✔ **Concept Check** Find and correct the error in the following:

$x - 4(x - 5) = x - 4x - 20$
$= -3x - 20$

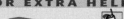
Objective Ⓐ *Simplify each expression. See Examples 1 through 5.*

1. $3(5 - 7)^4$

2. $7(3 - 8)^2$

3. $-3^2 + 2^3$

4. $-5^2 - 2^4$

5. $\dfrac{3.1 - (-1.4)}{-0.5}$

6. $\dfrac{4.2 - (-8.2)}{-0.4}$

7. $|3.6 - 7.2| + |3.6 + 7.2|$

8. $|8.6 - 1.9| - |2.1 + 5.3|$

9. $(-3)^2 + 2^3$

10. $(-15)^2 - 2^4$

11. $-8 \div 4 \cdot 2$

12. $-20 \div 5 \cdot 4$

13. $4[8 - (2 - 4)]$

14. $3[11 - (1 - 3)]$

15. $-8\left(-\dfrac{3}{4}\right) - 8$

16. $-10\left(-\dfrac{2}{5}\right) - 10$

17. $2 - [(7 - 6) + (9 - 19)]$

18. $8 - [(4 - 7) + (8 - 1)]$

19. $\dfrac{(-9 + 6)(-1^2)}{-2 - 2}$

20. $\dfrac{(-1 - 2)(-3^2)}{-6 - 3}$

21. $(\sqrt[3]{8})(-4) - (\sqrt{9})(-5)$

22. $(\sqrt[3]{27})(-5) - (\sqrt{25})(-3)$

23. $12 + \{6 - [5 - 2(-5)]\}$

24. $18 + \{9 - [1 - 6(-3)]\}$

25. $25 - [(3 - 5) + (14 - 18)]^2$

26. $10 - [(4 - 5)^2 + (12 - 14)]^4$

27. $\dfrac{(3 - \sqrt{9}) - (-5 - 1.3)}{-3}$

28. $\dfrac{-\sqrt{16} - (6 - 2.4)}{-2}$

29. $\dfrac{|3 - 9| - |-5|}{-3}$

30. $\dfrac{|-14| - |2 - 7|}{-15}$

31. $\dfrac{3(-2 + 1)}{5} - \dfrac{-7(2 - 4)}{1 - (-2)}$

32. $\dfrac{-1 - 2}{2(-3) + 10} - \dfrac{2(-5)}{-1(8) + 1}$

33. $\dfrac{\dfrac{1}{3} \cdot 9 - 7}{3 + \dfrac{1}{2} \cdot 4}$

34. $\dfrac{\dfrac{1}{5} \cdot 20 - 6}{10 + \dfrac{1}{4} \cdot 12}$

35. $3\{-2 + 5[1 - 2(-2 + 5)]\}$

36. $2\{-1 + 3[7 - 4(-10 + 12)]\}$

37. $\dfrac{-4\sqrt{80 + 1} + (-4)^2}{3^3 + |-2(3)|}$

38. $\dfrac{(-2)^4 + 3\sqrt{120 - 20}}{4^3 + |5(-1)|}$

39. $-150(3.25 - 1.68)$

40. $-290(9.61 - 6.27)$

41. $\left(\dfrac{5.6 - 8.4}{1.9 - 2.7}\right)^2$

42. $\left(\dfrac{9.4 - 10.8}{8.7 - 7.9}\right)^2$

Objective **B** *Evaluate each expression when x = 9 and y = −2. See Example 6.*

43. $9x - 6y$

44. $4x - 10y$

45. $-3y^2$

46. $-7y^2$

47. $\dfrac{\sqrt{x}}{y} - \dfrac{y}{x}$

48. $\dfrac{y}{2x} - \dfrac{\sqrt{x}}{3y}$

49. $\dfrac{3 + 2|x - y|}{x + 2y}$

50. $\dfrac{5 + 2|y - x|}{x + 6y}$

(*Hint:* Remember order of operations.)

51. $\dfrac{y^3 + \sqrt{x - 5}}{|4x - y|}$

52. $\dfrac{y^2 + \sqrt{x + 7}}{|3x - y|}$

Complete each table. See Example 7.

53. The algebraic expression $8 + 2y$ represents the perimeter of a rectangle with width 4 and length y.

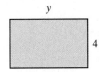

y

4

a. Complete the table by evaluating this expression at the given values of y.

Length	y	5	7	10	100
Perimeter	$8 + 2y$				

b. Use the results of the table in (a) to answer the following question. As the width of a rectangle remains the same and the length increases, does the perimeter increase or decrease? Explain how you arrived at your answer.

54. The algebraic expression πr^2 represents the area of a circle with radius r.

r

a. Complete the table by evaluating this expression at the given values of r. (Use 3.14 for π.)

Radius	r	2	3	7	10
Area	πr^2				

b. As the radius of a circle increases, does its area increase or decrease? Explain your answer.

55. The algebraic expression $\dfrac{100x + 5000}{x}$ represents the cost per bookshelf (in dollars) of producing x bookshelves.

a. Complete the table.

Number of Bookshelves	x	10	100	1000
Cost per Bookshelf	$\dfrac{100x + 5000}{x}$			

b. As the number of bookshelves manufactured increases, does the cost per bookshelf increase or decrease? Why do you think that this is so?

56. If C is degrees Celsius, the algebraic expression $1.8C + 32$ represents the equivalent temperature in degrees Fahrenheit.

a. Complete the table.

Degrees Celsius	C	−10	0	50
Degrees Fahrenheit	$1.8C + 32$			

b. As degrees Celsius increase, do degrees Fahrenheit increase or decrease? Explain your answer.

57. The average price for an ounce of gold in the United States during November 2004 was $439.39. The algebraic expression $439.39z$ gives the average cost of z ounces of gold during this period. Find the average cost if 8.4 ounces of gold had been purchased during this time. (*Source:* www.kitco.com)

58. On December 28, 2004, the velocity of the solar wind in Earth's upper atmosphere was 431.2 kilometers per second. At this speed, the algebraic expression $431.2t$ gives the total distance covered in t seconds. Find the distance covered by a proton traveling in the solar wind in 5 seconds. (*Source:* NOAA Space Environment Center)

Objective **C** *Simplify. See Examples 8 through 12.*

59. $6x + 2x$ **60.** $9y - 11y$ **61.** $19y - y$ **62.** $14x - x$

63. $9x - 8 - 10x$ **64.** $14x - 1 - 20x$ **65.** $-9 + 4x + 18 - 10x$ **66.** $5y - 14 + 7y - 20y$

67. $3a - 4b + a - 9b$ **68.** $11x - y + 11x - 6y$ **69.** $x - y + x - y$ **70.** $a - b + 3a - 3b$

71. $1.5x + 2.3 - 0.7x - 5.9$ **72.** $6.3y - 9.7 + 2.2y - 11.1$ **73.** $\frac{3}{4}b - \frac{1}{2} + \frac{1}{6}b - \frac{2}{3}$

74. $\frac{7}{8}a - \frac{11}{12} - \frac{1}{2}a + \frac{5}{6}$ **75.** $8ab - 4.6 - 11ab - 8.2$ **76.** $4mn - 6.01 - 6mn - 8.1$

Simplify. See Examples 13 through 16.

77. $2(3x + 7)$ **78.** $4(5y + 12)$ **79.** $5k - (3k - 10)$ **80.** $-11c - (4 - 2c)$

81. $(3x + 4) - (6x - 1)$ **82.** $(8 - 5y) - (4 - 3y)$ **83.** $3(x - 2) + x + 15$ **84.** $4(y + 3) - 7y + 1$

85. $-(n + 5) + (5n - 3)$ **86.** $-(8 - t) + (2t - 6)$ **87.** $4(6n - 3) - 3(8n + 4)$ **88.** $5(2z - 6) + 10(3 - z)$

89. $\frac{1}{4}(8x - 4) - \frac{1}{5}(20x - 6y)$ **90.** $\frac{1}{2}(10x - 2) - \frac{1}{6}(60x - 5y)$

91. $3x - 2(x - 5) + x$ **92.** $7n + 3(2n - 6) - n$

93. $\frac{1}{6}(24a - 18b) - \frac{1}{7}(7a - 21b - 2) - \frac{1}{5}$ **94.** $\frac{1}{3}(6x - 33y) - \frac{1}{8}(24x - 40y + 1) - \frac{1}{3}$

95. $5.7a + 1.7 - 3(2.1a - 0.6)$ **96.** $6.2b + 5.1 - 2(4.2b - 0.1)$

97. $-4[6(2t + 1) - (9 + 10t)]$ **98.** $-5[3(4x + 2) - (13 + 8x)]$ **99.** $-1.2(5.7x - 3.6) + 8.75x$

100. $5.8(-9.6 - 31.2y) - 18.65$ **101.** $8.1z + 7.3(z + 5.2) - 6.85$ **102.** $6.5y - 4.4(1.8y - 3.3) + 10.95$

Concept Extensions

Insert parentheses so that when simplified each expression is equal to the given number.

103. $2 + 7 \cdot 1 + 3$; 36 **104.** $6 - 5 \cdot 2 + 2$; -6

105. Write an algebraic expression that simplifies to $-3x - 1$.

106. Write an algebraic expression that simplifies to $-7x - 4$.

Find and correct the error in each expression.

107. $(3x + 2) - (5x + 7) = 3x + 2 - 5x + 7$
$$= -2x + 9$$

108. $2 + 3(7 - 6x) = 5(7 - 6x)$
$$= 35 - 30x$$

The following graph is called a broken-line graph, or simply a line graph. This particular graph shows the past, present, and future predicted U.S. population over 65. Just as with a bar graph, to find the population over 65 for a particular year, read the height of the corresponding point. To read the height, follow the point horizontally to the left until you reach the vertical axis. Use this graph to answer Exercises 109 through 114.

U.S. Population Over 65

Source: U.S. Census Bureau   * Projected

109. Estimate the population over 65 in the year 1970.

110. Estimate the predicted population over 65 in the year 2050.

111. Estimate the predicted population over 65 in the year 2030.

112. Estimate the population over 65 in the year 2000.

113. Is the population over 65 increasing as time passes or decreasing? Explain how you arrived at your answer.

114. The percent of Americans over 65 in 1950 was 8.1%. The percent of Americans over 65 in 2050 is expected to be 2.5 times the percent over 65 in 1950. Estimate the percent of Americans expected to be over age 65 in 2050.

Simplify. Round each result to the nearest ten-thousandth.

115. $\dfrac{-1.682 - 17.895}{(-7.102)(-4.691)}$

116. $\dfrac{(-5.161)(3.222)}{7.955 - 19.676}$

1.6 EXPONENTS AND SCIENTIFIC NOTATION

Objectives

A Use the Product Rule for Exponents.

B Simplify Expressions Raised to the Zero Power.

C Use the Quotient Rule for Exponents.

D Simplify Expressions Raised to Negative Powers.

E Simplify Exponential Expressions Containing Variables in the Exponent.

F Convert between Scientific Notation and Standard Notation.

Objective **A** Using the Product Rule

Recall from Section 1.4 that exponents may be used to write repeated factors in a more compact form. As we have seen in the previous sections, exponents can be used when the repeated factor is a number or a variable. For example,

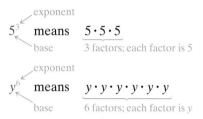

Expressions that contain exponents such as 5^3 and y^6 are called **exponential expressions.**

Exponential expressions can be multiplied, divided, added, subtracted, and themselves raised to powers. In this section, we review operations on exponential expressions.

We review multiplication first. To multiply x^2 by x^3, we use the definition of a^n:

$$x^2 \cdot x^3 = \underbrace{(x \cdot x)(x \cdot x \cdot x)}_{x \text{ is a factor 5 times.}}$$
$$= x^5$$

Notice that the result is exactly the same if we add the exponents.

$$x^2 \cdot x^3 = x^{2+3} = x^5$$

This suggests the following rule.

Product Rule for Exponents

If m and n are integers and a is a real number, then

$$a^m \cdot a^n = a^{m+n}$$

In other words, the *product* of exponential expressions with a common base is the common base raised to a power equal to the *sum* of the exponents of the factors.

EXAMPLES Use the product rule to simplify.

1. $2^2 \cdot 2^5 = 2^{2+5} = 2^7$

2. $x^7 \cdot x^3 = x^{7+3} = x^{10}$

3. $y \cdot y^2 \cdot y^4 = (y^1 \cdot y^2) \cdot y^4$
$= y^3 \cdot y^4$
$= y^7$

■ **Work Practice Problems 1–3**

EXAMPLES Use the product rule to simplify.

4. $(3x^6)(5x) = 3(5)x^6x^1 = 15x^7$ Use properties of multiplication to group like bases.

5. $(-2.4x^3p^2)(4xp^{10}) = -2.4(4)x^3x^1p^2p^{10} = -9.6x^4p^{12}$

■ **Work Practice Problems 4–5**

PRACTICE PROBLEMS 1–3

Use the product rule to simplify.
1. $5^2 \cdot 5^6$ **2.** $x^5 \cdot x^9$ **3.** $y \cdot y^4 \cdot y^3$

PRACTICE PROBLEMS 4–5

Use the product rule to simplify.
4. $(7y^5)(6y)$
5. $(-3x^2y^7)(5xy^6)$

Answers
1. 5^8,   **2.** x^{14},   **3.** y^8,   **4.** $42y^6$,
5. $-15x^3y^{13}$

Objective B Simplifying Expressions Raised to the Zero Power

The definition of a^n does not include the possibility that n might be 0. But if it did, then, by the product rule,

$$a^0 \cdot a^n = a^{0+n} = a^n = 1 \cdot a^n$$

From this, we reasonably define that $a^0 = 1$, as long as a does not equal 0.

Zero Exponent

If a does not equal 0, then $a^0 = 1$.

EXAMPLES Evaluate each expression.

6. $7^0 = 1$
7. $-7^0 = -7^0 = -1$ Without parentheses, only 7 is raised to the 0 power.
8. $(2x + 5)^0 = 1$
9. $2x^0 = 2(1) = 2$

■ **Work Practice Problems 6–9**

Objective C Using the Quotient Rule

To find quotients of exponential expressions, we again begin with the definition of a^n to simplify $\dfrac{x^9}{x^2}$:

$$\frac{x^9}{x^2} = \frac{x \cdot x \cdot x \cdot x \cdot x \cdot x \cdot x \cdot x \cdot x}{x \cdot x} = \frac{x}{x} \cdot \frac{x}{x} \cdot \frac{x^7}{1} = 1 \cdot 1 \cdot x^7 = x^7$$

(Assume for the next two sections that denominators containing variables are not 0.) Notice that the result is exactly the same if we subtract the exponents.

$$\frac{x^9}{x^2} = x^{9-2} = x^7$$

This suggests the following rule.

Quotient Rule for Exponents

If a is a nonzero real number and m and n are integers, then

$$\frac{a^m}{a^n} = a^{m-n}$$

In other words, the *quotient* of exponential expressions with a common base is the common base raised to a power equal to the *difference* of the exponents.

EXAMPLES Use the quotient rule to simplify.

10. $\dfrac{5^8}{5^2} = 5^{8-2} = 5^6$

11. $\dfrac{x^7}{x^4} = x^{7-4} = x^3$

12. $\dfrac{20x^6}{4x^5} = 5x^{6-5} = 5x^1$, or $5x$

13. $\dfrac{12y^{10}z^7}{14y^8z^7} = \dfrac{6}{7}y^{10-8} \cdot z^{7-7} = \dfrac{6}{7}y^2 \cdot z^0 = \dfrac{6}{7}y^2 \cdot 1 = \dfrac{6}{7}y^2$ or $\dfrac{6y^2}{7}$

■ **Work Practice Problems 10–13**

PRACTICE PROBLEMS 6–9

Evaluate each expression.
6. 8^0
7. -14^0
8. $(y - 3)^0$
9. $5x^0$

PRACTICE PROBLEMS 10–13

Use the quotient rule to simplify.
10. $\dfrac{6^{10}}{6^2}$ **11.** $\dfrac{y^6}{y^2}$
12. $\dfrac{36x^5}{9x}$ **13.** $\dfrac{10a^7b^9}{15a^5b^9}$

Answers
6. 1, **7.** −1, **8.** 1, **9.** 5, **10.** 6^8, **11.** y^4, **12.** $4x^4$, **13.** $\dfrac{2}{3}a^2$

Objective D Simplifying Expressions Raised to Negative Powers

When the exponent of the denominator is larger than the exponent of the numerator, applying the quotient rule gives a negative exponent. For example,

$$\frac{x^3}{x^5} = x^{3-5} = x^{-2}$$

However, using the definition of a^n gives us

$$\frac{x^3}{x^5} = \frac{x \cdot x \cdot x}{x \cdot x \cdot x \cdot x \cdot x} = 1 \cdot 1 \cdot 1 \cdot \frac{1}{x^2} = \frac{1}{x^2}$$

From this, we reasonably define $x^{-2} = \frac{1}{x^2}$ or, in general, $a^{-n} = \frac{1}{a^n}$.

Negative Exponents

If a is a real number other than 0 and n is a positive integer, then

$$a^{-n} = \frac{1}{a^n}$$

EXAMPLES Simplify and write with positive exponents only.

14. $5^{-2} = \dfrac{1}{5^2} = \dfrac{1}{25}$

15. $2x^{-3} = 2 \cdot \dfrac{1}{x^3} = \dfrac{2}{x^3}$ Without parentheses, only x is raised to the -3 power.

16. $(3x)^{-1} = \dfrac{1}{(3x)^1} = \dfrac{1}{3x}$ With parentheses, both 3 and x are raised to the -1 power.

17. $2^{-1} + 3^{-2} = \dfrac{1}{2^1} + \dfrac{1}{3^2} = \dfrac{1}{2} + \dfrac{1}{9} = \dfrac{9}{18} + \dfrac{2}{18} = \dfrac{11}{18}$

18. $\dfrac{1}{t^{-5}} = \dfrac{1}{\frac{1}{t^5}} = 1 \div \dfrac{1}{t^5} = 1 \cdot \dfrac{t^5}{1} = t^5$

■ **Work Practice Problems 14–18**

Helpful Hint

Notice that when a factor containing an exponent is moved from the numerator to the denominator or from the denominator to the numerator, the sign of its exponent changes.

$$x^{-3} = \frac{1}{x^3} \qquad 5^{-2} = \frac{1}{5^2} = \frac{1}{25}$$

$$\frac{1}{t^{-5}} = t^5 \qquad \frac{1}{2^{-3}} = 2^3 = 8$$

EXAMPLES Simplify and write using positive exponents only.

19. $\dfrac{m^5}{m^{15}} = m^{5-15} = m^{-10} = \dfrac{1}{m^{10}}$

20. $\dfrac{3^3}{3^6} = 3^{3-6} = 3^{-3} = \dfrac{1}{3^3} = \dfrac{1}{27}$

21. $x^{-9} \cdot x^2 = x^{-9+2} = x^{-7} = \dfrac{1}{x^7}$

Continued on next page

PRACTICE PROBLEMS 14–18

Simplify and write with positive exponents only.

14. 7^{-2}

15. $5x^{-4}$

16. $(2x)^{-1}$

17. $3^{-1} + 2^{-2}$

18. $\dfrac{1}{y^{-4}}$

PRACTICE PROBLEMS 19–25

Simplify and write using positive exponents only.

19. $\dfrac{x^3}{x^{10}}$ **20.** $\dfrac{4^2}{4^5}$ **21.** $y^{-10} \cdot y^3$

22. $\dfrac{q^5}{q^{-4}}$ **23.** $\dfrac{5^{-4}}{5^{-2}}$ **24.** $\dfrac{10x^{-8}y^5}{20xy^{-5}}$

25. $\dfrac{(4x^{-1})(x^5)}{x^7}$

Answers

14. $\dfrac{1}{49}$, **15.** $\dfrac{5}{x^4}$, **16.** $\dfrac{1}{2x}$, **17.** $\dfrac{7}{12}$,

18. y^4, **19.** $\dfrac{1}{x^7}$, **20.** $\dfrac{1}{64}$, **21.** $\dfrac{1}{y^7}$,

22. q^9, **23.** $\dfrac{1}{25}$, **24.** $\dfrac{y^{10}}{2x^9}$, **25.** $\dfrac{4}{x^3}$

22. $\dfrac{5p^4}{p^{-3}} = 5 \cdot p^{4-(-3)} = 5p^7$

23. $\dfrac{2^{-3}}{2^{-1}} = 2^{-3-(-1)} = 2^{-2} = \dfrac{1}{2^2} = \dfrac{1}{4}$

24. $\dfrac{2x^{-7}y^2}{10xy^{-5}} = \dfrac{x^{-7-1} \cdot y^{2-(-5)}}{5} = \dfrac{x^{-8}y^7}{5} = \dfrac{y^7}{5x^8}$

25. $\dfrac{(3x^{-3})(x^2)}{x^6} = \dfrac{3x^{-3+2}}{x^6} = \dfrac{3x^{-1}}{x^6} = 3x^{-1-6} = 3x^{-7} = \dfrac{3}{x^7}$

▥ **Work Practice Problems 19–25**

✔ **Concept Check** Find and correct the error in the following:

$$\dfrac{y^{-6}}{y^{-2}} = y^{-6-2} = y^{-8} = \dfrac{1}{y^8}$$

Objective E Simplifying with Variables in the Exponent

EXAMPLES Simplify. Assume that a and t are nonzero integers and that x is not 0.

26. $x^{2a} \cdot x^3 = x^{2a+3}$ Use the product rule.

27. $\dfrac{x^{2t-1}}{x^{t-5}} = x^{(2t-1)-(t-5)}$ Use the quotient rule.

$= x^{2t-1-t+5} = x^{t+4}$

▥ **Work Practice Problems 26–27**

Objective F Converting between Scientific Notation and Standard Notation

Very large and very small numbers occur frequently in nature. For example, the distance between Earth and the sun is approximately 150,000,000 kilometers. A helium atom has a diameter of 0.000000022 centimeters. It can be tedious to write these very large and very small numbers in standard notation like this. **Scientific notation** is a convenient shorthand notation for writing very large and very small numbers.

Helium atom
0.000 000 022 cm

Scientific Notation

A positive number is written in **scientific notation** if it is written as the product of a number a, where $1 \le a < 10$, and an integer power n of 10: $a \times 10^n$.

For example,

2.03×10^2 7.362×10^7 8.1×10^{-5}

Writing a Number in Scientific Notation

Step 1: Move the decimal point in the original number until the new number has a value between 1 and 10.

Step 2: Count the number of decimal places the decimal point was moved in Step 1. If the original number is 10 or greater, the count is positive. If the original number is less than 1, the count is negative.

Step 3: Write the product of the new number in Step 1 by 10 raised to an exponent equal to the count found in Step 2.

EXAMPLE 28 Write 730,000 in scientific notation.

Solution:

Step 1. Move the decimal point until the number is between 1 and 10.

730,000.

Step 2. The decimal point is moved 5 places and the original number is 10 or greater, so the count is positive 5.

Step 3. $730,000 = 7.3 \times 10^5$

▣ **Work Practice Problem 28**

EXAMPLE 29 Write 0.00000104 in scientific notation.

Solution:

Step 1. Move the decimal point until the number is between 1 and 10.

0.00000104

Step 2. The decimal point is moved 6 places and the original number is less than 1, so the count is −6.

Step 3. $0.00000104 = 1.04 \times 10^{-6}$

▣ **Work Practice Problem 29**

To write a scientific notation number in standard form, we reverse the preceding steps.

Writing a Scientific Notation Number in Standard Notation

Move the decimal point in the number the same number of places as the exponent on 10. If the exponent is positive, move the decimal point to the right. If the exponent is negative, move the decimal point to the left.

EXAMPLES Write each number in standard notation.

30. $7.7 \times 10^8 = 777,000,000$ Since the exponent is positive, move the decimal point 8 places to the right. Add zeros as needed.

31. $1.025 \times 10^{-3} = 0.001025$ Since the exponent is negative, move the decimal point 3 places to the left. Add zeros as needed.

▣ **Work Practice Problems 30–31**

✔ **Concept Check** Which of the following numbers have values that are less than 1?

a. 3.5×10^{-5} **b.** 3.5×10^5 **c.** -3.5×10^5 **d.** -3.5×10^{-5}

PRACTICE PROBLEM 28

Write 1,760,000 in scientific notation.

PRACTICE PROBLEM 29

Write 0.00028 in scientific notation.

PRACTICE PROBLEMS 30–31

Write each number in standard notation.

30. 8.6×10^7

31. 3.022×10^{-4}

Answers

28. 1.76×10^6,   **29.** 2.8×10^{-4},
30. 86,000,000,   **31.** 0.0003022

✔ **Concept Check Answer**

a, c, d

Multiply 5,000,000 by 700,000 on your calculator. The display should read $\boxed{3.5 \quad 12}$ or $\boxed{3.5 \text{ E } 12}$, which is the product written in scientific notation. Both these notations mean 3.5×10^{12}.

To enter a number written in scientific notation on a calculator, find the key marked $\boxed{\text{EE}}$. (On some calculators, this key may be marked $\boxed{\text{EXP}}$.)

To enter 7.26×10^{13}, press the keys

$\boxed{7.26} \quad \boxed{\text{EE}} \quad \boxed{13}$

The display will read $\boxed{7.26 \quad 13}$ or $\boxed{7.26\text{E}13}$.

Use your calculator to perform each indicated operation.

1. Multiply 3×10^{11} and 2×10^{32}.

2. Divide 6×10^{14} by 3×10^9.

3. Multiply 5.2×10^{23} and 7.3×10^4.

4. Divide 4.38×10^{41} by 3×10^{17}.

Mental Math

Write each expression without negative exponents.

1. $5x^{-1}y^{-2}$ **2.** $7xy^{-4}$ **3.** $a^2b^{-1}c^{-5}$ **4.** $a^{-4}b^2c^{-6}$ **5.** $\dfrac{y^{-2}}{x^{-4}}$ **6.** $\dfrac{x^{-7}}{z^{-3}}$

1.6 EXERCISE SET

Objective Ⓐ *Use the product rule to simplify each expression. See Examples 1 through 5.*

1. $4^2 \cdot 4^3$ **2.** $3^3 \cdot 3^5$ **3.** $x^5 \cdot x^3$ **4.** $a^2 \cdot a^9$

5. $m \cdot m^7 \cdot m^6$ **6.** $n \cdot n^{10} \cdot n^{12}$ **7.** $(4xy)(-5x)$ **8.** $(-7xy)(7y)$

9. $(-4x^3p^2)(4y^3x^3)$ **10.** $(-6a^2b^3)(-3ab^3)$

Objective Ⓑ *Evaluate each expression. See Examples 6 through 9.*

11. -8^0 **12.** $(-9)^0$ **13.** $(4x + 5)^0$ **14.** $(3x - 1)^0$

15. $-x^0$ **16.** $-5x^0$ **17.** $4x^0 + 5$ **18.** $8x^0 + 1$

Objective Ⓒ *Use the quotient rule to simplify. See Examples 10 through 13.*

19. $\dfrac{a^5}{a^2}$ **20.** $\dfrac{x^9}{x^4}$ **21.** $-\dfrac{26z^{11}}{2z^7}$ **22.** $-\dfrac{16x^5}{8x}$

23. $\dfrac{x^9y^6}{x^8y^6}$ **24.** $\dfrac{a^{12}b^2}{a^9b}$ **25.** $\dfrac{12x^4y^7}{9xy^5}$ **26.** $\dfrac{24a^{10}b^{11}}{10ab^3}$

27. $\dfrac{-36a^5b^7c^{10}}{6ab^3c^4}$ **28.** $\dfrac{49a^3bc^{14}}{-7abc^8}$

Objective D *Simplify and write using positive exponents only. See Examples 14 through 25.*

29. 4^{-2}

30. 2^{-3}

31. $(-3)^{-3}$

32. $(-6)^{-2}$

33. $\dfrac{x^7}{x^{15}}$

34. $\dfrac{z}{z^3}$

35. $5a^{-4}$

36. $10b^{-1}$

37. $\dfrac{x^{-7}}{y^{-2}}$

38. $\dfrac{p^{-13}}{q^{-3}}$

39. $\dfrac{x^{-2}}{x^5}$

40. $\dfrac{z^{-12}}{z^{10}}$

41. $\dfrac{8r^4}{2r^{-4}}$

42. $\dfrac{3s^3}{15s^{-3}}$

43. $\dfrac{x^{-9}x^4}{x^{-5}}$

44. $\dfrac{y^{-7}y}{y^8}$

45. $\dfrac{2a^{-6}b^2}{18ab^{-5}}$

46. $\dfrac{18ab^{-6}}{3a^{-3}b^6}$

47. $\dfrac{(24x^8)(x)}{20x^{-7}}$

48. $\dfrac{(30z^2)(z^5)}{55z^{-4}}$

Objectives A B C D **Mixed Practice** *Simplify and write using positive exponents only. See Examples 1 through 25.*

49. $-7x^3 \cdot 20x^9$

50. $-3y \cdot -9y^4$

51. $x^7 \cdot x^8 \cdot x$

52. $y^6 \cdot y \cdot y^9$

53. $2x^3 \cdot 5x^7$

54. $-3z^4 \cdot 10z^7$

55. $(5x)^0 + 5x^0$

56. $4y^0 - (4y)^0$

57. $\dfrac{z^{12}}{z^{15}}$

58. $\dfrac{x^{11}}{x^{20}}$

59. $3^0 - 3t^0$

60. $4^0 + 4x^0$

61. $\dfrac{y^{-3}}{y^{-7}}$

62. $\dfrac{y^{-6}}{y^{-9}}$

63. $4^{-1} + 3^{-2}$

64. $1^{-3} - 4^{-2}$

65. $3x^{-1}$

66. $(4x)^{-1}$

67. $\dfrac{r^4}{r^{-4}}$

68. $\dfrac{x^{-5}}{x^3}$

69. $\dfrac{x^{-7}y^{-2}}{x^2y^2}$

70. $\dfrac{a^{-5}b^7}{a^{-2}b^{-3}}$

71. $(-4x^2y)(3x^4)(-2xy^5)$

72. $(-6a^4b)(2b^3)(-3ab^6)$

73. $2^{-4} \cdot x$

74. $5^{-2} \cdot y$

75. $\dfrac{5^{17}}{5^{13}}$

76. $\dfrac{10^{25}}{10^{23}}$

77. $\dfrac{8^{-7}}{8^{-6}}$

78. $\dfrac{13^{-10}}{13^{-9}}$

79. $\dfrac{9^{-5}a^4}{9^{-3}a^{-1}}$

80. $\dfrac{11^{-9}b^3}{11^{-7}b^{-4}}$

81. $\dfrac{14x^{-2}yz^{-4}}{2xyz}$

82. $\dfrac{30x^{-7}yz^{-14}}{3xyz}$

Objective E *Simplify. Assume that variables in the exponents represent nonzero integers and that x, y, and z are not 0. See Examples 26 and 27.*

83. $x^5 \cdot x^{7a}$

84. $y^{2p} \cdot y^{9p}$

85. $\dfrac{x^{3t-1}}{x^t}$

86. $\dfrac{y^{4p-2}}{y^{3p}}$

87. $x^{4a} \cdot x^7$

88. $x^{9y} \cdot x^{-7y}$

89. $\dfrac{z^{6x}}{z^7}$

90. $\dfrac{y^6}{y^{4z}}$

91. $\dfrac{x^{3t} \cdot x^{4t-1}}{x^t}$

92. $\dfrac{z^{5x} \cdot z^{x-7}}{z^x}$

Objective **F** *Write each number in scientific notation. See Examples 28 and 29.*

93. 31,250,000   **94.** 678,000   **95.** 0.016   **96.** 0.007613   **97.** 67,413

98. 36,800,000   **99.** 0.0125   **100.** 0.00084   **101.** 0.000053   **102.** 98,700,000,000

Write each number in scientific notation.

103. The approximate distance between Jupiter and the sun is 778,300,000 kilometers, (*Source:* National Space Data Center)

104. For the 2004 Major League Baseball season, the World Series Champion Red Sox payroll was approximately $130,395,000. (*Source: The Boston Globe*)

105. The estimated world population in December 2004 was 6,404,000,000 (*Source:* U.S. Census Bureau)

106. Total revenues for Microsoft in fiscal year 2004 were $36,835,000,000. (*Source:* Microsoft Corporation)

107. Lake Mead, created from the Colorado River by the Hoover Dam, has a capacity of 124,000,000,000 cubic feet of water. (*Source:* U.S. Bureau of Reclamation)

108. The temperature of the core of the sun is about 27,000,000°F.

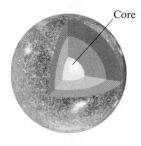

Core

109. A pulsar is a rotating neutron star that gives off sharp, regular pulses of radio waves. For one particular pulsar, the rate of pulses is every 0.001 second.

110. To convert from cubic inches to cubic meters, multiply by 0.0000164.

Write each number in standard notation. See Examples 30 and 31.

111. 3.6×10^{-9} **112.** 2.7×10^{-5} **113.** 9.3×10^{7} **114.** 6.378×10^{8} **115.** 1.278×10^{6}

116. 7.6×10^{4} **117.** 7.35×10^{12} **118.** 1.66×10^{-5} **119.** 4.03×10^{-7} **120.** 8.007×10^{8}

Write each number in standard notation.

121. The estimated world population in 1 A.D. was 3.0×10^{8}. (*Source: World Almanac and Book of Facts,* 2005)

122. There are 3.949×10^{6} miles of highways, roads, and streets in the United States. (*Source:* Bureau of Transportation Statistics)

123. In 2005, teenagers and children are expected to spend 4.9×10^9 dollars on purchases and transactions made online. (*Source:* Jupiter Research)

124. Each day, an estimated 1.2×10^9 beverages consumed throughout the world are Coca Cola products. (*Source:* Coca Cola)

Concept Extensions

125. Explain how to convert a number from standard notation to scientific notation.

126. Explain how to convert a number from scientific notation to standard notation.

127. Explain why $(-5)^0$ simplifies to 1 but -5^0 simplifies to -1.

128. Explain why both $4x^0 - 3y^0$ and $(4x - 3y)^0$ simplify to 1.

STUDY SKILLS BUILDER

What to Do the Day of an Exam?

Your first exam may be soon. On the day of an exam, don't forget to try the following:

- Allow yourself plenty of time to arrive.
- Read the directions on the test carefully.
- Read each problem carefully as you take your test. Make sure that you answer the question asked.
- Watch your time and pace yourself so that you may attempt each problem on your test.
- Check your work and answers.
- ***Do not turn your test in early.*** If you have extra time, spend it double-checking your work.

Good luck!

Answer the following questions based on your most recent mathematics exam, whenever that was.

1. How soon before class did you arrive?

2. Did you read the directions on the test carefully?

3. Did you make sure you answered the question asked for each problem on the exam?

4. Were you able to attempt each problem on your exam?

5. If your answer to question 4 is no, list reasons why.

6. Did you have extra time on your exam?

7. If your answer to question 6 is yes, describe how you spent that extra time.

A Use the Power Rules for Exponents.

B Use Exponent Rules and Definitions to Simplify Exponential Expressions.

C Simplify Exponential Expressions Containing Variables in the Exponent.

D Use Scientific Notation to Compute.

1.7 MORE WORK WITH EXPONENTS AND SCIENTIFIC NOTATION

Objective **A** Using the Power Rules

The volume of the cube shown whose side measures x^2 units is $(x^2)^3$ cubic units. To simplify an expression such as $(x^2)^3$, we use the definition of a^n:

$$(x^2)^3 = \underbrace{(x^2)(x^2)(x^2)}_{x^2 \text{ is a factor 3 times.}} = x^{2+2+2} = x^6$$

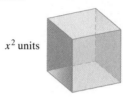

x^2 units

Notice that the result is exactly the same if the exponents are multiplied.

$$(x^2)^3 = x^{2\cdot3} = x^6$$

This suggests that an expression raised to a power that is then all raised to another power is equal to the original expression raised to the product of the powers. Two additional rules for exponents are given in the following box.

> ### Power Rule and Power of a Product or Quotient Rules for Exponents
>
> If a and b are real numbers and m and n are integers, then
>
> | | |
> |---|---|
> | $(a^m)^n = a^{m\cdot n}$ | Power rule |
> | $(ab)^m = a^m b^m$ | Power of a product |
> | $\left(\dfrac{a}{b}\right)^n = \dfrac{a^n}{b^n} \qquad (b \neq 0)$ | Power of a quotient |

EXAMPLES Use the power rule to simplify each expression. Write each answer using positive exponents only.

1. $(x^5)^7 = x^{5\cdot7} = x^{35}$

2. $(2^2)^3 = 2^{2\cdot3} = 2^6 = 64$

3. $(5^{-1})^2 = 5^{-1\cdot2} = 5^{-2} = \dfrac{1}{5^2} = \dfrac{1}{25}$

4. $(y^{-3})^{-4} = y^{-3(-4)} = y^{12}$

Work Practice Problems 1–4

EXAMPLES Use the power rules to simplify each expression. Write each answer using positive exponents only.

5. $(5x^2)^3 = 5^3 \cdot (x^2)^3 = 5^3 \cdot x^{2\cdot3} = 125x^6$

6. $\left(\dfrac{2}{3}\right)^3 = \dfrac{2^3}{3^3} = \dfrac{8}{27}$

7. $\left(\dfrac{3p^4}{q^5}\right)^2 = \dfrac{(3p^4)^2}{(q^5)^2} = \dfrac{3^2 \cdot (p^4)^2}{(q^5)^2} = \dfrac{9p^8}{q^{10}}$

Use the power rule to simplify each expression. Write each answer using positive exponents only.

1. $(y^2)^8$ **2.** $(3^3)^2$

3. $(6^2)^{-1}$ **4.** $(x^{-5})^{-7}$

Use the power rules to simplify each expression. Write each answer using positive exponents only.

5. $(3x^4)^3$ **6.** $\left(\dfrac{4}{5}\right)^2$

7. $\left(\dfrac{4m^5}{n^3}\right)^3$ **8.** $\left(\dfrac{2^{-1}}{y}\right)^{-3}$

9. $(a^{-4}b^3c^{-2})^6$

Answers

1. y^{16},   **2.** 729,   **3.** $\dfrac{1}{36}$,   **4.** x^{35},

5. $27x^{12}$,   **6.** $\dfrac{16}{25}$,   **7.** $\dfrac{64m^{15}}{n^9}$,   **8.** $8y^3$,

9. $\dfrac{b^{18}}{a^{24}c^{12}}$

8. $\left(\dfrac{2^{-3}}{y}\right)^{-2} = \dfrac{(2^{-3})^{-2}}{y^{-2}}$

$\qquad = \dfrac{2^6}{y^{-2}} = 64y^2$ Use the negative exponent rule.

9. $(x^{-5}y^2z^{-1})^7 = (x^{-5})^7 \cdot (y^2)^7 \cdot (z^{-1})^7$

$\qquad = x^{-35}y^{14}z^{-7} = \dfrac{y^{14}}{x^{35}z^7}$

🔲 **Work Practice Problems 5–9**

Objective B Using Exponent Rules to Simplify Expressions

In the next few examples, we practice the use of several of the rules and definitions for exponents. The following is a summary of these rules and definitions.

Summary of Rules for Exponents

If a and b are real numbers and m and n are integers, then

Product rule	$a^m \cdot a^n = a^{m+n}$	
Zero exponent	$a^0 = 1$	$(a \neq 0)$
Negative exponent	$a^{-n} = \dfrac{1}{a^n}$	$(a \neq 0)$
Quotient rule	$\dfrac{a^m}{a^n} = a^{m-n}$	$(a \neq 0)$
Power rule	$(a^m)^n = a^{m \cdot n}$	
Power of a product	$(ab)^m = a^m \cdot b^m$	
Power of a quotient	$\left(\dfrac{a}{b}\right)^n = \dfrac{a^n}{b^n}$	$(b \neq 0)$

EXAMPLES Simplify each expression. Write each answer using positive exponents only.

10. $(2x^0y^{-3})^{-2} = 2^{-2}(x^0)^{-2}(y^{-3})^{-2}$

$\qquad = 2^{-2}x^0y^6$

$\qquad = \dfrac{1(y^6)}{2^2}$ Write x^0 as 1.

$\qquad = \dfrac{y^6}{4}$

11. $\left(\dfrac{x^{-5}}{x^{-2}}\right)^{-3} = \dfrac{(x^{-5})^{-3}}{(x^{-2})^{-3}} = \dfrac{x^{15}}{x^6} = x^{15-6} = x^9$

12. $\left(\dfrac{2}{7}\right)^{-2} = \dfrac{2^{-2}}{7^{-2}} = \dfrac{7^2}{2^2} = \dfrac{49}{4}$

13. $\dfrac{5^{-2}x^{-3}y^{11}}{x^2y^{-5}} = 5^{-2}x^{-3-2}y^{11-(-5)} = 5^{-2}x^{-5}y^{16} = \dfrac{y^{16}}{5^2x^5} = \dfrac{y^{16}}{25x^5}$

🔲 **Work Practice Problems 10–13**

PRACTICE PROBLEMS 10–13

Simplify each expression. Write each answer using positive exponents only.

10. $(7xy^{-2})^{-2}$

11. $\left(\dfrac{y^{-7}}{y^{-10}}\right)^{-3}$

12. $\left(\dfrac{3}{5}\right)^{-2}$

13. $\dfrac{6^{-2}x^{-4}y^{10}}{x^2y^{-6}}$

Answers

10. $\dfrac{y^4}{49x^2}$,   **11.** $\dfrac{1}{y^9}$,   **12.** $\dfrac{25}{9}$,   **13.** $\dfrac{y^{16}}{36x^6}$

PRACTICE PROBLEMS 14-15

Simplify each expression. Write each answer using positive exponents only.

14. $\left(\dfrac{4a^3b^2}{b^{-6}c}\right)^{-2}$

15. $\left(\dfrac{4x^3}{3y^{-1}}\right)^3\left(\dfrac{y^{-2}}{3x^{-1}}\right)^{-1}$

EXAMPLES Simplify each expression. Write each answer using positive exponents only.

14. $\left(\dfrac{3x^2y}{y^{-9}z}\right)^{-2} = \left(\dfrac{3x^2y^{10}}{z}\right)^{-2} = \dfrac{3^{-2}x^{-4}y^{-20}}{z^{-2}} = \dfrac{z^2}{3^2x^4y^{20}} = \dfrac{z^2}{9x^4y^{20}}$

15. $\left(\dfrac{3a^2}{2x^{-1}}\right)^3\left(\dfrac{x^{-3}}{4a^{-2}}\right)^{-1} = \dfrac{27a^6}{8x^{-3}}\cdot\dfrac{x^3}{4^{-1}a^2}$

$= \dfrac{27\cdot4\cdot a^6\cdot x^3\cdot x^3}{8\cdot a^2} = \dfrac{27a^4x^6}{2}$

◼ **Work Practice Problems 14-15**

Objective C Simplifying with Variables in the Exponent

PRACTICE PROBLEMS 16-17

Simplify. Assume that m and n are integers and that x and y are not 0.

16. $x^{-n}(3x^n)^2$ **17.** $\dfrac{(y^{2m})^2}{y^{m-3}}$

EXAMPLES Simplify. Assume that a and b are integers and that x and y are not 0.

16. $x^{-b}(2x^b)^2 = x^{-b}2^2x^{2b} = 4x^{-b+2b} = 4x^b$

17. $\dfrac{(y^{3a})^2}{y^{a-6}} = \dfrac{y^{2(3a)}}{y^{a-6}} = \dfrac{y^{6a}}{y^{a-6}} = y^{6a-(a-6)} = y^{6a-a+6} = y^{5a+6}$

◼ **Work Practice Problems 16-17**

Objective D Using Scientific Notation to Compute

To perform operations on numbers written in scientific notation, we use the properties of exponents.

PRACTICE PROBLEMS 18-19

Perform each indicated operation. Write each answer in scientific notation.

18. $(9.6 \times 10^6)(4 \times 10^{-8})$

19. $\dfrac{4.2 \times 10^7}{7 \times 10^{-3}}$

EXAMPLES Perform each indicated operation. Write each answer in scientific notation.

18. $(8.1 \times 10^5)(5 \times 10^{-7}) = 8.1 \times 5 \times 10^5 \times 10^{-7}$

$= 40.5 \times 10^{-2}$

$= (4.05 \times 10^1) \times 10^{-2}$

$= 4.05 \times 10^{-1}$

19. $\dfrac{1.2 \times 10^4}{3 \times 10^{-2}} = \left(\dfrac{1.2}{3}\right)\left(\dfrac{10^4}{10^{-2}}\right) = 0.4 \times 10^{4-(-2)}$

$= 0.4 \times 10^6 = (4 \times 10^{-1}) \times 10^6 = 4 \times 10^5$

◼ **Work Practice Problems 18-19**

PRACTICE PROBLEM 20

Use scientific notation to simplify:

$\dfrac{3000 \times 0.000012}{400}$

EXAMPLE 20 Use scientific notation to simplify: $\dfrac{2000 \times 0.000021}{700}$

Solution:

$\dfrac{2000 \times 0.000021}{700} = \dfrac{(2 \times 10^3)(2.1 \times 10^{-5})}{7 \times 10^2} = \dfrac{2(2.1)}{7}\cdot\dfrac{10^3\cdot10^{-5}}{10^2}$

$= 0.6 \times 10^{-4}$

$= (6 \times 10^{-1}) \times 10^{-4}$

$= 6 \times 10^{-5}$

◼ **Work Practice Problem 20**

Answers

14. $\dfrac{c^2}{16a^6b^{16}}$, **15.** $\dfrac{64x^8y^5}{9}$, **16.** $9x^n$,

17. y^{3m+3}, **18.** 3.84×10^{-1},

19. 6×10^9, **20.** 9×10^{-5}

Mental Math

Simplify. See Examples 1 through 4.

1. $(x^4)^5$ **2.** $(5^6)^2$ **3.** $x^4 \cdot x^5$ **4.** $x^7 \cdot x^8$ **5.** $(y^6)^7$

6. $(x^3)^4$ **7.** $(z^4)^9$ **8.** $(z^3)^7$ **9.** $(z^{-6})^{-3}$ **10.** $(y^{-4})^{-2}$

1.7 EXERCISE SET

FOR EXTRA HELP

Student Solutions Manual   PH Math/Tutor Center   CD/Video for Review   MathXL MathXL®   MyMathLab MyMathLab

Objective A *Simplify. Write each answer using positive exponents only. See Examples 1 through 9.*

1. $(3^{-1})^2$ **2.** $(2^{-2})^2$ **3.** $(x^4)^{-9}$ **4.** $(y^7)^{-3}$

5. $(3x^2y^3)^2$ **6.** $(4x^3yz)^2$ **7.** $\left(\dfrac{2x^5}{y^{-3}}\right)^4$ **8.** $\left(\dfrac{3a^{-4}}{b^7}\right)^3$

9. $(2a^2bc^{-3})^{-6}$ **10.** $(6x^{-6}y^7z^0)^{-2}$ **11.** $\left(\dfrac{7}{8}\right)^3$ **12.** $\left(\dfrac{4}{3}\right)^2$

13. $(-2^{-2}y^{-1})^{-3}$ **14.** $(-4^{-6}y^{-6})^{-4}$

Objective B *Simplify. Write each answer using positive exponents only. See Examples 10 through 15.*

15. $\left(\dfrac{a^{-4}}{a^{-5}}\right)^{-2}$ **16.** $\left(\dfrac{x^{-9}}{x^{-4}}\right)^{-3}$ **17.** $\left(\dfrac{6p^6}{p^{12}}\right)^2$ **18.** $\left(\dfrac{4p^6}{p^9}\right)^3$

19. $(-8y^3xa^{-2})^{-3}$ **20.** $(-5y^0x^2a^3)^{-3}$ **21.** $\left(\dfrac{3}{4}\right)^{-3}$ **22.** $\left(\dfrac{5}{8}\right)^{-2}$

 23. $\left(\dfrac{2a^{-2}b^5}{4a^2b^7}\right)^{-2}$ **24.** $\left(\dfrac{5x^7y^4}{10x^3y^{-2}}\right)^{-3}$ **25.** $\left(\dfrac{x^{-2}y^{-2}}{a^{-3}}\right)^{-7}$ **26.** $\left(\dfrac{x^{-1}y^{-2}}{z^{-3}}\right)^{-5}$

Objectives A B Mixed Practice *Simplify. Write each answer using positive exponents only. See Examples 1 through 15.*

27. $(y^{-5})^2$ **28.** $(z^{-2})^{13}$ **29.** $(5^{-1})^3$ **30.** $(8^2)^{-1}$

31. $(x^7)^{-9}$ **32.** $(y^{-4})^5$ **33.** $\left(\dfrac{x^7y^{-3}}{z^{-4}}\right)^{-5}$ **34.** $\left(\dfrac{a^{-2}b^{-5}}{c^{-11}}\right)^{-6}$

35. $(4x^2)^2$ **36.** $(-8x^3)^2$ **37.** $\left(\dfrac{4^{-4}}{y^3x}\right)^{-2}$ **38.** $\left(\dfrac{7^{-3}}{ab^2}\right)^{-2}$

39. $\left(\dfrac{2x^{-3}}{y^{-1}}\right)^{-3}$ **40.** $\left(\dfrac{n^5}{2m^{-2}}\right)^{-4}$ **41.** $\dfrac{4^{-1}x^2yz}{x^{-2}yz^3}$ **42.** $\dfrac{8^{-2}x^{-3}y^{11}}{x^2y^{-5}}$

43. $\left(\dfrac{3x^5}{6x^4}\right)^4$ **44.** $\left(\dfrac{8^{-3}}{y^2}\right)^{-2}$ **45.** $\dfrac{(y^3)^{-4}}{y^3}$ **46.** $\dfrac{2(y^3)^{-3}}{y^{-3}}$

47. $\dfrac{3^{-2}a^{-5}b^6}{4^{-2}a^{-7}b^{-3}}$ **48.** $\dfrac{2^{-3}m^{-4}n^{-5}}{5^{-2}m^{-5}n}$ **49.** $(4x^6y^5)^{-2}(6x^4y^3)$ **50.** $(5x^2y^4)^{-2}(3x^9y^4)$

51. $x^6(x^6bc)^{-6}$ **52.** $y^2(y^2bx)^{-4}$ **53.** $\dfrac{2^{-3}x^2y^{-5}}{5^{-2}x^7y^{-1}}$ **54.** $\dfrac{7^{-1}a^{-3}b^5}{a^2b^{-2}}$

55. $\left(\dfrac{2x^2}{y^4}\right)^3\left(\dfrac{2x^5}{y}\right)^{-2}$ **56.** $\left(\dfrac{3z^{-2}}{y}\right)^2\left(\dfrac{9y^{-4}}{z^{-3}}\right)^{-1}$

Objective C *Simplify. Assume that variables in the exponents represent nonzero integers and that all other variables are not 0. See Examples 16 and 17.*

57. $(x^{3a+6})^3$

58. $(x^{2b+7})^2$

59. $\dfrac{x^{4a}(x^{4a})^3}{x}$

60. $\dfrac{x^{-5y+2}x^{2y}}{x}$

61. $(b^{5x-2})^2$

62. $(c^{2a+3})^3$

63. $\dfrac{(y^{2a})^8}{y^{a-3}}$

64. $\dfrac{(y^{4a})^7}{y^{2a-1}}$

65. $\left(\dfrac{2x^{3t}}{x^{2t-1}}\right)^4$

66. $\left(\dfrac{3y^{5a}}{y^{-a+1}}\right)^2$

67. $\dfrac{25x^{2a+b}y^{2a-b}}{5x^{a-b}y^{a+b}}$

68. $\dfrac{16x^{3a-b}y^{4a+b}}{2x^{a-2b}y^{a+3b}}$

Objective D *Perform each indicated operation. Write each answer in scientific notation. See Examples 18 through 20.*

69. $(5 \times 10^{11})(2.9 \times 10^{-3})$ **70.** $(3.6 \times 10^{-12})(6 \times 10^9)$ **71.** $(2 \times 10^5)^3$

72. $(3 \times 10^{-7})^3$

73. $\dfrac{3.6 \times 10^{-4}}{9 \times 10^2}$

74. $\dfrac{1.2 \times 10^9}{2 \times 10^{-5}}$

75. $\dfrac{0.0069}{0.023}$

76. $\dfrac{0.00048}{0.0016}$

77. $\dfrac{18,200 \times 100}{91,000}$

78. $\dfrac{0.0003 \times 0.0024}{0.0006 \times 20}$

79. $\dfrac{6000 \times 0.006}{0.009 \times 400}$

80. $\dfrac{0.00016 \times 300}{0.064 \times 100}$

81. $\dfrac{0.00064 \times 2000}{16,000}$

82. $\dfrac{0.00072 \times 0.003}{0.00024}$

83. $\dfrac{66,000 \times 0.001}{0.002 \times 0.003}$

84. $\dfrac{0.0007 \times 11,000}{0.001 \times 0.0001}$

85. $\dfrac{9.24 \times 10^{15}}{(2.2 \times 10^{-2})(1.2 \times 10^{-5})}$

86. $\dfrac{(2.6 \times 10^{-3})(4.8 \times 10^{-4})}{1.3 \times 10^{-12}}$

Solve.

87. A computer can add two numbers in about 10^{-8} second. Express in scientific notation how long it would take this computer to do this task 200,000 times.

88. To convert from square inches to square meters, multiply by 6.452×10^{-4}. The area of the following square is 4×10^{-2} square inches. Convert this area to square meters.

4×10^{-2} sq in.

89. To convert from cubic inches to cubic meters, multiply by 1.64×10^{-5}. A grain of salt is in the shape of a cube. If an average size of a grain of salt is 3.8×10^{-6} cubic inches, convert this volume to cubic meters.

Concept Extensions

90. Each side of the cube shown is $\dfrac{2x^{-2}}{y}$ meters. Find its volume.

91. The lot shown is in the shape of a parallelogram with base $\dfrac{3x^{-1}}{y^{-3}}$ feet and height $5x^{-7}$ feet. Find its area.

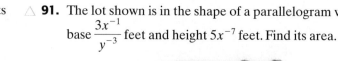

$\dfrac{2x^{-2}}{y}$ m

$5x^{-7}$ ft

$\dfrac{3x^{-1}}{y^{-3}}$ ft

92. The density D of an object is equivalent to the quotient of its mass M and volume V. Thus $D = \dfrac{M}{V}$. Express in scientific notation the density of an object whose mass is 500,000 pounds and whose volume is 250 cubic feet.

93. The density of ordinary water is 3.12×10^{-2} tons per cubic foot. The volume of water in the largest of the Great Lakes, Lake Superior, is 4.269×10^{14} cubic feet. Use the formula $D = \dfrac{M}{V}$ (see Exercise 92) to find the mass (in tons) of the water in Lake Superior. Express your answer in scientific notation. (*Source:* National Ocean Service)

94. Is there a number a such that $a^{-1} = a^{1}$? If so, give the value of a.

95. Is there a number a such that a^{-2} is a negative number? If so, give the value of a.

96. Explain whether 0.4×10^{-5} is written in scientific notation.

97. The estimated population of the United States in 2004 was 2.95×10^{8} people. The land area of the United States is 3.536×10^{6} square miles. Find the population density (number of people per square mile) for the United States in 2004. Round to the nearest whole. (*Source:* U.S. Census Bureau)

98. In October 2004, the value of goods and services imported into the United States was $\$1.535 \times 10^{11}$. The estimated population of the United States in 2004 was 2.95×10^{8} people. Find the average value of imports per person in the United States for October 2004. Round to the nearest dollar. (*Sources:* U.S. Census Bureau, Bureau of Economic Analysis)

99. In 2004, the population of Japan was 1.276×10^{8} people. At the same time, the population of Oceania (including the countries of Australia, New Zealand, Fiji, etc.) was 3.3×10^{7} people. How many times greater was the population of Japan than the population of Oceania? Round to the nearest tenth. (*Source:* Population Reference Bureau)

STUDY SKILLS BUILDER

Are You Prepared for a Test on Chapter 1?

Below I have listed some common trouble areas for students in Chapter 1. After studying for your test—but before taking your test—read these.

- Do you remember the meaning of a negative exponent?

$$7^{-2} = \frac{1}{7^2} = \frac{1}{49}$$

- Don't forget the order of operations and the distributive property.

$$7 - 3(2x - 6y) + 5 = 7 - 6x + 18y + 5 \quad \text{Use the distributive}$$
$$\uparrow \qquad \qquad \qquad \text{property.}$$
$$\text{Notice the sign.}$$

$$= -6x + 18y + 12 \quad \text{Combine like terms.}$$

- Don't forget the difference between $(-3)^{-2}$ and -3^{-2}.

$$(-3)^{-2} = \frac{1}{(-3)^2} = \frac{1}{9}$$

$$-3^{-2} = -1 \cdot 3^{-2} = \frac{-1}{3^2} = \frac{-1}{9} \text{ or } -\frac{1}{9}$$

- Remember that

$$\frac{0}{8} = 0 \text{ while } \frac{8}{0} \text{ is undefined.}$$

- Don't forget the difference between reciprocal and opposite.

The opposite of $-\dfrac{3}{5}$ is $\dfrac{3}{5}$.

The reciprocal of $-\dfrac{3}{5}$ is $-\dfrac{5}{3}$.

Remember: This is simply a checklist of common trouble areas. For a review of Chapter 1, see the Highlights and Chapter Review at the end of Chapter 1.

CHAPTER 1 Group Activity

Geometry Investigations

Sections 1.2 through 1.6

Recall that the perimeter of a figure is the distance around the outside of the figure. For a rectangle with length l and width w, the perimeter of the rectangle is given by the expression $2l + 2w$.

Area is a measure of the surface of a region. For example, we measure a plot of land or the floor space of a home by area. For a rectangle with length l and width w, the area of the rectangle is given by the expression lw.

A circular cylinder can be formed by rolling a rectangle into a tube. The surface area of the cylinder (excluding the two ends of the cylinder) is the same as the area of the rectangle used to form the cylinder. Recall that volume is a measure of the space inside a three-dimensional region. The volume of a circular cylinder with height h and radius r is given by the expression $\pi r^2 h$.

Group Activity

1. Work together to discover whether two rectangles with the same perimeter always have the same area. Explain your results. Give examples.

2. Do figures with the same surface area always have the same volume? To see, take two $8\frac{1}{2}$-by-11-inch sheets of paper and construct two cylinders using the following figures as a guide. Verify that both cylinders have the same surface area. Measure the height and radius of each resulting cylinder. Then find the volume of each cylinder to the nearest tenth of a cubic inch. Explain your results.

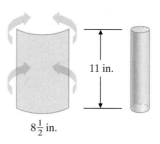

$8\frac{1}{2}$ in.

Cylinder 1

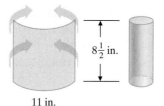

11 in.

Cylinder 2

Chapter 1 Vocabulary Check

Fill in each blank with one of the words or phrases listed below.

distributive	absolute value	inequality	algebraic expression
real	opposite	commutative	exponent
reciprocals	associative	whole	variable

1. A(n) _____ is formed by numbers and variables connected by the operations of addition, subtraction, multiplication, division, raising to powers, and/or taking roots.
2. The _____ of a number a is $-a$.
3. $3(x - 6) = 3x - 18$ by the _____ property.
4. The _____ of a number is the distance between that number and 0 on the number line.
5. A(n) _____ is a shorthand notation for repeated multiplication of the same factor.
6. A letter that represents a number is called a _____.
7. The symbols $<$ and $>$ are called _____ symbols.
8. If a is not 0, then a and $\dfrac{1}{a}$ are called _____ .
9. $A + B = B + A$ by the _____ property of addition.
10. $(A + B) + C = A + (B + C)$ by the _____ property of addition.
11. The numbers $0, 1, 2, 3, \dots$ are called _____ numbers.
12. If a number corresponds to a point on the number line, we know that number is a ____ number.

Helpful Hint

Are you preparing for your test? Don't forget to take the Chapter 1 Test on page 67. Then check your answers at the back of the text and use the Chapter Test Prep Video CD to see the fully worked-out solutions to any of the exercises you want to review.

1 Chapter Highlights

DEFINITIONS AND CONCEPTS	**EXAMPLES**
Section 1.2 Algebraic Expressions and Sets of Numbers	

Letters that represent numbers are called **variables.**	$x, \quad a, \quad m, \quad y$
An **algebraic expression** is formed by numbers and variables connected by the operations of addition, subtraction, multiplication, division, raising to powers, or taking roots.	$7y, \quad -3, \quad \dfrac{x^2 - 9}{-2} + 14x, \quad \sqrt{3} + \sqrt{m}$
Natural numbers: $\{1, 2, 3, \dots\}$	Given the set $\left\{-9.6, -5, -\sqrt{2}, 0, \dfrac{2}{5}, 101\right\}$, list the elements that belong to the set of
Whole numbers: $\{0, 1, 2, 3, \dots\}$	Natural numbers 101
Integers: $\{\dots, -3, -2, -1, 0, 1, 2, 3, \dots\}$	Whole numbers $0, 101$
Each listing of three dots is called an **ellipsis.**	Integers $-5, 0, 101$
The objects of a set are called its members or **elements.**	
Set builder notation describes the elements of a set but does not list them.	Real numbers $-9.6, -5, -\sqrt{2}, 0, \dfrac{2}{5}, 101$
Real numbers: $\{x \mid x \text{ corresponds to a point on the number line}\}$	Rational numbers $-9.6, -5, 0, \dfrac{2}{5}, 101$
Rational numbers: $\left\{\dfrac{a}{b} \,\middle|\, a \text{ and } b \text{ are integers and } b \neq 0\right\}$	Irrational numbers $-\sqrt{2}$
Irrational numbers: $\{x \mid x \text{ is a real number and } x \text{ is not a rational number}\}$	Write the set $\{x \mid x \text{ is an integer between } -2 \text{ and } 5\}$ in roster form.
	$\{-1, 0, 1, 2, 3, 4\}$
If all the elements of set A are also in set B, we say that set A is a **subset** of set B.	The set of integers is a subset of the set of rational numbers.

DEFINITIONS AND CONCEPTS	**EXAMPLES**
Section 1.3 Equations, Inequalities, and Properties of Real Numbers	

SYMBOLS

$=$	is equal to
$\neq$	is not equal to
$>$	is greater than
$<$	is less than
$\geq$	is greater than or equal to
$\leq$	is less than or equal to

$$-5 = -5$$
$$-5 \neq -3$$
$$1.7 > 1.2$$
$$-1.7 < -1.2$$
$$\frac{5}{3} \geq \frac{5}{3}$$
$$-\frac{1}{2} \leq \frac{1}{2}$$

IDENTITY

$$a + 0 = a \qquad 0 + a = a$$
$$a \cdot 1 = a \qquad 1 \cdot a = a$$

$$3 + 0 = 3 \qquad 0 + 3 = 3$$
$$-1.8 \cdot 1 = -1.8 \qquad 1 \cdot -1.8 = -1.8$$

INVERSE

$$a + (-a) = 0 \qquad -a + a = 0$$
$$a \cdot \frac{1}{a} = 1 \qquad \frac{1}{a} \cdot a = 1$$

$$7 + (-7) = 0 \qquad -7 + 7 = 0$$
$$5 \cdot \frac{1}{5} = 1 \qquad \frac{1}{5} \cdot 5 = 1$$

COMMUTATIVE

$$a + b = b + a$$
$$a \cdot b = b \cdot a$$

$$x + 7 = 7 + x$$
$$9 \cdot y = y \cdot 9$$

ASSOCIATIVE

$$(a + b) + c = a + (b + c)$$
$$(a \cdot b) \cdot c = a \cdot (b \cdot c)$$

$$(3 + 1) + 10 = 3 + (1 + 10)$$
$$(3 \cdot 1) \cdot 10 = 3 \cdot (1 \cdot 10)$$

DISTRIBUTIVE

$$a(b + c) = ab + ac$$

$$6(x + 5) = 6 \cdot x + 6 \cdot 5$$
$$= 6x + 30$$

Section 1.4 Operations on Real Numbers	

ABSOLUTE VALUE

$$|a| = \begin{cases} a \text{ if } a \text{ is } 0 \text{ or a positive number} \\ -a \text{ if } a \text{ is a negative number} \end{cases}$$

$$|3| = 3, \quad |0| = 0, \quad |-7.2| = 7.2$$

ADDING REAL NUMBERS

1. To add two numbers with the same sign, add their absolute values and attach their common sign.

2. To add two numbers with different signs, subtract the smaller absolute value from the larger absolute value and attach the sign of the number with the larger absolute value.

$$\frac{2}{7} + \frac{1}{7} = \frac{3}{7}$$
$$-5 + (-2.6) = -7.6$$
$$-18 + 6 = -12$$
$$20.8 + (-10.2) = 10.6$$

SUBTRACTING REAL NUMBERS

$$a - b = a + (-b)$$

$$18 - 21 = 18 + (-21) = -3$$

MULTIPLYING AND DIVIDING REAL NUMBERS

The product or quotient of two numbers with the same sign is positive.

$$(-8)(-4) = 32 \qquad \frac{-8}{-4} = 2$$
$$8 \cdot 4 = 32 \qquad \frac{8}{4} = 2$$

DEFINITIONS AND CONCEPTS	**EXAMPLES**

Section 1.4   Operations on Real Numbers (*continued*)

The product or quotient of two numbers with different signs is negative.	$-17 \cdot 2 = -34 \qquad \dfrac{-14}{2} = -7$ $4(-1.6) = -6.4 \qquad \dfrac{22}{-2} = -11$
A natural number **exponent** is a shorthand notation for repeated multiplication of the same factor.	$3^4 = 3 \cdot 3 \cdot 3 \cdot 3 = 81$
The notation $\sqrt{a}$ is used to denote the **positive,** or **principal square root** of a nonnegative number a. $\quad \sqrt{a} = b$ if $b^2 = a$ and b is positive Also, $\quad \sqrt[3]{a} = b$ if $b^3 = a$ $\quad \sqrt[4]{a} = b$ if $b^4 = a$ and b is positive.	$\sqrt{49} = 7$ $\sqrt[3]{64} = 4$ $\sqrt[4]{16} = 2$

Section 1.5   Order of Operations and Algebraic Expressions

ORDER OF OPERATIONS Simplify expressions using the order that follows. **1.** If grouping symbols such as parentheses are present, simplify expressions within those first, starting with the innermost set. **2.** Evaluate exponential expressions, roots, and absolute values. **3.** Multiply or divide in order from left to right. **4.** Add or subtract in order from left to right. To **evaluate** an algebraic expression containing variables, substitute the given numbers for the variables and simplify. The result is called the **value** of the expression.	Simplify: $\dfrac{42 - 2(3^2 - \sqrt{16})}{-8}$ $\dfrac{42 - 2(3^2 - \sqrt{16})}{-8} = \dfrac{42 - 2(9 - 4)}{-8}$ $= \dfrac{42 - 2(5)}{-8}$ $= \dfrac{42 - 10}{-8}$ $= \dfrac{32}{-8} = -4$ Evaluate: $2.7x$ when $x = 3$ $2.7x = 2.7(3)$ $\quad\ \ = 8.1$

Section 1.6   Exponents and Scientific Notation

PRODUCT RULE $a^m \cdot a^n = a^{m+n}$	$x^2 \cdot x^3 = x^5$
ZERO EXPONENT $a^0 = 1 \quad (a \neq 0)$	$7^0 = 1, (-10)^0 = 1$
QUOTIENT RULE $\dfrac{a^m}{a^n} = a^{m-n} \quad (a \neq 0)$	$\dfrac{y^{10}}{y^4} = y^{10-4} = y^6$
NEGATIVE EXPONENT $a^{-n} = \dfrac{1}{a^n} \quad (a \neq 0)$	$3^{-2} = \dfrac{1}{3^2} = \dfrac{1}{9}, \dfrac{x^{-5}}{x^{-7}} = x^{-5-(-7)} = x^2$
A positive number is written in **scientific notation** if it is written as the product of a number a, where $1 \leq a < 10$, and an integer power of 10: $a \times 10^n$.	Numbers written in scientific notation: $\quad 568{,}000 = 5.68 \times 10^5$ $\quad 0.0002117 = 2.117 \times 10^{-4}$

DEFINITIONS AND CONCEPTS	EXAMPLES
Section 1.7 More Work with Exponents and Scientific Notation	

POWER RULES

$$(a^m)^n = a^{m \cdot n}$$

$$(ab)^m = a^m b^m$$

$$\left(\frac{a}{b}\right)^n = \frac{a^n}{b^n}$$

$$(7^8)^2 = 7^{16}$$

$$(2y)^3 = 2^3 y^3 = 8y^3$$

$$\left(\frac{5x^{-3}}{x^2}\right)^{-2} = \frac{5^{-2}x^6}{x^{-4}}$$

$$= 5^{-2} \cdot x^{6-(-4)}$$

$$= \frac{x^{10}}{5^2}, \text{ or } \frac{x^{10}}{25}$$

1 CHAPTER REVIEW

(1.2) *Write each as an expression. Use x to represent the number.*

1. The quotient of a number and seven

2. The product of a number and seven

3. Four times the sum of a number and ten

4. The difference of three times a number and nine

Write each set in roster form.

5. $\{x \mid x$ is an odd integer between -2 and $4\}$

6. $\{x \mid x$ is an even integer between -3 and $7\}$

7. $\{x \mid x$ is a negative whole number$\}$

8. $\{x \mid x$ is a natural number that is not a rational number$\}$

9. $\{x \mid x$ is a whole number greater than $5\}$

10. $\{x \mid x$ is an integer less than $3\}$

Determine whether each statement is true or false if $A = \{6, 10, 12\}$, $B = \{5, 9, 11\}$, $C = \{\ldots, -3, -2, -1, 0, 1, 2, 3, \ldots\}$, $D = \{2, 4, 6, \ldots, 16\}$, $E = \{x \mid x$ is a rational number$\}$, $F = \{\ \}$, $G = \{x \mid x$ is an irrational number$\}$, and $H = \{x \mid x$ is a real number$\}$.

11. $10 \in D$

12. $59 \in B$

13. $\sqrt{169} \notin A$

14. $0 \notin F$

15. $\pi \in E$

16. $\pi \in H$

17. $\sqrt{4} \in G$

18. $-9 \in C$

List the elements of the set $\left\{5, -\frac{2}{3}, \frac{8}{2}, \sqrt{9}, 0.3, \sqrt{7}, 1\frac{5}{8}, -1, \pi\right\}$ that are also elements of each given set.

19. Whole numbers

20. Natural numbers

21. Rational numbers

22. Irrational numbers

23. Real numbers

24. Integers

(1.3) *Write each statement as an equation.*

25. Twelve is the product of x and negative 4.

26. The sum of n and twice n is negative fifteen.

27. Four times the sum of y and three is -1.

28. The difference of t and five, multiplied by six is four.

29. Seven subtracted from z is six.

30. Ten less than the product of x and nine is five.

31. The difference of x and 5 is the same as 12.

32. The opposite of four is equal to the product of y and seven.

33. Two-thirds is equal to twice the sum of n and one-fourth.

34. The sum of t and six amounts to negative twelve.

Find the opposite, or additive inverse, of each number.

35. $-\dfrac{3}{4}$ **36.** 0.6   **37.** 0   **38.** -1

Find the reciprocal, or multiplicative inverse, of each number.

39. $-\dfrac{3}{4}$ **40.** $\dfrac{1}{5}$ **41.** 0   **42.** -1

Name each property illustrated.

43. $(M + 5) + P = M + (5 + P)$

44. $5(3x - 4) = 15x - 20$

45. $(-4) + 4 = 0$

46. $(3 + x) + 7 = 7 + (3 + x)$

47. $(XY)Z = X(YZ)$

48. $\left(-\dfrac{3}{5}\right) \cdot \left(-\dfrac{5}{3}\right) = 1$

49. $T \cdot 1 = T$

50. $(x + y) + z = (y + x) + z$

51. $A + 0 = A$

52. $8 \cdot 1 = 8$

Complete each equation using the given property.

53. $5(x - 3z) = $ _____ Distributive property

54. $(7 + y) + (3 + x) = $ _____ Commutative property

55. $0 = $ _____ Additive inverse property

56. $1 = $ _____ Multiplicative inverse property

57. $[(3.4)(0.7)]5 = $ _____ Associative property

58. $7 = $ _____ Additive identity property

Insert $<, >,$ or $=$ to make each statement true.

59. -9 -12 **60.** 0 -6 **61.** -3 -1 **62.** 7 $|-7|$

63. -5 $-(-5)$ **64.** $-(-2)$ -2

Use the distributive property to multiply.

65. $-3(2x - 7y)$

66. $-9(10a + 4b)$

67. $\dfrac{1}{2}(18m - 8n + 1)$

68. $\dfrac{1}{3}(12x - 33y + 2)$

(1.4) *Simplify.*

69. $-7 + 3$

70. $-10 + (-25)$

71. $5(-0.4)$

72. $(-3.1)(-0.1)$

73. $-7 - (-15)$

74. $9 - (-4.3)$

75. $\sqrt{16} - 2^3$

76. $\sqrt[3]{27} - 5^2$

77. $(-24) \div 0$

78. $0 \div (-45)$

79. $(-36) \div (-9)$

80. $60 \div (-12)$

81. $-\dfrac{4}{5} - \left(-\dfrac{2}{3}\right)$

82. $\dfrac{5}{4} - \left(-2\dfrac{3}{4}\right)$

△ **83.** Determine the unknown fractional part.

84. The Bertha Rogers gas well in Washita County, Oklahoma, is the deepest well in the United States. From the surface, this now-capped well extends 31,441 feet into the earth. The elevation of the nearby Cordell Municipal Airport is 1589 feet above sea level. Assuming that the surface elevation of the well is the same as at the Cordell Municipal Airport, find the elevation relative to sea level of the *bottom* of the Bertha Rogers gas well. *(Sources:* U.S. Geological Survey, Oklahoma Department of Transportation)

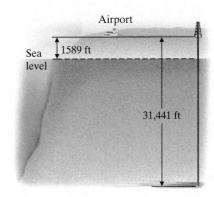

(1.5) *Simplify.*

85. $-5 + 7 - 3 - (-10)$

86. $8 - (-3) + (-4) + 6$

87. $3(4 - 5)^4$

88. $6(7 - 10)^2$

89. $\left(-\dfrac{8}{15}\right) \cdot \left(-\dfrac{2}{3}\right)^2$

90. $\left(-\dfrac{3}{4}\right)^2 \cdot \left(-\dfrac{10}{21}\right)$

91. $-\dfrac{6}{15} \div \dfrac{8}{25}$

92. $\dfrac{4}{9} \div \left(-\dfrac{8}{45}\right)$

93. $-\dfrac{3}{8} + 3(2) \div 6$

94. $5(-2) - (-3) - \dfrac{1}{6} + \dfrac{2}{3}$

95. $|2^3 - 3^2| - |5 - 7|$

96. $|5^2 - 2^2| + |9 \div (-3)|$

97. $(2^3 - 3^2) - (5 - 7)$

98. $(5^2 - 2^2) + [9 \div (-3)]$

99. $\dfrac{(8 - 10)^3 - (-4)^2}{2 + 8(2) \div 4}$

100. $\dfrac{(2 + 4)^2 + (-1)^5}{12 \div 2 \cdot 3 - 3}$

101. $\dfrac{(4 - 9) + 4 - 9}{10 - 12 \div 4 \cdot 8}$

102. $\dfrac{3 - 7 - (7 - 3)}{15 + 30 \div 6 \cdot 2}$

103. $\dfrac{\sqrt{25}}{4 + 3 \cdot 7}$

104. $\dfrac{\sqrt{64}}{24 - 8 \cdot 2}$

△ *The algebraic expression $2\pi r$ represents the circumference of (distance around) a circle of radius r.*

△ **105.** Complete the table by evaluating the expression at the given values of r. (Use 3.14 for π.)

Radius	r	1	10	100
Circumference				

106. As the radius of a circle increases, does the circumference of the circle increase or decrease?

Simplify.

107. $14x - 3 - 11x - 10$

108. $81y + 19 - y - 20$

109. $7a - 3(2a - y) + 4y - 6$

110. $9b - 4(8b - x) + 9x - 10$

111. $\frac{1}{5}(15m - 5n) - (3m - 5) + 2n$

112. $-(2x - 9) + \frac{1}{4}(8x + 4y) - 13$

(1.6) *Evaluate.*

113. $(-2)^2$

114. $(-3)^4$

115. -2^2

116. -3^4

117. 8^0

118. -9^0

119. -4^{-2}

120. $(-4)^{-2}$

Simplify each expression. Write each answer with positive exponents only. Assume that variables in the exponents represent nonzero integers and that all other variables are not 0.

121. $-xy^2 \cdot y^3 \cdot xy^2z$

122. $(-4xy)(-3xy^2b)$

123. $a^{-14} \cdot a^5$

124. $\dfrac{a^{16}}{a^{17}}$

125. $\dfrac{x^{-7}}{x^4}$

126. $\dfrac{9a(a^{-3})}{18a^{15}}$

127. $\dfrac{y^{6p-3}}{y^{6p+2}}$

128. $(3x^{2a+b}y^{-3b})^2$

Write each number in scientific notation.

129. 36,890,000

130. 0.000362

Write each number without exponents.

131. 1.678×10^{-6}

132. 4.1×10^5

(1.7) *Simplify. Write each answer with positive exponents only.*

133. $(8^5)^3$

134. $\left(\dfrac{a}{4}\right)^2$

135. $(3x)^3$

136. $(-4x)^{-2}$

137. $\left(\dfrac{6x}{5}\right)^2$ **138.** $(8^6)^{-3}$ **139.** $\left(\dfrac{4}{3}\right)^{-2}$ **140.** $(-2x^3)^{-3}$

141. $\left(\dfrac{8p^6}{4p^4}\right)^{-2}$ **142.** $(-3x^{-2}y^2)^3$ **143.** $\left(\dfrac{x^{-5}y^{-3}}{z^3}\right)^{-5}$ **144.** $\dfrac{4^{-1}x^3yz}{x^{-2}yx^4}$

145. $(5xyz)^{-4}(x^{-2})^{-3}$ **146.** $\dfrac{2(3yz)^{-3}}{y^{-3}}$

Simplify each expression.

147. $x^{4a}(3x^{5a})^3$ **148.** $\dfrac{4y^{3x-3}}{2y^{2x+4}}$

Mixed Review

Complete the table.

	Number	Opposite of Number	Reciprocal of Number
149.	$-\dfrac{3}{4}$		
150.		-5	

Simplify. If necessary, write answers with positive exponents only.

151. $-2\left(5x + \dfrac{1}{2}\right) + 7.1$ **152.** $\sqrt{36} \div 2 \cdot 3$

153. $-\dfrac{7}{11} - \left(-\dfrac{1}{11}\right)$ **154.** $10 - (-1) + (-2) + 6$

155. $\left(-\dfrac{2}{3}\right)^3 \div \dfrac{10}{9}$ **156.** $\dfrac{(3-5)^2 + (-1)^3}{1 + 2(3 - (-1))^2}$

157. $\dfrac{1}{3}(9x - 3y) - (4x - 1) + 4y$ **158.** -5^{-2}

159. $\left(\dfrac{5x^7}{10x^{-3}}\right)^{-3}$ **160.** $(-5a^{-2}bc)^{-2} \cdot (3a^{-3}b^2c^3)^2$

1 CHAPTER TEST

 Remember to use the Chapter Test Prep Video CD to see the fully worked-out solutions to any of the exercises you want to review.

Determine whether each statement is true or false.

1. $-2.3 > -2.33$

2. $-6^2 = (-6)^2$

3. $(-2)(-3)(0) = \dfrac{(-4)}{0}$

4. Write the set in roster form.
$\{x \mid x \text{ is a whole number less than 2}\}$

5. All natural numbers are integers.

6. All rational numbers are integers.

Simplify.

7. $5 - 12 \div 3(2)$

8. $(4 - 9)^3 - |-4 - 6|^2$

9. $[3|4 - 5|^5 - (-9)] \div (-6)$

10. $\dfrac{6(7 - 9)^3 + (-2)}{(-2)(-5)(-5)}$

11. $\dfrac{(4 - \sqrt{16}) - (-7 - 20)}{-2(1 - 4)^2}$

Evaluate each expression when $q = 4$, $r = -2$, and $t = 1$.

12. $q^2 - r^2$

13. $\dfrac{5t - 3q}{3r - 1}$

14. The algebraic expression $5.75x$ represents the total cost for x adults to attend the theater.
  a. Complete the table that follows.

Adults	x	1	3	10	20
Total Cost	$5.75x$				

  b. As the number of adults increases, does the total cost increase or decrease?

Write each statement as an equation.

15. Three times the quotient of n and five is the opposite of n.

16. Twenty is equal to six subtracted from twice x.

17. Negative two is equal to x divided by the sum of x and five.

Answers

1. _____

2. _____

3. _____

4. _____

5. _____

6. _____

7. _____

8. _____

9. _____

10. _____

11. _____

12. _____

13. _____

14. **a.** see table

   b. _____

15. _____

16. _____

17. _____

18. _____

19. _____

20. _____

21. _____

22. _____

23. _____

24. _____

25. _____

26. _____

27. _____

28. _____

29. _____

30. _____

31. _____

32. _____

33. _____

34. _____

Name each property illustrated.

18. $6(x - 4) = 6x - 24$

19. $(4 + x) + z = 4 + (x + z)$

20. $(-7) + 7 = 0$

21. Find the reciprocal and opposite of $-\dfrac{7}{11}$.

Simplify.

22. $9x + 12y - 3x - 6.2 + 20$

23. $\dfrac{1}{3}(15x - 27y) - 2(3x - y) - 4y$

Simplify. Write answers using positive exponents only.

24. $(-9x)^{-2}$

25. $\dfrac{6^{-1}a^2b^{-3}}{3^{-2}a^{-5}b^2}$

26. $\left(\dfrac{-xy^{-5}z}{xy^3}\right)^{-5}$

27. $(-6a^{-5}b^{12})^{-2}(3ab^5)^2$

28. $\dfrac{(x^w)^2}{(x^{4w})^{-2}}$

Write each number in scientific notation.

29. 630,000,000

30. 0.01200

31. Write 5.0×10^{-6} without exponents.

Use scientific notation to find the quotient. Express the quotient in scientific notation.

32. $\dfrac{(0.00012)(144{,}000)}{0.0003}$

33. $\dfrac{(0.0024)(0.00012)}{0.00032}$

34. In fiscal year 2004, the United States Postal Service handled approximately 848,600,000 pieces of Priority Mail. Write this number in scientific notation. (*Source*: United States Postal Service)

2

Equations, Inequalities, and Problem Solving

Mathematics is a tool for solving problems in such diverse fields as transportation, engineering, economics, medicine, business, and biology. We solve problems using mathematics by modeling real-world phenomena with mathematical equations or inequalities. Our ability to solve problems using mathematics, then, depends in part on our ability to solve equations and inequalities. In this chapter, we solve linear equations and inequalities in one variable and graph their solutions on number lines.

The federal Bureau of Labor Statistics (BLS) has issued its projections for job growth in the United States between 2000 and 2012. These projections are based on employer surveys distributed and collected by BLS. In Exercise Set 2.2, Exercises 37 and 38, you will find the actual increase in the number of some jobs as well as the percent increase in the number of jobs.

Occupations with the Largest Predicted Job Growth 2000-2012

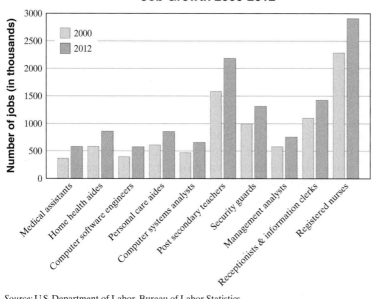

Source: U.S. Department of Labor, Bureau of Labor Statistics

Objectives

A Decide Whether a Number Is a Solution of an Equation.

B Solve Linear Equations Using Properties of Equality.

C Solve Linear Equations That Can Be Simplified by Combining Like Terms.

D Solve Linear Equations Containing Fractions or Decimals.

E Recognize Identities and Equations with No Solution.

Objective A Deciding Whether a Number Is a Solution of an Equation

An **equation** is a statement that two expressions are equal. To solve problems, we need to be able to solve equations. In this section, we will solve a special type of equation called a *linear equation in one variable.*

Linear equations model many real-life problems. For example, we can use a linear equation to calculate the increase in households (in millions) with digital cameras.

With the help of your computer, digital cameras allow you to see your pictures and make copies immediately, send them in e-mail or use them on a Web page. Numbers of households with these cameras for various years are shown in the graph below.

Households with a Digital Camera

Source: Internet Search

To find the increase in households from 2002 to 2003, for example, we can use the equation below.

In words:	Increase in households	is	households in 2003	minus	households in 2002
Translate:	x	$=$	31.5	$-$	22.8

Since our variable x (increase in households) is by itself on one side of the equation, we can find the value of x by simplifying the right side.

$$x = 8.7$$

The increase in households with digital cameras from 2002 to 2003 is $8.7 million.

The **equation,** $x = 31.5 - 22.8$, like every other equation, is a statement that two expressions are equal. Oftentimes, the unknown variable is not by itself on one side of the equation. In these cases, we will use properties of equality to write equivalent equations so that a solution many be found. This is called **solving the equation.** In this section, we concentrate on solving equations such as this one, called **linear equations** in one variable. Linear equations are also called **first-degree equations** since the exponent on the variable is 1.

Linear Equations in One Variable

$$3x = -15 \qquad 7 - y = 3y \qquad 4n - 9n + 6 = 0 \qquad z = -2$$

Linear Equation in One Variable

A **linear equation in one variable** is an equation that can be written in the form

$$ax + b = c$$

where a, b, and c are real numbers and $a \neq 0$.

When a variable in an equation is replaced by a number and the resulting equation is true, then that number is called a **solution** of the equation. For example, 1 is a solution of the equation $3x + 4 = 7$ since $3(1) + 4 = 7$ is a true statement. But 2 is not a solution of this equation since $3(2) + 4 = 7$ is *not* a true statement. The **solution set** of an equation is the set of solutions of the equation. For example, the solution set of $3x + 4 = 7$ is $\{1\}$.

EXAMPLE 1 Determine whether -15 is a solution of $x - 9 = -24$.

Solution: We replace x with -15 and see whether a true statement results.

$$x - 9 = -24$$
$$-15 - 9 \stackrel{?}{=} -24 \quad \text{Replace } x \text{ with } -15.$$
$$-24 = -24 \quad \text{True}$$

Since a true statement results, -15 is a solution.

🔲 **Work Practice Problem 1**

EXAMPLE 2 Determine whether 5 is a solution of $2x - 3 = x + 3$.

Solution:

$$2x - 3 = x + 3$$
$$2 \cdot 5 - 3 \stackrel{?}{=} 5 + 3 \quad \text{Replace } x \text{ with 5.}$$
$$7 = 8 \quad \text{False}$$

Since a false statement results, 5 is not a solution.

🔲 **Work Practice Problem 2**

Objective B Using the Properties of Equality

To **solve an equation** is to find the solution set of an equation. Equations with the same solution set are called **equivalent equations.** For example,

$$3x + 4 = 7 \qquad 3x = 3 \qquad x = 1$$

are equivalent equations because they all have the same solution set, namely, $\{1\}$. To solve an equation in x, we start with the given equation and write a series of simpler equivalent equations until we obtain an equation of the form

$$x = \textbf{number}$$

PRACTICE PROBLEM 1

Determine whether -7 is a solution of $14 - x = 21$.

PRACTICE PROBLEM 2

Determine whether 8 is a solution of $x - 10 = 2x - 14$.

To write equivalent equations, we use two important properties.

> ### Addition Property of Equality
>
> If a, b, and c, are real numbers, then
>
> $$a = b \quad \text{and} \quad a + c = b + c$$
>
> are equivalent equations.
>
> ### Multiplication Property of Equality
>
> If $c \neq 0$, then
>
> $$a = b \quad \text{and} \quad ac = bc$$
>
> are equivalent equations.

The **addition property of equality** guarantees that the same number may be added to both sides of an equation and the result is an equivalent equation. Recall that we define subtraction in terms of addition.

$$7 - 10 = 7 + (-10) = -3$$

This means that the addition property also says we can *subtract* the same number from both sides and the result is an equivalent equation.

The **multiplication property of equality** guarantees that both sides of an equation may be multiplied by the same nonzero number and the result is an equivalent equation. Recall that we define division in terms of multiplication. This means that the multiplication property also says we can *divide* both sides by the same nonzero number and the result is an equivalent equation.

For example, to solve $2x + 5 = 9$, we use the addition and multiplication properties of equality to get x alone—that is, to write an equivalent equation of the form

$$x = \text{number}$$

We will do this in the next example.

PRACTICE PROBLEM 3

Solve: $3x + 6 = 21$

EXAMPLE 3 Solve: $2x + 5 = 9$

Solution: First we use the addition property of equality and subtract 5 from both sides.

$$2x + 5 = 9$$
$$2x + 5 - 5 = 9 - 5 \quad \text{Subtract 5 from both sides.}$$
$$2x = 4 \quad \text{Simplify.}$$

Now we use the multiplication property of equality and divide both sides by 2.

$$\frac{2x}{2} = \frac{4}{2} \quad \text{Divide both sides by 2.}$$
$$x = 2 \quad \text{Simplify.}$$

Check: To check, we replace x in the original equation with 2.

$$2x + 5 = 9 \quad \text{Original equation}$$
$$2(2) + 5 \stackrel{?}{=} 9 \quad \text{Replace } x \text{ with 2.}$$
$$4 + 5 \stackrel{?}{=} 9$$
$$9 = 9 \quad \text{True}$$

The solution set is $\{2\}$.

■ **Work Practice Problem 3**

Answer

3. $\{5\}$

EXAMPLE 4 Solve: $0.6 = 2 - 3.5c$

Solution: We use both the addition property and the multiplication property of equality.

$$0.6 = 2 - 3.5c$$

$$0.6 - 2 = 2 - 3.5c - 2 \quad \text{Subtract 2 from both sides.}$$

$$-1.4 = -3.5c \quad \text{Simplify.}$$

$$\frac{-1.4}{-3.5} = \frac{-3.5c}{-3.5} \quad \text{Divide both sides by } -3.5.$$

$$0.4 = c \quad \text{Simplify } \frac{-1.4}{-3.5}.$$

Check:

$$0.6 = 2 - 3.5c$$

$$0.6 \stackrel{?}{=} 2 - 3.5(0.4) \quad \text{Replace } c \text{ with } 0.4.$$

$$0.6 \stackrel{?}{=} 2 - 1.4 \quad \text{Multiply.}$$

$$0.6 = 0.6 \quad \text{True}$$

The solution set is $\{0.4\}$.

■ **Work Practice Problem 4**

PRACTICE PROBLEM 4

Solve: $4.5 = 3 + 2.5x$

> **Helpful Hint**
>
> Don't forget that
> $$0.4 = c \text{ and } c = 0.4$$
> are equivalent equations. We may solve an equation so that the variable is alone on either side of the equation.

Objective ⓒ Solving Linear Equations by Combining Like Terms

Often, an equation can be simplified by removing any grouping symbols and combining any like terms.

EXAMPLE 5 Solve: $-4x - 1 + 5x = 9x + 3 - 7x$

Solution: First we simplify both sides of this equation by combining like terms. Then, let's get variable terms on the same side of the equation by using the addition property of equality to subtract $2x$ from both sides. Next, we use this same property to add 1 to both sides of the equation.

$$-4x - 1 + 5x = 9x + 3 - 7x$$

$$x - 1 = 2x + 3 \quad \text{Combine like terms.}$$

$$x - 1 - 2x = 2x + 3 - 2x \quad \text{Subtract } 2x \text{ from both sides.}$$

$$-x - 1 = 3 \quad \text{Simplify.}$$

$$-x - 1 + 1 = 3 + 1 \quad \text{Add 1 to both sides.}$$

$$-x = 4 \quad \text{Simplify.}$$

Notice that this equation is not solved for x since we have $-x$, or $-1x$, not x. To get x alone, we divide both sides by -1.

$$\frac{-x}{-1} = \frac{4}{-1} \quad \text{Divide both sides by } -1.$$

$$x = -4 \quad \text{Simplify.}$$

Check to see that the solution set is $\{-4\}$.

■ **Work Practice Problem 5**

If an equation contains parentheses, we use the distributive property to remove them.

PRACTICE PROBLEM 5

Solve:
$-2x + 2 - 3x = 8x + 20 - 7x$

Answers

4. $\{0.6\}$, **5.** $\{-3\}$

PRACTICE PROBLEM 6

Solve: $4(x - 2) = 6x - 10$

EXAMPLE 6 Solve: $3(x - 3) = 5x - 9$

Solution: First we use the distributive property.

$$3(x - 3) = 5x - 9$$
$$3x - 9 = 5x - 9 \quad \text{Use the distributive property.}$$

Next we get variable terms on the same side of the equation by using the addition property of equality.

$$3x - 9 - 5x = 5x - 9 - 5x \quad \text{Subtract } 5x \text{ from both sides.}$$
$$-2x - 9 = -9 \quad \text{Simplify.}$$
$$-2x - 9 + 9 = -9 + 9 \quad \text{Add 9 to both sides.}$$
$$-2x = 0 \quad \text{Simplify.}$$
$$\frac{-2x}{-2} = \frac{0}{-2} \quad \text{Divide both sides by } -2.$$
$$x = 0$$

Check to see that $\{0\}$ is the solution set.

Work Practice Problem 6

Objective **D** **Solving Linear Equations Containing Fractions or Decimals**

If an equation contains fractions, we can first clear the equation of fractions by multiplying both sides of the equation by the *least common denominator* (LCD) of all fractions in the equation.

PRACTICE PROBLEM 7

Solve: $\dfrac{x}{6} - \dfrac{x}{8} = \dfrac{1}{8}$

EXAMPLE 7 Solve: $\dfrac{y}{3} - \dfrac{y}{4} = \dfrac{1}{6}$

Solution: First we clear the equation of fractions by multiplying both sides of the equation by 12, the LCD of the denominators 3, 4, and 6.

$$\frac{y}{3} - \frac{y}{4} = \frac{1}{6}$$
$$12\left(\frac{y}{3} - \frac{y}{4}\right) = 12\left(\frac{1}{6}\right) \quad \text{Multiply both sides by the LCD, 12.}$$
$$12\left(\frac{y}{3}\right) - 12\left(\frac{y}{4}\right) = 2 \quad \text{Use the distributive property.}$$
$$4y - 3y = 2 \quad \text{Simplify.}$$
$$y = 2 \quad \text{Simplify.}$$

Check: To check, we replace y with 2 in the original equation.

$$\frac{y}{3} - \frac{y}{4} = \frac{1}{6} \quad \text{Original equation}$$
$$\frac{2}{3} - \frac{2}{4} \stackrel{?}{=} \frac{1}{6} \quad \text{Replace } y \text{ with 2.}$$
$$\frac{8}{12} - \frac{6}{12} \stackrel{?}{=} \frac{1}{6} \quad \text{Write fractions with the LCD.}$$
$$\frac{2}{12} \stackrel{?}{=} \frac{1}{6} \quad \text{Subtract.}$$
$$\frac{1}{6} = \frac{1}{6} \quad \text{True}$$

Since a true statement results, the solution set is $\{2\}$.

Work Practice Problem 7

Answers

6. $\{1\}$, 7. $\{3\}$

As a general guideline, the following steps may be used to solve a linear equation in one variable.

Solving a Linear Equation in One Variable

Step 1: Clear the equation of fractions or decimals by multiplying both sides of the equation by an appropriate nonzero number.

Step 2: Use the distributive property to remove grouping symbols such as parentheses.

Step 3: Combine like terms on each side of the equation.

Step 4: Use the addition property of equality to rewrite the equation as an equivalent equation, with variable terms on one side and numbers on the other side.

Step 5: Use the multiplication property of equality to get the variable alone.

Step 6: Check the proposed solution in the original equation.

EXAMPLE 8 Solve: $\dfrac{x + 5}{2} + \dfrac{1}{2} = \dfrac{1}{8}(15x + 3)$

Solution: To begin, we multiply both sides of the equation by 8, the LCD of 2 and 8. This will clear the equation of fractions.

$$8\left(\frac{x+5}{2} + \frac{1}{2}\right) = 8\left[\frac{1}{8}(15x + 3)\right] \quad \text{Multiply both sides by 8.}$$

$$8\left(\frac{x+5}{2}\right) + 8\left(\frac{1}{2}\right) = 8 \cdot \frac{1}{8}(15x + 3) \quad \text{Use the distributive property.}$$

$$4(x + 5) + 4 = 15x + 3 \quad \text{Multiply.}$$

$$4x + 20 + 4 = 15x + 3 \quad \text{Use the distributive property to remove parentheses.}$$

$$4x + 24 = 15x + 3 \quad \text{Combine like terms.}$$

$$4x - 15x = 3 - 24 \quad \text{Subtract } 15x \text{ and 24 from both sides.}$$

$$-11x = -21 \quad \text{Simplify.}$$

$$\frac{-11x}{-11} = \frac{-21}{-11} \quad \text{Divide both sides by } -11.$$

$$x = \frac{21}{11} \quad \text{Simplify.}$$

To check, verify that replacing x with $\dfrac{21}{11}$ makes the original equation true. The solution set is $\left\{\dfrac{21}{11}\right\}$.

Work Practice Problem 8

If an equation contains decimals, you may want to first clear the equation of decimals by multiplying by an appropriate power of 10.

PRACTICE PROBLEM 8

Solve:

$$\frac{x - 1}{3} + \frac{2}{3} = x - \frac{2x + 3}{9}$$

Helpful Hint

When we multiply both sides of an equation by a number, the distributive property tells us that each term of the equation is multiplied by the number.

Answer

8. $\left\{\dfrac{3}{2}\right\}$

PRACTICE PROBLEM 9

Solve:

$0.2x + 0.1 = 0.12x - 0.06$

EXAMPLE 9 Solve: $0.3x + 0.1 = 0.27x - 0.02$

Solution: To clear this equation of decimals, we multiply both sides of the equation by 100. Recall that multiplying a number by 100 moves its decimal point two places to the right.

$$100(0.3x + 0.1) = 100(0.27x - 0.02)$$

$$100(0.3x) + 100(0.1) = 100(0.27x) - 100(0.02) \quad \text{Use the distributive property.}$$

$$30x + 10 = 27x - 2 \quad \text{Multiply.}$$

$$30x - 27x = -2 - 10 \quad \text{Subtract } 27x \text{ and } 10 \text{ from both sides.}$$

$$3x = -12 \quad \text{Simplify.}$$

$$\frac{3x}{3} = \frac{-12}{3} \quad \text{Divide both sides by 3.}$$

$$x = -4 \quad \text{Simplify.}$$

Check to see that the solution set is $\{-4\}$.

🔲 **Work Practice Problem 9**

✔**Concept Check** Explain what is wrong with the following:

$$3x - 5 = 16$$
$$3x = 11$$
$$\frac{3x}{3} = \frac{11}{3}$$
$$x = \frac{11}{3}$$

Objective **E** **Recognizing Identities and Equations with No Solution**

So far, each linear equation that we have solved has had a single solution. We will now look at two other types of equations: *contradictions* and *identities*.

An equation in one variable that has no solution is called a **contradiction,** and an equation in one variable that has every number (for which the equation is defined) as a solution is called an **identity.** The next examples show how to recognize contradictions and identities.

PRACTICE PROBLEM 10

Solve: $5x - 1 = 5(x + 3)$

EXAMPLE 10 Solve: $3x + 5 = 3(x + 2)$

Solution: First we use the distributive property to remove parentheses.

$$3x + 5 = 3(x + 2)$$
$$3x + 5 = 3x + 6 \quad \text{Use the distributive property.}$$
$$3x + 5 - 3x = 3x + 6 - 3x \quad \text{Subtract } 3x \text{ from both sides.}$$
$$5 = 6 \quad \text{False.}$$

The equation $5 = 6$ is a false statement no matter what value the variable x might have. Thus the original equation has no solution. Its solution set is written either as $\{ \ \}$ or $\varnothing$. This equation is a contradiction.

🔲 **Work Practice Problem 10**

Answers

9. $\{-2\}$, 10. $\varnothing$

✔ **Concept Check Answer**

$$3x - 5 = 16$$
$$3x = 21$$
$$x = 7$$

Therefore the correct solution set is $\{7\}$.

Helpful Hint

A solution set of $\{0\}$ and a solution set of $\{ \ \}$ are not the same. The solution set $\{0\}$ means 1 solution, 0. The solution set $\{ \ \}$ means no solution.

EXAMPLE 11 Solve: $6x - 4 = 2 + 6(x - 1)$

Solution: First we use the distributive property to remove parentheses.

$6x - 4 = 2 + 6(x - 1)$

$6x - 4 = 2 + 6x - 6$ Use the distributive property.

$6x - 4 = 6x - 4$ Combine the terms.

At this point we might notice that both sides of the equation are the same, so replacing x by any real number gives a true statement. Thus the solution set of this equation is the set of real numbers, and the equation is an identity. Continuing to "solve" $6x - 4 = 6x - 4$, we eventually arrive at the same conclusion.

$6x - 4 + 4 = 6x - 4 + 4$ Add 4 to both sides.

$6x = 6x$ Simplify.

$6x - 6x = 6x - 6x$ Subtract $6x$ from both sides.

$0 = 0$ True

Since $0 = 0$ is a true statement for every value of x, the solution set is the set of all real numbers, which can be written as $\{x \mid x \text{ is a real number}\}$. The equation is called an identity.

■ **Work Practice Problem 11**

Helpful Hint

For linear equations, *any* false statement such as $5 = 6$, $0 = 1$, or $-2 = 2$ informs us that the original equation has no solution. Also, *any* true statement such as $0 = 0$, $2 = 2$, or $-5 = -5$ informs us that the original equation is an identity.

PRACTICE PROBLEM 11

Solve:
$-4(x - 1) = -4x - 9 + 13$

Answer
11. $\{x \mid x \text{ is a real number}\}$

Mental Math

Solve each equation.

1. $3x = 18$ **2.** $2x = 60$ **3.** $x - 7 = 10$ **4.** $x - 2 = 15$

5. $\dfrac{x}{2} = 4$ **6.** $\dfrac{x}{3} = 5$ **7.** $x + 1 = 11$ **8.** $x + 4 = 20$

2.1 EXERCISE SET

FOR EXTRA HELP

Student Solutions Manual   PH Math/Tutor Center   CD/Video for Review   Math XL MathXL®   MyMathLab MyMathLab

Objective A *Determine whether each number is a solution of the given equation. See Examples 1 and 2.*

1. $-24;\ \dfrac{x}{-6} = 4$    **2.** $15;\ \dfrac{x}{-3} = -5$ **3.** $-3;\ x - 17 = 20$

4. $-8;\ x - 10 = -2$ **5.** $-2;\ 5 + 3x = -1$ **6.** $-1;\ 6 - 2x = 4$

7. $5;\ x - 7 = x + 2$ **8.** $5;\ x - 1 = x - 1$ **9.** $5;\ 4(x - 3) = 12$

10. $12;\ 5(x - 6) = 30$ **11.** $-8;\ 4x - 2 = 5x + 6$ **12.** $2;\ 7x + 1 = 6x - 1$

Objective B *Solve each equation and check. See Examples 3 and 4.*

13. $-5x = -30$ **14.** $-2x = 18$ **15.** $-10 = x + 12$

16. $-25 = y + 30$ **17.** $x - 2.8 = 1.9$ **18.** $y - 8.6 = -6.3$

19. $5x - 4 = 26 + 2x$ **20.** $5y - 3 = 11 + 3y$ **21.** $-4.1 - 7z = 3.6$

22. $10.3 - 6x = -2.3$ **23.** $5y + 12 = 2y - 3$ **24.** $4x + 14 = 6x + 8$

Objective C *Solve each equation and check. See Examples 5 and 6.*

25. $3x - 4 - 5x = x + 4 + x$ **26.** $13x - 15x + 8 = 4x + 2 - 24$

27. $8x - 5x + 3 = x - 7 + 10$ **28.** $6 + 3x + x = -x + 8 - 26 + 24$

29. $5x + 12 = 2(2x + 7)$ **30.** $2(4x + 3) = 7x + 5$

31. $3(x - 6) = 5x$ **32.** $6x = 4(x - 5)$

33. $-2(5y - 1) - y = -4(y - 3)$ **34.** $-4(3n - 2) - n = -11(n - 1)$

Objective D *Solve each equation and check. See Examples 7 through 9.*

35. $\dfrac{x}{2} + \dfrac{x}{3} = \dfrac{3}{4}$ **36.** $\dfrac{x}{2} + \dfrac{x}{5} = \dfrac{5}{4}$ **37.** $\dfrac{3t}{4} - \dfrac{t}{2} = 1$ **38.** $\dfrac{4r}{5} - 7 = \dfrac{r}{10}$

39. $\dfrac{n-3}{4} + \dfrac{n+5}{7} = \dfrac{5}{14}$

40. $\dfrac{2+h}{9} + \dfrac{h-1}{3} = \dfrac{1}{3}$

41. $0.6x - 10 = 1.4x - 14$

42. $0.3x + 2.4 = 0.1x + 4$

43. $\dfrac{3x-1}{9} + x = \dfrac{3x+1}{3} + 4$

44. $\dfrac{2z+7}{8} - 2 = z + \dfrac{z-1}{2}$

45. $1.5(4 - x) = 1.3(2 - x)$

46. $2.4(2x + 3) = -0.1(2x + 3)$

Objective **E** *Solve each equation. See Examples 10 and 11.*

47. $4(n + 3) = 2(6 + 2n)$

48. $6(4n + 4) = 8(3 + 3n)$

49. $3(x + 1) + 5 = 3x + 2$

50. $4(x + 2) + 4 = 4x - 8$

51. $2(x - 8) + x = 3(x - 6) + 2$

52. $5(x - 4) + x = 6(x - 2) - 8$

53. $4(x + 5) = 3(x - 4) + x$

54. $9(x - 2) = 8(x - 3) + x$

Objectives **B** **C** **D** **E** **Mixed Practice** *Solve each equation. See Examples 3 through 11.*

55. $\dfrac{3}{8} + \dfrac{b}{3} = \dfrac{5}{12}$

56. $\dfrac{a}{2} + \dfrac{7}{4} = 5$

57. $x - 10 = -6x - 10$

58. $4x - 7 = 2x - 7$

59. $5(x - 2) + 2x = 7(x + 4) - 38$

60. $3x + 2(x + 4) = 5(x + 1) + 3$

61. $y + 0.2 = 0.6(y + 3)$

62. $-(w + 0.2) = 0.3(4 - w)$

63. $\dfrac{1}{4}(a + 2) = \dfrac{1}{6}(5 - a)$

64. $\dfrac{1}{3}(8 + 2c) = \dfrac{1}{5}(3c - 5)$

65. $2y + 5(y - 4) = 4y - 2(y - 10)$

66. $9c - 3(6 - 5c) = c - 2(3c + 9)$

67. $6x - 2(x - 3) = 4(x + 1) + 4$

68. $10x - 2(x + 4) = 8(x - 2) + 6$

69. $\dfrac{m-4}{3} - \dfrac{3m-1}{5} = 1$

70. $\dfrac{n+1}{8} - \dfrac{2-n}{3} = \dfrac{5}{6}$

71. $8x - 12 - 3x = 9x - 7$

72. $10y - 18 - 4y = 12y - 13$

73. $-(3x - 5) - (2x - 6) + 1 = -5(x - 1) - (3x + 2) + 3$

74. $-4(2x - 3) - (10x + 7) - 2 = -(12x - 5) - (4x + 9) - 1$

75. $\dfrac{1}{3}(y + 4) + 6 = \dfrac{1}{4}(3y - 1) - 2$

76. $\dfrac{1}{5}(2y - 1) - 2 = \dfrac{1}{2}(3y - 5) + 3$

77. $2[7 - 5(1 - n)] + 8n = -16 + 3[6(n + 1) - 3n]$

78. $3[8 - 4(n - 2)] + 5n = -20 + 2[5(1 - n) - 6n]$

Review

Translate each phrase into an expression. Use the variable x to represent each unknown number. See Section 1.2.

79. The quotient of 8 and a number

80. The sum of 8 and a number

81. The product of 8 and a number

82. The difference of 8 and a number

83. Five subtracted from twice a number

84. Two more than three times a number

Concept Extensions

Find the error for each proposed solution. Then correct the proposed solution. See the Concept Check in this section.

85.
$$2x + 19 = 13$$
$$2x = 32$$
$$\frac{2x}{2} = \frac{32}{2}$$
$$x = 16$$

86.
$$-3(x - 4) = 10$$
$$-3x - 12 = 10$$
$$-3x = 22$$
$$\frac{-3x}{-3} = \frac{22}{-3}$$
$$x = -\frac{22}{3}$$

87.
$$9x + 1.6 = 4x + 0.4$$
$$5x = 1.2$$
$$\frac{5x}{5} = \frac{1.2}{5}$$
$$x = 0.24$$

88.
$$\frac{x}{3} + 7 = \frac{5x}{3}$$
$$x + 7 = 5x$$
$$7 = 4x$$
$$\frac{7}{4} = \frac{4x}{4}$$
$$\frac{7}{2} = x$$

89. **a.** Simplify the expression $4(x + 1) + 1$.
b. Solve the equation $4(x + 1) + 1 = -7$.
c. Explain the difference between solving an equation for a variable and simplifying an expression.

90. Explain why the multiplication property of equality does not include multiplying both sides of an equation by 0. (*Hint:* Write down a false statement and then multiply both sides by 0. Is the result true or false? What does this mean?)

91. In your own words, explain why the equation $x + 7 = x + 6$ has no solution while the solution set of the equation $x + 7 = x + 7$ contains all real numbers.

92. In your own words, explain why the equation $x = -x$ has one solution—namely, 0—while the solution set of the equation $x = x$ is all real numbers.

Find the value of K such that the equations are equivalent.

93. $3.2x + 4 = 5.4x - 7$
$3.2x = 5.4x + K$

94. $-7.6y - 10 = -1.1y + 12$
$-7.6y = -1.1y + K$

95. $\frac{7}{11}x + 9 = \frac{3}{11}x - 14$
$\frac{7}{11}x = \frac{3}{11}x + K$

96. $\frac{x}{6} + 4 = \frac{x}{3}$
$x + K = 2x$

97. Write a linear equation in x whose only solution is 5.

98. Write an equation in x that has no solution.

Solve and check.

 99. $-9.112y = -47.537304$

100. $2.86z - 8.1258 = -3.75$

101. $x(x - 6) + 7 = x(x + 1)$

102. $7x^2 + 2x - 3 = 6x(x + 4) + x^2$

STUDY SKILLS BUILDER

Have You Decided to Complete This Course Successfully?

Ask yourself if one of your current goals is to complete this course successfully.

If it is not a goal of yours, ask yourself why? One common reason is fear of failure. Amazingly enough, fear of failure alone can be strong enough to keep many of us from doing our best in any endeavor.

Another common reason is that you simply haven't taken the time to make successfully completing this course one of your goals. How do you do this? Start by writing this goal in your mathematics notebook. Then list steps you will take to ensure success. A great first step is to read or reread Section 1.1 and make a commitment to try the suggestions in that section.

Good luck, and don't forget that a positive attitude will make a big difference.

Let's see how you are doing.

1. Have you decided to make "successfully completing this course" a goal of yours? If no, please list reasons why this has not happened. Study your list and talk to your instructor about this.

2. If your answer to question 1 is yes, take a moment and list in your notebook further specific goals that will help you achieve this major goal of successfully completing this course. (For example, "My goal this semester is not to miss any of my mathematics classes.")

3. Rate your commitment to this course with a number between 1 and 5. Use the diagram below to help.

High Commitment	Average Commitment	Not Committed at all
5 4	3	2 1

4. If you have rated your personal commitment level (from the exercise above) as a 1, 2, or 3, list the reasons why this is so. Then determine whether it is possible to increase your commitment level to a 4 or 5.

2.2 AN INTRODUCTION TO PROBLEM SOLVING

Objective **A** Writing Algebraic Expressions

In order to prepare for problem solving, we practice writing algebraic expressions that can be simplified.

Our first example involves consecutive integers and perimeter. Recall that *consecutive integers* are integers that follow one another in order. Study the examples of consecutive integers, consecutive even integers, and consecutive odd integers and their representations.

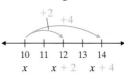

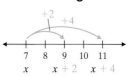

> **Helpful Hint**
> You may want to begin this section by studying key words and phrases and their translations in Sections 1.2 Objective B and 1.3 Objective A.

PRACTICE PROBLEM 1

Write the following as algebraic expressions. Then simplify.

a. The sum of three consecutive even integers, if x is the first even integer.

b. The perimeter of the triangle with sides of length x, $2x + 1$, and $4x$.

EXAMPLE 1 Write the following as algebraic expressions. Then simplify.

a. The sum of three consecutive integers, if x is the first consecutive integer.

b. The perimeter of the triangle with sides of length x, $5x$, and $6x - 3$.

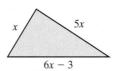

Solution:

a. Recall that if x is the first integer, then the next consecutive integer is 1 more, or $x + 1$, and the next consecutive integer is 1 more than $x + 1$, or $x + 1 + 1$, or $x + 2$.

In words:	first integer	plus	next consecutive integer	plus	next consecutive integer
Translate:	x	$+$	$(x + 1)$	$+$	$(x + 2)$

Then $x + (x + 1) + (x + 2) = x + x + 1 + x + 2$

$\qquad\qquad\qquad\qquad\quad = 3x + 3$ Simplify by combining like terms.

b. The perimeter of a triangle is the sum of the lengths of the sides.

In words:	side	+	side	+	side
Translate:	x	$+$	$5x$	$+$	$(6x - 3)$

Then $x + 5x + (6x - 3) = x + 5x + 6x - 3$

$\qquad\qquad\qquad\qquad\quad = 12x - 3$ Simplify.

Work Practice Problem 1

Answers

1. a. $3x + 6$, **b.** $7x + 1$

82

EXAMPLE 2 **Writing Algebraic Expressions Representing Metropolitan Regions**

The most populous metropolitan region in the United States is New York City, although it is only the third most populous metropolitan region in the world. Tokyo is the most populous metropolitan region, followed by Mexico City. Mexico City's population is 0.4 million more than New York, and Tokyo's is twice that of New York decreased by 1.6 million. Write the sum of the populations of these three metropolitan regions as an algebraic expression. Let x be the population of New York (in millions). (*Source:* United Nations, Department of Economic and Social Affairs)

Solution:

If x = the population of New York (in millions) then

$x + 0.4$ = the population of Mexico City (in millions) and

$2x - 1.6$ = the population of Tokyo (in millions)

In words:

population of New York	+	population of Mexico City	+	population of Tokyo

Translate: x + $(x + 0.4)$ + $(2x - 1.6)$

Then $x + (x + 0.4) + (2x - 1.6) = x + x + 2x + 0.4 - 1.6$

$$= 4x - 1.2$$

In Exercise 29, we will find the actual populations of these cities.

■ **Work Practice Problem 2**

Objective **B** **Solving Problems**

Our main purpose for studying algebra is to solve problems. The following problem-solving strategy will be used throughout this text and may also be used to solve real-life problems that occur outside the mathematics classroom.

General Strategy for Problem Solving

1. UNDERSTAND the problem. During this step, become comfortable with the problem. Some ways of doing this are:

   Read and reread the problem.

   Choose a variable to represent the unknown.

   Construct a drawing if necessary.

   Propose a solution and check. Pay careful attention to how you check your proposed solution. This will help when writing an equation to model the problem.

2. TRANSLATE the problem into an equation.

3. SOLVE the equation.

4. INTERPRET the results: If possible, check to see whether your answer is reasonable. Then *check* the proposed solution in the stated problem and *state* your conclusion.

PRACTICE PROBLEM 2

Audience ratings for three new online radio services for an average week in October 2004 were reported by comScore Arbitron Online Radio Ratings. These new online radio stations were AOL® radio Network, Yahoo!®'s LAUNCHcast, and Microsoft's MSN Radio and windowsMedia.com. AOL® radio Network had 1359.2 thousand listeners more than the number of listeners Microsoft's MSN Radio and windowsMedia.com had, and Yahoo!®'s LAUNCHcast had 230.1 thousand listeners more than four times the number of listeners Microsoft's MSN Radio and windowsMedia.com had. Write the sum of the listeners of these three online radio services as an algebraic expression. Let x be the number of Microsoft's MSN Radio and windowsMedia.com subscribers. (In Exercise 30, we will find the actual number of listeners for each online radio service) (*Source:* comScore Networks, Inc.)

Answer

2. $6x + 1589.3$

Let's review this strategy by solving a problem involving unknown numbers.

PRACTICE PROBLEM 3

One number is three times the first number. A third number is 50 more than the first number. If their sum is 235, find the three numbers.

The purpose of guessing a solution is not to guess correctly but to gain confidence and to help understand the problem and how to model it.

EXAMPLE 3 **Finding Unknown Numbers**

Find three numbers such that the second number is 3 more than twice the first number, and the third number is four times the first number. The sum of the three numbers is 164.

Solution:

1. UNDERSTAND the problem. First let's read and reread the problem and then propose a solution. For example, if the first number is 25, then the second number is 3 more than twice 25, or 53. The third number is four times 25, or 100. The sum of 25, 53, and 100 is 178, not the required sum, but we have gained some valuable information about the problem. First, we know that the first number is less than 25 since our guess led to a sum greater than the required sum. Also, we have gained some information as to how to model the problem.

   Next let's assign a variable and use this variable to represent any other unknown quantities. If we let

   $$x = \text{the first number, then}$$

   $$2x + 3 = \text{the second number}$$

   3 more than
   twice the second number

   $$4x = \text{the third number}$$

2. TRANSLATE the problem into an equation. To do so, we use the fact that the sum of the numbers is 164. First let's write this relationship in words and then translate to an equation.

In words:	first number	added to	second number	added to	third number	is	164
	↓	↓	↓	↓	↓	↓	↓
Translate:	x	$+$	$(2x + 3)$	$+$	$4x$	$=$	164

3. SOLVE the equation.

   $$x + (2x + 3) + 4x = 164$$

   $$x + 2x + 4x + 3 = 164 \quad \text{Remove parentheses.}$$

   $$7x + 3 = 164 \quad \text{Combine like terms.}$$

   $$7x = 161 \quad \text{Subtract 3 from both sides.}$$

   $$x = 23 \quad \text{Divide both sides by 7.}$$

4. INTERPRET. Here, we *check* our work and *state* the solution. Recall that if the first number $x = 23$, then the second number $2x + 3 = 2 \cdot 23 + 3 = 49$ and the third number $4x = 4 \cdot 23 = 92$.

Check: Is the second number 3 more than twice the first number? Yes, since 3 more than twice 23 is $46 + 3$, or 49. Also, their sum, $23 + 49 + 92 = 164$, is the required sum.

State: The three numbers are 23, 49, and 92.

☐ **Work Practice Problem 3**

Many of today's rates and statistics are given as percents. Interest rates, tax rates, nutrition labeling, and percent of households in a given category are just a few examples. Before we practice solving problems containing percents, let's take a moment to review the meaning of percent and how to find a percent of a number.

The word *percent* means *per hundred,* and the symbol % is used to denote percent. This means that 23% is 23 per one hundred, or $\dfrac{23}{100}$. Also,

$$41\% = \frac{41}{100} = 0.41$$

Answer

3. 37, 87, and 111

To find a percent of a number, we multiply.

$$16\% \text{ of } 25 = 16\% \cdot 25 = 0.16 \cdot 25 = 4$$

Thus, 16% of 25 is 4.

✔**Concept Check** Suppose you are finding 112% of a number x. Which of the following is a correct description of the result? Explain.

a. The result is less than x.

b. The result is equal to x.

c. The result is greater than x.

Next, we solve a problem containing a percent.

PRACTICE PROBLEM 4

EXAMPLE 4 **Finding the Original Price of a Computer**

Suppose that a computer store just announced an 8% decrease in the price of a particular computer model. If this computer sells for $2162 after the decrease, find its original price.

The price of a home was just decreased by 6%. If the decreased price is $83,660, find the original price of the home.

Solution:

1. UNDERSTAND. Read and reread the problem. Recall that a percent decrease means a percent of the original price. Let's guess that the original price of the computer is $2500. The amount of decrease is then 8% of $2500, or $(0.08)($2500) = 200. This means that the new price of the computer is the original price minus the decrease, or $2500 - $200 = 2300. Our guess is incorrect, but we now have an idea of how to model this problem. In our model, we will let x = the original price of the computer.

2. TRANSLATE.

In words:	original price of computer	minus	8% of original price	is	new price
	↓	↓	↓	↓	↓
Translate:	x	−	$0.08x$	=	2162

3. SOLVE the equation.

$$x - 0.08x = 2162$$
$$0.92x = 2162 \qquad \text{Combine like terms.}$$
$$x = \frac{2162}{0.92} = 2350 \qquad \text{Divide both sides by 0.92.}$$

Continued on next page

Answer
4. $89,000

✔ **Concept Check Answer**
c; the result is greater than x

4. INTERPRET.

Check: The amount $2350 is a reasonable price for a computer. If the original price of the computer was $2350, the new price is

$$\$2350 - (0.08)(\$2350) = \$2350 - \$188$$
$$= \$2162 \quad \text{The given new price}$$

State: The original price of the computer was $2350.

⬛ **Work Practice Problem 4**

PRACTICE PROBLEM 5

A rectangle has a perimeter of 106 meters. Its length is 5 meters more than twice its width. Find the length and the width of the rectangle.

△ **EXAMPLE 5** **Finding the Lengths of a Triangle's Sides**

A pennant in the shape of an isosceles triangle is to be constructed for the Slidell High School Athletic Club and sold at a fund-raiser. The company manufacturing the pennant charges according to perimeter, and the athletic club has determined that a perimeter of 149 centimeters should make a nice profit. If each equal side of the triangle is twice the length of the third side, increased by 12 centimeters, find the lengths of the sides of the triangular pennant.

Solution:

1. UNDERSTAND. Read and reread the problem. Recall that the perimeter of a triangle is the distance around. Let's guess that the third side of the triangular pennant is 20 centimeters. This means that each equal side is twice 20 centimeters, increased by 12 centimeters, or $2(20) + 12 = 52$ centimeters.

This gives a perimeter of $20 + 52 + 52 = 124$ centimeters. Our guess is incorrect, but we now have a better understanding of how to model this problem. Now we let

$$x = \text{the third side of the triangle, then}$$
$$2x + 12 = \text{the first side}$$
$$2x + 12 = \text{the second side}$$

2. TRANSLATE.

In words: first side + second side + third side = 149

Translate: $(2x + 12) + (2x + 12) + x = 149$

3. SOLVE the equation.

$$(2x + 12) + (2x + 12) + x = 149$$
$$2x + 12 + 2x + 12 + x = 149 \quad \text{Remove parentheses.}$$
$$5x + 24 = 149 \quad \text{Combine like terms.}$$
$$5x = 125 \quad \text{Subtract 24 from both sides.}$$
$$x = 25 \quad \text{Divide both sides by 5.}$$

Answer

5. length: 16 m, width: 37 m

4. INTERPRET. If the third side is 25 centimeters, then the first side is 2(25) + 12 = 62 centimeters and the second side is 62 centimeters also.

Check: The first and second sides are each twice 25 centimeters increased by 12 centimeters or 62 centimeters. Also, the perimeter is 25 + 62 + 62 = 149 centimeters, the required perimeter.

State: The dimensions of the triangle are 25 centimeters, 62 centimeters, and 62 centimeters.

▣ **Work Practice Problem 5**

EXAMPLE 6 **Finding Consecutive Integers**

Kelsey Ohleger was helping her friend Benji Burnstine study for an exam. Kelsey told Benji that her three latest art history quiz scores happened to be three consecutive even integers whose sum is 264. Help Benji find the scores.

Solution:

1. **UNDERSTAND.** Read and reread the problem. Since we are looking for consecutive even integers, let

   x = the first integer. Then

   $x + 2$ = the next consecutive even integer.

   $x + 4$ = " " " " "

2. **TRANSLATE.**

   In words: | first integer | + | next even integer | + | next even integer | = | 264 |

   Translate: x + $(x + 2)$ + $(x + 4)$ = 264

3. **SOLVE.**

   $x + (x + 2) + (x + 4) = 264$

   $3x + 6 = 264$ Combine like terms.

   $3x = 258$ Subtract 6 from both sides.

   $x = 86$ Divide both sides by 3.

4. **INTERPRET.** If $x = 86$, then $x + 2 = 86 + 2$ or 88, and $x + 4 = 86 + 4$ or 90.

Check: The numbers 86, 88, and 90 are three consecutive even integers. Their sum is 264, the required sum.

State: Kelsey's art history quiz scores are 86, 88, and 90.

▣ **Work Practice Problem 6**

PRACTICE PROBLEM 6

Find three consecutive integers whose sum is 378.

Answer

6. 125, 126, and 127

2.2

EXERCISE SET

FOR EXTRA HELP

  Student Solutions Manual

PH Math/Tutor Center

CD/Video for Review

 MathXL®

 MyMathLab

Objective 🅐 *Write the following as algebraic expressions. Then simplify. See Examples 1 and 2.*

△ **1.** The perimeter of the square with side length *y*.

2. The perimeter of the rectangle with length *x* and width $x - 5$.

3. The sum of three consecutive integers if the first is *z*.

4. The sum of three consecutive odd integers if the first integer is *x*.

5. The total amount of money (in cents) in *x* nickels, $(x + 3)$ dimes, and $2x$ quarters. (*Hint:* the value of a nickel is 5 cents, the value of a dime is 10 cents, and the value of a quarter is 25 cents)

6. The total amount of money (in cents) in *y* quarters, $7y$ dimes, and $(2y - 1)$ nickels. (Use the hint for Exercise 5.)

△ **7.** A piece of land along Bayou Liberty is to be fenced and subdivided as shown so that each rectangle has the same dimensions. Express the total amount of fencing needed as an algebraic expression in *x*.

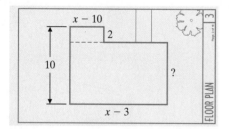

8. A flooded piece of land near the Mississippi River in New Orleans is to be surveyed and divided into 4 rectangles of equal dimension. Express the total amount of fencing needed as an algebraic expression in *y*.

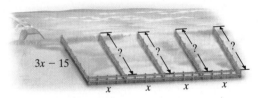

△ **9.** Write the perimeter of the floor plan shown as an algebraic expression in *x*.

10. Write the perimeter of the floor plan shown as an algebraic expression in *x*.

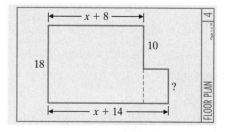

Objective 🅑 *Solve. See Example 3.*

11. Four times the difference of a number and 2 is the same as 2 increased by four times the number plus twice the number. Find the number.

12. Twice the sum of a number and 3 is the same as five times the number minus 1 minus four times the number. Find the number.

13. One number is five times a first number. A third number is 100 more than the first number. If the sum of the three numbers is 415, find the numbers.

14. One number is 6 less than a first number. A third number is twice the first number. If the sum of the three numbers is 306, find the numbers.

Solve. See Examples 4 through 6.

15. The United States consists of 2271 million acres of land. Approximately 29% of this land is federally owned. Find the number of acres that are not federally owned. (*Source:* U.S. General Services Administration)

16. The state of Nevada contains the most federally owned acres of land in the United States. If 90% of the state's 70 million acres of land is federally owned, find the number of acres that are not federally owned. (*Source:* U.S. General Services Administration)

17. In 2004, a total of 3456 earthquakes occurred in the United States. Of these, 91% were minor tremors with magnitudes of 3.9 or less on the Richter scale. How many minor earthquakes occurred in the United States in 2004? Round to the nearest whole. (*Source:* U.S. Geological Survey National Earthquake Information Center)

18. Of the 1376 tornadoes that occurred in the United States during 2003, 39.5% occurred during the month of May. How many tornadoes occurred during May 2003? Round to the nearest whole. (*Source:* Storm Prediction Center)

19. In a recent survey, 15% of online shoppers in the United States say that they prefer to do business only with large, well-known retailers. In a group of 1500 online shoppers, how many are willing to do business with any size retailers? (*Source:* Inc.com)

20. On average 12.8% of American men eat a commercially prepared lunch 5 times or more per week. As of 2003, San Francisco has a male population of 394,828. How many San Francisco men would you expect do not eat a commercially prepared lunch 5 times or more per week? Round to the nearest whole. (*Source:* National Restaurant Association, U.S. Census Bureau)

The following graph is called a circle graph or a pie chart. The circle represents a whole, or in this case, 100%. This particular graph shows the number of minutes per day that people use e-mail at work. Use this graph to answer Exercises 21 through 24.

Time Spent on E-mail at Work

Source: Pew Internet & American Life Project

21. What percent of e-mail users at work spend less than 15 minutes on e-mail per day?

22. Among e-mail users at work, what is the most common time spent on e-mail per day?

23. If it were estimated that a large company has 4633 employees, how many of these would you expect to be using e-mail more than 3 hours per day?

24. If it were estimated that a medium size company has 250 employees, how many of these would you expect to be using e-mail between 2 and 3 hours per day?

25. INVESCO Field at Mile High, home to the Denver Broncos, has 11,675 more seats than Heinz Field, home to the Pittsburgh Steelers. Together, these two stadiums can seat a total of 140,575 NFL fans. How many seats does each stadium have? (*Sources:* Denver Broncos, Pittsburgh Steelers)

26. For the 2001 Major League Baseball season, the opening day payroll for the Minnesota Twins was $46,718,000 less than the opening day payroll for the Colorado Rockies. The total of the opening day payrolls for these two teams was $95,418,000. What was the opening day payroll for each team? (*Source:* Associated Press)

27. The perimeter of the triangle in Example 1b in this section is 483 feet. Find the length of each side.

28. The perimeter of the triangle in Practice Problem 1b in this section is 130.5 meters. Find its dimensions.

29. The sum of the populations of the metropolitan regions of New York, Tokyo and Mexico City is 72 million. Use this information and Example 2 in this section to find the population of each metropolitan region. (*Source:* United Nations Department of Economic and Social Affairs)

30. The total number of listeners to the online radio services AOL® radio Network, Yahoo!®'s LAUNCHcast, and Microsoft's MSN Radio and windowsMedia.com during October 2004 was 4137.5 thousand. Use this information and Practice Problem 2 to find the number of listeners for each online radio service. (*Source:* comScore Networks, Inc.)

Use the diagrams to find the unknown measures of angles or lengths of sides. Recall that the sum of the angle measures of a triangle is 180°.

31.

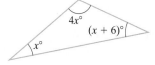

32.

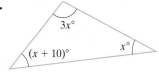

33.

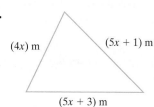

Perimeter is 102 meters.

34.

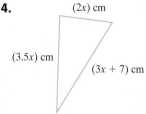

Perimeter is 75 centimeters.

35.

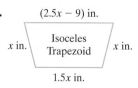

Perimeter is 99 inches.

36.

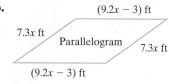

Perimeter is 324 feet.

Many companies and government agencies predict the growth or decline of occupations. The following data are based on information from the U.S. Department of Labor. Notice that the first table is in increase in number of jobs (in thousands) and the second table is in percent increase in number of jobs.

37. Use the table to find the actual number of jobs for each occupation.

Occupation	Increase in Number of Jobs (in thousands) from 2000 to 2012
Security guards	$2x - 51$
Home health aides	$\frac{3}{2}x + 3$
Computer systems analysts	x
Total	780 thousand

38. Use the table to find the actual percent increase in number of jobs for each occupation.

Occupation	Precent Increase in Number of Jobs from 2000 to 2012
Computer software engineers	$\frac{3}{2}x + 1$
Management analysts	x
Receptionist and information clerks	$x - 1$
Total	105%

Solve.

39. The occupations of postsecondary teachers, registered nurses, and medical assistants are among the ten with the largest growth from 2000 to 2012. (See the Chapter 2 opener.) The number of postsecondary teacher jobs will grow 173 thousand more than twice the number of medical assistant jobs. The number of registered nurse jobs will grow 22 thousand less than three times the number of medical assistant jobs. If the total growth of these three jobs is predicted to be 1441 thousand, find the predicted growth of each job.

40. The occupations of telephone operators, fishers, and sewing machine operators are among the ten with the largest job decline from 2000 to 2012, according to the U.S. Department of Labor. The number of telephone operators will decline 8 thousand more than twice the number of fishers. The number of sewing machine operators will decline 1 thousand less than 10 times the number of telephone operators. If the total decline of these three jobs is predicted to be 137 thousand, find the predicted decline of each job.

41. The B767-300ER aircraft has 88 more seats than the B737-200 aircraft. The F-100 has 32 fewer seats than the B737-200 aircraft. If their total number of seats is 413, find the number of seats for each aircraft. (*Source:* Air Transport Association of America)

42. The governor of Delaware makes $36,000 less per year than the governor of Connecticut. The governor of New Jersey makes $7000 more per year than the Connecticut governor. If the total of these salaries is $421,000, find the salary of each governor. (*Source:* World Almanac, 2005)

43. A new fax machine was recently purchased for an office in Hopedale for $464.40 including tax. If the tax rate in Hopedale is 8%, find the price of the fax machine before tax.

44. A premedical student at a local university was complaining that she had just paid $158.60 for her human anatomy book, including tax. Find the price of the book before taxes if the tax rate at this university is 9%.

45. Google, Inc. is a worldwide search engine. On December 28, 2004, shares of Google stock closed at $192.76 per share. This represents a 92.1% increase in stock price from the closing price on August 19, 2004, the day when Google stock was first traded on the NASDAQ exchange. Find the closing price on August 19, 2004. (*Source:* Yahoo! Finance)

46. In 2004, the population of Morocco was 32.2 million. This represented an increase in population of 1.6% from a year earlier. What was the population of Morocco in 2003? Round to the nearest tenth of a million. (*Source:* Population Reference Bureau)

47. The sum of three consecutive integers is 228. Find the integers.

48. The sum of three consecutive odd integers is 327. Find the integers.

Recall that the sum of the angle measures of a triangle is 180°.

△ **49.** Find the measures of the angles of a triangle if the measure of one angle is twice the measure of a second angle and the third angle measures 3 times the second angle decreased by 12.

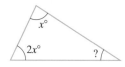

△ **50.** Find the angles of an isoceles triangle whose two base angles are equal and whose third angle is 10° less than three times a base angle.

51. Two frames are needed with the same perimeter: one frame in the shape of a square and one in the shape of an equilateral triangle. Each side of the triangle is 6 centimeters longer than each side of the square. Find the dimensions of each frame. (An equilateral triangle has sides that are the same length.)

52. Two frames are needed with the same perimeter: one frame in the shape of a square and one in the shape of a regular pentagon. Each side of the square is 7 inches longer than each side of the pentagon. Find the dimensions of each frame. (A regular polygon has sides that are the same length.)

53. In 2004, the population of South Africa was 46,900,000 people. From 2004 to 2050, South Africa's population is expected to decrease by 11%. Find the expected population of South Africa in 2050. (*Source:* Population Reference Bureau)

54. Dana, an auto parts supplier headquartered in Toledo, Ohio, recently announced it would be cutting 11,000 jobs worldwide. This is equivalent to 15% of Dana's workforce. Find the size of Dana's workforce prior to this round of job layoffs. Round to the nearest whole. (*Source:* Dana Corporation)

55. The zip codes of three Nevada locations—Fallon, Fernley, and Gardnerville Ranchos—are three consecutive even integers. If twice the first integer added to the third is 268,222, find each zip code.

56. During a recent year, the average SAT scores in math for the states of Alabama, Louisiana, and Michigan were 3 consecutive integers. If the sum of the first integer, second integer, and three times the third integer is 2637, find each score.

Recall that two angles are complements of each other if their sum is 90°. Two angles are supplements of each other if their sum is 180°. Find the measure of each angle.

57. One angle is three times its supplement increased by 20°. Find the measures of the two supplementary angles.

58. One angle is twice its complement increased by 30°. Find the measure of the two complementary angles.

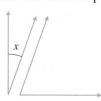

59. Incandescent, fluorescent, and halogen bulbs are lasting longer today than ever before. On average, the number of bulb hours for a fluorescent bulb is 25 times the number of bulb hours for a halogen bulb. The number of bulb hours for an incandescent bulb is 2500 less than the halogen bulb. If the total number of bulb hours for the three types of bulbs is 105,500, find the number of bulb hours for each type. (*Source: Popular Science* magazine)

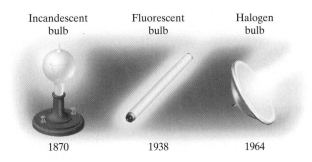

Incandescent bulb	Fluorescent bulb	Halogen bulb
1870	1938	1964

60. Taiwan, Luxembourg, and Italy are the top three countries that have the greatest penetration rate (number of cellular subscriptions per 100 inhabitants) of cellular subscribers in the world. Taiwan has a 9.0 greater penetration rate than Italy. Luxembourg has a 4.3 greater penetration rate than Italy. (*Source: World Almanac and Book of Facts*, 2005)

   a. If the sum of the penetration rates is 318.7, find the number of inhabitants per 100 in each of these countries with cellular subscriptions.

   b. Explain, in your own words, how the penetration rate of cellular subscribers in these three countries is greater than the number of inhabitants.

61. During the 2004 Major League Baseball season, the numbers of home runs hit by Derek Jeter of the New York Yankees, Mark Mulder of the Oakland Athletics, and Geoff Jenkins of the Milwaukee Brewers were three consecutive odd integers. Of these three players, Jenkins hit the most home runs and Jeter hit the fewest home runs. The total number of home runs hit by these three players was 75. How many home runs did each player hit during the 2004 season? (*Source:* Major League Baseball)

62. In the 2004 Summer Olympics, France won more gold medals than Italy, who won more gold medals than Korea. If the total number of gold medals won by these three countries is three consecutive integers whose sum is 30, find the number of gold medals won by each. (*Source:* Athens Olympic Committee)

Review

Find the value of each expression for the given values. See Section 1.5.

63. $4ab - 3bc$; $a = -5$, $b = -8$, and $c = 2$

64. $ab + 6bc$; $a = 0$, $b = -1$, and $c = 9$

65. $n^2 - m^2$; $n = -3$ and $m = -8$

66. $2n^2 + 3m^2$; $n = -2$ and $m = 7$

67. $P + PRT$; $P = 3000$, $R = 0.0325$, and $T = 2$

68. $\frac{1}{3}lwh$; $l = 37.8$, $w = 5.6$, and $h = 7.9$

Concept Extensions

69. For Exercise 38, the percents have a sum of 105%. Is this possible? Why or why not?

70. In your own words, explain the differences in the tables for Exercises 37 and 38.

71. Choose five occupations from the chapter opener graph and define these occupations.

72. Find an angle such that its supplement is equal to twice its complement increased by 50°.

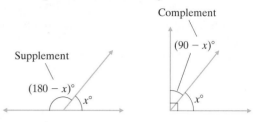

73. The average annual number of cigarettes smoked by an American adult continues to decline. For the years 1990–2002, the equation $y = -72.8x + 2843.5$ approximates this data. Here, x is the number of years after 1990 and y is the average annual number of cigarettes smoked.

   a. If this trend continues, find the year in which the average annual number of cigarettes smoked is 0. To do this, let $y = 0$ and solve for x.

   b. Predict the average annual number of cigarettes smoked by an American adult in 2010. To do so, let $x = 20$ (since $2010 - 1990 = 20$) and find y.

   c. Use the result of part b to predict the average *daily* number of cigarettes smoked by an American adult in 2010. Round to the nearest whole. Do you think this number represents the average daily number of cigarettes smoked by an adult? Why or why not?

74. Determine whether there are three consecutive integers such that their sum is three times the second integer.

75. Determine whether there are two consecutive odd integers such that 7 times the first exceeds 5 times the second by 54.

To break even in a manufacturing business, income or revenue R must equal the cost of production C. Use this information to answer Exercises 76 through 79.

76. The cost C to produce x number of skateboards is $C = 100 + 20x$. The skateboards are sold wholesale for $24 each, so revenue R is given by $R = 24x$. Find how many skateboards the manufacturer needs to produce and sell to break even. (*Hint:* Set the cost expression equal to the revenue expression and solve for x.)

77. The revenue R from selling x number of computer boards is given by $R = 60x$, and the cost C of producing them is given by $C = 50x + 5000$. Find how many boards must be sold to break even. Find how much money is needed to produce the break-even number of boards.

78. In your own words, explain what happens if a company makes and sells fewer products than the break-even number.

79. In your own words, explain what happens if more products than the break-even number are made and sold.

2.3 FORMULAS AND PROBLEM SOLVING

Objective **A** Solving Formulas for Specified Variables

Solving problems that we encounter in the real world sometimes requires us to express relationships among measured quantities. A **formula** is an equation that describes a known relationship among measured phenomena, such as time, area, and gravity. Some examples of formulas follow.

Formula	Meaning
$I = Prt$	Interest = principal · rate · time
$A = lw$	Area of a rectangle = length · width
$d = rt$	Distance = rate · time
$C = 2\pi r$	Circumference of a circle = $2 \cdot \pi \cdot$ radius
$V = lwh$	Volume of a rectangular solid = length · width · height

Other formulas are listed on the inside front cover of this text. Notice that the formula for the volume of a rectangular solid, $V = lwh$, is solved for V since V is by itself on one side of the equation with no Vs on the other side of the equation. Suppose that the volume of a rectangular solid is known as well as its width and its length, and we wish to find its height. One way to find its height is to begin by solving the formula $V = lwh$ for h.

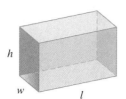

EXAMPLE 1 Solve $V = lwh$ for h.

Solution: To solve $V = lwh$ for h, we want to get h alone on one side of the equation. To do so, we divide both sides of the equation by lw.

$$V = lwh$$

$$\frac{V}{lw} = \frac{lw\ h}{lw} \quad \text{Divide both sides by } lw.$$

$$\frac{V}{lw} = h \text{ or } h = \frac{V}{lw} \quad \text{Simplify.}$$

Thus we see that to find the height of a rectangular solid, we divide its volume by the product of its length and its width.

Work Practice Problem 1

The following steps may be used to solve formulas and equations for a specified variable.

PRACTICE PROBLEM 1
Solve $I = Prt$ for P.

Answer

1. $P = \dfrac{I}{rt}$

Solving an Equation for a Specified Variable

Step 1: Clear the equation of fractions or decimals by multiplying each side of the equation by an appropriate nonzero number.

Step 2: Use the distributive property to remove grouping symbols such as parentheses.

Step 3: Combine like terms on each side of the equation.

Step 4: Use the addition property of equality to rewrite the equation as an equivalent equation with terms containing the specified variable on one side and all other terms on the other side.

Step 5: Use the distributive property and the multiplication property of equality to get the specified variable alone.

PRACTICE PROBLEM 2

Solve $2y + 5x = 10$ for y.

EXAMPLE 2 Solve $3y - 2x = 7$ for y.

Solution: This is a linear equation in two variables. Often an equation such as this is solved for y to reveal some properties about the graph of this equation, which we will learn more about in Chapter 3. Since there are no fractions or grouping symbols, we begin with Step 4 and get the term containing the specified variable y on one side by adding $2x$ to both sides of the equation.

$$3y - 2x = 7$$
$$3y - 2x + 2x = 7 + 2x \quad \text{Add } 2x \text{ to both sides.}$$
$$3y = 7 + 2x$$

To solve for y, we divide both sides by 3.

$$\frac{3y}{3} = \frac{7 + 2x}{3} \quad \text{Divide both sides by 3.}$$

$$y = \frac{7 + 2x}{3} \quad \text{or} \quad y = \frac{7}{3} + \frac{2}{3}x$$

Work Practice Problem 2

PRACTICE PROBLEM 3

Solve $A = \frac{1}{2}(B + b)h$ for B.

EXAMPLE 3 Solve $A = \frac{1}{2}(B + b)h$ for b.

Solution: Since this formula for finding the area of a trapezoid contains fractions, we begin by multiplying both sides of the equation by the LCD, 2.

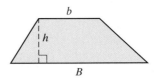

Helpful Hint

Remember that we may get the specified variable alone on either side of the equation.

$$A = \frac{1}{2}(B + b)h$$

$$2 \cdot A = 2 \cdot \frac{1}{2}(B + b)h \qquad \text{Multiply both sides by 2.}$$

$$2A = (B + b)h \qquad \text{Simplify.}$$

$$2A = Bh + bh \qquad \text{Use the distributive property.}$$

$$2A - Bh = bh \qquad \text{Get the term containing } b \text{ alone by subtracting } Bh \text{ from both sides.}$$

$$\frac{2A - Bh}{h} = \frac{bh}{h} \qquad \text{Divide both sides by } h.$$

$$\frac{2A - Bh}{h} = b \quad \text{or} \quad b = \frac{2A - Bh}{h} \quad \text{or} \quad b = \frac{2A}{h} - B$$

Work Practice Problem 3

Answers

2. $y = \dfrac{10 - 5x}{2}$ or $y = 5 - \dfrac{5}{2}x$,

3. $B = \dfrac{2A - bh}{h}$ or $B = \dfrac{2A}{h} - b$

Objective B Using Formulas to Solve Problems

In this section, we also solve problems that can be modeled by known formulas. We use the same problem-solving steps that were introduced in the previous section.

Formulas are very useful in problem solving. For example, the compound interest formula

$$A = P\left(1 + \frac{r}{n}\right)^{nt}$$

is used by banks to compute the amount A in an account that pays compound interest. The variable P represents the principal or amount invested in the account, r is the annual rate of interest, t is the time in years, and n is the number of times compounded per year.

EXAMPLE 4 **Finding the Amount in a Savings Account**

Marial Callier just received an inheritance of $10,000 and plans to place all the money in a savings account that pays 5% compounded quarterly to help her son go to college in 3 years. How much money will be in the account in 3 years?

Solution:

1. UNDERSTAND. Read and reread the problem. The appropriate formula needed to solve this problem is the compound interest formula

$$A = P\left(1 + \frac{r}{n}\right)^{nt}$$

Make sure that you understand the meaning of all the variables in this formula:

A = amount in the account after t years

P = principal or amount invested

t = time in years

r = annual rate of interest

n = number of times compounded per year

2. TRANSLATE. Use the compound interest formula and let $P = \$10,000$, $r = 5\% = 0.05$, $t = 3$ years, and $n = 4$ since the account is compounded quarterly, or 4 times a year.

Formula: $A = P\left(1 + \dfrac{r}{n}\right)^{nt}$

Substitute: $A = 10,000\left(1 + \dfrac{0.05}{4}\right)^{4 \cdot 3}$

3. SOLVE. We simplify the right side of the equation.

$$A = 10,000\left(1 + \frac{0.05}{4}\right)^{4 \cdot 3}$$

$A = 10,000(1.0125)^{12}$ Simplify $1 + \dfrac{0.05}{4}$ and write $4 \cdot 3$ as 12.

$A \approx 10,000(1.160754518)$ Approximate $(1.0125)^{12}$.

$A \approx 11,607.55$ Multiply and round to two decimal places.

4. INTERPRET.

Check: Repeat your calculations to make sure that you made no error. Notice that $11,607.55 is a reasonable amount to have in the account after 3 years.

State: In 3 years, the account will contain $11,607.55.

▨ Work Practice Problem 4

PRACTICE PROBLEM 4

If $5000 is invested in an account paying 4% compounded monthly, determine how much money will be in the account in 2 years. Use the formula from Example 4.

Answer

4. $5415.71

CALCULATOR EXPLORATIONS Graphing

To solve Example 4, we approximated the expression

$$10{,}000\left(1 + \frac{0.05}{4}\right)^{4 \cdot 3}.$$

Use the keystrokes (right) to evaluate this expression using a graphing calculator (or graphing utility, or grapher). Notice the use of parentheses.

```
10000(1+(.05/4))
^(4*3)
        11607.54518
```

Mental Math

Solve each equation for the specified variable. See Examples 1 through 3.

1. $2x + y = 5$ for y

2. $7x - y = 3$ for y

3. $a - 5b = 8$ for a

4. $7r + s = 10$ for s

5. $5j + k - h = 6$ for k

6. $w - 4y + z = 0$ for z

2.3 EXERCISE SET

FOR EXTRA HELP

Student Solutions Manual   PH Math/Tutor Center   CD/Video for Review   MathXL®   MyMathLab

Objective **A** *Solve each equation for the specified variable. See Examples 1 through 3.*

1. $d = rt$ for t

2. $W = gh$ for g

3. $I = Prt$ for r

4. $V = lwh$ for h

5. $P = a + b + c$ for c

6. $a^2 + b^2 = c^2$ for b^2

7. $9x - 4y = 16$ for y

8. $2x + 3y = 17$ for y

9. $P = 2l + 2w$ for l

10. $P = 2l + 2w$ for w

11. $E = I(r + R)$ for r

12. $A = P(1 + rt)$ for t

13. $5x + 4y = 20$ for y

14. $-9x - 5y = 18$ for y

15. $S = 2LW + 2LH + 2WH$ for H

16. $S = 2\pi r^2 + 2\pi rh$ for h

17. $C = 2\pi r$ for r

18. $A = \pi r^2$ for π

19. $C = \frac{5}{9}(F - 32)$ for F

20. $F = \frac{9}{5}C + 32$ for C

98

Objective **B** *Solve. Round all dollar amounts to two decimal places. See Example 4.*

21. Complete the table and find the balance A if $3500 is invested at an annual rate of 3% for 10 years and compounded n times a year.

n	1	2	4	12	365
A					

22. Complete the table and find the balance A if $5000 is invested at an annual rate of 6% for 15 years and compounded n times a year.

n	1	2	4	12	365
A					

23. A principal of $6000 is invested in an account paying an annual rate of 4%. Find the amount in the account after 5 years if the account is compounded
   a. semiannually.
   b. quarterly.
   c. monthly.

24. A principal of $25,000 is invested in an account paying an annual rate of 5%. Find the amount in the account after 2 years if the account is compounded
   a. semiannually.
   b. quarterly.
   c. monthly.

25. One day's high temperature in Phoenix, Arizona, was recorded as 104°F. Write 104°F as degrees Celsius. [Use the formula $C = \dfrac{5}{9}(F - 32)$.]

26. One year's low temperature in Nome, Alaska, was recorded as −15°C. Write −15°C as degrees Fahrenheit. (Use the formula $F = \dfrac{9}{5}C + 32$.)

27. Omaha, Nebraska, is about 90 miles from Lincoln, Nebraska. Irania Schmidt must go to the law library in Lincoln to get a document for the law firm she works for. Find how long it takes her to drive *round-trip* if she averages 50 mph.

28. It took the Selby family $5\dfrac{1}{2}$ hours round-trip to drive from their house to their beach house 154 miles away. Find their average speed.

29. A package of floor tiles contains 24 one-foot-square tiles. Find how many packages should be bought to cover a square ballroom floor whose side measures 64 feet.

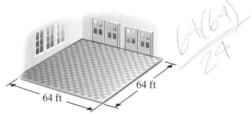

64 ft
64 ft

30. One-foot-square ceiling tiles are sold in packages of 50. Find how many packages must be bought for a rectangular ceiling 18 feet by 12 feet.

31. If the area of a triangular kite is 18 square feet and its base is 4 feet, find the height of the kite.

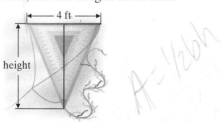

4 ft
height

32. Bryan, Eric, Mandy, and Melissa would like to go to Disneyland in 3 years. The total cost should be $4500. If each invests $1000 in a savings account paying 5.5% interest, compounded semiannually, will they have enough in 3 years?

33. A gallon of latex paint can cover 500 square feet. Find how many gallon containers of paint should be bought to paint two coats on the walls of a rectangular room whose dimensions are 14 feet by 16 feet. (Assume 8-foot ceilings and disregard any openings such as windows or doors.

34. A gallon of enamel paint can cover 300 square feet. Find how many gallon containers of paint should be bought to paint three coats on a wall measuring 21 feet by 8 feet.

35. A portion of the external tank of the Space Shuttle *Endeavour* is a liquid hydrogen tank. If the ends of the tank are hemispheres, find the volume of the tank. To do so, answer parts **(a)** through **(c)**. (*Source:* NASA/Kennedy Space Center)

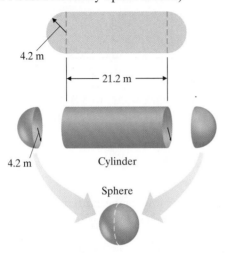

4.2 m

21.2 m

4.2 m   Cylinder

Sphere

a. Find the volume of the cylinder shown. Round to two decimal places.
b. Find the volume of the sphere shown. Round to two decimal places.
c. Add the results of parts **(a)** and **(b)**. This sum is the approximate volume of the tank.

36. A different kind of space exploration was launched March 2, 2004. The European Space Agency's *Rosetta* spacecraft is the first to undertake the long-term exploration of a comet, Comet 67P/Churyumov-Gerasimenko, at close quarters. It will take more than 10 years for the spacecraft to reach the comet, arriving in 2014. During its mission, *Rosetta* will travel a total of 490,883,093 miles in 125 months. Find the average speed of the spacecraft in miles per hour. Round to the nearest whole mile per hour. (*Hint:* Convert 125 months to hours using 1 month = 30 days and then use the formula $d = rt$.) (*Source:* European Space Agency)

37. In 1945, Arthur C. Clarke, a scientist and science-fiction writer, predicted that an artificial satellite placed at a height of 22,248 miles directly above the equator would orbit the globe at the same speed with which Earth was rotating. This belt along the equator is known as the Clarke belt. Use the formula for the circumference of a circle and approximate the "length" of the Clarke belt. Round to the nearest whole mile. (*Hint:* Recall that the radius of Earth is approximately 4000 miles.)

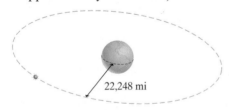

22,248 mi

38. The *Endeavour* Space Shuttle has a cargo bay that is in the shape of a cylinder whose length is 18.3 meters and whose diameter is 4.6 meters. Find its volume.

39. The deepest hole in the ocean floor is beneath the Pacific Ocean and is called Hole 504B. It is located off the coast of Ecuador. Scientists are drilling it to learn more about Earth's history. Currently, the hole is in the shape of a cylinder whose volume is approximately 3800 cubic feet and whose length is 1.3 miles. Find the radius of the hole to the nearest hundredth of a foot. (*Hint:* Make sure the same units of measurement are used.)

40. The deepest man-made hole is called the Kola Superdeep Borehole. It is approximately 8 miles deep and is located near a small Russian town in the Arctic Circle. If it takes 7.5 hours to remove the drill from the bottom of the hole, find the rate that the drill can be retrieved in feet per second. Round to the nearest tenth. (*Hint:* Write 8 miles as feet, 7.5 hours as seconds, and then use the formula $d = rt$.)

△ **41.** Eartha is the world's largest globe. It is located at the headquarters of DeLorme, a mapmaking company in Yarmouth, Maine. Eartha is 41.125 feet in diameter. Find its exact circumference (distance around) and then approximate its circumference using 3.14 for π. (*Source:* DeLorme)

△ **42.** Eartha is in the shape of a sphere. Its radius is about 20.6 feet. Approximate Eartha's volume to the nearest cubic foot. Use the approximation 3.14 for π. (*Source:* DeLorme)

The calorie count of a serving of food can be computed based on its composition of carbohydrate, fat, and protein. The calorie count C for a serving of food can be computed using the formula $C = 4h + 9f + 4p$, where h is the number of grams of carbohydrate contained in the serving, f is the number of grams of fat contained in the serving, and p is the number of grams of protein contained in the serving.

43. Solve this formula for f, the number of grams of fat contained in a serving of food.

44. Solve this formula for h, the number of grams of carbohydrate contained in a serving of food.

45. A serving of cashews contains 14 grams of fat, 7 grams of carbohydrate, and 6 grams of protein. How many calories are in this serving of cashews?

46. A serving of chocolate candies contains 9 grams of fat, 30 grams of carbohydrate, and 2 grams of protein. How many calories are in this serving of chocolate candies?

47. A serving of raisins contains 130 calories and 31 grams of carbohydrate. If raisins are a fat-free food, how much protein is provided by this serving of raisins?

48. A serving of yogurt contains 120 calories, 21 grams of carbohydrate, and 5 grams of protein. How much fat is provided by this serving of yogurt? Round to the nearest tenth of a gram.

Review

Determine which numbers in the set $\{-3, -2, -1, 0, 1, 2, 3\}$ are solutions of each inequality. See Sections 1.3 and 2.1.

49. $x < 0$ **50.** $x > 1$ **51.** $x + 5 \leq 6$ **52.** $x - 3 \geq -7$

53. In your own words, explain what real numbers are solutions of $x < 0$.

54. In your own words, explain what real numbers are solutions of $x > 1$.

Concept Extensions

55. Solar System distances are so great that units other than miles or kilometers are often used. For example, the astronomical unit (AU) is the average distance between Earth and the sun, or 92,900,000 miles. Use this information to convert each planet's distance in miles from the sun to astronomical units. Round to three decimal places. (*Source:* National Space Science Data Center)

Planet	Miles from the Sun	AU from the Sun	Planet	Miles from the Sun	AU from the Sun
Mercury	36 million		Saturn	886.1 million	
Venus	67.2 million		Uranus	1783 million	
Earth	92.9 million		Neptune	2793 million	
Mars	141.5 million		Pluto	3670 million	
Jupiter	483.3 million				

56. An orbit such as Clarke's belt in Exercise 37 is called a geostationary orbit. In your own words, why do you think that communications satellites are placed in geostationary orbits?

57. How much do you think it costs each American to build a Space Shuttle? Write down your estimate. The Space Shuttle *Endeavour* was completed in 1992 and cost approximately $1.7 billion. If the population of the United States in 1992 was 250 million, find the cost per person to build the *Endeavour*. How close was your estimate?

58. Find *how much interest* $10,000 earns in 2 years in a certificate of deposit account paying 8.5% interest compounded quarterly.

59. If you are investing money in a savings account paying a rate of *r*, which account should you choose—an account compounded 4 times a year or 12 times a year? Explain your choice.

60. To borrow money at a rate of *r*, which loan plan should you choose—one compounding 4 times a year or 12 times a year? Explain your choice.

61. The Drake Equation is a formula used to estimate the number of technological civilizations that might exist in our own Milky Way Galaxy. The Drake Equation is given as $N = R^* \times f_p \times n_e \times f_l \times f_i \times f_c \times L$. Solve the Drake Equation for the variable n_e. (*Note:* Descriptions of the meaning of each variable in this equation, as well as Drake Equation calculators, exist online. For more information, try doing a Web search on "Drake Equation.")

STUDY SKILLS BUILDER

How Are Your Homework Assignments Going?

It is very important in mathematics to keep up with homework. Why? Many concepts build on each other. Often your understanding of a day's concepts depends on an understanding of the previous day's material.

Remember that completing your homework assignment involves a lot more than attempting a few of the problems assigned.

To complete a homework assignment, remember these four things:

- Attempt all of it.
- Check it.
- Correct it.
- If needed, ask questions about it.

Take a moment and review your completed homework assignments. Answer the questions below based on this review.

1. Approximate the fraction of your homework you have attempted.

2. Approximate the fraction of your homework you have checked (if possible).

3. If you are able to check your homework, have you corrected it when errors have been found?

4. When working homework, if you do not understand a concept, what do you do?

2.4 LINEAR INEQUALITIES AND PROBLEM SOLVING

Objectives

A Use Interval Notation.

B Solve Linear Inequalities Using the Addition Property of Inequality.

C Solve Linear Inequalities Using the Multiplication Property of Inequality.

D Solve Linear Inequalities Using Both Properties of Inequality.

E Solve Problems That Can Be Modeled by Linear Inequalities.

Relationships among measurable quantities are not always described by equations. For example, suppose that a salesperson earns a base of $600 per month plus a commission of 20% of sales. Suppose we want to find the minimum amount of sales needed to receive a total income of *at least* $1500 per month. Here, the phrase "at least" implies that an income of $1500 *or more* is acceptable. In symbols, we can write

income $\geq$ 1500

This is an example of an inequality, which we will solve in Example 12.

A *linear inequality* in one variable is similar to a linear equation in one variable except that the equality symbol is replaced with an inequality symbol, such as $<, >, \leq,$ or $\geq$.

Linear Inequalities in One Variable

$$3x + 5 \geq 4 \qquad 2y < 0 \qquad 3(x - 4) > 5x \qquad \frac{x}{3} \leq 5$$

| is greater than or equal to | is less than | is greater than | is less than or equal to |

Linear Inequality in One Variable

A **linear inequality in one variable** is an inequality that can be written in the form

$$ax + b < c$$

where a, b, and c are real numbers and $a \neq 0$.

In this section, when we make definitions, state properties, or list steps about an inequality containing the symbol $<$, we mean that the definition, property, or steps apply to an inequality containing the symbols $>$, $\leq$, and $\geq$ also.

Objective **A** Using Interval Notation

A **solution** of an inequality is a value of the variable that makes the inequality a true statement. The **solution set** of an inequality is the set of all solutions. Notice that the solution set of the inequality $x > 2$, for example, contains all numbers greater than 2. Its graph is an interval on the number line since an infinite number of values satisfy the variable. If we use open/closed-circle notation, the graph of $\{x \mid x > 2\}$ looks like:

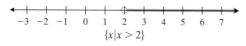

$\{x \mid x > 2\}$

In this text, a different graphing notation will be used to help us understand **interval notation.** Instead of an open circle, we use a parenthesis; instead of a closed circle, we use a bracket. With this new notation, the graph of $\{x \mid x > 2\}$ now looks like:

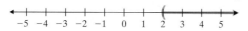

and can be represented in interval notation as $(2, \infty)$. The symbol ∞ is read "infinity" and indicates that the interval includes *all* numbers greater than 2. The left parenthesis indicates that 2 *is not* included in the interval. Using a left bracket, [, would indicate that 2 *is* included in the interval. The following table shows three

103

equivalent ways to describe an interval: in set notation, as a graph, and in interval notation.

Set Notation	Graph	Interval Notation
$\{x \mid x < a\}$	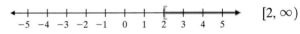	$(-\infty, a)$
$\{x \mid x > a\}$		(a, ∞)
$\{x \mid x \le a\}$		$(-\infty, a]$
$\{x \mid x \ge a\}$		$[a, \infty)$
$\{x \mid a < x < b\}$		(a, b)
$\{x \mid a \le x \le b\}$		$[a, b]$
$\{x \mid a < x \le b\}$		$(a, b]$
$\{x \mid a \le x < b\}$		$[a, b)$
$\{x \mid x \text{ is a real number}\}$		$(-\infty, \infty)$

Helpful Hint

Notice that a parenthesis is always used to enclose ∞ and $-\infty$.

PRACTICE PROBLEMS 1–3

Graph each set on a number line and then write it in interval notation.

1. $\{x \mid x > -3\}$

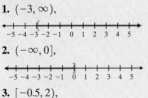

2. $\{x \mid x \le 0\}$

3. $\{x \mid -0.5 \le x < 2\}$

EXAMPLES Graph each set on a number line and then write it in interval notation.

1. $\{x \mid x \ge 2\}$ $[2, \infty)$

2. $\{x \mid x < -1\}$ $(-\infty, -1)$

3. $\{x \mid 0.5 < x \le 3\}$ $(0.5, 3]$

Work Practice Problems 1–3

✔**Concept Check** Explain what is wrong with writing the interval $(5, \infty]$.

Objective B Using the Addition Property of Inequality

To solve a linear inequality, we use a process similar to the one used to solve a linear equation. We use properties of inequalities to write equivalent inequalities until the variable is alone on one side of the inequality.

Answers

1. $(-3, \infty)$,

2. $(-\infty, 0]$,

3. $[-0.5, 2)$,

✔ **Concept Check Answer**

should be $(5, \infty)$ since a parenthesis is always used to enclose ∞

Addition Property of Inequality

If a, b, and c are real numbers, then

$$a < b \quad \text{and} \quad a + c < b + c$$

are equivalent inequalities.

In other words, we may add the same real number to both sides of an inequality, and the resulting inequality will have the same solution set. This property also allows us to subtract the same real number from both sides.

EXAMPLE 4 Solve: $x - 2 < 5$. Graph the solution set.

Solution:

$$x - 2 < 5$$
$$x - 2 + 2 < 5 + 2 \quad \text{Add 2 to both sides.}$$
$$x < 7 \quad \text{Simplify.}$$

The solution set is $\{x \mid x < 7\}$, which in interval notation is $(-\infty, 7)$. The graph of the solution set is

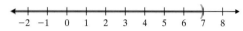

🔲 **Work Practice Problem 4**

Helpful Hint

In Example 4, the solution set is $\{x \mid x < 7\}$. This means that *all* numbers less than 7 are solutions. For example, 6.9, 0, $-\pi$, 1, and -56.7 are solutions, just to name a few. To see this, replace x in $x - 2 < 5$ with each of these numbers and see that the result is a true inequality.

EXAMPLE 5 Solve: $4x - 2 < 5x$. Graph the solution set.

Solution: To get x alone on one side of the inequality, we subtract $4x$ from both sides.

$$4x - 2 < 5x$$
$$4x - 2 - 4x < 5x - 4x \quad \text{Subtract } 4x \text{ from both sides.}$$
$$-2 < x \quad \text{or} \quad x > -2 \quad \text{Simplify.}$$

The solution set is $\{x \mid x > -2\}$, which in interval notation is $(-2, \infty)$. The graph is

🔲 **Work Practice Problem 5**

EXAMPLE 6 Solve: $3x + 4 \geq 2x - 6$. Graph the solution set.

Solution:

$$3x + 4 \geq 2x - 6$$
$$3x + 4 - 2x \geq 2x - 6 - 2x \quad \text{Subtract } 2x \text{ from both sides.}$$
$$x + 4 \geq -6 \quad \text{Combine like terms.}$$
$$x + 4 - 4 \geq -6 - 4 \quad \text{Subtract 4 from both sides.}$$
$$x \geq -10 \quad \text{Simplify}$$

The solution set is $\{x \mid x \geq -10\}$, which in interval notation is $[-10, \infty)$. The graph of the solution set is

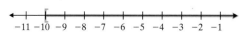

🔲 **Work Practice Problem 6**

PRACTICE PROBLEM 4

Solve: $x + 3 < 1$. Graph the solution set.

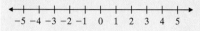

PRACTICE PROBLEM 5

Solve: $3x - 4 < 4x$. Graph the solution set.

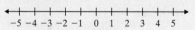

Helpful Hint
Don't forget that $-2 < x$ means the same as $x > -2$.

PRACTICE PROBLEM 6

Solve: $5x - 1 \geq 4x + 4$. Graph the solution set.

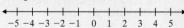

Answers

4. $\{x \mid x < -2\}, (-\infty, -2),$

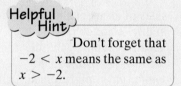

5. $\{x \mid x > -4\}, (-4, -\infty),$

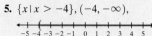

6. $\{x \mid x \geq 5\}, [5, \infty),$

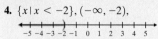

Objective ⒞ Using the Multiplication Property of Inequality

Next, we introduce and use the multiplication property of inequality to solve linear inequalities. To understand this property, let's start with the true statement $-3 < 7$ and multiply both sides by 2.

$$-3 < 7$$
$$2(-3) < 2(7) \quad \text{Multiply both sides by 2.}$$
$$-6 < 14 \quad \text{True}$$

The statement remains true.

Notice what happens if both sides of $-3 < 7$ are multiplied by -2.

$$-3 < 7$$
$$-2(-3) < -2(7) \quad \text{Multiply both sides by 2.}$$
$$6 < -14 \quad \text{False}$$

The inequality $6 < -14$ is a false statement. However, *if the direction of the inequality sign is reversed,* the result is

$$6 > -14 \quad \text{True}$$

These examples suggest the following property.

Multiplication Property of Inequality

If a, b, and c are real numbers and c is **positive,** then

$$a < b \text{ and } ac < bc$$

are equivalent inequalities.

If a, b, and c are real numbers and c is **negative,** then

$$a < b \text{ and } ac > bc$$

are equivalent inequalities.

In other words, we may multiply both sides of an inequality by the same positive real number, and the result is an equivalent inequality. We may also multiply both sides of an inequality by the same *negative number* and *reverse the direction of the inequality symbol,* and the result is an equivalent inequality. The multiplication property holds for division also since division is defined in terms of multiplication.

> **Helpful Hint**
> Whenever both sides of an inequality are multiplied or divided by a negative number, the direction of the inequality symbol *must be* reversed to form an equivalent inequality.

PRACTICE PROBLEM 7

Solve: $\frac{1}{6}x \le \frac{2}{3}$. Graph the solution set.

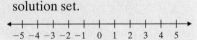

EXAMPLE 7 Solve: $\frac{1}{4}x \le \frac{3}{2}$. Graph the solution set.

Solution:

$$\frac{1}{4}x \le \frac{3}{2}$$
$$4 \cdot \frac{1}{4}x \le 4 \cdot \frac{3}{2} \quad \text{Multiply both sides by 4.}$$
$$x \le 6 \quad \text{Simplify.}$$

> **Helpful Hint**
> The inequality symbol is the same since we are multiplying by a *positive* number.

The solution set is $\{x \mid x \le 6\}$, which in interval notation is $(-\infty, 6]$. The graph of this solution set is

▣ **Work Practice Problem 7**

Answer

7. $\{x \mid x \le 4\}$, $(-\infty, 4]$,

EXAMPLE 8 Solve: $-2.3x < 6.9$. Graph the solution set.

Solution:

$$-2.3x < 6.9$$

$$\frac{-2.3x}{-2.3} > \frac{6.9}{-2.3}$$ Divide both sides by -2.3 and reverse the inequality symbol.

$$x > -3$$ Simplify.

Helpful Hint

The inequality symbol is *reversed* since we divided by a *negative* number.

The solution set is $\{x \mid x > -3\}$, which is $(-3, \infty)$ in interval notation. The graph of the solution set is

Work Practice Problem 8

✔ **Concept Check** In which of the following inequalities must the inequality symbol be reversed during the solution process? (If necessary, assume that variable terms are moved to the left and constants are moved to the right.)

a. $-2x > 7$ **c.** $-x + 4 + 3x < 5$

b. $2x - 3 > 10$ **d.** $-x + 4 < 5$

Objective D Using Both Properties of Inequality

Many problems require us to use both properties of inequality. To solve linear inequalities in general, we follow steps similar to those for solving linear equations.

Solving a Linear Inequality in One Variable

Step 1: Clear the equation of fractions or decimals by multiplying both sides of the inequality by an appropriate nonzero number.

Step 2: Use the distributive property to remove grouping symbols such as parentheses.

Step 3: Combine like terms on each side of the inequality.

Step 4: Use the addition property of inequality to rewrite the inequality as an equivalent inequality with variable terms on one side and numbers on the other side.

Step 5: Use the multiplication property of inequality to get the variable alone on one side of the inequality.

EXAMPLE 9 Solve: $-(x - 3) + 2 \le 3(2x - 5) + x$. Graph the solution set.

Solution:

$$-(x - 3) + 2 \le 3(2x - 5) + x$$

$$-x + 3 + 2 \le 6x - 15 + x$$ Apply the distributive property.

$$5 - x \le 7x - 15$$ Combine like terms.

$$5 - x + x \le 7x - 15 + x$$ Add x to both sides.

$$5 \le 8x - 15$$ Combine like terms.

$$5 + 15 \le 8x - 15 + 15$$ Add 15 to both sides.

$$20 \le 8x$$ Combine like terms.

Continued on next page

Continued on next page

PRACTICE PROBLEM 8

Solve: $-1.1x < 5.5$. Graph the solution set.

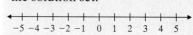

PRACTICE PROBLEM 9

Solve:

$-(2x - 6) \le 4(2x - 4) + 2.$ Write the solution in interval notation.

Answers

8. $\{x \mid x > -5\}, (-5, \infty)$,

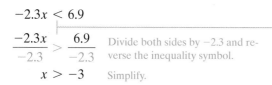

9. $[2, \infty)$

✔ **Concept Check Answer**

a, d

In Example 9, don't forget that $\frac{5}{2} \leq x$ means the same as $x \geq \frac{5}{2}$.

$$\frac{20}{8} \leq \frac{8x}{8} \qquad \text{Divide both sides by 8.}$$

$$\frac{5}{2} \leq x, \quad \text{or} \quad x \geq \frac{5}{2} \qquad \text{Simplify.}$$

The solution set written in interval notation is $\left[\frac{5}{2}, \infty\right)$ and its graph is

$$\frac{5}{2}$$

$\leftarrow\!\!+\!\!\!-\!\!+\!\!\!-\!\!+\!\!\!-\!\!+\!\!\!-\!\!+\!\!\!-\!\!+\!\!\!-\!\![\!\!+\!\!\!-\!\!+\!\!\!-\!\!+\!\!\!\rightarrow$
$\quad -5 \;\; -4 \;\; -3 \;\; -2 \;\; -1 \quad 0 \quad 1 \quad 2 \;\; 3 \quad 4 \quad 5$

◼ **Work Practice Problem 9**

PRACTICE PROBLEM 10

Solve: $\frac{3}{4}(x + 2) \geq x - 6$.

Write the solution set in interval notation.

EXAMPLE 10 Solve: $\frac{2}{5}(x - 6) \geq x - 1$. Write the solution set in interval notation.

Solution:

$$\frac{2}{5}(x - 6) \geq x - 1$$

$$5\left[\frac{2}{5}(x - 6)\right] \geq 5(x - 1) \qquad \text{Multiply both sides by 5 to eliminate fractions.}$$

$$2(x - 6) \geq 5(x - 1)$$

$$2x - 12 \geq 5x - 5 \qquad \text{Use the distributive property.}$$

$$-3x - 12 \geq -5 \qquad \text{Subtract } 5x \text{ from both sides.}$$

$$-3x \geq 7 \qquad \text{Add 12 to both sides.}$$

$$\frac{-3x}{-3} \leq \frac{7}{-3} \qquad \text{Divide both sides by } -3 \text{ and reverse the inequality symbol.}$$

$$x \leq -\frac{7}{3} \qquad \text{Simplify.}$$

The solution set is $\left(-\infty, -\frac{7}{3}\right]$.

◼ **Work Practice Problem 10**

PRACTICE PROBLEM 11

Solve: $5(x - 3) < 5x + 2$.

EXAMPLE 11 Solve: $2(x + 3) > 2x + 1$.

Solution:

$$2(x + 3) > 2x + 1$$

$$2x + 6 > 2x + 1 \qquad \text{Distribute on the left side.}$$

$$2x + 6 - 2x > 2x + 1 - 2x \qquad \text{Subtract } 2x \text{ from both sides.}$$

$$6 > 1 \qquad \text{Simplify.}$$

$6 > 1$ is a true statement for all values of x, so this inequality and the original inequality are true for all numbers. The solution set is $\{x \mid x \text{ is a real number}\}$, or $(-\infty, \infty)$ in interval notation, and its graph is

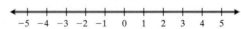

◼ **Work Practice Problem 11**

Answers

10. $(-\infty, 30]$,

11. $(-\infty, \infty)$

Objective E Linear Inequalities and Problem Solving

Problems containing words such as "at least," "at most," "between," "no more than," and "no less than" usually indicate that an inequality is to be solved instead of an equation. In solving applications involving linear inequalities, we use the same four-step strategy as when we solved applications involving linear equations.

EXAMPLE 12 Calculating Income with Commission

A salesperson earns $600 per month plus a commission of 20% of sales. Find the minimum amount of sales needed to receive a total income of at least $1500 per month.

Solution:

1. UNDERSTAND. Read and reread the problem. Let

   x = amount of sales

2. TRANSLATE. Since the income is to be at least $1500, this means, we want the income to be greater than or equal to $1500. To write an inequality, notice that the salesperson's income consists of $600 plus a commission (20% of sales).

   In words: 600 + commission of 20% of sales ≥ 1500

   Translate: 600 + 0.20x ≥ 1500

3. SOLVE the inequality for x.

$$600 + 0.20x \geq 1500$$
$$600 + 0.20x - 600 \geq 1500 - 600$$
$$0.20x \geq 900$$
$$x \geq 4500$$

4. INTERPRET.

Check: The income for sales of $4500 is

   $600 + 0.20(4500)$, or 1500

Thus, if sales are greater than or equal to $4500, income is greater than or equal to $1500.

State: The minimum amount of sales needed for the salesperson to earn at least $1500 per month is $4500.

◼ Work Practice Problem 12

EXAMPLE 13 Finding the Annual Consumption

In the United States, the annual consumption of cigarettes is declining. The consumption c in billions of cigarettes per year since 1990 can be approximated by the formula $c = -9.2t + 527.33$, where t is the number of years after 1990. Use this formula to predict the first year that the consumption of cigarettes will be less than 200 billion per year. (*Source:* U.S. Department of Agriculture Economic Research Service.)

Continued on next page

PRACTICE PROBLEM 12

A salesperson earns $1000 a month plus a commission of 15% of sales. Find the minimum amount of sales needed to receive a total income of at least $4000 per month.

PRACTICE PROBLEM 13

Use the formula given in Example 13 to predict when the consumption of cigarettes will be less than 100 billion per year.

Answers
12. $20,000, **13.** after the year 2037

Solution:

1. UNDERSTAND. Read and reread the problem. To become familiar with the given formula, let's find the cigarette consumption after 20 years, which would be the year 1990 + 20 or 2010. To do so, we substitute 20 for t in the given formula.

    $c = -9.2(20) + 527.33 = 343.33$

    Thus, in 2010, we predict cigarette consumption to be about 343.3 billion. Variables have already been assigned in the given formula. For review, they are

    $c =$ the annual consumption of cigarettes in the United States in billions of cigarettes

    $t =$ the number of years after 1990

2. TRANSLATE. We are looking for the first year that the consumption of cigarettes c *is less than 200*. Since we are finding years t, we substitute the expression in the formula given for c, or

    $-9.2t + 527.33 < 200$

3. SOLVE the inequality.

    $$-9.2t + 527.33 < 200$$
    $$-9.2t < -327.33$$
    $$t > \text{approximately } 35.58$$

4. INTERPRET.

Check: Substitute a number greater than 35.58 and see that c is less than 200.

State: The annual consumption of cigarettes will be less than 200 billion more than 35.58 years after 1990, or in approximately 36 + 1990 = 2026.

Work Practice Problem 13

Mental Math

Solve each inequality.

1. $x - 2 < 4$
2. $x - 1 > 6$
3. $x + 5 \geq 15$
4. $x + 1 \leq 8$

5. $3x > 12$
6. $5x < 20$
7. $\dfrac{x}{2} \leq 1$
8. $\dfrac{x}{4} \geq 2$

2.4 EXERCISE SET

Objective A *Graph the solution set of each inequality on a number line and then write it in interval notation. See Examples 1 through 3.*

1. $\{x \mid x < -3\}$

2. $\{x \mid x > 5\}$

3. $\{x \mid x \geq 0.3\}$

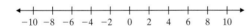

4. $\{x \mid x < -0.2\}$

5. $\{x \mid -7 \leq x\}$

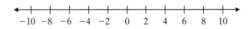

6. $\{x \mid -7 \geq x\}$

7. $\{x \mid -2 < x < 5\}$

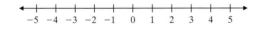

8. $\{x \mid -5 \leq x \leq -1\}$

9. $\{x \mid 5 \geq x > -1\}$

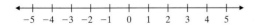

10. $\{x \mid -3 > x \geq -7\}$

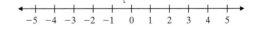

Objective B *Solve. Graph the solution set and write it in interval notation. See Examples 4 through 6.*

11. $x - 7 \geq -9$

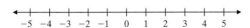

12. $x + 2 \leq -1$

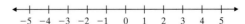

13. $7x < 6x + 1$

14. $11x < 10x + 5$

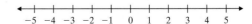

15. $8x - 7 \leq 7x - 5$

16. $7x - 1 \geq 6x - 1$

111

Objective 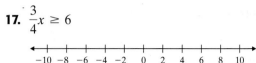 *Solve. Graph the solution set and then write it in interval notation. See Examples 7 and 8.*

17. $\frac{3}{4}x \geq 6$

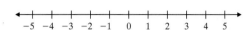

18. $\frac{5}{6}x \geq 5$

19. $5x < -23.5$

20. $4x > -11.2$

21. $-3x \geq 9$

22. $-4x \geq 8$

Objective D *Solve. Write the solution set using interval notation. See Examples 9 through 11.*

23. $-2x + 7 \geq 9$

24. $8 - 5x \leq 23$

25. $15 + 2x \geq 4x - 7$

26. $20 + x < 6x - 15$

27. $4(2x + 1) > 4$

28. $6(2 - 3x) \geq 12$

29. $3(x - 5) < 2(2x - 1)$

30. $5(x + 4) \leq 4(2x + 3)$

31. $\frac{5x + 1}{7} - \frac{2x - 6}{4} \geq -4$

32. $\frac{1 - 2x}{3} + \frac{3x + 7}{7} > 1$

33. $-3(2x - 1) < -4[2 + 3(x + 2)]$

34. $-2(4x + 2) > -5[1 + 2(x - 1)]$

Objectives B C D **Mixed Practice** *Solve. Write the solution set using interval notation. See Examples 4 through 11.*

35. $x + 9 < 3$

36. $x - 9 < -12$

37. $-x < -4$

38. $-x > -2$

39. $-7x \leq 3.5$

40. $-6x \leq 4.2$

41. $\frac{1}{2} + \frac{2}{3} \geq \frac{x}{6}$

42. $\frac{3}{4} - \frac{2}{3} \geq \frac{x}{6}$

43. $-5x + 4 \leq -4(x - 1)$

44. $-6x + 2 < -3(x + 4)$

45. $\frac{3}{4}(x - 7) \geq x + 2$

46. $\frac{4}{5}(x + 1) \leq x + 1$

47. $0.8x + 0.6x \geq 4.2$

48. $0.7x - x > 0.45$

49. $4(x - 6) + 2x - 4 \geq 3(x - 7) + 10x$

50. $7(2x + 3) + 4x \leq 7 + 5(3x - 4) + x$

51. $14 - (5x - 6) \geq -6(x + 1) - 5$

52. $13y - (9y + 2) \leq 5(y - 6) + 10$

53. $\frac{1}{2}(3x - 4) \leq \frac{3}{4}(x - 6) + 1$

54. $\frac{2}{3}(x + 3) < \frac{1}{6}(2x - 8) + 2$

55. $0.4(4x - 3) < 1.2(x + 2)$

56. $0.2(8x - 2) < 1.2(x - 3)$

57. $\dfrac{2}{5}x - \dfrac{1}{4} \le \dfrac{3}{10}x - \dfrac{4}{5}$

58. $\dfrac{7}{12}x - \dfrac{1}{3} \le \dfrac{3}{8}x - \dfrac{5}{6}$

Objective **E** *Solve. See Examples 12 and 13. For Exercises 59 through 66,* **a.** *answer with an inequality, and* **b.** *in your own words, explain the meaning of your answer to part a.*

Exercises 59 and 60 are written to help you get started.

59. Shureka Washburn has scores of 72, 67, 82, and 79 on her algebra tests.
   a. Use an inequality to find the scores she must make on the final exam to pass the course with an average of 77 or higher, given that the final exam counts as two tests.
   b. In your own words explain the meaning of your answer to part a.

60. In a Winter Olympics 5000-meter speed-skating event, Hans Holden scored times of 6.85, 7.04, and 6.92 minutes on his first three trials.
   a. Use an inequality to find the times he can score on his last trial so that his average time is under 7.0 minutes.
   b. In your own words, explain the meaning of your answer to part a.

61. A small plane's maximum takeoff weight is 2000 pounds or less. Six passengers weigh an average of 160 pounds each. Use an inequality to find the luggage and cargo weights the plane can carry.

62. A shopping mall parking garage charges $1 for the first half-hour and 60 cents for each additional half-hour. Use an inequality to find how long you can park if you have only $4.00 in cash.

63. A clerk must use the elevator to move boxes of paper. The elevator's maximum weight limit is 1500 pounds. If each box of paper weighs 66 pounds and the clerk weighs 147 pounds, use an inequality to find the number of whole boxes she can move on the elevator at one time.

64. To mail an envelope first class, the U.S. Post Office charges 37 cents for the first ounce and 23 cents per ounce for each additional ounce. Use an inequality to find the number of whole ounces that can be mailed for no more than $4.00

65. Northeast Telephone Company offers two billing plans for local calls.

Plan 1: $25 per month for unlimited calls
Plan 2: $13 per month plus $0.06 per call

Use an inequality to find the number of monthly calls for which plan 1 is more economical than plan 2.

66. A car rental company offers two subcompact rental plans.

Plan A: $36 per day and unlimited mileage
Plan B: $24 per day plus $0.15 per mile

Use an inequality to find the number of daily miles for which plan A is more economical than plan B.

67. At room temperature, glass used in windows actually has some properties of a liquid. It has a very slow, viscous flow. (Viscosity is the property of a fluid that resists internal flow. For example, lemonade flows more easily than fudge syrup. Fudge syrup has a higher viscosity than lemonade.) Glass does not become a true liquid until temperatures are greater than or equal to 500°C. Find the Fahrenheit temperatures for which glass is a liquid. (Use the formula $F = \dfrac{9}{5}C + 32$.)

68. Stibnite is a silvery white mineral with a metallic luster. It is one of the few minerals that melts easily in a match flame or at temperatures of approximately 977°F or greater. Find the Celsius temperatures for which stibnite melts. [Use the formula $C = \dfrac{5}{9}(F - 32)$.]

69. Although beginning salaries vary greatly according to your field of study, the equation

$$s = 651.2t + 27{,}821$$

can be used to approximate and to predict average beginning salaries for candidates with bachelor's degrees. The variable s is the starting salary and t is the number of years after 1989.

   a. Approximate when beginning salaries for candidates will be greater than $35,000.

   b. Determine the year you plan to graduate from college. Use this year to find the corresponding value of t and approximate your beginning salary.

70. a. Use the formula in Example 13 to estimate the years that the consumption of cigarettes will be less than 50 billion per year.

   b. Use your answer to part a to describe the limitations of your answer.

The average consumption per person per year of whole milk w can be approximated by the equation

$$w = -1.9t + 70.1$$

where t is the number of years after 1996 and w is measured in pounds. The average consumption of skim milk s per person per year can be approximated by the equation

$$s = -0.8t + 34.3$$

where t is the number of years after 1996 and s is measured in pounds. The consumption of whole milk is shown on the graph in blue and the consumption of skim milk is shown on the graph in red. Use this information to answer Exercises 71 through 79.

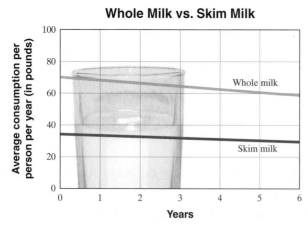

Whole Milk vs. Skim Milk

Source: Based on data from U.S. Department of Agriculture, Economic Research Service

71. Is the consumption of whole milk increasing or decreasing over time? Explain how you arrived at your answer.

72. Is the consumption of skim milk increasing or decreasing over time? Explain how you arrived at your answer.

73. Predict the consumption of whole milk in 2010. (*Hint:* Find the value of t that corresponds to 2010.)

74. Predict the consumption of skim milk in 2010. (*Hint:* Find the value of t that corresponds to 2010.)

75. Determine when the consumption of whole milk will be less than 50 pounds per person per year.

76. For 1996 through 2002, the consumption of whole milk was greater than the consumption of skim milk. Explain how this can be determined from the graph.

77. Both lines have a negative slope, that is, the amount of each type of milk consumed per person per year is decreasing as time goes on. However, the amount of whole milk being consumed is decreasing faster than the amount of skim milk being consumed. Explain how this could be.

78. Do you think it is possible that the consumption of whole milk will eventually be the same as the consumption of skim milk? Explain your answer.

79. The consumption of skim milk will be greater than as the consumption of whole milk when $s > w$.
   a. Find when this will occur by substituting the given equivalent expression for w and the given equivalent expression for s and solving for t.
   b. Estimate to the nearest whole the first year when this will occur.

Review

List or describe the integers that make both inequalities true.

80. $x < 5$ and $x > 1$

81. $x \geq 0$ and $x \leq 7$

82. $x \geq -2$ and $x \geq 2$

83. $x < 6$ and $x < -5$

Solve each equation for x.

84. $2x - 6 = 4$

85. $3x - 12 = 3$

86. $-x + 7 = 5x - 6$

87. $-5x - 4 = -x - 4$

Concept Extensions

In which inequality must the inequality symbol be reversed during the solution process? If necessary, assume that variable terms are moved to the left and constants to the right.

88. $3x > -14$

89. $-3x \leq 14$

90. $-3x < -14$

91. $-x - 17 \geq -23$

92. Write an inequality whose solution set is $\{x \mid x \leq 2\}$.

93. Solve: $2x - 3 = 5$

94. Solve: $2x - 3 < 5$

95. Solve: $2x - 3 > 5$

96. Read the equations and inequalities for Exercises 93, 94, and 95 and their solutions. In your own words, write down your thoughts.

Solve each inequality.

97. $4(x - 1) \geq 4x - 8$

98. $3x + 1 < 3(x - 2)$

99. $7x < 7(x - 2)$

100. $8(x + 3) \leq 7(x + 5) + x$

101. When graphing the solution set of an inequality, explain how you know whether to use a parenthesis or a bracket.

102. Explain what is wrong with the interval notation $(-6, -\infty)$.

103. Explain how solving a linear inequality is similar to solving a linear equation.

104. Explain how solving a linear inequality is different from solving a linear equation.

INTEGRATED REVIEW

Sections 2.1–2.4

Linear Equations and Inequalities

Solve each equation or inequality.

1. $-4x = 20$

2. $-4x < 20$

3. $\dfrac{3x}{4} \geq 2$

4. $5x + 3 \geq 2 + 4x$

5. $6(y - 4) = 3(y - 8)$

6. $-4x \leq \dfrac{2}{5}$

7. $-3x \geq \dfrac{1}{2}$

8. $5(y + 4) = 4(y + 5)$

9. $7x < 7(x - 2)$

10. $\dfrac{-5x + 11}{2} \leq 7$

11. $-5x + 1.5 = -19.5$

12. $-5x + 4 = -26$

13. $5 + 2x - x = -x + 3 - 14$

14. $12x + 14 < 11x - 2$

116

15. $\dfrac{x}{5} - \dfrac{x}{4} = \dfrac{x-2}{2}$

16. $12x - 12 = 8(x - 1)$

17. $2(x - 3) > 70$

18. $-3x - 4.7 = 11.8$

19. $-2(b - 4) - (3b - 1) = 5b + 3$

20. $8(x + 3) < 7(x + 5) + x$

21. $\dfrac{3t + 1}{8} = \dfrac{5 + 2t}{7} + 2$

22. $4(x - 6) - x = 8(x - 3) - 5x$

23. $\dfrac{x + 3}{12} + \dfrac{x - 5}{15} < \dfrac{2}{3}$

24. $\dfrac{y}{3} + \dfrac{y}{5} = \dfrac{y + 3}{10}$

25. $5(x - 6) + 2x > 3(2x - 1) - 4$

26. $14(x - 1) - 7x \le 2(3x - 6) + 4$

27. $\dfrac{1}{4}(3x + 2) - x \ge \dfrac{3}{8}(x - 5) + 2$

28. $\dfrac{1}{3}(x - 10) - 4x > \dfrac{5}{6}(2x + 1) - 1$

15. _____

16. _____

17. _____

18. _____

19. _____

20. _____

21. _____

22. _____

23. _____

24. _____

25. _____

26. _____

27. _____

28. _____

2.5 SETS AND COMPOUND INEQUALITIES

Two inequalities joined by the words **and** or **or** are called **compound inequalities.**

Compound Inequalities

$x + 3 < 8$ and $x > 2$

$\dfrac{2x}{3} \geq 5$ or $-x + 10 < 7$

Objective **A** Finding the Intersection of Two Sets

The solution set of a compound inequality formed by the word **and** is the *intersection* of the solution sets of the two inequalities. We use the symbol ∩ to denote "intersection."

Intersection of Two Sets

The **intersection** of two sets, A and B, is the set of all elements common to both sets. A intersect B is denoted by

$A \cap B$

PRACTICE PROBLEM 1

Find the intersection:

$\{1, 2, 3, 4, 5\} \cap \{3, 4, 5, 6\}$

EXAMPLE 1 Find the intersection: $\{2, 4, 6, 8\} \cap \{3, 4, 5, 6\}$

Solution: The numbers 4 and 6 are in both sets. The intersection is $\{4, 6\}$.

■ Work Practice Problem 1

Objective **B** Solving Compound Inequalities Containing *"and"*

A value is a **solution** of a compound inequality formed by the word **and** if it is a solution of *both* inequalities. For example, the solution set of the compound inequality $x \leq 5$ and $x \geq 3$ contains all values of x that make the inequality $x \leq 5$ a true statement **and** the inequality $x \geq 3$ a true statement. The first graph shown below is the graph of $x \leq 5$, the second graph is the graph of $x \geq 3$, and the third graph shows the intersection of the two graphs. The third graph is the graph of $x \leq 5$ **and** $x \geq 3$.

$\{x \mid x \leq 5\}$ $(-\infty, 5]$

$\{x \mid x \geq 3\}$ $[3, \infty)$

$\{x \mid x \leq 5 \text{ and } x \geq 3\}$,
also $\{x \mid 3 \leq x \leq 5\}$ $[3, 5]$
(see below)

Since $x \geq 3$ is the same as $3 \leq x$, the compound inequality $3 \leq x$ and $x \leq 5$ can be written in a more compact form as $3 \leq x \leq 5$. The solution set $\{x \mid 3 \leq x \leq 5\}$ includes all numbers that are greater than or equal to 3 and at the same time less than or equal to 5.

In interval notation, the set $\{x \mid x \leq 5 \text{ and } x \geq 3\}$ or $\{x \mid 3 \leq x \leq 5\}$ is written as $[3, 5]$.

Don't forget that some compound inequalities containing "and" can be written in a more compact form.

Compound Inequality	Compact Form	Interval Notation
$2 \leq x$ and $x \leq 6$	$2 \leq x \leq 6$	$[2, 6]$

Graph:

EXAMPLE 2 Solve: $x - 7 < 2$ and $2x + 1 < 9$

Solution: First we solve each inequality separately.

$$x - 7 < 2 \quad and \quad 2x + 1 < 9$$
$$x < 9 \quad and \quad 2x < 8$$
$$x < 9 \quad and \quad x < 4$$

Now we can graph the two intervals on two number lines and find their intersection.

$\{x | x < 9\}$ $(-\infty, 9)$

$\{x | x < 4\}$ $(-\infty, 4)$

$\{x | x < 9 \ and \ x < 4\}$ $(-\infty, 4)$
$= \{x | x < 4\}$

The solution set is $(-\infty, 4)$.

Work Practice Problem 2

EXAMPLE 3 Solve: $2x \geq 0$ and $4x - 1 \leq -9$

Solution: First we solve each inequality separately.

$$2x \geq 0 \quad and \quad 4x - 1 \leq -9$$
$$x \geq 0 \quad and \quad 4x \leq -8$$
$$x \geq 0 \quad and \quad x \leq -2$$

Now we can graph the two intervals and find their intersection.

$\{x | x \geq 0\}$ $[0, \infty)$

$\{x | x \leq -2\}$ $(-\infty, -2]$

$\{x | x \geq 0 \ and$
$x \leq -2\} = \varnothing$

There is no number that is greater than or equal to 0 **and** less than or equal to -2. The solution set is $\varnothing$.

Work Practice Problem 3

Example 3 shows that some compound inequalities have no solution. Also, some have all real numbers as solutions.

To solve a compound inequality like $2 < 4 - x < 7$, we get x alone in the middle. Since a compound inequality is really two inequalities in one statement, we must perform the same operation to all three parts of the inequality.

PRACTICE PROBLEM 2

Solve: $x + 5 < 9$ and $3x - 1 < 2$

PRACTICE PROBLEM 3

Solve: $4x \geq 0$ and $2x + 4 \leq 2$

Answers
2. $(-\infty, 1)$, **3.** $\varnothing$

PRACTICE PROBLEM 4

Solve: $5 < 1 - x < 9$

Helpful Hint

Don't forget to reverse both inequality symbols.

EXAMPLE 4 Solve: $2 < 4 - x < 7$

Solution: To get x alone, we first subtract 4 from all three parts.

$$2 < 4 - x < 7$$
$$2 - 4 < 4 - x - 4 < 7 - 4 \quad \text{Subtract 4 from all three parts.}$$
$$-2 < -x < 3 \quad \text{Simplify.}$$
$$\frac{-2}{-1} > \frac{-x}{-1} > \frac{3}{-1} \quad \text{Divide all three parts by } -1 \text{ and reverse the inequality symbols.}$$
$$2 > x > -3$$

This is equivalent to $-3 < x < 2$, and its graph is shown.

The solution set in interval notation is $(-3, 2)$.

🖢 **Work Practice Problem 4**

PRACTICE PROBLEM 5

Solve: $-2 < \frac{3}{4}x + 2 \le 5$

EXAMPLE 5 Solve: $-1 \le \frac{2}{3}x + 5 < 2$

Solution: First we clear the inequality of fractions by multiplying all three parts by the LCD, 3.

$$-1 \le \frac{2}{3}x + 5 < 2$$
$$3(-1) \le 3\left(\frac{2}{3}x + 5\right) < 3(2) \quad \text{Multiply all three parts by the LCD, 3.}$$
$$-3 \le 2x + 15 < 6 \quad \text{Use the distributive property and multiply.}$$
$$-3 - 15 \le 2x + 15 - 15 < 6 - 15 \quad \text{Subtract 15 from all three parts.}$$
$$-18 \le 2x < -9 \quad \text{Simplify.}$$
$$\frac{-18}{2} \le \frac{2x}{2} < \frac{-9}{2} \quad \text{Divide all three parts by 2.}$$
$$-9 \le x < -\frac{9}{2} \quad \text{Simplify.}$$

The graph of the solution is shown.

The solution set in interval notation is $\left[-9, -\frac{9}{2}\right)$.

🖢 **Work Practice Problem 5**

Objective 🄲 **Finding the Union of Two Sets**

The solution set of a compound inequality formed by the word **or** is the **union** of the solution sets of the two inequalities. We use the symbol $\cup$ to denote "union."

Helpful Hint

The word "either" in this definition means "one or the other or both."

Union of Two Sets

The **union** of two sets, A and B, is the set of elements that belong to *either* of the sets. A union B is denoted by

$$A \cup B$$

Answers

4. $(-8, -4)$, 5. $\left(-\frac{16}{3}, 4\right]$

EXAMPLE 6 Find the union: $\{2, 4, 6, 8\} \cup \{3, 4, 5, 6\}$

Solution: The numbers that are in either set are $\{2, 3, 4, 5, 6, 8\}$. This set is the union.

◻ **Work Practice Problem 6**

PRACTICE PROBLEM 6

Find the union:
$\{1, 2, 3, 4, 5\} \cup \{3, 4, 5, 6\}$

Objective ◻ **Solving Compound Inequalities Containing "or"**

A value of x is a solution of a compound inequality formed by the word **or** if it is a solution of **either** inequality. For example, the solution set of the compound inequality $x \leq 1$ **or** $x \geq 3$ contains all numbers that make the inequality $x \leq 1$ a true statement **or** the inequality $x \geq 3$ a true statement. In other words, the solution of such an inequality is the *union* of the solutions of the individual inequalities.

$\{x \mid x \leq 1\}$ $(-\infty, 1]$

$\{x \mid x \geq 3\}$ $[3, \infty)$

$\{x \mid x \leq 1$ or $x \geq 3\}$ $(-\infty, 1] \cup [3, \infty)$

In interval notation, the set $\{x \mid x \leq 1 \text{ or } x \geq 3\}$ is written as $(-\infty, 1] \cup [3, \infty)$.

EXAMPLE 7 Solve: $5x - 3 \leq 10 \text{ or } x + 1 \geq 5$

Solution: First we solve each inequality separately.

$$5x - 3 \leq 10 \quad or \quad x + 1 \geq 5$$
$$5x \leq 13 \quad or \quad x \geq 4$$
$$x \leq \frac{13}{5} \quad or \quad x \geq 4$$

Now we can graph each interval and find their union.

$\left\{ x \mid x \leq \dfrac{13}{5} \right\}$ $\left(-\infty, \dfrac{13}{5}\right]$

$\{x \mid x \geq 4\}$ $[4, \infty)$

$\left\{ x \mid x \leq \dfrac{13}{5} \text{ or } x \geq 4 \right\}$

$\left(-\infty, \dfrac{13}{5}\right] \cup [4, \infty)$

The solution set is $\left(-\infty, \dfrac{13}{5}\right] \cup [4, \infty)$.

◻ **Work Practice Problem 7**

PRACTICE PROBLEM 7

Solve:
$3x - 2 \geq 10 \text{ or } x - 6 \leq -4$

Answers
6. $\{1, 2, 3, 4, 5, 6\}$,
7. $(-\infty, 2] \cup [4, \infty)$

Solve: $x - 7 \leq -1$ or $2x - 6 \geq 2$

EXAMPLE 8 Solve: $-2x - 5 < -3$ *or* $6x < 0$

Solution: First we solve each inequality separately.

$$-2x - 5 < -3 \quad or \quad 6x < 0$$
$$-2x < 2 \quad or \quad x < 0$$
$$x > -1 \quad or \quad x < 0$$

Now we can graph each interval and find their union.

$\{x \mid x > -1\}$ $(-1, \infty)$

$\{x \mid x < 0\}$ $(-\infty, 0)$

$\{x \mid x > -1 \text{ or } x < 0\}$ $(-\infty, \infty)$
= all real numbers

The solution set is $(-\infty, \infty)$.

Work Practice Problem 8

✔ **Concept Check** Which of the following is *not* a correct way to represent the set of all numbers between -3 and 5?

a. $\{x \mid -3 < x < 5\}$
b. $-3 < x$ or $x < 5$
c. $(-3, 5)$
d. $x > -3$ and $x < 5$

Answer

8. $(-\infty, \infty)$

✔ **Concept Check Answer**

b is not correct

Objectives Ⓐ Ⓒ **Mixed Practice** *If* $A = \{x | x$ *is an even integer*$\}$, $B = \{x | x$ *is an odd integer*$\}$, $C = \{2, 3, 4, 5\}$, *and* $D = \{4, 5, 6, 7\}$, *list the elements of each set. See Examples 1 and 6.*

1. $C \cup D$

2. $C \cap D$

3. $A \cap D$

4. $A \cup D$

5. $A \cup B$

6. $A \cap B$

7. $B \cap D$

8. $B \cup D$

9. $B \cup C$

10. $B \cap C$

11. $A \cap C$

12. $A \cup C$

Objective Ⓑ *Solve each compound inequality. Graph the two inequalities on the first two number lines and the solution set on the third number line. See Examples 2 and 3.*

13. $x < 1$ *and* $x > -3$

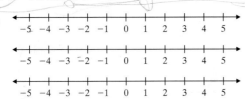

14. $x \leq 0$ *and* $x \geq -2$

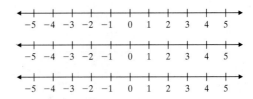

15. $x \leq -3$ *and* $x \geq -2$

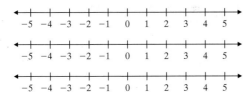

16. $x < 2$ *and* $x > 4$

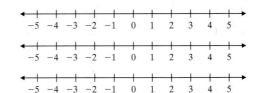

17. $x < -1$ *and* $x < 1$

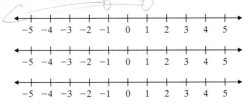

18. $x \geq -4$ *and* $x > 1$

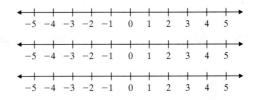

Solve each compound inequality. Write solutions in interval notation. See Examples 2 and 3.

19. $x + 1 \geq 7$ *and* $3x - 1 \geq 5$

20. $x + 2 \geq 3$ *and* $5x - 1 \geq 9$

21. $4x + 2 \leq -10$ *and* $2x \leq 0$

22. $2x + 4 > 0$ *and* $4x > 0$

23. $-2x < -8$ *and* $x - 5 < 5$

24. $-7x \leq -21$ *and* $x - 20 \leq -15$

Solve each compound inequality. See Examples 4 and 5.

25. $5 < x - 6 < 11$

26. $-2 \leq x + 3 \leq 0$

27. $-2 \leq 3x - 5 \leq 7$

28. $1 < 4 + 2x < 7$

29. $1 \le \frac{2}{3}x + 3 \le 4$

30. $-2 < \frac{1}{2}x - 5 < 1$

31. $-5 \le \frac{-3x + 1}{4} \le 2$

32. $-4 \le \frac{-2x + 5}{3} \le 1$

Objective D *Solve each compound inequality. Graph the two given inequalities on the first two number lines and the solution set on the third number line. See Examples 7 and 8.*

33. $x < 4 \; or \; x < 5$

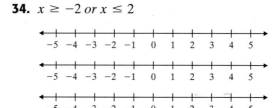

34. $x \ge -2 \; or \; x \le 2$

35. $x \le -4 \; or \; x \ge 1$

36. $x < 0 \; or \; x < 1$

37. $x > 0 \; or \; x < 3$

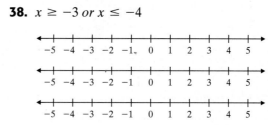

38. $x \ge -3 \; or \; x \le -4$

Solve each compound inequality. Write answers in interval notation. See Examples 7 and 8.

39. $-2x \le -4 \; or \; 5x - 20 \ge 5$

40. $-5x \le 10 \; or \; 3x - 5 \ge 1$

41. $x + 4 < 0 \; or \; 6x > -12$

42. $x + 9 < 0 \; or \; 4x > -12$

43. $3(x - 1) < 12 \; or \; x + 7 > 10$

44. $5(x - 1) \ge -5 \; or \; 5 - x \le 11$

Objectives B D **Mixed Practice** *Solve each compound inequality. Write solutions in interval notation.*

45. $x < \frac{2}{3} \; and \; x > -\frac{1}{2}$

46. $x < \frac{5}{7} \; and \; x < 1$

47. $x < \frac{2}{3} \; or \; x > -\frac{1}{2}$

48. $x < \frac{5}{7} \; or \; x < 1$

49. $0 \le 2x - 3 \le 9$

50. $3 < 5x + 1 < 11$

51. $\frac{1}{2} < x - \frac{3}{4} < 2$

52. $\frac{2}{3} < x + \frac{1}{2} < 4$

53. $x + 3 \ge 3 \; and \; x + 3 \le 2$

54. $2x - 1 \ge 3 \; and \; -x > 2$

55. $3x \ge 5 \; or \; -\frac{5}{8}x - 6 > 1$

56. $\frac{3}{8}x + 1 \le 0 \; or \; -2x < -4$

57. $0 < \dfrac{5 - 2x}{3} < 5$

58. $-2 < \dfrac{-2x - 1}{3} < 2$

59. $-6 < 3(x - 2) \le 8$

60. $-5 < 2(x + 4) < 8$

61. $-x + 5 > 6$ and $1 + 2x \le -5$

62. $5x \le 0$ and $-x + 5 < 8$

63. $3x + 2 \le 5$ or $7x > 29$

64. $-x < 7$ or $3x + 1 < -20$

65. $5 - x > 7$ and $2x + 3 \ge 13$

66. $-2x < -6$ or $1 - x > -2$

67. $-\dfrac{1}{2} \le \dfrac{4x - 1}{6} < \dfrac{5}{6}$

68. $-\dfrac{1}{2} \le \dfrac{3x - 1}{10} < \dfrac{1}{2}$

69. $\dfrac{1}{15} < \dfrac{8 - 3x}{15} < \dfrac{4}{5}$

70. $-\dfrac{1}{4} < \dfrac{6 - x}{12} < -\dfrac{1}{6}$

71. $0.3 < 0.2x - 0.9 < 1.5$

72. $-0.7 \le 0.4x + 0.8 < 0.5$

Review

Evaluate. See Section 1.5.

73. $|-7| - |19|$

74. $|-7 - 19|$

75. $-(-6) - |-10|$

76. $|-4| - (-4) + |-20|$

Find by inspection all values for x that make each equation true.

77. $|x| = 7$

78. $|x| = 5$

79. $|x| = 0$

80. $|x| = -2$

Concept Extensions

Use the graph to answer Exercises 81 and 82.

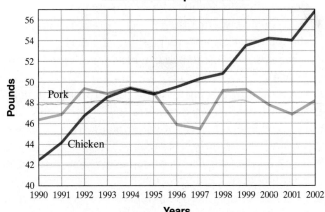

U.S. Consumption of Pork and Chicken per Person

Pounds / Years

Source: Based on data from Economic Research Service, U.S. Department of Agriculture, *Agricultural Outlook*, October 2001

81. For what years was the consumption of pork greater than 48 pounds per person *and* the consumption of chicken greater than 48 pounds per person?

82. For what years was the consumption of pork less than 45 pounds per person *or* the consumption of chicken greater than 55 pounds per person?

83. In your own words, describe how to find the union of two sets.

84. In your own words, describe how to find the intersection of two sets.

To solve a compound inequality such as $x - 6 < 3x < 2x + 5$, we solve

$$x - 6 < 3x \quad \text{and} \quad 3x < 2x + 5$$

Use this information to solve the inequalities in Exercises 85 through 88.

85. $x - 6 < 3x < 2x + 5$

86. $2x - 3 < 3x + 1 < 4x - 5$

87. $x + 3 < 2x + 1 < 4x + 6$

88. $-3(x - 2) \le 3 - 2x \le 10 - 3x$

The formula for converting Fahrenheit temperatures to Celsius temperatures is $C = \dfrac{5}{9}(F - 32)$. Use this formula for Exercises 89 and 90.

89. During a recent year, the temperatures in Chicago ranged from $-29°$ to $35°$ Celsius. Use a compound inequality to convert these temperatures to Fahrenheit temperatures.

90. In Oslo, the average temperature ranges from $-10°$ to $18°$ Celsius. Use a compound inequality to convert these temperatures to the Fahrenheit scale.

Solve.

91. Christian D'Angelo has scores of 68, 65, 75, and 78 on his algebra tests. Use a compound inequality to find the scores he can make on his final exam to receive a C in the course. The final exam counts as two tests, and a C is received if the final course average is from 70 to 79.

92. Wendy Wood has scores of 80, 90, 82, and 75 on her chemistry tests. Use a compound inequality to find the range of scores she can make on her final exam to receive a B in the course. The final exam counts as two tests, and a B is received if the final course average is from 80 to 89.

 STUDY SKILLS BUILDER

Organizing a Notebook

It's never too late to get organized. If you need ideas about organizing a notebook for your mathematics course, try some of these:

- Use a spiral or ring binder notebook with pockets and use it for mathematics only.
- Start each page by writing the book's section number you are working on at the top.
- When your instructor is lecturing, take notes. *Always* include any examples your instructor works for you.
- Place your worked-out homework exercises in your notebook immediately after the lecture notes from that section. This way, a section's worth of material is together.
- Homework exercises: Attempt all assigned homework. For odd-numbered exercises, you are not through until you check your answers against the back of the book. Correct any exercises with incorrect answers. You may want to place a "?" by any homework exercises or notes that you need to ask questions about. Also, consider placing a "!" by any notes or exercises you feel are important.

- Place graded quizzes in the pockets of your notebook. If you are using a binder, you can place your quizzes in a special section of your binder.

Let's check your notebook organization by answering the following questions.

1. Do you have a spiral or ring binder notebook for your mathematics course only?
2. Have you ever had to flip through several sheets of notes and work in your mathematics notebook to determine what section's work you are in?
3. Are you now writing the textbook's section number at the top of each notebook page?
4. Have you ever lost or had trouble finding a graded quiz or test?
5. Are you now placing all your graded work in a dedicated place in your notebook?
6. Are you attempting all of your homework and placing all of your work in your notebook?
7. Are you checking and correcting your homework in your notebook? If not, why not?
8. Are you writing in your notebook the examples your instructor works for you in class?

2.6 ABSOLUTE VALUE EQUATIONS AND INEQUALITIES

Objectives

A Solve Absolute Value Equations.

B Solve Absolute Value Inequalities.

In Chapter 1, we defined the absolute value of a number as its distance from 0 on a number line.

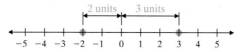

$|-2| = 2$ and $|3| = 3$

In this section, we concentrate on solving equations and inequalities containing the absolute value of a variable or a variable expression. Examples of absolute value equations and inequalities are

$$|x| = 3 \quad -5 \geq |2y + 7| \quad |z - 6.7| = |3z + 1.2| \quad |x - 3| > 7$$

Absolute value equations and inequalities are extremely useful in data analysis, especially for calculating acceptable measurement error and errors that result from the way numbers are sometimes represented in computers.

Objective **A** Solving Absolute Value Equations

To begin, let's solve a few absolute value equations by inspection.

EXAMPLE 1 Solve: $|x| = 3$

Solution: The solution set of this equation will contain all numbers whose distance from 0 is 3 units. Two numbers are 3 units away from 0 on the number line: 3 and -3.

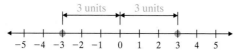

Check: To check, let $x = 3$ and $x = -3$ in the original equation.

$|x| = 3$	$|x| = 3$
$|3| \overset{?}{=} 3$ Let $x = 3$.	$|-3| \overset{?}{=} 3$ Let $x = -3$.
$3 = 3$ True	$3 = 3$ True

Both solutions check. Thus the solution set of the equation $|x| = 3$ is $\{3, -3\}$.

🖳 **Work Practice Problem 1**

PRACTICE PROBLEM 1

Solve: $|y| = 5$

EXAMPLE 2 Solve: $|x| = -2$

Solution: The absolute value of a number is never negative, so this equation has no solution. The solution set is $\{\ \}$ or $\varnothing$.

🖳 **Work Practice Problem 2**

PRACTICE PROBLEM 2

Solve: $|p| = -4$

EXAMPLE 3 Solve: $|y| = 0$

Solution: We are looking for all numbers whose distance from 0 is zero units. The only number is 0. The solution set is $\{0\}$.

🖳 **Work Practice Problem 3**

PRACTICE PROBLEM 3

Solve: $|x| = 0$

Answers

1. $\{-5, 5\}$, **2.** $\varnothing$, **3.** $\{0\}$

From the above examples, we have the following.

Absolute Value Property

To solve $|X| = a$,

If a is positive, then solve $X = a$ or $X = -a$.

If a is 0, then $X = 0$.

If a is negative, the equation $|X| = a$ has no solution.

> **Helpful Hint**
>
> For the equation $|X| = a$ in the box above, X can be a single variable or a variable expression.

When we are solving absolute value equations, if $|X|$ is not alone on one side of the equation we first use properties of equality to get $|X|$ alone.

PRACTICE PROBLEM 4

Solve: $|4x + 2| = 6$.

EXAMPLE 4 Solve: $|5w + 3| = 7$

Solution: Here the expression inside the absolute value bars is $5w + 3$. If we think of the expression $5w + 3$ as X in the absolute value property, we see that $|X| = 7$ is equivalent to

$$X = 7 \quad \text{or} \quad X = -7$$

Then substitute $5w + 3$ for X, and we have

$$5w + 3 = 7 \quad \text{or} \quad 5w + 3 = -7$$

Solve these two equations for w.

$$
\begin{array}{rcl}
5w + 3 = 7 & \text{or} & 5w + 3 = -7 \\
5w = 4 & \text{or} & 5w = -10 \\
w = \dfrac{4}{5} & \text{or} & w = -2
\end{array}
$$

Check: To check, let $w = -2$ and then $w = \dfrac{4}{5}$ in the original equation.

Let $w = -2$ Let $w = \dfrac{4}{5}$

$$|5(-2) + 3| = 7 \qquad\qquad \left|5\left(\dfrac{4}{5}\right) + 3\right| = 7$$

$$|-10 + 3| = 7 \qquad\qquad |4 + 3| = 7$$

$$|-7| = 7 \qquad\qquad\qquad |7| = 7$$

$$7 = 7 \quad \text{True} \qquad\qquad 7 = 7 \quad \text{True}$$

Both solutions check, and the solution set is $\left\{-2, \dfrac{4}{5}\right\}$.

▣ **Work Practice Problem 4**

PRACTICE PROBLEM 5

Solve: $\left|\dfrac{x}{3} + 4\right| = 1$

EXAMPLE 5 Solve: $\left|\dfrac{x}{2} - 1\right| = 11$

Solution: $\left|\dfrac{x}{2} - 1\right| = 11$ is equivalent to

$$
\begin{array}{rcll}
\dfrac{x}{2} - 1 = 11 & \text{or} & \dfrac{x}{2} - 1 = -11 \\[2mm]
2\left(\dfrac{x}{2} - 1\right) = 2(11) & \text{or} & 2\left(\dfrac{x}{2} - 1\right) = 2(-11) & \text{Clear fractions.} \\[2mm]
x - 2 = 22 & \text{or} & x - 2 = -22 & \text{Apply the} \\
 & & & \text{distributive property.} \\
x = 24 & \text{or} & x = -20 &
\end{array}
$$

The solution set is $\{-20, 24\}$.

▣ **Work Practice Problem 5**

Answers

4. $\{1, -2\}$,   **5.** $\{-9, -15\}$

Don't forget that to use the absolute value property you must first make sure that the absolute value expression is alone on one side of the equation.

Helpful Hint

If the equation has a single absolute value expression containing variables, get the absolute value expression alone. Then use the absolute value property.

EXAMPLE 6 Solve: $|2x - 1| + 5 = 6$

Solution: We want the absolute value expression alone on one side of the equation, so we begin by subtracting 5 from both sides. Then we use the absolute value property.

$$|2x - 1| + 5 = 6$$

| $|2x - 1| = 1$ | Subtract 5 from both sides. |

| $2x - 1 = 1$ or $2x - 1 = -1$ | Use the absolute value property. |

$$2x = 2 \quad \text{or} \quad 2x = 0$$

| $x = 1$ or $x = 0$ | Solve. |

The solution set is $\{0, 1\}$.

▣ **Work Practice Problem 6**

Given two absolute value expressions, we might ask, when are the absolute values of two expressions equal? To see the answer, notice that

$$|2| = |2| \quad |-2| = |-2| \quad |-2| = |2| \quad |2| = |-2|$$

same   same   opposites   opposites

Two absolute value expressions are equal when the expressions inside the absolute value bars are equal to or are opposites of each other. In otherwords,

---

To solve $|X| = |Y|$, solve $X = Y$ or $X = -Y$.

EXAMPLE 7 Solve: $|3x + 2| = |5x - 8|$

Solution: This equation is true if the expressions inside the absolute value bars are equal to or are opposites of each other.

$$3x + 2 = 5x - 8 \quad \text{or} \quad 3x + 2 = -(5x - 8)$$

Next we solve each equation.

$$3x + 2 = 5x - 8 \quad \text{or} \quad 3x + 2 = -5x + 8$$
$$-2x + 2 = -8 \quad \text{or} \quad 8x + 2 = 8$$
$$-2x = -10 \quad \text{or} \quad 8x = 6$$
$$x = 5 \quad \text{or} \quad x = \frac{3}{4}$$

Check to see that replacing x with 5 or with $\frac{3}{4}$ results in a true statement.

The solution set is $\left\{ \frac{3}{4}, 5 \right\}$.

▣ **Work Practice Problem 7**

PRACTICE PROBLEM 6
Solve: $|4x + 2| + 1 = 7$

PRACTICE PROBLEM 7
Solve: $|4x - 5| = |3x + 5|$

Answers
6. $\{1, -2\}$, **7.** $\{0, 10\}$

PRACTICE PROBLEM 8

Solve: $|x + 2| = |4 - x|$

EXAMPLE 8 Solve: $|x - 3| = |5 - x|$

Solution:

$$
\begin{array}{lll}
x - 3 = 5 - x & \text{or} & x - 3 = -(5 - x) \\
2x - 3 = 5 & \text{or} & x - 3 = -5 + x \\
2x = 8 & \text{or} & x - 3 - x = -5 + x - x \\
x = 4 & \text{or} & -3 = -5 \qquad \text{False}
\end{array}
$$

Recall from Section 2.1 that when an equation simplifies to a false statement, the equation has no solution. Thus the only solution for the original absolute value equation is 4, and the solution set is $\{4\}$.

▇ **Work Practice Problem 8**

✔**Concept Check** True or false? Absolute value equations always have two solutions. Explain your answer.

Objective **B** **Solving Absolute Value Inequalities**

To begin, let's solve a few absolute value inequalities by inspection.

PRACTICE PROBLEM 9

Solve $|x| < 4$ using a number line.

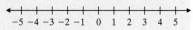

EXAMPLE 9 Solve $|x| < 2$ using a number line.

Solution: The solution set contains all numbers whose distance from 0 is less than 2 units on the number line.

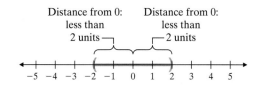

The solution set is $\{x \mid -2 < x < 2\}$, or $(-2, 2)$ in interval notation.

▇ **Work Practice Problem 9**

PRACTICE PROBLEM 10

Solve $|x| \geq 5$ using a number line.

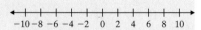

EXAMPLE 10 Solve $|x| \geq 3$ using a number line.

Solution: The solution set contains all numbers whose distance from 0 is 3 or more units. Thus the graph of the solution set contains 3 and all points to the right of 3 on the number line or -3 and all points to the left of -3 on the number line.

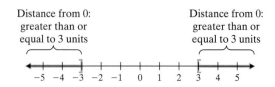

This solution set is $\{x \mid x \leq -3 \text{ or } x \geq 3\}$. In interval notation, the solution set is $(-\infty, -3] \cup [3, \infty)$, since **or** means union.

▇ **Work Practice Problem 10**

The following box summarizes solving absolute value equations and inequalities.

Answers

8. $\{1\}$,

9.

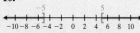

10.

✔ **Concept Check Answer**

false; answers may vary

Solving Absolute Value Equations and Inequalities

If a is a positive number,

To solve $|X| = a$, solve $X = a$ or $X = -a$.

To solve $|X| = |Y|$, solve $X = Y$ or $X = -Y$.

To solve $|X| < a$, solve $-a < X < a$.

To solve $|X| > a$, solve $X < -a$ or $X > a$.

EXAMPLE 11 Solve: $|x - 3| > 7$

Solution: Since 7 is positive, to solve $|x - 3| > 7$, we solve the compound inequality $x - 3 < -7$ or $x - 3 > 7$.

$$x - 3 < -7 \quad \text{or} \quad x - 3 > 7$$
$$x < -4 \quad \text{or} \quad \qquad x > 10 \qquad \text{Add 3 to both sides.}$$

The solution set is $\{x \mid x < -4 \text{ or } x > 10\}$ or $(-\infty, -4) \cup (10, \infty)$ in interval notation. Its graph is shown.

🔲 **Work Practice Problem 11**

Let's remember the differences between solving absolute value equations and inequalities by solving an absolute value equation.

EXAMPLE 12 Solve: $|x + 1| = 6$

Solution: This is an equation, so we solve

$$x + 1 = 6 \quad \text{or} \quad x + 1 = -6$$
$$x = 5 \quad \text{or} \quad \qquad x = -7$$

The solution set is $\{-7, 5\}$. Its graph is shown.

🔲 **Work Practice Problem 12**

Notice that the next example is an absolute value inequality.

EXAMPLE 13 Solve: $|x - 6| \leq 2$

Solution: To solve $|x - 6| \leq 2$, we solve

$$-2 \leq x - 6 \leq 2$$
$$-2 + 6 \leq x - 6 + 6 \leq 2 + 6 \qquad \text{Add 6 to all three parts.}$$
$$4 \leq x \leq 8 \qquad\qquad\qquad \text{Simplify.}$$

The solution set is $\{x \mid 4 \leq x \leq 8\}$, or $[4, 8]$ in interval notation. Its graph is shown.

🔲 **Work Practice Problem 13**

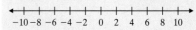
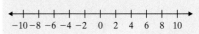
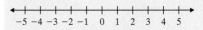
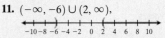
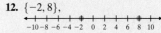

> **Helpful Hint**
> As with absolute value equations, before using an absolute value inequality property, get an absolute value expression alone on one side of the inequality.

PRACTICE PROBLEM 14

Solve: $|2x - 5| + 2 \leq 9$

EXAMPLE 14 Solve: $|5x + 1| + 1 \leq 10$

Solution: First we get the absolute value expression alone by subtracting 1 from both sides.

$$|5x + 1| + 1 \leq 10$$
$$|5x + 1| \leq 10 - 1 \quad \text{Subtract 1 from both sides.}$$
$$|5x + 1| \leq 9 \quad \text{Simplify.}$$

Since 9 is positive, to solve $|5x + 1| \leq 9$, we solve

$$-9 \leq 5x + 1 \leq 9$$
$$-9 - 1 \leq 5x + 1 - 1 \leq 9 - 1 \quad \text{Subtract 1 from all three parts.}$$
$$-10 \leq 5x \leq 8 \quad \text{Simplify.}$$
$$-2 \leq x \leq \frac{8}{5} \quad \text{Divide all three parts by 5.}$$

The solution set is $\left[-2, \frac{8}{5}\right]$.

■ **Work Practice Problem 14**

The next few examples are special cases of absolute value inequalities.

PRACTICE PROBLEM 15

Solve: $|x| < -1$

EXAMPLE 15 Solve: $|x| \leq -3$

Solution: The absolute value of a number is never negative. Thus it will then never be less than or equal to -3. The solution set is $\{ \ \}$ or $\varnothing$.

■ **Work Practice Problem 15**

PRACTICE PROBLEM 16

Solve: $|x + 1| \geq -3$

EXAMPLE 16 Solve: $|x - 1| > -2$

Solution: The absolute value of a number is always nonnegative. Thus it will always be greater than -2. The solution set contains all real numbers, or $(-\infty, \infty)$.

■ **Work Practice Problem 16**

✔ **Concept Check** Without taking any solution steps, how do you know that the absolute value inequality $|3x - 2| > -9$ has a solution? What is its solution?

Answers

14. $[-1, 6]$, **15.** $\varnothing$, **16.** $(-\infty, \infty)$

✔ **Concept Check Answer**

$(-\infty, \infty)$ since the absolute value is always nonnegative

Mental Math

Match each absolute value equation or inequality with an equivalent statement.

1. $|2x + 1| = 3$

2. $|2x + 1| \leq 3$

3. $|2x + 1| < 3$

4. $|2x + 1| \geq 3$

5. $|2x + 1| > 3$

a. $2x + 1 > 3 \text{ or } 2x + 1 < -3$

b. $2x + 1 \geq 3 \text{ or } 2x + 1 \leq -3$

c. $-3 < 2x + 1 < 3$

d. $2x + 1 = 3 \text{ or } 2x + 1 = -3$

e. $-3 \leq 2x + 1 \leq 3$

2.6 EXERCISE SET

FOR EXTRA HELP

Student Solutions Manual | PH Math/Tutor Center | CD/Video for Review | MathXL® | MyMathLab

Objective A *Solve. See Examples 1 through 6.*

1. $|x| = 7$

2. $|y| = 15$

3. $|x| = -4$

4. $|x| = -20$

5. $|3x| = 12.6$

6. $|6n| = 12.6$

7. $3|x| - 5 = 7$

8. $5|x| - 12 = 8$

9. $-6|x| + 44 = -10$

10. $-4|x| + 18 = -22$

11. $|x - 9| = 14$

12. $|x + 2| = 8$

13. $|2x - 5| = 9$

14. $|6 + 2n| = 4$

15. $\left|\dfrac{x}{2} - 3\right| = 1$

16. $\left|\dfrac{n}{3} + 2\right| = 4$

17. $|z| + 4 = 9$

18. $|x| + 1 = 3$

19. $|3x| + 5 = 14$

20. $|2x| - 6 = 4$

21. $\left|\dfrac{4x - 6}{3}\right| = 6$

22. $\left|\dfrac{2x + 1}{5}\right| = 7$

23. $|2x| = 0$

24. $|7z| = 0$

25. $|4n + 1| + 10 = 4$

26. $|3z - 2| + 8 = 1$

27. $3|x - 1| + 19 = 23$

28. $5|x + 1| - 1 = 3$

Solve. See Examples 7 and 8.

29. $|5x - 7| = |3x + 11|$

30. $|9y + 1| = |6y + 4|$

31. $|z + 8| = |z - 3|$

32. $|2x - 5| = |2x + 5|$

33. $|2y - 3| = |9 - 4y|$

34. $|5z - 1| = |7 - z|$

35. $\left|\dfrac{3}{4}x - 2\right| = \left|\dfrac{1}{4}x + 6\right|$

36. $\left|\dfrac{2}{3}x - 5\right| = \left|\dfrac{1}{3}x + 4\right|$

37. $|2x - 6| = |10 - 2x|$

38. $|4n + 5| = |4n + 3|$

39. $|x + 4| = |7 - x|$

40. $|8 - y| = |y + 2|$

41. $\left|\dfrac{2x + 1}{5}\right| = \left|\dfrac{3x - 7}{3}\right|$

42. $\left|\dfrac{5x - 1}{2}\right| = \left|\dfrac{4x + 5}{6}\right|$

43. $|5x + 1| = |4x - 7|$

44. $|3 + 6n| = |4n + 11|$

Objective **B** *Solve. Graph the solution set. See Examples 9 through 16.*

45. $|x| \le 4$

46. $|x| < 6$

47. $|x| > 3$

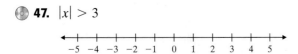

48. $|y| \ge 4$

49. $|x + 3| < 2$

50. $|x + 4| < 6$

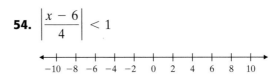

51. $|y - 6| \ge 7$

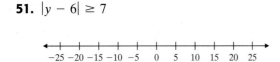

52. $|x - 3| \ge 10$

53. $\left|\dfrac{x + 2}{3}\right| < 1$

54. $\left|\dfrac{x - 6}{4}\right| < 1$

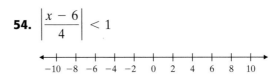

55. $|x| + 7 \le 12$

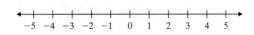

56. $|x| + 6 \le 7$

57. $|2x + 3| \le 0$

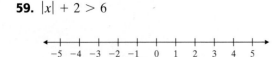

58. $|7x + 1| \le 0$

59. $|x| + 2 > 6$

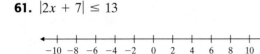

60. $|x| - 1 > 3$

61. $|2x + 7| \le 13$

62. $|5x - 3| \le 18$

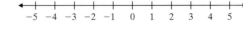

63. $|8 - 3x| < 5$

64. $|7 - 4x| < 5$

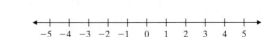

65. $|x + 10| \geq 14$

66. $|x - 9| \geq 2$

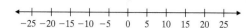

67. $|2x - 7| \leq 11$

68. $|5x + 2| < 8$

69. $|x| > -4$

70. $|x| \leq -7$

71. $6 + |4x - 1| \leq 9$

72. $-3 + |5x - 2| \leq 4$

73. $|6x - 8| - 7 > -3$

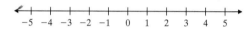

74. $|10 + 3x| - 2 > -1$

75. $|5x + 3| < -6$

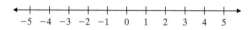

76. $|4 + 9x| \geq -6$

77. $\left| \dfrac{x + 6}{3} \right| > 2$

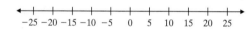

78. $\left| \dfrac{7 + x}{2} \right| \geq 4$

Objectives Ⓐ Ⓑ **Mixed Practice** *Solve each equation or inequality for x. See Examples 1 through 16.*

79. $|x| = 13$

80. $|x| > 13$

81. $|x| < 13$

82. $|3x| = 12$

83. $|x| + 12 = 9$

84. $|x| - 4 = -9$

85. $2|x| - 9 \leq 11$

86. $4|x| - 2 \geq 6$

87. $|2x - 3| = 7$

88. $|5 - 6x| = 29$

89. $|x - 5| \geq 12$

90. $|x + 4| \geq 20$

91. $|9 + 4x| = 0$

92. $|9 + 4x| \geq 0$

93. $|2x + 1| - 7 < -4$

94. $-11 + |5x - 3| \geq -8$

95. $\left| \dfrac{1}{3}x + 1 \right| > 5$

96. $\left| \dfrac{1}{4}x - 2 \right| < 1$

97. $|3x - 5| + 4 = 5$

98. $|5x - 1| + 7 = 11$

99. $|6x + 11| = -1$

100. $|4x - 4| = -3$

101. $\left|\dfrac{1 - 2x}{3}\right| = 6$

102. $\left|\dfrac{6 - 3x}{4}\right| = 5$

103. $\left|\dfrac{3x - 5}{6}\right| > 5$

104. $\left|\dfrac{4x - 7}{5}\right| < 2$

105. $|6x - 3| = |4x + 5|$

106. $|3x + 1| = |4x + 10|$

107. $\left|\dfrac{1 + 3n}{4}\right| = |-4|$

108. $\left|\dfrac{5x + 2}{2}\right| = |-6|$

Review

The circle graph shows the types of cheese produced in the United States in 2003. Use this graph to answer Exercises 109 through 112. See Section 2.2.

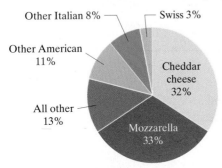

Cheese Production Percent by Type, 2003

Other Italian 8% — Swiss 3%

Other American 11%

Cheddar cheese 32%

All other 13%

Mozzarella 33%

Source: National Agricultural Statistics Service, U.S. Department of Agriculture

109. In 2003, cheddar cheese made up what percent of U. S. cheese production?

110. Which cheese had the highest U.S. production in 2003?

111. A circle contains 360°. Find the number of degrees found in the 8% sector for other Italian cheese.

112. In 2003, the total production of cheese in the United States was 8,598,000,000 pounds. Find the amount of Swiss cheese produced during that year.

Concept Extensions

Without going through a solution procedure, determine the solution of each absolute value equation or inequality.

113. $|x - 7| = -4$

114. $|x - 7| < -4$

115. $|x - 7| > -4$

116. $\left|\dfrac{3x - 2}{7}\right| \geq -7$

117. Write an absolute value equation representing all numbers x whose distance from 0 is 5 units.

118. Write an absolute value equation representing all numbers x whose distance from 0 is 2 units.

119. Write an absolute value inequality representing all numbers x whose distance from 0 is less than 7 units.

120. Write an absolute value inequality representing all numbers x whose distance from 0 is greater than 4 units.

121. Write $-5 \leq x \leq 5$ as an equivalent inequality containing an absolute value. Explain your answer.

122. Write $x > 1$ or $x < -1$ as an equivalent inequality containing an absolute value. Explain your answer.

The expression $|x_T - x|$ is defined to be the absolute error in x, where x_T is the true value of a quantity and x is the measured value or value as stored in a computer.

123. If the true value of a quantity is 3.5 and the absolute error must be less than 0.05, find the acceptable measured values.

124. If the true value of a quantity is 0.2 and the approximate value stored in a computer is $\dfrac{51}{256}$, find the absolute error.

THE BIGGER PICTURE   Solving Equations and Inequalities

This is a special feature that will be repeated and expanded throughout this text. It is very important for you to be able to recognize and solve different types of equations and inequalities. To help you do this, we will begin an outline below and continually expand this outline as different equations and inequalities are introduced. Although suggestions will be given, this outline should be in your own words and you should include at least "how to recognize" and "how to begin to solve" under each letter heading.

For example:

Solving Equations and Inequalities

I. Equations

A. Linear equations: Power on variable is 1 and there are no variables in denominator. (Section 2.1)

$$5(x - 2) = \frac{4(2x + 1)}{3}$$

$$3 \cdot 5(x - 2) = \cancel{3} \cdot \frac{4(2x + 1)}{\cancel{3}}$$ Multiply both sides by the LCD, 3.

$$15(x - 2) = 4(2x + 1)$$ Simplify.

$$15x - 30 = 8x + 4$$ Multiply.

$$7x = 34$$ Add 30 and subtract $8x$ from both sides.

$$x = \frac{34}{7}$$ Divide both sides by 7.

B. Absolute Value Equations: Equation contains the absolute value of a variable expression. (Section 2.6)

$$|3x - 1| - 12 = -4$$ Absolute value equation.

$$|3x - 1| = 8$$ Isolate absolute value.

$$3x - 1 = 8 \quad \text{or} \quad 3x - 1 = -8$$

$$3x = 9 \quad \text{or} \quad 3x = -7$$

$$x = 3 \quad \text{or} \quad x = -\frac{7}{3}$$

$$|x - 5| = |x + 1|$$ Absolute value equation.

$$x - 5 = x + 1 \quad \text{or} \quad x - 5 = -(x + 1)$$

$$\underbrace{-5 = 1}_{\text{No solution}} \quad \text{or} \quad x - 5 = -x - 1$$

$$\text{or} \quad 2x = 4$$

$$x = 2$$

II. Inequalities

A. Linear Inequalities: Inequality sign and power on x is 1 and there are no variables in denominator. (Section 2.4)

$$-3(x + 2) \geq 6$$ Linear inequality

$$-3x - 6 \geq 6$$ Multiply.

$$-3x \geq 12$$ Add 6 to both sides.

$$\frac{-3x}{-3} \leq \frac{12}{-3}$$ Divide both sides by -3, and *change* the direction of the inequality sign.

$$x \leq -4 \text{ or } (-\infty, -4]$$

B. Compound Inequalities: Two inequality signs or 2 inequalities separated by "and" or "or." *Or* means *union* and *and* means *intersection*. (Section 2.5)

$$x \leq 3 \text{ and } x < -7 \qquad x \leq 3 \text{ or } x < -7$$

$$(-\infty, -7) \qquad\qquad (-\infty, 3]$$

C. Absolute Value Inequalities: Inequality with absolute value bars about variable expression. (Section 2.6)

$$|x - 5| - 8 < -2 \qquad\qquad |2x + 1| \geq 17$$

$$|x - 5| < 6 \qquad\qquad 2x + 1 \geq 17 \text{ or } 2x + 1 \leq -17$$

$$-6 < x - 5 < 6 \qquad\qquad 2x \geq 16 \text{ or } \quad 2x \leq -18$$

$$-1 < x < 11 \qquad\qquad x \geq 8 \text{ or } \quad x \leq -9$$

$$(-1, 11) \qquad\qquad (-\infty, -9] \cup [8, \infty)$$

Solve.

1. $9x - 14 = 11x + 2$

2. $|x - 4| = 17$

3. $x - 1 \leq 5 \text{ or } 3x - 2 \leq 10$

4. $-x < 7 \text{ and } 4x \leq 20$

5. $|x - 2| = |x + 15|$

6. $9y - 6y + 1 = 4y + 10 - y + 3$

7. $1.5x - 3 = 1.2x - 18$

8. $\dfrac{7x + 1}{8} - 3 = x + \dfrac{2x + 1}{4}$

9. $|5x + 2| - 10 \leq -3$

10. $|x + 11| > 2$

11. $|9x + 2| - 1 = 24$

12. $\left|\dfrac{3x - 1}{2}\right| = |2x + 5|$

CHAPTER 2   Group Activity

Nutrition Labels

Since 1994, the Food and Drug Administration (FDA) of the Department of Health and Human Services and the Food Safety and Inspection Service of the U.S. Department of Agriculture (USDA) have required nutrition labels like the one below on most food packaging. The labels were designed to help consumers make healthful food choices by giving standardized nutrition information.

Nutrition Facts

| Serving Size | (36g) |
| Servings Per Container | 6 |

Amount Per Serving

| Calories 150 | Calories from Fat 50 |

	% Daily Value*
Total Fat 6g	**9%**
Saturated Fat 2g	**9%**
Cholesterol 0mg	**0%**
Sodium 80mg	**3%**
Potassium 40mg	**1%**
Total Carbohydrate 24g	**8%**
Dietary Fiber 0g	**0%**
Sugars 11g	
Protein 2g	

Vitamin A	25%	Vitamin C	25%
Calcium	50%	Iron	25%
Vitamin D	25%	Vitamin E	25%
Thiamin	25%	Riboflavin	25%
Niacin	25%	Vitamin B₆	25%
Folate	25%	Vitamin B₁₂	25%
Biotin	25%	Pant. Acid	25%
Phosphorus	30%	Iodine	25%
Magnesium	25%	Zinc	25%
Copper	25%		

* Percent Daily Values are based on a 2,000 calorie diet. Your daily values may be higher or lower depending on your calorie needs:

One key feature of this labeling is the column of % Daily Value figures. Most of these values are based on a 2000-calorie diet. Critics complain that this diet applies to only a small segment of the population and should be more versatile. However, FDA and USDA officials responded that these are based on 2000 calories as a guideline only to help consumers gauge the relative amount of a nutrient contained by a food product. For instance, a food with 10 grams of saturated fat per serving could be mistaken for a food low in saturated fat. However, in a 2000-calorie diet, 10 grams of saturated fat represents 50% of the allowable daily intake. This percentage signals to consumers that this product is relatively high in saturated fat. Similarly, a food with 125 mg of sodium per serving could be mistaken for a high-sodium food. However, because a person should ingest no more than 2400 mg of sodium per day, a corresponding % Daily Value figure of about 5% conveys to a consumer that a serving of food with 125 mg of sodium is relatively low in sodium.

The % Daily Value figures that depend on the number of calories consumed per day include total fat, saturated fat, carbohydrate, protein, and dietary fiber. The % Daily Value figures for nutrients such as cholesterol, sodium, and potassium do not depend on calories.

For diets that include more or less than 2000 calories, the daily allowable amount of a nutrient that depends on calories can be figured with the following guidelines:

- The daily allowable amount of total fat is based on 30% of calories.
- The daily allowable amount of saturated fat is based on 10% of calories.
- The daily allowable amount of carbohydrate is based on 60% of calories.
- The daily allowable amount of protein is based on 10% of calories.
- The daily allowable amount of dietary fiber is based on 11.5 grams of fiber per 1000 calories.

Additionally, each gram of protein and carbohydrate contains 4 calories. A gram of fat contains 9 calories.

The above information can be used to calculate % Daily Value figures for diets with other calorie levels. For instance, the daily allowable amount of total fat in a 2200-calorie diet is $0.30(2200) = 660$ calories from fat. For the nutrition label shown, one serving contains 6 grams of fat or $6(9) = 54$ calories from fat. The % Daily Value for total fat that one serving of this food provides in a 2200-calorie diet is $54 \div 660 \approx 0.08$ or 8%.

1. Calculate the daily allowable amounts of total fat, saturated fat, carbohydrate, protein, and dietary fiber in 1500-calorie, 1800-calorie, 2500-calorie, and 2800-calorie diets. Summarize your results in a table.

2. Choose five different food products having Nutrition Facts labels. For each product, use the information given on the label to calculate the % Daily Value figures for total fat, saturated fat, carbohydrate, protein, and dietary fiber for (a) a 1500-calorie diet, (b) an 1800-calorie diet, (c) a 2500-calorie diet, and (d) a 2800-calorie diet. Create a chart showing your results for each food product.

3. Use the data given in the food product nutrition labels used in Question 2 to estimate the daily allowable amounts of cholesterol, sodium, and potassium. Recall that the allowable amounts for these nutrients do not depend on calorie intake.

Chapter 2 Vocabulary Check

Fill in each blank with one of the words or phrases listed below.

contradiction	linear inequality in one variable	compound inequality	solution
absolute value	consecutive integers	identity	union
formula	linear equation in one variable	intersection	

1. The statement "$x < 5$ or $x > 7$" is called a(n) _____.
2. An equation in one variable that has no solution is called a(n) _____.
3. The _____ of two sets is the set of all elements common to both sets.
4. The _____ of two sets is the set of all elements that belong to either of the sets.
5. An equation in one variable that has every number (for which the equation is defined) as a solution is called a(n) _____.
6. The equation $d = rt$ is also called a(n) _____.
7. A number's distance from 0 is called its _____.
8. When a variable in an equation is replaced by a number and the resulting equation is true, then that number is called a(n) _____ of the equation.
9. The integers 17, 18, 19 are examples of _____.
10. The statement $5x - 0.2 < 7$ is an example of a(n) _____.
11. The statement $5x - 0.2 = 7$ is an example of a(n) _____.

> **Helpful Hint**
>
> Are you preparing for your test? Don't forget to take the Chapter 2 Test on page 149. Then check your answers at the back of the text and use the Chapter Test Prep Video CD to see the fully worked-out solutions to any of the exercises you want to review.

2 Chapter Highlights

DEFINITIONS AND CONCEPTS	**EXAMPLES**
Section 2.1 Linear Equations in One Variable	

An **equation** is a statement that two expressions are equal.	$5 = 5$ $7x + 2 = -14$ $3(x-1)^2 = 9x^2 - 6$
A **linear equation in one variable** is an equation that can be written in the form $ax + b = c$, where a, b, and c are real numbers and $a \neq 0$.	$7x + 2 = -14$ $x = -3$ $5(2y - 7) = -2(8y - 1)$
A **solution** of an equation is a value for the variable that makes the equation a true statement.	Determine whether -1 is a solution of $$3(x - 1) = 4x - 2$$ $$3(-1 - 1) \overset{?}{=} 4(-1) - 2$$ $$3(-2) \overset{?}{=} -4 - 2$$ $$-6 = -6 \quad \text{True}$$ Thus, -1 is a solution.
Equivalent equations have the same solution.	$x - 12 = 14$ and $x = 26$ are equivalent equations.
The **addition property of equality** guarantees that the same number may be added to (or subtracted from) both sides of an equation and the result is an equivalent equation.	Solve: $-3x - 2 = 10$ $$-3x - 2 + 2 = 10 + 2 \quad \text{Add 2 to both sides.}$$ $$-3x = 12$$
The **multiplication property of equality** guarantees that both sides of an equation may be multiplied by (or divided by) the same nonzero number and the result is an equivalent equation.	$$\frac{-3x}{-3} = \frac{12}{-3} \quad \text{Divide both sides by } -3.$$ $$x = -4$$

continued

DEFINITIONS AND CONCEPTS	**EXAMPLES**

Section 2.1   Linear Equations in One Variable (*continued*)

SOLVING A LINEAR EQUATION IN ONE VARIABLE

Solve: $x - \dfrac{x - 2}{6} = \dfrac{x - 7}{3} + \dfrac{2}{3}$

Step 1. Clear the equation of fractions and decimals.

1. $6\left(x - \dfrac{x - 2}{6}\right) = 6\left(\dfrac{x - 7}{3} + \dfrac{2}{3}\right)$ Multiply both sides by 6.

$6x - (x - 2) = 2(x - 7) + 2(2)$ Use the distributive property.

Step 2. Remove grouping symbols such as parentheses.

2. $6x - x + 2 = 2x - 14 + 4$

Step 3. Simplify by combining like terms.

3. $5x + 2 = 2x - 10$

Step 4. Write variable terms on one side and numbers on the other side by using the addition property of equality.

4. $5x + 2 - 2 = 2x - 10 - 2$ Subtract 2 from both sides.

$5x = 2x - 12$

$5x - 2x = 2x - 12 - 2x$ Subtract $2x$ from both sides.

$3x = -12$

Step 5. Get the variable alone by using the multiplication property of equality.

5. $\dfrac{3x}{3} = \dfrac{-12}{3}$ Divide both sides by 3.

$x = -4$

Step 6. Check the proposed solution in the original equation.

6. $-4 - \dfrac{-4 - 2}{6} \overset{?}{=} \dfrac{-4 - 7}{3} + \dfrac{2}{3}$ Replace x with -4 in the original equation.

$-4 - \dfrac{-6}{6} \overset{?}{=} \dfrac{-11}{3} + \dfrac{2}{3}$

$-4 - (-1) \overset{?}{=} \dfrac{-9}{3}$

$-3 = -3$ True

Section 2.2   An Introduction to Problem Solving

PROBLEM-SOLVING STRATEGY

1. UNDERSTAND the problem.

Colorado is shaped like a rectangle whose length is about 1.3 times its width. If the perimeter of Colorado is 2070 kilometers, find its dimensions.

1. Read and reread the problem. Guess a solution and check your guess.

Let x = width of Colorado in kilometers. Then $1.3x$ = length of Colorado in kilometers.

1.3x

2. TRANSLATE the problem.

2. In words:

twice the length	+	twice the width	=	perimeter

Translate: $2(1.3x)$ + $2x$ = 2070

DEFINITIONS AND CONCEPTS	**EXAMPLES**

Section 2.2   An Introduction to Problem Solving (*continued*)

3. SOLVE the equation.

4. INTERPRET the results.

3.
$$2.6x + 2x = 2070$$
$$4.6x = 2070$$
$$x = 450$$

4. If $x = 450$ kilometers, then $1.3x = 1.3(450) = 585$ kilometers. *Check:* The perimeter of a rectangle whose width is 450 kilometers and length is 585 kilometers is $2(450) + 2(585) = 2070$ kilometers, the required perimeter. *State:* The dimensions of Colorado are 450 kilometers by 585 kilometers.

Section 2.3   Formulas and Problem Solving

An equation that describes a known relationship among quantities is called a **formula.**

To solve a formula or equation for a specified variable, use the steps for solving an equation. Treat the specified variable as the only variable of the equation.

$A = \pi r^2$ (area of a circle)

$I = Prt$ (interest = principal · rate · time)

Solve $A = 2HW + 2LW + 2LH$ for H.

$A - 2LW = 2HW + 2LH$	Subtract $2LW$ from both sides.
$A - 2LW = H(2W + 2L)$	Use the distributive property.
$\dfrac{A - 2LW}{2W + 2L} = \dfrac{H(2W + 2L)}{2W + 2L}$	Divide both sides by $2W + 2L$.
$\dfrac{A - 2LW}{2W + 2L} = H$	Simplify.

Section 2.4   Linear Inequalities and Problem Solving

A **linear inequality in one variable** is an inequality that can be written in the form $ax + b < c$, where a, b, and c are real numbers and $a \neq 0$. (The inequality symbols $>$, $\leq$, and $\geq$ also apply here.)

The **addition property of inequality** guarantees that the same number may be added to (or subtracted from) both sides of an inequality, and the resulting inequality will have the same solution set.

The **multiplication property of inequality** guarantees that both sides of an inequality may be multiplied by (or divided by) the same **positive** number, and the resulting inequality will have the same solution set. We may also multiply (or divide) both sides of an inequality by the same **negative** number and **reverse the direction of the inequality symbol,** and the result will be an inequality with the same solution set.

$5x - 2 \leq -7 \quad 3y > 1 \quad \dfrac{z}{7} < -9(z - 3)$

$x - 9 \leq -16$	
$x - 9 + 9 \leq -16 + 9$	Add 9 to both sides.
$x \leq -7$	

$6x < -66$	Divide both sides by 6. Do not reverse the direction of the inequality symbol.
$\dfrac{6x}{6} < \dfrac{-66}{6}$	
$x < -11$	
$-6x < -66$	Divide both sides by -6. Reverse the direction of the inequality symbol.
$\dfrac{-6x}{-6} > \dfrac{-66}{-6}$	
$x > 11$	

(continued)

DEFINITIONS AND CONCEPTS	**EXAMPLES**

Section 2.4   Linear Inequalities and Problem Solving (*continued*)

SOLVING A LINEAR INEQUALITY IN ONE VARIABLE	Solve: $\dfrac{3}{7}(x - 4) \geq x + 2$
Step 1. Clear the equation of fractions and decimals.	**1.** $7\left[\dfrac{3}{7}(x - 4)\right] \geq 7(x + 2)$ Multiply both sides by 7.
	$3(x - 4) \geq 7(x + 2)$
Step 2. Remove grouping symbols such as parentheses.	**2.** $3x - 12 \geq 7x + 14$ Use the distributive property.
Step 3. Simplify by combining like terms.	
Step 4. Write variable terms on one side and numbers on the other side using the addition property of inequality.	**4.** $-4x - 12 \geq 14$ Subtract $7x$ from both sides. $-4x \geq 26$ Add 12 to both sides.
Step 5. Get the variable alone using the multiplication property of inequality.	$\dfrac{-4x}{-4} \leq \dfrac{26}{-4}$ Divide both sides by -4. Reverse the direction of the inequality symbol. $x \leq -\dfrac{13}{2}$

Section 2.5   Sets and Compound Inequalities

Two inequalities joined by the words **and** or **or** are called **compound inequalities.**	$x - 7 \leq 4$ and $x \geq -21$ $2x + 7 > x - 3$ or $5x + 2 > -3$
The solution set of a compound inequality formed by the word **and** is the **intersection** $\cap$ of the solution sets of the two inequalities.	Solve: $x < 5$ and $x < 3$
	$\{x \mid x < 5\}$ $(-\infty, 5)$
	$\{x \mid x < 3\}$ $(-\infty, 3)$
The solution set of a compound inequality formed by the word **or** is the **union** $\cup$ of the solution sets of the two inequalities.	$\{x \mid x < 3$ and $x < 5\}$ $(-\infty, 3)$
	Solve: $x - 2 \geq -3$ or $2x \leq -4$ $x \geq -1$ or $x \leq -2$
	$\{x \mid x \geq -1\}$ $[-1, \infty)$
	$\{x \mid x \leq -2\}$ $(-\infty, -2]$
	$\{x \mid x \leq -2$ or $x \geq -1\}$ $(-\infty, -2]$ $\cup [-1, \infty)$

DEFINITIONS AND CONCEPTS	**EXAMPLES**
Section 2.6 Absolute Value Equations and Inequalities	

If a is a positive number, then $|X| = a$ is equivalent to $X = a$ or $X = -a$.	Solve: $|5y - 1| - 7 = 4$

$$|5y - 1| = 11 \quad \text{Add 7 to both sides.}$$

$$5y - 1 = 11 \quad \text{or} \quad 5y - 1 = -11$$

$$5y = 12 \quad \text{or} \qquad 5y = -10 \quad \text{Add 1 to both sides.}$$

$$y = \frac{12}{5} \quad \text{or} \qquad 5y = -2 \quad \text{Divide both sides by 5.}$$

The solution set is $\left\{ -2, \dfrac{12}{5} \right\}$.

If a is negative, then $|X| = a$ has no solution.

Solve: $\left| \dfrac{x}{2} - 7 \right| = -1$

The solution set is $\{\ \}$, or $\varnothing$.

To solve $|X| = |Y|$, solve $X = Y$ or $X = -Y$.

Solve: $|x - 7| = |2x + 1|$

$$x - 7 = 2x + 1 \quad \text{or} \quad x - 7 = -(2x + 1)$$

$$x = 2x + 8 \quad \text{or} \quad x - 7 = -2x - 1$$

$$-x = 8 \qquad \text{or} \qquad x = -2x + 6$$

$$x = -8 \qquad \text{or} \qquad 3x = 6$$

$$x = 2$$

The solution set is $\{-8, 2\}$.

If a is a positive number, then to solve $|X| < a$, solve $-a < X < a$.

Solve: $|y - 5| \le 3$

$$-3 \le y - 5 \le 3$$

$$-3 + 5 \le y - 5 + 5 \le 3 + 5 \quad \text{Add 5 to all three parts.}$$

$$2 \le y \le 8$$

The solution set is $[2, 8]$.

If a is a positive number, then to solve $|X| > a$, solve $X < -a$ or $X > a$.

Solve: $\left| \dfrac{x}{2} - 3 \right| > 7$

$$\frac{x}{2} - 3 < -7 \quad \text{or} \quad \frac{x}{2} - 3 > 7 \quad \text{Multiply both sides by 2.}$$

$$x - 6 < -14 \quad \text{or} \quad x - 6 > 14$$

$$x < -8 \quad \text{or} \qquad x > 20 \quad \text{Add 6 to both sides.}$$

The solution set is $(-\infty, -8) \cup (20, \infty)$.

Are You Prepared for a Test on Chapter 2?

Below I have listed some common trouble areas for students in Chapter 2. After studying for your test—but before taking your test—read these.

- Remember to reverse the direction of the inequality symbol when multiplying or dividing both sides of an inequality by a negative number.

$$-11x < 33$$
$$\frac{-11x}{-11} > \frac{33}{-11} \quad \text{Direction of arrow is reversed.}$$
$$x > -3$$

- Remember the differences when solving absolute value equations and inequalities.

$|x + 1| = 3$	$|x + 1| < 3$	$|x + 1| > 3$
$x + 1 = 3$ or $x + 1 = -3$	$-3 < x + 1 < 3$	$x + 1 < -3$ or $x + 1 > 3$
$x = 2$ or $x = -4$	$-3 - 1 < x < 3 - 1$	$x < -4$ or $x > 2$
$\{2, -4\}$	$-4 < x < 2$	$(-\infty, -4) \cup (2, \infty)$
	$(-4, 2)$	

- Remember that an equation is not solved for a specified variable unless the variable is alone on one side of an equation *and* the other side contains *no* specified variables.

$$y = 10x + 6 - y \quad \text{Equation is } not \text{ solved for } y.$$
$$2y = 10x + 6 \quad \text{Add } y \text{ to both sides.}$$
$$y = 5x + 3 \quad \text{Divide both sides by 2.}$$

Remember: This is simply a checklist of common trouble areas. For a review of Chapter 2, see the Highlights and Chapter Review at the end of this chapter.

2 CHAPTER REVIEW

(2.1) *Solve each linear equation.*

1. $4(x - 5) = 2x - 14$

2. $x + 7 = -2(x + 8)$

3. $3(2y - 1) = -8(6 + y)$

4. $-(z + 12) = 5(2z - 1)$

5. $n - (8 + 4n) = 2(3n - 4)$

6. $4(9v + 2) = 6(1 + 6v) - 10$

7. $0.3(x - 2) = 1.2$

8. $1.5 = 0.2(c - 0.3)$

9. $-4(2 - 3x) = 2(3x - 4) + 6x$

10. $6(m - 1) + 3(2 - m) = 0$

11. $6 - 3(2g + 4) - 4g = 5(1 - 2g)$

12. $20 - 5(p + 1) + 3p = -(2p - 15)$

13. $\frac{x}{3} - 4 = x - 2$

14. $\frac{9}{4}y = \frac{2}{3}y$

15. $\frac{3n}{8} - 1 = 3 + \frac{n}{6}$

16. $\frac{z}{6} + 1 = \frac{z}{2} + 2$

17. $\dfrac{y}{4} - \dfrac{y}{2} = -8$

18. $\dfrac{2x}{3} - \dfrac{8}{3} = x$

19. $\dfrac{b - 2}{3} = \dfrac{b + 2}{5}$

20. $\dfrac{2t - 1}{3} = \dfrac{3t + 2}{15}$

21. $\dfrac{2(t + 1)}{3} = \dfrac{2(t - 1)}{3}$

22. $\dfrac{3a - 3}{6} = \dfrac{4a + 1}{15} + 2$

(2.2) *Solve.*

23. Twice the difference of a number and 3 is the same as 1 added to three times the number. Find the number.

24. One number is 5 more than another number. If the sum of the numbers is 285, find the numbers.

25. Find 40% of 130.

26. Find 1.5% of 8.

27. In 2003, the average annual earnings for a worker with a bachelor's degree were $60,665. This represents a 17.45% increase over the average annual earnings for a worker with a bachelor's degree in 2000. Find the average annual earnings for a worker with a bachelor's degree in 2000. Round to the nearest whole dollar. (*Source:* U.S. Census Bureau)

28. Find four consecutive integers such that twice the first subtracted from the sum of the other three integers is 16.

29. Determine whether there are two consecutive odd integers such that 5 times the first exceeds 3 times the second by 54.

△ **30.** The length of a rectangular playing field is 5 meters less than twice its width. If 230 meters of fencing goes around the field, find the dimensions of the field.

31. A car rental company charges $19.95 per day for a compact car plus 12 cents per mile for every mile over 100 miles driven per day. If Mr. Woo's bill for 2 days use is $46.86, find how many miles he drove.

32. The cost C of producing x number of scientific calculators is given by $C = 4.50x + 3000$ and the revenue R from selling them is given by $R = 16.50x$. Find the number of calculators that must be sold to break even. (Recall that to break even, revenue = cost.)

(2.3) *Solve each equation for the specified variable.*

△ **33.** $V = LWH$ for W

△ **34.** $C = 2\pi r$ for r

35. $5x - 4y = -12$ for y

36. $5x - 4y = -12$ for x

37. $y - y_1 = m(x - x_1)$ for m

38. $y - y_1 = m(x - x_1)$ for x

39. $E = I(R + r)$ for r

40. $S = vt + gt^2$ for g

41. $T = gr + gvt$ for g

42. $I = Prt + P$ for P

43. A principal of $3000 is invested in an account paying an annual percentage rate of 3%. Find the amount (to the nearest cent) in the account after 7 years if the amount is compounded
  a. semiannually.
  b. weekly.

44. The high temperature in Slidell, Louisiana, one day was 90° Fahrenheit. Convert this temperature to degrees Celsius.

△ **45.** Angie Applegate has a photograph for which the length is 2 inches longer than the width. If she increases each dimension by 4 inches, the area is increased by 88 square inches. Find the original dimensions.

△ **46.** One-square-foot floor tiles come 24 to a package. Find how many packages are needed to cover a rectangular floor 18 feet by 21 feet.

(2.4) *Solve each linear inequality. Write your answers in interval notation.*

47. $3(x - 5) > -(x + 3)$

48. $-2(x + 7) \geq 3(x + 2)$

49. $4x - (5 + 2x) < 3x - 1$

50. $3(x - 8) < 7x + 2(5 - x)$

51. $24 \geq 6x - 2(3x - 5) + 2x$

52. $\dfrac{x}{3} + \dfrac{1}{2} > \dfrac{2}{3}$

53. $x + \dfrac{3}{4} < -\dfrac{x}{2} + \dfrac{9}{4}$

54. $\dfrac{x - 5}{2} \leq \dfrac{3}{8}(2x + 6)$

Solve.

55. George Boros can pay his housekeeper $15 per week to do his laundry, or he can have the laundromat do it at a cost of 50 cents per pound for the first 10 pounds and 40 cents for each additional pound. Use an inequality to find the weight at which it is more economical to use the housekeeper than the laundromat.

56. Ceramic firing temperatures usually range from 500° to 1000° Fahrenheit. Use a compound inequality to convert this range to the Celsius scale. Round to the nearest degree.

57. In the Olympic gymnastics competition, Nana must average a score of 9.65 to win the silver medal. Seven of the eight judges have reported scores of 9.5, 9.7, 9.9, 9.7, 9.7, 9.6, and 9.5. Use an inequality to find the minimum score that Nana must receive from the last judge to win the silver medal.

58. Carol would like to pay cash for a car when she graduates from college and estimates that she can afford a car that costs between $4000 and $8000. She has saved $500 so far and plans to earn the rest of the money by working the next two summers. If Carol plans to save the same amount each summer, use a compound inequality to find the range of money she must save each summer to buy the car.

(2.5) *Solve each inequality. Write your answers in interval notation.*

59. $1 \leq 4x - 7 \leq 3$

60. $-2 \leq 8 + 5x < -1$

61. $-3 < 4(2x - 1) < 12$

62. $-6 < x - (3 - 4x) < -3$

63. $\dfrac{1}{6} < \dfrac{4x - 3}{3} \leq \dfrac{4}{5}$

64. $x \leq 2$ and $x > -5$

65. $3x - 5 > 6$ or $-x < -5$

(2.6) *Solve each absolute value equation.*

66. $|x - 7| = 9$

67. $|8 - x| = 3$

68. $|2x + 9| = 9$

69. $|-3x + 4| = 7$

70. $|3x - 2| + 6 = 10$

71. $5 + |6x + 1| = 5$

72. $-5 = |4x - 3|$

73. $|5 - 6x| + 8 = 3$

74. $-8 = |x - 3| - 10$

75. $\left|\dfrac{3x - 7}{4}\right| = 2$

76. $|6x + 1| = |15 + 4x|$

Solve each absolute value inequality. Graph the solution set and write it in interval notaion.

77. $|5x - 1| < 9$

78. $|6 + 4x| \geq 10$

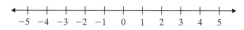

79. $|3x| - 8 > 1$

80. $9 + |5x| < 24$

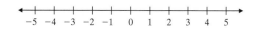

81. $|6x - 5| \leq -1$

82. $\left|3x + \dfrac{2}{5}\right| \geq 4$

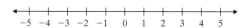

83. $\left|\dfrac{x}{3} + 6\right| - 8 > -5$

84. $\left|\dfrac{4(x - 1)}{7}\right| + 10 < 2$

Mixed Review

Solve. If an inequality, write your answer in interval notation.

85. $\dfrac{x - 2}{5} + \dfrac{x + 2}{2} = \dfrac{x + 4}{3}$

86. $\dfrac{2z - 3}{4} - \dfrac{4 - z}{2} = \dfrac{z + 1}{3}$

87. China, USA, and France are predicted to be the top tourist destinations by 2020. In this year, USA is predicted to have 9 million more tourists than France, and China is predicted to have 44 million more tourists than France. If the total number of tourists predicted for these three countries is 332 million, find the number predicted for each country in 2020.

△ **88.** $A = \dfrac{h}{2}(B + b)$ for B

△ **89.** $V = \dfrac{1}{3}\pi r^2 h$ for h

△ **90.** Determine which container holds more ice cream, an 8 inch $\times$ 5 inch $\times$ 3 inch box or a cylinder with radius of 3 inches and height of 6 inches.

91. Erasmos Gonzalez left Los Angeles at 11 a.m. and drove nonstop to San Diego, 130 miles away. If he arrived at 1:15 p.m., find his average speed, rounded to the nearest mile per hour.

92. $48 + x \geq 5(2x + 4) - 2x$

93. $\dfrac{3(x - 2)}{5} > \dfrac{-5(x - 2)}{3}$

94. $0 \leq \dfrac{2(3x + 4)}{5} \leq 3$

95. $x \leq 2$ or $x > -5$

96. $-2x \leq 6$ and $-2x + 3 < -7$

97. $|7x| - 26 = -5$

98. $\left| \dfrac{9 - 2x}{5} \right| = -3$

99. $|x - 3| = |7 + 2x|$

100. $|6x - 5| \geq -1$

101. $\left| \dfrac{4x - 3}{5} \right| < 1$

2 CHAPTER TEST

Remember to use the Chapter Test Prep Video CD to see the fully worked-out solutions to any of the exercises you want to review.

Solve each equation.

1. $8x + 14 = 5x + 44$

2. $9(x + 2) = 5[11 - 2(2 - x) + 3]$

3. $3(y - 4) + y = 2(6 + 2y)$

4. $7n - 6 + n = 2(4n - 3)$

5. $\dfrac{7w}{4} + 5 = \dfrac{3w}{10} + 1$

6. $\dfrac{z + 7}{9} + 1 = \dfrac{2z + 1}{6}$

7. $|6x - 5| - 3 = -2$

8. $|8 - 2t| = -6$

9. $|2x - 3| = |4x + 5|$

10. $|x - 5| = |x + 2|$

Solve each equation for the specified variable.

11. $3x - 4y = 8$ for y

12. $S = gt^2 + gvt$ for g

13. $F = \dfrac{9}{5}C + 32$ for C

Solve each inequality. Write your answers in interval notation.

14. $3(2x - 7) - 4x > -(x + 6)$

15. $\dfrac{3x - 2}{3} - \dfrac{5x + 1}{4} \geq 0$

16. $-3 < 2(x - 3) \leq 4$

17. $|3x + 1| > 5$

18. $|x - 5| - 4 < -2$

19. $x \geq 5$ and $x \geq 4$

1. _____

2. _____

3. _____

4. _____

5. _____

6. _____

7. _____

8. _____

9. _____

10. _____

11. _____

12. _____

13. _____

14. _____

15. _____

16. _____

17. _____

18. _____

19. _____

149

20. _____

21. _____

22. _____

23. _____

24. _____

25. _____

26. _____

27. _____

28. _____

20. $x \geq 5$ or $x \geq 4$

21. $-1 \leq \dfrac{2x - 5}{3} < 2$

22. $6x + 1 > 5x + 4$ or $1 - x > -4$

23. Find 12% of 80.

Solve.

24. In 2003, Ford sold 6,720,000 new vehicles worldwide. This represents a 9.48% decrease over the number of new vehicles sold by Ford in 2000. Use this information to find the number of new vehicles sold by Ford in 2000. Round to the nearest thousand. (*Source:* Ford Motor Company)

△ **25.** A circular dog pen has a circumference of 78.5 feet. Approximate π by 3.14 and estimate how many hunting dogs could be safely kept in the pen if each dog needs at least 60 square feet of room.

26. In 2006, the number of people employed as database administrators, computer support specialists, and all other computer scientists is expected to be 461,000 in the United States. This represents a 118% increase over the number of people employed in these occupations in 1996. Find the number of database administrators, computer support specialists, and all other computer scientists employed in 1996. (*Source:* U.S. Bureau of Labor Statistics)

27. Find the amount of money in an account after 10 years if a principal of $2500 is invested at 3.5% interest compounded quarterly. (Round to the nearest cent.)

28. The top three states where international travelers spend the most money are Florida, California, and New York. International travelers spent $4 billion more money in California than New York, and in Florida they spent $1 billion less than twice the amount spent in New York. If total international spending in these three states is $39 billion, find the amount spent in each state. (*Source:* Travel Industry Asso. of America)

Chapters 1–2

1. Write the set $\{x \mid x$ is a natural number greater than $100\}$ in roster form.

2. List the elements in each set.
  a. $\{x \mid x$ is an integer between -3 and $5\}$
  b. $\{x \mid x$ is a whole number between 3 and $5\}$

Write each sentence as an equation.

3. The sum of x and 5 is 20.

4. Subtract 19 from x, and the difference is the product of 4 and x.

5. The quotient of z and 9 amounts to 9 plus z.

6. Insert $<, >,$ or $=$ between each pair of numbers to form a true statement.
  a. $-3 \quad -5$
  b. $\dfrac{-12}{-4} \quad 3$
  c. $0 \quad -2$

Find the opposite or additive inverse of each number.

7. 8

8. 0

9. $-\dfrac{1}{5}$

10. -7.3

Add.

11. $-3 + (-11)$

12. $-20.2 + 7.8$

13. $-10 + 15$

14. $-\dfrac{1}{2} + \dfrac{7}{2}$

15. $-\dfrac{2}{3} + \dfrac{3}{7}$

16. Subtract: $1.7 - 8.9$

Divide, if possible.

17. $\dfrac{20}{-4}$

18. $\dfrac{30}{0}$

19. $\dfrac{0}{-8}$

20. $-\dfrac{3}{4} \div \dfrac{9}{4}$

21. $\dfrac{-10}{-80}$

22. Multiply: $-\dfrac{3}{4}\left(-\dfrac{4}{7}\right)$

23. Simplify: $3 + 2 \cdot 30$

Answers

1. _____
2. a. _____
   b. _____
3. _____
4. _____
5. _____
6. a. _____
   b. _____
   c. _____
7. _____
8. _____
9. _____
10. _____
11. _____
12. _____
13. _____
14. _____
15. _____
16. _____
17. _____
18. _____
19. _____
20. _____
21. _____
22. _____
23. _____

151

24. _____

25. _____

26. _____

27. _____

28. _____

29. _____

30. _____

31. _____

32. _____

33. _____

34. _____

35. _____

36. _____

37. _____

38. _____

39. _____

40. a. _____

    b. _____

41. _____

42. _____

43. _____

44. _____

45. _____

46. _____

47. _____

48. _____

49. _____

50. _____

24. Simplify: $\dfrac{\sqrt{9+40}-2^2}{20 \div 2 \cdot 2}$

Simplify by multiplying (if possible), then combining like terms.

25. $3x - 5x + 4$

26. $-3(4+y) + \dfrac{1}{3}(6-12y)$

27. $y + 3y$

28. $6 + 2(7x - 1) + 4$

Use the quotient rule to simplify. Write each answer using positive exponents only.

29. $\dfrac{x^7}{x^4}$

30. $\dfrac{a^{-2} \cdot b^{-8}}{b^{-7}}$

31. $\dfrac{20x^6}{4x^5}$

32. $\dfrac{(-3)^9 z^7 \cdot z^2}{(-3)^{11} z^{-11}}$

Simplify each expression. Write each answer using positive exponents only.

33. $\left(\dfrac{3x^2 y}{y^{-9} z}\right)^{-2}$

34. $(-4a^{-3}b^{-4})(3a^{-2}b)^2$

35. $\left(\dfrac{3a^2}{2x^{-1}}\right)^3 \left(\dfrac{x^{-3}}{4a^{-2}}\right)^{-1}$

Solve.

36. $11.2 = 1.2 - 5x$

37. $2x + 5 = 9$

38. $2x + 1.5 = -0.2 + 1.6x$

39. $6x - 4 = 2 + 6(x - 1)$

40. Write the following as algebraic expressions. Then simplify.
  a. The sum of three consecutive integers if x is the first consecutive integers.
  b. The perimeter of a square with side length $3x + 1$.

41. Suppose that a computer store just announced an 8% decrease in the price of a particular computer model. If this computer sells for $2162 after the decrease, find its original price.

42. Find two numbers such that the second number is 2 more than three times the first number and the difference of the two numbers is 24.

43. Solve $V = lwh$ for h.

44. Solve $7x - 4y = 10$ for x.

45. Solve $\dfrac{1}{4}x \le \dfrac{3}{2}$. Graph the solution set.

Solve.

46. $2(7x - 1) - 5x > -(-7x) + 4$

47. $-2x - 5 < -3 \ or \ 6x < 0$

48. $4(x + 1) - 3 < 4x + 1$

49. $|x| = 3$

50. $|x + 3| = |7 - x|$

3

Graphs and Functions

The linear equations we explored in Chapter 2 are statements about a single variable. This chapter examines statements about two variables: linear equations and inequalities in two variables. We focus particularly on graphs of these equations and inequalities, which lead to the notion of relation and to the notion of function, perhaps the single most important and useful concept in all of mathematics.

At the beginning of the 20th century, there were approximately 237,600 students enrolled in the 977 institutions of higher education in the United States. At that time, only 19% of bachelor's degree recipients were women. By the year 2010, the projected 4700 degree-granting colleges and universities in the United States will have an estimated 17,490,000 students. Of these, roughly 59% of bachelor's degree recipients are expected to be women. The phenomenal growth of colleges and universities can also be seen in the average tuition costs at these institutions of higher learning. For instance, the average annual tuition at a private four-year college or university has increased from $1809 in 1970 to $15,380 in 2000, an increase of about 750%! In Exercise 34 of Section 3.3, we will use a linear equation to predict future annual tuition costs.

Highest Level of Education Attained by Persons 25 Years of Age and Older

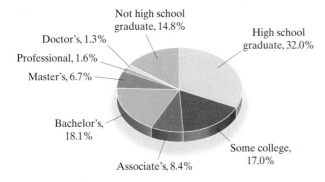

- Not high school graduate, 14.8%
- High school graduate, 32.0%
- Doctor's, 1.3%
- Professional, 1.6%
- Master's, 6.7%
- Bachelor's, 18.1%
- Associate's, 8.4%
- Some college, 17.0%

Total persons age 25 and over = 186.9 million

Source: U.S. Department of Commerce, Census Bureau, Current Population Survey (CPS), March 2004. unpublished data.

Note: Detail may not sum to 100.0% because of rounding.

A Plot Ordered Pairs on a
   Rectangular Coordinate System.

B Determine Whether an Ordered
   Pair of Numbers Is a Solution of
   an Equation in Two Variables.

C Graph Linear Equations.

D Graph Vertical and Horizontal
   Lines.

3.1 GRAPHING LINEAR EQUATIONS

Graphs are widely used today in newspapers, magazines, and all forms of newsletters as a quick way to display data. A few examples of graphs are shown here.

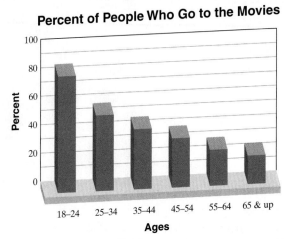

Percent of People Who Go to the Movies

Source: TELENATION/Market Facts, Inc.

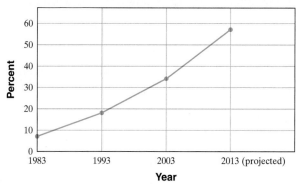

Percent of Sales Completed Using Cards*

Source: The Nilson Report
* These include credit or debit cards, prepaid cards and EBT
  (electronic benefits transfer) cards.

To help us understand how to read these graphs, we will review their basis—the rectangular coordinate system.

Objective A Plotting Ordered Pairs on a Rectangular Coordinate System

One way to locate points on a plane is by using a **rectangular coordinate system,** which is also called a **Cartesian coordinate system** after its inventor, René Descartes (1596–1650). A rectangular coordinate system consists of two number lines that intersect at right angles at their 0 coordinates. We position these axes on paper such that one number line is horizontal and the other number line is then vertical. The horizontal number line is called the *x*-axis (or the axis of the **abscissa**), and the vertical number line is called the *y*-axis (or the axis of the **ordinate**). The point of intersection of these axes is named the **origin.**

Notice that the axes divide the plane into four regions. These regions are called **quadrants.** The top-right region is quadrant I. Quadrants II, III, and IV are numbered counterclockwise from the first quadrant as shown. The x-axis and the y-axis are not in any quadrant.

Each point in the plane can be located, or **plotted,** by describing its position in terms of distances along each axis from the origin. An **ordered pair,** represented by the notation (x, y), records these distances.

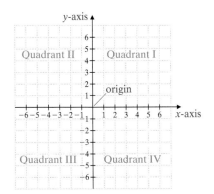 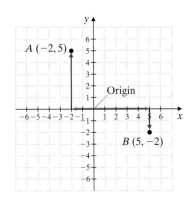

For example, the location of point A in the figure on the right above is described as 2 units to the left of the origin along the x-axis and 5 units upward parallel to the y-axis. Thus, we identify point A with the ordered pair $(-2, 5)$. Notice that the *order* of these numbers is *critical*. The x-value -2 is called the **x-coordinate** and is associated with the x-axis. The y-value 5 is called the **y-coordinate** and is associated with the y-axis.

Compare the location of point A with the location of point B, which corresponds to the ordered pair $(5, -2)$. The x-coordinate 5 indicates that we move 5 units to the right of the origin along the x-axis. The y-coordinate -2 indicates that we move 2 units down parallel to the y-axis. Point A is in a different position than point B. Two ordered pairs are considered equal and correspond to the same point if and only if their x-coordinates are equal and their y-coordinates are equal.

Keep in mind that *each ordered pair corresponds to exactly one point in the coordinate plane and that each point in the plane corresponds to exactly one ordered pair.* Thus, we may refer to the ordered pair (x, y) as the **point** (x, y).

EXAMPLE 1 Plot each ordered pair on a rectangular coordinate system and name the quadrant in which the point is located.

a. $(2, -1)$ **b.** $(0, 5)$ **c.** $(-3, 5)$

d. $(-2, 0)$ **e.** $\left(-\frac{1}{2}, -4\right)$ **f.** $(1.5, 1.5)$

Solution: The six points are graphed as shown in the figure.

a. $(2, -1)$ lies in quadrant IV.
b. $(0, 5)$ is not in any quadrant.
c. $(-3, 5)$ lies in quadrant II.
d. $(-2, 0)$ is not in any quadrant.
e. $\left(-\frac{1}{2}, -4\right)$ is in quadrant III.
f. $(1.5, 1.5)$ is in quadrant I.

Work Practice Problem 1

PRACTICE PROBLEM 1

Plot each ordered pair on a rectangular coordinate system and name the quadrant in which the point is located.

a. $(3, -2)$ **b.** $(3.5, 4.5)$
c. $(0, 3)$ **d.** $(-4, 1)$
e. $(-1, 0)$ **f.** $\left(-2\frac{1}{2}, -3\right)$

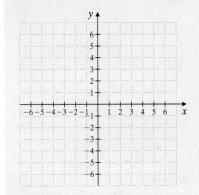

Answers
1.
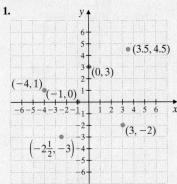

a. quadrant IV, **b.** quadrant I,
c. not in any quadrant, **d.** quadrant II,
e. not in any quadrant, **f.** quadrant III

Notice that the y-coordinate of any point on the x-axis is 0. For example, the point with coordinates $(-2, 0)$ lies on the x-axis. Also, the x-coordinate of any point on the y-axis is 0. For example, the point with coordinates $(0, 5)$ lies on the y-axis. A point on an axis is called a **quadrantal** point.

✔**Concept Check** Which of the following correctly describes the location of the point $(3, -6)$ in a rectangular coordinate system?

a. 3 units to the left of the y-axis and 6 units above the x-axis
b. 3 units above the x-axis and 6 units to the left of the y-axis
c. 3 units to the right of the y-axis and 6 units below the x-axis
d. 3 units below the x-axis and 6 units to the right of the y-axis

Many types of real-world data occur in pairs. For example, the data pairs below are for Walt Disney World ticket prices for the years shown. The graph of paired data, such as the one below, is called a **scatter diagram.** Such diagrams are used to look for patterns and relationships in paired data.

PRACTICE PROBLEM 2

Create a scatter diagram for the given paired data.

3-Point Baskets Made by Lisa Leslie of the Los Angeles Sparks	
WNBA Season	3-Point Baskets Made
2000	7
2001	22
2002	12
2003	12
2004	6
(*Source:* WNBA)	

3-Point Baskets by Lisa Leslie

Paired Data	
Year, x	Price (in dollars), y
0	35.90
1	37.10
2	38.00
3	39.22
4	40.81
5	42.14
6	44.52
7	46.64
8	48.76
9	50.88
10	52.00
11	54.75
12	59.75

Note: The notation $0 \leftrightarrow$ 1992 means that year 0 corresponds to the year 1992, 1 corresponds to 1993, and so on.

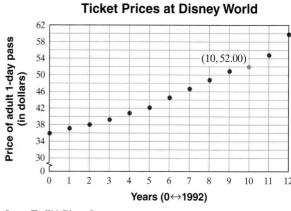

Ticket Prices at Disney World

(10, 52.00)

Source: The Walt Disney Company

Helpful Hint
Notice, for example, the paired data (10, 52.00) and its corresponding plotted point, both in blue.

Answer

2.

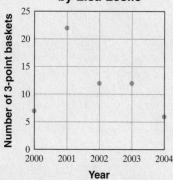

3-Point Baskets by Lisa Leslie

EXAMPLE 2 Create a scatter diagram for the given paired data.

Price of a Big Mac in Russia	
Year	Price (in U.S. dollars)
2001	1.21
2002	1.32
2003	1.43
2004	1.45
(*Sources: The Economist,* McDonald's)	

✔ **Concept Check Answer**
c

Solution: To graph the paired data in the table, we use the first column for the *x*- (or horizontal) axis and the second column for the *y*- (or vertical) axis.

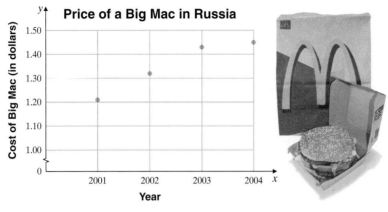

Work Practice Problem 2

Objective B Determining Whether an Ordered Pair of Numbers Is a Solution of an Equation

A **solution** of an equation in two variables consists of two numbers that make the equation true. These two numbers can be written as an ordered pair of numbers. Unless we are told otherwise, we will assume that variable values are written as ordered pairs in alphabetical order (that is, *x* first and then *y*).

EXAMPLE 3 Determine whether $(0, -12)$, $(1, 9)$, and $(2, -6)$ are solutions of the equation $3x - y = 12$.

Solution: To check each ordered pair, we replace *x* with the *x*-coordinate and *y* with the *y*-coordinate and see whether a true statement results.

Let $x = 0$ and $y = -12$.
$$3x - y = 12$$
$$3(0) - (-12) \overset{?}{=} 12$$
$$0 + 12 \overset{?}{=} 12$$
$$12 = 12 \quad \text{True}$$

Let $x = 1$ and $y = 9$.
$$3x - y = 12$$
$$3(1) - 9 \overset{?}{=} 12$$
$$3 - 9 \overset{?}{=} 12$$
$$-6 = 12 \quad \text{False}$$

Let $x = 2$ and $y = -6$.
$$3x - y = 12$$
$$3(2) - (-6) \overset{?}{=} 12$$
$$6 + 6 \overset{?}{=} 12$$
$$12 = 12 \quad \text{True}$$

We see that $(1, 9)$ is not a solution but both $(2, -6)$ and $(0, -12)$ are solutions.

Work Practice Problem 3

Objective C Graphing Linear Equations

As we saw in Example 3, some linear equations have more than one ordered pair solution. In fact, the equation $3x - y = 12$ has an infinite number of ordered pair solutions. Since it is impossible to list all solutions, we visualize them by graphing them.

A few more ordered pairs that satisfy $3x - y = 12$ are $(4, 0)$, $(3, -3)$, $(5, 3)$, and $(1, -9)$. These ordered pair solutions, along with the ordered pair solutions from Example 3, are plotted on the following graph. The graph of $3x - y = 12$ is the single line containing these points. Every ordered pair solution of the equation

PRACTICE PROBLEM 3

Determine whether $(0, -6)$, $(1, 4)$, and $(-1, -4)$ are solutions of the equation $2x + y = -6$.

Answer

3. yes; no; yes

corresponds to a point on this line, and every point on this line corresponds to an ordered pair solution.

x	y	3x − y = 12
5	3	$3 \cdot 5 - 3 = 12$
4	0	$3 \cdot 4 - 0 = 12$
3	−3	$3 \cdot 3 - (-3) = 12$
2	−6	$3 \cdot 2 - (-6) = 12$
1	−9	$3 \cdot 1 - (-9) = 12$
0	−12	$3 \cdot 0 - (-12) = 12$

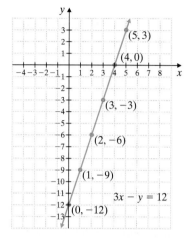

The equation $3x - y = 12$ is called a *linear equation in two variables*, and *the graph of every linear equation in two variables is a line.*

Linear Equation in Two Variables

A **linear equation in two variables** is an equation that can be written in the form

$$Ax + By = C$$

where A, B, and C are real numbers, and A and B are not both 0. This form is called **standard form** and the graph of a linear equation in two variables is a line.

Some examples of linear equations in standard form are:

$$3x - y = 12$$
$$-2.1x + 5.6y = 0$$

Helpful Hint

Remember that in a linear equation in standard form, all of the variable terms are on one side of the equation and the constant is on the other side.

Recall from geometry that a line is determined by two points. This means that to graph a linear equation in two variables, just two solutions are needed. We will find a third solution, just to check our work. To find ordered pair solutions of linear equations in two variables, we can choose an x-value and find its corresponding y-value, or we can choose a y-value and find its corresponding x-value. The number 0 is often a convenient value to choose for x and also for y.

EXAMPLE 4 Graph: $y = -2x + 3$

Solution: This is a linear equation. (In standard form it is $2x + y = 3$.) Find three ordered pair solutions, and plot the ordered pairs. The line through the plotted points is the graph. Since the equation is solved for y, let's choose three x-values. Let's let x be 0, 2, and then −1 to find our three ordered pair solutions.

Let $x = 0$	Let $x = 2$	Let $x = -1$
$y = -2x + 3$	$y = -2x + 3$	$y = -2x + 3$
$y = -2 \cdot 0 + 3$	$y = -2 \cdot 2 + 3$	$y = -2(-1) + 3$
$y = 3$ Simplify.	$y = -1$ Simplify.	$y = 5$ Simplify.

PRACTICE PROBLEM 4

Graph: $y = 4x - 3$

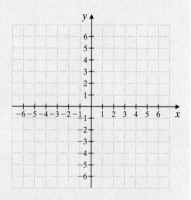

Answer

4.

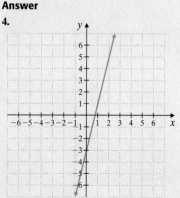

The three ordered pairs $(0, 3)$, $(2, -1)$ and $(-1, 5)$ are listed in the table and the graph is shown.

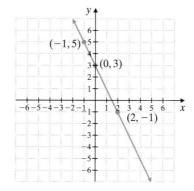

x	y
0	3
2	−1
−1	5

■ **Work Practice Problem 4**

Helpful Hint

Since the equation $y = -2x + 3$ is solved for y, we choose x-values for finding points. This way, we simply need to evaluate an expression to find the y-value, as shown.

Notice that the graph crosses the x-axis at the point $\left(\frac{3}{2}, 0\right)$. This point is called the **x-intercept.** (You may sometimes see just the number $\frac{3}{2}$ called the x-intercept.) This graph also crosses the y-axis at the point $(0, 3)$. This point is called the **y-intercept.** (You may also see just the number 3 called the y-intercept.)

Since every point on the x-axis has an y-value of 0, we can find the x-intercept of a graph by letting $y = 0$ and solving for x. Also, every point on the y-axis has an x-value of 0. To find the y-intercept, we let $x = 0$ and solve for y.

Finding x- and y-Intercepts

To find an x-intercept, let $y = 0$ and solve for x.
To find a y-intercept, let $x = 0$ and solve for y.

We will study intercepts further in Section 3.3.

One way to graph a linear equation in two variables is to find x- and y-intercepts. This provides at most two points, so make sure a third solution point is found to check.

EXAMPLE 5 Find the intercepts and graph: $3x + 4y = -12$

Solution: To find the y-intercept, we let $x = 0$ and solve for y. To find the x-intercept, we let $y = 0$ and solve for x. Let's let $x = 0$, $y = 0$, and then let $x = 2$ to find our third check point.

Let $x = 0$.	Let $y = 0$.	Let $x = 2$.
$3x + 4y = -12$	$3x + 4y = -12$	$3x + 4y = -12$
$3 \cdot 0 + 4y = -12$	$3x + 4 \cdot 0 = -12$	$3 \cdot 2 + 4y = -12$
$4y = -12$	$3x = -12$	$6 + 4y = -12$
$y = -3$	$x = -4$	$4y = -18$
		$y = -\dfrac{18}{4} = -4\dfrac{1}{2}$
$(0, -3)$	$(-4, 0)$	$\left(2, -4\dfrac{1}{2}\right)$

Continued on next page

PRACTICE PROBLEM 5

Find the intercepts and graph:
$2x + 3y = -6$

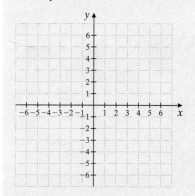

Answer

5.

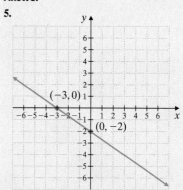

The ordered pairs are $(0, -3)$, $(-4, 0)$, and $\left(2, -4\frac{1}{2}\right)$. We plot these points to obtain the graph shown.

x	y	
0	-3	← y-intercept
-4	0	← x-intercept
2	$-4\frac{1}{2}$	

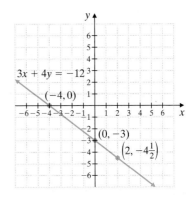

■ **Work Practice Problem 5**

Graph: $y = \dfrac{1}{3}x$

Solution: To graph, we find ordered pair solutions, plot the ordered pairs, and draw a line through the plotted points. We will choose x-values and substitute in the equation. To avoid fractions, we choose x-values that are multiples of 3. To find the intercepts, we will also let $x = 0$ and solve for y, then let $y = 0$ and solve for x.

$y = \dfrac{1}{3}x$

If $x = 6$, then $y = \dfrac{1}{3}(6)$, or 2.

If $x = -3$, then $y = \dfrac{1}{3}(-3)$, or -1.

If $x = 0$, then $y = \dfrac{1}{3}(0)$, or 0.

If $y = 0$, then $0 = \dfrac{1}{3}x$, or $0 = x$.

Multiply both sides of $0 = \dfrac{1}{3}x$ by 3.

x	y
-3	-1
0	0
6	2

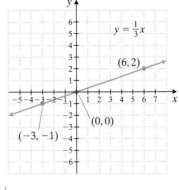

This graph crosses the x-axis at $(0, 0)$ and the y-axis at $(0, 0)$. This means that the x-intercept is $(0, 0)$ and that the y-intercept is $(0, 0)$. This happens when the graph passes through the origin.

■ **Work Practice Problem 6**

Objective D Graphing Vertical and Horizontal Lines

The equation $x = c$, where c is a real number constant, is a linear equation in two variables because it can be written in the form $x + 0y = c$. The graph of this equation is a vertical line, as shown in the next example.

PRACTICE PROBLEM 6

Graph: $y = \dfrac{1}{4}x$

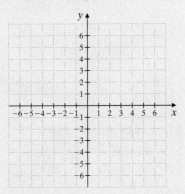

Helpful Hint

Notice that by using multiples of 3 for x, we avoid fractions.

Answer

6.

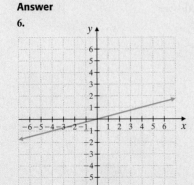

EXAMPLE 7 Graph: $x = 2$

Solution: The equation $x = 2$ can be written as $x + 0y = 2$. Notice that for any y-value chosen, x is 2. No other value for x satisfies $x + 0y = 2$. Any ordered pair whose x-coordinate is 2 is a solution to $x + 0y = 2$ because 2 added to 0 times any value of y is $2 + 0$, or 2. We will use the ordered pairs $(2, 3)$, $(2, 0)$ and $(2, -3)$ to graph $x = 2$.

x	y
2	3
2	0
3	-3

x-intercept →

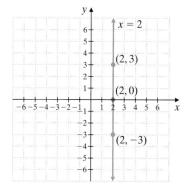

The graph is a vertical line with x-intercept $(2, 0)$. It has no y-intercept because x is never 0.

◻ **Work Practice Problem 7**

EXAMPLE 8 Graph: $y = -3$

Solution: The equation $y = -3$ can be written as $0x + y = -3$. For any x-value chosen, y is -3. If we choose 4, 0, and -2 as x-values, the ordered pair solutions are $(4, -3)$, $(0, -3)$, and $(-2, -3)$. We will use these ordered pairs to graph $y = -3$.

x	y
4	-3
0	-3
-2	-3

← y-intercept

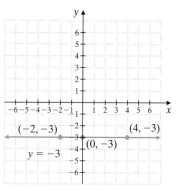

The graph is a horizontal line with y-intercept $(0, -3)$. It has no x-intercept because y is never 0.

◻ **Work Practice Problem 8**

From Examples 7 and 8, we have the following generalization.

Graphing Vertical and Horizontal Lines

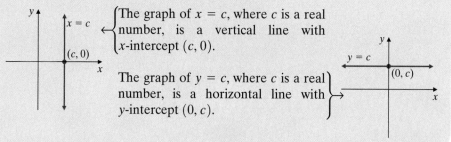

The graph of $x = c$, where c is a real number, is a vertical line with x-intercept $(c, 0)$.

The graph of $y = c$, where c is a real number, is a horizontal line with y-intercept $(0, c)$.

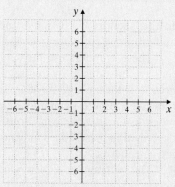
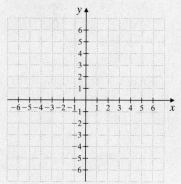
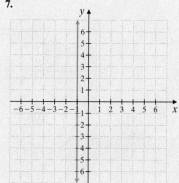
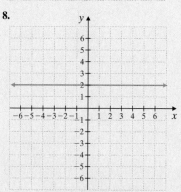

CALCULATOR EXPLORATIONS Graphing

In this section, we begin a study of graphing calculators and graphing software packages for computers.

These graphers use the same point-plotting technique that we introduced in this section. The advantage of this graphing technology is, of course, that graphing calculators and computers can find and plot ordered pair solutions much faster than we can. Note, however, that the features described in these boxes may not be available on all graphing calculators.

The rectangular screen where a portion of the rectangular coordinate system is displayed is called a **window.** We call it a **standard window** for graphing when both the x- and y-axes display coordinates between -10 and 10. This information is often displayed in the window menu on a graphing calculator as

    Xmin = -10
    Xmax = 10
     Xscl = 1   The scale on the x-axis is one unit per tick mark.
    Ymin = -10
    Ymax = 10
     Yscl = 1   The scale on the y-axis is one unit per tick mark.

To use a graphing calculator to graph the equation $y = -5x + 4$, press the $\boxed{Y=}$ key and enter the keystrokes

$\boxed{(-)}$ $\boxed{5}$ $\boxed{X}$ $\boxed{+}$ $\boxed{4}$

↑
(Check your owner's manual to make sure the "negative" key is pressed here and not the "subtraction" key.)

The top row should now read $Y_1 = -5x + 4$. Next press the $\boxed{GRAPH}$ key, and the display should look like this:

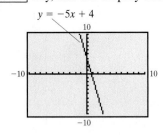

Use a standard window and graph each linear equation. (Unless otherwise stated, we will use a standard window when graphing.)

1. $y = 6x - 1$

2. $y = 3x - 2$

3. $y = -3.2x + 7.9$

4. $y = -x + 5.85$

5. $y = \dfrac{1}{4}x - \dfrac{2}{3}$ $\left(\text{Parentheses may need to be inserted around } \dfrac{1}{4}.\right)$

6. $y = \dfrac{2}{3}x - \dfrac{1}{5}$ $\left(\text{Parentheses may need to be inserted around } \dfrac{2}{3}.\right)$

Mental Math

Determine the coordinates of each point on the graph.

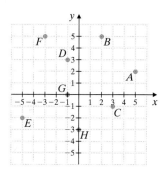

1. Point A

2. Point B

3. Point C

4. Point D

5. Point E

6. Point F

7. Point G

8. Point H

3.1 EXERCISE SET

Objective Ⓐ *Plot each ordered pair on a rectangular coordinate system and name the quadrant (or axis) in which the point lies. See Example 1.*

1. $(3, 2)$

$(-5, 3)$

$\left(5\frac{1}{2}, -4\right)$

$(0, 3.5)$

$(-2, -4)$

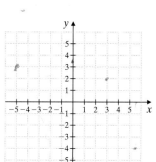

2. $(2, -1)$

$(-3, -1)$

$\left(-2, 6\frac{1}{3}\right)$

$(-2, 4)$

$(-4.2, 0)$

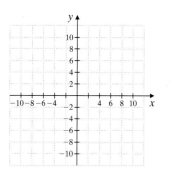

Given that x is a positive number and y is a positive number, determine the quadrant (or axis) in which each point lies. See Example 1.

3. $(x, -y)$

4. $(-x, y)$

5. $(x, 0)$

6. $(0, -y)$

7. $(-x, -y)$

8. $(0, 0)$

Create a scatter diagram for the given paired data. See Example 2.

9. Domestic Airline Revenues in the United States

Year	Revenue (in millions of dollars)
1999	582
2000	610
2001	570
2002	560
2003	593
(*Source: The World Almanac*)	

10. Median Age at Retirement for U.S. Men

Year	Age (in years)
1950	67
1960	65
1970	63
1980	63
1990	63
2000	62
(*Source:* U.S. Bureau of Labor Statistics)	

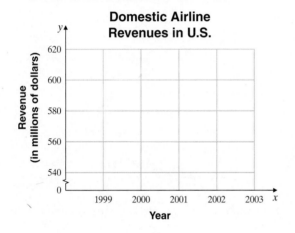

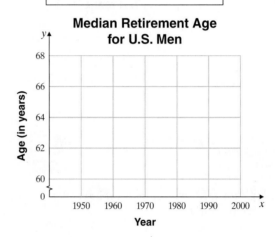

Objective **B** *Determine whether each ordered pair is a solution of the given equation. See Example 3.*

11. $y = 3x - 5$; $(0, 5)$, $(-1, -8)$

12. $y = -2x + 7$; $(1, 5)$, $(-2, 3)$

13. $-6x + 5y = -6$; $(1, 0)$, $\left(2, \dfrac{6}{5}\right)$

14. $5x - 3y = 9$; $(0, 3)$, $\left(\dfrac{12}{5}, -1\right)$

15. $y = -3$; $(1, -3)$, $(-3, 6)$

16. $y = 2$; $(2, 5)$, $(0, 2)$

Objectives **C** **D** **Mixed Practice** *Graph each linear equation. See Examples 4 through 8.*

17. $x - 2y = 4$

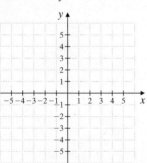

18. $y - 2x = 4$

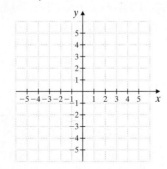

19. $3x + 2y = 6$

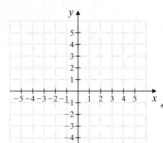

20. $2x + 4y = 8$

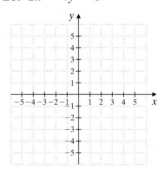

21. $x = 4$

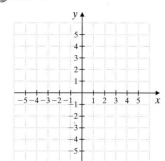

22. $y = 5$

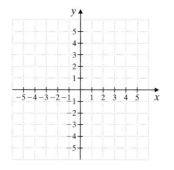

23. $x - 3y = 6$

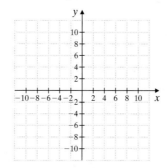

24. $x - 4y = 8$

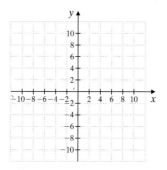

25. $y = 3x$

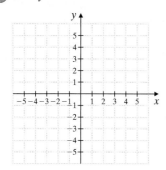

26. $y = -4x$

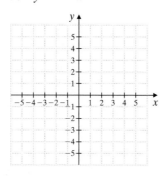

27. $y = -2$

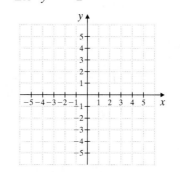

28. $x = -3$

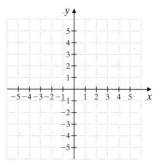

29. $4x + 5y = 15$

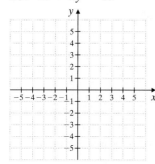

30. $2x + 3y = 9$

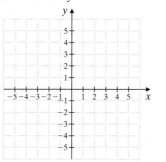

31. $5y = x - 10$

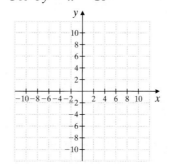

32. $3y = x - 3$

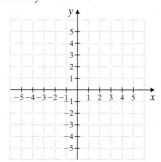

33. $x = \dfrac{1}{2}$

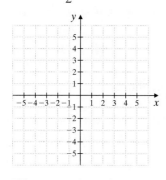

34. $y = -\dfrac{5}{2}$

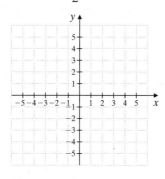

35. $y = 0.5x$

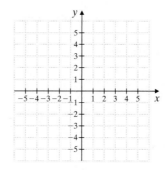

36. $x = 0.5y$

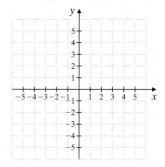

37. $y = -4x + 1$

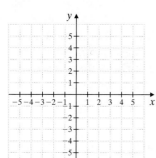

38. $y = -3x + 1$

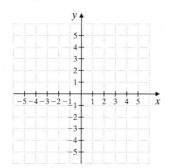

39. $2y - 6 = 0$

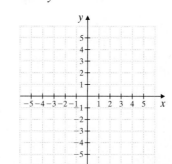

40. $3x + 6 = 0$

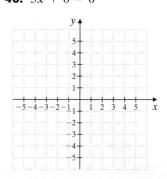

41. $y = -\dfrac{2}{3}x - 4$

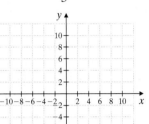

42. $y = -\dfrac{3}{2}x + 6$

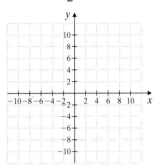

43. $5x - 2y = -10$

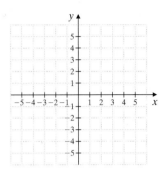

44. $5x - 3y = -15$

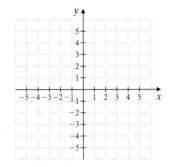

Review

Simplify. See Section 1.5.

45. $\dfrac{-6 - 3}{2 - 8}$

46. $\dfrac{4 - 5}{-1 - 0}$

47. $\dfrac{-8 - (-2)}{-3 - (-2)}$

48. $\dfrac{12 - 3}{10 - 9}$

49. $\dfrac{0 - 6}{5 - 0}$

50. $\dfrac{2 - 2}{3 - 5}$

Concept Extensions

Solve. See the Concept Check in this section.

51. Which correctly describes the location of the point $(-1, 5.3)$ in a rectangular coordinate system?

   a. 1 unit to the right of the y-axis and 5.3 units above the x-axis

   b. 1 unit to the left of the y-axis and 5.3 units above the x-axis

   c. 1 unit to the left of the y-axis and 5.3 units below the x-axis

   d. 1 unit to the right of the y-axis and 5.3 units below the x-axis

52. Which correctly describes the location of the point $\left(0, -\dfrac{3}{4}\right)$ in a rectangular coordinate system?

   a. on the x-axis and $\dfrac{3}{4}$ unit to the left of the y-axis

   b. on the x-axis and $\dfrac{3}{4}$ unit to the right of the y-axis

   c. on the y-axis and $\dfrac{3}{4}$ unit above the x-axis

   d. on the y-axis and $\dfrac{3}{4}$ unit below the x-axis

For Exercises 53 through 56, match each description with the graph that best illustrates it.

53. Moe worked 40 hours per week until the fall semester started. He quit and didn't work again until he worked 60 hours a week during the holiday season starting mid-December.

54. Kawana worked 40 hours a week for her father during the summer. She slowly cut back her hours to not working at all during the fall semester. During the holiday season in December, she started working again and increased her hours to 60 hours per week.

55. Wendy worked from July through February, never quitting. She worked between 10 and 30 hours per week.

56. Bartholomew worked from July through February, never quitting. He worked between 10 and 40 hours per week. During the month of December, he worked 40 hours per week.

A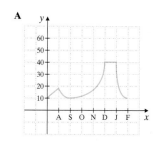

B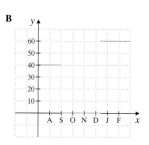

C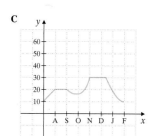

D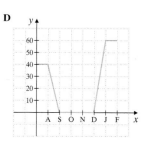

This broken line graph shows hourly minimum wages and the years it increased. Use this graph for Exercises 57 through 60.

The Federal Hourly Minimum Wage

$6.65
$6.15
$5.65
$5.15
$4.75
$4.25
$3.80
$3.35
$3.10

Hourly minimum wage (in dollars)

1980 82 84 86 88 90 92 94 96 98 00 02 04 06*

Year

* Proposed by Congress

57. What was the first year that the minimum hourly wage rose above $4.00?

58. What was the first year that the minimum hourly wage rose above $5.00?

59. Why do you think that this graph is shaped the way it is?

60. The federal hourly minimum wage started in 1938 at $0.25. How much will it have increased by in 2007?

61. The perimeter y of a rectangle whose width is a constant 3 inches and whose length is x inches is given by the equation

x inches

$$y = 2x + 6$$

3 inches

a. Draw a graph of this equation in the first quadrant only.
b. Read from the graph the perimeter y of a rectangle whose length x is 4 inches.

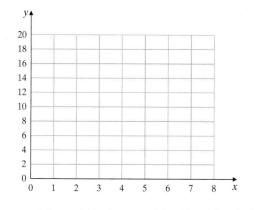

62. The distance y traveled in a train moving at a constant speed of 50 miles per hour is given by the equation

$$y = 50x$$

where x is the time in hours traveled.

a. Draw a graph of this equation in the first quadrant only.
b. Read from the graph the distance y traveled after 6 hours.

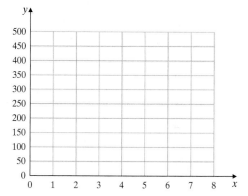

*For income tax purposes, Jason Verges, owner of Copy Services, uses a method called **straight-line depreciation** to show the loss in value of a copy machine he recently purchased. Jason assumes that he can use the machine for 7 years. The following graph shows the value of the machine over the years. Use this graph to answer the following questions.*

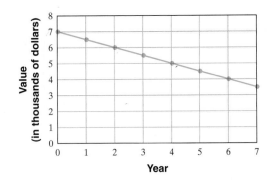

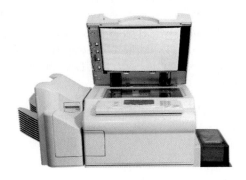

63. What was the purchase price of the copy machine?

64. What is the depreciated value of the machine in 7 years?

65. What loss in value occurred during the first year?

66. What loss in value occurred during the second year?

67. Why do you think that this method depreciating is called straight-line depreciation?

68. Why is the line tilted downward?

69. Broyhill Furniture found that it takes 2 hours to manufacture each table for one of its special dining room sets. Each chair takes 3 hours to manufacture. A total of 1500 hours is available to produce tables and chairs of this style. The linear equation that models this situation is $2x + 3y = 1500$, where x represents the number of tables produced and y the number of chairs produced.

   a. Complete the ordered pair solution $(0, \quad)$ of this equation. Describe the manufacturing situation to which this solution corresponds.

   b. Complete the ordered pair solution $(\quad, 0)$ for this equation. Describe the manufacturing situation to which this solution corresponds.

   c. If 50 tables are produced, find the greatest number of chairs the company can make.

70. While manufacturing two different camera models, Kodak found that the basic model costs $55 to produce, whereas the deluxe model costs $75. The weekly budget for those two models is limited to $33,000 in production costs. The linear equation that models this situation is $55x + 75y = 33,000$, where x represents the number of basic models and y the number of deluxe models.

   a. Complete the ordered pair solution $(0, \quad)$ of this equation. Describe the manufacturing situation to which this solution corresponds.

   b. Complete the ordered pair solution $(\quad, 0)$ of this equation. Describe the manufacturing situation to which this solution corresponds.

   c. If 350 deluxe models are produced, find the greatest number of basic models that can be made in one week.

71. On the same set of axes, graph $y = 2x$, $y = 2x - 5$, and $y = 2x + 5$. What patterns do you see in these graphs?

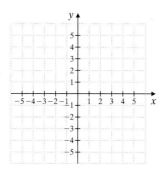

 72. Explain why we generally use three points to graph a line, when only two points are needed.

 73. Discuss whether a vertical line ever has a y-intercept.

 74. Discuss whether a horizontal line ever has an x-intercept.

Use a grapher to verify the graph of each exercise.

75. Exercise 25

76. Exercise 26

77. Exercise 37

78. Exercise 38

STUDY SKILLS BUILDER

How Well Do You Know Your Textbook?

The questions below will determine whether you are familiar with your textbook. For help, see Section 1.1 in this text.

1. What does the 💿 mean?

2. What does the ✎ mean?

3. What does the △ mean?

4. Where can you find a review for each chapter? What answers to this review can be found in the back of your text?

5. Each chapter contains an overview of the chapter along with examples. What is this feature called?

6. Each chapter contains a review of vocabulary. What is this feature called?

7. There is a CD in your text. What content is contained on this CD?

8. What is the location of the section that is entirely devoted to study skills?

9. There are Practice Problems that are contained in the margin of the text. What are they and how can they be used?

3.2 THE SLOPE OF A LINE

You may have noticed by now that different lines often tilt differently. It is very important in many fields to be able to measure and compare the tilt, or **slope,** of lines. For example, a wheelchair ramp with a slope of $\frac{1}{12}$ means that the ramp rises 1 foot for every 12 horizontal feet. A road with a slope or grade of 8% $\left(\text{or } \frac{8}{100}\right)$ means that the road rises 8 feet for every 100 horizontal feet.

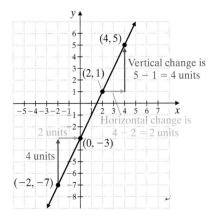

We measure the slope of a line as a ratio of **vertical change** to **horizontal change.** Slope is usually designated by the letter m.

Objective **A** Finding Slope from Two Points

Suppose that we want to measure the slope of the following line.

The vertical change between *both* pairs of points on the line is 4 units per horizontal change of 2 units. Then

$$\text{slope } m = \frac{\text{change in } y \text{ (vertical change)}}{\text{change in } x \text{ (horizontal change)}} = \frac{4}{2} = 2$$

Notice that slope is a **rate of change** between points. A slope of 2 or $\frac{2}{1}$ means that between pairs of points on the line, the rate of change is a vertical change of 2 units per horizontal change of 1 unit.

In general, consider the line that passes through the points (x_1, y_1) and (x_2, y_2). (The notation x_1 is read "x-sub-one.") The vertical change, or **rise,** between these points is the difference in the y-coordinates: $y_2 - y_1$. The horizontal change, or **run,** between the points is the difference of the x-coordinates: $x_2 - x_1$.

Slope of a Line

Given a line passing through points (x_1, y_1) and (x_2, y_2) the **slope** m of the line is

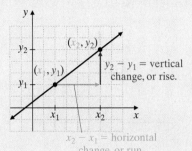

$$m = \frac{\text{rise}}{\text{run}} = \frac{y_2 - y_1}{x_2 - x_1}$$

as long as $x_2 \neq x_1$.

$y_2 - y_1 = $ vertical change, or rise.

$x_2 - x_1 = $ horizontal change, or run.

✔**Concept Check** In the definition of slope, we state that $x_2 \neq x_1$. Explain why.

EXAMPLE 1 Find the slope of the line containing the points $(0, 3)$ and $(2, 5)$. Graph the line.

Solution: We use the slope formula. It does not matter which point we call (x_1, y_1) and which point we call (x_2, y_2). We'll let $(x_1, y_1) = (0, 3)$ and $(x_2, y_2) = (2, 5)$.

$$m = \frac{y_2 - y_1}{x_2 - x_1}$$

$$= \frac{5 - 3}{2 - 0} = \frac{2}{2} = 1$$

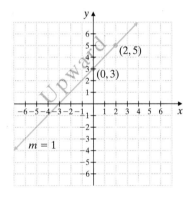

Notice in this example that the slope is *positive* and that the graph of the line containing $(0, 3)$ and $(2, 5)$ moves *upward*—that is, the y-values increase—as we go from left to right.

▨ **Work Practice Problem 1**

Helpful Hint
          The slope of a line is the same no matter which 2 points of a line you choose to calculate slope. The line in Example 1 also contains the point $(-3, 0)$. Below, we calculate the slope of the line using $(0, 3)$ as (x_1, y_1) and $(-3, 0)$ as (x_2, y_2).

$$m = \frac{y_2 - y_1}{x_2 - x_1} = \frac{0 - 3}{-3 - 0} = \frac{-3}{-3} = 1$$ Same slope as found in Example 1.

PRACTICE PROBLEM 1

Find the slope of the line containing the points $(-1, -2)$ and $(2, 5)$. Graph the line.

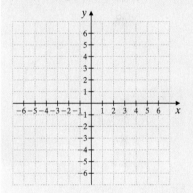

Answer

1. $m = \dfrac{7}{3}$,

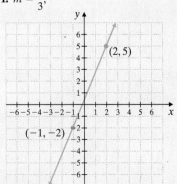

✔ **Concept Check Answer**

so that the denominator is never 0

PRACTICE PROBLEM 2

Find the slope of the line containing the points $(1, -1)$ and $(-2, 4)$. Graph the line.

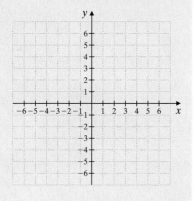

EXAMPLE 2 Find the slope of the line containing the points $(5, -4)$ and $(-3, 3)$. Graph the line.

Solution: We use the slope formula, and let $(x_1, y_1) = (5, -4)$ and $(x_2, y_2) = (-3, 3)$.

$$m = \frac{y_2 - y_1}{x_2 - x_1}$$

$$= \frac{3 - (-4)}{-3 - 5} = \frac{7}{-8} = -\frac{7}{8}$$

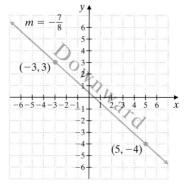

Notice in this example that the slope is negative and that the graph of the line through $(5, -4)$ and $(-3, 3)$ moves downward—that is, the y-values decrease—as we go from left to right.

■ **Work Practice Problem 2**

> **Helpful Hint**
>
> When we are trying to find the slope of a line through two given points, it makes no difference which given point is called (x_1, y_1) and which is called (x_2, y_2). Once an x-coordinate is called x_1, however, make sure its corresponding y-coordinate is called y_1.

✔ **Concept Check** Find and correct the error in the following calculation of slope of the line containing the points $(12, 2)$ and $(4, 7)$.

$$m = \frac{12 - 4}{2 - 7} = \frac{8}{-5} = -\frac{8}{5}$$

PRACTICE PROBLEM 3

Find the slope of the line $y = 2x + 4$.

Objective B Finding Slope from an Equation

As we have seen, the slope of a line is defined by two points on the line. Thus, if we know the equation of a line, we can find its slope.

Answers

2. $m = -\dfrac{5}{3}$,

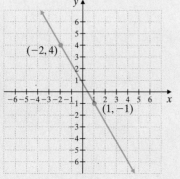

3. $m = 2$

✔ Concept Check Answer

$m = \dfrac{2 - 7}{12 - 4} = \dfrac{-5}{8} = -\dfrac{5}{8}$

EXAMPLE 3 Find the slope of the line $y = 3x + 2$.

Solution: We must find two points on the line defined by $y = 3x + 2$ to find its slope. We will let $x = 0$ and then $x = 1$ to find the required points.

Let $x = 0$.

$y = 3x + 2$

$y = 3 \cdot 0 + 2$

$y = 2$

Let $x = 1$.

$y = 3x + 2$

$y = 3 \cdot 1 + 2$

$y = 5$

Now we use the points $(0, 2)$ and $(1, 5)$ to find the slope. We'll let (x_1, y_1) be $(0, 2)$ and (x_2, y_2) be $(1, 5)$. Then

$$m = \frac{y_2 - y_1}{x_2 - x_1} = \frac{5 - 2}{1 - 0} = \frac{3}{1} = 3$$

■ **Work Practice Problem 3**

Analyzing the results of Example 3, you may notice a striking pattern:

The slope of $y = 3x + 2$ is 3, the same as the coefficient of x.

The y-coordinate of the y-intercept $(0, 2)$, the same as the constant term.

We have just illustrated, not proved, an amazing pattern with linear equations in two variables. When a linear equation is written in the form $y = mx + b$, m is the slope of the line and $(0, b)$ is its y-intercept. The form $y = mx + b$ is appropriately called the *slope-intercept form*.

Slope-Intercept Form

When a linear equation in two variables is written in **slope-intercept form,**

$$\underset{\downarrow}{\text{slope}} \quad \underset{\downarrow}{y\text{-intercept is } (0, b)}$$

$$y = mx + b$$

then m is the slope of the line and $(0, b)$ is the y-intercept of the line.

EXAMPLE 4 Find the slope and the y-intercept of the line $3x - 4y = 4$.

Solution: We write the equation in slope-intercept form by solving for y.

$$3x - 4y = 4$$

$$-4y = -3x + 4 \qquad \text{Subtract } 3x \text{ from both sides.}$$

$$\frac{-4y}{-4} = \frac{-3x}{-4} + \frac{4}{-4} \qquad \text{Divide both sides by } -4.$$

$$y = \frac{3}{4}x - 1 \qquad \text{Simplify.}$$

The coefficient of x, $\frac{3}{4}$, is the slope, and $(0, -1)$, is the y-intercept.

Work Practice Problem 4

The graphs of $y = \frac{1}{2}x + 1$ and $y = 5x + 1$ are shown below. The graph of $y = \frac{1}{2}x + 1$ has a slope of $\frac{1}{2}$ and the graph of $y = 5x + 1$ has a slope of 5.

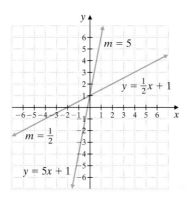

Notice that the line with the slope of 5 is steeper than the line with the slope of $\frac{1}{2}$. This is true in general for positive slopes.

For a line with positive slope m, as m increases the line becomes steeper.

PRACTICE PROBLEM 4

Find the slope and the y-intercept of the line $2x - 4y = 8$.

Objective C Finding Slopes of Horizontal and Vertical Lines

Next we find the slopes of two special types of lines: vertical lines and horizontal lines.

Find the slope of the line $x = 3$.

EXAMPLE 5 Find the slope of the line $x = -5$.

Solution: Recall that the graph of $x = -5$ is a vertical line with x-intercept $(-5, 0)$. To find the slope, we find two ordered pair solutions of $x = -5$. Of course, solutions of $x = -5$ must have an x-value of -5. We will let $(x_1, y_1) = (-5, 0)$ and $(x_2, y_2) = (-5, 4)$. Then

$$m = \frac{y_2 - y_1}{x_2 - x_1} = \frac{4 - 0}{-5 - (-5)} = \frac{4}{0}$$

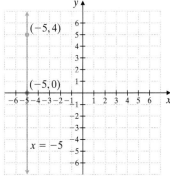

Since $\frac{4}{0}$ is undefined, we say that the slope of the vertical line $x = -5$ is undefined.

▣ **Work Practice Problem 5**

Find the slope of the line $y = -3$.

EXAMPLE 6 Find the slope of the line $y = 2$.

Solution: Recall that the graph of $y = 2$ is a horizontal line with y-intercept $(0, 2)$. To find the slope, we find two points on the line, such as $(0, 2)$ and $(1, 2)$, and use these points to find the slope.

$$m = \frac{2 - 2}{1 - 0} = \frac{0}{1} = 0$$

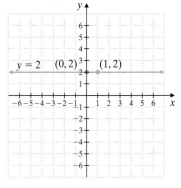

The slope of the horizontal line $y = 2$ is 0.

▣ **Work Practice Problem 6**

From the above examples, we have the following generalization.

Slopes of Vertical and Horizontal Lines

The slope of any vertical line is undefined.
The slope of any horizontal line is 0.

Helpful Hint

Slope of 0 and undefined slope are not the same. Vertical lines have un-defined slope, whereas horizontal lines have slope of 0.

The following four graphs summarize the overall appearance of lines with posi-tive, negative, zero, and undefined slopes.

Increasing line, positive slope   Decreasing line, negative slope   Horizontal line, zero-slope   Vertical line, undefined slope

Objective D Finding the Slope of a Line Given the Graph of a Line

Now that we know the appearance of lines with positive, negative, zero, and unde-fined slopes, let's practice finding the slope of a line given its graph.

EXAMPLE 7 Find the slope of the line graphed.

Solution: The two points shown have coordinates $(2, 1)$ and $(0, -4)$. Thus,

$$m = \frac{-4 - 1}{0 - 2} = \frac{-5}{-2} = \frac{5}{2}$$

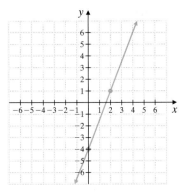

As a quick check, notice that the line goes upward from left to right, so its slope is positive.

The slope of the line is $\frac{5}{2}$.

Work Practice Problem 7

Objective E Comparing Slopes of Parallel and Perpendicular Lines

Slopes of lines can help us determine whether lines are parallel. Parallel lines are dis-tinct lines with the same steepness, so it follows that they have the same slope.

PRACTICE PROBLEM 7

Find the slope of the line graphed.

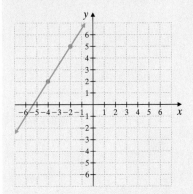

Answer

7. $m = \frac{3}{2}$

Parallel Lines

Two nonvertical lines are parallel if they have the same slope and different *y*-intercepts.

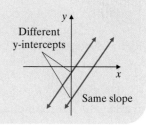

How do the slopes of perpendicular lines compare? (Two lines intersecting at right angles are called **perpendicular lines.**) Suppose that a line has a slope of $\frac{a}{b}$. If the line is rotated 90°, the rise and run are now switched, except that the run is now negative. This means that the new slope is $-\frac{b}{a}$. Notice that

$$\left(\frac{a}{b}\right) \cdot \left(-\frac{b}{a}\right) = -1$$

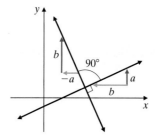

This is how we tell whether two lines are perpendicular.

Perpendicular Lines

Two nonvertical lines are perpendicular if the product of their slopes is −1.

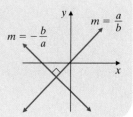

In other words, two nonvertical lines are perpendicular if the slope of one is the negative reciprocal of the slope of the other.

PRACTICE PROBLEM 8

Determine whether the two lines are parallel, perpendicular, or neither.

$$2x + 5y = 1$$
$$4x + 10y = 3$$

EXAMPLE 8 Determine whether the two lines are parallel, perpendicular, or neither.

$$3x + 7y = 4$$
$$6x + 14y = 7$$

Solution: We find the slope of each line by solving each equation for *y*.

$$3x + 7y = 4 \qquad\qquad 6x + 14y = 7$$
$$7y = -3x + 4 \qquad\qquad 14y = -6x + 7$$
$$\frac{7y}{7} = \frac{-3x}{7} + \frac{4}{7} \qquad\qquad \frac{14y}{14} = \frac{-6x}{14} + \frac{7}{14}$$
$$y = -\frac{3}{7}x + \frac{4}{7} \qquad\qquad y = -\frac{3}{7}x + \frac{1}{2}$$

slope   y-intercept $\left(0, \frac{4}{7}\right)$ slope   y-intercept $\left(0, \frac{1}{2}\right)$

Answer

8. parallel

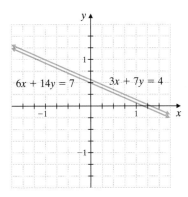

The slopes of both lines are $-\dfrac{3}{7}$. The y-intercepts are different. Therefore, the lines are parallel.

🖱 **Work Practice Problem 8**

EXAMPLE 9 Determine whether the two lines are parallel, perpendicular, or neither.

$$-x + 3y = 2$$
$$2x + 6y = 5$$

Solution: When we solve each equation for y, we have

$$-x + 3y = 2 \qquad\qquad 2x + 6y = 5$$
$$3y = x + 2 \qquad\qquad 6y = -2x + 5$$
$$\frac{3y}{3} = \frac{x}{3} + \frac{2}{3} \qquad\qquad \frac{6y}{6} = \frac{-2x}{6} + \frac{5}{6}$$
$$y = \frac{1}{3}x + \frac{2}{3} \qquad\qquad y = -\frac{1}{3}x + \frac{5}{6}$$

slope y-intercept $\left(0, \dfrac{2}{3}\right)$ slope y-intercept $\left(0, \dfrac{5}{6}\right)$

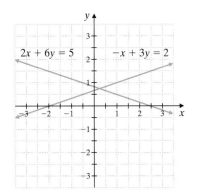

The slopes are not the same and their product is not -1 $\left[\left(\dfrac{1}{3}\right) \cdot \left(-\dfrac{1}{3}\right) = -\dfrac{1}{9}\right]$.

Therefore, the lines are neither parallel nor perpendicular.

🖱 **Work Practice Problem 9**

✔**Concept Check** What is *different* about the equations of two parallel lines?

PRACTICE PROBLEM 9

Determine whether the two lines are parallel, perpendicular, or neither.

$$x - 4y = 3$$
$$3x + 12y = 7$$

Answer

9. neither parallel nor perpendicular

✔**Concept Check Answer**

y-intercepts are different

CALCULATOR EXPLORATIONS Graphing

It is possible to use a grapher to sketch the graph of more than one equation on the same set of axes. For example, let's graph the equations $y = 2x - 3$ and $y = 2x + 5$ on the same set of axes.

To graph on the same set of axes, press the $\boxed{Y=}$ key and enter the equations on the first two lines.

$$Y_1 = 2x - 3$$
$$Y_2 = 2x + 5$$

Then press the $\boxed{\text{GRAPH}}$ key as usual. The screen should look like this:

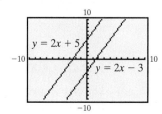

Notice the slopes and y-intercepts of the graphs. Since their slopes are the same and they have different y-intercepts, we have parallel lines, as shown.

Graph each pair of equations on the same set of axes. Describe the similarities and differences in their graphs.

1. $y = 3x, y = 3x + 4$

2. $y = 5x, y = 5x - 2$

3. $y = -\dfrac{2}{3}x + 1, y = -\dfrac{2}{3} + 6$

4. $y = -\dfrac{1}{4}x - 3, y = -\dfrac{1}{4}x + 6$

5. $y = 4.61x - 1.86, y = 4.61x + 2.11$

6. $y = 3.78x + 1.92, y = 3.78x + 8.08$

Mental Math

Determine whether a line with the given slope slants upward, downward, horizontally, or vertically from left to right.

1. $m = \dfrac{7}{6}$

2. $m = -3$

3. $m = 0$

4. m is undefined.

3.2 EXERCISE SET

Objectives Ⓐ Ⓓ **Mixed Practice** *Find the slope of the line containing each pair of points. See Examples 1, 2, and 7.*

1. $(3, 2), (8, 11)$

2. $(1, 6), (7, 11)$

3. $(3, 1), (1, 8)$

4. $(2, 9), (6, 4)$

5. $(-2, 8), (4, 3)$

6. $(3, 7), (-2, 11)$

7. $(-2, -6), (4, -4)$

8. $(-3, -4), (-1, 6)$

9. $(-3, -1), (-12, 11)$

10. $(3, -1), (-6, 5)$

11. $(-2, 5), (3, 5)$

12. $(4, 2), (4, 0)$

13.

14.

15.

16.

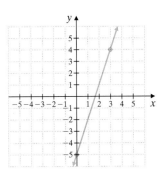
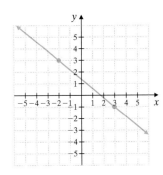
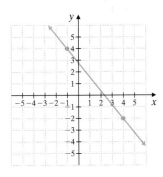

Find each slope. See Examples 1 and 2.

17. Find the pitch, or slope, of the roof shown.

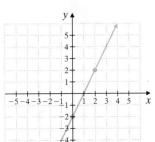

18. Upon takeoff, a Delta Airlines jet climbs to 3 miles as it passes over 25 miles of land below it. Find the slope of its climb.

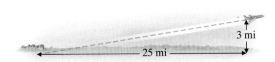

19. Find the grade, or slope, of the road shown.

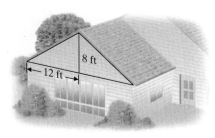

20. Driving down Bald Mountain in Wyoming, Bob Dean finds that he descends 1600 feet in elevation by the time he is 2.5 miles (horizontally) away from the high point on the mountain road. Find the slope of his descent. (*Hint:* 1 mile = 5280 feet.)

Objectives Ⓑ Ⓒ **Mixed Practice** *Find the slope and the y-intercept of each line. See Examples 3 through 6.*

21. $y = -x + 5$

22. $y = x + 2$

23. $y = 5x - 2$

24. $y = -2x + 6$

25. $2x + y = 7$

26. $-5x + y = 10$

27. $2x - 3y = 10$

28. $-3x - 4y = 6$

29. $x = 4$

30. $x = 7.2$

31. $y = -2$

32. $y = -3.6$

33. $y = \frac{1}{2}x$

34. $y = -\frac{1}{4}x$

35. $3y + 8 = x$

36. $2y - 7 = x$

37. $-6x + 5y = 30$

38. $4x - 7y = 28$

39. $y = 7x$

40. $y = \frac{1}{7}x$

41. $x + 2 = 0$

42. $y - 7 = 0$

Match each graph with its equation.

43. $y = 2x + 3$

44. $y = 2x - 3$

45. $y = -2x + 3$

46. $y = -2x - 3$

A
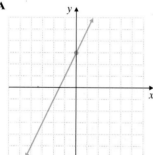

B
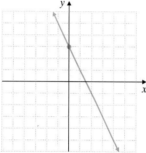

C
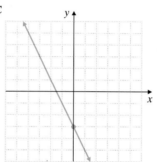

D
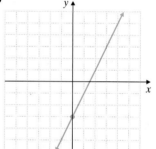

Two lines are graphed on each set of axes. For each graph, determine whether l_1 or l_2 has the greater slope.

47.
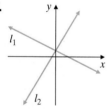

48.
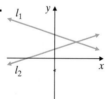

49.
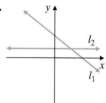

50.
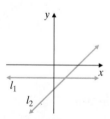

51.
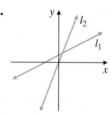

52.
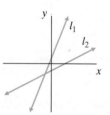

Objective **E** *Determine whether each pair of lines is parallel, perpendicular, or neither. See Examples 8 and 9.*

53. $y = -3x + 6$
$y = 3x + 5$

54. $y = 5x - 6$
$y = 5x + 2$

55. $y = -9x + 3$
$y = \frac{3}{2}x - 7$

56. $y = 2x - 12$
$y = \frac{1}{2}x - 6$

57. $2x - y = -10$
$2x + 4y = 2$

58. $-2x + 3y = 1$
$3x + 2y = 12$

59. $y = 12x + 6$
$y = 12x - 2$

60. $y = 5x + 8$
$y = \frac{1}{5}x - 8$

61. $-4x + 2y = 5$

$2x - y = 7$

62. $8x - 11y = 15$

$11x + 8y = 1$

63. $7x - y = 1$

$x + 7y = 3$

64. $x + 4y = 7$

$-2x - 8y = 0$

65. Line 1 goes through $(0, 5)$ and $(10, 0)$.

Line 2 goes through $(0, -10)$ and $(5, 0)$.

66. Line 1 goes through $(7, 0)$ and $(0, -21)$.

Line 2 goes through $(21, 0)$ and $(0, 7)$.

67. Line 1 goes through $(-3, 8)$ and $(1, -9)$.

Line 2 goes through $(-2, -1)$ and $(-1, 16)$.

68. Line 1 goes through $(-4, 10)$ and $(2, -3)$.

Line 2 goes through $(-3, 8)$ and $(2, -5)$.

Review

Solve. See Section 2.6.

69. $|x - 3| = 6$

70. $|x + 2| < 4$

71. $|2x + 5| > 3$

72. $|5x| = 10$

73. $|3x - 4| \leq 2$

74. $|7x - 2| \geq 5$

Concept Extensions

Determine whether each slope calculation is correct or incorrect. If incorrect, then correct the calculation. See the Concept Check in this section.

75. $(-2, 6)$ and $(7, -14)$

$$m = \frac{-14 - 6}{7 - 2} = \frac{-20}{5} = -4$$

76. $(-1, 4)$ and $(-3, 9)$

$$m = \frac{9 - 4}{-3 - 1} = \frac{5}{-4} \text{ or } -\frac{5}{4}$$

77. $(-8, -10)$ and $(-11, -5)$

$$m = \frac{-10 - (-5)}{-8 - (-11)} = \frac{-5}{3} \text{ or } -\frac{5}{3}$$

78. $(0, -4)$ and $(-6, -6)$

$$m = \frac{0 - (-6)}{-4 - (-6)} = \frac{6}{2} = 3$$

79. Find the slope of a line parallel to the line

$y = -\frac{7}{2}x - 6$.

80. Find the slope of a line parallel to the line $y = x$.

81. Find the slope of a line perpendicular to the line

$y = -\frac{7}{2}x - 6$.

82. Find the slope of a line perpendicular to the line

$y = x$.

83. Find the slope of a line parallel to the line

$5x - 2y = 6$.

84. Find the slope of a line parallel to the line

$-3x + 4y = 10$.

85. Find the slope of a line perpendicular to the line

$5x - 2y = 6$.

The following graph shows the altitude of a seagull in flight over a time period of 30 seconds. Use this graph to answer Exercises 86 through 89.

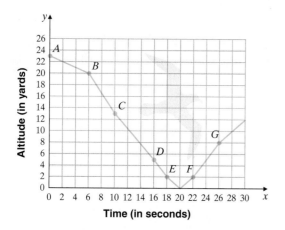

86. Find the coordinates of point *B*.

87. Find the coordinates of point *C*.

88. Find the rate of change of altitude between points *B* and *C*. (Recall that the rate of change between points is the slope between points. This rate of change will be in yards per second.)

89. Find the rate of change of altitude (in yards per second) between points *F* and *G*.

90. Professional plumbers suggest that a sewer pipe should be sloped 0.25 inch for every foot. Find the recommended slope for a sewer pipe. (*Source: Rules of Thumb* by Tom Parker, Houghton Mifflin Company)

91. Explain whether two lines, both with positive slopes, can be perpendicular.

92. Explain how merely looking at a line can tell us whether its slope is negative, positive, undefined, or zero.

93. Each line on the graph has negative slope.

    a. Find the slope of each line.

    b. Use the results of part (a) to fill in the blank: For lines with negative slopes, the steeper line has the _____ (greater/lesser) slope.

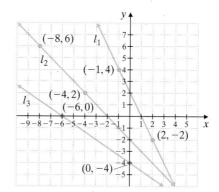

94. a. On a single screen of a graphing calculator, graph $y = \frac{1}{2}x + 1$, $y = x + 1$, and $y = 2x + 1$. Notice the change in slope for each graph.

b. On a single screen of a graphing calculator, graph $y = -\frac{1}{2}x + 1$, $y = -x + 1$, and $y = -2x + 1$. Notice the change in slope for each graph.

c. Determine whether the following statement is true or false for slope m of a given line: As $|m|$ becomes greater, the line becomes steeper.

3.3 THE SLOPE-INTERCEPT FORM

Objectives

A Graph a Line Using Its Slope and y-Intercept.

B Use the Slope-Intercept Form to Write an Equation of the Line.

C Interpret the Slope-Intercept Form in an Application.

Objective **A** Graphing a Line Using Its Slope and y-Intercept

In the last section, we learned that the slope-intercept form of a linear equation is $y = mx + b$. Recall that when an equation is written in this form, the slope of the line is the same as the coefficient m of x. Also, the y-intercept of the line is $(0, b)$. For example, the slope of the line defined by $y = 2x + 3$ is 2 and its y-intercept is $(0, 3)$.

We may also use the slope-intercept form to graph a linear equation.

EXAMPLE 1 Graph: $y = \dfrac{1}{4}x - 3$

Solution: Recall that the slope of the graph of $y = \dfrac{1}{4}x - 3$ is $\dfrac{1}{4}$ and the y-intercept is $(0, -3)$. To graph the line, we first plot the y-intercept $(0, -3)$. To find another point on the line, we recall that slope is $\dfrac{rise}{run} = \dfrac{1}{4}$. We may then plot another point by starting at $(0, -3)$, rising 1 unit up, and then running 4 units to the right. We are now at the point $(4, -2)$. The graph of $y = \dfrac{1}{4}x - 3$ is the line through points $(0, -3)$ and $(4, -2)$, as shown.

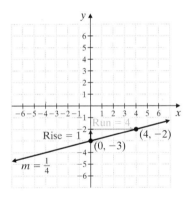

Work Practice Problem 1

EXAMPLE 2 Graph: $2x + y = 3$

Solution: First, we solve the equation for y to write it in slope-intercept form. In slope-intercept form, the equation is $y = -2x + 3$. Next we plot the y-intercept $(0, 3)$. To find another point on the line, we use the slope -2, which can be written as $\dfrac{rise}{run} = \dfrac{-2}{1}$. We start at $(0, 3)$ and move vertically 2 units down, since the numerator of the slope is -2; then we move horizontally 1 unit to the right since the denominator of the slope is 1. We arrive at the point $(1, 1)$. The line through $(1, 1)$ and $(0, 3)$ will have the required slope of -2.

Continued on next page

PRACTICE PROBLEM 1

Graph: $y = \dfrac{2}{3}x + 1$

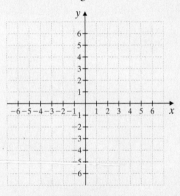

PRACTICE PROBLEM 2

Graph: $3x + y = -2$

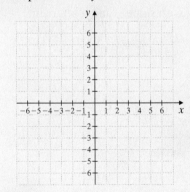

Answers

1.

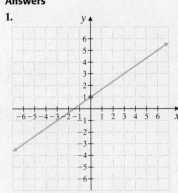

2. See page 184.

183

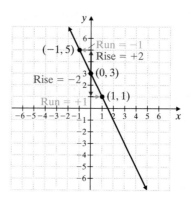

The slope -2 can also be written as $\dfrac{2}{-1}$, so to find another point for Example 2 we could start at $(0, 3)$ and move 2 units up and then 1 unit left. We would stop at the point $(-1, 5)$. The line through $(-1, 5)$ and $(0, 3)$ will have the required slope and will be the same line as shown previously through $(1, 1)$ and $(0, 3)$.

◾ **Work Practice Problem 2**

Objective **B** Using the Slope-Intercept Form to Write an Equation

Given the slope and y-intercept of a line, we may write its equation as well as graph the line.

PRACTICE PROBLEM 3

Write an equation of the line with slope $\dfrac{1}{7}$ and y-intercept $(0, -5)$.

EXAMPLE 3   Write an equation of the line with y-intercept $(0, -7)$ and slope of $\dfrac{2}{3}$.

Solution: We are given the slope and the y-intercept. We let $m = \dfrac{2}{3}$ and $b = -7$, and write the equation in slope-intercept form, $y = mx + b$.

$$y = mx + b$$
$$y = \frac{2}{3}x + (-7) \quad \text{Let } m = \frac{1}{4} \text{ and } b = -7.$$
$$y = \frac{2}{3}x - 7 \quad \text{Simplify.}$$

Notice that the graph of this equation will have slope $\dfrac{2}{3}$ and y-intercept $(0, -7)$ as desired.

◾ **Work Practice Problem 3**

✔**Concept Check**   What is wrong with the following equation of a line with y-intercept $(0, 4)$ and slope 2?

$$y = 4x + 2$$

Objective **C** Interpreting the Slope-Intercept Form

Recall from Section 3.1 the graph of an adult one-day pass price for Disney World. Notice that the graph resembles the graph of a line. Often, businesses depend on equations that "closely fit" lines like this one to model the data and predict future trends. For example, by a method called least squares regression, the linear equation

Answers

2.

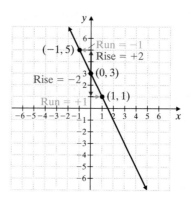

3. $y = \dfrac{1}{7}x - 5$

$y = 1.883x + 34.12$ approximates the data shown, where x is the number of years since 1992 and y is the ticket price for that year.

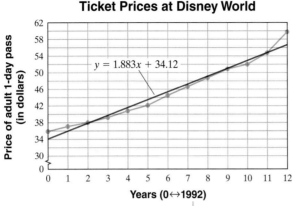

Ticket Prices at Disney World

Source: The Walt Disney Company

Helpful Hint

Don't forget—the notation $0 \leftrightarrow 1992$ means that the number 0 corresponds to the year 1992, 1 corresponds to the year 1993, and so on.

EXAMPLE 4 **Predicting Future Prices**

The adult one-day pass price y for Disney World is given by

$$y = 1.883x + 34.12$$

where x is the number of years since 1992.

a. Use this equation to predict the ticket price for 2008.
b. What does the slope of this equation mean?
c. What does the y-intercept of this equation mean?

Solution:

a. To predict the price of a pass in 2008, we need to find y when x is 17. (Since the year 1992 corresponds to $x = 0$, the year 2008 corresponds to $x = 2008 - 1992 = 16$).

$$y = 1.883x + 34.12$$
$$= 1.883(16) + 34.12 \quad \text{Let } x = 16.$$
$$= 64.248$$

We predict that in the year 2008 the price of an adult one-day pass to Disney World will be about $64.25

b. The slope of $y = 1.883x + 34.12$ is 1.883. We can think of this number as $\dfrac{\text{rise}}{\text{run}}$ or $\dfrac{1.883}{1}$. This means that the ticket price increases on the average by $1.883 each year.

c. The y-intercept of $y = 1.883x + 34.12$ is 34.12. Notice that it corresponds to the point of the graph $(0, 34.12)$

year price

This means that at year $x = 0$, or 1992, the ticket price was about $34.12.

🖥 **Work Practice Problem 4**

PRACTICE PROBLEM 4

For the period 1980 through 2020, the number of people y age 85 or older living in the United States is given by the equation $y = 110{,}520x + 2{,}127{,}400$, where x is the number of years since 1980. (*Source:* Based on data and estimates from the U.S. Bureau of the Census)

a. Estimate the number of people age 85 or older living in the United States in 2010.

b. What does the slope of this equation mean?

c. What does the y-intercept of this equation mean?

Answers

a. 5,443,000, **b.** The number of people age 85 or older in the United States increases at a rate of 110,520 per year, **c.** At year $x = 0$, or 1980, there were 2,127,400 people age 85 or older in the United States.

You may have noticed by now that to use the $\boxed{Y =}$ key on a grapher to graph an equation, the equation must be solved for y.

Graph each equation by first solving the equation for y.

1. $x = 3.5y$

2. $-2.7y = x$

3. $5.78x + 2.31y = 10.98$

4. $-7.22x + 3.89y = 12.57$

5. $y - x = 3.78$

6. $3y - 5x = 6x - 4$

7. $y - 5.6x = 7.7x + 1.5$

8. $y + 2.6x = -3.2$

Mental Math

Find the slope and the y-intercept of each line.

1. $y = -4x + 12$

2. $y = \dfrac{2}{3}x - \dfrac{7}{2}$

3. $y = 5x$

4. $y = -x$

5. $y = \dfrac{1}{2}x + 6$

6. $y = -\dfrac{2}{3}x + 5$

3.3 EXERCISE SET

 Student Solutions Manual   PH Math/Tutor Center   CD/Video for Review   Math XL MathXL®   MyMathLab MyMathLab

Objective **A** *Graph each line passing through the given point with the given slope. See Examples 1 and 2.*

1. Through $(1, 3)$ with slope $\dfrac{3}{2}$

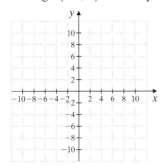

2. Through $(-2, -4)$ with slope $\dfrac{2}{5}$

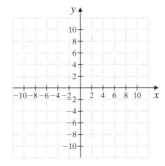

3. Through $(0, 0)$ with slope 5

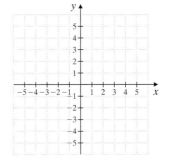

4. Through $(-5, 2)$ with slope 2

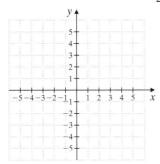

5. Through $(0, 7)$ with slope -1

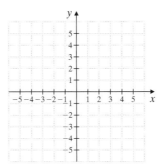

6. Through $(3, 0)$ with slope -3

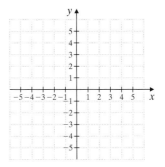

Graph each linear equation using the slope and y-intercept. See Examples 1 and 2.

7. $y = -2x$

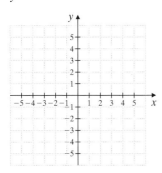

8. $y = 2x$

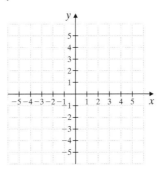

9. $y = -2x + 3$

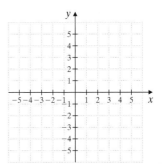

10. $y = 2x + 6$

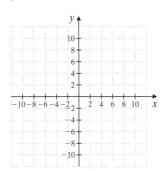

11. $y = \dfrac{1}{2}x$

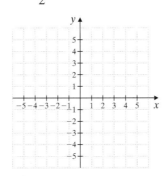

12. $y = \dfrac{1}{3}x$

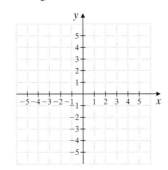

13. $y = \frac{1}{2}x - 4$

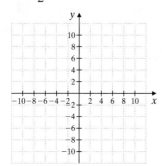

14. $y = \frac{1}{3}x - 2$

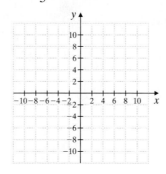

15. $7x - 2y = 10$

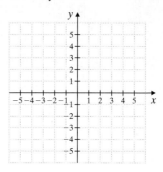

16. $8x - 3y = 9$

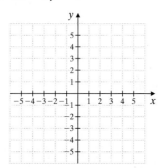

17. $x + 2y = 8$

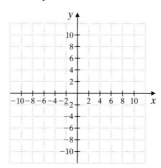

18. $x - 3y = 3$

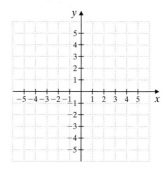

Match each equation with its graph. See Examples 1 and 2.

19. $y = \frac{1}{2}x - 1$

20. $y = \frac{1}{4}x - 1$

21. $y = -2x + 3$

22. $y = -3x + 3$

A

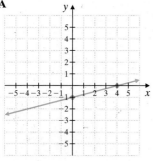

B

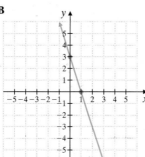

C

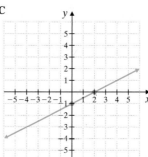

D

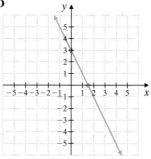

Objective B *Use the slope-intercept form of a linear equation to write the equation of each line with the given slope and y-intercept. See Example 3.*

23. Slope -1; y-intercept $(0, 1)$

24. Slope $\frac{1}{2}$; y-intercept $(0, -6)$

25. Slope 2; y-intercept $\left(0, \frac{3}{4}\right)$

26. Slope -3; y-intercept $\left(0, -\frac{1}{5}\right)$

27. Slope $\frac{2}{7}$; y-intercept $(0, 0)$

28. Slope $-\frac{4}{5}$; y-intercept $(0, 0)$

Objective **C** *Solve. See Example 4.*

29. The annual average income y of an American man with an associate's degree is given by the linear equation $y = 1983x + 42{,}972$, where x is the number of years after 1999. (*Source:* Based on data from U.S. Bureau of the Census, 1999–2004)

   a. Find the average income of an American man with an associate's degree in 2000.
   b. Find and interpret the slope of the equation.
   c. Find and interpret the y-intercept of the equation.

30. The annual average income y of an American woman with a bachelor's degree is given by the equation $y = 4174.9x + 42{,}173$, where x is the number of years after 1999. (*Source:* Based on data from U.S. Bureau of the Census, 1999–2004)

   a. Find the average income of an American woman with a bachelor's degree in 2000.
   b. Find and interpret the slope of the equation.
   c. Find and interpret the y-intercept of the equation.

31. One of the top ten occupations in terms of job growth in the next few years is expected to be network systems and data communications analysts. The number of people y, in thousands, employed as network systems and data communications analysts in the United States can be estimated by the linear equation $53x - 5y = -930$, where x is the number of years after 2002. (*Source:* Based on projections from the U.S. Bureau of Labor Statistics, 2002–2012)

   a. Find the slope and the y-intercept of the linear equation.
   b. What does the slope mean in this context?

   c. What does the y-intercept mean in this context?

32. One of the fastest-growing occupations over the next few years is expected to be medical assistant. The number of people y, in thousands, employed as medical assistants in the United States can be estimated by the linear equation $43x - 2y = -730$, where x is the number of years after 2002. (*Source:* Based on projections from the U.S. Bureau of Labor Statistics, 2002–2012)

   a. Find the slope and the y-intercept of the linear equation.
   b. What does the slope mean in this context?

   c. What does the y-intercept mean in this context?

33. The number of cellular subscribers y (in millions) in the United States can be estimated by the linear equation $y = 17.7x + 44.9$, where x is the number of years after 1997. (*Source:* Cellular Telecommunications Industry Association)

   a. Use this equation to estimate the number of cellular telephone subscribers in the United States in 2004.

   b. Use this equation to predict in what year the number of cellular telephone subscribers in the United States will exceed 220 million. (*Hint:* Let $y = 220$ and solve for x.)
   c. Use this equation to estimate the number of cellular telephone subscribers in the present year. Do you have a personal cell phone? Do your friends?

34. The yearly cost of undergraduate tuition and required fees for attending a public two-year college full-time can be estimated by the linear equation $y = 170.5x + 1401$, where x is the number of years after 2000 and y is the total cost in dollars. (*Source:* Based on data from the College Board, 2000–2004).

   a. Use this equation to approximate the yearly cost of attending a public two-year college in 2010.
   b. Use this equation to predict in what year the yearly cost of tuition and required fees will exceed $4000. (*Hint:* Let $y = 4000$ and solve for x)
   c. Use this equation to approximate the yearly cost of attending a two-year college in the present year. If you attend a two-year college, is this amount greater or less than the amount currently charged by the college you attend?

Review

Simplify and solve for y. See Section 2.3.

35. $y - 2 = 5(x + 6)$ **36.** $y - 0 = -3[x - (-10)]$ **37.** $y - (-1) = 2(x - 0)$ **38.** $y - 9 = -8[x - (-4)]$

Concept Extensions

Determine whether each equation is correct or incorrect. If incorrect, then correct. See the Concept Check in this section.

39. The equation of the line with y-intercept $(0, 1.7)$ and slope 3 is $y = 1.7x + 3$.

40. The equation of the line with slope $-\frac{1}{9}$ and y-intercept $(0, -5)$ is $y = -\frac{1}{9}x + 5$.

41. The equation of the line with slope -5 and y-intercept $\left(0, -\frac{1}{3}\right)$ is $y = -5x - \frac{1}{3}$.

 42. In your own words, explain how to graph an equation using its slope and y-intercept.

43. Suppose that the revenue of a company has increased at a steady rate of $42,000 per year since 1995. Also the company's revenue in 1995 was $2,900,000. Write an equation that describes the company's revenue since 1995.

44. Suppose that a bird dives off a 500-foot cliff and descends at a rate of 7 feet per second. Write an equation that describes the bird's height at any time x.

STUDY SKILLS BUILDER

Are You Organized?

Have you ever had trouble finding a completed assignment? When it's time to study for a test, are your notes neat and organized? Have you ever had trouble reading your own mathematics handwriting? (Be honest—I have.)

When any of these things happen, it's time to get organized. Here are a few suggestions:

Write your notes and complete your homework assignment in a notebook with pockets (spiral or ring binder.) Take class notes in this notebook, and then follow the notes with your completed homework assignment. When you receive graded papers or handouts, place them in the notebook pocket so that you will not lose them.

Remember to mark (possibly with an exclamation point) any note(s) that seem extra important to you. Also remember to mark (possibly with a question mark) any notes or homework that you are having trouble with. Don't forget to see your instructor or a math tutor to help you with the concepts or exercises that you are having trouble understanding.

Also, if you are having trouble reading your own handwriting, *slow down* and write your mathematics work clearly!

Exercises

1. Have you been completing your assignments on time?

2. Have you been correcting any exercises you may be having difficulty with?

3. If you are having trouble with a mathematical concept or correcting any homework exercises, have you visited your instructor, a tutor, or your campus math lab?

4. Are you taking lecture notes in your mathematics course? (By the way, these notes should include all worked-out examples solved by your instructor.)

5. Is your mathematics course material (handouts, graded papers, lecture notes) organized?

6. If your answer to Exercise 5 is no, take a moment and review your course material. List at least two ways that you might better organize it. Then read the Study Skills Builder on organizing a notebook in Chapter 2.

3.4 MORE EQUATIONS OF LINES

Objectives

A Use the Point-Slope Form to Write the Equation of a Line.

B Write Equations of Vertical and Horizontal Lines.

C Write Equations of Parallel and Perpendicular Lines.

D Use the Point-Slope Form in Real-World Applications.

Objective **A** Using the Point-Slope Form to Write an Equation

When the slope of a line and a point on the line are known, the equation of the line can also be found. To do this, we use the slope formula to write the slope of a line that passes through points (x, y), and (x_1, y_1). We have

$$m = \frac{y - y_1}{x - x_1}$$

We multiply both sides of this equation by $x - x_1$ to obtain

$$y - y_1 = m(x - x_1)$$

This form is called the *point-slope form* of the equation of a line.

Point-Slope Form of the Equation of a Line

The **point-slope form** of the equation of a line is

$$y - y_1 = m(x - x_1)$$

where m is the slope of the line and (x_1, y_1) is a point on the line.

EXAMPLE 1 Write an equation of the line with slope -3 and containing the point $(1, -5)$. Write the equation in slope-intercept form, $y = mx + b$.

Solution: Because we know the slope and a point on the line, we use the point-slope form with $m = -3$ and $(x_1, y_1) = (1, -5)$.

$$y - y_1 = m(x - x_1) \quad \text{Point-slope form}$$
$$y - (-5) = -3(x - 1) \quad \text{Let } m = -3 \text{ and } (x_1, y_1) = (1, -5).$$
$$y + 5 = -3x + 3 \quad \text{Use the distributive property.}$$
$$y = -3x - 2 \quad \text{Write in slope-intercept form (solved for } y\text{).}$$

In slope-intercept form, the equation is $y = -3x - 2$.

 Work Practice Problem 1

PRACTICE PROBLEM 1

Write an equation of the line with slope -2 and containing the point $(2, -4)$. Write the equation in slope-intercept form, $y = mx + b$.

Helpful Hint

Remember, "slope-intercept form" means the equation is "solved for y."

EXAMPLE 2 Write an equation of the line through points $(4, 0)$ and $(-4, -5)$. Write the equation in standard form $Ax + By = C$.

Solution: First we find the slope of the line.

$$m = \frac{-5 - 0}{-4 - 4} = \frac{-5}{-8} = \frac{5}{8}$$

Next we make use of the point-slope form. We replace (x_1, y_1) by either $(4, 0)$ or $(-4, -5)$ in the point-slope equation. We will choose the point $(4, 0)$. The line through $(4, 0)$ with slope $\frac{5}{8}$ is as follows.

PRACTICE PROBLEM 2

Write an equation of the line through points $(3, 0)$ and $(-2, 4)$. Write the equation in standard form, $Ax + By = C$.

Answers
1. $y = -2x$, **2.** $4x + 5y = 12$

Continued on next page

$$y - y_1 = m(x - x_1) \quad \text{Point-slope form}$$

$$y - 0 = \frac{5}{8}(x - 4) \quad \text{Let } m = \frac{5}{8} \text{ and } (x_1, y_1) = (4, 0).$$

Let's multiply through by 8 so that the coefficients are integers and are less tedious to work with.

$$8(y - 0) = 8 \cdot \frac{5}{8}(x - 4)$$

$$8y = 5(x - 4) \qquad \text{Simplify.}$$

$$8y = 5x - 20 \qquad \text{Multiply.}$$

$$-5x + 8y = -20 \qquad \text{Write in standard form.}$$

If we multiply both sides of $-5x + 8y = -20$ by -1, we have an equivalent equation in standard form, also. Both $-5x + 8y = -20$ and $5x - 8y = 20$ are acceptable.

🔲 **Work Practice Problem 2**

Objective B Writing Equations of Vertical and Horizontal Lines

A few special types of linear equations are those whose graphs are vertical and horizontal lines.

PRACTICE PROBLEM 3

Write an equation of the horizontal line containing the point $(-1, 6)$.

EXAMPLE 3 Write an equation of the horizontal line containing the point $(2, 3)$.

Solution: Recall, from Section 3.1, that a horizontal line has an equation of the form $y = b$. Since the line contains the point $(2, 3)$, the equation is $y = 3$.

🔲 **Work Practice Problem 3**

PRACTICE PROBLEM 4

Write an equation of the line containing the point $(4, 7)$ with undefined slope.

EXAMPLE 4 Write an equation of the line containing the point $(2, 3)$ with undefined slope.

Solution:

Since the line has undefined slope, the line must be vertical. A vertical line has an equation of the form $x = c$, and since the line contains the point $(2, 3)$, the equation is $x = 2$.

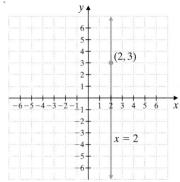

🔲 **Work Practice Problem 4**

Objective C Writing Equations of Parallel and Perpendicular Lines

Next, we write equations of parallel and perpendicular lines.

Answers

3. $y = 6$, **4.** $x = 4$

EXAMPLE 5 Write an equation of the line containing the point $(4, 4)$ and parallel to the line $2x + y = -6$. Write the equation in slope-intercept form, $y = mx + b$.

Solution: Because the line we want to find is *parallel* to the line $2x + y = -6$, the two lines must have equal slopes. So we first find the slope of $2x + y = -6$ by solving the equation for y to write it in the form $y = mx + b$. Here $y = -2x - 6$, so the slope is -2.

Now we use the point-slope form to write the equation of a line through $(4, 4)$ with slope -2.

$$y - y_1 = m(x - x_1)$$
$$y - 4 = -2(x - 4) \quad \text{Let } m = -2, x_1 = 4, \text{ and } y_1 = 4.$$
$$y - 4 = -2x + 8 \quad \text{Use the distributive property.}$$
$$y = -2x + 12$$

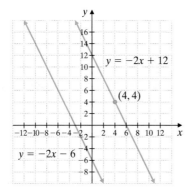

The equation, $y = -2x - 6$, and the new equation, $y = -2x + 12$, have the same slope but different y-intercepts so their graphs are parallel. Also, the graph of $y = -2x + 12$ contains the point $(4, 4)$, as desired.

Work Practice Problem 5

EXAMPLE 6 Write an equation of the line containing the point $(-2, 1)$ and perpendicular to the line $3x + 5y = 4$. Write the equation in slope-intercept form, $y = mx + b$.

Solution: First we find the slope of $3x + 5y = 4$ by solving the equation for y.

$$5y = -3x + 4$$
$$y = -\frac{3}{5}x + \frac{4}{5}$$

The slope of the given line is $-\frac{3}{5}$. A line perpendicular to this line will have a slope that is the negative reciprocal of $-\frac{3}{5}$, or $\frac{5}{3}$. We use the point-slope form to write an equation of a new line through $(-2, 1)$ with slope $\frac{5}{3}$.

$$y - 1 = \frac{5}{3}[x - (-2)]$$
$$y - 1 = \frac{5}{3}(x + 2) \quad \text{Simplify.}$$
$$y - 1 = \frac{5}{3}x + \frac{10}{3} \quad \text{Use the distributive property.}$$
$$y = \frac{5}{3}x + \frac{13}{3} \quad \text{Add 1 to both sides.}$$

Continued on next page

Write an equation of the line containing the point $(-1, 2)$ and parallel to the line $3x + y = 5$. Write the equation in slope-intercept form $y = mx + b$.

Write an equation of the line containing the point $(3, 4)$ and perpendicular to the line $2x + 4y = 5$. Write the equation in slope-intercept form, $y = mx + b$.

Answers

5. $y = -3x - 1$, **6.** $y = 2x - 2$

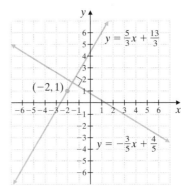

The equation $y = -\dfrac{3}{5}x + \dfrac{4}{5}$ and the new equation $y = \dfrac{5}{3}x + \dfrac{13}{3}$ have negative reciprocal slopes, so their graphs are perpendicular. Also, the graph of $y = \dfrac{5}{3}x + \dfrac{13}{3}$ contains the point $(-2, 1)$, as desired.

▣ **Work Practice Problem 6**

Objective D Using the Point-Slope Form in Applications

The point-slope form of an equation is very useful for solving real-world problems.

EXAMPLE 7 Predicting Sales

Southern Star Realty is an established real estate company that has enjoyed constant growth in sales since 1996. In 2000 the company sold 250 houses, and in 2004 the company sold 330 houses. Use these figures to predict the number of houses this company will sell in 2008.

Solution:

1. UNDERSTAND. Read and reread the problem. Then let

    x = the number of years after 1996 and

    y = the number of houses sold in the year corresponding to x

    The information provided than gives the ordered pairs $(4, 250)$ and $(8, 330)$. To better visualize the sales of Southern Star Realty, we graph the line that passes through the points $(4, 250)$ and $(8, 330)$.

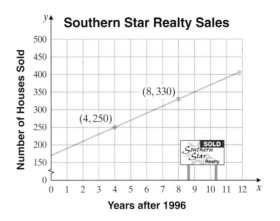

2. TRANSLATE. We write the equation of the line that passes through the points $(4, 250)$ and $(8, 330)$. To do so, we first find the slope of the line.

    $$m = \frac{330 - 250}{8 - 4} = \frac{80}{4} = 20$$

Then using the point-slope form to write the equation, we have

$$y - y_1 = m(x - x_1)$$

$$y - 250 = 20(x - 4) \quad \text{Let } m = 20 \text{ and } (x_1, y_1) = (4, 250).$$

$$y - 250 = 20x - 80 \quad \text{Multiply.}$$

$$y = 20x + 170 \quad \text{Add 250 to both sides.}$$

3. SOLVE. To predict the number of houses sold in 2008, we use $y = 20x + 170$ and complete the ordered pair (12,), since $2008 - 1996 = 12$.

$$y = 20(12) + 170$$

$$y = 410$$

4. INTERPRET.

Check: Verify that the point $(12, 410)$ is a point on the line graphed in Step 1.

State: Southern Star Realty should expect to sell 410 houses in 2008.

 Work Practice Problem 7

CALCULATOR EXPLORATIONS   Graphing

Many graphing calculators have a TRACE feature. This feature allows you to trace along a graph and see the corresponding x- and y-coordinates appear on the screen. Use this feature for the following exercises.

Graph each equation and then use the TRACE feature to complete each ordered pair solution. (Many times the tracer will not show the exact x- or y-value asked for. In each case, trace as closely as you can to the given x- or y-coordinate and approximate the other, unknown coordinate to one decimal place.)

1. $y = 2.3x + 6.7$;
   $x = 5.1, y = ?$

2. $y = -4.8x + 2.9$;
   $x = -1.8, y = ?$

3. $y = -5.9x - 1.6$;
   $x = ?, y = 7.2$

4. $y = 0.4x - 8.6$;
   $x = ?, y = -4.4$

5. $y = 5.2x - 3.3$;
   $x = 2.3, y = ?$
   $x = ?, y = 36$

6. $y = -6.2x - 8.3$;
   $x = 3.2, y = ?$
   $x = ?, y = 12$

Mental Math

Find the slope of and a point on the line described by each equation.

1. $y - 4 = -2(x - 1)$

2. $y - 6 = -3(x - 4)$

3. $y - 0 = \dfrac{1}{4}(x - 2)$

4. $y - 1 = -\dfrac{2}{3}(x - 0)$

5. $y + 2 = 5(x - 3)$

6. $y - 7 = 4(x + 6)$

3.4 EXERCISE SET

FOR EXTRA HELP

Student Solutions Manual   PH Math/Tutor Center   CD/Video for Review   Math XL MathXL®   MyMathLab MyMathLab

Objective Ⓐ *Write an equation of each line with the given slope and containing the given point. Write the equation in the slope-intercept form $y = mx + b$. See Example 1.*

 1. Slope 3; through $(1, 2)$

2. Slope 4; through $(5, 1)$

3. Slope -2; through $(1, -3)$

4. Slope -4; through $(2, -4)$

5. Slope $\dfrac{1}{2}$; through $(-6, 2)$

6. Slope $\dfrac{2}{3}$; through $(-9, 4)$

7. Slope $-\dfrac{9}{10}$; through $(-3, 0)$

8. Slope $-\dfrac{1}{5}$; through $(4, -6)$

Write an equation of the line passing through the given points. Write the equation in standard form $Ax + By = C$. See Example 2.

9. $(2, 0)$ and $(4, 6)$

10. $(3, 0)$ and $(7, 8)$

11. $(-2, 5)$ and $(-6, 13)$

12. $(7, -4)$ and $(2, 6)$

13. $(-2, -4)$ and $(-4, -3)$

14. $(-9, -2)$ and $(-3, 10)$

 15. $(-3, -8)$ and $(-6, -9)$

16. $(8, -3)$ and $(4, -8)$

17. $\left(\dfrac{3}{5}, \dfrac{4}{10}\right)$ and $\left(-\dfrac{1}{5}, \dfrac{7}{10}\right)$

18. $\left(\dfrac{1}{2}, -\dfrac{1}{4}\right)$ and $\left(\dfrac{3}{2}, \dfrac{3}{4}\right)$

Objective Ⓑ *Write an equation of each line. See Examples 3 and 4.*

19. Vertical; through $(2, 6)$

20. Slope 0; through $(-2, -4)$

21. Horizontal; through $(-3, 1)$

22. Vertical; through $(4, 7)$

23. Undefined slope; through $(0, 5)$

24. Horizontal; through $(0, 5)$

196

Objective [C] *Write an equation of each line. Write the equation in the form* $x = a$, $y = b$, *or* $y = mx + b$. *See Examples 5 and 6.*

25. Through $(3, 8)$; parallel to $y = 4x - 2$

26. Through $(1, 5)$; parallel to $y = 3x - 4$

27. Through $(2, -5)$; perpendicular to $3y = x - 6$

28. Through $(-4, 8)$; perpendicular to $2x - 3y = 1$

29. Through $(1, 4)$; parallel to $y = 7$

30. Through $(-2, 6)$; perpendicular to $y = 7$

31. Through $(-2, -3)$; parallel to $3x + 2y = 5$

32. Through $(-2, -3)$; perpendicular to $3x + 2y = 5$

33. Through $(-1, -5)$; perpendicular to $x = 3$

34. Through $(4, -6)$; parallel to $x = -2$

35. Through $(-1, 5)$; perpendicular to $x - 4y = 4$

36. Through $(2, -3)$; perpendicular to $x - 5y = 10$

Objectives [A] [B] [C] **Mixed Practice** *Find the equation of each line. Write the equation in standard form unless indicated otherwise. See Examples 1 through 6.*

37. Slope 2; through $(-2, 3)$

38. Slope 3; through $(-4, 2)$

39. Through $(1, 6)$ and $(5, 2)$; use slope-intercept form.

40. Through $(2, 9)$ and $(8, 6)$; use slope-intercept form.

41. With slope $-\dfrac{1}{2}$; y-intercept $\left(0, \dfrac{3}{8}\right)$; use slope-intercept form.

42. With slope -4; y-intercept $\left(0, \dfrac{2}{9}\right)$; use slope-intercept form.

43. Through $(-7, -4)$ and $(0, -6)$

44. Through $(2, -8)$ and $(-4, -3)$

45. Slope $-\dfrac{4}{3}$; through $(-5, 0)$

46. Slope $-\dfrac{3}{5}$; through $(4, -1)$

47. Vertical line; through $(-2, -10)$

48. Horizontal line; through $(1, 0)$

49. Through $(6, -2)$; parallel to the line $2x + 4y = 9$

50. Through $(8, -3)$; parallel to the line $6x + 2y = 5$

51. Slope 0; through $(-9, 12)$

52. Undefined slope; through $(10, -8)$

53. Through $(6, 1)$; parallel to the line $8x - y = 9$

54. Through $(3, 5)$; perpendicular to the line $2x - y = 8$

55. Through $(5, -6)$; perpendicular to $y = 9$

56. Through $(-3, -5)$; parallel to $y = 9$

57. Through $(2, -8)$ and $(-6, -5)$; use slope-intercept form.

58. Through $(-4, -2)$ and $(-6, 5)$; use slope-intercept form.

Objective **D** *Solve. See Example 7.*

59. A rock is dropped from the top of a 400-foot building. After 1 second, the rock is traveling 32 feet per second. After 3 seconds, the rock is traveling 96 feet per second. Let y be the rate of descent and x be the number of seconds since the rock was dropped.

a. Write a linear equation that relates time x to rate y. [*Hint:* Use the ordered pairs $(1, 32)$ and $(3, 96)$.]

b. Use this equation to determine the rate of travel of the rock 4 seconds after it was dropped.

60. The Whammo Company has learned that by pricing a newly released Frisbee at $6, sales will reach 2000 per day. Raising the price to $8 will cause the sales to fall to 1500 per day. Assume that the ratio of change in price to change in daily sales is constant, and let x be the price of the Frisbee and y be number of sales.

a. Find the linear equation that models the price–sales relationship for this Frisbee. [*Hint:* The line must pass through $(6, 2000)$ and $(8, 1500)$.]

b. Use this equation to predict the daily sales of Frisbees if the price is set at $7.50.

61. A fruit company recently released a new applesauce. By the end of its first year, profits on this product amounted to $30,000. The anticipated profit for the end of the fourth year is $66,000. The ratio of change in time to change in profit is constant. Let x be years and y be profit.

a. Write a linear equation that relates profit and time.

b. Use this equation to predict the company's profit at the end of the seventh year.

c. Predict when the profit should reach $126,000.

62. The Pool Fun Company has learned that, by pricing a newly released Fun Noodle at $3, sales will reach 10,000 Fun Noodles per day during the summer. Raising the price to $5 will cause the sales to fall to 8000 Fun Noodles per day. Let x be price and y be the number sold.

a. Assume that the relationship between sales price and number of Fun Noodles sold is linear and write an equation describing this relationship.

b. Use this equation to predict the daily sales of Fun Noodles if the price is $3.50.

63. In 2000, the median price of an existing home in the United States was $142,200. In 2003, the median price of an existing home was $170,000. Let y be the median price of an existing home in the year x, where $x = 0$ represents the year 2000. (*Source:* National Association of REALTORS)

a. Write a linear equation that models the median existing home price in terms of the year x. [*Hint:* The line must pass through the point $(0, 142,200)$ and $(3, 170,000)$.]

b. Use this equation to predict the median existing home price for 2008.

64. The number of commercial aircraft delivered to customers by Boeing in 1997 was 374. In 2003, Boeing delivered a total of 281 commercial airplanes to customers. Let y be the number of Boeing commercial aircraft delivered to customers in the year x, where $x = 0$ represents 1997. (*Source:* The Boeing Company)

a. Write a linear equation that models the number of Boeing commercial aircraft delivered to customers in terms of the year x. [*Hint:* The line must pass through $(0, 374)$ and $(6, 281)$.]

b. Use this information to predict the number of commercial aircraft Boeing delivers to customers in 2007.

65. The number of basic cable TV subscribers in the United States for 2001 was approximately 70.0 million. In 2004, there were 70.3 million basic cable TV subscribers. Let y be the number of basic cable TV subscribers, in millions in year x, where $x = 0$ represents 2001. (*Source:* Cisco Systems)

a. Write a linear equation that models the number of basic cable TV subscribers (in the millions) in the year x.

b. Use this equation to estimate the number of basic cable TV subscribers for the year 2008.

66. The number of people employed in the United States as registered nurses was 2284 thousand in 2002. By 2012, this number is expected to rise to 2908 thousand. Let y be the number of registered nurses (in thousands) employed in the United States in the year x, where $x = 0$ represents 2002. (*Source:* U.S. Bureau of Labor Statistics)

a. Write a linear equation that models the number of people (in thousands) employed as registered nurses in year x.

b. Use this equation to estimate the number of people who will be employed as registered nurses in 2009.

Review

Complete each ordered pair for the given equation. See Section 3.1.

67. $y = 7x + 3; (4, \quad)$

68. $y = 2x - 6; (2, \quad)$

69. $y = 4.2x; (-2, \quad)$

70. $y = -1.3x; (6, \quad)$

71. $y = x^2 + 2x + 1; (1, \quad)$

72. $y = x^2 - 6x + 4; (0, \quad)$

Concept Extensions

Find an equation of each line graphed. Write the equation in standard form.

73.
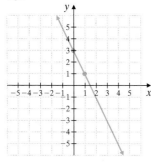

74.
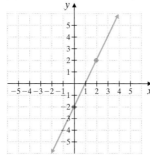

75.
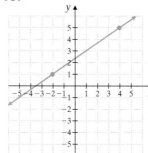

76.
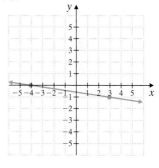

Answer true or false.

77. A vertical line is always perpendicular to a horizontal line.

78. A vertical line is always parallel to another vertical line.

79. Describe how to check to see if the graph of $2x - 4y = 7$ passes through the points $(1.4, -1.05)$ and $(0, -1.75)$. Then follow your directions and check these points.

Use a grapher with a TRACE feature to see the results of each exercise.

80. Exercise 25: Graph the equation and verify that it passes through $(3, 8)$ and is parallel to $y = 4x - 2$.

81. Exercise 26: Graph the equation and verify that it passes through $(1, 5)$ and is parallel to $y = 3x - 4$.

Linear Equations in Two Variables

Below is a review of equations of lines.

Forms of Linear Equations

$Ax + By = C$	**Standard form** of a linear equation. A and B are not both 0.
$y = mx + b$	**Slope-intercept form** of a linear equation. The slope is m, and the y-intercept is $(0, b)$.
$y - y_1 = m(x - x_1)$	**Point-slope form** of a linear equation. The slope is m, and (x_1, y_1) is a point on the line.
$y = c$	**Horizontal line**. The slope is 0, and the y-intercept is $(0, c)$.
$x = c$	**Vertical line**. The slope is undefined and the x-intercept is $(c, 0)$.

Parallel and Perpendicular Lines

Nonvertical parallel lines have the same slope.
The product of the slopes of two nonvertical perpendicular lines is -1.

Graph each linear equation.

1. $y = -2x$

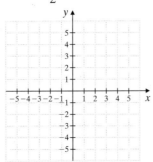

2. $3x - 2y = 6$

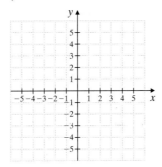

3. $x = -1\dfrac{1}{2}$

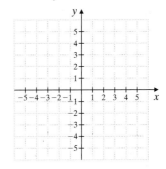

4. $y = 1.5$

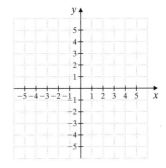

Find the slope of the line containing each pair of points.

5. $(-2, -5), (3, -5)$

6. $(5, 2), (0, 5)$

Answers

1. see graph

2. see graph

3. see graph

4. see graph

5. _____

6. _____

Find the slope and y-intercept of each line.

7. $y = 3x - 5$

8. $5x - 2y = 7$

Determine whether each pair of lines is parallel, perpendicular, or neither.

9. $y = 8x - 6$
$y = 8x + 6$

10. $y = \dfrac{2}{3}x + 1$
$2y + 3x = 1$

Find the equation of each line. Write the equation in the form $x = a$, $y = b$, or $y = mx + b$.

11. Through $(1, 6)$ and $(5, 2)$

12. Vertical line; through $(-2, -10)$

13. Horizontal line; through $(1, 0)$

14. Through $(2, -8)$ and $(-6, -5)$

15. Through $(-2, 4)$ with slope -5

16. Slope -4; y-intercept $\left(0, \dfrac{1}{3}\right)$

17. Slope $\dfrac{1}{2}$; y-intercept $(0, -1)$

18. Through $\left(\dfrac{1}{2}, 0\right)$ with slope 3

19. Through $(-1, -5)$; parallel to
$3x - y = 5$

20. Through $(0, 4)$; perpendicular to
$4x - 5y = 10$

21. Through $(2, -3)$; perpendicular to
$4x + y = \dfrac{2}{3}$

22. Through $(-1, 0)$; parallel to
$5x + 2y = 2$

23. Undefined slope; through $(-1, 3)$

24. $m = 0$; through $(-1, 3)$

7. _____

8. _____

9. _____

10. _____

11. _____

12. _____

13. _____

14. _____

15. _____

16. _____

17. _____

18. _____

19. _____

20. _____

21. _____

22. _____

23. _____

24. _____

Objectives

[A] Graph Linear Inequalities.

[B] Graph the Intersection or Union of Two Linear Inequalities.

3.5 GRAPHING LINEAR INEQUALITIES

Objective [A] Graphing Linear Inequalities

Recall that the graph of a linear equation in two variables is the graph of all ordered pairs that satisfy the equation and that the graph is a line. Here we graph linear inequalities in two variables; that is, we graph all the ordered pairs that satisfy the inequality.

If the equal sign in a linear equation in two variables is replaced with an inequality symbol, the result is a **linear inequality in two variables.**

Examples of Linear Inequalities in Two Variables

$$x + y < 3 \qquad 2x - 4y \geq -3$$
$$4x > 2 \qquad y \leq 5$$

To graph the linear inequality $x + y < 3$, we first graph the related equation $x + y = 3$. The resulting **boundary line** contains all ordered pairs whose coordinates add up to 3. The line separates the plane into two regions called **half-planes.** All points above the boundary line $x + y = 3$ have coordinates that satisfy the inequality $x + y > 3$, and all points below the line have coordinates that satisfy the inequality $x + y < 3$.

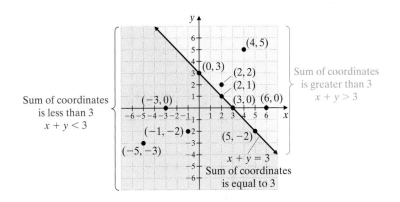

The graph, or **solution region,** for $x + y < 3$, then, is the half-plane below the boundary line and is shown shaded in the figure below. The boundary line is shown dashed since it is not a part of the solution region. The ordered pairs on this line satisfy $x + y = 3$, but not $x + y < 3$.

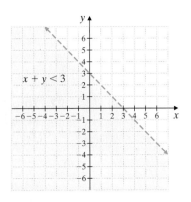

The following steps may be used to graph linear inequalities in two variables.

Graphing a Linear Inequality in Two Variables

Step 1: Graph the boundary line found by replacing the inequality sign with an equal sign. If the inequality sign is $<$ or $>$, graph a dashed line (indicating that points on the line are not solutions of the inequality). If the inequality sign is $\le$ or $\ge$, graph a solid line (indicating that points on the line are solutions of the inequality).

Step 2: Choose a **test point** *not on the boundary line* and substitute the coordinates of this test point into the *original inequality*.

Step 3: If a true statement is obtained in Step 2, shade the half-plane that contains the test point. If a false statement is obtained, shade the half-plane that does not contain the test point.

EXAMPLE 1 Graph: $2x - y < 6$

Solution: The boundary line for this inequality is the graph of $2x - y = 6$. We graph a dashed boundary line because the inequality symbol is $<$. Next we choose a test point on either side of the boundary line. The point $(0, 0)$ is not on the boundary line, so we use this point. Replacing x with 0 and y with 0 in the *original inequality* $2x - y < 6$ leads to the following:

$$2x - y < 6$$
$$2(0) - 0 < 6 \quad \text{Let } x = 0 \text{ and } y = 0.$$
$$0 < 6 \quad \text{True}$$

Because $(0, 0)$ satisfies the inequality, so does every point on the same side of the boundary line as $(0, 0)$. We shade the half-plane that contains $(0, 0)$, as shown. Every point in the shaded half-plane satisfies the original inequality.

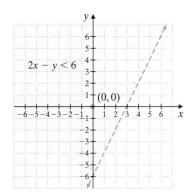

■ **Work Practice Problem 1**

EXAMPLE 2 Graph: $3x \ge y$

Solution: The boundary line is the graph of $3x = y$. We graph a solid boundary line because the inequality symbol is $\ge$. We choose a test point not on the boundary line to determine which half-plane contains points that satisfy the inequality. Let's choose $(0, 1)$ as our test point.

$$3x \ge y$$
$$3(0) \ge 1 \quad \text{Let } x = 0 \text{ and } y = 1.$$
$$0 \ge 1 \quad \text{False}$$

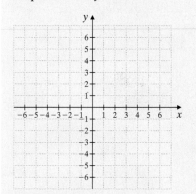

Continued on next page

PRACTICE PROBLEM 1

Graph: $x + 3y > 4$

PRACTICE PROBLEM 2

Graph: $x \le 2y$

Answers

1.

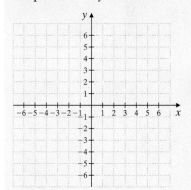

2. See page 204.

This point does not satisfy the inequality, so the correct half-plane is on the opposite side of the boundary line from $(0, 1)$. The graph of $3x \geq y$ is the boundary line together with the shaded region, as shown.

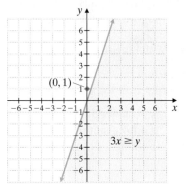

■ **Work Practice Problem 2**

✔ **Concept Check** If a point on the boundary line is included in the solution of an inequality in two variables, should the graph of the boundary line be solid or dashed?

Objective **B** **Graphing Intersections and Unions**

The intersections and the unions of linear inequalities can also be graphed, as shown in the next two examples.

EXAMPLE 3 Graph the intersection of $x \geq 1$ and $y \geq 2x - 1$.

Solution: First we graph each inequality. The intersection of the two graphs is all points common to both regions, as shown by the *heaviest* shading in the third graph.

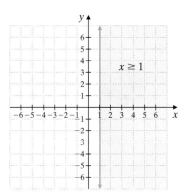 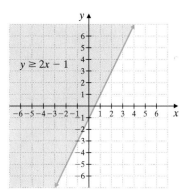

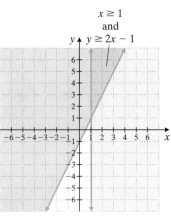

■ **Work Practice Problem 3**

PRACTICE PROBLEM 3

Graph the intersection of $x \leq 2$ and $y \geq x + 1$.

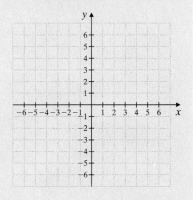

Answers

2.

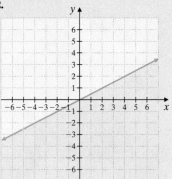

3.

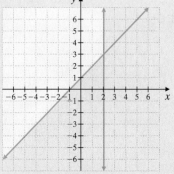

✔ **Concept Check Answer**

Solid

EXAMPLE 4 Graph the union of $2x + y \geq -8$ or $y \leq -2$.

Solution: First we graph each inequality. The union of the two inequalities is both shaded regions, including the solid boundary lines, as shown in the third graph.

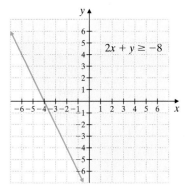

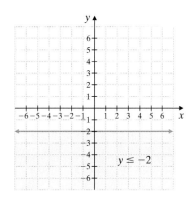

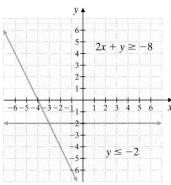

■ **Work Practice Problem 4**

PRACTICE PROBLEM 4

Graph the union of $x + 2y \leq 4$ or $y \geq -1$.

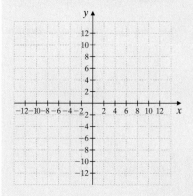

Answer
4.

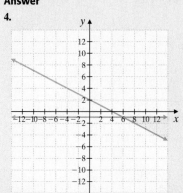

Objective **A** *Graph each inequality. See Examples 1 and 2.*

1. $x < 2$

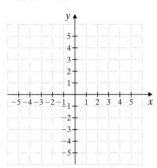

2. $x > -3$

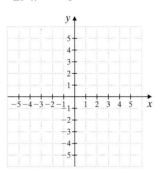

3. $x - y \geq 7$

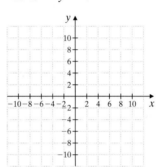

4. $3x + y \leq 1$

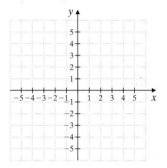

5. $3x + y > 6$

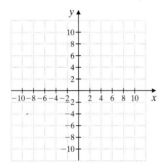

6. $2x + y > 2$

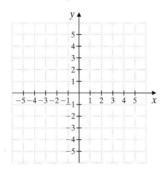

7. $y \leq -2x$

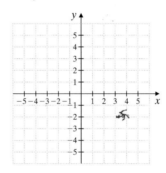

8. $y \leq 3x$

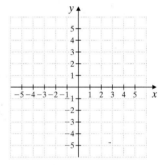

9. $2x + 4y \geq 8$

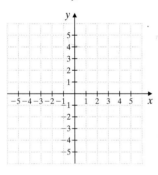

10. $2x + 6y \leq 12$

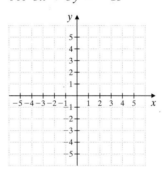

11. $5x + 3y > -15$

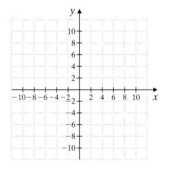

12. $2x + 5y < -20$

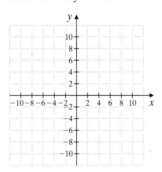

13. $y \leq -4.5$

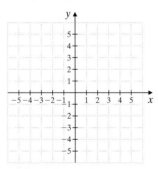

14. $y \geq -2\frac{1}{3}$

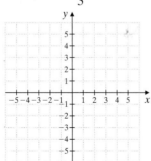

15. $x > 4y$

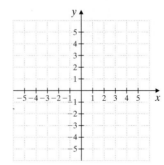

16. $x < -3y$

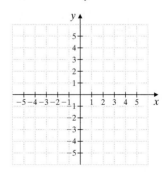

Objective **B** *Graph each union or intersection. See Examples 3 and 4.*

17. The intersection of $x \geq 3$ and $y \leq -2$

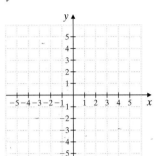

18. The union of $x \geq 3$ or $y \leq -2$

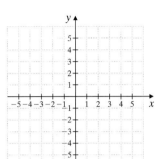

19. The union of $x \leq -2$ or $y \geq 4$

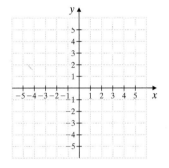

20. The intersection of $x \leq -2$ and $y \geq 4$

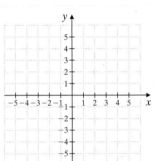

21. The intersection of $x - y < 3$ and $x > 4$

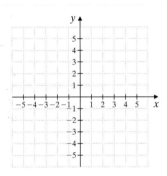

22. The intersection of $x + y \leq 1$ and $y \leq -1$

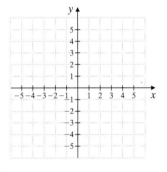

23. The union of $x + y \leq 3$ or $x - y \geq 5$

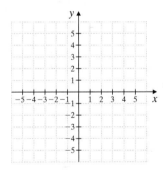

24. The union of $x - y \leq 3$ or $x + y > -1$

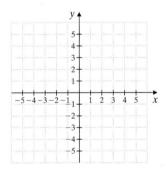

25. The union of $x - y \geq 2$ or $y < 5$

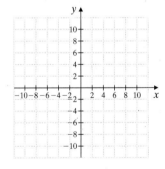

26. The union of $x - y < 3$ or $x > 4$

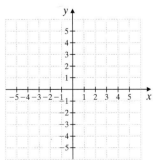

27. The intersection of $2x > y$ and $3x - 9y > 12$

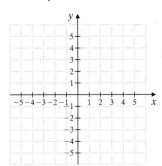

28. The intersection of $y \geq x$ and $2x - 4y \geq 6$

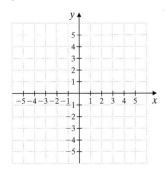

Match each inequality with its graph.

29. $y \le 2x + 3$ **30.** $y < 2x + 3$ **31.** $y > 2x + 3$ **32.** $y \ge 2x + 3$

A 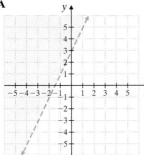   B 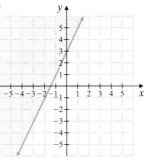   C 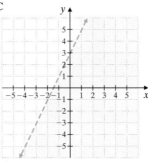   D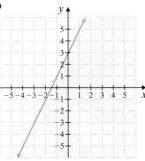

Review

Determine whether the given ordered pair is a solution of both equations. See Section 3.1.

33. $(3, -1); x - y = 4$
$x + 2y = 1$

34. $(0, 2); x + 3y = 6$
$4x - y = -2$

35. $(-4, 0); 3x + 2y = -12$
$x = 4y$

36. $(-5, 2); x + y = -3$
$2x - y = -8$

Concept Extensions

37. Explain when a dashed boundary line should be used in the graph of an inequality.

38. Explain why, after the boundary line is sketched, we test a point on either side of this boundary in the original inequality.

Solve.

39. Chris-Craft manufactures boats out of Fiberglas and wood. Fiberglas hulls require 2 hours of work, whereas wood hulls require 4 hours of work. Employees work at most 40 hours a week. The following inequalities model these restrictions, where x represents the number of Fiberglas hulls produced and y represents the number of wood hulls produced.

$$x \ge 0$$
$$y \ge 0$$
$$2x + 4y \le 40$$

Graph the intersection of these inequalities.

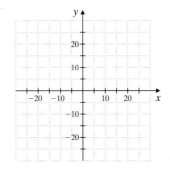

40. Rheem Abo-Zahrah decides that she will study at most 20 hours every week and that she must work at least 10 hours every week. Let x represent the hours studying and y represent the hours working. Write two inequalities that model this situation and graph their intersection.

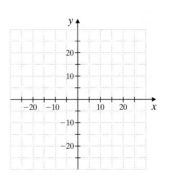

3.6 INTRODUCTION TO FUNCTIONS

Objectives

A Define Relation, Domain, and Range.

B Identify Functions.

C Use the Vertical Line Test for Functions.

D Use Function Notation.

E Graph a Linear Function.

Objective A Defining Relation, Domain, and Range

Equations in two variables, such as $y = 2x + 1$, describe **relations** between x-values and y-values. For example, if $x = 1$, then this equation describes how to find the y-value related to $x = 1$. In words, the equation $y = 2x + 1$ says that twice the x-value increased by 1 gives the corresponding y-value. The x-value of 1 corresponds to the y-value of $2(1) + 1 = 3$ for this equation, and we have the ordered pair $(1, 3)$. In other words, for the relationship (or relation) between x and y defined by $y = 2x + 1$, the x-value 1 is paired with the y-value 3.

There are other ways of describing relations or correspondences between two numbers or, in general, a set of first components (sometimes called the set of *inputs*) and a set of second components (sometimes called the set of *outputs*). For example,

First Set: Input	Correspondence	Second Set: Output
People in a certain city	Each person's age	The set of nonnegative integers

A few examples of ordered pairs from this relation might be (Ana, 4); (Bob, 36); (Trey, 21); and so on.

Below are just a few other ways of describing relations between two sets and the ordered pairs that they generate.

First Set: Input   **Second Set: Output**

Correspondence

Ordered Pairs

$(a, 3), (c, 3), (e, 1)$

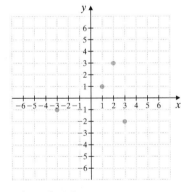

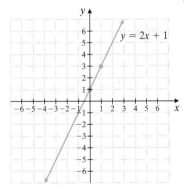

Ordered Pairs

$(-3, -1), (1, 1), (2, 3), (3, -2)$

Some Ordered Pairs

$(1, 3), (0, 1),$ and so on

Relation, Domain, and Range

A **relation** is a set of ordered pairs.

The **domain** of the relation is the set of all first components of the ordered pairs.

The **range** of the relation is the set of all second components of the ordered pairs.

209

For example, the domain for our middle relation on the previous page is $\{a, c, e\}$ and the range is $\{1, 3\}$. Notice that the range does not include the element 2 of the second set. This is because no element of the first set is assigned to this element. If a relation is defined in terms of x- and y-values, we will agree that the domain corresponds to x-values and that the range corresponds to y-values that are paired with x-values.

> **Helpful Hint**
>
> Remember that the range only includes elements that are paired with domain values. For
>
> | First Set: Input | Correspondence | Second Set: Output |
> |---|---|---|
>
> 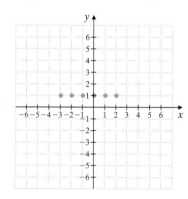, the range is $\{a\}$.

PRACTICE PROBLEMS 1–3

Determine the domain and range of each relation.
1. $\{(1, 6), (2, 8), (0, 3), (0, -2)\}$
2.

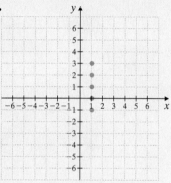

3.

Input: States	Output: Number of Congressional Representatives

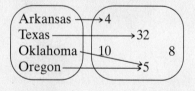

EXAMPLES Determine the domain and range of each relation.

1. $\{(2, 3), (2, 4), (0, -1), (3, -1)\}$
   The domain is the set of all first coordinates of the ordered pairs, $\{2, 0, 3\}$. The range is the set of all second coordinates, $\{3, 4, -1\}$.

2.

> **Helpful Hint**
> Equivalent domain or range elements that occur more than once need only to be listed once.

The relation is $\{(-3, 1), (-2, 1), (-1, 1), (0, 1), (1, 1), (2, 1)\}$.
The domain is $\{-3, -2, -1, 0, 1, 2\}$.
The range is $\{1\}$.

3.

Input: Cities	Output: Population (in thousands)

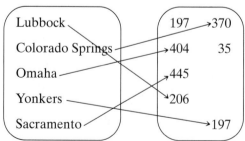

The domain is the set of inputs {Lubbock, Colorado Springs, Omaha, Yonkers, Sacramento}. The range is the numbers in the set of outputs that correspond to elements in the set of inputs {370, 404, 445, 206, 197}.

Work Practice Problems 1–3

Answers
1. domain: $\{1, 2, 0\}$; range: $\{6, 8, 3, -2\}$,
2. domain: $\{1\}$; range: $\{-1, 0, 1, 2, 3\}$,
3. domain: {Arkansas, Texas, Oklahoma, Oregon}; range: $\{4, 5, 32\}$

Objective B Identifying Functions

Now we consider a special kind of relation called a *function*.

> ### Function
>
> A **function** is a relation in which each first component in the ordered pairs corresponds to *exactly one* second component.

EXAMPLES Determine whether each relation is also a function.

4. $\{(-2, 5), (2, 7), (-3, 5), (9, 9)\}$

   Although the ordered pairs $(-2, 5)$ and $(-3, 5)$ have the same y-value, each x-value is assigned to only one y-value, so this set of ordered pairs is a function.

5.

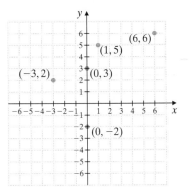

The x-value 0 is assigned to two y-values, -2 and 3, in this graph, so this relation is not a function.

6.

Input	Correspondence	Output
People in a certain city	Each person's age	The set of nonnegative integers

This relation is a function because although two different people may have the same age, each person has only one age. This means that each element in the first set is assigned to only one element in the second set.

Work Practice Problems 4–6

✔ **Concept Check** Explain why a function can contain both the ordered pairs $(1, 3)$ and $(2, 3)$ but not both $(3, 1)$ and $(3, 2)$.

Recall that an equation such as $y = 2x + 1$ is a relation since this equation defines a set of ordered pair solutions.

EXAMPLE 7 Determine whether the relation $y = 2x + 1$ is also a function.

Solution: The relation $y = 2x + 1$ is a function if each x-value corresponds to just one y-value. For each x-value substituted in the equation $y = 2x + 1$, the multiplication and addition performed gives a single result, so only one y-value will be associated with each x-value. Thus, $y = 2x + 1$ is a function.

Work Practice Problem 7

EXAMPLE 8 Determine whether the relation $x = y^2$ is also a function.

Solution: In $x = y^2$, if $y = 3$, then $x = 9$. Also, if $y = -3$, then $x = 9$. In other words, we have the ordered pairs $(9, 3)$ and $(9, -3)$. Since the x-value 9 corresponds to two y-values, 3 and -3, $x = y^2$ is not a function.

Work Practice Problem 8

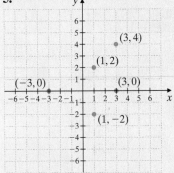

PRACTICE PROBLEMS 9–13

Use the vertical line test to determine which are graphs of functions.

9.

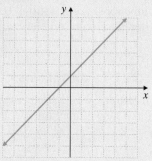

10.

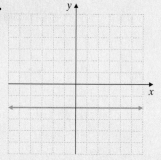

11.

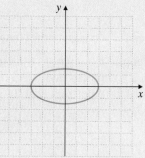

12.

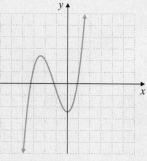

13.

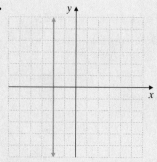

Answers

9. function,   **10.** function,
11. not a function,   **12.** function,
13. not a function

Objective **C** Using the Vertical Line Test

As we have seen, not all relations are functions. Consider the graphs of $y = 2x + 1$ and $x = y^2$ shown next. On the graph of $y = 2x + 1$, notice that each x-value corresponds to only one y-value. Recall from Example 7 that $y = 2x + 1$ is a function.

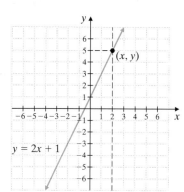

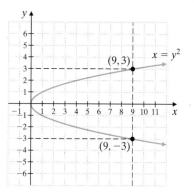

On the graph of $x = y^2$, the x-value 9, for example, corresponds to two y-values, 3 and –3, as shown by the vertical line. Recall from Example 8 that $x = y^2$ is not a function.

Graphs can be used to help determine whether a relation is also a function by the following **vertical line test.**

Vertical Line Test

If no vertical line can be drawn so that it intersects a graph more than once, the graph is the graph of a function. If such a line can be drawn, the graph is not that of a function.

EXAMPLES Use the vertical line test to determine which are graphs of functions.

9.

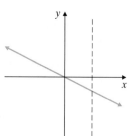

10.

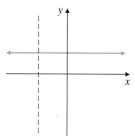

11.

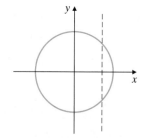

This is the graph of a function since no vertical line will intersect this graph more than once.

This is the graph of a function.

This is not the graph of a function. Note that vertical lines can be drawn that intersect the graph in two points.

12.

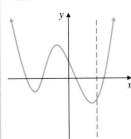

13.

This is the graph of a function.

This is not the graph of a function. A vertical line can be drawn that intersects this line at every point.

■ **Work Practice Problems 9–13**

✔**Concept Check** Determine which equations represent functions. Explain your answer.

a. $y = 14$ **b.** $x = -5$ **c.** $x + y = 6$

Objective D Using Function Notation

Many times letters such as $f, g,$ and h are used to name functions. To denote that y is a function of x, we can write

$$y = f(x)$$

This means that **y is a function of x** or that y *depends on x.* For this reason, y is called the **dependent variable** and x the **independent variable.** The notation $f(x)$ is read "f of x" and is called **function notation.**

For example, to use function notation with the function $y = 4x + 3$, we write $f(x) = 4x + 3$. The notation $f(1)$ means to replace x with 1 and find the resulting y- or function value. Since

$$f(x) = 4x + 3$$

then

$$f(1) = 4(1) + 3 = 7$$

This means that when $x = 1$, y or $f(x) = 7$. The corresponding ordered pair is $(1, 7)$. Here, the input is 1 and the output is $f(1)$ or 7. Now let's find $f(2), f(0),$ and $f(-1)$.

$f(x) = 4x + 3$	$f(x) = 4x + 3$	$f(x) = 4x + 3$
$f(2) = 4(2) + 3$	$f(0) = 4(0) + 3$	$f(-1) = 4(-1) + 3$
$= 8 + 3$	$= 0 + 3$	$= -4 + 3$
$= 11$	$= 3$	$= -1$

Ordered Pairs:

$(2, 11)$ $(0, 3)$ $(-1, -1)$

Helpful Hint

Note that $f(x)$ is a special symbol in mathematics used to denote a function. The symbol $f(x)$ is read "f of x." It does *not* mean $f \cdot x$ (f times x).

PRACTICE PROBLEMS 14–17

Find each function value.

14. If $g(x) = 4x + 5$, find $g(0)$.

15. If $g(x) = 4x + 5$, find $g(-5)$.

16. If $f(x) = 3x^2 - x + 2$, find $f(2)$.

17. If $f(x) = 3x^2 - x + 2$, find $f(-1)$.

EXAMPLES Find each function value.

14. If $g(x) = 3x - 2$, find $g(1)$.
$$g(1) = 3(1) - 2 = 1$$

15. If $g(x) = 3x - 2$, find $g(0)$.
$$g(0) = 3(0) - 2 = -2$$

16. If $f(x) = 7x^2 - 3x + 1$, find $f(1)$.
$$f(1) = 7(1)^2 - 3(1) + 1 = 5$$

17. If $f(x) = 7x^2 - 3x + 1$, find $f(-2)$.
$$f(-2) = 7(-2)^2 - 3(-2) + 1 = 35$$

▣ **Work Practice Problems 14–17**

✔**Concept Check** Suppose $y = f(x)$ and we are told that $f(3) = 9$. Which is not true?

a. When $x = 3$, $y = 9$.

b. A possible function is $f(x) = x^2$.

c. A point on the graph of the function is $(3, 9)$.

d. A possible function is $f(x) = 2x + 4$.

Objective ▣ Graphing Linear Functions

Recall that the graph of a linear equation in two variables is a line, and a line that is not vertical will always pass the vertical line test. Thus, *all linear equations are functions except those whose graphs are vertical lines.* We call such functions *linear functions.*

> ### Linear Function
>
> A **linear function** is a function that can be written in the form
> $$f(x) = mx + b$$

EXAMPLE 18 Graph the function $f(x) = 2x + 1$.

Solution: Since $y = f(x)$, we can replace $f(x)$ with y and graph as usual. The graph of $y = 2x + 1$ has slope 2 and y-intercept $(0, 1)$. Its graph is shown.

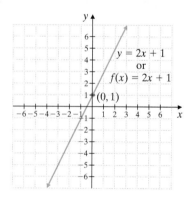

▣ **Work Practice Problem 18**

PRACTICE PROBLEM 18

Graph the function $f(x) = 3x - 2$.

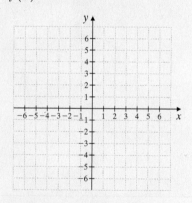

Answers

14. $g(0) = 5$, **15.** $g(-5) = -15$,
16. $f(2) = 12$, **17.** $f(-1) = 6$,
18.

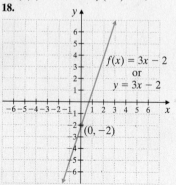

✔ **Concept Check Answer**

d

3.6 EXERCISE SET

 Student Solutions Manual   PH Math/Tutor Center CD/Video for Review Math XL MathXL®   MyMathLab MyMathLab

Objectives Ⓐ Ⓑ **Mixed Practice** *Find the domain and the range of each relation. Also determine whether the relation is a function. See Examples 1 through 6.*

1. $\{(-1, 7), (0, 6), (-2, 2), (5, 6)\}$

2. $\{(4, 9), (-4, 9), (2, 3), (10, -5)\}$

3. $\{(-2, 4), (6, 4), (-2, -3), (-7, -8)\}$

4. $\{(6, 6), (5, 6), (5, -2), (7, 6)\}$

 5. $\{(1, 1), (1, 2), (1, 3), (1, 4)\}$

6. $\{(1, 1), (2, 1), (3, 1), (4, 1)\}$

7. $\left\{ \left(\frac{3}{2}, \frac{1}{2}\right), \left(1\frac{1}{2}, -7\right), \left(0, \frac{4}{5}\right) \right\}$

8. $\{(\pi, 0), (0, \pi), (-2, 4), (4, -2)\}$

9. $\{(-3, -3), (0, 0), (3, 3)\}$

10. $\left\{ \left(\frac{1}{2}, \frac{1}{4}\right), \left(0, \frac{7}{8}\right), (0.5, \pi) \right\}$

11.

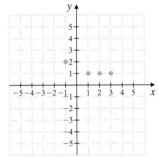

12.

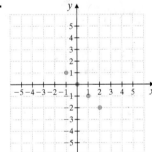

13. Input: **Output:**
    States   **Number of Congressional**
                      Representatives

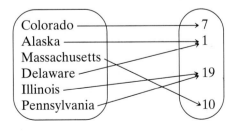

14. Input: **Output:**
    Animal   **Average Life Span**
                      (in years)

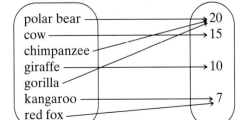

15. Input: Degrees Fahrenheit   **Output:** Degrees Celsius

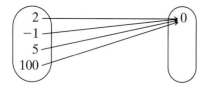

16. Input: Words   **Output:** Number of Letters

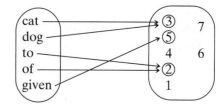

17. Input: **Output:**

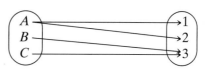

18. Input: **Output:**

Determine whether each relation is a function. See Examples 4 through 6.

19. First Set: Input   **Correspondence**   **Second Set: Output**

Class of algebra students   Grade average   Set of nonnegative numbers

20. First Set: Input   **Correspondence**   **Second Set: Output**

People in New Orleans (population 500,000)   Birthdate   Days of the year

Determine whether each relation is also a function. See Examples 7 and 8.

21. $y = x + 1$

22. $y = x - 1$

23. $x = 2y^2$

24. $y = x^2$

25. $y - x = 7$

26. $2x - 3y = 9$

Objective *Use the vertical line test to determine whether each graph is the graph of a function. See Examples 9 through 13.*

27.

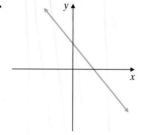

28.

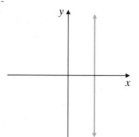

29.

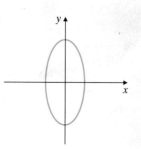

30.
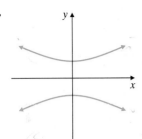

31.
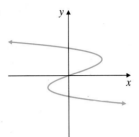

32.
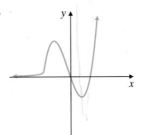

33.
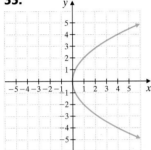

34.
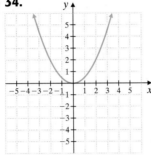

35.
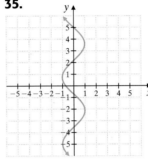

36.
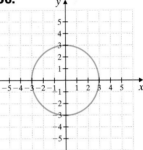

37.
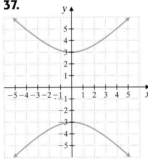

38.
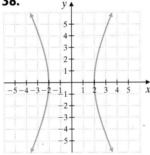

39.
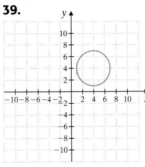

40.
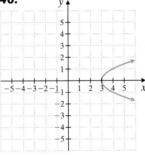

41.
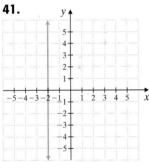

42.
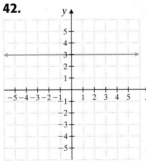

43.
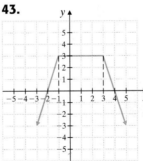

44.
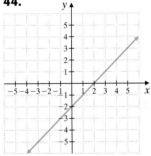

Objective D *If* $f(x) = 3x + 3$, $g(x) = 4x^2 - 6x + 3$, *and* $h(x) = 5x^2 - 7$, *find each function value.*
See Examples 14 through 17.

45. $f(4)$

46. $f(-1)$

47. $h(-3)$

48. $h(0)$

49. $g(2)$

50. $g(1)$

51. $g(0)$

52. $h(-2)$

For each function, find the indicated values. See Examples 14 through 17.

53. $f(x) = \frac{1}{2}x;$

   a. $f(0)$
   b. $f(2)$
   c. $f(-2)$

54. $g(x) = -\frac{1}{3}x;$

   a. $g(0)$
   b. $g(-1)$
   c. $g(3)$

55. $f(x) = -5;$

   a. $f(2)$
   b. $f(0)$
   c. $f(606)$

56. $h(x) = 7;$

   a. $h(7)$
   b. $h(542)$
   c. $h\left(-\frac{3}{4}\right)$

The function $A(r) = \pi r^2$ may be used to find the area of a circle if we are given its radius. Use this function to answer Exercises 57 and 58.

△ **57.** Find the area of a circle whose radius is 5 centimeters. (Do not approximate π.)

△ **58.** Find the area of a circular garden whose radius is 8 feet. (Do not approximate π.)

The function $V(x) = x^3$ may be used to find the volume of a cube if we are given the length x of a side. Use this function to answer Exercises 59 and 60.

△ **59.** Find the volume of a cube whose side is 14 inches.

△ **60.** Find the volume of a die whose side is 1.7 centimeters.

Forensic scientists use the following functions to find the height of a woman if they are given the height of her femur bone (f) or her tibia bone (t) in centimeters.

$$H(f) = 2.59f + 47.24$$
$$H(t) = 2.72t + 61.28$$

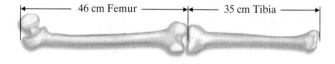

Use these functions to answer Exercises 61 and 62.

61. Find the height of a woman whose femur measures 46 centimeters.

62. Find the height of a woman whose tibia measures 35 centimeters.

The dosage in milligrams D of Ivermectin, a heartworm preventive, for a dog who weighs x pounds is given by

$$D(x) = \frac{136}{25}x$$

Use this function to answer Exercises 63 and 64.

63. Find the proper dosage for a dog that weighs 30 pounds.

64. Find the proper dosage for a dog that weighs 50 pounds.

65. The per capita consumption (in pounds) of all poultry in the United States is given by the function $C(x) = 2.28x + 94.86$, where x is the number of years since 2001. (*Source:* Based on actual and estimated data from the Economic Research Service, U.S. Dept. of Agriculture, 2001–2005)

   a. Find and interpret $C(2)$.
   b. Estimate the per capita consumption of all poultry in the United States in 2007.

66. The amount of money (in billions of dollars) spent by U.S. biotechnology companies on research and development annually is represented by the function $R(x) = 1.84x + 8.57$, where x is the number of years since 1997. (*Source:* Based on data from The Biotechnology Industry Organization)

   a. Find and interpret $R(7)$.
   b. Estimate the amount of money spent on biotechnology research and development in 2000.

3.7 FIND DOMAINS AND RANGES FROM GRAPHS AND GRAPHING PIECEWISE-DEFINED FUNCTIONS

Objectives

A Find the Domain and Range from a Graph.

B Graph Piecewise-Defined Functions.

Objective A Finding the Domain and Range from a Graph

Recall from Section 3.6 that the

domain of a relation is the set of all first components of the ordered pairs of the relation and the

range of a relation is the set of all second components of the ordered pairs of the relation.

In this section we use the graph of a relation to find its domain and range. Let's use interval notation to write these domains and ranges. Remember, we use a parenthesis to indicate that a number is not part of the domain and we use a bracket to indicate that a number if part of the domain. Of course, as usual, parentheses are placed about infinity symbols indicating that we approach but never reach infinity.

To find the domain of a function (or relation) from its graph, recall that on the rectangular coordinate system, "domain" is the set of first components of the ordered pairs, so this means what x-values are graphed. Similarly, "range" is the set of second components of the ordered pairs, so this means what y-values are graphed.

EXAMPLES Find the domain and range of each relation.

1.

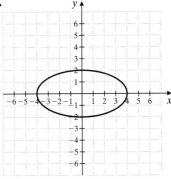

2.

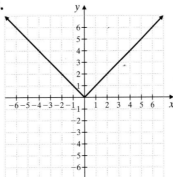

3.

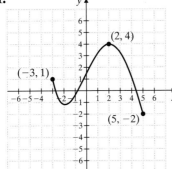

4.

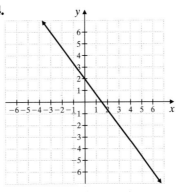

Continued on next page

PRACTICE PROBLEMS 1–4

Find the domain and range of each relation.

1.

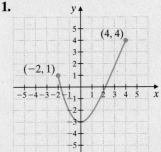

2.

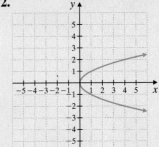

3.

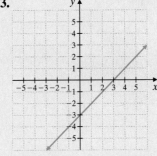

Answers

1. domain: $[-2, 4]$; range: $[-3, 4]$,

2. domain: $[0, \infty)$; range: $(-\infty, \infty)$,

3. domain: $(-\infty, \infty)$; range: $(-\infty, \infty)$,

4.

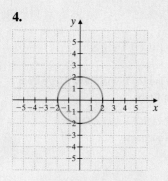

Solution: Notice that the graphs for Examples 1, 2, and 4 are graphs of functions because each passes the vertical line test.

1.

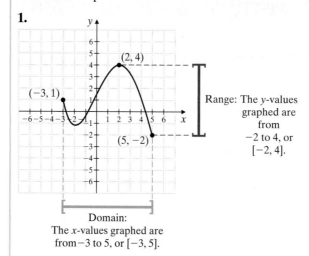

Range: The y-values graphed are from −2 to 4, or [−2, 4].

Domain:
The x-values graphed are from −3 to 5, or [−3, 5].

2.

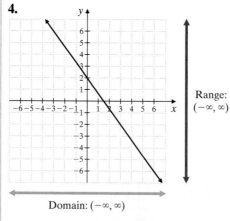

Range: [0, ∞)

Domain: (−∞, ∞)

3.

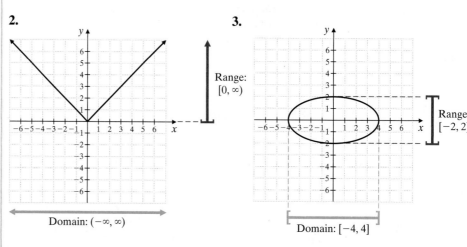

Range: [−2, 2]

Domain: [−4, 4]

4.

Range: (−∞, ∞)

Domain: (−∞, ∞)

■ **Work Practice Problems 1–4**

Objective B Graphing Piecewise-Defined Functions

In the last section we graphed functions. There are many special functions. In fact, sometimes a function is defined by two or more expressions. The equation to use depends upon the value of x. Before we actually graph these piecewise-defined functions, let's practice finding function values.

Answer
4. domain: [−2, 2]; range: [−2, 2]

EXAMPLE 5 Evaluate $f(2), f(-6)$, and $f(0)$ for the function

$$f(x) = \begin{cases} 2x + 3 & \text{if } x \le 0 \\ -x - 1 & \text{if } x > 0 \end{cases}$$

Then write your results in ordered-pair form.

Solution: Take a moment and study this function. It is a single function defined by two expressions depending on the value of x. From above, if $x \le 0$, use $f(x) = 2x + 3$. If $x > 0$, use $f(x) = 3x - 1$. Thus

$f(2) = -(2) - 1$
    $= -3$ since $2 > 0$
$f(2) = -3$
Ordered pairs: $(2, -3)$

$f(-6) = 2(-6) + 3$
    $= -9$ since $-6 \le 0$
$f(-6) = -9$
    $(-6, -9)$

$f(0) = 2(0) + 3$
    $= 3$ since $0 \le 0$
$f(0) = 3$
    $(0, 3)$

◾ **Work Practice Problem 5**

Now, let's graph a piecewise-defined function.

EXAMPLE 6 Graph $f(x) = \begin{cases} 2x + 3 & \text{if } x \le 0 \\ -x - 1 & \text{if } x > 0 \end{cases}$

Solution: Let's graph each piece.

If $x \le 0$,
$f(x) = 2x + 3$

Values ≤ 0
x	$f(x) = 2x + 3$
0	3 Closed circle
-1	1
-2	-1

If $x > 0$,
$f(x) = -x - 1$

Values > 0
x	$f(x) = -x - 1$
1	-2
2	-3
3	-4

The graph of the first part of $f(x)$ listed will look like a ray with a closed-circle endpoint at $(0, 3)$. The graph of the second part of $f(x)$ listed will look like a ray with an open-circle endpoint. To find the exact location of the open-circle endpoint, use $f(x) = -x - 1$ and find $f(0)$. Since $f(0) = -0 - 1 = -1$, we graph the second table and place an open circle at $(0, -1)$.

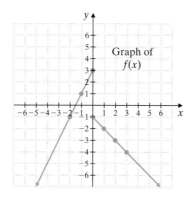

Graph of $f(x)$

Notice that this graph is the graph of a function because it passes the vertical line test. The domain of this function is $(-\infty, \infty)$ and the range is $(-\infty, 3]$.

◾ **Work Practice Problem 6**

In the next section, we shall graph piecewise-defined functions whose pieces are not necessarily pieces of lines.

PRACTICE PROBLEM 5

Evaluate $f(-4), f(3)$, and $f(0)$ for the function

$$f(x) = \begin{cases} 3x + 4 & \text{if } x < 0 \\ -x + 2 & \text{if } x \ge 0 \end{cases}$$

Then write your results in ordered-pair solution form.

PRACTICE PROBLEM 6

Graph

$$f(x) = \begin{cases} 3x + 4 & \text{if } x < 0 \\ -x + 2 & \text{if } x \ge 0 \end{cases}$$

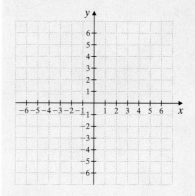

Answers

5. $f(-4) = -8$; $f(3) = -1$; $f(0) = 2$; $(-4, -8); (3, -1); (0, 2)$,

6.

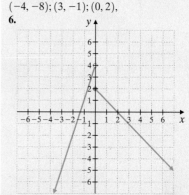

3.7 EXERCISE SET

Objective **A** *Find the domain and the range of each relation. See Examples 1 through 4.*

1.

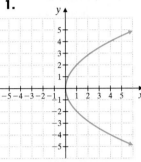

2.

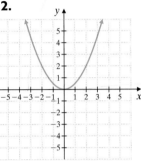

3.

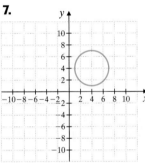

4.

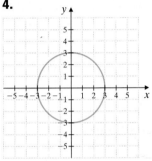

5.

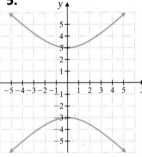

6.

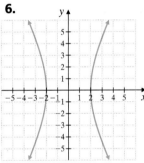

7.

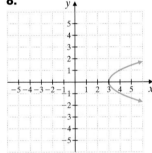

8.

9.

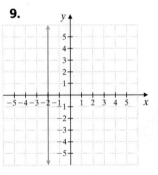

10.

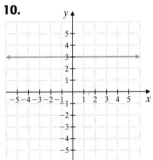

11.

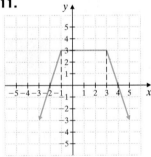

12.

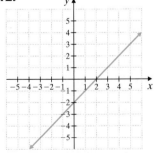

13.

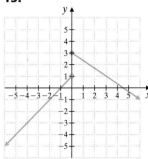

14.

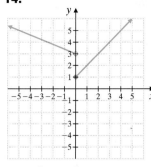

15.

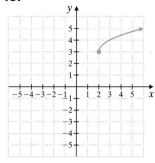

16.

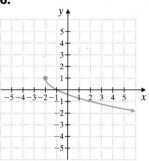

Objective **B** *Graph each piecewise-defined function. See Examples 5 and 6.*

17. $f(x) = \begin{cases} 2x & \text{if} \quad x < 0 \\ x + 1 & \text{if} \quad x \geq 0 \end{cases}$

18. $f(x) = \begin{cases} 3x & \text{if} \quad x < 0 \\ x + 2 & \text{if} \quad x \geq 0 \end{cases}$

19. $f(x) = \begin{cases} 4x + 5 & \text{if} \quad x \leq 0 \\ \dfrac{1}{4}x + 2 & \text{if} \quad x > 0 \end{cases}$

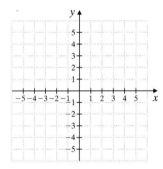

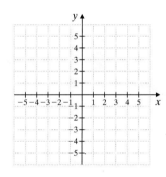

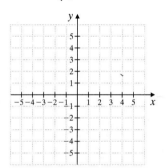

20. $f(x) = \begin{cases} 5x + 4 & \text{if } x \leq 0 \\ \dfrac{1}{3}x - 1 & \text{if } x > 0 \end{cases}$

21. $g(x) = \begin{cases} -x & \text{if} \quad x \leq 1 \\ 2x + 1 & \text{if} \quad x > 1 \end{cases}$

22. $g(x) = \begin{cases} 3x - 1 & \text{if} \quad x \leq 2 \\ -x & \text{if} \quad x > 2 \end{cases}$

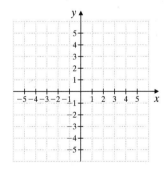

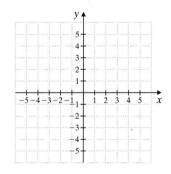

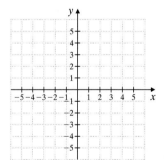

23. $f(x) = \begin{cases} 5 & \text{if} \quad x < -2 \\ 3 & \text{if} \quad x \geq -2 \end{cases}$

24. $f(x) = \begin{cases} 4 & \text{if} \quad x < -3 \\ -2 & \text{if} \quad x \geq -3 \end{cases}$

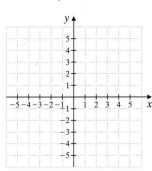

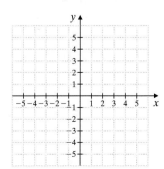

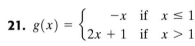

Objectives **A** **B** **Mixed Practice** *Graph each piecewise-defined function. Use the graph to determine the domain and range of the function. See Examples 1 through 6.*

25. $f(x) = \begin{cases} -2x & \text{if } x \le 0 \\ 2x + 1 & \text{if } x > 0 \end{cases}$

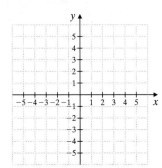

26. $g(x) = \begin{cases} -3x & \text{if } x \le 0 \\ 3x + 2 & \text{if } x > 0 \end{cases}$

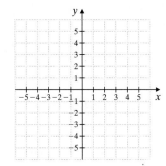

27. $h(x) = \begin{cases} 5x - 5 & \text{if } x < 2 \\ -x + 3 & \text{if } x \ge 2 \end{cases}$

28. $f(x) = \begin{cases} 4x - 4 & \text{if } x < 2 \\ -x + 1 & \text{if } x \ge 2 \end{cases}$

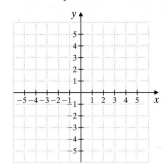

29. $f(x) = \begin{cases} x + 3 & \text{if } x < -1 \\ -2x + 4 & \text{if } x \ge -1 \end{cases}$

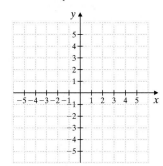

30. $h(x) = \begin{cases} x + 2 & \text{if } x < 1 \\ 2x + 1 & \text{if } x \ge 1 \end{cases}$

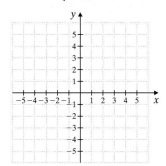

31. $g(x) = \begin{cases} -2 & \text{if } x \le 0 \\ -4 & \text{if } x \ge 1 \end{cases}$

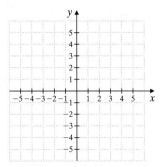

32. $f(x) = \begin{cases} -1 & \text{if } x \le 0 \\ -3 & \text{if } x \ge 2 \end{cases}$

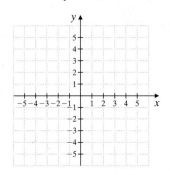

Review

Match each equation with its graph.

33. $y = -1$

A

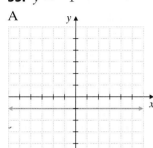

34. $x = -1$

B

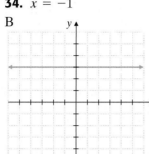

35. $x = 3$

C

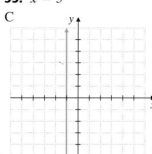

36. $y = 3$

D

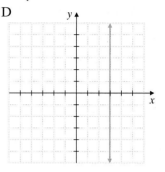

Concept Extensions

37. Draw a graph whose domain is $(-\infty, 5]$ and whose range is $[2, \infty)$.

38. In your own words, describe how to graph a piecewise-defined function.

39. Graph: $f(x) = \begin{cases} -\dfrac{1}{2}x & \text{if } x \le 0 \\ x + 1 & \text{if } 0 < x \le 2 \\ 2x - 1 & \text{if } x > 2 \end{cases}$

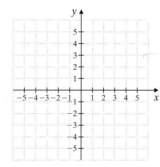

3.8 SHIFTING AND REFLECTING GRAPHS OF FUNCTIONS

In this section, we take common graphs and learn how more complicated graphs are actually formed by shifting and reflecting these common graphs. These shifts and reflections are called transformations, and it is possible to combine transformations. A knowledge of these transformations will help you simplify future graphs.

Objective **A** Graphing Common Equations

Let's begin with the graphs of four common functions.

First, let's graph the linear function $f(x) = x$, or $y = x$. Ordered-pair solutions of these graphs consist of ordered pairs whose x- and y-values are the same.

x	y or $f(x) = x$
-3	-3
0	0
1	1
4	4

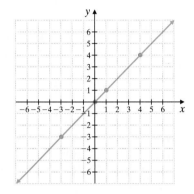

Next, let's graph the nonlinear function $f(x) = x^2$ or $y = x^2$.

This equation is not linear because the x^2 term does not allow us to write it in the form $Ax + By = C$. Its graph is not a line. We begin by finding ordered pair solutions. Because this graph is solved for $f(x)$, or y, we choose x-values and find corresponding $f(x)$, or y-values.

If $x = -3$, then $y = (-3)^2$, or 9.

If $x = -2$, then $y = (-2)^2$, or 4.

If $x = -1$, then $y = (-1)^2$, or 1.

If $x = 0$, then $y = 0^2$, or 0.

If $x = 1$, then $y = 1^2$, or 1.

If $x = 2$, then $y = 2^2$, or 4.

If $x = 3$, then $y = 3^2$, or 9.

x	$f(x)$ or y
-3	9
-2	4
-1	1
0	0
1	1
2	4
3	9

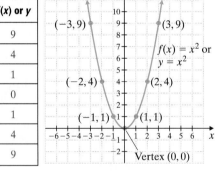

Study the table a moment and look for patterns. Notice that the ordered pair solution $(0, 0)$ contains the smallest y-value because any other x-value squared will give a positive result. This means that the point $(0, 0)$ will be the lowest point on the graph. Also notice that all other y-values correspond to two different x-values. For example, $3^2 = 9$ and also $(-3)^2 = 9$. This means that the graph will be a mirror image of itself across the y-axis. Connect the plotted points with a smooth curve to sketch its graph.

This curve is given a special name, a **parabola.** We will study more about parabolas in later chapters.

Next, let's graph another nonlinear function $f(x) = |x|$ or $y = |x|$.

This is not a linear equation since it cannot be written in the form $Ax + By = C$. Its graph is not a line. Because we do not know the shape of this graph, we find many ordered pair solutions. We will choose x-values and substitute to find corresponding y-values.

If $x = -3$, then $y = |-3|$, or 3.
If $x = -2$, then $y = |-2|$, or 2.
If $x = -1$, then $y = |-1|$, or 1.
If $x = 0$, then $y = |0|$, or 0.
If $x = 1$, then $y = |1|$, or 1.
If $x = 2$, then $y = |2|$, or 2.
If $x = 3$, then $y = |3|$, or 3.

x	y
-3	3
-2	2
-1	1
0	0
1	1
2	2
3	3

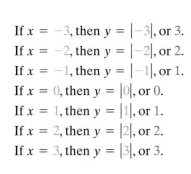

Again, study the table of values for a moment and notice any patterns.

From the plotted ordered pairs, we see that the graph of this absolute value equation is V-shaped.

Finally, a fourth common function, $f(x) = \sqrt{x}$ or $y = \sqrt{x}$. For this graph, you need to recall basic facts about square roots and use your calculator to approximate some square roots to help locate points. Recall also that the square root of a negative number is not a real number, so be careful when finding your domain.

Now let's graph the square root function $f(x) = \sqrt{x}$, or $y = \sqrt{x}$.

To graph, we identify the domain, evaluate the function for several values of x, plot the resulting points, and connect the points with a smooth curve. Since $\sqrt{x}$ represents the nonnegative square root of x, the domain of this function is the set of all nonnegative numbers, $\{x | x \geq 0\}$, or $[0, \infty)$. We have approximated $\sqrt{3}$ below to help us locate the point corresponding to $(3, \sqrt{3})$.

If $x = 0$, then $y = \sqrt{0}$, or 0.
If $x = 1$, then $y = \sqrt{1}$, or 1.
If $x = 3$, then $y = \sqrt{3}$, or 1.7.
If $x = 4$, then $y = \sqrt{4}$, or 2.
If $x = 9$, then $y = \sqrt{9}$, or 3.

x	$f(x) = \sqrt{x}$
0	0
1	1
3	$\sqrt{3} \approx 1.7$
4	2
9	3

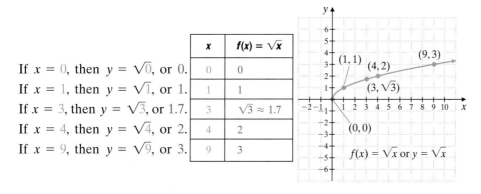

Notice that the graph of this function passes the vertical line test, as expected.

Below is a summary of our four common graphs. Take a moment and study these graphs. Your success in the rest of this section depends on your knowledge of these graphs.

Common Graphs

$f(x) = x$

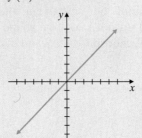

$f(x) = x^2$

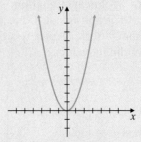

$f(x) = \sqrt{x}$

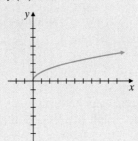

$f(x) = |x|$

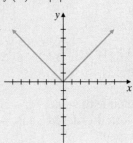

1.

2.

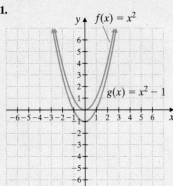

3.

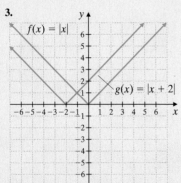

Objective B **Vertical and Horizontal Shifting**

Your knowledge of the slope-intercept form, $f(x) = mx + b$, will help you understand simple shifting of transformations such as vertical shifts. For example, what is the difference between the graphs of $f(x) = x$ and $g(x) = x + 3$?

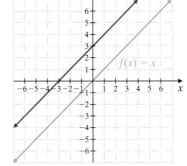

$$f(x) = x \qquad\qquad g(x) = x + 3$$
slope, $m = 1$ slope, $m = 1$
y-intercept is $(0, 0)$ y-intercept is $(0, 3)$

Notice that the graph of $g(x) = x + 3$ is the same as the graph of $f(x) = x$, but moved upward 3 units. This is an example of a **vertical shift** and is true for graphs in general.

Vertical Shifts (Upward and Downward)
Let k be a Positive Number

Graph of	Same As	Moved
$g(x) = f(x) + k$	$f(x)$	k units upward
$g(x) = f(x) - k$	$f(x)$	k units downward

EXAMPLES Without plotting points, sketch the graph of each pair of functions on the same set of axes.

1. $f(x) = x^2$ and $g(x) = x^2 + 2$

2. $f(x) = \sqrt{x}$ and $g(x) = \sqrt{x} - 3$

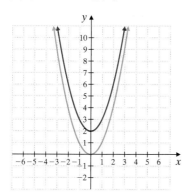

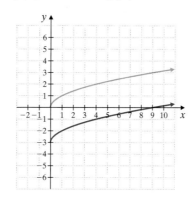

🔲 **Work Practice Problems 1–2**

A horizontal shift to the left or right may be slightly more difficult to understand. Let's graph $g(x) = |x - 2|$ and compare it with $f(x) = |x|$.

EXAMPLE 3 Without plotting points, sketch the graphs of $f(x) = |x|$ and $g(x) = |x - 2|$ on the same set of axes.

x	$f(x) = |x|$	$g(x) = |x - 2|$
−3	3	5
−2	2	4
−1	1	3
0	0	2
1	1	1
2	2	0
3	3	1

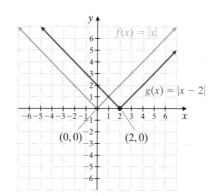

The graph of $g(x) = |x - 2|$ is the same as the graph of $f(x) = |x|$, but moved 2 units to the right. This is an example of a **horizontal shift** and is true for graphs in general.

Horizontal Shift (To the Left or Right)
Let h be a Positive Number

Graph of	Same as	Moved
$g(x) = f(x - h)$	$f(x)$	h units to the right
$g(x) = f(x + h)$	$f(x)$	h units to the left

Helpful Hint

Notice that $f(x - h)$ corresponds to a shift to the right and $f(x + h)$ corresponds to a shift to the left.

🔲 **Work Practice Problem 3**

Vertical and horizontal shifts can be combined.

PRACTICE PROBLEMS 1–2

Without plotting points, sketch the graphs of each pair of functions on the same set of axes.
1. $f(x) = x^2$ and $g(x) = x^2 - 1$

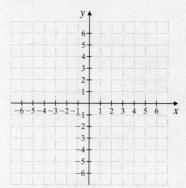

2. $f(x) = \sqrt{x}$ and
$g(x) = \sqrt{x} + 2$

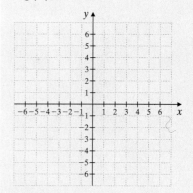

PRACTICE PROBLEM 3

Without plotting points, sketch the graphs of $f(x) = |x|$ and $g(x) = |x + 2|$ on the same set of axes.

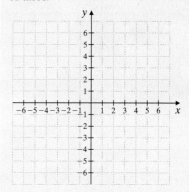

Answers
1–3. See page 230.

PRACTICE PROBLEM 4

Sketch the graphs of $f(x) = x^2$ and $g(x) = (x + 2)^2 - 1$ on the same set of axes.

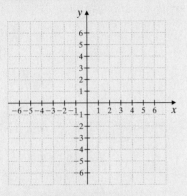

PRACTICE PROBLEM 5

Sketch the graph of $h(x) = -(x - 3)^2 + 2$.

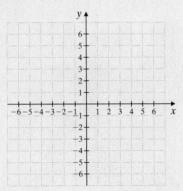

Answers

4.

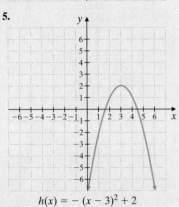

5.

$h(x) = -(x - 3)^2 + 2$

EXAMPLE 4 Sketch the graphs of $f(x) = x^2$ and $g(x) = (x - 2)^2 + 1$ on the same set of axes.

Solution: The graph of $g(x)$ is the same as the graph of $f(x)$ shifted 2 units to the right and 1 unit up.

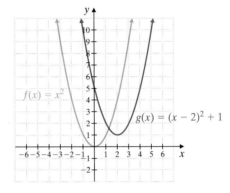

◉ **Work Practice Problem 4**

Objective C Reflecting Graphs

Another type of transformation is called a **reflection.** In this section, we will study reflections (mirror images) about the x-axis only. For example, take a moment and study these two graphs. The graph of $g(x) = -x^2$ can be found, as usual, by plotting points.

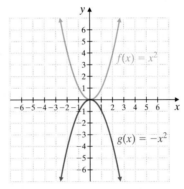

Reflection about the x-axis

The graph of $g(x) = -f(x)$ is the graph of $f(x)$ reflected about the x-axis.

EXAMPLE 5 Sketch the graph of $h(x) = -|x - 3| + 2$.

Solution: The graph of $h(x) = -|x - 3| + 2$ is the same as the graph of $f(x) = |x|$ reflected about the x-axis, then moved three units to the right and two units upward.

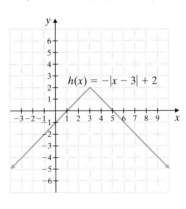

◉ **Work Practice Problem 5**

There are other transformations, such as stretching that won't be covered in this section. For a review of this transformation, see the Appendix.

Mental Math

Objective **A** *Match each equation with its graph.*

1. $y = \sqrt{x}$

2. $y = x^2$

3. $y = x$

4. $y = |x|$

a.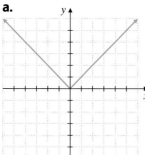

b.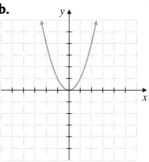

c.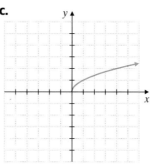

d.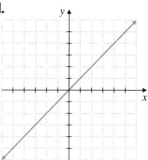

Objectives **A** **B** **Mixed Practice** *Sketch the graph of function. See Examples 1 through 4.*

1. $f(x) = |x| + 3$

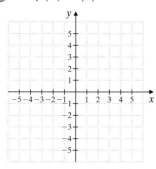

2. $f(x) = |x| - 2$

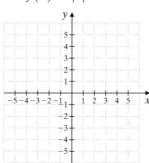

3. $f(x) = \sqrt{x} - 2$

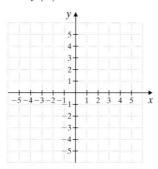

4. $f(x) = \sqrt{x} + 3$

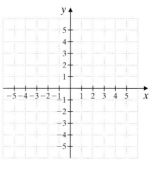

5. $f(x) = |x - 4|$

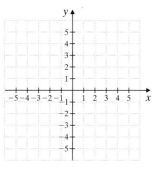

6. $f(x) = |x + 3|$

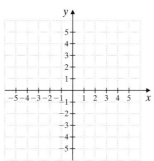

7. $f(x) = \sqrt{x + 2}$

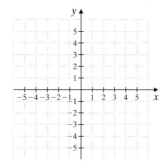

8. $f(x) = \sqrt{x - 2}$

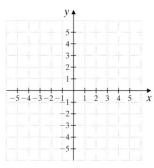

233

9. $y = (x - 4)^2$

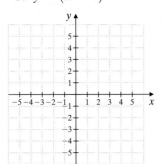

10. $y = (x + 4)^2$

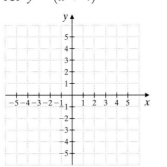

11. $f(x) = x^2 + 4$

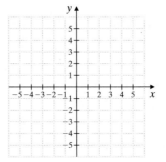

12. $f(x) = x^2 - 4$

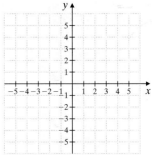

13. $f(x) = \sqrt{x - 2} + 3$

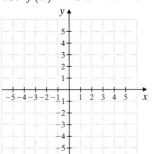

14. $f(x) = \sqrt{x - 1} + 3$

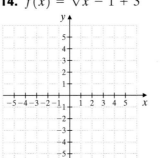

15. $f(x) = |x - 1| + 5$

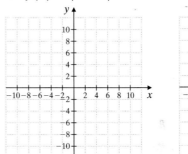

16. $f(x) = |x - 3| + 2$

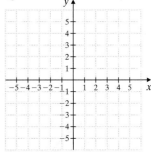

17. $f(x) = \sqrt{x + 1} + 1$

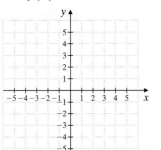

18. $f(x) = \sqrt{x + 3} + 2$

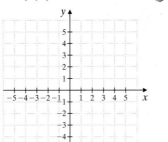

19. $f(x) = |x + 3| - 1$

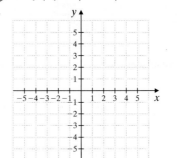

20. $f(x) = |x + 1| - 4$

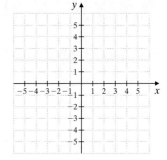

21. $g(x) = (x - 1)^2 - 1$

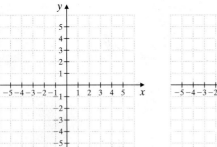

22. $h(x) = (x + 2)^2 + 2$

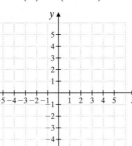

23. $f(x) = (x + 3)^2 - 2$

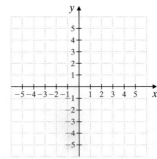

24. $f(x) = (x + 2)^2 + 4$

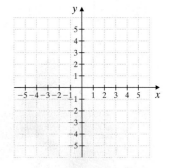

Objectives Ⓐ Ⓑ Ⓒ **Mixed Practice** *Sketch the graph of each function.*

25. $f(x) = -(x - 1)^2$

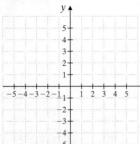

26. $g(x) = -(x + 2)^2$

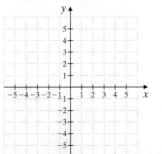

27. $h(x) = -\sqrt{x} + 3$

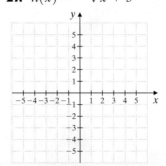

28. $f(x) = -\sqrt{x + 3}$

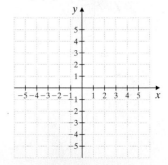

29. $h(x) = -|x + 2| + 3$ **30.** $g(x) = -|x + 1| + 1$ **31.** $f(x) = (x - 3) + 2$ **32.** $f(x) = (x - 1) + 4$

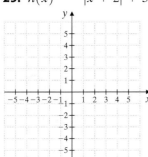   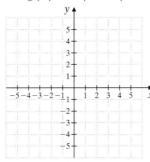   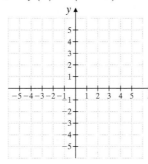   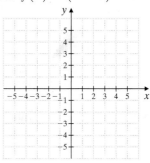

Review

Simplify. See Sections 1.5 through 1.7.

33. $-3x^4 \cdot 5x^4$ **34.** $-3x^4 + 5x^4$ **35.** $y^7 \cdot y^{11}$ **36.** $8(y^7 + y^{11})$

Concept Extensions

Mixed Practice (*Sections 3.7, 3.8*) *Write the domain and range of each graphed function in this section.*

37. Exercise 13   **38.** Exercise 14

39. Exercise 29   **40.** Exercise 30

Without graphing, find the domain of each function.

41. $f(x) = 5\sqrt{x - 20} + 1$ **42.** $g(x) = -3\sqrt{x + 5}$ **43.** $h(x) = 5|x - 20| + 1$

44. $f(x) = -3|x + 5.7|$ **45.** $g(x) = 9 - \sqrt{x + 103}$ **46.** $h(x) = \sqrt{x - 17} - 3$

Sketch the graph of each piecewise-defined function. Write the domain and range of each function.

47. $f(x) = \begin{cases} |x| & \text{if } x \le 0 \\ x^2 & \text{if } x > 0 \end{cases}$

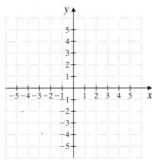

48. $f(x) = \begin{cases} x^2 & \text{if } x < 0 \\ \sqrt{x} & \text{if } x \ge 0 \end{cases}$

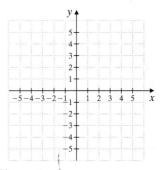

49. $g(x) = \begin{cases} |x - 2| & \text{if } x < 0 \\ -x^2 & \text{if } x \ge 0 \end{cases}$

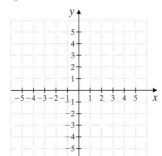

50. $g(x) = \begin{cases} -|x + 1| - 1 & \text{if } x < -2 \\ \sqrt{x + 2} - 4 & \text{if } x \ge -2 \end{cases}$

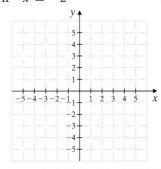

CHAPTER 3   Group Activity

Linear Modeling

As we saw in Sections 3.3 and 3.4, businesses often depend on equations that "closely fit" data. To *model* the data means to find an equation that describes the relationship between the paired data of two variables, such as time in years and profit. A model that accurately summarizes the relationship between two variables can be used to replace a potentially lengthy listing of the raw data. An accurate model might also be used to predict future trends by answering questions such as "If the trend seen in our company's performance in the last several years continues, what level of profit can we reasonably expect in 3 years?"

There are several ways to find a linear equation that models a set of data. If only two ordered pair data points are involved, an exact equation that contains both points can be found using the methods of Section 3.4. When more than two ordered pair data points are involved, it may be impossible to find a linear equation that contains all of the data points. In this case, the graph of the **best fit equation** should have a majority of the plotted ordered pair data points on the graph or close to it. In statistics, a technique called least squares regression is used to determine an equation that best fits a set of data. Various graphing utilities have built-in capabilities for finding an equation (called a regression equation) that best fits a set of ordered pair data points. Regression capabilities are often found with a graphing utility's statistics features.* A best fit equation can also be estimated using an algebraic method, which is outlined in the Group Activity below. In either case, a useful first step when finding a linear equation that models a set of data is creating a scatter diagram of the ordered pair data points to verify that a linear equation is an appropriate model.

Group Activity

Windows operating system is one of the well known products of the Microsoft Corporation. This company develops, manufactures, licenses and supports a wide range of software products for various computing devices. They also provide the MSN network of Internet products and services, and even the Xbox video game system. The table shows Microsoft's total revenues (in billions) for the years 1997 through 2004. Use the table along with your answers to the questions below to find a linear equation $y = mx + b$ that represents total revenue y (in billions) for Microsoft Corporation, where x represents the number of years after 1997.

Year	1997	1998	1999	2000	2001	2002
Total Microsoft Revenues (in billions)	20	23	25	28	32	37

(*Source:* Microsoft Corporation, Reuters Business Reports)

1. Create a scatter diagram of the paired data given in the table. Does a linear model seem appropriate for the data?

2. Use a straightedge to draw on your graph what appears to be the line that "best fits" the data you plotted.

3. Estimate the coordinates of two points that fall on your best fit line. Use these points to find the equation of the line that passes through both points.

4. Use this equation to find the value of y for $x = 11$. Interpret the meaning of this pair of data.

5. How could this equation be useful to accountants who work at Microsoft?

6. Compare your group's linear equation with other groups' equations. Are they the same or different? Explain why.

7. (Optional) Enter the data from the table into a graphing utility and use the linear regression feature to find a linear equation that models the data. Compare this equation with the one you found in Question 3. How are they alike or different?

8. (Optional) Using corporation annual reports or articles from magazines or newspapers, search for a set of business-related data that could be modeled with a linear equation. Explain how modeling this data could be useful to a business. Then find the best fit equation for the data.

*To find out more about using a graphing utility to find a regression equation, consult the user's manual for your graphing utility.

Chapter 3 Vocabulary Check

Fill in each blank with one of the words or phrases listed below.

relation	line	function	standard	parallel
slope-intercept	*x*	*y*	range	domain
point-slope	perpendicular	linear inequality	slope	linear function

1. A _____ is a set of ordered pairs.
2. The graph of every linear equation in two variables is a _____.
3. The statement $-x + 2y > 0$ is called a _____ in two variables.
4. _____ form of linear equation in two variables is $Ax + By = C$.
5. The _____ of a relation is the set of all second components of the ordered pairs of the relation.
6. _____ lines have the same slope and different *y*-intercepts.
7. _____ form of a linear equation in two variables is $y = mx + b$.
8. A _____ is a relation in which each first component in the ordered pairs corresponds to exactly one second component.
9. In the equation $y = 4x - 2$, the coefficient of *x* is the _____ of its corresponding graph.
10. Two lines are _____ if the product of their slopes is -1.
11. To find the *x*-intercept of a linear equation, let __ $= 0$ and solve for the other variable.
12. The _____ of a relation is the set of all first components of the ordered pairs of the relation.
13. A _____ is a function that can be written in the form $f(x) = mx + b$.
14. To find the *y*-intercept of a linear equation, let __ $= 0$ and solve for the other variable.
15. The equation $y - 8 = -5(x + 1)$ is written in _____ form.

Helpful Hint

Are you preparing for your test? Don't forget to take the Chapter 3 Test on page 250. Then check your answers at the back of the text and use the Chapter Test Prep Video CD to see the fully worked-out solutions to any of the exercises you want to review.

3 Chapter Highlights

DEFINITIONS AND CONCEPTS	**EXAMPLES**
Section 3.1 Graphing Linear Equations	

The **rectangular coordinate system,** or **Cartesian coordinate system,** consists of a vertical and a horizontal number line on a plane intersecting at their 0 coordinate. The vertical number line is called the **y-axis,** and the horizontal number line is called **x-axis.** The point of intersection of the axes is called the **origin.**

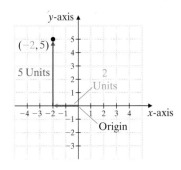

To **plot** or **graph** an ordered pair means to find its corresponding point on a rectangular coordinate system.

An ordered pair is a **solution** of an equation in two variables if replacing the variables by the corresponding coordinates results in a true statement.

Plot or graph the ordered pair $(-2, 5)$.
Start at the origin. Move 2 units to the left along the *x*-axis, then 5 units upward parallel to the *y*-axis.
Determine whether $(-2, 3)$ is a solution of $3x + 2y = 0$.
$$3(-2) + 2(3) = 0$$
$$-6 + 6 = 0$$
$$0 = 0 \quad \text{True}$$
$(-2, 3)$ is a solution.

continued

DEFINITIONS AND CONCEPTS	**EXAMPLES**

Section 3.1   Graphing Linear Equations (*continued*)

A **linear equation in two variables** is an equation that can be written in the form $Ax + By = C$, where A, B, and C are real numbers and A and B are not both 0. The form $Ax + By = C$ is called **standard form.**

$y = -2x + 5$, $x = 7$,
$y - 3 = 0$, $6x - 4y = 10$
$6x - 4y = 10$ is in standard form.

The **graph of a linear equation** in two variables is a line. To graph a linear equation in two variables, find three ordered pair solutions. Plot the solution points, and draw the line connecting the points.

Graph: $3x + y = -6$

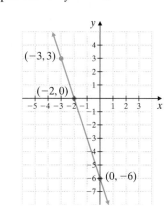

x	y
0	−6
−2	0
−3	3

To find an x-intercept, let $y = 0$ and solve for x.
To find a y-intercept, let $x = 0$ and solve for y.

The graph of $x = c$ is a vertical line with x-intercept c.
The graph of $y = c$ is a horizontal line with y-intercept c.

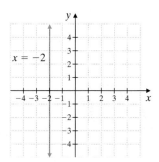

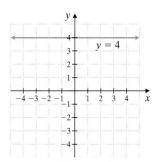

Section 3.2   The Slope of a Line

The **slope** m of the line through (x_1, y_1) and (x_2, y_2) is given by

$$m = \frac{y_2 - y_1}{x_2 - x_1}$$

as long as $x_2 \neq x_1$

Find the slope of the line through $(-1, 7)$ and $(-2, -3)$.

$$m = \frac{y_2 - y_1}{x_2 - x_1} = \frac{-3 - 7}{-2 - (-1)} = \frac{-10}{-1} = 10$$

The **slope-intercept form** of a linear equation is

$$y = mx + b$$

where m is the slope of the line and $(0, b)$ is the y-intercept.

Find the slope and y-intercept of $-3x + 2y = -8$.

$$2y = 3x - 8$$

$$\frac{2y}{2} = \frac{3x}{2} - \frac{8}{2}$$

$$y = \frac{3}{2}x - 4$$

The slope of the line is $\frac{3}{2}$, and the y-intercept is $(0, -4)$.

DEFINITIONS AND CONCEPTS	**EXAMPLES**

Section 3.2   The Slope of a Line (*continued*)

The slope of a horizontal line is 0. The slope of a vertical line is undefined. Parallel lines have the same slope.	The slope of $y = -2$ is 0. The slope of $x = 5$ is undefined. 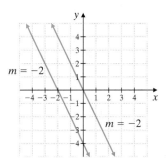
If the product of the slopes of two nonvertical lines is -1, then the lines are perpendicular.	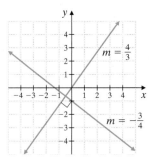

Section 3.3   The Slope-Intercept Form

We can use the slope-intercept form to write an equation of a line given its slope and y-intercept.	Write an equation of the line with y-intercept $(0, -1)$ and slope $\dfrac{2}{3}$. $$y = mx + b$$ $$y = \frac{2}{3}x - 1$$

Section 3.4   More Equations of Lines

The **point-slope form** of the equation of a line is $$y - y_1 = m(x - x_1)$$ where m is the slope of the line and (x_1, y_1) is a point on the line.	Find an equation of the line with slope 2 containing the point $(1, -4)$. Write the equation in standard form: $Ax + By = C$. $$y - y_1 = m(x - x_1)$$ $$y - (-4) = 2(x - 1)$$ $$y + 4 = 2x - 2$$ $$-2x + y = -6 \quad \text{Standard form}$$

DEFINITIONS AND CONCEPTS	**EXAMPLES**

Section 3.5   Graphing Linear Inequalities

If the equal sign in a linear equation in two variables is replaced with an inequality symbol, the result is a **linear inequality in two variables.**	$x \leq -5y$, $y \geq 2$, $3x - 2y > 7$, $x < -5$
GRAPHING A LINEAR INEQUALITY	Graph: $2x - 4y > 4$
Step 1. Graph the **boundary line** by graphing the related equation. Draw a solid line if the inequality symbol is $\leq$ or $\geq$. Draw a dashed line if the inequality symbol is $<$ or $>$.	**1.** Graph $2x - 4y = 4$. Draw a dashed line because the inequality symbol is $>$.
Step 2. Choose a **test point** not on the line. Substitute its coordinates into the original inequality.	**2.** Check the test point $(0, 0)$ in the inequality $2x - 4y > 4$. $2 \cdot 0 - 4 \cdot 0 > 4$ Let $x = 0$ and $y = 0$. $0 > 4$ False
Step 3. If the resulting inequality is true, shade the **half-plane** that contains the test point. If the inequality is not true, shade the half-plane that does not contain the test point.	**3.** The inequality is false, so shade the half-plane that does not contain $(0, 0)$. 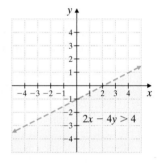

Section 3.6   Introduction to Functions

A **relation** is a set of ordered pairs. The **domain** of the relation is the set of all first components of the ordered pairs. The **range** of the relation is the set of all second components of the ordered pairs.	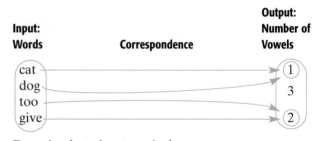 Domain: {cat, dog, too, give} Range: {1, 2}
A **function** is a relation in which each element of the first set corresponds to exactly one element of the second set.	The previous relation is a function. Each word contains one exact number of vowels.
VERTICAL LINE TEST	Find the domain and the range of the relation. Also determine whether the relation is a function.
If no vertical line can be drawn so that it intersects a graph more than once, the graph is the graph of a function. If such a line can be drawn, the graph is not that of a function.	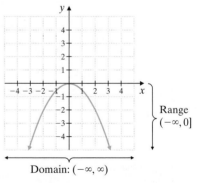

By the vertical line test, this is the graph of a function.

DEFINITIONS AND CONCEPTS	**EXAMPLES**

Section 3.6   Introduction to Functions (*continued*)

The symbol $f(x)$ means **function of x** and is called **function notation.**

If $f(x) = 2x^2 - 5$, find $f(-3)$.

$$f(-3) = 2(-3)^2 - 5 = 2(9) - 5 = 13$$

A **linear function** is a function that can be written in the form

$$f(x) = mx + b$$

To graph a linear function, use the slope and y-intercept.

$$f(x) = -3, g(x) = 5x, h(x) = -\frac{1}{3}x - 7$$

Graph: $f(x) = -2x$
(or $y = -2x + 0$)

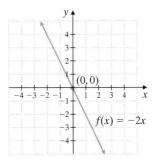

The slope is $\frac{2}{-1}$.
The y-intercept is $(0,0)$.

Section 3.7   Find Domains and Ranges from Graphs and Graphing Piecewise-Defined Functions

To find the domain of a function (or relation) from its graph, recall that on the rectangular coordinate system, "domain" means what x-values are graphed. Similarly, "range" means what y-values are graphed.

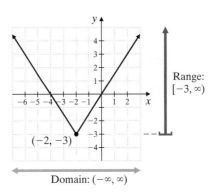

Section 3.8   Shifting and Reflecting Graphs of Functions

Vertical shifts (upward and downward) let k be a positive number.

The graph of $h(x) = -|x - 3| + 1$ is the same as the graph of $f(x) = |x|$, reflected about the x-axis, shifted 3 units right, then 1 unit up.

Graph of	**Same as**	**Moved**
$g(x) = f(x) + k$	$f(x)$	k units upward
$g(x) = f(x) + (-k)$	$f(x)$	k units downward

Horizontal shift (to the left or right) let h be a positive number.

Graph of	**Same as**	**Moved**
$g(x) = f(x - h)$	$f(x)$	h units to the right
$g(x) = f(x + h)$	$f(x)$	h units to the left

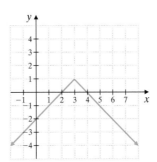

Reflection about the x-axis
The graph of $g(x) = -f(x)$ is the graph of $f(x)$ reflected about the x-axis.

Are You Prepared for a Test on Chapter 3?

Below I have listed some common trouble areas for students in Chapter 3. After studying for your test—but before taking your test—read these.

- Don't forget that the graph of an ordered pair is a *single* point in the rectangular coordinate plane.
- Remember that the slope of a horizontal line is 0 while a vertical line has undefined slope or no slope.
- For a linear equation such as $2y = 3x - 6$, the slope is not the coefficient of x unless the equation is solved for y. Solving this equation for y, we have $y = \frac{3}{2}x - 3$. The slope is $\frac{3}{2}$ and the y-intercept is $(0, -3)$.
- Parallel lines have the same slope while perpendicular lines have negative reciprocal slopes.

Slope	Parallel line	Perpendicular line
$m = 6$	$m = 6$	$m = -\frac{1}{6}$
$m = -\frac{2}{3}$	$m = -\frac{2}{3}$	$m = \frac{3}{2}$

- Don't forget that the statement $f(2) = 3$ corresponds to the ordered pair $(2, 3)$.

Remember: this is simply a checklist of common trouble areas. For a review of Chapter 3, see the Highlights and Chapter Review at the end of this chapter.

3 CHAPTER REVIEW

(3.1) *Plot the points and name the quadrant or axis in which each point lies.*

1. $A(2, -1), B(-2, 1), C(0, 3), D(-3, -5)$

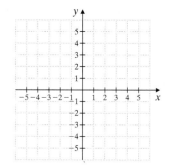

2. $A(-3, 4), B(4, -3), C(-2, 0), D(-4, 1)$

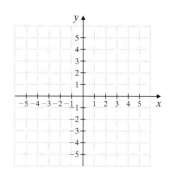

Create a scatter diagram for the given paired data.

3. Per Person Consumption of Caloric Sweeteners in the United States

Year	Per Person Consumption (pounds per person)
1998	149
1999	151
2000	149
2001	147
2002	146
2003	147

4. U.S. Armed Forces Active Duty Personnel

Year	Armed Forces (in millions)
2000	1
2001	1.4
2002	0.8
2003	1.5
2004	1.5

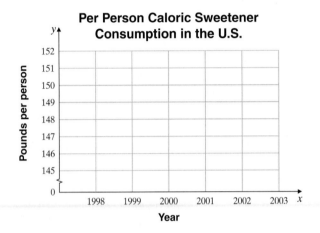

Per Person Caloric Sweetener Consumption in the U.S.

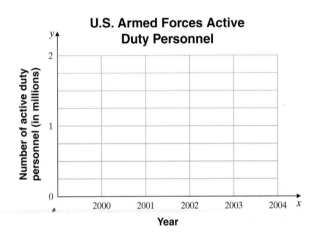

U.S. Armed Forces Active Duty Personnel

Determine whether each ordered pair is a solution to the given equation.

5. $7x - 8y = 56; (0, 56), (8, 0)$

6. $-2x + 5y = 10; (-5, 0), (1, 1)$

7. $x = 13; (13, 5), (13, 13)$

8. $y = 2; (7, 2), (2, 7)$

Graph each linear equation.

9. $3x - y = 3$

10. $2x - y = 4$

11. $4x + 5y = 20$

12. $y = 5$

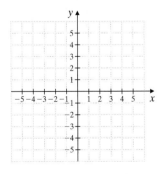

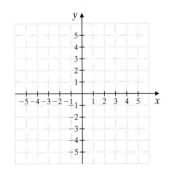

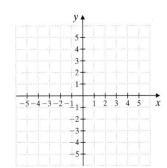

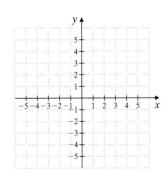

13. $x = -2$

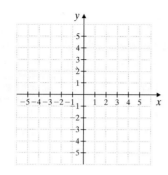

14. $y = \dfrac{1}{3}x$

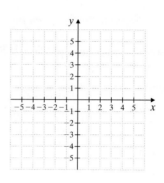

(3.2) *Find the slope of the line through each pair of points.*

15. $(2, 8)$ and $(6, -4)$

16. $(-3, 9)$ and $(5, 13)$

17. $(-7, -4)$ and $(-3, 6)$

18. $(7, -2)$ and $(-5, 7)$

Determine the slope of each line.

19.

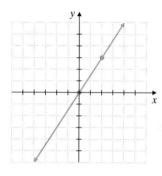

20.

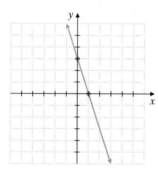

21.

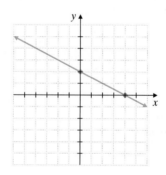

22.

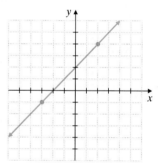

Two lines are graphed on each set of axes. Determine whether l_1 or l_2 has the greater slope.

23.

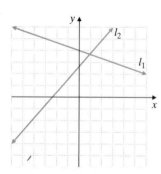

24.

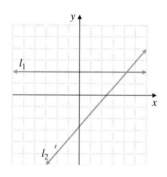

25.

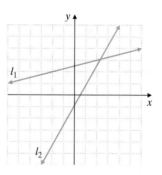

26.

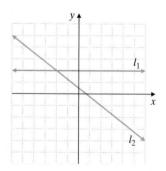

Find the slope and y-intercept of each line.

27. $y = -3x + \dfrac{1}{2}$

28. $y = 2x + 4$

29. $6x - 15y = 20$

30. $4x + 14y = 21$

Find the slope of each line.

31. $y - 3 = 0$

32. $x = -5$

Determine whether each pair of lines is parallel, perpendicular, or neither.

33. $y = -2x + 6$
$\quad\ y = 2x - 1$

34. $-x + 3y = 2$
$\quad\ 6x - 18y = 3$

35. $y = \dfrac{3}{4}x + 1$

$\qquad y = -\dfrac{4}{3}x + 1$

36. $x - 2y = 6$
$\quad\ 4x + y = 8$

(3.3) *Graph each line passing through the given point with the given slope.*

37. Through $(2, -3)$ with slope $\dfrac{2}{3}$

38. Through $(1, -4)$ with slope $\dfrac{1}{2}$

39. Through $(0, 1)$ with slope 2

40. Through $(-2, 0)$ with slope -3

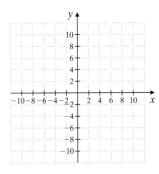
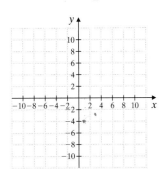
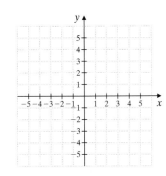
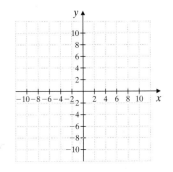

Graph each linear equation using the slope and y-intercept.

41. $y = -x + 1$

42. $y = 4x - 3$

43. $3x - y = 6$

44. $y = -5x$

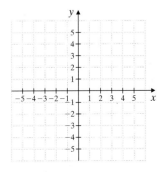
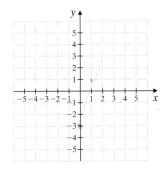
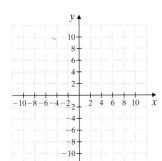
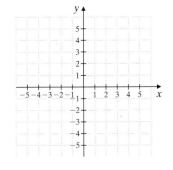

45. The cost C, in dollars, of renting a minivan for a day is given by the linear equation $C = 0.3x + 42$, where x is number of miles driven.

   a. Find the cost of renting the minivan for a day and driving it 150 miles.

   b. Find and interpret the slope of this equation.

   c. Find and interpret the y-intercept of this equation.

(3.4) *Write an equation of the line satisfying each set of conditions.*

46. Horizontal; through $(3, -1)$

47. Slope undefined; through $(-4, -3)$

Write the equation of the line satisfying each set of conditions. Write the equation in the form $y = mx + b$.

48. Through $(-3, 5)$; slope 3

49. Through $(-6, -1)$ and $(-4, -2)$

50. Through $(2, -6)$; parallel to $y = -2x + 3$

51. Through $(-6, -1)$; perpendicular to $4x + 3y = 5$

52. The value of an automobile bought in 2000 continues to decrease as time passes. Two years after the car was bought, it was worth \$14,300; four years after it was bought it was worth \$10,200.
   a. Assuming that this relationship between the number of years past 2000 and the value of the car is linear, write an equation describing this relationship. [*Hint:* Use ordered pairs of the form (years past 2000, value of the automobile)]
   b. Use this equation to estimate the value of the automobile in 2006.

53. The value of a building bought in 1990 continues to increase as time passes. Seven years after the building was bought, it was worth \$210,000; 12 years after it was bought, it was worth \$270,000.
   a. Assuming that this relationship between the number of years past 1990 and the value of the building is linear, write an equation describing this relationship.
   b. Use this equation to estimate the value of the building in 2008.

(3.5) *Graph each linear inequality.*

54. $3x + y > 4$

55. $\frac{1}{2}x - y < 2$

56. $5x - 2y \leq 9$

57. $y < 1$

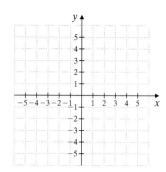 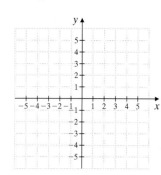 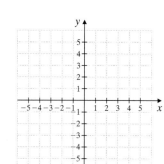 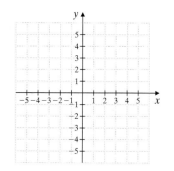

58. $x > -2$

59. Graph the union of $y > 2x + 3$ or $x \leq -3$.

60. Graph the intersection of $2x < 3y + 8$ and $y \geq -2$.

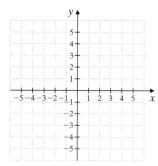

(3.6) *Find the domain and range for each relation. Then determine whether the relation is also a function.*

61. $\left\{ \left(-\frac{1}{2}, \frac{3}{4} \right), (6, 0.65), (0, -12), (25, 25) \right\}$

62. $\left\{ \left(\frac{3}{4}, -\frac{1}{2} \right), (0.65, 6), (-12, 0), (25, 25) \right\}$

63.

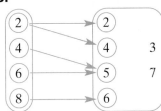

64.

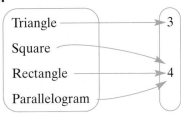

65.

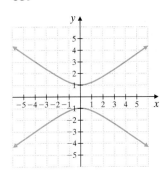

66.

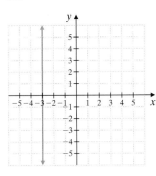

67.

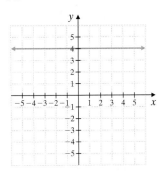

68.

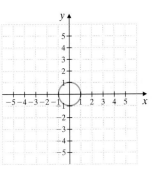

If $f(x) = x - 5$, $g(x) = -3x$, and $h(x) = 2x^2 - 6x + 1$, find each function value.

69. $f(2)$

70. $g(0)$

71. $g(-6)$

72. $h(-1)$

73. $h(1)$

74. $f(5)$

The function $J(x) = 2.54x$ may be used to calculate the weight of an object on Jupiter (J) given its weight on Earth (x).

75. If a person weighs 150 pounds on Earth, find the equivalent weight on Jupiter.

76. A 2000-pound probe on Earth weighs how many pounds on Jupiter?

Graph each linear function.

77. $f(x) = x + 2$

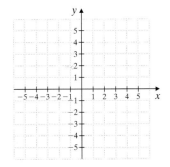

78. $f(x) = -\dfrac{1}{2}x + 3$

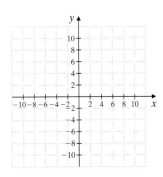

(3.7) *Find the domain and range of each relation.*

79.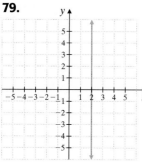

80.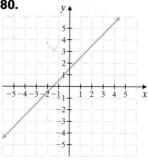

81.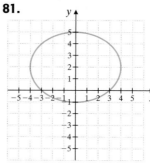

82.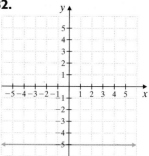

Graph each function.

83. $f(x) = \begin{cases} -3x & \text{if } x < 0 \\ x - 3 & \text{if } x \geq 0 \end{cases}$

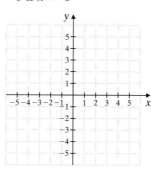

84. $g(x) = \begin{cases} -\dfrac{1}{5}x & \text{if } x \leq -1 \\ -4x + 2 & \text{if } x > -1 \end{cases}$

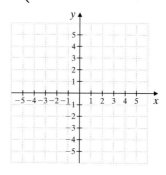

(3.8) *Graph each function.*

85. $y = \sqrt{x} - 4$

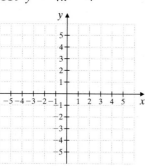

86. $f(x) = \sqrt{x - 4}$

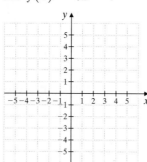

87. $g(x) = |x - 2| - 2$

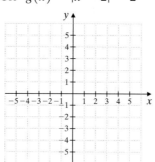

88. $h(x) = -(x + 3)^2 - 1$

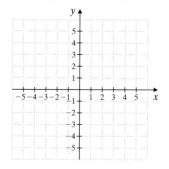

Mixed Review

Graph each linear equation or inequality.

89. $3x - 2y = -9$

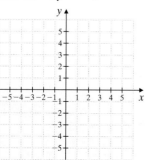

90. $x = -4y$

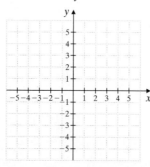

91. $3y \geq x$

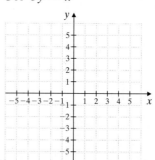

Write an equation of the line satisfying each set of conditions. If possible, write the equation in the form $y = mx + b$.

92. Vertical; through $(-2, -4)$

93. Slope 0; through $(2, 5)$

94. Slope 2; through $(5, -2)$

95. Through $(-5, 3)$ and $(-4, -8)$

96. Through $(-4, -2)$; parallel to $y = -\dfrac{3}{2}x + 1$

97. Through $(-4, 5)$; perpendicular to $2x - 3y = 6$

Find the domain and range of each relation.

98.

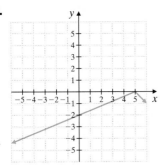

99.

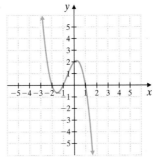

Graph each piecewise-defined function.

100. $f(x) = \begin{cases} x - 2 & \text{if } x \le 0 \\ -\dfrac{x}{3} & \text{if } x \ge 3 \end{cases}$

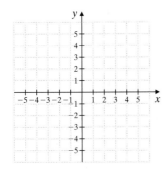

101. $g(x) = \begin{cases} 4x - 3 & \text{if } x \le 1 \\ 2x & \text{if } x > 1 \end{cases}$

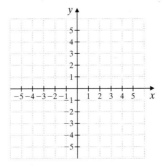

Graph each function.

102. $f(x) = \sqrt{x - 2}$

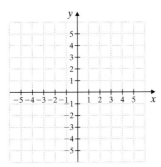

103. $f(x) = |x + 1| - 3$

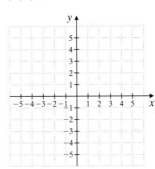

CHAPTER TEST

Remember to use the Chapter Test Prep Video CD to see the fully worked-out solutions to any of the exercises you want to review.

1. Plot the points, and name the quadrant in which each is located: $A(6, -2), B(4, 0), C(-1, 6)$.

2. Create a scatter diagram for the paired data.

1. see graph

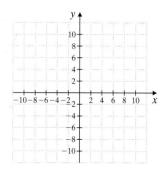

2. see graph

Krispy Kreme Donut Locations	
Year	Number of Krispy Kreme Donut Locations
2000	144
2001	174
2002	218
2003	276
2004	360

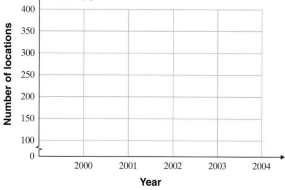

3. see graph

4. see graph

Graph each linear equation.

3. $-3x + y = -3$

4. $2x - 3y = -6$

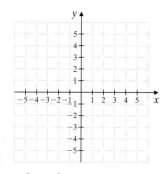

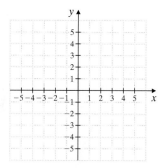

5. see graph

6. see graph

5. $4x + 6y = 8$

6. $y = -3$

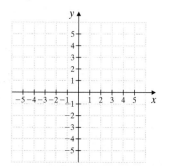

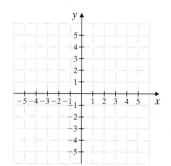

250

7. Find the slope of the line that passes through $(5, -8)$ and $(-7, 10)$.

8. Find the slope and the y-intercept of the line $3x + 12y = 8$.

Match each equation with its graph.

9. $f(x) = 3x + 1$

10. $f(x) = 3x - 2$

11. $f(x) = 3x + 2$

12. $f(x) = 3x - 5$

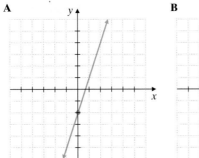
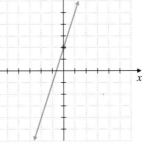

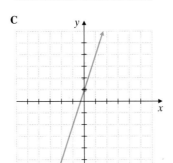
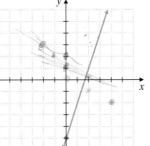

Find an equation of the line satisfying each set of conditions. Write the equations in the form $x = a$, $y = b$, or $y = mx + b$. For Exercises 16 and 17, write the equations in standard form, $Ax + By = C$.

13. Horizontal; through $(2, -8)$

14. Vertical; through $(-4, -3)$

15. Perpendicular to $x = 5$; through $(3, -2)$

16. Through $(4, -1)$; slope -3

17. Through $(0, -2)$; slope 5

18. Through $(4, -2)$ and $(6, -3)$

19. Through $(-1, 2)$; perpendicular to $3x - y = 4$

20. Parallel to $2y + x = 3$; through $(3, -2)$

21. _____

21. Line L_1 has the equation $2x - 5y = 8$. Line L_2 passes through the points $(1, 4)$ and $(-1, -1)$. Determine whether these lines are parallel, perpendicular, or neither.

22. For the 2004 major league baseball season, the following linear equation describes the relationship between a team's payroll x (in millions of dollars) and the number of games y that team won during the regular season.

$$y = 0.232x + 63.953$$

22. a. _____

Round to the nearest whole. (*Sources:* Based on data from *The Boston Globe* and Major League Baseball)

a. According to this equation, how many games would have been won during the 2004 season by a team with a payroll of $60 million?

b. _____

b. The St. Louis Cardinals had a payroll of $93 million in 2004. According to this equation, how many games would they have won during the season?

c. According to this equation, what payroll would have been necessary in 2004 to have won 95 games during the season?

c. _____

d. Find and interpret the slope of the equation.

Graph each inequality.

23. $x \leq -4$

24. $2x - y > 5$

d. _____

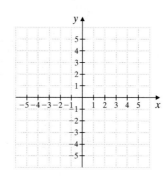

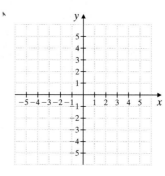

23. see graph

25. The intersection of $2x + 4y < 6$ and $y \leq -4$

24. see graph

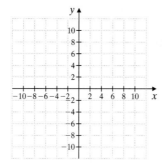

25. see graph

Find the domain and range of each relation. Also determine whether the relation is a function.

26. _____

26.

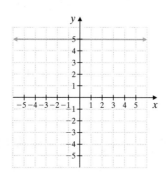

27.

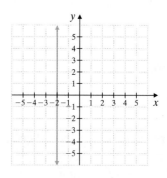

27. _____

28.

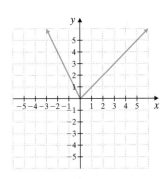

29.

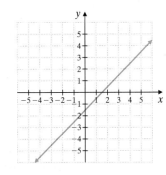

Graph each function. For Exercises 30 and 32, state the domain and the range of the function.

30. $f(x) = \begin{cases} -\dfrac{1}{2}x \text{ if } x \le 0 \\ 2x - 3 \text{ if } x > 0 \end{cases}$

31. $f(x) = (x - 4)^2$

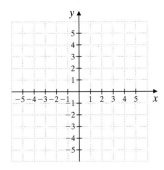

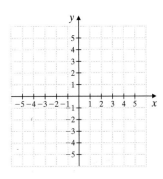

32. $g(x) = -|x + 2| - 1$

33. $h(x) = \sqrt{x} - 1$

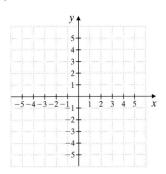

28. _____

29. _____

30. see graph _____

31. see graph _____

32. see graph _____

33. see graph _____

Answers

Write each set in roster form.

1. $\{x|x \text{ is a whole number between } 1 \text{ and } 6\}$

2. $\{x|x \text{ is an integer between } -2 \text{ and } 3\}$

3. $\{x|x \text{ is a natural number greater than } 100\}$

4. Multiply or divide.

    a. $\dfrac{-42}{-6}$

    b. $\dfrac{0}{14}$

    c. $-1(-5)(-2)$

Find the reciprocal, or multiplicative inverse, of each number.

5. -9

6. 10.2

7. $\dfrac{7}{4}$

8. Evaluate $2x^2$ for

    a. $x = 7$

    b. $x = -7$

Find each absolute value.

9. $|3|$

10. $\left|-\dfrac{7}{15}\right|$

11. $-|2.7|$

12. $-|-102.86|$

13. Simplify: $3 - [6(4 - 6) + 2(5 - 9)]$

14. Simplify $-2 + 3[5 - (7 - 10)]$.

Use the product rule to simplify.

15. $2^2 \cdot 2^5$

16. $-7x^9 \cdot \dfrac{3}{14}x^4$

17. $y \cdot y^2 \cdot y^4$

18. $(-2ab)(-0.3a^7b^{10})(8.1b^2)$

Answers

1. _____

2. _____

3. _____

4. a. _____

  b. _____

  c. _____

5. _____

6. _____

7. _____

8. a. _____

  b. _____

9. _____

10. _____

11. _____

12. _____

13. _____

14. _____

15. _____

16. _____

17. _____

18. _____

Use the power rules to simplify. Write each result using positive exponents only.

19. $(5x^2)^3$

20. $(3^{-2}y^4)^2$

21. $\left(\dfrac{2^{-3}}{y}\right)^{-2}$

22. $\left(\dfrac{5a^3}{b^{-2}c^5}\right)^{-3}$

23. Solve: $0.6 = 2 - 3.5c$.

24. Solve: $5(x - 7) = 4x - 35 + x$.

25. Find three numbers such that the second number is 3 more than twice the first number and the third number is four times the first number. The sum of the three numbers is 164.

26. Find 3 consecutive odd integers whose sum is 213.

27. Marial Callier just received an inheritance of \$10,000 and plans to place all the money in a savings account that pays 5% compounded quarterly to help her son go to college in 3 years. How much money will be in the account in 3 years?

28. Add.
  a. $-4 + (-3)$
  b. $\dfrac{1}{2} - \left(-\dfrac{1}{3}\right)$
  c. $7 - 20$

Graph each set on a number line and then write it in interval notation.

29. $\{x | x \geq 2\}$

30. $\left\{x \,\middle|\, x < -\dfrac{3}{2}\right\}$

31. $\{x | 0.5 < x \leq 3\}$

Solve.

32. $3x + 10 > \dfrac{5}{2}(x - 1)$

33. $x - 7 < 2 \text{ and } 2x + 1 < 9$

34. $x - 2 < 6 \text{ or } 3x + 1 > 1$

35. $|x - 3| = |5 - x|$

36. $|5x - 2| - 7 = -4$

37. $|x - 3| > 7$

19. _____

20. _____

21. _____

22. _____

23. _____

24. _____

25. _____

26. _____

27. _____

28. a. _____

    b. _____

    c. _____

29. _____

30. _____

31. _____

32. _____

33. _____

34. _____

35. _____

36. _____

37. _____

38. _____

39. see graph _____

40. _____

41. _____

42. _____

43. _____

44. _____

45. _____

46. _____

47. _____

48. _____

49. _____

50. _____

38. $|-x + 8| - 2 \leq 8$

39. Find the intercepts and graph:
$3x + 4y = 12$

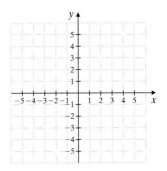

40. Find the slope and y-intercept of $7x + 2y = 10$.

41. Find the slope of the line $y = 3x + 2$.

42. Find the slope of the line through $(-1, 6)$ and $(0, 9)$.

43. Write an equation of the line with y-intercept $(0, -7)$ and slope of $\frac{2}{3}$.

44. Write an equation of the line through $(-2, 5)$ and $(-4, 7)$ in standard form.

45. Find an equation of the horizontal line containing the point $(2, 3)$.

46. Find an equation of the vertical line through $\left(-2, -\dfrac{3}{4}\right)$.

Find the following.

47. If $f(x) = 7x^2 - 3x + 1$, find $f(1)$.

48. Find the slope of the line defined by $f(x) = -2.3x - 6$.

49. If $g(x) = 3x - 2$, find $g(0)$.

50. Determine whether the graph below is the graph of a function.

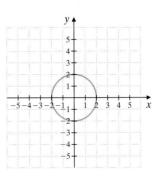

4

Systems of Equations and Inequalities

In this chapter, two or more equations in two or more variables are solved simultaneously. Such a collection of equations is called a *system of equations*. Systems of equations are good mathematical models for many real-world problems because these problems may involve several related patterns. We will study various methods for solving systems of equations and will conclude with a look at systems of inequalities.

"Please be kind and rewind." It finally happened in 2001 (or 2002, depending on who you believe). What happened? The number of DVD players sold in the United States was greater than the number of VCR decks sold. A trip to your local movie rental store will at least convince you that rentals of DVDs are increasing while rentals of VCR tapes are decreasing. The graph below compares the money spent on VCR decks with the money spent on DVD players. In Exercise 55 on page 313, we will find the year in which the amount of money spent on sales of VCR decks equals the amount of money spent on DVD players.

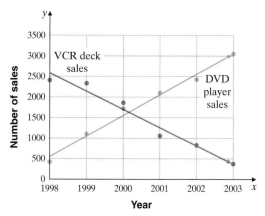

(*Source:* Consumer Electronics Association, data used from 1998–2003)

A Determine Whether an Ordered Pair is a Solution of a System of Two Linear Equations.

B Solve a System of Two Equations by Graphing.

C Solve a System Using Substitution.

D Solve a System Using Elimination.

4.1 SOLVING SYSTEMS OF LINEAR EQUATIONS IN TWO VARIABLES

Recall from Chapter 3 that the graph of a linear equation in two variables is a line. Two or more linear equations form a **system of linear equations.** Some examples of systems of linear equations in two variables follow.

$$\begin{cases} x - 2y = -7 \\ 3x + y = 0 \end{cases} \qquad \begin{cases} x = 5 \\ x + \dfrac{y}{2} = 9 \end{cases} \qquad \begin{cases} x - 3 = 2y + 6 \\ y = 1 \end{cases}$$

Objective **A** Determining Whether an Ordered Pair Is a Solution

Recall that a solution of an equation in two variables is an ordered pair (x, y) that makes the equation true. A **solution of a system** of two equations in two variables is an ordered pair (x, y) that makes both equations true.

PRACTICE PROBLEM 1

Determine whether the ordered pair $(4, 1)$ is a solution of the system.

$$\begin{cases} x - y = 3 \\ 2x - 3y = 5 \end{cases}$$

EXAMPLE 1 Determine whether the ordered pair $(-1, 1)$ is a solution of the system.

$$\begin{cases} -x + y = 2 \\ 2x - y = -3 \end{cases}$$

Solution: We replace x with -1 and y with 1 in each equation.

$$\begin{aligned} -x + y &= 2 && \text{First equation} \\ -(-1) + (1) &\stackrel{?}{=} 2 && \text{Let } x = -1 \text{ and } y = 1. \\ 1 + 1 &\stackrel{?}{=} 2 \\ 2 &= 2 && \text{True} \end{aligned}$$

$$\begin{aligned} 2x - y &= -3 && \text{Second equation} \\ 2(-1) - (1) &\stackrel{?}{=} -3 && \text{Let } x = -1 \text{ and } y = 1. \\ -2 - 1 &\stackrel{?}{=} -3 \\ -3 &= -3 && \text{True} \end{aligned}$$

Since $(-1, 1)$ makes both equations true, it is a solution.

Work Practice Problem 1

PRACTICE PROBLEM 2

Determine whether the ordered pair $(-3, 3)$ is a solution of the system.

$$\begin{cases} 3x - y = -12 \\ x - y = 0 \end{cases}$$

EXAMPLE 2 Determine whether the ordered pair $(-2, 3)$ is a solution of the system.

$$\begin{cases} 5x + 3y = -1 \\ x - y = 1 \end{cases}$$

Solution: We replace x with -2 and y with 3 in each equation.

$$\begin{aligned} 5x + 3y &= -1 && \text{First equation} \\ 5(-2) + 3(3) &\stackrel{?}{=} -1 && \text{Let } x = -2 \text{ and } y = 3. \\ -10 + 9 &\stackrel{?}{=} -1 \\ -1 &= -1 && \text{True} \end{aligned}$$

Answers

1. yes, a solution, **2.** no, not a solution

$$x - y = 1 \quad \text{Second equation}$$
$$(-2) - (3) \stackrel{?}{=} 1 \quad \text{Let } x = -2 \text{ and } y = 3.$$
$$-5 = 1 \quad \text{False}$$

Since the ordered pair $(-2, 3)$ does not make both equations true, it is not a solution of the system.

🔲 **Work Practice Problem 2**

Objective B **Solving a System by Graphing**

The graph of each linear equation in a system is a line. Each point on each line corresponds to an ordered pair solution of its equation. If the lines intersect, the point of intersection lies on both lines and corresponds to an ordered pair solution of both equations. In other words, the point of intersection corresponds to an ordered pair solution of the system. Therefore, we can estimate the solutions of a system by graphing the equations on the same rectangular coordinate system and estimating the coordinates of any points of intersection.

EXAMPLE 3 Solve the system by graphing.

$$\begin{cases} x + y = 2 \\ 3x - y = -2 \end{cases}$$

Solution: First we graph the linear equations on the same rectangular coordinate system. These lines intersect at one point as shown. The coordinates of the point of intersection appear to be $(0, 2)$. We check this estimated solution by replacing x with 0 and y with 2 in *both* equations.

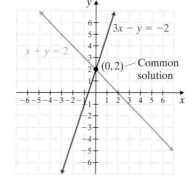

$$x + y = 2 \quad \text{First equation}$$
$$0 + 2 \stackrel{?}{=} 2 \quad \text{Let } x = 0 \text{ and } y = 2.$$
$$2 = 2 \quad \text{True}$$

$$3x - y = -2 \quad \text{Second equation}$$
$$3(0) - 2 \stackrel{?}{=} -2 \quad \text{Let } x = 0 \text{ and } y = 2.$$
$$-2 = -2 \quad \text{True}$$

The ordered pair $(0, 2)$ is the solution of the system. A system that has at least one solution, such as this one, is said to be **consistent.**

🔲 **Work Practice Problem 3**

EXAMPLE 4 Solve the system by graphing.

$$\begin{cases} x - 2y = 4 \\ x \quad\;\; = 2y \end{cases}$$

Solution: We graph each linear equation.

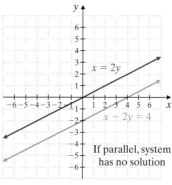

If parallel, system has no solution

Continued on next page

Helpful Hint

Reading values from graphs may not be accurate. Until a proposed solution is checked in both equations of the system, we can only assume that we have *estimated* a solution.

PRACTICE PROBLEMS 3–4

Solve each system by graphing. If the system has just one solution, estimate the solution.

3. $\begin{cases} x - y = 2 \\ x + 3y = 6 \end{cases}$

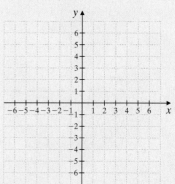

4. $\begin{cases} y = -3x \\ 6x + 2y = 4 \end{cases}$

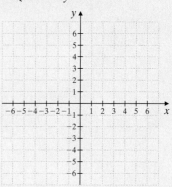

Answers

3. $(3, 1)$,

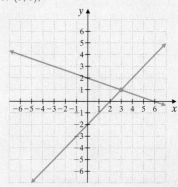

4. See page 260.

The lines appear to be parallel. To be sure, let's write each equation in slope-intercept form, $y = mx + b$. To do so, we solve for y.

$x - 2y = 4$	First equation	$x = 2y$	Second equation
$-2y = -x + 4$	Subtract x from both sides.	$\frac{1}{2}x = y$	Divide both sides by 2.
$y = \frac{1}{2}x - 2$	Divide both sides by -2.	$y = \frac{1}{2}x$	

The graphs of these equations have the same slope, $\frac{1}{2}$, but different y-intercepts, so these lines are parallel. Therefore, the system has no solution since the equations have no common solution (there are no intersection points). A system that has no solution is said to be **inconsistent.**

◻ **Work Practice Problem 4**

PRACTICE PROBLEM 5

Solve the system by graphing.

$$\begin{cases} -2x + y = 1 \\ 4x - 2y = -2 \end{cases}$$

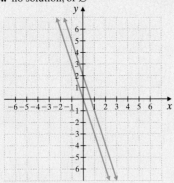

> **Helpful Hint**
> • If a system of equations has *at least one solution*, the system is *consistent*.
> • If a system of equations has *no solution*, the system is *inconsistent*.

The pairs of equations in Examples 3 and 4 are called independent because their graphs differ. In Example 5, we see an example of dependent equations.

✔ **Concept Check** How can you tell just by looking at the following system that it has no solution?

$$\begin{cases} y = 3x + 5 \\ y = 3x - 7 \end{cases}$$

Answers

4. no solution, or ∅

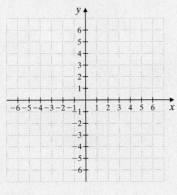

5. $\{(x, y) | -2x + y = 1\}$

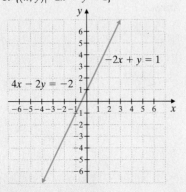

EXAMPLE 5 Solve the system by graphing.

$$\begin{cases} 2x + 4y = 10 \\ x + 2y = 5 \end{cases}$$

Solution: We graph each linear equation. We see that the graphs of the equations are the same line. To confirm this, notice that if both sides of the second equation are multiplied by 2, the result is the first equation. This means that the equations have identical solutions. Any ordered pair solution of one equation satisfies the other equation also. These equations are said to be **dependent equations.** The solution set of the system is $\{(x, y) | x + 2y = 5\}$ or, equivalently, $\{(x, y) | 2x + 4y = 10\}$ since the lines describe identical ordered pairs. Written the second way, the solution set is read "the set of all ordered pairs (x, y), such that $2x + 4y = 10$." There are an infinite number of solutions to this system.

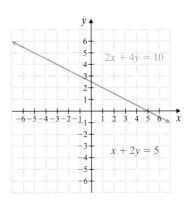

◻ **Work Practice Problem 5**

✔ **Concept Check Answer**

answers may vary

> **Helpful Hint**
> • If the graphs of two equations *differ*, they are *independent* equations.
> • If the graphs of two equations are the *same*, they are *dependent* equations.

✔**Concept Check** How can you tell just by looking at the following system that it has infinitely many solutions?

$$\begin{cases} x + y = 5 \\ 2x + 2y = 10 \end{cases}$$

We can summarize the information discovered in Examples 3 through 5 as follows.

Possible Solutions to Systems of Two Linear Equations

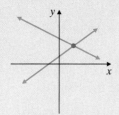

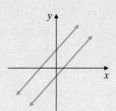

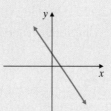

One solution: consistent system; independent equations

No solution: inconsistent system; independent equations

Infinite number of solutions: consistent system; dependent equations

---

Objective C Solving a System Using Substitution

Graphing the equations of a system by hand is often a good method for finding approximate solutions of a system, but it is not a reliable method for finding exact solutions. To find an exact solution, we need to use *algebra*. One *algebraic* method is called the **substitution method.**

EXAMPLE 6 Use the substitution method to solve the system:

$$\begin{cases} 2x + 4y = -6 & \text{First equation} \\ x = 2y - 5 & \text{Second equation} \end{cases}$$

Solution: In the second equation, we are told that x is equal to $2y - 5$. Since they are equal, we can *substitute* $2y - 5$ for x in the first equation. This will give us an equation in one variable, which we can solve for y.

$$2x + 4y = -6 \qquad \text{First equation}$$

$$2(\overbrace{2y - 5}) + 4y = -6 \qquad \text{Substitute } 2y - 5 \text{ for } x.$$

$$4y - 10 + 4y = -6$$

$$8y = 4$$

$$y = \frac{4}{8} = \frac{1}{2} \qquad \text{Solve for } y.$$

The y-coordinate of the solution is $\frac{1}{2}$. To find the x-coordinate, we replace y with $\frac{1}{2}$ in the second equation,

$$x = 2y - 5$$

$$x = 2y - 5$$

$$x = 2\left(\frac{1}{2}\right) - 5 = 1 - 5 = -4$$

The ordered pair solution is $\left(-4, \frac{1}{2}\right)$. Check to see that $\left(-4, \frac{1}{2}\right)$ satisfies both equations of the system.

■ **Work Practice Problem 6**

PRACTICE PROBLEM 6

Use the substitution method to solve the system:

$$\begin{cases} 6x - 4y = 10 \\ y = 3x - 3 \end{cases}$$

Answer

6. $\left(\frac{1}{3}, -2\right)$

✔ **Concept Check Answer**

answers may vary

The substitution method is summarized below. Feel free to use these steps.

Solving a System of Two Equations Using the Substitution Method

Step 1: Solve one of the equations for one of its variables.

Step 2: Substitute the expression for the variable found in Step 1 into the other equation.

Step 3: Find the value of one variable by solving the equation from Step 2.

Step 4: Find the value of the other variable by substituting the value found in Step 3 into the equation from Step 1.

Step 5: Check the ordered pair solution in *both* original equations.

PRACTICE PROBLEM 7

Use the substitution method to solve the system:

$$\begin{cases} -\dfrac{x}{2} + \dfrac{y}{4} = \dfrac{1}{2} \\ \dfrac{x}{2} + \dfrac{y}{2} = -\dfrac{1}{8} \end{cases}$$

EXAMPLE 7 Use the substitution method to solve the system:

$$\begin{cases} -\dfrac{x}{6} + \dfrac{y}{2} = \dfrac{1}{2} \\ \dfrac{x}{3} - \dfrac{y}{6} = -\dfrac{3}{4} \end{cases}$$

Solution: First we multiply each equation by its least common denominator to clear the system of fractions. We multiply the first equation by 6 and the second equation by 12.

$$\begin{cases} 6\left(-\dfrac{x}{6} + \dfrac{y}{2} \right) = 6\left(\dfrac{1}{2} \right) \\ 12\left(\dfrac{x}{3} - \dfrac{y}{6} \right) = 12\left(-\dfrac{3}{4} \right) \end{cases} \quad \text{simplifies to} \quad \begin{cases} -x + 3y = 3 & \text{First equation} \\ 4x - 2y = -9 & \text{Second equation} \end{cases}$$

We now solve the first equation for x so that we may substitute our findings into the second equation.

$$-x + 3y = 3 \quad \text{First equation}$$
$$3y - 3 = x \quad \text{Solve for } x.$$

Next we replace x with $3y - 3$ in the second equation.

$$4x - 2y = -9 \quad \text{Second equation}$$
$$4(3y - 3) - 2y = -9$$
$$12y - 12 - 2y = -9$$
$$10y = 3$$
$$y = \dfrac{3}{10} \quad \text{Solve for } y.$$

> **Helpful Hint**
> To avoid tedious fractions, solve for a variable whose coefficient is 1 or −1, if possible.

The y-coordinate is $\dfrac{3}{10}$. To find the x-coordinate, we replace y with $\dfrac{3}{10}$ in the equation $x = 3y - 3$. Then

$$x = 3\left(\dfrac{3}{10} \right) - 3 = \dfrac{9}{10} - 3 = \dfrac{9}{10} - \dfrac{30}{10} = -\dfrac{21}{10}$$

The ordered pair solution is $\left(-\dfrac{21}{10}, \dfrac{3}{10} \right)$. Check to see that this solution satisfies both original equations.

■ **Work Practice Problem 7**

Answer

7. $\left(-\dfrac{3}{4}, \dfrac{1}{2} \right)$

Helpful Hint

If a system of equations contains equations with fractions, the first step is to clear the equations of fractions.

Objective D Solving a System Using Elimination

The **elimination method,** or **addition method,** is a second algebraic technique for solving systems of equations. For this method, we rely on a version of the addition property of equality, which states that "equals added to equals are equal."

If $A = B$ and $C = D$ then $A + C = B + D$

EXAMPLE 8 Use the elimination method to solve the system:

$$\begin{cases} x - 5y = -12 & \text{First equation} \\ -x + y = 4 & \text{Second equation} \end{cases}$$

Solution: Since the left side of each equation is equal to the right side, we add equal quantities by adding the left sides of the equations and the right sides of the equations. This sum gives us an equation in one variable, y, which we can solve for y.

$$\begin{array}{r} x - 5y = -12 \quad \text{First equation} \\ \underline{-x + y = 4} \quad \text{Second equation} \\ -4y = -8 \quad \text{Add.} \\ y = 2 \quad \text{Solve for } y. \end{array}$$

The y-coordinate of the solution is 2. To find the corresponding x-coordinate, we replace y with 2 in either original equation of the system. Let's use the second equation.

$$\begin{array}{r} -x + y = 4 \quad \text{Second equation} \\ -x + 2 = 4 \quad \text{Let } y = 2. \\ -x = 2 \\ x = -2 \end{array}$$

The ordered pair solution is $(-2, 2)$. Check to see that $(-2, 2)$ satisfies both equations of the system.

◙ **Work Practice Problem 8**

The steps below summarize the elimination method.

Solving a System of Two Linear Equations Using the Elimination Method

Step 1: Rewrite each equation in standard form, $Ax + By = C$.

Step 2: If necessary, multiply one or both equations by some nonzero number so that the coefficient of one variable in one equation is the opposite of the coefficient of that variable in the other equation.

Step 3: Add the equations. Your chosen variable should be eliminated.

Step 4: Find the value of the remaining variable by solving the equation from Step 3.

Step 5: Find the value of the other variable by substituting the value found in Step 4 into either original equation.

Step 6: Check the proposed ordered pair solution in *both* original equations.

PRACTICE PROBLEM 8

Use the elimination method to solve the system:

$$\begin{cases} 3x - y = 1 \\ 4x + y = 6 \end{cases}$$

Answer
8. $(1, 2)$

PRACTICE PROBLEM 9

Use the elimination method to solve the system:

$$\begin{cases} \dfrac{x}{3} + 2y = -1 \\ x + 6y = 2 \end{cases}$$

EXAMPLE 9 Use the elimination method to solve the system:

$$\begin{cases} 3x + \dfrac{y}{2} = 2 \\ 6x + y = 5 \end{cases}$$

Solution: If we add the two equations, the sum will still be an equation in two variables. Notice, however, that if we multiply both sides of the first equation by -2, the coefficients of x in the two equations will be opposites. Then

$$\begin{cases} -2\left(3x + \dfrac{y}{2}\right) = -2(2) \\ 6x + y = 5 \end{cases} \quad \text{simplifies to} \quad \begin{cases} -6x - y = -4 \\ 6x + y = 5 \end{cases}$$

Now we can add the left sides and add the right sides.

$$\begin{array}{r} -6x - y = -4 \\ 6x + y = 5 \\ \hline 0 = 1 \quad \text{False} \end{array}$$

The resulting equation, $0 = 1$, is false for all values of y or x. Thus, the system has no solution. The solution set is $\{\ \}$ or $\varnothing$. This system is inconsistent, and the graphs of the equations are parallel lines.

■ **Work Practice Problem 9**

PRACTICE PROBLEM 10

Use the elimination method to solve the system:

$$\begin{cases} 2x - 5y = 6 \\ 3x - 4y = 9 \end{cases}$$

EXAMPLE 10 Use the elimination method to solve the system:

$$\begin{cases} 3x - 2y = 10 \\ 4x - 3y = 15 \end{cases}$$

Solution: To eliminate y, our first step is to multiply both sides of the first equation by 3 and both sides of the second equation by -2. Then

$$\begin{cases} 3(3x - 2y) = 3(10) \\ -2(4x - 3y) = -2(15) \end{cases} \quad \text{simplifies to} \quad \begin{cases} 9x - 6y = 30 \\ -8x + 6y = -30 \end{cases}$$

Next we add the left sides and add the right sides.

$$\begin{array}{r} 9x - 6y = 30 \\ -8x + 6y = -30 \\ \hline x \qquad = 0 \end{array}$$

To find y, we let $x = 0$ in either equation of the system

$$\begin{aligned} 3x - 2y &= 10 \quad \text{First equation} \\ 3(0) - 2y &= 10 \quad \text{Let } x = 0. \\ -2y &= 10 \\ y &= -5 \end{aligned}$$

The ordered pair solution is $(0, -5)$. Check to see that $(0, -5)$ satisfies both equations.

■ **Work Practice Problem 10**

Answers

9. no solution or $\varnothing$, **10.** $(3, 0)$

EXAMPLE 11 Use the elimination method to solve the system:

$$\begin{cases} -5x - 3y = 9 \\ 10x + 6y = -18 \end{cases}$$

Solution: To eliminate x, our first step is to multiply both sides of the first equation by 2. Then

$$\begin{cases} 2(-5x - 3y) = 2(9) \\ 10x + 6y = -18 \end{cases} \text{ simplifies to } \begin{cases} -10x - 6y = 18 \\ 10x + 6y = -18 \end{cases}$$

Next we add the equations.

$$\begin{array}{r} -10x - 6y = 18 \\ \underline{10x + 6y = -18} \\ 0 = 0 \end{array}$$

The resulting equation, $0 = 0$, is true for all possible values of y or x. Notice in the original system that if both sides of the first equation are multiplied by -2, the result is the second equation. This means that the two equations are equivalent. They have the same solution set and there are an infinite number of solutions. Thus, the equations of this system are dependent, and the solution set of the system is

$$\{(x, y)|-5x - 3y = 9\} \qquad \text{or, equivalently,} \qquad \{(x, y)|10x + 6y = -18\}$$

▨ **Work Practice Problem 11**

Helpful Hint

Remember that not all ordered pairs are solutions of the system in Example 11. Only the infinite number of ordered pairs that satisfy $-5x - 3y = 9$ or, equivalently, $10x + 6y = -18$.

PRACTICE PROBLEM 11

Use the elimination method to solve the system:

$$\begin{cases} 4x - 7y = 10 \\ -8x + 14y = -20 \end{cases}$$

Answer

11. $\{(x, y)|4x - 7y = 10\}$

▦ **CALCULATOR EXPLORATIONS** Graphing

We may use a grapher to approximate solutions of systems of equations by graphing both equations on the same set of axes and approximating any points of intersection. For example, let's approximate the solution of the system

$$\begin{cases} y = -2.6x + 5.6 \\ y = 4.3x - 4.9 \end{cases}$$

We use a standard window and graph the equations on a single screen.

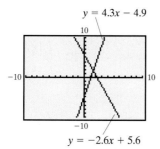

The two lines intersect. To approximate the point of intersection, we trace to the point of intersection and use an INTERSECT feature of the grapher, a ZOOM IN feature of the grapher, or redefine the window to [0, 3] by [0, 3]. If we redefine the window to [0, 3] by [0, 3], the screen should look like the following:

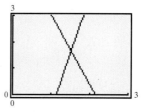

By tracing along the curves, we can see that the point of intersection has an x-value between 1.5 and 1.532. We can continue to zoom and trace or redefine the window until the coordinates of the point of intersection can be determined to the nearest hundredth. The approximate point of intersection is (1.52, 1.64).

Solve each system of equations. Approximate each solution to two decimal places.

1. $y = -1.65x + 3.65$
$y = 4.56x - 9.44$

2. $y = 7.61x + 3.48$
$y = -1.26x - 6.43$

3. $2.33x - 4.72y = 10.61$
$5.86x - 6.22y = -8.89$

4. $-7.89x - 5.68y = 3.26$
$-3.65x + 4.98y = 11.77$

Mental Math

Match each graph with the solution of the corresponding system.

A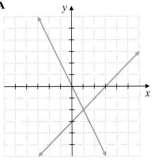

B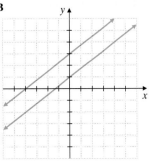

C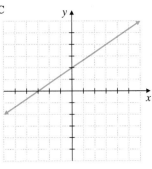

D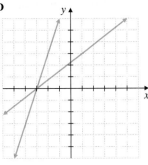

1. No solution

2. Infinite number of solutions

3. $(1, -2)$

4. $(-3, 0)$

FOR EXTRA HELP

Student Solutions Manual   PH Math/Tutor Center   CD/Video for Review   Math XL   MathXL®   MyMathLab   MyMathLab

4.1 EXERCISE SET

Objective **A** *Determine whether the given ordered pair is a solution of the system. See Examples 1 and 2.*

1. $\begin{cases} x - y = 3 \\ 2x - 4y = 8 \end{cases}$ $(2, -1)$

2. $\begin{cases} x - y = -4 \\ 2x + 10y = 4 \end{cases}$ $(-3, 1)$

3. $\begin{cases} 2x - 3y = -9 \\ 4x + 2y = -2 \end{cases}$ $(3, 5)$

4. $\begin{cases} 2x - 5y = -2 \\ 3x + 4y = 4 \end{cases}$ $(4, 2)$

5. $\begin{cases} y = -5x \\ x = -2 \end{cases}$ $(-2, 10)$

6. $\begin{cases} y = 6 \\ x = -2y \end{cases}$ $(-12, 6)$

7. $\begin{cases} 3x + 7y = -19 \\ -6x = 5y + 8 \end{cases}$ $\left(\dfrac{2}{3}, -3\right)$

8. $\begin{cases} 4x + 5y = -7 \\ -8x = 3y - 1 \end{cases}$ $\left(\dfrac{3}{4}, -2\right)$

Objective **B** *Solve each system by graphing. See Examples 3 through 5.*

9. $\begin{cases} x + y = 1 \\ x - 2y = 4 \end{cases}$

10. $\begin{cases} 2x - y = 8 \\ x + 3y = 11 \end{cases}$

11. $\begin{cases} 2y - 4 = 0 \\ x + 2y = 5 \end{cases}$

12. $\begin{cases} 4x - y = 6 \\ x - y = 0 \end{cases}$

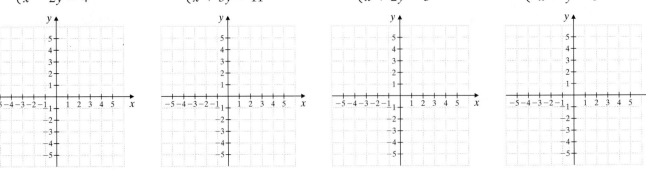

266

13. $\begin{cases} 3x - y = 4 \\ 6x - 2y = 4 \end{cases}$

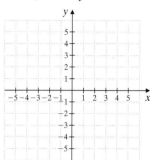

14. $\begin{cases} -x + 3y = 6 \\ 3x - 9y = 9 \end{cases}$

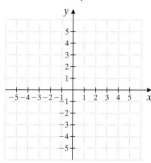

15. $\begin{cases} y = -3x \\ 2x - y = -5 \end{cases}$

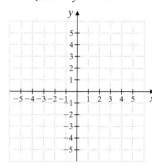

16. $\begin{cases} y = -2x \\ -3x + y = 10 \end{cases}$

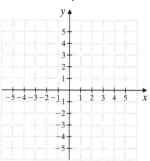

Objective **C** *Use the substitution method to solve each system of equations. See Examples 6 and 7.*

17. $\begin{cases} x + y = 10 \\ y = 4x \end{cases}$

18. $\begin{cases} 5x + 2y = -17 \\ x = 3y \end{cases}$

19. $\begin{cases} 4x - y = 9 \\ 2x + 3y = -27 \end{cases}$

20. $\begin{cases} 3x - y = 6 \\ -4x + 2y = -8 \end{cases}$

21. $\begin{cases} \dfrac{1}{2}x + \dfrac{3}{4}y = -\dfrac{1}{4} \\ \dfrac{3}{4}x - \dfrac{1}{4}y = 1 \end{cases}$

22. $\begin{cases} \dfrac{2}{5}x + \dfrac{1}{5}y = -1 \\ x + \dfrac{2}{5}y = -\dfrac{8}{5} \end{cases}$

23. $\begin{cases} x = -3y + 4 \\ 3x + 9y = 12 \end{cases}$

24. $\begin{cases} x = 3y - 1 \\ 2x - 6y = -2 \end{cases}$

Objective **D** *Use the elimination method to solve each system of equations. See Examples 8 through 11.*

25. $\begin{cases} 2x - 4y = 0 \\ x + 2y = 5 \end{cases}$

26. $\begin{cases} 2x - 3y = 0 \\ 2x + 6y = 3 \end{cases}$

27. $\begin{cases} 5x + 2y = 1 \\ x - 3y = 7 \end{cases}$

28. $\begin{cases} 6x - y = -5 \\ 4x - 2y = 6 \end{cases}$

29. $\begin{cases} 5x - 2y = 27 \\ -3x + 5y = 18 \end{cases}$

30. $\begin{cases} 3x + 4y = 2 \\ 2x + 5y = -1 \end{cases}$

31. $\begin{cases} 3x = 5y + 11 \\ 2x = 6y + 2 \end{cases}$

32. $\begin{cases} 6x = 3y - 3 \\ 4x = -5y - 9 \end{cases}$

33. $\begin{cases} x - 2y = 4 \\ 2x - 4y = 4 \end{cases}$

34. $\begin{cases} -x + 3y = 6 \\ 3x - 9y = 9 \end{cases}$

35. $\begin{cases} 3x + y = 1 \\ 2y = 2 - 6x \end{cases}$

36. $\begin{cases} y = 2x - 5 \\ 8x - 4y = 20 \end{cases}$

Objectives **C** **D** **Mixed Practice** *Solve each system of equations.*

37. $\begin{cases} 2x + 5y = 8 \\ 6x + y = 10 \end{cases}$

38. $\begin{cases} x - 4y = -5 \\ -3x - 8y = 0 \end{cases}$

39. $\begin{cases} x + y = 1 \\ x - 2y = 4 \end{cases}$

40. $\begin{cases} 2x - y = 8 \\ x + 3y = 11 \end{cases}$

41. $\begin{cases} \dfrac{1}{3}x + y = \dfrac{4}{3} \\ -\dfrac{1}{4}x - \dfrac{1}{2}y = -\dfrac{1}{4} \end{cases}$

42. $\begin{cases} \dfrac{3}{4}x - \dfrac{1}{2}y = -\dfrac{1}{2} \\ x + y = -\dfrac{3}{2} \end{cases}$

43. $\begin{cases} 4x + 2y = 5 \\ 2x + y = -1 \end{cases}$

44. $\begin{cases} 3x + 6y = 15 \\ 2x + 4y = 3 \end{cases}$

45. $\begin{cases} 10y - 2x = 1 \\ 5y = 4 - 6x \end{cases}$

46. $\begin{cases} 3x + 4y = 0 \\ 7x = 3y \end{cases}$

47. $\begin{cases} \dfrac{3}{4}x + \dfrac{5}{2}y = 11 \\ \dfrac{1}{16}x - \dfrac{3}{4}y = -1 \end{cases}$

48. $\begin{cases} \dfrac{2}{3}x + \dfrac{1}{4}y = -\dfrac{3}{2} \\ \dfrac{1}{2}x - \dfrac{1}{4}y = -2 \end{cases}$

49. $\begin{cases} x = 3y + 2 \\ 5x - 15y = 10 \end{cases}$

50. $\begin{cases} x = 7y - 21 \\ 2x - 14y = -42 \end{cases}$

51. $\begin{cases} \dfrac{x}{3} + y = \dfrac{4}{3} \\ -x + 2y = 11 \end{cases}$

52. $\begin{cases} \dfrac{x}{8} - \dfrac{y}{2} = 1 \\ \dfrac{x}{3} - y = 2 \end{cases}$

53. $\begin{cases} 2x = 6 \\ y = 5 - x \end{cases}$

54. $\begin{cases} x = 3y + 4 \\ -y = 5 \end{cases}$

55. $\begin{cases} \dfrac{x+5}{2} = \dfrac{6-4y}{3} \\ \dfrac{3x}{5} = \dfrac{21-7y}{10} \end{cases}$

56. $\begin{cases} \dfrac{y}{5} = \dfrac{8-x}{2} \\ x = \dfrac{2y-8}{3} \end{cases}$

57. $\begin{cases} 4x - 7y = 7 \\ 12x - 21y = 24 \end{cases}$

58. $\begin{cases} 2x - 5y = 12 \\ -4x + 10y = 20 \end{cases}$

59. $\begin{cases} \dfrac{2}{3}x - \dfrac{3}{4}y = -1 \\ -\dfrac{1}{6}x + \dfrac{3}{8}y = 1 \end{cases}$

60. $\begin{cases} \dfrac{1}{2}x - \dfrac{1}{3}y = -3 \\ \dfrac{1}{8}x + \dfrac{1}{6}y = 0 \end{cases}$

61. $\begin{cases} 2x - y = -1 \\ y = -2x \end{cases}$

62. $\begin{cases} 4y - x = -1 \\ x = -2y \end{cases}$

63. $\begin{cases} 0.7x - 0.2y = -1.6 \\ 0.2x - y = -1.4 \end{cases}$

64. $\begin{cases} -0.7x + 0.6y = 1.3 \\ 0.5x - 0.3y = -0.8 \end{cases}$

65. $\begin{cases} 4x - 1.5y = 10.2 \\ 2x + 7.8y = -25.68 \end{cases}$

66. $\begin{cases} x - 3y = -5.3 \\ 6.3x + 6y = 3.96 \end{cases}$

Review

Determine whether the given replacement values make each equation true or false. See Section 2.1.

67. $3x - 4y + 2z = 5$;
$x = 1, y = 2,$ and $z = 5$

68. $x + 2y - z = 7$;
$x = 2, y = -3,$ and $z = 3$

69. $-x - 5y + 3z = 15$;
$x = 0, y = -1,$ and $z = 5$

70. $-4x + y - 8z = 4$;
$x = 1, y = 0,$ and $z = -1$

Add the equations. See this section.

71. $\begin{cases} 3x + 2y - 5z = 10 \\ -3x + 4y + z = 15 \end{cases}$

72. $\begin{cases} x + 4y - 5z = 20 \\ 2x - 4y - 2z = -17 \end{cases}$

73. $\begin{cases} 10x + 5y + 6z = 14 \\ -9x + 5y - 6z = -12 \end{cases}$

74. $\begin{cases} -9x - 8y - z = 31 \\ 9x + 4y - z = 12 \end{cases}$

Concept Extensions

Without graphing, determine whether each system has one solution, no solution, or an infinite number of solutions. See the Concept Checks in this section.

75. $\begin{cases} y = 2x - 5 \\ y = 2x + 1 \end{cases}$

76. $\begin{cases} y = 3x - \dfrac{1}{2} \\ y = -2x + \dfrac{1}{5} \end{cases}$

77. $\begin{cases} x + y = 3 \\ 5x + 5y = 15 \end{cases}$

78. $\begin{cases} y = 5x - 2 \\ y = -\dfrac{1}{5}x - 2 \end{cases}$

79. Can a system consisting of two linear equations have exactly two solutions? Explain why or why not.

80. Suppose the graph of the equations in a system of two equations in two variables consists of a circle and a line. Discuss the possible number of solutions for this system.

The concept of supply and demand is used often in business. In general, as the unit price of a commodity increases, the demand for that commodity decreases. Also, as a commodity's unit price increases, the manufacturer normally increases the supply. The point where supply is equal to demand is called the equilibrium point. The following shows the graph of a demand equation and the graph of a supply equation for previously rented DVDs. The x-axis represents the number of DVDs in thousands, and the y-axis represents the cost of a DVD. Use this graph to answer Exercises 81 through 84.

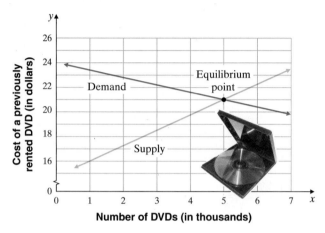

81. Find the number of DVDs and the price per DVD when supply equals demand.

82. When x is between 3 and 4, is supply greater than demand or is demand greater than supply?

83. When x is greater than 6, is supply greater than demand or is demand greater than supply?

84. For what x-values are the y-values corresponding to the supply equation greater than the y-values corresponding to the demand equation?

The revenue equation for a certain brand of toothpaste is $y = 2.5x$, where x is the number of tubes of toothpaste sold and y is the total income for selling x tubes. The cost equation is $y = 0.9x + 3000$, where x is the number of tubes of toothpaste manufactured and y is the cost of producing x tubes. The following set of axes shows the graph of the cost and revenue equations. Use this graph for Exercises 85 through 90.

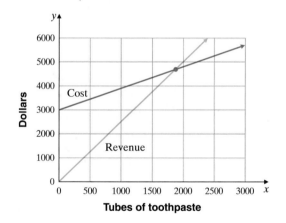

85. Find the coordinates of the point of intersection, or break-even point, by solving the system
$$\begin{cases} y = 2.5x \\ y = 0.9x + 3000 \end{cases}$$

86. Explain the meaning of the y-value of the point of intersection.

87. If the company sells 2000 tubes of toothpaste, does the company make money or lose money?

88. If the company sells 1000 tubes of toothpaste, does the company make money or lose money?

89. For what x-values will the company make a profit? (*Hint:* For what x-values is the revenue graph "higher" than the cost graph?)

90. For what x-values will the company lose money? (*Hint:* For what x-values is the revenue graph "lower" than the cost graph?)

91. Write a system of two linear equations in x and y that has the ordered pair solution $(2, 5)$.

92. Which method would you use to solve the system?
$$\begin{cases} 5x - 2y = 6 \\ 2x + 3y = 5 \end{cases}$$

Explain your choice.

93. The amount y of red meat consumed per person in the United States (in pounds) in the year x can be modeled by the linear equation $y = -0.56x + 114$. The amount y of all poultry consumed per person in the United States (in pounds) in the year x can be modeled by the linear equation $y = 0.98x + 67$. In both models, $x = 0$ represents the year 1997. (*Source:* Based on data and forecasts from the Economic Research Service, U.S. Department of Agriculture)

   a. What does the slope of each equation tell you about the patterns of red meat and poultry consumption in the United States?

   b. Solve this system of equations. (Round your final results to the nearest whole numbers.)

   c. Explain the meaning of your answer to part (b).

94. The amount of U.S. federal government income y (in billions of dollars) for the fiscal year x, from 2000 through 2004 ($x = 0$ represents 2000), can be modeled by the linear equation $y = -48.2x + 2025.4$. The amount of U.S. federal government expenditures y (in billions of dollars) for the same period can be modeled by the linear equation $y = 99x + 2165.5$. Did expenses ever equal income during this period? If so, in what year? (*Source:* Based on data from Financial Management Service, U.S. Department of the Treasury, 1995–2000)

Solve each system. To do so you may want to let $a = \dfrac{1}{x}$ (if x is in the denominator) and let $b = \dfrac{1}{y}$ (if y is in the denominator.)

95. $\begin{cases} \dfrac{1}{x} + y = 12 \\ \dfrac{3}{x} - y = 4 \end{cases}$

96. $\begin{cases} x + \dfrac{2}{y} = 7 \\ 3x + \dfrac{3}{y} = 6 \end{cases}$

97. $\begin{cases} \dfrac{1}{x} + \dfrac{1}{y} = 5 \\ \dfrac{1}{x} - \dfrac{1}{y} = 1 \end{cases}$

98. $\begin{cases} \dfrac{2}{x} + \dfrac{3}{y} = 5 \\ \dfrac{5}{x} - \dfrac{3}{y} = 2 \end{cases}$

99. $\begin{cases} \dfrac{2}{x} + \dfrac{3}{y} = -1 \\ \dfrac{3}{x} - \dfrac{2}{y} = 18 \end{cases}$

100. $\begin{cases} \dfrac{3}{x} - \dfrac{2}{y} = -18 \\ \dfrac{2}{x} + \dfrac{3}{y} = 1 \end{cases}$

101. $\begin{cases} \dfrac{2}{x} - \dfrac{4}{y} = 5 \\ \dfrac{1}{x} - \dfrac{2}{y} = \dfrac{3}{2} \end{cases}$

102. $\begin{cases} \dfrac{5}{x} + \dfrac{7}{y} = 1 \\ -\dfrac{10}{x} - \dfrac{14}{y} = 0 \end{cases}$

STUDY SKILLS BUILDER

Are You Getting All the Mathematics Help That You Need?

Remember that, in addition to your instructor, there are many places to get help with your mathematics course. For example,

- This text has an accompanying video lesson for every section and worked out solutions to every Chapter Test exercise on video.
- The back of the book contains answers to odd-numbered exercises and selected solutions.
- A student *Solutions Manual* is available that contains worked-out solutions to odd-numbered exercises as well as solutions to every exercise in the Integrated Reviews, Chapter Reviews, Chapter Tests, and Cumulative Reviews.

- Don't forget to check with your instructor for other local resources available to you, such as a tutor center.

Exercises

1. List items you find helpful in the text and all student supplements to this text.

2. List all the campus help that is available to you for this course.

3. List any help (besides the textbook) from Exercises 1 and 2 above that you are using.

4. List any help (besides the textbook) that you feel you should try.

5. Write a goal for yourself that includes trying anything you listed in Exercise 4 during the next week.

4.2 SOLVING SYSTEMS OF LINEAR EQUATIONS IN THREE VARIABLES

Objective

[A] Solve a System of Three Linear Equations in Three Variables.

In this section, we solve systems of linear equations in three variables. We call the equation $3x - y + z = -15$, for example, a **linear equation in three variables** since there are three variables and each variable is raised only to the power 1. A solution of this equation is an **ordered triple (x, y, z)** that makes the equation a true statement.

For example, the ordered triple $(2, 0, -21)$ is a solution of $3x - y + z = -15$ since replacing x with 2, y with 0, and z with -21 yields the true statement

$$3(2) - 0 + (-21) = -15$$

The graph of this equation is a plane in three-dimensional space, just as the graph of a linear equation in two variables is a line in two-dimensional space.

Although we will not discuss the techniques for graphing equations in three variables, visualizing the possible patterns of intersecting planes gives us insight into the possible patterns of solutions of a system of three three-variable linear equations. There are four possible patterns.

1. Three planes have a single point in common. This point represents the single solution of the system. This system is **consistent.**

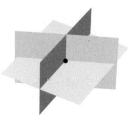

2. Three planes intersect at no point common to all three. This system has no solution. A few ways that this can occur are shown. This system is **inconsistent.**

3. Three planes intersect at all the points of a single line. The system has infinitely many solutions. This system is **consistent.**

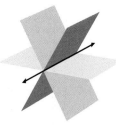

4. Three planes coincide at all points on the plane. The system is **consistent,** and the equations are **dependent.**

Objective A Solving a System of Three Linear Equations in Three Variables

Just as with systems of two equations in two variables, we can use the elimination or substitution method to solve a system of three equations in three variables. To use the elimination method, we eliminate a variable and obtain a system of two equations in two variables. Then we use the methods we learned in the previous section to solve the system of two equations.

PRACTICE PROBLEM 1

Solve the system:

$$\begin{cases} 2x - y + 3z = 13 \\ x + y - z = -2 \\ 3x + 2y + 2z = 13 \end{cases}$$

EXAMPLE 1 Solve the system:

$$\begin{cases} 3x - y + z = -15 & \text{Equation (1)} \\ x + 2y - z = 1 & \text{Equation (2)} \\ 2x + 3y - 2z = 0 & \text{Equation (3)} \end{cases}$$

Solution: Let's add equations (1) and (2) to eliminate z.

$$\begin{aligned} 3x - y + z &= -15 \\ \underline{x + 2y - z} &= \underline{1} \\ 4x + y &= -14 \quad \text{Equation (4)} \end{aligned}$$

Helpful Hint

Make sure you add two other equations besides equations (1) and (2) and *also* **eliminate the same variable.** You will see why as you follow this example.

Next we add two *other* equations and *eliminate z again*. To do so, we multiply both sides of equation (1) by 2 and add the resulting equation to equation (3). Then

$$\begin{cases} 2(3x - y + z) = 2(-15) \\ 2x + 3y - 2z = 0 \end{cases} \text{simplifies to} \begin{cases} 6x - 2y + 2z = -30 \\ \underline{2x + 3y - 2z = 0} \\ 8x + y = -30 \end{cases}$$
$$\text{Equation (5)}$$

We now have two equations (4 and 5) in the same two variables. This means we can solve equations (4) and (5) for x and y. To solve by elimination, we multiply both sides of equation (4) by -1 and add the resulting equation to equation (5). Then

$$\begin{cases} -1(4x + y) = -1(-14) \\ 8x + y = -30 \end{cases} \text{simplifies to} \begin{cases} -4x - y = 14 \\ \underline{8x + y = -30} \\ 4x = -16 \quad \text{Add the equations.} \\ x = -4 \quad \text{Solve for } x. \end{cases}$$

We now replace x with -4 in equation (4) or (5).

$$\begin{aligned} 4x + y &= -14 \quad \text{Equation (4)} \\ 4(-4) + y &= -14 \quad \text{Let } x = -4. \\ y &= 2 \quad \text{Solve for } y. \end{aligned}$$

Finally, we replace x with -4 and y with 2 in equation (1), (2), or (3).

$$\begin{aligned} x + 2y - z &= 1 \quad \text{Equation (2)} \\ -4 + 2(2) - z &= 1 \quad \text{Let } x = -4 \text{ and } y = 2. \\ -4 + 4 - z &= 1 \\ -z &= 1 \\ z &= -1 \end{aligned}$$

The ordered triple solution is $(-4, 2, -1)$. To check, we let $x = -4$, $y = 2$, and $z = -1$ in *all three* original equations of the system.

Equation (1)

$$\begin{aligned} 3x - y + z &= -15 \\ 3(-4) - 2 + (-1) &\stackrel{?}{=} -15 \\ -12 - 2 - 1 &\stackrel{?}{=} -15 \\ -15 &= -15 \quad \text{True} \end{aligned}$$

Equation (2)

$$\begin{aligned} x + 2y - z &= 1 \\ -4 + 2(2) - (-1) &\stackrel{?}{=} 1 \\ -4 + 4 + 1 &\stackrel{?}{=} 1 \\ 1 &= 1 \quad \text{True} \end{aligned}$$

Answer

1. $(1, 1, 4)$

Equation (3)

$$2x + 3y - 2z = 0$$
$$2(-4) + 3(2) - 2(-1) \stackrel{?}{=} 0$$
$$-8 + 6 + 2 \stackrel{?}{=} 0$$
$$0 = 0 \quad \text{True}$$

All three statements are true, so the ordered triple solution is $(-4, 2, -1)$.

■ **Work Practice Problem 1**

EXAMPLE 2 Solve the system:

$$\begin{cases} 2x - 4y + 8z = 2 & (1) \\ -x - 3y + z = 11 & (2) \\ x - 2y + 4z = 0 & (3) \end{cases}$$

Solution: Add equations (2) and (3) to eliminate x, and the new equation is

$$-5y + 5z = 11 \quad (4)$$

To eliminate x again, we multiply both sides of equation (2) by 2 and add the resulting equation to equation (1). Then

$$\begin{cases} 2x - 4y + 8z = 2 \\ 2(-x - 3y + z) = 2(11) \end{cases} \text{ simplifies to } \begin{cases} 2x - 4y + 8z = 2 \\ \underline{-2x - 6y + 2z = 22} \\ -10y + 10z = 24 \quad (5) \end{cases}$$

Next we solve for y and z using equations (4) and (5). To do so, we multiply both sides of equation (4) by -2 and add the resulting equation to equation (5).

$$\begin{cases} -2(-5y + 5z) = -2(11) \\ -10y + 10z = 24 \end{cases} \text{ simplifies to } \begin{cases} 10y - 10z = -22 \\ \underline{-10y + 10z = 24} \\ 0 = 2 \quad \text{False} \end{cases}$$

Since the statement is false, this system is inconsistent and has no solution. The solution set is the empty set $\{\ \}$ or $\varnothing$.

■ **Work Practice Problem 2**

The elimination method is summarized next.

Solving a System of Three Linear Equations by the Elimination Method

Step 1: Write each equation in standard form, $Ax + By + Cz = D$.

Step 2: Choose a pair of equations and use them to eliminate a variable.

Step 3: Choose any other pair of equations and eliminate the *same variable* as in Step 2.

Step 4: Two equations in two variables should be obtained from Step 2 and Step 3. Use methods from Section 4.1 to solve this system for both variables.

Step 5: To solve for the third variable, substitute the values of the variables found in Step 4 into any of the original equations containing the third variable.

Step 6: Check the ordered triple solution in *all three* original equations.

PRACTICE PROBLEM 2

Solve the system:

$$\begin{cases} 2x + 4y - 2z = 3 \\ -x + y - z = 6 \\ x + 2y - z = 1 \end{cases}$$

Helpful Hint

Make sure you read closely and follow Step 3.

Answer

2. $\varnothing$

✔**Concept Check** In the system

$$\begin{cases} x + y + z = 6 & \text{Equation (1)} \\ 2x - y + z = 3 & \text{Equation (2)} \\ x + 2y + 3z = 14 & \text{Equation (3)} \end{cases}$$

equations (1) and (2) are used to eliminate y. Which action could be used to finish solving? Why?

a. Use (1) and (2) to eliminate z.
b. Use (2) and (3) to eliminate y.
c. Use (1) and (3) to eliminate x.

PRACTICE PROBLEM 3

Solve the system:

$$\begin{cases} 3x + 2y = -1 \\ 6x - 2z = 4 \\ y - 3z = 2 \end{cases}$$

EXAMPLE 3 Solve the system:

$$\begin{cases} 2x + 4y = 1 & (1) \\ 4x - 4z = -1 & (2) \\ y - 4z = -3 & (3) \end{cases}$$

Solution: Notice that equation (2) has no term containing the variable y. Let's eliminate y using equations (1) and (3). We multiply both sides of equation (3) by -4 and add the resulting equation to equation (1). Then

$$\begin{cases} 2x + 4y = 1 \\ -4(y - 4z) = -4(-3) \end{cases} \text{ simplifies to } \begin{cases} 2x + 4y = 1 \\ -4y + 16z = 12 \\ \hline 2x + 16z = 13 \quad (4) \end{cases}$$

Next we solve for z using equations (4) and (2). We multiply both sides of equation (4) by -2 and add the resulting equation to equation (2).

$$\begin{cases} -2(2x + 16z) = -2(13) \\ 4x - 4z = -1 \end{cases} \text{ simplifies to } \begin{cases} -4x - 32z = -26 \\ 4x - 4z = -1 \\ \hline -36z = -27 \\ z = \dfrac{3}{4} \end{cases}$$

Now we replace z with $\dfrac{3}{4}$ in equation (3) and solve for y.

$$y - 4\left(\dfrac{3}{4}\right) = -3 \quad \text{Let } z = \dfrac{3}{4} \text{ in equation (3).}$$
$$y - 3 = -3$$
$$y = 0$$

Finally, we replace y with 0 in equation (1) and solve for x.

$$2x + 4(0) = 1 \quad \text{Let } y = 0 \text{ in equation (1).}$$
$$2x = 1$$
$$x = \dfrac{1}{2}$$

The ordered triple solution is $\left(\dfrac{1}{2}, 0, \dfrac{3}{4}\right)$. Check to see that this solution satisfies *all three* equations of the system.

■ **Work Practice Problem 3**

Answer

3. $\left(\dfrac{1}{3}, -1, -1\right)$

✔ **Concept Check Answer**

b; answers may vary

EXAMPLE 4 Solve the system:

$$\begin{cases} x - 5y - 2z = 6 & (1) \\ -2x + 10y + 4z = -12 & (2) \\ \dfrac{1}{2}x - \dfrac{5}{2}y - z = 3 & (3) \end{cases}$$

Solution: We multiply both sides of equation (3) by 2 to eliminate fractions, and we multiply both sides of equation (2) by $-\dfrac{1}{2}$ so that the coefficient of x is 1. The resulting system is then

$$\begin{cases} x - 5y - 2z = 6 & (1) \\ x - 5y - 2z = 6 & \text{Multiply (2) by } -\dfrac{1}{2}. \\ x - 5y - 2z = 6 & \text{Multiply (3) by 2.} \end{cases}$$

All three resulting equations are identical, and therefore equations (1), (2), and (3) are all equivalent. There are infinitely many solutions of this system. The equations are dependent. The solution set can be written as $\{(x, y, z)|x - 5y - 2z = 6\}$.

◻ **Work Practice Problem 4**

As mentioned earlier, we can also use the substitution method to solve a system of linear equations in three variables.

EXAMPLE 5 Solve the system:

$$\begin{cases} x - 4y - 5z = 35 & (1) \\ x - 3y = 0 & (2) \\ -y + z = -55 & (3) \end{cases}$$

Solution: Notice in equations (2) and (3) that a variable is missing. Also notice that both equations contain the variable y. Let's use the substitution method by solving equation (2) for x and equation (3) for z and substituting the results in equation (1).

$$x - 3y = 0 \qquad (2)$$
$$x = 3y \qquad \text{Solve equation (2) for } x.$$
$$-y + z = -55 \qquad (3)$$
$$z = y - 55 \qquad \text{Solve equation (3) for } z.$$

Now substitute $3y$ for x and $y - 55$ for z in equation (1).

$$x - 4y - 5z = 35 \qquad (1)$$
$$3y - 4y - 5(y - 55) = 35 \qquad \text{Let } x = 3y \text{ and } z = y - 55.$$
$$3y - 4y - 5y + 275 = 35 \qquad \text{Use the distributive property and multiply.}$$
$$-6y + 275 = 35 \qquad \text{Combine like terms.}$$
$$-6y = -240 \qquad \text{Subtract 275 from both sides.}$$
$$y = 40 \qquad \text{Solve.}$$

To find x, recall that $x = 3y$ and substitute 40 for y. Then $x = 3y$ becomes $x = 3 \cdot 40 = 120$. To find z, recall that $z = y - 55$ and also substitute 40 for y. Then $z = y - 55$ becomes $z = 40 - 55 = -15$. The solution is $(120, 40, -15)$.

◻ **Work Practice Problem 5**

PRACTICE PROBLEM 4

Solve the system:

$$\begin{cases} x - 3y + 4z = 2 \\ -2x + 6y - 8z = -4 \\ \dfrac{1}{2}x - \dfrac{3}{2}y + 2z = 1 \end{cases}$$

PRACTICE PROBLEM 5

Solve the system:

$$\begin{cases} 2x + 5y - 3z = 30 & (1) \\ x + y = -3 & (2) \\ 2x - z = 0 & (3) \end{cases}$$

(*Hint:* Equations (2) and (3) each contain the variable x and have a variable missing.)

Helpful Hint

Do not forget to distribute.

Objective A *Solve.*

1. Choose the equation(s) that has $(-1, 3, 1)$ as a solution.

   a. $x + y + z = 3$ **b.** $-x + y + z = 5$
   c. $-x + y + 2z = 0$ **d.** $x + 2y - 3z = 2$

2. Choose the equation(s) that has $(2, 1, -4)$ as a solution.

   a. $x + y + z = -1$ **b.** $x - y - z = -3$
   c. $2x - y + z = -1$ **d.** $-x - 3y - z = -1$

3. Use the result of Exercise 1 to determine whether $(-1, 3, 1)$ is a solution of the system below. Explain your answer.
$$\begin{cases} x + y + z = 3 \\ -x + y + z = 5 \\ x + 2y - 3z = 2 \end{cases}$$

4. Use the result of Exercise 2 to determine whether $(2, 1, -4)$ is a solution of the system below. Explain your answer.
$$\begin{cases} x + y + z = -1 \\ x - y - z = -3 \\ 2x - y + z = -1 \end{cases}$$

Mixed Practice *Solve each system. See Examples 1 through 5.*

5. $\begin{cases} x - y + z = -4 \\ 3x + 2y - z = 5 \\ -2x + 3y - z = 15 \end{cases}$

6. $\begin{cases} x + y - z = -1 \\ -4x - y + 2z = -7 \\ 2x - 2y - 5z = 7 \end{cases}$

7. $\begin{cases} x + y = 3 \\ 2y = 10 \\ 3x + 2y - 3z = 1 \end{cases}$

8. $\begin{cases} 5x = 5 \\ 2x + y = 4 \\ 3x + y - 4z = -15 \end{cases}$

9. $\begin{cases} 2x + 2y + z = 1 \\ -x + y + 2z = 3 \\ x + 2y + 4z = 0 \end{cases}$

10. $\begin{cases} 2x - 3y + z = 5 \\ x + y + z = 0 \\ 4x + 2y + 4z = 4 \end{cases}$

11. $\begin{cases} x - 2y + z = -5 \\ -3x + 6y - 3z = 15 \\ 2x - 4y + 2z = -10 \end{cases}$

12. $\begin{cases} 3x + y - 2z = 2 \\ -6x - 2y + 4z = -2 \\ 9x + 3y - 6z = 6 \end{cases}$

13. $\begin{cases} 4x - y + 2z = 5 \\ 2y + z = 4 \\ 4x + y + 3z = 10 \end{cases}$

14. $\begin{cases} 5y - 7z = 14 \\ 2x + y + 4z = 10 \\ 2x + 6y - 3z = 30 \end{cases}$

15. $\begin{cases} x + 5z = 0 \\ 5x + y = 0 \\ y - 3z = 0 \end{cases}$

16. $\begin{cases} x - 5y = 0 \\ x - z = 0 \\ -x + 5z = 0 \end{cases}$

17. $\begin{cases} 6x - 5z = 17 \\ 5x - y + 3z = -1 \\ 2x + y = -41 \end{cases}$

18. $\begin{cases} x + 2y = 6 \\ 7x + 3y + z = -33 \\ x - z = 16 \end{cases}$

19. $\begin{cases} x + y + z = 8 \\ 2x - y - z = 10 \\ x - 2y - 3z = 22 \end{cases}$

20. $\begin{cases} 5x + y + 3z = 1 \\ x - y + 3z = -7 \\ -x + y = 1 \end{cases}$

21. $\begin{cases} x + 2y - z = 5 \\ 6x + y + z = 7 \\ 2x + 4y - 2z = 5 \end{cases}$

22. $\begin{cases} 4x - y + 3z = 10 \\ x + y - z = 5 \\ 8x - 2y + 6z = 10 \end{cases}$

23. $\begin{cases} 2x - 3y + z = 2 \\ x - 5y + 5z = 3 \\ 3x + y - 3z = 5 \end{cases}$

24. $\begin{cases} 4x + y - z = 8 \\ x - y + 2z = 3 \\ 3x - y + z = 6 \end{cases}$

25. $\begin{cases} -2x - 4y + 6z = -8 \\ x + 2y - 3z = 4 \\ 4x + 8y - 12z = 16 \end{cases}$

26. $\begin{cases} -6x + 12y + 3z = -6 \\ 2x - 4y - z = 2 \\ -x + 2y + \dfrac{z}{2} = -1 \end{cases}$

27. $\begin{cases} 2x + 2y - 3z = 1 \\ y + 2z = -14 \\ 3x - 2y = -1 \end{cases}$

28. $\begin{cases} 7x + 4y = 10 \\ x - 4y + 2z = 6 \\ y - 2z = -1 \end{cases}$

29. $\begin{cases} x + 2y - z = 5 \\ -3x - 2y - 3z = 11 \\ 4x + 4y + 5z = -18 \end{cases}$

30. $\begin{cases} 3x - 3y + z = -1 \\ 3x - y - z = 3 \\ -6x + 4y + 3z = -8 \end{cases}$

31. $\begin{cases} \dfrac{3}{4}x - \dfrac{1}{3}y + \dfrac{1}{2}z = 9 \\ \dfrac{1}{6}x + \dfrac{1}{3}y - \dfrac{1}{2}z = 2 \\ \dfrac{1}{2}x - y + \dfrac{1}{2}z = 2 \end{cases}$

32. $\begin{cases} \dfrac{1}{3}x - \dfrac{1}{4}y + z = -9 \\ \dfrac{1}{2}x - \dfrac{1}{3}y - \dfrac{1}{4}z = -6 \\ x - \dfrac{1}{2}y - z = -8 \end{cases}$

Review

Solve. See Section 2.1.

33. $2(x - 1) - 3x = x - 12$

34. $7(2x - 1) + 4 = 11(3x - 2)$

35. $-y - 5(y + 5) = 3y - 10$

36. $z - 3(z + 7) = 6(2z + 1)$

Solve. See Section 2.2.

37. The sum of two numbers is 45 and one number is twice the other. Find the numbers.

38. The difference between two numbers is 5. Twice the smaller number added to five times the larger number is 53. Find the numbers.

Concept Extensions

39. Write a single linear equation in three variables that has $(-1, 2, -4)$ as a solution. (There are many possibilities.) Explain the process you used to write an equation.

40. Write a system of three linear equations in three variables that has $(2, 1, 5)$ as a solution. (There are many possibilities.) Explain the process you used to write an equation.

41. Write a system of linear equations in three variables that has the solution $(-1, 2, -4)$. Explain the process you used to write your system.

42. When solving a system of three equations in three unknowns, explain how to determine that a system has no solution.

43. The fraction $\dfrac{1}{24}$ can be written as the following sum:

$$\frac{1}{24} = \frac{x}{8} + \frac{y}{4} + \frac{z}{3}$$

where the numbers x, y, and z are solutions of

$$\begin{cases} x + y + z = 1 \\ 2x - y + z = 0 \\ -x + 2y + 2z = -1 \end{cases}$$

Solve the system and see that the sum of the fractions is $\dfrac{1}{24}$.

44. The fraction $\dfrac{1}{18}$ can be written as the following sum:

$$\frac{1}{18} = \frac{x}{2} + \frac{y}{3} + \frac{z}{9}$$

where the numbers x, y, and z are solutions of

$$\begin{cases} x + 3y + z = -3 \\ -x + y + 2z = -14 \\ 3x + 2y - z = 12 \end{cases}$$

Solve the system and see that the sum of the fractions is $\dfrac{1}{18}$.

Solving systems involving more than three variables can be accomplished with methods similar to those encountered in this section. Apply what you already know to solve each system of equations in four variables.

45. $\begin{cases} x + y \quad\;\; - w = 0 \\ \quad\;\; y + 2z + w = 3 \\ x \quad\quad - z \quad\quad = 1 \\ 2x - y \quad\quad\;\; - w = -1 \end{cases}$

46. $\begin{cases} 5x + 4y \quad\quad\quad = 29 \\ \quad\quad y + z - w = -2 \\ 5x \quad\quad + z \quad\quad = 23 \\ \quad\quad y - z + w = 4 \end{cases}$

47. $\begin{cases} x + y + z + w = 5 \\ 2x + y + z + w = 6 \\ x + y + z \quad\quad = 2 \\ x + y \quad\quad\quad = 0 \end{cases}$

48. $\begin{cases} 2x \quad\quad - z \quad\quad = -1 \\ \quad\; y + z + \;\; w = 9 \\ \quad\; y \quad\quad - 2w = -6 \\ x + y \quad\quad\quad = 3 \end{cases}$

49. Write a system of three linear equations in three variables that are dependent equations.

50. What is the solution to the system in Exercise 49?

4.3 SYSTEMS OF LINEAR EQUATIONS AND PROBLEM SOLVING

Objectives

A Solve Problems That Can Be Modeled by a System of Two Linear Equations.

B Solve Problems with Cost and Revenue Functions.

C Solve Problems That Can Be Modeled by a System of Three Linear Equations.

Objective **A** Solving Problems Modeled by Systems of Two Equations

Thus far, we have solved problems by writing one-variable equations and solving for the variable. Some of these problems can be solved, perhaps more easily, by writing a system of equations, as illustrated in this section.

EXAMPLE 1 Predicting Equal Consumption of Red Meat and Poultry

America's consumption of red meat has decreased most years since 1980 while consumption of poultry has increased. The function $y = -0.71x + 125.6$ approximates the annual pounds of red meat consumed per capita, where x is the number of years since 1980. The function $y = 1.56x + 39.7$ approximates the annual pounds of poultry consumed per capita, where x is also the number of years since 1980. If this trend continues, determine the year when the annual consumption of red meat and poultry are equal.

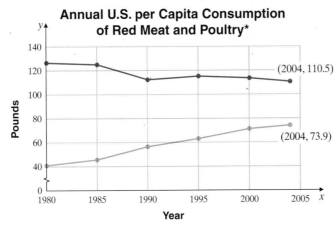

Annual U.S. per Capita Consumption of Red Meat and Poultry*

(2004, 110.5)

(2004, 73.9)

Source: USDA Economic Research Service

* Excludes shipments to Puerto Rico and other U.S. possessions

Solution:

1. **UNDERSTAND.** Read and reread the problem and guess a year. Let's guess the year 2010. This year is 30 years since 1980, so $x = 30$. Now let $x = 30$ in each given function.

   Red meat: $y = -0.71x + 125.6 = -0.71(30) + 125.6 = 104.3$ pounds

   Poultry: $y = 1.56x + 39.7 = 1.56(30) + 39.7 = 86.5$ pounds

   Since the projected pounds in 2010 for red meat and poultry are not the same, we guessed incorrectly, but we do have a better understanding of the problem, and we know that the year will be later than 2010.

2. **TRANSLATE.** We are already given the system of equations.

3. **SOLVE.** We want to know the year x in which pounds y are the same, so we solve the system:

$$\begin{cases} y = -0.71x + 125.6 \\ y = 1.56x + 39.7 \end{cases}$$

Continued on next page

PRACTICE PROBLEM 1

Read Example 1. If we use the years 2000, 2001, 2002, and 2004 only to write functions approximating the consumption of red meat and poultry, we have the following.

Red meat: $y = -0.7x + 113.5$

Poultry: $y = 2.4x + 66.3$

where x is years since 2000 and y is pounds per year consumed.

a. Assuming this trend continues, predict the year when consumption of red meat and poultry will be the same. Round to the nearest year.

b. Does your answer differ from the answer to Example 1? Why or why not.

Answers

1. a. 2015, **b.** yes; answers may vary

Since both equations are solved for y, one way to solve is to use the substitution method.

$$y = -0.71x + 125.6 \quad \text{First equation}$$

$$1.56x + 39.7 = -0.71x + 125.6 \quad \text{Let } y = 1.56x + 39.7.$$

$$2.27x = 85.9$$

$$x = \frac{85.9}{2.27} \approx 37.84$$

4. INTERPRET. Since we are only asked to find the year, we need only solve for x.

Check: To check, see whether $x \approx 37.84$ gives approximately the same number of pounds of red meat and poultry.

Red meat: $y = -0.71x + 125.6 = -0.71(37.84) + 125.6 \approx 98.73$ pounds
Poultry: $y = 1.56x + 39.7 = 1.56(37.84) + 39.7 \approx 98.73$ pounds

Since we rounded the number of years, the number of pounds do differ slightly. They differ only by 0.0032, so we can assume that we solved correctly.

State: The consumption of red meat and poultry will be the same about 37.84 years after 1980, or 2017.84. Thus, in the year 2017, we predict the consumption will be the same.

⬛ **Work Practice Problem 1**

PRACTICE PROBLEM 2

A first number is 7 greater than a second number. Twice the first number is 4 more than three times the second. Find the numbers.

EXAMPLE 2 **Finding Unknown Numbers**

A first number is 4 less than a second number. Four times the first number is 6 more than twice the second. Find the numbers.

Solution:

1. UNDERSTAND. Read and reread the problem and guess a solution. If one number is 10 and this is 4 less than a second number, the second number is 14. Four times the first number is $4(10)$, or 40. This is not equal to 6 more than twice the second number, which is $2(14) + 6$ or 34. Although we guessed incorrectly, we now have a better understanding of the problem.

Since we are looking for two numbers, we will let

x = first number

y = second number

2. TRANSLATE. Since we have assigned two variables to this problem, we will translate the given facts into two equations. For the first statement we have

In words:	the first number	is	4 less than second number
	↓	↓	↓
Translate:	x	$=$	$y - 4$

Next we translate the second statement into an equation.

In words:	four times the first number	is	6 more than twice the second number
	↓	↓	↓
Translate:	$4x$	$=$	$2y + 6$

3. SOLVE. Now we solve the system

$$\begin{cases} x = y - 4 \\ 4x = 2y + 6 \end{cases}$$

Answer

2. 17 and 10

Since the first equation expresses x in terms of y, we will use substitution. We substitute $y - 4$ for x in the second equation and solve for y.

$$4x = 2y + 6 \quad \text{Second equation}$$

$$4(y - 4) = 2y + 6 \quad \text{Let } x = y - 4.$$
$$4y - 16 = 2y + 6$$
$$2y = 22$$
$$y = 11$$

Now we replace y with 11 in the equation $x = y - 4$ and solve for x. Then $x = y - 4$ becomes $x = 11 - 4 = 7$. The ordered pair solution of the system is $(7, 11)$.

4. **INTERPRET.** Since the solution of the system is $(7, 11)$, the first number we are looking for is 7 and the second number is 11.

Check: Notice that 7 *is* 4 less than 11, and 4 times 7 *is* 6 more than twice 11. The proposed numbers, 7 and 11, are correct.

State: The numbers are 7 and 11.

⬛ **Work Practice Problem 2**

EXAMPLE 3 **Finding the Rate of Speed**

Two cars leave Indianapolis, one traveling east and the other west. After 3 hours they are 297 miles apart. If one car is traveling 5 mph faster than the other, what is the speed of each?

Solution:

1. **UNDERSTAND.** Read and reread the problem. Let's guess a solution and use the formula $d = r \cdot t$ to check. Suppose the faster car is traveling at a rate of 55 mph. This means that the other car is traveling at a rate of 50 mph since we are told that one car is traveling 5 mph faster than the other. To find the distance apart after 3 hours, we will first find the distance traveled by each car. One car's distance is rate $\cdot$ time $= 55(3) = 165$ miles. The other car's distance is rate $\cdot$ time $= 50(3) = 150$ miles. Since one car is traveling east and the other west, their distance apart is the sum of their distances, or 165 miles $+$ 150 miles $=$ 315 miles. Although this distance apart is not the required distance of 297 miles, we now have a better understanding of the problem.

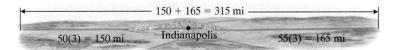

Let's model the problem with a system of equations. We will let

$x =$ speed of one car

$y =$ speed of the other car

We summarize the information on the following chart. Both cars have traveled 3 hours. Since distance $=$ rate $\cdot$ time, their distances are $3x$ and $3y$ miles, respectively.

	Rate	·	Time	=	Distance
One Car	x		3		$3x$
Other Car	y		3		$3y$

Continued on next page

---

2. TRANSLATE. We can now translate the stated conditions into two equations.

In words:

one car's distance	added to	the other car's distance	is	297
↓	↓	↓	↓	↓

Translate: $3x$ + $3y$ =   297

In words:

one car's speed	is	5 mph faster than the other
↓	↓	↓

Translate: x = $y + 5$

3. SOLVE. Now we solve the system.

$$\begin{cases} 3x + 3y = 297 \\ x = y + 5 \end{cases}$$

Again, the substitution method is appropriate. We replace x with $y + 5$ in the first equation and solve for y.

$$3x + 3y = 297 \quad \text{\small First equation}$$
$$3(y + 5) + 3y = 297 \quad \text{\small Let } x = y + 5.$$
$$3y + 15 + 3y = 297$$
$$6y = 282$$
$$y = 47$$

To find x, we replace y with 47 in the equation $x = y + 5$. Then $x = 47 + 5 = 52$. The ordered pair solution of the system is $(52, 47)$.

4. INTERPRET. The solution $(52, 47)$ means that the cars are traveling at 52 mph and 47 mph, respectively.

Check: Notice that one car is traveling 5 mph faster than the other. Also, if one car travels 52 mph for 3 hours, the distance is $3(52) = 156$ miles. The other car traveling for 3 hours at 47 mph travels a distance of $3(47) = 141$ miles. The sum of the distances $156 + 141$ is 297 miles, the required distance.

State: The cars are traveling at 52 mph and 47 mph.

🔲 **Work Practice Problem 3**

Helpful Hint
  Don't forget to attach units, if appropriate.

PRACTICE PROBLEM 4

One solution contains 20% acid and a second solution contains 50% acid. How many ounces of each solution should be mixed in order to have 60 ounces of a 30% acid solution?

Answer

4. 40 oz of 20% solution; 20 oz of 50% solution

EXAMPLE 4 **Mixing Solutions**

Lynn Pike, a pharmacist, needs 70 liters of a 50% alcohol solution. She has available a 30% alcohol solution and an 80% alcohol solution. How many liters of each solution should she mix to obtain 70 liters of a 50% alcohol solution?

Solution:

1. UNDERSTAND. Read and reread the problem. Next, guess the solution. Suppose that we need 20 liters of the 30% solution. Then we need $70 - 20 = 50$ liters of the 80% solution. To see if this gives us 70 liters of a 50% alcohol solution, let's find the amount of pure alcohol in each solution.

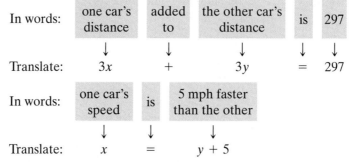

number of liters	×	alcohol strength	=	amount of pure alcohol
↓		↓		↓
20 liters	×	0.30	=	6 liters
50 liters	×	0.80	=	40 liters
70 liters	×	0.50	=	35 liters

Since 6 liters + 40 liters = 46 liters and not 35 liters, our guess is incorrect, but we have gained some insight as to how to model and check this problem.

We will let

x = amount of 30% solution, in liters

y = amount of 80% solution, in liters

and use a table to organize the given data.

	Number of Liters	Alcohol Strength	Amount of Pure Alcohol
30% Solution	x	30%	$0.30x$
80% Solution	y	80%	$0.80y$
50% Solution Needed	70	50%	$(0.50)(70)$

2. **TRANSLATE.** We translate the stated conditions into two equations.

In words:	amount of 30% solution	+	amount of 80% solution	=	70
	↓		↓		↓
Translate:	x	+	y	=	70

In words:	amount of pure alcohol in 30% solution	+	amount of pure alcohol in 80% solution	=	amount of pure alcohol in 50% solution
	↓		↓		↓
Translate:	$0.30x$	+	$0.80y$	=	$(0.50)(70)$

3. **SOLVE.** Now we solve the system

$$\begin{cases} x + y = 70 \\ 0.30x + 0.80y = (0.50)(70) \end{cases}$$

To solve this system, we use the elimination method. We multiply both sides of the first equation by -3 and both sides of the second equation by 10. Then

$$\begin{cases} -3(x + y) = -3(70) \\ 10(0.30x + 0.80y) = 10(0.50)(70) \end{cases} \text{ simplifies to } \begin{cases} -3x - 3y = -210 \\ \underline{3x + 8y = 350} \\ 5y = 140 \\ y = 28 \end{cases}$$

Now we replace y with 28 in the equation $x + y = 70$ and find that $x + 28 = 70$, or $x = 42$. The ordered pair solution of the system is $(42, 28)$.

4. **INTERPRET.** The solution $(42, 28)$ means that 42 liters of the 30% solution and 28 liters of the 80% solution.

Check: We check the solution in the same way that we checked our guess.

State: The pharmacist needs to mix 42 liters of 30% solution and 28 liters of 80% solution to obtain 70 liters of 50% solution.

◼ **Work Practice Problem 4**

✔ *Concept Check* Suppose you mix an amount of 25% acid solution with an amount of 60% acid solution. You then calculate the acid strength of the resulting acid mixture. For which of the following results should you suspect an error in your calculation? Why?

a. 14% **b.** 32% **c.** 55%

Objective B Solving Problems with Cost and Revenue Functions

Recall that businesses are often computing cost and revenue functions or equations to predict sales, to determine whether prices need to be adjusted, and to see whether the company is making or losing money. Recall also that the value at which revenue equals cost is called the break-even point. When revenue is less than cost, the company is losing money; when revenue is greater than cost, the company is making money.

EXAMPLE 5 Finding a Break-Even Point

A manufacturing company recently purchased $3000 worth of new equipment to create new personalized stationery to sell to its customers. The cost of producing a package of personalized stationery is $3.00, and it is sold for $5.50. Find the number of packages that must be sold for the company to break even.

Solution:

1. UNDERSTAND. Read and reread the problem.

   Notice that the cost to the company will include a one-time cost of $3000 for the equipment and then $3.00 per package produced. The revenue will be $5.50 per package sold.

   To model this problem, we will let

   x = number of packages of personalized stationery

   $C(x)$ = total cost for producing x packages of stationery

   $R(x)$ = total revenue for selling x packages of stationery

2. TRANSLATE. The revenue equation is

In words:	revenue for selling x packages of stationery	=	price per package	·	number of packages
	↓		↓		↓
Translate:	$R(x)$	=	5.5	·	x

   The cost equation is

In words:	cost for producing x packages of stationery	=	cost per package	·	number of packages	+	cost for equipment
	↓		↓		↓		↓
Translate:	$C(x)$	=	3	·	x	+	3000

3. SOLVE. Since the break-even point is when $R(x) = C(x)$, we solve the equation $5.5x = 3x + 3000$.

$$5.5x = 3x + 3000$$
$$2.5x = 3000 \qquad \text{Subtract } 3x \text{ from both sides.}$$
$$x = 1200 \qquad \text{Divide both sides by } 2.5.$$

4. INTERPRET.

Check: To see whether the break-even point occurs when 1200 packages are produced and sold, we check to see if revenue equals cost when $x = 1200$. When $x = 1200$,

$$R(x) = 5.5x = 5.5(1200) = 6600$$
$$C(x) = 3x + 3000 = 3(1200) + 3000 = 6600$$

Since $R(1200) = C(1200) = 6600$, the break-even point is 1200.

State: The company must sell 1200 packages of stationery to break even. The graph of this system is shown.

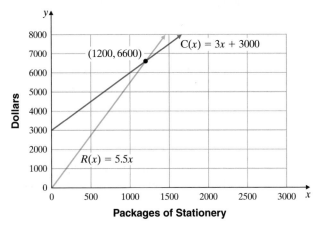

Work Practice Problem 5

Objective C Solving Problems Modeled by Systems of Three Equations

To introduce problem solving with systems of three linear equations in three variables, we solve a problem about triangles.

EXAMPLE 6 Finding Angle Measures

The measure of the largest angle of a triangle is 80° more than the measure of the smallest angle, and the measure of the remaining angle is 10° more than the measure of the smallest angle. Find the measure of each angle.

Solution:

1. UNDERSTAND. Read and reread the problem. Recall that the sum of the measures of the angles of a triangle is 180°. Then guess a solution. If the smallest angle measures 20°, the measure of the largest angle is 80° more, or 20° + 80° = 100°. The measure of the remaining angle is 10° more than the measure of the smallest angle, or 20° + 10° = 30°. The sum of these three angles is 20° + 100° + 30° = 150°, not the required 180°. We now know that the measure of the smallest angle is greater than 20°.

Continued on next page

PRACTICE PROBLEM 6

The measure of the largest angle of a triangle is 80° more than the measure of the smallest angle, and the measure of the remaining angle is 40° more than the measure of the smallest angle. Find the measure of each angle.

Answer

6. 20°; 60°; 100°

To model this problem we will let

x = degree measure of the smallest angle

y = degree measure of the largest angle

z = degree measure of the remaining angle

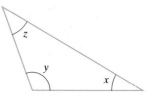

2. TRANSLATE. We translate the given information into three equations.

In words: | the sum of the measures | = | 180
Translate: | $x + y + z$ | = | 180

In words: | the largest angle | is | 80 more than the smallest angle
Translate: | y | = | $x + 80$

In words: | the remaining angle | is | 10 more than the smallest angle
Translate: | z | = | $x + 10$

3. SOLVE. We solve the system

$$\begin{cases} x + y + z = 180 \\ y = x + 80 \\ z = x + 10 \end{cases}$$

Since y and z are both expressed in terms of x, we will solve using the substitution method. We substitute $y = x + 80$ and $z = x + 10$ in the first equation. Then

$$x + y + z = 180 \quad \text{First equation}$$

$$x + (x + 80) + (x + 10) = 180 \quad \text{Let } y = x + 80 \text{ and } z = x + 10.$$
$$3x + 90 = 180$$
$$3x = 90$$
$$x = 30$$

Then $y = x + 80 = 30 + 80 = 110$, and $z = x + 10 = 30 + 10 = 40$. The ordered triple solution is $(30,\ 110,\ 40)$.

4. INTERPRET.

Check: Notice that $30° + 40° + 110° = 180°$. Also, the measure of the largest angle, $110°$, is $80°$ more than the measure of the smallest angle, $30°$. The measure of the remaining angle, $40°$, is $10°$ more than the measure of the smallest angle, $30°$.

State: The angles measure $30°, 110°,$ and $40°$.

▨ **Work Practice Problem 6**

Objective **A** *Solve. See Examples 1 through 4.*

1. One number is two more than a second number. Twice the first is 4 less than 3 times the second. Find the numbers.

2. Three times one number minus a second is 8, and the sum of the numbers is 12. Find the numbers.

3. The U.S.A. has the world's only "large deck" aircraft carriers which can hold up to 72 aircraft. The Enterprise class carrier is longest in length while the Nimitz class carrier is the second longest. The total length of these two carriers is 2193 feet while the difference of their lengths is only 9 feet. (*Source: USA Today,* May, 2001)

  a. Find the length of each class carrier.
  b. If a football field has a length of 100 yards, determine the length of the Enterprise class carrier in terms of number of football fields.

4. The rate of growth of participation (age 6 and older) in sports featured in the X-Games is surpassing that for older sports such as football and baseball. The most popular X-Game sport is in-line roller skating, followed by skateboarding. In 2003, the total number of participants in both sports was 30.3 million. If the number of participants in in-line skating was 3 million less than twice the number of participants in skateboarding, find the number of participants in each sport. (*Source:* April 2004 Sporting Goods Manufacturers Association Sports Participation Topline Report)

5. A Delta 727 traveled 560 mph with the wind and 480 mph against the wind. Find the speed of the plane in still air and the speed of the wind.

6. Terry Watkins can row about 10.6 kilometers in 1 hour downstream and 6.8 kilometers upstream in 1 hour. Find how fast he can row in still water, and find the speed of the current.

7. Find how many quarts of 4% butterfat milk and 1% butterfat milk should be mixed to yield 60 quarts of 2% butterfat milk.

8. A pharmacist needs 500 milliliters of a 20% phenobarbital solution but has only 5% and 25% phenobarbital solutions available. Find how many milliliters of each she should mix to get the desired solution.

9. In recent years, the United Kingdom was the most popular host country for U.S. students traveling abroad to study. Italy was the second most popular destination. A total of 50,642 students visited one of the two countries. If 12,770 more U.S. students studied in the United Kingdom than in Italy, how many students studied abroad in each country? (*Source:* Institute of International Education, Open Doors 2004)

10. Harvard University and Cornell University are each known for their excellent libraries, and each is participating with Google to put their collections into Google's searchable database. In 2003, Harvard libraries contained 584,531 more printed volumes than twice the total number of printed volumes in the libraries of Cornell. Together these two great libraries house 22,479,758 printed volumes. Find the number of printed volumes in each library. (*Source:* Harvard libraries, Cornell libraries)

11. Karen Karlin bought some large frames for $15 each and some small frames for $8 each at a closeout sale. If she bought 22 frames for $239, find how many of each type she bought.

12. Hilton University Drama Club sold 311 tickets for a play. Student tickets cost 50 cents each; nonstudent tickets cost $1.50. If total receipts were $385.50, find how many tickets of each type were sold.

13. One number is two less than a second number. Twice the first is 4 more than 3 times the second. Find the numbers.

14. Twice one number plus a second number is 42, and the first number minus the second number is -6. Find the numbers.

15. In the United States, the percent of women using the Internet is increasing faster than the percent of men. For the years 1998–2003, the function $y = 5.6x + 20.7$ can be used to estimate the percent of females using the Internet while the function $y = 4.8x + 29.4$ can be used to estimate the percent of males. For both functions, x is the number of years since 1998. If this trend continues, predict the year in which the percent of females using the Internet is equal to the percent of males. (*Source:* Pew Internet & American Life Project)

16. The percent of car vehicle sales is decreasing while the percent of light trucks (pickups, sport-utility vehicles, and minivans) is increasing. For the years 1999 through 2003, the function $y = -1.2x + 52.6$ can be used to estimate the percent of vehicle sales being cars while the function $y = 1.2x + 47.4$ can be used to estimate the percent of vehicle sales being light trucks. For both functions, x is the number of years since 1999. (*Source:* Based on data from the Bureau of Transportation Statistics, U.S. Department of Transportation)

    a. If this trend continues, predict the year in which the percent of car sales equals the percent of light truck sales.

    b. Before the actual 2001 vehicle sales data was published, USA today predicted that light truck sales would likely be greater than car sales in the year 2001. Does your prediction from part a agree with this statement? (*Source:* USA Today and Autodata)

17. An office supply store in San Diego sells 7 writing tablets and 4 pens for $6.40. Also, 2 tablets and 19 pens cost $5.40. Find the price of each.

18. A Candy Barrel shop manager mixes M&M's worth $2.00 per pound with trail mix worth $1.50 per pound. Find how many pounds of each she should use to get 50 pounds of a party mix worth $1.80 per pound.

19. An airplane takes 3 hours to travel a distance of 2160 miles with the wind. The return trip takes 4 hours against the wind. Find the speed of the plane in still air and the speed of the wind.

20. Two cyclists start at the same point and travel in opposite directions. One travels 4 mph faster than the other. In 4 hours they are 112 miles apart. Find how fast each is traveling.

21. Two of the major job categories defined by the United States Department of Labor are manufacturing jobs and jobs in the service sector. Jobs in the manufacturing sector have decreased nearly every year since the 1960s. During the same time period, service sector jobs have been steadily increasing. For the years from 1970 through 2003, the function $y = -0.513x + 33.5$ approximates the percent of jobs in the U.S. economy which are manufacturing jobs, while the function $y = 0.56x + 18.4$ approximates the percent of jobs in the United States economy which are service sector jobs. (*Source:* Based on data from the United States Department of Labor)

    a. Explain how the decrease in manufacturing jobs can be verified by the given function, while the increase of service sector jobs can be verified by their given function.

    b. Based on this information, determine the year when the percent of manufacturing jobs and the percent of service sector jobs were the same.

22. The annual U.S. per capita consumption of cheddar cheese has remained about the same since the early 1990s while the consumption of mozzarella cheese has increased. For the years 1997–2003, the function $y = 0.07x + 9.5$ approximates the annual U.S. per capita consumption of cheddar cheese in pounds and the function $y = 0.3x + 8.3$ approximates the annual U.S. per capita consumption of mozzarella cheese in pounds. For both functions, x is the number of years after 1997. Based on this information, determine the year in which the pounds of cheddar cheese consumed equaled the pounds of mozzarella cheese consumed.

△ **23.** The perimeter of a triangle is 93 centimeters. If two sides are equally long and the third side is 9 centimeters longer than the others, find the lengths of the three sides.

24. Jack Reinholt, a car salesman, has a choice of two pay arrangements: a weekly salary of $200 plus 5% commission on sales, or a straight 15% commission. Find the amount of weekly sales for which Jack's earnings are the same regardless of the pay arrangement.

25. Hertz car rental agency charges $25 daily plus 10 cents per mile. Budget charges $20 daily plus 25 cents per mile. Find the daily mileage for which the Budget charge for the day is twice that of the Hertz charge for the day.

26. Carroll Blakemore, a drafting student, bought three templates and a pencil one day for $6.45. Another day he bought two pads of paper and four pencils for $7.50. If the price of a pad of paper is three times the price of a pencil, find the price of each type of item.

△ **27.** In the figure, line l and line m are parallel lines cut by transversal t. Find the values of x and y.

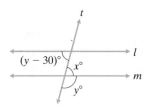

△ **28.** Find the values of x and y in the following isosceles triangle.

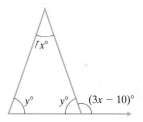

Objective B *Given the cost function $C(x)$ and the revenue function $R(x)$, find the number of units x that must be sold to break even. See Example 5.*

29. $C(x) = 30x + 10{,}000 \quad R(x) = 46x$

30. $C(x) = 12x + 15{,}000 \quad R(x) = 32x$

31. $C(x) = 1.2x + 1500 \quad R(x) = 1.7x$

32. $C(x) = 0.8x + 900 \quad R(x) = 2x$

33. $C(x) = 75x + 160{,}000 \quad R(x) = 200x$

34. $C(x) = 105x + 70{,}000 \quad R(x) = 245x$

35. The planning department of Abstract Office Supplies has been asked to determine whether the company should introduce a new computer desk next year. The department estimates that $6000 of new manufacturing equipment will need to be purchased and that the cost of constructing each desk will be $200. The department also estimates that the revenue from each desk will be $450.
   a. Determine the revenue function $R(x)$ from the sale of x desks.
   b. Determine the cost function $C(x)$ for manufacturing x desks.
   c. Find the break-even point.

36. Baskets, Inc., is planning to introduce a new woven basket. The company estimates that $500 worth of new equipment will be needed to manufacture this new type of basket and that it will cost $15 per basket to manufacture. The company also estimates that the revenue from each basket will be $31.
   a. Determine the revenue function $R(x)$ from the sale of x baskets.
   b. Determine the cost function $C(x)$ for manufacturing x baskets.
   c. Find the break-even point. Round up to the nearest whole basket.

Objective **C** *Solve. See Example 6.*

37. Rabbits in a lab are to be kept on a strict daily diet that includes 30 grams of protein, 16 grams of fat, and 24 grams of carbohydrates. The scientist has only three food mixes available with the following grams of nutrients per unit.

	Protein	Fat	Carbohydrate
Mix A	4	6	3
Mix B	6	1	2
Mix C	4	1	12

Find how many units of each mix are needed daily to meet each rabbit's dietary need.

38. Gerry Gundersen mixes different solutions with concentrations of 25%, 40%, and 50% to get 200 liters of a 32% solution. If he uses twice as much of the 25% solution as of the 40% solution, find how many liters of each kind he uses.

△ **39.** The perimeter of a quadrilateral (four-sided polygon) is 29 inches. The longest side is twice as long as the shortest side. The other two sides are equally long and are 2 inches longer than the shortest side. Find the length of all four sides.

△ **40.** The measure of the largest angle of a triangle is 90° more than the measure of the smallest angle, and the measure of the remaining angle is 30° more than the measure of the smallest angle. Find the measure of each angle.

41. The sum of three numbers is 40. One number is five more than a second number. It is also twice the third. Find the numbers.

42. The sum of the digits of a three-digit number is 15. The tens-place digit is twice the hundreds-place digit, and the ones-place digit is 1 less than the hundreds-place digit. Find the three-digit number.

43. In 2004, the WNBA's top scorer was Lauren Jackson of the Seattle Storm. She scored a total of 738 points during the regular season. The number of two-point field goals Jackson made was 12 more than four times the number of three-point field goals she made. The number of free throws (each worth one point) she made was seventy-eight fewer than the number of two-point field goals she made. Find how many free throws, two-point field goals, and three-point field goals Lauren Jackson made during the 2004 regular WNBA season. (*Source:* Women's National Basketball Association)

44. During the 2003–2004 regular NBA season, the top-scoring player was Kevin Garnett of the Minnesota Timberwolves. Garnett scored a total of 1987 points during the regular season. The number of free throws (each worth one point) he made was five more than thirty-three times the number of three-point field goals he made. The number of two-point field goals that Garnett made was fifty-seven more than twice the number of free throws he made. How many free throws, three-point field goals, and two-point field goals did Kevin Garnett make during the 2003–2004 NBA season? (*Source:* National Basketball Association)

△ **45.** Find the values of x, y, and z in the following triangle.

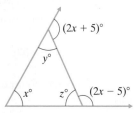

△ **46.** The sum of the measures of the angles of a quadrilateral is 360°. Find the values of x, y, and z in the following quadrilateral.

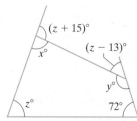

Review

Multiply both sides of equation (1) by 2, and add the resulting equation to equation (2). See Section 4.2.

47. $3x - y + z = 2$ (1)
$-x + 2y + 3z = 6$ (2)

48. $2x + y + 3z = 7$ (1)
$-4x + y + 2z = 4$ (2)

Multiply both sides of equation (1) by −3, and add the resulting equation to equation (2). See Section 4.2.

49. $x + 2y - z = 0$ (1)
$3x + y - z = 2$ (2)

50. $2x - 3y + 2z = 5$ (1)
$x - 9y + z = -1$ (2)

Concept Extensions

51. The number of personal bankruptcy petitions filed in the United States has been on the rise since the 1980s, but a recent sharper rise has occurred. In 2003, the number of petitions filed was 132,466 less than twice the number of bankruptcy petitions filed in 1993. This is equivalent to an increase of 764,765 petitions filed from 1993 to 2003. Find how many personal bankruptcy petitions were filed in each year. (*Source:* Based on data from the Administrative Office of the United States Courts)

52. In 2003, the median weekly earnings for male janitors was $71 more than the median weekly earnings for female janitors. The median weekly earnings for female janitors was 0.83 times that of their male counterparts. Also in 2003, the median weekly earnings for female pharmacists was $169 less than the median weekly incomes for male pharmacists. The median weekly earning for male pharmacists was 1.12 times that of their female counterparts. (*Source:* Based on data from the U.S. Bureau of Labor Statistics)

   a. Find the median weekly earnings for female janitors in the United States in 2003. (Round to the nearest dollar.)

   b. Find the median weekly earnings for female pharmacists in the United States in 2003. (Round to the nearest dollar.)

   c. Of the four groups of workers described in the problem, which group makes the greatest weekly earnings? Which group makes the least weekly earnings?

53. Find the values of a, b, and c such that the equation $y = ax^2 + bx + c$ has ordered pair solutions $(1, 6)$, $(-1, -2)$, and $(0, -1)$. To do so, substitute each ordered pair solution into the equation. Each time, the result is an equation in three unknowns: a, b, and c. Then solve the resulting system of three linear equations in three unknowns, a, b, and c.

54. Find the values of a, b, and c such that the equation $y = ax^2 + bx + c$ has ordered pair solutions $(1, 2)$, $(2, 3)$, and $(-1, 6)$. (*Hint:* See Exercise 53.)

55. Data (x, y) for the total number (in thousands) of college-bound students who took the ACT assessment in the year x are $(3, 925)$, $(6, 1019)$, and $(11, 1171)$, where $x = 3$ represents 1996 and $x = 6$ represents 1999. Find the values a, b, and c such that the equation $y = ax^2 + bx + c$ models these data. According to your model, how many students will take the ACT in 2009? (*Source:* ACT, Inc.)

56. Monthly normal rainfall data (x, y) for Portland, Oregon, are $(4, 2.47)$, $(7, 0.6)$, $(8, 1.1)$, where x represents time in months (with $x = 1$ representing January) and y represents rainfall in inches. Find the values of a, b, and c rounded to 2 decimal places such that the equation $y = ax^2 + bx + c$ models this data. According to your model, how much rain should Portland expect during September? (*Source:* National Climatic Data Center)

INTEGRATED REVIEW

Sections 4.1–4.3

Systems of Linear Equations

The graphs of various systems of equations are shown. Match each graph with the solution of its corresponding system.

A 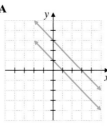 **B** 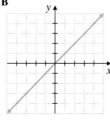 **C** 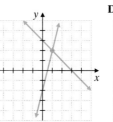 **D**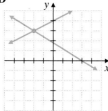

1. Solution: $(1, 2)$

2. Solution: $(-2, 3)$

3. No solution

4. Infinite number of solutions

Solve each system by elimination or substitution.

5. $\begin{cases} x + y = 4 \\ y = 3x \end{cases}$

6. $\begin{cases} x - y = -4 \\ y = 4x \end{cases}$

7. $\begin{cases} x + y = 1 \\ x - 2y = 4 \end{cases}$

8. $\begin{cases} 2x - y = 8 \\ x + 3y = 11 \end{cases}$

9. $\begin{cases} 2x + 5y = 8 \\ 6x + y = 10 \end{cases}$

10. $\begin{cases} \dfrac{1}{8}x - \dfrac{1}{2}y = -\dfrac{5}{8} \\ -3x - 8y = 0 \end{cases}$

11. $\begin{cases} 4x - 7y = 7 \\ 12x - 21y = 24 \end{cases}$

12. $\begin{cases} 2x - 5y = 3 \\ -4x + 10y = -6 \end{cases}$

13. $\begin{cases} y = \dfrac{1}{3}x \\ 5x - 3y = 4 \end{cases}$

14. $\begin{cases} y = \dfrac{1}{4}x \\ 2x - 4y = 3 \end{cases}$

15. $\begin{cases} x + y = 2 \\ -3y + z = -7 \\ 2x + y - z = -1 \end{cases}$

16. $\begin{cases} y + 2z = -3 \\ x - 2y = 7 \\ 2x - y + z = 5 \end{cases}$

17. $\begin{cases} 2x + 4y - 6z = 3 \\ -x + y - z = 6 \\ x + 2y - 3z = 1 \end{cases}$

18. $\begin{cases} x - y + 3z = 2 \\ -2x + 2y - 6z = -4 \\ 3x - 3y + 9z = 6 \end{cases}$

19. $\begin{cases} x + y - 4z = 5 \\ x - y + 2z = -2 \\ 3x + 2y + 4z = 18 \end{cases}$

20. $\begin{cases} 2x - y + 3z = 2 \\ x + y - 6z = 0 \\ 3x + 4y - 3z = 6 \end{cases}$

21. A first number is 8 less than a second number. Twice the first number is 11 more than the second number. Find the numbers.

△**22.** The sum of the measures of the angles of a quadrilateral is 360°. The two smallest angles of the quadrilateral have the same measure. The third angle measures 30° more than the measure of one of the smallest angles and the fourth angle measures 50° more than the measure of one of the smallest angles. Find the measure of each angle.

4.4 SOLVING SYSTEMS OF EQUATIONS USING MATRICES

By now, you may have noticed that the solution of a system of equations depends on the coefficients of the equations in the system and not on the variables. In this section, we introduce how to solve a system of equations using a **matrix.**

Objective A Using Matrices to Solve a System of Two Equations

A **matrix** (plural: **matrices**) is a rectangular array of numbers. The following are examples of matrices.

$$\begin{bmatrix} 1 & 0 \\ 0 & 1 \end{bmatrix} \quad \begin{bmatrix} 2 & 1 & 3 & -1 \\ 0 & -1 & 4 & 5 \\ -6 & 2 & 1 & 0 \end{bmatrix} \quad \begin{bmatrix} a & b & c \\ d & e & f \end{bmatrix}$$

The numbers aligned horizontally in a matrix are in the same **row.** The numbers aligned vertically are in the same **column.**

Row 1 →
Row 2 →
$$\begin{bmatrix} 2 & 1 & 0 \\ -1 & 6 & 2 \end{bmatrix}$$
Column 1   Column 2   Column 3

This matrix has 2 rows and 3 columns. It is called a 2×3 (read "two by three") matrix.

To see the relationship between systems of equations and matrices, study the example below.

System of Equations (in Standard Form)

$$\begin{cases} 2x - 3y = 6 & \text{Equation 1} \\ x + y = 0 & \text{Equation 2} \end{cases}$$

Corresponding Matrix

$$\begin{bmatrix} 2 & -3 & \vdots & 6 \\ 1 & 1 & \vdots & 0 \end{bmatrix} \begin{matrix} \text{Row 1} \\ \text{Row 2} \end{matrix}$$

> **Helpful Hint**
> Before writing the corresponding matrix associated with a system of equations, make sure that the equations are written in standard form.

Notice that the rows of the matrix correspond to the equations in the system. The coefficients of the variables are placed to the left of a vertical dashed line. The constants are placed to the right. Each of these numbers in the matrix is called an **element.**

The method of solving systems by matrices is to write this matrix as an equivalent matrix from which we can easily identify the solution. Two matrices are equivalent if they represent systems that have the same solution set. The following **row operations** can be performed on matrices, and the result is an equivalent matrix.

Elementary Row Operations

1. Any two rows in a matrix may be interchanged.

2. The elements of any row may be multiplied (or divided) by the same nonzero number.

3. The elements of any row may be multiplied (or divided) by a nonzero number and added to their corresponding elements in any other row.

> **Helpful Hint**
> Notice that these *row* operations are the same operations that we can perform on *equations* in a system.

293

To solve a system of two equations in x and y by matrices, write the corresponding matrix associated with the system. Then use elementary row operations to write equivalent matrices until you have a matrix of the form

$$\begin{bmatrix} 1 & a & | & b \\ 0 & 1 & | & c \end{bmatrix},$$

where a, b, and c are constants. Why? If a matrix associated with a system of equations is in this form, we can easily solve for x and y. For example,

Matrix | **System of Equations**

$$\begin{bmatrix} 1 & 2 & | & -3 \\ 0 & 1 & | & 5 \end{bmatrix} \text{ corresponds to } \begin{cases} 1x + 2y = -3 \\ 0x + 1y = 5 \end{cases} \text{ or } \begin{cases} x + 2y = -3 \\ y = 5 \end{cases}$$

In the second equation, we have $y = 5$. Substituting this in the first equation, we have $x + 2(5) = -3$ or $x = -13$. The solution of the system is the ordered pair $(-13, 5)$.

PRACTICE PROBLEM 1

Use matrices to solve the system:

$$\begin{cases} x + 2y = -4 \\ 2x - 3y = 13 \end{cases}$$

EXAMPLE 1 Use matrices to solve the system:

$$\begin{cases} x + 3y = 5 \\ 2x - y = -4 \end{cases}$$

Solution: The corresponding matrix is $\begin{bmatrix} 1 & 3 & | & 5 \\ 2 & -1 & | & -4 \end{bmatrix}$. We use elementary row

operations to write an equivalent matrix that looks like $\begin{bmatrix} 1 & a & | & b \\ 0 & 1 & | & c \end{bmatrix}$.

For the matrix given, the element in the first row, first column is already 1, as desired. Next we write an equivalent matrix with a 0 below the 1. To do this, we multiply row 1 by -2 and add to row 2. *We will change only row 2.*

$$\begin{bmatrix} 1 & 3 & | & 5 \\ -2(1) + 2 & -2(3) + (-1) & | & -2(5) + (-4) \end{bmatrix} \text{ simplifies to}$$

Row 1 Row 2   Row 1 Row 2   Row 1 Row 2
element element   element element   element element

$$\begin{bmatrix} 1 & 3 & | & 5 \\ 0 & -7 & | & -14 \end{bmatrix}$$

Now we change the -7 to a 1 by use of an elementary row operation. We divide row 2 by -7, then

$$\begin{bmatrix} 1 & 3 & | & 5 \\ \frac{0}{-7} & \frac{-7}{-7} & | & \frac{-14}{-7} \end{bmatrix} \text{ simplifies to } \begin{bmatrix} 1 & 3 & | & 5 \\ 0 & 1 & | & 2 \end{bmatrix}$$

This last matrix corresponds to the system

$$\begin{cases} x + 3y = 5 \\ y = 2 \end{cases}$$

Thus we know that y is 2. To find x, we let $y = 2$ in the first equation, $x + 3y = 5$.

$x + 3y = 5$ First equation
$x + 3(2) = 5$ Let $y = 2$.
$x = -1$

The ordered pair solution is $(-1, 2)$. Check to see that this ordered pair satisfies both original equations.

Work Practice Problem 1

Answer

1. $(2, -3)$

EXAMPLE 2 Use matrices to solve the system:

$$\begin{cases} 2x - y = 3 \\ 4x - 2y = 5 \end{cases}$$

Solution: The corresponding matrix is $\begin{bmatrix} 2 & -1 & | & 3 \\ 4 & -2 & | & 5 \end{bmatrix}$. To get 1 in the row 1, column 1 position, we divide the elements of row 1 by 2.

$$\begin{bmatrix} \frac{2}{2} & -\frac{1}{2} & | & \frac{3}{2} \\ 4 & -2 & | & 5 \end{bmatrix} \text{ simplifies to } \begin{bmatrix} 1 & -\frac{1}{2} & | & \frac{3}{2} \\ 4 & -2 & | & 5 \end{bmatrix}$$

To get 0 under the 1, we multiply the elements of row 1 by -4 and add the new elements to the elements of row 2.

$$\begin{bmatrix} 1 & -\frac{1}{2} & | & \frac{3}{2} \\ -4(1)+4 & -4\left(-\frac{1}{2}\right)-2 & | & -4\left(\frac{3}{2}\right)+5 \end{bmatrix} \begin{matrix} \text{simplifies} \\ \text{to} \end{matrix} \begin{bmatrix} 1 & -\frac{1}{2} & | & \frac{3}{2} \\ 0 & 0 & | & -1 \end{bmatrix}$$

The corresponding system is $\begin{cases} x - \frac{1}{2}y = \frac{3}{2} \\ 0 = -1 \end{cases}$. The equation $0 = -1$ is false for all y or x values; hence the system is inconsistent and has no solution. The solution set is $\varnothing$ or $\{\ \}$.

■ **Work Practice Problem 2**

✔ **Concept Check** Consider the system

$$\begin{cases} 2x - 3y = 8 \\ x + 5y = -3 \end{cases}$$

What is wrong with its corresponding matrix shown below?

$$\begin{bmatrix} 2 & -3 & | & 8 \\ 0 & 5 & | & -3 \end{bmatrix}$$

Objective B Using Matrices to Solve a System of Three Equations

To solve a system of three equations in three variables using matrices, we will write the corresponding matrix in the equivalent form

$$\begin{bmatrix} 1 & a & b & | & d \\ 0 & 1 & c & | & e \\ 0 & 0 & 1 & | & f \end{bmatrix}$$

EXAMPLE 3 Use matrices to solve the system:

$$\begin{cases} x + 2y + z = 2 \\ -2x - y + 2z = 5 \\ x + 3y - 2z = -8 \end{cases}$$

Solution: The corresponding matrix is $\begin{bmatrix} 1 & 2 & 1 & | & 2 \\ -2 & -1 & 2 & | & 5 \\ 1 & 3 & -2 & | & -8 \end{bmatrix}$.

Continued on next page

PRACTICE PROBLEM 2

Use matrices to solve the system:

$$\begin{cases} -3x + y = 0 \\ -6x + 2y = 2 \end{cases}$$

PRACTICE PROBLEM 3

Use matrices to solve the system:

$$\begin{cases} x + 3y + z = 5 \\ -3x + y - 3z = 5 \\ x + 2y - 2z = 9 \end{cases}$$

Answers

2. no solution, **3.** $(1, 2, -2)$

✔ **Concept Check Answer**

matrix should be $\begin{bmatrix} 2 & -3 & | & 8 \\ 1 & 5 & | & -3 \end{bmatrix}$

Our goal is to write an equivalent matrix with 1s along the diagonal (see the numbers in red) and 0s below the 1s. The element in row 1, column 1 is already 1. Next we get 0s for each element in the rest of column 1. To do this, first we multiply the elements of row 1 by 2 and add the new elements to row 2. Also, we multiply the elements of row 1 by -1 and add the new elements to the elements of row 3. *We do not change row 1.* Then

$$\left[\begin{array}{ccc|c} 1 & 2 & 1 & 2 \\ 2(1)-2 & 2(2)-1 & 2(1)+2 & 2(2)+5 \\ -1(1)+1 & -1(2)+3 & -1(1)-2 & -1(2)-8 \end{array}\right] \text{ simplifies to}$$

$$\left[\begin{array}{ccc|c} 1 & 2 & 1 & 2 \\ 0 & 3 & 4 & 9 \\ 0 & 1 & -3 & -10 \end{array}\right]$$

We continue down the diagonal and use elementary row operations to get 1 where the element 3 is now. To do this, we interchange rows 2 and 3.

$$\left[\begin{array}{ccc|c} 1 & 2 & 1 & 2 \\ 0 & 3 & 4 & 9 \\ 0 & 1 & -3 & -10 \end{array}\right] \quad \text{is equivalent to} \quad \left[\begin{array}{ccc|c} 1 & 2 & 1 & 2 \\ 0 & 1 & -3 & -10 \\ 0 & 3 & 4 & 9 \end{array}\right]$$

Next we want the new row 3, column 2 element to be 0. We multiply the elements of row 2 by -3 and add the result to the elements of row 3.

$$\left[\begin{array}{ccc|c} 1 & 2 & 1 & 2 \\ 0 & 1 & -3 & -10 \\ -3(0)+0 & -3(1)+3 & -3(-3)+4 & -3(-10)+9 \end{array}\right] \text{ simplifies to}$$

$$\left[\begin{array}{ccc|c} 1 & 2 & 1 & 2 \\ 0 & 1 & -3 & -10 \\ 0 & 0 & 13 & 39 \end{array}\right]$$

Finally, we divide the elements of row 3 by 13 so that the final diagonal element is 1.

$$\left[\begin{array}{ccc|c} 1 & 2 & 1 & 2 \\ 0 & 1 & -3 & -10 \\ \frac{0}{13} & \frac{0}{13} & \frac{13}{13} & \frac{39}{13} \end{array}\right] \quad \text{simplifies to} \quad \left[\begin{array}{ccc|c} 1 & 2 & 1 & 2 \\ 0 & 1 & -3 & -10 \\ 0 & 0 & 1 & 3 \end{array}\right]$$

This matrix corresponds to the system

$$\begin{cases} x + 2y + z = 2 \\ y - 3z = -10 \\ z = 3 \end{cases}$$

We identify the z-coordinate of the solution as 3. Next we replace z with 3 in the second equation and solve for y.

$$y - 3z = -10 \quad \text{Second equation}$$
$$y - 3(3) = -10 \quad \text{Let } z = 3.$$
$$y = -1$$

To find x, we let $z = 3$ and $y = -1$ in the first equation.

$$x + 2y + z = 2 \quad \text{First equation}$$
$$x + 2(-1) + 3 = 2 \quad \text{Let } z = 3 \text{ and } y = -1.$$
$$x = 1$$

The ordered triple solution is $(1, -1, 3)$. Check to see that it satisfies all three equations in the original system.

◼ **Work Practice Problem 3**

4.4 EXERCISE SET

Objective Ⓐ *Use matrices to solve each system of linear equations. See Example 1.*

1. $\begin{cases} x + y = 1 \\ x - 2y = 4 \end{cases}$

2. $\begin{cases} 2x - y = 8 \\ x + 3y = 11 \end{cases}$

 3. $\begin{cases} x + 3y = 2 \\ x + 2y = 0 \end{cases}$

4. $\begin{cases} 4x - y = 5 \\ 3x - 3y = 6 \end{cases}$

Use matrices to solve each system of linear equations. See Example 2.

5. $\begin{cases} x - 2y = 4 \\ 2x - 4y = 4 \end{cases}$

6. $\begin{cases} -x + 3y = 6 \\ 3x - 9y = 9 \end{cases}$

7. $\begin{cases} 3x - 3y = 9 \\ 2x - 2y = 6 \end{cases}$

8. $\begin{cases} 9x - 3y = 6 \\ -18x + 6y = -12 \end{cases}$

Objective Ⓑ *Use matrices to solve each system of linear equations. See Example 3.*

9. $\begin{cases} x + y = 3 \\ 2y = 10 \\ 3x + 2y - 4z = 12 \end{cases}$

10. $\begin{cases} 5x = 5 \\ 2x + y = 4 \\ 3x + y - 5z = -15 \end{cases}$

11. $\begin{cases} 2y - z = -7 \\ x + 4y + z = -4 \\ 5x - y + 2z = 13 \end{cases}$

12. $\begin{cases} 4y + 3z = -2 \\ 5x - 4y = 1 \\ -5x + 4y + z = -3 \end{cases}$

Objectives Ⓐ Ⓑ **Mixed Practice** *Solve each system of linear equations using matrices. See Examples 1 through 3.*

13. $\begin{cases} x - 4 = 0 \\ x + y = 1 \end{cases}$

14. $\begin{cases} 3y = 6 \\ x + y = 7 \end{cases}$

15. $\begin{cases} x + y + z = 2 \\ 2x - z = 5 \\ 3y + z = 2 \end{cases}$

16. $\begin{cases} x + 2y + z = 5 \\ x - y - z = 3 \\ y + z = 2 \end{cases}$

17. $\begin{cases} 5x - 2y = 27 \\ -3x + 5y = 18 \end{cases}$

18. $\begin{cases} 4x - y = 9 \\ 2x + 3y = -27 \end{cases}$

19. $\begin{cases} 4x - 7y = 7 \\ 12x - 21y = 24 \end{cases}$

20. $\begin{cases} 2x - 5y = 12 \\ -4x + 10y = 20 \end{cases}$

21. $\begin{cases} 4x - y + 2z = 5 \\ 2y + z = 4 \\ 4x + y + 3z = 10 \end{cases}$

22. $\begin{cases} 5y - 7z = 14 \\ 2x + y + 4z = 10 \\ 2x + 6y - 3z = 30 \end{cases}$

23. $\begin{cases} 4x + y + z = 3 \\ -x + y - 2z = -11 \\ x + 2y + 2z = -1 \end{cases}$

24. $\begin{cases} x + y + z = 9 \\ 3x - y + z = -1 \\ -2x + 2y - 3z = -2 \end{cases}$

Review

Determine whether each graph is the graph of a function. See Section 3.6.

25.

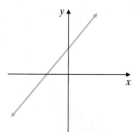

26.

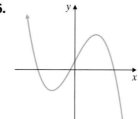

27.

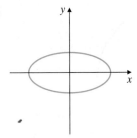

28.

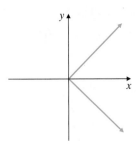

Concept Extensions

Solve. See the Concept Check in this section.

29. For the system $\begin{cases} x \quad + z = 7 \\ \quad y + 2z = -6, \\ 3x - y \quad = 0 \end{cases}$ which is the correct corresponding matrix?

a. $\begin{bmatrix} 1 & 1 & | & 7 \\ 1 & 2 & | & -6 \\ 3 & -1 & | & 0 \end{bmatrix}$ **b.** $\begin{bmatrix} 1 & 0 & 1 & | & 7 \\ 1 & 2 & 0 & | & -6 \\ 3 & -1 & 0 & | & 0 \end{bmatrix}$ **c.** $\begin{bmatrix} 1 & 0 & 1 & | & 7 \\ 0 & 1 & 2 & | & -6 \\ 3 & -1 & 0 & | & 0 \end{bmatrix}$

30. For the system $\begin{cases} x - 6 = 0 \\ 2x - 3y = 1 \end{cases}$, which is the correct corresponding matrix?

a. $\begin{bmatrix} 1 & -6 & | & 0 \\ 2 & -3 & | & 1 \end{bmatrix}$ **b.** $\begin{bmatrix} 1 & 0 & | & 6 \\ 2 & -3 & | & 1 \end{bmatrix}$ **c.** $\begin{bmatrix} 1 & 0 & | & -6 \\ 2 & -3 & | & 1 \end{bmatrix}$

31. The percent y of U.S. households that owned a black-and-white television set between the years 1980 and 1993 can be modeled by the linear equation $2.3x + y = 52$, where x represents the number of years after 1980. Similarly, the percent y of U.S. households that owned a microwave oven during this same period can be modeled by the linear equation $-5.4x + y = 14$. (*Source:* Based on data from the Energy Information Administration, U.S. Department of Energy)

a. The data used to form these two models were incomplete. It is impossible to tell from the data the year in which the percent of households owning black-and-white television sets was the same as the percent of households owning microwave ovens. Use matrix methods to estimate the year in which this occurred.

b. Did more households own black-and-white television sets or microwave ovens in 1980? In 1993? What trends do these models show? Does this seem to make sense? Why or why not?

c. According to the models, when will the percent of households owning black-and-white television sets reach 0%?

d. Do you think your answer to part **c** is accurate? Why or why not?

32. The most popular amusement park in the world (according to annual attendance) is Tokyo Disneyland, whose yearly attendance in thousands can be approximated by the equation $y = 1201x + 16,507$ where x is the number of years after 2000. In second place is Walt Disney World's Magic Kingdom, whose yearly attendance, in thousands, can be approximated by $y = -616x + 15,400$. Find the last year when attendance in Magic Kingdom was greater than attendance in Tokyo Disneyland. (*Source:* Amusement Business)

33. For the system $\begin{cases} 2x - 3y = 8 \\ x + 5y = -3 \end{cases}$, explain what is wrong with writing the corresponding matrix as $\begin{bmatrix} 2 & 3 & | & 8 \\ 0 & 5 & | & -3 \end{bmatrix}$.

STUDY SKILLS BUILDER

Are You Satisfied with Your Performance on a Particular Quiz or Exam?

If not, don't forget to analyze your quiz or exam and look for common errors. Were most of your errors a result of:

- *Carelessness?* Did you turn in your quiz or exam before the allotted time expired? If so, resolve next time to use the entire time allotted. Any extra time can be spent checking your work.

- *Running out of time?* If so, make a point to better manage your time on your next quiz or exam. Try completing any questions that you are unsure of last and delay checking your work until all questions have been answered.

- *Not understanding a concept?* If so, review that concept and correct your work. Try to understand how this happened so that you make sure it doesn't happen before the next quiz or exam.

- *Test conditions?* When studying for a quiz or exam, make sure you place yourself in conditions similar to test conditions. For example, before your next quiz or exam, use a few sheets of blank paper and take a sample test without the aid of your notes or text.

(See your instructor or use the Chapter Test at the end of each chapter.)

Exercises

1. Have you corrected all your previous quizzes and exams?

2. List any errors you have found common to two or more of your graded papers.

3. Is one of your common errors not understanding a concept? If so, are you making sure you understand all the concepts for the next quiz or exam?

4. Is one of your common errors making careless mistakes? If so, are you now taking all the time allotted to check over your work so that you can minimize the number of careless mistakes?

5. Are you satisfied with your grades thus far on quizzes and tests?

6. If your answer to Exercise 5 is no, are there any more suggestions you can make to your instructor or yourself to help? If so, list them here and share these with your instructor.

4.5 SYSTEMS OF LINEAR INEQUALITIES

Objective **A** Graphing Systems of Linear Inequalities

In Section 3.5 we solved linear inequalities in two variables. Just as two linear equations make a system of linear equations, two linear inequalities make a **system of linear inequalities.** Systems of inequalities are very important in a process called linear programming. Many businesses use linear programming to find the most profitable way to use limited resources such as employees, machines, or buildings.

A **solution of a system of linear inequalities** is an ordered pair that satisfies each inequality in the system. The set of all such ordered pairs is the solution set of the system. Graphing this set gives us a picture of the solution set. We can graph a system of inequalities by graphing each inequality in the system and identifying the region of overlap.

PRACTICE PROBLEM 1

Graph the solutions of the system:

$$\begin{cases} 2x \le y \\ x + 4y \ge 4 \end{cases}$$

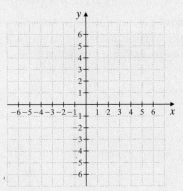

Graphing the Solutions of a System of Linear Inequalities

Step 1: Graph each inequality in the system on the same set of axes.

Step 2: The solutions of the system are the points common to the graphs of all the inequalities in the system.

EXAMPLE 1 Graph the solutions of the system: $\begin{cases} 3x \ge y \\ x + 2y \le 8 \end{cases}$

Solution: We begin by graphing each inequality on the *same* set of axes. The graph of the solutions of the system is the region contained in the graphs of both inequalities. In other words, it is their intersection.

First let's graph $3x \ge y$. The boundary line is the graph of $3x = y$. We sketch a solid boundary line since the inequality $3x \ge y$ means $3x > y$ or $3x = y$. The test point $(1, 0)$ satisfies the inequality, so we shade the half-plane that includes $(1, 0)$.

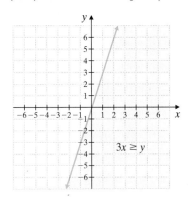

$3x \ge y$

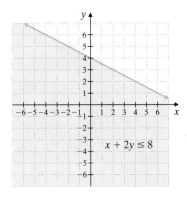

$x + 2y \le 8$

Answers

1.

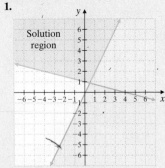

2.

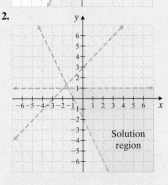

Next we sketch a solid boundary line $x + 2y = 8$ on the same set of axes. The test point $(0, 0)$ satisfies the inequality $x + 2y \le 8$, so we shade the half-plane that includes $(0, 0)$. (For clarity, the graph of $x + 2y \le 8$ is shown here on a separate set of axes.) An ordered pair solution of the system must satisfy both inequalities. These solutions are points that lie in both shaded regions. The solution of the system is the darkest shaded region. This solution includes parts of both boundary lines.

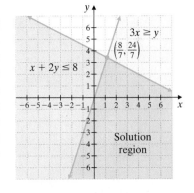

▢ Work Practice Problem 1

17. $\begin{cases} 3x - 4y \geq -6 \\ 2x + y \leq 7 \\ y \geq -3 \end{cases}$

18. $\begin{cases} 4x - y \geq -2 \\ 2x + 3y \leq -8 \\ y \geq -5 \end{cases}$

19. $\begin{cases} 2x + y \leq 5 \\ x \leq 3 \\ x \geq 0 \\ y \geq 0 \end{cases}$

20. $\begin{cases} 3x + y \leq 4 \\ x \leq 4 \\ x \geq 0 \\ y \geq 0 \end{cases}$

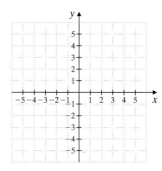
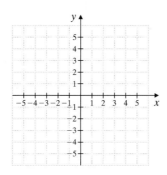
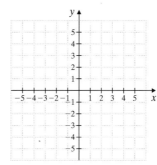
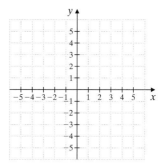

Match each system of inequalities to the corresponding graph.

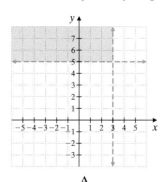
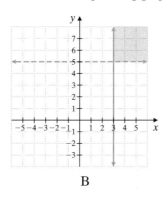
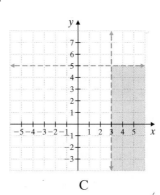
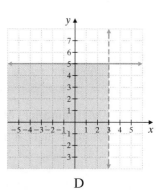

A	B	C	D

21. $\begin{cases} y < 5 \\ x > 3 \end{cases}$

22. $\begin{cases} y > 5 \\ x < 3 \end{cases}$

23. $\begin{cases} y \leq 5 \\ x < 3 \end{cases}$

24. $\begin{cases} y > 5 \\ x \geq 3 \end{cases}$

Review

Evaluate each expression. See Section 1.7.

25. $(-3)^2$

26. $(-5)^3$

27. $\left(\dfrac{2}{3}\right)^2$

28. $\left(\dfrac{3}{4}\right)^3$

Perform each indicated operation. See Section 1.4.

29. $(-2)^2 - (-3) + 2(-1)$ **30.** $5^2 - 11 + 3(-5)$

31. $8^2 + (-13) - 4(-2)$ **32.** $(-12)^2 + (-1)(2) - 6$

Concept Extensions

Solve. See the Concept Check in this section.

33. Describe the solution of the system: $\begin{cases} y \leq 3 \\ y \geq 3 \end{cases}$

34. Describe the solution of the system: $\begin{cases} x \leq 5 \\ x \leq 3 \end{cases}$

35. Explain how to decide which region to shade to show the solution region of the following system.

$$\begin{cases} x \geq 3 \\ y \geq -2 \end{cases}$$

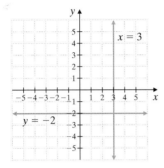

36. Tony Noellert budgets his time at work today. Part of the day he can write bills; the rest of the day he can use to write purchase orders. The total time available is at most 8 hours. Less than 3 hours is to be spent writing bills.

a. Write a system of inequalities to describe the situation. (Let x = hours available for writing bills and y = hours available for writing purchase orders.)

b. Graph the solutions of the system.

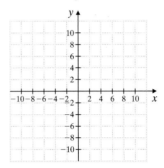

CHAPTER 4   Group Activity

Another Mathematical Model

Sometimes mathematical models other than linear models are appropriate for data. Suppose that an equation of the form $y = ax^2 + bx + c$ is an appropriate model for the ordered pairs $(x_1, y_1), (x_2, y_2),$ and (x_3, y_3). Then it is necessary to find the values of a, b, and c such that the given ordered pairs are solutions of the equation $y = ax^2 + bx + c$. To do so, substitute each ordered pair into the equation. Each time, the result is an equation in three unknowns: a, b, and c. Solving the resulting system of three linear equation in three unknowns will give the required values of a, b, and c.

1. The table gives the total beef supply (in billions of pounds) in the United States in each of the years listed.

a. Write the data as ordered pairs of the form (x, y), where y is the beef supply (in billions of pounds) in the year $x(x = 0$ represents 1999).

b. Find the values of a, b, and c such that the equation $y = ax^2 + bx + c$ models this data.

c. Verify that the model you found in part (b) gives each of the ordered pair solutions from part (a).

d. According to the model, what was the U.S. beef supply in 2002?

Total U.S. Beef Supply	
Year	**Beef Supply (billions of pounds)**
1999	29.8
2001	29.9
2003	30.1

(*Source:* Economic Research Service, U.S. Department of Agriculture)

2. The table gives Porsche sales figures for each of the years listed.

a. Write the data as ordered pairs of the form (x, y), where y is Porsche sales in the year $x(x = 0$ represents 1999).

b. Find the values of a, b, and c such that the equation $y = ax^2 + bx + c$ models this data.

c. According to the model, what was the total Porsche sales in 2002?

Total Porsche Sales	
Year	**Number of Porsche Vehicles Produced**
1999	43,982
2001	54,586
2003	66,803

(*Source:* Automotive Intelligence, www.autointell.com)

3. a. Make up an equation of the form $y = ax^2 + bx + c$.

b. Find three ordered pair solutions of the equation.

c. Without revealing your equation from part (a), exchange lists of ordered pair solutions with another group.

d. Use the method described above to find the values of a, b, and c such that the equation $y = ax^2 + bx + c$ has the ordered pair solutions you received from the other group.

e. Check with the other group to see if your equation from part (d) is the correct one.

Chapter 4 Vocabulary Check

Fill in each blank with one of the words or phrases listed below.

matrix   consistent   system of equations

solution   inconsistent   square

1. Two or more linear equations in two variables form a _____ .
2. A _____ of a system of two equations in two variables is an ordered pair that makes both equations true.
3. A(n) _____ system of equations has at least one solution.
4. If a matrix has the same number of rows and columns, it is called a _____ matrix.
5. A(n) _____ system of equations has no solution.
6. A _____ is a rectangular array of numbers.

Helpful Hint

Are you preparing for your test? Don't forget to take the Chapter 4 Test on page 314. Then check your answers at the back of the text and use the Chapter Test Prep Video CD to see the fully worked-out solutions to any of the exercises you want to review.

4 Chapter Highlights

DEFINITIONS AND CONCEPTS	**EXAMPLES**

Section 4.1   Solving Systems of Linear Equations in Two Variables

A **system of linear equations** consists of two or more linear equations.	$\begin{cases} x - 3y = 6 \\ y = \dfrac{1}{2}x \end{cases} \qquad \begin{cases} x + 2y - z = 1 \\ 3x - y + 4z = 0 \\ 5y + z = 6 \end{cases}$
A **solution** of a system of two equations in two variables is an ordered pair (x, y) that makes both equations true.	Determine whether $(2, -5)$ is a solution of the system. $\begin{cases} x + y = -3 \\ 2x - 3y = 19 \end{cases}$ Replace x with 2 and y with -5 in both equations. $\begin{array}{cc} x + y = -3 & 2x - 3y = 19 \\ 2 + (-5) \stackrel{?}{=} -3 & 2(2) - 3(-5) \stackrel{?}{=} 19 \\ -3 = -3 \quad \text{True} & 4 + 15 \stackrel{?}{=} 19 \\ & 19 = 19 \quad \text{True} \end{array}$ $(2, -5)$ is a solution of the system.
Geometrically, a solution of a system in two variables is a point of intersection of the graphs of the equations.	Solve by graphing: $\begin{cases} y = 2x - 1 \\ x + 2y = 13 \end{cases}$ 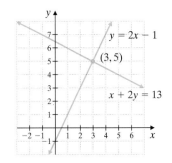

continued

DEFINITIONS AND CONCEPTS	**EXAMPLES**

Section 4.1   Solving Systems of Linear Equations in Two Variables (*continued*)

A system of equations with at least one solution is a **consistent system.** A system that has no solution is an **inconsistent system.**

If the graphs of two linear equations are identical, the equations are **dependent.**

If their graphs are different, the equations are **independent.**

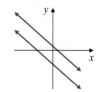

One solution:
Independent equations
Consistent system

No solution:
Independent equations
Inconsistent system

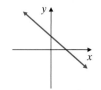

Infinite number of solutions:
Dependent equations
Consistent system

SOLVING A SYSTEM OF LINEAR EQUATIONS BY THE SUBSTITUTION METHOD

Step 1. Solve one equation for a variable.

Step 2. Substitute the expression for the variable into the other equation.

Step 3. Solve the equation from Step 2 to find the value of one variable.

Step 4. Substitute the value from Step 3 in either original equation to find the value of the other variable.

Step 5. Check the solution in both equations.

Solve by substitution:

$$\begin{cases} y = x + 2 \\ 3x - 2y = -5 \end{cases}$$

Since the first equation is solved for y, substitute $x + 2$ for y in the second equation.

$$\begin{aligned} 3x - 2y &= -5 \quad \text{Second equation} \\ 3x - 2(x + 2) &= -5 \quad \text{Let } y = x + 2. \\ 3x - 2x - 4 &= -5 \\ x - 4 &= -5 \quad \text{Simplify.} \\ x &= -1 \quad \text{Add 4.} \end{aligned}$$

To find y, let $x = -1$ in $y = x + 2$, so $y = -1 + 2 = 1$. The solution $(-1, 1)$ checks.

SOLVING A SYSTEM OF LINEAR EQUATIONS BY THE ELIMINATION METHOD

Step 1. Rewrite each equation in standard form, $Ax + By = C$.

Step 2. Multiply one or both equations by a nonzero number so that the coefficients of a variable are opposites.

Step 3. Add the equations.

Step 4. Find the value of the remaining variable by solving the resulting equation.

Step 5. Substitute the value from Step 4 into either original equation to find the value of the other variable.

Step 6. Check the solution in both equations.

Solve by elimination:

$$\begin{cases} x - 3y = -3 \\ -2x + y = 6 \end{cases}$$

Multiply both sides of the first equation by 2.

$$\begin{aligned} 2x - 6y &= -6 \\ \underline{-2x + y = 6} \\ -5y &= 0 \quad \text{Add.} \\ y &= 0 \quad \text{Divide by } -5. \end{aligned}$$

To find x, let $y = 0$ in an original equation.

$$\begin{aligned} x - 3y &= -3 \\ x - 3 \cdot 0 &= -3 \\ x &= -3 \end{aligned}$$

The solution $(-3, 0)$ checks.

DEFINITIONS AND CONCEPTS	**EXAMPLES**

Section 4.2   Solving Systems of Linear Equations in Three Variables

A **solution** of an equation in three variables x, y, and z is an **ordered triple** (x, y, z) that makes the equation a true statement.

Verify that $(-2, 1, 3)$ is a solution of $2x + 3y - 2z = -7$. Replace x with -2, y with 1, and z with 3.

$$2(-2) + 3(1) - 2(3) \stackrel{?}{=} -7$$
$$-4 + 3 - 6 \stackrel{?}{=} -7$$
$$-7 = -7 \quad \text{True}$$

$(-2, 1, 3)$ is a solution.

SOLVING A SYSTEM OF THREE LINEAR EQUATIONS BY THE ELIMINATION METHOD

Step 1. Write each equation in standard form, $Ax + By + Cz = D$.

Step 2. Choose a pair of equations and use them to eliminate a variable.

Step 3. Choose any other pair of equations and eliminate the same variable.

Step 4. Solve the system of two equations in two variables from Steps 2 and 3.

Step 5. Solve for the third variable by substituting the values of the variables from Step 4 into any of the original equations.

Step 6. Check the solution in all three original equations.

Solve:
$$\begin{cases} 2x + y - z = 0 & (1) \\ x - y - 2z = -6 & (2) \\ -3x - 2y + 3z = -22 & (3) \end{cases}$$

1. Each equation is written in standard form.

2.
$$\begin{array}{l} 2x + y - z = 0 \quad (1) \\ \underline{x - y - 2z = -6} \quad (2) \\ 3x \quad\quad - 3z = -6 \quad (4) \quad \text{Add.} \end{array}$$

3. Eliminate y from equations (1) and (3) also.
$$\begin{array}{l} 4x + 2y - 2z = 0 \quad\quad \text{Multiply equation (1) by 2.} \\ \underline{-3x - 2y + 3z = -22} \quad (3) \\ x \quad\quad + z = -22 \quad (5) \quad \text{Add.} \end{array}$$

4. Solve.
$$\begin{cases} 3x - 3z = -6 & (4) \\ x + z = -22 & (5) \end{cases}$$

$$\begin{array}{l} x - z = -2 \quad \text{Divide equation (4) by 3.} \\ \underline{x + z = -22} \quad (5) \\ 2x \quad\quad = -24 \\ x \quad\quad = -12 \end{array}$$

To find z, use equation (5).
$$x + z = -22$$
$$-12 + z = -22$$
$$z = -10$$

5. To find y, use equation (1).
$$2x + y - z = 0$$
$$2(-12) + y - (-10) = 0$$
$$-24 + y + 10 = 0$$
$$y = 14$$

6. The solution $(-12, 14, -10)$ checks.

DEFINITIONS AND CONCEPTS	EXAMPLES

Section 4.3   Systems of Linear Equations and Problem Solving

	Two numbers have a sum of 11. Twice one number is 3 less than 3 times the other. Find the numbers.
1. UNDERSTAND the problem.	**1.** Read and reread. x = one number y = other number
2. TRANSLATE.	**2.** In words: 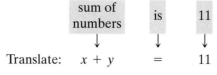 Translate: $x + y$ $=$ 11 In words: 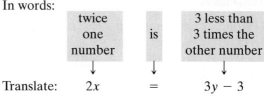 Translate: $2x$ $=$ $3y - 3$
3. SOLVE.	**3.** Solve the system: $\begin{cases} x + y = 11 \\ 2x = 3y - 3 \end{cases}$ In the first equation, $x = 11 - y$. Substitute into the other equation. $$2x = 3y - 3$$ $$2(11 - y) = 3y - 3$$ $$22 - 2y = 3y - 3$$ $$-5y = -25$$ $$y = 5$$ Replace y with 5 in the equation $x = 11 - y$. Then $x = 11 - 5 = 6$. The solution is $(6, 5)$.
4. INTERPRET.	**4.** *Check:* See that $6 + 5 = 11$ is the required sum and that twice 6 is 3 times 5 less 3. *State:* The numbers are 6 and 5.

Section 4.4   Solving Systems of Equations Using Matrices

A **matrix** is a rectangular array of numbers.	$\begin{bmatrix} -7 & 0 & 3 \\ 1 & 2 & 4 \end{bmatrix}$ $\begin{bmatrix} a & b & c \\ d & e & f \\ g & h & i \end{bmatrix}$
The **matrix** corresponding to a system is composed of the coefficients of the variables and the constants of the system.	The matrix corresponding to the system $\begin{cases} x - y = 1 \\ 2x + y = 11 \end{cases}$ is $\begin{bmatrix} 1 & -1 & 1 \\ 2 & 1 & 11 \end{bmatrix}$

DEFINITIONS AND CONCEPTS	**EXAMPLES**

Section 4.4   Solving Systems of Equations Using Matrices (*continued*)

The following **row operations** can be performed on matrices, and the result is an equivalent matrix.

Elementary row operations:

1. Interchange any two rows.

2. Multiply (or divide) the elements of one row by the same nonzero number.

3. Multiply (or divide) the elements of one row by the same nonzero number and add them to their corresponding elements in any other row.

Use matrices to solve: $\begin{cases} x - y = 1 \\ 2x + y = 11 \end{cases}$

The corresponding matrix is

$$\begin{bmatrix} 1 & -1 & | & 1 \\ 2 & 1 & | & 11 \end{bmatrix}$$

Use row operations to write an equivalent matrix with 1s along the diagonal and 0s below each 1 in the diagonal. Multiply row 1 by -2 and add to row 2. Change row 2 only.

$$\begin{bmatrix} 1 & -1 & | & 1 \\ -2(1) + 2 & -2(-1) + 1 & | & -2(1) + 11 \end{bmatrix}$$

simplifies to $\begin{bmatrix} 1 & -1 & | & 1 \\ 0 & 3 & | & 9 \end{bmatrix}$

Divide row 2 by 3.

$$\begin{bmatrix} 1 & -1 & | & 1 \\ \dfrac{0}{3} & \dfrac{3}{3} & | & \dfrac{9}{3} \end{bmatrix} \text{ simplifies to } \begin{bmatrix} 1 & -1 & | & 1 \\ 0 & 1 & | & 3 \end{bmatrix}$$

This matrix corresponds to the system

$$\begin{cases} x - y = 1 \\ y = 3 \end{cases}$$

Let $y = 3$ in the first equation.

$$x - 3 = 1$$
$$x = 4$$

The ordered pair solution is $(4, 3)$.

Section 4. 5   Systems of Linear Inequalities

A **system of linear inequalities** consists of two or more linear inequalities.

To graph a system of inequalities, graph each inequality in the system. The overlapping region is the solution of the system.

$$\begin{cases} x - y \geq 3 \\ y \leq -2x \end{cases}$$

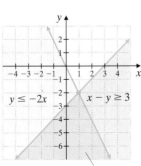

Solution region

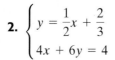

(4.1) *Solve each system of equations in two variables by each method: (a) graphing, (b) substitution, and (c) elimination.*

1. $\begin{cases} 3x + 10y = 1 \\ x + 2y = -1 \end{cases}$

2. $\begin{cases} y = \dfrac{1}{2}x + \dfrac{2}{3} \\ 4x + 6y = 4 \end{cases}$

3. $\begin{cases} 2x - 4y = 22 \\ 5x - 10y = 15 \end{cases}$

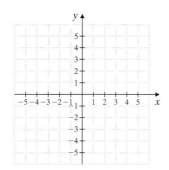
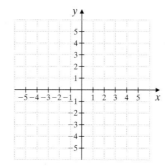
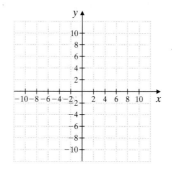

4. $\begin{cases} 3x - 6y = 12 \\ 2y = x - 4 \end{cases}$

5. $\begin{cases} \dfrac{1}{2}x - \dfrac{3}{4}y = -\dfrac{1}{2} \\ \dfrac{1}{8}x + \dfrac{3}{4}y = \dfrac{19}{8} \end{cases}$

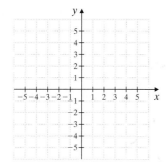
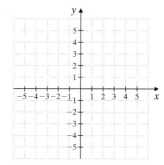

6. The revenue equation for a certain style of backpack is $y = 32x$, where x is the number of backpacks sold and y is the income in dollars for selling x backpacks. The cost equation for these units is $y = 15x + 25{,}500$, where x is the number of backpacks manufactured and y is the cost in dollars for manufacturing x backpacks. Find the number of units to be sold for the company to break even. (*Hint:* Solve the system of equations formed by the two given equations.)

(4.2) *Solve each system of equations in three variables.*

7. $\begin{cases} x \quad\;\; + z = 4 \\ 2x - y \quad\;\; = 4 \\ x + y - z = 0 \end{cases}$

8. $\begin{cases} 2x + 5y \quad\;\; = 4 \\ x - 5y + z = -1 \\ 4x \quad\quad - z = 11 \end{cases}$

9. $\begin{cases} \quad\;\; 4y + 2z = 5 \\ 2x + 8y \quad\;\; = 5 \\ 6x + \quad\;\; 4z = 1 \end{cases}$

10. $\begin{cases} 5x + 7y \quad\quad = 9 \\ \quad\quad 14y - z = 28 \\ 4x \quad\quad + 2z = -4 \end{cases}$

11. $\begin{cases} 3x - 2y + 2z = 5 \\ -x + 6y + z = 4 \\ 3x + 14y + 7z = 20 \end{cases}$

12. $\begin{cases} x + 2y + 3z = 11 \\ \quad\quad y + 2z = 3 \\ 2x \quad\quad + 2z = 10 \end{cases}$

13. $\begin{cases} 7x - 3y + 2z = 0 \\ 4x - 4y - z = 2 \\ 5x + 2y + 3z = 1 \end{cases}$

14. $\begin{cases} x - 3y - 5z = -5 \\ 4x - 2y + 3z = 13 \\ 5x + 3y + 4z = 22 \end{cases}$

(4.3) *Use systems of equations to solve.*

15. The sum of three numbers is 98. The sum of the first and second is two more than the third number, and the second is four times the first. Find the numbers.

16. One number is three times a second number, and twice the sum of the numbers is 168. Find the numbers.

17. Two cars leave Chicago, one traveling east and the other west. After 4 hours they are 492 miles apart. If one car is traveling 7 mph faster than the other, find the speed of each.

△ **18.** The foundation for a rectangular Hardware Warehouse has a length three times the width and is 296 feet around. Find the dimensions of the building.

19. James Callahan has available a 10% alcohol solution and a 60% alcohol solution. Find how many liters of each solution he should mix to make 50 liters of a 40% alcohol solution.

20. An employee at See's Candy Store needs a special mixture of candy. She has creme-filled chocolates that sell for $3.00 per pound, chocolate-covered nuts that sell for $2.70 per pound, and chocolate-covered raisins that sell for $2.25 per pound. She wants to have twice as many raisins as nuts in the mixture. Find how many pounds of each she should use to make 45 pounds worth $2.80 per pound.

21. Chris Kringler has $2.77 in her coin jar—all in pennies, nickels, and dimes. If she has 53 coins in all and four more nickels than dimes, find how many of each type of coin she has.

22. If $10,000 and $4000 are invested such that $1250 in interest is earned in one year, and if the rate of interest on the larger investment is 2% more than that on the smaller investment, find the rates of interest.

△ **23.** The perimeter of an isosceles (two sides equal) triangle is 73 centimeters. If the unequal side is 7 centimeters longer than the two equal sides, find the lengths of the three sides.

24. The sum of three numbers is 295. One number is five more than a second and twice the third. Find the numbers.

(4.4) *Use matrices to solve each system.*

25. $\begin{cases} 3x + 10y = 1 \\ x + 2y = -1 \end{cases}$

26. $\begin{cases} 3x - 6y = 12 \\ 2y = x - 4 \end{cases}$

27. $\begin{cases} 3x - 2y = -8 \\ 6x + 5y = 11 \end{cases}$

28. $\begin{cases} 6x - 6y = -5 \\ 10x - 2y = 1 \end{cases}$

29. $\begin{cases} 3x - 6y = 0 \\ 2x + 4y = 5 \end{cases}$

30. $\begin{cases} 5x - 3y = 10 \\ -2x + y = -1 \end{cases}$

31. $\begin{cases} 0.2x - 0.3y = -0.7 \\ 0.5x + 0.3y = 1.4 \end{cases}$

32. $\begin{cases} 3x + 2y = 8 \\ 3x - y = 5 \end{cases}$

33. $\begin{cases} x + z = 4 \\ 2x - y = 0 \\ x + y - z = 0 \end{cases}$

34. $\begin{cases} 2x + 5y = 4 \\ x - 5y + z = -1 \\ 4x - z = 11 \end{cases}$

35. $\begin{cases} 3x - y = 11 \\ x + 2z = 13 \\ y - z = -7 \end{cases}$

36. $\begin{cases} 5x + 7y + 3z = 9 \\ 14y - z = 28 \\ 4x + 2z = -4 \end{cases}$

37. $\begin{cases} 7x - 3y + 2z = 0 \\ 4x - 4y - z = 2 \\ 5x + 2y + 3z = 1 \end{cases}$

38. $\begin{cases} x + 2y + 3z = 14 \\ y + 2z = 3 \\ 2x - 2z = 10 \end{cases}$

(4.5) *Graph the solution of each system of linear inequalities.*

39. $\begin{cases} y \geq 2x - 3 \\ y \leq -2x + 1 \end{cases}$

40. $\begin{cases} y \leq -3x - 3 \\ y \leq 2x + 7 \end{cases}$

41. $\begin{cases} x + 2y > 0 \\ x - y \leq 6 \end{cases}$

42. $\begin{cases} x - 2y \geq 7 \\ x + y \leq -5 \end{cases}$

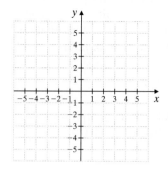
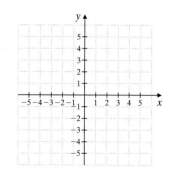
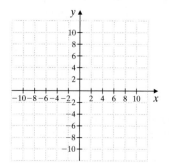
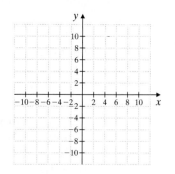

43. $\begin{cases} 3x - 2y \le 4 \\ 2x + y \ge 5 \\ y \le 4 \end{cases}$

44. $\begin{cases} 4x - y \le 0 \\ 3x - 2y \ge -5 \\ y \ge -4 \end{cases}$

45. $\begin{cases} x + 2y \le 5 \\ x \le 2 \\ x \ge 0 \\ y \ge 0 \end{cases}$

46. $\begin{cases} x + 3y \le 7 \\ y \le 5 \\ x \ge 0 \\ y \ge 0 \end{cases}$

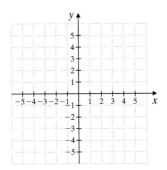
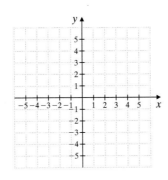
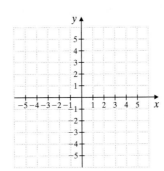
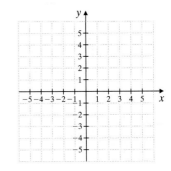

Mixed Review

Solve each system.

47. $\begin{cases} y = x - 5 \\ y = -2x + 2 \end{cases}$

48. $\begin{cases} \dfrac{2}{5}x + \dfrac{3}{4}y = 1 \\ x + 3y = -2 \end{cases}$

49. $\begin{cases} 5x - 2y = 10 \\ x = \dfrac{2}{5}y + 2 \end{cases}$

50. $\begin{cases} x - 4y = 4 \\ \dfrac{1}{8}x - \dfrac{1}{2}y = 3 \end{cases}$

51. $\begin{cases} x - 3y + 2z = 0 \\ 9y - z = 22 \\ 5x + 3z = 10 \end{cases}$

52. One number is five less than three times a second number. If the sum of the numbers is 127, find the numbers.

53. The perimeter of a triangle is 126 units. The length of one side is twice the length of the shortest side. The length of the third side is fourteen more than the length of the shortest side. Find the lengths of the sides of the triangles.

54. Graph the solution of the system: $\begin{cases} y \le 3x - \dfrac{1}{2} \\ 3x + 4y \ge 6 \end{cases}$

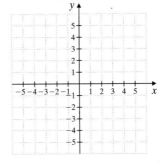

55. In the United States, the consumer spending on VCR decks is decreasing while the spending on DVD players is increasing. For the years 1998–2003, the function $y = -443x + 2584$ estimates the millions of dollars spent on purchasing VCR decks while the function $y = 500x + 551$ estimates the millions of dollars spent on purchasing DVD players. For both functions, x is the number of years since 1998. Use these equations to determine the year in which the amount of money spent on VCR decks equals the amount of money spent on DVD players. (*Source:* Consumer Electronics Asso.)

4 CHAPTER TEST

 Remember to use the Chapter Test Prep Video CD to see the fully worked-out solutions to any of the exercises you want to review.

Solve each system of equations graphically and then solve by the elimination method or the substitution method.

1. $\begin{cases} 2x - y = -1 \\ 5x + 4y = 17 \end{cases}$

2. $\begin{cases} 7x - 14y = 5 \\ x = 2y \end{cases}$

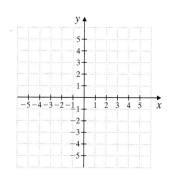

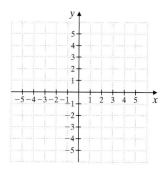

Solve each system.

3. $\begin{cases} 4x - 7y = 29 \\ 2x + 5y = -11 \end{cases}$

4. $\begin{cases} 15x + 6y = 15 \\ 10x + 4y = 10 \end{cases}$

5. $\begin{cases} 2x - 3y = 4 \\ 3y + 2z = 2 \\ x - z = -5 \end{cases}$

6. $\begin{cases} 3x - 2y - z = -1 \\ 2x - 2y = 4 \\ 2x - 2z = -12 \end{cases}$

7. $\begin{cases} \dfrac{x}{2} + \dfrac{y}{4} = -\dfrac{3}{4} \\ x + \dfrac{3}{4}y = -4 \end{cases}$

Use matrices to solve each system.

8. $\begin{cases} x - y = -2 \\ 3x - 3y = -6 \end{cases}$

9. $\begin{cases} x + 2y = -1 \\ 2x + 5y = -5 \end{cases}$

10. $\begin{cases} x - y - z = 0 \\ 3x - y - 5z = -2 \\ 2x + 3y = -5 \end{cases}$

11. A motel in New Orleans charges $90 per day for double occupancy and $80 per day for single occupancy. If 80 rooms are occupied for a total of $6930, how many rooms of each kind are occupied?

12. The research department of a company that manufactures children's fruit drinks is experimenting with a new flavor. A 17.5% fructose solution is needed, but only 10% and 20% solutions are available. How many gallons of a 10% fructose solution should be mixed with a 20% fructose solution to obtain 20 gallons of a 17.5% fructose solution?

Answers

1. _____
2. _____
3. _____
4. _____
5. _____
6. _____
7. _____
8. _____
9. _____
10. _____
11. _____
12. _____

13. A company that manufactures boxes recently purchased $2000 worth of new equipment to offer gift boxes to its customers. The cost of producing a package of gift boxes is $1.50 and it is sold for $4.00. Find the number of packages that must be sold for the company to break even.

14. The measure of the largest angle of a triangle is three less than 5 times the measure of the smallest angle. The measure of the remaining angle is 1 less than twice the measure of the smallest angle. Find the measure of each angle.

13. _____

14. _____

Graph the solutions of each system of linear inequalities.

15. $\begin{cases} 2y - x \geq 1 \\ x + y \geq -4 \\ y \leq 2 \end{cases}$

15. see graph _____

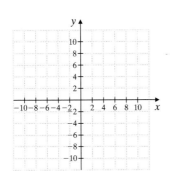

Answers

1. _____

2. _____

3. _____

4. _____

5. _____

6. _____

7. _____

8. _____

9. _____

10. _____

11. _____

12. _____

13. _____

14. _____

15. _____

16. a. _____

  b. _____

17. _____

18. _____

19. _____

20. _____

21. _____

22. _____

23. _____

24. _____

25. _____

26. _____

27. _____

28. _____

Determine whether each statement is true or false.

1. $7 \notin \{1, 2, 3\}$

2. $6 \in \{1, 2, 3\}$

3. $\dfrac{1}{5}$ is an irrational number.

4. Every integer is a real number.

Insert $<$, $>$, or $=$ between each pair of numbers to form a true statement.

5. $-5 \quad 5$

6. $-21 \quad -23$

7. $-2.5 \quad -2.1$

8. $-\dfrac{3}{3} \quad -\dfrac{7.2}{7.2}$

9. $\dfrac{2}{3} \quad \dfrac{3}{4}$

10. $-\dfrac{1}{11} \quad \dfrac{1}{11}$

Find each absolute value.

11. $\left| -\dfrac{1}{7} \right|$

12. $|-7.8|$

13. $-|-8|$

14. $-\left| -\dfrac{3}{8} \right|$

15. Simplify: $\dfrac{|-2|^3 + 1}{-7 - \sqrt{4}}$

16. Subtract.
   a. $-7 - (-2)$
   b. $14 - 38$

Simplify and write with positive exponents only.

17. $(3x)^{-1}$

18. 9^{-2}

19. $2^{-1} + 3^{-2}$

20. $(x^2)^3 \cdot x^{10}$

21. Determine whether -15 is a solution of $x - 9 = -24$.

22. Solve: $8y - 14 = 6y - 14$

23. Suppose that a computer store just announced an 8% decrease in the price of a particular computer model. If this computer sells for $2162 after the decrease, find its original price.

24. A quadrilateral has 4 angles whose sum is 360°. In a particular quadrilateral, two angles have the same measure. A third angle is 10° more than the measure of one of the equal angles, and the fourth angle is half the measure of one of the equal angles. Find the measures of the angles.

△ **25.** Solve $A = \dfrac{1}{2}(B + b)h$ for b.

26. Solve:
$$2(m - 6) - m = 4(m - 3) - 3m$$

27. Solve: $4x - 2 < 5x$

28. Solve: $5(2x - 1) > -5$

29. Find the intersection:
$\{2, 4, 6, 8, \} \cap \{3, 4, 5, 6\}$

30. Find the union:
$\{2, 4, 6, 8\} \cup \{3, 4, 5, 6\}$

31. Solve: $|y| = 0$

32. Solve: $|x - 5| = 4$

33. Graph: $y = -3$

34. Solve: $|2x + 1| > 5$

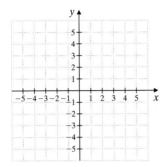

35. Find the slope and the y-intercept of the line $3x - 4y = 4$.

36. Name the quadrant or axis each point is located.
   a. $(-1, -5)$ **b.** $(4, -2)$ **c.** $(0, 2)$

37. Write an equation of the line with slope -3 containing the point $(1, -5)$. Write the equation in slope-intercept form.

38. If $f(x) = 3x^2$, find the following.
   a. $f(5)$ **b.** $f(-2)$

39. Determine whether the relation $y = 2x + 1$ is also a function.

40. Find the slope of the line containing $(-2, 6)$ and $(0, 9)$.

41. Graph the intersection of $x \geq 1$ and $y \geq 2x - 1$.

42. Find an equation of the line through $(-2, 6)$ perpendicular to
$$f(x) = \frac{1}{2}x - \frac{1}{3}.$$

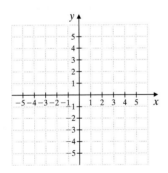

43. Use the elimination method to solve the system: $\begin{cases} 3x - 2y = 10 \\ 4x - 3y = 15 \end{cases}$

44. Solve the system: $\begin{cases} 5x + y = -2 \\ 4x - 2y = -10 \end{cases}$

45. Solve the system:
$$\begin{cases} 2x - 4y + 8z = 2 \\ -x - 3y + z = 11 \\ x - 2y + 4z = 0 \end{cases}$$

46. Solve the system:
$$\begin{cases} x - 2y + z = 0 \\ 3x - y - 2z = -15 \\ 2x - 3y + 3z = 7 \end{cases}$$

47. A first number is 4 less than a second number. Four times the first number is 6 more than twice the second. Find the numbers.

48. Solve the system:
$$\begin{cases} -6x + 8y = 0 \\ 9x - 12y = 2 \end{cases}$$

29. _____

30. _____

31. _____

32. _____

33. see graph _____

34. _____

35. _____

36. a. _____

     b. _____

     c. _____

37. _____

38. a. _____

     b. _____

39. _____

40. _____

41. see graph _____

42. _____

43. _____

44. _____

45. _____

46. _____

47. _____

48. _____

5

Polynomials and Polynomial Functions

Linear equations are important for solving problems. They are not sufficient, however, to solve all problems. Many real-world phenomena are modeled by polynomials. In the first portion of this chapter we will study operations on polynomials. We then look at how polynomials can be used in problem solving.

Mount Rushmore, the world's greatest mountain carving, rises in splendor amid South Dakota's Black Hills. This epic sculpture of four towering presidents represents the first 150 years of our American history. The president chosen to embody this era were George Washington, Thomas Jefferson, Abraham Lincoln, and Theodore Roosevelt. These 60-foot-high faces, 500 feet up, look out over a wilderness of pine, spruce, birch, and aspen.

The carving of Mount Rushmore, created by Sculptor Gutzon Borglum, was begun on August 10, 1927, and spanned a period of fourteen years. Work continued on the project until the death of Gutzon Borglum in 1941. No carving has been done on the mountain since that time, and none is planned in the future. The project cost close to one million dollars and today, millions of visitors come to see it.

In Exercise 80, Section 5.1, we will use a polynomial function to find the height of an object thrown from the top of Mount Rushmore.

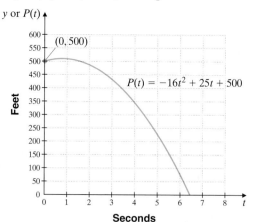

Graph of $P(t) = -16t^2 + 25t + 500$, the height of an object above the ground at time t, if the object is thrown upward from the top of Washington's head on Mount Rushmore, with an initial velocity of 25 feet per second.

5.1 POLYNOMIAL FUNCTIONS AND ADDING AND SUBTRACTING POLYNOMIALS

Objectives

- **A** Define Term, Constant, Polynomial, Monomial, Binomial, Trinomial, Degree of a Term, and Degree of a Polynomial.
- **B** Combine Like Terms.
- **C** Add Polynomials.
- **D** Subtract Polynomials.
- **E** Evaluate Polynomial Functions.

Objective **A** Defining a Polynomial and Related Terms

A **term** is a number or the product of a number and one or more variables raised to powers. The **numerical coefficient,** or simply the **coefficient,** is the numerical factor of a term.

Term	Numerical Coefficient of Term
$-1.2x^5$	-1.2
x^3y	1
$-z$	-1
2	2
$\dfrac{x^9}{7}\left(\text{or } \dfrac{1}{7}x^9\right)$	$\dfrac{1}{7}$

If a term contains only a number, it is called a **constant term,** or simply a **constant.**

A **polynomial** is a finite sum of terms in which all variables are raised to nonnegative integer powers and no variables appear in any denominator.

Polynomials	Not Polynomials	
$4x^5y + 7xz$	$5x^{-3} + 2x$	Negative integer exponent
$-5x^3 + 2x + \dfrac{2}{3}$	$\dfrac{6}{x^2} - 5x + 1$	Variable in denominator

A polynomial that contains only one variable is called a **polynomial in one variable.** For example, $3x^2 - 2x + 7$ is a **polynomial in x.** This polynomial in x is written in **descending order** since the terms are listed in descending order of the variable's exponents. (The term 7 can be thought of as $7x^0$.) The following examples are polynomials in one variable written in descending order:

$$4x^3 - \frac{7}{8}x^2 + 5 \qquad y^2 - 4.7 \qquad \frac{8a^4}{11} - 7a^2 + \frac{4a}{3}$$

A **monomial** is a polynomial consisting of one term. A **binomial** is a polynomial consisting of two terms. A **trinomial** is a polynomial consisting of three terms.

Monomials	Binomials	Trinomials
ax^2	$x + y$	$x^2 + 4xy + y^2$
$-3x$	$6y^2 - 2.9$	$-x^4 + 3x^3 + 0.1$
4	$\dfrac{5}{7}z^3 - 2z$	$8y^2 - 2y - \dfrac{10}{17}$

By definition, all monomials, binomials, and trinomials are also polynomials.

Each term of a polynomial has a **degree.**

> **Helpful Hint**
> We will write answers that are polynomials in one variable in descending order.

Degree of a Term

The **degree of a term** is the sum of the exponents on the *variables* contained in the term.

PRACTICE PROBLEMS 1–5

Determine the degree of each term.
1. $2x^3$ **2.** 7^2x^4
3. x **4.** $15xy^2z^4$
5. 9.1

EXAMPLES Determine the degree of each term.

1. $3x^2$ The exponent on x is 2, so the degree of the term is 2.
2. -2^3x^5 The exponent on x is 5, so the degree of the term is 5. (Recall that the degree is the sum of the exponents on *only* the *variables*).
3. y The degree of y, or y^1, is 1.
4. $12x^2yz^3$ The degree is the sum of the exponents on the variables, or $2 + 1 + 3 = 6$.
5. 5.27 The degree of 5.27, which can be written as $5.27x^0$, is 0.

■ **Work Practice Problems 1–5**

From the preceding examples, we can say that the degree of a constant is 0. Also, the term 0 has no degree.

Each polynomial also has a degree.

Degree of a Polynomial

The **degree of a polynomial** is the greatest degree of any of its terms.

PRACTICE PROBLEMS 6–8

Determine the degree of each polynomial and indicate whether the polynomial is a monomial, binomial, or trinomial.
6. $4x^5 + 7x^3 - 1$
7. $-2xy^2z$
8. $y^3 + 6y$

EXAMPLES Determine the degree of each polynomial and also indicate whether the polynomial is a monomial, binomial, or trinomial.

	Polynomial	Degree	Classification
6.	$7x^3 - \dfrac{3}{4}x + 2$	3	Trinomial
7.	$-xyz$	$1 + 1 + 1 = 3$	Monomial
8.	$x^4 - 16.5$	4	Binomial

■ **Work Practice Problems 6–8**

PRACTICE PROBLEM 9

Determine the degree of the polynomial
$7x^2y - 6x^2yz + 2 - 4y^3$.

EXAMPLE 9 Determine the degree of the polynomial
$$3xy + x^2y^2 - 5x^2 - 6.7$$

Solution: The degree of each term is
$$3xy + x^2y^2 - 5x^2 - 6.7$$
$$\downarrow \quad\quad \downarrow \quad\quad \downarrow \quad\quad \downarrow$$
Degree: $\quad 2 \quad\quad 4 \quad\quad 2 \quad\quad 0$

The greatest degree of any term is 4, so the degree of this polynomial is 4.

■ **Work Practice Problem 9**

Objective **B** Combining Like Terms

Before we add polynomials, recall from Section 1.5 that terms are considered to be **like terms** if they contain exactly the same variables raised to exactly the same powers.

Like Terms	Unlike Terms
$-5x^2, -x^2$	$4x^2, 3x$
$7xy^3z, -2xzy^3$	$12x^2y^3, -2xy^3$

To simplify a polynomial, we **combine like terms** by using the distributive property. For example, by the distributive property.

$$5x + 7x = (5 + 7)x = 12x$$

Answers
1. 3, **2.** 4, **3.** 1, **4.** 7, **5.** 0,
6. 5; trinomial, **7.** 4; monomial,
8. 3; binomial, **9.** 4

EXAMPLES Simplify each polynomial by combining like terms.

10. $-12x^2 + 7x^2 - 6x = (-12 + 7)x^2 - 6x = -5x^2 - 6x$

11. $3xy - 2x + 5xy - x = 3xy + 5xy - 2x - x$
$$= (3 + 5)xy + (-2 - 1)x$$
$$= \underbrace{8xy} - \underbrace{3x}$$

■ **Work Practice Problems 10–11**

Helpful Hint
These two terms are unlike terms. They cannot be combined.

PRACTICE PROBLEMS 10–11

Simplify each polynomial by combining like terms.
10. $10x^3 - 12x^3 - 3x$
11. $-6ab + 2a + 12ab - a$

Objective C Adding Polynomials

Now we have reviewed the skills we need to add polynomials.

Adding Polynomials

To add polynomials, combine all like terms.

EXAMPLE 12 Add $11x^3 - 12x^2 + x - 3$ and $x^3 - 10x + 5$.

Solution:

$(11x^3 - 12x^2 + x - 3) + (x^3 - 10x + 5)$
$= 11x^3 + x^3 - 12x^2 + x - 10x - 3 + 5$ Group like terms.
$= 12x^3 - 12x^2 - 9x + 2$ Combine like terms.

■ **Work Practice Problem 12**

PRACTICE PROBLEM 12

Add $14x^4 - 6x^3 + x^2 - 6$ and $x^3 - 5x^2 + 1$.

Sometimes it is more convenient to add polynomials vertically. To do this, we line up like terms beneath one another and then add like terms.

EXAMPLE 13 Add $11x^3 - 12x^2 + x - 3$ and $x^3 - 10x + 5$ vertically.

Solution:

$$\begin{array}{r} 11x^3 - 12x^2 + x - 3 \\ \underline{x^3 \qquad - 10x + 5} \\ 12x^3 - 12x^2 - 9x + 2 \end{array}$$ Line up like terms.
   Combine like terms.

This example is the same as Example 12, only here we added vertically.

■ **Work Practice Problem 13**

PRACTICE PROBLEM 13

Add $10y^3 - y^2 + 4y - 11$ and $y^3 - 4y^2 + 3y$ vertically.

EXAMPLE 14 Add: $(7x^3y - xy^3 + 11) + (6x^3y - 4)$

Solution: To add these polynomials, we remove the parentheses and group like terms.

$(7x^3y - xy^3 + 11) + (6x^3y - 4)$
$= 7x^3y - xy^3 + 11 + 6x^3y - 4$ Remove parentheses.
$= 7x^3y + 6x^3y - xy^3 + 11 - 4$ Group like terms.
$= 13x^3y - xy^3 + 7$ Combine like terms.

■ **Work Practice Problem 14**

PRACTICE PROBLEM 14

Add:
$(4x^2y - xy^2 + 5) + (-6x^2y - 1)$

Objective D Subtracting Polynomials

The definition of subtraction of real numbers can be extended to apply to polynomials. To subtract a number, we add its opposite:

$$a - b = a + (-b)$$

Answers
10. $-2x^3 - 3x$, **11.** $6ab + a$,
12. $14x^4 - 5x^3 - 4x^2 - 5$,
13. $11y^3 - 5y^2 + 7y - 11$,
14. $-2x^2y - xy^2 + 4$

Likewise, to subtract a polynomial we add its opposite. In other words, if P and Q are polynomials, then

$$P - Q = P + (-Q)$$

The polynomial $-Q$ is the **opposite,** or **additive inverse,** of the polynomial Q. We can find $-Q$ by changing the sign of each term of Q.

✔ **Concept Check**   Which polynomial is the opposite of $16x^3 - 5x + 7$?

a. $-16x^3 - 5x + 7$ b. $-16x^3 + 5x - 7$

c. $16x^3 + 5x + 7$ d. $-16x^3 + 5x + 7$

Subtracting Polynomials

To subtract polynomials, change the signs of the terms of the polynomial being subtracted and then add.

Review the example below.

To subtract, change the signs; then add.

$$(3x^2 + 4x - 7) - (3x^2 - 2x - 5) = (3x^2 + 4x - 7) + (-3x^2 + 2x + 5)$$

$$= 3x^2 + 4x - 7 - 3x^2 + 2x + 5$$

$$= 6x - 2 \qquad \text{Combine like terms.}$$

PRACTICE PROBLEM 15

Subtract:
$(7x^4 - 8x^2 + x)$
$\quad - (-9x^4 + x^2 - 18)$

EXAMPLE 15   Subtract: $(12z^5 - 12z^3 + z) - (-3z^4 + z^3 + 12z)$

Solution: First we change the sign of each term of the second polynomial, and then we add the result to the first polynomial.

$$(12z^5 - 12z^3 + z) - (-3z^4 + z^3 + 12z)$$

$$= 12z^5 - 12z^3 + z + 3z^4 - z^3 - 12z \qquad \text{Change signs and add.}$$

$$= 12z^5 + 3z^4 - 12z^3 - z^3 + z - 12z \qquad \text{Group like terms; write in descending order.}$$

$$= 12z^5 + 3z^4 - 13z^3 - 11z \qquad \text{Combine like terms.}$$

▩ **Work Practice Problem 15**

PRACTICE PROBLEM 16

Subtract
$(2y^4 + 4y) - (6y^4 + 7y^3 - 3y)$
vertically.

EXAMPLE 16   Subtract $(10x^3 - 7x^2) - (4x^3 - 3x^2 + 2)$ vertically.

Solution: To subtract these polynomials, we add the opposite of the second polynomial to the first one.

$$\begin{array}{ll} 10x^3 - 7x^2 \\ \underline{-(4x^3 - 3x^2 + 2)} \end{array} \quad \text{is equivalent to} \quad \begin{array}{l} 10x^3 - 7x^2 \\ \underline{-4x^3 + 3x^2 - 2} \\ 6x^3 - 4x^2 - 2 \quad \text{Add.} \end{array}$$

▩ **Work Practice Problem 16**

Answers

15. $16x^4 - 9x^2 + x + 18$,
16. $-4y^4 - 7y^3 + 7y$

✔ **Concept Check Answers**

b;

With parentheses removed, the expression should be
$\quad 7z - 5 - 3z + 4 = 4z - 1$

✔ **Concept Check**   Why is the following subtraction incorrect?

$$(7z - 5) - (3z - 4)$$

$$= 7z - 5 - 3z - 4$$

$$= 4z - 9$$

EXAMPLE 17 Subtract $4x^3y^2 - 3x^2y^2 + 2y^2$ from $10x^3y^2 - 7x^2y^2$.

Solution: Notice the order of the expressions, and then write "Subtract $4x^3y^2 - 3x^2y^2 + 2y^2$ from $10x^3y^2 - 7x^2y^2$" as a mathematical expression. (For example, if we subtract 2 from 8, we would write $8 - 2 = 6$.)

$$(10x^3y^2 - 7x^2y^2) - (4x^3y^2 - 3x^2y^2 + 2y^2)$$
$$= 10x^3y^2 - 7x^2y^2 - 4x^3y^2 + 3x^2y^2 - 2y^2 \quad \text{Remove parentheses.}$$
$$= 6x^3y^2 - 4x^2y^2 - 2y^2 \quad \text{Combine like terms.}$$

◼ **Work Practice Problem 17**

PRACTICE PROBLEM 17

Subtract $3a^2b^3 - 4ab^2 + 6a$ from $7a^2b^3 - ab^2$.

Objective E Evaluating Polynomial Functions

Recall function notation first introduced in Section 3.6. At times it is convenient to use function notation to represent polynomials. For example, we may write $P(x)$ to represent the polynomial $3x^4 - 2x^2 - 5$. In symbols, we would write

$$P(x) = 3x^4 - 2x^2 - 5$$

This function is called a **polynomial function** because the expression $3x^4 - 2x^2 - 5$ is a polynomial.

> **Helpful Hint**
>
> Recall that the symbol $P(x)$ *does not mean* P times x. It is a special symbol used to denote a function.

EXAMPLES If $P(x) = 3x^4 - 2x^2 - 5$, find each function value.

18. $P(1) = 3(1)^4 - 2(1)^2 - 5 = -4$ Let $x = 1$ in the function $P(x)$.

19. $P(-2) = 3(-2)^4 - 2(-2)^2 - 5$ Let $x = -2$ in the function $P(x)$.
$$= 3(16) - 2(4) - 5$$
$$= 35$$

◼ **Work Practice Problems 18–19**

PRACTICE PROBLEMS 18–19

If $P(x) = 5x^4 - 3x^2 + 7$, find each function value.
18. $P(2)$
19. $P(-1)$

Many real-world phenomena are modeled by polynomial functions. If the polynomial function model is given, we can often find the solution of a problem by evaluating the function at a certain value.

EXAMPLE 20 Finding the Height of an Object

The Millau Viaduct, at 1125 feet, is the highest bridge in the world and overlooks the river Tarn in France. An object is dropped from the top of this bridge. Neglecting air resistance, the height of the object at time t seconds is given by the polynomial function $P(t) = -16t^2 + 1125$. Find the height of the object when $t = 1$ second and when $t = 8$ seconds.

Solution: To find the height of the object at 1 second, we find $P(1)$.

$$P(t) = -16t^2 + 1125$$
$$P(1) = -16(1)^2 + 1125$$
$$P(1) = 1109$$

Continued on next page

PRACTICE PROBLEM 20

Use the polynomial function in Example 20 to find the height of the object when $t = 3$ seconds and $t = 7$ seconds.

Answers
17. $4a^2b^3 + 3ab^2 - 6a$,
18. $P(2) = 75$,   **19.** $P(-1) = 9$,
20. at 3 sec, height is 981 ft; at 7 sec, height is 341 ft

When $t = 1$ second, the height of the object is 1109 feet.

To find the height of the object at 8 seconds, we find $P(8)$.

$$P(t) = -16t^2 + 1125$$
$$P(8) = -16(8)^2 + 1125$$
$$P(8) = -1024 + 1125$$
$$P(8) = 101$$

When $t = 8$ seconds, the height of the object is 101 feet. Notice that as time t increases, the height of the object decreases.

Work Practice Problem 20

CALCULATOR EXPLORATIONS   Graphing

A graphing calculator may be used to check addition and subtraction of polynomials in one variable. For example, to check the polynomial subtraction statement

$$(3x^2 - 6x + 9) - (x^2 - 5x + 6) = 2x^2 - x + 3$$

graph both

$$Y_1 = (3x^2 - 6x + 9) - (x^2 - 5x + 6) \quad \text{Left side of equation}$$

and

$$Y_2 = 2x^2 - x + 3 \quad \text{Right side of equation}$$

on the same screen and see that their graphs coincide. (*Note:* If the graphs do not coincide, we can be sure that a mistake has been made either in combining polynomials or in calculator keystrokes. However, if the graphs appear to coincide, we cannot be sure that our work is correct. This is because it is possible for the graphs to differ so slightly that we do not notice it.)

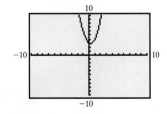

The graphs of Y_1 and Y_2 are shown. The graphs appear to coincide, so the subtraction statement

$$(3x^2 - 6x + 9) - (x^2 - 5x + 6) = 2x^2 - x + 3$$

appears to be correct.

Perform each indicated operation. Then use the procedure described above to check your work.

1. $(2x^2 + 7x + 6) + (x^3 - 6x^2 - 14)$

2. $(-14x^3 - x + 2) + (-x^3 + 3x^2 + 4x)$

3. $(1.8x^2 - 6.8x - 1.7) - (3.9x^2 - 3.6x)$

4. $(-4.8x^2 + 12.5x - 7.8) - (3.1x^2 - 7.8x)$

5. $(1.29x - 5.68) + (7.69x^2 - 2.55x + 10.98)$

6. $(-0.98x^2 - 1.56x + 5.57) + (4.36x - 3.71)$

Mental Math

Add or subtract as indicated.

1. $7x + 3x$ **2.** $8x - 2x$ **3.** $14y - 9y$ **4.** $14y + 9y$ **5.** $3z - 12z$ **6.** $2z - 6z$

5.1 EXERCISE SET

FOR EXTRA HELP

Student Solutions Manual   PH Math/Tutor Center   CD/Video for Review   MathXL®   MyMathLab

Objective Ⓐ *Find the degree of each term. See Examples 1 through 5.*

1. 4   **2.** 7   **3.** $5x^2$ **4.** $-z^3$ **5.** $-3xy^2$

6. $12x^3z$ **7.** -8^7y^3 **8.** $-9^{11}y^5$ **9.** $3.78ab^3c^5$ **10.** $9.11r^2st^{12}$

Find the degree of each polynomial and indicate whether the polynomial is a monomial, binomial, trinomial, or none of these. See Examples 6 through 9.

11. $6x + 0.3$ **12.** $7x - 0.8$

13. $3x^2 - 2x + 5$ **14.** $5x^2 - 3x - 2$

15. -3^4xy^2 **16.** -7^5abc

17. $x^2y - 4xy^2 + 5x + y^4$ **18.** $-2x^2y - 3y^2 + 4x + y^5$

Objective Ⓑ *Simplify each polynomial by combining like terms. See Examples 10 and 11.*

19. $5y + y - 6y^2 - y^2$ **20.** $-x + 3x - 4x^2 - 9x^2$

21. $4x + 7x - 3x^4$ **22.** $-8y + 9y + 4y^6$

23. $4x^2y + 2x - 3x^2y - \dfrac{1}{2} - 7x$ **24.** $-8xy^2 + 4x - x + 2xy^2 - \dfrac{11}{15}$

Objective Ⓒ *Add. See Examples 12 through 14.*

25. $(9y^2 + y - 8) + (9y^2 - y - 9)$ **26.** $(x^2 + 4x - 7) + (8x^2 + 9x - 7)$

27. $(x^2 + xy - y^2)$ and $(2x^2 - 4xy + 7y^2)$ **28.** $(6x^2 + 5x + 7)$ and $(x^2 + 6x - 3)$

29. $x^2 - 6x + 3$
     $+ \quad (2x + 5)$
     ―――――――――

30. $-2x^2 + 3x - 9$
     $+ \quad (2x - 3)$
     ―――――――――

31. $(7x^3y - 4xy + 8) + (5x^3y + 4xy + 8x)$ **32.** $(9xyz + 4x - y) + (-9xyz - 3x + y + 2)$

325

33. $(0.6x^3 + 1.2x^2 - 4.5x + 9.1) + (3.9x^3 - x^2 + 0.7x)$ **34.** $(9.3y^2 - y + 12.8) + (2.6y^2 + 4.4y - 8.9)$

Objective D *Subtract. See Examples 15 through 17.*

35. $(9y^2 - 7y + 5) - (8y^2 - 7y + 2)$ **36.** $(2x^2 + 3x + 12) - (20x^2 - 5x - 7)$

37. Subtract $(6x^2 - 3x)$ from $(4x^2 + 2x)$.   **38.** Subtract $(8y^2 + 4x)$ from $(y^2 + x)$.

39. $\quad 6y^2 - 6y + 4$
$\quad -(-y^2 + 6y + 7)$

40. $\quad -4x^3 + 4x^2 - 4x$
$\quad -(2x^3 - 2x^2 + 3x)$

41. $(9x^3 - 2x^2 + 4x - 7) - (2x^3 - 6x^2 - 4x + 3)$ **42.** $(3x^2 + 6xy + 3y^2) - (8x^2 - 6xy - y^2)$

43. Subtract $\left(y^2 + 4yx + \dfrac{1}{7}\right)$ from $\left(-19y^2 + 7yx + \dfrac{1}{7}\right)$.   **44.** Subtract $\left(13x^2 + x^2y - \dfrac{1}{4}\right)$ from $\left(3x^2 - 4x^2y - \dfrac{1}{4}\right)$.

Objectives B C D **Mixed Practice** *Perform indicated operations and simplify. See Examples 10 through 17.*

45. $(-3x + 8) + (-3x^2 + 3x - 5)$ **46.** $(-5y^2 - 2y + 4) + (3y + 7)$

47. $(5y^4 - 7y^2 + x^2 - 3) + (-3y^4 + 2y^2 + 4)$ **48.** $(8x^4 - 14x^2 + x + 6) + (-12x^6 - 21x^4 - 9x^2)$

49. $(4x^2 - 6x + 2) - (-x^2 + 3x + 5)$ **50.** $(7x^2 + x + 1) - (6x^2 + x - 1)$

51. $(5x^2 + x + 9) - (2x^2 - 9)$ **52.** $(4x - 4) - (-x - 4)$

53. $(5x - 11) + (-x - 2)$ **54.** $(3x^2 - 2x) + (-5x^2 - 9x)$

55. $(3x^3 - b + 2a - 6) + (-4x^3 + b + 6a - 6)$ **56.** $(9y^3 - a + 7b - 3) + (-2y^3 + a + 6b - 8)$

57. $(14ab - 10a^2b + 6b^2) - (18a^2 - 20a^2b - 6b^2)$ **58.** $(13x^2 - 26x^2y^2 + 4) - (19x^2 + x^2y^2 - 11)$

59. $\quad 3x^2 + 15x + 8$
$\quad +(2x^2 + 7x + 8)$

60. $\quad 9x^2 + 9x - 4$
$\quad +(7x^2 - 3x - 4)$

61. $(7x^2 - 5) + (-3x^2 - 2) - (4x^2 - 7)$ **62.** $(9y^2 - 3) + (-4y^2 + 1) - (5y^2 - 2)$

63. $(-3 + 4x^2 + 7xy^2) + (2x^3 - x^2 + xy^2)$ **64.** $(-3x^2y + 4) + (-7x^2y - 8y)$

65. $\quad 3x^2 - 4x + 8$
$\quad -\quad (5x - 7)$

66. $\quad -3x^2 - 4x + 8$
$\quad -\quad (5x + 12)$

67. Subtract $(3x + 7)$ from the sum of $(7x^2 + 4x + 9)$ and $(8x^2 + 7x - 8)$.

68. Subtract $(9x + 8)$ from the sum of $(3x^2 - 2x - x^3 + 2)$ and $(5x^2 - 8x - x^3 + 4)$.

69. $\left(\dfrac{2}{3}x^2 - \dfrac{1}{6}x + \dfrac{5}{6}\right) - \left(\dfrac{1}{3}x^2 + \dfrac{5}{6}x - \dfrac{1}{6}\right)$

70. $\left(\dfrac{3}{16}x^2 + \dfrac{5}{8}x - \dfrac{1}{4}\right) - \left(\dfrac{5}{16}x^2 - \dfrac{3}{8}x + \dfrac{3}{4}\right)$

Objective **E** *If $P(x) = x^2 + x + 1$ and $Q(x) = 5x^2 - 1$, find each function value. See Examples 18 and 19.*

71. $P(7)$ **72.** $Q(4)$ **73.** $Q(-10)$ **74.** $P(-4)$ **75.** $P(0)$ **76.** $Q(0)$

Solve. See Example 20.

The surface area of a rectangular box is given by the polynomial

$$2HL + 2LW + 2HW$$

and is measured in square units. In business, surface area is often calculated to help determine cost of materials.

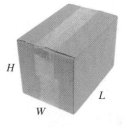

H

L

W

77. A rectangular box is to be constructed to hold a new camcorder. The box is to have dimensions 5 inches by 4 inches by 9 inches. Find the surface area of the box.

78. Suppose it has been determined that a box of dimensions 4 inches by 4 inches by 8.5 inches can be used to contain the camcorder in Exercise 77. Find the surface area of this box and calculate the square inches of material saved by using this box instead of the box in Exercise 77.

79. A projectile is fired upward from the ground with an initial velocity of 300 feet per second. Neglecting air resistance, the height of the projectile at any time t can be described by the polynomial function $P(t) = -16t^2 + 300t$. Find the height of the projectile at each given time.

   a. $t = 1$ second
   b. $t = 2$ seconds
   c. $t = 3$ seconds
   d. $t = 4$ seconds
   e. Explain why the height increases and then decreases as time passes.
   f. Approximate (to the nearest second) how long before the object hits the ground.

80. An object is thrown upward with an initial velocity of 25 feet per second from the top of Washington's head on Mount Rushmore. The height of the object above the ground at any time t can be described by the polynomial function $P(t) = -16t^2 + 25t + 500$. Find the height of the projectile at each given time.

   a. $t = 1$ second
   b. $t = 3$ seconds
   c. $t = 5$ seconds
   d. approximate (to the nearest second) how long before the object hits the ground.

81. The polynomial function $P(x) = 45x - 100{,}000$ models the relationship between the number of computer briefcases x that a company sells and the profit the company makes, $P(x)$. Find $P(4000)$, the profit from selling 4000 computer briefcases.

82. The total cost (in dollars) for MCD, Inc., Manufacturing Company to produce x blank audio-cassette tapes per week is given by the polynomial function $C(x) = 0.8x + 10{,}000$. Find the total cost of producing 20,000 tapes per week.

83. The total revenues (in dollars) for MCD, Inc., Manufacturing Company to sell x blank audiocassette tapes per week is given by the polynomial function $R(x) = 2x$. Find the total revenue from selling 20,000 tapes per week.

Review

Use the distributive property to multiply. See Section 1.3.

84. $5(3x - 2)$

85. $-7(2z - 6y)$

86. $-2(x^2 - 5x + 6)$

87. $5(-3y^2 - 2y + 7)$

88. In business, profit equals revenue minus cost, or $P(x) = R(x) - C(x)$. Find the profit function for MCD, Inc. by subtracting the functions given in Exercises 82 and 83.

Concept Extensions

Solve. See the Concept Checks in this section.

89. Which polynomial(s) is the opposite of $8x - 6$?
   a. $-(8x - 6)$
   b. $8x + 6$
   c. $-8x + 6$
   d. $-8x - 6$

90. Which polynomial(s) is the opposite of $-y^5 + 10y^3 - 2.3$?
   a. $y^5 + 10y^3 + 2.3$
   b. $-y^5 - 10y^3 - 2.3$
   c. $y^5 + 10y^3 - 2.3$
   d. $y^5 - 10y^3 + 2.3$

91. Correct the subtraction.
$$(12x - 1.7) - (15x + 6.2) = 12x - 1.7 - 15x + 6.2$$
$$= -3x + 4.5$$

92. Correct the addition.
$$(12x - 1.7) + (15x + 6.2) = 12x - 1.7 + 15x + 6.2$$
$$= 27x + 7.9$$

93. Write a function, $P(x)$, so that $P(0) = 7$.

94. Write a function, $R(x)$, so that $R(1) = 2$.

95. In your own words, describe how to find the degree of a term.

96. In your own words, describe how to find the degree of a polynomial.

97. The function $f(x) = 0.014x^2 + 0.043x + 0.584$ can be used to approximate the amazing growth of the number of Web logs (blogs) appearing on the Internet from June 2003 through December 2004, where x is the number of months after June 2003 and y is the number of new Web logs (in millions). Round answers to the nearest tenth of a million. (*Source:* Clickz.com)

   a. Approximate the number of Web logs on the Internet in June 2003.

   b. Approximate the number of Web logs on the Internet in June 2004.

   c. Use this function to estimate the number of Web logs on the Internet in June 2005.

   d. From parts (a), (b), and (c), determine whether the number of Web logs on the Internet is increasing at a steady rate. Explain why or why not.

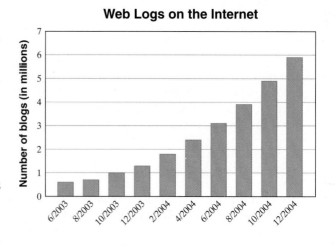

Web Logs on the Internet

98. The function $f(x) = -1.6x^2 + 9.6x + 67.6$ can be used to approximate the number of Americans enrolled in health maintenance organizations (HMOs) during the period 1997–2003, where x is the number of years after 1997 and $f(x)$ is the number of Americans in millions. Round answers to the nearest tenth of a million. (*Source:* Based on data from *Health, United States, 2004,* National Center for Health Statistics)

  a. Approximate the number of Americans enrolled in HMOs in 1997.
  b. Approximate the number of Americans enrolled in HMOs in 2003.
  c. Use the function to predict the number of Americans enrolled in HMOs in 2006.
  d. From parts (a), (b), and (c), determine whether the number of American enrolled in HMOs is changing at a steady rate. Explain why or why not.

If $P(x) = 3x + 3$, $Q(x) = 4x^2 - 6x + 3$, and $R(x) = 5x^2 - 7$, find each function.

99. $P(x) + Q(x)$

100. $Q(x) - R(x)$

101. If $P(x) = 2x - 3$, find $P(a)$, $P(-x)$, and $P(x + h)$.

Perform each indicated operation.

102. $(8x^{2y} - 7x^y + 3) + (-4x^{2y} + 9x^y - 14)$

103. $(14z^{5x} + 3z^{2x} + z) - (2z^{5x} - 10z^{2x} + 3z)$

Find each perimeter.

△ **104.**

$(x + y)$ units

$(3x^2 - x + 2y)$ units

△ **105.**

$(5xy - y^2)$ units

$(3xy + 4y^2)$ units

$(2x^2 - xy + 7y^2)$ units

Objectives

PRACTICE PROBLEMS 1–2

Multiply.

1. $(7y^2)(4y^5)$

2. $(-a^2b^3c)(10ab^2c^{12})$

PRACTICE PROBLEMS 3–5

Multiply.

3. $4x(3x - 2)$

4. $-2y^3(5y^2 - 2y + 6)$

5. $-a^2b(4a^3 - 2ab + b^2)$

Answers

1. $28y^7$, **2.** $-10a^3b^5c^{13}$, **3.** $12x^2 - 8x$,
4. $-10y^5 + 4y^4 - 12y^3$,
5. $-4a^5b + 2a^3b^2 - a^2b^3$

✔ **Concept Check Answer**

$4x(x - 5) + 2x$
$= 4x(x) + 4x(-5) + 2x$
$= 4x^2 - 20x + 2x$
$= 4x^2 - 18x$

330

5.2 MULTIPLYING POLYNOMIALS

Objective **A** Multiplying Any Two Polynomials

Properties of real numbers and exponents are used continually in the process of multiplying polynomials. To multiply monomials, for example, we apply the commutative and associative properties of real numbers and the product rule for exponents.

EXAMPLES Multiply.

Group like bases and apply the product rule for exponents.

1. $(2x^3)(5x^6) = 2(5)(x^3)(x^6) = 10x^{3+6} = 10x^9$

2. $(7y^4z^4)(-xy^{11}z^5) = 7(-1)x(y^4y^{11})(z^4z^5) = -7xy^{4+11}z^{4+5} = -7xy^{15}z^9$

Work Practice Problems 1–2

> **Helpful Hint**
>
> See Sections 1.6 and 1.7 to review exponential expressions further.

To multiply a monomial by a polynomial other than a monomial, we use an expanded form of the distributive property.

$$a(b + c + d + \cdots + z) = ab + ac + ad + \cdots + az$$

Notice that the monomial a is multiplied by each term of the polynomial.

EXAMPLES Multiply.

3. $2x(5x - 4) = 2x(5x) + 2x(-4)$ Use the distributive property.

$\qquad = 10x^2 - 8x$ Multiply.

4. $-3x^2(4x^2 - 6x + 1) = -3x^2(4x^2) + (-3x^2)(-6x) + (-3x^2)(1)$

$\qquad = -12x^4 + 18x^3 - 3x^2$

5. $-xy(7x^2y + 3xy - 11) = -xy(7x^2y) + (-xy)(3xy) + (-xy)(-11)$

$\qquad = -7x^3y^2 - 3x^2y^2 + 11xy$

Work Practice Problems 3–5

✔ **Concept Check** Find the error:

$4x(x - 5) + 2x$
$= 4x(x) + 4x(-5) + 4x(2x)$
$= 4x^2 - 20x + 8x^2$
$= 12x^2 - 20x$

To Multiply any two polynomials, we can use the following.

Multiplying Two Polynomials

To multiply any two polynomials, use the distributive property and multiply each term of one polynomial by each term of the other polynomial. Then combine any like terms.

EXAMPLE 6 Multiply: $(x + 3)(2x + 5)$

Solution: We multiply each term of $(x + 3)$ by $(2x + 5)$.

$$(x + 3)(2x + 5) = x(2x + 5) + 3(2x + 5) \quad \text{Use the distributive property.}$$
$$= 2x^2 + 5x + 6x + 15 \quad \text{Use the distributive property again.}$$
$$= 2x^2 + 11x + 15 \quad \text{Combine like terms.}$$

▧ **Work Practice Problem 6**

EXAMPLE 7 Multiply: $(2x - 3)(5x^2 - 6x + 7)$

Solution: We multiply each term of $(2x - 3)$ by each term of $(5x^2 - 6x + 7)$.

$$(2x - 3)(5x^2 - 6x + 7) = 2x(5x^2 - 6x + 7) + (-3)(5x^2 - 6x + 7)$$
$$= 10x^3 - 12x^2 + 14x - 15x^2 + 18x - 21$$
$$= 10x^3 - 27x^2 + 32x - 21 \quad \text{Combine like terms.}$$

▧ **Work Practice Problem 7**

Sometimes polynomials are easier to multiply vertically, in the same way we multiply real numbers. When multiplying vertically, we line up like terms in the **partial products** vertically. This makes combining like terms easier.

EXAMPLE 8 Multiply vertically: $(4x^2 + 7)(x^2 + 2x + 8)$

Solution:

$$
\begin{array}{r}
x^2 + 2x + 8 \\
4x^2 + 7 \\
\hline
7x^2 + 14x + 56 \quad 7(x^2 + 2x + 8) \\
4x^4 + 8x^3 + 32x^2 \quad\quad\quad 4x^2(x^2 + 2x + 8) \\
\hline
4x^4 + 8x^3 + 39x^2 + 14x + 56 \quad \text{Combine like terms.}
\end{array}
$$

▧ **Work Practice Problem 8**

Objective **B** Multiplying Binomials

When multiplying a binomial by a binomial, we can follow a special order for multiplying terms, called the **FOIL** order. The letters of FOIL stand for "First–**O**uter–**I**nner–**L**ast." To illustrate this method, let's multiply $(2x - 3)$ by $(3x + 1)$.

Multiply the **First** terms of each binomial. $(2x - 3)(3x + 1)$ F $2x(3x) = 6x^2$

Multiply the **O**uter terms of each binomial. $(2x - 3)(3x + 1)$ O $2x(1) = 2x$

Multiply the **I**nner terms of each binomial. $(2x - 3)(3x + 1)$ I $-3(3x) = -9x$

Multiply the **L**ast terms of each binomial. $(2x - 3)(3x + 1)$ L $-3(1) = -3$

Combine like terms.

$$6x^2 + 2x - 9x - 3 = 6x^2 - 7x - 3$$

PRACTICE PROBLEM 6
Multiply: $(x + 2)(3x + 1)$

PRACTICE PROBLEM 7
Multiply:
$$(5x - 1)(2x^2 - x + 4)$$

PRACTICE PROBLEM 8
Multiply vertically:
$$(3x^2 + 5)(x^2 - 6x + 1)$$

Answers
6. $3x^2 + 7x + 2$,
7. $10x^3 - 7x^2 + 21x - 4$,
8. $3x^4 - 18x^3 + 8x^2 - 30x + 5$

PRACTICE PROBLEM 9

Use the FOIL order to multiply $(x - 7)(x + 5)$.

Helpful Hint

The FOIL *order* is simply that. It is an *order* you may choose to use when multiplying two binomials.

PRACTICE PROBLEMS 10–11

Use the FOIL order to multiply.

10. $(4x - 3)(x - 6)$
11. $(6x^2 + 5y)(2x^2 - y)$

EXAMPLE 9 Use the FOIL order to multiply $(x - 1)(x + 2)$.

Solution:

First · Outer · Inner · Last

$$(x - 1)(x + 2) = x \cdot x + x \cdot 2 + (-1)x + (-1)(2)$$
$$= x^2 + 2x - x - 2$$
$$= x^2 + x - 2 \qquad \text{Combine like terms.}$$

Work Practice Problem 9

EXAMPLES Use the FOIL order to multiply.

First · Outer · Inner · Last

10. $(2x - 7)(3x - 4) = 2x(3x) + 2x(-4) + (-7)(3x) + (-7)(-4)$
$$= 6x^2 - 8x - 21x + 28$$
$$= 6x^2 - 29x + 28$$

F · O · I · L

11. $(3x^2 + y)(5x^2 - 2y) = 15x^4 - 6x^2y + 5x^2y - 2y^2$
$$= 15x^4 - x^2y - 2y^2$$

Work Practice Problems 10–11

Objective C Squaring Binomials

The **square of a binomial** is a special case of the product of two binomials. By the FOIL order for multiplying two binomials, we have

$$(a + b)^2 = (a + b)(a + b)$$

F · O · I · L

$$= a^2 + ab + ba + b^2$$
$$= a^2 + 2ab + b^2$$

We can visualize this product geometrically by analyzing areas.

Area of square in the margin: $(a + b)^2$
Sum of areas of smaller rectangles: $a^2 + 2ab + b^2$
Thus, $(a + b)^2 = a^2 + 2ab + b^2$

The same pattern occurs for the square of a difference. In general, we have the following.

Square of a Binomial

$$(a + b)^2 = a^2 + 2ab + b^2 \qquad (a - b)^2 = a^2 - 2ab + b^2$$

In other words, a binomial squared is the sum of the first term squared, twice the product of both terms, and the second term squared.

Answers

9. $x^2 - 2x - 35$, **10.** $4x^2 - 27x + 18$,
11. $12x^4 + 4x^2y - 5y^2$

EXAMPLES Multiply.

$$(a + b)^2 = a^2 + 2 \cdot a \cdot b + b^2$$

12. $(x + 5)^2 = x^2 + 2 \cdot x \cdot 5 + 5^2 = x^2 + 10x + 25$
13. $(x - 9)^2 = x^2 - 2 \cdot x \cdot 9 + 9^2 = x^2 - 18x + 81$
14. $(3x + 2z)^2 = (3x)^2 + 2(3x)(2z) + (2z)^2 = 9x^2 + 12xz + 4z^2$
15. $(4m^2 - 3n)^2 = (4m^2)^2 - 2(4m^2)(3n) + (3n)^2 = 16m^4 - 24m^2n + 9n^2$

 Work Practice Problems 12–15

Helpful Hint

Note that $(a + b)^2 = a^2 + 2ab + b^2$, not $a^2 + b^2$. Also,
$(a - b)^2 = a^2 - 2ab + b^2$, not $a^2 - b^2$.

Objective D Multiplying the Sum and Difference of Two Terms

Another special product applies to the sum and difference of the same two terms. Multiply $(a + b)(a - b)$ to see a pattern.

$$(a + b)(a - b) = a^2 - ab + ba - b^2 = a^2 - b^2$$

Product of the Sum and Difference of Two Terms

$$(a + b)(a - b) = a^2 - b^2$$

In other words, the product of the sum and difference of the same two terms is the difference of the first term squared and the second term squared.

EXAMPLES Multiply.

$$(a + b) (a - b) = a^2 - b^2$$

16. $(x + 3)(x - 3) = x^2 - 3^2 = x^2 - 9$
17. $(4y - 1)(4y + 1) = (4y)^2 - 1^2 = 16y^2 - 1$
18. $(x^2 + 2y)(x^2 - 2y) = (x^2)^2 - (2y)^2 = x^4 - 4y^2$

19. $\left(3m^2 - \dfrac{1}{2}\right)\left(3m^2 + \dfrac{1}{2}\right) = (3m^2)^2 - \left(\dfrac{1}{2}\right)^2 = 9m^4 - \dfrac{1}{4}$

 Work Practice Problems 16–19

EXAMPLE 20 Multiply $[3 + (2a + b)]^2$.

Solution: Think of 3 as the first term and $(2a + b)$ as the second term, and apply the method for squaring a binomial.

$$[a + b]^2 = a^2 + 2(a) \cdot b + b^2$$

$$[3 + (2a + b)]^2 = 3^2 + 2(3)(2a + b) + (2a + b)^2$$
$$= 9 + 6(2a + b) + (2a + b)^2$$
$$= 9 + 12a + 6b + (2a)^2 + 2(2a)(b) + b^2 \quad \text{Square } (2a + b).$$
$$= 9 + 12a + 6b + 4a^2 + 4ab + b^2$$

 Work Practice Problem 20

PRACTICE PROBLEMS 12–15

Multiply.
12. $(x + 3)^2$
13. $(y - 6)^2$
14. $(2x + 5y)^2$
15. $(6a^2 - 2b)^2$

PRACTICE PROBLEMS 16–19

Multiply.
16. $(x + 4)(x - 4)$
17. $(3m - 6)(3m + 6)$
18. $(a^2 + 5y)(a^2 - 5y)$
19. $\left(4y^2 - \dfrac{1}{3}\right)\left(4y^2 + \dfrac{1}{3}\right)$

PRACTICE PROBLEM 20

Multiply:
$$[2 - (3x + y)]^2$$

Answers
12. $x^2 + 6x + 9$, **13.** $y^2 - 12y + 36$,
14. $4x^2 + 20xy + 25y^2$,
15. $36a^4 - 24a^2b + 4b^2$,
16. $x^2 - 16$, **17.** $9m^2 - 36$,
18. $a^4 - 25y^2$, **19.** $16y^4 - \dfrac{1}{9}$,
20. $4 - 12x - 4y + 9x^2 + 6xy + y^2$

PRACTICE PROBLEM 21

Multiply:

$[(2x + 3y) - 2][(2x + 3y) + 2]$

EXAMPLE 21 Multiply: $[(5x - 2y) - 1][(5x - 2y) + 1]$

Solution: We can think of $(5x - 2y)$ as the first term and 1 as the second term. Then we can apply the method for the product of the sum and difference of two terms.

$$\overbrace{[(5x - 2y)}^{a} \overbrace{- 1]}^{- b}\overbrace{[(5x - 2y)}^{a} \overbrace{+ 1]}^{+ b} = \overbrace{(5x - 2y)^2}^{a^2} \overbrace{- 1^2}^{- b^2}$$

$$= (5x)^2 - 2(5x)(2y) + (2y)^2 - 1 \qquad \text{Square } (5x - 2y).$$

$$= 25x^2 - 20xy + 4y^2 - 1$$

Work Practice Problem 21

Objective E Multiplying Three or More Polynomials

To multiply three or more polynomials, more than one method may be needed.

PRACTICE PROBLEM 22

Multiply:

$(y - 2)(y + 2)(y^2 - 4)$

EXAMPLE 22 Multiply: $(x - 3)(x + 3)(x^2 - 9)$

Solution: We multiply the first two binomials, the sum and difference of two terms. Then we multiply the resulting two binomials, the square of a binomial.

$$(x - 3)(x + 3)(x^2 - 9) = (x^2 - 9)(x^2 - 9) \qquad \text{Multiply } (x - 3)(x + 3)$$

$$= (x^2 - 9)^2$$

$$= x^4 - 18x^2 + 81 \qquad \text{Square } (x^2 - 9).$$

Work Practice Problem 22

Objective F Evaluating Polynomial Functions

Our work in multiplying polynomials is often useful in evaluating polynomial functions.

PRACTICE PROBLEM 23

If $f(x) = x^2 - 6x + 1$, find $f(b - 1)$

EXAMPLE 23 If $f(x) = x^2 + 5x - 2$, find $f(a + 1)$.

Solution: To find $f(a + 1)$, replace x with the expression $a + 1$ in the polynomial function $f(x)$.

$$f(x) = x^2 + 5x - 2$$

$$f(a + 1) = (a + 1)^2 + 5(a + 1) - 2$$

$$= a^2 + 2a + 1 + 5a + 5 - 2$$

$$= a^2 + 7a + 4$$

Work Practice Problem 23

Answers

21. $4x^2 + 12xy + 9y^2 - 4$,

22. $y^4 - 8y^2 + 16$, **23.** $b^2 - 8b + 8$

In the previous section, we used a graphing calculator to check addition and subtraction of polynomials in one variable. In this section, we use the same method to check multiplication of polynomials in one variable. For example, to see that

$$(x - 2)(x + 1) = x^2 - x - 2$$

graph both $Y_1 = (x - 2)(x + 1)$ and $Y_2 = x^2 - x - 2$ on the same screen and see whether their graphs coincide.

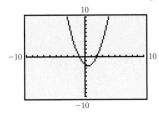

By tracing along both graphs, we see that the graphs of Y_1 and Y_2 appear to coincide, and thus $(x - 2)(x + 1) = x^2 - x - 2$ appears to be correct.

Multiply. Then use a graphing calculator to check the results.

1. $(x + 4)(x - 4)$
2. $(x + 3)(x + 3)$
3. $(3x - 7)^2$
4. $(5x - 2)^2$
5. $(5x + 1)(x^2 - 3x - 2)$
6. $(7x + 4)(2x^2 + 3x - 5)$

5.2 EXERCISE SET

FOR EXTRA HELP

Student Solutions Manual   PH Math/Tutor Center   CD/Video for Review   Math XL   MyMathLab
MathXL®   MyMathLab

Objective A *Multiply. See Examples 1 through 8.*

1. $(-4x^3)(3x^2)$ **2.** $(-6a)(4a)$ **3.** $(8.6a^4b^5c)(10ab^3c^2)$ **4.** $(7.1xy^2z^{11})(10xy^7z)$

5. $3x(4x + 7)$ **6.** $5x(6x - 4)$ **7.** $-6xy(4x + y)$ **8.** $-8y(6xy + 4x)$

9. $-4ab(xa^2 + ya^2 - 3)$ **10.** $-6b^2z(z^2a + baz - 3b)$ **11.** $(x - 3)(2x + 4)$

12. $(y + 5)(3y - 2)$ **13.** $(2x + 3)(x^3 - x + 2)$ **14.** $(a + 2)(3a^2 - a + 5)$

15. $\begin{array}{r} 3x - 2 \\ \times\ 5x + 1 \\ \hline \end{array}$ **16.** $\begin{array}{r} 2z - 4 \\ \times\ 6z - 2 \\ \hline \end{array}$ **17.** $\begin{array}{r} 3m^2 + 2m - 1 \\ \times\ \quad\quad 5m + 2 \\ \hline \end{array}$ **18.** $\begin{array}{r} 2x^2 - 3x - 4 \\ \times\ \quad\quad 4x + 5 \\ \hline \end{array}$

19. $-6a^2b^2(5a^2b^2 - 6a - 6b)$ **20.** $7x^2y^3(-3ax - 4xy + z)$

Objective B *Multiply the binomials. See Examples 9 through 11.*

21. $(x - 3)(x + 4)$ **22.** $(c - 3)(c + 1)$ **23.** $(5x - 8y)(2x - y)$ **24.** $(2n - 9m)(n - 7m)$

25. $\left(4x + \dfrac{1}{3}\right)\left(4x - \dfrac{1}{2}\right)$ **26.** $\left(4y - \dfrac{1}{3}\right)\left(3y - \dfrac{1}{8}\right)$ **27.** $(5x^2 - 2y^2)(x^2 - 3y^2)$ **28.** $(4x^2 - 5y^2)(x^2 - 2y^2)$

335

Objectives Ⓒ Ⓓ *Use special products to multiply. See Examples 12 through 21.*

29. $(x + 4)^2$ **30.** $(x - 5)^2$ **31.** $(6y - 1)(6y + 1)$ **32.** $(7x - 9)(7x + 9)$

33. $(3x - y)^2$ **34.** $(4x + z)^2$ **35.** $(7ab + 3c)(7ab - 3c)$ **36.** $(3xy - 2b)(3xy + 2b)$

37. $\left(3x + \dfrac{1}{2}\right)\left(3x - \dfrac{1}{2}\right)$ **38.** $\left(2x - \dfrac{1}{3}\right)\left(2x + \dfrac{1}{3}\right)$

39. $[3 + (4b + 1)]^2$ **40.** $[5 - (3b - 3)]^2$

41. $[(2s - 3) - 1][(2s - 3) + 1]$ **42.** $[(2y + 5) + 6][(2y + 5) - 6]$

Objective Ⓔ *Multiply. See Example 22.*

43. $(x + y)(2x - 1)(x + 1)$ **44.** $(z + 2)(z - 3)(2z + 1)$ **45.** $(x - 2)^4$

46. $(x - 1)^4$ **47.** $(x - 5)(x + 5)(x^2 + 25)$ **48.** $(x + 3)(x - 3)(x^2 + 9)$

Objectives Ⓐ Ⓑ Ⓒ Ⓓ Ⓔ **Mixed Practice** *Multiply. See Examples 1 through 22.*

49. $-8a^2b(3b^2 - 5b + 20)$ **50.** $-9xy^2(3x^2 - 2x + 10)$ **51.** $(6x + 1)^2$

52. $(4x + 7)^2$ **53.** $(5x^3 + 2y)(5x^3 - 2y)$ **54.** $(3x^4 + 2y)(3x^4 - 2y)$

55. $(2x^3 + 5)(5x^2 + 4x + 1)$ **56.** $(3y^3 - 1)(3y^3 - 6y + 1)$ **57.** $(3x^2 + 2x - 1)^2$

58. $(4x^2 + 4x - 4)^2$ **59.** $(3x - 1)(x + 3)$ **60.** $(5d - 3)(d + 6)$

61. $(3x^4 + 1)(3x^2 + 5)$ **62.** $(4x^3 - 5)(5x^2 + 6)$ **63.** $(3x + 1)^2$

64. $(4x + 6)^2$ **65.** $(3b - 6y)(3b + 6y)$ **66.** $(2x - 4y)(2x + 4y)$

67. $(7x - 3)(7x + 3)$ **68.** $(4x + 1)(4x - 1)$ **69.** $\begin{array}{r} 3x^2 + 4x - 4 \\ \times \quad 3x + 6 \\ \hline \end{array}$

70. $\begin{array}{r} 6x^2 + 2x - 1 \\ \times \quad 3x - 6 \\ \hline \end{array}$ **71.** $(4x^2 - 2x + 5)(3x + 1)$ **72.** $(5x^2 - x - 2)(2x - 1)$

73. $[(xy + 4) - 6]^2$

74. $[(2a^2 + 4) - 1]^2$

75. $(11a^2 + 1)(2a + 1)$

76. $(13x^2 + 1)(3x + 1)$

77. $\left(\dfrac{2}{3}n - 2\right)\left(\dfrac{1}{2}n - 9\right)$

78. $\left(\dfrac{3}{5}y - 6\right)\left(\dfrac{1}{3}y - 10\right)$

79. $(3x + 1)(3x - 1)(2y + 5x)$

80. $(2a + 1)(2a - 1)(6a + 7b)$

Objective **F** *If* $f(x) = x^2 - 3x$, *find the following. See Example 23.*

81. $f(a)$

82. $f(c)$

83. $f(a + h)$

84. $f(a + 5)$

85. $f(b - 2)$

86. $f(a - b)$

Review

Simplify. See Section 1.6.

87. $\dfrac{6x^3}{3x}$

88. $\dfrac{4x^7}{x^2}$

89. $\dfrac{20a^3b^5}{18ab^2}$

90. $\dfrac{15x^7y^2}{6xy^2}$

91. $\dfrac{8m^4n}{12mn}$

92. $\dfrac{6n^6p}{8np}$

Concept Extensions

Solve. See the Concept Check in this section.

93. Find the error: $7y(3z - 2) + 1$
$= 21yz - 14y + 7y$
$= 21yz - 7y$

94. Find the error: $2x + 3x(12 - x)$
$= 5x(12 - x)$
$= 60x - 5x^2$

95. Explain how to multiply a polynomial by a polynomial.

96. Explain why $(3x + 2)^2$ does not equal $9x^2 + 4$.

Multiply. Assume that variables represent positive integers.

97. $5x^2y^n(6y^{n+1} - 2)$

98. $-3yz^n(2y^3z^{2n} - 1)$

99. $(x^a + 5)(x^{2a} - 3)$

100. $(x^a + y^{2b})(x^a - y^{2b})$

101. Perform each indicated operation.

     a. $(3x + 5) + (3x + 7)$

     b. $(3x + 5)(3x + 7)$

     c. Explain the difference between the two problems.

102. Explain when the FOIL method can be used to multiply polynomials.

If $R(x) = x + 5$, $Q(x) = x^2 - 2$, and $P(x) = 5x$, find each function.

103. $P(x) \cdot R(x)$

104. $P(x) \cdot Q(x)$

If $f(x) = x^3 - 2x^2$, find each function value.

105. $f(a)$

106. $f(a + h)$

Find the area of each shaded region.

△ **107.**

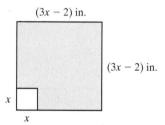

(3x − 2) in.

(3x − 2) in.

x

x

△ **108.**

$x - 7$

$2x$ x

△ **109.** Find the area of the circle. Do not approximate π.

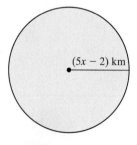

(5x − 2) km

△ **110.** Find the volume of the cylinder. Do not approximate π.

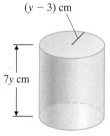

(y − 3) cm

7y cm

 STUDY SKILLS BUILDER

How Are You Doing?

If you haven't done so yet, take a few moments and think about how you are doing in this course. Are you working toward your goal of successfully completing this course? Is your performance on homework, quizzes, and tests satisfactory? If not, you might want to see your instructor to see if he/she has any suggestions on how you can improve your performance. Reread Section 1.1 for ideas on places to get help with your mathematics course.

Answer the following.

1. List any textbook supplements you are using to help you through this course.

2. List any campus resources you are using to help you through this course.

3. Write a short paragraph describing how you are doing in your mathematics course.

4. If improvement is needed, list ways that you can work toward improving your situation as described in Exercise 3.

5.3 DIVIDING POLYNOMIALS AND SYNTHETIC DIVISION

Objectives

A Divide a Polynomial by a Monomial.

B Divide by a Polynomial.

C Use Synthetic Division.

Now that we have added, subtracted, and multiplied polynomials, we will learn how to divide them.

Objective A Dividing a Polynomial by a Monomial

Recall the following addition fact for fractions with a common denominator:

$$\frac{a}{c} + \frac{b}{c} = \frac{a+b}{c}$$

If $a, b,$ and c are monomials, we can read this equation from right to left and gain insight into how to divide a polynomial by a monomial.

> ### Dividing a Polynomial by a Monomial
>
> To divide a polynomial by a monomial, divide each term in the polynomial by the monomial.
>
> $$\frac{a+b}{c} = \frac{a}{c} + \frac{b}{c}, \quad c \neq 0$$

EXAMPLE 1 Divide $10x^3 - 5x^2 + 20x$ by $5x$.

Solution: We divide each term of $10x^3 - 5x^2 + 20x$ by $5x$ and simplify.

$$\frac{10x^3 - 5x^2 + 20x}{5x} = \frac{10x^3}{5x} - \frac{5x^2}{5x} + \frac{20x}{5x} = 2x^2 - x + 4$$

To check, see that (quotient)(divisor) = dividend, or

$$(2x^2 - x + 4)(5x) = 10x^3 - 5x^2 + 20x$$

■ **Work Practice Problem 1**

EXAMPLE 2 Divide: $\dfrac{3x^5y^2 - 15x^3y - x^2y - 6x}{x^2y}$

Solution: We divide each term in the numerator by x^2y.

$$\frac{3x^5y^2 - 15x^3y - x^2y - 6x}{x^2y} = \frac{3x^5y^2}{x^2y} - \frac{15x^3y}{x^2y} - \frac{x^2y}{x^2y} - \frac{6x}{x^2y}$$

$$= 3x^3y - 15x - 1 - \frac{6}{xy}$$

■ **Work Practice Problem 2**

Objective B Dividing by a Polynomial

To divide a polynomial by a polynomial other than a monomial, we use **long division.** Polynomial long division is similar to long division of real numbers. We review long division of real numbers by dividing 7 into 296.

$$
\begin{array}{r}
42 \\
\text{Divisor: } 7\overline{)296} \\
-28 \quad 4(7) = 28 \\
\hline
16 \quad \text{Subtract and bring down the next digit in the dividend.} \\
-14 \quad 2(7) = 14 \\
\hline
2 \quad \text{Subtract. The remainder is 2.}
\end{array}
$$

PRACTICE PROBLEM 1

Divide $16y^3 - 8y^2 + 6y$ by $2y$.

PRACTICE PROBLEM 2

Divide:

$$\frac{9a^3b^3 - 6a^2b^2 + a^2b - 4a}{a^2b}$$

Answers

1. $8y^2 - 4y + 3,$

2. $9ab^2 - 6b + 1 - \dfrac{4}{ab}$

339

The quotient is

$$42\frac{2}{7} \quad \begin{array}{l}\text{remainder}\\ \text{divisor}\end{array}$$

To check, notice that $42(7) + 2 = 296$, which is the dividend. This same division process can be applied to polynomials, as shown next.

PRACTICE PROBLEM 3
Divide $6x^2 + 11x - 2$ by $x + 2$.

EXAMPLE 3 Divide $2x^2 - x - 10$ by $x + 2$.

Solution: $2x^2 - x - 10$ is the dividend, and $x + 2$ is the divisor.

Step 1: Divide $2x^2$ by x.

$$x + 2\overline{)2x^2 - x - 10} \overset{2x}{}$$

$\dfrac{2x^2}{x} = 2x$, so $2x$ is the first term of the quotient.

Step 2: Multiply $2x(x + 2)$.

$$\begin{array}{r} 2x \\ x + 2\overline{)2x^2 - x - 10} \\ 2x^2 + 4x \end{array}$$

Multiply: $2x(x + 2)$. Like terms are lined up vertically.

Step 3: Subtract $(2x^2 + 4x)$ from $(2x^2 - x - 10)$ by changing the signs of $(2x^2 + 4x)$ and adding.

$$\begin{array}{r} 2x \\ x + 2\overline{)2x^2 - x - 10} \\ \cancel{+}2x^2 \cancel{+} 4x \\ \hline -5x \end{array}$$

Step 4: Bring down the next term, -10, and start the process over.

$$\begin{array}{r} 2x \\ x + 2\overline{)2x^2 - x - 10} \\ \cancel{+}2x^2 \cancel{+} 4x \downarrow \\ \hline -5x - 10 \end{array}$$

Step 5: Divide $-5x$ by x.

$$\begin{array}{r} 2x - 5 \\ x + 2\overline{)2x^2 - x - 10} \\ \cancel{+}2x^2 \cancel{+} 4x \\ \hline -5x - 10 \end{array}$$

$\dfrac{-5x}{x} = -5$ so -5 is the second term of the quotient.

Step 6: Multiply $-5(x + 2)$.

$$\begin{array}{r} 2x - 5 \\ x + 2\overline{)2x^2 - x - 10} \\ \cancel{+}2x^2 \cancel{+} 4x \\ \hline -5x - 10 \\ -5x - 10 \end{array}$$

Multiply: $-5(x + 2)$. Like terms are lined up vertically.

Step 7: Subtract by changing signs of $-5x - 10$ and adding.

$$\begin{array}{r} 2x - 5 \\ x + 2\overline{)2x^2 - x - 10} \\ \cancel{+}2x^2 \cancel{+} 4x \\ \hline -5x - 10 \\ \cancel{+}5x \cancel{+} 10 \\ \hline 0 \end{array}$$

Subtract.

Remainder.

Answer

3. $6x - 1$

Then $\dfrac{2x^2 - x - 10}{x + 2} = 2x - 5$. There is no remainder.

Check this result by multiplying $2x - 5$ by $x + 2$, the divisor. Their product is $(2x - 5)(x + 2) = 2x^2 - x - 10$, the dividend.

🔲 **Work Practice Problem 3**

EXAMPLE 4 Divide: $(6x^2 - 19x + 12) \div (3x - 5)$

Solution:

$$
\begin{array}{r}
2x \\
3x - 5 \overline{)6x^2 - 19x + 12} \\
\underline{6x^2 - 10x} \qquad\downarrow \\
-9x + 12
\end{array}
$$

Divide: $\dfrac{6x^2}{3x} = 2x$

Multiply: $2x(3x - 5)$

Subtract: $(6x^2 - 19x) - (6x^2 - 10x) = -9x$

Bring down the next term, $+12$.

$$
\begin{array}{r}
2x - 3 \\
3x - 5 \overline{)6x^2 - 19x + 12} \\
\underline{6x^2 - 10x} \\
-9x + 12 \\
\underline{-9x + 15} \\
-3
\end{array}
$$

Divide: $\dfrac{-9x}{3x} = -3$

Multiply: $-3(3x - 5)$

Subtract: $(-9x + 12) - (-9x + 15) = -3$

Check:

divisor · quotient + remainder

$$(3x - 5)(2x - 3) + (-3) = 6x^2 - 19x + 15 - 3$$
$$= 6x^2 - 19x + 12 \qquad \text{The dividend}$$

The division checks, so

$$\dfrac{6x^2 - 19x + 12}{3x - 5} = 2x - 3 + \dfrac{-3}{3x - 5}$$

Helpful Hint   This fraction is the remainder over the divisor.

🔲 **Work Practice Problem 4**

EXAMPLE 5 Divide: $(7x^3 + 16x^2 + 2x - 1) \div (x + 4)$.

Solution:

$$
\begin{array}{r}
7x^2 - 12x + 50 \\
x + 4 \overline{)7x^3 + 16x^2 + 2x - 1} \\
\underline{7x^3 + 28x^2} \\
-12x^2 + 2x \\
\underline{-12x^2 + 48x} \\
50x - 1 \\
\underline{50x + 200} \\
-201
\end{array}
$$

Divide $\dfrac{7x^3}{x} = 7x^2$.

$7x^2(x + 4)$

Subtract. Bring down $2x$.

$\dfrac{-12x^2}{x} = -12x$, a term of the quotient.

$-12x(x + 4)$

Subtract. Bring down -1.

$\dfrac{50x}{x} = 50$, a term of the quotient.

$50(x + 4)$.

Subtract.

Thus, $\dfrac{7x^3 + 16x^2 + 2x - 1}{x + 4} = 7x^2 - 12x + 50 + \dfrac{-201}{x + 4}$ or

$$7x^2 - 12x + 50 - \dfrac{201}{x + 4}.$$

🔲 **Work Practice Problem 5**

PRACTICE PROBLEM 4

Divide:
$(10x^2 - 17x + 5) \div (5x - 1)$

PRACTICE PROBLEM 5

Divide:
$(5x^3 - 4x^2 + 3x - 4) \div (x - 2)$.

Answers

4. $2x - 3 + \dfrac{2}{5x - 1}$,

5. $5x^2 + 6x + 15 + \dfrac{26}{x - 2}$

PRACTICE PROBLEM 6

Divide $3x^4 + 4x^2 - 6x + 1$ by $x^2 + 1$.

EXAMPLE 6 Divide $2x^3 + 3x^4 - 8x + 6$ by $x^2 - 1$.

Solution: Before dividing, we write terms in descending order of powers of x. Also, we represent any "missing powers" by the product of 0 and the variable raised to the missing power. There is no x^2-term in the dividend, so we include $0x^2$ to represent the missing term. Also, there is no x term in the divisor, so we include $0x$ in the divisor.

$$
\begin{array}{r}
3x^2 + 2x + 3 \\
x^2 + 0x - 1 \overline{)\ 3x^4 + 2x^3 + 0x^2 - 8x + 6} \\
\underline{3x^4 \ * \ 0x^3 \ \not{-} \ 3x^2} \\
2x^3 + 3x^2 - 8x \\
\underline{2x^3 \ * \ 0x^2 \ \not{-} \ 2x} \\
3x^2 - 6x + 6 \\
\underline{3x^2 \ * \ 0x \ \not{-} \ 3} \\
-6x + 9
\end{array}
$$

$\dfrac{3x^4}{x^2} = 3x^2$

$3x^2(x^2 + 0x - 1)$
Subtract. Bring down $-8x$.
$2x^3/x^2 = 2x$, a term of the quotient
$2x(x^2 + 0x - 1)$
Subtract. Bring down 6.
$3x^2/x^2 = 3$, a term of the quotient
$3(x^2 + 0x - 1)$
Subtract.

> The division process is finished when the degree of the remainder polynomial is less than the degree of the divisor.

Thus,

$$\frac{3x^4 + 2x^3 - 8x + 6}{x^2 - 1} = 3x^2 + 2x + 3 + \frac{-6x + 9}{x^2 - 1}$$

▢ **Work Practice Problem 6**

PRACTICE PROBLEM 7

Divide $64x^3 - 27$ by $4x - 3$.

EXAMPLE 7 Divide $27x^3 + 8$ by $3x + 2$.

Solution: We replace the missing terms in the dividend with $0x^2$ and $0x$.

$$
\begin{array}{r}
9x^2 - 6x + 4 \\
3x + 2 \overline{)\ 27x^3 + \ 0x^2 + 0x + 8} \\
\underline{27x^3 \ * \ 18x^2} \\
-18x^2 + \ 0x \\
\underline{\not{-}18x^2 \ * \ 12x} \\
12x + 8 \\
\underline{12x \ * \ 8}
\end{array}
$$

$9x^2(3x + 2)$
Subtract. Bring down $0x$.
$-6x(3x + 2)$
Subtract. Bring down 8.
$4(3x + 2)$

Helpful Hint

The degree of the remainder polynomial (1) is the same as the degree of the divisor (1), so we continue the division process.

Thus, $\dfrac{27x^3 + 8}{3x + 2} = 9x^2 - 6x + 4.$

▢ **Work Practice Problem 7**

✔ **Concept Check** In a division problem, the divisor is $4x^3 - 5$. The division process can be stopped when which of these possible remainder polynomials is reached?

a. $2x^4 + x^2 - 3$ **b.** $x^3 - 5^2$ **c.** $4x^2 + 25$

Objective C Using Synthetic Division

When a polynomial is to be divided by a binomial of the form $x - c$, a shortcut process called **synthetic division** may be used. On the left is an example of long division, and on the right is the same example showing the coefficients of the variables only.

Answers

6. $3x^2 + 1 - \dfrac{6x}{x^2 + 1}$,

7. $16x^2 + 12x + 9$

✔ **Concept Check Answer**

c

$$
\begin{array}{r}
2x^2 + 5x + 2 \\
x - 3\overline{)2x^3 - x^2 - 13x + 1} \\
\underline{2x^3 - 6x^2} \\
5x^2 - 13x \\
\underline{5x^2 - 15x} \\
2x + 1 \\
\underline{2x - 6} \\
7
\end{array}
\qquad
\begin{array}{r}
2 \quad 5 \quad 2 \\
1 - 3\overline{)2 - 1 - 13 + 1} \\
\underline{2 - 6} \\
5 - 13 \\
\underline{5 - 15} \\
2 + 1 \\
\underline{2 - 6} \\
7
\end{array}
$$

Notice that as long as we keep coefficients of powers of x in the same column, we can perform division of polynomials by performing algebraic operations on the coefficients only. This shorter process of dividing with coefficients only in a special format is called synthetic division. To find $(2x^3 - x^2 - 13x + 1) \div (x - 3)$ by synthetic division, follow the next example.

EXAMPLE 8   Use synthetic division to divide $2x^3 - x^2 - 13x + 1$ by $x - 3$.

Solution: To use synthetic division, the divisor must be in the form $x - c$. Since we are dividing by $x - 3$, c is 3. We write down 3 and the coefficients of the dividend.

c

$$
\begin{array}{c|cccc}
3 & 2 & -1 & -13 & 1 \\
& & & & \\
\hline
& 2 & & &
\end{array}
$$

Next, draw a line and bring down the first coefficient of the dividend.

$$
\begin{array}{c|cccc}
3 & 2 & -1 & -13 & 1 \\
& & 6 & & \\
\hline
& 2 & & &
\end{array}
$$

Multiply $3 \cdot 2$ and write down the product, 6.

$$
\begin{array}{c|cccc}
3 & 2 & -1 & -13 & 1 \\
& & 6 & & \\
\hline
& 2 & 5 & &
\end{array}
$$

Add $-1 + 6$. Write down the sum, 5.

$$
\begin{array}{c|cccc}
3 & 2 & -1 & -13 & 1 \\
& & 6 & 15 & \\
\hline
& 2 & 5 & 2 &
\end{array}
$$

$3 \cdot 5 = 15$
$-13 + 15 = 2$

$$
\begin{array}{c|cccc}
3 & 2 & -1 & -13 & 1 \\
& & 6 & 15 & 6 \\
\hline
& 2 & 5 & 2 & 7
\end{array}
$$

$3 \cdot 2 = 6$
$1 + 6 = 7$

The quotient is found in the bottom row. The numbers 2, 5, and 2 are the coefficients of the quotient polynomial, and the number 7 is the remainder. The degree of the quotient polynomial is one less than the degree of the dividend. In our example, the degree of the dividend is 3, so the degree of the quotient polynomial is 2. As we found when we performed the long division, the quotient is

$2x^2 + 5x + 2$,   remainder 7

or

$2x^2 + 5x + 2 + \dfrac{7}{x - 3}$

Work Practice Problem 8

PRACTICE PROBLEM 8

Use synthetic division to divide $3x^3 - 2x^2 + 5x + 4$ by $x - 2$.

Answer

8. $3x^2 + 4x + 13 + \dfrac{30}{x - 2}$

When using synthetic division, if there are missing powers of the variable, insert 0s as coefficients.

PRACTICE PROBLEM 9

Use synthetic division to divide $x^4 + 3x^3 - 5x + 4$ by $x + 1$.

EXAMPLE 9   Use synthetic division to divide $x^4 - 2x^3 - 11x^2 + 34$ by $x + 2$.

Solution: The divisor is $x + 2$, which in the form $x - c$ is $x - (-2)$. Thus, c is -2. There is no x-term in the dividend, so we insert a coefficient of 0. The dividend coefficients are $1, -2, -11, 0$, and 34.

$$
\begin{array}{r|rrrrr}
-2 & 1 & -2 & -11 & 0 & 34 \\
   &   & -2 & 8 & 6 & -12 \\
\hline
   & 1 & -4 & -3 & 6 & 22
\end{array}
$$

The dividend is a fourth-degree polynomial, so the quotient polynomial is a third-degree polynomial. The quotient is $x^3 - 4x^2 - 3x + 6$ with a remainder of 22. Thus,

$$
\frac{x^4 - 2x^3 - 11x^2 + 34}{x + 2} = x^3 - 4x^2 - 3x + 6 + \frac{22}{x + 2}
$$

Work Practice Problem 9

Helpful Hint

Before dividing by long division or by synthetic division, write the dividend in descending order of variable exponents. Any "missing powers" of the variable must be represented by 0 times the variable raised to the missing power.

✔ **Concept Check**   Which division problems are candidates for the synthetic division process?

a. $(3x^2 + 5) \div (x + 4)$
b. $(x^3 - x^2 + 2) \div (3x^3 - 2)$
c. $(y^4 + y - 3) \div (x^2 + 1)$
d. $x^5 \div (x - 5)$

Answer

9. $x^3 + 2x^2 - 2x - 3 + \dfrac{7}{x + 1}$

✔ **Concept Check Answer**

a and d

57. A board of length $(3x^4 + 6x^2 - 18)$ meters is to be cut into three pieces of the same length. Find the length of each piece.

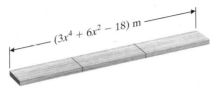

58. The perimeter of a regular hexagon is given to be $(12x^5 - 48x^3 + 3)$ miles. Find the length of each side.

59. If the area of the rectangle is $(15x^2 - 29x - 14)$ square inches, and its length is $(5x + 2)$ inches, find its width.

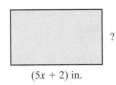

?

$(5x + 2)$ in.

60. If the area of a parallelogram is $(2x^2 - 17x + 35)$ square centimeters and its base is $(2x - 7)$ centimeters, find its height.

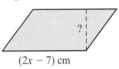

$(2x - 7)$ cm

61. Find $P(1)$ for the polynomial function $P(x) = 3x^3 + 2x^2 - 4x + 3$. Next, divide $3x^3 + 2x^2 - 4x + 3$ by $x - 1$. Compare the remainder with $P(1)$.

62. Find $P(-2)$ for the polynomial function $P(x) = x^3 - 4x^2 - 3x + 5$. Next, divide $x^3 - 4x^2 - 3x + 5$ by $x + 2$. Compare the remainder with $P(-2)$.

63. Find $P(-3)$ for the polynomial $P(x) = 5x^4 - 2x^2 + 3x - 6$. Next, divide $5x^4 - 2x^2 + 3x - 6$ by $x + 3$. Compare the remainder with $P(-3)$.

64. Find $P(2)$ for the polynomial function $P(x) = -4x^4 + 2x^3 - 6x + 3$. Next, divide $-4x^4 + 2x^3 - 6x + 3$ by $x - 2$. Compare the remainder with $P(2)$.

65. Write down any patterns you noticed from Exercises 61–64.

66. Explain how to check polynomial long division.

Divide.

67. $\left(x^4 + \dfrac{2}{3}x^3 + x \right) \div (x - 1)$

68. $\left(3x^4 - x - x^3 + \dfrac{1}{2} \right) \div (2x - 1)$

For each given $f(x)$ and $g(x)$, find $\dfrac{f(x)}{g(x)}$. Also find any x-values that are not in the domain of $\dfrac{f(x)}{g(x)}$. (Note: Since $g(x)$ is in the denominator, $g(x)$ cannot be 0).

69. $f(x) = 25x^2 - 5x + 30; g(x) = 5x$

70. $f(x) = 12x^4 - 9x^3 + 3x - 1; g(x) = 3x$

71. $f(x) = 7x^4 - 3x^2 + 2; g(x) = x - 2$

72. $f(x) = 2x^3 - 4x^2 + 1; g(x) = x + 3$

73. Try performing the following division without changing the order of the terms. Describe why this makes the process more complicated. Then perform the division again after putting the terms in the dividend in descending order of exponents.

$$\frac{4x^2 - 12x - 12 + 3x^3}{x - 2}$$

74. Explain how to divide a polynomial in x by $(x - c)$ using synthetic division.

75. Explain an advantage of using synthetic division instead of long division.

76. Dell, Inc., is a provider of products and services required for customers worldwide to build their information-technology and Internet infrastructures. Dell's annual revenues since 1999 can be modeled by the polynomial function $R(x) = 0.4x^3 - 3.6x^2 + 11.6x + 17.9$, where $R(x)$ is revenue in billions of dollars and x is the number of years since 1999. Dell's profit can be modeled by the function $P(x) = 0.2x^3 - 1x^2 + 1.4x + 1.4$, where $P(x)$ is the profit in billions of dollars and x is the number of years since 1999. (*Source:* Based on data from Dell, Inc.)

a. Suppose that a market analyst has found the model $P(x)$ and another analyst at the same firm has found the model $R(x)$. The analysts have been asked by their manager to work together to find a model for Dell's profit margin. The analysts know that a company's profit margin is the ratio of its profit to its revenue. Describe how these two analysts could collaborate to find a function $m(x)$ that models Dell's net profit margin based on the work they have done independently.

b. Without actually finding $m(x)$, give a general description of what you would expect the form of the answer to be.

5.4 THE GREATEST COMMON FACTOR AND FACTORING BY GROUPING

OBJECTIVES

A Factor out the Greatest Common Factor of a Polynomials Terms.

B Factor Polynomials by Grouping.

Factoring is the reverse process of multiplying. It is the process of writing a polynomial as a product

$$6x^2 + 13x - 5 = (3x - 1)(2x + 5)$$

with "factoring" labeled above and "multiplying" labeled below.

In the next few sections, we review techniques for factoring polynomials.

Objective **A** Factoring Out the Greatest Common Factor

To factor a polynomial, we first **factor out** the greatest common factor of its terms, using the distributive property. The **greatest common factor** (GCF) of the terms of a polynomial is the product of the GCF of the numerical coefficients and the GCF of each common variable.

Let's find the GCF of $20x^3y$, $10x^2y^2$, and $35x^3$.

The GCF of the numerical coefficients 20, 10, and 35 is 5, the largest integer that is a factor of each integer. The GCF of the variable factors x^3, x^2, and x^3 is x^2 because x^2 is the largest factor common to all three powers of x. The variable y is not a common factor because it does not appear in all three monomials. The GCF is thus

$$5 \cdot x^2, \quad \text{or} \quad 5x^2$$

EXAMPLE 1 Factor: $8x + 4$

Solution: The greatest common factor of the terms $8x$ and 4 is 4.

$$8x + 4 = 4 \cdot 2x + 4 \cdot 1 \quad \text{Factor out 4 from each term.}$$
$$= 4(2x + 1) \quad \text{Use the distributive property.}$$

The factored form of $8x + 4$ is $4(2x + 1)$. To check, multiply $4(2x + 1)$ to see that the product is $8x + 4$.

■ **Work Practice Problem 1**

PRACTICE PROBLEM 1

Factor: $9x + 3$

EXAMPLES Factor.

2. $6x^2 + 3x^3 = 3x^2 \cdot 2 + 3x^2 \cdot x$ The GCF of 6 and 3 is 3 and the GCF of x^2 and x^3 is x^2. Thus, the GCF of the terms is $3x^2$.

$$= 3x^2(2 + x) \quad \text{Use the distributive property.}$$

3. $3y + 1$ There is no common factor other than 1.

4. $17x^3y^2 - 34x^4y^2 = 17x^3y^2 \cdot 1 - 17x^3y^2 \cdot 2x$ Factor out the greatest common factor, $17x^3y^2$.

$$= 17x^3y^2(1 - 2x) \quad \text{Use the distributive property.}$$

■ **Work Practice Problems 2-4**

PRACTICE PROBLEMS 2-4

Factor each polynomial.
2. $20y^2 - 4y^3$
3. $6a - 7$
4. $6a^4b^2 - 3a^2b^2$

Helpful Hint

If the greatest common factor happens to be one of the terms in the polynomial, a factor of 1 will remain for this term when the greatest common factor is factored out. For example, in the polynomial $21x^2 + 7x$, the greatest common factor of $21x^2$ and $7x$ is $7x$, so

$$21x^2 + 7x = 7x(3x) + 7x(1) = 7x(3x + 1)$$

Answers
1. $3(3x + 1)$, 2. $4y^2(5 - y)$,
3. $6a - 7$, 4. $3a^2b^2(2a^2 - 1)$

349

PRACTICE PROBLEM 5

Factor: $-2x^2y - 4xy + 10y$

✔**Concept Check** Which factorization of $12x^2 + 9x - 3$ is correct?

a. $3(4x^2 + 3x + 1)$ **b.** $3(4x^2 + 3x - 1)$

c. $3(4x^2 + 3x - 3)$ **d.** $3(4x^2 + 3x)$

EXAMPLE 5 Factor: $-3x^3y + 2x^2y - 5xy$

Solution: Two possibilities are shown for factoring this polynomial. First, the common factor xy is factored out.

$$-3x^3y + 2x^2y - 5xy = xy(-3x^2 + 2x - 5)$$

Also, the common factor $-xy$ can be factored out as shown.

$$-3x^3y + 2x^2y - 5xy = (-xy)(3x^2) + (-xy)(-2x) + (-xy)(5)$$
$$= -xy(3x^2 - 2x + 5)$$

Both of these are correct.

◻ **Work Practice Problem 5**

PRACTICE PROBLEM 6

Factor: $3(x + 7) + 5y(x + 7)$

EXAMPLE 6 Factor: $2(x - 5) + 3a(x - 5)$

Solution: The greatest common factor is the binomial factor $(x - 5)$.

$$2(x - 5) + 3a(x - 5) = (x - 5)(2 + 3a)$$

◻ **Work Practice Problem 6**

PRACTICE PROBLEM 7

Factor:

$6a(2a + 3b) - (2a + 3b)$

EXAMPLE 7 Factor: $7x(x^2 + 5y) - (x^2 + 5y)$

Solution:

$$7x(x^2 + 5y) - (x^2 + 5y) = 7x(x^2 + 5y) - 1(x^2 + 5y)$$
$$= (x^2 + 5y)(7x - 1)$$

Helpful Hint

Notice that we wrote $-(x^2 + 5y)$ as $-1(x^2 + 5y)$ to aid in factoring.

◻ **Work Practice Problem 7**

Objective B **Factoring by Grouping**

Sometimes it is possible to factor a polynomial by grouping the terms of the polynomial and looking for common factors in each group. This method of factoring is called **factoring by grouping.**

PRACTICE PROBLEM 8

Factor: $xy - 5y + 3x - 15$

EXAMPLE 8 Factor: $ab - 6a + 2b - 12$

Solution:

$$ab - 6a + 2b - 12 = (ab - 6a) + (2b - 12) \quad \text{Group pairs of terms.}$$
$$= a(b - 6) + 2(b - 6) \quad \text{Factor each binomial.}$$
$$= (b - 6)(a + 2) \quad \text{Factor out the greatest common factor, } (b - 6).$$

To check, multiply $(b - 6)$ and $(a + 2)$ to see that the product is $ab - 6a + 2b - 12$.

◻ **Work Practice Problem 8**

Answers
5. $-2y(x^2 + 2x - 5)$ or $2y(-x^2 - 2x + 5)$,
6. $(x + 7)(3 + 5y)$,
7. $(2a + 3b)(6a - 1)$,
8. $(x - 5)(y + 3)$

✔ **Concept Check Answer**

b

> **Helpful Hint**
>
> Notice that the polynomial $a(b - 6) + 2(b - 6)$ is *not* in factored form. It is a *sum*, not a *product*. The factored form is $(b - 6)(a + 2)$.

EXAMPLE 9 Factor: $x^3 + 5x^2 + 3x + 15$

Solution:

$$\begin{aligned} x^3 + 5x^2 + 3x + 15 &= (x^3 + 5x^2) + (3x + 15) && \text{Group pairs of terms.} \\ &= x^2(x + 5) + 3(x + 5) && \text{Factor each binomial.} \\ &= (x + 5)(x^2 + 3) && \text{Factor out the common} \\ &&& \text{factor, } (x + 5). \end{aligned}$$

🔲 **Work Practice Problem 9**

EXAMPLE 10 Factor: $m^2n^2 + m^2 - 2n^2 - 2$

Solution:

$$\begin{aligned} m^2n^2 + m^2 - 2n^2 - 2 &= (m^2n^2 + m^2) + (-2n^2 - 2) && \text{Group pairs of terms.} \\ &= m^2(n^2 + 1) - 2(n^2 + 1) && \text{Factor each binomial.} \\ &= (n^2 + 1)(m^2 - 2) && \text{Factor out the common} \\ &&& \text{factor, } (n^2 + 1). \end{aligned}$$

🔲 **Work Practice Problem 10**

EXAMPLE 11 Factor: $xy + 2x - y - 2$

Solution:

$$\begin{aligned} xy + 2x - y - 2 &= (xy + 2x) + (-y - 2) && \text{Group pairs of terms.} \\ &= x(y + 2) - 1(y + 2) && \text{Factor each binomial.} \\ &= (y + 2)(x - 1) && \text{Factor out the common factor, } y + 2. \end{aligned}$$

🔲 **Work Practice Problem 11**

PRACTICE PROBLEM 9
Factor: $y^3 + 6y^2 + 4y + 24$

PRACTICE PROBLEM 10
Factor: $a^2b^2 + a^2 - 3b^2 - 3$

PRACTICE PROBLEM 11
Factor: $ab + 5a - b - 5$

Answers
9. $(y + 6)(y^2 + 4)$,
10. $(b^2 + 1)(a^2 - 3)$,
11. $(b + 5)(a - 1)$

Mental Math

Find the greatest common factor of each list of monomials.

1. $6, 12$

2. $9, 27$

3. $15x, 10$

4. $9x, 12$

5. $13x, 2x$

6. $4y, 5y$

7. $7x, 14x$

8. $8z, 4z$

5.4 EXERCISE SET

FOR EXTRA HELP

Student Solutions Manual PH Math/Tutor Center CD/Video for Review Math XL MathXL® MyMathLab MyMathLab

Objective Ⓐ *Factor out the greatest common factor. See Examples 1 through 7.*

1. $18x - 12$

2. $21x + 14$

3. $4y^2 - 16xy^3$

4. $3z - 21xz^4$

5. $6x^5 - 8x^4 + 2x^3$

6. $9x + 3x^2 - 6x^3$

7. $8a^3b^3 - 4a^2b^2 + 4ab + 16ab^2$

8. $12a^3b - 6ab + 18ab^2 - 18a^2b$

9. $6(x + 3) + 5a(x + 3)$

10. $2(x - 4) + 3y(x - 4)$

11. $2x(z + 7) + (z + 7)$

12. $9x(y - 2) + (y - 2)$

13. $3x(6x^2 + 5) - 2(6x^2 + 5)$

14. $4x(2y^2 + 3) - 5(2y^2 + 3)$

Objective Ⓑ *Factor each polynomial by grouping. See Examples 8 through 11.*

15. $ab + 3a + 2b + 6$

16. $ab + 2a + 5b + 10$

17. $ac + 4a - 2c - 8$

18. $bc + 8b - 3c - 24$

19. $2xy - 3x - 4y + 6$

20. $12xy - 18x - 10y + 15$

21. $12xy - 8x - 3y + 2$

22. $20xy - 15x - 4y + 3$

Objectives Ⓐ Ⓑ **Mixed Practice** *Factor each polynomial. See Examples 1 through 11.*

23. $6x^3 + 9$

24. $6x^2 - 8$

25. $x^3 + 3x^2$

26. $x^4 - 4x^3$

27. $8a^3 - 4a$

28. $12b^4 + 3b^2$

29. $-20x^2y + 16xy^3$

30. $-18xy^3 + 27x^4y$

31. $10a^2b^3 + 5ab^2 - 15ab^3$

32. $10ef - 20e^2f^3 + 30e^3f$

33. $9abc^2 + 6a^2bc - 6ab + 3bc$

34. $4a^2b^2c - 6ab^2c - 4ac + 8a$

35. $4x(y - 2) - 3(y - 2)$

36. $8y(z + 8) - 3(z + 8)$

37. $6xy + 10x + 9y + 15$

38. $15xy + 20x + 6y + 8$

39. $xy + 3y - 5x - 15$

40. $xy + 4y - 3x - 12$

41. $6ab - 2a - 9b + 3$

42. $16ab - 8a - 6b + 3$

43. $12xy + 18x + 2y + 3$

44. $20xy + 8x + 5y + 2$

45. $2m(n - 8) - (n - 8)$

46. $3a(b - 4) - (b - 4)$

47. $15x^3y^2 - 18x^2y^2$

48. $12x^4y^2 - 16x^3y^3$

49. $2x^2 + 3xy + 4x + 6y$

50. $3x^2 + 12x + 4xy + 16y$

51. $5x^2 + 5xy - 3x - 3y$

52. $4x^2 + 2xy - 10x - 5y$

53. $x^3 + 3x^2 + 4x + 12$ **54.** $x^3 + 4x^2 + 3x + 12$ **55.** $x^3 - x^2 - 2x + 2$ **56.** $x^3 - 2x^2 - 3x + 6$

Review

Find each product by using the FOIL order of multiplying binomials. See Section 5.2.

57. $(x + 2)(x - 5)$

58. $(x - 7)(x - 1)$

59. $(x + 3)(x + 2)$

60. $(x - 4)(x + 2)$

61. $(y - 3)(y - 1)$

62. $(s + 8)(s + 10)$

Concept Extensions

Solve. See the Concept Check in this section.

63. Which factorization of $10x^2 - 2x - 2$ is correct?
  a. $2(5x^2 - x + 1)$
  b. $2(5x^2 - x)$
  c. $2(5x^2 - x - 2)$
  d. $2(5x^2 - x - 1)$

$2(5x^2 - x - 1)$

64. Which factorization of $x^4 + 5x^3 - x^2$ is correct?
  a. $-1(x^4 + 5x^3 + x^2)$
  b. $x^2(x^2 + 5x^3 - x^2)$
  c. $x^2(x^2 + 5x - 1)$
  d. $5x^2(x^2 + 5x - 5)$

△ **65.** The area of the material needed to manufacture a tin can is given by the polynomial $2\pi r^2 + 2\pi rh$, where the radius is r and height is h. Factor this expression.

66. To estimate the cost of a new product, one expression used by the production department is $4\pi r^2 + \frac{4}{3}\pi r^3$. Write an equivalent expression by factoring $4\pi r^2$ from both terms.

67. At the end of T years, the amount of money A in a savings account earning simple interest from an initial investment of $5600 at rate r is given by the formula $A = 5600 + 5600rt$. Write an equivalent equation by factoring the expression $5600 + 5600rt$.

△ **68.** An open-topped box has a square base and a height of 10 inches. If each of the bottom edges of the box has length x inches, find the amount of material needed to construct the box. Write the answer in factored form.

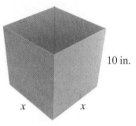

10 in.

x $\quad$ x

69. When $3x^2 - 9x + 3$ is factored, the result is $3(x^2 - 3x + 1)$. Explain why it is necessary to include the term 1 in this factored form.

70. Construct a trinomial whose greatest common factor is $5x^2y^3$.

71. A factored polynomial can be in many forms. For example, a factored form of $xy - 3x - 2y + 6$ is $(x - 2)(y - 3)$. Which of the following is not a factored form of $xy - 3x - 2y + 6$?
   a. $(2 - x)(3 - y)$ $\qquad$ **b.** $(-2 + x)(-3 + y)$
   c. $(y - 3)(x - 2)$ $\qquad$ **d.** $(-x + 2)(-y + 3)$

72. Consider the following sequence of algebraic steps:
$$x^3 - 6x^2 + 2x - 10 = (x^3 - 6x^2) + (2x - 10)$$
$$= x^2(x - 6) + 2(x - 5)$$

Explain whether the final result is the factored form of the original polynomial.

73. Which factorization of $12x^2 + 9x + 3$ is correct?
   a. $3(4x^2 + 3x + 1)$
   b. $3(4x^2 + 3x - 1)$
   c. $3(4x^2 + 3x - 3)$
   d. $3(4x^2 + 3x)$

74. The amount E of voltage in an electrical circuit is given by the formula
$$IR_1 + IR_2 = E$$
Write an equivalent equation by factoring the expression $IR_1 + IR_2$.

75. At the end of T years, the amount of money A in a savings account earning simple interest from an initial investment of P dollars at rate R is given by the formula
$$A = P + PRT$$
Write an equivalent equation by factoring the expression $P + PRT$.

Factor out the greatest common factor. Assume that variables used as exponents represent positive integers.

76. $x^{3n} - 2x^{2n} + 5x^n$ $\qquad$ **77.** $3y^n + 3y^{2n} + 5y^{8n}$ $\qquad$ **78.** $6x^{8a} - 2x^{5a} - 4x^{3a}$ $\qquad$ **79.** $3x^{5a} - 6x^{3a} + 9x^{2a}$

80. An object is thrown upward from the ground with an initial velocity of 64 feet per second. The height $h(t)$ in feet of the object after t seconds is given by the polynomial function
$$h(t) = -16t^2 + 64t$$
   a. Write an equivalent factored expression for the function $h(t)$ by factoring $-16t^2 + 64t$.
   b. Find $h(1)$ by using
$$h(t) = -16t^2 + 64t$$
and then by using the factored form of $h(t)$.
   c. Explain why the values found in part (b) are the same.

81. An object is dropped from the gondola of a hot-air balloon at a height of 224 feet. The height $h(t)$ in feet of the object after t seconds is given by the polynomial function

$$h(t) = -16t^2 + 224$$

a. Write an equivalent factored expression for the function $h(t)$ by factoring $-16t^2 + 224$.

b. Find $h(2)$ by using

$$h(t) = -16t^2 + 224$$

and then by using the factored form of the function.

c. Explain why the values found in part (b) are the same.

224 ft

82. The polynomial function $f(x) = 35,000x^3 - 171,000x^2 + 190,000x + 320,000$ models the number of applications for trademark registrations for the years 1999–2003, where x represents the number of years after 1999 and $f(x)$ is the number of trademark registration applications. Write an equivalent expression for $f(x)$ by factoring the greatest common factor from the terms of $f(x) = 35,000x^3 - 171,000x^2 + 190,000x + 320,000$. (*Source:* International Trademark Association)

83. The percentage of public school instructional rooms with Internet access is increasing. The polynomial function $f(x) = -2x^2 + 22x + 28$ models the percent of public school instructional rooms with Internet access for the years 1997–2003, where x is the years after 1997 and $f(x)$ is the percent of Internet-connected instructional spaces. Write the equivalent expression for $f(x)$ by factoring the greatest common factor from the terms of $f(x) = -2x^2 + 22x + 28$. (*Source:* U.S. Department of Education)

A Factor Trinomials of the Form $x^2 + bx + c$.

B Factor Trinomials of the Form $ax^2 + bx + c$.
  a. Method 1—Trial and Check
  b. Method 2—Grouping

C Factor by Substitution.

5.5 FACTORING TRINOMIALS

Objective **A** Factoring Trinomials of the Form $x^2 + bx + c$

In the previous section, we used factoring by grouping to factor four-term polynomials. In this section, we present techniques for factoring trinomials. Since $(x - 2)(x + 5) = x^2 + 3x - 10$, we say that $(x - 2)(x + 5)$ is a factored form of $x^2 + 3x - 10$. Taking a close look at how $(x - 2)$ and $(x + 5)$ are multiplied suggests a pattern for factoring trinomials of the form $x^2 + bx + c$.

$$(x - 2)(x + 5) = x^2 + 3x - 10$$

The pattern for factoring is summarized next.

Factoring a Trinomial of the Form $x^2 + bx + c$

Find two numbers whose product is c and whose sum is b. The factored form of $x^2 + bx + c$ is

$$(x + \text{one number})(x + \text{other number})$$

PRACTICE PROBLEM 1
Factor: $x^2 + 8x + 15$

EXAMPLE 1 Factor: $x^2 + 10x + 16$

Solution: We look for two integers whose product is 16 and whose sum is 10. Since our integers must have a positive product and a positive sum, we look at only positive factors of 16.

Positive Factors of 16	Sum of Factors
1, 16	1 + 16 = 17
4, 4	4 + 4 = 8
2, 8	2 + 8 = 10 Correct pair

The correct pair of numbers is 2 and 8 because their product is 16 and their sum is 10. Thus,

$$x^2 + 10x + 16 = (x + 2)(x + 8)$$

Check: To check, see that $(x + 2)(x + 8) = x^2 + 10x + 16$.

☐ **Work Practice Problem 1**

PRACTICE PROBLEM 2
Factor: $x^2 - 10x + 24$

EXAMPLE 2 Factor: $x^2 - 12x + 35$

Solution: We need to find two integers whose product is 35 and whose sum is -12. Since our integers must have a positive product and a negative sum, we consider only negative factors of 35.

Negative Factors of 35	Sum of Factors
$-1, -35$	$-1 + (-35) = -36$
$-5, -7$	$-5 + (-7) = -12$ Correct pair

The numbers are -5 and -7.

$$x^2 - 12x + 35 = [x + (-5)][x + (-7)]$$
$$= (x - 5)(x - 7)$$

Check: To check, see that $(x - 5)(x - 7) = x^2 - 12x + 35$.

☐ **Work Practice Problem 2**

Answers
1. $(x + 5)(x + 3)$,
2. $(x - 4)(x - 6)$

EXAMPLE 3 Factor: $5x^3 - 30x^2 - 35x$

Solution: First we factor out the greatest common factor, $5x$.

$$5x^3 - 30x^2 - 35x = 5x(x^2 - 6x - 7)$$

Next we factor $x^2 - 6x - 7$ by finding two numbers whose product is -7 and whose sum is -6. The numbers are 1 and -7.

$$5x^3 - 30x^2 - 35x = 5x(x^2 - 6x - 7)$$
$$= 5x(x + 1)(x - 7)$$

🖿 **Work Practice Problem 3**

PRACTICE PROBLEM 3

Factor: $6x^3 + 24x^2 - 30x$

> **Helpful Hint**
>
> If the polynomial to be factored contains a common factor that is factored out, don't forget to include that common factor in the final factored form of the original polynomial.

EXAMPLE 4 Factor: $2n^2 - 38n + 80$

Solution: The terms of this polynomial have a greatest common factor of 2, which we factor out first.

$$2n^2 - 38n + 80 = 2(n^2 - 19n + 40)$$

Next we factor $n^2 - 19n + 40$ by finding two numbers whose product is 40 and whose sum is -19. Both numbers must be negative since their product is positive and their sum is negative. Possibilities are

$$-1 \text{ and } -40, \quad -2 \text{ and } -20, \quad -4 \text{ and } -10, \quad -5 \text{ and } -8$$

None of the pairs has a sum of -19, so no further factoring with integers is possible. The factored form of $2n^2 - 38n + 80$ is

$$2n^2 - 38n + 80 = 2(n^2 - 19n + 40)$$

🖿 **Work Practice Problem 4**

PRACTICE PROBLEM 4

Factor: $3y^2 + 6y + 6$

We call a polynomial such as $n^2 - 19n + 40$, which cannot be factored further, a **prime polynomial.**

Objective B Factoring Trinomials of the Form $ax^2 + bx + c$

Next, we factor trinomials of the form $ax^2 + bx + c$, where the coefficient a of x^2 is not 1. Don't forget that the first step in factoring any polynomial is to factor out the greatest common factor of its terms.

Method 1—Factoring $ax^2 + bx + c$ by Trial and Check

EXAMPLE 5 Factor: $2x^2 + 11x + 15$

Solution: Factors of $2x^2$ are $2x$ and x. Let's try these factors as first terms of the binomials.

$$2x^2 + 11x + 15 = (2x + \quad)(x + \quad)$$

PRACTICE PROBLEM 5

Factor: $3x^2 + 13x + 4$

Answers

3. $6x(x - 1)(x + 5)$, **4.** $3(y^2 + 2y + 2)$,
5. $(3x + 1)(x + 4)$

Continued on next page

Next we try combinations of factors of 15 until the correct middle term, $11x$, is obtained. We will try only positive factors of 15 since the coefficient of the middle term, 11, is positive. Positive factors of 15 are 1 and 15 and 3 and 5.

$$(2x + 1)(x + 15)$$
$$1x$$
$$\frac{30x}{31x} \quad \text{Incorrect middle term}$$

$$(2x + 15)(x + 1)$$
$$15x$$
$$\frac{2x}{17x} \quad \text{Incorrect middle term}$$

$$(2x + 3)(x + 5)$$
$$3x$$
$$\frac{10x}{13x} \quad \text{Incorrect middle term}$$

$$(2x + 5)(x + 3)$$
$$5x$$
$$\frac{6x}{11x} \quad \text{Correct middle term}$$

Thus, the factored form of $2x^2 + 11x + 15$ is $(2x + 5)(x + 3)$.

■ **Work Practice Problem 5**

Factoring a Trinomial of the Form $ax^2 + bx + c$

Step 1: Write all pairs of factors of ax^2.

Step 2: Write all pairs of factors of c, the constant term.

Step 3: Try various combinations of these factors until the correct middle term bx is found.

Step 4: If no combination exists, the polynomial is **prime.**

PRACTICE PROBLEM 6

Factor: $5x^2 + 13x - 6$

EXAMPLE 6 Factor: $3x^2 - x - 4$

Solution: Factors of $3x^2$: $3x \cdot x$

Factors of -4: $-1 \cdot 4, \quad 1 \cdot -4 \quad -2 \cdot 2, \quad 2 \cdot -2$

Let's try possible combinations of these factors.

$$(3x - 1)(x + 4)$$
$$-1x$$
$$\frac{12x}{11x} \quad \text{Incorrect middle term}$$

$$(3x + 4)(x - 1)$$
$$4x$$
$$\frac{-3x}{1x} \quad \text{Incorrect middle term}$$

$$(3x - 4)(x + 1)$$
$$-4x$$
$$\frac{3x}{-1x} \quad \text{Correct middle term}$$

Thus, $3x^2 - x - 4 = (3x - 4)(x + 1)$.

■ **Work Practice Problem 6**

Answer

6. $(x + 3)(5x - 2)$

> **Helpful Hint**
>
> A positive constant in a trinomial tells us to look for two numbers with the same sign. The sign of the coefficient of the middle term tells us whether the signs are both positive or both negative.
>
> both positive   same sign
>
> $2x^2 + 7x + 3 = (2x + 1)(x + 3)$
>
> both negative   same sign
>
> $2x^2 - 7x + 3 = (2x - 1)(x - 3)$
>
> A negative constant in a trinomial tells us to look for two numbers with opposite signs.
>
> opposite signs
>
> $2x^2 - 5x - 3 = (2x + 1)(x - 3)$
>
> opposite signs
>
> $2x^2 + 5x - 3 = (2x - 1)(x + 3)$

EXAMPLE 7 Factor: $12x^3y - 22x^2y + 8xy$

Solution: First we factor out the greatest common factor of the terms of this trinomial, $2xy$.

$$12x^3y - 22x^2y + 8xy = 2xy(6x^2 - 11x + 4)$$

Now we try to factor the trinomial $6x^2 - 11x + 4$.

Factors of $6x^2$: $2x \cdot 3x$, $6x \cdot x$

Let's try $2x$ and $3x$.

$$2xy(6x^2 - 11x + 4) = 2xy(2x + \quad)(3x + \quad)$$

The constant term, 4, is positive and the coefficient of the middle term, -11, is negative, so we factor 4 into negative factors only.

Negative factors of 4: $-4(-1)$, $-2(-2)$

Let's try -4 and -1.

$$2xy(2x - 4)(3x - 1)$$

    $-12x$
    $\underline{-2x}$
    $-14x$ Incorrect middle term

This combination cannot be correct because one of the factors, $(2x - 4)$, has a common factor of 2. This cannot happen if the polynomial $6x^2 - 11x + 4$ has no common factors.

Now let's try -1 and -4.

$$2xy(2x - 1)(3x - 4)$$

    $-3x$
    $\underline{-8x}$
    $-11x$ Correct middle term

Thus,

$$12x^3y - 22x^2y + 8xy = 2xy(2x - 1)(3x - 4)$$

If this combination had not worked, we would try -2 and -2 as factors of 4 and then $6x$ and x as factors of $6x^2$.

Work Practice Problem 7

PRACTICE PROBLEM 7

Factor: $24x^2y^2 - 42xy^2 + 9y^2$

> **Helpful Hint**
>
> If a trinomial has no common factor (other than 1), then none of its binomial factors will contain a common factor (other than 1).

Answer

7. $3y^2(2x - 3)(4x - 1)$

PRACTICE PROBLEM 8

Factor: $4x^2 + 28xy + 49y^2$

EXAMPLE 8 Factor: $16x^2 + 24xy + 9y^2$

Solution: No greatest common factor can be factored out of this trinomial.

Factors of $16x^2$: $16x \cdot x,$ $8x \cdot 2x,$ $4x \cdot 4x$

Factors of $9y^2$: $y \cdot 9y,$ $3y \cdot 3y$

We try possible combinations until the correct factorization is found.

$16x^2 + 24xy + 9y^2 = (4x + 3y)(4x + 3y)$ or $(4x + 3y)^2$

◻ Work Practice Problem 8

The trinomial $16x^2 + 24xy + 9y^2$ in Example 8 is an example of a **perfect square trinomial** since its factors are two identical binomials. In the next section, we examine a special method for factoring perfect square trinomials.

Method 2—Factoring $ax^2 + bx + c$ by Grouping

There is another method we can use when factoring trinomials of the form $ax^2 + bx + c$: Write the trinomial as a four-term polynomial, and then factor by grouping.

Factoring a Trinomial of the Form $ax^2 + bx + c$ by Grouping

Step 1: Find two numbers whose product is $a \cdot c$ and whose sum is b.

Step 2: Write the term bx as a sum by using the factors found in Step 1.

Step 3: Factor by grouping.

PRACTICE PROBLEM 9

Factor: $12x^2 + 11x + 2$

EXAMPLE 9 Factor: $6x^2 + 13x + 6$

Solution: In this trinomial, $a = 6, b = 13,$ and $c = 6$.

Step 1: Find two numbers whose product is $a \cdot c$, or $6 \cdot 6 = 36$, and whose sum is $b, 13$. The two numbers are 4 and 9.

Step 2: Write the middle term $13x$ as the sum $4x + 9x$.

$6x^2 + 13x + 6 = 6x^2 + 4x + 9x + 6$

Step 3: Factor $6x^2 + 4x + 9x + 6$ by grouping.

$(6x^2 + 4x) + (9x + 6) = 2x(3x + 2) + 3(3x + 2)$
$= (3x + 2)(2x + 3)$

◻ Work Practice Problem 9

✔ Concept Check Name one way that a factorization can be checked.

PRACTICE PROBLEM 10

Factor: $14x^2 - x - 3$

EXAMPLE 10 Factor: $18x^2 - 9x - 2$

Solution: In this trinomial, $a = 18, b = -9,$ and $c = -2$.

Step 1: Find two numbers whose product is $a \cdot c$ or $18(-2) = -36$ and whose sum is $b, -9$. The two numbers are -12 and 3.

Step 2: Write the middle term, $-9x$, as the sum $-12x + 3x$.

$18x^2 - 9x - 2 = 18x^2 - 12x + 3x - 2$

Step 3: Factor by grouping.

$(18x^2 - 12x) + (3x - 2) = 6x(3x - 2) + 1(3x - 2)$
$= (3x - 2)(6x + 1)$

◻ Work Practice Problem 10

Answers

8. $(2x + 7y)^2$, **9.** $(3x + 2)(4x + 1)$,
10. $(7x + 3)(2x - 1)$

✔ Concept Check Answer

Answers may vary. A sample is: By multiplying the factors to see that the product is the original polynomial.

Objective C Factoring by Substitution

A complicated-looking polynomial may be a simpler trinomial "in disguise." Revealing the simpler trinomial is possible by substitution.

EXAMPLE 11 Factor: $2(a + 3)^2 - 5(a + 3) - 7$

Solution: The quantity $(a + 3)$ is in two of the terms of this polynomial. If we *substitute* x for $(a + 3)$, the result is the following simpler trinomial.

$$2(a + 3)^2 - 5(a + 3) - 7 \quad \text{Original trinomial}$$
$$= 2(x)^2 - 5(x) - 7 \quad \text{Substitute } x \text{ for } (a + 3).$$

Now we can factor $2x^2 - 5x - 7$.

$$2x^2 - 5x - 7 = (2x - 7)(x + 1)$$

But the quantity in the original polynomial was $(a + 3)$, not x. Thus we need to reverse the substitution and replace x with $(a + 3)$.

$$(2x - 7)(x + 1) \quad \text{Factored expression}$$
$$= [2(a + 3) - 7][(a + 3) + 1] \quad \text{Substitute } (a + 3) \text{ for } x.$$
$$= (2a + 6 - 7)(a + 3 + 1) \quad \text{Remove inside parentheses.}$$
$$= (2a - 1)(a + 4) \quad \text{Simplify.}$$

Thus, $2(a + 3)^2 - 5(a + 3) - 7 = (2a - 1)(a + 4)$.

Work Practice Problem 11

EXAMPLE 12 Factor: $5x^4 + 29x^2 - 42$

Solution: Again, substitution may help us factor this polynomial more easily. We will let $y = x^2$, so $y^2 = (x^2)^2$, or x^4. Then

$$5x^4 + 29x^2 - 42$$
becomes
$$5y^2 + 29y - 42$$

which factors as

$$5y^2 + 29y - 42 = (5y - 6)(y + 7)$$

Now we replace y with x^2 to get

$$(5x^2 - 6)(x^2 + 7)$$

Work Practice Problem 12

PRACTICE PROBLEM 11

Factor: $3(z + 2)^2 - 19(z + 2) + 6$

PRACTICE PROBLEM 12

Factor: $14x^4 + 23x^2 + 3$

Answers
11. $(3z + 5)(z - 4)$,
12. $(2x^2 + 3)(7x^2 + 1)$

Mental Math

1. Find two numbers whose product is 10 and whose sum is 7.

2. Find two numbers whose product is 12 and whose sum is 8.

3. Find two numbers whose product is 24 and whose sum is 11.

4. Find two numbers whose product is 30 and whose sum is 13.

5.5 EXERCISE SET

FOR EXTRA HELP

Student Solutions Manual · PH Math/Tutor Center · CD/Video for Review · MathXL® · MyMathLab

Objective Ⓐ *Factor each trinomial. See Examples 1 through 4.*

 1. $x^2 + 9x + 18$

2. $x^2 + 9x + 20$

 3. $x^2 - 12x + 32$

4. $x^2 - 12x + 27$

5. $x^2 + 10x - 24$

6. $x^2 + 3x - 54$

 7. $x^2 - 2x - 24$

8. $x^2 - 9x - 36$

9. $3x^2 - 18x + 24$

10. $x^2y^2 + 4xy^2 + 3y^2$

11. $4x^2z + 28xz + 40z$

12. $5x^2 - 45x + 70$

13. $2x^2 - 24x - 64$

14. $3n^2 - 6n - 51$

Objective Ⓑ *Factor each trinomial. See Examples 5 through 9.*

15. $5x^2 + 16x + 3$

16. $3x^2 + 8x + 4$

17. $2x^2 - 11x + 12$

18. $3x^2 - 19x + 20$

19. $2x^2 + 25x - 20$

20. $6x^2 + 13x + 8$

21. $4x^2 - 12x + 9$

22. $25x^2 - 30x + 9$

23. $12x^2 + 10x - 50$

24. $12y^2 - 48y + 45$

25. $3y^4 - y^3 - 10y^2$

26. $2x^2z + 5xz - 12z$

27. $6x^3 + 8x^2 + 24x$

28. $18y^3 + 12y^2 + 2y$

29. $2x^2 - 5xy - 3y^2$

30. $6x^2 + 11xy + 4y^2$

31. $28y^2 + 22y + 4$

32. $24y^3 - 2y^2 - y$

33. $2x^2 + 15x - 27$

34. $3x^2 + 14x + 15$

Objective Ⓒ *Use substitution to factor each polynomial completely. See Examples 11 and 12.*

35. $x^4 + x^2 - 6$

36. $x^4 - x^2 - 20$

37. $(5x + 1)^2 + 8(5x + 1) + 7$

38. $(3x - 1)^2 + 5(3x - 1) + 6$

39. $x^6 - 7x^3 + 12$

40. $x^6 - 4x^3 - 12$

41. $(a + 5)^2 - 5(a + 5) - 24$

42. $(3c + 6)^2 + 12(3c + 6) - 28$

Objectives Ⓐ Ⓑ Ⓒ **Mixed Practice** *Factor each polynomial completely. See Examples 1 through 12.*

43. $x^2 - 24x - 81$

44. $x^2 - 48x - 100$

45. $x^2 - 15x - 54$

46. $x^2 - 15x + 54$

47. $3x^2 - 6x + 3$

48. $8x^2 - 8x + 2$

49. $3x^2 - 5x - 2$

50. $5x^2 - 14x - 3$

51. $8x^2 - 26x + 15$

52. $12x^2 - 17x + 6$

53. $18x^4 + 21x^3 + 6x^2$

54. $20x^5 + 54x^4 + 10x^3$

55. $x^2 + 8xz + 7z^2$

56. $a^2 - 2ab - 15b^2$

57. $x^2 - x - 12$

58. $x^2 + 4x - 5$

59. $3a^2 + 12ab + 12b^2$

60. $2x^2 + 16xy + 32y^2$

61. $x^2 + 4x + 5$

62. $x^2 + 6x + 8$

63. $2(x + 4)^2 + 3(x + 4) - 5$

64. $3(x + 3)^2 + 2(x + 3) - 5$

65. $6x^2 - 49x + 30$

66. $4x^2 - 39x + 27$

67. $x^4 - 5x^2 - 6$

68. $x^4 - 5x^2 + 6$

69. $6x^3 - x^2 - x$

70. $12x^3 + x^2 - x$

71. $12a^2 - 29ab + 15b^2$

72. $16y^2 + 6yx - 27x^2$

73. $9x^2 + 30x + 25$

74. $4x^2 + 6x + 9$

75. $3x^2y - 11xy + 8y$

76. $5xy^2 - 9xy + 4x$

77. $2x^2 + 2x - 12$

78. $3x^2 + 6x - 45$

79. $(x - 4)^2 + 3(x - 4) - 18$

80. $(x - 3)^2 - 2(x - 3) - 8$

81. $2x^6 + 3x^3 - 9$

82. $3x^6 - 14x^3 + 8$

83. $72xy^4 - 24xy^2z + 2xz^2$

84. $36xy^2 - 48xyz^2 + 16xz^4$

85. $2x^3y + 2x^2y - 12xy$

86. $3x^2y^3 + 6x^2y^2 - 45x^2y$

87. $x^2 + 6xy + 5y^2$

88. $x^2 + 6xy + 8y^2$

Review

Multiply. See Section 5.2.

89. $(x - 3)(x + 3)$

90. $(x - 4)(x + 4)$

91. $(2x + 1)^2$

92. $(3x + 5)^2$

93. $(x - 2)(x^2 + 2x + 4)$

94. $(y + 1)(y^2 - y + 1)$

Concept Extensions

95. Find all positive and negative integers b such that $x^2 + bx + 6$ is factorable.

96. Find all positive and negative integers b such that $x^2 + bx - 10$ is factorable.

97. The volume $V(x)$ of a box in terms of its height x is given by the function $V(x) = x^3 + 2x^2 - 8x$. Factor this expression for $V(x)$.

98. Based on your results from Exercise 97, find the length and width of the box if the height is 5 inches and the dimensions of the box are whole numbers.

99. Suppose that a movie is being filmed in New York City. An action shot requires an object to be thrown upward with an initial velocity of 80 feet per second off the top of 1 Madison Square Plaza, a height of 576 feet. The height $h(t)$ in feet of the object after t seconds is given by the function $h(t) = -16t^2 + 80t + 576$. (*Source: The World Almanac, 2001*)

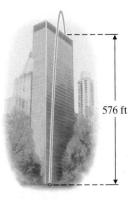

576 ft

   a. Find the height of the object at $t = 0$ seconds, $t = 2$ seconds, $t = 4$ seconds, and $t = 6$ seconds.

   b. Explain why the height of the object increases and then decreases as time passes.

   c. Factor the polynomial $-16t^2 + 80t + 576$.

100. Suppose that an object is thrown upward with an initial velocity of 64 feet per second off the edge of a 960-foot-cliff. The height $h(t)$ in feet of the object after t seconds is given by the function

$$h(t) = -16t^2 + 64t + 960$$

   a. Find the height of the object at $t = 0$ seconds, $t = 3$ seconds, $t = 6$ seconds, and $t = 9$ seconds.

   b. Explain why the height of the object increases and then decreases as time passes.

   c. Factor the polynomial $-16t^2 + 64t + 960$.

Factor. Assume that variables used as exponents represent positive integers.

101. $x^{2n} + 10x^n + 16$ **102.** $x^{2n} - 7x^n + 12$ **103.** $x^{2n} - 3x^n - 18$ **104.** $x^{2n} + 7x^n - 18$

105. $2x^{2n} + 11x^n + 5$ **106.** $3x^{2n} - 8x^n + 4$ **107.** $4x^{2n} - 12x^n + 9$ **108.** $9x^{2n} + 24x^n + 16$

Recall that a graphing calculator may be used to check addition, subtraction, and multiplication of polynomials. In the same manner, a graphing calculator may be used to check factoring of polynomials in one variable. For example, to see that

$$2x^3 - 9x^2 - 5x = x(2x + 1)(x - 5)$$

graph $Y_1 = 2x^3 - 9x^2 - 5x$ *and* $Y_2 = x(2x + 1)(x - 5)$. *Then trace along both graphs to see that they coincide. Factor the following and use this method to check your results.*

109. $x^4 + 6x^3 + 5x^2$ **110.** $x^3 + 6x^2 + 8x$ **111.** $30x^3 + 9x^2 - 3x$ **112.** $-6x^4 + 10x^3 - 4x^2$

STUDY SKILLS BUILDER

Is Your Notebook Still Organized?

It's never too late to organize your material in a course. Let's see how you are doing.

1. Are all your graded papers in one place in your math notebook or binder?

2. Flip through the pages of your notebook. Are your notes neat and readable?

3. Are your notes complete with no sections missing?

4. Are important notes marked in some way (like an exclamation point) so that you will know to review them before a quiz or test?

5. Are your assignments complete?

6. Do exercises that have given you trouble have a mark (like a question mark) so that you will remember to talk to your instructor or a tutor about them?

7. Describe your attitude toward this course.

8. List ways your attitude can improve and make a commitment to work on at least one of these during the next week.

5.6 FACTORING BY SPECIAL PRODUCTS

Objectives

A Factor a Perfect Square Trinomial.

B Factor the Difference of Two Squares.

C Factor the Sum or Difference of Two Cubes.

Objective **A** Factoring Perfect Square Trinomials

In the previous section, we considered a variety of ways to factor trinomials of the form $ax^2 + bx + c$. In Example 8, we factored $16x^2 + 24xy + 9y^2$ as

$$16x^2 + 24xy + 9y^2 = (4x + 3y)^2$$

Recall that we called $16x^2 + 24xy + 9y^2$ a perfect square trinomial because its factors are two identical binomials. A trinomial is a perfect square trinomial if it can be written so that its first term is the square of some quantity a, its last term is the square of some quantity b, and its middle term is twice the product of the quantities a and b.

The following special formulas can be used to factor perfect square trinomials.

> **Perfect Square Trinomials**
> $$a^2 + 2ab + b^2 = (a + b)^2$$
> $$a^2 - 2ab + b^2 = (a - b)^2$$

Notice that these equations are the same special products from Section 5.2 for the square of a binomial.

From

$$a^2 + 2ab + b^2 = (a + b)^2$$

we see that

$$16x^2 + 24xy + 9y^2 = (4x)^2 + 2(4x)(3y) + (3y)^2 = (4x + 3y)^2$$

EXAMPLE 1 Factor: $m^2 + 10m + 25$

Solution: Notice that the first term is a square: $m^2 = (m)^2$, the last term is a square: $25 = 5^2$, and $10m = 2 \cdot 5 \cdot m$.

This is a perfect square trinomial. Thus,

$$m^2 + 10m + 25 = m^2 + 2(m)(5) + 5^2 = (m + 5)^2$$

▢ **Work Practice Problem 1**

EXAMPLES Factor each trinomial.

2. $4x^2 + 4x + 1 = (2x)^2 + 2 \cdot 2x \cdot 1 + 1^2$ See whether it is a perfect square trinomial.
$$= (2x + 1)^2$$ Factor.

3. $9x^2 - 12x + 4 = (3x)^2 - 2(3x)(2) + 2^2$ See whether it is a perfect square trinomial.
$$= (3x - 2)^2$$ Factor.

▢ **Work Practice Problems 2–3**

EXAMPLE 4 Factor: $3a^2x - 12abx + 12b^2x$

Solution: The terms of this trinomial have a greatest common factor of $3x$, which we factor out first.

$$3a^2x - 12abx + 12b^2x = 3x(a^2 - 4ab + 4b^2)$$

The polynomial $a^2 - 4ab + 4b^2$ is a perfect square trinomial. Notice that the first term is a square: $a^2 = (a)^2$, the last term is a square: $4b^2 = (2b)^2$, and $4ab = 2(a)(2b)$. The factoring can now be completed as

$$3x(a^2 - 4ab + 4b^2) = 3x(a - 2b)^2$$

▢ **Work Practice Problem 4**

PRACTICE PROBLEM 1

Factor: $x^2 + 8x + 16$

PRACTICE PROBLEMS 2–3

Factor.

2. $9x^2 + 6x + 1$

3. $25x^2 - 20x + 4$

PRACTICE PROBLEM 4

Factor: $4x^3 - 32x^2y + 64xy^2$

Answers

1. $(x + 4)^2$,   **2.** $(3x + 1)^2$,

3. $(5x - 2)^2$,   **4.** $4x(x - 4y)^2$

Helpful Hint
If you recognize a trinomial as a perfect square trinomial, use the special formulas to factor. However, methods for factoring trinomials in general from Section 5.5 will also result in the correct factored form.

Objective B Factoring the Difference of Two Squares

We now factor special types of binomials, beginning with the **difference of two squares.** The special product pattern presented in Section 5.2 for the product of a sum and a difference of two terms is used again here. However, the emphasis is now on factoring rather than on multiplying.

Difference of Two Squares

$$a^2 - b^2 = (a + b)(a - b)$$

Notice that a binomial is a difference of two squares when it is the difference of the square of some quantity a and the square of some quantity b.

PRACTICE PROBLEMS 5–8

Factor:

5. $x^2 - 49$

6. $4y^2 - 81$

7. $12 - 3a^2$

8. $y^2 - \dfrac{1}{25}$

EXAMPLES Factor.

5. $\begin{aligned} x^2 - 9 &= x^2 - 3^2 \\ &= (x + 3)(x - 3) \end{aligned}$

6. $\begin{aligned} 16y^2 - 9 &= (4y)^2 - 3^2 \\ &= (4y + 3)(4y - 3) \end{aligned}$

7. $\begin{aligned} 50 - 8y^2 &= 2(25 - 4y^2) \qquad \text{Factor out the common factor of 2.} \\ &= 2[5^2 - (2y)^2] \\ &= 2(5 + 2y)(5 - 2y) \end{aligned}$

8. $\begin{aligned} x^2 - \dfrac{1}{4} &= x^2 - \left(\dfrac{1}{2}\right)^2 \\ &= \left(x + \dfrac{1}{2}\right)\left(x - \dfrac{1}{2}\right) \end{aligned}$

▣ **Work Practice Problems 5–8**

Helpful Hint
The sum of two squares whose greatest common factor is 1 usually cannot be factored by using real numbers.

The binomial $x^2 + 9$ is a **sum of two squares** and cannot be factored by using real numbers. *In general, except for factoring out a greatest common factor, the sum of two squares usually cannot be factored by using real numbers.*

PRACTICE PROBLEM 9

Factor: $a^4 - 81$

EXAMPLE 9 Factor: $p^4 - 16$

Solution:

$$\begin{aligned} p^4 - 16 &= (p^2)^2 - 4^2 \\ &= (p^2 + 4)(p^2 - 4) \end{aligned}$$

The binomial factor $p^2 + 4$ cannot be factored by using real numbers, but the binomial factor $p^2 - 4$ is a difference of squares.

$$(p^2 + 4)(p^2 - 4) = (p^2 + 4)(p + 2)(p - 2)$$

▣ **Work Practice Problem 9**

Answers

5. $(x + 7)(x - 7)$,

6. $(2y + 9)(2y - 9)$,

7. $3(2 + a)(2 - a)$,

8. $\left(y + \dfrac{1}{5}\right)\left(y - \dfrac{1}{5}\right)$,

9. $(a^2 + 9)(a + 3)(a - 3)$

✔**Concept Check** Is $(x - 4)(y^2 - 9)$ completely factored? Why or why not?

EXAMPLE 10 Factor: $(x + 3)^2 - 36$

Solution:

$$
\begin{aligned}
(x + 3)^2 - 36 &= (x + 3)^2 - 6^2 \\
&= [(x + 3) + 6][(x + 3) - 6] \quad \text{Factor as the difference of two squares.} \\
&= [x + 3 + 6][x + 3 - 6] \quad \text{Remove parentheses.} \\
&= (x + 9)(x - 3) \quad \text{Simplify.}
\end{aligned}
$$

🔲 **Work Practice Problem 10**

PRACTICE PROBLEM 10

Factor: $(x + 1)^2 - 9$

EXAMPLE 11 Factor: $x^2 + 4x + 4 - y^2$

Solution: Notice that the first three terms form a perfect square trinomial. Let's try factoring by grouping. To do so, let's group and factor the first three terms.

$$
\begin{aligned}
x^2 + 4x + 4 - y^2 &= (x^2 + 4x + 4) - y^2 \quad \text{Group the first three terms.} \\
&= (x + 2)^2 - y^2 \quad \text{Factor the perfect square trinomial.}
\end{aligned}
$$

This is not completely factored yet since we have a *difference*, not a *product*. Since $(x + 2)^2 - y^2$ is a difference of squares, we have

$$
\begin{aligned}
(x + 2)^2 - y^2 &= [(x + 2) + y][(x + 2) - y] \\
&= (x + 2 + y)(x + 2 - y)
\end{aligned}
$$

🔲 **Work Practice Problem 11**

PRACTICE PROBLEM 11

Factor: $a^2 + 2a + 1 - b^2$

Objective 🅲 **Factoring the Sum or Difference of Two Cubes**

Although the sum of two squares usually cannot be factored, the sum of two cubes, as well as the difference of two cubes, can be factored as follows.

Sum and Difference of Two Cubes

$$a^3 + b^3 = (a + b)(a^2 - ab + b^2)$$
$$a^3 - b^3 = (a - b)(a^2 + ab + b^2)$$

To check the first pattern, let's find the product of $(a + b)$ and $(a^2 - ab + b^2)$.

$$
\begin{aligned}
(a + b)(a^2 - ab + b^2) &= a(a^2 - ab + b^2) + b(a^2 - ab + b^2) \\
&= a^3 - a^2b + ab^2 + a^2b - ab^2 + b^3 \\
&= a^3 + b^3
\end{aligned}
$$

EXAMPLE 12 Factor: $x^3 + 8$

Solution: First we write the binomial in the form $a^3 + b^3$. Then we use the formula

$$a^3 + b^3 = (a + b)(a^2 - a \cdot b + b^2), \text{ where } a \text{ is } x \text{ and } b \text{ is } 2.$$

$$x^3 + 8 = x^3 + 2^3 = (x + 2)(x^2 - x \cdot 2 + 2^2)$$

Thus, $x^3 + 8 = (x + 2)(x^2 - 2x + 4)$

🔲 **Work Practice Problem 12**

PRACTICE PROBLEM 12

Factor: $z^3 + 27$

Answers
10. $(x - 2)(x + 4)$,
11. $(a + 1 + b)(a + 1 - b)$,
12. $(z + 3)(z^2 - 3z + 9)$

✔ **Concept Check Answer**
no; $(y^2 - 9)$ can be factored

PRACTICE PROBLEM 13

Factor: $x^3 + 64y^3$

EXAMPLE 13 Factor: $p^3 + 27q^3$

Solution:

$$p^3 + 27q^3 = p^3 + (3q)^3$$
$$= (p + 3q)[p^2 - (p)(3q) + (3q)^2]$$
$$= (p + 3q)(p^2 - 3pq + 9q^2)$$

🔲 **Work Practice Problem 13**

PRACTICE PROBLEM 14

Factor: $y^3 - 8$

EXAMPLE 14 Factor: $y^3 - 64$

Solution: This is a difference of cubes since $y^3 - 64 = y^3 - 4^3$.

From $a^3 - b^3 = (a - b)(a^2 + a \cdot b + b^2)$ we have that

$$y^3 - 4^3 = (y - 4)(y^2 + y \cdot 4 + 4^2)$$
$$= (y - 4)(y^2 + 4y + 16)$$

🔲 **Work Practice Problem 14**

Helpful Hint

When factoring sums or differences of cubes, be sure to notice the sign patterns.

$$\overset{\text{Same sign}}{x^3 + y^3} = (x + y)(x^2 - xy + y^2)$$

Opposite sign   Always positive

$$\overset{\text{Same sign}}{x^3 - y^3} = (x - y)(x^2 + xy + y^2)$$

Opposite sign

PRACTICE PROBLEM 15

Factor: $27a^2 - b^3a^2$

EXAMPLE 15 Factor: $125q^2 - n^3q^2$

Solution: First we factor out a common factor of q^2.

$$125q^2 - n^3q^2 = q^2(125 - n^3)$$
$$= q^2(5^3 - n^3)$$

Opposite sign   Positive

$$= q^2(5 - n)[5^2 + (5)(n) + (n^2)]$$
$$= q^2(5 - n)(25 + 5n + n^2)$$

Thus, $125q^2 - n^3q^2 = q^2(5 - n)(25 + 5n + n^2)$. The trinomial $25 + 5n + n^2$ cannot be factored further.

🔲 **Work Practice Problem 15**

Answers

13. $(x + 4y)(x^2 - 4xy + 16y^2)$,
14. $(y - 2)(y^2 + 2y + 4)$,
15. $a^2(3 - b)(9 + 3b + b^2)$

5.6 EXERCISE SET

Objective **A** *Factor. See Examples 1 through 4.*

1. $x^2 + 6x + 9$

2. $x^2 - 10x + 25$

3. $4x^2 - 12x + 9$

4. $25x^2 + 10x + 1$

5. $4a^2 + 12a + 9$

6. $9a^2 - 30a + 25$

7. $3x^2 - 24x + 48$

8. $2x^2 + 28x + 98$

9. $9y^2x^2 + 12yx^2 + 4x^2$

10. $4x^2y^3 - 4xy^3 + y^3$

11. $16x^2 - 56xy + 49y^2$

12. $81x^2 + 36xy + 4y^2$

Objective **B** *Factor. See Examples 5 through 11.*

13. $x^2 - 25$

14. $y^2 - 100$

15. $\dfrac{1}{9} - 4z^2$

16. $\dfrac{1}{16} - y^2$

17. $(y + 2)^2 - 49$

18. $(x - 1)^2 - z^2$

19. $64x^2 - 100$

20. $4x^2 - 36$

21. $(x + 2y)^2 - 9$

22. $(3x + y)^2 - 25$

23. $x^2 + 16x + 64 - x^4$

24. $x^2 + 20x + 100 - x^4$

Objective **C** *Factor. See Examples 12 through 15.*

25. $x^3 + 27$

26. $y^3 + 1$

27. $z^3 - 1$

28. $8 - x^3$

29. $m^3 + n^3$

30. $p^3 + 125q^3$

31. $x^3y^2 - 27y^2$

32. $a^3b + 8b^4$

33. $64q^2 - q^2p^3$

34. $8ab^3 + 27a^4$

35. $250y^3 - 16x^3$

36. $54y^3 - 128$

Objectives **A** **B** **C** **Mixed Practice** *Factor completely. See Examples 1 through 15.*

37. $x^2 - 12x + 36$

38. $x^2 - 18x + 81$

39. $18x^2y - 2y$

40. $12xy^2 - 108x$

41. $9x^2 - 49$

42. $25x^2 - 4$

43. $x^4 - 1$

44. $x^4 - 256$

45. $x^6 - y^3$

46. $x^3 - y^6$

47. $8x^3 + 27y^3$

48. $125x^3 + 8y^3$

49. $4x^2 + 4x + 1 - z^2$

50. $9y^2 + 12y + 4 - x^2$

51. $3x^6y^2 + 81y^2$

52. $x^2y^9 + x^2y^3$

53. $n^3 - \dfrac{1}{27}$

54. $p^3 + \dfrac{1}{125}$

55. $-16y^2 + 64$

56. $-12y^2 + 108$

57. $x^2 - 10x + 25 - y^2$

58. $x^2 - 18x + 81 - y^2$

59. $a^3b^3 + 125$

60. $x^3y^3 + 216$

61. $\dfrac{x^2}{25} - \dfrac{y^2}{9}$

62. $\dfrac{a^2}{4} - \dfrac{b^2}{49}$

63. $(x + y)^3 + 125$

64. $(r + s)^3 + 27$

Review

Solve each equation. See Section 2.1.

65. $x - 5 = 0$

66. $x + 7 = 0$

67. $3x + 1 = 0$

68. $5x - 15 = 0$

69. $-2x = 0$

70. $3x = 0$

71. $-5x + 25 = 0$

72. $-4x - 16 = 0$

Concept Extensions

Determine whether each polynomial is factored completely or not. See the Concept Check in this section.

73. $5x(x^2 - 4)$

74. $x^2y^2(x^3 - y^3)$

75. $7y(a^2 + a + 1)$

76. $9z(x^2 + 4)$

77. A manufacturer of metal washers needs to determine the cross-sectional area of each washer. If the outer radius of the washer is R and the radius of the hole is r, express the area of the washer as a polynomial. Factor this polynomial completely.

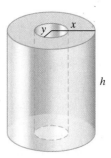

78. Express the area of the shaded region as a polynomial. Factor the polynomial completely.

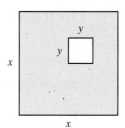

Express the volume of each solid as a polynomial. To do so, subtract the volume of the "hole" from the volume of the larger solid. Then factor the resulting polynomial.

79.

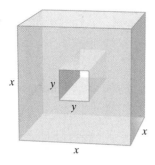

80.

Find the value of c that makes each trinomial a perfect square trinomial.

81. $x^2 + 6x + c$

82. $y^2 + 10y + c$

83. $m^2 - 14m + c$

84. $n^2 - 2n + c$

85. Factor $x^6 - 1$ completely, using the following methods from this chapter.
   a. Factor the expression by treating it as the difference of two squares, $(x^3)^2 - 1^2$.
   b. Factor the expression treating it as the difference of two cubes, $(x^2)^3 - 1^3$.
   c. Are the answers to parts **(a)** and **(b)** the same? Why or why not?

Factor. Assume that variables used as exponents represent positive integers.

86. $x^{2n} - 25$

87. $x^{2n} - 36$

88. $36x^{2n} - 49$

89. $25x^{2n} - 81$

90. $x^{4n} - 16$

91. $x^{4n} - 625$

Operations on Polynomials and Factoring Strategies

Operations on Polynomials

Perform each indicated operation.

1. $(-y^2 + 6y - 1) + (3y^2 - 4y - 10)$

2. $(5z^4 - 6z^2 + z + 1) - (7z^4 - 2z + 1)$

3. Subtract $(x - 5)$ from $(x^2 - 6x + 2)$

4. $(2x^2 + 6x - 5) + (5x^2 - 10x)$

5. $(5x - 3)^2$

6. $(5x^2 - 14x - 3) \div (5x + 1)$

7. $(2x^4 - 3x^2 + 5x - 2) \div (x + 2)$

8. $(4x - 1)(x^2 - 3x - 2)$

Factoring Strategies

The key to proficiency in factoring polynomials is to practice until you are comfortable with each technique. A strategy for factoring polynomials completely is given next.

Factoring a Polynomial

Step 1: Are there any common factors? If so, factor out the greatest common factor.

Step 2: How many terms are in the polynomial?

    a. If there are *two* terms, decide if one of the following formulas may be applied:

      i. Difference of two squares: $a^2 - b^2 = (a - b)(a + b)$

      ii. Difference of two cubes: $a^3 - b^3 = (a - b)(a^2 + ab + b^2)$

      iii. Sum of two cubes: $a^3 + b^3 = (a + b)(a^2 - ab + b^2)$

    b. If there are *three* terms, try one of the following:

      i. Perfect square trinomial: $a^2 + 2ab + b^2 = (a + b)^2$
$$a^2 - 2ab + b^2 = (a - b)^2$$

      ii. If not a perfect square trinomial, factor by using the methods presented in Section 5.5.

    c. If there are *four* or more terms, try factoring by grouping.

Step 3: See whether any factors in the factored polynomial can be factored further.

Factor completely.

9. $x^2 - 8x + 16 - y^2$

10. $12x^2 - 22x - 20$

11. $x^4 - x$

12. $(2x + 1)^2 - 3(2x + 1) + 2$

13. $14x^2y - 2xy$

14. $24ab^2 - 6ab$

15. $4x^2 - 16$

16. $9x^2 - 81$

17. $3x^2 - 8x - 11$

18. $5x^2 - 2x - 3$

19. $4x^2 + 8x - 12$

20. $6x^2 - 6x - 12$

21. $4x^2 + 36x + 81$

22. $25x^2 + 40x + 16$

23. $8x^3 + 125y^3$

24. $27x^3 - 64y^3$ **25.** $64x^2y^3 - 8x^2$ **26.** $27x^5y^4 - 216x^2y$

27. $(x + 5)^3 + y^3$ **28.** $(y - 1)^3 + 27x^3$

29. $(5a - 3)^2 - 6(5a - 3) + 9$ **30.** $(4r + 1)^2 + 8(4r + 1) + 16$

31. $7x^2 - 63x$ **32.** $20x^2 + 23x + 6$ **33.** $ab - 6a + 7b - 42$

34. $20x^2 - 220x + 600$ **35.** $x^4 - 1$ **36.** $15x^2 - 20x$

37. $10x^2 - 7x - 33$ **38.** $45m^3n^3 - 27m^2n^2$ **39.** $5a^3b^3 - 50a^3b$

40. $x^4 + x$ **41.** $16x^2 + 25$ **42.** $20x^3 + 20y^3$

43. $10x^3 - 210x^2 + 1100x$ **44.** $9y^2 - 42y + 49$ **45.** $64a^3b^4 - 27a^3b$

46. $y^4 - 16$ **47.** $2x^3 - 54$ **48.** $2sr + 10s - r - 5$

49. $3y^5 - 5y^4 + 6y - 10$ **50.** $64a^2 + b^2$ **51.** $100z^3 + 100$

52. $250x^4 - 16x$ **53.** $4b^2 - 36b + 81$ **54.** $2a^5 - a^4 + 6a - 3$

55. $(y - 6)^2 + 3(y - 6) + 2$ **56.** $(c + 2)^2 - 6(c + 2) + 5$

△ **57.** Express the area of the shaded region as a polynomial. Factor the polynomial completely.

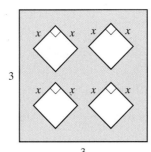

24.	_____
25.	_____
26.	_____
27.	_____
28.	_____
29.	_____
30.	_____
31.	_____
32.	_____
33.	_____
34.	_____
35.	_____
36.	_____
37.	_____
38.	_____
39.	_____
40.	_____
41.	_____
42.	_____
43.	_____
44.	_____
45.	_____
46.	_____
47.	_____
48.	_____
49.	_____
50.	_____
51.	_____
52.	_____
53.	_____
54.	_____
55.	_____
56.	_____
57.	_____

A Solve Polynomial Equations by Factoring.

B Solve Problems That Can Be Modeled by Polynomial Equations.

5.7 SOLVING EQUATIONS BY FACTORING AND SOLVING PROBLEMS

Objective **A** Solving Polynomial Equations by Factoring

In this section, your efforts to learn factoring will start to pay off. We use factoring to solve polynomial equations.

A **polynomial equation** is the result of setting two polynomials equal to each other. Examples are shown below.

$$3x^3 - 2x^2 = x^2 + 2x \qquad 2.6x + 7 = -1.3$$
$$-5x^2 - 5 = -9x^2 - 2x + 1$$

A polynomial equation is in **standard form** if one side of the equation is 0, as in the following examples.

$$3x^3 - 3x^2 - 2x + 1 = 0 \qquad 2.6x + 8.3 = 0$$
$$4x^2 + 2x - 6 = 0$$

The degree of a simplified polynomial equation in standard form is the same as the highest degree of any of its terms. A polynomial equation of degree 2 is also called a **quadratic equation.**

A solution of a polynomial equation in one variable is a value of the variable that makes the equation true. The method presented in this section for solving polynomial equations is called the **factoring method.** This method is based on the **zero-factor property.**

Zero-Factor Property

If a and b are real numbers and $a \cdot b = 0$, then $a = 0$ or $b = 0$. This property is true for three or more factors also.

In other words, if the product of two or more real numbers is zero, then at least one of the numbers must be zero.

PRACTICE PROBLEM 1

Solve: $(x - 3)(x + 5) = 0$

EXAMPLE 1 Solve: $(x + 2)(x - 6) = 0$

Solution: By the zero-factor property, $(x + 2)(x - 6) = 0$ only if

$$x + 2 = 0 \quad \text{or} \quad x - 6 = 0.$$

$x + 2 = 0 \quad \text{or} \quad x - 6 = 0$ Use the zero-factor property.

$\qquad x = -2 \qquad\qquad x = 6$ Solve each linear equation.

To check, let $x = -2$ and then let $x = 6$ in the original equation.

Let $x = -2$.	Let $x = 6$.
$(x + 2)(x - 6) = 0$	$(x + 2)(x - 6) = 0$
$(-2 + 2)(-2 - 6) = 0$	$(6 + 2)(6 - 6) = 0$
$(0)(-8) = 0$	$(8)(0) = 0$
$0 = 0$ True	$0 = 0$ True

Both -2 and 6 check, so the solution set is $\{-2, 6\}$.

☐ **Work Practice Problem 1**

Answer

1. $\{-5, 3\}$

EXAMPLE 2 Solve: $2x^2 + 9x - 5 = 0$

Solution: To use the zero-factor property, one side of the equation must be 0, and the other side must be in factored form.

$$2x^2 + 9x - 5 = 0$$
$$(2x - 1)(x + 5) = 0 \qquad\qquad \text{Factor.}$$
$$2x - 1 = 0 \quad \text{or} \quad x + 5 = 0 \qquad \text{Set each factor equal to 0.}$$
$$2x = 1 \qquad\qquad x = -5 \qquad \text{Solve each linear equation.}$$
$$x = \frac{1}{2}$$

To check, let $x = \dfrac{1}{2}$ in the original equation; then let $x = -5$ in the original equation. The solution set is $\left\{-5, \dfrac{1}{2}\right\}$.

🔲 **Work Practice Problem 2**

PRACTICE PROBLEM 2

Solve: $3x^2 + 5x - 2 = 0$

Solving a Polynomial Equation by Factoring

Step 1: Write the equation in standard form so that one side of the equation is 0.

Step 2: Factor the polynomial completely.

Step 3: Set each factor containing a variable equal to 0.

Step 4: Solve the resulting equations.

Step 5: Check each solution in the original equation.

Since it is not always possible to factor a polynomial, not all polynomial equations can be solved by factoring. Other methods of solving polynomial equations are presented in Chapter 8.

EXAMPLE 3 Solve: $x(2x - 7) = 4$

Solution: We first write the equation in standard form; then we factor.

$$x(2x - 7) = 4$$
$$2x^2 - 7x = 4 \qquad\qquad \text{Multiply. Write in standard form.}$$
$$2x^2 - 7x - 4 = 0$$
$$(2x + 1)(x - 4) = 0 \qquad\qquad \text{Factor.}$$
$$2x + 1 = 0 \quad \text{or} \quad x - 4 = 0 \qquad \text{Set each factor equal to 0.}$$
$$2x = -1 \qquad\qquad x = 4 \qquad \text{Solve.}$$
$$x = -\frac{1}{2}$$

Check both solutions in the original equation. The solution set is $\left\{-\dfrac{1}{2}, 4\right\}$.

🔲 **Work Practice Problem 3**

PRACTICE PROBLEM 3

Solve: $x(5x - 7) = -2$

Helpful Hint

To apply the zero-factor property, one side of the equation must be 0, and the other side of the equation must be factored. To solve the equation $x(2x - 7) = 4$, for example, you may *not* set each factor equal to 4.

Answers

2. $\left\{-2, \dfrac{1}{3}\right\}$, **3.** $\left\{\dfrac{2}{5}, 1\right\}$

PRACTICE PROBLEM 4

Solve:

$$2(x^2 + 5) + 10 = -2(x^2 + 10x) - 5$$

EXAMPLE 4 Solve: $3(x^2 + 4) + 5 = -6(x^2 + 2x) + 13$

Solution: We rewrite the equation so that one side is 0.

$$3(x^2 + 4) + 5 = -6(x^2 + 2x) + 13$$
$$3x^2 + 12 + 5 = -6x^2 - 12x + 13 \quad \text{Use the distributive property.}$$
$$9x^2 + 12x + 4 = 0 \quad \text{Rewrite the equation in standard form so that one side is 0.}$$
$$(3x + 2)(3x + 2) = 0 \quad \text{Factor.}$$
$$3x + 2 = 0 \quad \text{or} \quad 3x + 2 = 0 \quad \text{Set each factor equal to 0.}$$
$$3x = -2 \qquad 3x = -2 \quad \text{Solve each equation.}$$
$$x = -\frac{2}{3} \qquad x = -\frac{2}{3}$$

Check by substituting $-\frac{2}{3}$ into the original equation. The solution set is $\left\{-\frac{2}{3}\right\}$.

■ **Work Practice Problem 4**

If the equation contains fractions, we clear the equation of fractions as a first step.

PRACTICE PROBLEM 5

Solve: $2x^2 + \frac{5}{2}x = 3$

EXAMPLE 5 Solve: $2x^2 = \frac{17}{3}x + 1$

Solution:

$$2x^2 = \frac{17}{3}x + 1$$
$$3(2x^2) = 3\left(\frac{17}{3}x + 1\right) \quad \text{Clear the equation of fractions.}$$
$$6x^2 = 17x + 3 \quad \text{Use the distributive property.}$$
$$6x^2 - 17x - 3 = 0 \quad \text{Rewrite the equation in standard form so that one side is 0.}$$
$$(6x + 1)(x - 3) = 0 \quad \text{Factor.}$$
$$6x + 1 = 0 \quad \text{or} \quad x - 3 = 0 \quad \text{Set each factor equal to 0.}$$
$$6x = -1 \qquad x = 3 \quad \text{Solve each equation.}$$
$$x = -\frac{1}{6}$$

Check by substituting into the original equation. The solution set is $\left\{-\frac{1}{6}, 3\right\}$.

■ **Work Practice Problem 5**

PRACTICE PROBLEM 6

Solve: $x^3 = x^2 + 6x$

EXAMPLE 6 Solve: $x^3 = 4x$

Solution:

$$x^3 = 4x$$
$$x^3 - 4x = 0 \quad \text{Rewrite the equation in standard form so that one side is 0.}$$
$$x(x^2 - 4) = 0 \quad \text{Factor out the greatest common factor.}$$
$$x(x + 2)(x - 2) = 0 \quad \text{Factor the difference of squares.}$$
$$x = 0 \quad \text{or} \quad x + 2 = 0 \quad \text{or} \quad x - 2 = 0 \quad \text{Set each factor equal to 0.}$$
$$x = -2 \qquad x = 2 \quad \text{Solve each equation.}$$

Check by substituting into the original equation. The solution set is $\{-2, 0, 2\}$.

■ **Work Practice Problem 6**

Notice that the *third*-degree equation of Example 6 yielded *three* solutions.

Answers

4. $\left\{-\frac{5}{2}\right\}$, 5. $\left\{-2, \frac{3}{4}\right\}$, 6. $\{-2, 0, 3\}$

EXAMPLE 7 Solve: $x^3 - x = -5x^2 + 5$.

Solution: First, write the equation so that one side is 0.

$$x^3 - x + 5x^2 - 5 = 0$$
$$(x^3 - x) + (5x^2 - 5) = 0 \qquad \text{Factor by grouping.}$$
$$x(x^2 - 1) + 5(x^2 - 1) = 0$$
$$(x^2 - 1)(x + 5) = 0$$
$$(x + 1)(x - 1)(x + 5) = 0 \qquad \text{Factor the difference of squares.}$$
$$x + 1 = 0 \quad \text{or} \quad x - 1 = 0 \quad \text{or} \quad x + 5 = 0 \qquad \text{Set each factor equal to 0.}$$
$$x = -1 \quad \text{or} \qquad x = 1 \quad \text{or} \qquad x = -5 \qquad \text{Solve each equation.}$$

The solution set is $\{-5, -1, 1\}$. Check in the original equation.

■ **Work Practice Problem 7**

✔**Concept Check** Which solution strategies are incorrect? Why?
a. Solve $(y - 2)(y + 2) = 4$ by setting each factor equal to 4.
b. Solve $(x + 1)(x + 3) = 0$ by setting each factor equal to 0.
c. Solve $z^2 + 5z + 6 = 0$ by factoring $z^2 + 5z + 6$ and setting each factor equal to 0.
d. Solve $x^2 + 6x + 8 = 10$ by factoring $x^2 + 6x + 8$ and setting each factor equal to 0.

Objective B Solving Problems Modeled by Polynomial Equations

Some problems may be modeled by polynomial equations. To solve these problems, we use the same problem-solving steps that were introduced in Section 2.2. When solving these problems, keep in mind that a solution of an equation that models a problem is not always a solution to the problem. For example, a person's weight or the length of a side of a geometric figure is always a positive number. Discard solutions that do not make sense as solutions of the problem.

EXAMPLE 8 Finding the Return Time of a Rocket

An Alpha III model rocket is launched from the ground with an A8-3 engine. Without a parachute the height h in feet of the rocket at time t seconds is approximated by the equation

$$h = -16t^2 + 144t$$

Find how long it takes the rocket to return to the ground.

Solution:

1. UNDERSTAND. Read and reread the problem. The equation $h = -16t^2 + 144t$ models the height of the rocket. Familiarize yourself with this equation by finding a few values.

   When $t = 1$ second, the height of the rocket is

   $$h = -16(1)^2 + 144(1) = 128 \text{ feet}$$

   When $t = 2$ seconds, the height of the rocket is

   $$h = -16(2)^2 + 144(2) = 224 \text{ feet}$$

2. TRANSLATE. To find how long it takes the rocket to return to the ground, we want to know what value of t makes the height h equal to 0. That is, we want to solve for t when $h = 0$.

   $$-16t^2 + 144t = 0$$

Continued on next page

PRACTICE PROBLEM 7
Solve: $x^3 + 7x^2 = 4x + 28$

PRACTICE PROBLEM 8
A model rocket is launched from the ground. Its height h in feet at time t seconds is approximated by the equation

$$h = -16t^2 + 112t$$

Find how long it takes the rocket to return to the ground.

Answers
7. $\{-2, 2, -7\}$, **8.** 7 sec

✔ **Concept Check Answer**
a and d; the zero-factor property works only if one side of the equation is 0

3. SOLVE the quadratic equation by factoring.

$$-16t^2 + 144t = 0$$
$$-16t(t - 9) = 0$$
$$-16t = 0 \quad \text{or} \quad t - 9 = 0$$
$$t = 0 \qquad\qquad t = 9$$

4. INTERPRET. The height h is 0 feet at time 0 seconds (when the rocket is launched) and at time 9 seconds.

Check: See that the height of the rocket at 9 seconds equals 0.

$$h = -16(9)^2 + 144(9) = -1296 + 1296 = 0$$

State: The rocket returns to the ground 9 seconds after it is launched.

▣ **Work Practice Problem 8**

Some of the exercises at the end of this section make use of the **Pythagorean theorem.** Before we review this theorem, recall that a **right triangle** is a triangle that contains a 90° angle, or right angle. The **hypotenuse** of a right triangle is the side opposite the right angle and is the longest side of the triangle. The **legs** of a right triangle are the other sides of the triangle.

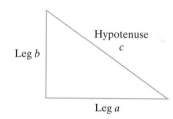

Pythagorean Theorem

In a right triangle, the sum of the squares of the lengths of the two legs is equal to the square of the length of the hypotenuse.

$$(\text{leg})^2 + (\text{leg})^2 = (\text{hypotenuse})^2 \quad \text{or} \quad a^2 + b^2 = c^2$$

PRACTICE PROBLEM 9 △

Find a right triangle whose sides are three consecutive even integers.

△ **EXAMPLE 9** **Using the Pythagorean Theorem**

While framing an addition to an existing home, Kim Menzies, a carpenter, used the Pythagorean theorem to determine whether a wall was "square"—that is, whether the wall formed a right angle with the floor. He used a triangle whose sides are three consecutive integers. Find a right triangle whose sides are three consecutive integers.

Solution:

1. UNDERSTAND. Read and reread the problem. Let x, $x + 1$, and $x + 2$ be three consecutive integers. Since these integers represent lengths of the sides of a right triangle, we have

$$x = \text{one leg},$$
$$x + 1 = \text{other leg, and}$$
$$x + 2 = \text{hypotenuse (longest side)}$$

Answer

9. 6, 8, and 10 units

2. TRANSLATE. By the Pythagorean theorem, we have

In words: $(\text{leg})^2 + (\text{leg})^2 = (\text{hypotenuse})^2$

$\qquad\qquad\quad \downarrow \qquad\quad \downarrow \qquad\qquad \downarrow$

Translate: $(x)^2 + (x + 1)^2 = (x + 2)^2$

3. SOLVE the equation.

$$x^2 + (x + 1)^2 = (x + 2)^2$$
$$x^2 + x^2 + 2x + 1 = x^2 + 4x + 4 \qquad \text{Multiply.}$$
$$2x^2 + 2x + 1 = x^2 + 4x + 4$$
$$x^2 - 2x - 3 = 0 \qquad \text{Write in standard form.}$$
$$(x - 3)(x + 1) = 0$$
$$x - 3 = 0 \quad \text{or} \quad x + 1 = 0$$
$$x = 3 \qquad\qquad x = -1$$

4. INTERPRET. Discard $x = -1$ since length cannot be negative. If $x = 3$, then $x + 1 = 4$ and $x + 2 = 5$.

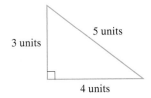

Check: To check, see that $(\text{leg})^2 + (\text{leg})^2 = (\text{hypotenuse})^2$.

$$3^2 + 4^2 = 5^2$$
$$9 + 16 = 25 \qquad \text{True}$$

State: The lengths of the sides of the right triangle are 3, 4, and 5 units. Kim used this information by marking off lengths of 3 and 4 feet on the floor and framing, respectively. If the diagonal length between these marks was 5 feet, the wall was square. If not, adjustments were made.

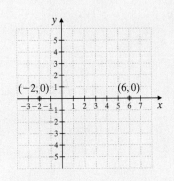

Work Practice Problem 9

Recall that to find the x-intercepts of the graph of a function, let $f(x) = 0$, or $y = 0$, and solve for x. This fact gives us a visual interpretation of the results of this section.

From Example 1, we know that the solutions of the equation $(x + 2)(x - 6) = 0$ are -2 and 6. These solutions give us important information about the related polynomial function $p(x) = (x + 2)(x - 6)$. We know that when x is -2 or when x is 6, the value of $p(x)$ is 0.

$$p(x) = (x + 2)(x - 6)$$
$$p(-2) = (-2 + 2)(-2 - 6) = (0)(-8) = 0$$
$$p(6) = (6 + 2)(6 - 6) = (8)(0) = 0$$

Thus, we know that $(-2, 0)$ and $(6, 0)$ are the x-intercepts of the graph of $p(x)$.

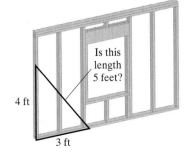

We also know that the graph of $p(x)$ does not cross the x-axis at any other point. For this reason, and the fact that $p(x) = (x + 2)(x - 6) = x^2 - 4x - 12$ has degree 2, we conclude that the graph of p must look something like one of these two graphs:

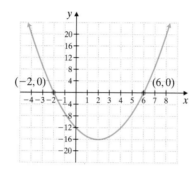 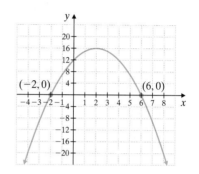

In a later chapter, we explore these graphs more fully. For the moment, know that the solutions of a polynomial equation are the x-intercepts of the graph of the related function and that the x-intercepts of the graph of a polynomial function are the solutions of the related polynomial equation. These values are also called **roots,** or **zeros,** of a polynomial function.

PRACTICE PROBLEM 10

Match each function with its graph.

$f(x) = (x + 1)(x - 3)$
$g(x) = (x - 5)(x - 1)(x + 1)$
$h(x) = x(x - 1)(x + 1)$

a.

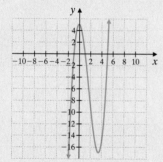

b.

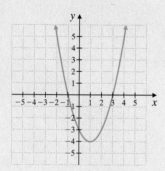

c.

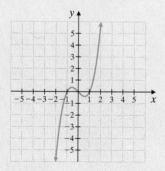

Answers

10. a. $g(x) = (x - 5)(x - 1)(x + 1)$,
   b. $f(x) = (x + 1)(x - 3)$,
   c. $h(x) = x(x - 1)(x + 1)$

EXAMPLE 10 Match Each Function with Its Graph

$f(x) = (x - 3)(x + 2)$ $g(x) = x(x + 2)(x - 2)$
$h(x) = (x - 2)(x + 2)(x - 1)$

A

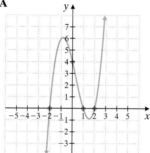

B

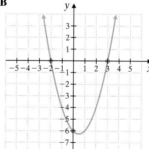

C

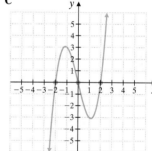

Solution: The graph of the function $f(x) = (x - 3)(x + 2)$ has two x-intercepts, $(3, 0)$ and $(-2, 0)$, because the equation $0 = (x - 3)(x + 2)$ has two solutions, 3 and -2.

The graph of $f(x)$ is graph B.

The graph of the function $g(x) = x(x + 2)(x - 2)$ has three x-intercepts, $(0, 0)$, $(-2, 0)$, and $(2, 0)$, because the equation $0 = x(x + 2)(x - 2)$ has three solutions, $0, -2$, and 2.

The graph of $g(x)$ is graph C.

The graph of the function $h(x) = (x - 2)(x + 2)(x - 1)$ has three x-intercepts, $(-2, 0)$, $(1, 0)$, and $(2, 0)$, because the equation $0 = (x - 2)(x + 2)(x - 1)$ has three solutions, $-2, 1$, and 2.

The graph of $h(x)$ is graph A.

Work Practice Problem 10

CALCULATOR EXPLORATIONS Graphing

We can use a graphing calculator to approximate real number solutions of any quadratic equation in standard form, whether the associated polynomial is factorable or not. For example, let's solve the quadratic equation $x^2 - 2x - 4 = 0$. The solutions of this equation will be the x-intercepts of the graph of the function $f(x) = x^2 - 2x - 4$. (Recall that to find x-intercepts, we let $f(x) = 0$, or $y = 0$.) When we use a standard window, the graph of this function looks like this.

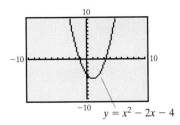

$y = x^2 - 2x - 4$

The graph appears to have one x-intercept between -2 and -1 and one between 3 and 4. To find the x-intercept between 3 and 4 to the nearest hundredth, we can use a zero feature, a Zoom feature, which magnifies a portion of the graph around the cursor, or we can redefine our window. If we redefine our window to

Xmin = 2   Ymin = -1
Xmax = 5   Ymax = 1
Xscl = 1   Yscl = 1

the resulting screen is

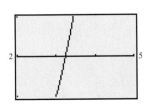

By using the Trace feature, we can now see that one of the intercepts is between 3.21 and 3.25. To approximate to the nearest hundredth, Zoom again or redefine the window to

Xmin = 3.2   Ymin = -0.1
Xmax = 3.3   Ymax = 0.1
Xscl = 1   Yscl = 1

If we use the Trace feature again, we see that, to the nearest hundredth, the x-intercept is 3.24. By repeating this process, we can approximate the other x-intercept to be -1.24.

To check, find $f(3.24)$ and $f(-1.24)$. Both of these values should be close to 0. (They will not be exactly 0 since we approximated these solutions.)

$$f(3.24) = 0.0176 \quad \text{and} \quad f(-1.24) = 0.0176$$

Solve each of these quadratic equations by graphing a related function and approximating the x-intercepts to the nearest thousandth.

1. $x^2 + 3x - 2 = 0$
2. $5x^2 - 7x + 1 = 0$
3. $2.3x^2 - 4.4x - 5.6 = 0$
4. $0.2x^2 + 6.2x + 2.1 = 0$
5. $0.09x^2 - 0.13x - 0.08 = 0$
6. $x^2 + 0.08x - 0.01 = 0$

Mental Math

Solve each equation for the variable.

1. $(x - 3)(x + 5) = 0$

2. $(y + 5)(y + 3) = 0$

3. $(z - 3)(z + 7) = 0$

4. $(c - 2)(c - 4) = 0$

5. $x(x - 9) = 0$

6. $w(w + 7) = 0$

5.7 EXERCISE SET

FOR EXTRA HELP

Student Solutions Manual   PH Math/Tutor Center   CD/Video for Review   MathXL®   MyMathLab

Objective Ⓐ *Solve each equation. See Example 1*

1. $(x + 3)(3x - 4) = 0$

2. $(5x + 1)(x - 2) = 0$

3. $3(2x - 5)(4x + 3) = 0$

4. $8(3x - 4)(2x - 7) = 0$

Solve each equation. See Examples 2 through 5.

5. $x^2 + 11x + 24 = 0$

6. $y^2 - 10y + 24 = 0$

7. $12x^2 + 5x - 2 = 0$

8. $3y^2 - y - 14 = 0$

9. $z^2 + 9 = 10z$

10. $n^2 + n = 72$

11. $x(5x + 2) = 3$

12. $n(2n - 3) = 2$

13. $x^2 - 6x = x(8 + x)$

14. $n(3 + n) = n^2 + 4n$

15. $\dfrac{z^2}{6} - \dfrac{z}{2} - 3 = 0$

16. $\dfrac{c^2}{20} - \dfrac{c}{4} + \dfrac{1}{5} = 0$

17. $\dfrac{x^2}{2} + \dfrac{x}{20} = \dfrac{1}{10}$

18. $\dfrac{y^2}{30} = \dfrac{y}{15} + \dfrac{1}{2}$

19. $\dfrac{4t^2}{5} = \dfrac{t}{5} + \dfrac{3}{10}$

20. $\dfrac{5x^2}{6} - \dfrac{7x}{2} + \dfrac{2}{3} = 0$

Solve each equation. See Examples 6 and 7.

21. $(x + 2)(x - 7)(3x - 8) = 0$

22. $(4x + 9)(x - 4)(x + 1) = 0$

23. $y^3 = 9y$

24. $n^3 = 16n$

25. $x^3 - x = 2x^2 - 2$

26. $m^3 = m^2 + 12m$

382

Solve each equation. This section of exercises contains some linear equations. See Examples 1 through 7.

27. $(2x + 7)(x - 10) = 0$

28. $(x + 4)(5x - 1) = 0$

29. $3x(x - 5) = 0$

30. $4x(2x + 3) = 0$

31. $x^2 - 2x - 15 = 0$

32. $x^2 + 6x - 7 = 0$

33. $12x^2 + 2x - 2 = 0$

34. $8x^2 + 13x + 5 = 0$

35. $w^2 - 5w = 36$

36. $x^2 + 32 = 12x$

37. $25x^2 - 40x + 16 = 0$

38. $9n^2 + 30n + 25 = 0$

39. $2r^3 + 6r^2 = 20r$

40. $-2t^3 = 108t - 30t^2$

41. $z(5z - 4)(z + 3) = 0$

42. $2r(r - 3)(5r + 4) = 0$

43. $2z(z + 6) = 2z^2 + 12z - 8$

44. $3c^2 - 8c + 2 = c(3c - 8)$

45. $(x - 1)(x + 4) = 24$

46. $(2x - 1)(x + 2) = -3$

47. $\dfrac{x^2}{4} - \dfrac{5}{2}x + 6 = 0$

48. $\dfrac{x^2}{18} + \dfrac{x}{2} + 1 = 0$

49. $y^2 + \dfrac{1}{4} = -y$

50. $\dfrac{x^2}{10} + \dfrac{5}{2} = x$

51. $y^3 + 4y^2 = 9y + 36$

52. $x^3 + 5x^2 = x + 5$

53. $2x^3 = 50x$

54. $m^5 = 36m^3$

55. $x^2 + (x + 1)^2 = 61$

56. $y^2 + (y + 2)^2 = 34$

57. $m^2(3m - 2) = m$

58. $x^2(5x + 3) = 26x$

59. $3x^2 = -x$

60. $y^2 = -5y$

61. $x(x - 3) = x^2 + 5x + 7$

62. $z^2 - 4z + 10 = z(z - 5)$

63. $3(t - 8) + 2t = 7 + t$

64. $7c - 2(3c + 1) = 5(4 - 2c)$

65. $-3(x - 4) + x = 5(3 - x)$

66. $-4(a + 1) - 3a = -7(2a - 3)$

Objective **B** *Solve. See Examples 8 and 9.*

67. One number exceeds another by five, and their product is 66, Find the numbers.

68. If the sum of two numbers is 4 and their product is $\dfrac{15}{4}$, find the numbers.

69. An electrician needs to run a cable from the top of a 60-foot tower to a transmitter box located 45 feet away from the base of the tower. Find how long he should cut the cable.

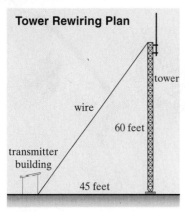

Tower Rewiring Plan

tower

wire

60 feet

transmitter building

45 feet

70. A stereo-system installer needs to run speaker wire above the ceiling along the two diagonals of a rectangular room whose dimensions are 40 feet by 75 feet. Find how much speaker wire she needs.

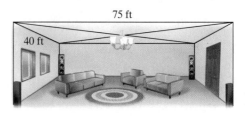

75 ft

40 ft

71. If the cost, $C(x)$, for manufacturing x units of a certain product is given by $C(x) = x^2 - 15x + 50$, find the number of units manufactured at a cost of $9500.

72. Determine whether any three consecutive integers represent the lengths of the sides of a right triangle.

73. The shorter leg of a right triangle is 3 centimeters less than the other leg. Find the length of the two legs if the hypotenuse is 15 centimeters.

74. The longer leg of a right triangle is 4 feet longer than the other leg. Find the length of the two legs if the hypotenuse is 20 feet.

75. Marie Mulroney has a rectangular board 12 inches by 16 inches around which she wants to put a uniform border of shells. If she has enough shells for a border whose area is 128 square inches, determine the width of the border.

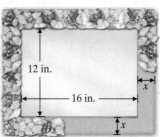

12 in.

16 in.

x

x

76. A gardener has a rose garden that measures 30 feet by 20 feet. He wants to put a uniform border of pine bark around the outside of the garden. Find how wide the border should be if he has enough pine bark to cover 336 square feet.

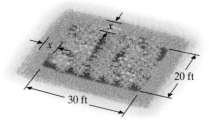

x

x

20 ft

30 ft

77. While hovering near the top of Ribbon Falls in Yosemite National Park at 1600 feet, a helicopter pilot accidentally drops his sunglasses. The height $h(t)$ of the sunglasses after t seconds is given by the polynomial function

$$h(t) = -16t^2 + 1600$$

When will the sunglasses hit the ground?

78. After t seconds, the height $h(t)$ of a model rocket launched from the ground into the air is given by the function

$$h(t) = -16t^2 + 80t$$

Find how long it takes the rocket to reach a height of 96 feet.

△ **79.** The floor of a shed has an area of 90 square feet. The floor is in the shape of a rectangle whose length is 3 feet less than twice the width. Find the length and the width of the floor of the shed.

△ **80.** A vegetable garden with an area of 200 square feet is to be fertilized. If the length of the garden is 1 foot less than three times the width, find the dimensions of the garden.

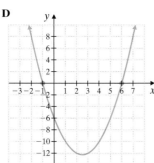

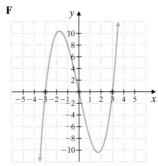

81. The function $W(x) = 0.5x^2$ gives the number of servings of wedding cake that can be obtained from a two-layer x-inch square wedding cake tier. What size square wedding cake tier is needed to serve 50 people? (*Source:* Based on data from the *Wilton 2000 Yearbook of Cake Decorating*)

82. Use the function in Exercise 81 to determine what size wedding cake tier is needed to serve 200 people.

Match each polynomial function with its graph (A–F). See Example 10.

83. $f(x) = (x - 2)(x + 5)$

84. $g(x) = (x + 1)(x - 6)$

85. $h(x) = x(x + 3)(x - 3)$

86. $F(x) = (x + 1)(x - 2)(x + 5)$

87. $G(x) = 2x^2 + 9x + 4$

88. $H(x) = 2x^2 - 7x - 4$

A

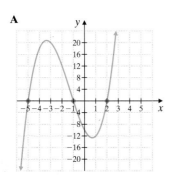

B

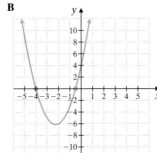

C

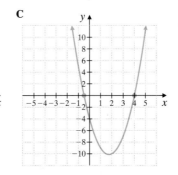

D

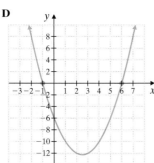

E

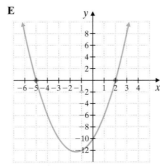

F

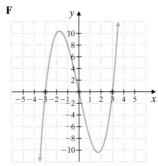

Review

Write the x- and y-intercepts for each graph. See Section 3.1.

89.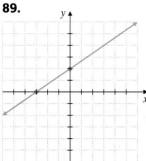

90.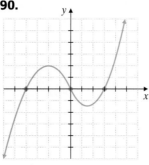

91.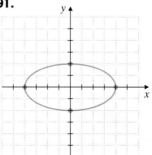

92.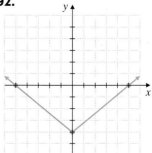

Concept Extensions

Each exercise contains an error. Find and correct the error. See the Concept Check in the section.

93. $(x - 5)(x + 2) = 0$
$x - 5 = 0$ or $x + 2 = 0$
$x = -5$ or $x = -2$

94. $(4x - 5)(x + 7) = 0$
$4x - 5 = 0$ or $x + 7 = 0$
$x = \dfrac{4}{5}$ or $x = -7$

95. $y(y - 5) = -6$
$y = -6$ or $y - 5 = -5$
$y = -7$ or $y = 0$

96. $3x^2 - 19x = 14$
$-16x = 14$
$x = -\dfrac{14}{16}$
$x = -\dfrac{7}{8}$

Solve.

97. $(x^2 + x - 6)(3x^2 - 14x - 5) = 0$

98. $(x^2 - 9)(x^2 + 8x + 16) = 0$

99. Explain how solving $2(x - 3)(x - 1) = 0$ differs from solving $2x(x - 3)(x - 1) = 0$.

100. Explain why the zero-factor property works for more than two numbers whose product is 0.

101. Is the following step correct? Why or why not?

$$x(x - 3) = 5$$
$$x = 5 \quad \text{or} \quad x - 3 = 5$$

Write a quadratic equation that has the given numbers as solutions.

102. 5, 3

103. 6, 7

 THE BIGGER PICTURE   Solving Equations and Inequalities

Continue the outline started in Section 2.6. Write how to recognize and how to solve quadratic equations by factoring.

Solving Equations and Inequalities

I. **Equations**

   A. **Linear equations** (Sec. 2.1)

   B. **Absolute value equations** (Sec. 2.6)

   C. **Quadratic equations:** Equation can be written in the standard form $ax^2 + bx + c = 0$, with $a \neq 0$.

   Solve: $2x^2 - 7x = 9$

$2x^2 - 7x - 9 = 0$	Write in standard form so that equation is equal to 0.
$(2x - 9)(x + 1) = 0$	Factor.
$2x - 9 = 0 \quad \text{or} \quad x + 1 = 0$	Set each factor equal to 0.
$x = \dfrac{9}{2} \quad \text{or} \quad x = -1$	Solve.

II. **Inequalities**

   A. **Linear inequalities** (Sec. 2.4)

   B. **Compound inequalities** (Sec. 2.5)

   C. **Absolute value inequalities** (Sec. 2.6)

Solve. If an inequality write your answer in interval notation.

1. $|7x - 3| = |5x + 9|$

2. $\left| \dfrac{x + 2}{5} \right| < 1$

3. $3(x - 6) + 2 = 9 + 5(3x - 1)$

4. $(x - 6)(2x + 3) = 0$

5. $|-3x + 10| \geq -2$

6. $|-2x - 5| = 11$

7. $x(x - 7) = 30$

8. $8x - 4 \geq 15x - 4$

CHAPTER 5   Group Activity

Finding the Largest Area

This activity may be completed by working in groups or individually.

A picture framer has a piece of wood that measures 1 inch wide by 50 inches long. She would like to make a picture frame with the largest possible interior area. Complete the following activity to help her determine the dimensions of the frame that she should use to achieve her goal.

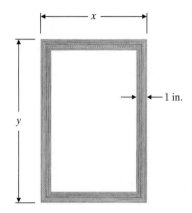

1. Use the situation shown in the figure to write an equation in x and y for the *outer* perimeter of the frame. (Remember that the outer perimeter will equal 50 inches.)

2. Use your equation from Question 1 to help you find the value of y for each value of x given in the table. Complete the y column of the table. (*Note:* The first two columns of the table give possible combinations for the outer dimensions of the frame.)

3. How is the interior width of the frame related to the exterior width of the frame? How is the interior height of the frame related to the exterior height of the frame? Use these relationships to complete the two columns of the table labeled "Interior Width" and "Interior Height."

4. Complete the last column of the table labeled "Interior Area" by using the columns of dimensions for the interior width and height.

5. From the table, what appears to be the largest interior area of the frame? Which exterior dimensions of the frame provide this area?

6. Use the patterns in the table to write an algebraic expression in terms of x for the interior width of the frame.

7. Use the patterns in the table to write an algebraic expression in terms of y for the interior height of the frame.

8. Use the perimeter equation from Question 1 to rewrite the algebraic expression for the interior height of the frame in terms of x.

Frame's Interior Dimensions				
x	**y**	**Interior Width**	**Interior Height**	**Interior Area**
2.0				
2.5				
3.0				
3.5				
4.0				
4.5				
5.0				
5.5				
6.0				
6.5				
7.0				
7.5				
8.0				
8.5				
9.0				
9.5				
10.0				
10.5				
11.0				
11.5				
12.0				
12.5				
13.0				
13.5				
14.0				
14.5				
15.0				

9. Find a function A that gives the interior area of the frame in terms of its exterior width x. (*Hint:* Study the patterns in the table. How could the expressions from Questions 6 and 8 be used to write this function?)

10. Graph the function A. Locate and label the point from the table that represents the maximum interior area. Describe the location of the point in relation to the rest of the graph.

Chapter 5 Vocabulary Check

Fill in each blank with one of the words or phrases listed below.

| quadratic equation | synthetic division | polynomial | FOIL | 0 | monomial |
| binomial | trinomial | degree of a polynomial | degree of a term | | factoring |

1. A _____ is a finite sum of terms in which all variables are raised to nonnegative integer powers and no variables appear in any denominator.
2. _____ is the process of writing a polynomial as a product.
3. The _____ is the sum of the exponents on the variables contained in the term.
4. A _____ is a polynomial with one term.
5. A _____ is a polynomial with three terms.
6. A polynomial equation of degree 2 is also called a _____.
7. The _____ is the largest degree of all of its terms.
8. A _____ is a polynomial with two terms.
9. If a and b are real numbers and $a \cdot b = $ ___, then $a = 0$ or $b = 0$.
10. The _____ order may be used when multiplying two binomials.
11. A shortcut method called _____ may be used to divide a polynomial by a binomial of the form $x - c$.

Helpful Hint

Are you preparing for your test? Don't forget to take the Chapter 5 Test on page 397. Then check your answers at the back of the text and use the Chapter Test Prep Video CD to see the fully worked-out solutions to any of the exercises you want to review.

5 Chapter Highlights

DEFINITIONS AND CONCEPTS	EXAMPLES
Section 5.1 Polynomial Functions and Adding and Subtracting Polynomials	

A **polynomial** is a finite sum of terms in which all variables have exponents raised to nonnegative integer powers and no variables appear in any denominator.	$1.3x^2$ Monomial $-\dfrac{1}{3}y + 5$ Binomial $6z^2 - 5z + 7$ Trinomial
To **add polynomials,** combine all like terms.	Add: $(3y^2x - 2yx + 11) + (-5y^2x - 7)$ $= -2y^2x - 2yx + 4$
To **subtract polynomials,** change the signs of the terms of the polynomial being subtracted; then add.	Subtract: $(-2z^3 - z + 1) - (3z^3 + z - 6)$ $= -2z^3 - z + 1 - 3z^3 - z + 6$ $= -5z^3 - 2z + 7$
A function P is a **polynomial function** if $P(x)$ is a polynomial.	For the polynomial function $$P(x) = -x^2 + 6x - 12$$ find $P(-2)$. $$P(-2) = -(-2)^2 + 6(-2) - 12 = -28$$

DEFINITIONS AND CONCEPTS	EXAMPLES

Section 5.2   Multiplying Polynomials

MULTIPLYING TWO POLYNOMIALS

Use the distributive property and multiply each term of one polynomial by each term of the other polynomial; then combine like terms.

Multiply.

$$(x^2 - 2x)(3x^2 - 5x + 1)$$
$$= 3x^4 - 5x^3 + x^2 - 6x^3 + 10x^2 - 2x$$
$$= 3x^4 - 11x^3 + 11x^2 - 2x$$

SPECIAL PRODUCTS

$$(a + b)^2 = a^2 + 2ab + b^2$$
$$(a - b)^2 = a^2 - 2ab + b^2$$
$$(a + b)(a - b) = a^2 - b^2$$

$$(3m + 2n)^2 = 9m^2 + 12mn + 4n^2$$
$$(z^2 - 5)^2 = z^4 - 10z^2 + 25$$
$$(7y + 1)(7y - 1) = 49y^2 - 1$$

The **FOIL order** may be used when multiplying two binomials.

Multiply.

$$(x^2 + 5)(2x^2 - 9)$$

$$\overset{\text{F}}{\downarrow} \quad \overset{\text{O}}{\downarrow} \quad \overset{\text{I}}{\downarrow} \quad \overset{\text{L}}{\downarrow}$$

$$= x^2(2x^2) + x^2(-9) + 5(2x^2) + 5(-9)$$
$$= 2x^4 - 9x^2 + 10x^2 - 45$$
$$= 2x^4 + x^2 - 45$$

Section 5.3   Dividing Polynomials and Synthetic Division

DIVIDING A POLYNOMIAL BY A MONOMIAL

Divide each term in the polynomial by the monomial.

$$\frac{12a^5b^3 - 6a^2b^2 + ab}{6a^2b^2}$$

$$= \frac{12a^5b^3}{6a^2b^2} - \frac{6a^2b^2}{6a^2b^2} + \frac{ab}{6a^2b^2}$$

$$= 2a^3b - 1 + \frac{1}{6ab}$$

DIVIDING A POLYNOMIAL BY A POLYNOMIAL OTHER THAN A MONOMIAL

Use **long division.**

Divide $2x^3 - x^2 - 8x - 1$ by $x - 2$.

$$\begin{array}{r} 2x^2 + 3x - 2 \\ x - 2 \overline{)\, 2x^3 - x^2 - 8x - 1} \\ \underline{2x^3 - 4x^2} \\ 3x^2 - 8x \\ \underline{3x^2 - 6x} \\ -2x - 1 \\ \underline{-2x + 4} \\ -5 \end{array}$$

The quotient is $2x^2 + 3x - 2 - \dfrac{5}{x - 2}$.

A shortcut method called **synthetic division** may be used to divide a polynomial by a binomial of the form $x - c$.

Use synthetic division to divide $2x^3 - x^2 - 8x - 1$ by $x - 2$.

$$\begin{array}{r|rrrr} 2 & 2 & -1 & -8 & -1 \\ & & 4 & 6 & -4 \\ \hline & 2 & 3 & -2 & -5 \end{array}$$

The quotient is $2x^2 + 3x - 2 - \dfrac{5}{x - 2}$.

DEFINITIONS AND CONCEPTS	**EXAMPLES**

Section 5.4   The Greatest Common Factor and Factoring by Grouping

The **greatest common factor** of the terms of a polynomial is the product of the greatest common factor of the numerical coefficients and the greatest common factor of the variable factors.

Factor: $14xy^3 - 2xy^2 = 2 \cdot 7 \cdot x \cdot y^3 - 2 \cdot x \cdot y^2$
The greatest common factor is $2 \cdot x \cdot y^2$, or $2xy^2$.

$$14xy^3 - 2xy^2 = 2xy^2(7y - 1)$$

FACTORING A POLYNOMIAL BY GROUPING

Group the terms so that each group has a common factor. Factor out these common factors. Then see if the new groups have a common factor.

Factor: $x^4y - 5x^3 + 2xy - 10$
$$= x^3(xy - 5) + 2(xy - 5)$$
$$= (xy - 5)(x^3 + 2)$$

Section 5.5   Factoring Trinomials

FACTORING $ax^2 + bx + c$

Step 1. Write all pairs of factors of ax^2.

Step 2. Write all pairs of factors of c.

Step 3. Try combinations of these factors until the middle term bx is found.

Factor: $28x^2 - 27x - 10$
Factors of $28x^2$: $28x$ and x, $2x$ and $14x$, $4x$ and $7x$.
Factors of -10: -2 and 5, 2 and -5, -10 and 1, 10 and -1.

$$28x^2 - 27x - 10 = (7x + 2)(4x - 5)$$

Section 5.6   Factoring by Special Products

PERFECT SQUARE TRINOMIAL

$$a^2 + 2ab + b^2 = (a + b)^2$$
$$a^2 - 2ab + b^2 = (a - b)^2$$

Factor.

$$25x^2 + 30x + 9 = (5x + 3)^2$$
$$49z^2 - 28z + 4 = (7z - 2)^2$$

DIFFERENCE OF TWO SQUARES

$$a^2 - b^2 = (a + b)(a - b)$$

Factor.

$$36x^2 - y^2 = (6x + y)(6x - y)$$

SUM AND DIFFERENCE OF TWO CUBES

$$a^3 + b^3 = (a + b)(a^2 - ab + b^2)$$
$$a^3 - b^3 = (a - b)(a^2 + ab + b^2)$$

Factor.

$$8y^3 + 1 = (2y + 1)(4y^2 - 2y + 1)$$
$$27p^3 - 64q^3 = (3p - 4q)(9p^2 + 12pq + 16q^2)$$

Section 5.7   Solving Equations by Factoring and Solving Problems

SOLVING A POLYNOMIAL EQUATION BY FACTORING

Step 1. Write the equation so that one side is 0.

Step 2. Factor the polynomial completely.

Step 3. Set each factor equal to 0.

Step 4. Solve the resulting equations.

Step 5. Check each solution.

Solve: $2x^3 - 5x^2 = 3x$
$$2x^3 - 5x^2 - 3x = 0$$
$$x(2x + 1)(x - 3) = 0$$
$x = 0$ or $2x + 1 = 0$ or $x - 3 = 0$
$x = 0$ $x = -\dfrac{1}{2}$ $x = 3$

Are You Prepared for a Test on Chapter 5?

Below I have listed some *common trouble areas* for students in Chapter 5. After studying for your test—but before taking your test—read these.

- Don't forget to watch your signs when subtracting polynomials.

$$(7x^3 - 6x^2 + 2x) - (9x^3 + 7x^2 - 20x - 20)$$
$$= 7x^3 - 6x^2 + 2x - 9x^3 - 7x^2 + 20x + 20$$
$$= -2x^3 - 13x^2 + 22x + 20$$

- Can you evaluate $P(-1)$ if $P(x) = -16x^2 + 2x$?

$$P(-1) = -16(-1)^2 + 2(-1)$$
$$= -16 \cdot 1 + (-2) = -18$$

- Don't forget how to square a binomial.

$$(3x + 5y)^2 = (3x)^2 + 2(3x)(5y) + (5y)^2$$
$$= 9x^2 + 30xy + 25y^2$$

- Remember that the first step in factoring a polynomial is to factor out any common factors. Also, always check to see if a factor can be factored further.

Factor:

$$4x^4 - 64 = 4(x^4 - 16)$$
$$= 4(x^2 + 4)(x^2 - 4)$$
$$= 4(x^2 + 4)(x + 2)(x - 2)$$

- When factoring the sum or difference of two cubes, it may be helpful to first write each term as a quantity cubed.

Factor:

$$27x^3 - 8y^3 = (3x)^3 - (2y)^3$$
$$= (3x - 2y)[(3x)^2 + (3x)(2y) + (2y)^2]$$
$$= (3x - 2y)(9x^2 + 6xy + 4y^2)$$

- Remember that to use the zero-factor property to solve a quadratic equation, one side of the equation must be 0 and the other side must be a factored polynomial.

Solve:

$$x(5x + 3) = 2 \quad \text{Cannot use zero-factor property.}$$
$$5x^2 + 3x - 2 = 0 \quad \text{Multiply and subtract 2 from both sides.}$$
$$(5x - 2)(x + 1) = 0 \quad \text{Now you can use zero-factor property.}$$
$$5x - 2 = 0 \quad \text{or} \quad x + 1 = 0 \quad \text{Set each factor equal to 0.}$$
$$x = \frac{2}{5} \quad \text{or} \qquad x = -1 \quad \text{Solve.}$$

Remember: This is simply a checklist of common trouble areas. For a review of Chapter 5, see the Highlights and Chapter Review at the end of this chapter.

---

5 CHAPTER REVIEW

(5.1) *Find the degree of each polynomial.*

1. $x^2y - 3xy^3z + 5x + 7y$

2. $3x + 2$

Simplify by combining like terms.

3. $4x + 8x - 6x^2 - 6x^2y$

4. $-8xy^3 + 4xy^3 - 3x^3y$

Add or subtract as indicated.

5. $(3x + 7y) + (4x^2 - 3x + 7) + (y - 1)$

6. $(4x^2 - 6xy + 9y^2) - (8x^2 - 6xy - y^2)$

7. $(3x^2 - 4b + 28) + (9x^2 - 30) - (4x^2 - 6b + 20)$

8. Add $(9xy + 4x^2 + 18)$ and $(7xy - 4x^3 - 9x)$.

9. Subtract $(x - 7)$ from the sum of $(3x^2y - 7xy - 4)$ and $(9x^2y + x)$.

392

10. $\begin{aligned} x^2 - 5x + 7 \\ -\quad\ (x + 4) \\ \hline \end{aligned}$

11. $\begin{aligned} x^3\quad + 2xy^2 - y \\ +\ (x - 4xy^2\quad - 7) \\ \hline \end{aligned}$

If $P(x) = 9x^2 - 7x + 8,$ *find each function value.*

12. $P(6)$

13. $P(-2)$

14. $P(-3)$

If $P(x) = 2x - 1$ *and* $Q(x) = x^2 + 2x - 5,$ *find each function value.*

15. $P(x) + Q(x)$

16. $2 \cdot P(x) - Q(x)$

△ **17.** Find the perimeter of the rectangle.

$(x^2y + 5)$ cm

$(2x^2y - 6x + 1)$ cm

(5.2) *Multiply.*

18. $-6x(4x^2 - 6x + 1)$

19. $-4ab^2(3ab^3 + 7ab + 1)$

20. $(x - 4)(2x + 9)$

21. $(-3xa + 4b)^2$

22. $(9x^2 + 4x + 1)(4x - 3)$

23. $(5x - 9y)(3x + 9y)$

24. $\left(x - \dfrac{1}{3}\right)\left(x + \dfrac{2}{3}\right)$

25. $(x^2 + 9x + 1)^2$

26. $(2x - 1)(x^2 + 2x - 5)$

Use special products to multiply.

27. $(3x - y)^2$

28. $(4x + 9)^2$

29. $(x + 3y)(x - 3y)$

30. $[4 + (3a - b)][4 - (3a - b)]$

△ **31.** Find the area of the rectangle.

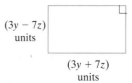

$(3y - 7z)$ units

$(3y + 7z)$ units

(5.3) *Divide.*

32. $(4xy + 2x^2 - 9) \div (4xy)$

33. $12xb^2 + 16xb^4$ by $4xb^3$

34. $(3x^4 - 25x^2 - 20) \div (x - 3)$

35. $(-x^2 + 2x^4 + 5x - 12) \div (x - 3)$

36. $(2x^4 - x^3 + 2x^2 - 3x + 1) \div \left(x - \dfrac{1}{2}\right)$

37. $(x^3 + 3x^2 - 2x + 2) \div \left(x - \dfrac{1}{2}\right)$

38. $(3x^4 + 5x^3 + 7x^2 + 3x - 2) \div (x^2 + x + 2)$

39. $(9x^4 - 6x^3 + 3x^2 - 12x - 30) \div (3x^2 - 2x - 5)$

Use synthetic division to find each quotient.

40. $(3x^3 + 12x - 4) \div (x - 2)$

41. $(4x^3 + 2x^2 - 4x - 2) \div \left(x + \dfrac{3}{2}\right).$

42. $(x^5 - 1) \div (x + 1)$

43. $(x^3 - 81) \div (x - 3)$

44. $(x^3 - x^2 + 3x^4 - 2) \div (x - 4)$

45. $(3x^4 - 2x^2 + 10) \div (x + 2)$

(5.4) *Factor out the greatest common factor.*

46. $16x^3 - 24x^2$

47. $36y - 24y^2$

48. $6ab^2 + 8ab - 4a^2b^2$

49. $14a^2b^2 - 21ab^2 + 7ab$

50. $6a(a + 3b) - 5(a + 3b)$

51. $4x(x - 2y) - 5(x - 2y)$

Factor by grouping.

52. $xy - 6y + 3x - 18$

53. $ab - 8b + 4a - 32$

54. $pq - 3p - 5q + 15$

55. $x^3 - x^2 - 2x + 2$

△**56.** A smaller square is cut from a larger rectangle. Write the area of the shaded region as a factored polynomial.

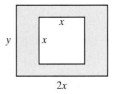

(5.5) *Factor each polynomial completely.*

57. $x^2 - 14x - 72$

58. $x^2 + 16x - 80$

59. $2x^2 - 18x + 28$

60. $3x^2 + 33x + 54$

61. $2x^3 - 7x^2 - 9x$

62. $3x^2 + 2x - 16$

63. $6x^2 + 17x + 10$

64. $15x^2 - 91x + 6$

65. $4x^2 + 2x - 12$

66. $9x^2 - 12x - 12$

67. $y^2(x + 6)^2 - 2y(x + 6)^2 - 3(x + 6)^2$

68. $(x + 5)^2 + 6(x + 5) + 8$

69. $x^4 - 6x^2 - 16$

70. $x^4 + 8x^2 - 20$

(5.6) *Factor each polynomial completely.*

71. $x^2 - 100$

72. $x^2 - 81$

73. $2x^2 - 32$

74. $6x^2 - 54$

75. $81 - x^4$

76. $16 - y^4$

77. $(y + 2)^2 - 25$

78. $(x - 3)^2 - 16$

79. $x^3 + 216$

80. $y^3 + 512$

81. $8 - 27y^3$

82. $1 - 64y^3$

83. $6x^4y + 48xy$

84. $2x^5 + 16x^2y^3$

85. $x^2 - 2x + 1 - y^2$

86. $x^2 - 6x + 9 - 4y^2$

87. $4x^2 + 12x + 9$

88. $16a^2 - 40ab + 25b^2$

△ **89.** The volume of the cylindrical shell is $\pi R^2 h - \pi r^2 h$ cubic units. Write this volume as a factored expression.

(5.7) *Solve each polynomial equation for the variable.*

90. $(3x - 1)(x + 7) = 0$

91. $3(x + 5)(8x - 3) = 0$

92. $5x(x - 4)(2x - 9) = 0$

93. $6(x + 3)(x - 4)(5x + 1) = 0$

94. $2x^2 = 12x$

95. $4x^3 - 36x = 0$

96. $(1 - x)(3x + 2) = -4x$

97. $2x(x - 12) = -40$

98. $3x^2 + 2x = 12 - 7x$

99. $2x^2 + 3x = 35$

100. $x^3 - 18x = 3x^2$

101. $19x^2 - 42x = -x^3$

102. $12x = 6x^3 + 6x^2$

103. $8x^3 + 10x^2 = 3x$

104. The sum of a number and twice its square is 105. Find the number.

△ **105.** The length of a rectangular piece of carpet is 2 meters less than 5 times its width. Find the dimensions of the carpet if its area is 16 square meters.

106. A scene from an adventure film calls for a stunt dummy to be dropped from above the second-story platform of the Eiffel Tower, a distance of 400 feet. Its height h in feet at the time t seconds is given by $h = -16t^2 + 400$. Determine how long before the stunt dummy reaches the ground.

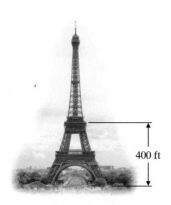

400 ft

Mixed Review

Perform the indicated operation.

107. $(x + 5)(3x^2 - 2x + 1)$

108. $(3x^2 + 4x - 1.2) - (5x^2 - x + 5.7)$

109. $(3x^2 + 4x - 1.2) + (5x^2 - x + 5.7)$

110. $\left(7ab - \dfrac{1}{2}\right)^2$

If $P(x) = -x^2 + x - 4$, find

111. $P(5)$

112. $P(-2)$

Factor each polynomial completely.

113. $12y^5 - 6y^4$

114. $x^2y + 4x^2 - 3y - 12$

115. $6x^2 - 34x - 12$

116. $y^2(4x + 3)^2 - 19y(4x + 3)^2 - 20(4x + 3)^2$

117. $4z^7 - 49z^5$

118. $5x^4 + 4x^2 - 9$

Solve each equation.

119. $8x^2 = 24x$

120. $x(x - 11) = 26$

5 CHAPTER TEST

Answers

Perform each indicated operation.

1. $(4x^3 - 3x - 4) - (9x^3 + 8x + 5)$ **2.** $-3xy(4x + y)$

3. $(3x + 4)(4x - 7)$ **4.** $(5a - 2b)(5a + 2b)$

5. $(6m + n)^2$ **6.** $(2x - 1)(x^2 - 6x + 4)$

7. $(4x^2y + 9x + z) \div (3xz)$

8. $(4x^5 - 2x^4 + 4x^2 - 6x + 3) \div (2x - 1)$

9. Use synthetic division to divide $4x^4 - 3x^3 + 2x^2 - x - 1$ by $x + 3$.

Factor each polynomial completely.

10. $16x^3y - 12x^2y^4$ **11.** $x^2 - 13x - 30$ **12.** $4y^2 + 20y + 25$

13. $6x^2 - 15x - 9$ **14.** $4x^2 - 25$ **15.** $x^3 + 64$

16. $3x^2y - 27y^3$ **17.** $16y^3 - 2$ **18.** $x^2y - 9y - 3x^2 + 27$

1. _____

2. _____

3. _____

4. _____

5. _____

6. _____

7. _____

8. _____

9. _____

10. _____

11. _____

12. _____

13. _____

14. _____

15. _____

16. _____

17. _____

18. _____

19. _____

20. _____

21. _____

22. _____

23. a. _____

    b. _____

    c. _____

    d. _____

Solve each equation for the variable.

19. $3n(7n - 20) = 96$

20. $(x + 2)(x - 2) = 5(x + 4)$

21. $2x^3 + 5x^2 = 8x + 20$

△ **22.** Write the area of the shaded region as a factored polynomial.

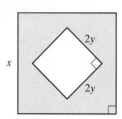

23. A pebble is hurled upward from the top of the 880-foot-tall Canada Trust Tower with an initial velocity of 96 feet per second. Neglecting air resistance, the height $h(t)$ in feet of the pebble after t seconds is given by the polynomial function

$$h(t) = -16t^2 + 96t + 880$$

  a. Find the height of the pebble when $t = 1$.

  b. Find the height of the pebble when $t = 5.1$.

  c. Factor the polynomial $-16t^2 + 96t + 880$.

  d. When will the pebble hit the ground?

Simplify.

1. $-2(x + 3) + 7$

2. $6z - 9z - \dfrac{1}{2} + z - \dfrac{3}{8}$

3. $7x + 3 - 5(x - 4)$

4. $7 + 2(9x - 1) - (1.7x + 5.8)$

5. Use scientific notation to simplify:
$$\dfrac{2000 \times 0.000021}{700}$$

6. Simplify each expression. Write the answers with positive exponents only.

a. $(4a^{-1}b^0)^{-3}$ **b.** $\left(\dfrac{a^{-6}}{a^{-8}}\right)^{-2}$

c. $\left(\dfrac{2}{3}\right)^{-3}$ **d.** $\dfrac{3^{-2}a^{-2}b^{12}}{a^4b^{-5}}$

7. Solve: $3(x - 3) = 5x - 9$

8. Solve: $0.3y + 2.4 = 0.1y + 4$

9. Solve $3y - 2x = 7$ for y.

10. A gallon of latex paint can cover 400 square feet. How many gallon containers of paint should be bought to paint two coats on each wall of a rectangular room whose dimensions are 14 feet by 18 feet? (Assume 8-foot ceilings.)

Solve. Write the solution set in interval notation.

11. $-(x - 3) + 2 \leq 3(2x - 5) + x$

12. $x + 2 < \dfrac{1}{4}(x - 7)$

13. $-1 \leq \dfrac{2}{3}x + 5 < 2$

14. Solve: $-\dfrac{1}{3} < \dfrac{3x + 1}{6} \leq \dfrac{1}{3}$

15. Solve: $|3x + 2| = |5x - 8|$

16. Solve: $|5x - 1| + 9 > 5$

1. _____

2. _____

3. _____

4. _____

5. _____

6. a. _____

b. _____

c. _____

d. _____

7. _____

8. _____

9. _____

10. _____

11. _____

12. _____

13. _____

14. _____

15. _____

16. _____

17. _____

18. _____

19. see graph _____

20. see graph _____

21. _____

22. _____

23. _____

24. _____

25. _____

26. _____

27. _____

28. a. _____

   b. _____

29. see graph _____

30. _____

31. _____

32. _____

33. _____

34. _____

35. _____

36. _____

17. Find the slope of the line $x = -5$.

18. Find the slope of the line, $f(x) = -2x - 3$.

19. Graph: $y = \dfrac{1}{4}x - 3$

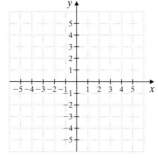

20. Graph: $y = 3x$

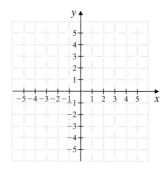

21. Write an equation of the line containing the point $(-2, 1)$ and perpendicular to the line $3x + 5y = 4$. Write the equation in slope-intercept form.

22. Find the equation of the vertical line containing the point $(-3, 2)$.

23. Determine the domain and range of the relation: $\{(2, 3), (2, 4), (0, -1), (3, -1)\}$

24. Find the equation of the line containing the point $(-2, 3)$ and slope of 0.

Find each function value.

25. If $g(x) = 3x - 2$, find $g(0)$.

26. If $f(x) = 3x^2 + 2x + 3$, find $f(-3)$.

27. If $f(x) = 7x^2 - 3x + 1$, find $f(1)$.

28. Multiply.

   a. $(3y^6)(4.7y^2)$ **b.** $(6a^3b^2)(-a^2bc^4)$

29. Graph: $2x - y < 6$

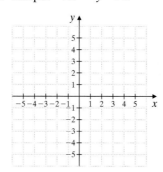

30. One solution contains 20% acid and a second solution contains 60% acid. How many ounces of each solution should be mixed in order to have 50 ounces of a 30% acid solution?

31. Use the substitution method to solve the system: $\begin{cases} 2x + 4y = -6 \\ x = 2y - 5 \end{cases}$

32. $\begin{cases} 4y = 8 \\ x + y = 7 \end{cases}$

33. Solve the system: $\begin{cases} 3x - y + z = -15 \\ x + 2y - z = 1 \\ 2x + 3y - 2z = 0 \end{cases}$

34. $\begin{cases} x + y + z = 0 \\ 2x - 3y + z = 5 \\ 2x + y + 2z = 2 \end{cases}$

35. Use matrices to solve the system: $\begin{cases} x + 3y = 5 \\ 2x - y = -4 \end{cases}$

36. Solve the system.

$\begin{cases} x + y - \dfrac{3}{2}z = \dfrac{1}{2} \\ -y - 2z = 14 \\ x - \dfrac{2}{3}y = -\dfrac{1}{3} \end{cases}$

37. Graph the solutions of the system

$$\begin{cases} x - y < 2 \\ x + 2y > -1 \\ y < 2 \end{cases}$$

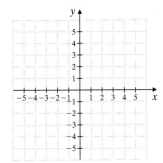

38. Write each number in scientific notation.
 a. 8,250,000
 b. 0.0000346

39. Determine the degree of the polynomial: $3xy + x^2y^2 - 5x^2 - 6.7$

40. Subtract $(5x^2 + 3x)$ from $(3x^2 - 2x)$.

Multiply.

41. $2x(5x - 4)$

42. $(2x - 1)(5x^2 - 6x + 2)$

43. $-xy(7x^2y + 3xy - 11)$

44. $\left(7x - \dfrac{1}{2}\right)^2$

45. Divide $10x^3 - 5x^2 + 20x$ by $5x$.

Factor.

46. $12x^3y - 3xy^3$

47. $x^2 + 10x + 16$

48. $5a^2 + 14a - 3$

49. Solve: $2x^2 + 9x - 5 = 0$

50. Solve: $3x^2 - 10x - 8 = 0$

37. see graph

38. a. _____

 b. _____

39. _____

40. _____

41. _____

42. _____

43. _____

44. _____

45. _____

46. _____

47. _____

48. _____

49. _____

50. _____

6

Rational Expressions

Dividing one polynomial by another in Section 5.3, we found quotients of polynomials. When the remainder part of the quotient was not 0, the remainder was a fraction, such as $\frac{x}{x^2 + 1}$. This fraction is not a polynomial, since it cannot be written as the sum of whole number powers. Instead, it is called a *rational expression*. In this chapter, we study these algebraic forms, the operations that can be performed on them, and the *rational functions* they generate.

"**P**lay Ball!" Each spring, this call resounds throughout America as another baseball season begins. Baseball, often referred to as "America's Pastime," was supposedly created by Abner Doubleday in Cooperstown, New York, in 1839. However, it appears that this quintessential American game has been around much longer than that. Just recently, a bylaw was uncovered from the musty records of a courthouse in the small town of Pittsfield, Massachusetts. The 1791 statute aims to protect windows in a "new" town meeting house by prohibiting anyone from playing "baseball" within 80 yards of the building.

Rational expressions can be found in formulas associated with baseball. In Exercise 60, Section 6.3, you will simplify the rational formula for calculating the earned run average (ERA) statistic for pitchers. The ERA determines the average number of earned runs scored against a pitcher in nine innings.

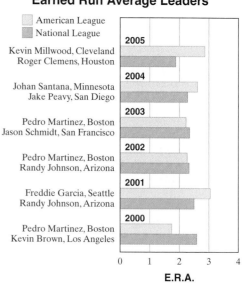

Earned Run Average Leaders

American League
National League

2005
Kevin Millwood, Cleveland
Roger Clemens, Houston

2004
Johan Santana, Minnesota
Jake Peavy, San Diego

2003
Pedro Martinez, Boston
Jason Schmidt, San Francisco

2002
Pedro Martinez, Boston
Randy Johnson, Arizona

2001
Freddie Garcia, Seattle
Randy Johnson, Arizona

2000
Pedro Martinez, Boston
Kevin Brown, Los Angeles

E.R.A.

6.1 RATIONAL FUNCTIONS AND MULTIPLYING AND DIVIDING RATIONAL EXPRESSIONS

Objectives

A Find the Domain of a Rational Function.

B Simplify Rational Expressions.

C Multiply Rational Expressions.

D Divide Rational Expressions.

E Use Rational Functions in Real-World Applications.

Recall that a *rational number*, or *fraction*, is a number that can be written as the quotient $\frac{p}{q}$ of two integers p and q as long as q is not 0. A **rational expression** is an expression that can be written as the quotient $\frac{P}{Q}$ of two polynomials P and Q as long as Q is not 0.

Examples of Rational Expressions

$$\frac{8x^3 + 7x^2 + 20}{2} \qquad \frac{5x^2 - 3}{x - 1} \qquad \frac{7x - 2}{x^2 - 2x - 15}$$

Rational expressions are sometimes used to describe functions. For example, we call the function $f(x) = \frac{x^2 + 2}{x - 3}$ a **rational function** since $\frac{x^2 + 2}{x - 3}$ is a rational expression.

Objective A Finding Domains of Rational Functions

As with fractions, a rational expression is **undefined** if the denominator is 0. If a variable in a rational expression is replaced with a number that makes the denominator 0, we say that the rational expression is **undefined** for this value of the variable. For example, the rational expression $\frac{x^2 + 2}{x - 3}$ is undefined when x is 3, because replacing x with 3 results in a denominator of 0. For this reason, we must exclude 3 from the domain of the function $f(x) = \frac{x^2 + 2}{x - 3}$.

The domain of f is then

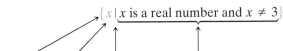

$$\{x \mid x \text{ is a real number and } x \neq 3\}$$

"The set of all x such that x is a real number and x is not equal to 3."
In this section, we will use this set builder notations to write domains.

Unless told otherwise, we assume that the domain of a function described by an equation is the set of all real numbers for which the equation is defined.

EXAMPLE 1 Find the domain of each rational function.

a. $f(x) = \dfrac{8x^3 + 7x^2 + 20}{2}$ **b.** $g(x) = \dfrac{5x^2 - 3}{x - 1}$

c. $f(x) = \dfrac{7x - 2}{x^2 - 2x - 15}$

Solution: The domain of each function will contain all real numbers except those values that make the denominator 0.

a. No matter what the value of x, the denominator of $f(x) = \dfrac{8x^3 + 7x^2 + 20}{2}$ is never 0, so the domain of f is $\{x \mid x \text{ is a real number}\}$.

b. To find the values of x that make the denominator of $g(x)$ equal to 0, we solve the equation "denominator = 0":

$$x - 1 = 0, \quad \text{or} \quad x = 1$$

The domain must exclude 1 since the rational expression is undefined when x is 1. The domain of g is $\{x \mid x \text{ is a real number and } x \neq 1\}$.

Continued on next page

PRACTICE PROBLEM 1

Find the domain of each rational function.

a. $f(x) = \dfrac{x^2 + 1}{x - 6}$

b. $g(x) = \dfrac{5x + 4}{x^2 - 3x - 10}$

c. $h(x) = \dfrac{x^2 - 9}{4}$

Answer

1. a. $\{x \mid x \text{ is a real number, } x \neq 6\}$,
b. $\{x \mid x \text{ is a real number, } x \neq 5, x \neq -2\}$,
c. $\{x \mid x \text{ is a real number}\}$

c. We find the domain by setting the denominator equal to 0.

$$x^2 - 2x - 15 = 0 \quad \text{\small Set the denominator equal to 0 and solve.}$$
$$(x - 5)(x + 3) = 0$$
$$x - 5 = 0 \quad \text{or} \quad x + 3 = 0$$
$$x = 5 \quad \text{or} \qquad x = -3$$

If x is replaced with 5 or with -3, the rational expression is undefined.

The domain of f is $\{x \mid x$ is a real number and $x \neq 5, x \neq -3\}$.

■ **Work Practice Problem 1**

Objective B Simplifying Rational Expressions

Recall that a fraction is in lowest terms or simplest form if the numerator and denominator have no common factors other than 1 (or -1). For example, $\frac{3}{13}$ is in lowest terms since 3 and 13 have no common factors other than 1 (or -1).

To **simplify** a rational expression, or to write it in lowest terms, we use a method similar to simplifying a fraction.

Recall that to simplify a fraction, we essentially "remove factors of 1." Our ability to do this comes from these facts:

- If $c \neq 0$, then $\dfrac{c}{c} = 1$. For example, $\dfrac{7}{7} = 1$ and $\dfrac{-8.65}{-8.65} = 1$.

- $n \cdot 1 = n$. For example, $-5 \cdot 1 = -5$, $126.8 \cdot 1 = 126.8$, and $\dfrac{a}{b} \cdot 1 = \dfrac{a}{b}, b \neq 0$.

In other words, we have the following:

$$\frac{a \cdot c}{b \cdot c} = \frac{a}{b} \cdot \frac{c}{c} = \frac{a}{b}$$

$$\text{\small Since } \tfrac{a}{b} \cdot 1 = \tfrac{a}{b}$$

Let's practice simplifying a fraction by simplifying $\dfrac{15}{65}$.

$$\frac{15}{65} = \frac{3 \cdot 5}{13 \cdot 5} = \frac{3}{13} \cdot \frac{5}{5} = \frac{3}{13} \cdot 1 = \frac{3}{13}$$

Let's use the same technique and simplify the rational expression $\dfrac{(x + 2)^2}{x^2 - 4}$.

$$\frac{(x + 2)^2}{x^2 - 4} = \frac{(x + 2)(x + 2)}{(x - 2)(x + 2)}$$

$$= \frac{(x + 2)}{(x - 2)} \cdot \frac{x + 2}{x + 2}$$

$$= \frac{x + 2}{x - 2} \cdot 1$$

$$= \frac{x + 2}{x - 2}$$

This means that the rational expression $\dfrac{(x + 2)^2}{x^2 - 4}$ has the same value as the rational expression $\dfrac{x + 2}{x - 2}$ for all values of x except 2 and -2. (Remember that when x is 2, the denominators of both rational expressions are 0 and that when x is -2, the original rational expression has a denominator of 0.)

As we simplify rational expressions, we will assume that the simplified rational expression is equivalent to the original rational expression for all real numbers except those for which either denominator is 0.

Just as for numerical fractions, we can use a shortcut notation. Remember that as long as exact factors in both the numerator and denominator are divided out, we are "removing a factor of 1." We can use the following notation:

$$\frac{(x+2)^2}{x^2-4} = \frac{(x+2)\;(x+2)}{(x-2)\;(x+2)} \qquad \text{A factor of 1 is identified by the shading.}$$

$$= \frac{x+2}{x-2} \qquad \text{"Remove" the factor of 1.}$$

In general, the following steps may be used to simplify rational expressions or to write a rational expression in lowest terms.

Simplifying or Writing a Rational Expression in Lowest Terms

Step 1: Completely factor the numerator and denominator of the rational expression.

Step 2: Divide out factors common to the numerator and denominator. (This is the same as "removing a factor of 1.")

For now, we assume that variables in a rational expression do not represent values that make the denominator 0.

EXAMPLES Simplify each rational expression.

2. $\dfrac{2x^2}{10x^3-2x^2} = \dfrac{2x^2 \cdot 1}{2x^2\,(5x-1)}$ Factor the numerator and denominator.

$$= 1 \cdot \frac{1}{5x-1} \qquad \text{Since } \frac{2x^2}{2x^2} = 1.$$

$$= \frac{1}{5x-1} \qquad \text{Simplest form.}$$

3. $\dfrac{9x^2+13x+4}{8x^2+x-7} = \dfrac{(9x+4)\;(x+1)}{(8x-7)\;(x+1)}$ Factor the numerator and denominator.

$$= \frac{9x+4}{8x-7} \cdot 1 \qquad \text{Since } \frac{x+1}{x+1} = 1.$$

$$= \frac{9x+4}{8x-7} \qquad \text{Simplest form.}$$

⬛ **Work Practice Problems 2–3**

EXAMPLES Simplify each rational expression.

4. $\dfrac{2+x}{x+2} = \dfrac{x+2}{x+2} = 1$ By the commutative property of addition, $2+x = x+2$.

5. $\dfrac{2-x}{x-2}$

The terms in the numerator of $\dfrac{2-x}{x-2}$ differ by sign from the terms of the denominator, so the polynomials are opposites of each other and the expression simplifies to -1. To see this, we factor out -1 from the numerator or the denominator.

Continued on next page

PRACTICE PROBLEMS 2–3

Simplify each rational expression.

2. $\dfrac{3y^3}{6y^4-3y^3}$

3. $\dfrac{5x^2+13x+6}{6x^2+7x-10}$

PRACTICE PROBLEMS 4–6

Simplify each rational expression.

4. $\dfrac{5+x}{x+5}$ **5.** $\dfrac{5-x}{x-5}$

6. $\dfrac{3-3x^2}{x^2+x-2}$

Answers

2. $\dfrac{1}{2y-1}$,   **3.** $\dfrac{5x+3}{6x-5}$,   **4.** 1,   **5.** -1,

6. $-\dfrac{3(x+1)}{x+2}$

If -1 is factored from the *numerator,* then

$$\frac{2 - x}{x - 2} = \frac{-1(-2 + x)}{x - 2} = \frac{-1\,(x - 2)}{x - 2} = \frac{-1}{1} = -1$$

If -1 is factored from the *denominator,* the result is the same.

$$\frac{2 - x}{x - 2} = \frac{2 - x}{-1(-x + 2)} = \frac{2 - x}{-1\,(2 - x)} = \frac{1}{-1} = -1$$

6. $\dfrac{18 - 2x^2}{x^2 - 2x - 3} = \dfrac{2(9 - x^2)}{(x + 1)(x - 3)}$ Factor.

$$= \frac{2(3 + x)(3 - x)}{(x + 1)(x - 3)}$$ Factor completely.

Notice the opposites $3 - x$ and $x - 3$. We write $3 - x$ as $-1(x - 3)$ and simplify.

$$\frac{2(3 + x)(3 - x)}{(x + 1)(x - 3)} = \frac{2(3 + x) \cdot -1\,(x - 3)}{(x + 1)\,(x - 3)} = -\frac{2(3 + x)}{x + 1}$$

Work Practice Problems 4–6

✔ **Concept Check** Which of the following expressions are equivalent to $\dfrac{x}{8 - x}$?

a. $\dfrac{-x}{x - 8}$ **b.** $\dfrac{-x}{8 - x}$ **c.** $\dfrac{x}{x - 8}$ **d.** $\dfrac{-x}{-8 + x}$

EXAMPLES Simplify each rational expression.

7. $\dfrac{x^3 + 8}{x + 2} = \dfrac{(x + 2)\,(x^2 - 2x + 4)}{x + 2}$ Factor the sum of the two cubes.

$$= x^2 - 2x + 4$$ Simplest form.

8. $\dfrac{2y^2 + 2}{y^3 - 5y^2 + y - 5} = \dfrac{2(y^2 + 1)}{(y^3 - 5y^2) + (y - 5)}$ Factor the numerator.

$$= \frac{2(y^2 + 1)}{y^2(y - 5) + 1(y - 5)}$$ Factor the denominator by grouping.

$$= \frac{2\,(y^2 + 1)}{(y - 5)\,(y^2 + 1)}$$

$$= \frac{2}{y - 5}$$ Simplest form.

Work Practice Problems 7–8

✔**Concept Check** Does $\dfrac{n}{n+2}$ simplify to $\dfrac{1}{2}$? Why or why not?

Objective ⓒ Multiplying Rational Expressions

Arithmetic operations on rational expressions are performed in the same way as they are on rational numbers. To multiply rational expressions, we multiply numerators and multiply denominators.

Multiplying Rational Expressions

The rule for multiplying rational expressions is

$$\frac{P}{Q} \cdot \frac{R}{S} = \frac{PR}{QS} \quad \text{as long as } Q \neq 0 \text{ and } S \neq 0.$$

To multiply rational expressions, you may use these steps:

Step 1: Completely factor each numerator and denominator.

Step 2: Use the rule above and multiply the numerators and the denominators.

Step 3: Simplify the product.

When we multiply rational expressions, notice that we factor each numerator and denominator first. This helps when we check to see whether the product is in simplest form.

EXAMPLES Multiply.

9. $\dfrac{3n+1}{2n} \cdot \dfrac{2n-4}{3n^2-2n-1} = \dfrac{3n+1}{2n} \cdot \dfrac{2(n-2)}{(3n+1)(n-1)}$ Factor.

$$= \frac{(3n+1) \cdot 2 \,(n-2)}{2 \, n \,(3n+1)\,(n-1)} \qquad \text{Multiply.}$$

$$= \frac{n-2}{n(n-1)} \qquad \text{Simplest form.}$$

10. $\dfrac{x^3-1}{-3x+3} \cdot \dfrac{15x^2}{x^2+x+1} = \dfrac{(x-1)(x^2+x+1)}{-3(x-1)} \cdot \dfrac{15x^2}{x^2+x+1}$ Factor.

$$= \frac{(x-1)(x^2+x+1)\cdot 3 \cdot 5x^2}{-1 \cdot 3(x-1)(x^2+x+1)} \qquad \text{Factor.}$$

$$= \frac{5x^2}{-1} = -5x^2 \qquad \text{Simplest form.}$$

▨ **Work Practice Problems 9–10**

Objective ⒟ Dividing Rational Expressions

Recall that two numbers are reciprocals of each other if their product is 1. Similarly, if $\dfrac{P}{Q}$ is a rational expression and $P \neq 0$, then $\dfrac{Q}{P}$ is its **reciprocal**, since

$$\frac{P}{Q} \cdot \frac{Q}{P} = \frac{P \cdot Q}{Q \cdot P} = 1$$

PRACTICE PROBLEMS 9–10

Multiply.

9. $\dfrac{2x-3}{5x} \cdot \dfrac{5x+5}{2x^2-x-3}$

10. $\dfrac{x^3+27}{-2x-6} \cdot \dfrac{4x^3}{x^2-3x+9}$

Answers

9. $\dfrac{1}{x}$,   **10.** $-2x^3$

✔ **Concept Check Answer**

no; answers may vary

The following are examples of expressions and their reciprocals.

Expression	Reciprocal
$\dfrac{3}{x}$	$\dfrac{x}{3}$
$\dfrac{2 + x^2}{4x - 3}$	$\dfrac{4x - 3}{2 + x^2}$
x^3	$\dfrac{1}{x^3}$
0	no reciprocal

Dividing Rational Expressions

The rule for dividing rational expressions is

$$\frac{P}{Q} \div \frac{R}{S} = \frac{P}{Q} \cdot \frac{S}{R} = \frac{PS}{QR} \quad \text{as long as } Q \neq 0, S \neq 0, \text{ and } R \neq 0.$$

To divide by a rational expression, use the rule above and multiply by its reciprocal. Then simplify if possible.

Notice that division of rational expressions is the same as for rational numbers.

EXAMPLES Divide.

11. $\dfrac{8m^2}{3m^2 - 12} \div \dfrac{40}{2 - m} = \dfrac{8m^2}{3m^2 - 12} \cdot \dfrac{2 - m}{40}$ Multiply by the reciprocal of the divisor.

$= \dfrac{8m^2(2 - m)}{3(m + 2)(m - 2) \cdot 40}$ Factor and multiply.

$= \dfrac{8\,m^2 \cdot -1\,(m - 2)}{3(m + 2)\,(m - 2) \cdot 8 \cdot 5}$ Write $(2 - m)$ as $-1(m - 2)$.

$= -\dfrac{m^2}{15(m + 2)}$ Simplify.

12. $\dfrac{18y^2 + 9y - 2}{24y^2 - 10y + 1} \div \dfrac{3y^2 + 17y + 10}{8y^2 + 18y - 5}$

$= \dfrac{18y^2 + 9y - 2}{24y^2 - 10y + 1} \cdot \dfrac{8y^2 + 18y - 5}{3y^2 + 17y + 10}$ Multiply by the reciprocal.

$= \dfrac{(6y - 1)\,(3y + 2) \cdot (4y - 1)\,(2y + 5)}{(6y - 1)\,(4y - 1) \cdot (3y + 2)\,(y + 5)}$ Factor.

$= \dfrac{2y + 5}{y + 5}$ Simplest form.

■ **Work Practice Problems 11–12**

PRACTICE PROBLEMS 11–12

Divide.

11. $\dfrac{12y^3}{5y^2 - 5} \div \dfrac{6}{1 - y}$

12. $\dfrac{8z^2 + 14z + 3}{20z^2 + z - 1}$

$\div \dfrac{2z^2 + 7z + 6}{35z^2 + 3z - 2}$

Helpful Hint

When dividing rational expressions, do not divide out common factors until the division problem is rewritten as a multiplication problem.

Answers

11. $-\dfrac{2y^3}{5(y + 1)}$, **12.** $\dfrac{7z + 2}{z + 2}$

EXAMPLE 13 Perform each indicated operation.

$$\frac{x^2 - 25}{(x + 5)^2} \cdot \frac{3x + 15}{4x} \div \frac{x^2 - 3x - 10}{x}$$

Solution:

$$\frac{x^2 - 25}{(x + 5)^2} \cdot \frac{3x + 15}{4x} \div \frac{x^2 - 3x - 10}{x}$$

$$= \frac{x^2 - 25}{(x + 5)^2} \cdot \frac{3x + 15}{4x} \cdot \frac{x}{x^2 - 3x - 10} \quad \text{To divide, multiply by the reciprocal.}$$

$$= \frac{(x + 5)\ (x - 5)}{(x + 5)\ (x + 5)} \cdot \frac{3\ (x + 5)}{4\ x} \cdot \frac{x}{(x - 5)\ (x + 2)}$$

$$= \frac{3}{4(x + 2)}$$

■ **Work Practice Problem 13**

Objective **E** **Applications with Rational Functions**

Rational functions are often used to model real-life situations. Don't forget to be aware of the domains of these functions. See the Graphing Calculator Explorations on page 410 for further domain exercises.

EXAMPLE 14 **Finding Unit Cost**

For the ICL Production Company, the rational function $C(x) = \dfrac{2.6x + 10,000}{x}$ describes the company's cost per disc of pressing x compact discs. Find the cost per disc for pressing:

a. 100 compact discs
b. 1000 compact discs

Solution:

a. $C(100) = \dfrac{2.6(100) + 10,000}{100} = \dfrac{10,260}{100} = 102.6$

The cost per disc for pressing 100 compact discs is $102.60.

b. $C(1000) = \dfrac{2.6(1000) + 10,000}{1000} = \dfrac{12,600}{1000} = 12.6$

The cost per disc for pressing 1000 compact discs is $12.60. Notice that as more compact discs are produced, the cost per disc decreases.

■ **Work Practice Problem 14**

PRACTICE PROBLEM 13

Perform each indicated operation.

$$\frac{(x + 3)^2}{x^2 - 9} \cdot \frac{2x - 6}{5x}$$
$$\div \frac{x^2 + 7x + 12}{x}$$

PRACTICE PROBLEM 14

A company's cost per book for printing x particular books is given by the rational function $C(x) = \dfrac{0.8x + 5000}{x}$.

Find the cost per book for printing:

a. 100 books
b. 1000 books

Answers

13. $\dfrac{2}{5(x + 4)}$,   **14. a.** $50.80,   **b.** $5.80

CALCULATOR EXPLORATIONS Graphing

Recall that since the rational expression $\dfrac{7x-2}{(x-2)(x+5)}$ is not defined when $x = 2$ or when $x = -5$, we say that the domain of the rational function $f(x) = \dfrac{7x-2}{(x-2)(x+5)}$ is all real numbers except 2 and -5. This domain can be written as $\{x \mid x$ is a real number and $x \neq 2, x \neq -5\}$. This means that the graph of f should not cross the vertical lines $x = 2$ and $x = -5$. The graph of f in *connected* mode is shown below. In connected mode the grapher tries to connect all dots of the graph so that the result is a smooth curve. This is what has happened in the graph. Notice that the graph appears to contain vertical lines at $x = 2$ and at $x = -5$. We know that this cannot happen because the function is not defined at $x = 2$ and at $x = -5$. We also know that this cannot happen because the graph of this function would not pass the vertical line test.

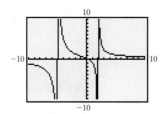

If we graph f in *dot* mode, the graph appears as below. In dot mode the grapher will not connect dots with a smooth curve. Notice that the vertical lines have disappeared, and we have a better picture of the graph. It actually appears more like the hand-drawn graph to its right. By using a TABLE feature, a CALCULATE VALUE feature, or by tracing, we can see that the function is not defined at $x = 2$ and at $x = -5$.

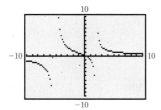

Note: Some calculator manufacturers now offer download-able operating systems that eliminate the need to use dot mode to graph rational functions.

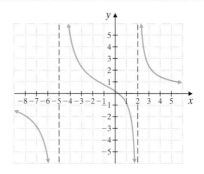

Find the domain of each rational function. Then graph each rational function and use the graph to confirm the domain.

1. $f(x) = \dfrac{5x}{x-6}$

2. $f(x) = \dfrac{x}{x+4}$

3. $f(x) = \dfrac{x+1}{x^2-4}$

4. $g(x) = \dfrac{5x}{x^2-9}$

5. $h(x) = \dfrac{x^2}{2x^2+7x-4}$

6. $f(x) = \dfrac{3x+2}{4x^2-19x-5}$

7. $g(x) = \dfrac{x^2+x+1}{5}$

8. $h(x) = \dfrac{x^2+25}{2}$

Mental Math

Multiply.

1. $\dfrac{x}{5} \cdot \dfrac{y}{2}$ **2.** $\dfrac{y}{6} \cdot \dfrac{z}{5}$ **3.** $\dfrac{2}{x} \cdot \dfrac{y}{3}$ **4.** $\dfrac{a}{5} \cdot \dfrac{7}{b}$ **5.** $\dfrac{m}{6} \cdot \dfrac{m}{6}$ **6.** $\dfrac{9}{x} \cdot \dfrac{8}{x}$

6.1 EXERCISE SET

Objective A *Find the domain of each rational expression. See Example 1.*

1. $f(x) = \dfrac{5x - 7}{4}$

2. $g(x) = \dfrac{4 - 3x}{2}$

3. $s(t) = \dfrac{t^2 + 1}{2t}$

4. $v(t) = -\dfrac{5t + t^2}{3t}$

5. $f(x) = \dfrac{3x}{7 - x}$

6. $f(x) = \dfrac{-4x}{-2 + x}$

7. $f(x) = \dfrac{x}{3x - 1}$

8. $g(x) = \dfrac{-2}{2x + 5}$

9. $R(x) = \dfrac{3 + 2x}{x^3 + x^2 - 2x}$

10. $h(x) = \dfrac{5 - 3x}{2x^2 - 14x + 20}$

11. $C(x) = \dfrac{x + 3}{x^2 - 4}$

12. $R(x) = \dfrac{5}{x^2 - 7x}$

Objective B *Simplify each rational expression. See Examples 2 through 8.*

13. $\dfrac{8x - 16x^2}{8x}$

14. $\dfrac{3x - 6x^2}{3x}$

15. $\dfrac{x^2 - 9}{3 + x}$

16. $\dfrac{x^2 - 25}{5 + x}$

17. $\dfrac{9y - 18}{7y - 14}$

18. $\dfrac{6y - 18}{2y - 6}$

19. $\dfrac{x^2 + 6x - 40}{x + 10}$

20. $\dfrac{x^2 - 8x + 16}{x - 4}$

21. $\dfrac{x - 9}{9 - x}$

22. $\dfrac{x - 4}{4 - x}$

23. $\dfrac{x^2 - 49}{7 - x}$

24. $\dfrac{x^2 - y^2}{y - x}$

25. $\dfrac{2x^2 - 7x - 4}{x^2 - 5x + 4}$

26. $\dfrac{3x^2 - 11x + 10}{x^2 - 7x + 10}$

27. $\dfrac{x^3 - 125}{2x - 10}$

28. $\dfrac{4x + 4}{x^3 + 1}$

29. $\dfrac{3x^2 - 5x - 2}{6x^3 + 2x^2 + 3x + 1}$

30. $\dfrac{2x^2 - x - 3}{2x^3 - 3x^2 + 2x - 3}$

31. $\dfrac{9x^2 - 15x + 25}{27x^3 + 125}$

32. $\dfrac{8x^3 - 27}{4x^2 + 6x + 9}$

Objective C *Multiply and simplify. See Examples 9 and 10.*

33. $\dfrac{2x - 4}{15} \cdot \dfrac{6}{2 - x}$

34. $\dfrac{10 - 2x}{7} \cdot \dfrac{14}{5x - 25}$

35. $\dfrac{18a - 12a^2}{4a^2 + 4a + 1} \cdot \dfrac{4a^2 + 8a + 3}{4a^2 - 9}$

36. $\dfrac{a - 5b}{a^2 + ab} \cdot \dfrac{b^2 - a^2}{10b - 2a}$

37. $\dfrac{9x + 9}{4x + 8} \cdot \dfrac{2x + 4}{3x^2 - 3}$

38. $\dfrac{2x^2 - 2}{10x + 30} \cdot \dfrac{12x + 36}{3x - 3}$

39. $\dfrac{2x^3 - 16}{6x^2 + 6x - 36} \cdot \dfrac{9x + 18}{3x^2 + 6x + 12}$

40. $\dfrac{x^2 - 3x + 9}{5x^2 - 20x - 105} \cdot \dfrac{x^2 - 49}{x^3 + 27}$

41. $\dfrac{a^3 + a^2b + a + b}{5a^3 + 5a} \cdot \dfrac{6a^2}{2a^2 - 2b^2}$

42. $\dfrac{4a^2 - 8a}{ab - 2b + 3a - 6} \cdot \dfrac{8b + 24}{3a + 6}$

43. $\dfrac{x^2 - 6x - 16}{2x^2 - 128} \cdot \dfrac{x^2 + 16x + 64}{3x^2 + 30x + 48}$

44. $\dfrac{2x^2 + 12x - 32}{x^2 + 16x + 64} \cdot \dfrac{x^2 + 10x + 16}{x^2 - 3x - 10}$

Objective D *Divide and simplify. See Examples 11 and 12.*

45. $\dfrac{2x}{5} \div \dfrac{6x + 12}{5x + 10}$

46. $\dfrac{7}{3x} \div \dfrac{14 - 7x}{18 - 9x}$

47. $\dfrac{a + b}{ab} \div \dfrac{a^2 - b^2}{4a^3b}$

48. $\dfrac{6a^2b^2}{a^2 - 4} \div \dfrac{3ab^2}{a - 2}$

49. $\dfrac{x^2 - 6x + 9}{x^2 - x - 6} \div \dfrac{x^2 - 9}{4}$

50. $\dfrac{x^2 - 4}{3x + 6} \div \dfrac{2x^2 - 8x + 8}{x^2 + 4x + 4}$

51. $\dfrac{x^2 - 6x - 16}{2x^2 - 128} \div \dfrac{x^2 + 10x + 16}{x^2 + 16x + 64}$

52. $\dfrac{a^2 - a - 6}{a^2 - 81} \div \dfrac{a^2 - 7a - 18}{4a + 36}$

53. $\dfrac{3x - x^2}{x^3 - 27} \div \dfrac{x}{x^2 + 3x + 9}$

54. $\dfrac{x^2 - 3x}{x^3 - 27} \div \dfrac{2x}{2x^2 + 6x + 18}$

55. $\dfrac{8b + 24}{3a + 6} \div \dfrac{ab - 2b + 3a - 6}{a^2 - 4a + 4}$

56. $\dfrac{2a^2 - 2b^2}{a^3 + a^2b + a + b} \div \dfrac{6a^2}{a^3 + a}$

Objectives B C D Mixed Practice *Perform each indicated operation. See Examples 2 through 13.*

57. $\dfrac{x^2 - 9}{4} \cdot \dfrac{x^2 - x - 6}{x^2 - 6x + 9}$

58. $\dfrac{x^2 - 4}{9} \cdot \dfrac{x^2 - 6x + 9}{x^2 - 5x + 6}$

59. $\dfrac{2x^2 - 4x - 30}{5x^2 - 40x - 75} \div \dfrac{x^2 - 8x + 15}{x^2 - 6x + 9}$

60. $\dfrac{4a + 36}{a^2 - 7a - 18} \div \dfrac{a^2 - a - 6}{a^2 - 81}$

61. Simplify: $\dfrac{r^3 + s^3}{r + s}$

62. Simplify: $\dfrac{m^3 - n^3}{m - n}$

63. $\dfrac{4}{x} \div \dfrac{3xy}{x^2} \cdot \dfrac{6x^2}{x^4}$

64. $\dfrac{4}{x} \cdot \dfrac{3xy}{x^2} \div \dfrac{6x^2}{x^4}$

65. $\dfrac{3x^2 - 5x - 2}{y^2 + y - 2} \cdot \dfrac{y^2 + 4y - 5}{12x^2 + 7x + 1} \div \dfrac{5x^2 - 9x - 2}{8x^2 - 2x - 1}$

66. $\dfrac{x^2 + x - 2}{3y^2 - 5y - 2} \cdot \dfrac{12y^2 + y - 1}{x^2 + 4x - 5} \div \dfrac{8y^2 - 6y + 1}{5y^2 - 9y - 2}$

Objective **E** *Find each function value. See Example 14.*

67. If $f(x) = \dfrac{x + 8}{2x - 1}$, find $f(2)$, $f(0)$, and $f(-1)$.

68. If $f(x) = \dfrac{x - 2}{-5 + x}$, find $f(-5)$, $f(0)$, and $f(10)$.

69. $g(x) = \dfrac{x^2 + 8}{x^3 - 25x}$; $g(3)$, $g(-2)$, $g(1)$

70. $s(t) = \dfrac{t^3 + 1}{t^2 + 1}$; $s(-1)$, $s(1)$, $s(2)$

71. The total revenue from the sale of a popular book is approximated by the rational function
$$R(x) = \frac{1000x^2}{x^2 + 4},$$ where x is the number of years since publication and $R(x)$ is the total revenue in millions of dollars.

  a. Find the total revenue at the end of the first year.
  b. Find the total revenue at the end of the second year.
  c. Find the revenue during the second year only.
  d. Find the domain of function R.

72. The function $f(x) = \dfrac{100,000x}{100 - x}$ models the cost in dollars for removing x percent of the pollutants from a bayou in which a nearby company dumped creosol.

  a. Find the cost of removing 20% of the pollutants from the bayou. [*Hint:* Find $f(20)$.]
  b. Find the cost of removing 60% of the pollutants and then 80% of the pollutants.
  c. Find $f(90)$, then $f(95)$, and then $f(99)$. What happens to the cost as x approaches 100%?
  d. Find the domain of function f.

Review

Perform each indicated operation. See Section 1.4.

73. $\dfrac{4}{5} + \dfrac{3}{5}$

74. $\dfrac{4}{10} - \dfrac{7}{10}$

75. $\dfrac{5}{28} - \dfrac{2}{21}$

76. $\dfrac{5}{13} + \dfrac{2}{7}$

77. $\dfrac{3}{8} + \dfrac{1}{2} - \dfrac{3}{16}$

78. $\dfrac{2}{9} - \dfrac{1}{6} + \dfrac{2}{3}$

Concept Extensions

Solve. For Exercises 79 and 80, see the first Concept Check in this section; for Exercises 81 and 82 see the second Concept Check.

79. Which of the expressions are equivalent to $\dfrac{x}{5-x}$?

   a. $\dfrac{-x}{5-x}$ **b.** $\dfrac{-x}{-5+x}$

   c. $\dfrac{x}{x-5}$ **d.** $\dfrac{-x}{x-5}$

80. Which of the expressions are equivalent to $\dfrac{-2+x}{x}$?

   a. $\dfrac{2-x}{-x}$ **b.** $-\dfrac{2-x}{x}$

   c. $\dfrac{x-2}{x}$ **d.** $\dfrac{x-2}{-x}$

81. Does $\dfrac{x}{x+5}$ simplify to $\dfrac{1}{5}$? Why or why not?

82. Does $\dfrac{x+7}{x}$ simplify to 7? Why or why not?

△ 83. Find the area of the rectangle.

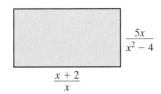

△ 84. Find the area of the triangle.

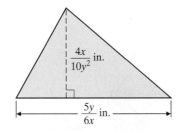

△ 85. A parallelogram has an area of $\dfrac{x^2+x-2}{x^3}$ square feet and a height of $\dfrac{x^2}{x-1}$ feet. Express the length of its base as a rational expression in x. (*Hint:* Since $A = b \cdot h$, then $b = \dfrac{A}{h}$ or $b = A \div h$.)

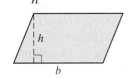

86. A lottery prize of $\dfrac{15x^3}{y^2}$ dollars is to be divided among $5x$ people. Express the amount of money each person is to receive as a rational expression in x and y.

87. In your own words explain how to simplify a rational expression.

88. In your own words, explain the difference between multiplying rational expressions and dividing rational expressions.

89. Decide whether each rational expression equals 1, −1, or neither.

   a. $\dfrac{x+5}{5+x}$ **b.** $\dfrac{x-5}{5-x}$

   c. $\dfrac{x+5}{x-5}$ **d.** $\dfrac{-x-5}{x+5}$

   e. $\dfrac{x-5}{-x+5}$ **f.** $\dfrac{-5+x}{x-5}$

90. In our definition of division for

$$\dfrac{P}{Q} \div \dfrac{R}{S}$$

we stated that $Q \neq 0$, $S \neq 0$, and $R \neq 0$. Explain why R cannot equal 0.

91. Find the polynomial in the second numerator such that the following statement is true.

$$\dfrac{x^2-4}{x^2-7x+10} \cdot \dfrac{?}{2x^2+11x+14} = 1$$

92. In your own words, explain how to find the domain of a rational function.

Simplify. Assume that no denominator is 0.

93. $\dfrac{p^x - 4}{4 - p^x}$

94. $\dfrac{3 + q^n}{q^n + 3}$

95. $\dfrac{x^n + 4}{x^{2n} - 16}$

96. $\dfrac{x^{2k} - 9}{3 + x^k}$

 STUDY SKILLS BUILDER

Are You Satisfied with Your Performance in this Course thus Far?

To see if there is room for improvement, answer these questions:

1. Am I attending all classes and arriving on time?

2. Am I working and checking my homework assignments on time?

3. Am I getting help (from my instructor or a campus learning resource lab) when I need it?

4. In addition to my instructor, am I using the text supplements that might help me?

5. Am I satisfied with my performance on quizzes and exams?

If you answered no to any of these questions, read or reread Section 1.1 for suggestions in these areas. Also, you might want to contact your instructor for additional feedback.

A Add or Subtract Rational Expressions with the Same Denominator.

B Find the Least Common Denominator (LCD) of Two or More Rational Expressions.

C Add and Subtract Rational Expressions with Different Denominators.

6.2 ADDING AND SUBTRACTING RATIONAL EXPRESSIONS

Objective **A** Adding or Subtracting Rational Expressions with the Same Denominator

We add or subtract rational expressions just as we add or subtract fractions.

Adding or Subtracting Rational Expressions with Common Denominators

If $\dfrac{P}{Q}$ and $\dfrac{R}{Q}$ are rational expressions, then

$$\frac{P}{Q} + \frac{R}{Q} = \frac{P+R}{Q} \quad \text{and} \quad \frac{P}{Q} - \frac{R}{Q} = \frac{P-R}{Q}$$

To add or subtract rational expressions with common denominators, add or subtract the numerators and write the sum or difference over the common denominator.

PRACTICE PROBLEMS 1–2

Add.

1. $\dfrac{9}{11x^4} + \dfrac{y}{11x^4}$ **2.** $\dfrac{x}{6} + \dfrac{7x}{6}$

EXAMPLES Add.

1. $\dfrac{5}{7z^2} + \dfrac{x}{7z^2} = \dfrac{5+x}{7z^2}$ Add the numerators and write the result over the common denominator.

2. $\dfrac{x}{4} + \dfrac{5x}{4} = \dfrac{x+5x}{4} = \dfrac{6x}{4} = \dfrac{3x}{2}$

■ **Work Practice Problems 1–2**

PRACTICE PROBLEMS 3–4

Subtract.

3. $\dfrac{x^2}{x+3} - \dfrac{9}{x+3}$

4. $\dfrac{a}{5b^3} - \dfrac{a+2}{5b^3}$

EXAMPLES Subtract.

3. $\dfrac{x^2}{x+7} - \dfrac{49}{x+7} = \dfrac{x^2-49}{x+7}$ Subtract the numerators and write the result over the common denominator.

$$= \frac{(x+7)(x-7)}{x+7}$$ Factor the numerator.

$$= x - 7$$ Simplify.

4. $\dfrac{x}{3y^2} - \dfrac{x+1}{3y^2} = \dfrac{x-(x+1)}{3y^2}$ Subtract the numerators.

$$= \frac{x-x-1}{3y^2}$$ Use the distributive property.

$$= -\frac{1}{3y^2}$$ Simplify.

Helpful Hint
Be sure to insert parentheses here so that the entire second numerator is subtracted.

■ **Work Practice Problems 3–4**

Answers

1. $\dfrac{9+y}{11x^4}$, **2.** $\dfrac{4x}{3}$, **3.** $x-3$, **4.** $-\dfrac{2}{5b^3}$

✔ Concept Check Answer

$\dfrac{3+2y}{y^2-1} - \dfrac{y+3}{y^2-1} = \dfrac{3+2y-y-3}{y^2-1}$

$= \dfrac{y}{y^2-1}$

✔ Concept Check Find and correct the error.

$$\frac{3+2y}{y^2-1} - \frac{y+3}{y^2-1}$$

$$= \frac{3+2y-y+3}{y^2-1}$$

$$= \frac{y+6}{y^2-1}$$

Objective B Finding the LCD of Rational Expressions

To add or subtract rational expressions with unlike, or different, denominators, we first write the rational expressions as equivalent rational expressions with common denominators.

The **least common denominator (LCD)** is usually the easiest common denominator to work with. The LCD of a list of rational expressions is a polynomial of least degree whose factors include all the factors of the denominators in the list.

The following steps can be used to find the LCD.

Finding the Least Common Denominator (LCD)

Step 1: Factor each denominator completely.

Step 2: The LCD is the product of all unique factors each raised to a power equal to the greatest number of times that the factor appears in any one factored denominator.

EXAMPLE 5 Find the LCD of the rational expressions in each list.

a. $\dfrac{2}{3x^5y^2}, \dfrac{3z}{5xy^3}$

b. $\dfrac{7}{z+1}, \dfrac{z}{z-1}$

c. $\dfrac{m-1}{m^2-25}, \dfrac{2m}{2m^2-9m-5}, \dfrac{7}{m^2-10m+25}$

d. $\dfrac{x}{x^2-4}, \dfrac{11}{6-3x}$

Solution:

a. First we factor each denominator.

$3x^5y^2 = 3 \cdot x^5 \cdot y^2$
$5xy^3 = 5 \cdot x \cdot y^3$
$\text{LCD} = 3 \cdot 5 \cdot x^5 \cdot y^3 = 15x^5y^3$

> **Helpful Hint**
> The greatest power of x is 5, so we have a factor of x^5. The greatest power of y is 3, so we have a factor of y^3.

b. The denominators $z+1$ and $z-1$ do not factor further.

$(z+1) = (z+1)$
$(z-1) = (z-1)$
$\text{LCD} = (z+1)(z-1)$

c. We first factor each denominator.

$m^2 - 25 = (m+5)(m-5)$
$2m^2 - 9m - 5 = (2m+1)(m-5)$
$m^2 - 10m + 25 = (m-5)(m-5)$
$\text{LCD} = (m+5)(2m+1)(m-5)^2$

d. We factor each denominator.

$x^2 - 4 = (x+2)(x-2)$

$6 - 3x = 3(2-x) = 3(-1)(x-2)$

$\text{LCD} = 3(-1)(x+2)(x-2)$
$= -3(x+2)(x-2)$

> **Helpful Hint**
> $(x-2)$ and $(2-x)$ are opposite factors. Notice that a -1 was factored from $(2-x)$ so that the factors are identical.

📖 **Work Practice Problem 5**

> **Helpful Hint**
>
> If opposite factors occur, do not use both in the LCD. Instead, factor -1 from one of the opposite factors so that the factors are then identical.

Objective C Adding or Subtracting Rational Expressions with Different Denominators

To add or subtract rational expressions with different denominators, we write each rational expression as an equivalent rational expression with the LCD as the denominator. To do this, we use the multiplication property of 1 and multiply each rational expression by a form of 1 so that each denominator becomes the LCD.

> ### Adding or Subtracting Rational Expressions with Different Denominators
>
> **Step 1:** Find the LCD of the rational expressions.
>
> **Step 2:** Write each rational expression as an equivalent rational expression whose denominator is the LCD found in Step 1.
>
> **Step 3:** Add or subtract numerators, and write the result over the common denominator.
>
> **Step 4:** Simplify the resulting rational expression.

PRACTICE PROBLEM 6

Add: $\dfrac{7}{a^3} + \dfrac{9}{2a^4}$

EXAMPLE 6 Add: $\dfrac{2}{x^2} + \dfrac{5}{3x^3}$

Solution: The LCD is $3x^3$. To write the first rational expression as an equivalent rational expression with denominator $3x^3$, we multiply by 1 in the form of $\dfrac{3x}{3x}$.

$$\frac{2}{x^2} + \frac{5}{3x^3} = \frac{2 \cdot 3x}{x^2 \cdot 3x} + \frac{5}{3x^3} \qquad \text{The second expression already has a denominator of } 3x^3.$$

$$= \frac{6x}{3x^3} + \frac{5}{3x^3}$$

$$= \frac{6x + 5}{3x^3} \qquad \text{Add the numerators.}$$

■ **Work Practice Problem 6**

PRACTICE PROBLEM 7

Add: $\dfrac{1}{x + 5} + \dfrac{6x}{x - 5}$

EXAMPLE 7 Add: $\dfrac{3}{x + 2} + \dfrac{2x}{x - 2}$

Solution: The LCD is the product of the two denominators: $(x + 2)(x - 2)$.

$$\frac{3}{x + 2} + \frac{2x}{x - 2} = \frac{3 \cdot (x - 2)}{(x + 2) \cdot (x - 2)} + \frac{2x \cdot (x + 2)}{(x - 2) \cdot (x + 2)} \qquad \text{Write equivalent rational expressions.}$$

$$= \frac{3x - 6}{(x + 2)(x - 2)} + \frac{2x^2 + 4x}{(x + 2)(x - 2)} \qquad \text{Multiply in the numerators.}$$

$$= \frac{3x - 6 + 2x^2 + 4x}{(x + 2)(x - 2)} \qquad \text{Add the numerators.}$$

$$= \frac{2x^2 + 7x - 6}{(x + 2)(x - 2)} \qquad \text{Simplify.}$$

■ **Work Practice Problem 7**

Answers

6. $\dfrac{14a + 9}{2a^4}$,   **7.** $\dfrac{6x^2 + 31x - 5}{(x + 5)(x - 5)}$

EXAMPLE 8 Subtract: $\dfrac{2x-6}{x-1} - \dfrac{4}{1-x}$

Solution: The LCD is either $x-1$ or $1-x$. To get a common denominator of $x-1$, we factor -1 from the denominator of the second rational expression.

$$\dfrac{2x-6}{x-1} - \dfrac{4}{1-x} = \dfrac{2x-6}{x-1} - \dfrac{4}{-1(x-1)} \quad \text{Write } 1-x \text{ as } -1(x-1).$$

$$= \dfrac{2x-6}{x-1} - \dfrac{-1\cdot 4}{x-1} \quad \text{Write } \dfrac{4}{-1(x-1)} \text{ as } \dfrac{-1\cdot 4}{x-1}.$$

$$= \dfrac{2x-6-(-4)}{x-1}$$

$$= \dfrac{2x-6+4}{x-1} \quad \text{Simplify.}$$

$$= \dfrac{2x-2}{x-1}$$

$$= \dfrac{2(x-1)}{x-1}$$

$$= 2$$

🔲 **Work Practice Problem 8**

EXAMPLE 9 Subtract: $\dfrac{5k}{k^2-4} - \dfrac{2}{k^2+k-2}$

Solution:

$$\dfrac{5k}{k^2-4} - \dfrac{2}{k^2+k-2} = \dfrac{5k}{(k+2)(k-2)} - \dfrac{2}{(k+2)(k-1)} \quad \begin{array}{l}\text{Factor each}\\ \text{denominator to}\\ \text{find the LCD.}\end{array}$$

The LCD is $(k+2)(k-2)(k-1)$. We write equivalent rational expressions with the LCD as the denominators.

$$\dfrac{5k}{(k+2)(k-2)} - \dfrac{2}{(k+2)(k-1)}$$

$$= \dfrac{5k\cdot(k-1)}{(k+2)(k-2)\cdot(k-1)} - \dfrac{2\cdot(k-2)}{(k+2)(k-1)\cdot(k-2)}$$

$$= \dfrac{5k^2-5k}{(k+2)(k-2)(k-1)} - \dfrac{2k-4}{(k+2)(k-2)(k-1)} \quad \begin{array}{l}\text{Multiply in the}\\ \text{numerators.}\end{array}$$

$$= \dfrac{5k^2-5k-2k+4}{(k+2)(k-2)(k-1)} \quad \begin{array}{l}\text{Subtract the}\\ \text{numerators.}\end{array}$$

$$= \dfrac{5k^2-7k+4}{(k+2)(k-2)(k-1)} \quad \text{Simplify.}$$

The numerator is a prime polynomial, so the expression cannot be simplified further.

🔲 **Work Practice Problem 9**

PRACTICE PROBLEM 8

Subtract: $\dfrac{3m-26}{m-6} - \dfrac{8}{6-m}$

PRACTICE PROBLEM 9

Subtract:

$\dfrac{2x}{x^2-9} - \dfrac{3}{x^2-4x+3}$

Answers

8. 3, **9.** $\dfrac{2x^2-5x-9}{(x+3)(x-3)(x-1)}$

PRACTICE PROBLEM 10

Add:

$$\frac{x + 1}{x^2 + x - 12} + \frac{2x - 1}{x^2 + 6x + 8}$$

EXAMPLE 10 Add: $\dfrac{2x - 1}{2x^2 - 9x - 5} + \dfrac{x + 3}{6x^2 - x - 2}$ Factor the denominators.

Solution:

$$\frac{2x - 1}{2x^2 - 9x - 5} + \frac{x + 3}{6x^2 - x - 2} = \frac{2x - 1}{(2x + 1)(x - 5)} + \frac{x + 3}{(2x + 1)(3x - 2)}$$

The LCD is $(2x + 1)(x - 5)(3x - 2)$.

$$= \frac{(2x - 1) \cdot (3x - 2)}{(2x + 1)(x - 5) \cdot (3x - 2)} + \frac{(x + 3) \cdot (x - 5)}{(2x + 1)(3x - 2) \cdot (x - 5)}$$

$$= \frac{6x^2 - 7x + 2}{(2x + 1)(x - 5)(3x - 2)} + \frac{x^2 - 2x - 15}{(2x + 1)(x - 5)(3x - 2)}$$ Multiply in the numerators.

$$= \frac{6x^2 - 7x + 2 + x^2 - 2x - 15}{(2x + 1)(x - 5)(3x - 2)}$$ Add the numerators.

$$= \frac{7x^2 - 9x - 13}{(2x + 1)(x - 5)(3x - 2)}$$ Simplify.

The numerator is a prime polynomial, so the expression cannot be simplified further.

■ **Work Practice Problem 10**

PRACTICE PROBLEM 11

Perform each indicated operation.

$$\frac{6}{x - 5} + \frac{x - 35}{x^2 - 5x} - \frac{2}{x}$$

EXAMPLE 11 Perform each indicated operation:

$$\frac{7}{x - 1} + \frac{10x}{x^2 - 1} - \frac{5}{x + 1}$$

Solution:

$$\frac{7}{x - 1} + \frac{10x}{x^2 - 1} - \frac{5}{x + 1} = \frac{7}{x - 1} + \frac{10x}{(x - 1)(x + 1)} - \frac{5}{x + 1}$$

The LCD is $(x - 1)(x + 1)$.

$$= \frac{7 \cdot (x + 1)}{(x - 1) \cdot (x + 1)} + \frac{10x}{(x - 1) \cdot (x + 1)} - \frac{5 \cdot (x - 1)}{(x + 1) \cdot (x - 1)}$$

$$= \frac{7x + 7}{(x - 1)(x + 1)} + \frac{10x}{(x - 1)(x + 1)} - \frac{5x - 5}{(x + 1)(x - 1)}$$ Multiply in the numerators.

$$= \frac{7x + 7 + 10x - 5x + 5}{(x - 1)(x + 1)}$$ Add and subtract the numerators.

$$= \frac{12x + 12}{(x - 1)(x + 1)}$$ Simplify.

$$= \frac{12(x + 1)}{(x - 1)(x + 1)}$$ Factor the numerator.

$$= \frac{12}{x - 1}$$ Simplify.

■ **Work Practice Problem 11**

Answers

10. $\dfrac{3x^2 - 4x + 5}{(x + 2)(x - 3)(x + 4)}$, 11. $\dfrac{5}{x}$

CALCULATOR EXPLORATIONS Graphing

A grapher can be used to support the results of operations on rational expressions. For example, to verify the result of Example 7, graph

$$Y_1 = \frac{3}{x + 2} + \frac{2x}{x - 2} \quad \text{and} \quad Y_2 = \frac{2x^2 + 7x - 6}{(x + 2)(x - 2)}$$

on the same set of axes. The graphs should be the same. Use a TABLE feature or a TRACE feature to see that this is true.

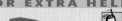
Objective A *Add or subtract as indicated. Simplify each answer. See Examples 1 through 4.*

1. $\dfrac{2}{xz^2} - \dfrac{5}{xz^2}$

2. $\dfrac{4}{x^2y} + \dfrac{2}{x^2y}$

3. $\dfrac{2}{x-2} + \dfrac{x}{x-2}$

4. $\dfrac{x}{5-x} + \dfrac{7}{5-x}$

5. $\dfrac{x^2}{x+2} - \dfrac{4}{x+2}$

6. $\dfrac{x^2}{x+6} - \dfrac{36}{x+6}$

7. $\dfrac{2x-6}{x^2+x-6} + \dfrac{3-3x}{x^2+x-6}$

8. $\dfrac{5x+2}{x^2+2x-8} + \dfrac{2-4x}{x^2+2x-8}$

9. $\dfrac{x-5}{2x} - \dfrac{x+5}{2x}$

10. $\dfrac{x+4}{4x} - \dfrac{x-4}{4x}$

Objective B *Find the LCD of the rational expressions in each list. See Example 5.*

11. $\dfrac{2}{7}, \dfrac{3}{5x}$

12. $\dfrac{4}{5y}, \dfrac{3}{4y^2}$

13. $\dfrac{3}{x}, \dfrac{2}{x+1}$

14. $\dfrac{5}{2x}, \dfrac{7}{2+x}$

15. $\dfrac{12}{x+7}, \dfrac{8}{x-7}$

16. $\dfrac{1}{2x-1}, \dfrac{8}{2x+1}$

17. $\dfrac{5}{3x+6}, \dfrac{2x}{2x-4}$

18. $\dfrac{2}{3a+9}, \dfrac{5}{5a-15}$

19. $\dfrac{2a}{a^2-b^2}, \dfrac{1}{a^2-2ab+b^2}$

20. $\dfrac{2a}{a^2+8a+16}, \dfrac{7a}{a^2+a-12}$

21. $\dfrac{x}{x^2-9}, \dfrac{5}{x}, \dfrac{7}{12-4x}$

22. $\dfrac{9}{x^2-25}, \dfrac{1}{50-10x}, \dfrac{6}{x}$

Objective C *Add or subtract as indicated. Simplify each answer. See Examples 6 and 7.*

23. $\dfrac{4}{3x} + \dfrac{3}{2x}$

24. $\dfrac{10}{7x} - \dfrac{5}{2x}$

25. $\dfrac{5}{2y^2} - \dfrac{2}{7y}$

26. $\dfrac{4}{11x^4} - \dfrac{1}{4x^2}$

27. $\dfrac{x-3}{x+4} - \dfrac{x+2}{x-4}$

28. $\dfrac{x-1}{x-5} - \dfrac{x+2}{x+5}$

29. $\dfrac{1}{x-5} + \dfrac{2x-19}{(x-5)(x+4)}$

30. $\dfrac{4x-2}{(x-5)(x+4)} - \dfrac{2}{x+4}$

Perform the indicated operation. If possible, simplify your answer. See Example 8.

31. $\dfrac{1}{a-b} + \dfrac{1}{b-a}$

32. $\dfrac{1}{a-3} - \dfrac{1}{3-a}$

421

33. $\dfrac{x+1}{1-x} + \dfrac{1}{x-1}$

34. $\dfrac{5}{1-x} - \dfrac{1}{x-1}$

35. $\dfrac{5}{x-2} + \dfrac{x+4}{2-x}$

36. $\dfrac{3}{5-x} + \dfrac{x+2}{x-5}$

Perform each indicated operation. If possible, simplify your answer. See Examples 6 through 10.

37. $\dfrac{y+1}{y^2-6y+8} - \dfrac{3}{y^2-16}$

38. $\dfrac{x+2}{x^2-36} - \dfrac{x}{x^2+9x+18}$

39. $\dfrac{x+4}{3x^2+11x+6} + \dfrac{x}{2x^2+x-15}$

40. $\dfrac{x+3}{5x^2+12x+4} + \dfrac{6}{x^2-x-6}$

41. $\dfrac{7}{x^2-x-2} + \dfrac{x}{x^2+4x+3}$

42. $\dfrac{a}{a^2+10a+25} + \dfrac{4}{a^2+6a+5}$

43. $\dfrac{x+4}{3x^2+11x+6} + \dfrac{x}{2x^2+x-15}$

44. $\dfrac{x+3}{5x^2+12x+4} + \dfrac{6}{x^2-x-6}$

45. $\dfrac{2}{a^2+2a+1} + \dfrac{3}{a^2-1}$

46. $\dfrac{9x+2}{3x^2-2x-8} + \dfrac{7}{3x^2+x-4}$

Objectives Ⓐ Ⓒ **Mixed Practice** *Add or subtract as indicated. If possible, simplify your answer. See Examples 1 through 10.*

47. $\dfrac{4}{3x^2y^3} + \dfrac{5}{3x^2y^3}$

48. $\dfrac{7}{2xy^4} + \dfrac{1}{2xy^4}$

49. $\dfrac{13x-5}{2x} - \dfrac{13x+5}{2x}$

50. $\dfrac{17x+4}{4x} - \dfrac{17x-4}{4x}$

51. $\dfrac{3}{2x+10} + \dfrac{8}{3x+15}$

52. $\dfrac{10}{3x-3} + \dfrac{1}{7x-7}$

53. $\dfrac{-2}{x^2-3x} - \dfrac{1}{x^3-3x^2}$

54. $\dfrac{-3}{2a+8} - \dfrac{8}{a^2+4a}$

55. $\dfrac{ab}{a^2-b^2} + \dfrac{b}{a+b}$

56. $\dfrac{x}{25-x^2} + \dfrac{2}{3x-15}$

57. $\dfrac{5}{x^2 - 4} - \dfrac{3}{x^2 + 4x + 4}$

58. $\dfrac{3z}{z^2 - 9} - \dfrac{2}{3 - z}$

59. $\dfrac{3x}{2x^2 - 11x + 5} + \dfrac{7}{x^2 - 2x - 15}$

60. $\dfrac{2x}{3x^2 - 13x + 4} + \dfrac{5}{x^2 - 2x - 8}$

Objective **C** *Perform each indicated operation. Simplify each answer. See Example 11.*

61. $\dfrac{2}{x + 1} - \dfrac{3x}{3x + 3} + \dfrac{1}{2x + 2}$

62. $\dfrac{5}{3x - 6} - \dfrac{x}{x - 2} + \dfrac{3 + 2x}{5x - 10}$

63. $\dfrac{3}{x + 3} + \dfrac{5}{x^2 + 6x + 9} - \dfrac{x}{x^2 - 9}$

64. $\dfrac{x + 2}{x^2 - 2x - 3} + \dfrac{x}{x - 3} - \dfrac{x}{x + 1}$

65. $\dfrac{x}{x^2 - 9} + \dfrac{3}{x^2 - 6x + 9} - \dfrac{1}{x + 3}$

66. $\dfrac{3}{x^2 - 9} - \dfrac{x}{x^2 - 6x + 9} + \dfrac{1}{x + 3}$

67. $\left(\dfrac{1}{x} + \dfrac{2}{3}\right) - \left(\dfrac{1}{x} - \dfrac{2}{3}\right)$

68. $\left(\dfrac{1}{2} + \dfrac{2}{x}\right) - \left(\dfrac{1}{2} - \dfrac{1}{x}\right)$

Review

Use the distributive property to multiply each expression. See Section 1.3.

69. $12\left(\dfrac{2}{3} + \dfrac{1}{6}\right)$

70. $14\left(\dfrac{1}{7} + \dfrac{3}{14}\right)$

71. $x^2\left(\dfrac{4}{x^2} + 1\right)$

72. $5y^2\left(\dfrac{1}{y^2} - \dfrac{1}{5}\right)$

Concept Extensions

Find and correct each error. See the Concept Check in this section.

73. $\dfrac{2x - 3}{x^2 + 1} - \dfrac{x - 6}{x^2 + 1} = \dfrac{2x - 3 - x - 6}{x^2 + 1}$
$= \dfrac{x - 9}{x^2 + 1}$

74. $\dfrac{7}{x + 7} - \dfrac{x + 3}{x + 7} = \dfrac{7 - x - 3}{(x + 7)^2}$
$= \dfrac{-x + 4}{(x + 7)^2}$

75. Find the perimeter and the area of the square.

$\dfrac{x}{x + 5}$ ft

76. Find the perimeter of the quadrilateral.

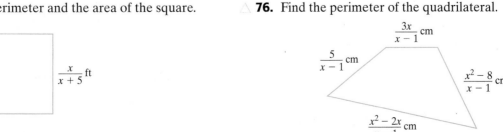

$\dfrac{3x}{x - 1}$ cm

$\dfrac{5}{x - 1}$ cm

$\dfrac{x^2 - 8}{x - 1}$ cm

$\dfrac{x^2 - 2x}{x - 1}$ cm

77. When is the LCD of two rational expressions equal to the product of their denominators? $\left(\textit{Hint:}\right.$ What is the LCD of $\dfrac{1}{x}$ and $\dfrac{7}{x+5}$? $\left.\right)$

78. When is the LCD of two rational expressions with different denominators equal to one of the denominators? $\left(\textit{Hint:}\right.$ What is the LCD of $\dfrac{3x}{x+2}$ and $\dfrac{7x+1}{(x+2)^3}$? $\left.\right)$

79. In your own words, explain how to add rational expressions with different denominators.

80. In your own words, explain how to multiply rational expressions.

81. In your own words, explain how to divide rational expressions.

82. In your own words, explain how to subtract rational expressions with different denominators.

Mixed Practice (Sections 6.1, 6.2) *Perform the indicated operation. If possible, simplify your answer.*

83. $\left(\dfrac{2}{3}-\dfrac{1}{x}\right)\cdot\left(\dfrac{3}{x}+\dfrac{1}{2}\right)$

84. $\left(\dfrac{2}{3}-\dfrac{1}{x}\right)\div\left(\dfrac{3}{x}+\dfrac{1}{2}\right)$

85. $\left(\dfrac{2a}{3}\right)^2\div\left(\dfrac{a^2}{a+1}-\dfrac{1}{a+1}\right)$

86. $\left(\dfrac{x+2}{2x}-\dfrac{x-2}{2x}\right)\cdot\left(\dfrac{5x}{4}\right)^2$

87. $\left(\dfrac{2x}{3}\right)^2\div\left(\dfrac{x}{3}\right)^2$

88. $\left(\dfrac{2x}{3}\right)^2\cdot\left(\dfrac{3}{x}\right)^2$

89. $\left(\dfrac{x}{x+1}-\dfrac{x}{x-1}\right)\div\dfrac{x}{2x+2}$

90. $\dfrac{x}{2x+2}\div\left(\dfrac{x}{x+1}+\dfrac{x}{x-1}\right)$

91. $\dfrac{4}{x}\cdot\left(\dfrac{2}{x+2}-\dfrac{2}{x-2}\right)$

92. $\dfrac{1}{x+1}\cdot\left(\dfrac{5}{x}+\dfrac{2}{x-3}\right)$

Perform each indicated operation. (Hint: First write each expression with positive exponents.)

93. $x^{-1}+(2x)^{-1}$

94. $y^{-1}+(4y)^{-1}$

95. $4x^{-2}-3x^{-1}$

96. $(4x)^{-2}-(3x)^{-1}$

Use a graphing calculator to support the results of each exercise.

97. Exercise 3

98. Exercise 4

6.3 SIMPLIFYING COMPLEX FRACTIONS

Objectives

A Simplify Complex Fractions by Simplifying the Numerator and Denominator and Then Dividing.

B Simplify Complex Fractions by Multiplying by the Least Common Denominator (LCD).

C Simplify Expressions with Negative Exponents.

A rational expression whose numerator, denominator, or both contain one or more rational expressions is called a **complex rational expression** or a **complex fraction.** Examples are

$$\dfrac{\frac{1}{a}}{\frac{b}{2}} \qquad \dfrac{\frac{x}{2y^2}}{\frac{6x-2}{9y}} \qquad \dfrac{x+\frac{1}{y}}{y+1}$$

The parts of a complex fraction are

$$\dfrac{\left.\dfrac{x}{y+2}\right\}}{\left.7+\dfrac{1}{y}\right\}}$$

$\leftarrow$ Numerator of complex fraction

$\leftarrow$ Main fraction bar

$\leftarrow$ Denominator of complex fraction

Our goal in this section is to simplify complex fractions. A complex fraction is simplified when it is in the form $\dfrac{P}{Q}$, where P and Q are polynomials that have no common factors. Two methods of simplifying complex fractions are introduced.

Objective **A** Method 1: Simplifying a Complex Fraction by Simplifying the Numerator and Denominator and Then Dividing

In the first method we study, we simplify complex fractions by simplifying and dividing.

Simplifying a Complex Fraction: Method 1

Step 1: Simplify the numerator and the denominator of the complex fraction so that each is a single fraction.

Step 2: Perform the indicated division by multiplying the numerator of the complex fraction by the reciprocal of the denominator of the complex fraction.

Step 3: Simplify if possible.

EXAMPLE 1 Simplify: $\dfrac{\frac{5x}{x+2}}{\frac{10}{x-2}}$

Solution:

$$\dfrac{\frac{5x}{x+2}}{\frac{10}{x-2}} = \frac{5x}{x+2} \cdot \frac{x-2}{10} \qquad \text{Multiply by the reciprocal of } \frac{10}{x-2}.$$

$$= \frac{5\,x(x-2)}{2\cdot 5\,(x+2)}$$

$$= \frac{x(x-2)}{2(x+2)} \qquad \text{Simplify.}$$

■ **Work Practice Problem 1**

PRACTICE PROBLEM 1

Use Method 1 to simplify:

$$\dfrac{\frac{6x}{x-5}}{\frac{12}{x+5}}$$

Answer

1. $\dfrac{x(x+5)}{2(x-5)}$

✔ **Concept Check** Which of the following are equivalent $\dfrac{\dfrac{1}{x}}{\dfrac{3}{y}}$?

a. $\dfrac{1}{x} \div \dfrac{3}{y}$ **b.** $\dfrac{1}{x} \cdot \dfrac{y}{3}$ **c.** $\dfrac{1}{x} \div \dfrac{y}{3}$

PRACTICE PROBLEM 2

Use Method 1 to simplify:

$$\dfrac{\dfrac{x}{y^2} - \dfrac{1}{y}}{\dfrac{y}{x^2} - \dfrac{1}{x}}$$

EXAMPLE 2 Simplify: $\dfrac{\dfrac{x}{y^2} + \dfrac{1}{y}}{\dfrac{y}{x^2} + \dfrac{1}{x}}$

Solution: First we simplify the numerator and the denominator of the complex fraction separately so that each is a single fraction.

$$\dfrac{\dfrac{x}{y^2} + \dfrac{1}{y}}{\dfrac{y}{x^2} + \dfrac{1}{x}} = \dfrac{\dfrac{x}{y^2} + \dfrac{1 \cdot y}{y \cdot y}}{\dfrac{y}{x^2} + \dfrac{1 \cdot x}{x \cdot x}} \qquad \begin{array}{l} \text{The LCD is } y^2. \\[6pt] \text{The LCD is } x^2. \end{array}$$

$$= \dfrac{\dfrac{x + y}{y^2}}{\dfrac{y + x}{x^2}} \qquad \begin{array}{l} \text{Add.} \\[18pt] \text{Add.} \end{array}$$

$$= \dfrac{x + y}{y^2} \cdot \dfrac{x^2}{y + x} \qquad \text{Multiply by the reciprocal of } \dfrac{y + x}{x^2}.$$

$$= \dfrac{x^2 \,(x + y)}{y^2 \,(y + x)}$$

$$= \dfrac{x^2}{y^2} \qquad \text{Simplify.}$$

⬛ **Work Practice Problem 2**

Objective **B** **Method 2: Simplifying a Complex Fraction by Multiplying the Numerator and Denominator by the LCD**

With this method, we multiply the numerator and the denominator of the complex fraction by the least common denominator (LCD) of all fractions in the complex fraction.

Simplifying a Complex Fraction: Method 2

Step 1: Multiply the numerator and the denominator of the complex fraction by the LCD of all the fractions in both the numerator and the denominator.

Step 2: Simplify.

PRACTICE PROBLEM 3

Use Method 2 to simplify:

$$\dfrac{\dfrac{6x}{x - 5}}{\dfrac{12}{x + 5}}$$

EXAMPLE 3 Simplify: $\dfrac{\dfrac{5x}{x + 2}}{\dfrac{10}{x - 2}}$

Solution: Notice we are reworking Example 1 using Method 2. The least common denominator of $\dfrac{5x}{x + 2}$ and $\dfrac{10}{x - 2}$ is $(x + 2)(x - 2)$. We multiply both the numerator, $\dfrac{5x}{x + 2}$, and the denominator, $\dfrac{10}{x - 2}$, by this LCD.

Answers

2. $-\dfrac{x^2}{y^2}$, 3. $\dfrac{x(x + 5)}{2(x - 5)}$

✔ **Concept Check Answer**

a and b

$$\dfrac{\dfrac{5x}{x+2}}{\dfrac{10}{x-2}} = \dfrac{\left(\dfrac{5x}{x+2}\right)\cdot(x+2)(x-2)}{\left(\dfrac{10}{x-2}\right)\cdot(x+2)(x-2)}$$

Multiply the numerator and denominator by the LCD.

$$= \dfrac{5x\cdot(x-2)}{2\cdot 5\cdot(x+2)}$$

Simplify.

$$= \dfrac{x(x-2)}{2(x+2)}$$

Simplify.

Work Practice Problem 3

Examples 1 and 3 are the same and simplify to the same rational expression. Regardless of what method you choose to use, the simplification is the same.

EXAMPLE 4 Simplify: $\dfrac{\dfrac{x}{y^2}+\dfrac{1}{y}}{\dfrac{y}{x^2}+\dfrac{1}{x}}$

Solution: The least common denominator of $\dfrac{x}{y^2}, \dfrac{1}{y}, \dfrac{y}{x^2},$ and $\dfrac{1}{x}$ is x^2y^2.

$$\dfrac{\dfrac{x}{y^2}+\dfrac{1}{y}}{\dfrac{y}{x^2}+\dfrac{1}{x}} = \dfrac{\left(\dfrac{x}{y^2}+\dfrac{1}{y}\right)\cdot x^2y^2}{\left(\dfrac{y}{x^2}+\dfrac{1}{x}\right)\cdot x^2y^2}$$

Multiply the numerator and denominator by the LCD.

$$= \dfrac{\dfrac{x}{y^2}\cdot x^2y^2 + \dfrac{1}{y}\cdot x^2y^2}{\dfrac{y}{x^2}\cdot x^2y^2 + \dfrac{1}{x}\cdot x^2y^2}$$

Use the distributive property.

$$= \dfrac{x^3 + x^2y}{y^3 + xy^2}$$

Simplify.

$$= \dfrac{x^2(x+y)}{y^2(y+x)}$$

Factor.

$$= \dfrac{x^2}{y^2}$$

Simplify.

Work Practice Problem 4

Helpful Hint

Just as for Examples 1 and 3, Examples 2 and 4 are the same and they simplify to the same rational expression. Note that regardless of what method you use, the result is the same.

Objective **C** **Simplifying Expressions with Negative Exponents**

Some expressions containing negative exponents can be written as complex fractions. To simplify these expressions, we first write them as equivalent expressions with positive exponents.

PRACTICE PROBLEM 4

Use Method 2 to simplify:

$$\dfrac{\dfrac{x}{y^2}-\dfrac{1}{y}}{\dfrac{y}{x^2}-\dfrac{1}{x}}$$

Answer

4. $-\dfrac{x^2}{y^2}$

PRACTICE PROBLEM 5

Simplify: $\dfrac{2x^{-1} + 3y^{-1}}{x^{-1} - 2y^{-1}}$

EXAMPLE 5 Simplify: $\dfrac{x^{-1} + 2xy^{-1}}{x^{-2} - x^{-2}y^{-1}}$

Solution: This fraction does not appear to be a complex fraction. However, if we write it by using only positive exponents we see that it is a complex fraction.

$$\frac{x^{-1} + 2xy^{-1}}{x^{-2} - x^{-2}y^{-1}} = \frac{\dfrac{1}{x} + \dfrac{2x}{y}}{\dfrac{1}{x^2} - \dfrac{1}{x^2 y}}$$

The LCD of $\dfrac{1}{x}, \dfrac{2x}{y}, \dfrac{1}{x^2}$, and $\dfrac{1}{x^2 y}$ is $x^2 y$. We multiply both the numerator and denominator by $x^2 y$.

$$\frac{\dfrac{1}{x} + \dfrac{2x}{y}}{\dfrac{1}{x^2} - \dfrac{1}{x^2 y}} = \frac{\left(\dfrac{1}{x} + \dfrac{2x}{y}\right) \cdot x^2 y}{\left(\dfrac{1}{x^2} - \dfrac{1}{x^2 y}\right) \cdot x^2 y}$$

$$= \frac{\dfrac{1}{x} \cdot x^2 y + \dfrac{2x}{y} \cdot x^2 y}{\dfrac{1}{x^2} \cdot x^2 y - \dfrac{1}{x^2 y} \cdot x^2 y} \qquad \text{Use the distributive property.}$$

$$= \frac{xy + 2x^3}{y - 1} \qquad \text{Simplify.}$$

$$\text{or } \frac{x(y + 2x^2)}{y - 1}$$

Work Practice Problem 5

PRACTICE PROBLEM 6

Simplify: $\dfrac{5 - 3x^{-1}}{2 + (3x)^{-1}}$

Helpful Hint

Don't forget that $(2x)^{-1} = \dfrac{1}{2x}$, but

$2x^{-1} = 2 \cdot \dfrac{1}{x} = \dfrac{2}{x}$.

EXAMPLE 6 Simplify: $\dfrac{(2x)^{-1} + 1}{2x^{-1} - 1}$

Solution: $\dfrac{(2x)^{-1} + 1}{2x^{-1} - 1} = \dfrac{\dfrac{1}{2x} + 1}{\dfrac{2}{x} - 1}$ \qquad Write using positive exponents.

$$= \frac{\left(\dfrac{1}{2x} + 1\right) \cdot 2x}{\left(\dfrac{2}{x} - 1\right) \cdot 2x} \qquad \text{The LCD of } \dfrac{1}{2x} \text{ and } \dfrac{2}{x} \text{ is } 2x.$$

$$= \frac{\dfrac{1}{2x} \cdot 2x + 1 \cdot 2x}{\dfrac{2}{x} \cdot 2x - 1 \cdot 2x} \qquad \text{Use distributive property.}$$

$$= \frac{1 + 2x}{4 - 2x} \qquad \text{Simplify.}$$

Work Practice Problem 6

Answers

5. $\dfrac{2y + 3x}{y - 2x}$, **6.** $\dfrac{15x - 9}{6x + 1}$ or $\dfrac{3(5x - 3)}{6x + 1}$

Objectives **A** **B** **Mixed Practice** *Simplify each complex fraction. See Examples 1 through 4.*

1. $\dfrac{1 + \dfrac{2}{5}}{2 + \dfrac{3}{5}}$

2. $\dfrac{2 + \dfrac{1}{7}}{3 - \dfrac{4}{7}}$

3. $\dfrac{\dfrac{4}{x - 1}}{\dfrac{x}{x - 1}}$

4. $\dfrac{\dfrac{x}{x + 2}}{\dfrac{2}{x + 2}}$

5. $\dfrac{1 - \dfrac{2}{x}}{x + \dfrac{4}{9x}}$

6. $\dfrac{5 - \dfrac{3}{x}}{x + \dfrac{2}{3x}}$

7. $\dfrac{\dfrac{10}{3x}}{\dfrac{5}{6x}}$

8. $\dfrac{\dfrac{15}{2x}}{\dfrac{5}{6x}}$

9. $\dfrac{\dfrac{4x^2 - y^2}{xy}}{\dfrac{2}{y} - \dfrac{1}{x}}$

10. $\dfrac{\dfrac{x^2 - 9y^2}{xy}}{\dfrac{1}{y} - \dfrac{3}{x}}$

11. $\dfrac{\dfrac{x + 1}{3}}{\dfrac{2x - 1}{6}}$

12. $\dfrac{\dfrac{x + 3}{12}}{\dfrac{4x - 5}{15}}$

13. $\dfrac{\dfrac{2}{x} + \dfrac{3}{x^2}}{\dfrac{4}{x^2} - \dfrac{9}{x}}$

14. $\dfrac{\dfrac{2}{x^2} + \dfrac{1}{x}}{\dfrac{4}{x^2} - \dfrac{1}{x}}$

15. $\dfrac{\dfrac{1}{x} + \dfrac{2}{x^2}}{x + \dfrac{8}{x^2}}$

16. $\dfrac{\dfrac{1}{y} + \dfrac{3}{y^2}}{y + \dfrac{27}{y^2}}$

17. $\dfrac{\dfrac{4}{5 - x} + \dfrac{5}{x - 5}}{\dfrac{2}{x} + \dfrac{3}{x - 5}}$

18. $\dfrac{\dfrac{3}{x - 4} - \dfrac{2}{4 - x}}{\dfrac{2}{x - 4} - \dfrac{2}{x}}$

19. $\dfrac{\dfrac{x + 2}{x} - \dfrac{2}{x - 1}}{\dfrac{x + 1}{x} + \dfrac{x + 1}{x - 1}}$

20. $\dfrac{\dfrac{5}{a + 2} - \dfrac{1}{a - 2}}{\dfrac{3}{2 + a} + \dfrac{6}{2 - a}}$

21. $\dfrac{\dfrac{2}{x} + 3}{\dfrac{4}{x^2} - 9}$

22. $\dfrac{2 + \dfrac{1}{x}}{4x - \dfrac{1}{x}}$

23. $\dfrac{1 - \dfrac{x}{y}}{\dfrac{x^2}{y^2} - 1}$

24. $\dfrac{1 - \dfrac{2}{x}}{x - \dfrac{4}{x}}$

25. $\dfrac{\dfrac{-2x}{x^2 - xy}}{\dfrac{y}{x^2}}$

26. $\dfrac{\dfrac{7y}{x^2 + xy}}{\dfrac{y^2}{x^2}}$

27. $\dfrac{\dfrac{2}{x} + \dfrac{1}{x^2}}{\dfrac{y}{x^2} + 1}$

28. $\dfrac{\dfrac{5}{x^2} - \dfrac{2}{x}}{\dfrac{1}{x} + 2}$

29. $\dfrac{\dfrac{x}{9} - \dfrac{1}{x}}{1 + \dfrac{3}{x}}$

30. $\dfrac{\dfrac{x}{4} - \dfrac{4}{x}}{1 - \dfrac{4}{x}}$

31. $\dfrac{\dfrac{x - 1}{x^2 - 4}}{1 + \dfrac{1}{x - 2}}$

429

32. $\dfrac{\dfrac{x-2}{x^2-9}}{1+\dfrac{1}{x-3}}$

33. $\dfrac{\dfrac{2}{x+5}+\dfrac{4}{x+3}}{\dfrac{3x+13}{x^2+8x+15}}$

34. $\dfrac{\dfrac{2}{x+2}+\dfrac{6}{x+7}}{\dfrac{4x+13}{x^2+9x+14}}$

Objective C *Simplify. See Examples 5 and 6.*

35. $\dfrac{x^{-1}}{x^{-2}+y^{-2}}$

36. $\dfrac{a^{-3}+b^{-1}}{a^{-2}}$

37. $\dfrac{2a^{-1}+3b^{-2}}{a^{-1}-b^{-1}}$

38. $\dfrac{x^{-1}+y^{-1}}{3x^{-2}+5y^{-2}}$

39. $\dfrac{1}{x-x^{-1}}$

40. $\dfrac{x^{-2}}{x+3x^{-1}}$

41. $\dfrac{a^{-1}+1}{a^{-1}-1}$

42. $\dfrac{a^{-1}-4}{4+a^{-1}}$

43. $\dfrac{3x^{-1}+(2y)^{-1}}{x^{-2}}$

44. $\dfrac{5x^{-2}-3y^{-1}}{x^{-1}+y^{-1}}$

45. $\dfrac{2a^{-1}+(2a)^{-1}}{a^{-1}+2a^{-2}}$

46. $\dfrac{a^{-1}+2a^{-2}}{2a^{-1}+(2a)^{-1}}$

47. $\dfrac{5x^{-1}+2y^{-1}}{x^{-2}y^{-2}}$

48. $\dfrac{x^{-2}y^{-2}}{5x^{-1}+2y^{-1}}$

49. $\dfrac{5x^{-1}-2y^{-1}}{25x^{-2}-.4y^{-2}}$

50. $\dfrac{3x^{-1}+3y^{-1}}{4x^{-2}-9y^{-2}}$

Review

Solve each equation for x. See Sections 2.1 and 5.7.

51. $7x+2=x-3$

52. $4-2x=17-5x$

53. $x^2=4x-4$

54. $5x^2+10x=15$

55. $\dfrac{x}{3}-5=13$

56. $\dfrac{2x}{9}+1=\dfrac{7}{9}$

Concept Extensions

Solve. See the Concept Check in the section.

57. Which of the following are equivalent to $\dfrac{\dfrac{x+1}{9}}{\dfrac{y-2}{5}}$?

   a. $\dfrac{x+1}{9}\div\dfrac{y-2}{5}$

   b. $\dfrac{x+1}{9}\cdot\dfrac{y-2}{5}$

   c. $\dfrac{x+1}{9}\cdot\dfrac{5}{y-2}$

58. Which of the following are equivalent to $\dfrac{\dfrac{a}{7}}{\dfrac{b}{13}}$?

   a. $\dfrac{a}{7}\cdot\dfrac{b}{13}$

   b. $\dfrac{a}{7}\div\dfrac{b}{13}$

   c. $\dfrac{a}{7}\div\dfrac{13}{b}$

   d. $\dfrac{a}{7}\cdot\dfrac{13}{b}$

59. When the source of a sound is traveling toward a listener, the pitch that the listener hears due to the Doppler effect is given by the complex rational compression $\dfrac{a}{1 - \dfrac{s}{770}}$, where a is the actual pitch of the sound and s is the speed of the sound source. Simplify this expression.

60. In baseball, the earned run average (ERA) statistic gives the average number of earned runs scored on a pitcher per game. It is computed with the following expression: $\dfrac{E}{\dfrac{I}{9}}$, where E is the number of earned runs scored on a pitcher and I is the total number of innings pitched by the pitcher. Simplify this expression.

61. Which of the following are equivalent to $\dfrac{\dfrac{1}{x}}{\dfrac{3}{y}}$?

a. $\dfrac{1}{x} \div \dfrac{3}{y}$ **b.** $\dfrac{1}{x} \cdot \dfrac{y}{3}$ **c.** $\dfrac{1}{x} \div \dfrac{y}{3}$

62. In your own words, explain one method for simplifying a complex fraction.

Simplify.

63. $\dfrac{\dfrac{2}{y^2} - \dfrac{5}{xy} - \dfrac{3}{x^2}}{\dfrac{2}{y^2} + \dfrac{7}{xy} + \dfrac{3}{x^2}}$

64. $\dfrac{\dfrac{2}{x^2} - \dfrac{1}{xy} - \dfrac{1}{y^2}}{\dfrac{1}{x^2} - \dfrac{3}{xy} + \dfrac{2}{y^2}}$

65. $\dfrac{1}{1 + (1 + x)^{-1}}$ $\dfrac{1}{1 + \dfrac{1}{1 + x}}$

66. $\dfrac{(x + 2)^{-1} + (x - 2)^{-1}}{(x^2 - 4)^{-1}}$

67. $\dfrac{x}{1 - \dfrac{1}{1 + \dfrac{1}{x}}}$

68. $\dfrac{x}{1 - \dfrac{1}{1 - \dfrac{1}{x}}}$

In the study of calculus, the difference quotient $\dfrac{f(a + h) - f(a)}{h}$ *is often found and simplified. Find and simplify this quotient for each function $f(x)$ by following steps **a** through **d**.*

a. *Find $(a + h)$.*
b. *Find $f(a)$.*
c. *Use steps **a** and **b** to find $\dfrac{f(a + h) - f(a)}{h}$*
d. *Simplify the result of step **c**.*

69. $f(x) = \dfrac{1}{x}$

70. $f(x) = \dfrac{5}{x}$

71. $\dfrac{3}{x+1}$

72. $\dfrac{2}{x^2}$

 STUDY SKILLS BUILDER

Are You Familiar with Your Textbook Supplements?

Below is a review of some of the student supplements available for additional study. Check to see if you are using the ones most helpful to you.

- Chapter Test Prep Videos on CD. This material is found with your textbook and is fully explained there. The CD contains video clip solutions to the Chapter Test exercises in this text and are excellent help when studying for chapter tests.

- Lecture Videos on CD-ROM. These video segments are keyed to each section of the text. The material is presented by me, Elayn Martin-Gay, and I have placed a ⊙ by the exercises in the text that I have worked on the video.

- The *Student Solutions Manual*. This contains worked out solutions to odd-numbered exercises as well as every exercise in the Integrated Reviews, Chapter Reviews, Chapter Tests, and Cumulative Reviews.

- Prentice Hall Tutor Center. Mathematic questions may be phoned, faxed, or emailed to this center.

- MyMathLab, MathXL, and Interact Math. These are computer and Internet tutorials. This supplement may already be available to you somewhere on campus, for example at your local learning resource lab. Take a moment and find the name and location of any such lab on campus.

   As usual, your instructor is your best source of information.

Let's see how you are doing with textbook supplements.

1. Name one way the Lecture Videos can be helpful to you.

2. Name one way the Chapter Test Prep Video can help you prepare for a chapter test.

3. List any textbook supplements that you have found useful.

4. Have you located and visited a learning resource lab located on your campus?

5. List the textbook supplements that are currently housed in your campus' learning resource lab.

6.4 SOLVING EQUATIONS CONTAINING RATIONAL EXPRESSIONS

Objective **A** Solving Equations Containing Rational Expressions

In this section, we solve rational equations. A *rational equation* is an equation containing at least one rational expression. Before beginning this section, make sure that you understand the difference between an *equation* and an *expression*. An **equation** contains an equal sign and an **expression** does not.

Equation	**Expression**
$\dfrac{x}{2} + \dfrac{x}{6} = \dfrac{2}{3}$ ↑ equal sign	$\dfrac{x}{2} + \dfrac{x}{6}$

Solving an Equation Containing Rational Expressions

To solve an *equation* containing rational expressions, first clear the equation of fractions by multiplying both sides of the equation by the LCD of all rational expressions. Then solve as usual.

> **Helpful Hint**
> The method described is for equations only. It may *not* be used for performing operations on expressions.

✔ **Concept Check**   True or false? Clearing fractions is valid when solving an equation and when simplifying rational expressions. Explain.

EXAMPLE 1   Solve: $\dfrac{4x}{5} + \dfrac{3}{2} = \dfrac{3x}{10}$

Solution: The LCD of $\dfrac{4x}{5}, \dfrac{3}{2}$, and $\dfrac{3x}{10}$ is 10. We multiply both sides of the equation by 10.

$$\frac{4x}{5} + \frac{3}{2} = \frac{3x}{10}$$

$$10\left(\frac{4x}{5} + \frac{3}{2}\right) = 10\left(\frac{3x}{10}\right) \quad \text{Multiply both sides by the LCD.}$$

$$10 \cdot \frac{4x}{5} + 10 \cdot \frac{3}{2} = 10 \cdot \frac{3x}{10} \quad \text{Use the distributive property.}$$

$$8x + 15 = 3x \quad \text{Simplify.}$$

$$15 = -5x \quad \text{Subtract } 8x \text{ from both sides.}$$

$$-3 = x \quad \text{Solve.}$$

We verify this solution by replacing x with -3 in the original equation.

Check: $\dfrac{4x}{5} + \dfrac{3}{2} = \dfrac{3x}{10}$

$$\frac{4(-3)}{5} + \frac{3}{2} \overset{?}{=} \frac{3(-3)}{10}$$

$$\frac{-12}{5} + \frac{3}{2} \overset{?}{=} \frac{-9}{10}$$

$$-\frac{24}{10} + \frac{15}{10} \overset{?}{=} -\frac{9}{10}$$

$$-\frac{9}{10} = -\frac{9}{10} \quad \text{True}$$

The solution set is $\{-3\}$.

■ **Work Practice Problem 1**

Answer

1. $\{-1\}$

✔ **Concept Check Answer**

false; answers may vary

433

The important difference about the equations in this section is that the denominator of a rational expression may contain a variable. Recall that a rational expression is undefined for values of the variable that make the denominator 0. If a proposed solution makes the denominator 0, then it must be rejected as a solution of the original equation. Such proposed solutions are called **extraneous solutions.**

PRACTICE PROBLEM 2

Solve: $\dfrac{5}{x} - \dfrac{3x + 6}{2x} = \dfrac{7}{2}$

EXAMPLE 2 Solve: $\dfrac{3}{x} - \dfrac{x + 21}{3x} = \dfrac{5}{3}$

Solution: The LCD of the denominators $x, 3x,$ and 3 is $3x.$ We multiply both sides by $3x.$

$$\frac{3}{x} - \frac{x + 21}{3x} = \frac{5}{3}$$

$$3x\left(\frac{3}{x} - \frac{x + 21}{3x}\right) = 3x\left(\frac{5}{3}\right) \qquad \text{Multiply both sides by the LCD.}$$

$$3x \cdot \frac{3}{x} - 3x \cdot \frac{x + 21}{3x} = 3x \cdot \frac{5}{3} \qquad \text{Use the distributive property.}$$

$$9 - (x + 21) = 5x \qquad \text{Simplify.}$$

$$9 - x - 21 = 5x$$

$$-12 = 6x$$

$$-2 = x \qquad \text{Solve.}$$

The proposed solution is $-2.$

Check: We check the proposed solution in the original equation.

$$\frac{3}{x} - \frac{x + 21}{3x} = \frac{5}{3}$$

$$\frac{3}{-2} - \frac{-2 + 21}{3(-2)} \stackrel{?}{=} \frac{5}{3}$$

$$-\frac{9}{6} + \frac{19}{6} \stackrel{?}{=} \frac{5}{3}$$

$$\frac{10}{6} = \frac{5}{3} \qquad \text{True}$$

The solution set is $\{-2\}.$

▣ **Work Practice Problem 2**

The following steps may be used to solve equations containing rational expressions.

To Solve an Equation Containing Rational Expressions

Step 1: Multiply both sides of the equation by the LCD of all rational expressions in the equation.

Step 2: Simplify both sides.

Step 3: Determine whether the equation is linear, quadratic, or higher degree and solve accordingly.

Step 4: Check the solution in the original equation.

Let's talk more about multiplying both sides of an equation by the LCD of the rational expressions in the equation. In Example 3 that follows, the LCD is $x - 2,$ so we will first multiply both sides of the equation by $x - 2.$ Recall that the multiplication property for equations allows us to multiply both sides of an equation by any

Answer

2. $\left\{\dfrac{2}{5}\right\}$

nonzero number. In other words, for Example 3 below, we may multiply both sides of the equation by $x - 2$ *as long as* $x - 2 \neq 0$ or as long as $x \neq 2$. Keep this in mind when solving these equations.

EXAMPLE 3 Solve: $\dfrac{x + 6}{x - 2} = \dfrac{2(x + 2)}{x - 2}$

Solution: First multiply both sides of the equation by the LCD, $x - 2$. (Remember, we can only do this if $x \neq 2$, so that we are not multiplying by 0.)

$$\dfrac{x + 6}{x - 2} = \dfrac{2(x + 2)}{x - 2}$$

$$(x - 2) \cdot \dfrac{x + 6}{x - 2} = (x - 2) \cdot \dfrac{2(x + 2)}{x - 2} \qquad \text{Multiply both sides by } x - 2.$$

$$x + 6 = 2(x + 2) \qquad \text{Simplify.}$$

$$x + 6 = 2x + 4 \qquad \text{Use the distributive property.}$$

$$2 = x \qquad \text{Solve.}$$

From above, we assumed that $x \neq 2$, so this equation has no solution. This will also show as we attempt to check this proposed solution.

Check: The proposed solution is 2. Notice that 2 makes the denominator 0 in the original equation. This can also be seen in a check. Check the proposed solution 2 in the original equation.

$$\dfrac{x + 6}{x - 2} = \dfrac{2(x + 2)}{x - 2}$$

$$\dfrac{2 + 6}{6 - 2} = \dfrac{2(2 + 2)}{2 - 2}$$

$$\dfrac{8}{0} = \dfrac{2(4)}{0}$$

The denominators are 0, so 2 is not a solution of the original equation. The solution set is $\varnothing$ or $\{\ \}$.

■ **Work Practice Problem 3**

EXAMPLE 4 Solve: $\dfrac{2x}{2x - 1} + \dfrac{1}{x} = \dfrac{1}{2x - 1}$

Solution: The LCD is $x(2x - 1)$. Multiply both sides by $x(2x - 1)$. By the distributive property, this is the same as multiplying each term by $x(2x - 1)$.

$$x(2x - 1) \cdot \dfrac{2x}{2x - 1} + x(2x - 1) \cdot \dfrac{1}{x} = x(2x - 1) \cdot \dfrac{1}{2x - 1}$$

$$x(2x) + (2x - 1) = x$$

$$2x^2 + 2x - 1 - x = 0$$

$$2x^2 + x - 1 = 0$$

$$(x + 1)(2x - 1) = 0$$

$$x + 1 = 0 \qquad \text{or} \qquad 2x - 1 = 0$$

$$x = -1 \qquad\qquad x = \dfrac{1}{2}$$

The number $\dfrac{1}{2}$ makes the denominator $2x - 1$ equal 0, so it is not a solution. The solution set is $\{-1\}$.

■ **Work Practice Problem 4**

PRACTICE PROBLEM 3

Solve: $\dfrac{x + 5}{x - 3} = \dfrac{2(x + 1)}{x - 3}$

PRACTICE PROBLEM 4

Solve: $\dfrac{3x}{3x - 1} + \dfrac{1}{x} = \dfrac{1}{3x - 1}$

Answers

3. $\varnothing$, **4.** $\{-1\}$

PRACTICE PROBLEM 5

Solve:

$$\frac{2x}{x - 4} + \frac{10 - 5x}{x^2 - 16} = \frac{x}{x + 4}$$

EXAMPLE 5 Solve: $\dfrac{2x}{x - 3} + \dfrac{6 - 2x}{x^2 - 9} = \dfrac{x}{x + 3}$

Solution: We factor the second denominator to find that the LCD is $(x + 3)(x - 3)$. We multiply both sides of the equation by $(x + 3)(x - 3)$. By the distributive property, this is the same as multiplying each term by $(x + 3)(x - 3)$.

$$\frac{2x}{x - 3} + \frac{6 - 2x}{x^2 - 9} = \frac{x}{x + 3}$$

$$(x + 3)(x - 3) \cdot \frac{2x}{x - 3} + (x + 3)(x - 3) \cdot \frac{6 - 2x}{(x + 3)(x - 3)}$$

$$= (x + 3)(x - 3)\left(\frac{x}{x + 3}\right)$$

$2x(x + 3) + (6 - 2x) = x(x - 3)$ Simplify.

$2x^2 + 6x + 6 - 2x = x^2 - 3x$ Use the distributive property.

Next we solve this quadratic equation by the factoring method. To do so, we first write the equation so that one side is 0.

$$x^2 + 7x + 6 = 0$$

$(x + 6)(x + 1) = 0$ Factor.

$x = -6$ or $x = -1$ Set each factor equal to 0 and solve.

Neither -6 nor -1 makes any denominator 0. Check to see that the solution set is $\{-6, -1\}$.

■ **Work Practice Problem 5**

PRACTICE PROBLEM 6

Solve:

$$\frac{3z}{3z^2 + 7z - 6} - \frac{1}{z} = \frac{4}{3z^2 - 2z}$$

EXAMPLE 6 Solve: $\dfrac{z}{2z^2 + 3z - 2} - \dfrac{1}{2z} = \dfrac{3}{z^2 + 2z}$

Solution: Factor the denominators to find that the LCD is $2z(z + 2)(2z - 1)$. Multiply both sides by the LCD. Remember, by using the distributive property, this is the same as multiplying each term by $2z(z + 2)(2z - 1)$.

$$\frac{z}{2z^2 + 3z - 2} - \frac{1}{2z} = \frac{3}{z^2 + 2z}$$

$$\frac{z}{(2z - 1)(z + 2)} - \frac{1}{2z} = \frac{3}{z(z + 2)}$$

$$2z(z + 2)(2z - 1) \cdot \frac{z}{(2z - 1)(z + 2)} - 2z(z + 2)(2z - 1) \cdot \frac{1}{2z}$$

$$= 2z(z + 2)(2z - 1) \cdot \frac{3}{z(z + 2)}$$ Apply the distributive property.

$2z(z) - (z + 2)(2z - 1) = 3 \cdot 2(2z - 1)$ Simplify.

$2z^2 - (2z^2 + 3z - 2) = 12z - 6$

$2z^2 - 2z^2 - 3z + 2 = 12z - 6$

$-3z + 2 = 12z - 6$

$-15z = -8$

$$z = \frac{8}{15}$$ Solve.

The proposed solution $\dfrac{8}{15}$ does not make any denominator 0; the solution set is $\left\{\dfrac{8}{15}\right\}$.

■ **Work Practice Problem 6**

Answers

5. $\{-5, -2\}$, **6.** $\left\{-\dfrac{6}{11}\right\}$

A graph can be helpful in visualizing solutions of equations. For example, to visualize the solution of the equation $\dfrac{3}{x} - \dfrac{x+21}{3x} = \dfrac{5}{3}$ in Example 2, the graph of the related rational function $f(x) = \dfrac{3}{x} - \dfrac{x+21}{3x}$ is shown. A solution of the equation is an x-value that corresponds to a y-value of $\dfrac{5}{3}$.

Notice that an x-value of -2 corresponds to a y-value of $\dfrac{5}{3}$. The solution of the equation is indeed -2 as shown in Example 2.

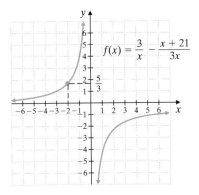

Mental Math

Determine whether each is an equation or an expression. Do not solve or simplify.

1. $\dfrac{x}{2} = \dfrac{3x}{5} + \dfrac{x}{6}$

2. $\dfrac{3x}{5} + \dfrac{x}{6}$

3. $\dfrac{x}{x-1} + \dfrac{2x}{x+1}$

4. $\dfrac{x}{x-1} + \dfrac{2x}{x+1} = 5$

5. $\dfrac{y+7}{2} = \dfrac{y+1}{6} + \dfrac{1}{y}$

6. $\dfrac{y+1}{6} + \dfrac{1}{y}$

6.4 EXERCISE SET

FOR EXTRA HELP

Student Solutions Manual   PH Math/Tutor Center   CD/Video for Review   MathXL®   MyMathLab

Objective A *Solve each equation. See Examples 1 and 2.*

1. $\dfrac{x}{2} - \dfrac{x}{3} = 12$

2. $x = \dfrac{x}{2} - 4$

3. $\dfrac{x}{3} = \dfrac{1}{6} + \dfrac{x}{4}$

4. $\dfrac{x}{2} = \dfrac{21}{10} - \dfrac{x}{5}$

5. $\dfrac{2}{x} + \dfrac{1}{2} = \dfrac{5}{x}$

6. $\dfrac{5}{3x} + 1 = \dfrac{7}{6}$

7. $\dfrac{x^2 + 1}{x} = \dfrac{5}{x}$

8. $\dfrac{x^2 - 14}{2x} = -\dfrac{5}{2x}$

Solve each equation. See Examples 3 through 6.

9. $\dfrac{x+5}{x+3} = \dfrac{2}{x+3}$

10. $\dfrac{x-7}{x-1} = \dfrac{11}{x-1}$

11. $\dfrac{5}{x-2} - \dfrac{2}{x+4} = \dfrac{-4}{x^2 + 2x - 8}$

12. $\dfrac{1}{x-1} + \dfrac{1}{x+1} = \dfrac{2}{x^2 - 1}$

13. $\dfrac{1}{x-1} = \dfrac{2}{x+1}$

14. $\dfrac{6}{x+3} = \dfrac{4}{x-3}$

15. $\dfrac{x^2 - 23}{2x^2 - 5x - 3} + \dfrac{2}{x-3} = \dfrac{-1}{2x+1}$

16. $\dfrac{4x^2 - 24x}{3x^2 - x - 2} + \dfrac{3}{3x+2} = \dfrac{-4}{x-1}$

17. $\dfrac{1}{x-4} - \dfrac{3x}{x^2 - 16} = \dfrac{2}{x+4}$

18. $\dfrac{3}{2x+3} - \dfrac{1}{2x-3} = \dfrac{4}{4x^2 - 9}$

19. $\dfrac{1}{x-4} = \dfrac{8}{x^2 - 16}$

20. $\dfrac{2}{x^2 - 4} = \dfrac{1}{2x - 4}$

21. $\dfrac{1}{x-2} - \dfrac{2}{x^2 - 2x} = 1$

22. $\dfrac{12}{3x^2 + 12x} = 1 - \dfrac{1}{x+4}$

Mixed Practice *Solve each equation. See Examples 1 through 6.*

23. $\dfrac{5}{x} = \dfrac{20}{12}$

24. $\dfrac{2}{x} = \dfrac{10}{5}$

25. $1 - \dfrac{4}{a} = 5$

26. $7 + \dfrac{6}{a} = 5$

27. $\dfrac{x^2 + 5}{x} - 1 = \dfrac{5(x+1)}{x}$

28. $\dfrac{x^2 + 6}{x} + 5 = \dfrac{2(x+3)}{x}$

29. $\dfrac{1}{2x} - \dfrac{1}{x+1} = \dfrac{1}{3x^2 + 3x}$

30. $\dfrac{2}{x-5} + \dfrac{1}{2x} = \dfrac{5}{3x^2 - 15x}$

31. $\dfrac{1}{x} - \dfrac{x}{25} = 0$

32. $\dfrac{x}{4} + \dfrac{5}{x} = 3$

33. $5 - \dfrac{2}{2y - 5} = \dfrac{3}{2y - 5}$

34. $1 - \dfrac{5}{y+7} = \dfrac{4}{y+7}$

438

35. $\dfrac{x-1}{x+2}=\dfrac{2}{3}$

36. $\dfrac{6x+7}{2x+9}=\dfrac{5}{3}$

37. $\dfrac{x+3}{x+2}=\dfrac{1}{x+2}$

38. $\dfrac{2x+1}{4-x}=\dfrac{9}{4-x}$

39. $\dfrac{1}{a-3}+\dfrac{2}{a+3}=\dfrac{1}{a^2-9}$

40. $\dfrac{12}{9-a^2}+\dfrac{3}{3+a}=\dfrac{2}{3-a}$

41. $\dfrac{64}{x^2-16}+1=\dfrac{2x}{x-4}$

42. $2+\dfrac{3}{x}=\dfrac{2x}{x+3}$

43. $\dfrac{-15}{4y+1}+4=y$

44. $\dfrac{36}{x^2-9}+1=\dfrac{2x}{x+3}$

45. $\dfrac{28}{x^2-9}+\dfrac{2x}{x-3}+\dfrac{6}{x+3}=0$

46. $\dfrac{x^2-20}{x^2-7x+12}=\dfrac{3}{x-3}+\dfrac{5}{x-4}$

47. $\dfrac{x+2}{x^2+7x+10}=\dfrac{1}{3x+6}-\dfrac{1}{x+5}$

48. $\dfrac{3}{2x-5}+\dfrac{2}{2x+3}=0$

Review

Write each sentence as an equation and solve. See Section 2.2.

49. Four more than 3 times a number is 19. Find the number.

50. The sum of two consecutive integers is 147. Find the integers.

51. The length of a rectangle is 5 inches more than the width. Its perimeter is 50 inches. Find the length and width.

52. The sum of a number and its reciprocal is $\dfrac{5}{2}$. Find the number and its reciprocal.

The following graph is from a recent survey of state and federal prisons. Use this histogram to answer Exercises 53 through 58.

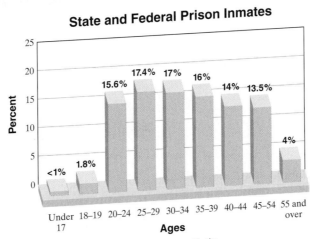

State and Federal Prison Inmates

Source: Bureau of Justice Statistics, U.S. Department of Justice

53. What percent of state and federal prison inmates are age 45 to 54?

54. What percent of state and federal prison inmates are 55 years old or older?

55. What age category shows the highest percent of prison inmates?

56. What percent of state and federal prison inmates are 20 to 34 years old?

57. At the end of 2003, there were 36,612 inmates under the jurisdiction of state and federal correctional authorities in the state of Louisiana. Approximately how many 25- to 29-year-old inmates would you expect to have been held in Louisiana at the end of 2003? Round to nearest whole. (*Source:* U.S. Bureau of Justice Statistics)

58. Use the data from Exercise 57 to answer the following.

    a. Approximate the number of 35- to 39-year-old inmates you might expect to have been held in Louisiana at the end of 2003. Round to the nearest whole.

    b. Is your answer to part a greater than or less than your answer to Exercise 57? Is this reasonable? Why or why not.

Concept Extensions

59. In your own words, explain the differences between equations and expressions.

60. In your own words, explain why it is necessary to check solutions to equations containing rational expressions.

61. The average cost of producing x game disks for a computer is given by the function $C(x) = 3.3 + \dfrac{5400}{x}$. Find the number of game disks that must be produced for the average cost to be \$5.10.

62. The average cost of producing x electric pencil sharpeners is given by the function $C(x) = 20 + \dfrac{4000}{x}$. Find the number of electric pencil sharpeners that must be produced for the average cost to be \$25.

Solve each equation. Begin by writing each equation with positive exponents only.

63. $x^{-2} - 19x^{-1} + 48 = 0$

64. $x^{-2} - 5x^{-1} - 36 = 0$

65. $p^{-2} + 4p^{-1} - 5 = 0$

66. $6p^{-2} - 5p^{-1} + 1 = 0$

Solve each equation. Round solutions to two decimal places.

67. $\dfrac{1.4}{x - 2.6} = \dfrac{-3.5}{x + 7.1}$

68. $\dfrac{-8.5}{x + 1.9} = \dfrac{5.7}{x - 3.6}$

69. $\dfrac{10.6}{y} - 14.7 = \dfrac{9.92}{3.2} + 7.6$

70. $\dfrac{12.2}{x} + 17.3 = \dfrac{9.6}{x} - 14.7$

Solve each equation by substitution. For example, to solve Exercise 71, first let $u = x - 1$. After substituting, we have $u^2 + 3u + 2 = 0$. Solve for u and then substitute back to solve for x.

71. $(x - 1)^2 + 3(x - 1) + 2 = 0$

72. $(4 - x)^2 - 5(4 - x) + 6 = 0$

73. $\left(\dfrac{3}{x - 1}\right)^2 + 2\left(\dfrac{3}{x - 1}\right) + 1 = 0$

74. $\left(\dfrac{5}{2 + x}\right)^2 + \left(\dfrac{5}{2 + x}\right) - 20 = 0$

Use a graphing calculator to verify the solution of each given exercise.

75. Exercise 23   **76.** Exercise 24   **77.** Exercise 35   **78.** Exercise 36

THE BIGGER PICTURE   Solving Equations and Inequalities

Continue the outline from Sections 2.6 and 5.7. Write how to recognize and how to solve exponential and logarithmic equations in your own words. For example:

Solving Equations and Inequalities

I. Equations

    A. Linear equations (Sec. 2.1)

    B. Absolute value equations (Sec. 2.6)

    C. Quadratic equations (Sec. 5.7)

    D. Equations with rational expressions: Equation contains rational expressions.

$$\frac{7}{x-1} + \frac{3}{x+1} = \frac{x+3}{x^2-1} \quad \text{Equation with rational expressions.}$$

$$\cancel{(x-1)}(x+1)\cdot\frac{7}{\cancel{x-1}} + (x-1)\cancel{(x+1)}\cdot\frac{3}{\cancel{x+1}} = \cancel{(x-1)}\cancel{(x+1)}\cdot\frac{x+3}{\cancel{(x-1)}\cancel{(x+1)}} \quad \text{Multiply by the LCD.}$$

$$7(x+1) + 3(x-1) = x+3 \quad \text{Simplify.}$$
$$7x + 7 + 3x - 3 = x + 3$$
$$9x = -1$$
$$x = -\frac{1}{9} \quad \text{Solve.}$$

II. Inequalities

    A. Linear inequalities (Sec. 2.4)

    B. Compound inequalities (Sec. 2.5)

    C. Absolute value inequalities (Sec. 2.7)

Solve. Write solutions to inequalities using interval notation.

1. $|-7x + 1| < 15$

2. $|-7x + 1| = 15$

3. $x^2 - 121 = 0$

4. $\dfrac{8}{x+2} - \dfrac{3}{x-1} = \dfrac{x+6}{x^2+x-2}$

5. $9x + 6 = 4x - 2$

6. $3x \leq 6$ or $-x \geq 5$

7. $3x \leq 6$ and $-x \geq 5$

8. $-9 \leq -3x + 21 < 0$

9. $\left|\dfrac{2x-1}{5}\right| > 7$

10. $15x^3 - 16x^2 = 7x$

Answers

1. _____

2. _____

3. _____

4. _____

5. _____

6. _____

7. _____

8. _____

9. _____

10. _____

11. _____

12. _____

13. _____

14. _____

15. _____

16. _____

17. _____

18. _____

19. _____

20. _____

Expressions and Equations Containing Rational Expressions

It is very important that you understand the difference between an expression and an equation containing rational expressions. An equation contains an equal sign; an expression does not.

Expression to be Simplified	**Equation to be Solved**

$$\frac{x}{2} + \frac{x}{6}$$

$$\frac{x}{2} + \frac{x}{6} = \frac{2}{3}$$

Write both rational expressions with the LCD, 6, as the denominator.

Multiply both sides by the LCD, 6.

$$\frac{x}{2} + \frac{x}{6} = \frac{x \cdot 3}{2 \cdot 3} + \frac{x}{6}$$

$$6\left(\frac{1}{2} + \frac{x}{6}\right) = 6\left(\frac{2}{3}\right)$$

$$= \frac{3x}{6} + \frac{x}{6}$$

$$3 + x = 4$$

$$= \frac{4x}{6} = \frac{2x}{3}$$

$$x = 1$$

Check to see that the solution set is {1}.

 Helpful Hint

Remember: Equations can be cleared of fractions, expressions cannot.

Perform each indicated operation and either simplify the expression, or solve the equation for the variable.

1. $\dfrac{x}{2} = \dfrac{1}{8} + \dfrac{x}{4}$

2. $\dfrac{x}{4} = \dfrac{3}{2} + \dfrac{x}{10}$

3. $\dfrac{1}{8} + \dfrac{x}{4}$

4. $\dfrac{3}{2} + \dfrac{x}{10}$

5. $\dfrac{4}{x+2} - \dfrac{2}{x-1}$

6. $\dfrac{5}{x-2} - \dfrac{10}{x+4}$

7. $\dfrac{4}{x+2} = \dfrac{2}{x-1}$

8. $\dfrac{5}{x-2} = \dfrac{10}{x+4}$

9. $\dfrac{2}{x^2-4} = \dfrac{1}{x+2} - \dfrac{3}{x-2}$

10. $\dfrac{3}{x^2-25} = \dfrac{1}{x+5} + \dfrac{2}{x-5}$

11. $\dfrac{5}{x^2-3x} + \dfrac{4}{2x-6}$

12. $\dfrac{5}{x^2-3x} \div \dfrac{4}{2x-6}$

13. $\dfrac{x-1}{x+1} + \dfrac{x+7}{x-1} = \dfrac{4}{x^2-1}$

14. $\left(1 - \dfrac{y}{x}\right) \div \left(1 - \dfrac{x}{y}\right)$

15. $\dfrac{a^2-9}{a-6} \cdot \dfrac{a^2-5a-6}{a^2-a-6}$

16. $\dfrac{2}{a-6} + \dfrac{3a}{a^2-5a-6} - \dfrac{a}{5a+5}$

17. $\dfrac{2x+3}{3x-2} = \dfrac{4x+1}{6x+1}$

18. $\dfrac{5x-3}{2x} = \dfrac{10x+3}{4x+1}$

19. $\dfrac{a}{9a^2-1} + \dfrac{2}{6a-2}$

20. $\dfrac{3}{4a-8} - \dfrac{a+2}{a^2-2a}$

21. $-\dfrac{3}{x^2} - \dfrac{1}{x} + 2 = 0$

22. $\dfrac{x}{2x+6} + \dfrac{5}{x^2-9}$

23. $\dfrac{x-8}{x^2-x-2} + \dfrac{2}{x-2}$

24. $\dfrac{x-8}{x^2-x-2} + \dfrac{2}{x-2} = \dfrac{3}{x+1}$

25. $\dfrac{3}{a} - 5 = \dfrac{7}{a} - 1$

26. $\dfrac{7}{3z-9} + \dfrac{5}{z}$

Use $\dfrac{x}{5} - \dfrac{x}{4} = \dfrac{1}{10}$ and $\dfrac{x}{5} - \dfrac{x}{4} + \dfrac{1}{10}$ for Exercises 27 and 28.

27. a. Which one above is an expression?
   b. Describe the first step to simplify this expression.
   c. Simplify the expression.

28. a. Which one above is an equation?
   b. Describe the first step to solve this equation.
   c. Solve the equation.

For each exercise, choose the correct statement.* Each figure represents a real number and no denominators are 0.

29. a. $\dfrac{\triangle + \square}{\triangle} = \square$

   b. $\dfrac{\triangle + \square}{\triangle} = 1 + \dfrac{\square}{\triangle}$

   c. $\dfrac{\triangle + \square}{\triangle} = \dfrac{\square}{\triangle}$

   d. $\dfrac{\triangle + \square}{\triangle} = 1 + \square$

   e. $\dfrac{\triangle + \square}{\triangle - \square} = -1$

30. a. $\dfrac{\triangle}{\square} + \dfrac{\square}{\triangle} = \dfrac{\triangle + \square}{\square + \triangle} = 1$

   b. $\dfrac{\triangle}{\square} + \dfrac{\square}{\triangle} = \dfrac{\triangle + \square}{\triangle\square}$

   c. $\dfrac{\triangle}{\square} + \dfrac{\square}{\triangle} = \triangle\triangle + \square\square$

   d. $\dfrac{\triangle}{\square} + \dfrac{\square}{\triangle} = \dfrac{\triangle\triangle + \square\square}{\square\triangle}$

   e. $\dfrac{\triangle}{\square} + \dfrac{\square}{\triangle} = \dfrac{\triangle\square}{\square\triangle} = 1$

31. a. $\dfrac{\triangle}{\square} \cdot \dfrac{\bigcirc}{\square} = \dfrac{\triangle\bigcirc}{\square}$

   b. $\dfrac{\triangle}{\square} \cdot \dfrac{\bigcirc}{\square} = \triangle\bigcirc$

   c. $\dfrac{\triangle}{\square} \cdot \dfrac{\bigcirc}{\square} = \dfrac{\triangle + \bigcirc}{\square + \square}$

   d. $\dfrac{\triangle}{\square} \cdot \dfrac{\bigcirc}{\square} = \dfrac{\triangle\bigcirc}{\square\square}$

32. a. $\dfrac{\triangle}{\square} \div \dfrac{\bigcirc}{\triangle} = \dfrac{\triangle\triangle}{\square\bigcirc}$

   b. $\dfrac{\triangle}{\square} \div \dfrac{\bigcirc}{\triangle} = \dfrac{\bigcirc\square}{\triangle\triangle}$

   c. $\dfrac{\triangle}{\square} \div \dfrac{\bigcirc}{\triangle} = \dfrac{\bigcirc}{\square}$

   d. $\dfrac{\triangle}{\square} \div \dfrac{\bigcirc}{\triangle} = \dfrac{\triangle + \triangle}{\square + \bigcirc}$

33. a. $\dfrac{\frac{\triangle + \square}{\bigcirc}}{\frac{\triangle}{\bigcirc}} = \square$

   b. $\dfrac{\frac{\triangle + \square}{\bigcirc}}{\frac{\triangle}{\bigcirc}} = \dfrac{\triangle\triangle + \triangle\square}{\bigcirc\bigcirc}$

   c. $\dfrac{\frac{\triangle + \square}{\bigcirc}}{\frac{\triangle}{\bigcirc}} = 1 + \square$

   d. $\dfrac{\frac{\triangle + \square}{\bigcirc}}{\frac{\triangle}{\bigcirc}} = \dfrac{\triangle + \square}{\triangle}$

21. _____

22. _____

23. _____

24. _____

25. _____

26. _____

27. a. _____

   b. _____

   c. _____

28. a. _____

   b. _____

   c. _____

29. _____

30. _____

31. _____

32. _____

33. _____

*My thanks to Kelly Champagne for permission to use her exercises for 29 through 33.

A Solve an Equation Containing Rational Expressions for a Specified Variable.

B Solve Number Problems by Writing Equations Containing Rational Expressions.

C Solve Problems Modeled by Proportions.

D Solve Problems About Work.

E Solve Problems About Distance, Rate, and Time.

6.5 RATIONAL EQUATIONS AND PROBLEM SOLVING

Objective **A** Solving Rational Equations for a Specified Variable

In Section 2.3, we solved equations for a specified variable. In this section, we continue practicing this skill by solving equations containing rational expressions for a specified variable. The steps given in Section 2.3 for solving equations for a specified variable are repeated here.

Solving an Equation for a Specified Variable

Step 1: Clear the equation of fractions or rational expressions by multiplying each side of the equation by the least common denominator (LCD) of all denominators in the equation.

Step 2: Use the distributive property to remove grouping symbols such as parentheses.

Step 3: Combine like terms on each side of the equation.

Step 4: Use the addition property of equality to rewrite the equation as an equivalent equation with terms containing the specified variable on one side and all other terms on the other side.

Step 5: Use the distributive property and the multiplication property of equality to get the specified variable alone.

PRACTICE PROBLEM 1

Solve $\dfrac{1}{x} + \dfrac{1}{y} = \dfrac{1}{z}$ for y.

EXAMPLE 1 Solve $\dfrac{1}{x} + \dfrac{1}{y} = \dfrac{1}{z}$ for x.

Solution: To clear this equation of fractions, we multiply both sides of the equation by xyz, the LCD of $\dfrac{1}{x}, \dfrac{1}{y},$ and $\dfrac{1}{z}$.

$$\frac{1}{x} + \frac{1}{y} = \frac{1}{z}$$

$$xyz\left(\frac{1}{x} + \frac{1}{y}\right) = xyz\left(\frac{1}{z}\right) \quad \text{Multiply both sides by } xyz.$$

$$xyz\left(\frac{1}{x}\right) + xyz\left(\frac{1}{y}\right) = xyz\left(\frac{1}{z}\right) \quad \text{Use the distributive property.}$$

$$yz + xz = xy \quad \text{Simplify.}$$

Notice the two terms that contain the specified variable, x.

Next, we subtract xz from both sides so that all terms containing the specified variable x are on one side of the equation and all other terms are on the other side.

$$yz = xy - xz$$

Now we use the distributive property to factor x from $xy - xz$ and then the multiplication property of equality to solve for x.

$$yz = x(y - z)$$

$$\frac{yz}{y - z} = x \quad \text{or} \quad x = \frac{yz}{y - z} \quad \text{Divide both sides by } y - z.$$

■ **Work Practice Problem 1**

Answer

1. $y = \dfrac{xz}{x - z}$

Objective B Solving Number Problems Modeled by Rational Equations

Problem solving sometimes involves modeling a described situation with an equation containing rational expressions. In Examples 2 through 5, we practice solving such problems and use the problem-solving steps first introduced in Section 2.2.

EXAMPLE 2 Finding an Unknown Number

If a certain number is subtracted from the numerator and added to the denominator of $\frac{9}{19}$, the new fraction is equivalent to $\frac{1}{3}$. Find the number.

Solution:

1. **UNDERSTAND** the problem. Read and reread the problem and try guessing the solution. For example, if the unknown number is 3, we have

$$\frac{9-3}{19+3} = \frac{6}{22} = \frac{3}{11} \neq \frac{1}{3}$$

Thus, $\frac{9-3}{19+3} \neq \frac{1}{3}$ and 3 is not the correct number. Remember that the purpose of this step is not to guess the correct solution but to gain an understanding of the problem posed.

   We will let n = the number to be subtracted from the numerator and added to the denominator.

2. **TRANSLATE** the problem.

In words:	when the number is subtracted from the numerator and added to the denominator of the fraction $\frac{9}{19}$	this is equivalent to	$\frac{1}{3}$
	↓	↓	↓
Translate:	$\dfrac{9-n}{19+n}$	=	$\dfrac{1}{3}$

3. **SOLVE** the equation for n.

$$\frac{9-n}{19+n} = \frac{1}{3}$$

To solve for n, we begin by multiplying both sides by the LCD, $3(19+n)$.

$$3(19+n) \cdot \frac{9-n}{19+n} = 3(19+n) \cdot \frac{1}{3} \qquad \text{Multiply both sides by the LCD.}$$
$$3(9-n) = 19+n \qquad \text{Simplify.}$$
$$27-3n = 19+n$$
$$8 = 4n$$
$$2 = n \qquad \text{Solve.}$$

4. **INTERPRET** the results.

Check: If we subtract 2 from the numerator and add 2 to the denominator of $\frac{9}{19}$, we have $\frac{9-2}{19+2} = \frac{7}{21} = \frac{1}{3}$, and the problem checks.

State: The unknown number is 2.

 Work Practice Problem 2

Objective C Solving Problems Modeled by Proportions

A **ratio** is the quotient of two numbers or two quantities. Since rational expressions are quotients of quantities, rational expressions are ratios, also. A **proportion** is a mathematical statement that two ratios are equal.

Let's review two methods for solving a proportion such as $\dfrac{x-3}{10} = \dfrac{7}{15}$. We can multiply both sides of the equation by the LCD, 30.

Multiply both sides by the LCD, 30.

$$30 \cdot \frac{x-3}{10} = 30 \cdot \frac{7}{15}$$
$$3(x-3) = 2 \cdot 7$$
$$3x - 9 = 14$$
$$3x = 23$$
$$x = \frac{23}{3}$$

We can also solve a proportion by setting cross products equal. Here, we are using the fact that if $\dfrac{a}{b} = \dfrac{c}{d}$, then $ad = bc$.

$$\frac{x-3}{10} = \frac{7}{15}$$

$$15(x-3) = 10 \cdot 7 \qquad \text{Set cross products equal.}$$
$$15x - 45 = 70 \qquad \text{Use the distributive property.}$$
$$15x = 115$$
$$x = \frac{115}{15} \quad \text{or} \quad \frac{23}{3}$$

A ratio of two different quantities is called a **rate.** For example $\dfrac{3 \text{ miles}}{2 \text{ hours}}$ or 1.5 miles/hour is a rate. The proportions we write to solve problems will sometimes include rates. When this happens, make sure that the rates contain units written in the same order.

PRACTICE PROBLEM 3

In the United States, 1 out of 50 homes is heated by wood. At this rate, how many homes in a community of 36,000 homes are heated by wood? (*Source:* 2000 Census Survey)

EXAMPLE 3

In the United States, 7 out of every 25 homes are heated by electricity. At this rate, how many homes in a community of 36,000 homes would you predict are heated by electricity? (*Source:* Census Survey)

Solution:

1. UNDERSTAND. Read and reread the problem. Try to estimate a reasonable solution. For example, since 7 is less than $\dfrac{1}{3}$ of 25, we might reason that the solution would be less than $\dfrac{1}{3}$ of 36,000 or 12,000.

   Let's let x = number of homes in the community heated by electricity

2. TRANSLATE.

   homes heated by electricity → $\dfrac{7}{25} = \dfrac{x}{36,000}$ ← homes heated by electricity

   total homes → ← total homes

3. SOLVE. To solve this proportion we will set cross products equal.

$$\frac{7}{25} = \frac{x}{36,000}$$

$$7 \cdot 36,000 = 25x$$
$$\frac{252,000}{25} = x$$
$$10,080 = x$$

Answer

3. 720 homes

4. INTERPRET.

Check: To check, replace x with 10,080 in the proportion and see that a true statement results. Notice that our answer is reasonable since it is less than 12,000 as we stated above.

State: We predict that 10,080 homes are heated by electricity.

■ **Work Practice Problem 3**

Objective D Solving Problems About Work

The following work example leads to an equation containing rational expressions.

EXAMPLE 4 **Calculating Work Hours**

Melissa Scarlatti can clean the house in 4 hours, whereas her husband, Zack, can do the same job in 5 hours. They have agreed to clean together so that they can finish in time to watch a movie on TV that starts in 2 hours. How long will it take them to clean the house together? Can they finish before the movie starts?

Solution:

1. Read and reread the problem. The key idea here is the relationship between the *time* (in hours) it takes to complete the job and the *part of the job* completed in 1 unit of time (1 hour). For example, if the *time* it takes Melissa to complete the job is 4 hours, the *part of the job* she can complete in 1 hour is $\frac{1}{4}$. Similarly, Zack can complete $\frac{1}{5}$ of the job in 1 hour.

   We will let t = the *time* in hours it takes Melissa and Zack to clean the house together. Then $\frac{1}{t}$ represents the *part of the job* they complete in 1 hour. We summarize the given information on a chart.

	Hours to Complete the Job	Part of Job Completed in 1 Hour
Melissa Alone	4	$\frac{1}{4}$
Zack Alone	5	$\frac{1}{5}$
Together	t	$\frac{1}{t}$

2. **TRANSLATE.**

In words:	part of job Melissa can complete in 1 hour	added to	part of job Zack can complete in 1 hour	is equal to	part of job they can complete together in 1 hour
	↓	↓	↓	↓	↓
Translate:	$\frac{1}{4}$	$+$	$\frac{1}{5}$	$=$	$\frac{1}{t}$

Continued on next page

PRACTICE PROBLEM 4
Greg Guillot can paint a room alone in 3 hours. His brother Phillip can do the same job alone in 5 hours. How long would it take them to paint the room if they work together?

Answer

4. $1\frac{7}{8}$ hr

3. SOLVE:

$$\frac{1}{4} + \frac{1}{5} = \frac{1}{t}$$

$$20t\left(\frac{1}{4} + \frac{1}{5}\right) = 20t\left(\frac{1}{t}\right) \qquad \text{Multiply both sides by the LCD, } 20t.$$

$$5t + 4t = 20$$

$$9t = 20$$

$$t = \frac{20}{9} \quad \text{or} \quad 2\frac{2}{9} \quad \text{Solve.}$$

4. INTERPRET.

Check: The proposed solution is $2\frac{2}{9}$. That is, Melissa and Zack would take $2\frac{2}{9}$ hours to clean the house together. This proposed solution is reasonable since $2\frac{2}{9}$ hours is more than half of Melissa's time and less than half of Zack's time. Check this solution in the originally stated problem.

State: Melissa and Zack can clean the house together in $2\frac{2}{9}$ hours. They cannot complete the job before the movie starts.

■ **Work Practice Problem 4**

Objective **E** **Solving Problems About Distance, Rate, and Time**

EXAMPLE 5 **Finding the Speed of a Current**

Steve Deitmer takes $1\frac{1}{2}$ times as long to go 72 miles upstream in his boat as he does to return. If the boat cruises at 30 mph in still water, what is the speed of the current?

Solution:

1. UNDERSTAND. Read and reread the problem. Guess a solution. Suppose that the current is 4 mph. The speed of the boat upstream is slowed down by the current: $30 - 4$, or 26 mph, and the speed of the boat downstream is speeded up by the current: $30 + 4$, or 34 mph. Next let's find out how long it takes to travel 72 miles upstream and 72 miles downstream. To do so, we use the formula

$$d = r \cdot t, \text{ or } \frac{d}{r} = t.$$

Upstream

$$\frac{d}{r} = t$$

$$\frac{72}{26} = t$$

$$2\frac{10}{13} = t$$

Downstream

$$\frac{d}{r} = t$$

$$\frac{72}{34} = t$$

$$2\frac{2}{17} = t$$

Since the time upstream $\left(2\frac{10}{13} \text{ hours}\right)$ is not $1\frac{1}{2}$ times the time downstream $\left(2\frac{2}{17} \text{ hours}\right)$, our guess is not correct. We do, however, have a better understanding of the problem.

We will let

$x =$ the speed of the current

$30 + x =$ the speed of the boat downstream

$30 - x =$ the speed of the boat upstream

This information is summarized in the following chart, where we use the formula $\frac{d}{t} = t$.

	Distance	Rate	Time $\left(\dfrac{d}{r}\right)$
Upstream	72	$30 - x$	$\dfrac{72}{30 - x}$
Downstream	72	$30 + x$	$\dfrac{72}{30 + x}$

2. TRANSLATE. Since the time spent traveling upstream is $1\frac{1}{2}$ times the time spent traveling downstream, we have

In words:

time upstream	is	$1\frac{1}{2}$	times	time downstream
↓	↓	↓	↓	↓

Translate:

$$\frac{72}{30 - x} \quad = \quad \frac{3}{2} \quad \cdot \quad \frac{72}{30 + x}$$

3. SOLVE. $\dfrac{72}{30 - x} = \dfrac{3}{2} \cdot \dfrac{72}{30 + x}$

First we multiply both sides by the LCD, $2(30 + x)(30 - x)$.

$$2(30 + x)(30 - x) \cdot \frac{72}{30 - x} = 2(30 + x)(30 - x)\left(\frac{3}{2} \cdot \frac{72}{30 + x}\right)$$

$$72 \cdot 2(30 + x) = 3 \cdot 72 \cdot (30 - x) \quad \text{Simplify.}$$

$$2(30 + x) = 3(30 - x) \quad \text{Divide both sides by 72.}$$

$$60 + 2x = 90 - 3x \quad \text{Use the distributive property.}$$

$$5x = 30$$

$$x = 6 \quad \text{Solve.}$$

4. INTERPRET.

Check: Check the proposed solution of 6 mph in the originally stated problem.

State: The current's speed is 6 mph.

■ **Work Practice Problem 5**

6.5 EXERCISE SET

FOR EXTRA HELP

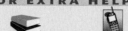
Student Solutions Manual

PH Math/Tutor Center

CD/Video for Review

Math XL
MathXL®

MyMathLab
MyMathLab

Objective **A** *Solve each equation for the specified variable. See Example 1.*

1. $F = \dfrac{9}{5}C + 32$ for C

2. $V = \dfrac{1}{3}\pi r^2 h$ for h

3. $Q = \dfrac{A - I}{L}$ for I

4. $P = 1 - \dfrac{C}{S}$ for S

5. $\dfrac{1}{R} = \dfrac{1}{R_1} + \dfrac{1}{R_2}$ for R

6. $\dfrac{1}{R} = \dfrac{1}{R_1} + \dfrac{1}{R_2}$ for R_1

7. $S = \dfrac{n(a + L)}{2}$ for n

8. $S = \dfrac{n(a + L)}{2}$ for a

9. $A = \dfrac{h(a + b)}{2}$ for b

10. $A = \dfrac{h(a + b)}{2}$ for h

11. $\dfrac{P_1 V_1}{T_1} = \dfrac{P_2 V_2}{T_2}$ for T_2

12. $H = \dfrac{kA(T_1 - T_2)}{L}$ for T_2

13. $f = \dfrac{f_1 f_2}{f_1 + f_2}$ for f_2

14. $I = \dfrac{E}{R + r}$ for r

15. $\lambda = \dfrac{2L}{n}$ for L

16. $S = \dfrac{a_1 - a_n r}{1 - r}$ for a_1

17. $\dfrac{\theta}{\omega} = \dfrac{2L}{c}$ for c

18. $F = \dfrac{-GMm}{r^2}$ for M

Objective **B** *Solve. See Example 2.*

19. The sum of a number and 5 times its reciprocal is 6. Find the number(s).

20. The quotient of a number and 9 times its reciprocal is 1. Find the number(s).

21. If a number is added to the numerator of $\dfrac{12}{41}$ and twice the number is added to the denominator of $\dfrac{12}{41}$, the resulting fraction is equivalent to $\dfrac{1}{3}$. Find the number.

22. If a number is subtracted from the numerator of $\dfrac{13}{8}$ and added to the denominator of $\dfrac{13}{8}$, the resulting fraction is equivalent to $\dfrac{2}{5}$. Find the number.

Objective **C** *Solve. See Example 3.*

23. An Arabian camel can drink 15 gallons of water in 10 minutes. At this rate, how much water can the camel drink in 3 minutes? (*Source:* Grolier, Inc.)

24. An Arabian camel can travel 20 miles in 8 hours, carrying a 300-pound load on its back. At this rate, how far can the camel travel in 10 hours? (*Source:* Grolier, Inc.)

25. In 2004, 10.7 out of every 100 Coast Guard personnel were women. If there were 40,151 total Coast Guard personnel on active duty, estimate the number of women. Round to the nearest whole. (*Source: The World Almanac, 2005*)

26. In 2004, 47 out of every 50 Marines personnel were men. If there were 175,202 total Marine personnel on active duty, estimate the number of men. Round to the nearest whole. (*Source: The World Almanac, 2005*)

Objective D *Solve. See Example 4.*

27. An experienced roofer can roof a house in 26 hours. A beginning roofer needs 39 hours to complete the same job. Find how long it takes for the two to do the job together.

28. Alan Cantrell can word process a research paper in 6 hours. With Steve Isaac's help, the paper can be processed in 4 hours. Find how long it takes Steve to word process the paper alone.

29. Three postal workers can sort a stack of mail in 20 minutes, 30 minutes, and 60 minutes, respectively. Find how long it takes them to sort the mail if all three work together.

30. A new printing press can print newspapers twice as fast as the old one can. The old one can print the afternoon edition in 4 hours. Find how long it takes to print the afternoon edition if both printers are operating.

Objective E *Solve. See Example 5.*

31. Mattie Evans drove 150 miles in the same amount of time that it took a turbopropeller plane to travel 600 miles. The speed of the plane was 150 mph faster than the speed of the car. Find the speed of the plane.

32. An F-100 plane and a Toyota truck leave the same town at sunrise and head for a town 450 miles away. The speed of the plane is three times the speed of the truck, and the plane arrives 6 hours ahead of the truck. Find the speed of the truck.

33. The speed of Lazy River's current is 5 mph. If a boat travels 20 miles downstream in the same time that it takes to travel 10 miles upstream, find the speed of the boat in still water.

34. The speed of a boat in still water is 24 mph. If the boat travels 54 miles upstream in the same time that it takes to travel 90 miles downstream, find the speed of the current.

Objectives B C D E **Mixed Practice** *Solve.*

35. The sum of the reciprocals of two consecutive integers is $-\frac{15}{56}$. Find the two integers.

36. The sum of the reciprocals of two consecutive odd integers is $\frac{20}{99}$. Find the two integers.

37. One hose can fill a goldfish pond in 45 minutes, and two hoses can fill the same pond in 20 minutes. Find how long it takes the second hose alone to fill the pond.

38. If Sarah Clark can do a job in 5 hours and Dick Belli and Sarah working together can do the same job in 2 hours, find how long it takes Dick to do the job alone.

39. Two trains going in opposite directions leave at the same time. One train travels 15 mph faster than the other. In 6 hours the trains are 630 miles apart. Find the speed of each.

40. The speed of a bicyclist is 10 mph faster than the speed of a walker. If the bicyclist travels 26 miles in the same amount of time that the walker travels 6 miles, find the speed of the bicyclist.

41. A giant tortoise can travel 0.17 miles in 1 hour. At this rate, how long would it take the tortoise to travel 1 mile? Round to the nearest tenth of an hour. (*Source: The World Almanac*)

42. A black mamba snake can travel 88 feet in 3 seconds. At this rate, how long does it take to travel 300 feet (the length of a football field)? Round to the nearest tenth of a second. (*Source: The World Almanac, 2002*).

43. Moo Dairy has three machines to fill half-gallon milk cartons. The machines can fill the daily quota in 5 hours, 6 hours, and 7.5 hours, respectively. Find how long it takes to fill the daily quota if all three machines are running.

44. The inlet pipe of an oil tank can fill the tank in 1 hour, 30 minutes. The outlet pipe can empty the tank in 1 hour. Find how long it takes to empty a full tank if both pipes are open.

45. A plane flies 465 miles with the wind and 345 miles against the wind in the same length of time. If the speed of the wind is 20 mph, find the speed of the plane in still air.

46. Two rockets are launched. The first travels at 9000 mph. Fifteen minutes later the second is launched at 10,000 mph. Find the distance at which both rockets are an equal distance from Earth.

47. Two joggers, one averaging 8 mph and one averaging 6 mph, start from a designated initial point. The slower jogger arrives at the end of the run a half-hour after the other jogger. Find the distance of the run.

48. A semi truck travels 300 miles through the flatland in the same amount of time that it travels 180 miles through the Great Smoky Mountains. The rate of the truck is 20 miles per hour slower in the mountains than in the flatland. Find both the flatland rate and mountain rate.

49. The denominator of a fraction is 1 more than the numerator. If both the numerator and the denominator are decreased by 3, the resulting fraction is equivalent to $\frac{4}{5}$. Find the fraction.

50. The numerator of a fraction is 4 less than the denominator. If both the numerator and the denominator are increased by 2, the resulting fraction is equivalent to $\frac{2}{3}$. Find the fraction.

51. In 2 minutes, a conveyor belt can move 300 pounds of recyclable aluminum from the delivery truck to a storage area. A smaller belt can move the same quantity of cans the same distance in 6 minutes. If both belts are used, find how long it takes to move the cans to the storage area.

52. Gary Marcus and Tony Alva work at Lombardo's Pipe and Concrete. Mr. Lombardo is preparing an estimate for a customer. He knows that Gary can lay a slab of concrete in 6 hours. Tony can lay the same size slab in 4 hours. If both work on the job and the cost of labor is $45.00 per hour, determine what the labor estimate should be.

53. The world record for the largest white bass caught is held by Ronald Sprouse of Virginia. The bass weighed 6 pounds 13 ounces. If Ronald rows to his favorite fishing spot 9 miles downstream in the same amount of time that he rows 3 miles upstream and if the current is 6 mph, find how long it takes him to cover the 12 miles.

54. An amateur cyclist training for a road race rode the first 20-mile portion of his workout at a constant rate. For the 16-mile cooldown portion of his workout, he reduced his speed by 2 miles per hour. Each portion of the workout took equal time. Find the cyclist's rate during the first portion and his rate during the cooldown portion.

55. Smith Engineering is in the process of reviewing the salaries of their surveyors. During this review, the company found that an experienced surveyor can survey a roadbed in 4 hours. An apprentice surveyor needs 5 hours to survey the same stretch of road. If the two work together, find how long it takes them to complete the job.

56. Mr. Dodson can paint his house by himself in four days. His son will need an additional day to complete the job if he works by himself. If they work together, find how long it takes to paint the house.

57. An experienced bricklayer can construct a small wall in 3 hours. An apprentice can complete the job in 6 hours. Find how long it takes if they work together.

58. A marketing manager travels 1080 miles in a corporate jet and then an additional 240 miles by car. If the car ride takes 1 hour longer, and if the rate of the jet is 6 times the rate of the car, find the time the manager travels by jet and find the time she travels by car.

Review

Solve each equation for x. See Section 2.1.

59. $\dfrac{x}{5} = \dfrac{x+2}{3}$

60. $\dfrac{x}{4} = \dfrac{x+3}{6}$

61. $\dfrac{x-3}{2} = \dfrac{x-5}{6}$

62. $\dfrac{x-6}{4} = \dfrac{x-2}{5}$

Concept Extensions

Calculating body-mass index (BMI) is a way to gauge whether a person should lose weight. Doctors recommend that body-mass index values fall between 19 and 25. The formula for body-mass index B is $B = \dfrac{705w}{h^2}$, where w is weight in pounds and h is height in inches. Use this formula to answer Exercises 63 and 64.

63. A patient is 5 ft 8 in. tall. What should his or her weight be to have a body-mass index of 25? Round to the nearest whole pound.

64. A doctor recorded a body-mass index of 47 on a patient's chart. Later, a nurse notices that the doctor recorded the patient's weight as 240 pounds but neglected to record the patient's height. Explain how the nurse can use the information from the chart to find the patient's height. Then find the height.

In physics, when the source of a sound is traveling toward an observer, the relationship between the actual pitch a of the sound and the pitch h that the observer hears due to the Doppler effect is described by the formula $h = \dfrac{a}{1 - \dfrac{s}{770}}$, where s is the speed of the sound source in miles per hour. Use this formula to answer Exercise 65.

65. An emergency vehicle has a single-tone siren with the pitch of the musical note E. As it approaches an observer standing by the road, the vehicle is traveling 50 mph. Is the pitch that the observer hears due to the Doppler effect lower or higher than the actual pitch? To which musical note is the pitch that the observer hears closest?

Pitch of an Octave of Musical Notes in Hertz (Hz)	
Note	**Pitch**
Middle C	261.63
D	293.66
E	329.63
F	349.23
G	392.00
A	440.00
B	493.88

Note: Greater numbers indicate higher pitches (acoustically).

(*Source:* American Standards Association)

In electronics, the relationship among the resistances R_1 and R_2 of two resistors wired in a parallel circuit and their combined resistance R is described by the formula $\dfrac{1}{R} = \dfrac{1}{R_1} + \dfrac{1}{R_2}$. Use this formula to solve Exercises 66 through 68.

66. If the combined resistance is 2 ohms and one of the two resistances is 3 ohms, find the other resistance.

67. Find the combined resistance of two resistors of 12 ohms each when they are wired in a parallel circuit.

68. The relationship among resistance of two resistors wired in a parallel circuit and their combined resistance may be extended to three resistors of resistances R_1, R_2, and R_3. Write an equation you think may describe the relationship, and use it to find the combined resistance if R_1 is 5, R_2 is 6, and R_3 is 2.

6.6 VARIATION AND PROBLEM SOLVING

Objectives

A Solve Problems Involving Direct Variation.

B Solve Problems Involving Inverse Variation.

C Solve Problems Involving Joint Variation.

D Solve Problems Involving Combined Variation.

Objective **A** Solving Problems Involving Direct Variation

A very familiar example of **direct variation** is the relationship of the circumference C of a circle to its radius r. The formula $C = 2\pi r$ expresses that the circumference is always 2π times the radius. In other words, C is always a constant multiple (2π) of r. Because it is, we say that *C varies directly as r*, that *C varies directly with r*, or that *C is directly proportional to r*.

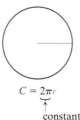

$$C = 2\pi r$$
constant

Direct Variation

y varies directly as x, or **y is directly proportional to x,** if there is a nonzero constant k such that

$$y = kx$$

The number k is called the **constant of variation** or the **constant of proportionality.**

In the above definition, the relationship described between x and y is a linear one. In other words, the graph of $y = kx$ is a line. The slope of the line is k, and the line passes through the origin.

For example, the graph of the direct variation equation $C = 2\pi r$ is shown. The horizontal axis represents the radius r, and the vertical axis is the circumference C. From the graph we can read that when the radius is 6 units, the circumference is approximately 38 units. Also, when the circumference is 45 units, the radius is between 7 and 8 units. Notice that as the radius increases, the circumference increases.

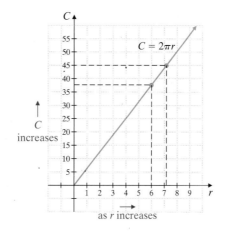

EXAMPLE 1 Suppose that y varies directly as x. If y is 5 when x is 30, find the constant of variation and the direct variation equation.

Solution: Since y varies directly as x, we write $y = kx$. If $y = 5$ when $x = 30$, we have that

$$y = kx$$
$$5 = k(30) \qquad \text{Replace } y \text{ with 5 and } x \text{ with 30.}$$
$$\frac{1}{6} = k \qquad \text{Solve for } k.$$

The constant of variation is $\frac{1}{6}$.

After finding the constant of variation k, the direct variation equation can be written as $y = \frac{1}{6}x$.

Work Practice Problem 1

PRACTICE PROBLEM 1

Suppose that y varies directly as x. If y is 24 when x is 8, find the constant of variation and the direct variation equation.

Answer

1. $k = 3$; $y = 3x$

455

Copyright 2007 Pearson Education, Inc.

PRACTICE PROBLEM 2

Use Hooke's law as stated in Example 2. If a 56-pound weight attached to a spring stretches the spring 8 inches, find the distance that an 85-pound weight attached to the spring stretches the spring.

EXAMPLE 2 **Using Direct Variation and Hooke's Law**

Hooke's law states that the distance a spring stretches is directly proportional to the weight attached to the spring. If a 40-pound weight attached to the spring stretches the spring 5 inches, find the distance that a 65-pound weight attached to the spring stretches the spring.

Solution:

1. UNDERSTAND. Read and reread the problem. Notice that we are given that the distance a spring stretches is *directly proportional* to the weight attached. We let

    d = the distance stretched.

    w = the weight attached

    The constant of variation is represented by k.

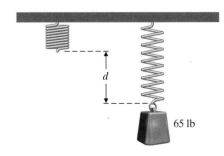

65 lb

2. TRANSLATE. Because d is directly proportional to w, we write

    $$d = kw$$

3. SOLVE. When a weight of 40 pounds is attached, the spring stretches 5 inches. That is, when $w = 40$, $d = 5$.

    $5 = k(40)$ Replace d with 5 and w with 40.

    $\dfrac{1}{8} = k$ Solve for k.

    Now when we replace k with $\dfrac{1}{8}$ in the equation $d = kw$, we have

    $$d = \frac{1}{8}w$$

    To find the stretch when a weight of 65 pounds is attached, we replace w with 65 to find d.

    $$d = \frac{1}{8}(65)$$

    $$= \frac{65}{8} = 8\frac{1}{8} \text{ or } 8.125$$

4. INTERPRET.

    Check: Check the proposed solution of 8.125 inches in the original problem.

    State: The spring stretches 8.125 inches when a 65-pound weight is attached.

■ **Work Practice Problem 2**

Objective **B** **Solving Problems Involving Inverse Variation**

When y is proportional to the *reciprocal* of another variable x, we say that y *varies inversely as* x, or that y *is inversely proportional to* x. An example of the **inverse variation** relationship is the relationship between the pressure that a gas exerts and

Answer

2. $12\frac{1}{7}$ in.

the volume of its container. As the volume of a container decreases, the pressure of the gas it contains increases.

Inverse Variation

y varies inversely as x, or **y is inversely proportional to x,** if there is a nonzero constant k such that

$$y = \frac{k}{x}$$

The number k is called the **constant of variation** or the **constant of proportionality.**

Notice that $y = \dfrac{k}{x}$ is an equation containing a rational expression. Its graph for $k > 0$ and $x > 0$ is shown. From the graph, we can see that as x increases, y decreases.

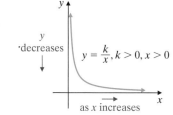

$y = \dfrac{k}{x}, k > 0, x > 0$

as x increases

EXAMPLE 3 Suppose that u varies inversely as w. If u is 3 when w is 5, find the constant of variation and the inverse variation equation.

Solution: Since u varies inversely as w, we have $u = \dfrac{k}{w}$. We let $u = 3$ and $w = 5$, and we solve for k.

$$u = \frac{k}{w}$$

$$3 = \frac{k}{5} \qquad \text{Let } u = 3 \text{ and } w = 5.$$

$$15 = k \qquad \text{Multiply both sides by 5.}$$

The constant of variation k is 15. This gives the inverse variation equation

$$u = \frac{15}{w}$$

■ **Work Practice Problem 3**

EXAMPLE 4 **Using Inverse Variation and Boyle's Law**

Boyle's law says that if the temperature stays the same, the pressure P of a gas is inversely proportional to the volume V. If a cylinder in a steam engine has a pressure of 960 kilopascals when the volume is 1.4 cubic meters, find the pressure when the volume increases to 2.5 cubic meters.

Continued on next page

PRACTICE PROBLEM 3

Suppose that y varies inversely as x. If y is 6 when x is 3, find the constant of variation and the inverse variation equation.

PRACTICE PROBLEM 4

The speed r at which one needs to drive in order to travel a constant distance is inversely proportional to the time t. A fixed distance can be driven in 5 hours at a rate of 24 mph. Find the rate needed to drive the same distance in 4 hours.

Answers

3. $k = 18; y = \dfrac{18}{x}$, **4.** 30 mph

Solution:

1. UNDERSTAND. Read and reread the problem. Notice that we are given that the pressure of a gas is *inversely proportional* to the volume. We will let $P =$ the pressure and $V =$ the volume. The constant of variation is represented by k.

2. TRANSLATE. Because P is inversely proportional to V, we write

$$P = \frac{k}{V}$$

When $P = 960$ kilopascals, the volume $V = 1.4$ cubic meters. We use this information to find k.

$$960 = \frac{k}{1.4} \quad \text{Let } P = 960 \text{ and } V = 1.4.$$

$$1344 = k \quad \text{Multiply both sides by 1.4.}$$

Thus, the value of k is 1344. Replacing k with 1344 in the variation equation, we have

$$P = \frac{1344}{V}$$

Next we find P when V is 2.5 cubic meters.

3. SOLVE.

$$P = \frac{1344}{2.5} \quad \text{Let } V = 2.5$$

$$= 537.6$$

4. INTERPRET. *Check* the proposed solution in the original problem.

State: When the volume is 2.5 cubic meters, the pressure is   kilopascals.

◼ **Work Practice Problem 4**

Objective ◉ **Solving Problems Involving Joint Variation**

Sometimes the ratio of a variable to the product of many other variables is constant. For example, the ratio of distance traveled to the product of speed and time traveled is constantly 1:

$$\frac{d}{rt} = 1 \quad \text{or} \quad d = rt$$

Such a relationship is called **joint variation.**

Joint Variation

If the ratio of a variable y to the product of two or more variables is constant, then y **varies jointly as,** or **is jointly proportional to,** the other variables. If

$$y = kxz$$

then the number k is the **constant of variation** or the **constant of proportionality.**

✔**Concept Check** Which type of variation is represented by the equation $xy = 8$? Explain.

a. Direct variation

b. Inverse variation

c. Joint variation

Solve. See Example 7.

39. The maximum weight that a rectangular beam can support varies jointly as its width and the square of its height and inversely as its length. If a beam $\frac{1}{2}$ foot wide, $\frac{1}{3}$ foot high, and 10 feet long can support 12 tons find how much a similar beam can support if the beam is $\frac{2}{3}$ foot wide, $\frac{1}{2}$ foot high, and 16 feet long.

40. The number of cars manufactured on an assembly line at a General Motors plant varies jointly as the number of workers and the time they work. If 200 workers can produce 60 cars in 2 hours, find how many cars 240 workers should be able to make in 3 hours.

41. The volume of a cone varies jointly as its height and the square of its radius. If the volume of a cone is 32π cubic inches when the radius is 4 inches and the height is 6 inches, find the volume of a cone when the radius is 3 inches and the height is 5 inches.

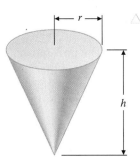

42. When a wind blows perpendicularly against a flat surface, its force is jointly proportional to the surface area and the speed of the wind. A sail whose surface area is 12 square feet experiences a 20-pound force when the wind speed is 10 miles per hour. Find the force on an 8-square-foot sail if the wind speed is 12 miles per hour.

43. The intensity of light (in foot-candles) varies inversely as the square of x, the distance in feet from the light source. The intensity of light 2 feet from the source is 80 foot-candles. How far away is the source if the intensity of light is 5 foot-candles?

44. The horsepower that can be safely transmitted to a shaft varies jointly as the shaft's angular speed of rotation (in revolutions per minute) and the cube of its diameter. A 2-inch shaft making 120 revolutions per minute safely transmits 40 horsepower. Find how much horsepower can be safely transmitted by a 3-inch shaft making 80 revolutions per minute.

Objectives Ⓐ Ⓑ Ⓒ Ⓓ **Mixed Practice** *Write an equation to describe each variation. Use k for the constant of proportionality. See Examples 1 through 7.*

45. y varies directly as x

46. p varies directly as q

47. a varies inversely as b

48. y varies inversely as x

49. y varies jointly as x and z

50. y varies jointly as q, r, and t

51. y varies inversely as x^3

52. y varies inversely as a^4

53. y varies directly as x and inversely as p^2

54. y varies directly as a^5 and inversely as b

Review

Find the exact circumference and area of each circle. See the inside cover for a list of geometric formulas.

55.

6 cm

56.

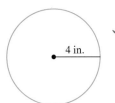

4 in.

57.

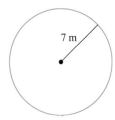

7 m

58.

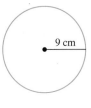

9 cm

Find the slope of the line containing each pair of points. See Section 3.2.

59. $(3, 6), (-2, 6)$ **60.** $(-5, -2), (0, 7)$ **61.** $(4, -1), (5, -2)$ **62.** $(2, 1), (2, -3)$

Concept Extensions

Solve. See the Concept Check in this section. Choose the type of variation that each equation represents.
a. *Direct variation*   **b.** *Inverse variation*   **c.** *Joint variation*

63. $y = \dfrac{2}{3}x$ **64.** $y = \dfrac{0.6}{x}$ **65.** $y = 9ab$ **66.** $xy = \dfrac{2}{11}$

67. The volume of a cylinder varies jointly as the height and the square of the radius. If the height is halved and the radius is doubled, determine what happens to the volume.

68. The horsepower to drive a boat varies directly as the cube of the speed of the boat. If the speed of the boat is to double, determine the corresponding increase in horsepower required.

69. Suppose that y varies directly as x^2. If x is doubled, what is the effect on y?

70. Suppose that y varies directly as x. If x is doubled, what is the effect on y?

 CHAPTER 6   Group Activity

Fastest-Growing Occupations

We reviewed fastest-growing occupations originally in Chapter 2. In this chapter, let's study this important data in terms of percents.

According to U.S. Bureau of Labor Statistics projections, the careers listed in the following table will be the top ten fastest-growing jobs into the next decade, according to the percent increase in the number of jobs.

Employment (in thousands)			
Occupation	**2002**	**2012**	**% Change**
Medical assistants	365	579	
Computer system analysts	468	653	
Computer software engineers	394	573	
Personal and home care aides	608	854	
Registered nurses	2,284	2,908	
Management analysts	577	753	
Postsecondary teachers	1,581	2,184	
Security guards	995	1,313	
Receptionists and information clerks	1,100	1,425	
Home health aides	580	859	

What do all these fast-growing occupations have in common? They all require knowledge of math! For some careers, such as management analysts, registered nurses, and computer software engineers, the ways math is used on the job may be obvious. For other occupations, the use of math may not be quite as apparent. However, tasks common to many jobs, like filling in a time sheet, writing up an expense or mileage report, planning a budget, figuring a bill, ordering supplies, and even making a work schedule, all require math.

Group Activity

1. Find the percent change in the number of jobs available from 2002 to 2012 for each occupation in the list.

2. Rank these top-ten occupations according to percent growth, from greatest to least.

3. Which occupation will be the fastest growing during this period?

4. How many occupations will have more than 50% more positions in 2012 than in 2002?

5. Which of the listed occupations will be the slowest growing during this period?

Chapter 6 Vocabulary Check

Fill in each blank with one of the words or phrases listed below.

rational expression   equation   complex fraction   opposites   directly

least common denominator   expression   inversely   jointly

1. A rational expression whose numerator, denominator, or both contain one or more rational expressions is called a

_____ .

2. In the equation $y = kx$, y varies _____ as x.

3. In the equation $y = \dfrac{k}{x}$, y varies _____ as x.

4. The _____ of a list of rational expressions is a polynomial of least degree whose factors include the denominator factors in the list.

5. In the equation $y = kxz$, y varies _____ as x and z.

6. The expressions $(x - 5)$ and $(5 - x)$ are called _____ .

7. A _____ is an expression that can be written as the quotient of $\dfrac{P}{Q}$ of two polynomials P and Q as long as Q is not 0.

8. Which is an expression and which is an equation? An example

of an _____ is $\dfrac{2}{x} + \dfrac{2}{x^2} = 7$ and an example of an _____

is $\dfrac{2}{x} + \dfrac{5}{x^2}$.

> **Helpful Hint**
>
> Are you preparing for your test? Don't forget to take the Chapter 6 Test on page 472. Then check your answers at the back of the text and use the Chapter Test Prep Video CD to see the fully worked-out solutions to any of the exercises you want to review.

 # Chapter Highlights

DEFINITIONS AND CONCEPTS	**EXAMPLES**
Section 6.1 Rational Functions and Multiplying and Dividing Rational Expressions	

A **rational expression** is the quotient $\dfrac{P}{Q}$ of two polynomials P and Q, as long as Q is not 0.

$$\frac{2x - 6}{7}, \frac{t^2 - 3t + 5}{t - 1}$$

SIMPLIFYING A RATIONAL EXPRESSION

Step 1. Completely factor the numerator and the denominator.

Step 2. Divide out common factors.

Simplify.

$$\frac{2x^2 + 9x - 5}{x^2 - 25} = \frac{(2x - 1)\,(x + 5)}{(x - 5)\,(x + 5)}$$

$$= \frac{2x - 1}{x - 5}$$

MULTIPLYING RATIONAL EXPRESSIONS

Step 1. Completely factor numerators and denominators.

Step 2. Multiply the numerators and multiply the denominators.

Step 3. Simplify the product.

Multiply: $\dfrac{x^3 + 8}{12x - 18} \cdot \dfrac{14x^2 - 21x}{x^2 + 2x}$

$$= \frac{(x + 2)(x^2 - 2x + 4)}{6(2x - 3)} \cdot \frac{7x(2x - 3)}{x(x + 2)}$$

$$= \frac{7(x^2 - 2x + 4)}{6}$$

DIVIDING RATIONAL EXPRESSIONS

Multiply the first rational expression by the reciprocal of the second rational expression.

Divide: $\dfrac{x^2 + 6x + 9}{5xy - 5y} \div \dfrac{x + 3}{10y}$

$$= \frac{(x + 3)(x + 3)}{5y(x - 1)} \cdot \frac{2 \cdot 5y}{x + 3}$$

$$= \frac{2(x + 3)}{x - 1}$$

A **rational function** is a function described by a rational expression.

$$f(x) = \frac{2x - 6}{7}, h(t) = \frac{t^2 - 3t + 5}{t - 1}$$

DEFINITIONS AND CONCEPTS	**EXAMPLES**

Section 6.2   Adding and Subtracting Rational Expressions

ADDING OR SUBTRACTING RATIONAL EXPRESSIONS

Step 1. Find the LCD.

Step 2. Write each rational expression as an equivalent rational expression whose denominator is the LCD.

Step 3. Add or subtract numerators and write the sum or difference over the common denominator.

Step 4. Simplify the result.

Subtract: $\dfrac{3}{x+2} - \dfrac{x+1}{x-3}$

$$= \frac{3 \cdot (x-3)}{(x+2) \cdot (x-3)} - \frac{(x+1) \cdot (x+2)}{(x-3) \cdot (x+2)}$$

$$= \frac{3(x-3) - (x+1)(x+2)}{(x+2)(x-3)}$$

$$= \frac{3x - 9 - (x^2 + 3x + 2)}{(x+2)(x-3)}$$

$$= \frac{3x - 9 - x^2 - 3x - 2}{(x+2)(x-3)}$$

$$= \frac{-x^2 - 11}{(x+2)(x-3)}$$

Section 6.3   Simplifying Complex Fractions

Method 1: Simplify the numerator and the denominator so that each is a single fraction. Then perform the indicated division and simplify if possible.

Simplify: $\dfrac{\dfrac{x+2}{x}}{x - \dfrac{4}{x}}$

Method 1: $\dfrac{\dfrac{x+2}{x}}{\dfrac{x \cdot x}{1 \cdot x} - \dfrac{4}{x}} = \dfrac{\dfrac{x+2}{x}}{\dfrac{x^2 - 4}{x}}$

$$= \frac{x+2}{x} \cdot \frac{x}{(x+2)(x-2)} = \frac{1}{x-2}$$

Method 2: Multiply the numerator and the denominator of the complex fraction by the LCD of the fractions in both the numerator and the denominator. Then simplify if possible.

Method 2: $\dfrac{\left(\dfrac{x+2}{x}\right) \cdot x}{\left(x - \dfrac{4}{x}\right) \cdot x} = \dfrac{x+2}{x \cdot x - \dfrac{4}{x} \cdot x}$

$$= \frac{x+2}{x^2 - 4} = \frac{x+2}{(x+2)(x-2)} = \frac{1}{x-2}$$

Section 6.4   Solving Equations Containing Rational Expressions

SOLVING AN EQUATION CONTAINING RATIONAL EXPRESSIONS

Multiply both sides of the equation by the LCD of all rational expressions. Then use the distributive property and simplify. Solve the resulting equation and then check each proposed solution to see whether it makes any denominator 0. Discard any solutions that do.

Solve: $x - \dfrac{3}{x} = \dfrac{1}{2}$

$2x\left(x - \dfrac{3}{x}\right) = 2x\left(\dfrac{1}{2}\right)$ The LCD is $2x$.

$2x \cdot x - 2x\left(\dfrac{3}{x}\right) = 2x\left(\dfrac{1}{2}\right)$ Distribute.

$2x^2 - 6 = x$

$2x^2 - x - 6 = 0$ Subtract x from both sides.

$(2x+3)(x-2) = 0$ Factor.

$x = -\dfrac{3}{2}$ or $x = 2$

Both $-\dfrac{3}{2}$ and 2 check. The solution set is $\left\{2, -\dfrac{3}{2}\right\}$.

DEFINITIONS AND CONCEPTS	**EXAMPLES**
Section 6.5 Rational Equations and Problem Solving	

SOLVING AN EQUATION FOR A SPECIFIED VARIABLE

Treat the specified variable as the only variable of the equation and solve as usual.

Solve for x.

$$A = \frac{2x + 3y}{5}$$

$5A = 2x + 3y$ Multiply both sides by 5.

$5A - 3y = 2x$ Subtract $3y$ from both sides.

$$\frac{5A - 3y}{2} = x$$ Divide both sides by 2.

SOLVING A PROBLEM THAT INVOLVES A RATIONAL EQUATION

Jeanee and David Dillon volunteer every year to clean a strip of Lake Ponchartrain beach. Jeanee can clean all the trash in this area of beach in 6 hours; David takes 5 hours. Find how long it will take them to clean the area of beach together.

1. UNDERSTAND.

1. Read and reread the problem. Let x = time in hours that it takes Jeanee and David to clean the beach together.

	Hours to Complete	Part Completed in 1 Hour
Jeanee Alone	6	$\frac{1}{6}$
David Alone	5	$\frac{1}{5}$
Together	x	$\frac{1}{x}$

2. TRANSLATE.

2. In words:

part Jeanee can complete in 1 hour	+	part David can complete in 1 hour	=	part they can complete together in 1 hour
$\downarrow$		$\downarrow$		$\downarrow$

Translate:

$$\frac{1}{6} \quad + \quad \frac{1}{5} \quad = \quad \frac{1}{x}$$

3. SOLVE.

3. $\dfrac{1}{6} + \dfrac{1}{5} = \dfrac{1}{x}$

$5x + 6x = 30$ Multiply both sides by $30x$.

$11x = 30$

$$x = \frac{30}{11} \quad \text{or} \quad 2\frac{8}{11}$$

4. INTERPRET.

4. *Check* and then *state*. Together, they can clean the beach in $2\frac{8}{11}$ hours.

DEFINITIONS AND CONCEPTS	**EXAMPLES**
Section 6.6 Variation and Problem Solving	

y **varies directly** as *x*, or *y* is **directly proportional** to *x*, if there is a nonzero constant *k* such that $$y = kx$$	The circumference of a circle *C* varies directly as its radius *r*. $$C = \underbrace{2\pi}_{k} r$$
y **varies inversely** as *x*, or *y* is **inversely proportional** to *x*, if there is a nonzero constant *k* such that $$y = \frac{k}{x}$$	Pressure *P* varies inversely with volume *V*. $$P = \frac{k}{V}$$
y **varies jointly** as *x* and *z*, or *y* is **jointly proportional** to *x* and *z*, if there is a nonzero constant *k* such that $$y = kxz$$	The lateral surface area *S* of a cylinder varies jointly as its radius *r* and height *h*. $$S = \underbrace{2\pi}_{k} rh$$

6 CHAPTER REVIEW

(6.1) *Find the domain for each rational function.*

1. $f(x) = \dfrac{3 - 5x}{7}$

2. $g(x) = \dfrac{2x + 4}{11}$

3. $F(x) = \dfrac{-3x^2}{x - 5}$

4. $h(x) = \dfrac{4x}{3x - 12}$

5. $f(x) = \dfrac{x^3 + 2}{x^2 + 8x}$

6. $G(x) = \dfrac{20}{3x^2 - 48}$

Write each rational expression in lowest terms.

7. $\dfrac{x - 12}{12 - x}$

8. $\dfrac{5x - 15}{25x - 75}$

9. $\dfrac{2x}{2x^2 - 2x}$

10. $\dfrac{x + 7}{x^2 - 49}$

11. $\dfrac{2x^2 + 4x - 30}{x^2 + x - 20}$

12. The average cost of manufacturing *x* bookcases is given by the rational function.

$$C(x) = \frac{35x + 4200}{x}$$

a. Find the average cost per bookcase of manufacturing 50 bookcases.

b. Find the average cost per bookcase of manufacturing 100 bookcases.

c. As the number of bookcases increases, does the average cost per bookcase increase or decrease? (See parts (a) and (b).)

Perform each indicated operation. Write your answers in lowest terms.

13. $\dfrac{4-x}{5} \cdot \dfrac{15}{2x-8}$

14. $\dfrac{x^2-6x+9}{2x^2-18} \cdot \dfrac{4x+12}{5x-15}$

15. $\dfrac{a-4b}{a^2+ab} \cdot \dfrac{b^2-a^2}{8b-2a}$

16. $\dfrac{x^2-x-12}{2x^2-32} \cdot \dfrac{x^2+8x+16}{3x^2+21x+36}$

17. $\dfrac{4x+8y}{3} \div \dfrac{5x+10y}{9}$

18. $\dfrac{x^2-25}{3} \div \dfrac{x^2-10x+25}{x^2-x-20}$

19. $\dfrac{a-4b}{a^2+ab} \div \dfrac{20b-5a}{b^2-a^2}$

20. $\dfrac{3x+3}{x-1} \div \dfrac{x^2-6x-7}{x^2-1}$

21. $\dfrac{2x-x^2}{x^3-8} \div \dfrac{x^2}{x^2+2x+4}$

22. $\dfrac{5x-15}{3-x} \cdot \dfrac{x+2}{10x+20} \cdot \dfrac{x^2-9}{x^2-x-6}$

(6.2) *Find the LCD of the rational expressions in each list.*

23. $\dfrac{5}{4x^2y^5}, \dfrac{3}{10x^2y^4}, \dfrac{x}{6y^4}$

24. $\dfrac{5}{2x}, \dfrac{7}{x-2}$

25. $\dfrac{3}{5x}, \dfrac{2}{x-5}$

26. $\dfrac{1}{5x^3}, \dfrac{4}{x^2+3x-28}, \dfrac{11}{10x^2-30x}$

Perform each indicated operation. Write your answers in lowest terms.

27. $\dfrac{4}{x-4} + \dfrac{x}{x-4}$

28. $\dfrac{4}{3x^2} + \dfrac{2}{3x^2}$

29. $\dfrac{1}{x-2} - \dfrac{1}{4-2x}$

30. $\dfrac{1}{10-x} + \dfrac{x-1}{x-10}$

31. $\dfrac{x}{9-x^2} - \dfrac{2}{5x-15}$

32. $2x+1 - \dfrac{1}{x-3}$

33. $\dfrac{2}{a^2-2a+1} + \dfrac{3}{a^2-1}$

34. $\dfrac{x}{9x^2+12x+16} - \dfrac{3x+4}{27x^3-64}$

Perform each indicated operation. Write your answers in lowest terms.

35. $\dfrac{2}{x-1} - \dfrac{3x}{3x-3} + \dfrac{1}{2x-2}$

△ **36.** Find the perimeter of the heptagon (a polygon with seven sides).

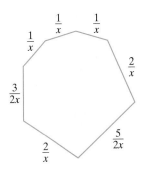

(6.3) *Simplify each complex fraction.*

37. $\dfrac{1 - \dfrac{3x}{4}}{2 + \dfrac{x}{4}}$

38. $\dfrac{\dfrac{x^2}{15}}{\dfrac{x + 1}{5x}}$

39. $\dfrac{2 - \dfrac{3}{2x}}{x - \dfrac{2}{5x}}$

40. $\dfrac{1 + \dfrac{x}{y}}{\dfrac{x^2}{y^2} - 1}$

41. $\dfrac{\dfrac{5}{x} + \dfrac{1}{xy}}{\dfrac{3}{x^2}}$

42. $\dfrac{\dfrac{x}{3} - \dfrac{3}{x}}{1 + \dfrac{3}{x}}$

43. $\dfrac{\dfrac{1}{x - 1} + 1}{\dfrac{1}{x + 1} - 1}$

44. $\dfrac{\dfrac{x - 3}{x + 3} + \dfrac{x + 3}{x - 3}}{\dfrac{x - 3}{x + 3} - \dfrac{x + 3}{x - 3}}$

(6.4) *Solve each equation.*

45. $\dfrac{3}{x} + \dfrac{1}{3} = \dfrac{5}{x}$

46. $\dfrac{2x + 3}{5x - 9} = \dfrac{3}{2}$

47. $\dfrac{1}{x - 2} - \dfrac{3x}{x^2 - 4} = \dfrac{2}{x + 2}$

48. $\dfrac{7}{x} - \dfrac{x}{7} = 0$

Solve each equation or perform each indicated operation. Simplify.

49. $\dfrac{5}{x^2 - 7x} + \dfrac{4}{2x - 14}$

50. $3 - \dfrac{5}{x} - \dfrac{2}{x^2} = 0$

51. $\dfrac{4}{3 - x} - \dfrac{7}{2x - 6} + \dfrac{5}{x}$

(6.5) *Solve each equation for the specified variable.*

52. $A = \dfrac{h(a + b)}{2}$ for a

53. $\dfrac{1}{R} = \dfrac{1}{R_1} + \dfrac{1}{R_2}$ for R_2

54. $I = \dfrac{E}{R + r}$ for R

55. $A = P + Prt$ for r

56. $H = \dfrac{kA(T_1 - T_2)}{L}$ for A

Solve.

57. The sum of a number and twice its reciprocal is 3. Find the number(s).

58. If a number is added to the numerator of $\dfrac{3}{7}$, and twice that number is added to the denominator of $\dfrac{3}{7}$, the result is equivalent to $\dfrac{10}{21}$. Find the number.

59. Three boys can paint a fence in 4 hours, 5 hours, and 6 hours, respectively. Find how long it will take all three boys to paint the fence.

60. If Sue Katz can type a certain number of mailing labels in 6 hours and Tom Neilson and Sue working together can type the same number of mailing labels in 4 hours, find how long it takes Tom alone to type the mailing labels.

61. The speed of a Ranger boat in still water is 32 mph. If the boat travels 72 miles upstream in the same time that it takes to travel 120 miles downstream, find the current of the stream.

62. The speed of a jogger is 3 mph faster than the speed of a walker. If the jogger travels 14 miles in the same amount of time that the walker travels 8 miles, find the speed of the walker.

(6.6) *Solve each variation problem.*

63. A is directly proportional to B. If $A = 6$ when $B = 14$, find A when $B = 21$.

64. According to Boyle's law, the pressure exerted by a gas is inversely proportional to the volume, as long as the temperature stays the same. If a gas exerts a pressure of 1250 kilopascals when the volume is 2 cubic meters, find the volume when the pressure is 800 kilopascals.

Mixed Review

For expressions, perform the indicated operation and/or simplify. For equations, solve the equation for the unknown variable.

65. $\dfrac{22x + 8}{11x + 4}$

66. $\dfrac{xy - 3x + 2y - 6}{x^2 + 4x + 4}$

67. $\dfrac{2}{5x} \div \dfrac{4 - 18x}{6 - 27x}$

68. $\dfrac{7x + 28}{2x + 4} \div \dfrac{x^2 + 2x - 8}{x^2 - 2x - 8}$

69. $\dfrac{5a^2 - 20}{a^3 + 2a^2 + a + 2} \div \dfrac{7a}{a^3 + a}$

70. $\dfrac{4a + 8}{5a^2 - 20} \cdot \dfrac{3a^2 - 6a}{a + 3} \div \dfrac{2a^2}{5a + 15}$

71. $\dfrac{7}{2x} + \dfrac{5}{6x}$

72. $\dfrac{x - 2}{x + 1} - \dfrac{x - 3}{x - 1}$

73. $\dfrac{2x + 1}{x^2 + x - 6} + \dfrac{2 - x}{x^2 + x - 6}$

74. $\dfrac{2}{x^2 - 16} - \dfrac{3x}{x^2 + 8x + 16} + \dfrac{3}{x + 4}$

75. $\dfrac{\dfrac{1}{x} - \dfrac{2}{3x}}{\dfrac{5}{2x} - \dfrac{1}{3}}$

76. $\dfrac{2}{1 - \dfrac{2}{x}}$

77. $\dfrac{\dfrac{x^2 + 5x - 6}{4x + 3}}{\dfrac{(x + 6)^2}{8x + 6}}$

78. $\dfrac{\dfrac{3}{x - 1} - \dfrac{2}{1 - x}}{\dfrac{2}{x - 1} - \dfrac{2}{x}}$

79. $4 + \dfrac{8}{x} = 8$

80. $\dfrac{x - 2}{x^2 - 7x + 10} = \dfrac{1}{5x - 10} - \dfrac{1}{x - 5}$

81. The denominator of a fraction is 2 more than the numerator. If the numerator is decreased by 3 and the denominator is increased by 5, the resulting fraction is equivalent to $\dfrac{2}{3}$. Find the fraction.

82. The sum of the reciprocals of two consecutive even integers is $-\dfrac{9}{40}$. Find the two integers.

83. The inlet pipe of a water tank can fill the tank in 2 hours and 30 minutes. The outlet pipe can empty the tank in 2 hours. Find how long it takes to empty a full tank if both pipes are open.

84. Timmy Garnica drove 210 miles in the same amount of time that it took a DC-10 jet to travel 1715 miles. The speed of the jet was 430 mph faster than the speed of the car. Find the speed of the jet.

85. Two Amtrak trains traveling on parallel tracks leave Tucson at the same time. In 6 hours the faster train is 382 miles from Tucson and the trains are 112 miles apart. Find how fast each train is traveling.

86. C is inversely proportional to D. If $C = 12$ when $D = 8$, find C when $D = 24$.

△**87.** The surface area of a sphere varies directly as the square of its radius. If the surface area is 36π square inches when the radius is 3 inches, find the surface area when the radius is 4 inches.

6 CHAPTER TEST

 Remember to use the Chapter Test Prep Video CD to see the fully worked-out solutions to any of the exercises you want to review.

Find the domain of each rational function.

1. $f(x) = \dfrac{5x^2}{1 - x}$

2. $g(x) = \dfrac{9x^2 - 9}{x^2 + 4x + 3}$

Write each rational expression in lowest terms.

3. $\dfrac{7x - 21}{24 - 8x}$

4. $\dfrac{x^2 - 4x}{x^2 + 5x - 36}$

5. $\dfrac{x^3 - 8}{x - 2}$

Perform the indicated operation. If possible, simplify your answer.

6. $\dfrac{2x^3 + 16}{6x^2 + 12x} \cdot \dfrac{5}{x^2 - 2x + 4}$

7. $\dfrac{5}{4x^3} + \dfrac{7}{4x^3}$

8. $\dfrac{3x^2 - 12}{x^2 + 2x - 8} \div \dfrac{6x + 18}{x + 4}$

9. $\dfrac{4x - 12}{2x - 9} \div \dfrac{3 - x}{4x^2 - 81} \cdot \dfrac{x + 3}{5x + 15}$

10. $\dfrac{3 + 2x}{10 - x} + \dfrac{13 + x}{x - 10}$

11. $\dfrac{2x^2 + 7}{2x^4 - 18x^2} - \dfrac{6x + 7}{2x^4 - 18x^2}$

12. $\dfrac{3}{x^2 - x - 6} + \dfrac{2}{x^2 - 5x + 6}$

13. $\dfrac{5}{x - 7} - \dfrac{2x}{3x - 21} + \dfrac{x}{2x - 14}$

14. $\dfrac{3x}{5} \cdot \left(\dfrac{5}{x} - \dfrac{5}{2x} \right)$

Simplify each complex fraction.

15. $\dfrac{\dfrac{5}{x} - \dfrac{7}{3x}}{\dfrac{9}{8x} - \dfrac{1}{x}}$

16. $\dfrac{\dfrac{x^2 - 5x + 6}{x + 3}}{\dfrac{x^2 - 4x + 4}{x^2 - 9}}$

Solve each equation for x.

17. $\dfrac{5x + 3}{3x - 7} = \dfrac{19}{7}$

18. $\dfrac{x}{x - 4} = 3 - \dfrac{4}{x - 4}$

19. $\dfrac{3}{x + 2} - \dfrac{1}{5x} = \dfrac{2}{5x^2 + 10x}$

20. $\dfrac{x^2 + 8}{x} - 1 = \dfrac{2(x + 4)}{x}$

Answers

1. _____

2. _____

3. _____

4. _____

5. _____

6. _____

7. _____

8. _____

9. _____

10. _____

11. _____

12. _____

13. _____

14. _____

15. _____

16. _____

17. _____

18. _____

19. _____

20. _____

21. Solve for x: $\dfrac{x+b}{a} = \dfrac{4x-7a}{b}$

22. The product of one more than a number and twice the reciprocal of the number is $\dfrac{12}{5}$. Find the number.

23. if Jan can weed the garden in 2 hours and her husband can weed it in 1 hour and 30 minutes, find how long, it takes them to weed the garden together.

24. Suppose that W is inversely proportional to V. If $W = 20$ when $V = 12$, find W when $V = 15$.

25. Suppose that Q is jointly proportional to R and the square of S. If $Q = 24$ when $R = 3$ and $S = 4$, find Q when $R = 2$ and $S = 3$.

26. When an anvil is dropped into a gorge, the speed with which it strikes the ground is directly proportional to the square root of the distance it falls. An anvil that falls 400 feet hits the ground at a speed of 160 feet per second. Find the height of a cliff over the gorge if a dropped anvil hits the ground at a speed of 128 feet per second.

27. In baseball, a pitcher's earned run average statistic A is computed as $A = \dfrac{E}{\frac{I}{9}}$, where E is the number of earned runs scored on the pitcher and I is the total number of innings he or she pitched. During the 2004 season, pitcher Jake Peavy of the San Diego Padres led the Major Leagues with an earned run average of 2.27. Peavy pitched a total of 166.1 innings during the season. How many earned runs were scored on Jake Peavy? Round to the nearest whole. (*Source:* Major League Baseball)

21. _____

22. _____

23. _____

24. _____

25. _____

26. _____

27. _____

1.

2. a.

   b.

   c.

   d.

3.

4.

5.

6.

7.

8.

9.

10.

11.

12. a.

    b.

    c.

13.

14.

15.

16.

17.

18.

1. Determine whether the statement is true or false.

$3 \in \{x \mid x \text{ is a natural number}\}$

2. Translate each phrase to an algebraic expression. Use the variable x to represent each unknown number.

   a. One third subtracted from a number.

   b. Six less than five times a number.

   c. Three more than eight times a number

   d. The quotient of seven and the difference of two and a number.

Evaluate each expression.

3. -7^0

4. $(-5xy^7)(-0.2xy^{-5})$

5. $(2x + 5)^0$

6. $\dfrac{(3a^{-2}b)^{-2}}{(2ab^{-3})^{-3}}$

7. Solve: $2x \geq 0$ *and* $4x - 1 \leq -9$

8. Solve: $\left|\dfrac{2x - 1}{3}\right| + 6 = 3$

9. Solve: $|w + 3| = 7$

10. Solve: $\left|\dfrac{3(x - 1)}{4}\right| \geq 2$

11. Determine whether $(0, -12)$, $(1, 9)$, and $(2, -6)$ are solutions of the equation $3x - y = 12$.

12. If $f(x) = -x^2 + 3x - 2$, find

   a. $f(0)$ **b.** $f(-3)$

   c. $f\left(\dfrac{1}{3}\right)$

13. Determine whether the two lines are parallel, perpendicular, or neither.

   $3x + 7y = 4$

   $6x + 14y = 7$

14. Solve $\dfrac{x}{7} + \dfrac{x}{5} = \dfrac{12}{5}$

15. Write an equation of the line through points $(4, 0)$ and $(-4, -5)$. Write the equation in standard form $Ax + By = C$.

16. Find an equation of the line with slope $\dfrac{1}{2}$ containing the point $(-1, 3)$. Use function notation to write the equation.

Find the domain and range of each relation.

17.

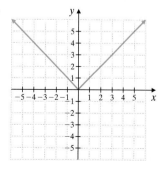

18.

19.

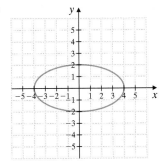

20.

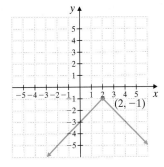

(2, −1)

21. Use the substitution method to solve the system:
$$\begin{cases} -\dfrac{x}{6} + \dfrac{y}{2} = \dfrac{1}{2} \\ \dfrac{x}{3} - \dfrac{y}{6} = -\dfrac{3}{4} \end{cases}$$

22. Use the substitution method to solve the system.
$$\begin{cases} -2x + 3y = 6 \\ 3x - y = 5 \end{cases}$$

23. Solve the system:
$$\begin{cases} 2x + 4y \quad\;\; = 1 \\ 4x \quad\quad - 4z = -1 \\ \quad\quad y - 4z = -3 \end{cases}$$

24. Solve the system:
$$\begin{cases} 2x - 2y + 4z = 6 \\ -4x - y + z = -8 \\ 3x - y + z = 6 \end{cases}$$

25. Two cars leave Indianapolis, one traveling east and the other west. After 3 hours they are 297 miles apart. If one car is traveling 5 mph faster than the other, what is the speed of each?

26. Kernersville office supply sold three reams of paper and two boxes of manila folders for $21.90. Also, five reams of paper and one box of manila folders cost $24.25. Find the price of a ream of paper and a box of manila folders.

27. Use matrices to solve the system:
$$\begin{cases} x + 2y + z = 2 \\ -2x - y + 2z = 5 \\ x + 3y - 2z = -8 \end{cases}$$

28. Use matrices to solve the system.
$$\begin{cases} x + y + z = 9 \\ 2x - 2y + 3z = 2 \\ -3x + y - z = 1 \end{cases}$$

29. Simplify the expression. Write the answer using positive exponents only.
$$\left(\dfrac{3x^2y}{y^{-9}z}\right)^{-2}$$

30. Simplify the following. Write answers with positive exponents.
a. $2^{-2} + 3^{-1}$ **b.** $-6a^0$
c. $\dfrac{x^{-5}}{x^{-2}}$

31. Graph the solutions of the system:
$$\begin{cases} 3x \geq y \\ x + 2y \leq 8 \end{cases}$$

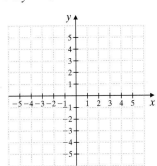

32. Simplify each. Assume that a and b are integers and that x and y are not 0.
a. $3x^{4a}(4x^{-a})^2$
b. $\dfrac{(y^{4b})^3}{y^{2b-3}}$

19. _____

20. _____

21. _____

22. _____

23. _____

24. _____

25. _____

26. _____

27. _____

28. _____

29. _____

30. a. _____

  b. _____

  c. _____

31. see graph

32. a. _____

  b. _____

Perform indicated operations, if necessary. Then simplify by combining like terms.

33. $-12x^2 + 7x^2 - 6x$

34. Subtract $(2x - 7)$ from $2x^2 + 8x - 3$

35. $3xy - 2x + 5xy - x$

36. $7 + 2(x - 7y) + \dfrac{1}{2} - \dfrac{1}{4}(8x - 16y)$

37. Multiply vertically:
$(4x^2 + 7)(x^2 + 2x + 8)$

38. Multiply: $[4 + (3x - y)]^2$

39. Divide: $2x^2 - x - 10$ by $x + 2$.

40. Factor: $xy + 2x - 5y - 10$

41. Factor: $-3x^3y + 2x^2y - 5xy$

42. Factor: $6x^2 - x - 35$

43. Factor: $x^2 + 10x + 16$

44. Factor: $4x^2 - 4x + 1 - 9y^2$

45. Factor: $p^4 - 16$

46. For the graph of $f(x)$, answer the following:
   a. Find the domain and range.
   b. List the x- and y-intercepts.
   c. Find the coordinates of the point with the greatest y-value.
   d. Find the coordinates of the point with the least y-value.
   e. List the x-values whose y-values are equal to 0.
   f. List the x-values whose y-values are less than 0.
   g. Find the solutions of $f(x) = 0$.

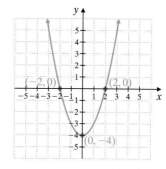

47. Solve: $(x + 2)(x - 6) = 0$

48. Solve: $2x(3x + 1)(x - 3) = 0$

49. Simplify: $\dfrac{\dfrac{5x}{x+2}}{\dfrac{10}{x-2}}$

50. Solve: $\dfrac{3x - 4}{2x} = -\dfrac{8}{x}$.

33. _____
34. _____
35. _____
36. _____
37. _____
38. _____
39. _____
40. _____
41. _____
42. _____
43. _____
44. _____
45. _____
46. a. _____
   b. _____
   c. _____
   d. _____
   e. _____
   f. _____
   g. _____
47. _____
48. _____
49. _____
50. _____

7

Rational Exponents, Radicals, and Complex Numbers

In this chapter, radical notation is reviewed, and then rational exponents are introduced. As the name implies, rational exponents are exponents that are rational numbers. We present an interpretation of rational exponents that is consistent with the meaning and rules already established for integer exponents, and we present two forms of notation for roots: radical and exponent. We conclude this chapter with complex numbers, a natural extension of the real number system.

What is a zorb? Simply put, a zorb is a large inflated ball within a ball, and zorbing is a recreational activity which may involve rolling down a hill strapped in a zorb. Zorbing started in New Zealand (as well as bungee jumping) and was invented by Andrew Akers and Dwane van der Sluis. The first site was set up in New Zealand's North Island. This downhill course has a length of about 490 feet and you can reach speeds of up to 20 mph.

An example of a course can be seen in the diagram below and you can see the mathematics involved. In Section 7.3, Exercise 99, you will calculate the outer radius of a zorb, which would certainly be closely associated with the cost of production.

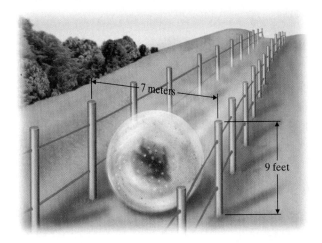

7 meters

9 feet

A Find Square Roots.

B Approximate Roots Using a Calculator.

C Find Cube Roots.

D Find *n*th Roots.

E Find $\sqrt[n]{a^n}$ when *a* Is Any Real Number.

F Find Function Values of Radical Functions.

PRACTICE PROBLEMS 1–3

Find the square roots of each number.

1. 36 **2.** 81 **3.** −16

7.1 RADICAL EXPRESSIONS AND RADICAL FUNCTIONS

Objective **A** Finding Square Roots

Recall from Section 1.4 that to find a *square root* of a number *a*, we find a number that was squared to get *a*.

Square Root

The number *b* is a **square root** of *a* if $b^2 = a$.

EXAMPLES Find the real square roots of each number.

1. 25   Since $5^2 = 25$ and $(-5)^2 = 25$, the square roots of 25 are 5 and −5.

2. 49   Since $7^2 = 49$ and $(-7)^2 = 49$, the square roots of 49 are 7 and −7.

3. −4   There is no real number whose square is −4. The number −4 has no real number square root.

Work Practice Problems 1–3

Recall that we denote the *nonnegative*, or *principal, square root* with the **radical sign:**

$$\sqrt{25} = 5$$

We denote the *negative square root* with the **negative radical sign:**

$$-\sqrt{25} = -5$$

An expression containing a radical sign is called a **radical expression.** An expression within, or "under," a radical sign is called a **radicand.**

radical expression: $\overset{\text{radical sign}}{\sqrt{a}}_{\text{radicand}}$

Principal and Negative Square Roots

The **principal square root** of a nonnegative number *a* is its nonnegative square root. The principal square root is written as $\sqrt{a}$. The **negative square root** of *a* is written as $-\sqrt{a}$.

EXAMPLES Find each square root. Assume that all variables represent nonnegative real numbers.

4. $\sqrt{36} = 6$ because $6^2 = 36$.

5. $\sqrt{0} = 0$ because $0^2 = 0$.

6. $\sqrt{\dfrac{4}{49}} = \dfrac{2}{7}$ because $\left(\dfrac{2}{7}\right)^2 = \dfrac{4}{49}$.

7. $\sqrt{0.25} = 0.5$ because $(0.5)^2 = 0.25$.

8. $\sqrt{x^6} = x^3$ because $(x^3)^2 = x^6$.

Answers

1. 6, −6, **2.** 9, −9,
3. no real number square root

9. $\sqrt{9x^{10}} = 3x^5$ because $(3x^5)^2 = 9x^{10}$.

10. $-\sqrt{81} = -9$. The negative in front of the radical indicates the negative square root of 81.

11. $\sqrt{-81}$ is not a real number.

 Work Practice Problems 4–11

Helpful Hint
──
• Remember: $\sqrt{0} = 0$.
• Don't forget that the square root of a negative number is not a real number. For example,

$\sqrt{-9}$ is not a real number

because there is no real number that when multiplied by itself would give a product of -9. In Section 7.7, we will see what kind of a number $\sqrt{-9}$ is.

Objective B Approximating Roots

Recall that numbers such as 1, 4, 9, and 25 are called **perfect squares,** since $1 = 1^2, 4 = 2^2, 9 = 3^2$, and $25 = 5^2$. Square roots of perfect square radicands simplify to rational numbers. What happens when we try to simplify a root such as $\sqrt{3}$? Since 3 is not a perfect square, $\sqrt{3}$ is not a rational number. It is called an **irrational number,** and we can find a decimal **approximation** of it. To find decimal approximations, we can use a calculator. For example, an approximation for $\sqrt{3}$ is

$$\sqrt{3} \approx 1.732$$
↑
approximation symbol

To see if the approximation is reasonable, notice that since

$1 < 3 < 4$, then
$\sqrt{1} < \sqrt{3} < \sqrt{4}$, or
$1 < \sqrt{3} < 2$.

We found $\sqrt{3} \approx 1.732$, a number between 1 and 2, so our result is reasonable.

EXAMPLE 12

Use a calculator or the appendix to approximate $\sqrt{20}$. Round the approximation to three decimal places and check to see that your approximation is reasonable.

Solution:

$$\sqrt{20} \approx 4.472$$

Is this reasonable? Since $16 < 20 < 25$, then $\sqrt{16} < \sqrt{20} < \sqrt{25}$, or $4 < \sqrt{20} < 5$. The approximation is between 4 and 5 and is thus reasonable.

 Work Practice Problem 12

Objective C Finding Cube Roots

Finding roots can be extended to other roots such as cube roots. For example, since $2^3 = 8$, we call 2 the *cube root* of 8. In symbols, we write

$$\sqrt[3]{8} = 2$$

PRACTICE PROBLEMS 4–11

Find each square root. Assume that all variables represent non-negative real numbers.

4. $\sqrt{25}$ **5.** $\sqrt{0}$

6. $\sqrt{\dfrac{9}{25}}$ **7.** $\sqrt{0.36}$

8. $\sqrt{x^{10}}$ **9.** $\sqrt{36x^6}$

10. $-\sqrt{25}$ **11.** $\sqrt{-25}$

PRACTICE PROBLEM 12

Use a calculator or the appendix to approximate $\sqrt{30}$. Round the approximation to three decimal places and check to see that your approximation is reasonable.

Answers
4. 5, **5.** 0, **6.** $\dfrac{3}{5}$, **7.** 0.6, **8.** x^5,
9. $6x^3$, **10.** -5, **11.** not a real number,
12. 5.477

Cube Root

The **cube root** of a real number a is written as $\sqrt[3]{a}$, and

$$\sqrt[3]{a} = b \quad \text{only if} \quad b^3 = a$$

From this definition, we have

$\sqrt[3]{64} = 4$ since $4^3 = 64$

$\sqrt[3]{-27} = -3$ since $(-3)^3 = -27$

$\sqrt[3]{x^3} = x$ since $x^3 = x^3$

Notice that, unlike with square roots, *it is possible to have a negative radicand when finding a cube root.* This is so because the *cube* of a negative number is a negative number. Therefore, the *cube root* of a negative number is a negative number.

PRACTICE PROBLEMS 13–17

Find each cube root.

13. $\sqrt[3]{0}$ **14.** $\sqrt[3]{-8}$

15. $\sqrt[3]{\dfrac{1}{64}}$ **16.** $\sqrt[3]{x^9}$

17. $\sqrt[3]{-64x^6}$

EXAMPLES Find each cube root.

13. $\sqrt[3]{1} = 1$ because $1^3 = 1$.

14. $\sqrt[3]{-64} = -4$ because $(-4)^3 = -64$.

15. $\sqrt[3]{\dfrac{8}{125}} = \dfrac{2}{5}$ because $\left(\dfrac{2}{5}\right)^3 = \dfrac{8}{125}$.

16. $\sqrt[3]{x^6} = x^2$ because $(x^2)^3 = x^6$.

17. $\sqrt[3]{-8x^9} = -2x^3$ because $(-2x^3)^3 = -8x^9$.

💻 **Work Practice Problems 13–17**

Objective D Finding *n*th Roots

Just as we can raise a real number to powers other than 2 or 3, we can find roots other than square roots and cube roots. In fact, we can find the **nth root** of a number, where n is any natural number. In symbols, the nth root of a is written as $\sqrt[n]{a}$, where n is called the **index.** The index 2 is usually omitted for square roots.

> **Helpful Hint**
>
> If the index is even, such as in $\sqrt{\ }, \sqrt[4]{\ }, \sqrt[6]{\ }$, and so on, the radicand must be nonnegative for the root to be a real number. For example,
>
> $\sqrt[4]{16} = 2$, but $\sqrt[4]{-16}$ is not a real number,
>
> $\sqrt[6]{64} = 2$, but $\sqrt[6]{-64}$ is not a real number.
>
> If the index is odd, such as in $\sqrt[3]{\ }, \sqrt[5]{\ }$, and so on, the radicand may be any real number. For example,
>
> $\sqrt[3]{64} = 4$ and $\sqrt[3]{-64} = -4$,
>
> $\sqrt[5]{32} = 2$ and $\sqrt[5]{-32} = -2$.

Answers

13. 0,   **14.** -2,   **15.** $\dfrac{1}{4}$,   **16.** x^3,

17. $-4x^2$

✔ **Concept Check Answer**

b

✔**Concept Check**   Which one is not a real number?

a. $\sqrt[3]{-15}$ **b.** $\sqrt[4]{-15}$ **c.** $\sqrt[5]{-15}$ **d.** $\sqrt{(-15)^2}$

EXAMPLES Find each root.

18. $\sqrt[4]{81} = 3$ because $3^4 = 81$ and 3 is positive.

19. $\sqrt[5]{-243} = -3$ because $(-3)^5 = -243$.

20. $-\sqrt{25} = -5$ because -5 is the opposite of $\sqrt{25}$.

21. $\sqrt[4]{-81}$ is not a real number. There is no real number that, when raised to the fourth power, is -81.

22. $\sqrt[3]{64x^3} = 4x$ because $(4x)^3 = 64x^3$.

▣ **Work Practice Problems 18–22**

Objective **E** Finding $\sqrt[n]{a^n}$ when *a* Is Any Real Number

Recall that the notation $\sqrt{a^2}$ indicates the positive square root of a^2 only. For example,

$$\sqrt{(-5)^2} = \sqrt{25} = 5$$

When variables are present in the radicand and it is *unclear whether the variable represents a positive number or a negative number,* absolute value bars are sometimes needed to ensure that the result is a positive number. For example,

$$\sqrt{x^2} = |x|$$

This ensures that the result is positive. This same situation may occur when the index is any *even* positive integer. When the index is any *odd* positive integer, absolute value bars are not necessary.

Finding $\sqrt[n]{a^n}$

If *n* is an *even* positive integer, then $\sqrt[n]{a^n} = |a|$.
If *n* is an *odd* positive integer, then $\sqrt[n]{a^n} = a$.

EXAMPLES Simplify. Assume that the variables represent any real number.

23. $\sqrt{(-3)^2} = |-3| = 3$ When the index is even, the absolute value bars ensure that the result is not negative.

24. $\sqrt{x^2} = |x|$

25. $\sqrt[4]{(x-2)^4} = |x-2|$

26. $\sqrt[3]{(-5)^3} = -5$ Absolute value bars are not needed when the index is odd.

27. $\sqrt[5]{(2x-7)^5} = 2x - 7$

28. $\sqrt{25x^2} = 5|x|$

29. $\sqrt{x^2 + 2x + 1} = \sqrt{(x+1)^2} = |x+1|$

▣ **Work Practice Problems 23–29**

Objective **F** Finding Function Values

Functions of the form

$$f(x) = \sqrt[n]{x}$$

are called **radical functions.** Recall that the domain of a function in *x* is the set of all possible replacement values of *x*. This means that if *n* is even, the domain is the set of all nonnegative numbers, or $\{x \mid x \geq 0\}$. If *n* is odd, the domain is the set of all real numbers. Keep this in mind as we find function values. In Chapter 10, we will graph these functions and discuss their domains further.

PRACTICE PROBLEMS 30–33

If $f(x) = \sqrt{x + 2}$ and $g(x) = \sqrt[3]{x - 1}$, find each function value.

30. $f(7)$ **31.** $g(9)$

32. $f(0)$ **33.** $g(10)$

EXAMPLES If $f(x) = \sqrt{x - 4}$ and $g(x) = \sqrt[3]{x + 2}$, find each function value.

30. $f(8) = \sqrt{8 - 4} = \sqrt{4} = 2$ **31.** $f(6) = \sqrt{6 - 4} = \sqrt{2}$

32. $g(-1) = \sqrt[3]{-1 + 2} = \sqrt[3]{1} = 1$ **33.** $g(1) = \sqrt[3]{1 + 2} = \sqrt[3]{3}$

■ **Work Practice Problems 30–33**

Notice that for the function $f(x) = \sqrt{x - 4}$, the domain includes all real numbers that make the radicand ≥ 0. To see what numbers these are, solve $x - 4 \geq 0$ and find that $x \geq 4$. The domain is $\{x \mid x \geq 4\}$ or $[4, \infty)$.

The domain of the cube root function $g(x) = \sqrt[3]{x + 2}$ is the set of real numbers or $(-\infty, \infty)$.

See Chapter 9 for further discussions of domains.

Recall from Chapter 3 that the graph of $f(x) = \sqrt{x}$ is

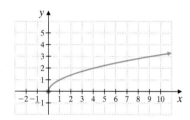

PRACTICE PROBLEM 34

Graph the cube root function $h(x) = \sqrt[3]{x} + 2$.

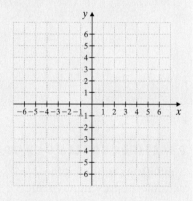

Let's now graph the function $f(x) = \sqrt[3]{x}$.

EXAMPLE 34 Graph the function $f(x) = \sqrt[3]{x}$.

Solution: To graph, we identify the domain, plot points, and connect the points with a smooth curve. The domain of this function is the set of all real numbers. The table comes from the function values obtained earlier. We have approximated $\sqrt[3]{6}$ and $\sqrt[3]{-6}$ for graphing purposes.

x	$f(x) = \sqrt[3]{x}$
0	0
1	1
−1	−1
6	$\sqrt[3]{6} \approx 1.8$
−6	$\sqrt[3]{-6} \approx -1.8$
8	2
−8	−2

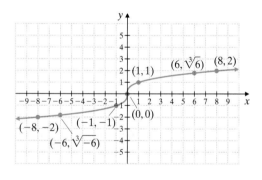

The graph of this function passes the vertical line test, as expected.

■ **Work Practice Problem 34**

Answers

30. 3,   **31.** 2,   **32.** $\sqrt{2}$,   **33.** $\sqrt[3]{9}$,

34.

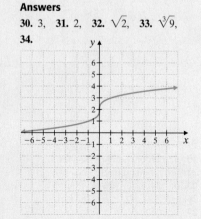

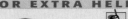

Objective **A** *Find the real square roots of each number. See Examples 1 through 3.*

1. 4
2. 9
3. -25

4. -49
5. 100
6. 64

Find each square root. Assume that all variables represent nonnegative real numbers. See Examples 4 through 11.

7. $\sqrt{100}$
8. $\sqrt{400}$
9. $\sqrt{\dfrac{1}{4}}$
10. $\sqrt{\dfrac{9}{25}}$
11. $\sqrt{0.0001}$
12. $\sqrt{0.04}$

13. $-\sqrt{36}$
14. $-\sqrt{9}$
15. $\sqrt{x^{10}}$
16. $\sqrt{x^{16}}$
17. $\sqrt{16y^6}$
18. $\sqrt{64y^{20}}$

Objective **B** *Use a calculator to approximate each square root to three decimal places. Check to see that each approximation is reasonable. See Example 12.*

19. $\sqrt{7}$
20. $\sqrt{11}$
21. $\sqrt{38}$
22. $\sqrt{56}$
23. $\sqrt{200}$
24. $\sqrt{300}$

Objective **C** *Find each cube root. See Examples 13 through 17.*

25. $\sqrt[3]{64}$
26. $\sqrt[3]{27}$
27. $\sqrt[3]{\dfrac{1}{8}}$
28. $\sqrt[3]{\dfrac{27}{64}}$
29. $\sqrt[3]{-1}$

30. $\sqrt[3]{-125}$
31. $\sqrt[3]{x^{12}}$
32. $\sqrt[3]{x^{15}}$
33. $\sqrt[3]{-27x^9}$
34. $\sqrt[3]{-64x^6}$

Objective **D** *Find each root. Assume that all variables represent nonnegative real numbers. See Examples 18 through 22.*

35. $-\sqrt[4]{16}$
36. $\sqrt[5]{-243}$
37. $\sqrt[4]{-16}$
38. $\sqrt{-16}$
39. $\sqrt[5]{-32}$

40. $\sqrt[5]{-1}$
41. $\sqrt[5]{x^{20}}$
42. $\sqrt[4]{x^{20}}$
43. $\sqrt[6]{64x^{12}}$
44. $\sqrt[5]{-32x^{15}}$

45. $\sqrt{81x^4}$
46. $\sqrt[4]{81x^4}$
47. $\sqrt[4]{256x^8}$
48. $\sqrt{256x^8}$

Objective **E** *Simplify. Assume that the variables represent any real number. See Examples 23 through 29.*

49. $\sqrt{(-8)^2}$
50. $\sqrt{(-7)^2}$
51. $\sqrt[3]{(-8)^3}$
52. $\sqrt[5]{(-7)^5}$
53. $\sqrt{4x^2}$

54. $\sqrt[4]{16x^4}$
55. $\sqrt[3]{x^3}$
56. $\sqrt[5]{x^5}$
57. $\sqrt[4]{(x-2)^4}$
58. $\sqrt[6]{(2x-1)^6}$

59. $\sqrt{x^2 + 4x + 4}$
(*Hint:* Factor the polynomial first.)

60. $\sqrt{x^2 - 8x + 16}$
(*Hint:* Factor the polynomial first.)

Objectives Ⓐ Ⓑ Ⓒ Ⓓ **Mixed Practice** *Simplify each radical. Assume that all variables represent positive real numbers.*

61. $-\sqrt{121}$

62. $-\sqrt[3]{125}$

63. $\sqrt[3]{8x^3}$

64. $\sqrt{16x^8}$

65. $\sqrt{y^{12}}$

66. $\sqrt[3]{y^{12}}$

67. $\sqrt{25a^2b^{20}}$

68. $\sqrt{9x^4y^6}$

69. $\sqrt[3]{-27x^{12}y^9}$

70. $\sqrt[3]{-8a^{21}y^6}$

71. $\sqrt[4]{a^{16}b^4}$

72. $\sqrt[4]{x^8y^{12}}$

73. $\sqrt[5]{-32x^{10}y^5}$

74. $\sqrt[5]{-243z^{15}}$

75. $\sqrt{\dfrac{25}{49}}$

76. $\sqrt{\dfrac{4}{81}}$

77. $\sqrt{\dfrac{x^2}{4y^2}}$

78. $\sqrt{\dfrac{y^{10}}{9x^6}}$

79. $-\sqrt[3]{\dfrac{z^{21}}{27x^3}}$

80. $-\sqrt[3]{\dfrac{64a^3}{b^9}}$

81. $\sqrt[4]{\dfrac{x^4}{16}}$

82. $\sqrt[4]{\dfrac{y^4}{81x^4}}$

Objective Ⓕ *If $f(x) = \sqrt{2x+3}$ and $g(x) = \sqrt[3]{x-8}$, find each function value. See Examples 30 through 33.*

83. $f(0)$

84. $g(0)$

85. $g(7)$

86. $f(-1)$

87. $g(-19)$

88. $f(3)$

89. $f(2)$

90. $g(1)$

Identify the domain and then graph each function. See Example 34.

91. $f(x) = \sqrt[3]{x} + 1$

92. $f(x) = \sqrt[3]{x} - 2$

93. $g(x) = \sqrt[3]{x-1}$

94. $g(x) = \sqrt[3]{x+1}$

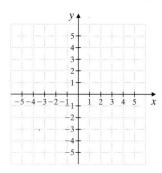
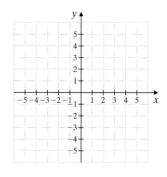
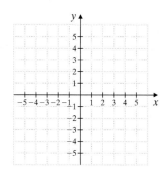
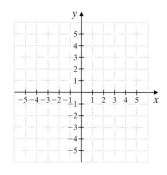

Review

Simplify each exponential expression. See Sections 1.6 and 1.7.

95. $(-2x^3y^2)^5$

96. $(4y^6z^7)^3$

97. $(-3x^2y^3z^5)(20x^5y^7)$

98. $(-14a^5bc^2)(2abc^4)$

99. $\dfrac{7x^{-1}y}{14(x^5y^2)^{-2}}$

100. $\dfrac{(2a^{-1}b^2)^3}{(8a^2b)^{-2}}$

Which of the following are not real numbers? See the Concept Check in this section.

101. $\sqrt{-17}$

102. $\sqrt[3]{-17}$

103. $\sqrt[10]{-17}$

104. $\sqrt[15]{-17}$

Concept Extensions

105. Explain why $\sqrt{-64}$ is not a real number.

106. Explain why $\sqrt[3]{-64}$ is a real number.

For Exercises 107 through 110, do not use a calculator.

107. $\sqrt{160}$ is closest to
   a. 10   **b.** 13   **c.** 20   **d.** 40

108. $\sqrt{1000}$ is closest to
   a. 10   **b.** 30   **c.** 100   **d.** 500

109. The perimeter of the triangle is closest to
   a. 12   **b.** 18   **c.** 66   **d.** 132

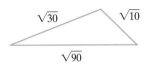

110. The length of the bent wire is closest to
   a. 5   **b.** $\sqrt{28}$ **c.** 7   **d.** 14

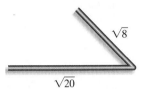

The Mosteller formula for calculating adult body surface area is $B = \sqrt{\dfrac{hw}{3131}}$, where B is an individual's body surface area in square meters, h is the individual's height in inches, and w is the individual's weight in pounds. Use this information to answer Exercises 111 and 112. Round answers to 2 decimal places.

111. Find the body surface area of an individual who is 66 inches tall and who weighs 135 pounds.

112. Find the body surface area of an individual who is 74 inches tall and who weighs 225 pounds.

113. Suppose that a friend tells you that $\sqrt{13} \approx 5.7$. Without a calculator, how can you convince your friend that he or she must have made an error?

114. Escape velocity is the minimum speed that an object must reach to escape a planet's pull of gravity. Escape velocity v is given by the equation $v = \sqrt{\dfrac{2Gm}{r}}$, where m is the mass of the planet, r is its radius, and G is the universal gravitational constant, which has a value of $G = 6.67 \times 10^{-11}\, \text{m}^3/\text{kg} \cdot \text{s}^2$. The mass of Earth is 5.97×10^{24} kg and its radius is 6.37×10^6 m. Use this information to find the escape velocity for Earth. Round to the nearest whole number. (*Source:* National Space Science Data Center)

Objective [A] Understanding $a^{1/n}$

So far in this text, we have not defined expressions with rational exponents such as $3^{1/2}$, $x^{2/3}$, and $-9^{-1/4}$. We will define these expressions so that the rules for exponents shall apply to these rational exponents as well.

Suppose that $x = 5^{1/3}$. Then

$$x^3 = (5^{1/3})^3 = 5^{1/3 \cdot 3} = 5^1 \text{ or } 5$$

$\quad\quad \llcorner$ using rules $\uparrow$
for exponents

Since $x^3 = 5$, then x is the number whose cube is 5, or $x = \sqrt[3]{5}$. Notice that we also know that $x = 5^{1/3}$. This means that

$$5^{1/3} = \sqrt[3]{5}$$

Definition of $a^{1/n}$

If n is a positive integer greater than 1 and $\sqrt[n]{a}$ is a real number, then

$$a^{1/n} = \sqrt[n]{a}$$

Notice that the denominator of the rational exponent corresponds to the index of the radical.

EXAMPLES Use radical notation to rewrite each expression. Simplify if possible.

1. $4^{1/2} = \sqrt{4} = 2$
2. $64^{1/3} = \sqrt[3]{64} = 4$
3. $x^{1/4} = \sqrt[4]{x}$
4. $-9^{1/2} = -\sqrt{9} = -3$
5. $(81x^8)^{1/4} = \sqrt[4]{81x^8} = 3x^2$
6. $5y^{1/3} = 5\sqrt[3]{y}$

◼ **Work Practice Problems 1-6**

Objective [B] Understanding $a^{m/n}$

As we expand our use of exponents to include $\dfrac{m}{n}$, we define their meaning so that rules for exponents still hold true. For example, by properties of exponents,

$$8^{2/3} = (8^{1/3})^2 = (\sqrt[3]{8})^2 \quad \text{or} \quad 8^{2/3} = (8^2)^{1/3} = \sqrt[3]{8^2}$$

Definition of $a^{m/n}$

If m and n are positive integers greater than 1 with $\dfrac{m}{n}$ in simplest form, then

$$a^{m/n} = \sqrt[n]{a^m} = (\sqrt[n]{a})^m$$

as long as $\sqrt[n]{a}$ is a real number.

Notice that the denominator n of the rational exponent corresponds to the index of the radical. The numerator m of the rational exponent indicates that the base is to be raised to the mth power. This means that

$$8^{2/3} = \sqrt[3]{8^2} = \sqrt[3]{64} = 4 \quad \text{or} \quad 8^{2/3} = (\sqrt[3]{8})^2 = 2^2 = 4$$

Helpful Hint

Most of the time, $(\sqrt[n]{a})^m$ will be easier to calculate than $\sqrt[n]{a^m}$.

EXAMPLES Use radical notation to rewrite each expression. Simplify if possible.

7. $4^{3/2} = (\sqrt{4})^3 = 2^3 = 8$

8. $-16^{3/4} = -(\sqrt[4]{16})^3 = -(2)^3 = -8$

9. $(-27)^{2/3} = (\sqrt[3]{-27})^2 = (-3)^2 = 9$

10. $\left(\dfrac{1}{9}\right)^{3/2} = \left(\sqrt{\dfrac{1}{9}}\right)^3 = \left(\dfrac{1}{3}\right)^3 = \dfrac{1}{27}$

11. $(4x - 1)^{3/5} = \sqrt[5]{(4x - 1)^3}$

 Work Practice Problems 7–11

Helpful Hint

The *denominator* of a rational exponent is the index of the corresponding radical. For example, $x^{1/5} = \sqrt[5]{x}$, and $z^{2/3} = \sqrt[3]{z^2}$ or $z^{2/3} = (\sqrt[3]{z})^2$.

Objective **C** Understanding $a^{-m/n}$

The rational exponents we have given meaning to exclude negative rational numbers. To complete the set of definitions, we define $a^{-m/n}$.

Definition of $a^{-m/n}$

$$a^{-m/n} = \dfrac{1}{a^{m/n}}$$

as long as $a^{m/n}$ is a nonzero real number.

EXAMPLES Write each expression with a positive exponent. Then simplify.

12. $16^{-3/4} = \dfrac{1}{16^{3/4}} = \dfrac{1}{(\sqrt[4]{16})^3} = \dfrac{1}{2^3} = \dfrac{1}{8}$

13. $(-27)^{-2/3} = \dfrac{1}{(-27)^{2/3}} = \dfrac{1}{(\sqrt[3]{-27})^2} = \dfrac{1}{(-3)^2} = \dfrac{1}{9}$

 Work Practice Problems 12–13

Helpful Hint

If an expression contains a negative rational exponent, you may want to first write the expression with a positive exponent, then interpret the rational exponent. Notice that the sign of the base is not affected by the sign of its exponent. For example,

$$9^{-3/2} = \frac{1}{9^{3/2}} = \frac{1}{(\sqrt{9})^3} = \frac{1}{27}$$

Also,

$$(-27)^{-1/3} = \frac{1}{(-27)^{1/3}} = -\frac{1}{3}$$

✔ **Concept Check** Which one is correct?

a. $-8^{2/3} = \frac{1}{4}$ **b.** $8^{-2/3} = -\frac{1}{4}$ **c.** $8^{-2/3} = -4$ **d.** $-8^{-2/3} = -\frac{1}{4}$

Objective D Using Rules for Exponents

It can be shown that the properties of integer exponents hold for rational exponents. By using these properties and definitions, we can now simplify expressions that contain rational exponents. These rules are repeated here for review.

Summary of Exponent Rules

If m and n are rational numbers, and a, b, and c are numbers for which the expressions below exist, then

Product rule for exponents:	$a^m \cdot a^n = a^{m+n}$
Power rule for exponents:	$(a^m)^n = a^{m \cdot n}$
Power rules for products and quotients:	$(ab)^n = a^n b^n$ and
	$\left(\dfrac{a}{c}\right)^n = \dfrac{a^n}{c^n}, \quad c \neq 0$
Quotient rule for exponents:	$\dfrac{a^m}{a^n} = a^{m-n}, \quad a \neq 0$
Zero exponent:	$a^0 = 1, \quad a \neq 0$
Negative exponent:	$a^{-n} = \dfrac{1}{a^n}, \quad a \neq 0$

PRACTICE PROBLEMS 14–18

Use the properties of exponents to simplify.

14. $x^{1/3}x^{1/4}$ **15.** $\dfrac{9^{2/5}}{9^{12/5}}$

16. $y^{-3/10} \cdot y^{6/10}$ **17.** $(11^{2/9})^3$

18. $\dfrac{(3x^{2/3})^3}{x^2}$

Answers

14. $x^{7/12}$,   **15.** $\dfrac{1}{81}$,   **16.** $y^{3/10}$,

17. $11^{2/3}$,   **18.** 27

✔ **Concept Check Answer**

d

EXAMPLES Use the properties of exponents to simplify.

14. $x^{1/2}x^{1/3} = x^{1/2+1/3} = x^{3/6+2/6} = x^{5/6}$ Use the product rule.

15. $\dfrac{7^{1/3}}{7^{4/3}} = 7^{1/3-4/3} = 7^{-3/3} = 7^{-1} = \dfrac{1}{7}$ Use the quotient rule.

16. $y^{-4/7} \cdot y^{6/7} = y^{-4/7+6/7} = y^{2/7}$ Use the product rule.

17. $(5^{3/8})^4 = 5^{3/8 \cdot 4} = 5^{12/8} = 5^{3/2}$ Use the power rule.

18. $\dfrac{(2x^{2/5})^5}{x^2} = \dfrac{2^5(x^{2/5})^5}{x^2}$ Use the power rule.

$\qquad = \dfrac{32x^2}{x^2}$ Simplify.

$\qquad = 32x^{2-2}$ Use the quotient rule.

$\qquad = 32x^0$ Simplify.

$\qquad = 32 \cdot 1 \quad$ or $\quad 32$ Substitute 1 for x^0.

Work Practice Problems 14–18

Objective E Using Rational Exponents to Simplify Radical Expressions

We can simplify some radical expressions by first writing the expression with rational exponents. Use the properties of exponents to simplify, and then convert back to radical notation.

EXAMPLES Use rational exponents to simplify. Assume that all variables represent positive real numbers.

19. $\sqrt[8]{x^4} = x^{4/8}$ Write with rational exponents.

      $= x^{1/2}$ Simplify the exponent.

      $= \sqrt{x}$ Write with radical notation.

20. $\sqrt[6]{25} = 25^{1/6}$ Write with rational exponents.

      $= (5^2)^{1/6}$ Write 25 as 5^2.

      $= 5^{2/6}$ Use the power rule.

      $= 5^{1/3}$ Simplify the exponent.

      $= \sqrt[3]{5}$ Write with radical notation.

21. $\sqrt[6]{r^2 s^4} = (r^2 s^4)^{1/6}$ Write with rational exponents.

      $= r^{2/6} s^{4/6}$ Use the power rule.

      $= r^{1/3} s^{2/3}$ Simplify the exponents.

      $= (r s^2)^{1/3}$ Use $a^n b^n = (ab)^n$.

      $= \sqrt[3]{r s^2}$ Write with radical notation.

◼ **Work Practice Problems 19–21**

EXAMPLES Use rational exponents to write as a single radical.

22. $\sqrt{x} \cdot \sqrt[4]{x} = x^{1/2} \cdot x^{1/4} = x^{1/2 + 1/4}$

      $= x^{3/4} = \sqrt[4]{x^3}$

23. $\dfrac{\sqrt{x}}{\sqrt[3]{x}} = \dfrac{x^{1/2}}{x^{1/3}} = x^{1/2 - 1/3} = x^{3/6 - 2/6}$

      $= x^{1/6} = \sqrt[6]{x}$

24. $\sqrt[3]{3} \cdot \sqrt{2} = 3^{1/3} \cdot 2^{1/2}$ Write with rational exponents.

      $= 3^{2/6} \cdot 2^{3/6}$ Write the exponents so that they have the same denominator.

      $= (3^2 \cdot 2^3)^{1/6}$ Use $a^n b^n = (ab)^n$.

      $= \sqrt[6]{3^2 \cdot 2^3}$ Write with radical notation.

      $= \sqrt[6]{72}$ Multiply $3^2 \cdot 2^3$.

◼ **Work Practice Problems 22–24**

PRACTICE PROBLEMS 19–21

Use rational exponents to simplify. Assume that all variables represent positive real numbers.

19. $\sqrt[10]{y^5}$ **20.** $\sqrt[4]{9}$

21. $\sqrt[9]{a^6 b^3}$

PRACTICE PROBLEMS 22–24

Use rational exponents to write as a single radical.

22. $\sqrt{y} \cdot \sqrt[3]{y}$ **23.** $\dfrac{\sqrt[3]{x}}{\sqrt[4]{x}}$

24. $\sqrt{5} \cdot \sqrt[3]{2}$

Answers

19. $\sqrt{y}$,   **20.** $\sqrt{3}$,   **21.** $\sqrt[3]{a^2 b}$,

22. $\sqrt[6]{y^5}$,   **23.** $\sqrt[12]{x}$,   **24.** $\sqrt[6]{500}$

7.2 EXERCISE SET

FOR EXTRA HELP

Student Solutions Manual

PH Math/Tutor Center

CD/Video for Review

Math XL
MathXL®

MyMathLab
MyMathLab

Objective **A** *Use radical notation to rewrite each expression. Simplify if possible. See Examples 1 through 6.*

1. $49^{1/2}$ **2.** $64^{1/3}$ **3.** $27^{1/3}$ **4.** $8^{1/3}$ **5.** $\left(\dfrac{1}{16}\right)^{1/4}$ **6.** $\left(\dfrac{1}{64}\right)^{1/2}$

7. $169^{1/2}$ **8.** $81^{1/4}$ **9.** $2m^{1/3}$ **10.** $(2m)^{1/3}$ **11.** $(9x^4)^{1/2}$ **12.** $(16x^8)^{1/2}$

13. $(-27)^{1/3}$ **14.** $-64^{1/2}$ **15.** $-16^{1/4}$ **16.** $(-32)^{1/5}$

Objective **B** *Use radical notation to rewrite each expression. Simplify if possible. See Examples 7 through 11.*

17. $16^{3/4}$ **18.** $4^{5/2}$ **19.** $(-64)^{2/3}$ **20.** $(-8)^{4/3}$ **21.** $(-16)^{3/4}$ **22.** $(-9)^{3/2}$

23. $(2x)^{3/5}$ **24.** $2x^{3/5}$ **25.** $(7x+2)^{2/3}$ **26.** $(x-4)^{3/4}$ **27.** $\left(\dfrac{16}{9}\right)^{3/2}$ **28.** $\left(\dfrac{49}{25}\right)^{3/2}$

Objective **C** *Write with positive exponents. Simplify if possible. See Examples 12 and 13.*

29. $8^{-4/3}$ **30.** $64^{-2/3}$ **31.** $(-64)^{-2/3}$ **32.** $(-8)^{-4/3}$ **33.** $(-4)^{-3/2}$ **34.** $(-16)^{-5/4}$

35. $x^{-1/4}$ **36.** $y^{-1/6}$ **37.** $\dfrac{1}{a^{-2/3}}$ **38.** $\dfrac{1}{n^{-8/9}}$ **39.** $\dfrac{5}{7x^{-3/4}}$ **40.** $\dfrac{2}{3y^{-5/7}}$

Objective **D** *Use the properties of exponents to simplify each expression. Write with positive exponents. See Examples 14 through 18.*

41. $a^{2/3}a^{5/3}$ **42.** $b^{9/5}b^{8/5}$ **43.** $x^{-2/5}\cdot x^{7/5}$ **44.** $y^{4/3}\cdot y^{-1/3}$ **45.** $3^{1/4}\cdot 3^{3/8}$

46. $5^{1/2}\cdot 5^{1/6}$ **47.** $\dfrac{y^{1/3}}{y^{1/6}}$ **48.** $\dfrac{x^{3/4}}{x^{1/8}}$ **49.** $(4u^2)^{3/2}$ **50.** $(32^{1/5}x^{2/3})^3$

51. $\dfrac{b^{1/2}b^{3/4}}{-b^{1/4}}$ **52.** $\dfrac{a^{1/4}a^{-1/2}}{a^{2/3}}$ **53.** $\dfrac{(x^3)^{1/2}}{x^{7/2}}$ **54.** $\dfrac{y^{11/3}}{(y^5)^{1/3}}$ **55.** $\dfrac{(3x^{1/4})^3}{x^{1/12}}$

56. $\dfrac{(2x^{1/5})^4}{x^{3/10}}$

57. $\dfrac{(y^3z)^{1/6}}{y^{-1/2}z^{1/3}}$

58. $\dfrac{(m^2n)^{1/4}}{m^{-1/2}n^{5/8}}$

59. $\dfrac{(x^3y^2)^{1/4}}{(x^{-5}y^{-1})^{-1/2}}$

60. $\dfrac{(a^{-2}b^3)^{1/8}}{(a^{-3}b)^{-1/4}}$

Objective **E** *Use rational exponents to simplify each radical. Assume that all variables represent positive real numbers. See Examples 19 through 21.*

61. $\sqrt[6]{x^3}$

62. $\sqrt[9]{a^3}$

63. $\sqrt[6]{4}$

64. $\sqrt[4]{36}$

65. $\sqrt[4]{16x^2}$

66. $\sqrt[8]{4y^2}$

67. $\sqrt[4]{(x+3)^2}$

68. $\sqrt[8]{(y+1)^4}$

69. $\sqrt[8]{x^4y^4}$

70. $\sqrt[9]{y^6z^3}$

71. $\sqrt[12]{a^8b^4}$

72. $\sqrt[10]{a^5b^5}$

Use rational expressions to write as a single radical expression. See Examples 22 through 24.

73. $\sqrt[3]{y}\cdot\sqrt[5]{y^2}$

74. $\sqrt[3]{y^2}\cdot\sqrt[6]{y}$

75. $\dfrac{\sqrt[3]{b^2}}{\sqrt[4]{b}}$

76. $\dfrac{\sqrt[4]{a}}{\sqrt[5]{a}}$

77. $\sqrt[3]{x}\cdot\sqrt[4]{x}\cdot\sqrt[8]{x^3}$

78. $\sqrt[6]{y}\cdot\sqrt[3]{y}\cdot\sqrt[5]{y^2}$

79. $\dfrac{\sqrt[3]{a^2}}{\sqrt[6]{a}}$

80. $\dfrac{\sqrt[5]{b^2}}{\sqrt[10]{b^3}}$

81. $\sqrt{3}\cdot\sqrt[3]{4}$

82. $\sqrt[3]{5}\cdot\sqrt{2}$

83. $\sqrt[5]{7}\cdot\sqrt[3]{y}$

84. $\sqrt[4]{5}\cdot\sqrt[3]{x}$

85. $\sqrt{5r}\cdot\sqrt[3]{s}$

86. $\sqrt[3]{b}\cdot\sqrt[5]{4a}$

Review

Write each integer as a product of two integers such that one of the factors is a perfect square. For example, write 18 as $9\cdot2$, because 9 is a perfect square.

87. 75

88. 20

89. 48

90. 45

Write each integer as a product of two integers such that one of the factors is a perfect cube. For example, write 24 as $8\cdot3$, because 8 is a perfect cube.

91. 16

92. 56

93. 54

94. 80

Concept Extensions

Basal metabolic rate (BMR) is the number of calories per day a person needs to maintain life. A person's basal metabolic rate $B(w)$ in calories per day can be estimated with the function $B(w) = 70w^{3/4}$, where w is the person's weight in kilograms. Use this information to answer Exercises 95 and 96.

95. Estimate the BMR for a person who weighs 60 kilograms. Round to the nearest calorie. (*Note:* 60 kilograms is approximately 132 pounds.)

96. Estimate the BMR for a person who weighs 90 kilograms. Round to the nearest calorie. (*Note:* 90 kilograms is approximately 198 pounds.)

The number of cellular telephone subscriptions in the United States from 1999 through 2004 can be modeled by the function $f(x) = 12x^{7/6}$, where y is the number of cellular telephone subscriptions in millions, x years after 1994. (Source: Based on data from the Cellular Telecommunications & Internet Association, 1985–2004.) Use this information to answer Exercises 97 and 98.

97. Use this model to estimate the number of cellular telephone subscriptions in 2004. Round to the nearest tenth of a million.

98. Predict the number of cellular telephone subscriptions in 2009. Round to the nearest tenth of a million.

99. Explain how writing x^{-7} with positive exponents is similar to writing $x^{-1/4}$ with positive exponents.

100. Explain how writing $2x^{-5}$ with positive exponents is similar to writing $2x^{-3/4}$ with positive exponents.

Fill in each box with the correct expression.

101. $\square \cdot a^{2/3} = a^{3/3}$, or a

102. $\square \cdot x^{1/8} = x^{4/8}$, or $x^{1/2}$

103. $\dfrac{\square}{x^{-2/5}} = x^{3/5}$

104. $\dfrac{\square}{x^{-3/4}} = y^{4/4}$, or y

Use a calculator to write a four-decimal-place approximation of each number.

105. $8^{1/4}$

106. $18^{3/5}$

107. In physics, the speed of a wave traveling over a stretched string with tension t and density u is given by the expression $\dfrac{\sqrt{t}}{\sqrt{u}}$. Write this expression with rational exponents.

108. In electronics, the angular frequency of oscillations in a certain type of circuit is given by the expression $(LC)^{-1/2}$. Use radical notation to write this expression.

7.3 SIMPLIFYING RADICAL EXPRESSIONS

Objectives

Objective A Using the Product Rule

It is possible to simplify some radicals that do not evaluate to rational numbers. To do so, we use a product rule and a quotient rule for radicals. To discover the product rule, notice the following pattern:

$$\sqrt{9} \cdot \sqrt{4} = 3 \cdot 2 = 6$$
$$\sqrt{9 \cdot 4} = \sqrt{36} = 6$$

Since both expressions simplify to 6, it is true that

$$\sqrt{9} \cdot \sqrt{4} = \sqrt{9 \cdot 4}$$

This pattern suggests the following product rule for radicals.

Product Rule for Radicals

If $\sqrt[n]{a}$ and $\sqrt[n]{b}$ are real numbers, then

$$\sqrt[n]{a} \cdot \sqrt[n]{b} = \sqrt[n]{ab}$$

Notice that the product rule is the relationship $a^{1/n} \cdot b^{1/n} = (ab)^{1/n}$ stated in radical notation.

EXAMPLES Use the product rule to multiply.

1. $\sqrt{3} \cdot \sqrt{5} = \sqrt{3 \cdot 5} = \sqrt{15}$
2. $\sqrt{21} \cdot \sqrt{x} = \sqrt{21x}$
3. $\sqrt[3]{4} \cdot \sqrt[3]{2} = \sqrt[3]{4 \cdot 2} = \sqrt[3]{8} = 2$
4. $\sqrt[4]{5} \cdot \sqrt[4]{2x^3} = \sqrt[4]{5 \cdot 2x^3} = \sqrt[4]{10x^3}$
5. $\sqrt{\dfrac{2}{a}} \cdot \sqrt{\dfrac{b}{3}} = \sqrt{\dfrac{2}{a} \cdot \dfrac{b}{3}} = \sqrt{\dfrac{2b}{3a}}$

■ Work Practice Problems 1–5

Objective B Using the Quotient Rule

To discover the quotient rule for radicals, notice the following pattern:

$$\sqrt{\dfrac{4}{9}} = \dfrac{2}{3}$$

$$\dfrac{\sqrt{4}}{\sqrt{9}} = \dfrac{2}{3}$$

Since both expressions simplify to $\dfrac{2}{3}$, it is true that

$$\sqrt{\dfrac{4}{9}} = \dfrac{\sqrt{4}}{\sqrt{9}}$$

Objectives

A Use the Product Rule for Radicals.

B Use the Quotient Rule for Radicals.

C Simplify Radicals.

D Use the Distance and Midpoint Formula.

PRACTICE PROBLEMS 1–5

Use the product rule to multiply.

1. $\sqrt{2} \cdot \sqrt{13}$ 2. $\sqrt{17} \cdot \sqrt{y}$
3. $\sqrt[3]{2} \cdot \sqrt[3]{32}$ 4. $\sqrt[4]{6} \cdot \sqrt[4]{3x^2}$
5. $\sqrt{\dfrac{3}{x}} \cdot \sqrt{\dfrac{y}{2}}$

Answers

1. $\sqrt{26}$, 2. $\sqrt{17y}$, 3. 4,
4. $\sqrt[4]{18x^2}$, 5. $\sqrt{\dfrac{3y}{2x}}$

493

This pattern suggests the following quotient rule for radicals.

Quotient Rule for Radicals

If $\sqrt[n]{a}$ and $\sqrt[n]{b}$ are real numbers and $\sqrt[n]{b}$ is not zero, then

$$\sqrt[n]{\frac{a}{b}} = \frac{\sqrt[n]{a}}{\sqrt[n]{b}}$$

Notice that the quotient rule is the relationship $\left(\dfrac{a}{b}\right)^{1/n} = \dfrac{a^{1/n}}{b^{1/n}}$ stated in radical notation. We can use the quotient rule to simplify radical expressions by reading the rule from left to right or to divide radicals by reading the rule from right to left.

For example.

$$\sqrt{\frac{x}{16}} = \frac{\sqrt{x}}{\sqrt{16}} = \frac{\sqrt{x}}{4} \qquad \text{Using } \sqrt[n]{\frac{a}{b}} = \frac{\sqrt[n]{a}}{\sqrt[n]{b}}$$

$$\frac{\sqrt{50}}{\sqrt{2}} = \sqrt{\frac{50}{2}} = \sqrt{25} = 5 \qquad \text{Using } \frac{\sqrt[n]{a}}{\sqrt[n]{b}} = \sqrt[n]{\frac{a}{b}}$$

Note: *For the remainder of this chapter, we will assume that variables represent positive real numbers. If this is so, we need not insert absolute value bars when we simplify even roots.*

PRACTICE PROBLEMS 6–9

Use the quotient rule to simplify. Assume that all variables represent positive real numbers.

6. $\sqrt{\dfrac{9}{25}}$ **7.** $\sqrt{\dfrac{y}{36}}$

8. $\sqrt[3]{\dfrac{27}{64}}$ **9.** $\sqrt[5]{\dfrac{7}{32x^5}}$

EXAMPLES Use the quotient rule to simplify.

6. $\sqrt{\dfrac{25}{49}} = \dfrac{\sqrt{25}}{\sqrt{49}} = \dfrac{5}{7}$

7. $\sqrt{\dfrac{x}{9}} = \dfrac{\sqrt{x}}{\sqrt{9}} = \dfrac{\sqrt{x}}{3}$

8. $\sqrt[3]{\dfrac{8}{27}} = \dfrac{\sqrt[3]{8}}{\sqrt[3]{27}} = \dfrac{2}{3}$

9. $\sqrt[4]{\dfrac{3}{16y^4}} = \dfrac{\sqrt[4]{3}}{\sqrt[4]{16y^4}} = \dfrac{\sqrt[4]{3}}{2y}$

🔲 **Work Practice Problems 6–9**

Objective C Simplifying Radicals

Both the product and quotient rules can be used to simplify a radical. If the product rule is read from right to left, we have that $\sqrt[n]{ab} = \sqrt[n]{a} \cdot \sqrt[n]{b}$. We use this to simplify the following radicals.

PRACTICE PROBLEM 10

Simplify: $\sqrt{18}$

EXAMPLE 10 Simplify: $\sqrt{50}$

Solution: We factor 50 such that one factor is the largest perfect square that divides 50. The largest perfect square factor of 50 is 25, so we write 50 as $25 \cdot 2$ and use the product rule for radicals to simplify.

$$\sqrt{50} = \sqrt{25 \cdot 2} = \sqrt{25} \cdot \sqrt{2} = 5\sqrt{2}$$

⌐ the largest perfect square factor of 50

Helpful Hint

Don't forget that, for example, $5\sqrt{2}$ means $5 \cdot \sqrt{2}$.

🔲 **Work Practice Problem 10**

Answers

6. $\dfrac{3}{5}$, **7.** $\dfrac{\sqrt{y}}{6}$, **8.** $\dfrac{3}{4}$, **9.** $\dfrac{\sqrt[5]{7}}{2x}$,

10. $3\sqrt{2}$

EXAMPLES Simplify.

11. $\sqrt[3]{24} = \sqrt[3]{8 \cdot 3} = \sqrt[3]{8} \cdot \sqrt[3]{3} = 2\sqrt[3]{3}$
        └ the largest perfect cube factor of 24

12. $\sqrt{26}$ The largest perfect square factor of 26 is 1, so $\sqrt{26}$ cannot be simplified further.

13. $\sqrt[4]{32} = \sqrt[4]{16 \cdot 2} = \sqrt[4]{16} \cdot \sqrt[4]{2} = 2\sqrt[4]{2}$
          └ the largest 4th power factor of 32

☐ **Work Practice Problems 11–13**

After simplifying a radical such as a square root, always check the radicand to see that it contains no other perfect square factors. It may, if the largest perfect square factor of the radicand was not originally recognized. For Example,

$$\sqrt{200} = \sqrt{4 \cdot 50} = \sqrt{4} \cdot \sqrt{50} = 2\sqrt{50}$$

Notice that the radicand 50 still contains the perfect square factor 25. This is because 4 is not the largest perfect square factor of 200. We continue as follows:

$$2\sqrt{50} = 2\sqrt{25 \cdot 2} = 2 \cdot \sqrt{25} \cdot \sqrt{2} = 2 \cdot 5 \cdot \sqrt{2} = 10\sqrt{2}$$

The radical is now simplified since 2 contains no perfect square factors (other than 1).

Helpful Hint

To recognize the largest perfect power factors of a radicand, it will help if you are familiar with some perfect powers. A few are listed below.

Perfect Squares $1, 4, 9, 16, 25, 36, 49, 64, 81, 100, 121, 144$
                              $1^2 \; 2^2 \; 3^2 \; 4^2 \; 5^2 \; 6^2 \; 7^2 \; 8^2 \; 9^2 \; 10^2 \; 11^2 \; 12^2$

Perfect Cubes $1, 8, 27, 64, 125$
                              $1^3 \; 2^3 \; 3^3 \; \; 4^3 \; \; 5^3$

Perfect 4th powers $1, 16, 81, 256$
                              $1^4 \; 2^4 \; \; 3^4 \; \; 4^4$

Helpful Hint

We say that a radical of the form $\sqrt[n]{a}$ is simplified when the radicand a contains no factors that are perfect nth powers (other than 1 or −1).

EXAMPLES Simplify. Assume that all variables represent positive real numbers.

14. $\sqrt{25x^3} = \sqrt{25 \cdot x^2 \cdot x}$ Find the largest perfect square factor.

        $= \sqrt{25 \cdot x^2} \cdot \sqrt{x}$ Use the product rule.

        $= 5x\sqrt{x}$ Simplify.

15. $\sqrt[3]{54x^6y^8} = \sqrt[3]{27 \cdot 2 \cdot x^6 \cdot y^6 \cdot y^2}$ Factor the radicand and identify perfect cube factors.

              $= \sqrt[3]{27 \cdot x^6 \cdot y^6 \cdot 2y^2}$

              $= \sqrt[3]{27 \cdot x^6 \cdot y^6} \cdot \sqrt[3]{2y^2}$ Use the product rule.

              $= 3x^2y^2\sqrt[3]{2y^2}$ Simplify.

Continued on next page

PRACTICE PROBLEMS 11–13

Simplify.

11. $\sqrt[3]{40}$ **12.** $\sqrt{14}$

13. $\sqrt[4]{162}$

PRACTICE PROBLEMS 14–16

Simplify. Assume that all variables represent positive real numbers.

14. $\sqrt{49a^5}$ **15.** $\sqrt[3]{24x^9y^7}$

16. $\sqrt[4]{16z^9}$

Answers

11. $2\sqrt[3]{5}$, **12.** $\sqrt{14}$, **13.** $3\sqrt[4]{2}$,

14. $7a^2\sqrt{a}$, **15.** $2x^3y^2\sqrt[3]{3y}$,

16. $2z^2\sqrt[4]{z}$

16. $\sqrt[4]{81z^{11}} = \sqrt[4]{81 \cdot z^8 \cdot z^3}$ Factor the radicand and identify perfect 4th power factors.

$\qquad = \sqrt[4]{81 \cdot z^8} \cdot \sqrt[4]{z^3}$ Use the product rule.

$\qquad = 3z^2\sqrt[4]{z^3}$ Simplify.

▣ **Work Practice Problems 14–16**

PRACTICE PROBLEMS 17–20

Use the quotient rule to divide. Then simplify if possible. Assume that all variables represent positive real numbers.

17. $\dfrac{\sqrt{75}}{\sqrt{3}}$ **18.** $\dfrac{\sqrt{80y}}{3\sqrt{5}}$

19. $\dfrac{5\sqrt[3]{162x^8}}{\sqrt[3]{3x^2}}$ **20.** $\dfrac{3\sqrt[4]{243x^9y^6}}{\sqrt[4]{x^{-3}y}}$

EXAMPLES Use the quotient rule to divide. Then simplify if possible. Assume that all variables represent positive real numbers.

17. $\dfrac{\sqrt{20}}{\sqrt{5}} = \sqrt{\dfrac{20}{5}}$ Use the quotient rule.

$\qquad = \sqrt{4}$ Simplify.

$\qquad = 2$ Simplify.

18. $\dfrac{\sqrt{50x}}{2\sqrt{2}} = \dfrac{1}{2} \cdot \sqrt{\dfrac{50x}{2}}$ Use the quotient rule.

$\qquad = \dfrac{1}{2} \cdot \sqrt{25x}$ Simplify.

$\qquad = \dfrac{1}{2} \cdot \sqrt{25} \cdot \sqrt{x}$ Factor 25x.

$\qquad = \dfrac{1}{2} \cdot 5 \cdot \sqrt{x}$ Simplify.

$\qquad = \dfrac{5}{2}\sqrt{x}$

19. $\dfrac{7\sqrt[3]{48y^4}}{\sqrt[3]{2y}} = 7\sqrt[3]{\dfrac{48y^4}{2y}} = 7\sqrt[3]{24y^3} = 7\sqrt[3]{8 \cdot y^3 \cdot 3}$

$\qquad = 7\sqrt[3]{8 \cdot y^3} \cdot \sqrt[3]{3} = 7 \cdot 2y\sqrt[3]{3} = 14y\sqrt[3]{3}$

20. $\dfrac{2\sqrt[4]{32a^8b^6}}{\sqrt[4]{a^{-1}b^2}} = 2\sqrt[4]{\dfrac{32a^8b^6}{a^{-1}b^2}} = 2\sqrt[4]{32a^9b^4} = 2\sqrt[4]{16 \cdot a^8 \cdot b^4 \cdot 2 \cdot a}$

$\qquad = 2\sqrt[4]{16 \cdot a^8 \cdot b^4} \cdot \sqrt[4]{2 \cdot a} = 2 \cdot 2a^2b \cdot \sqrt[4]{2a} = 4a^2b\sqrt[4]{2a}$

▣ **Work Practice Problems 17–20**

✔**Concept Check** Find and correct the error:

$$\dfrac{\sqrt[3]{27}}{\sqrt{9}} \bcancel{= \sqrt[3]{\dfrac{27}{9}} = \sqrt[3]{3}}$$

Objective **D** Using the Distance and Midpoint Formulas

Now that we know how to simplify radicals, we can derive and use the distance formula. The midpoint formula is often confused with the distance formula, so to clarify both, we will also review the midpoint formula.

The Cartesian coordinate system helps us visualize a distance between points. To find the distance between two points, we use the distance formula, which is derived from the Pythagorean theorem.

To find the distance d between two points (x_1, y_1) and (x_2, y_2), draw vertical and horizontal lines so that a right triangle is formed, as shown. Notice that the length of leg a is $x_2 - x_1$ and that

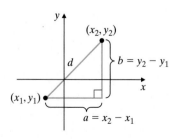

Answers

17. 5, **18.** $\dfrac{4}{3}\sqrt{y}$,

19. $15x^2\sqrt[3]{2}$, **20.** $9x^3y\sqrt[4]{3y}$

✔ **Concept Check Answer**

$\dfrac{\sqrt[3]{27}}{\sqrt{9}} = \dfrac{3}{3} = 1$

the length of leg b is $y_2 - y_1$. Thus, the Pythagorean theorem tells us that

$$d^2 = a^2 + b^2$$

or

$$d^2 = (x_2 - x_1)^2 + (y_2 - y_1)^2$$

or

$$d = \sqrt{(x_2 - x_1)^2 + (y_2 - y_1)^2}$$

This formula gives us the distance between any two points on the real plane.

Distance Formula

The distance d between two points (x_1, y_1) and (x_2, y_2) is given by

$$d = \sqrt{(x_2 - x_1)^2 + (y_2 - y_1)^2}$$

EXAMPLE 21 Find the distance between $(2, -5)$ and $(1, -4)$. Give an exact distance and a three-decimal-place approximation.

Solution: To use the distance formula, it makes no difference which point we call (x_1, y_1) and which point we call (x_2, y_2). We will let $(x_1, y_1) = (2, -5)$ and $(x_2, y_2) = (1, -4)$.

$$\begin{aligned}
d &= \sqrt{(x_2 - x_1)^2 + (y_2 - y_1)^2} \\
&= \sqrt{(1 - 2)^2 + [-4 - (-5)]^2} \\
&= \sqrt{(-1)^2 + (1)^2} \\
&= \sqrt{1 + 1} \\
&= \sqrt{2} \approx 1.414
\end{aligned}$$

The distance between the two points is exactly $\sqrt{2}$ units, or approximately 1.414 units.

■ **Work Practice Problem 21**

The **midpoint** of a line segment is the **point** located exactly halfway between the two endpoints of the line segment. On the following graph, the point M is the midpoint of line segment PQ. Thus, the distance between M and P equals the distance between M and Q. *Note:* We usually need no knowledge of roots to calculate the midpoint of a line segment. We review midpoint here only because it is often confused with the distance between two points.

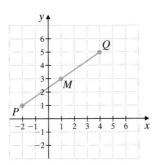

The x-coordinate of M is at half the distance between the x-coordinates of P and Q, and the y-coordinate of M is at half the distance between the y-coordinates of P and Q. That is, the x-coordinate of M is the average of the x-coordinates of P and Q; the y-coordinate of M is the average of the y-coordinates of P and Q.

PRACTICE PROBLEM 21

Find the distance between $(-1, 3)$ and $(-2, 6)$. Give an exact distance and a three-decimal-place approximation.

Answer

21. $\sqrt{10} \approx 3.162$

Midpoint Formula

The midpoint of the line segment whose endpoints are (x_1, y_1) and (x_2, y_2) is the point with coordinates

$$\left(\frac{x_1 + x_2}{2}, \frac{y_1 + y_2}{2} \right)$$

PRACTICE PROBLEM 22

Find the midpoint of the line segment that joins points $P(-2, 5)$ and $Q(4, -6)$.

EXAMPLE 22 Find the midpoint of the line segment that joins points $P(-3, 3)$ and $Q(1, 0)$.

Solution: To use the midpoint formula, it makes no difference which point we call (x_1, y_1) and which point we call (x_2, y_2). We will let $(x_1, y_1) = (-3, 3)$ and $(x_2, y_2) = (1, 0)$.

$$\text{midpoint} = \left(\frac{x_1 + x_2}{2}, \frac{y_1 + y_2}{2} \right)$$
$$= \left(\frac{-3 + 1}{2}, \frac{3 + 0}{2} \right)$$
$$= \left(\frac{-2}{2}, \frac{3}{2} \right)$$
$$= \left(-1, \frac{3}{2} \right)$$

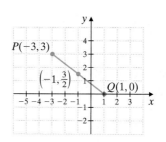

The midpoint of the segment is $\left(-1, \frac{3}{2} \right)$.

🔲 **Work Practice Problem 22**

> **Helpful Hint**
>
> The distance between two points is a distance. The midpoint of a line segment is the point halfway between the endpoints of the segment.
>
> distance—measured in units
>
> midpoint—it is a point

Answer

22. $\left(1, -\frac{1}{2} \right)$

7.3 EXERCISE SET

FOR EXTRA HELP

Student Solutions Manual

PH Math/Tutor Center

CD/Video for Review

MathXL®

MyMathLab

Objective A *Use the product rule to multiply. Assume that all variables represent positive real numbers. See Examples 1 through 5.*

1. $\sqrt{7} \cdot \sqrt{2}$

2. $\sqrt{11} \cdot \sqrt{10}$

3. $\sqrt[4]{8} \cdot \sqrt[4]{2}$

4. $\sqrt[4]{27} \cdot \sqrt[4]{3}$

5. $\sqrt[3]{4} \cdot \sqrt[3]{9}$

6. $\sqrt[3]{10} \cdot \sqrt[3]{5}$

7. $\sqrt{2} \cdot \sqrt{3x}$

8. $\sqrt{3y} \cdot \sqrt{5x}$

9. $\sqrt{\dfrac{7}{x}} \cdot \sqrt{\dfrac{2}{y}}$

10. $\sqrt{\dfrac{6}{m}} \cdot \sqrt{\dfrac{n}{5}}$

11. $\sqrt[4]{4x^3} \cdot \sqrt[4]{5}$

12. $\sqrt[4]{ab^2} \cdot \sqrt[4]{27ab}$

Objective B *Use the quotient rule to simplify. Assume that all variables represent positive real numbers. See Examples 6 through 9.*

13. $\sqrt{\dfrac{6}{49}}$

14. $\sqrt{\dfrac{8}{81}}$

15. $\sqrt{\dfrac{2}{49}}$

16. $\sqrt{\dfrac{5}{121}}$

17. $\sqrt[4]{\dfrac{x^3}{16}}$

18. $\sqrt[4]{\dfrac{y}{81x^4}}$

19. $\sqrt[3]{\dfrac{4}{27}}$

20. $\sqrt[3]{\dfrac{3}{64}}$

21. $\sqrt[4]{\dfrac{8}{x^8}}$

22. $\sqrt[4]{\dfrac{a^3}{81}}$

23. $\sqrt[3]{\dfrac{2x}{81y^{12}}}$

24. $\sqrt[3]{\dfrac{3}{8x^6}}$

25. $\sqrt{\dfrac{x^2 y}{100}}$

26. $\sqrt{\dfrac{y^2 z}{400}}$

27. $\sqrt{\dfrac{5x^2}{169y^2}}$

28. $\sqrt{\dfrac{y^{10}}{225x^6}}$

29. $-\sqrt[3]{\dfrac{z^7}{125x^3}}$

30. $-\sqrt[3]{\dfrac{1000a}{b^9}}$

Objective C *Simplify. Assume that all variables represent positive real numbers. See Examples 10 through 16.*

31. $\sqrt{32}$

32. $\sqrt{27}$

33. $\sqrt[3]{192}$

34. $\sqrt[3]{108}$

35. $5\sqrt{75}$

36. $3\sqrt{8}$

37. $\sqrt{24}$

38. $\sqrt{20}$

39. $\sqrt{100x^5}$

40. $\sqrt{64y^9}$

41. $\sqrt[3]{16y^7}$

42. $\sqrt[3]{64y^9}$

43. $\sqrt[4]{a^8 b^7}$ **44.** $\sqrt[5]{32 z^{12}}$ **45.** $\sqrt{y^5}$ **46.** $\sqrt[3]{y^5}$ ⊙ **47.** $\sqrt{25 a^2 b^3}$ **48.** $\sqrt{9 x^5 y^7}$

49. $\sqrt[5]{-32 x^{10} y}$ **50.** $\sqrt[5]{-243 z^9}$ **51.** $\sqrt[3]{50 x^{14}}$ **52.** $\sqrt[3]{40 y^{10}}$ **53.** $-\sqrt{32 a^8 b^7}$ **54.** $-\sqrt{20 a b^6}$

55. $\sqrt{9 x^7 y^9}$ **56.** $\sqrt{12 r^9 s^{12}}$ ⊙ **57.** $\sqrt[3]{125 r^9 s^{12}}$ **58.** $\sqrt[3]{8 a^6 b^9}$

Use the quotient rule to divide. Then simplify if possible. Assume that all variables represent positive real numbers. See Examples 17 through 20.

⊙ **59.** $\dfrac{\sqrt{14}}{\sqrt{7}}$

60. $\dfrac{\sqrt{45}}{\sqrt{9}}$

61. $\dfrac{\sqrt[3]{24}}{\sqrt[3]{3}}$

62. $\dfrac{\sqrt[3]{10}}{\sqrt[3]{2}}$

63. $\dfrac{5\sqrt[4]{48}}{\sqrt[4]{3}}$

64. $\dfrac{7\sqrt[4]{162}}{\sqrt[4]{2}}$

⊙ **65.** $\dfrac{\sqrt{x^5 y^3}}{\sqrt{xy}}$

66. $\dfrac{\sqrt{a^7 b^6}}{\sqrt{a^3 b^2}}$

67. $\dfrac{8\sqrt[3]{54 m^7}}{\sqrt[3]{2m}}$

68. $\dfrac{\sqrt[3]{128 x^3}}{-3\sqrt[3]{2x}}$

69. $\dfrac{3\sqrt{100 x^2}}{2\sqrt{2 x^{-1}}}$

70. $\dfrac{\sqrt{270 y^2}}{5\sqrt{3 y^{-4}}}$

71. $\dfrac{\sqrt[4]{96 a^{10} b^3}}{\sqrt[4]{3 a^2 b^3}}$

72. $\dfrac{\sqrt[5]{64 x^{10} y^3}}{\sqrt[5]{2 x^3 y^{-7}}}$

Objective Ⓓ *Find the distance between each pair of points. Give an exact distance and a three-decimal-place approximation. See Example 21.*

73. $(5, 1)$ and $(8, 5)$

74. $(2, 3)$ and $(14, 8)$

⊙ **75.** $(-3, 2)$ and $(1, -3)$

76. $(3, -2)$ and $(-4, 1)$

77. $(0, -\sqrt{2})$ and $(\sqrt{3}, 0)$

78. $(-\sqrt{5}, 0)$ and $(0, \sqrt{7})$

79. $(1.7, -3.6)$ and $(-8.6, 5.7)$

80. $(9.6, 2.5)$ and $(-1.9, -3.7)$

Find the midpoint of each line segment whose endpoints are given. See Example 22.

81. $(6, -8); (2, 4)$ **82.** $(3, 9); (7, 11)$ ⬤ **83.** $(-2, -1); (-8, 6)$ **84.** $(-3, -4); (6, -8)$

85. $\left(\dfrac{1}{2}, \dfrac{3}{8}\right); \left(-\dfrac{3}{2}, \dfrac{5}{8}\right)$ **86.** $\left(-\dfrac{2}{5}, \dfrac{7}{15}\right); \left(-\dfrac{2}{5}, -\dfrac{4}{15}\right)$ **87.** $(\sqrt{2}, 3\sqrt{5}); (\sqrt{2}, -2\sqrt{5})$ **88.** $(\sqrt{8}, -\sqrt{12}); (3\sqrt{2}, 7\sqrt{3})$

Review

Perform each indicated operation. See Sections 1.5 and 5.2.

89. $6x + 8x$ **90.** $(6x)(8x)$ **91.** $(2x + 3)(x - 5)$ **92.** $(2x + 3) + (x - 5)$

93. $9y^2 - 8y^2$ **94.** $(9y^2)(-8y^2)$ **95.** $(x - 4)^2$ **96.** $(2x + 1)^2$

Concept Extensions

Find and correct the error. See the Concept Check in this section.

97. $\dfrac{\sqrt[3]{64}}{\sqrt{64}} = \sqrt[3]{\dfrac{64}{64}} = \sqrt[3]{1} = 1$

98. $\dfrac{\sqrt[4]{16}}{\sqrt{4}} = \sqrt[4]{\dfrac{16}{4}} = \sqrt[4]{4}$

99. The formula for the radius r of a sphere with surface area A is given by $r = \sqrt{\dfrac{A}{4\pi}}$. Calculate the radius of a standard zorb whose outside surface area is 32.17 sq m. Round to the nearest tenth. (*Source:* Zorb, Ltd.)

100. Before Mount Vesuvius, a volcano in Italy, erupted violently in 79 A.D., its height was 4190 feet. Vesuvius was roughly cone shaped, and its base had a radius of approximately 25,200 feet. Use the formula $A = \pi r \sqrt{r^2 + h^2}$ for the surface area A of a cone with radius r and height h to approximate the surface area of this volcano before it erupted. (*Source:* Global Volcanism Network)

101. The owner of Knightime Video has determined that the demand equation for renting older released tapes is $F(x) = 0.6\sqrt{49 - x^2}$, where x is the price in dollars per two-day rental and $F(x)$ is the number of times the video is demanded per week.

  a. Approximate to one decimal place the demand per week of an older released video if the rental price is $3 per two-day rental.

  b. Approximate to one decimal place the demand per week of an older released video if the rental price is $5 per two-day rental.

  c. Explain how the owner of the video store can use this equation to predict the number of copies of each tape that should be in stock.

7.4 ADDING, SUBTRACTING, AND MULTIPLYING RADICAL EXPRESSIONS

Objective **A** Adding or Subtracting Radical Expressions

We have learned that the sum or difference of like terms can be simplified. To simplify these sums or differences, we use the distributive property. For example,

$$2x + 3x = (2 + 3)x = 5x$$

The distributive property can also be used to add *like radicals*.

Like Radicals

Radicals with the same index and the same radicand are **like radicals.** The example below shows how to use the distributive property to simplify an expression containing like radicals.

$$2\sqrt{7} + 3\sqrt{7} = (2 + 3)\sqrt{7} = 5\sqrt{7}$$

Like radicals

Helpful Hint

The expression

$$5\sqrt{7} - 3\sqrt{6}$$

does not contain like radicals and cannot be simplified further.

PRACTICE PROBLEMS 1–3

Add or subtract as indicated.

1. $5\sqrt{15} + 2\sqrt{15}$
2. $9\sqrt[3]{2y} - 15\sqrt[3]{2y}$
3. $6\sqrt{10} - 3\sqrt[3]{10}$

EXAMPLES Add or subtract as indicated.

1. $4\sqrt{11} + 8\sqrt{11} = (4 + 8)\sqrt{11} = 12\sqrt{11}$
2. $5\sqrt[3]{3x} - 7\sqrt[3]{3x} = (5 - 7)\sqrt[3]{3x} = -2\sqrt[3]{3x}$
3. $2\sqrt{7} + 2\sqrt[3]{7}$ This expression cannot be simplified since $2\sqrt{7}$ and $2\sqrt[3]{7}$ do not contain like radicals.

Work Practice Problems 1–3

PRACTICE PROBLEMS 4–8

Add or subtract as indicated. Assume that all variables represent positive real numbers.

4. $\sqrt{50} + 5\sqrt{18}$
5. $\sqrt[3]{24} - 4\sqrt[3]{192} + \sqrt[3]{3}$
6. $\sqrt{20x} - 6\sqrt{16x} + \sqrt{45x}$
7. $\sqrt[4]{32} + \sqrt{32}$
8. $\sqrt[3]{8y^5} + \sqrt[3]{27y^5}$

✔**Concept Check** True or false:

$$\sqrt{a} + \sqrt{b} = \sqrt{a + b}$$

Explain.

When adding or subtracting radicals, always check first to see whether any radicals can be simplified.

Answers

1. $7\sqrt{15}$, 2. $-6\sqrt[3]{2y}$,
3. $6\sqrt{10} - 3\sqrt[3]{10}$, 4. $20\sqrt{2}$,
5. $-13\sqrt[3]{3}$, 6. $5\sqrt{5x} - 24\sqrt{x}$,
7. $2\sqrt[4]{2} + 4\sqrt{2}$, 8. $5y\sqrt[3]{y^2}$

✔ **Concept Check Answer**
false; answers may vary

502

EXAMPLES Add or subtract as indicated. Assume that all variables represent positive real numbers.

4. $\sqrt{20} + 2\sqrt{45} = \sqrt{4 \cdot 5} + 2\sqrt{9 \cdot 5}$ — Factor 20 and 45.

$\qquad = \sqrt{4} \cdot \sqrt{5} + 2 \cdot \sqrt{9} \cdot \sqrt{5}$ — Use the product rule.

$\qquad = 2 \cdot \sqrt{5} + 2 \cdot 3 \cdot \sqrt{5}$ — Simplify $\sqrt{4}$ and $\sqrt{9}$.

$\qquad = 2\sqrt{5} + 6\sqrt{5}$ — Add like radicals.

$\qquad = 8\sqrt{5}$

5. $\sqrt[3]{54} - 5\sqrt[3]{16} + \sqrt[3]{2}$

$= \sqrt[3]{27} \cdot \sqrt[3]{2} - 5 \cdot \sqrt[3]{8} \cdot \sqrt[3]{2} + \sqrt[3]{2}$ Factor and use the product rule.

$= 3 \cdot \sqrt[3]{2} - 5 \cdot 2 \cdot \sqrt[3]{2} + \sqrt[3]{2}$ Simplify $\sqrt[3]{27}$ and $\sqrt[3]{8}$.

$= 3\sqrt[3]{2} - 10\sqrt[3]{2} + \sqrt[3]{2}$ Write $5 \cdot 2$ as 10.

$= -6\sqrt[3]{2}$ Combine like radicals.

6. $\sqrt{27x} - 2\sqrt{9x} + \sqrt{72x}$

$= \sqrt{9} \cdot \sqrt{3x} - 2 \cdot \sqrt{9} \cdot \sqrt{x} + \sqrt{36} \cdot \sqrt{2x}$ Factor and use the product rule.

$= 3 \cdot \sqrt{3x} - 2 \cdot 3 \cdot \sqrt{x} + 6 \cdot \sqrt{2x}$ Simplify $\sqrt{9}$ and $\sqrt{36}$.

$= 3\sqrt{3x} - 6\sqrt{x} + 6\sqrt{2x}$ Write $2 \cdot 3$ as 6.

7. $\sqrt[3]{98} + \sqrt{98} = \sqrt[3]{98} + \sqrt{49} \cdot \sqrt{2}$ Factor and use the product rule.

$= \sqrt[3]{98} + 7\sqrt{2}$ No further simplification is possible.

8. $\sqrt[3]{48y^4} + \sqrt[3]{6y^4} = \sqrt[3]{8y^3} \cdot \sqrt[3]{6y} + \sqrt[3]{y^3} \cdot \sqrt[3]{6y}$ Factor and use the product rule.

$= 2y\sqrt[3]{6y} + y\sqrt[3]{6y}$ Simplify $\sqrt[3]{8y^3}$ and $\sqrt[3]{y^3}$.

$= 3y\sqrt[3]{6y}$ Combine like radicals.

> **Helpful Hint**
> None of these terms contain like radicals. We can simplify no further.

■ **Work Practice Problems 4–8**

EXAMPLES Add or subtract as indicated. Assume that all variables represent positive real numbers.

9. $\dfrac{\sqrt{45}}{4} - \dfrac{\sqrt{5}}{3} = \dfrac{3\sqrt{5}}{4} - \dfrac{\sqrt{5}}{3}$ To subtract, notice that the LCD is 12.

$= \dfrac{3\sqrt{5} \cdot 3}{4 \cdot 3} - \dfrac{\sqrt{5} \cdot 4}{3 \cdot 4}$ Write each expression as an equivalent expression with a denominator of 12.

$= \dfrac{9\sqrt{5}}{12} - \dfrac{4\sqrt{5}}{12}$ Multiply factors in the numerators and the denominators.

$= \dfrac{5\sqrt{5}}{12}$ Subtract.

10. $\sqrt[3]{\dfrac{7x}{8}} + 2\sqrt[3]{7x} = \dfrac{\sqrt[3]{7x}}{\sqrt[3]{8}} + 2\sqrt[3]{7x}$ Use the quotient rule for radicals.

$= \dfrac{\sqrt[3]{7x}}{2} + 2\sqrt[3]{7x}$ Simplify.

$= \dfrac{\sqrt[3]{7x}}{2} + \dfrac{2\sqrt[3]{7x} \cdot 2}{2}$ Write each expression as an equivalent expression with a denominator of 2.

$= \dfrac{\sqrt[3]{7x}}{2} + \dfrac{4\sqrt[3]{7x}}{2}$

$= \dfrac{5\sqrt[3]{7x}}{2}$ Add.

■ **Work Practice Problems 9–10**

PRACTICE PROBLEMS 9–10

Add or subtract as indicated. Assume that all variables represent positive real numbers.

9. $\dfrac{\sqrt{75}}{9} - \dfrac{\sqrt{3}}{2}$

10. $\sqrt[3]{\dfrac{5x}{27}} + 4\sqrt[3]{5x}$

Objective B Multiplying Radical Expressions

We can multiply radical expressions by using many of the same properties used to multiply polynomial expressions. For instance, to multiply $\sqrt{2}(\sqrt{6} - 3\sqrt{2})$, we use the distributive property and multiply $\sqrt{2}$ by each term inside the parentheses.

$$\sqrt{2}(\sqrt{6} - 3\sqrt{2}) = \sqrt{2}(\sqrt{6}) - \sqrt{2}(3\sqrt{2}) \quad \text{Use the distributive property.}$$
$$= \sqrt{2 \cdot 6} - 3\sqrt{2 \cdot 2}$$
$$= \sqrt{2 \cdot 2 \cdot 3} - 3 \cdot 2 \quad \text{Use the product rule for radicals.}$$
$$= 2\sqrt{3} - 6$$

PRACTICE PROBLEM 11

Multiply: $\sqrt{2}(6 + \sqrt{10})$

EXAMPLE 11 Multiply: $\sqrt{3}(5 + \sqrt{30})$

Solution:

$$\sqrt{3}(5 + \sqrt{30}) = \sqrt{3}(5) + \sqrt{3}(\sqrt{30})$$
$$= 5\sqrt{3} + \sqrt{3 \cdot 30}$$
$$= 5\sqrt{3} + \sqrt{3 \cdot 3 \cdot 10}$$
$$= 5\sqrt{3} + 3\sqrt{10}$$

Work Practice Problem 11

PRACTICE PROBLEMS 12–15

Multiply. Assume that all variables represent positive real numbers.

12. $(\sqrt{3} - \sqrt{5})(\sqrt{2} + 7)$
13. $(\sqrt{5y} + 2)(\sqrt{5y} - 2)$
14. $(\sqrt{3} - 7)^2$
15. $(\sqrt{x + 1} + 2)^2$

EXAMPLES Multiply. Assume that all variables represent positive real numbers.

12. $(\sqrt{5} - \sqrt{6})(\sqrt{7} + 1) = \overset{\text{First}}{\sqrt{5} \cdot \sqrt{7}} + \overset{\text{Outer}}{\sqrt{5} \cdot 1} - \overset{\text{Inner}}{\sqrt{6} \cdot \sqrt{7}} - \overset{\text{Last}}{\sqrt{6} \cdot 1}$

$$= \sqrt{35} + \sqrt{5} - \sqrt{42} - \sqrt{6} \quad \begin{array}{l}\text{Using the FOIL order.}\\ \text{Simplify.}\end{array}$$

13. $(\sqrt{2x} + 5)(\sqrt{2x} - 5) = (\sqrt{2x})^2 - 5^2 \quad \begin{array}{l}\text{Multiply the sum and difference of}\\ \text{two terms: } (a + b)(a - b) = a^2 - b^2\end{array}$

$$= 2x - 25$$

14. $(\sqrt{3} - 1)^2 = (\sqrt{3})^2 - 2 \cdot \sqrt{3} \cdot 1 + 1^2 \quad \begin{array}{l}\text{Square the binomial:}\\ (a - b)^2 = a^2 - 2ab + b^2\end{array}$

$$= 3 - 2\sqrt{3} + 1$$
$$= 4 - 2\sqrt{3} \quad \begin{array}{l}\text{Square the binomial:}\\ (a + b)^2 = a^2 + 2ab + b^2\end{array}$$

15. $\underset{a}{(\sqrt{x - 3}} + \underset{b}{5)^2} = \underset{a^2}{(\sqrt{x - 3})^2} + \underset{+\,2\,\cdot}{2} \cdot \underset{a}{\sqrt{x - 3}} \cdot \underset{\cdot\, b\, +\, b^2}{5 + 5^2}$

$$= x - 3 + 10\sqrt{x - 3} + 25 \quad \text{Simplify.}$$
$$= x + 22 + 10\sqrt{x - 3} \quad \text{Combine like terms.}$$

Work Practice Problems 12–15

Answers

11. $6\sqrt{2} + 2\sqrt{5}$,
12. $\sqrt{6} + 7\sqrt{3} - \sqrt{10} - 7\sqrt{5}$,
13. $5y - 4$, **14.** $52 - 14\sqrt{3}$,
15. $x + 5 + 4\sqrt{x + 1}$

Mental Math

Simplify. Assume that all variables represent positive real numbers.

1. $2\sqrt{3} + 4\sqrt{3}$

2. $5\sqrt{7} + 3\sqrt{7}$

3. $8\sqrt{x} - 5\sqrt{x}$

4. $3\sqrt{y} + 10\sqrt{y}$

5. $7\sqrt[3]{x} + 5\sqrt[3]{x}$

6. $8\sqrt[3]{z} - 2\sqrt[3]{z}$

7. $(\sqrt{3})^2$

8. $(\sqrt{4x+1})^2$

7.4 EXERCISE SET

Objective A *Add or subtract as indicated. Assume that all variables represent positive real numbers. See Examples 1 through 10.*

1. $\sqrt{8} - \sqrt{32}$

2. $\sqrt{27} - \sqrt{75}$

3. $2\sqrt{2x^3} + 4x\sqrt{8x}$

4. $3\sqrt{45x^3} + x\sqrt{5x}$

 5. $2\sqrt{50} - 3\sqrt{125} + \sqrt{98}$

6. $4\sqrt{32} - \sqrt{18} + 2\sqrt{128}$

7. $\sqrt[3]{16x} - \sqrt[3]{54x}$

8. $2\sqrt[3]{3a^4} - 3a\sqrt[3]{81a}$

9. $\sqrt{9b^3} - \sqrt{25b^3} + \sqrt{49b^3}$

10. $\sqrt{4x^7} + 9x^2\sqrt{x^3} - 5x\sqrt{x^5}$

11. $\dfrac{5\sqrt{2}}{3} + \dfrac{2\sqrt{2}}{5}$

12. $\dfrac{\sqrt{3}}{2} + \dfrac{4\sqrt{3}}{3}$

13. $\sqrt[3]{\dfrac{11}{8}} - \dfrac{\sqrt[3]{11}}{6}$

14. $\dfrac{2\sqrt[3]{4}}{7} - \dfrac{\sqrt[3]{4}}{14}$

15. $\dfrac{\sqrt{20x}}{9} + \sqrt{\dfrac{5x}{9}}$

16. $\dfrac{3x\sqrt{7}}{5} + \sqrt{\dfrac{7x^2}{100}}$

17. $7\sqrt{9} - 7 + \sqrt{3}$

18. $\sqrt{16} - 5\sqrt{10} + 7$

19. $2 + 3\sqrt{y^2} - 6\sqrt{y^2} + 5$

20. $3\sqrt{7} - \sqrt[3]{x} + 4\sqrt{7} - 3\sqrt[3]{x}$

21. $3\sqrt{108} - 2\sqrt{18} - 3\sqrt{48}$

22. $-\sqrt{75} + \sqrt{12} - 3\sqrt{3}$

23. $-5\sqrt[3]{625} + \sqrt[3]{40}$

24. $-2\sqrt[3]{108} - \sqrt[3]{32}$

25. $\sqrt{9b^3} - \sqrt{25b^3} + \sqrt{16b^3}$

26. $\sqrt{4x^7y^5} + 9x^2\sqrt{x^3y^5} - 5xy\sqrt{x^5y^3}$

27. $5y\sqrt{8y} + 2\sqrt{50y^3}$

28. $3\sqrt{8x^2y^3} - 2x\sqrt{32y^3}$

29. $\sqrt[3]{54xy^3} - 5\sqrt[3]{2xy^3} + y\sqrt[3]{128x}$

30. $2\sqrt[3]{24x^3y^4} + 4x\sqrt[3]{81y^4}$

31. $6\sqrt[3]{11} + 8\sqrt{11} - 12\sqrt{11}$

32. $3\sqrt[3]{5} + 4\sqrt{5}$

33. $-2\sqrt[4]{x^7} + 3\sqrt[4]{16x^7}$

34. $6\sqrt[3]{24x^3} - 2\sqrt[3]{81x^3} - x\sqrt[3]{3}$

35. $\dfrac{4\sqrt{3}}{3} - \dfrac{\sqrt{12}}{3}$

36. $\dfrac{\sqrt{45}}{10} + \dfrac{7\sqrt{5}}{10}$

37. $\dfrac{\sqrt[3]{8x^4}}{7} + \dfrac{3x\sqrt[3]{x}}{7}$

38. $\dfrac{\sqrt[4]{48}}{5x} - \dfrac{2\sqrt[4]{3}}{10x}$

39. $\sqrt{\dfrac{28}{x^2}} + \sqrt{\dfrac{7}{4x^2}}$

40. $\dfrac{\sqrt{99}}{5x} - \sqrt{\dfrac{44}{x^2}}$

41. $\sqrt[3]{\dfrac{16}{27}} - \dfrac{\sqrt[3]{54}}{6}$

42. $\dfrac{\sqrt[3]{3}}{10} + \sqrt[3]{\dfrac{24}{125}}$

43. $-\dfrac{\sqrt[3]{2x^4}}{9} + \sqrt[3]{\dfrac{250x^4}{27}}$

44. $\dfrac{\sqrt[3]{y^5}}{8} + \dfrac{5y\sqrt[3]{y^2}}{4}$

45. Find the perimeter of the trapezoid.

2√12 in., 3√3 in., √12 in., 2√27 in.

46. Find the perimeter of the triangle.

√8 m, √32 m, √45 m

Objective B *Multiply. Then simplify if possible. Assume that all variables represent positive real numbers. See Examples 11 through 15.*

47. $\sqrt{7}(\sqrt{5} + \sqrt{3})$

48. $\sqrt{5}(\sqrt{15} - \sqrt{35})$

49. $(\sqrt{5} - \sqrt{2})^2$

50. $(3x - \sqrt{2})(3x - \sqrt{2})$

51. $\sqrt{3x}(\sqrt{3} - \sqrt{x})$

52. $\sqrt{5y}(\sqrt{y} + \sqrt{5})$

53. $(2\sqrt{x} - 5)(3\sqrt{x} + 1)$

54. $(8\sqrt{y} + z)(4\sqrt{y} - 1)$

55. $(\sqrt[3]{a} - 4)(\sqrt[3]{a} + 5)$

56. $(\sqrt[3]{a} + 2)(\sqrt[3]{a} + 7)$

57. $6(\sqrt{2} - 2)$

58. $\sqrt{5}(6 - \sqrt{5})$

59. $\sqrt{2}(\sqrt{2} + x\sqrt{6})$

60. $\sqrt{3}(\sqrt{3} - 2\sqrt{5x})$

61. $(2\sqrt{7} + 3\sqrt{5})(\sqrt{7} - 2\sqrt{5})$

62. $(\sqrt{6} - 4\sqrt{2})(3\sqrt{6} + 1)$

63. $(\sqrt{x} - y)(\sqrt{x} + y)$

64. $(3\sqrt{x} + 2)(\sqrt{3x} - 2)$

65. $(\sqrt{3} + x)^2$

66. $(\sqrt{y} - 3x)^2$

67. $(\sqrt{5x} - 3\sqrt{2})(\sqrt{5x} - 3\sqrt{3})$

68. $(5\sqrt{3x} - \sqrt{y})(4\sqrt{x} + 1)$ **69.** $(\sqrt[3]{4} + 2)(\sqrt[3]{2} - 1)$ **70.** $(\sqrt[3]{3} + \sqrt[3]{2})(\sqrt[3]{9} - \sqrt[3]{4})$

71. $(\sqrt[3]{x} + 1)(\sqrt[3]{x} - 4\sqrt{x} + 7)$ **72.** $(\sqrt[3]{3x} + 3)(\sqrt[3]{2x} - 3x - 1)$

73. $(\sqrt{x - 1} + 5)^2$ **74.** $(\sqrt{3x + 1} + 2)^2$

75. $(\sqrt{2x + 5} - 1)^2$ **76.** $(\sqrt{x - 6} - 7)^2$

Review

Factor each numerator and denominator. Then simplify if possible. See Section 6.1.

77. $\dfrac{2x - 14}{2}$ **78.** $\dfrac{8x - 24y}{4}$ **79.** $\dfrac{7x - 7y}{x^2 - y^2}$ **80.** $\dfrac{x^3 - 8}{4x - 8}$

81. $\dfrac{6a^2b - 9ab}{3ab}$ **82.** $\dfrac{14r - 28r^2s^2}{7rs}$ **83.** $\dfrac{-4 + 2\sqrt{3}}{6}$ **84.** $\dfrac{-5 + 10\sqrt{7}}{5}$

Concept Extensions

△ **85.** Find the perimeter and area of the rectangle.

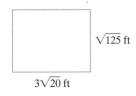

△ **86.** Find the area and perimeter of the trapezoid. (*Hint:* The area of a trapezoid is the product of half the height $6\sqrt{3}$ meters and the sum of the bases $2\sqrt{63}$ and $7\sqrt{7}$ meters.)

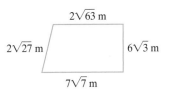

87. a. Add: $\sqrt{3} + \sqrt{3}$
  b. Multiply: $\sqrt{3} \cdot \sqrt{3}$
  c. Describe the differences in parts a and b.

88. Multiply: $(\sqrt{2} + \sqrt{3} - 1)^2$

STUDY SKILLS BUILDER

Have You Decided to Successfully Complete this Course?

Hopefully by now, one of your current goals is to successfully complete this course.

If it is not a goal of yours, ask yourself why? One common reason is fear of failure. Amazingly enough, fear of failure alone can be strong enough to keep many of us from doing our best in any endeavor. Another common reason is that you simply haven't taken the time to make successfully completing this course one of your goals.

Anytime you are registered for a course, successfully completing this course should probably be a goal. How do you do this? Start by writing this goal in your mathematics notebook. Then list steps you will take to ensure success. A great first step is to read or reread Section 1.1 and make a commitment to try the suggestions in this section.

Good luck and don't forget that a positive attitude will make a big difference.

Let's see how you are doing.

1. Have you made the decision to make "successfully completing this course" a goal of yours? If not, please list reasons that this has not happened. Study your list and talk to your instructor about this.

2. If your answer to Exercise 1 is yes, take a moment and list, in your notebook, further specific goals that will help you achieve this major goal of successfully completing this course. (For example, my goal this semester is not to miss any of my mathematics classes.)

3. Rate your commitment to this course with a number between 1 and 5. Use the diagram below to help.

High Commitment		Average Commitment		Not Committed at All
5	4	3	2	1

4. If you have rated your personal commitment level (from the exercise above) as a 1, 2, or 3, list the reasons why this is so. Then determine whether it is possible to increase your commitment level to a 4 or 5.

7.5 RATIONALIZING NUMERATORS AND DENOMINATORS OF RADICAL EXPRESSIONS

Objectives

Ⓐ Rationalize Denominators.

Ⓑ Rationalize Denominators Having Two Terms.

Ⓒ Rationalize Numerators.

Objective Ⓐ Rationalizing Denominators

Often in mathematics it is helpful to write a radical expression such as $\dfrac{\sqrt{3}}{\sqrt{2}}$ either without a radical in the denominator or without a radical in the numerator. The process of writing this expression as an equivalent expression but without a radical in the denominator is called **rationalizing the denominator.** To rationalize the denominator of $\dfrac{\sqrt{3}}{\sqrt{2}}$, we multiply the numerator and the denominator by $\sqrt{2}$. Recall that this is the same as multiplying by $\dfrac{\sqrt{2}}{\sqrt{2}}$, which simplifies to 1.

$$\frac{\sqrt{3}}{\sqrt{2}} = \frac{\sqrt{3}\cdot\sqrt{2}}{\sqrt{2}\cdot\sqrt{2}} = \frac{\sqrt{6}}{\sqrt{4}} = \frac{\sqrt{6}}{2}$$

EXAMPLE 1 Rationalize the denominator of $\dfrac{2}{\sqrt{5}}$.

Solution: To rationalize the denominator, we multiply the numerator and denominator by a factor that makes the radicand in the denominator a perfect square.

$$\frac{2}{\sqrt{5}} = \frac{2\cdot\sqrt{5}}{\sqrt{5}\cdot\sqrt{5}} = \frac{2\sqrt{5}}{5} \quad \text{The denominator is now rationalized.}$$

▣ **Work Practice Problem 1**

EXAMPLE 2 Rationalize the denominator of $\dfrac{2\sqrt{16}}{\sqrt{9x}}$.

Solution: First we simplify the radicals; then we rationalize the denominator.

$$\frac{2\sqrt{16}}{\sqrt{9x}} = \frac{2(4)}{\sqrt{9}\cdot\sqrt{x}} = \frac{8}{3\sqrt{x}}$$

To rationalize the denominator, we multiply the numerator and the denominator by $\sqrt{x}$.

$$\frac{8}{3\sqrt{x}} = \frac{8\cdot\sqrt{x}}{3\sqrt{x}\cdot\sqrt{x}} = \frac{8\sqrt{x}}{3x}$$

▣ **Work Practice Problem 2**

EXAMPLE 3 Rationalize the denominator of $\sqrt[3]{\dfrac{1}{2}}$.

Solution: $\sqrt[3]{\dfrac{1}{2}} = \dfrac{\sqrt[3]{1}}{\sqrt[3]{2}} = \dfrac{1}{\sqrt[3]{2}}$

Now we rationalize the denominator. Since $\sqrt[3]{2}$ is a cube root, we want to multiply by a value that will make the radicand 2 a perfect cube. If we multiply by $\sqrt[3]{2^2}$, we get $\sqrt[3]{2^3} = 2$. Thus,

$$\frac{1\cdot\sqrt[3]{2^2}}{\sqrt[3]{2}\cdot\sqrt[3]{2^2}} = \frac{\sqrt[3]{4}}{\sqrt[3]{2^3}} = \frac{\sqrt[3]{4}}{2} \quad \text{Multiply numerator and denominator by } \sqrt[3]{2^2} \text{ and then simplify.}$$

▣ **Work Practice Problem 3**

PRACTICE PROBLEM 1

Rationalize the denominator of $\dfrac{7}{\sqrt{2}}$.

PRACTICE PROBLEM 2

Rationalize the denominator of $\dfrac{2\sqrt{9}}{\sqrt{16y}}$.

PRACTICE PROBLEM 3

Rationalize the denominator of $\sqrt[3]{\dfrac{2}{25}}$.

Answers

1. $\dfrac{7\sqrt{2}}{2}$, 2. $\dfrac{3\sqrt{y}}{2y}$, 3. $\dfrac{\sqrt[3]{10}}{5}$

✔**Concept Check** Determine by which number both the numerator and denominator should be multiplied to rationalize the denominator of the radical expression.

a. $\dfrac{1}{\sqrt[3]{7}}$ **b.** $\dfrac{1}{\sqrt[4]{8}}$

PRACTICE PROBLEM 4

Rationalize the denominator of $\sqrt{\dfrac{5m}{11n}}$. Assume that all variables represent positive real numbers.

EXAMPLE 4 Rationalize the denominator of $\sqrt{\dfrac{7x}{3y}}$. Assume that all variables represent positive real numbers.

Solution:

$$\sqrt{\frac{7x}{3y}} = \frac{\sqrt{7x}}{\sqrt{3y}}$$ Use the quotient rule. No radical may be simplified further.

$$= \frac{\sqrt{7x} \cdot \sqrt{3y}}{\sqrt{3y} \cdot \sqrt{3y}}$$ Multiply numerator and denominator by $\sqrt{3y}$ so that the radicand in the denominator is a perfect square.

$$= \frac{\sqrt{21xy}}{3y}$$ Use the product rule in the numerator and denominator. Remember that $\sqrt{3y} \cdot \sqrt{3y} = 3y$.

▣ **Work Practice Problem 4**

PRACTICE PROBLEM 5

Rationalize the denominator of $\dfrac{\sqrt[5]{a^2}}{\sqrt[5]{32b^{12}}}$. Assume that all variables represent positive real numbers.

EXAMPLE 5 Rationalize the denominator of $\dfrac{\sqrt[4]{x}}{\sqrt[4]{81y^5}}$. Assume that all variables represent positive real numbers.

Solution: First we simplify each radical if possible.

$$\frac{\sqrt[4]{x}}{\sqrt[4]{81y^5}} = \frac{\sqrt[4]{x}}{\sqrt[4]{81y^4} \cdot \sqrt[4]{y}}$$ Use the product rule in the denominator.

$$= \frac{\sqrt[4]{x}}{3y\sqrt[4]{y}}$$ Write $\sqrt[4]{81y^4}$ as $3y$.

$$= \frac{\sqrt[4]{x} \cdot \sqrt[4]{y^3}}{3y\sqrt[4]{y} \cdot \sqrt[4]{y^3}}$$ Multiply numerator and denominator by $\sqrt[4]{y^3}$ so that the radicand in the denominator is a perfect 4th power.

$$= \frac{\sqrt[4]{xy^3}}{3y\sqrt[4]{y^4}}$$ Use the product rule in the numerator and denominator.

$$= \frac{\sqrt[4]{xy^3}}{3y^2}$$ In the denominator, $\sqrt[4]{y^4} = y$ and $3y \cdot y = 3y^2$.

▣ **Work Practice Problem 5**

Objective ⬛ **Rationalizing Denominators Having Two Terms**

Remember the product of the sum and difference of two terms?

$$(a + b)(a - b) = a^2 - b^2$$

These two expressions are called **conjugates** of each other.

Answers

4. $\dfrac{\sqrt{55mn}}{11n}$, **5.** $\dfrac{\sqrt[5]{a^2b^3}}{2b^3}$

✔ **Concept Check Answers**

a. $\sqrt[3]{7^2}$ or $\sqrt[3]{49}$, **b.** $\sqrt[4]{2}$

To rationalize a denominator that is a sum or difference of two terms, we use conjugates. To see how and why this works, let's rationalize the denominator of the expression $\dfrac{5}{\sqrt{3} - 2}$. To do so, we multiply both the numerator and the denominator by $\sqrt{3} + 2$, the *conjugate* of the denominator $\sqrt{3} - 2$, and see what happens.

$$\frac{5}{\sqrt{3} - 2} = \frac{5(\sqrt{3} + 2)}{(\sqrt{3} - 2)(\sqrt{3} + 2)}$$

$$= \frac{5(\sqrt{3} + 2)}{(\sqrt{3})^2 - 2^2} \qquad \text{Multiply the sum and difference of two terms:}$$
$$\text{} \qquad \qquad \qquad (a + b)(a - b) = a^2 - b^2.$$

$$= \frac{5(\sqrt{3} + 2)}{3 - 4}$$

$$= \frac{5(\sqrt{3} + 2)}{-1}$$

$$= -5(\sqrt{3} + 2) \quad \text{or} \quad -5\sqrt{3} - 10$$

Notice in the denominator that the product of $(\sqrt{3} - 2)$ and its conjugate, $(\sqrt{3} + 2)$, is -1. In general, the product of an expression and its conjugate will contain no radical terms. This is why, when rationalizing a denominator or a numerator containing two terms, we multiply by its conjugate. Examples of conjugates are

$$\sqrt{a} - \sqrt{b} \qquad \text{and} \qquad \sqrt{a} + \sqrt{b}$$

$$x + \sqrt{y} \qquad \text{and} \qquad x - \sqrt{y}$$

EXAMPLE 6 Rationalize the denominator of $\dfrac{2}{3\sqrt{2} + 4}$.

Solution: We multiply the numerator and the denominator by the conjugate of $3\sqrt{2} + 4$.

$$\frac{2}{3\sqrt{2} + 4} = \frac{2(3\sqrt{2} - 4)}{(3\sqrt{2} + 4)(3\sqrt{2} - 4)}$$

$$= \frac{2(3\sqrt{2} - 4)}{(3\sqrt{2})^2 - 4^2} \qquad \text{Multiply the sum and difference}$$
$$\text{} \qquad \qquad \qquad \text{of two terms: } (a + b)(a - b) = a^2 - b^2.$$

$$= \frac{2(3\sqrt{2} - 4)}{18 - 16} \qquad \text{Write } (3\sqrt{2})^2 \text{ as } 9 \cdot 2 \text{ or } 18 \text{ and } 4^2 \text{ as } 16.$$

$$= \frac{2(3\sqrt{2} - 4)}{2} = 3\sqrt{2} - 4$$

Work Practice Problem 6

As we saw in Example 6, it is often helpful to leave a numerator in factored form to help determine whether the expression can be simplified.

PRACTICE PROBLEM 6

Rationalize the denominator of $\dfrac{3}{2\sqrt{5} + 1}$.

Answer

6. $\dfrac{3(2\sqrt{5} - 1)}{19}$

PRACTICE PROBLEM 7

Rationalize the denominator

of $\dfrac{\sqrt{5}+3}{\sqrt{3}-\sqrt{2}}$.

EXAMPLE 7 Rationalize the denominator of $\dfrac{\sqrt{6}+2}{\sqrt{5}-\sqrt{3}}$.

Solution: We multiply the numerator and the denominator by the conjugate of $\sqrt{5}-\sqrt{3}$.

$$\frac{\sqrt{6}+2}{\sqrt{5}-\sqrt{3}} = \frac{(\sqrt{6}+2)(\sqrt{5}+\sqrt{3})}{(\sqrt{5}-\sqrt{3})(\sqrt{5}+\sqrt{3})}$$

$$= \frac{\sqrt{6}\sqrt{5}+\sqrt{6}\sqrt{3}+2\sqrt{5}+2\sqrt{3}}{(\sqrt{5})^2-(\sqrt{3})^2}$$

$$= \frac{\sqrt{30}+\sqrt{18}+2\sqrt{5}+2\sqrt{3}}{5-3}$$

$$= \frac{\sqrt{30}+3\sqrt{2}+2\sqrt{5}+2\sqrt{3}}{2}$$

🔲 **Work Practice Problem 7**

PRACTICE PROBLEM 8

Rationalize the denominator

of $\dfrac{3}{2-\sqrt{x}}$. Assume that all

variables represent positive real numbers.

EXAMPLE 8 Rationalize the denominator of $\dfrac{2\sqrt{m}}{3\sqrt{x}+\sqrt{m}}$. Assume that all variables represent positive real numbers.

Solution: We multiply by the conjugate of $3\sqrt{x}+\sqrt{m}$ to eliminate the radicals from the denominator.

$$\frac{2\sqrt{m}}{3\sqrt{x}+\sqrt{m}} = \frac{2\sqrt{m}(3\sqrt{x}-\sqrt{m})}{(3\sqrt{x}+\sqrt{m})(3\sqrt{x}-\sqrt{m})} = \frac{6\sqrt{mx}-2m}{(3\sqrt{x})^2-(\sqrt{m})^2}$$

$$= \frac{6\sqrt{mx}-2m}{9x-m}$$

🔲 **Work Practice Problem 8**

Objective **C** **Rationalizing Numerators**

As mentioned earlier, it is also often helpful to write an expression such as $\dfrac{\sqrt{3}}{\sqrt{2}}$ as an equivalent expression without a radical in the numerator. This process is called **rationalizing the numerator.** To rationalize the numerator of $\dfrac{\sqrt{3}}{\sqrt{2}}$, we multiply the numerator and the denominator by $\sqrt{3}$.

$$\frac{\sqrt{3}}{\sqrt{2}} = \frac{\sqrt{3}\cdot\sqrt{3}}{\sqrt{2}\cdot\sqrt{3}} = \frac{\sqrt{9}}{\sqrt{6}} = \frac{3}{\sqrt{6}}$$

PRACTICE PROBLEM 9

Rationalize the numerator

of $\dfrac{\sqrt{18}}{\sqrt{75}}$.

EXAMPLE 9 Rationalize the numerator of $\dfrac{\sqrt{7}}{\sqrt{45}}$.

Solution: First we simplify $\sqrt{45}$.

$$\frac{\sqrt{7}}{\sqrt{45}} = \frac{\sqrt{7}}{\sqrt{9\cdot 5}} = \frac{\sqrt{7}}{3\sqrt{5}}$$

Next we rationalize the numerator by multiplying the numerator and the denominator by $\sqrt{7}$.

$$\frac{\sqrt{7}}{3\sqrt{5}} = \frac{\sqrt{7}\cdot\sqrt{7}}{3\sqrt{5}\cdot\sqrt{7}} = \frac{7}{3\sqrt{5}\cdot 7} = \frac{7}{3\sqrt{35}}$$

🔲 **Work Practice Problem 9**

Answers

7. $\sqrt{15}+\sqrt{10}+3\sqrt{3}+3\sqrt{2}$,

8. $\dfrac{6+3\sqrt{x}}{4-x}$, **9.** $\dfrac{6}{5\sqrt{6}}$

EXAMPLE 10 Rationalize the numerator of $\dfrac{\sqrt[3]{2x^2}}{\sqrt[3]{5y}}$.

Solution:

$$\dfrac{\sqrt[3]{2x^2}}{\sqrt[3]{5y}} = \dfrac{\sqrt[3]{2x^2} \cdot \sqrt[3]{2^2x}}{\sqrt[3]{5y} \cdot \sqrt[3]{2^2x}}$$ Multiply the numerator and denominator by $\sqrt[3]{2^2x}$ so that the radicand in the numerator is a perfect cube.

$$= \dfrac{\sqrt[3]{2^3x^3}}{\sqrt[3]{5y \cdot 2^2x}}$$ Use the product rule in the numerator and denominator.

$$= \dfrac{2x}{\sqrt[3]{20xy}}$$ Simplify.

■ **Work Practice Problem 10**

Just as for denominators, to rationalize a numerator that is a sum or difference of two terms, we use conjugates.

EXAMPLE 11 Rationalize the numerator of $\dfrac{\sqrt{x} + 2}{5}$. Assume that all variables represent positive real numbers.

Solution: We multiply the numerator and the denominator by the conjugate of $\sqrt{x} + 2$, the numerator.

$$\dfrac{\sqrt{x} + 2}{5} = \dfrac{(\sqrt{x} + 2)(\sqrt{x} - 2)}{5(\sqrt{x} - 2)}$$ Multiply by $\sqrt{x} - 2$, the conjugate of $\sqrt{x} + 2$.

$$= \dfrac{(\sqrt{x})^2 - 2^2}{5(\sqrt{x} - 2)}$$ $(a + b)(a - b) = a^2 - b^2$.

$$= \dfrac{x - 4}{5(\sqrt{x} - 2)}$$

■ **Work Practice Problem 11**

PRACTICE PROBLEM 10

Rationalize the numerator of $\dfrac{\sqrt[3]{3a}}{\sqrt[3]{7b}}$.

PRACTICE PROBLEM 11

Rationalize the numerator of $\dfrac{\sqrt{x} + 5}{3}$. Assume that all variables represent positive real numbers.

Answers

10. $\dfrac{3a}{\sqrt[3]{63a^2b}}$,   11. $\dfrac{x - 25}{3(\sqrt{x} - 5)}$

Mental Math

Find the conjugate of each expression.

1. $\sqrt{2} + x$

2. $\sqrt{3} + y$

3. $5 - \sqrt{a}$

4. $6 - \sqrt{b}$

5. $7\sqrt{4} + 8\sqrt{x}$

6. $9\sqrt{2} - 6\sqrt{y}$

7.5 EXERCISE SET

Objective A *Rationalize each denominator. Assume that all variables represent positive real numbers. See Examples 1 through 5.*

1. $\dfrac{\sqrt{2}}{\sqrt{7}}$

2. $\dfrac{\sqrt{3}}{\sqrt{2}}$

3. $\sqrt{\dfrac{1}{5}}$

4. $\sqrt{\dfrac{1}{2}}$

5. $\dfrac{4}{\sqrt[3]{3}}$

6. $\dfrac{6}{\sqrt[3]{9}}$

7. $\dfrac{3}{\sqrt{8x}}$

8. $\dfrac{5}{\sqrt{27a}}$

9. $\dfrac{3}{\sqrt[3]{4x^2}}$

10. $\dfrac{5}{\sqrt[3]{3y}}$

11. $\dfrac{9}{\sqrt{3a}}$

12. $\dfrac{x}{\sqrt{5}}$

13. $\dfrac{3}{\sqrt[3]{2}}$

14. $\dfrac{5}{\sqrt[3]{9}}$

15. $\dfrac{2\sqrt{3}}{\sqrt{7}}$

16. $\dfrac{-5\sqrt{2}}{\sqrt{11}}$

17. $\sqrt{\dfrac{2x}{5y}}$

18. $\sqrt{\dfrac{13a}{2b}}$

19. $\sqrt[3]{\dfrac{3}{5}}$

20. $\sqrt[3]{\dfrac{7}{10}}$

21. $\sqrt{\dfrac{3x}{50}}$

22. $\sqrt{\dfrac{11y}{45}}$

23. $\dfrac{1}{\sqrt{12z}}$

24. $\dfrac{1}{\sqrt{32x}}$

25. $\dfrac{\sqrt[3]{2y^2}}{\sqrt[3]{9x^2}}$

26. $\dfrac{\sqrt[3]{3x}}{\sqrt[3]{4y^4}}$

27. $\sqrt[4]{\dfrac{16}{9x^7}}$

28. $\sqrt[5]{\dfrac{32}{m^6 n^{13}}}$

29. $\dfrac{5a}{\sqrt[5]{8a^9 b^{11}}}$

30. $\dfrac{9y}{\sqrt[4]{4y^9}}$

Objective B *Rationalize each denominator. Assume that all variables represent positive real numbers. See Examples 6 through 8.*

31. $\dfrac{6}{2 - \sqrt{7}}$

32. $\dfrac{3}{\sqrt{7} - 4}$

33. $\dfrac{-7}{\sqrt{x} - 3}$

34. $\dfrac{-8}{\sqrt{y} + 4}$

35. $\dfrac{\sqrt{2} - \sqrt{3}}{\sqrt{2} + \sqrt{3}}$

36. $\dfrac{\sqrt{3} + \sqrt{4}}{\sqrt{2} + \sqrt{3}}$

37. $\dfrac{\sqrt{a}+1}{2\sqrt{a}-\sqrt{b}}$

38. $\dfrac{2\sqrt{a}-3}{2\sqrt{a}-\sqrt{b}}$

39. $\dfrac{8}{1+\sqrt{10}}$

40. $\dfrac{-3}{\sqrt{6}-2}$

41. $\dfrac{\sqrt{x}}{\sqrt{x}+\sqrt{y}}$

42. $\dfrac{2\sqrt{a}}{2\sqrt{x}-\sqrt{y}}$

43. $\dfrac{2\sqrt{3}+\sqrt{6}}{4\sqrt{3}-\sqrt{6}}$

44. $\dfrac{4\sqrt{5}+\sqrt{2}}{2\sqrt{5}-\sqrt{2}}$

Objective C *Rationalize each numerator. Assume that all variables represent positive real numbers. See Examples 9 and 10.*

45. $\sqrt{\dfrac{5}{3}}$

46. $\sqrt{\dfrac{3}{2}}$

47. $\sqrt{\dfrac{18}{5}}$

48. $\sqrt{\dfrac{12}{7}}$

49. $\dfrac{\sqrt{4x}}{7}$

50. $\dfrac{\sqrt{3x^5}}{6}$

51. $\dfrac{\sqrt[3]{5y^2}}{\sqrt[3]{4x}}$

52. $\dfrac{\sqrt[3]{4x}}{\sqrt[3]{z^4}}$

53. $\sqrt{\dfrac{2}{5}}$

54. $\sqrt{\dfrac{3}{7}}$

55. $\dfrac{\sqrt{2x}}{11}$

56. $\dfrac{\sqrt{y}}{7}$

57. $\sqrt[3]{\dfrac{7}{8}}$

58. $\sqrt[3]{\dfrac{25}{2}}$

59. $\dfrac{\sqrt[3]{3x^5}}{10}$

60. $\sqrt[3]{\dfrac{9y}{7}}$

61. $\sqrt{\dfrac{18x^4y^6}{3z}}$

62. $\sqrt{\dfrac{8x^5y}{2z}}$

63. When rationalizing the denominator of $\dfrac{\sqrt{5}}{\sqrt{7}}$, explain why both the numerator and the denominator must be multiplied by $\sqrt{7}$.

64. When rationalizing the numerator of $\dfrac{\sqrt{5}}{\sqrt{7}}$, explain why both the numerator and the denominator must be multiplied by $\sqrt{5}$.

Rationalize each numerator. Assume that all variables represent positive real numbers. See Example 11.

65. $\dfrac{2-\sqrt{11}}{6}$

66. $\dfrac{\sqrt{15}+1}{2}$

67. $\dfrac{2-\sqrt{7}}{-5}$

68. $\dfrac{\sqrt{5}+2}{\sqrt{2}}$

69. $\dfrac{\sqrt{x}+3}{\sqrt{x}}$

70. $\dfrac{5+\sqrt{2}}{\sqrt{2x}}$

71. $\dfrac{\sqrt{x}+1}{\sqrt{x}-1}$

72. $\dfrac{\sqrt{x}+\sqrt{y}}{\sqrt{x}-\sqrt{y}}$

Review

Solve each equation. See Sections 2.1 and 5.7.

73. $2x - 7 = 3(x - 4)$ **74.** $9x - 4 = 7(x - 2)$ **75.** $(x - 6)(2x + 1) = 0$

76. $(y + 2)(5y + 4) = 0$ **77.** $x^2 - 8x = -12$ **78.** $x^3 = x$

Concept Extensions

△ **79.** The formula of the radius r of a sphere with surface area A is

$$r = \sqrt{\frac{A}{4\pi}}$$

Rationalize the denominator of the radical expression in this formula.

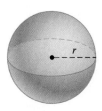

△ **80.** The formula for the radius r of a cone with height 7 centimeters and volume V is

$$r = \sqrt{\frac{3V}{7\pi}}$$

Rationalize the numerator of the radical expression in this formula.

7 cm

81. Explain why rationalizing the denominator does not change the value of the original expression.

82. Explain why rationalizing the numerator does not change the value of the original expression.

Determine the smallest number both the numerator and denominator should by multiplied by to rationalize the denominator of the radical expression. See the Concept Check in this section.

83. $\dfrac{9}{\sqrt[3]{5}}$ **84.** $\dfrac{5}{\sqrt{27}}$

Radicals and Rational Exponents

Throughout this review, assume that all variables represent positive real numbers.
Find each root.

1. $\sqrt{81}$

2. $\sqrt[3]{-8}$

3. $\sqrt[4]{\dfrac{1}{16}}$

4. $\sqrt{x^6}$

5. $\sqrt[3]{y^9}$

6. $\sqrt{4y^{10}}$

7. $\sqrt[5]{-32y^5}$

8. $\sqrt[4]{81b^{12}}$

Use radical notation to rewrite each expression. Simplify if possible.

9. $36^{1/2}$

10. $(3y)^{1/4}$

11. $64^{-2/3}$

12. $(x+1)^{3/5}$

Use the properties of exponents to simplify each expression. Write with positive exponents.

13. $y^{-1/6} \cdot y^{7/6}$

14. $\dfrac{(2x^{1/3})^4}{x^{5/6}}$

15. $\dfrac{x^{1/4}x^{3/4}}{x^{-1/4}}$

16. $4^{1/3} \cdot 4^{2/5}$

Use rational exponents to simplify each radical.

17. $\sqrt[3]{8x^6}$

18. $\sqrt[12]{a^9b^6}$

Use rational exponents to write each as a single radical expression.

19. $\sqrt[4]{x} \cdot \sqrt{x}$

20. $\sqrt{5} \cdot \sqrt[3]{2}$

Simplify.

21. $\sqrt{40}$

22. $\sqrt[4]{16x^7y^{10}}$

23. $\sqrt[3]{54x^4}$

24. $\sqrt[5]{-64b^{10}}$

Answers

1. _____
2. _____
3. _____
4. _____
5. _____
6. _____
7. _____
8. _____
9. _____
10. _____
11. _____
12. _____
13. _____
14. _____
15. _____
16. _____
17. _____
18. _____
19. _____
20. _____
21. _____
22. _____
23. _____
24. _____

25. _____

26. _____

27. _____

28. _____

29. _____

30. _____

31. _____

32. _____

33. _____

34. _____

35. _____

36. _____

37. _____

38. _____

39. _____

40. _____

Multiply or divide. Then simplify if possible.

25. $\sqrt{5} \cdot \sqrt{x}$

26. $\sqrt[3]{8x} \cdot \sqrt[3]{8x^2}$

27. $\dfrac{\sqrt{98y^6}}{\sqrt{2y}}$

28. $\dfrac{\sqrt[4]{48a^9b^3}}{\sqrt[4]{ab^3}}$

Perform each indicated operation.

29. $\sqrt{20} - \sqrt{75} + 5\sqrt{7}$

30. $\sqrt[3]{54y^4} - y\sqrt[3]{16y}$

31. $\sqrt{3}(\sqrt{5} - \sqrt{2})$

32. $(\sqrt{7} + \sqrt{3})^2$

33. $(2x - \sqrt{5})(2x + \sqrt{5})$

34. $(\sqrt{x+1} - 1)^2$

Rationalize each denominator.

35. $\sqrt{\dfrac{7}{3}}$

36. $\dfrac{5}{\sqrt[3]{2x^2}}$

37. $\dfrac{\sqrt{3} - \sqrt{7}}{2\sqrt{3} + \sqrt{7}}$

Rationalize each numerator.

38. $\sqrt{\dfrac{7}{3}}$

39. $\sqrt[3]{\dfrac{9y}{11}}$

40. $\dfrac{\sqrt{x} - 2}{\sqrt{x}}$

RADICAL EQUATIONS AND PROBLEM SOLVING

Objective **A** **Solving Equations That Contain Radical Expressions**

In this section, we present techniques to solve equations containing radical expressions such as

$$\sqrt{2x - 3} = 9$$

We use the power rule to help us solve these radical equations.

Power Rule

If both sides of an equation are raised to the same power, *all* solutions of the original equation are *among* the solutions of the new equation.

This property *does not* say that raising both sides of an equation to a power yields an equivalent equation. A solution of the new equation *may* or *may not* be a solution of the original equation. Thus, *each solution of the new equation must be checked* to make sure it is a solution of the original equation. Recall that a proposed solution that is not a solution of the original equation is called an extraneous solution.

EXAMPLE 1 Solve: $\sqrt{2x - 3} = 9$

Solution: We use the power rule to square both sides of the equation to eliminate the radical.

$$\sqrt{2x - 3} = 9$$
$$(\sqrt{2x - 3})^2 = 9$$
$$2x - 3 = 81$$
$$2x = 84$$
$$x = 42$$

Now we check the solution in the original equation.

Check: $\sqrt{2x - 3} = 9$

$$\sqrt{2(42) - 3} \stackrel{?}{=} 9 \quad \text{Let } x = 42.$$
$$\sqrt{84 - 3} \stackrel{?}{=} 9$$
$$\sqrt{81} \stackrel{?}{=} 9$$
$$9 = 9 \quad \text{True}$$

The solution checks, so we conclude that the solution set is $\{42\}$.

Work Practice Problem 1

To solve a radical equation, first isolate a radical on one side of the equation.

PRACTICE PROBLEM 1

Solve: $\sqrt{3x - 2} = 5$

Answer
1. $\{9\}$

PRACTICE PROBLEM 2
Solve: $\sqrt{9x - 2} - 2x = 0$

EXAMPLE 2   Solve: $\sqrt{-10x - 1} + 3x = 0$

Solution: First we isolate the radical on one side of the equation. To do this, we subtract $3x$ from both sides.

$$\sqrt{-10x - 1} + 3x = 0$$
$$\sqrt{-10x - 1} + 3x - 3x = 0 - 3x$$
$$\sqrt{-10x - 1} = -3x$$

Next we use the power rule to eliminate the radical.

$$(\sqrt{-10x - 1})^2 = (-3x)^2$$
$$-10x - 1 = 9x^2$$

Since this is a quadratic equation, we can set the equation equal to 0 and try to solve by factoring.

$$9x^2 + 10x + 1 = 0$$
$$(9x + 1)(x + 1) = 0 \qquad \text{Factor.}$$
$$9x + 1 = 0 \quad \text{or} \quad x + 1 = 0 \qquad \text{Set each factor equal to 0.}$$
$$x = -\frac{1}{9} \qquad\qquad x = -1$$

Check: Let $x = -\frac{1}{9}$.   Let $x = -1$.

$$\sqrt{-10x - 1} + 3x = 0 \qquad\qquad \sqrt{-10x - 1} + 3x = 0$$

$$\sqrt{-10\left(-\frac{1}{9}\right) - 1} + 3\left(-\frac{1}{9}\right) \stackrel{?}{=} 0 \qquad \sqrt{-10(-1) - 1} + 3(-1) \stackrel{?}{=} 0$$

$$\sqrt{\frac{10}{9} - \frac{9}{9}} - \frac{3}{9} \stackrel{?}{=} 0 \qquad\qquad \sqrt{10 - 1} - 3 \stackrel{?}{=} 0$$

$$\sqrt{\frac{1}{9}} - \frac{1}{3} \stackrel{?}{=} 0 \qquad\qquad\qquad \sqrt{9} - 3 \stackrel{?}{=} 0$$

$$\frac{1}{3} - \frac{1}{3} = 0 \quad \text{True} \qquad\qquad 3 - 3 = 0 \quad \text{True}$$

Both solutions check. The solution set is $\left\{-\frac{1}{9}, -1\right\}$.

⬛ **Work Practice Problem 2**

The following steps may be used to solve a radical equation.

Solving a Radical Equation

Step 1: Isolate one radical on one side of the equation.

Step 2: Raise each side of the equation to a power equal to the index of the radical and simplify.

Step 3: If the equation still contains a radical term, repeat Steps 1 and 2. If not, solve the equation.

Step 4: Check all proposed solutions in the original equation.

PRACTICE PROBLEM 3
Solve: $\sqrt[3]{x - 5} + 2 = 1$

EXAMPLE 3   Solve: $\sqrt[3]{x + 1} + 5 = 3$

Solution: First we isolate the radical by subtracting 5 from both sides of the equation.

$$\sqrt[3]{x + 1} + 5 = 3$$
$$\sqrt[3]{x + 1} = -2$$

Answers

2. $\left\{\frac{1}{4}, 2\right\}$,   **3.** $\{4\}$

Next we raise both sides of the equation to the third power to eliminate the radical.

$$(\sqrt[3]{x + 1})^3 = (-2)^3$$
$$x + 1 = -8$$
$$x = -9$$

The solution checks in the original equation, so the solution set is $\{-9\}$.

🖳 **Work Practice Problem 3**

EXAMPLE 4 Solve: $\sqrt{4 - x} = x - 2$

Solution:
$$\sqrt{4 - x} = x - 2$$
$$(\sqrt{4 - x})^2 = (x - 2)^2$$
$$4 - x = x^2 - 4x + 4$$
$$x^2 - 3x = 0 \qquad \text{Write the quadratic equation in standard form.}$$
$$x(x - 3) = 0 \qquad \text{Factor.}$$
$$x = 0 \quad \text{or} \quad x - 3 = 0 \qquad \text{Set each factor equal to 0.}$$
$$x = 3$$

Check:

$$\sqrt{4 - x} = x - 2 \qquad\qquad \sqrt{4 - x} = x - 2$$
$$\sqrt{4 - 0} \overset{?}{=} 0 - 2 \quad \text{Let } x = 0. \qquad \sqrt{4 - 3} \overset{?}{=} 3 - 2 \quad \text{Let } x = 3.$$
$$2 = -2 \quad \text{False} \qquad\qquad\qquad 1 = 1 \quad \text{True}$$

The proposed solution 3 checks, but 0 does not. Since 0 is an extraneous solution, the solution set is $\{3\}$.

🖳 **Work Practice Problem 4**

Helpful Hint

In Example 4, notice that $(x - 2)^2 = x^2 - 4x + 4$. Make sure binomials are squared correctly.

✔**Concept Check** How can you immediately tell that the equation $\sqrt{2y + 3} = -4$ has no real solution?

EXAMPLE 5 Solve: $\sqrt{2x + 5} + \sqrt{2x} = 3$

Solution: We get one radical alone by subtracting $\sqrt{2x}$ from both sides.

$$\sqrt{2x + 5} + \sqrt{2x} = 3$$
$$\sqrt{2x + 5} = 3 - \sqrt{2x}$$

Now we use the power rule to begin eliminating the radicals. First we square both sides.

$$(\sqrt{2x + 5})^2 = (3 - \sqrt{2x})^2$$
$$2x + 5 = 9 - 6\sqrt{2x} + 2x \quad \text{Multiply: } (3 - \sqrt{2x})(3 - \sqrt{2x})$$

Continued on next page

PRACTICE PROBLEM 4
Solve: $\sqrt{9 + x} = x + 3$

PRACTICE PROBLEM 5
Solve: $\sqrt{3x + 1} + \sqrt{3x} = 2$

Answers
4. $\{0\}$, 5. $\left\{\dfrac{3}{16}\right\}$

✔ **Concept Check Answer**
answers may vary

There is still a radical in the equation, so we get the radical alone again. Then we square both sides.

$$2x + 5 = 9 - 6\sqrt{2x} + 2x$$

$$6\sqrt{2x} = 4 \qquad \text{Get the radical alone.}$$

$$(6\sqrt{2x})^2 = 4^2 \qquad \text{Square both sides of the equation to eliminate the radical.}$$

$$36(2x) = 16$$

$$72x = 16 \qquad \text{Multiply.}$$

$$x = \frac{16}{72} \qquad \text{Solve.}$$

$$x = \frac{2}{9} \qquad \text{Simplify.}$$

The proposed solution $\frac{2}{9}$ checks in the original equation. The solution set is $\left\{\frac{2}{9}\right\}$.

■ **Work Practice Problem 5**

Helpful Hint

Make sure expressions are squared correctly. In Example 5, we squared $(3 - \sqrt{2x})$ as

$$(3 - \sqrt{2x})^2 = (3 - \sqrt{2x})(3 - \sqrt{2x})$$

$$= 3 \cdot 3 - 3\sqrt{2x} - 3\sqrt{2x} + \sqrt{2x} \cdot \sqrt{2x}$$

$$= 9 - 6\sqrt{2x} + 2x$$

✔ **Concept Check**   What is wrong with the following solution?

$$\sqrt{2x + 5} + \sqrt{4 - x} = 8$$

$$(\sqrt{2x + 5} + \sqrt{4 - x})^2 = 8^2$$

$$(2x + 5) + (4 - x) = 64$$

$$x + 9 = 64$$

$$x = 55$$

Objective B Using the Pythagorean Theorem

Recall that the Pythagorean theorem states that in a right triangle, the length of the hypotenuse squared equals the sum of the lengths of each of the legs squared.

Pythagorean Theorem

If a and b are the lengths of the legs of a right triangle and c is the length of the hypotenuse, then $a^2 + b^2 = c^2$.

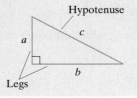

Hypotenuse

c

a

b

Legs

✔ **Concept Check Answer**

From the second line of the solution to the third line of the solution, the left side of the equation is squared incorrectly.

EXAMPLE 6 Find the length of the unknown leg of the right triangle.

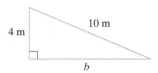

Solution: In the formula $a^2 + b^2 = c^2$, c is the hypotenuse. Here, $c = 10$, the length of the hypotenuse, and $a = 4$. We solve for b. Then $a^2 + b^2 = c^2$ becomes

$$4^2 + b^2 = 10^2$$
$$16 + b^2 = 100$$
$$b^2 = 84 \quad \text{Subtract 16 from both sides.}$$

Recall from Section 7.1 our definition of square root that if $b^2 = a$, then b is a square root of a. Since b is a length and thus is positive, we have that

$$b = \sqrt{84} = \sqrt{4 \cdot 21} = 2\sqrt{21}$$

The unknown leg of the triangle is exactly $2\sqrt{21}$ meters long. Using a calculator, this is approximately 9.2 meters.

Work Practice Problem 6

EXAMPLE 7 **Calculating Placement of a Wire**

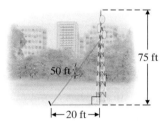

A 50-foot supporting wire is to be attached to a 75-foot antenna. Because of surrounding buildings, sidewalks, and roadways, the wire must be anchored exactly 20 feet from the base of the antenna.

a. How high from the base of the antenna must the wire be attached?

b. Local regulations require that a supporting wire be attached at a height no less than $\frac{3}{5}$ of the total height the antenna. From part (a), have local regulations been met?

Solution:

1. UNDERSTAND. Read and reread the problem. From the diagram we notice that a right triangle is formed with hypotenuse 50 feet and one leg 20 feet. We let $x =$ the height from the base of the antenna to the attached wire.

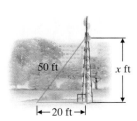

Continued on next page

PRACTICE PROBLEM 6

Find the length of the unknown leg of the right triangle.

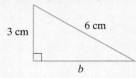

PRACTICE PROBLEM 7

A furniture upholsterer wishes to cut a strip from a piece of fabric that is 45 inches by 45 inches. The strip must be cut on the bias of the fabric. What is the longest strip that can be cut? Give an exact answer and a two-decimal-place approximation.

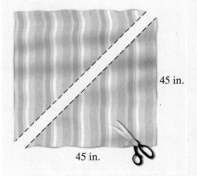

Answers
6. $3\sqrt{3}$ cm, **7.** $45\sqrt{2}$ in. ≈ 63.64 in.

2. TRANSLATE. We'll use the Pythagorean theorem.

$$a^2 + b^2 = c^2$$
$$20^2 + x^2 = 50^2 \quad a = 20, c = 50$$

3. SOLVE.

$$20^2 + x^2 = 50^2$$
$$400 + x^2 = 2500$$
$$x^2 = 2100 \quad \text{Subtract 400 from both sides.}$$
$$x = \sqrt{2100}$$
$$= 10\sqrt{21}$$

4. INTERPRET. *Check* the work and *state* the solution.

    a. The wire is attached exactly $10\sqrt{21}$ feet from the base of the pole, or approximately 45.8 feet.

    b. The supporting wire must be attached at a height no less than $\frac{3}{5}$ of the total height of the antenna. This height is $\frac{3}{5}$ (75 feet), or 45 feet.

    Since we know from part (a) that the wire is to be attached at a height of approximately 45.8 feet, local regulations have been met.

◾ **Work Practice Problem 7**

📇 CALCULATOR EXPLORATIONS Graphing

We can use a graphing calculator to solve radical equations. For example, to use a graphing calculator to approximate the solutions of the equation solved in Example 4, we graph the following:

$$Y_1 = \sqrt{4 - x} \quad \text{and} \quad Y_2 = x - 2$$

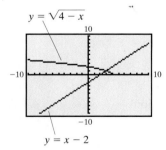

The *x*-value of the point of intersection is the solution. Use the INTERSECT feature or the ZOOM and TRACE features of your graphing calculator to see that the solution is 3.

Use a graphing calculator to solve each radical equation. Round all solutions to the nearest hundredth.

1. $\sqrt{x + 7} = x$

2. $\sqrt{3x + 5} = 2x$

3. $\sqrt{2x + 1} = \sqrt{2x + 2}$

4. $\sqrt{10x - 1} = \sqrt{-10x + 10} - 1$

5. $1.2x = \sqrt{3.1x + 5}$

6. $\sqrt{1.9x^2 - 2.2} = -0.8x + 3$

Objective Ⓐ *Solve. See Examples 1 and 2.*

1. $\sqrt{2x} = 4$

2. $\sqrt{3x} = 3$

3. $\sqrt{x - 3} = 2$

4. $\sqrt{x + 1} = 5$

5. $\sqrt{2x} = -4$

6. $\sqrt{5x} = -5$

7. $\sqrt{4x - 3} - 5 = 0$

8. $\sqrt{x - 3} - 1 = 0$

9. $\sqrt{2x - 3} - 2 = 1$

10. $\sqrt{3x + 3} - 4 = 8$

Solve. See Example 3.

11. $\sqrt[3]{6x} = -3$

12. $\sqrt[3]{4x} = -2$

13. $\sqrt[3]{x - 2} - 3 = 0$

14. $\sqrt[3]{2x - 6} - 4 = 0$

Solve. See Examples 4 and 5.

15. $\sqrt{13 - x} = x - 1$

16. $\sqrt{2x - 3} = 3 - x$

17. $x - \sqrt{4 - 3x} = -8$

18. $2x + \sqrt{x + 1} = 8$

19. $\sqrt{y + 5} = 2 - \sqrt{y - 4}$

20. $\sqrt{x + 3} + \sqrt{x - 5} = 3$

21. $\sqrt{x - 3} + \sqrt{x + 2} = 5$

22. $\sqrt{2x - 4} - \sqrt{3x + 4} = -2$

Solve. See Examples 1 through 5.

23. $\sqrt{3x - 2} = 5$

24. $\sqrt{5x - 4} = 9$

25. $-\sqrt{2x} + 4 = -6$

26. $-\sqrt{3x + 9} = -12$

27. $\sqrt{3x + 1} + 2 = 0$

28. $\sqrt{3x + 1} - 2 = 0$

29. $\sqrt[4]{4x + 1} - 2 = 0$

30. $\sqrt[4]{2x - 9} - 3 = 0$

31. $\sqrt{3x + 4} = 5$

32. $\sqrt{3x + 9} = 12$

33. $\sqrt[3]{6x - 3} - 3 = 0$

34. $\sqrt[3]{3x} + 4 = 7$

35. $\sqrt[3]{2x - 3} - 2 = -5$

36. $\sqrt[3]{x - 4} - 5 = -7$

37. $\sqrt{x + 4} = \sqrt{2x - 5}$

38. $\sqrt{3y + 6} = \sqrt{7y - 6}$

39. $x - \sqrt{1 - x} = -5$

40. $x - \sqrt{x - 2} = 4$

41. $\sqrt[3]{-6x - 1} = \sqrt[3]{-2x - 5}$

42. $x + \sqrt{x + 5} = 7$

43. $\sqrt{5x - 1} - \sqrt{x + 2} = 3$

44. $\sqrt{2x - 1} - 4 = -\sqrt{x - 4}$

45. $\sqrt{2x - 1} = \sqrt{1 - 2x}$

46. $\sqrt{7x - 4} = \sqrt{4 - 7x}$

47. $\sqrt{3x + 4} - 1 = \sqrt{2x + 1}$

48. $\sqrt{x - 2} + 3 = \sqrt{4x + 1}$

49. $\sqrt{y + 3} - \sqrt{y - 3} = 1$

50. $\sqrt{x + 1} - \sqrt{x - 1} = 2$

Objective **B** *Find the length of the unknown side of each triangle. See Example 6.*

△ **51.**
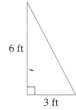
6 ft
3 ft

△ **52.**
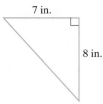
7 in.
8 in.

△ **53.**
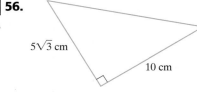
3 m
7 m

△ **54.**

4 cm
7 cm

Find the length of the unknown side of each triangle. Give the exact length and a one-decimal-place approximation. See Example 6.

⌨ **55.**
△
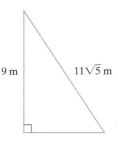
9 m
$11\sqrt{5}$ m

⌨ **56.**
△
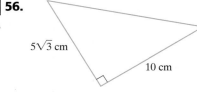
$5\sqrt{3}$ cm
10 cm

⌨ **57.**
△

7 mm
7.2 mm

⌨ **58.**
△
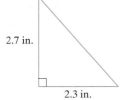
2.7 in.
2.3 in.

Solve. Give exact answers and two-decimal-place approximations where appropriate. See Example 7.

⌨ **59.** A wire is needed to support a vertical pole 15 feet high. The cable will be anchored to a stake 8 feet from the base of the pole. How much cable is needed?

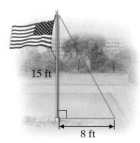
15 ft
8 ft

⌨ **60.** The tallest structure in the United States is a TV tower in Blanchard, North Dakota. Its height is 2063 feet. A 2382-foot length of wire is to be used as a guy wire attached to the top of the tower. Approximate to the nearest foot how far from the base of the tower the guy wire must be anchored. (*Source:* U.S. Geological Survey)

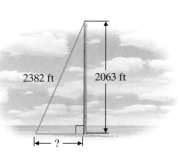
2382 ft
2063 ft
?

△ **61.** A spotlight is mounted on the eaves of a house 12 feet above the ground. A flower bed runs between the house and the sidewalk, so the closest the ladder can be placed to the house is 5 feet. How long a ladder is needed so that an electrician can reach the place where the light is mounted?

12 ft

5 ft

62. A wire is to be attached to support a telephone pole. Because of surrounding buildings, sidewalks, and roadway, the wire must be anchored exactly 15 feet from the base of the pole. Telephone company workers have only 30 feet of cable, and 2 feet of that must be used to attach the cable to the pole and to the stake on the ground. How high from the base of the pole can the wire be attached?

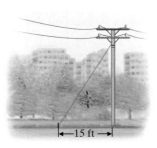

←15 ft→

△ **63.** The radius of the moon is 1080 miles. Use the formula for the radius r of a sphere given its surface area A.

$$r = \sqrt{\frac{A}{4\pi}}$$

to find the surface area of the moon. Round to the nearest square mile. (*Source:* National Space Science Data Center)

64. Police departments find it very useful to be able to approximate driving speeds in skidding accidents. If the road surface is wet concrete, the function $S(x) = \sqrt{10.5x}$ is used, where $S(x)$ is the speed of the car in miles per hour and x is the distance skidded in feet. Find how fast a car was moving if it skidded 280 feet on wet concrete.

65. The formula $v = \sqrt{2gh}$ relates the velocity v, in feet per second, of an object after it falls h feet accelerated by gravity g, in feet per second squared. If g is approximately 32 feet per second squared, find how far an object has fallen if its velocity is 80 feet per second.

66. Two tractors are pulling a tree stump from a field. If two forces A and B pull at right angles (90°) to each other, the size of the resulting force R is given by the formula $R = \sqrt{A^2 + B^2}$. If tractor A is exerting 600 pounds of force and the resulting force is 850 pounds, find how much force tractor B is exerting.

?

600 lb

In psychology, it has been suggested that the number S of nonsense syllables that a person can repeat consecutively depends on his or her IQ score I according to the equation $S = 2\sqrt{I} - 9$.

67. Use this relationship to estimate the IQ of a person who can repeat 11 nonsense syllables consecutively.

68. Use this relationship to estimate the IQ of a person who can repeat 15 nonsense syllables consecutively.

*The **period** of a pendulum is the time it takes for the pendulum to make one full back-and-forth swing. The period of a pendulum depends on the length of the pendulum. The formula for the period P, in seconds, is $P = 2\pi\sqrt{\dfrac{l}{32}}$, where l is the length of the pendulum in feet. Use this formula for Exercises 69 through 74.*

69. Find the period of a pendulum whose length is 2 feet. Give an exact answer and a two-decimal-place approximation.

70. Klockit sells a 43-inch lyre pendulum. Find the period of this pendulum. Round your answer to 2 decimal places. (*Hint:* First convert inches to feet.)

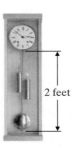

2 feet

71. Find the length of a pendulum whose period is 4 seconds. Round your answer to 2 decimal places.

72. Find the length of a pendulum whose period is 3 seconds. Round your answer to 3 decimal places.

73. Study the relationship between period and pendulum length in Exercises 69 through 72 and make a conjecture about this relationship.

74. Galileo experimented with pendulums. He supposedly made conjectures about pendulums of equal length with different bob weights. Try this experiment. Make two pendulums 3 feet long. Attach a heavy weight (lead) to one and a light weight (a cork) to the other. Pull both pendulums back the same angle measure and release. Make a conjecture from your observations.

If the three lengths of the sides of a triangle are known, Heron's formula can be used to find its area. If a, b, and c are the three lengths of the sides, Heron's formula *for area is*

$$A = \sqrt{s(s - a)(s - b)(s - c)}$$

where s is half the perimeter of the triangle, or $s = \dfrac{1}{2}(a + b + c)$. Use this formula to find the area of each triangle. Give an exact answer and then a 2-decimal place approximation.

△ **75.**

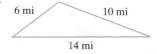

6 mi   10 mi

14 mi

△ **76.**

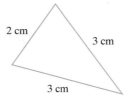

2 cm   3 cm

3 cm

77. Describe when Heron's formula might be useful.

78. In your own words, explain why you think S in *Heron's formula* is called the *semiperimeter.*

The maximum distance $D(h)$ in kilometers that a person can see from a height h kilometers above the ground is given by the function $D(h) = 111.7\sqrt{h}$. Use this function for Exercises 79 and 80. Round your answers to two decimal places.

79. Find the height that would allow a person to see 80 kilometers.

80. Find the height that would allow a person to see 40 kilometers.

Review

Simplify. See Section 6.3.

81. $\dfrac{\dfrac{x}{6}}{\dfrac{2x}{3} + \dfrac{1}{2}}$

82. $\dfrac{\dfrac{1}{y} + \dfrac{4}{5}}{\dfrac{-3}{20}}$

83. $\dfrac{\dfrac{z}{5} + \dfrac{1}{10}}{\dfrac{z}{20} - \dfrac{z}{5}}$

84. $\dfrac{\dfrac{1}{y} + \dfrac{1}{x}}{\dfrac{1}{y} - \dfrac{1}{x}}$

Concept Extensions

85. Solve: $\sqrt{\sqrt{x + 3} + \sqrt{x}} = \sqrt{3}$

86. Explain why proposed solutions of radical equations must be checked.

87. Find the error in the following solution and correct. See the Concept Check in this section.

$$\sqrt{5x - 1} + 4 = 7$$
$$(\sqrt{5x - 1} + 4)^2 = 7^2$$
$$5x - 1 + 16 = 49$$
$$5x = 34$$
$$x = \frac{34}{5}$$

88. Consider the equations $\sqrt{2x} = 4$ and $\sqrt[3]{2x} = 4$.
  a. Explain the difference in solving these equations.
  b. Explain the similarity in solving these equations.

89. The cost $C(x)$ in dollars per day to operate a small delivery service is given by $C(x) = 80\sqrt[3]{x} + 500$, where x is the number of deliveries per day. In July, the manager decides that it is necessary to keep delivery costs below $1620.00. Find the greatest number of deliveries this company can make per day and still keep overhead below $1620.00.

THE BIGGER PICTURE   Solving Equations and Inequalities

Continue your outline from Sections 2.6, 5.7, and 6.4. Write how to recognize and how to solve exponential and logarithmic equations in your own words. For example:

Solving Equations and Inequalities

I. Equations

   A. Linear equations (Sec. 2.1)

   B. Absolute value equations (Sec. 2.6)

   C. Quadratic and higher degree equations (Sec. 5.7)

   D. Equations with rational expressions (Sec. 6.4)

   E. Equations with radicals: Equation contains at least one root of a variable expression.

$\sqrt{5x + 10} - 2 = x$	Radical equation.
$\sqrt{5x + 10} = x + 2$	Isolate the radical.
$(\sqrt{5x + 10})^2 = (x + 2)^2$	Square both sides.
$5x + 10 = x^2 + 4x + 4$	Simplify.
$0 = x^2 - x - 6$	Write in standard form.
$0 = (x - 3)(x + 2)$	Factor.
$x - 3 = 0 \quad \text{or} \quad x + 2 = 0$	Set each factor equal to 0.
$x = 3 \quad \text{or} \quad x = -2$	Solve.

Both solutions check.

II. Inequalities

   A. Linear inequalities (Sec. 2.4)

   B. Compound inequalities (Sec. 2.5)

   C. Absolute value inequalities (Sec. 2.6)

Solve. Write inequality solutions in interval notation.

1. $\dfrac{x}{4} + \dfrac{x + 18}{20} = \dfrac{x - 5}{5}$

2. $|3x - 5| = 10$

3. $2x^2 - x = 45$

4. $-6 \le -5x - 1 \le 10$

5. $4(x - 1) + 3x > 1 + 2(x - 6)$

6. $\sqrt{x} + 14 = x - 6$

7. $x \ge 10$ or $-x < 5$

8. $\sqrt{3x - 1} + 4 = 1$

9. $|x - 2| > 15$

10. $5x - 4[x - 2(3x + 1)] = 25$

7.7 COMPLEX NUMBERS

Objectives

A Write Square Roots of Negative Numbers in the Form *bi*.

B Add or Subtract Complex Numbers.

C Multiply Complex Numbers.

D Divide Complex Numbers.

E Raise *i* to Powers.

Objective **A** Writing Numbers in the Form *bi*

Our work with radical expressions has excluded expressions such as $\sqrt{-16}$ because $\sqrt{-16}$ is not a real number; there is no real number whose square is -16. In this section, we discuss a number system that includes roots of negative numbers. This number system is the **complex number system,** and it includes the set of real numbers as a subset. The complex number system allows us to solve equations such as $x^2 + 1 = 0$ that have no real number solutions. The set of complex numbers includes the *imaginary unit.*

Imaginary Unit

The **imaginary unit,** written i, is the number whose square is -1. That is,

$$i^2 = -1 \quad \text{and} \quad i = \sqrt{-1}$$

To write the square root of a negative number in terms of i, we use the property that if a is a positive number, then

$$\sqrt{-a} = \sqrt{-1} \cdot \sqrt{a}$$
$$= i \cdot \sqrt{a}$$

Using i, we can write $\sqrt{-16}$ as

$$\sqrt{-16} = \sqrt{-1 \cdot 16} = \sqrt{-1} \cdot \sqrt{16} = i \cdot 4 \text{ or } 4i$$

EXAMPLES Write using i notation.

1. $\sqrt{-36} = \sqrt{-1 \cdot 36} = \sqrt{-1} \cdot \sqrt{36} = i \cdot 6 \text{ or } 6i$

2. $\sqrt{-5} = \sqrt{-1(5)} = \sqrt{-1} \cdot \sqrt{5} = i\sqrt{5}$

> **Helpful Hint**
> Since $\sqrt{5}i$ can easily be confused with $\sqrt{5i}$, we write $\sqrt{5}i$ as $i\sqrt{5}$.

3. $-\sqrt{-20} = -\sqrt{-1 \cdot 20} = -\sqrt{-1} \cdot \sqrt{4 \cdot 5} = -i \cdot 2\sqrt{5} = -2i\sqrt{5}$

🔲 **Work Practice Problems 1–3**

The product rule for radicals does not necessarily hold true for imaginary numbers. *To multiply square roots of negative numbers, first we write each number in terms of the imaginary unit i.* For example, to multiply $\sqrt{-4}$ and $\sqrt{-9}$, we first write each number in the form bi:

$$\sqrt{-4} \cdot \sqrt{-9} = 2i(3i) = 6i^2 = 6(-1) = -6 \quad \text{Correct.}$$

Make sure you notice that the product rule does not work for this example. In other words, $\sqrt{-4} \cdot \sqrt{-9} = \sqrt{(-4)(-9)} = \sqrt{36} = 6$ is incorrect!

EXAMPLES Multiply or divide as indicated.

4. $\sqrt{-3} \cdot \sqrt{-5} = i\sqrt{3}(i\sqrt{5}) = i^2\sqrt{15} = -1\sqrt{15} = -\sqrt{15}$

5. $\sqrt{-36} \cdot \sqrt{-1} = 6i(i) = 6i^2 = 6(-1) = -6$

6. $\sqrt{8} \cdot \sqrt{-2} = 2\sqrt{2}(i\sqrt{2}) = 2i(\sqrt{2}\sqrt{2}) = 2i(2) = 4i$

7. $\dfrac{\sqrt{-125}}{\sqrt{5}} = \dfrac{i\sqrt{125}}{\sqrt{5}} = i\sqrt{25} = 5i$

🔲 **Work Practice Problems 4–7**

PRACTICE PROBLEMS 1–3

Write using i notation.

1. $\sqrt{-25}$

2. $\sqrt{-17}$

3. $-\sqrt{-50}$

PRACTICE PROBLEMS 4–7

Multiply or divide as indicated.

4. $\sqrt{-3} \cdot \sqrt{-7}$

5. $\sqrt{-25} \cdot \sqrt{-1}$

6. $\sqrt{27} \cdot \sqrt{-3}$ **7.** $\dfrac{\sqrt{-8}}{\sqrt{2}}$

Answers

1. $5i$, **2.** $i\sqrt{17}$, **3.** $-5i\sqrt{2}$,
4. $-\sqrt{21}$, **5.** -5, **6.** $9i$, **7.** $2i$

Now that we have practiced working with the imaginary unit, we define *complex numbers*.

Complex Numbers

A **complex number** is a number that can be written in the form $a + bi$, where a and b are real numbers.

Notice that the set of real numbers is a subset of the complex numbers since any real number can be written in the form of a complex number. For example,

$$16 = 16 + 0i$$

In general, a complex number $a + bi$ is a real number if $b = 0$. Also, a complex number is called an **imaginary number** if $a = 0$. For example.

$$3i = 0 + 3i \quad \text{and} \quad i\sqrt{7} = 0 + i\sqrt{7}$$

are imaginary numbers.

The following diagram shows the relationship between complex numbers and their subsets.

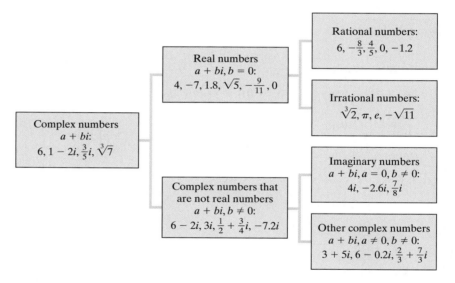

✔**Concept Check** True or false? Every complex number is also a real number.

Objective B Adding or Subtracting Complex Numbers

Two complex numbers $a + bi$ and $c + di$ are equal if and only if $a = c$ and $b = d$. Complex numbers can be added or subtracted by adding or subtracting their real parts and then adding or subtracting their imaginary parts.

Sum or Difference of Complex Numbers

If $a + bi$ and $c + di$ are complex numbers, then their sum is

$$(a + bi) + (c + di) = (a + c) + (b + d)i$$

Their difference is

$$(a + bi) - (c + di) = a + bi - c - di = (a - c) + (b - d)i$$

✔ **Concept Check Answer**

false

EXAMPLES Add or subtract as indicated.

8. $(2 + 3i) + (-3 + 2i) = (2 - 3) + (3 + 2)i = -1 + 5i$

9. $5i - (1 - i) = 5i - 1 + i$
$$= -1 + (5 + 1)i$$
$$= -1 + 6i$$

10. $(-3 - 7i) - (-6) = -3 - 7i + 6$
$$= (-3 + 6) - 7i$$
$$= 3 - 7i$$

📖 **Work Practice Problems 8–10**

PRACTICE PROBLEMS 8–10

Add or subtract as indicated.

8. $(5 + 2i) + (4 - 3i)$

9. $6i - (2 - i)$

10. $(-2 - 4i) - (-3)$

Objective C Multiplying Complex Numbers

To multiply two complex numbers of the form $a + bi$, we multiply as though they were binomials. Then we use the relationship $i^2 = -1$ to simplify.

EXAMPLES Multiply.

11. $-7i \cdot 3i = -21i^2$
$$= -21(-1) \quad \text{Replace } i^2 \text{ with } -1.$$
$$= 21$$

12. $3i(2 - i) = 3i \cdot 2 - 3i \cdot i \quad \text{Use the distributive property.}$
$$= 6i - 3i^2 \qquad \text{Multiply.}$$
$$= 6i - 3(-1) \quad \text{Replace } i^2 \text{ with } -1.$$
$$= 6i + 3$$
$$= 3 + 6i \qquad \text{Use the FOIL order. (First, Outer, Inner, Last)}$$

13. $(2 - 5i)(4 + i) = 2(4) + 2(i) - 5i(4) - 5i(i)$
$$\qquad\qquad\quad \text{F} \qquad \text{O} \qquad \text{I} \qquad \text{L}$$
$$= 8 + 2i - 20i - 5i^2$$
$$= 8 - 18i - 5(-1) \quad i^2 = -1$$
$$= 8 - 18i + 5$$
$$= 13 - 18i$$

14. $(2 - i)^2 = (2 - i)(2 - i)$
$$= 2(2) - 2(i) - 2(i) + i^2$$
$$= 4 - 4i + (-1) \quad i^2 = -1$$
$$= 3 - 4i$$

15. $(7 + 3i)(7 - 3i) = 7(7) - 7(3i) + 3i(7) - 3i(3i)$
$$= 49 - 21i + 21i - 9i^2$$
$$= 49 - 9(-1) \quad i^2 = -1$$
$$= 49 + 9$$
$$= 58$$

📖 **Work Practice Problems 11–15**

Notice that if you add, subtract, or multiply two complex numbers, the result is a complex number.

Objective D Dividing Complex Numbers

From Example 15, notice that the product of $7 + 3i$ and $7 - 3i$ is a real number. These two complex numbers are called *complex conjugates* of one another. In general, we have the following definition.

PRACTICE PROBLEMS 11–15

Multiply.

11. $-5i \cdot 3i$

12. $-2i(6 - 2i)$

13. $(3 - 4i)(6 + i)$

14. $(1 - 2i)^2$

15. $(6 + 5i)(6 - 5i)$

Answers

8. $9 - i$, **9.** $-2 + 7i$, **10.** $1 - 4i$,

11. 15, **12.** $-4 - 12i$, **13.** $22 - 21i$,

14. $-3 - 4i$, **15.** 61

Complex Conjugates

The complex numbers $(a + bi)$ and $(a - bi)$ are called **complex conjugates** of each other, and

$$(a + bi)(a - bi) = a^2 + b^2$$

To see that the product of a complex number $a + bi$ and its conjugate $a - bi$ is the real number $a^2 + b^2$, we multiply:

$$(a + bi)(a - bi) = a^2 - abi + abi - b^2i^2$$
$$= a^2 - b^2(-1)$$
$$= a^2 + b^2$$

We will use complex conjugates to divide by a complex number.

PRACTICE PROBLEM 16

Divide and write in the form $a + bi$: $\dfrac{3 + i}{2 - 3i}$

EXAMPLE 16 Divide and write in the form $a + bi$: $\dfrac{2 + i}{1 - i}$

Solution: We multiply the numerator and the denominator by the complex conjugate of $1 - i$ to eliminate the imaginary number in the denominator.

$$\frac{2 + i}{1 - i} = \frac{(2 + i)(1 + i)}{(1 - i)(1 + i)}$$
$$= \frac{2(1) + 2(i) + 1(i) + i^2}{1^2 - i^2}$$
$$= \frac{2 + 3i - 1}{1 + 1}$$
$$= \frac{1 + 3i}{2} = \frac{1}{2} + \frac{3}{2}i$$

▣ **Work Practice Problem 16**

PRACTICE PROBLEM 17

Divide and write in the form $a + bi$: $\dfrac{6}{5i}$

EXAMPLE 17 Divide and write in the form $a + bi$: $\dfrac{7}{3i}$

Solution: We multiply the numerator and the denominator by the conjugate of $3i$. Note that $3i = 0 + 3i$, so its conjugate is $0 - 3i$ or $-3i$.

$$\frac{7}{3i} = \frac{7(-3i)}{(3i)(-3i)} = \frac{-21i}{-9i^2} = \frac{-21i}{-9(-1)} = \frac{-21i}{9} = \frac{-7i}{3} = -\frac{7}{3}i$$

▣ **Work Practice Problem 17**

Objective 🄴 Finding Powers of *i*

We can use the fact that $i^2 = -1$ to simplify i^3 and i^4.

$$i^3 = i^2 \cdot i = (-1)i = -i$$
$$i^4 = i^2 \cdot i^2 = (-1) \cdot (-1) = 1$$

We continue this process and use the fact that $i^4 = 1$ and $i^2 = -1$ to simplify i^5 and i^6.

$$i^5 = i^4 \cdot i = 1 \cdot i = i$$
$$i^6 = i^4 \cdot i^2 = 1 \cdot (-1) = -1$$

Answers

16. $\dfrac{3}{13} + \dfrac{11}{13}i$,   **17.** $-\dfrac{6}{5}i$

If we continue finding powers of i, we generate the following pattern. Notice that the values i, -1, $-i$, and 1 repeat as i is raised to higher and higher powers.

$i^1 = i$	$i^5 = i$	$i^9 = i$
$i^2 = -1$	$i^6 = -1$	$i^{10} = -1$
$i^3 = -i$	$i^7 = -i$	$i^{11} = -i$
$i^4 = 1$	$i^8 = 1$	$i^{12} = 1$

This pattern allows us to find other powers of i. To do so, we will use the fact that $i^4 = 1$ and rewrite a power of i in terms of i^4. For example,

$$i^{22} = i^{20} \cdot i^2 = (i^4)^5 \cdot i^2 = 1^5 \cdot (-1) = 1 \cdot (-1) = -1$$

EXAMPLES Find each power of i.

18. $i^7 = i^4 \cdot i^3 = 1(-i) = -i$

19. $i^{20} = (i^4)^5 = 1^5 = 1$

20. $i^{46} = i^{44} \cdot i^2 = (i^4)^{11} \cdot i^2 = 1^{11}(-1) = -1$

21. $i^{-12} = \dfrac{1}{i^{12}} = \dfrac{1}{(i^4)^3} = \dfrac{1}{(1)^3} = \dfrac{1}{1} = 1$

◾ **Work Practice Problems 18–21**

PRACTICE PROBLEMS 18–21

Find the powers of i.

18. i^{11} **19.** i^{40}

20. i^{50} **21.** i^{-10}

Answers

18. $-i$, **19.** 1, **20.** -1, **21.** -1

Mental Math

Simplify. See Examples 1 through 3.

1. $\sqrt{-81}$ **2.** $\sqrt{-49}$ **3.** $\sqrt{-7}$ **4.** $\sqrt{-3}$

5. $-\sqrt{16}$ **6.** $-\sqrt{4}$ **7.** $\sqrt{-64}$ **8.** $\sqrt{-100}$

7.7 EXERCISE SET

FOR EXTRA HELP

 Student Solutions Manual   PH Math/Tutor Center   CD/Video for Review   Math XL MathXL®   MyMathLab MyMathLab

Objective A *Write using i notation. See Examples 1 through 3.*

1. $\sqrt{-24}$ **2.** $\sqrt{-32}$ **3.** $-\sqrt{-36}$ **4.** $-\sqrt{-121}$

5. $8\sqrt{-63}$ **6.** $4\sqrt{-20}$ **7.** $-\sqrt{54}$ **8.** $\sqrt{-63}$

Multiply or divide as indicated. See Examples 4 through 7.

9. $\sqrt{-2} \cdot \sqrt{-7}$ **10.** $\sqrt{-11} \cdot \sqrt{-3}$ **11.** $\sqrt{-5} \cdot \sqrt{-10}$

12. $\sqrt{-2} \cdot \sqrt{-6}$ **13.** $\sqrt{16} \cdot \sqrt{-1}$ **14.** $\sqrt{3} \cdot \sqrt{-27}$ **15.** $\dfrac{\sqrt{-9}}{\sqrt{3}}$

16. $\dfrac{\sqrt{49}}{\sqrt{-10}}$ **17.** $\dfrac{\sqrt{-80}}{\sqrt{-10}}$ **18.** $\dfrac{\sqrt{-40}}{\sqrt{-8}}$

Objective B *Add or subtract as indicated. Write your answers in the form a + bi. See Examples 8 through 10.*

19. $(4 - 7i) + (2 + 3i)$ **20.** $(2 - 4i) - (2 - i)$ **21.** $(6 + 5i) - (8 - i)$

22. $(8 - 3i) + (-8 + 3i)$ **23.** $6 - (8 + 4i)$ **24.** $(9 - 4i) - 9$

25. $(6 - 3i) - (4 - 2i)$ **26.** $(-2 - 4i) - (6 - 8i)$ **27.** $(5 - 6i) - 4i$

28. $(6 - 2i) + 7i$ **29.** $(2 + 4i) + (6 - 5i)$ **30.** $(5 - 3i) + (7 - 8i)$

Objective C *Multiply. Write your answers in the form a + bi. See Examples 11 through 15.*

31. $6i \cdot 2i$ **32.** $5i \cdot 7i$ **33.** $-9i \cdot 7i$

34. $-6i \cdot 4i$ **35.** $-10i \cdot -4i$ **36.** $-2i \cdot -11i$

37. $6i(2 - 3i)$ **38.** $5i(4 - 7i)$ **39.** $-3i(-1 + 9i)$

40. $-5i(-2 + i)$ **41.** $(4 + i)(5 + 2i)$ **42.** $(3 + i)(2 + 4i)$

43. $(\sqrt{3} + 2i)(\sqrt{3} - 2i)$ **44.** $(\sqrt{5} - 5i)(\sqrt{5} + 5i)$ **45.** $(4 - 2i)^2$

46. $(6 - 3i)^2$ **47.** $(6 - 2i)(3 + i)$ **48.** $(2 - 4i)(2 - i)$

49. $(1 - i)(1 + i)$ **50.** $(6 + 2i)(6 - 2i)$ **51.** $(9 + 8i)^2$

52. $(4 + 7i)^2$ **53.** $(1 - i)^2$ **54.** $(2 - 2i)^2$

Objective **D** *Divide. Write your answers in the form a + bi. See Examples 16 and 17.*

55. $\dfrac{4}{i}$ **56.** $\dfrac{5}{6i}$ **57.** $\dfrac{7}{4 + 3i}$ **58.** $\dfrac{9}{1 - 2i}$

59. $\dfrac{6i}{1 - 2i}$ **60.** $\dfrac{3i}{5 + i}$ **61.** $\dfrac{3 + 5i}{1 + i}$ **62.** $\dfrac{6 + 2i}{4 - 3i}$

63. $\dfrac{4 - 5i}{2i}$ **64.** $\dfrac{6 + 8i}{3i}$ **65.** $\dfrac{16 + 15i}{-3i}$ **66.** $\dfrac{2 - 3i}{-7i}$

67. $\dfrac{2}{3 + i}$ **68.** $\dfrac{5}{3 - 2i}$ **69.** $\dfrac{2 - 3i}{2 + i}$ **70.** $\dfrac{6 + 5i}{6 - 5i}$

Objective **E** *Find each power of i. See Examples 18 through 21.*

71. i^8 **72.** i^{10} **73.** i^{21} i **74.** i^{15} **75.** i^{11} **76.** i^{40}

77. i^{-6} **78.** i^{-9} **79.** $(2i)^6$ **80.** $(5i)^4$ **81.** $(-3i)^5$ **82.** $(-2i)^7$

Review

Thirty people were recently polled about the average monthly balance in their checking account. The results of this poll are shown in the bar graph. Use this graph to answer Exercises 83 through 88. See Section 1.2.

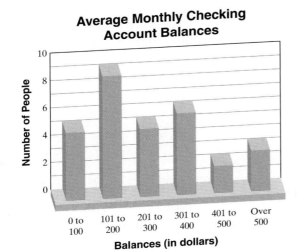

Average Monthly Checking Account Balances

83. How many people polled reported an average checking balance of $201 to $300?

84. How many people polled reported an average checking balance of $0 to $100?

85. How many people polled reported an average checking balance of $200 or less?

86. How many people polled reported an average checking balance of $301 or more?

87. What percent of people polled reported an average checking balance of $201 to $300?

88. What percent of people polled reported an average checking balance of 0 to $100?

Concept Extensions

Write each expression in the form a + bi.

89. $i^3 + i^4$ **90.** $i^8 - i^7$ **91.** $i^6 + i^8$ **92.** $i^4 + i^{12}$ **93.** $2 + \sqrt{-9}$

94. $5 - \sqrt{-16}$ **95.** $\dfrac{6 + \sqrt{-18}}{3}$ **96.** $\dfrac{4 - \sqrt{-8}}{2}$ **97.** $\dfrac{5 - \sqrt{-75}}{10}$

98. Describe how to find the conjugate of a complex number.

99. Explain why the product of a complex number and its complex conjugate is a real number.

Simplify.

100. $(8 - \sqrt{-3}) - (2 + \sqrt{-12})$ **101.** $(8 - \sqrt{-4}) - (2 + \sqrt{-16})$

102. Determine whether $2i$ is a solution of $x^2 + 4 = 0$.

103. Determine whether $-1 + i$ is a solution of $x^2 + 2x = -2$.

CHAPTER 7   Group Activity

Heron of Alexandria

Heron (also Hero) was a Greek mathematician and engineer. He lived and worked in Alexandria, Egypt, around 75 A.D. During his prolific work life, Heron developed a rotary steam engine called an aeolipile, a surveying tool called a dioptra, as well as a wind organ and a fire engine. As an engineer, he must have had the need to approximate square roots because he described an iterative method for doing so in his work *Metrica*. Heron's method for approximating a square root can be summarized as follows:

> Suppose that x is not a perfect square and a^2 is the nearest perfect square to x. For a rough estimate of the value of $\sqrt{x}$, find the value of $y_1 = \frac{1}{2}\left(a + \frac{x}{a}\right)$. This estimate can be improved by calculating a second estimate using the first estimate y_1 in place of a: $y_2 = \frac{1}{2}\left(y_1 + \frac{x}{y_1}\right)$.
>
> Repeating this process several times will give more and more accurate estimates of $\sqrt{x}$.

Critical Thinking

1. **a.** Which perfect square is closest to 80?

   b. Use Heron's method for approximating square roots to calculate the first estimate of the square root of 80.

   c. Use the first estimate of the square root of 80 to find a more refined second estimate.

   d. Use a calculator to find the actual value of the square root of 80. List all digits shown on your calculator's display.

   e. Compare the actual value from part (d) to the values of the first and second estimates. What do you notice?

   f. How many iterations of this process are necessary to get an estimate that differs no more than one digit from the actual value recorded in part (d)?

2. Repeat Question 1 for finding an estimate of the square root of 30.

3. Repeat Question 1 for finding an estimate of the square root of 4572.

4. Why would this iterative method have been important to people of Heron's era? Would you say that this method is as important today? Why or why not?

STUDY SKILLS BUILDER

Are You Prepared for a Test on Chapter 7?

Below I have listed some common trouble areas for students in Chapter 7. After studying for your test, but before taking your test, read these.

- Remember how to convert an expression with rational expressions to one with radicals and one with radicals to one with rational expressions.

$$7^{2/3} = \sqrt[3]{7^2} \text{ or } (\sqrt[3]{7})^2$$
$$\sqrt[5]{4^3} = 4^{3/5}$$

- Remember the difference between $\sqrt{x} + \sqrt{x}$ and $\sqrt{x} \cdot \sqrt{x}, x > 0$.

$$\sqrt{x} + \sqrt{x} = 2\sqrt{x}$$
$$\sqrt{x} \cdot \sqrt{x} = x$$

- Don't forget the difference between rationalizing the denominator of $\sqrt{\dfrac{2}{x}}$ and rationalizing the denominator of $\dfrac{\sqrt{2}}{\sqrt{x} + 1}, x > 0$.

$$\sqrt{\frac{2}{x}} = \frac{\sqrt{2}}{\sqrt{x}} = \frac{\sqrt{2}\cdot}{\sqrt{x}\cdot} = \frac{\sqrt{2x}}{x}$$

$$\frac{\sqrt{2}}{\sqrt{x} + 1} = \frac{\sqrt{2}(\sqrt{x} - 1)}{(\sqrt{x} + 1)(\sqrt{x} - 1)} = \frac{\sqrt{2}(\sqrt{x} - 1)}{x - 1}$$

- Remember that the midpoint of a segment is a *point*. The x-coordinate is the average of the x-coordinates of the endpoints of the segment and the y-coordinate is the average of the y-coordinates of the endpoints of the segment.

  The midpoint of the segment joining $(-1, 5)$ and $(3, 4)$ is $\left(\dfrac{-1 + 3}{2}, \dfrac{5 + 4}{2}\right)$ or $\left(1, \dfrac{9}{2}\right)$.

- Remember that the distance formula gives the *distance* between two points.

  The distance between $(-1, 5)$ and $(3, 4)$ is
$$\sqrt{(3 - (-1))^2 + (4 - 5)^2} = \sqrt{4^2 + (-1)^2}$$
$$= \sqrt{16 + 1} = \sqrt{17} \text{ units}$$

Remember: This is simply a checklist of common trouble areas. For a review of Chapter 7, see the Highlights and Chapter Review at the end of this chapter.

Chapter 7 Vocabulary Check

Fill in each blank with one of the words or phrases listed below.

index	rationalizing	conjugate	principal square root	cube root	midpoint
complex number	like radicals	radicand	imaginary unit		distance

1. The _____ of $\sqrt{3} + 2$ is $\sqrt{3} - 2$.
2. The _____ of a nonnegative number a is written as $\sqrt{a}$.
3. The process of writing a radical expression as an equivalent expression but without a radical in the denominator is called _____ the denominator.
4. The _____, written i, is the number whose square is -1.
5. The _____ of a number is written as $\sqrt[3]{a}$.
6. In the notation $\sqrt[n]{a}$, n is called the _____ and a is called the _____.
7. Radicals with the same index and the same radicand are called _____.
8. A _____ is a number that can be written in the form $a + bi$, where a and b are real numbers.
9. The _____ formula is $d = \sqrt{(x_2 - x_1)^2 + (y_2 - y_1)^2}$.
10. The _____ formula is $\left(\dfrac{x_1 + x_2}{2}, \dfrac{y_1 + y_2}{2} \right)$.

Helpful Hint

Are you preparing for your test? Don't forget to take the Chapter 7 Test on page 547. Then check your answers at the back of the text and use the Chapter Test Prep Video CD to see the fully worked-out solutions to any of the exercises you want to review.

7 Chapter Highlights

DEFINITIONS AND CONCEPTS	EXAMPLES
Section 7.1 Radical Expressions and Radical Functions	

DEFINITIONS AND CONCEPTS	EXAMPLES
The **positive,** or **principal, square root** of a nonnegative number a is written as $\sqrt{a}$. $\quad \sqrt{a} = b$ only if $b^2 = a$ and $b \geq 0$ The **negative square root** of a is written as $-\sqrt{a}$. The **cube root** of a real number a is written as $\sqrt[3]{a}$. $\quad \sqrt[3]{a} = b$ only if $b^3 = a$ If n is an even positive integer, then $\sqrt[n]{a^n} = \lvert a \rvert$. If n is an odd positive integer, then $\sqrt[n]{a^n} = a$. A **radical function** in x is a function defined by an expression containing a root of x.	$\sqrt{36} = 6 \qquad \sqrt{\dfrac{9}{100}} = \dfrac{3}{10}$ $-\sqrt{36} = -6 \qquad -\sqrt{0.04} = -0.2$ $\sqrt[3]{27} = 3 \qquad \sqrt[3]{-\dfrac{1}{8}} = -\dfrac{1}{2}$ $\sqrt[3]{y^6} = y^2 \qquad \sqrt[3]{64x^9} = 4x^3$ $\sqrt{(-3)^2} = \lvert -3 \rvert = 3$ $\sqrt[3]{(-7)^3} = -7$ If $f(x) = \sqrt{x} + 2$, $\quad f(1) = \sqrt{1} + 2 = 1 + 2 = 3$ $\quad f(3) = \sqrt{3} + 2 \approx 3.73$

DEFINITIONS AND CONCEPTS	**EXAMPLES**

Section 7.2   Rational Exponents

$a^{1/n} = \sqrt[n]{a}$ if $\sqrt[n]{a}$ is a real number.

If m and n are positive integers greater than 1 with $\dfrac{m}{n}$ in lowest terms and $\sqrt[n]{a}$ is a real number, then

$$a^{m/n} = (a^{1/n})^m = (\sqrt[n]{a})^m$$

$a^{-m/n} = \dfrac{1}{a^{m/n}}$ as long as $a^{m/n}$ is a nonzero number.

Exponent rules are true for rational exponents.

$81^{1/2} = \sqrt{81} = 9$
$(-8x^3)^{1/3} = \sqrt[3]{-8x^3} = -2x$
$4^{5/2} = (\sqrt{4})^5 = 2^5 = 32$
$27^{2/3} = (\sqrt[3]{27})^2 = 3^2 = 9$
$16^{-3/4} = \dfrac{1}{16^{3/4}} = \dfrac{1}{(\sqrt[4]{16})^3} = \dfrac{1}{2^3} = \dfrac{1}{8}$
$x^{2/3} \cdot x^{-5/6} = x^{2/3-5/6} = x^{-1/6} = \dfrac{1}{x^{1/6}}$
$(8^{14})^{1/7} = 8^2 = 64$
$\dfrac{a^{4/5}}{a^{-2/5}} = a^{4/5-(-2/5)} = a^{6/5}$

Section 7.3   Simplifying Radical Expressions

PRODUCT AND QUOTIENT RULES

If $\sqrt[n]{a}$ and $\sqrt[n]{b}$ are real numbers,

$$\sqrt[n]{a} \cdot \sqrt[n]{b} = \sqrt[n]{a \cdot b}$$

$$\dfrac{\sqrt[n]{a}}{\sqrt[n]{b}} = \sqrt[n]{\dfrac{a}{b}}, \text{provided } \sqrt[n]{b} \neq 0$$

A radical of the form $\sqrt[n]{a}$ is **simplified** when a contains no factors that are perfect nth powers.

Multiply or divide as indicated:

$$\sqrt{11} \cdot \sqrt{3} = \sqrt{33}$$

$$\dfrac{\sqrt[3]{40x}}{\sqrt[3]{5x}} = \sqrt[3]{8} = 2$$

$$\sqrt{40} = \sqrt{4 \cdot 10} = 2\sqrt{10}$$

$$\sqrt{36x^5} = \sqrt{36x^4 \cdot x} = 6x^2\sqrt{x}$$

$$\sqrt[3]{24x^7y^3} = \sqrt[3]{8x^6y^3 \cdot 3x} = 2x^2y\sqrt[3]{3x}$$

DISTANCE FORMULA

The distance d between two points (x_1, y_1) and (x_2, y_2) is given by

$$d = \sqrt{(x_2 - x_1)^2 + (y_2 - y_1)^2}$$

Find the distance between points $(-1, 6)$ and $(-2, -4)$. Let $(x_1, y_1) = (-1, 6)$ and $(x_2, y_2) = (-2, -4)$.

$$d = \sqrt{(x_2 - x_1)^2 + (y_2 - y_1)^2}$$

$$= \sqrt{(-2 - (-1))^2 + (-4 - 6)^2}$$

$$= \sqrt{1 + 100} = \sqrt{101}$$

MIDPOINT FORMULA

The midpoint of the line segment whose endpoints are (x_1, y_1) and (x_2, y_2) is the point with coordinates

$$\left(\dfrac{x_1 + x_2}{2}, \dfrac{y_1 + y_2}{2} \right)$$

Find the midpoint of the line segment whose endpoints are $(-1, 6)$ and $(-2, -4)$.

$$\left(\dfrac{-1 + (-2)}{2}, \dfrac{6 + (-4)}{2} \right)$$

The midpoint is $\left(-\dfrac{3}{2}, 1 \right)$.

Section 7.4   Adding, Subtracting, and Multiplying Radical Expressions

Radicals with the same index and the same radicand are **like radicals.**

The distributive property can be used to add like radicals.

Radical expressions are multiplied by using many of the same properties used to multiply polynomials.

$$5\sqrt{6} + 2\sqrt{6} = (5 + 2)\sqrt{6} = 7\sqrt{6}$$

$$-\sqrt[3]{3x} - 10\sqrt[3]{3x} + 3\sqrt[3]{10x}$$

$$= (-1 - 10)\sqrt[3]{3x} + 3\sqrt[3]{10x}$$

$$= -11\sqrt[3]{3x} + 3\sqrt[3]{10x}$$

DEFENITIONS AND CONCEPTS	EXAMPLES

Section 7.4   Adding, Subtracting, and Multiplying Radical Expressions (*continued*)

Multiply:

$$(\sqrt{5} - \sqrt{2x})(\sqrt{2} + \sqrt{2x})$$

$$= \sqrt{10} + \sqrt{10x} - \sqrt{4x} - 2x$$

$$= \sqrt{10} + \sqrt{10x} - 2\sqrt{x} - 2x$$

$$(2\sqrt{3} - \sqrt{8x})(2\sqrt{3} + \sqrt{8x})$$

$$= 4(3) - 8x = 12 - 8x$$

Section 7.5   Rationalizing Numerators and Denominators of Radical Expressions

The **conjugate** of $a + b$ is $a - b$.

The process of writing the denominator of a radical expression without a radical is called **rationalizing the denominator.**

The conjugate of $\sqrt{7} + \sqrt{3}$ is $\sqrt{7} - \sqrt{3}$.

Rationalize each denominator:

$$\frac{\sqrt{5}}{\sqrt{3}} = \frac{\sqrt{5} \cdot \sqrt{3}}{\sqrt{3} \cdot \sqrt{3}} = \frac{\sqrt{15}}{3}$$

$$\frac{6}{\sqrt{7} + \sqrt{3}} = \frac{6(\sqrt{7} - \sqrt{3})}{(\sqrt{7} + \sqrt{3})(\sqrt{7} - \sqrt{3})}$$

$$= \frac{6(\sqrt{7} - \sqrt{3})}{7 - 3}$$

$$= \frac{6(\sqrt{7} - \sqrt{3})}{4} = \frac{3(\sqrt{7} - \sqrt{3})}{2}$$

The process of writing the numerator of a radical expression without a radical is called **rationalizing the numerator.**

Rationalize each numerator:

$$\frac{\sqrt[3]{9}}{\sqrt[3]{5}} = \frac{\sqrt[3]{9} \cdot \sqrt[3]{3}}{\sqrt[3]{5} \cdot \sqrt[3]{3}} = \frac{\sqrt[3]{27}}{\sqrt[3]{15}} = \frac{3}{\sqrt[3]{15}}$$

$$\frac{\sqrt{9} + \sqrt{3x}}{12} = \frac{(\sqrt{9} + \sqrt{3x})(\sqrt{9} - \sqrt{3x})}{12(\sqrt{9} - \sqrt{3x})}$$

$$= \frac{9 - 3x}{12(\sqrt{9} - \sqrt{3x})}$$

$$= \frac{3(3 - x)}{3 \cdot 4(3 - \sqrt{3x})} = \frac{3 - x}{4(3 - \sqrt{3x})}$$

Section 7.6   Radical Equations and Problem Solving

SOLVING A RADICAL EQUATION

Step 1. Write the equation so that one radical is by itself on one side of the equation.

Step 2. Raise each side of the equation to a power equal to the index of the radical and simplify.

Step 3. If the equation still contains a radical, repeat Steps 1 and 2. If not, solve the equation.

Step 4. Check all proposed solutions in the original equation.

Solve: $x = \sqrt{4x + 9} + 3$

1. $x - 3 = \sqrt{4x + 9}$

2. $(x - 3)^2 = (\sqrt{4x + 9})^2$

$x^2 - 6x + 9 = 4x + 9$

3. $x^2 - 10x = 0$

$x(x - 10) = 0$

$x = 0$ or $x = 10$

4. The proposed solution 10 checks, but 0 does not. The solution set is $\{10\}$.

DEFINITIONS AND CONCEPTS	EXAMPLES
Section 7.7 Complex Numbers	

A **complex number** is a number that can be written in the form $a + bi$, where a and b are real numbers.

$$i^2 = -1 \quad \text{and} \quad i = \sqrt{-1}$$

Simplify: $\sqrt{-9}$

$$\sqrt{-9} = \sqrt{-1 \cdot 9} = \sqrt{-1} \cdot \sqrt{9} = i \cdot 3, \text{ or } 3i$$

Complex Numbers   Written in Form $a + bi$

12	$12 + 0i$
$-5i$	$0 + (-5)i$
$-2 - 3i$	$-2 + (-3)i$

Multiply:

$$\sqrt{-3} \cdot \sqrt{-7} = i\sqrt{3} \cdot i\sqrt{7}$$
$$= i^2\sqrt{21}$$
$$= -\sqrt{21}$$

To add or subtract complex numbers, add or subtract their real parts and then add or subtract their imaginary parts.

To multiply complex numbers, multiply as though they were binomials.

Perform each indicated operation.

$$(-3 + 2i) - (7 - 4i) = -3 + 2i - 7 + 4i$$
$$= -10 + 6i$$

$$(-7 - 2i)(6 + i) = -42 - 7i - 12i - 2i^2$$
$$= -42 - 19i - 2(-1)$$
$$= -42 - 19i + 2$$
$$= -40 - 19i$$

The complex numbers $(a + bi)$ and $(a - bi)$ are called **complex conjugates.**

The complex conjugate of

$$(3 + 6i) \text{ is } (3 - 6i).$$

Their product is a real number:

$$(3 - 6i)(3 + 6i) = 9 - 36i^2$$
$$= 9 - 36(-1) = 9 + 36 = 45$$

To divide complex numbers, multiply the numerator and the denominator by the conjugate of the denominator.

Divide:

$$\frac{4}{2 - i} = \frac{4(2 + i)}{(2 - i)(2 + i)}$$
$$= \frac{4(2 + i)}{4 - i^2}$$
$$= \frac{4(2 + i)}{4 - (-1)}$$
$$= \frac{8 + 4i}{5} = \frac{8}{5} + \frac{4}{5}i$$

7

(7.1) *Find each root. Assume that all variables represent positive real numbers.*

1. $\sqrt{81}$ *9*

2. $\sqrt[4]{81}$ *3*

3. $\sqrt[3]{-8}$ *−2*

4. $\sqrt[4]{-16}$ *∅*

5. $-\sqrt{\dfrac{1}{49}}$ *−1/7*

6. $\sqrt{x^{64}}$ *x³²*

7. $-\sqrt{36}$ *−6*

8. $\sqrt[3]{64}$ *4*

9. $\sqrt[3]{-a^6 b^9}$ *−a² b³*

10. $\sqrt{16a^4 b^{12}}$ *4a²b⁶*

11. $\sqrt[5]{32a^5 b^{10}}$ *2ab²*

12. $\sqrt[5]{-32x^{15}y^{20}}$ *−2x³y⁴*

13. $\sqrt{\dfrac{x^{12}}{36y^2}}$ *x⁶/6y*

14. $\sqrt[3]{\dfrac{27y^3}{z^{12}}}$ *3y/z⁴*

Simplify. Use absolute value bars when necessary.

15. $\sqrt{x^2}$ *x*

16. $\sqrt[4]{(x^2 - 4)^4}$ *(x−2)(x+2)*

17. $\sqrt[3]{(-27)^3}$ *−27*

18. $\sqrt[5]{(-5)^5}$ *−5*

19. $-\sqrt[5]{x^5}$ *−x*

20. $\sqrt[4]{16(2y + z)^4}$ *2(2y+z) 4y+2z*

21. $\sqrt{25(x - y)^2}$ *5x−5y*

22. $\sqrt[5]{y^5}$ *y*

23. $\sqrt[6]{x^6}$ *x*

24. If $f(x) = \sqrt{x} + 3$, find $f(0)$ and $f(9)$.   *f(0)=3 f(9)=6*

25. If $g(x) = \sqrt[3]{x} - 3$, find $g(11)$ and $g(20)$.   *g(11)=2 g(20)= 3√17*

(7.2) *Evaluate.*

26. $\left(\dfrac{1}{81}\right)^{1/4}$ *4√1/81 = 1/3*

27. $\left(-\dfrac{1}{27}\right)^{1/3}$

28. $(-27)^{-1/3}$

29. $(-64)^{-1/3}$

30. $-9^{3/2}$

31. $64^{-1/3}$

32. $(-25)^{5/2}$

33. $\left(\dfrac{25}{49}\right)^{-3/2}$

34. $\left(\dfrac{8}{27}\right)^{-2/3}$

35. $\left(-\dfrac{1}{36}\right)^{-1/4}$

Write with rational exponents.

36. $\sqrt[3]{x^2}$

37. $\sqrt[5]{5x^2 y^3}$

Write using radical notation.

38. $y^{4/5}$

39. $5(xy^2 z^5)^{1/3}$

40. $(x + 2y)^{-1/2}$

Simplify each expression. Assume that all variables represent positive real numbers. Write with only positive exponents.

41. $a^{1/3} a^{4/3} a^{1/2}$

42. $\dfrac{b^{1/3}}{b^{4/3}}$

43. $(a^{1/2} a^{-2})^3$

44. $(x^{-3} y^6)^{1/3}$

45. $\left(\dfrac{b^{3/4}}{a^{-1/2}}\right)^8$

46. $\dfrac{x^{1/4} x^{-1/2}}{x^{2/3}}$

47. $\left(\dfrac{49c^{5/3}}{a^{-1/4} b^{5/6}}\right)^{-1}$

48. $a^{-1/4}(a^{5/4} - a^{9/4})$

Use a calculator and write a three-decimal-place approximation of each number.

49. $\sqrt{20}$

50. $\sqrt[3]{-39}$

51. $\sqrt[4]{726}$

52. $56^{1/3}$

53. $-78^{3/4}$

54. $105^{-2/3}$

Use rational exponents to write each as a single radical.

55. $\sqrt[3]{2} \cdot \sqrt{7}$

56. $\sqrt[3]{3} \cdot \sqrt[4]{x}$

(7.3) *Perform each indicated operation and then simplify if possible. Assume that all variables represent positive real numbers.*

57. $\sqrt{3} \cdot \sqrt{8}$

58. $\sqrt[3]{7y} \cdot \sqrt[3]{x^2 z}$

59. $\dfrac{\sqrt{44x^3}}{\sqrt{11x}}$

60. $\dfrac{\sqrt[4]{a^6 b^{13}}}{\sqrt[4]{a^2 b}}$

Simplify.

61. $\sqrt{60}$

62. $-\sqrt{75}$

63. $\sqrt[3]{162}$

64. $\sqrt[3]{-32}$

65. $\sqrt{36x^7}$

66. $\sqrt[3]{24a^5 b^7}$

67. $\sqrt{\dfrac{p^{17}}{121}}$

68. $\sqrt[3]{\dfrac{y^5}{27x^6}}$

69. $\sqrt[4]{\dfrac{xy^6}{81}}$

70. $\sqrt{\dfrac{2x^3}{49y^4}}$

△ **71.** The formula for the radius r of a circle of area A is

$$r = \sqrt{\dfrac{A}{\pi}}$$

    a. Find the exact radius of a circle whose area is 25 square meters.

🖩 **b.** Approximate to two decimal places the radius of a circle whose area is 104 square inches.

Find the distance between each pair of points. Give an exact value and a three-decimal-place approximation.

72. $(-6, 3)$ and $(8, 4)$

73. $(-4, -6)$ and $(-1, 5)$

74. $(-1, 5)$ and $(2, -3)$

75. $(-\sqrt{2}, 0)$ and $(0, -4\sqrt{6})$

76. $(-\sqrt{5}, -\sqrt{11})$ and $(-\sqrt{5}, -3\sqrt{11})$

🖩 **77.** $(7.4, -8.6)$ and $(-1.2, 5.6)$

Find the midpoint of each line segment whose endpoints are given.

78. $(2, 6); (-12, 4)$

79. $(-6, -5); (-9, 7)$

80. $(4, -6); (-15, 2)$

81. $\left(0, -\dfrac{3}{8}\right); \left(\dfrac{1}{10}, 0\right)$

82. $\left(\dfrac{3}{4}, -\dfrac{1}{7}\right); \left(-\dfrac{1}{4}, -\dfrac{3}{7}\right)$

83. $(\sqrt{3}, -2\sqrt{6})$ and $(\sqrt{3}, -4\sqrt{6})$

(7.4) *Perform each indicated operation. Assume that all variables represent positive real numbers.*

84. $\sqrt{20} + \sqrt{45} - 7\sqrt{5}$

85. $x\sqrt{75x} - \sqrt{27x^3}$

86. $\sqrt[3]{128} + \sqrt[3]{250}$

87. $3\sqrt[4]{32a^5} - a\sqrt[4]{162a}$

88. $\dfrac{5}{\sqrt{4}} + \dfrac{\sqrt{3}}{3}$

89. $\sqrt{\dfrac{8}{x^2}} - \sqrt{\dfrac{50}{16x^2}}$

90. $2\sqrt{50} - 3\sqrt{125} + \sqrt{98}$

91. $2a\sqrt[4]{32b^5} - 3b\sqrt[4]{162a^4 b} + \sqrt[4]{2a^4 b^5}$

Multiply and then simplify if possible. Assume that all variables represent positive real numbers.

92. $\sqrt{3}(\sqrt{27} - \sqrt{3})$

93. $(\sqrt{x} - 3)^2$

94. $(\sqrt{5} - 5)(2\sqrt{5} + 2)$

95. $(2\sqrt{x} - 3\sqrt{y})(2\sqrt{x} + 3\sqrt{y})$

96. $(\sqrt{a} + 3)(\sqrt{a} - 3)$

97. $(\sqrt[3]{a} + 2)^2$

98. $(\sqrt[3]{5x} + 9)(\sqrt[3]{5x} - 9)$

99. $(\sqrt[3]{a} + 4)(\sqrt[3]{a^2} - 4\sqrt[3]{a} + 16)$

(7.5) *Rationalize each denominator. Assume that all variables represent positive real numbers.*

100. $\dfrac{3}{\sqrt{7}}$

101. $\sqrt{\dfrac{x}{12}}$

102. $\dfrac{5}{\sqrt[3]{4}}$

103. $\sqrt{\dfrac{24x^5}{3y^2}}$

104. $\sqrt[3]{\dfrac{15x^6y^7}{z^2}}$

105. $\sqrt[4]{\dfrac{81}{8x^{10}}}$

106. $\dfrac{3}{\sqrt{y}-2}$

107. $\dfrac{\sqrt{2}-\sqrt{3}}{\sqrt{2}+\sqrt{3}}$

Rationalize each numerator. Assume that all variables represent positive real numbers.

108. $\dfrac{\sqrt{11}}{3}$

109. $\sqrt{\dfrac{18}{y}}$

110. $\dfrac{\sqrt[3]{9}}{7}$

111. $\sqrt{\dfrac{24x^5}{3y^2}}$

112. $\sqrt[3]{\dfrac{xy^2}{10z}}$

113. $\dfrac{\sqrt{x}+5}{-3}$

(7.6) *Solve each equation.*

114. $\sqrt{y-7}=5$

115. $\sqrt{2x}+10=4$

116. $\sqrt[3]{2x-6}=4$

117. $\sqrt{x+6}=\sqrt{x+2}$

118. $2x-5\sqrt{x}=3$

119. $\sqrt{x+9}=2+\sqrt{x-7}$

Find each unknown length.

△ **120.**

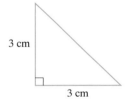

△ **121.**

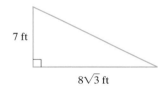

△ **122.** Craig and Daniel Cantwell want to determine the distance *x* across a pond on their property. They are able to measure the distances shown on the following diagram. Find how wide the lake is at the crossing point indicated by the triangle to the nearest tenth of a foot.

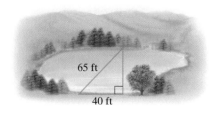

△ **123.** Andrea Roberts, a pipefitter, needs to connect two underground pipelines that are offset by 3 feet, as pictured in the diagram. Neglecting the joints needed to join the pipes, find the length of the shortest possible connecting pipe rounded to the nearest hundredth of a foot.

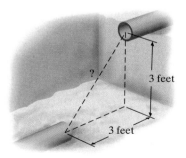

(7.7) *Perform each indicated operation and simplify. Write the results in the form a + bi.*

124. $\sqrt{-8}$

125. $-\sqrt{-6}$

126. $\sqrt{-4}+\sqrt{-16}$

127. $\sqrt{-2}\cdot\sqrt{-5}$

128. $(12-6i)+(3+2i)$

129. $(-8-7i)-(5-4i)$

130. $(2i)^6$ **131.** $-3i(6 - 4i)$ **132.** $(3 + 2i)(1 + i)$ **133.** $(2 - 3i)^2$

134. $(\sqrt{6} - 9i)(\sqrt{6} + 9i)$ **135.** $\dfrac{2 + 3i}{2i}$ **136.** $\dfrac{1 + i}{-3i}$

Mixed Review

Simplify. Use absolute value bars when necessary.

137. $\sqrt[3]{x^3}$ **138.** $\sqrt{(x + 2)^2}$

Simplify. Assume that all variables represent positive real numbers. If necessary, write answers with positive exponents only.

139. $-\sqrt{100}$ **140.** $\sqrt[3]{-x^{12}y^3}$ **141.** $\sqrt[4]{\dfrac{y^{20}}{16x^{12}}}$ **142.** $9^{1/2}$

143. $64^{-1/2}$ **144.** $\left(\dfrac{27}{64}\right)^{-2/3}$ **145.** $\dfrac{(x^{2/3}x^{-3})^3}{x^{-1/2}}$ **146.** $\sqrt{200x^9}$

147. $\sqrt{\dfrac{3n^3}{121m^{10}}}$ **148.** $3\sqrt{20} - 7x\sqrt[3]{40} + 3\sqrt[3]{5x^3}$

149. $(2\sqrt{x} - 5)^2$ **150.** Find the distance between $(-3, 5)$ and $(-8, 9)$.

151. Find the midpoint of the line segment joining $(-3, 8)$ and $(11, 24)$.

Rationalize each denominator.

152. $\dfrac{7}{\sqrt{13}}$ **153.** $\dfrac{2}{\sqrt{x} + 3}$

Solve.

154. $\sqrt{x} + 2 = x$

7 CHAPTER TEST

 Remember to use the Chapter Test Prep Video CD to see the fully worked-out solutions to any of the exercises you want to review.

Raise to the power or find the root. Assume that all variables represent positive real numbers. Write with only positive exponents.

1. $\sqrt{216}$

2. $-\sqrt[4]{x^{64}}$

3. $\left(\dfrac{1}{125}\right)^{1/3}$

4. $\left(\dfrac{1}{125}\right)^{-1/3}$

5. $\left(\dfrac{8x^3}{27}\right)^{2/3}$

6. $\sqrt[3]{-a^{18}b^9}$

7. $\left(\dfrac{64c^{4/3}}{a^{-2/3}b^{5/6}}\right)^{1/2}$

8. $a^{-2/3}(a^{5/4} - a^3)$

Find each root. Use absolute value bars when necessary.

9. $\sqrt[4]{(4xy)^4}$

10. $\sqrt[3]{(-27)^3}$

Rationalize each denominator. Assume that all variables represent positive real numbers.

11. $\sqrt{\dfrac{9}{y}}$

12. $\dfrac{4 - \sqrt{x}}{4 + 2\sqrt{x}}$

13. $\sqrt[3]{\dfrac{8}{9x}}$

14. Rationalize the numerator of $\dfrac{\sqrt{6} + x}{8}$ and simplify.

Perform each indicated operation. Assume that all variables represent positive real numbers.

15. $\sqrt{125x^3} - 3\sqrt{20x^3}$

16. $\sqrt{3}(\sqrt{16} - \sqrt{2})$

17. $(\sqrt{x} + 1)^2$

18. $(\sqrt{2} - 4)(\sqrt{3} + 1)$

19. $(\sqrt{5} + 5)(\sqrt{5} - 5)$

▦ *Use a calculator to approximate each number to three decimal places.*

20. $\sqrt{561}$

21. $386^{-2/3}$

Answers

1. _____

2. _____

3. _____

4. _____

5. _____

6. _____

7. _____

8. _____

9. _____

10. _____

11. _____

12. _____

13. _____

14. _____

15. _____

16. _____

17. _____

18. _____

19. _____

20. _____

21. _____

22. _____

23. _____

24. _____

25. _____

26. _____

27. _____

28. _____

29. _____

30. _____

31. _____

32. _____

33. _____

34. _____

35. _____

36. _____

37. _____

38. _____

Solve.

22. $x = \sqrt{x - 2} + 2$

23. $\sqrt{x^2 - 7} + 3 = 0$

24. $\sqrt{x + 5} = \sqrt{2x - 1}$

Perform each indicated operation and simplify. Write the results in the form a + bi.

25. $\sqrt{-2}$

26. $-\sqrt{-8}$

27. $(12 - 6i) - (12 - 3i)$

28. $(6 - 2i)(6 + 2i)$

29. $(4 + 3i)^2$

30. $\dfrac{1 + 4i}{1 - i}$

△ **31.** Find x.

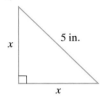

5 in.

x

x

32. If $g(x) = \sqrt{x + 2}$, find $g(0)$ and $g(23)$.

33. Find the distance between the points $(-6, 3)$ and $(-8, -7)$.

34. Find the distance between the points $(-2\sqrt{5}, \sqrt{10})$ and $(-\sqrt{5}, 4\sqrt{10})$.

35. Find the midpoint of the line segment whose endpoints are $(-2, -5)$ and $(-6, 12)$.

36. Find the midpoint of the line segment whose endpoints are $\left(-\dfrac{2}{3}, -\dfrac{1}{5}\right)$ and $\left(-\dfrac{1}{3}, \dfrac{4}{5}\right)$.

Solve.

37. The function $V = \sqrt{2.5r}$ can be used to estimate the maximum safe velocity, V, in miles per hour, at which a car can travel if it is driven along a curved road with a *radius of curvature, r,* in feet. To the nearest whole number, find the maximum safe speed if a cloverleaf exit on an expressway has a radius of curvature of 300 feet.

38. Use the formula from Exercise 37 to find the radius of curvature if the safe velocity is 30 miles per hour.

Simplify. Assume that a and b are integers and that x and y are not 0.

1. $x^{-b}(2x^b)^2$

2. $\left(-\dfrac{1}{3}x^a\right)\left(-\dfrac{2}{5}x^a\right)$

3. $\dfrac{(y^{3a})^2}{y^{a-6}}$

4. $\dfrac{2.8y^{7a}}{-2y^{-2a}}$

5. Solve: $|x + 1| = 6$

6. Solve: $|3x - 2| + 5 = 5$

7. Write an equation of the line containing the point $(4, 4)$ and parallel to the line $2x + y = -6$. Write the equation in slope-intercept form.

8. Solve: $\dfrac{a - 1}{2} + a = 2 - \dfrac{2a + 7}{8}$

9. Use the elimination method to solve the system: $\begin{cases} x - 5y = -12 \\ -x + y = 4 \end{cases}$

10. Find the slope of $y = -3$.

11. Solve the system:
$$\begin{cases} x - 5y - 2z = 6 \\ -2x + 10y + 4z = -12 \\ \dfrac{1}{2}x - \dfrac{5}{2}y - z = 3 \end{cases}$$

12. Use the substitution method to solve the system.
$$\begin{cases} \dfrac{x}{6} - \dfrac{y}{2} = 1 \\ \dfrac{x}{3} - \dfrac{y}{4} = 2 \end{cases}$$

13. Use matrices to solve the system:
$$\begin{cases} 2x - y = 3 \\ 4x - 2y = 5 \end{cases}$$

14. Simplify each expression.
  a. $2(x - 3) + (5x + 3)$
  b. $4(3x + 2) - 3(5x - 1)$
  c. $7x + 2(x - 7) - 3x$

15. Add:
$(7x^3y - xy^3 + 11) + (6x^3y - 4)$

16. Use scientific notation to simplify and write the answer in scientific notation.
$$\dfrac{0.0000035 \times 4000}{0.28}$$

If $P(x) = 3x^4 - 2x^2 - 5$, find the following.

17. $P(1)$

18. $P(0)$

19. $P(-2)$

20. $P\left(\dfrac{1}{2}\right)$

Multiply.

21. $(x + 5)^2$

22. $(y - 2)(3y + 4)$

Answers

1. _____
2. _____
3. _____
4. _____
5. _____
6. _____
7. _____
8. _____
9. _____
10. _____
11. _____
12. _____
13. _____
14. a. _____
   b. _____
   c. _____
15. _____
16. _____
17. _____
18. _____
19. _____
20. _____
21. _____
22. _____

23. _____

24. _____

25. _____

26. _____

27. _____

28. _____

29. _____

30. _____

31. _____

32. _____

33. _____

34. _____

35. _____

36. _____

37. _____

38. a. _____

   b. _____

39. _____

40. _____

41. _____

42. _____

23. $(4m^2 - 3n)^2$

24. $(3y - 1)(2y^2 + 3y - 1)$

25. Use synthetic division to divide $2x^3 - x^2 - 13x + 1$ by $x - 3$.

26. Use synthetic division to divide $4y^3 - 12y^2 - y + 12$ by $y - 3$.

Factor.

27. $ab - 6a + 2b - 12$

28. Factor: $x^3 - x^2 + 4x - 4$

29. $2n^2 - 38n + 80$

30. $2x^3 - 16$

31. $16x^2 + 24xy + 9y^2$

32. $x^2 + 2x + 1 - y^2$

33. $50 - 8y^2$

34. $x^4 - 1$

35. Solve: $x(2x - 7) = 4$

36. Divide $x^3 - 2x^2 + 3x - 6$ by $x - 2$.

37. Solve: $x^3 = 4x$

38. Simplify each complex fraction.

a. $\dfrac{\dfrac{y - 2}{16}}{\dfrac{2y + 3}{12}}$

b. $\dfrac{\dfrac{x}{16} - \dfrac{1}{x}}{1 - \dfrac{4}{x}}$

Simplify each rational expression.

39. $\dfrac{2 + x}{x + 2}$

40. $\dfrac{a^3 - 8}{2 - a}$

41. $\dfrac{2x^2}{10x^3 - 2x^2}$

42. $\dfrac{3a^2 - 3}{a^3 + 5a^2 - a - 5}$

43. Add: $\dfrac{5}{7z^2} + \dfrac{x}{7z^2}$

44. Perform the indicated operations.

   a. $\dfrac{3}{xy^2} - \dfrac{2}{3x^2y}$ **b.** $\dfrac{5x}{x+3} - \dfrac{2x}{x-3}$

   c. $\dfrac{x}{x-2} - \dfrac{5}{2-x}$

45. Solve: $\dfrac{3}{x} - \dfrac{x+21}{3x} = \dfrac{5}{3}$

46. Solve: $\dfrac{28}{9-a^2} = \dfrac{2a}{a-3} + \dfrac{6}{a+3}$

Write each expression with a positive exponent, and then simplify.

47. $16^{-3/4}$

48. $81^{-3/4}$

49. $(-27)^{-2/3}$

50. $(-125)^{-2/3}$

43. _____

44. a. _____

   b. _____

   c. _____

45. _____

46. _____

47. _____

48. _____

49. _____

50. _____

8

Quadratic Equations and Functions

An important part of algebra is learning to model and solve problems. Often, the model of a problem is a quadratic equation or a function containing a second-degree polynomial. In this chapter, we continue the work from Chapter 5, solving polynomial equations in one variable by factoring. Two other methods of solving quadratic equations are analyzed in this chapter, with methods of solving nonlinear inequalities in one variable and the graphs of quadratic functions.

America, as a society, is highly dependent on the automobile. There is a mystique about the "open road" that has been enhanced by popular songs and movies. Each year, more cars join the traffic flow, often instead leading to gridlock. Automobiles have a major impact on the environment. As the U.S. Environmental Protection Agency has stated, "driving a private car is probably a typical citizen's most 'polluting' daily activity."

Each year, the United States produces only about 10% of the world's petroleum but consumes about 26% of the world's total production. Cars and light trucks are the single largest users of petroleum, consuming about 43% of the total.

In Section 8.2, Exercise 89, we use a quadratic equation to examine the gasoline supply in the United States.

(*Source:* Environmental Protection Agency, Energy Information Administration)

Amount of Time It Takes in Car Commuting to Work

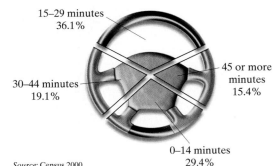

15–29 minutes
36.1%

45 or more minutes
15.4%

30–44 minutes
19.1%

0–14 minutes
29.4%

Source: Census 2000

8.1 SOLVING QUADRATIC EQUATIONS BY COMPLETING THE SQUARE

Objectives

A Use the Square Root Property to Solve Quadratic Equations.

B Write Perfect Square Trinomials.

C Solve Quadratic Equations by Completing the Square.

D Use Quadratic Equations to Solve Problems.

Objective A Using the Square Root Property

In Chapter 5, we solved quadratic equations by factoring. Recall that a **quadratic**, or **second-degree, equation** is an equation that can be written in the form $ax^2 + bx + c = 0$, where a, b, and c are real numbers and a is not 0. To solve a quadratic equation such as $x^2 = 9$ by factoring, we use the zero-factor property. To use the zero-factor property, the equation must first be written in the standard form $ax^2 + bx + c = 0$.

$$x^2 = 9$$
$$x^2 - 9 = 0 \qquad \text{Subtract 9 from both sides to write in standard form.}$$
$$(x + 3)(x - 3) = 0 \qquad \text{Factor.}$$
$$x + 3 = 0 \quad \text{or} \quad x - 3 = 0 \qquad \text{Set each factor equal to 0.}$$
$$x = -3 \qquad\qquad x = 3 \qquad \text{Solve.}$$

The solution set is $\{-3, 3\}$, the positive and negative square roots of 9.

Not all quadratic equations can be solved by factoring, so we need to explore other methods. Notice that the solutions of the equation $x^2 = 9$ are two numbers whose square is 9:

$$3^2 = 9 \qquad \text{and} \qquad (-3)^2 = 9$$

Thus, we can solve the equation $x^2 = 9$ by taking the square root of both sides. Be sure to include both $\sqrt{9}$ and $-\sqrt{9}$ as solutions since both $\sqrt{9}$ and $-\sqrt{9}$ are numbers whose square is 9.

$$x^2 = 9$$
$$x = \pm\sqrt{9} \qquad \text{The notation } \pm\sqrt{9} \text{ (read as "plus or minus } \sqrt{9}\text{") indicates the pair of}$$
$$x = \pm 3 \qquad \text{numbers } +\sqrt{9} \text{ and } -\sqrt{9}.$$

This illustrates the square root property.

> **Helpful Hint**
>
> The notation ± 3, for example, is read as "plus or minus 3." It is a shorthand notation for the pair of numbers $+3$ and -3.

Square Root Property

If b is a real number and if $a^2 = b$, then $a = \pm\sqrt{b}$.

EXAMPLE 1 Use the square root property to solve $x^2 = 50$.

Solution:

$$x^2 = 50$$
$$x = \pm\sqrt{50} \qquad \text{Use the square root property.}$$
$$x = \pm 5\sqrt{2} \qquad \text{Simplify the radical.}$$

Continued on next page

PRACTICE PROBLEM 1

Use the square root property to solve $x^2 = 45$.

Answer

1. $\{3\sqrt{5}, -3\sqrt{5}\}$

553

Check: Let $x = 5\sqrt{2}$.   Let $x = -5\sqrt{2}$.

$$x^2 = 50 \qquad\qquad\qquad x^2 = 50$$
$$(5\sqrt{2})^2 \stackrel{?}{=} 50 \qquad\qquad (-5\sqrt{2})^2 \stackrel{?}{=} 50$$
$$25 \cdot 2 \stackrel{?}{=} 50 \qquad\qquad 25 \cdot 2 \stackrel{?}{=} 50$$
$$50 = 50 \quad \text{True} \qquad\qquad 50 = 50 \quad \text{True}$$

The solution set is $\{5\sqrt{2}, -5\sqrt{2}\}$.

▣ **Work Practice Problem 1**

PRACTICE PROBLEM 2

Use the square root property to solve $5x^2 = 55$.

EXAMPLE 2 Use the square root property to solve $2x^2 = 14$.

Solution: First we get the squared variable alone on one side of the equation.

$$2x^2 = 14$$
$$x^2 = 7 \qquad \text{Divide both sides by 2.}$$
$$x = \pm\sqrt{7} \qquad \text{Use the square root property.}$$

Check: Let $x = \sqrt{7}$.   Let $x = -\sqrt{7}$.

$$2x^2 = 14 \qquad\qquad\qquad 2x^2 = 14$$
$$2(\sqrt{7})^2 \stackrel{?}{=} 14 \qquad\qquad 2(-\sqrt{7})^2 \stackrel{?}{=} 14$$
$$2 \cdot 7 \stackrel{?}{=} 14 \qquad\qquad 2 \cdot 7 \stackrel{?}{=} 14$$
$$14 = 14 \quad \text{True} \qquad\qquad 14 = 14 \quad \text{True}$$

The solution set is $\{\sqrt{7}, -\sqrt{7}\}$.

▣ **Work Practice Problem 2**

PRACTICE PROBLEM 3

Use the square root property to solve $(x + 2)^2 = 18$.

EXAMPLE 3 Use the square root property to solve $(x + 1)^2 = 12$.

Solution: $(x + 1)^2 = 12$

$$x + 1 = \pm\sqrt{12} \qquad \text{Use the square root property.}$$
$$x + 1 = \pm 2\sqrt{3} \qquad \text{Simplify the radical.}$$
$$x = -1 \pm 2\sqrt{3} \qquad \text{Subtract 1 from both sides.}$$

> **Helpful Hint**
>
> Don't forget that $-1 \pm 2\sqrt{3}$, for example, means $-1 + 2\sqrt{3}$ and $-1 - 2\sqrt{3}$. In other words, the equation in Example 3 has two solutions.

Check: Below is a check for $-1 + 2\sqrt{3}$. The check for $-1 - 2\sqrt{3}$ is almost the same and is left for you to do on your own.

$$(x + 1)^2 = 12$$
$$(-1 + 2\sqrt{3} + 1)^2 \stackrel{?}{=} 12$$
$$(2\sqrt{3})^2 \stackrel{?}{=} 12$$
$$4 \cdot 3 \stackrel{?}{=} 12$$
$$12 = 12 \quad \text{True}$$

The solution set is $\{-1 + 2\sqrt{3}, -1 - 2\sqrt{3}\}$.

▣ **Work Practice Problem 3**

Answers

2. $\{\sqrt{11}, -\sqrt{11}\}$,

3. $\{-2 + 3\sqrt{2}, -2 - 3\sqrt{2}\}$

EXAMPLE 4 Use the square root property to solve $(2x - 5)^2 = -16$.

Solution: $(2x - 5)^2 = -16$

$2x - 5 = \pm\sqrt{-16}$ Use the square root property.

$2x - 5 = \pm 4i$ Simplify the radical.

$2x = 5 \pm 4i$ Add 5 to both sides.

$x = \dfrac{5 \pm 4i}{2}$ Divide both sides by 2.

Check each proposed solution in the original equation to see that the solution set is $\left\{\dfrac{5 + 4i}{2}, \dfrac{5 - 4i}{2}\right\}$.

■ **Work Practice Problem 4**

✔ **Concept Check** How do you know just by looking that $(x - 2)^2 = -81$ has complex solutions?

Objective B Writing Perfect Square Trinomials

Notice from Examples 3 and 4 that, if we write a quadratic equation so that one side is the square of a binomial, we can solve by using the square root property. To write the square of a binomial, we must have a perfect square trinomial. Recall that a perfect square trinomial is a trinomial that can be factored into two identical binomial factors, that is, as a binomial squared.

Perfect Square Trinomials	Factored Form
$x^2 + 8x + 16$	$(x + 4)^2$
$x^2 - 6x + 9$	$(x - 3)^2$
$x^2 + 3x + \dfrac{9}{4}$	$\left(x + \dfrac{3}{2}\right)^2$

Notice that for each perfect square trinomial, *the constant term of the trinomial is the square of half the coefficient of the x-term.* For example,

$x^2 + 8x + 16$ $x^2 - 6x + 9$

$\dfrac{1}{2}(8) = 4$ and $4^2 = 16$ $\dfrac{1}{2}(-6) = -3$ and $(-3)^2 = 9$

EXAMPLE 5 Add the proper constant to $x^2 + 6x$ so that the result is a perfect square trinomial. Then factor.

Solution: We add the square of half the coefficient of x.

$x^2 + 6x + 9 \quad = \quad (x + 3)^2$ In factored form

$\dfrac{1}{2}(6) = 3$ and $3^2 = 9$

■ **Work Practice Problem 5**

PRACTICE PROBLEM 4

Use the square root property to solve $(3x - 1)^2 = -4$.

PRACTICE PROBLEM 5

Add the proper constant to $x^2 + 12x$ so that the result is a perfect square trinomial. Then factor.

Answers

4. $\left\{\dfrac{1 - 2i}{3}, \dfrac{1 + 2i}{3}\right\}$,

5. $x^2 + 12x + 36 = (x + 6)^2$

✔ **Concept Check Answer**

answers may vary

Add the proper constant to $y^2 - 5y$ so that the result is a perfect square trinomial. Then factor.

EXAMPLE 6   Add the proper constant to $x^2 - 3x$ so that the result is a perfect square trinomial. Then factor.

Solution: We add the square of half the coefficient of x.

$$x^2 - 3x + \frac{9}{4} \quad = \left(x - \frac{3}{2} \right)^2 \quad \text{In factored form}$$

$$\frac{1}{2}(-3) = -\frac{3}{2} \text{ and } \left(-\frac{3}{2} \right)^2 = \frac{9}{4}$$

▣ **Work Practice Problem 6**

Objective ⓒ **Solving by Completing the Square**

The process of writing a quadratic equation so that one side is a perfect square trinomial is called **completing the square.** We will use this process in the next examples.

PRACTICE PROBLEM 7

Solve $x^2 + 8x = 1$ by completing the square.

EXAMPLE 7   Solve $p^2 + 2p = 4$ by completing the square.

Solution: First we add the square of half the coefficient of p to both sides so that the resulting trinomial will be a perfect square trinomial. The coefficient of p is 2.

$$\frac{1}{2}(2) = 1 \qquad \text{and} \qquad 1^2 = 1$$

Now we add 1 to both sides of the original equation.

$$p^2 + 2p = 4$$
$$p^2 + 2p + 1 = 4 + 1 \quad \text{Add 1 to both sides.}$$
$$(p + 1)^2 = 5 \qquad \text{Factor the trinomial; simplify the right side.}$$

We may now use the square root property and solve for p.

$$p + 1 = \pm\sqrt{5} \qquad \text{Use the square root property.}$$
$$p = -1 \pm \sqrt{5} \quad \text{Subtract 1 from both sides.}$$

Don't forget that there are two solutions: $-1 + \sqrt{5}$ and $-1 - \sqrt{5}$. The solution set is $\{-1 + \sqrt{5}, -1 - \sqrt{5}\}$.

▣ **Work Practice Problem 7**

PRACTICE PROBLEM 8

Solve $y^2 - 5y + 2 = 0$ by completing the square.

EXAMPLE 8   Solve $m^2 - 7m - 1 = 0$ by completing the square.

Solution: First we add 1 to both sides of the equation so that the left side has no constant term. We can then add the constant term on both sides that will make the left side a perfect square trinomial.

$$m^2 - 7m - 1 = 0$$
$$m^2 - 7m = 1$$

Now we find the constant term that makes the left side a perfect square trinomial by squaring half the coefficient of m. We add this constant to both sides of the equation.

$$\frac{1}{2}(-7) = -\frac{7}{2} \qquad \text{and} \qquad \left(-\frac{7}{2} \right)^2 = \frac{49}{4}$$

Answers

6. $y^2 - 5y + \frac{25}{4} = \left(y - \frac{5}{2} \right)^2$,

7. $\{-4 - \sqrt{17}, -4 + \sqrt{17}\}$,

8. $\left\{ \dfrac{5 - \sqrt{17}}{2}, \dfrac{5 + \sqrt{17}}{2} \right\}$

$$m^2 - 7m + \frac{49}{4} = 1 + \frac{49}{4} \qquad \text{Add } \frac{49}{4} \text{ to both sides of the equation.}$$

$$\left(m - \frac{7}{2}\right)^2 = \frac{53}{4} \qquad \text{Factor the perfect square trinomial and simplify the right side.}$$

$$m - \frac{7}{2} = \pm\sqrt{\frac{53}{4}} \qquad \text{Use the square root property.}$$

$$m = \frac{7}{2} \pm \frac{\sqrt{53}}{2} \qquad \text{Add } \frac{7}{2} \text{ to both sides and simplify } \sqrt{\frac{53}{4}}.$$

$$m = \frac{7 \pm \sqrt{53}}{2} \qquad \text{Simplify.}$$

The solution set is $\left\{ \dfrac{7 + \sqrt{53}}{2}, \dfrac{7 - \sqrt{53}}{2} \right\}$.

■ **Work Practice Problem 8**

The following steps may be used to solve a quadratic equation such as $ax^2 + bx + c = 0$ by completing the square. This method may be used whether or not the polynomial $ax^2 + bx + c$ is factorable.

Solving a Quadratic Equation in x by Completing the Square

Step 1: If the coefficient of x^2 is 1, go to Step 2. Otherwise, divide both sides of the equation by the coefficient of x^2.

Step 2: Get all variable terms alone on one side of the equation.

Step 3: Complete the square for the resulting binomial by adding the square of half of the coefficient of x to both sides of the equation.

Step 4: Factor the resulting perfect square trinomial and write it as the square of a binomial.

Step 5: Use the square root property to solve for x.

EXAMPLE 9 Solve $4x^2 - 24x + 41 = 0$ by completing the square.

Solution: First we divide both sides of the equation by 4 so that the coefficient of x^2 is 1.

$$4x^2 - 24x + 41 = 0$$

Step 1: $\quad x^2 - 6x + \dfrac{41}{4} = 0 \qquad$ Divide both sides of the equation by 4.

Step 2: $\quad x^2 - 6x = -\dfrac{41}{4} \qquad$ Subtract $\dfrac{41}{4}$ from both sides.

Since $\dfrac{1}{2}(-6) = -3$ and $(-3)^2 = 9$, we add 9 to both sides of the equation.

Step 3: $\quad x^2 - 6x + 9 = -\dfrac{41}{4} + 9 \qquad$ Add 9 to both sides.

Step 4: $\quad (x - 3)^2 = -\dfrac{41}{4} + \dfrac{36}{4} \qquad$ Factor the perfect square trinomial.

$$(x - 3)^2 = -\dfrac{5}{4}$$

Continued on next page

PRACTICE PROBLEM 9

Solve $2x^2 - 2x + 7 = 0$ by completing the square.

Answer

9. $\left\{ \dfrac{1 + i\sqrt{13}}{2}, \dfrac{1 - i\sqrt{13}}{2} \right\}$

Step 5: $x - 3 = \pm\sqrt{-\dfrac{5}{4}}$ Use the square root property.

$x - 3 = \pm\dfrac{i\sqrt{5}}{2}$ Simplify the radical.

$x = 3 \pm \dfrac{i\sqrt{5}}{2}$ Add 3 to both sides.

$= \dfrac{6}{2} \pm \dfrac{i\sqrt{5}}{2}$ Find a common denominator.

$= \dfrac{6 \pm i\sqrt{5}}{2}$ Simplify.

The solution set is $\left\{\dfrac{6 + i\sqrt{5}}{2}, \dfrac{6 - i\sqrt{5}}{2}\right\}$.

■ **Work Practice Problem 9**

Objective D Solving Problems Modeled by Quadratic Equations

Recall the **simple interest** formula $I = Prt$, where I is the interest earned, P is the principal, r is the rate of interest, and t is time. If \$100 is invested at a simple interest rate of 5% annually, at the end of 3 years the total interest I earned is

$$I = P \cdot r \cdot t$$

or

$$I = 100 \cdot 0.05 \cdot 3 = \$15$$

and the new principal is

$$\$100 + \$15 = \$115$$

Most of the time, the interest computed on money borrowed or money deposited is **compound interest.** Unlike simple interest, compound interest is computed on original principal *and* on interest already earned. To see the difference between simple interest and compound interest, suppose that \$100 is invested at a rate of 5% compounded annually. To find the total amount of money at the end of 3 years, we calculate as follows:

$$I = P \cdot r \cdot t$$

First year: Interest $= \$100 \cdot 0.05 \cdot 1 = \$5.00$
New principal $= \$100.00 + \$5.00 = \$105.00$

Second year: Interest $= \$105.00 \cdot 0.05 \cdot 1 = \$5.25$
New principal $= 105.00 + \$5.25 = \$110.25$

Third year: Interest $= \$110.25 \cdot 0.05 \cdot 1 \approx \$5.51$
New principal $= \$110.25 + \$5.51 = \$115.76$

At the end of the third year, the total compound interest earned is \$15.76, whereas the total simple interest earned is \$15.

It is tedious to calculate compound interest as we did above, so we use a compound interest formula. The formula for calculating the total amount of money when interest is compounded annually is

$$A = P(1 + r)^t$$

where P is the original investment, r is the interest rate per compounding period, and t is the number of periods. For example, the amount of money A at the end of 3 years if \$100 is invested at 5% compounded annually is

$$A = \$100(1 + 0.05)^3 \approx 100(1.1576) = \$115.76$$

as we previously calculated.

EXAMPLE 10 **Finding an Interest Rate**

Find the interest rate r if \$2000 compounded annually grows to \$2420 in 2 years.

Solution:

1. **UNDERSTAND** the problem. For this example, make sure that you understand the formula for compounding interest annually.

2. **TRANSLATE.** We substitute the given values into the formula.

$$A = P(1 + r)^t$$
$$2420 = 2000(1 + r)^2 \quad \text{Let } A = 2420, P = 2000, \text{ and } t = 2.$$

3. **SOLVE.** We now solve the equation for r.

$$2420 = 2000(1 + r)^2$$

$$\frac{2420}{2000} = (1 + r)^2 \qquad \text{Divide both sides by 2000.}$$

$$\frac{121}{100} = (1 + r)^2 \qquad \text{Simplify the fraction.}$$

$$\pm\sqrt{\frac{121}{100}} = 1 + r \qquad \text{Use the square root property.}$$

$$\pm\frac{11}{10} = 1 + r \qquad \text{Simplify.}$$

$$-1 \pm \frac{11}{10} = r$$

$$-\frac{10}{10} \pm \frac{11}{10} = r$$

$$\frac{1}{10} = r \quad \text{or} \quad -\frac{21}{10} = r$$

4. **INTERPRET.** The rate cannot be negative, so we reject $-\frac{21}{10}$.

Check: $\frac{1}{10} = 0.10 = 10\%$ per year. If we invest \$2000 at 10% compounded annually, in 2 years the amount in the account would be $2000(1 + 0.10)^2 = 2420$ dollars, the desired amount.

State: The interest rate is 10% compounded annually.

■ **Work Practice Problem 10**

PRACTICE PROBLEM 10

Use the formula from Example 10 to find the interest rate r if \$1600 compounded annually grows to \$1764 in 2 years.

Answer

10. 5%

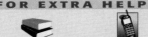
Objective A *Use the square root property to solve each equation. See Examples 1 through 4.*

1. $x^2 = 16$

2. $x^2 = 49$

3. $x^2 - 7 = 0$

4. $x^2 - 11 = 0$

5. $x^2 = 18$

6. $y^2 = 20$

7. $3z^2 - 30 = 0$

8. $2x^2 = 4$

9. $(x + 5)^2 = 9$

10. $(y - 3)^2 = 4$

11. $(z - 6)^2 = 18$

12. $(y + 4)^2 = 27$

13. $(2x - 3)^2 = 8$

14. $(4x + 9)^2 = 6$

15. $x^2 + 9 = 0$

16. $x^2 + 4 = 0$

17. $x^2 - 6 = 0$

18. $y^2 - 10 = 0$

19. $2z^2 + 16 = 0$

20. $3p^2 + 36 = 0$

21. $(3x - 1)^2 = -16$

22. $(4y + 2)^2 = -25$

23. $(z + 7)^2 = 5$

24. $(x + 10)^2 = 11$

25. $(x + 3)^2 + 8 = 0$

26. $(y - 4)^2 + 18 = 0$

Objective B *Add the proper constant to each binomial so that the resulting trinomial is a perfect square trinomial. Then factor the trinomial. See Examples 5 and 6.*

27. $x^2 + 16x$

28. $y^2 + 2y$

29. $z^2 - 12z$

30. $x^2 - 8x$

31. $p^2 + 9p$

32. $n^2 + 5n$

33. $r^2 - r$

34. $p^2 - 7p$

Objective **C** *Solve each equation by completing the square. See Examples 7 through 9.*

35. $x^2 + 8x = -15$

36. $y^2 + 6y = -8$

37. $x^2 + 6x + 2 = 0$

38. $x^2 - 2x - 2 = 0$

39. $x^2 + x - 1 = 0$

40. $x^2 + 3x - 2 = 0$

41. $x^2 + 2x - 5 = 0$

42. $x^2 - 6x + 3 = 0$

43. $y^2 + y - 7 = 0$

44. $x^2 - 7x - 1 = 0$

45. $x^2 + 8x + 1 = 0$

46. $x^2 - 10x + 2 = 0$

47. $3p^2 - 12p + 2 = 0$

48. $2x^2 + 14x - 1 = 0$

49. $2x^2 + 7x = 4$

50. $3x^2 - 4x = 4$

51. $3y^2 + 6y - 4 = 0$

52. $2y^2 + 12y + 3 = 0$

53. $y^2 + 2y + 2 = 0$

54. $x^2 + 4x + 6 = 0$

55. $2a^2 + 8a = -12$

56. $3x^2 + 12x = -14$

57. $2x^2 - x + 6 = 0$

58. $4x^2 - 2x + 5 = 0$

59. $x^2 + 10x + 28 = 0$

60. $y^2 + 8y + 18 = 0$

61. $z^2 + 3z - 4 = 0$

62. $y^2 + y - 2 = 0$

63. $2x^2 - 4x + 3 = 0$

64. $9x^2 - 36x = -40$

65. $3x^2 + 3x = 5$

66. $5y^2 - 15y = 1$

Objective D *Use the formula $A = P(1 + r)^t$ to solve Exercises 67 through 70. See Example 10.*

67. Find the rate r at which $3000 grows to $4320 in 2 years.

68. Find the rate r at which $800 grows to $882 in 2 years.

69. Find the rate r at which $810 grows to approximately $1000 in 2 years.

70. Find the rate r at which $2000 grows to $2880 in 2 years.

Neglecting air resistance, the distance $s(t)$ in feet traveled by a freely falling object is given by the function $s(t) = 16t^2$, where t is time in seconds. Use this formula to solve Exercises 71 through 74. Round answers to two decimal places.

71. The Petronas Towers in Kuala Lumpur, built in 1997, are the tallest buildings in Malaysia. Each tower is 1483 feet tall. How long would it take an object to fall to the ground from the top of one of the towers? (*Source:* Council on Tall Buildings and Urban Habitat, Lehigh University)

72. The height of the Chicago Beach Tower Hotel, built in 1998 in Dubai, United Arab Emirates, is 1053 feet. How long would it take an object to fall to the ground from the top of the building? (*Source:* Council on Tall Buildings and Urban Habitat, Lehigh University)

73. The Rogun Dam in Tajikistan (part of the former USSR that borders Afghanistan) is the tallest dam in the world at 1100 feet. How long would it take an object to fall from the top to the base of the dam? (*Source:* U.S. Committee on Large Dams of the International Commission on Large Dams)

74. The Hoover Dam, located on the Colorado River on the border of Nevada and Arizona near Las Vegas, is 725 feet tall. How long would it take an object to fall from the top to the base of the dam? (*Source:* U.S. Committee on Large Dams of the International Commission on Large Dams)

Solve.

△ **75.** The area of a square room is 225 square feet. Find the dimensions of the room.

△ **76.** The area of a circle is 36π square inches. Find the radius of the circle.

△ **77.** An isosceles right triangle has legs of equal length. If the hypotenuse is 20 centimeters long, find the length of each leg.

Review

Simplify each expression. See Section 7.5.

78. $\dfrac{6 + 4\sqrt{5}}{2}$ **79.** $\dfrac{10 - 20\sqrt{3}}{2}$ **80.** $\dfrac{3 - 9\sqrt{2}}{6}$ **81.** $\dfrac{12 - 8\sqrt{7}}{16}$

Evaluate $\sqrt{b^2 - 4ac}$ for each set of values. See Section 7.3.

82. $a = 2, b = 4, c = -1$ **83.** $a = 1, b = 6, c = 2$ **84.** $a = 3, b = -1, c = -2$ **85.** $a = 1, b = -3, c = -1$

Concept Extensions

Without solving, determine whether the solutions of each equation are real numbers or complex, but not real numbers. See the Concept Check in this section.

86. $(x + 1)^2 = -1$

87. $(y - 5)^2 = -9$

88. $3z^2 = 10$

89. $4x^2 = 17$

90. $(2y - 5)^2 + 7 = 3$

91. $(3m + 2)^2 + 4 = 1$

92. In your own words, what is the difference between simple interest and compound interest?

93. If you are depositing money in an account that pays 4%, would you prefer the interest to be simple or compound? Explain your answer.

94. If you are borrowing money at a rate of 10%, would you prefer the interest to be simple or compound? Explain your answer.

Find two possible missing terms so that each is a perfect square trinomial.

95. $x^2 + \;\blacksquare\; + 16$

96. $y^2 + \;\blacksquare\; + 9$

A common equation used in business is a demand equation. It expresses the relationship between the unit price of some commodity and the quantity demanded. For Exercises 97 and 98, p represents the unit price and x represents the quantity demanded in thousands.

97. A manufacturing company has found that the demand equation for a certain type of scissors is given by the equation $p = -x^2 + 47$. Find the demand for the scissors if the price is \$11 per pair.

98. Acme, Inc., sells desk lamps and has found that the demand equation for a certain style of desk lamp is given by the equation $p = -x^2 + 15$. Find the demand for the desk lamp if the price is \$7 per lamp.

 ## STUDY SKILLS BUILDER

Learning New Terms?

By now, you have encountered many new terms. It's never too late to make a list of new terms and review them frequently. Remember that placing these new terms (including page references) on 3 × 5 index cards might help you later when you're preparing for a quiz.

Answer the following.

1. How do new terms stand out in this text so that they can be found?

2. Name one way placing a word and its definition on a 3 × 5 card might be helpful.

8.2 SOLVING QUADRATIC EQUATIONS BY USING THE QUADRATIC FORMULA

Objectives

A Solve Quadratic Equations by Using the Quadratic Formula.

B Determine the Number and Type of Solutions of a Quadratic Equation by Using the Discriminant.

C Solve Problems Modeled by Quadratic Equations.

Objective **A** Solving Equations by Using the Quadratic Formula

Any quadratic equation can be solved by completing the square. Since the same sequence of steps is repeated each time we complete the square, let's complete the square for a general quadratic equation, $ax^2 + bx + c = 0$. By doing so, we will find a pattern for the solutions of a quadratic equation known as the **quadratic formula.**

Recall that to complete the square for an equation such as $ax^2 + bx + c = 0$, $a \neq 0$, we first divide both sides by the coefficient of x^2.

$$ax^2 + bx + c = 0$$

$$x^2 + \frac{b}{a}x + \frac{c}{a} = 0 \qquad \text{Divide both sides by } a, \text{ the coefficient of } x^2.$$

$$x^2 + \frac{b}{a}x = -\frac{c}{a} \qquad \text{Subtract the constant } \frac{c}{a} \text{ from both sides.}$$

Next we find the square of half $\dfrac{b}{a}$, the coefficient of x.

$$\frac{1}{2}\left(\frac{b}{a}\right) = \left(\frac{b}{2a}\right) \qquad \text{and} \qquad \left(\frac{b}{2a}\right)^2 = \frac{b^2}{4a^2}$$

Now we add this result to both sides of the equation.

$$x^2 + \frac{b}{a}x + \frac{b^2}{4a^2} = -\frac{c}{a} + \frac{b^2}{4a^2} \qquad \text{Add } \frac{b^2}{4a^2} \text{ to both sides.}$$

$$x^2 + \frac{b}{a}x + \frac{b^2}{4a^2} = \frac{-c \cdot 4a}{a \cdot 4a} + \frac{b^2}{4a^2} \qquad \text{Find a common denominator on the right side.}$$

$$x^2 + \frac{b}{a}x + \frac{b^2}{4a^2} = \frac{b^2 - 4ac}{4a^2} \qquad \text{Simplify the right side.}$$

$$\left(x + \frac{b}{2a}\right)^2 = \frac{b^2 - 4ac}{4a^2} \qquad \text{Factor the perfect square trinomial on the left side.}$$

$$x + \frac{b}{2a} = \pm\sqrt{\frac{b^2 - 4ac}{4a^2}} \qquad \text{Use the square root property.}$$

$$x + \frac{b}{2a} = \pm\frac{\sqrt{b^2 - 4ac}}{2a} \qquad \text{Simplify the radical.}$$

$$x = -\frac{b}{2a} \pm \frac{\sqrt{b^2 - 4ac}}{2a} \qquad \text{Subtract } \frac{b}{2a} \text{ from both sides.}$$

$$x = \frac{-b \pm \sqrt{b^2 - 4ac}}{2a} \qquad \text{Simplify.}$$

The resulting equation identifies the solutions of the general quadratic equation in standard form and is called the quadratic formula. It can be used to solve any equation written in standard form $ax^2 + bx + c = 0$ as long as a is not 0.

Quadratic Formula

A quadratic equation written in the form $ax^2 + bx + c = 0$, $a \neq 0$, has the solutions

$$x = \frac{-b \pm \sqrt{b^2 - 4ac}}{2a}$$

PRACTICE PROBLEM 1

Solve: $2x^2 + 9x + 10 = 0$

EXAMPLE 1 Solve: $3x^2 + 16x + 5 = 0$

Solution: This equation is in standard form with $a = 3$, $b = 16$, and $c = 5$. We substitute these values into the quadratic formula.

$$x = \frac{-b \pm \sqrt{b^2 - 4ac}}{2a} \qquad \text{Quadratic formula}$$

$$= \frac{-16 \pm \sqrt{16^2 - 4(3)(5)}}{2(3)} \qquad \text{Let } a = 3, b = 16, \text{ and } c = 5.$$

$$= \frac{-16 \pm \sqrt{256 - 60}}{6}$$

$$= \frac{-16 \pm \sqrt{196}}{6} = \frac{-16 \pm 14}{6}$$

$$x = \frac{-16 + 14}{6} = -\frac{1}{3} \quad \text{or} \quad x = \frac{-16 - 14}{6} = -\frac{30}{6} = -5$$

The solution set is $\left\{ -\frac{1}{3}, -5 \right\}$.

🔲 **Work Practice Problem 1**

> **Helpful Hint**
>
> To replace a, b, and c correctly in the quadratic formula, write the quadratic equation in standard form, $ax^2 + bx + c = 0$.

PRACTICE PROBLEM 2

Solve: $2x^2 - 6x - 1 = 0$

EXAMPLE 2 Solve: $2x^2 - 4x = 3$

Solution: First we write the equation in standard form by subtracting 3 from both sides.

$$2x^2 - 4x - 3 = 0$$

Now $a = 2$, $b = -4$, and $c = -3$. We substitute these values into the quadratic formula.

$$x = \frac{-b \pm \sqrt{b^2 - 4ac}}{2a}$$

$$= \frac{-(-4) \pm \sqrt{(-4)^2 - 4(2)(-3)}}{2(2)}$$

$$= \frac{4 \pm \sqrt{16 + 24}}{4}$$

$$= \frac{4 \pm \sqrt{40}}{4}$$

$$= \frac{4 \pm 2\sqrt{10}}{4}$$

$$= \frac{2\left(2 \pm \sqrt{10}\right)}{2 \cdot 2}$$

$$= \frac{2 \pm \sqrt{10}}{2}$$

The solution set is $\left\{ \dfrac{2 + \sqrt{10}}{2}, \dfrac{2 - \sqrt{10}}{2} \right\}$.

🔲 **Work Practice Problem 2**

Answers

1. $\left\{ -\dfrac{5}{2}, -2 \right\}$,

2. $\left\{ \dfrac{3 + \sqrt{11}}{2}, \dfrac{3 - \sqrt{11}}{2} \right\}$

✔**Concept Check** For the quadratic equation $x^2 = 7$, which substitution is correct?

a. $a = 1, b = 0,$ and $c = -7$ **b.** $a = 1, b = 0,$ and $c = 7$
c. $a = 0, b = 0,$ and $c = 7$ **d.** $a = 1, b = 1,$ and $c = -7$

Helpful Hint

To simplify the expression $\dfrac{4 \pm 2\sqrt{10}}{4}$ in Example 2, note that we factored 2 out of both terms of the numerator *before* simplifying.

$$\frac{4 \pm 2\sqrt{10}}{4} = \frac{2(2 \pm \sqrt{10})}{2 \cdot 2} = \frac{2 \pm \sqrt{10}}{2}$$

EXAMPLE 3 Solve: $\dfrac{1}{4}m^2 - m + \dfrac{1}{2} = 0$

Solution: We could use the quadratic formula with $a = \dfrac{1}{4}, b = -1,$ and $c = \dfrac{1}{2}$. Instead, let's find a simpler, equivalent, standard-form equation whose coefficients are not fractions.

First we multiply both sides of the equation by 4 to clear the fractions.

$$4\left(\frac{1}{4}m^2 - m + \frac{1}{2}\right) = 4 \cdot 0$$

$$m^2 - 4m + 2 = 0 \qquad \text{Simplify.}$$

Now we can substitute $a = 1, b = -4,$ and $c = 2$ into the quadratic formula and simplify.

$$m = \frac{-(-4) \pm \sqrt{(-4)^2 - 4(1)(2)}}{2(1)}$$

$$= \frac{4 \pm \sqrt{16 - 8}}{2}$$

$$= \frac{4 \pm \sqrt{8}}{2} = \frac{4 \pm 2\sqrt{2}}{2} = \frac{2(2 \pm \sqrt{2})}{2} = 2 \pm \sqrt{2}$$

The solution set is $\{2 + \sqrt{2}, 2 - \sqrt{2}\}$.

■ **Work Practice Problem 3**

PRACTICE PROBLEM 3

Solve: $\dfrac{1}{6}x^2 - \dfrac{1}{3}x - 1 = 0$

EXAMPLE 4 Solve: $p = -3p^2 - 3$

Solution: The equation in standard form is $3p^2 + p + 3 = 0.$ Thus, $a = 3, b = 1,$ and $c = 3$ in the quadratic formula.

$$p = \frac{-1 \pm \sqrt{1^2 - 4(3)(3)}}{2(3)} = \frac{-1 \pm \sqrt{1 - 36}}{6}$$

$$= \frac{-1 \pm \sqrt{-35}}{6} = \frac{-1 \pm i\sqrt{35}}{6}$$

The solution set is $\left\{\dfrac{-1 + i\sqrt{35}}{6}, \dfrac{-1 - i\sqrt{35}}{6}\right\}$.

■ **Work Practice Problem 4**

PRACTICE PROBLEM 4

Solve: $x = -4x^2 - 4$

Answers

3. $\{1 + \sqrt{7}, 1 - \sqrt{7}\},$

4. $\left\{\dfrac{-1 - 3i\sqrt{7}}{8}, \dfrac{-1 + 3i\sqrt{7}}{8}\right\}$

✔ **Concept Check Answer**

a

✔ **Concept Check** What is the first step in solving $-3x^2 = 5x - 4$ using the quadratic formula?

Objective **B** Using the Discriminant

In the quadratic formula $x = \dfrac{-b \pm \sqrt{b^2 - 4ac}}{2a}$, the radicand $b^2 - 4ac$ is called the **discriminant** because when we know its value, we can **discriminate** among the possible number and type of solutions of a quadratic equation. Possible values of the discriminant and their meanings are summarized next.

Discriminant

The following table relates the discriminant $b^2 - 4ac$ of a quadratic equation of the form $ax^2 + bx + c = 0$ with the number and type of solutions of the equation.

$b^2 - 4ac$	Number and Type of Solutions
Positive	Two real solutions
Zero	One real solution
Negative	Two complex but not real solutions

PRACTICE PROBLEM 5

Use the discriminant to determine the number and type of solutions of $x^2 + 4x + 4 = 0$.

EXAMPLE 5 Use the discriminant to determine the number and type of solutions of $x^2 + 2x + 1$.

Solution: In $x^2 + 2x + 1 = 0$, $a = 1$, $b = 2$, and $c = 1$. Thus,

$$b^2 - 4ac = 2^2 - 4(1)(1) = 0$$

Since $b^2 - 4ac = 0$, this quadratic equation has one real solution.

▣ **Work Practice Problem 5**

PRACTICE PROBLEM 6

Use the discriminant to determine the number and type of solutions of $5x^2 + 7 = 0$.

EXAMPLE 6 Use the discriminant to determine the number and type of solutions of $3x^2 + 2 = 0$.

Solution: In this equation, $a = 3$, $b = 0$, and $c = 2$. Then

$$b^2 - 4ac = 0^2 - 4(3)(2) = -24$$

Since $b^2 - 4ac$ is negative, this quadratic equation has two complex but not real solutions.

▣ **Work Practice Problem 6**

PRACTICE PROBLEM 7

Use the discriminant to determine the number and type of solutions of $3x^2 - 2x - 2 = 0$.

EXAMPLE 7 Use the discriminant to determine the number and type of solutions of $2x^2 - 7x - 4 = 0$.

Solution: In this equation, $a = 2$, $b = -7$, and $c = -4$. Then

$$b^2 - 4ac = (-7)^2 - 4(2)(-4) = 81$$

Since $b^2 - 4ac$ is positive, this quadratic equation has two real solutions.

▣ **Work Practice Problem 7**

Answers

5. one real solution,
6. two complex but not real solutions,
7. two real solutions

✔ **Concept Check Answer**

Write the equation in standard form.

The discriminant helps us determine the number and type of solutions of a quadratic equation, $ax^2 + bx + c = 0$. Recall from Chapter 5 that the solutions of this equation are the same as the x-intercepts of its related graph $f(x) = ax^2 + bx + c$. This means that the discriminant of $ax^2 + bx + c = 0$ also tells us the number of x-intercepts for the graph of $f(x) = ax^2 + bx + c$, or, equivalently, $y = ax^2 + bx + c$.

Graph of $f(x) = ax^2 + bx + c$ or $y = ax^2 + bx + c$

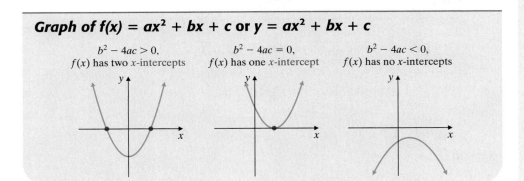

| $b^2 - 4ac > 0$, | $b^2 - 4ac = 0$, | $b^2 - 4ac < 0$, |
| $f(x)$ has two x-intercepts | $f(x)$ has one x-intercept | $f(x)$ has no x-intercepts |

Objective **C** Solving Problems Modeled by Quadratic Equations

The quadratic formula is useful in solving problems that are modeled by quadratic equations.

△ **EXAMPLE 8** **Calculating Distance Saved**

At a local university, students often leave the sidewalk and cut across the lawn to save walking distance. Given the diagram below of a favorite place to cut across the lawn, approximate to the nearest foot how many feet of walking distance a student saves by cutting across the lawn instead of walking on the sidewalk.

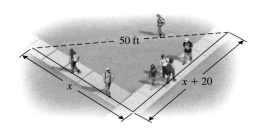

Solution:

1. UNDERSTAND. Read and reread the problem. You may want to review the Pythagorean theorem.
2. TRANSLATE. By the Pythagorean theorem, we have

    In words: $(\text{leg})^2 + (\text{leg})^2 = (\text{hypotenuse})^2$

    Translate: $x^2 + (x + 20)^2 = 50^2$

3. SOLVE. Use the quadratic formula to solve.

    $x^2 + x^2 + 40x + 400 = 2500$ Square $(x + 20)$ and 50.

    $2x^2 + 40x - 2100 = 0$ Write the equation in standard form.

Continued on next page

PRACTICE PROBLEM 8

Given the diagram below, approximate to the nearest foot how many feet of walking distance a person saves by cutting across the lawn instead of walking on the sidewalk.

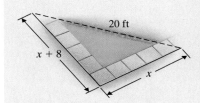

Answer

8. 8 ft

We can use the quadratic formula right now with $a = 2, b = 40,$ and $c = -2100.$ Instead, just as in Example 3, you may want to find a simpler, equivalent equation by dividing both sides of the equation by 2.

$$x^2 + 20x - 1050 = 0 \quad \text{Divide by 2.}$$

Here, $a = 1, b = 20,$ and $c = -1050.$ By the quadratic formula,

$$x = \frac{-20 \pm \sqrt{20^2 - 4(1)(-1050)}}{2 \cdot 1}$$

$$= \frac{-20 \pm \sqrt{400 + 4200}}{2} = \frac{-20 \pm \sqrt{4600}}{2}$$

$$= \frac{-20 \pm \sqrt{100 \cdot 46}}{2} = \frac{-20 \pm 10\sqrt{46}}{2}$$

$$= -10 \pm 5\sqrt{46} \qquad \qquad \text{Simplify.}$$

Check:

4. INTERPRET. We check our calculations from the quadratic formula. The length of a side of a triangle can't be negative, so we reject $-10 - 5\sqrt{46}.$ Since $-10 + 5\sqrt{46} \approx 24$ feet, the walking distance along the sidewalk is

$$x + (x + 20) \approx 24 + (24 + 20) = 68 \text{ feet.}$$

State: A person saves about $68 - 50$ or 18 feet of walking distance by cutting across the lawn.

▣ **Work Practice Problem 8**

PRACTICE PROBLEM 9

How long after the object in Example 9 is thrown will it be 100 feet from the ground? Round to the nearest tenth of a second.

EXAMPLE 9 **Calculating Landing Time**

An object is thrown upward from the top of a 200-foot cliff with a velocity of 12 feet per second. The height above ground h in feet of the object after t seconds is

$$h = -16t^2 + 12t + 200$$

How long after the object is thrown will it strike the ground? Round to the nearest tenth of a second.

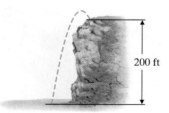

200 ft

Solution:

1. UNDERSTAND. Read and reread the problem.
2. TRANSLATE. Since we want to know when the object strikes the ground, we want to know when the height $h = 0,$ or

$$0 = -16t^2 + 12t + 200$$

3. SOLVE. First we divide both sides of the equation by $-4.$

$$0 = 4t^2 - 3t - 50 \quad \text{Divide both sides by } -4.$$

Answer
9. 1.7 sec

Here, $a = 4$, $b = -3$, and $c = -50$. By the quadratic formula,

$$t = \frac{-(-3) \pm \sqrt{(-3)^2 - 4(4)(-50)}}{2 \cdot 4}$$

$$= \frac{3 \pm \sqrt{9 + 800}}{8}$$

$$= \frac{3 \pm \sqrt{809}}{8}$$

Check:

4. INTERPRET. We check our calculations from the quadratic formula. Since the time won't be negative, we reject the proposed solution $\dfrac{3 - \sqrt{809}}{8}$.

State: The time it takes for the object to strike the ground is exactly $\dfrac{3 + \sqrt{809}}{8}$ seconds ≈ 3.9 seconds.

◼ **Work Practice Problem 9**

▦ **CALCULATOR EXPLORATIONS** Graphing

In Section 5.7, we showed how we can use a grapher to approximate real number solutions of a quadratic equation written in standard form. We can also use a grapher to solve a quadratic equation when it is not written in standard form. For example, to solve $(x + 1)^2 = 12$, the quadratic equation in Example 3 of Section 8.1, we graph the following on the same set of axes. We use Xmin $= -10$, Xmax $= 10$, Ymin $= -13$, and Ymax $= 13$.

$$Y_1 = (x + 1)^2 \quad \text{and} \quad Y_2 = 12$$

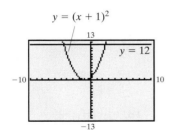

We use the INTERSECT feature or the ZOOM and TRACE features to locate the points of intersection of the graphs. The x-values of these points are the solutions of $(x + 1)^2 = 12$. The solutions, rounded to two decimal points, are 2.46 and -4.46.

Check to see that these numbers are approximations of the exact solutions, $-1 \pm 2\sqrt{3}$.

Use a grapher to solve each quadratic equation. Round all solutions to the nearest hundredth.

1. $x(x - 5) = 8$
2. $x(x + 2) = 5$
3. $x^2 + 0.5x = 0.3x + 1$
4. $x^2 - 2.6x = -2.2x + 3$
5. Use a grapher to solve $(2x - 5)^2 = -16$, (Example 4, Section 8.1) using the window

   Xmin $= -20$
   Xmax $= 20$
   Xscl $= 1$
   Ymin $= -20$
   Ymax $= 20$
   Yscl $= 1$

   Explain the results. Compare your results with the solution found in Example 4 of Section 8.1.
6. What are the advantages and disadvantages of using a grapher to solve quadratic equations?

Mental Math

Identify the values of a, b, and c in each quadratic equation.

1. $x^2 + 3x + 1 = 0$

2. $2x^2 - 5x - 7 = 0$

3. $7x^2 - 4 = 0$

4. $x^2 + 9 = 0$

5. $6x^2 - x = 0$

6. $5x^2 + 3x = 0$

8.2 EXERCISE SET

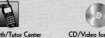

Objective A *Use the quadratic formula to solve each equation. These equations have real number solutions only. See Examples 1 through 3.*

1. $m^2 + 5m - 6 = 0$

2. $p^2 + 11p - 12 = 0$

3. $2y = 5y^2 - 3$

4. $5x^2 - 3 = 14x$

5. $x^2 - 6x + 9 = 0$

6. $y^2 + 10y + 25 = 0$

7. $x^2 + 7x + 4 = 0$

8. $y^2 + 5y + 3 = 0$

9. $8m^2 - 2m = 7$

10. $11n^2 - 9n = 1$

11. $3m^2 - 7m = 3$

12. $x^2 - 13 = 5x$

13. $\frac{1}{2}x^2 - x - 1 = 0$

14. $\frac{1}{6}x^2 + x + \frac{1}{3} = 0$

15. $\frac{2}{5}y^2 + \frac{1}{5}y = \frac{3}{5}$

16. $\frac{1}{8}x^2 + x = \frac{5}{2}$

17. $\frac{1}{3}y^2 - y - \frac{1}{6} = 0$

18. $\frac{1}{2}y^2 = y + \frac{1}{2}$

19. $x^2 + 5x = -2$

20. $y^2 - 8 = 4y$

21. $(m + 2)(2m - 6) = 5(m - 1) - 12$

22. $7p(p - 2) + 2(p + 4) = 3$

Mixed Practice *Use the quadratic formula to solve each equation. These equations have real solutions and complex, but not real, solutions. See Examples 1 through 4.*

23. $x^2 + 6x + 13 = 0$

24. $x^2 + 2x + 2 = 0$

25. $(x + 5)(x - 1) = 2$

26. $x(x + 6) = 2$

27. $6 = -4x^2 + 3x$

28. $9x^2 + x + 2 = 0$

29. $\dfrac{x^2}{3} - x = \dfrac{5}{3}$

30. $\dfrac{x^2}{2} - 3 = -\dfrac{9}{2}x$

31. $10y^2 + 10y + 3 = 0$

32. $3y^2 + 6y + 5 = 0$

33. $x(6x + 2) = 3$

34. $x(7x + 1) = 2$

35. $\dfrac{2}{5}y^2 + \dfrac{1}{5}y + \dfrac{3}{5} = 0$

36. $\dfrac{1}{8}x^2 + x + \dfrac{5}{2} = 0$

37. $\dfrac{1}{2}y^2 = y - \dfrac{1}{2}$

38. $\dfrac{2}{3}x^2 - \dfrac{20}{3}x = -\dfrac{100}{6}$

39. $(n - 2)^2 = 2n$

40. $\left(p - \dfrac{1}{2}\right)^2 = \dfrac{p}{2}$

Objective **B** *Use the discriminant to determine the number and types of solutions of each equation. See Examples 5 through 7.*

41. $x^2 - 5 = 0$

42. $x^2 - 7 = 0$

43. $4x^2 + 12x = -9$

44. $9x^2 + 1 = 6x$

45. $3x = -2x^2 + 7$

46. $3x^2 = 5 - 7x$

47. $6 = 4x - 5x^2$

48. $8x = 3 - 9x^2$

49. $9x - 2x^2 + 5 = 0$

50. $5 - 4x + 12x^2 = 0$

Objective **C** *Solve. See Examples 8 and 9.*

△ **51.** Nancy, Thelma, and John Varner live on a corner lot. Often, neighborhood children cut across their lot to save walking distance. Given the diagram below, approximate to the nearest foot how many feet of walking distance children save by cutting across their property instead of walking around the lot.

△ **52.** Given the diagram below, approximate to the nearest foot how many feet of walking distance a person saves by cutting across the lawn instead of walking on the sidewalk.

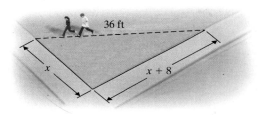

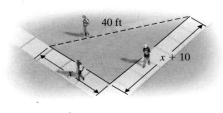

△ **53.** The hypotenuse of an isosceles right triangle is 2 centimeters longer than either of its legs. Find the exact length of each side. (*Hint:* An isosceles right triangle is a right triangle whose legs are the same length.)

△ **54.** The hypotenuse of an isosceles right triangle is one meter longer than either of its legs. Find the length of each side.

△ **55.** Bailey Wilson's rectangular dog pen for her Irish setter must have an area of 400 square feet. Also, the length must be 10 feet longer than the width. Find the dimensions of the pen.

△ **56.** An entry in the Peach Festival Poster Contest must be rectangular and have an area of 1200 square inches. Furthermore, its length must be 20 inches longer than its width. Find the dimensions each entry must have.

△ **57.** A holding pen for cattle must be square and have a diagonal length of 100 meters.
   a. Find the length of a side of the pen.
   b. Find the area of the pen.

△ **58.** A rectangle is three times longer than it is wide. It has a diagonal of length 50 centimeters.
   a. Find the dimensions of the rectangle.
   b. Find the perimeter of the rectangle.

▦ △ **59.** The heaviest reported door in the world is the 708.6 ton radiation shield door in the National Institute for Fusion Science at Toki, Japan. If the height of the door is 1.1 feet longer than its width, and its front area (neglecting depth) is 1439.9 square feet, find its width and height. [Interesting note: the door is 6.6 feet thick.] (*Source: Guiness World Records*)

▦ △ **60.** Christi and Robbie Wegmann are constructing a rectangular stained glass window whose length is 7.3 inches longer than its width. If the area of the window is 569.9 square inches, find its width and length.

△ **61.** The base of a triangle is four more than twice its height. If the area of the triangle is 42 square centimeters, find its base and height.

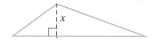

△ **62.** If a point B divides a line segment such that the smaller portion is to the larger portion as the larger is to the whole, the whole is the length of the *golden ratio*.

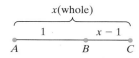

The golden ratio was thought by the Greeks to be the most pleasing to the eye, and many of their buildings contained numerous examples of the golden ratio. The value of the golden ratio is the positive solution of the following equation.

$$\text{(smaller)} \quad \frac{x-1}{1} = \frac{1}{x} \quad \text{(larger)}$$
$$\text{(larger)} \qquad\qquad\quad \text{(whole)}$$

Find this value.

The Wollomombi Falls in Australia have a height of 1100 feet. A pebble is thrown upward from the top of the falls with an initial velocity of 20 feet per second. The height of the pebble h in feet after t seconds is given by the equation $h = -16t^2 + 20t + 1100$. Use this equation for Exercises 63 and 64.

63. How long after the pebble is thrown will it hit the ground? Round to the nearest tenth of a second.

64. How long after the pebble is thrown will it be 550 feet from the ground? Round to the nearest tenth of a second.

A ball is thrown downward from the top of a 180-foot building with an initial velocity of 20 feet per second. The height of the ball h in feet after t seconds is given by the equation $h = -16t^2 - 20t + 180$. Use this equation to answer Exercises 65 and 66.

65. How long after the ball is thrown will it strike the ground? Round the result to the nearest tenth of a second.

66. How long after the ball is thrown will it be 50 feet from the ground? Round the result to the nearest tenth of a second.

Review

Solve each equation. See Sections 6.4 and 7.6.

67. $\sqrt{5x - 2} = 3$

68. $\sqrt{y + 2} + 7 = 12$

69. $\dfrac{1}{x} + \dfrac{2}{5} = \dfrac{7}{x}$

70. $\dfrac{10}{z} = \dfrac{5}{z} - \dfrac{1}{3}$

Factor. See Section 5.5 and 5.6.

71. $x^4 + x^2 - 20$

72. $2y^4 + 11y^2 - 6$

73. $z^4 - 13z^2 + 36$

74. $x^4 - 1$

Concept Extensions

For each quadratic equation, choose the correct substitution for a, b, and c in the standard form $ax^2 + bx + c = 0$.

75. $x^2 = -10$
   a. $a = 1, b = 0, c = -10$
   b. $a = 1, b = 0, c = 10$
   c. $a = 0, b = 1, c = -10$
   d. $a = 1, b = 1, c = 10$

76. $x^2 + 5 = -x$
   a. $a = 1, b = 5, c = -1$
   b. $a = 1, b = -1, c = 5$
   c. $a = 1, b = 5, c = 1$
   d. $a = 1, b = 1, c = 5$

77. Solve Exercise 1 by factoring. Explain the result.

78. Solve Exercise 2 by factoring. Explain the result.

Use the quadratic formula and a calculator to approximate each solution to the nearest tenth.

79. $2x^2 - 6x + 3 = 0$

80. $3.6x^2 + 1.8x - 4.3 = 0$.

The graph shows the daily low temperatures for one week in New Orleans, Louisiana. Use this graph to answer Exercises 81 through 84.

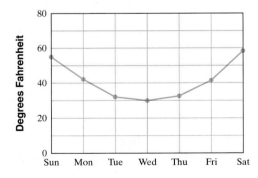

81. Which day of the week shows the greatest decrease in the low temperature?

82. Which day of the week shows the greatest increase in the low temperature?

83. Which day of the week had the lowest temperature?

84. Use the graph to estimate the low temperature on Thursday.

Notice that the shape of the temperature graph for Exercises 81 through 84 is similar to a parabola (see Section 3.8). In fact, this graph can be approximated by the quadratic function $f(x) = 3x^2 - 18x + 57$, where $f(x)$ is the temperature in degrees Fahrenheit and x is the number of days from Sunday. Use this function to answer Exercises 85 and 86.

85. Use the given quadratic function to approximate the low temperature on Thursday. Does your answer agree with the graph?

86. Use the given function and the quadratic formula to find when the low temperature was 35°F. [*Hint:* Let $f(x) = 35$ and solve for x.] Round your answer to one decimal place and interpret your result. Does your answer agree with the graph above?

87. Procter & Gamble's net earnings can be modeled by the quadratic function $f(x) = 213x^2 - 66.4x + 4689$, where $f(x)$ is net earnings in millions of dollars and x is the number of years after 2001. (*Source:* Based on data from The Procter & Gamble Company, 2001–2004)

   a. Find Procter & Gamble's net earnings in 2004.

   b. If the trend described by the model continues, predict the year after 2001 in which Procter & Gamble's net earnings will be $9682 million.

88. The number of inmates in custody in U.S. prisons and jails can be modeled by the quadratic function $p(x) = -18x^2 + 298x + 988$, where $p(x)$ is the number of inmates in thousands and x is the number of years after 1995. (*Source:* Based on data from the Bureau of Justice Statistics, U.S. Department of Justice, 1995–2004)

   a. Find the number of prison and jail inmates in the United States in 1998.

   b. Find the number of prison and jail inmates in the United States in 2000.

   c. If the trend described by the model continues, find a year in which the number of prisoners is 2,212,000 (that is, 2212 thousand).

89. The average total daily supply y of motor gasoline (in thousands of barrels per day) in the United States for the period 2000–2003 can be approximated by the equation $y = -13x^2 + 221x + 8476$, where x is the number of years after 2000. (*Source:* Based on data from the Energy Information Administration)

   a. Find the average total daily supply of motor gasoline in 2001.

   b. According to this model, in what year will the average total daily supply of motor gasoline be 9334 thousand barrels per day?

 90. The relationship between body weight and the Recommended Dietary Allowance (RDA) for vitamin A in children up to age 10 is modeled by the quadratic equation $y = 0.149x^2 - 4.475x + 406.478$, where y is the RDA for vitamin A in micrograms for a child whose weight is x pounds. (*Source:* Based on data from the Food and Nutrition Board, National Academy of Sciences—Institute of Medicine, 1989)

   a. Determine the vitamin A requirements of a child who weighs 35 pounds.

   b. What is the weight of a child whose RDA of vitamin A is 600 micrograms? Round your answer to the nearest pound.

91. Use a grapher to solve Exercise 79.

92. Use a grapher to solve Exercise 80.

📖 STUDY SKILLS BUILDER

How Well Do You Know Your Textbook?

Let's check to see whether you are familiar with your textbook yet. For help, see Section 1.1 in this text.

1. What does the ⊚ icon mean?

2. What does the ＼ icon mean?

3. What does the △ icon mean?

4. Where can you find a review for each chapter? What answers to this review can be found in the back of your text?

5. Each chapter contains an overview of the chapter along with examples. What is this feature called?

6. Each chapter contains a review of vocabulary. What is this feature called?

7. There are free CDs in your text. What content is contained on these CDs?

8. What is the location of the section that is entirely devoted to study skills?

9. There are Practice Problems that are contained in the margin of the text. What are they and how can they be used?

8.3 SOLVING EQUATIONS BY USING QUADRATIC METHODS

Objective **A** Solving Equations That Are Quadratic in Form

In this section, we discuss various types of equations that can be solved in part by using the methods for solving quadratic equations.

Once each equation is simplified, you may want to use these steps when deciding what method to use to solve the quadratic equation.

Solving a Quadratic Equation

Step 1: If the equation is in the form $(ax + b)^2 = c$, use the square root property and solve. If not, go to Step 2.

Step 2: Write the equation in standard form by setting it equal to 0: $ax^2 + bx + c = 0$.

Step 3: Try to solve the equation by the factoring method. If not possible, go to Step 4.

Step 4: Solve the equation by the quadratic formula.

The first example is a radical equation that becomes a quadratic equation once we square both sides.

PRACTICE PROBLEM 1

Solve: $x - \sqrt{x - 1} - 3 = 0$

EXAMPLE 1 Solve: $x - \sqrt{x} - 6 = 0$

Solution: Recall that to solve a radical equation, we first get the radical alone on one side of the equation. Then we square both sides.

$$x - 6 = \sqrt{x} \qquad \text{Add } \sqrt{x} \text{ to both sides.}$$
$$x^2 - 12x + 36 = x \qquad \text{Square both sides.}$$
$$x^2 - 13x + 36 = 0 \qquad \text{Set the equation equal to 0.}$$
$$(x - 9)(x - 4) = 0 \qquad \text{Factor.}$$
$$x - 9 = 0 \quad \text{or} \quad x - 4 = 0 \qquad \text{Set each factor equal to 0.}$$
$$x = 9 \qquad\qquad x = 4 \qquad \text{Solve.}$$

Check: Let $x = 9$.   Let $x = 4$.

$$x - \sqrt{x} - 6 = 0 \qquad\qquad x - \sqrt{x} - 6 = 0$$
$$9 - \sqrt{9} - 6 \overset{?}{=} 0 \qquad\qquad 4 - \sqrt{4} - 6 \overset{?}{=} 0$$
$$9 - 3 - 6 \overset{?}{=} 0 \qquad\qquad 4 - 2 - 6 \overset{?}{=} 0$$
$$0 = 0 \quad \text{True} \qquad\qquad -4 = 0 \quad \text{False}$$

The solution set is $\{9\}$.

■ **Work Practice Problem 1**

Answer

1. $\{5\}$

EXAMPLE 2 Solve: $\dfrac{3x}{x-2} - \dfrac{x+1}{x} = \dfrac{6}{x(x-2)}$

Solution: In this equation, x cannot be either 2 or 0 because these values cause denominators to equal zero. To solve for x, we first multiply both sides of the equation by $x(x-2)$ to clear the fractions. By the distributive property, this means that we multiply each term by $x(x-2)$.

$$x(x-2)\left(\frac{3x}{x-2}\right) - x(x-2)\left(\frac{x+1}{x}\right) = x(x-2)\left[\frac{6}{x(x-2)}\right]$$

$$3x^2 - (x-2)(x+1) = 6 \quad \text{Simplify.}$$
$$3x^2 - (x^2 - x - 2) = 6 \quad \text{Multiply.}$$
$$3x^2 - x^2 + x + 2 = 6$$
$$2x^2 + x - 4 = 0 \quad \text{Simplify.}$$

This equation cannot be factored using integers, so we solve by the quadratic formula.

$$x = \frac{-1 \pm \sqrt{1^2 - 4(2)(-4)}}{2\cdot 2} \quad \text{Let } a = 2, b = 1, \text{ and } c = -4, \text{ in the quadratic formula.}$$

$$= \frac{-1 \pm \sqrt{1 + 32}}{4} \quad \text{Simplify.}$$

$$= \frac{-1 \pm \sqrt{33}}{4}$$

Neither proposed solution will make the denominators 0.

The solution set is $\left\{ \dfrac{-1 + \sqrt{33}}{4}, \dfrac{-1 - \sqrt{33}}{4} \right\}$.

■ **Work Practice Problem 2**

EXAMPLE 3 Solve: $p^4 - 3p^2 - 4 = 0$

Solution: First we factor the trinomial.

$$p^4 - 3p^2 - 4 = 0$$
$$(p^2 - 4)(p^2 + 1) = 0 \quad \text{Factor.}$$
$$(p - 2)(p + 2)(p^2 + 1) = 0 \quad \text{Factor further.}$$
$$p - 2 = 0 \quad \text{or} \quad p + 2 = 0 \quad \text{or} \quad p^2 + 1 = 0 \quad \text{Set each factor equal to 0 and solve.}$$
$$p = 2 \qquad\qquad p = -2 \qquad\qquad p^2 = -1$$
$$p = \pm\sqrt{-1} = \pm i$$

The solution set is $\{2, -2, i, -i\}$.

■ **Work Practice Problem 3**

✔ **Concept Check**

a. True or false? The maximum number of solutions that a quadratic equation can have is 2.

b. True or false? The maximum number of solutions that an equation in quadratic form can have is 2.

PRACTICE PROBLEM 2

Solve:

$$\frac{2x}{x-1} - \frac{x+2}{x} = \frac{5}{x(x-1)}$$

PRACTICE PROBLEM 3

Solve: $x^4 - 5x^2 - 36 = 0$

Answers

2. $\left\{ \dfrac{1 + \sqrt{13}}{2}, \dfrac{1 - \sqrt{13}}{2} \right\}$

3. $\{3, -3, 2i, -2i\}$

✔ **Concept Check Answers**

a. true, b. false

PRACTICE PROBLEM 4

Solve:
$(x + 4)^2 - (x + 4) - 6 = 0$

EXAMPLE 4 Solve: $(x - 3)^2 - 3(x - 3) - 4 = 0$

Solution: Notice that the quantity $(x - 3)$ is repeated in this equation. Sometimes it is helpful to substitute a variable (in this case other than x) for the repeated quantity. We will let $u = x - 3$. Then

$$(x - 3)^2 - 3(x - 3) - 4 = 0$$

becomes

$$u^2 - 3u - 4 = 0 \quad \text{Let } x - 3 = u.$$
$$(u - 4)(u + 1) = 0 \quad \text{Factor.}$$

To solve, we use the zero-factor property.

$$u - 4 = 0 \quad \text{or} \quad u + 1 = 0 \quad \text{Set each factor equal to 0.}$$
$$u = 4 \qquad\qquad u = -1 \quad \text{Solve.}$$

To find values of x, we substitute back. That is, we substitute $x - 3$ for u.

$$x - 3 = 4 \quad \text{or} \quad x - 3 = -1$$
$$x = 7 \qquad\qquad x = 2$$

Both 2 and 7 check. The solution is $\{2, 7\}$.

Helpful Hint
When using substitution, don't forget to substitute back to the original variable.

Work Practice Problem 4

PRACTICE PROBLEM 5

Solve: $x^{2/3} - 7x^{1/3} + 10 = 0$

EXAMPLE 5 Solve: $x^{2/3} - 5x^{1/3} + 6 = 0$

Solution: The key to solving this equation is recognizing that $x^{2/3} = (x^{1/3})^2$. We replace $x^{1/3}$ with m so that

$$(x^{1/3})^2 - 5x^{1/3} + 6 = 0$$

becomes

$$m^2 - 5m + 6 = 0$$

Now we solve by factoring.

$$m^2 - 5m + 6 = 0$$
$$(m - 3)(m - 2) = 0 \qquad \text{Factor.}$$
$$m - 3 = 0 \quad \text{or} \quad m - 2 = 0 \quad \text{Set each factor equal to 0.}$$
$$m = 3 \qquad\qquad m = 2$$

Since $m = x^{1/3}$, we have

$$x^{1/3} = 3 \qquad \text{or} \quad x^{1/3} = 2$$
$$x = 3^3 = 27 \quad \text{or} \qquad x = 2^3 = 8$$

Both 8 and 27 check. The solution set is $\{8, 27\}$.

Work Practice Problem 5

Helpful Hint
Example 3 can be solved using substitution also. Think of $p^4 - 3p^2 - 4 = 0$ as

$$(p^2)^2 - 3p^2 - 4 = 0 \qquad \text{Then let } x = p^2, \text{ and solve and substitute back.}$$
$$\text{The solution set will be the same.}$$
$$x^2 - 3x - 4 = 0$$

Answers
4. $\{-1, -6\}$,   5. $\{8, 125\}$

Objective **B** Solving Problems That Lead to Quadratic Equations

The next example is a work problem. This problem is modeled by a rational equation that simplifies to a quadratic equation.

EXAMPLE 6 **Finding Work Time**

Together, an experienced word processor and an apprentice word processor can create a document in 6 hours. Alone, the experienced word processor can process the document 2 hours faster than the apprentice word processor can. Find the time in which each person can create the document alone.

Solution:

1. UNDERSTAND. Read and reread the problem. The key idea here is the relationship between the *time* (hours) it takes to complete the job and the *part of the job* completed in one unit of time (hour). For example, because they can complete the job together in 6 hours, the *part of the job* they can complete in 1 hour is $\frac{1}{6}$. We let

   x = the *time* in hours it takes the apprentice word processor to complete the job alone, and

   $x - 2$ = the *time* in hours it takes the experienced word processor to complete the job alone

   We can summarize in a chart the information discussed.

	Total Hours to Complete Job	Part of Job Completed in 1 Hour
Apprentice Word Processor	x	$\frac{1}{x}$
Experienced Word Processor	$x - 2$	$\frac{1}{x - 2}$
Together	6	$\frac{1}{6}$

2. TRANSLATE.

   In words:

part of job completed by apprentice word processor in 1 hour	added to	part of job completed by experienced word processor in 1 hour	is equal to	part of job completed together in 1 hour
↓	↓	↓	↓	↓

   Translate: $\frac{1}{x}$ $+$ $\frac{1}{x - 2}$ $=$ $\frac{1}{6}$

   Continued on next page

Continued on next page

PRACTICE PROBLEM 6

Together, Karen and Doug Lewis can clean a strip of beach in 5 hours. Alone, Karen can clean the strip of beach one hour faster than Doug. Find the time that each person can clean the strip of beach alone. Give an exact answer and a one-decimal-place approximation.

Answer

6. Doug: $\dfrac{11 + \sqrt{101}}{2} \approx 10.5$ hr;

   Karen: $\dfrac{9 + \sqrt{101}}{2} \approx 9.5$ hr

3. SOLVE.

$$\frac{1}{x} + \frac{1}{x-2} = \frac{1}{6}$$

$$6x(x-2)\left(\frac{1}{x} + \frac{1}{x-2}\right) = 6x(x-2)\cdot\frac{1}{6}$$ Multiply both sides by the LCD $6x(x-2)$.

$$6x(x-2)\cdot\frac{1}{x} + 6x(x-2)\cdot\frac{1}{x-2} = 6x(x-2)\cdot\frac{1}{6}$$ Use the distributive property.

$$6(x-2) + 6x = x(x-2)$$

$$6x - 12 + 6x = x^2 - 2x$$

$$0 = x^2 - 14x + 12$$

Now we can substitute $a = 1, b = -14,$ and $c = 12$ into the quadratic formula and simplify.

$$x = \frac{-(-14) \pm \sqrt{(-14)^2 - 4(1)(12)}}{2\cdot 1} = \frac{14 \pm \sqrt{148}}{2}$$

Using a calculator or a square root table, we see that $\sqrt{148} \approx 12.2$ rounded to one decimal place. Thus,

$$x \approx \frac{14 \pm 12.2}{2}$$

$$x \approx \frac{14 + 12.2}{2} = 13.1 \quad \text{or} \quad x \approx \frac{14 - 12.2}{2} = 0.9$$

4. INTERPRET.

Check: If the apprentice word processor completes the job alone in 0.9 hours, the experienced word processor completes the job alone in $x - 2 = 0.9 - 2 = -1.1$ hours. Since this is not possible, we reject the solution of 0.9. The approximate solution thus is 13.1 hours.

State: The apprentice word processor can complete the job alone in approximately 13.1 hours, and the experienced word processor can complete the job alone in approximately

$$x - 2 = 13.1 - 2 = 11.1 \text{ hours}$$

◼ **Work Practice Problem 6**

PRACTICE PROBLEM 7

A family drives 500 miles to the beach for a vacation. The return trip was made at a speed that was 10 miles per hour faster. The total traveling time was $18\frac{1}{3}$ hours. Find the speed to the beach and the return speed.

Answer

7. 50 mph to the beach; 60 mph returning

EXAMPLE 7 **Finding Driving Speeds**

Beach and Fargo are about 400 miles apart. A salesperson travels from Fargo to Beach one day at a certain speed. She returns to Fargo the next day and drives 10 mph faster. Her total travel time was $14\frac{2}{3}$ hours. Find her speed to Beach and the return speed to Fargo.

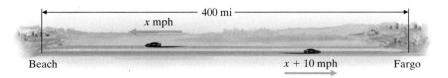

Beach   400 mi x mph $x + 10$ mph   Fargo

Solution:

1. UNDERSTAND. Read and reread the problem. Let

$$x = \text{the speed to Beach, so}$$
$$x + 10 = \text{the return speed to Fargo}$$

Then organize the given information in a table.

	Distance	=	Rate	·	Time	
To Beach	400		x		$\dfrac{400}{x}$	← distance ← rate
Return to Fargo	400		$x + 10$		$\dfrac{400}{x + 10}$	← distance ← rate

Helpful Hint

Since $d = rt$, then $t = \dfrac{d}{r}$. The time column was completed using $\dfrac{d}{r}$.

2. TRANSLATE.

In words:

time to Beach	+	return time to Fargo	=	$14\dfrac{2}{3}$ hours

Translate:

$$\frac{400}{x} \quad + \quad \frac{400}{x + 10} \quad = \quad \frac{44}{3}$$

3. SOLVE.

$$\frac{400}{x} + \frac{400}{x + 10} = \frac{44}{3}$$

This next step is optional. Notice that all three numerators in our equation are divisible by 4. To keep the numbers in our equation as simple as possible, we will take a step and divide through by 4.

$$\frac{100}{x} + \frac{100}{x + 10} = \frac{11}{3} \qquad \text{Divide both sides by 4.}$$

$$3x(x + 10)\left(\frac{100}{x} + \frac{100}{x + 10}\right) = 3x(x + 10) \cdot \frac{11}{3} \qquad \begin{array}{l}\text{Multiply both sides by the LCD,}\\ 3x(x + 10).\end{array}$$

$$3x(x + 10)\frac{100}{x} + 3x(x + 10)\frac{100}{x + 10} = 3x(x + 10) \cdot \frac{11}{3} \qquad \begin{array}{l}\text{Use the distribu-}\\ \text{tive property.}\end{array}$$

$$3(x + 10)100 + 3x(100) = x(x + 10)11$$

$$300x + 3000 + 300x = 11x^2 + 110x$$

$$0 = 11x^2 - 490x - 3000 \qquad \begin{array}{l}\text{Set equation}\\ \text{equal to 0.}\end{array}$$

$$0 = (11x + 60)(x - 50) \qquad \text{Factor.}$$

$$11x + 60 = 0 \qquad \text{or} \qquad x - 50 = 0 \qquad \text{Set each factor equal to 0.}$$

$$x = -\frac{60}{11} = -5\frac{5}{11} \qquad x = 50$$

4. INTERPRET.

Check: The speed is not negative, so it's not $-5\dfrac{5}{11}$. The number 50 does check.

State: The speed to Beach was 50 miles per hour and the return speed to Fargo was 60 miles per hour.

■ **Work Practice Problem 7**

8.3 EXERCISE SET

FOR EXTRA HELP

 Student Solutions Manual
 PH Math/Tutor Center
 CD/Video for Review
Math XL / MathXL®
MyMathLab / MyMathLab

Objective Ⓐ *Solve. See Example 1.*

1. $2x = \sqrt{10 + 3x}$

2. $3x = \sqrt{8x + 1}$

3. $x - 2\sqrt{x} = 8$

4. $x - \sqrt{2x} = 4$

5. $\sqrt{9x} = x + 2$

6. $\sqrt{16x} = x + 3$

Solve. See Example 2.

7. $\dfrac{2}{x} + \dfrac{3}{x - 1} = 1$

8. $\dfrac{6}{x^2} = \dfrac{3}{x + 1}$

9. $\dfrac{3}{x} + \dfrac{4}{x + 2} = 2$

10. $\dfrac{5}{x - 2} + \dfrac{4}{x + 2} = 1$

11. $\dfrac{7}{x^2 - 5x + 6} = \dfrac{2x}{x - 3} - \dfrac{x}{x - 2}$

12. $\dfrac{11}{2x^2 + x - 15} = \dfrac{5}{2x - 5} - \dfrac{x}{x + 3}$

Solve. See Example 3.

13. $p^4 - 16 = 0$

14. $z^4 = 81$

15. $z^4 - 13z^2 + 36 = 0$

16. $x^4 + 2x^2 - 3 = 0$

17. $4x^4 + 11x^2 = 3$

18. $9x^4 + 5x^2 - 4 = 0$

Solve. See Examples 4 and 5.

19. $x^{2/3} - 3x^{1/3} - 10 = 0$

20. $x^{2/3} + 2x^{1/3} + 1 = 0$

21. $(5n + 1)^2 + 2(5n + 1) - 3 = 0$

22. $(m - 6)^2 + 5(m - 6) + 4 = 0$

23. $2x^{2/3} - 5x^{1/3} = 3$

24. $3x^{2/3} + 11x^{1/3} = 4$

25. $1 + \dfrac{2}{3t - 2} = \dfrac{8}{(3t - 2)^2}$

26. $2 - \dfrac{7}{x + 6} = \dfrac{15}{(x + 6)^2}$

27. $20x^{2/3} - 6x^{1/3} - 2 = 0$

28. $4x^{2/3} + 16x^{1/3} = -15$

Mixed Practice *Solve. See Examples 1 through 5.*

29. $a^4 - 5a^2 + 6 = 0$

30. $x^4 - 12x^2 + 11 = 0$

31. $\dfrac{2x}{x - 2} + \dfrac{x}{x + 3} = \dfrac{-5}{x + 3}$

32. $\dfrac{5}{x - 3} + \dfrac{x}{x + 3} = \dfrac{19}{x^2 - 9}$

33. $(p + 2)^2 = 9(p + 2) - 20$

34. $2(4m - 3)^2 - 9(4m - 3) = 5$

35. $2x = \sqrt{11x + 3}$

36. $4x = \sqrt{2x + 3}$

37. $x^{2/3} - 8x^{1/3} + 15 = 0$

38. $x^{2/3} - 2x^{1/3} - 8 = 0$

39. $x - \sqrt{19 - 2x} - 2 = 0$

40. $x - \sqrt{17 - 4x} - 3 = 0$

41. $2x^{2/3} + 3x^{1/3} - 2 = 0$

42. $6x^{2/3} - 25x^{1/3} - 25 = 0$

43. $(t + 3)^2 - 2(t + 3) - 8 = 0$

44. $(2n - 3)^2 - 7(2n - 3) + 12 = 0$

45. $x - \sqrt{x} = 2$

46. $x - \sqrt{3x} = 6$

47. $\dfrac{x}{x - 1} + \dfrac{1}{x + 1} = \dfrac{2}{x^2 - 1}$

48. $\dfrac{x}{x - 5} + \dfrac{5}{x + 5} = \dfrac{-1}{x^2 - 25}$

49. $p^4 - p^2 - 20 = 0$

50. $x^4 - 10x^2 + 9 = 0$

51. $1 = \dfrac{4}{x - 7} + \dfrac{5}{(x - 7)^2}$

52. $3 + \dfrac{1}{2p + 4} = \dfrac{10}{(2p + 4)^2}$

53. $27y^4 + 15y^2 = 2$

54. $8z^4 + 14z^2 = -5$

Objective **B** *Solve. See Examples 6 and 7.*

55. A jogger ran 3 miles, decreased her speed by 1 mile per hour and then ran another 4 miles. If her total time jogging was $1\dfrac{3}{5}$ hours, find her speed for each part of her run.

56. Mark Keaton's workout consists of jogging for 3 miles, and then riding his bike for 5 miles at a speed 4 miles per hour faster than he jogs. If his total workout time is 1 hour, find his jogging speed and his biking speed.

57. A Chinese restaurant in Mandeville, Louisiana, has a large goldfish pond around the restaurant. Suppose that an inlet pipe and a hose together can fill the pond in 8 hours. The inlet pipe alone can complete the job in one hour less time than the hose alone. Find the time that the hose can complete the job alone and the time that the inlet pipe can complete the job alone. Round each to the nearest tenth of an hour.

58. A water tank on a farm in Flatonia, Texas, can be filled with a large inlet pipe and a small inlet pipe in 3 hours. The large inlet pipe alone can fill the tank in 2 hours less time than the small inlet pipe alone. Find the time to the nearest tenth of an hour each pipe can fill the tank alone.

59. Roma Sherry drove 330 miles from her home town to Tucson. During her return trip, she was able to increase her speed by 11 miles per hour. If her return trip took 1 hour less time, find her original speed and her speed returning home.

60. A salesperson drove to Portland, a distance of 300 miles. During the last 80 miles of his trip, heavy rainfall forced him to decrease his speed by 15 miles per hour. If his total driving time was 6 hours, find his original speed and his speed during the rainfall.

61. Bill Shaughnessy and his son Billy can clean the house together in 4 hours. When the son works alone, it takes him an hour longer to clean than it takes his dad alone. Find how long to the nearest tenth of an hour it takes the son to clean alone.

62. Together, Noodles and Freckles eat a 50-pound bag of dog food in 30 days. Noodles by herself eats a 50-pound bag in 2 weeks less time than Freckles does by himself. How many days to the nearest whole day would a 50-pound bag of dog food last Freckles?

63. The product of a number and 4 less than the number is 96. Find the number.

64. A whole number increased by its square is two more than twice itself. Find the number.

△ **65.** Suppose that we want to make an open box from a square sheet of cardboard by cutting out squares from each corner as shown and then folding along the dotted lines. If the box is to have a volume of 300 cubic inches, find the original dimensions of the sheet of cardboard.

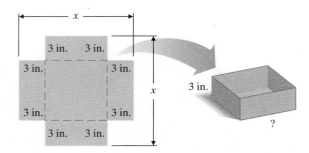

△ **66.** Suppose that we want to make an open box from a square sheet of cardboard by cutting out squares from each corner as shown and then folding along the dotted lines. If the box is to have a volume of 128 cubic inches, find the original dimensions of the sheet of cardboard. (*Hint:* Use Exercise 65 parts (a), (b), and (c) to help you.)

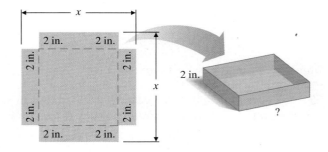

a. The ? in the drawing to the left will be the length (and also the width) of the box as shown in the drawing to the right. Represent this length in terms of x.

b. Use the formula for volume of a box, $V = l \cdot w \cdot h$, to write an equation in x.

c. Solve the equation for x and give the dimensions of the sheet of cardboard. Check your solution.

△ **67.** A sprinkler that sprays water in a circular pattern is to be used to water a square garden. If the area of the garden is 920 square feet, find the smallest whole number *radius* that the sprinkler can be adjusted to so that the entire garden is watered.

△ **68.** Suppose that a square field has an area of 6270 square feet. See Exercise 67 and find a new sprinkler radius.

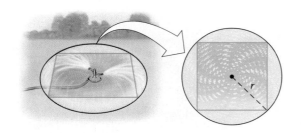

Review

Solve each inequality. See Section 2.4.

69. $\dfrac{5x}{3} + 2 \leq 7$

70. $\dfrac{2x}{3} + \dfrac{1}{6} \geq 2$

71. $\dfrac{y-1}{15} > -\dfrac{2}{5}$

72. $\dfrac{z-2}{12} < \dfrac{1}{4}$

Concept Extensions

Solve.

73. $y^3 + 9y - y^2 - 9 = 0$

74. $x^3 + x - 3x^2 - 3 = 0$

75. $x^{-2} - x^{-1} - 6 = 0$

76. $y^{-2} - 8y^{-1} + 7 = 0$

77. $2x^3 = -54$

78. $y^3 - 216 = 0$

79. Write a polynomial equation that has three solutions: 2, 5, and −7.

80. Write a polynomial equation that has three solutions: 0, 2*i*, and −2*i*.

81. During the 2004 Champ Car Grand Prix of Cleveland, held at Burke Lakefront Airport in Cleveland, Ohio, Bruno Junqueira posted the fastest speed, but Sebastien Bourdais won the race. The track is 11,119.68 feet long. Junqueira's fastest speed was 0.1159 feet per second faster than Bourdais's fastest speed. Traveling at these fastest speeds, Bourdais would have taken 0.036 seconds longer than Junqueira to complete a lap. (*Source:* Based on data from Championship Auto Racing Teams, Inc.)

   a. Find Bruno Junqueira's fastest speed during the race. Round to three decimal places.

   b. Find Sebastien Bourdais's fastest speed during the race. Round to three decimal places.

   c. Convert Bourdais's speed to miles per hour. Round to three decimal places.

82. Use a grapher to solve Exercise 29. Compare the solution with the solution from Exercise 29. Explain any differences.

1. _____

2. _____

3. _____

4. _____

5. _____

6. _____

7. _____

8. _____

9. _____

10. _____

11. _____

12. _____

13. _____

14. _____

15. _____

16. _____

17. _____

18. _____

INTEGRATED REVIEW　Sections 8.1–8.3

Summary on Solving Quadratic Equations

Use the square root property to solve each equation.

1. $x^2 - 10 = 0$

2. $3x^2 + 24 = 0$

3. $(x - 1)^2 = 8$

4. $(2x + 5)^2 = 12$

Solve each equation by completing the square.

5. $x^2 + 2x - 12 = 0$

6. $x^2 - 12x + 11 = 0$

7. $4x^2 + 12x = 8$

8. $16y^2 + 16y = 1$

Use the quadratic formula to solve each equation.

9. $2x^2 - 4x + 1 = 0$

10. $\frac{1}{2}x^2 + 3x + 2 = 0$

11. $x^2 + 4x = -7$

12. $5x^2 + 6x = -3$

Solve each equation. Use a method of your choice.

13. $x^2 + 3x + 6 = 0$

14. $2x^2 + 18 = 0$

15. $x^2 + 17x = 0$

16. $4x^2 - 2x - 3 = 0$

17. $(x - 2)^2 = 27$

18. $\frac{1}{2}x^2 - 2x + \frac{1}{2} = 0$

19. $3x^2 + 2x = 8$

20. $2x^2 = -5x - 1$

21. $x(x - 2) = 5$

22. $x^2 - 31 = 0$

23. $4x^2 - 48 = 0$

24. $5x^2 + 55 = 0$

25. $x(x + 5) = 66$

26. $5x^2 + 6x - 2 = 0$

27. $2x^2 + 3x = 1$

28. $x - \sqrt{13 - 3x} - 3 = 0$

29. $\dfrac{5x}{x - 2} - \dfrac{x + 1}{x} = \dfrac{3}{x(x - 2)}$

△ **30.** The diagonal of a square room measures 20 feet. Find the exact length of a side of the room. Then approximate the length to the nearest tenth of a foot.

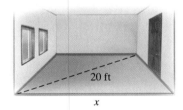

20 ft

x

31. Diane Gray and Lucy Hoag together can prepare a crawfish boil for a large party in 4 hours. Lucy alone can complete the job in 2 hours less time than Diane alone. Find the time that each person can prepare the crawfish boil alone. Round each time to the nearest tenth of an hour.

32. Kraig Blackwelder exercises at Total Body Gym. On the treadmill, he runs 5 miles, then increases his speed by 1 mile per hour and runs an additional 2 miles. If his total time on the treadmill is $1\frac{1}{3}$ hours, find his speed during each part of his run.

19. _____

20. _____

21. _____

22. _____

23. _____

24. _____

25. _____

26. _____

27. _____

28. _____

29. _____

30. _____

31. _____

32. _____

A Solve Polynomial Inequalities of Degree 2 or Greater.

B Solve Inequalities That Contain Rational Expressions with Variables in the Denominator.

8.4 NONLINEAR INEQUALITIES IN ONE VARIABLE

Objective A Solving Polynomial Inequalities

Just as we can solve linear inequalities in one variable, we can also solve quadratic and higher-degree inequalities in one variable. Let's begin with quadratic inequalities. A **quadratic inequality** is an inequality that can be written so that one side is a quadratic expression and the other side is 0. Here are examples of quadratic inequalities in one variable. Each is written in **standard form.**

$$x^2 - 10x + 7 \leq 0 \qquad 3x^2 + 2x - 6 > 0$$
$$2x^2 + 9x - 2 < 0 \qquad x^2 - 3x + 11 \geq 0$$

A solution of a quadratic inequality in one variable is a value of the variable that makes the inequality a true statement.

The value of an expression such as $x^2 - 3x - 10$ will sometimes be positive, sometimes negative, and sometimes 0, depending on the value substituted for x. To solve the inequality $x^2 - 3x - 10 < 0$, we look for all values of x that make the expression $x^2 - 3x - 10$ **less than 0,** or **negative.** To understand how we find these values, we'll study the graph of the quadratic function $y = x^2 - 3x - 10$.

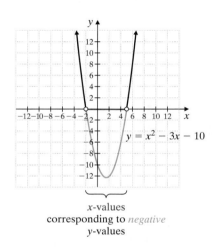

x-values
corresponding to *negative*
y-values

Notice that the x-values for which y or $x^2 - 3x - 10$ is positive are separated from the x-values for which y or $x^2 - 3x - 10$ is negative by the values for which y or $x^2 - 3x - 10$ is 0, the x-intercepts. Thus, the solution set of $x^2 - 3x - 10 < 0$ consists of all real numbers from -2 to 5 or, in interval notation, $(-2, 5)$.

It is not necessary to graph $y = x^2 - 3x - 10$ to solve the related inequality $x^2 - 3x - 10 < 0$. Instead, we can draw a number line representing the x-axis and keep the following in mind: *A region on the number line for which the value of $x^2 - 3x - 10$ is positive is separated from a region on the number line for which the value of $x^2 - 3x - 10$ is negative by a value for which the expression is 0.*

Let's find these values for which the expression is 0 by solving the related equation, $x^2 - 3x - 10 = 0$.

$$x^2 - 3x - 10 = 0$$
$$(x - 5)(x + 2) = 0 \qquad \text{Factor.}$$
$$x - 5 = 0 \quad \text{or} \quad x + 2 = 0 \qquad \text{Set each factor equal to 0.}$$
$$x = 5 \qquad\qquad x = -2 \qquad \text{Solve.}$$

These two numbers -2 and 5 divide the number line into three regions. We will call the regions A, B, and C. These regions are important because if the value of $x^2 - 3x - 10$ is negative when a number from a region is substituted for x, then $x^2 - 3x - 10$ is negative when any number in that region is substituted for x. Similarly, if the value of $x^2 - 3x - 10$ is positive when a number from a region is substituted for x, then $x^2 - 3x - 10$ is positive when any number in that region is substituted for x.

To see whether the inequality $x^2 - 3x - 10 < 0$ is true or false in each region, we choose a test point from each region and substitute its value for x in the inequality $x^2 - 3x - 10 < 0$. If the resulting inequality is true, the region containing the test point is a solution region.

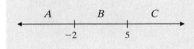

Region	Test Point Value	$(x - 5)(x + 2) < 0$	Result
A	-3	$(-8)(-1) < 0$	False
B	0	$(-5)(2) < 0$	True
C	6	$(1)(8) < 0$	False

The values in region B satisfy the inequality. The numbers -2 and 5 are not included in the solution set since the inequality symbol is $<$. The solution set is $(-2, 5)$, and its graph is shown.

$$
\begin{array}{ccc}
A & B & C \\
F \quad -2 \quad & T & \quad 5 \quad F
\end{array}
$$

EXAMPLE 1 Solve: $(x + 3)(x - 3) > 0$

Solution: First we solve the related equation, $(x + 3)(x - 3) = 0$.

$$(x + 3)(x - 3) = 0$$
$$x + 3 = 0 \quad \text{or} \quad x - 3 = 0$$
$$x = -3 \qquad\qquad x = 3$$

The two numbers -3 and 3 separate the number line into three regions, A, B, and C.

Now we substitute the value of a test point from each region. If the test value satisfies the inequality, every value in the region containing the test value is a solution.

Region	Test Point Value	$(x + 3)(x - 3) > 0$	Result
A	-4	$(-1)(-7) > 0$	True
B	0	$(3)(-3) > 0$	False
C	4	$(7)(1) > 0$	True

The points in regions A and C satisfy the inequality. The numbers -3 and 3 are not included in the solution since the inequality symbol is $>$. The solution set is $(-\infty, -3) \cup (3, \infty)$, and its graph is shown.

$$
\begin{array}{ccc}
A & B & C \\
T \quad -3 \quad & F & \quad 3 \quad T
\end{array}
$$

 Work Practice Problem 1

PRACTICE PROBLEM 1

Solve: $(x - 2)(x + 4) > 0$

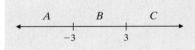

Answer

1. $(-\infty, -4) \cup (2, \infty)$

The steps below may be used to solve a polynomial inequality of degree 2 or greater.

> ### Solving a Polynomial Inequality of Degree 2 or Greater
>
> **Step 1:** Write the inequality in standard form and then solve the related equation.
>
> **Step 2:** Separate the number line into regions with the solutions from Step 1.
>
> **Step 3:** For each region, choose a test point and determine whether its value satisfies the *original inequality*.
>
> **Step 4:** The solution set includes the regions whose test point value is a solution. If the inequality symbol is $\leq$ or $\geq$, the values from Step 1 are solutions; if $<$ or $>$, they are not.

✔ **Concept Check** When choosing a test point in Step 4, why would the solutions from Step 2 not make good choices for test points?

PRACTICE PROBLEM 2

Solve: $x^2 - 6x \leq 0$

EXAMPLE 2 Solve: $x^2 - 4x \leq 0$

Solution: First we solve the related equation, $x^2 - 4x = 0$.

$$x^2 - 4x = 0$$
$$x(x - 4) = 0$$
$$x = 0 \quad \text{or} \quad x = 4$$

The numbers 0 and 4 separate the number line into three regions, A, B, and C.

$$\overset{A \qquad\quad B \qquad\quad C}{\underset{0 \qquad\qquad 4}{\longleftrightarrow}}$$

We check a test value in each region in the original inequality. Values in region B satisfy the inequality. The numbers 0 and 4 are included in the solution since the inequality symbol is $\leq$. The solution set is $[0, 4]$, and its graph is shown.

$$\overset{A \qquad\quad B \qquad\quad C}{\underset{F \quad 0 \quad T \quad 4 \quad F}{\longleftrightarrow}}$$

◼ **Work Practice Problem 2**

PRACTICE PROBLEM 3

Solve:
$(x - 2)(x + 1)(x + 5) \leq 0$

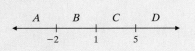

EXAMPLE 3 Solve: $(x + 2)(x - 1)(x - 5) \leq 0$

Solution: First we solve $(x + 2)(x - 1)(x - 5) = 0$. By inspection, we see that the solutions are $-2, 1$ and 5. They separate the number line into four regions, A, B, C, and D. Next we check test points from each region.

Region	Test Point Value	$(x + 2)(x - 1)(x - 5) \leq 0$	Result
A	-3	$(-1)(-4)(-8) \leq 0$	True
B	0	$(2)(-1)(-5) \leq 0$	False
C	2	$(4)(1)(-3) \leq 0$	True
D	6	$(8)(5)(1) \leq 0$	False

The solution set is $(-\infty, -2] \cup [1, 5]$, and its graph is shown. We include the numbers -2, 1, and 5 because the inequality symbol is $\leq$.

$$\overset{A \qquad B \qquad C \qquad D}{\underset{T \; -2 \quad F \quad 1 \quad T \quad 5 \quad F}{\longleftrightarrow}}$$

◼ **Work Practice Problem 3**

Answers

2. $[0, 6]$, **3.** $(-\infty, -5] \cup [-1, 2]$

✔ **Concept Check Answer**

The solutions found in Step 2 have a value of 0 in the original inequality.

Objective B Solving Rational Inequalities

Inequalities containing rational expressions with variables in the denominator are solved by using a similar procedure. Notice as we solve an example that unlike quadratic inequalities, we must also consider values for which the rational inequality is undefined. Why? As usual, these values may not be solution values for the inequality.

EXAMPLE 4 Solve: $\dfrac{x + 2}{x - 3} \le 0$

Solution: First we find all values that make the denominator equal to 0. To do this, we solve $x - 3 = 0$, or $x = 3$.

Next, we solve the related equation, $\dfrac{x + 2}{x - 3} = 0$.

$\dfrac{x + 2}{x - 3} = 0$ Multiply both sides by the LCD, $x - 3$.

$x + 2 = 0$

$x = -2$

Now we place these numbers on a number line and proceed as before, checking test point values in the original inequality.

Choose -3 from region A.	**Choose 0 from region B.**	**Choose 4 from region C.**
$\dfrac{x + 2}{x - 3} \le 0$	$\dfrac{x + 2}{x - 3} \le 0$	$\dfrac{x + 2}{x - 3} \le 0$
$\dfrac{-3 + 2}{-3 - 3} \le 0$	$\dfrac{0 + 2}{0 - 3} \le 0$	$\dfrac{4 + 2}{4 - 3} \le 0$
$\dfrac{-1}{-6} \le 0$	$-\dfrac{2}{3} \le 0$ True	$6 \le 0$ False
$\dfrac{1}{6} \le 0$ False		

The solution set is $[-2, 3)$. This interval includes -2 because -2 satisfies the original inequality. This interval does not include 3 because 3 would make the denominator 0.

Work Practice Problem 4

The steps below may be used to solve a rational inequality with variables in the denominator.

Solving a Rational Inequality

Step 1: Solve for values that make all denominators 0.

Step 2: Solve the related equation.

Step 3: Separate the number line into regions with the solutions from Steps 1 and 2.

Step 4: For each region, choose a test point and determine whether its value satisfies the *original inequality*.

Step 5: The solution set includes the regions whose test point value is a solution. Check whether to include values from Step 2. Be sure *not* to include values that make any denominator 0.

PRACTICE PROBLEM 4

Solve: $\dfrac{x - 3}{x + 5} \le 0$

Answer

4. $(-5, 3]$

PRACTICE PROBLEM 5

Solve: $\dfrac{3}{x-2} < 2$

EXAMPLE 5 Solve: $\dfrac{5}{x+1} < -2$

Solution: First we find values for x that make the denominator equal to 0.

$$x + 1 = 0$$
$$x = -1$$

Next we solve $\dfrac{5}{x+1} = -2$.

$$(x+1) \cdot \frac{5}{x+1} = (x+1) \cdot -2 \qquad \text{Multiply both sides by the LCD, } x+1.$$
$$5 = -2x - 2 \qquad \text{Simplify.}$$
$$7 = -2x$$
$$-\frac{7}{2} = x$$

We use these two solutions to divide a number line into three regions and choose test points. Only a test point value from region B satisfies the *original inequality*. The solution set is $\left(-\dfrac{7}{2}, -1\right)$, and its graph is shown.

■ **Work Practice Problem 5**

Answer

5. $(-\infty, 2) \cup \left(\dfrac{7}{2}, \infty\right)$

8.4 EXERCISE SET

Student Solutions Manual PH Math/Tutor Center CD/Video for Review Math XL MathXL® MyMathLab MyMathLab

Objective A *Solve. See Examples 1 through 3.*

1. $(x + 1)(x + 5) > 0$

2. $(x + 1)(x + 5) \leq 0$

3. $(x - 3)(x + 4) \leq 0$

4. $(x + 4)(x - 1) > 0$

5. $x^2 + 8x + 15 \geq 0$

6. $x^2 - 7x + 10 \leq 0$

7. $3x^2 + 16x < -5$

8. $2x^2 - 5x < 7$

9. $(x - 6)(x - 4)(x - 2) > 0$

10. $(x - 6)(x - 4)(x - 2) \leq 0$

11. $x(x - 1)(x + 4) \leq 0$

12. $x(x - 6)(x + 2) > 0$

13. $(x^2 - 9)(x^2 - 4) > 0$

14. $(x^2 - 16)(x^2 - 1) \leq 0$

Objective B *Solve. See Examples 4 and 5.*

15. $\dfrac{x + 7}{x - 2} < 0$

16. $\dfrac{x - 5}{x - 6} > 0$

17. $\dfrac{5}{x + 1} > 0$

18. $\dfrac{3}{y - 5} < 0$

19. $\dfrac{x + 1}{x - 4} \geq 0$

20. $\dfrac{x + 1}{x - 4} \leq 0$

21. $\dfrac{3}{x - 2} < 4$

22. $\dfrac{-2}{y + 3} > 2$

23. $\dfrac{x^2 + 6}{5x} \geq 1$

24. $\dfrac{y^2 + 15}{8y} \leq 1$

25. $\dfrac{x + 2}{x - 3} < 1$

26. $\dfrac{x - 1}{x + 4} > 2$

Objectives Ⓐ Ⓑ **Mixed Practice** *Solve each inequality. Write the solution set in interval notation.*

27. $(2x - 3)(4x + 5) \leq 0$ **28.** $(6x + 7)(7x - 12) > 0$ ◉ **29.** $x^2 > x$

30. $x^2 < 25$ ◉ **31.** $\dfrac{x}{x - 10} < 0$ **32.** $\dfrac{x + 10}{x - 10} > 0$

33. $(2x - 8)(x + 4)(x - 6) \leq 0$ **34.** $(3x - 12)(x + 5)(2x - 3) \geq 0$ **35.** $6x^2 - 5x \geq 6$

36. $12x^2 + 11x \leq 15$ **37.** $\dfrac{x - 5}{x + 4} \geq 0$ **38.** $\dfrac{x - 3}{x + 2} \leq 0$

39. $\dfrac{-1}{x - 1} > -1$ **40.** $\dfrac{4}{y + 2} < -2$ **41.** $4x^3 + 16x^2 - 9x - 36 > 0$

42. $x^3 + 2x^2 - 4x - 8 < 0$ **43.** $x^4 - 26x^2 + 25 \geq 0$ **44.** $16x^4 - 40x^2 + 9 \leq 0$

45. $\dfrac{x(x + 6)}{(x - 7)(x + 1)} \geq 0$ **46.** $\dfrac{(x - 2)(x + 2)}{(x + 1)(x - 4)} \leq 0$ **47.** $\dfrac{x}{x + 4} \leq 2$

48. $\dfrac{4x}{x - 3} \geq 5$ **49.** $(2x - 7)(3x + 5) > 0$ **50.** $(4x - 9)(2x + 5) < 0$

51. $\dfrac{z}{z - 5} \geq 2z$ **52.** $\dfrac{p}{p + 4} \leq 3p$ **53.** $\dfrac{(x + 1)^2}{5x} > 0$

54. $\dfrac{(2x - 3)^2}{x} < 0$

Review

Fill in each table so that each ordered pair is a solution of the given function. See Section 3.6.

55. $f(x) = x^2$

x	y
0	
1	
−1	
2	
−2	

56. $f(x) = 2x^2$

x	y
0	
1	
−1	
2	
−2	

57. $f(x) = -x^2$

x	y
0	
1	
−1	
2	
−2	

58. $f(x) = -3x^2$

x	y
0	
1	
−1	
2	
−2	

Concept Extensions

59. Explain why $\dfrac{x + 2}{x - 3} > 0$ and $(x + 2)(x - 3) > 0$ have the same solution sets.

60. Explain why $\dfrac{x + 2}{x - 3} \geq 0$ and $(x + 2)(x - 3) \geq 0$ do not have the same solution sets.

Find all numbers that satisfy each statement.

61. A number minus its reciprocal is less than zero. Find the numbers.

62. Twice a number, added to its reciprocal is nonnegative. Find the numbers.

63. The total profit $P(x)$ for a company producing x thousand units is given by the function $P(x) = -2x^2 + 26x - 44$. Find the values of x for which the company makes a profit. [*Hint:* The company makes a profit when $P(x) > 0$.]

64. A projectile is fired straight up from the ground with an initial velocity of 80 feet per second. Its height $s(t)$ in feet at any time t in seconds is given by the function $s(t) = -16t^2 + 80t$. Find the interval of time for which the height of the projectile is greater than 96 feet.

Solve each inequality, then use a graphing calculator to check.

65. $x^2 - x - 56 > 0$

66. $x^2 - 4x - 5 < 0$

THE BIGGER PICTURE   Solving Equations and Inequalities

Continue your outline from Sections 2.6, 5.7, 6.4, and 7.6. Write how to recognize and how to solve exponential and logarithmic equations in your own words. For example:

Solving Equations and Inequalities

I. Equations

  A. Linear equations (Sec. 2.1)

  B. Absolute value equations (Sec. 2.6)

  C. Quadratic and higher degree equations (Sec. 5.7, 8.1, 8.2, 8.3)

Solving by Factoring:	Solving by the Quadratic Formula:
$2x^2 - 7x = 9$	$2x^2 + x - 2 = 0$
$2x^2 - 7x - 9 = 0$	$a = 2, b = 1, c = -2$
$(2x - 9)(x + 1) = 0$	$x = \dfrac{-1 \pm \sqrt{(1)^2 - 4(2)(-2)}}{2 \cdot 2}$
$2x - 9 = 0$	
or $\quad x + 1 = 0$	$x = \dfrac{-1 \pm \sqrt{17}}{4}$
$x = \dfrac{9}{2}$ or $\quad x = -1$	

  D. Equations with rational expressions (Sec. 6.4)

  E. Equations with radicals (Sec. 7.6)

II. Inequalities

  A. Linear inequalities (Sec. 2.4)

  B. Compound inequalities (Sec. 2.5)

  C. Absolute value inequalities (Sec. 2.6)

  D. Nonlinear inequalities

Polynomial Inequality	Rational Inequality with variable in denominator
$x^2 - x < 6$	
$x^2 - x - 6 < 0$	$\dfrac{x - 5}{x + 1} \geq 0$
$(x - 3)(x + 2) < 0$	

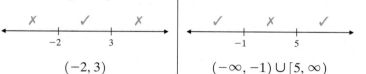

$(-2, 3)$	$(-\infty, -1) \cup [5, \infty)$

Solve. Write solutions to inequalities in interval notation.

1. $|x - 8| = |2x + 1|$

2. $0 < -x + 7 < 3$

3. $\sqrt{3x - 11} + 3 = x$

4. $x(3x + 1) = 1$

5. $\dfrac{x + 2}{x - 7} \leq 0$

6. $x(x - 6) + 4 = x^2 - 2(3 - x)$

7. $x(5x - 36) = -7$

8. $2x^2 - 4 \geq 7x$

9. $\left|\dfrac{x - 7}{3}\right| > 5$

10. $2(x - 5) + 4 < 1 + 7(x - 5) - x$

8.5 QUADRATIC FUNCTIONS AND THEIR GRAPHS

Objectives

A Graph Quadratic Functions of the Form $f(x) = x^2 + k$.

B Graph Quadratic Functions of the Form $f(x) = (x - h)^2$.

C Graph Quadratic Functions of the Form $f(x) = (x - h)^2 + k$.

D Graph Quadratic Functions of the Form $f(x) = ax^2$.

E Graph Quadratic Functions of the Form $f(x) = a(x - h)^2 + k$.

Objective **A** Graphing $f(x) = x^2 + k$

We first graphed the quadratic function $f(x) = x^2$ in Section 3.8. In that section, we discovered that the graph of a quadratic function is a parabola opening upward or downward. Now, as we continue our study, we will discover more details about quadratic functions and their graphs.

First, let's recall the definition of a *quadratic function*.

Quadratic Function

A **quadratic function** is a function that can be written in the form $f(x) = ax^2 + bx + c$, where a, b, and c are real numbers and $a \neq 0$.

Notice that equations of the form $y = ax^2 + bx + c$, where $a \neq 0$, also define quadratic functions since y is a function of x or $y = f(x)$.

Recall that if $a > 0$, the parabola opens upward and if $a < 0$, the parabola opens downward. Also, the vertex of a parabola is the lowest point if the parabola opens upward and the highest point if the parabola opens downward. The axis of symmetry is the vertical line that passes through the vertex.

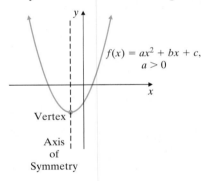

EXAMPLE 1 Graph $f(x) = x^2$ and $g(x) = x^2 + 3$ on the same set of axes.

Solution: First we construct a table of values for f and plot the points. Notice that for each x-value, the corresponding value of $g(x)$ must be 3 more than the corresponding value of $f(x)$ since $f(x) = x^2$ and $g(x) = x^2 + 3$. In other words, the graph of $g(x) = x^2 + 3$ is the same as the graph of $f(x) = x^2$ shifted upward 3 units. The axis of symmetry for both graphs is the y-axis.

x	$f(x) = x^2$	$g(x) = x^2 + 3$
-2	4	7
-1	1	4
0	0	3
1	1	4
2	4	7

Each y-value is increased by 3.

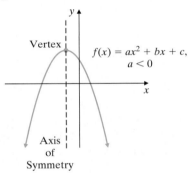

▣ **Work Practice Problem 1**

In general, we have the following properties.

PRACTICE PROBLEM 1

Graph $f(x) = x^2$ and $g(x) = x^2 + 4$ on the same set of axes.

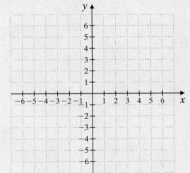

Answer

1.

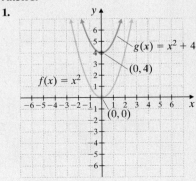

599

Graphing the Parabola Defined by $f(x) = x^2 + k$

If k is positive, the graph of $f(x) = x^2 + k$ is the graph of $y = x^2$ shifted upward k units.

If k is negative, the graph of $f(x) = x^2 + k$ is the graph of $y = x^2$ shifted downward $|k|$ units.

The vertex is $(0, k)$, and the axis of symmetry is the y-axis.

PRACTICE PROBLEMS 2–3

Graph each function.

2. $F(x) = x^2 + 1$

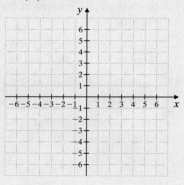

3. $g(x) = x^2 - 2$

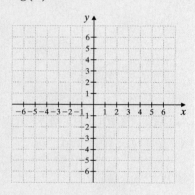

Answers

2.

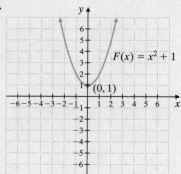

3.

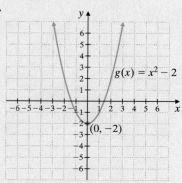

EXAMPLES Graph each function.

2. $F(x) = x^2 + 2$

The graph of $F(x) = x^2 + 2$ is obtained by shifting the graph of $y = x^2$ upward 2 units.

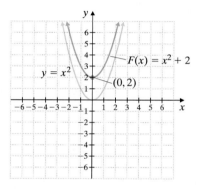

3. $g(x) = x^2 - 3$

The graph of $g(x) = x^2 - 3$ is obtained by shifting the graph of $y = x^2$ downward 3 units.

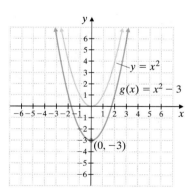

■ **Work Practice Problems 2–3**

Objective B Graphing $f(x) = (x - h)^2$

Now we will graph functions of the form $f(x) = (x - h)^2$.

EXAMPLE 4 Graph $f(x) = x^2$ and $g(x) = (x - 2)^2$ on the same set of axes.

Solution: By plotting points, we see that for each x-value, the corresponding value of $g(x)$ is the same as the value of $f(x)$ when the x-value is increased by 2. Thus, the graph of $g(x) = (x - 2)^2$ is the graph of $f(x) = x^2$ shifted to the right 2 units. The axis of symmetry for the graph of $g(x) = (x - 2)^2$ is also shifted 2 units to the right and is the line $x = 2$.

x	$f(x) = x^2$	x	$g(x) = (x - 2)^2$
-2	4	0	4
-1	1	1	1
0	0	2	0
1	1	3	1
2	4	4	4

Each x-value increased by 2 corresponds to same y-value.

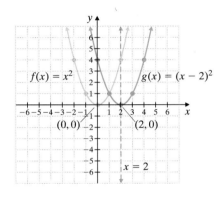

■ **Work Practice Problem 4**

In general, we have the following properties.

Graphing the Parabola Defined by $f(x) = (x - h)^2$

If h is positive, the graph of $f(x) = (x - h)^2$ is the graph of $y = x^2$ shifted to the right h units.

If h is negative, the graph of $f(x) = (x - h)^2$ is the graph of $y = x^2$ shifted to the left $|h|$ units.

The vertex is $(h, 0)$, and the axis of symmetry is the vertical line $x = h$.

EXAMPLES Graph each function.

5. $G(x) = (x - 3)^2$

The graph of $G(x) = (x - 3)^2$ is obtained by shifting the graph of $y = x^2$ to the right 3 units.

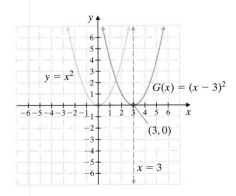

Continued on next page

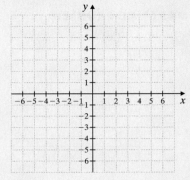
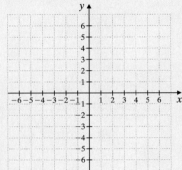
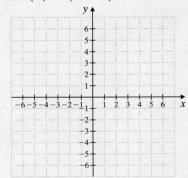
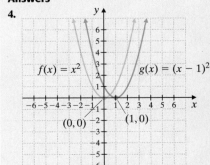

6. $F(x) = (x + 1)^2$

The equation $F(x) = (x + 1)^2$ can be written as $F(x) = [x - (-1)]^2$. The graph of $F(x) = [x - (-1)]^2$ is obtained by shifting the graph of $y = x^2$ to the left 1 unit.

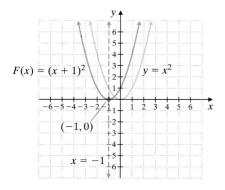

■ **Work Practice Problems 5–6**

PRACTICE PROBLEM 7

Graph: $F(x) = (x - 2)^2 + 3$

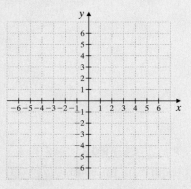

Answers

5.

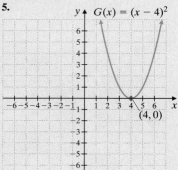

6.

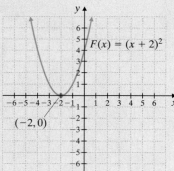

7.

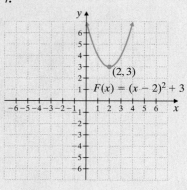

Objective **C** **Graphing $f(x) = (x - h)^2 + k$**

As we will see in graphing functions of the form $f(x) = (x - h)^2 + k$, it is possible to combine vertical and horizontal shifts.

> **Graphing the Parabola Defined by $f(x) = (x - h)^2 + k$**
>
> The parabola has the same shape as $y = x^2$.
> The vertex is (h, k), and the axis of symmetry is the vertical line $x = h$.

EXAMPLE 7 Graph: $F(x) = (x - 3)^2 + 1$

Solution: The graph of $F(x) = (x - 3)^2 + 1$ is the graph of $y = x^2$ shifted 3 units to the right and 1 unit up. The vertex is then $(3, 1)$, and the axis of symmetry is $x = 3$. A few ordered pair solutions are plotted to aid in graphing.

x	$F(x) = (x - 3)^2 + 1$
1	5
2	2
4	2
5	5

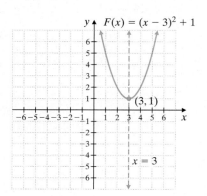

■ **Work Practice Problem 7**

Objective D Graphing $f(x) = ax^2$

Next, we discover the change in the shape of the graph when the coefficient of x^2 is not 1.

EXAMPLE 8 Graph $f(x) = x^2$, $g(x) = 3x^2$, and $h(x) = \dfrac{1}{2}x^2$ on the same set of axes.

Solution: Comparing the table of values, we see that for each x-value, the corresponding value of $g(x)$ is triple the corresponding value of $f(x)$. Similarly, the value of $h(x)$ is half the value of $f(x)$.

x	$f(x) = x^2$
-2	4
-1	1
0	0
1	1
2	4

x	$g(x) = 3x^2$
-2	12
-1	3
0	0
1	3
2	12

x	$h(x) = \dfrac{1}{2}x^2$
-2	2
-1	$\dfrac{1}{2}$
0	0
1	$\dfrac{1}{2}$
2	2

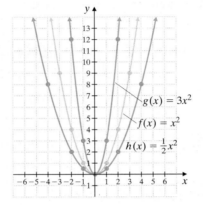

The result is that the graph of $g(x) = 3x^2$ is narrower than the graph of $f(x) = x^2$ and the graph of $h(x) = \dfrac{1}{2}x^2$ is wider. The vertex for each graph is $(0, 0)$, and the axis of symmetry is the y-axis.

Work Practice Problem 8

Graphing the Parabola Defined by $f(x) = ax^2$

If a is positive, the parabola opens upward, and if a is negative, the parabola opens downward.

If $|a| > 1$, the graph of the parabola is narrower than the graph of $y = x^2$.

If $|a| < 1$, the graph of the parabola is wider than the graph of $y = x^2$.

EXAMPLE 9 Graph: $f(x) = -2x^2$

Solution: Because $a = -2$, a negative value, this parabola opens downward. Since $|-2| = 2$ and $2 > 1$, the parabola is narrower than the graph of $y = x^2$. The vertex is $(0, 0)$, and the axis of symmetry is the y-axis. We verify this by plotting a few points.

Continued on next page

PRACTICE PROBLEM 8

Graph $f(x) = x^2$, $g(x) = 2x^2$, and $h(x) = \dfrac{1}{3}x^2$ on the same set of axes.

Answer

8.

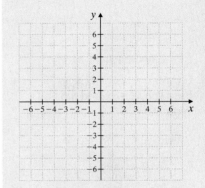

PRACTICE PROBLEM 9

Graph: $f(x) = -3x^2$

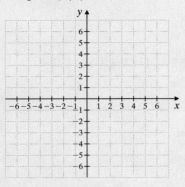

PRACTICE PROBLEM 10

Graph: $f(x) = 2(x + 3)^2 - 4$. Find the vertex and axis of symmetry.

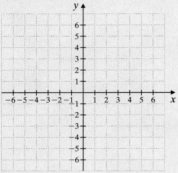

Answers

9.

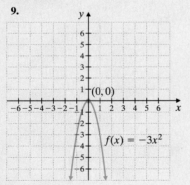

10.

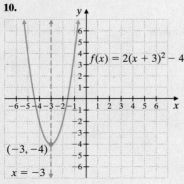

x	$f(x) = -2x^2$
−2	−8
−1	−2
0	0
1	−2
2	−8

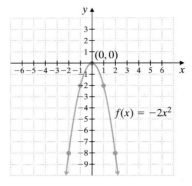

Work Practice Problem 9

Objective ☐ **Graphing $f(x) = a(x - h)^2 + k$**

Now we will see the shape of the graph of a quadratic function of the form $f(x) = a(x - h)^2 + k$.

EXAMPLE 10 Graph: $g(x) = \dfrac{1}{2}(x + 2)^2 + 5$. Find the vertex and the axis of symmetry.

Solution: The function $g(x) = \dfrac{1}{2}(x + 2)^2 + 5$ may be written as $g(x) = \dfrac{1}{2}[x - (-2)]^2 + 5$. Thus, this graph is the same as the graph of $y = x^2$ shifted 2 units to the left and 5 units upward and widened because a is $\dfrac{1}{2}$. The vertex is $(-2, 5)$, and the axis of symmetry is $x = -2$. We plot a few points to verify.

x	$g(x) = \dfrac{1}{2}(x + 2)^2 + 5$
−4	7
−3	$5\dfrac{1}{2}$
−2	5
−1	$5\dfrac{1}{2}$
0	7

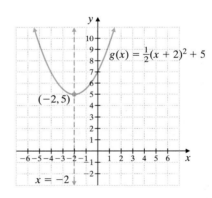

Work Practice Problem 10

In general, the following holds.

Graphing a Quadratic Function

The graph of a quadratic function written in the form $f(x) = a(x - h)^2 + k$ is a parabola with vertex (h, k).

If $a > 0$, the parabola opens upward.

If $a < 0$, the parabola opens downward. The axis of symmetry is the line whose equation is $x = h$.

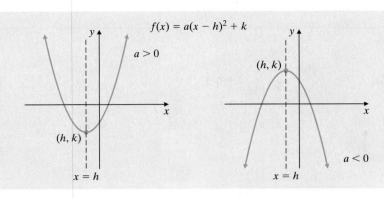

✔ Concept Check

Which description of the graph of $f(x) = -0.35(x + 3)^2 - 4$ is correct?

a. The graph opens downward and has its vertex at $(-3, 4)$.

b. The graph opens upward and has its vertex at $(-3, 4)$.

c. The graph opens downward and has its vertex at $(-3, -4)$.

d. The graph is narrower than the graph of $y = x^2$.

✔ **Concept Check Answer**

c

📟 **CALCULATOR EXPLORATIONS** Graphing

Use a graphing calculator to graph the first function of each pair. Then use its graph to predict the graph of the second function. Check your prediction by graphing both on the same set of axes. See this section and Section 3.8.

1. $F(x) = \sqrt{x}; G(x) = \sqrt{x} + 1$

2. $g(x) = x^3; H(x) = x^3 - 2$

3. $H(x) = |x|; f(x) = |x - 5|$

4. $h(x) = x^3 + 2; g(x) = (x - 3)^3 + 2$

5. $f(x) = |x + 4|; F(x) = |x + 4| + 3$

6. $G(x) = \sqrt{x} - 2; g(x) = \sqrt{x - 4} - 2$

Mental Math

State the vertex of the graph of each quadratic function.

1. $f(x) = x^2$

2. $f(x) = -5x^2$

3. $g(x) = (x - 2)^2$

4. $g(x) = (x + 5)^2$

5. $f(x) = 2x^2 + 3$

6. $h(x) = x^2 - 1$

7. $g(x) = (x + 1)^2 + 5$

8. $h(x) = (x - 10)^2 - 7$

8.5 EXERCISE SET

FOR EXTRA HELP

Student Solutions Manual   PH Math/Tutor Center   CD/Video for Review   Math XL MathXL®   MyMathLab MyMathLab

Objectives **A** **B** **Mixed Practice** *Graph each quadratic function. Label the vertex and sketch and label the axis of symmetry. See Examples 1 through 6.*

1. $f(x) = x^2 - 1$

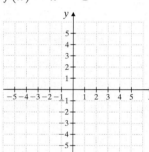

2. $h(x) = x^2 + 3$

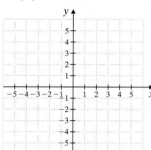

3. $f(x) = (x - 5)^2$

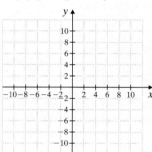

4. $g(x) = (x + 5)^2$

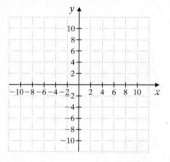

5. $h(x) = x^2 + 5$

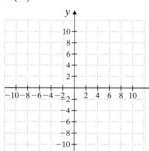

6. $h(x) = x^2 - 4$

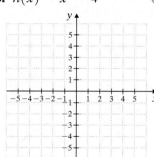

 7. $h(x) = (x + 2)^2$

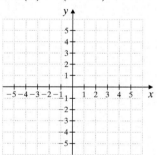

8. $H(x) = (x - 1)^2$

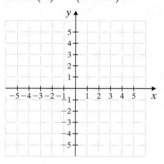

9. $g(x) = x^2 + 7$

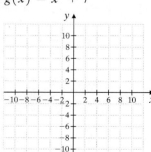

10. $f(x) = x^2 - 2$

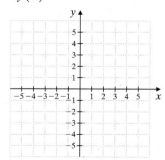

11. $G(x) = (x + 3)^2$

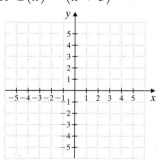

12. $f(x) = (x - 6)^2$

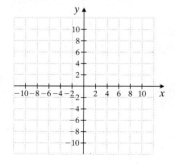

Objective C *Graph each quadratic function. Label the vertex and sketch and label the axis of symmetry. See Example 7.*

13. $f(x) = (x - 2)^2 + 5$

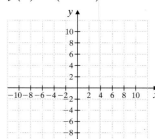

14. $g(x) = (x - 6)^2 + 1$

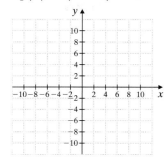

15. $h(x) = (x + 1)^2 + 4$

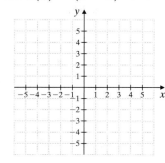

16. $G(x) = (x + 3)^2 + 3$

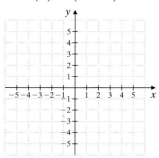

17. $g(x) = (x + 2)^2 - 5$

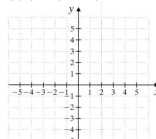

18. $h(x) = (x + 4)^2 - 6$

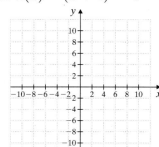

19. $h(x) = (x - 3)^2 + 2$

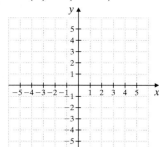

20. $F(x) = (x - 2)^2 - 3$

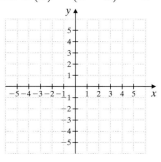

Objective D *Graph each quadratic function. Label the vertex and sketch and label the axis of symmetry.*
See Examples 8 and 9.

21. $g(x) = -x^2$

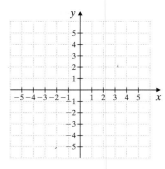

22. $f(x) = 5x^2$

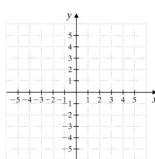

23. $h(x) = \frac{1}{3}x^2$

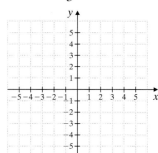

24. $g(x) = -3x^2$

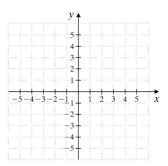

25. $H(x) = 2x^2$

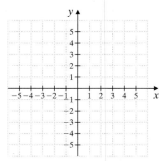

26. $f(x) = -\frac{1}{4}x^2$

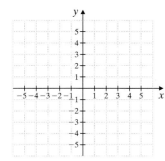

27. $F(x) = -4x^2$

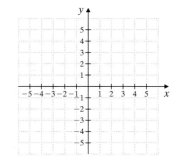

28. $G(x) = \frac{1}{5}x^2$

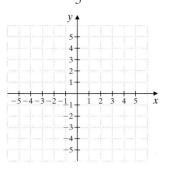

Objective **E** *Graph each quadratic function. Label the vertex and sketch and label the axis of symmetry. See Example 10.*

29. $f(x) = 10(x + 4)^2 - 6$

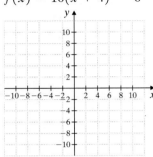

30. $g(x) = 4(x - 4)^2 + 2$

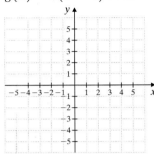

31. $h(x) = -3(x + 3)^2 + 1$

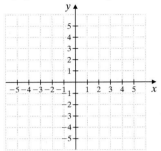

32. $f(x) = -(x - 2)^2 - 6$

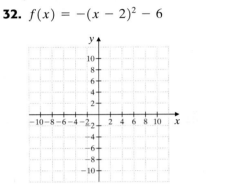

33. $H(x) = \dfrac{1}{2}(x - 6)^2 - 3$

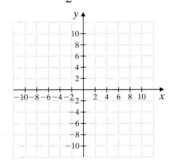

34. $G(x) = \dfrac{1}{5}(x + 4)^2 + 3$

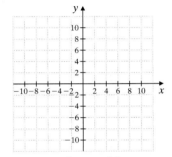

35. $f(x) = -(x - 1)^2$

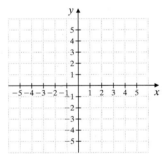

36. $f(x) = 2(x + 3)^2$

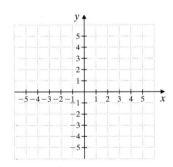

37. $F(x) = \left(x + \dfrac{1}{2}\right)^2 - 2$

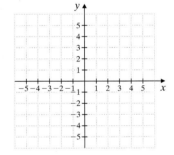

38. $H(x) = \left(x + \dfrac{1}{4}\right)^2 - 3$

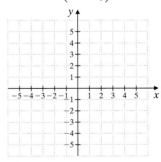

39. $F(x) = -x^2 + 2$

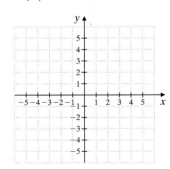

40. $G(x) = 3x^2 + 1$

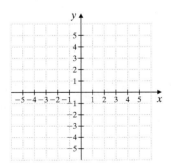

Review

Add the proper constant to each binomial so that the resulting trinomial is a perfect square trinomial. See Section 8.1.

41. $x^2 + 8x$

42. $y^2 + 4y$

43. $z^2 - 16z$

44. $x^2 - 10x$

45. $y^2 + y$

46. $z^2 - 3z$

Concept Extensions

Write the equation of the parabola that has the same shape as $f(x) = 5x^2$ but with each given vertex. Call each function $g(x)$.

47. $(2, 3)$ 　　　　**48.** $(1, 6)$ 　　　　**49.** $(-3, 6)$ 　　　　**50.** $(4, -1)$

Recall from Section 3.8 that the shifting properties covered in this section apply to the graphs of all functions. Given the accompanying graph of $y = f(x)$, graph each function.

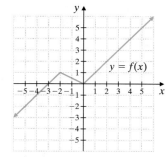

51. $y = f(x) + 1$ 　　　　　　**52.** $y = f(x) - 2$ 　　　　　　**53.** $y = f(x - 3)$

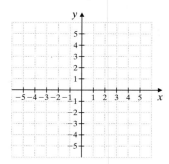 　　 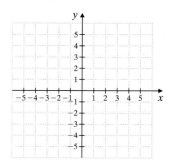 　　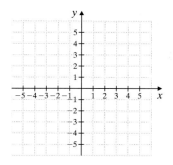

54. $y = f(x + 3)$ 　　　　　　**55.** $y = f(x + 2) + 2$ 　　　　　　**56.** $y = f(x - 1) + 1$

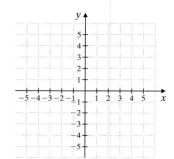 　　 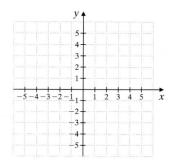 　　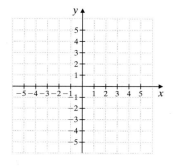

Solve. See the Concept Check in this section.

57. Which description of $f(x) = -213(x - 0.1)^2 + 3.6$ is correct?

Graph Opens	Vertex
a. upward	$(0.1, 3.6)$
b. upward	$(-213, 3.6)$
c. downward	$(0.1, 3.6)$
d. downward	$(-0.1, 3.6)$

58. Which description of $f(x) = 5\left(x + \dfrac{1}{2}\right)^2 + \dfrac{1}{2}$ is correct?

Graph Opens	Vertex
a. upward	$\left(\dfrac{1}{2}, \dfrac{1}{2}\right)$
b. upward	$\left(-\dfrac{1}{2}, \dfrac{1}{2}\right)$
c. downward	$\left(\dfrac{1}{2}, -\dfrac{1}{2}\right)$
d. downward	$\left(-\dfrac{1}{2}, -\dfrac{1}{2}\right)$

Objectives

A Write Quadratic Functions in the Form $y = a(x - h)^2 + k$.

B Derive a Formula for Finding the Vertex of a Parabola.

C Find the Minimum or Maximum Value of a Quadratic Function.

Objective **A** Writing Quadratic Functions in the Form $y = a(x - h)^2 + k$

We know that the graph of a quadratic function is a parabola. If a quadratic function is written in the form

$$f(x) = a(x - h)^2 + k$$

we can easily find the vertex (h, k) and graph the parabola. To write a quadratic function in this form, we need to complete the square. (See Section 8.1 for a review of completing the square.)

EXAMPLE 1 Graph: $f(x) = x^2 - 4x - 12$. Find the vertex and any intercepts.

Solution: The graph of this quadratic function is a parabola. To find the vertex of the parabola, we complete the square on the binomial $x^2 - 4x$. To simplify our work, we let $f(x) = y$.

$$y = x^2 - 4x - 12 \quad \text{Let } f(x) = y.$$
$$y + 12 = x^2 - 4x \quad \text{Add 12 to both sides to get the } x\text{-variable terms alone.}$$

Now we add the square of half of -4 to both sides.

$$\frac{1}{2}(-4) = -2 \quad \text{and} \quad (-2)^2 = 4$$

$$y + 12 + 4 = x^2 - 4x + 4 \quad \text{Add 4 to both sides.}$$
$$y + 16 = (x - 2)^2 \quad \text{Factor the trinomial.}$$
$$y = (x - 2)^2 - 16 \quad \text{Subtract 16 from both sides.}$$
$$f(x) = (x - 2)^2 - 16 \quad \text{Replace } y \text{ with } f(x).$$

From this equation, we can see that the vertex of the parabola is $(2, -16)$, a point in quadrant IV, and the axis of symmetry is the line $x = 2$.

Notice that $a = 1$. Since $a > 0$, the parabola opens upward. This parabola opening upward with vertex $(2, -16)$ will have two x-intercepts.

To find the x-intercepts, we let $f(x)$ or $y = 0$.

$$0 = x^2 - 4x - 12$$
$$0 = (x - 6)(x + 2)$$
$$0 = x - 6 \quad \text{or} \quad 0 = x + 2$$
$$6 = x \quad\quad\quad\quad -2 = x$$

The two x-intercepts are $(6, 0)$ and $(-2, 0)$. To find the y-intercept, we let $x = 0$.

$$f(0) = 0^2 - 4 \cdot 0 - 12 = -12$$

The y-intercept is $(0, -12)$. The sketch of $f(x) = x^2 - 4x - 12$ is shown.

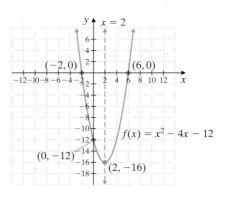

■ **Work Practice Problem 1**

PRACTICE PROBLEM 1

Graph: $f(x) = x^2 - 4x - 5$. Find the vertex and any intercepts.

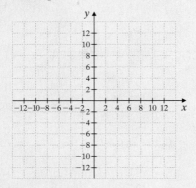

Answer

1. vertex: $(2, -9)$; x-intercepts: $(-1, 0)$, $(5, 0)$; y-intercept: $(0, -5)$

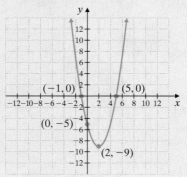

EXAMPLE 2 Graph: $f(x) = 3x^2 + 3x + 1$. Find the vertex and any intercepts.

Solution: We replace $f(x)$ with y and complete the square on x to write the equation in the form $y = a(x - h)^2 + k$.

$$y = 3x^2 + 3x + 1 \quad \text{Replace } f(x) \text{ with } y.$$
$$y - 1 = 3x^2 + 3x \qquad \text{Get the } x\text{-variable terms alone.}$$

Next we factor 3 from the terms $3x^2 + 3x$ so that the coefficient of x^2 is 1.

$$y - 1 = 3(x^2 + x) \quad \text{Factor out 3.}$$

The coefficient of x is 1. Then $\frac{1}{2}(1) = \frac{1}{2}$ and $\left(\frac{1}{2}\right)^2 = \frac{1}{4}$. Since we are adding $\frac{1}{4}$

inside the parentheses, we are really adding $3\left(\frac{1}{4}\right)$, so we *must* add $3\left(\frac{1}{4}\right)$ to the left side.

$$y - 1 + 3\left(\frac{1}{4}\right) = 3\left(x^2 + x + \frac{1}{4}\right)$$

$$y - \frac{1}{4} = 3\left(x + \frac{1}{2}\right)^2 \qquad \text{Simplify the left side and factor the right side.}$$

$$y = 3\left(x + \frac{1}{2}\right)^2 + \frac{1}{4} \quad \text{Add } \frac{1}{4} \text{ to both sides.}$$

$$f(x) = 3\left(x + \frac{1}{2}\right)^2 + \frac{1}{4} \quad \text{Replace } y \text{ with } f(x).$$

Then $a = 3, h = -\frac{1}{2}$, and $k = \frac{1}{4}$. This means that the parabola opens upward with

vertex $\left(-\frac{1}{2}, \frac{1}{4}\right)$ and that the axis of symmetry is the line $x = -\frac{1}{2}$. This parabola has no x-intercepts since the vertex is in the second quadrant and it opens upward.

To find the y-intercept, we let $x = 0$. Then

$$f(0) = 3(0)^2 + 3(0) + 1 = 1$$

We use the vertex, axis of symmetry, and y-intercept to graph the parabola.

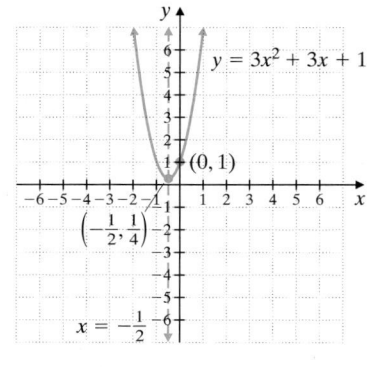

■ **Work Practice Problem 2**

PRACTICE PROBLEM 2

Graph: $f(x) = 2x^2 + 2x + 5$. Find the vertex and any intercepts.

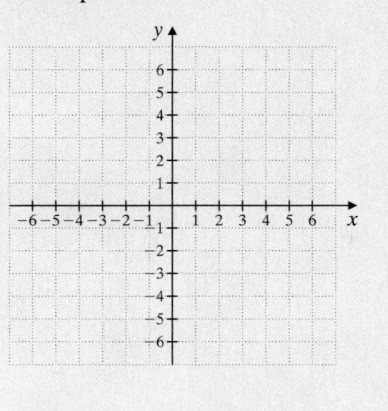

Answer

2. vertex: $\left(-\frac{1}{2}, \frac{9}{2}\right)$; y-intercept: $(0, 5)$

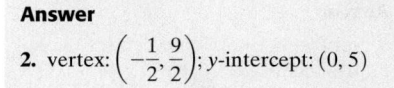
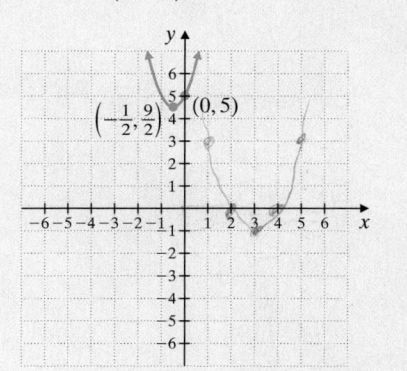

Helpful Hint

Parabola Opens Upward
Vertex in quadrant I or II: no x-intercepts
Vertex in quadrant III or IV: 2 x-intercepts

Parabola Opens Downward
Vertex in quadrant I or II: 2 x-intercepts
Vertex in quadrant III or IV: no x-intercepts

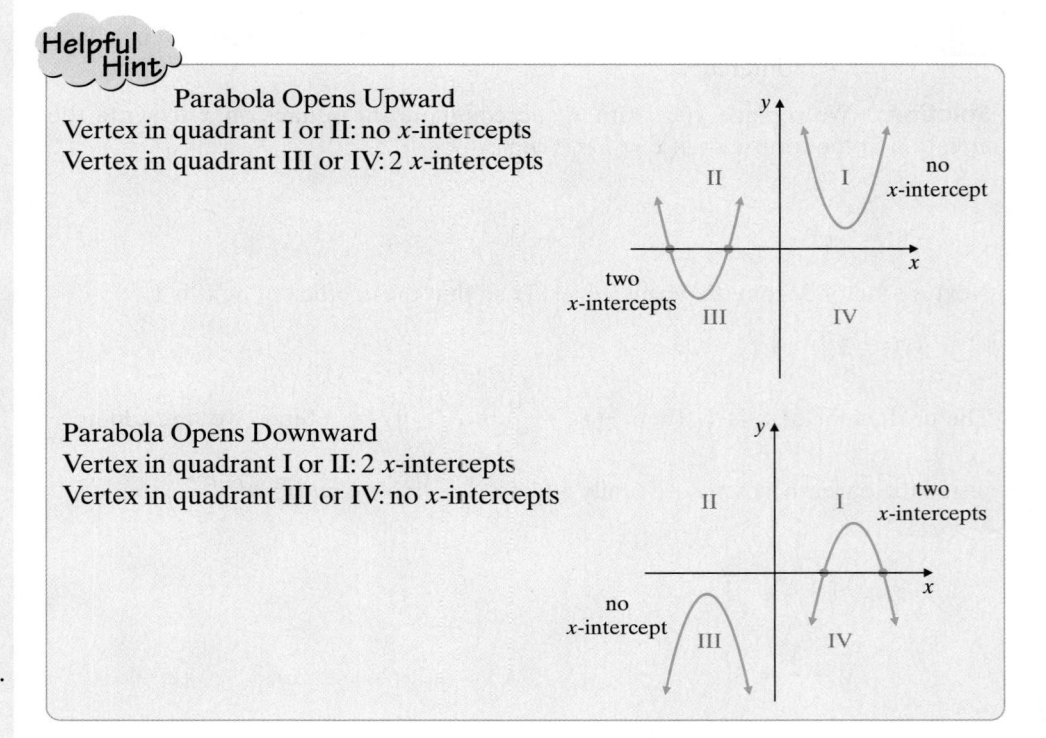

PRACTICE PROBLEM 3

Graph: $f(x) = -x^2 - 2x + 8$.
Find the vertex and any intercepts.

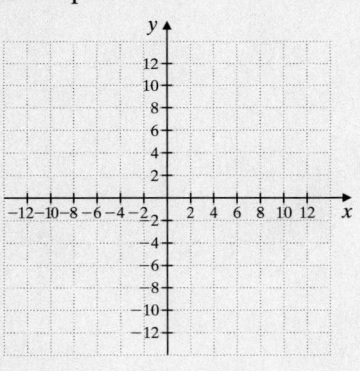

Helpful Hint

This can be written as $f(x) = -1[x - (-1)]^2 + 4$.
Notice that the vertex is $(-1, 4)$.

Answer

3. vertex: $(-1, 9)$; x-intercepts: $(-4, 0)$, $(2, 0)$; y-intercept: $(0, 8)$

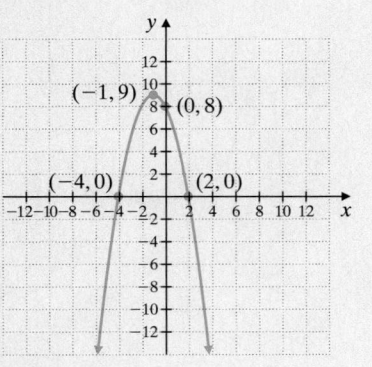

EXAMPLE 3 Graph: $f(x) = -x^2 - 2x + 3$. Find the vertex and any intercepts.

Solution: We write $f(x)$ in the form $a(x - h)^2 + k$ by completing the square. First we replace $f(x)$ with y.

$$f(x) = -x^2 - 2x + 3$$
$$y = -x^2 - 2x + 3$$
$$y - 3 = -x^2 - 2x \qquad \text{Subtract 3 from both sides to get the } x\text{-variable terms alone.}$$
$$y - 3 = -1(x^2 + 2x) \qquad \text{Factor } -1 \text{ from the terms } -x^2 - 2x.$$

The coefficient of x is 2. Then $\frac{1}{2}(2) = 1$ and $1^2 = 1$. We add 1 to the right side inside the parentheses and add $-1(1)$ to the left side.

$$y - 3 - 1(1) = -1(x^2 + 2x + 1)$$
$$y - 4 = -1(x + 1)^2 \qquad \text{Simplify the left side and factor the right side.}$$
$$y = -1(x + 1)^2 + 4 \qquad \text{Add 4 to both sides.}$$
$$\underbrace{f(x) = -1(x + 1)^2 + 4} \qquad \text{Replace } y \text{ with } f(x).$$

Since $a = -1$, the parabola opens downward with vertex $(-1, 4)$ and axis of symmetry $x = -1$.

To find the x-intercepts, we let y or $f(x) = 0$ and solve for x.

$$f(x) = -x^2 - 2x + 3$$
$$0 = -x^2 - 2x + 3 \qquad \text{Let } f(x) = 0.$$

Now we divide both sides by -1 so that the coefficient of x^2 is 1. (If you prefer, you may factor -1 from the trinomial on the right side.)

$$\frac{0}{-1} = \frac{-x^2}{-1} - \frac{2x}{-1} + \frac{3}{-1} \qquad \text{Divide both sides by } -1.$$

$$0 = x^2 + 2x - 3 \qquad \text{Simplify.}$$

$$0 = (x + 3)(x - 1) \qquad \text{Factor.}$$

$$x + 3 = 0 \quad \text{or} \quad x - 1 = 0 \qquad \text{Set each factor equal to 0.}$$

$$x = -3 \qquad\qquad x = 1 \qquad \text{Solve.}$$

The x-intercepts are $(-3, 0)$ and $(1, 0)$.

To find the y-intercept, we let $x = 0$ and solve for y. Then

$$f(0) = -0^2 - 2(0) + 3 = 3$$

Thus, $(0, 3)$ is the y-intercept. We use these points to graph the parabola.

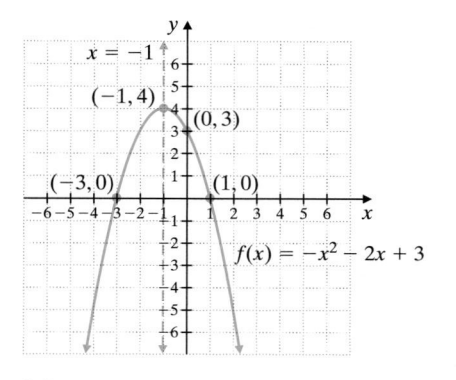

■ **Work Practice Problem 3**

Objective **B** Deriving a Formula for Finding the Vertex

As you have seen in previous examples, it can sometimes be tedious to find the vertex of a parabola by completing the square. There is a formula for finding the vertex of a parabola. Now that we have practiced completing the square, we will show that the x-coordinate of the vertex of the graph of $f(x)$ or $y = ax^2 + bx + c$ can be found by the formula $x = \dfrac{-b}{2a}$. To do so, we complete the square on x and write the equation in the form $y = (x - h)^2 + k$.

First we get the x-variable terms alone by subtracting c from both sides.

$$y = ax^2 + bx + c$$

$$y - c = ax^2 + bx$$

$$y - c = a\left(x^2 + \frac{b}{a}x \right) \qquad \text{Factor } a \text{ from the terms } ax^2 + bx.$$

Now we add the square of half of $\dfrac{b}{a}$, or $\left(\dfrac{b}{2a} \right)^2 = \dfrac{b^2}{4a^2}$, to the right side inside the parentheses. Because of the factor a, what we really added is $a\left(\dfrac{b^2}{4a^2} \right)$ and this must be added to the left side as well.

$$y - c + a\left(\frac{b^2}{4a^2} \right) = a\left(x^2 + \frac{b}{a}x + \frac{b^2}{4a^2} \right)$$

$$y - c + \frac{b^2}{4a} = a\left(x + \frac{b}{2a} \right)^2 \qquad \text{Simplify the left side and factor the right}$$

$$y = a\left(x + \frac{b}{2a} \right)^2 + c - \frac{b^2}{4a} \qquad \text{side. Add } c \text{ to both sides and subtract } \frac{b^2}{4a} \text{ from both sides.}$$

Compare this form with $f(x)$ or $y = a(x - h)^2 + k$ and see that h is $\dfrac{-b}{2a}$, which means that the x-coordinate of the vertex of the graph of $f(x) = ax^2 + bx + c$ is $\dfrac{-b}{2a}$.

Let's use the vertex formula below to find the vertex of the parabola we graphed in Example 1.

Vertex Formula

The graph of $f(x) = ax^2 + bx + c$, when $a \neq 0$, is a parabola with vertex

$$\left(\dfrac{-b}{2a}, f\left(\dfrac{-b}{2a}\right)\right)$$

PRACTICE PROBLEM 4

Find the vertex of the graph of $f(x) = x^2 - 4x - 5$. Compare your result with the result of Practice Problem 1.

EXAMPLE 4 Find the vertex of the graph of $f(x) = x^2 - 4x - 12$.

Solution: In the quadratic function $f(x) = x^2 - 4x - 12$, notice that $a = 1$, $b = -4$, and $c = -12$.

$$\dfrac{-b}{2a} = \dfrac{-(-4)}{2(1)} = 2$$

The x-value of the vertex is 2. To find the corresponding $f(x)$ or y-value, find $f(2)$. Then

$$f(2) = 2^2 - 4(2) - 12 = 4 - 8 - 12 = -16$$

The vertex is $(2, -16)$. These results agree with our findings in Example 1.

◾ **Work Practice Problem 4**

Objective C Finding Minimum and Maximum Values

The quadratic function whose graph is a parabola that opens upward has a minimum value, and the quadratic function whose graph is a parabola that opens downward has a maximum value. The $f(x)$- or y-value of the vertex is the minimum or maximum value of the function.

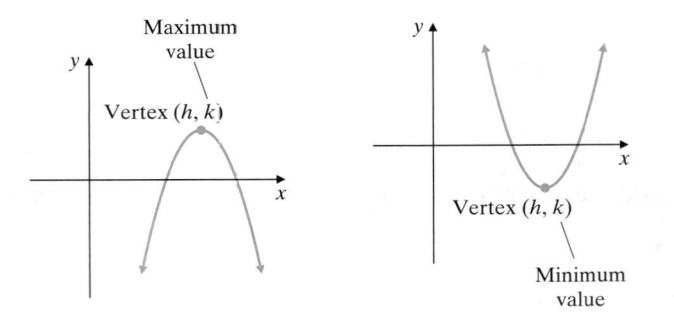

✔ **Concept Check** Without making any calculations, tell whether the graph of $f(x) = 7 - x - 0.3x^2$ has a maximum value or a minimum value. Explain your reasoning.

Answer

4. $(2, -9)$

✔ **Concept Check Answer**

$f(x)$ has a maximum value since it opens downward.

EXAMPLE 5 Finding Maximum Height

A rock is thrown upward from the ground. Its height in feet above ground after t seconds is given by the function $f(t) = -16t^2 + 20t$. Find the maximum height of the rock and the number of seconds it took for the rock to reach its maximum height.

Solution:

1. UNDERSTAND. The maximum height of the rock is the largest value of $f(t)$. Since the function $f(t) = -16t^2 + 20t$ is a quadratic function, its graph is a parabola. It opens downward since $-16 < 0$. Thus, the maximum value of $f(t)$ is the $f(t)$- or y-value of the vertex of its graph.

2. TRANSLATE. To find the vertex (h, k), we notice that for $f(t) = -16t^2 + 20t$, $a = -16$, $b = 20$, and $c = 0$. We will use these values and the vertex formula

$$\left(\frac{-b}{2a}, f\left(\frac{-b}{2a}\right)\right)$$

3. SOLVE. $h = \dfrac{-b}{2a} = \dfrac{-20}{-32} = \dfrac{5}{8}$

$$f\left(\frac{5}{8}\right) = -16\left(\frac{5}{8}\right)^2 + 20\left(\frac{5}{8}\right) = -16\left(\frac{25}{64}\right) + \frac{25}{2} = -\frac{25}{4} + \frac{50}{4} = \frac{25}{4}$$

4. INTERPRET. The graph of $f(t)$ is a parabola opening downward with vertex $\left(\frac{5}{8}, \frac{25}{4}\right)$. This means that the rock's maximum height is $\dfrac{25}{4}$ feet, or $6\dfrac{1}{4}$ feet, which was reached in $\dfrac{5}{8}$ second.

■ **Work Practice Problem 5**

Objectives **A** **B** **Mixed Practice** *Find the vertex of the graph of each quadratic function by completing the square or using the vertex formula. See Examples 1 through 4.*

1. $f(x) = x^2 + 8x + 7$ **2.** $f(x) = x^2 + 6x + 5$ **3.** $f(x) = -x^2 + 10x + 5$ **4.** $f(x) = -x^2 - 8x + 2$

5. $f(x) = 5x^2 - 10x + 3$ **6.** $f(x) = -3x^2 + 6x + 4$ **7.** $f(x) = -x^2 + x + 1$ **8.** $f(x) = x^2 - 9x + 8$

Match each function with its graph. See Examples 1 through 4.

A.

$(-1, -4)$

B.

$(1, -4)$

C.

$(-2, -1)$

D.

$(2, -1)$

9. $f(x) = x^2 - 4x + 3$ **10.** $f(x) = x^2 + 2x - 3$ **11.** $f(x) = x^2 - 2x - 3$ **12.** $f(x) = x^2 + 4x + 3$

Find the vertex of the graph of each quadratic function. Determine whether the graph opens upward or downward, find any intercepts, and graph the function. See Examples 1 through 4.

13. $f(x) = x^2 + 4x - 5$ **14.** $f(x) = x^2 + 2x - 3$ **15.** $f(x) = -x^2 + 2x - 1$

16. $f(x) = -x^2 + 4x - 4$

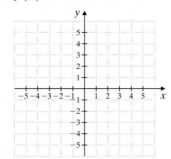

17. $f(x) = x^2 - 4$

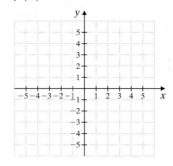

18. $f(x) = x^2 - 1$

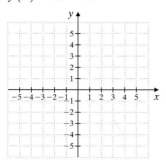

19. $f(x) = 4x^2 + 4x - 3$

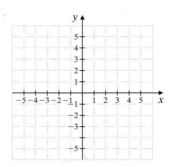

20. $f(x) = 2x^2 - x - 3$

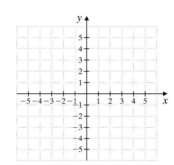

21. $f(x) = \frac{1}{2}x^2 + 4x + \frac{15}{2}$

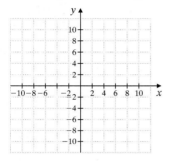

22. $f(x) = \frac{1}{5}x^2 + 2x + \frac{9}{5}$

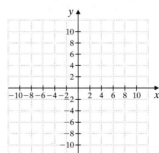

23. $f(x) = x^2 - 4x + 5$

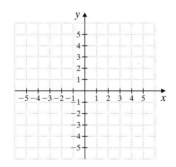

24. $f(x) = x^2 - 6x + 11$

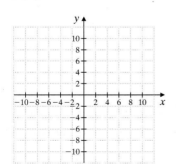

25. $f(x) = 2x^2 + 4x + 5$ **26.** $f(x) = 3x^2 + 12x + 16$ **27.** $f(x) = -2x^2 + 12x$ **28.** $f(x) = -4x^2 + 8x$

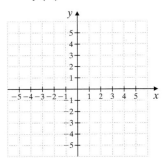 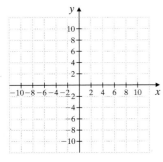 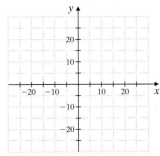 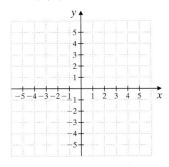

Objective C *Solve. See Example 5.*

29. If a projectile is fired straight upward from the ground with an initial speed of 96 feet per second, then its height h in feet after t seconds is given by the function $h(t) = -16t^2 + 96t$. Find the maximum height of the projectile.

30. If Rheam Gaspar throws a ball upward with an initial speed of 32 feet per second, then its height h in feet after t seconds is given by the function $h(t) = -16t^2 + 32t$. Find the maximum height of the ball.

31. The cost C in dollars of manufacturing x bicycles at Holladay's Production Plant is given by the function $C(x) = 2x^2 - 800x + 92,000$.

  a. Find the number of bicycles that must be manufactured to minimize the cost.

  b. Find the minimum cost.

32. The Utah Ski Club sells calendars to raise money. The profit P, in cents, from selling x calendars is given by the function $P(x) = 360x - x^2$.

  a. Find how many calendars must be sold to maximize profit.

  b. Find the maximum profit.

33. Find two numbers whose sum is 60 and whose product is as large as possible. [*Hint:* Let x and $60 - x$ be the two positive numbers. Their product can be described by the function $f(x) = x(60 - x)$.]

34. Find two numbers whose sum is 11 and whose product is as large as possible. (Use the hint for Exercise 33.)

35. Find two numbers whose difference is 10 and whose product is as small as possible. (Use the hint for Exercise 33.)

36. Find two numbers whose difference is 8 and whose product is as small as possible.

△ **37.** The length and width of a rectangle must have a sum of 40. Find the dimensions of the rectangle that will have the maximum area. (Use the hint for Exercise 33.)

△ **38.** The length and width of a rectangle must have a sum of 50. Find the dimensions of the rectangle that will have maximum area.

Review

Find the vertex of the graph of each function. See Section 8.5.

39. $f(x) = x^2 + 2$ **40.** $f(x) = (x - 3)^2$ **41.** $g(x) = (x + 2)^2$ **42.** $h(x) = x^2 - 3$

43. $f(x) = (x + 5)^2 + 2$ **44.** $f(x) = 2(x - 3)^2 + 2$ **45.** $f(x) = 3(x - 4)^2 + 1$ **46.** $f(x) = (x + 1)^2 + 4$

Concept Extensions

Without calculating, tell whether each graph has a minimum value or a maximum value. See the Concept Check in the section.

47. $f(x) = 2x^2 - 5$

48. $g(x) = -7x^2 + x + 1$

49. $F(x) = 3 - \frac{1}{2}x^2$

50. $G(x) = 3 - \frac{1}{2}x + 0.8x^2$

Find the vertex of the graph of each quadratic function. Determine whether the graph opens upward or downward, find the y-intercept, approximate the x-intercepts to one decimal place, and graph the function.

51. $f(x) = x^2 + 10x + 15$

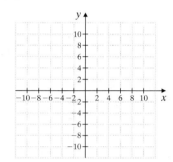

52. $f(x) = 2x^2 + 4x - 1$

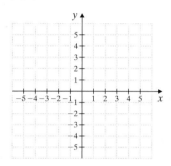

Use a graphing calculator to verify the graph of each exercise.

53. Exercise 21.

54. Exercise 22.

Find the maximum or minimum value of each function. Approximate to two decimal places.

55. $f(x) = 2.3x^2 - 6.1x + 3.2$

56. $f(x) = 7.6x^2 + 9.8x - 2.1$

57. The number of inmates in custody in U.S. prisons and jails can be modeled by the quadratic function $p(x) = -18x^2 + 298x + 988$, where $p(x)$ is the number of inmates in thousands and x is the number of years after 1995. (*Source:* Based on data from the Bureau of Justice Statistics, U.S. Department of Justice, 1995–2004.)

a. Will this function have a maximum or minimum. How can you tell?

b. According to this model, in what year will the number of prison and jail inmates in custody in the United States be at its maximum or minimum?

c. What is the maximum/minimum number of inmates predicted?

58. Methane is a gas produced by landfills, natural gas systems, and coal mining that contributes to the greenhouse effect and global warming. Projected methane emissions in the United States can be modeled by the quadratic function

$$f(x) = -0.072x^2 + 1.93x + 173.9$$

where $f(x)$ is the amount of methane produced in million metric tons and x is the number of years after 2000. (*Source:* Based on data from the U.S. Environmental Protection Agency, 2000–2020)

a. According to this model, what will U.S. emissions of methane be in 2009?

b. Will this function have a maximum or a minimum? How can you tell?

c. In what year will methane emissions in the United States be at their maximum/minimum? Round to the nearest whole year.

d. What is the level of methane emissions for that year? (Use your rounded answer from part c.)

CHAPTER 8   Group Activity

Recognizing Linear and Quadratic Models

This activity may be completed by working in groups or individually.

We have seen in this and previous chapters that data can be modeled by both linear models and quadratic models. However, when we are given a set of data to model, how can we tell if a linear or quadratic model is appropriate? The best answer requires looking at a scatter diagram of the data. If the plotted data points fall roughly on a line, a linear model is usually the better choice. If the plotted data points seem to fall on a definite curve or if a maximum or minimum point is apparent, a quadratic model is usually the better choice.

One of the sets of data shown in the tables is best modeled by a linear function and one is best modeled by a quadratic function. In each case, the variable *x* represents the number of years after 2000.

Hummer Vehicle Sales				
Year	2001	2002	2003	2004
x	1	2	3	4
Number of Hummers Sold, *y*	768	19,581	35,259	29,346

(*Source:* General Motors)

Number of Domestic Wal-Mart Stores and Supercenters					
Year	2000	2001	2002	2003	2004
x	0	1	2	3	4
Number of Stores, *y*	2522	2624	2713	2826	2949

(*Source:* Wal-Mart Stores, Inc.)

1. Make a scatter diagram for each set of data. Which type of model should be used for each set of data?

2. For the set of data that you have determined to be linear, find a linear function that fits the data points. Explain the method that you used.

3. For the set of data that you have determined to be quadratic, identify the point on your scatter diagram that appears to be the vertex of the parabola. Use the coordinates of this vertex in the quadratic model $f(x) = a(x - h)^2 + k$.

4. Solve for the remaining unknown constant in the quadratic model by substituting the coordinates for another data point into the function. Write the final form of the quadratic model for this data set.

5. Use your models to estimate the number of Hummers sold and the number of domestic Wal-Mart stores and supercenters in 2006.

6. (Optional) For each set of data, enter the data from the table into a graphing calculator and use either the linear regression feature or the quadratic regression feature to find an appropriate function that models the data.* Compare these functions with the ones you found by hand. How are they alike or different?

*To find out more about using your graphing calculator to find a regression equation, consult your user's manual.

Chapter 8 Vocabulary Check

Fill in each blank with one of the words or phrases listed below.

quadratic formula	quadratic	discriminant	$\pm\sqrt{b}$
completing the square	quadratic inequality		
(h, k) $(0, k)$ $(h, 0)$	$\dfrac{-b}{2a}$		

1. The _____ helps us know find the number and type of solutions of a quadratic equation.

2. If $a^2 = b$, then $a =$ _____.

3. The graph of $f(x) = ax^2 + bx + c$, where a is not 0, is a parabola whose vertex has an x-value of ____.

4. A(n) _____ is an inequality that can be written so that one side is a quadratic expression and the other side is 0.

5. The process of writing a quadratic equation so that one side is a perfect square trinomial is called _____.

6. The graph of $f(x) = x^2 + k$ has vertex _____.

7. The graph of $f(x) = (x - h)^2$ has vertex _____.

8. The graph of $f(x) = (x - h)^2 + k$ has vertex _____.

9. The formula $x = \dfrac{-b \pm \sqrt{b^2 - 4ac}}{2a}$ is called the _____.

10. A _____ equation is one that can be written in the form $ax^2 + bx + c = 0$, where $a, b,$ and c are real numbers and a is not 0.

Helpful Hint

Are you preparing for your test? Don't forget to take the Chapter 8 Test on page 629. Then check your answers at the back of the text and use the Chapter Test Prep Video CD to see the fully worked-out solutions to any of the exercises you want to review.

8 Chapter Highlights

DEFINITIONS AND CONCEPTS	**EXAMPLES**
Section 8.1 Solving Quadratic Equations by Completing the Square	

SQUARE ROOT PROPERTY If b is a real number and if $a^2 = b$, then $a = \pm\sqrt{b}$.	Solve: $(x + 3)^2 = 14$ $x + 3 = \pm\sqrt{14}$ $x = -3 \pm \sqrt{14}$
SOLVING A QUADRATIC EQUATION IN x BY COMPLETING THE SQUARE **Step 1.** If the coefficient of x^2 is not 1, divide both sides of the equation by the coefficient of x^2. **Step 2.** Get the variable terms alone. **Step 3.** Complete the square by adding the square of half of the coefficient of x to both sides. **Step 4.** Write the resulting trinomial as the square of a binomial. **Step 5.** Use the square root property.	Solve: $3x^2 - 12x - 18 = 0$ **1.** $x^2 - 4x - 6 = 0$ **2.** $\quad x^2 - 4x = 6$ **3.** $\dfrac{1}{2}(-4) = -2$ and $(-2)^2 = 4$ $\quad x^2 - 4x + 4 = 6 + 4$ **4.** $(x - 2)^2 = 10$ **5.** $x - 2 = \pm\sqrt{10}$ $\quad\quad x = 2 \pm \sqrt{10}$

DEFINITIONS AND CONCEPTS	**EXAMPLES**

Section 8.2   Solving Quadratic Equations by Using the Quadratic Formula

QUADRATIC FORMULA

A quadratic equation written in the form $ax^2 + bx + c = 0$ has solutions

$$x = \frac{-b \pm \sqrt{b^2 - 4ac}}{2a}$$

Solve: $x^2 - x - 3 = 0$

$$a = 1, b = -1, c = -3$$

$$x = \frac{-(-1) \pm \sqrt{(-1)^2 - 4(1)(-3)}}{2 \cdot 1}$$

$$x = \frac{1 \pm \sqrt{13}}{2}$$

Section 8.3   Solving Equations by Using Quadratic Methods

Substitution is often helpful in solving an equation that contains a repeated variable expression.

Solve: $(2x + 1)^2 - 5(2x + 1) + 6 = 0$

Let $m = 2x + 1$. Then

$$\begin{aligned} m^2 - 5m + 6 &= 0 & &\text{Let} \\ (m - 3)(m - 2) &= 0 & &m = 2x + 1. \\ m = 3 \quad &\text{or} \quad m = 2 \\ 2x + 1 = 3 \quad &\quad 2x + 1 = 2 & &\text{Substitute} \\ x = 1 \quad &\quad x = \frac{1}{2} & &\text{back.} \end{aligned}$$

Section 8.4   Nonlinear Inequalities in One Variable

SOLVING A POLYNOMIAL INEQUALITY

Step 1. Write the inequality in standard form and solve the related equation.

Step 2. Use solutions from Step 1 to separate the number line into regions.

Step 3. Use a test point to determine whether values in each region satisfy the original inequality.

Step 4. Write the solution set as the union of regions whose test point values are solutions.

Solve: $x^2 \geq 6x$

1. $x^2 - 6x \geq 0$

2. $x^2 - 6x = 0$

$$x(x - 6) = 0$$
$$x = 0 \quad \text{or} \quad x = 6$$

3.

4.

Region	Test Point Value	$x^2 \geq 6x$	Result
A	-2	$(-2)^2 \geq 6(-2)$	True
B	1	$1^2 \geq 6(1)$	False
C	7	$7^2 \geq 6(7)$	True

5.

The solution set is $(-\infty, 0] \cup [6, \infty)$.

SOLVING A RATIONAL INEQUALITY

Step 1. Solve for values that make all denominators 0.

Step 2. Solve the related equation.

Solve: $\dfrac{6}{x - 1} < -2$

1. $x - 1 = 0$ Set the denominator equal to 0.

$$x = 1$$

2. $\dfrac{6}{x - 1} = -2$

$$6 = -2(x - 1) \quad \text{Multiply by } (x - 1).$$
$$6 = -2x + 2$$
$$4 = -2x$$
$$-2 = x$$

DEFINITIONS AND CONCEPTS	**EXAMPLES**

Section 8.4   Nonlinear Inequalities in One Variable (*continued*)

Step 3. Use solutions from Steps 1 and 2 to separate the number line into regions.

Step 4. Use a test point to determine whether values in each region satisfy the original inequality.

Step 5. Write the solution set as the union of regions whose test point value is a solution.

3.

$$A \qquad B \qquad C$$

$$\overset{-2}{} \qquad \overset{1}{}$$

4. Only a test value from region B satisfies the original inequality.

5.

$$\overset{(}{-2} \qquad \overset{)}{1}$$

The solution set is $(-2, 1)$.

Section 8.5   Quadratic Functions and Their Graphs

GRAPHING A QUADRATIC FUNCTION

The graph of a quadratic function written in the form $f(x) = a(x - h)^2 + k$ is a parabola with vertex (h, k).
If $a > 0$, the parabola opens upward.
If $a < 0$, the parabola opens downward.
The axis of symmetry is the line whose equation is $x = h$.

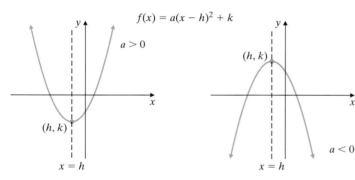

Graph: $g(x) = 3(x - 1)^2 + 4$
The graph is a parabola with vertex $(1, 4)$ and axis of symmetry $x = 1$. Since $a = 3$ is positive, the graph opens upward.

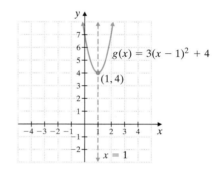

Section 8.6   Further Graphing of Quadratic Functions

The graph of $f(x) = ax^2 + bx + c, a \neq 0$, is a parabola with vertex

$$\left(\frac{-b}{2a}, f\left(\frac{-b}{2a} \right) \right)$$

Graph: $f(x) = x^2 - 2x - 8$. Find the vertex and x- and y-intercepts.

$$\frac{-b}{2a} = \frac{-(-2)}{2 \cdot 1} = 1$$

$$f(1) = 1^2 - 2(1) - 8 = -9$$

The vertex is $(1, -9)$.

$$0 = x^2 - 2x - 8$$
$$0 = (x - 4)(x + 2)$$
$$x = 4 \quad \text{or} \quad x = -2$$

continued

DEFINITIONS AND CONCEPTS	EXAMPLES

Section 8.6   Further Graphing of Quadratic Functions (*continued*)

The *x*-intercepts are $(4, 0)$ and $(-2, 0)$.

$$f(0) = 0^2 - 2 \cdot 0 - 8 = -8$$

The *y*-intercept is $(0, -8)$.

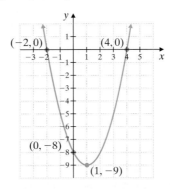

 STUDY SKILLS BUILDER

Are You Prepared for a Test on Chapter 8?

Below I have listed some common trouble areas for students in Chapter 8. After studying for your test—but before taking your test—read these.

- Don't forget that to solve a quadratic equation such as $x^2 + 6x = 1$, by completing the square, add the square of half of 6 to *both* sides.

$$x^2 + 6x = 1$$
$$x^2 + 6x + 9 = 1 + 9 \qquad \text{Add 9 to both sides.}$$
$$(x + 3)^2 = 10 \qquad \left(\frac{1}{2}(6) = 3 \text{ and } 3^2 = 9\right)$$
$$x + 3 = \pm\sqrt{10}$$
$$x = -3 \pm \sqrt{10}$$

- Remember to write a quadratic equation in standard form $(ax^2 + bx + c = 0)$ before using the quadratic formula of solve.

$$x(4x - 1) = 1$$
$$4x^2 - x - 1 = 0 \qquad \text{Write in standard form.}$$
$$x = \frac{-(-1) \pm \sqrt{(-1)^2 - 4(4)(-1)}}{2 \cdot 4} \qquad \begin{array}{l}\text{Use the quadratic}\\ \text{formula with } a = 4,\\ b = -1, \text{ and } c = -1.\end{array}$$
$$x = \frac{1 \pm \sqrt{17}}{8} \qquad \text{Simplify.}$$

- Review the steps for solving a quadratic equation in general on page 578.

- Don't forget how to graph a quadratic function in the form $f(x) = a(x - h)^2 + k$. The graph of $f(x) = -2(x - 3)^2 - 1$

opens downward / narrower   shift 3 units right   shift 1 units down

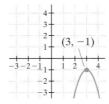

Remember: This is simply a checklist of common trouble areas. For a review of Chapter 8, see the Highlights and Chapter Review at the end of this chapter.

(8.1) *Solve by factoring.*

1. $x^2 - 15x + 14 = 0$

2. $7a^2 = 29a + 30$

Use the square root property to solve each equation.

3. $4m^2 = 196$

4. $(5x - 2)^2 = 2$

Solve by completing the square.

5. $z^2 + 3z + 1 = 0$

6. $(2x + 1)^2 = x$

7. If P dollars are invested, the formula $A = P(1 + r)^2$ gives the amount A in an account paying interest rate r compounded annually after 2 years. Find the interest rate r such that \$2500 increases to \$2717 in 2 years. Round the result to the nearest hundredth of a percent.

8. Two ships leave a port at the same time and travel at the same speed. One ship is traveling due north and the other due east. In a few hours, the ships are 150 miles apart. How many miles has each ship traveled? Give an exact answer and a one-decimal-place approximation.

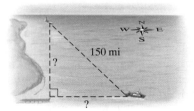

(8.2) *If the discriminant of a quadratic equation has the given value, determine the number and type of solutions of the equation.*

9. -8

10. 48

11. 100

12. 0

Use the quadratic formula to solve each equation.

13. $x^2 - 16x + 64 = 0$

14. $x^2 + 5x = 0$

15. $2x^2 + 3x = 5$

16. $6x^2 + 7 = 5x$

17. $9a^2 + 4 = 2a$

18. $(2x - 3)^2 = x$

19. Cadets graduating from military school usually toss their hats high into the air at the end of the ceremony. One cadet threw his hat so that its distance $d(t)$ in feet above the ground t seconds after it was thrown was $d(t) = -16t^2 + 30t + 6$.

     a. Find the distance above the ground of the hat 1 second after it was thrown.

     b. Find the time it took that hat to hit the ground. Give an exact time and a one-decimal-place approximation.

20. The hypotenuse of an isosceles right triangle is 6 centimeters longer than either of the legs. Find the length of the legs. (*Hint:* Don't forget that an isosceles triangle has two sides of equal length.)

(8.3) *Solve each equation.*

21. $x^3 = 27$

22. $y^3 = -64$

23. $\dfrac{5}{x} + \dfrac{6}{x-2} = 3$

24. $x^4 - 21x^2 - 100 = 0$

25. $5(x+3)^2 - 19(x+3) = 4$

26. $x^{2/3} - 6x^{1/3} + 5 = 0$

27. $a^6 - a^2 = a^4 - 1$

28. $y^{-2} + y^{-1} = 20$

29. Two postal workers, Jerome Grant and Tim Bozik, can sort a stack of mail in 5 hours. Working alone, Tim can sort the mail in 1 hour less time than Jerome can. Find the time that each postal worker can sort the mail alone. Round the result to one decimal place.

30. A negative number decreased by its reciprocal is $-\dfrac{24}{5}$. Find the number.

(8.4) *Solve each inequality for x. Write each solution set in interval notation.*

31. $2x^2 - 50 \le 0$

32. $\dfrac{1}{4}x^2 < \dfrac{1}{16}$

33. $(x^2 - 16)(x^2 - 1) > 0$

34. $\dfrac{x-5}{x-6} < 0$

35. $\dfrac{(4x+3)(x-5)}{x(x+6)} > 0$

36. $(x+5)(x-6)(x+2) \le 0$

37. $x^3 + 3x^2 - 25x - 75 > 0$

38. $\dfrac{x^2+4}{3x} \le 1$

39. $\dfrac{(5x+6)(x-3)}{x(6x-5)} < 0$

(8.5) *Graph each function. Label the vertex and the axis of symmetry of each graph.*

40. $f(x) = x^2 - 4$ **41.** $g(x) = x^2 + 7$ **42.** $H(x) = 2x^2$ **43.** $h(x) = -\dfrac{1}{3}x^2$

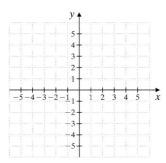   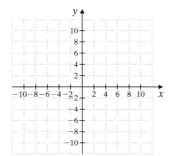   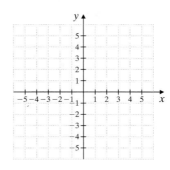   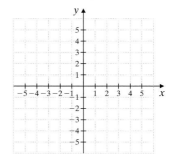

44. $F(x) = (x - 1)^2$ **45.** $G(x) = (x + 5)^2$ **46.** $f(x) = (x - 4)^2 - 2$ **47.** $f(x) = -3(x - 1)^2 + 1$

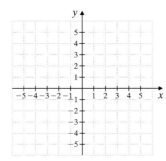   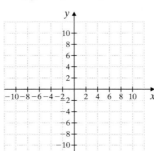   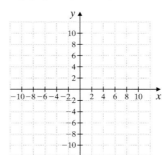   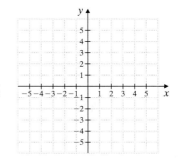

(8.6) *Graph each function. Find the vertex and any intercepts of each graph.*

48. $f(x) = x^2 + 10x + 25$ **49.** $f(x) = -x^2 + 6x - 9$

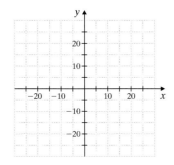   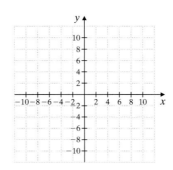

50. $f(x) = 4x^2 - 1$ **51.** $f(x) = -5x^2 + 5$

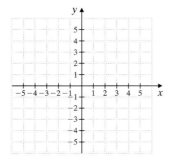   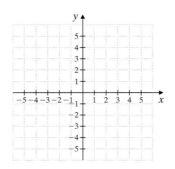

52. Find the vertex of the graph of $f(x) = -3x^2 - 5x + 4$. Determine whether the graph opens upward or downward, find the y-intercept, approximate the x-intercepts to one decimal place, and graph the function.

 53. The function $h(t) = -16t^2 + 120t + 300$ gives the height in feet of a projectile fired from the top of a building at t seconds.

    a. When will the object reach a height of 350 feet? Round your answer to one decimal place.

    b. Explain why part (a) has two answers.

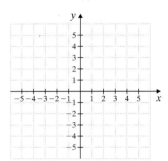

54. Find two numbers whose sum is 420 and whose product is as large as possible.

Mixed Review

Solve each equation.

55. $x^2 - x - 30 = 0$

56. $10x^2 = 3x + 4$

57. $9y^2 = 36$

58. $(9n + 1)^2 = 9$

59. $x^2 + x + 7 = 0$

60. $(3x - 4)^2 = 10x$

61. $x^2 + 11 = 0$

62. $(5a - 2)^2 - a = 0$

63. $\dfrac{7}{8} = \dfrac{8}{x^2}$

64. $x^{2/3} - 6x^{1/3} = -8$

65. $(2x - 3)(4x + 5) \geq 0$

66. $\dfrac{x(x + 5)}{4x - 3} \geq 0$

67. $\dfrac{3}{x - 2} > 2$

 68. The total amount of passenger traffic at Phoenix Sky Harbor International Airport in Phoenix, Arizona, during the period 1960 through 2000 can be modeled by the equation $y = 26x^2 - 136x + 917$, where y is the number of passengers enplaned and deplaned in thousands and x is the number of years after 1960. (*Source:* Based on data from The City of Phoenix Aviation Department, 1960–2000)

    a. Estimate the passenger traffic at Phoenix Sky Harbor International Airport in 2000.

    b. According to this model, in what year will passenger traffic at Phoenix Sky Harbor International Airport reach 60,000,000 passengers?

CHAPTER TEST

 Remember to use the Chapter Test Prep Video CD to see the fully worked-out solutions to any of the exercises you want to review.

Solve each equation.

1. $5x^2 - 2x = 7$

2. $(x + 1)^2 = 10$

3. $m^2 - m + 8 = 0$

4. $u^2 - 6u + 2 = 0$

5. $7x^2 + 8x + 1 = 0$

6. $y^2 - 3y = 5$

7. $\dfrac{4}{x + 2} + \dfrac{2x}{x - 2} = \dfrac{6}{x^2 - 4}$

8. $x^4 - 8x^2 - 9 = 0$

9. $x^6 + 1 = x^4 + x^2$

10. $(x + 1)^2 - 15(x + 1) + 56 = 0$

Solve by completing the square.

11. $x^2 - 6x = -2$

12. $2a^2 + 5 = 4a$

Solve each inequality. Write each solution set in interval notation.

13. $2x^2 - 7x > 15$

14. $(x^2 - 16)(x^2 - 25) \geq 0$

15. $\dfrac{5}{x + 3} < 1$

16. $\dfrac{7x - 14}{x^2 - 9} \leq 0$

Graph each function. Label the vertex for each graph.

17. $f(x) = 3x^2$

18. $G(x) = -2(x - 1)^2 + 5$

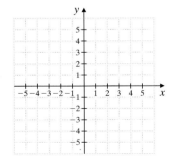

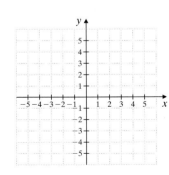

1. _____

2. _____

3. _____

4. _____

5. _____

6. _____

7. _____

8. _____

9. _____

10. _____

11. _____

12. _____

13. _____

14. _____

15. _____

16. _____

17. see graph

18. see graph

Graph each function. Find and label the vertex, y-intercept, and x-intercepts (if any) for each graph.

19. see graph

20. see graph

21. _____

22. _____

23. a. _____

b. _____

19. $h(x) = x^2 - 4x + 4$

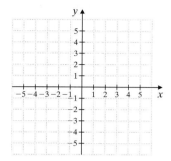

20. $F(x) = 2x^2 - 8x + 9$

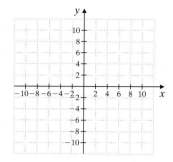

△ **21.** Given the diagram shown, approximate to the nearest foot how many feet of walking distance a person saves by cutting across the lawn instead of walking on the sidewalk.

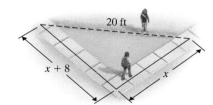

20 ft

$x + 8$ x

22. Dave and Sandy Hartranft can paint a room together in 4 hours. Working alone, Dave can paint the room in 2 hours less time than Sandy can. Find how long it takes Sandy to paint the room alone. Give an exact answer and a two-decimal-place approximation.

23. A stone is thrown upward from a bridge. The stone's height $s(t)$ in feet, above the water t seconds after the stone is thrown is given by the function $s(t) = -16t^2 + 32t + 256$.
   a. Find the maximum height of the stone.
   b. Find the time it takes the stone to hit the water. Round to the nearest hundredth of a second.

256 ft

1. Write 730,000 in scientific notation.

2. Write 2.068×10^{-3} in standard form.

Answers

3. Solve: $|2x - 1| + 5 = 6$

4. Solve: $|3x - 2| = -5$

5. Determine whether the relation $x = y^2$ is also a function.

6. Find the vertex and any intercepts of $f(x) = x^2 + x - 12$.

7. Use the elimination method to solve

the system: $\begin{cases} 3x + \dfrac{y}{2} = 2 \\ 6x + y = 5 \end{cases}$

8. Use the elimination method to solve the system.
$\begin{cases} -6x + y = 5 \\ 4x - 2y = 6 \end{cases}$

9. Subtract vertically:
$(10x^3 - 7x^2) - (4x^3 - 3x^2 + 2)$

10. Simplify: $\dfrac{x^2 - 4x + 4}{2 - x}$

Multiply.

11. $(2x - 7)(3x - 4)$

12. $(4a - 3)(7a - 2)$

13. $(3x^2 + y)(5x^2 - 2y)$

14. $(2a + b)(3a - 5b)$

15. Divide $2x^3 + 3x^4 - 8x + 6$ by $x^2 - 1$.

16. Simplify. Use positive exponents to write each answer.
  a. $(a^{-2}bc^3)^{-3}$
  b. $\left(\dfrac{a^{-4}b^2}{c^3}\right)^{-2}$
  c. $\left(\dfrac{3a^8b^2}{12a^5b^5}\right)^{-2}$

17. Factor: $6x^2 + 3x^3$

18. $9x^3 + 27x^2 - 15x$

Answers
1.
2.
3.
4.
5.
6.
7.
8.
9.
10.
11.
12.
13.
14.
15.
a. b. c.
16.
17.
18.

19. _____

20. _____

21. _____

22. _____

23. _____

24. _____

25. _____

26. _____

27. _____

28. _____

29. _____

30. _____

31. _____

32. _____

33. _____

34. _____

35. _____

36. _____

37. _____

38. _____

39. _____

40. _____

41. _____

19. Factor: $2(a + 3)^2 - 5(a + 3) - 7$

20. $2x(3y - 2) - 5(3y - 2)$

21. Factor: $3a^2x - 12abx + 12b^2x$

22. $2xy + 6x - y - 3$

23. Solve: $2x^2 = \dfrac{17}{3}x + 1$

24. Solve: $2(a^2 + 2) - 8 = -2a(a - 2) - 5$

Multiply.

25. $\dfrac{3n + 1}{2n} \cdot \dfrac{2n - 4}{3n^2 - 2n - 1}$

26. $(\sqrt{3} - 4)(2\sqrt{3} + 2)$

27. $\dfrac{x^3 - 1}{-3x + 3} \cdot \dfrac{15x^2}{x^2 + x + 1}$

28. Subtract: $\dfrac{3}{x^2 - 4} - \dfrac{1 - x}{x^2 - 4}$

29. Add: $\dfrac{2}{x^2} + \dfrac{5}{3x^3}$

30. Subtract: $\dfrac{a + 1}{a^2 - 6a + 8} - \dfrac{3}{16 - a^2}$

31. Simplify: $\dfrac{\dfrac{x}{y^2} + \dfrac{1}{y}}{\dfrac{y}{x^2} + \dfrac{1}{x}}$

32. Simplify: $\dfrac{(2a)^{-1} + b^{-1}}{a^{-1} + (2b)^{-1}}$

33. Solve: $\dfrac{2x}{x - 3} + \dfrac{6 - 2x}{x^2 - 9} = \dfrac{x}{x + 3}$

34. Solve: $\dfrac{x + 3}{x^2 + 5x + 6} = \dfrac{3}{2x + 4} - \dfrac{1}{x + 3}$

35. Solve: $\dfrac{1}{x} + \dfrac{1}{y} = \dfrac{1}{z}$ for x

36. Divide $x^3 - 3x^2 - 10x + 24$ by $x + 3$.

△ **37.** The lateral surface area of a cylinder varies jointly as its radius and height. Express surface area S in terms of radius r and height h.

38. Suppose that y varies inversely as x. If y is 8 when x is 24, find the constant of variation and the variation equation.

Find each cube root.

39. $\sqrt[3]{-64}$

40. $\sqrt[3]{24a^7b^9}$

41. $\sqrt[3]{\dfrac{8}{125}}$

42. Simplify and rationalize: $\dfrac{2\sqrt{16}}{\sqrt{9x}}$

Multiply.

43. $\sqrt[3]{4} \cdot \sqrt[3]{2}$

44. $(\sqrt{5} - x)^2$

45. $\sqrt{\dfrac{2}{a}} \cdot \sqrt{\dfrac{b}{3}}$

46. $(\sqrt{a} + b)(\sqrt{a} - b)$

47. Simplify: $\sqrt[3]{54} - 5\sqrt[3]{16} + \sqrt[3]{2}$

48. Solve: $\sqrt{x - 2} = \sqrt{4x + 1} - 3$

49. Add, then simplify: $\sqrt[3]{\dfrac{7x}{8}} + 2\sqrt[3]{7x}$

50. Simplify: $\sqrt[4]{48x^9}$

51. Solve $p^2 + 2p = 4$ by completing the square.

52. Use the square root property to solve $(y - 1)^2 = 24$.

53. Use the quadratic formula to solve $2x^2 - 4x = 3$.

54. Use the quadratic formula to solve. $m^2 = 4m + 8$

55. An object is thrown upward from the top of a 200-foot cliff with a velocity of 12 feet per second. The height above ground h in feet of the object after t seconds is $h = -16t^2 + 12t + 200$. How long after the object is thrown will it strike the ground? Round to the nearest tenth of a second.

56. Mr. Briley can roof his house in 24 hours. His son can roof the same house in 40 hours. If they work together, how long will it take to roof the house?

42. _____

43. _____

44. _____

45. _____

46. _____

47. _____

48. _____

49. _____

50. _____

51. _____

52. _____

53. _____

54. _____

55. _____

56. _____

9

Exponential and Logarithmic Functions

In this chapter, we discuss two closely-related types of functions: exponential and logarithmic functions. These functions are vital in applications in economics, finance, engineering, the sciences, education, and other fields. Models of tumor growth and learning curves are two examples of the uses of exponential and logarithmic functions.

Twice a day, every day of the year, weather balloons are released from about 1100 locations worldwide, including at least 92 from National Weather Service sites.

Weather balloons carry instrument packages, called *radiosondes*, high into the atmosphere to gather essential data needed to forecast the weather. Temperature, humidity, and air pressure are measured at various altitudes and transmitted via radio waves to a receiving station. Radio navigation supplies wind speed and direction at each altitude. Scientists use computers to display the data in a picture, called a *Skew-T diagram*. The balloons are filled with helium and expand to many times its original size, often to 30.48 km (about 20 miles). They eventually burst, and a parachute attached to the radiosonde package opens and conveys it gently back to earth. In Exercise 39 on page 660 we will explore the changing effect of atmospheric pressure on weather balloons. (*Source:* National Weather Service Forecast Office—El Paso)

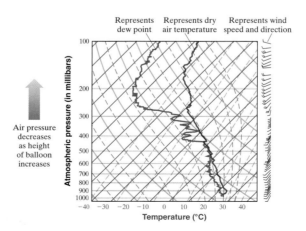

9.1 THE ALGEBRA OF FUNCTIONS

Objective A Adding, Subtracting, Multiplying, and Dividing Functions

Objectives

A Add, Subtract, Multiply, and Divide Functions.

B Compose Functions.

As we have seen in earlier chapters, it is possible to add, subtract, multiply, and divide functions. Although we have not stated them as such, the sums, differences, products, and quotients of functions are themselves functions. For example, if $f(x) = 3x$ and $g(x) = x + 1$, their product, $f(x) \cdot g(x) = 3x(x + 1) = 3x^2 + 3x$, is a new function. We can use the notation $(f \cdot g)(x)$ to denote this new function. Finding the sum, difference, product, and quotient of functions to generate new functions is called the **algebra of functions.**

Algebra of Functions

Let f and g be functions. New functions from f and g are defined as follows:

Sum	$(f + g)(x) = f(x) + g(x)$
Difference	$(f - g)(x) = f(x) - g(x)$
Product	$(f \cdot g)(x) = f(x) \cdot g(x)$
Quotient	$\left(\dfrac{f}{g}\right)(x) = \dfrac{f(x)}{g(x)}, \ g(x) \neq 0$

EXAMPLE 1 If $f(x) = x - 1$ and $g(x) = 2x - 3$, find the following.

a. $(f + g)(x)$

b. $(f - g)(x)$

c. $(f \cdot g)(x)$

d. $\left(\dfrac{f}{g}\right)(x)$

Solution: Use the algebra of functions and replace $f(x)$ by $x - 1$ and $g(x)$ by $2x - 3$. Then simplify.

a. $(f + g)(x) = f(x) + g(x)$
$$= (x - 1) + (2x - 3)$$
$$= 3x - 4$$

b. $(f - g)(x) = f(x) - g(x)$
$$= (x - 1) - (2x - 3)$$
$$= x - 1 - 2x + 3$$
$$= -x + 2$$

c. $(f \cdot g)(x) = f(x) \cdot g(x)$
$$= (x - 1)(2x - 3)$$
$$= 2x^2 - 5x + 3$$

d. $\left(\dfrac{f}{g}\right)(x) = \dfrac{f(x)}{g(x)} = \dfrac{x - 1}{2x - 3}$, where $x \neq \dfrac{3}{2}$

■ **Work Practice Problem 1**

PRACTICE PROBLEM 1

If $f(x) = x + 3$ and $g(x) = 3x - 1$, find

a. $(f + g)(x)$

b. $(f - g)(x)$

c. $(f \cdot g)(x)$

d. $\left(\dfrac{f}{g}\right)(x)$

There is an interesting but not surprising relationship between the graphs of functions and the graphs of their sum, difference, product, and quotient. For example, the graph of $(f + g)$ can be found by adding the graph of f to the graph of g. We add two graphs by adding corresponding y-values.

Answers

1. a. $4x + 2$, **b.** $-2x + 4$,

c. $3x^2 + 8x - 3$,

d. $\dfrac{x + 3}{3x - 1}$ where $x \neq \dfrac{1}{3}$

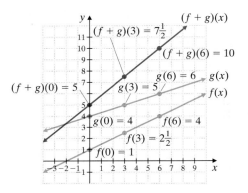

Objective B Composition of Functions

Another way to combine functions is called **function composition.** To understand this new way of combining functions, study the tables below. They show degrees Fahrenheit converted to equivalent degrees Celsius, and then degrees Celsius converted to equivalent degrees Kelvin. (The Kelvin scale is a temperature scale devised by Lord Kelvin in 1848.)

x = Degrees Fahrenheit (Input)	−31	−13	32	68	149	212
C(x) = Degrees Celsius (Output)	−35	−25	0	20	65	100

C(x) = Degrees Celsius (Input)	−35	−25	0	20	65	100
K(C(x)) = Kelvins (Output)	238.15	248.15	273.15	293.15	338.15	373.15

Suppose that we want a table that shows a direct conversion from degrees Fahrenheit to kelvins. In other words, suppose that a table is needed that shows kelvins as a function of degrees Fahrenheit. This can easily be done because in the tables, the output of the first table is the same as the input of the second table. The new table is as follows.

x = Degrees Fahrenheit (Input)	−31	−13	32	68	149	212
K(C(x)) = Kelvins (Output)	238.15	248.15	273.15	293.15	338.15	373.15

Since the output of the first table is used as the input of the second table, we write the new function as $K(C(x))$. The new function is formed from the composition of the other two functions. The mathematical symbol for this composition is $(K \circ C)(x)$. Thus, $(K \circ C)(x) = K(C(x))$.

It is possible to find an equation for the composition of the two functions $C(x)$ and $K(x)$. In other words, we can find a function that converts degrees Fahrenheit directly to kelvins. The function $C(x) = \dfrac{5}{9}(x - 32)$ converts degrees Fahrenheit to degrees Celsius, and the function $K(C(x)) = C(x) + 273.15$ converts degrees Celsius to kelvins. Thus,

$$(K \circ C)(x) = K(C(x)) = K\left(\frac{5}{9}(x - 32)\right) = \frac{5}{9}(x - 32) + 273.15$$

In general, the notation $f(g(x))$ means "f composed with g" and can be written as $(f \circ g)(x)$. Also $g(f(x))$, or $(g \circ f)(x)$, means "g composed with f."

Composite Functions

The composition of functions f and g is

$$(f \circ g)(x) = f(g(x))$$

Helpful Hint

$(f \circ g)(x)$ does not mean the same as $(f \cdot g)(x)$.

$(f \circ g)(x) = f(g(x))$ while $(f \cdot g)(x) = f(x) \cdot g(x)$

EXAMPLE 2 If $f(x) = x^2$ and $g(x) = x + 3$, find each composition.

a. $(f \circ g)(2)$ and $(g \circ f)(2)$
b. $(f \circ g)(x)$ and $(g \circ f)(x)$

Solution:

a. $(f \circ g)(2) = f(g(2))$
$\qquad\qquad\quad = f(5)$ Since $g(x) = x + 3$, then $g(2) = 2 + 3 = 5$.
$\qquad\qquad\quad = 5^2 = 25$
$\quad (g \circ f)(2) = g(f(2))$
$\qquad\qquad\quad = g(4)$ Since $f(x) = x^2$, then $f(2) = 2^2 = 4$.
$\qquad\qquad\quad = 4 + 3 = 7$

b. $(f \circ g)(x) = f(g(x))$
$\qquad\qquad\quad = f(x + 3)$ Replace $g(x)$ with $x + 3$.
$\qquad\qquad\quad = (x + 3)^2$ $f(x + 3) = (x + 3)^2$
$\qquad\qquad\quad = x^2 + 6x + 9$ Square $(x + 3)$.
$\quad (g \circ f)(x) = g(f(x))$
$\qquad\qquad\quad = g(x^2)$ Replace $f(x)$ with x^2.
$\qquad\qquad\quad = x^2 + 3$ $g(x^2) = x^2 + 3$

▣ **Work Practice Problem 2**

EXAMPLE 3 If $f(x) = |x|$ and $g(x) = x - 2$, find each composition.

a. $(f \circ g)(x)$
b. $(g \circ f)(x)$

Solution:

a. $(f \circ g)(x) = f(g(x)) = f(x - 2) = |x - 2|$
b. $(g \circ f)(x) = g(f(x)) = g(|x|) = |x| - 2$

▣ **Work Practice Problem 3**

Helpful Hint

In Examples 2 and 3, notice that $(g \circ f)(x) \neq (f \circ g)(x)$. In general, $(g \circ f)(x)$ *may* or *may not* equal $(f \circ g)(x)$.

PRACTICE PROBLEM 2

If $f(x) = x^2$ and $g(x) = 2x + 1$, find each composition.
a. $(f \circ g)(3)$ and $(g \circ f)(3)$
b. $(f \circ g)(x)$ and $(g \circ f)(x)$

PRACTICE PROBLEM 3

If $f(x) = \sqrt{x}$ and $g(x) = x + 1$, find each composition.
a. $(f \circ g)(x)$
b. $(g \circ f)(x)$

Answers
2. a. $49; 19,$ **b.** $4x^2 + 4x + 1; 2x^2 + 1,$
3. a. $\sqrt{x + 1},$ **b.** $\sqrt{x} + 1$

PRACTICE PROBLEM 4

If $f(x) = 2x, g(x) = x + 5$, and $h(x) = |x|$, write each function as a composition of f, g, or h.

a. $F(x) = |x + 5|$

b. $G(x) = 2x + 5$

EXAMPLE 4   If $f(x) = 5x, g(x) = x - 2$, and $h(x) = \sqrt{x}$, write each function as a composition with f, g, or h.

a. $F(x) = \sqrt{x - 2}$

b. $G(x) = 5x - 2$

Solution:

a. Notice the order in which the function F operates on an input value x. First, 2 is subtracted from x, and then the square root of that result is taken. This means that $F(x) = (h \circ g)(x)$. To check, we find $(h \circ g)(x)$.

$$(h \circ g)(x) = h(g(x)) = h(x - 2) = \sqrt{x - 2}$$

b. Notice the order in which the function G operates on an input value x. First, x is multiplied by 5, and then 2 is subtracted from the result. This means that $G(x) = (g \circ f)(x)$. To check, we find $(g \circ f)(x)$.

$$(g \circ f)(x) = g(f(x)) = g(5x) = 5x - 2$$

■ **Work Practice Problem 4**

▦ **CALCULATOR EXPLORATIONS**   Graphing

If $f(x) = \frac{1}{2}x + 2$ and $g(x) = \frac{1}{3}x^2 + 4$, then

$$(f + g)(x) = f(x) + g(x)$$

$$= \left(\frac{1}{2}x + 2\right) + \left(\frac{1}{3}x^2 + 4\right)$$

$$= \frac{1}{3}x^2 + \frac{1}{2}x + 6$$

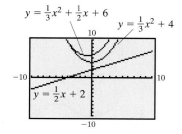

To visualize this addition of functions with a grapher, graph

$$Y_1 = \frac{1}{2}x + 2, Y_2 = \frac{1}{3}x^2 + 4, \quad \text{and} \quad Y_3 = \frac{1}{3}x^2 + \frac{1}{2}x + 6$$

Use a TABLE feature to verify that for a given x value, $Y_1 + Y_2 = Y_3$. For example, verify that when $x = 0$, $Y_1 = 2, Y_2 = 4$ and $Y_3 = 2 + 4 = 6$.

Answer

4. a. $(h \circ g)(x)$,   **b.** $(g \circ f)(x)$

9.1 EXERCISE SET

FOR EXTRA HELP

Student Solutions Manual

PH Math/Tutor Center

CD/Video for Review

MathXL®

MyMathLab

Objective Ⓐ *For the functions f and g, find* **a.** $(f + g)(x)$, **b.** $(f - g)(x)$, **c.** $(f \cdot g)(x)$, *and* **d.** $\left(\dfrac{f}{g}\right)(x)$. *See Example 1.*

1. $f(x) = x - 7$; $g(x) = 2x + 1$

2. $f(x) = x + 4$; $g(x) = 5x - 2$

⊙ **3.** $f(x) = x^2 + 1$; $g(x) = 5x$

4. $f(x) = x^2 - 2$; $g(x) = 3x$

5. $f(x) = \sqrt{x}$; $g(x) = x + 5$

6. $f(x) = \sqrt[3]{x}$; $g(x) = x - 3$

7. $f(x) = -3x$; $g(x) = 5x^2$

8. $f(x) = 4x^3$; $g(x) = -6x$

Objective Ⓑ *If* $f(x) = x^2 - 6x + 2$, $g(x) = -2x$, *and* $h(x) = \sqrt{x}$, *find each composition. See Example 2.*

9. $(f \circ g)(2)$

10. $(h \circ f)(-2)$

⊙ **11.** $(g \circ f)(-1)$

12. $(f \circ h)(1)$

13. $(g \circ h)(0)$

14. $(h \circ g)(0)$

Find $(f \circ g)(x)$ *and* $(g \circ f)(x)$. *See Examples 2 and 3.*

⊙ **15.** $f(x) = x^2 + 1$; $g(x) = 5x$

16. $f(x) = x - 3$; $g(x) = x^2$

17. $f(x) = 2x - 3$; $g(x) = x + 7$

18. $f(x) = x + 10$; $g(x) = 3x + 1$

19. $f(x) = x^3 + x - 2$; $g(x) = -2x$

20. $f(x) = -4x$; $g(x) = x^3 + x^2 - 6$

21. $f(x) = |x|; g(x) = 10x - 3$

22. $f(x) = |x|; g(x) = 14x - 8$

23. $f(x) = \sqrt{x}; g(x) = -5x + 2$

24. $f(x) = 7x - 1; g(x) = \sqrt[3]{x}$

If $f(x) = 3x$, $g(x) = \sqrt{x}$, and $h(x) = x^2 + 2$, write each function as a composition with f, g, or h. See Example 4.

25. $H(x) = \sqrt{x^2 + 2}$

26. $G(x) = \sqrt{3x}$

27. $F(x) = 9x^2 + 2$

28. $H(x) = 3x^2 + 6$

29. $G(x) = 3\sqrt{x}$

30. $F(x) = x + 2$

Find $f(x)$ and $g(x)$ so that the given function $h(x) = (f \circ g)(x)$.

31. $h(x) = (x + 2)^2$

32. $h(x) = |x - 1|$

33. $h(x) = \sqrt{x + 5} + 2$

34. $h(x) = (3x + 4)^2 + 3$

35. $h(x) = \dfrac{1}{2x - 3}$

36. $h(x) = \dfrac{1}{x + 10}$

Review

Solve each equation for y. See Section 2.3.

37. $x = y + 2$

38. $x = y - 5$

39. $x = 3y$

40. $x = -6y$

41. $x = -2y - 7$

42. $x = 4y + 7$

Concept Extensions

43. Business people are concerned with cost functions, revenue functions, and profit functions. Recall that the profit $P(x)$ obtained from selling x units of a product is equal to the revenue $R(x)$ from selling the x units minus the cost $C(x)$ of manufacturing the x units. Write an equation expressing this relationship among $C(x)$, $R(x)$, and $P(x)$.

44. Suppose the revenue $R(x)$ for x units of a product can be described by $R(x) = 25x$, and the cost $C(x)$ can be described by $C(x) = 50 + x^2 + 4x$. Find the profit $P(x)$ for x units.

45. If you are given $f(x)$ and $g(x)$, explain in your own words how to find $(f \circ g)(x)$, and then how to find $(g \circ f)(x)$.

46. Given $f(x)$ and $g(x)$, describe in your own words the difference between $(f \circ g)(x)$ and $(f \cdot g)(x)$.

STUDY SKILLS BUILDER

Tips for Studying for an Exam

To prepare for an exam, try the following study techniques.

- Start the study process days before your exam.
- Make sure that you are up-to-date on your assignments.
- If there is a topic that you are unsure of, use one of the many resources that are available to you. For example,

  See your instructor.

  Visit a learning resource center on campus.

  Read the textbook material and examples on the topic.

  View a video on the topic.

- Reread your notes and carefully review the Chapter Highlights at the end of any chapter.
- Work the review exercises at the end of the chapter. Check your answers and correct any mistakes. If you have trouble, use a resource listed above.
- Find a quiet place to take the Chapter Test found at the end of the chapter. Do not use any resources when taking this sample test. This way, you will have a clear indication of how prepared you are for your exam.

Check your answers and make sure that you correct any missed exercises.

- Get lots of rest the night before the exam. It's hard to show how well you know the material if your brain is foggy from lack of sleep.

Good luck and keep a positive attitude.

Let's see how you did on your last exam.

1. How many days before your last exam did you start studying?
2. Were you up-to-date on your assignments at that time or did you need to catch up on assignments?
3. List the most helpful text supplement (if you used one).
4. List the most helpful campus supplement (if you used one).
5. List your process for preparing for a mathematics test.
6. Was this process helpful? In other words, were you satisfied with your performance on your exam?
7. If not, what changes can you make in your process that will make it more helpful to you?

9.2 INVERSE FUNCTIONS

Objectives

A Determine Whether a Function Is a One-to-One Function.

B Use the Horizontal Line Test to Decide Whether a Function Is a One-to-One Function.

C Find the Inverse of a Function.

D Find the Equation of the Inverse of a Function.

E Graph Functions and Their Inverses.

In the next sections, we begin a study of two new functions: exponential and logarithmic functions. As we learn more about these functions, we will discover that they share a special relation to each other; they are inverses of each other.

Before we study these functions, we need to learn about inverses. We begin by defining one-to-one functions.

Objective **A** Determining Whether a Function Is One-to-One

Study the following table.

Degrees Fahrenheit (Input)	-31	-13	32	68	149	212
Degrees Celsius (Output)	-35	-25	0	20	65	100

Recall that since each degrees Fahrenheit (input) corresponds to exactly one degrees Celsius (output), this table of inputs and outputs does describe a function. Also notice that each output corresponds to a different input. This type of function is given a special name—a *one-to-one function*.

Does the set $f = \{(0, 1), (2, 2), (-3, 5), (7, 6)\}$ describe a one-to-one function? It is a function since each x-value corresponds to a unique y-value. For this particular function f, each y-value corresponds to a unique x-value. Thus, this function is also a one-to-one function.

One-to-One Function

For a **one-to-one function,** each x-value (input) corresponds to only one y-value (output) and each y-value (output) corresponds to only one x-value (input).

EXAMPLES Determine whether each function described is one-to-one.

1. $f = \{(6, 2), (5, 4), (-1, 0), (7, 3)\}$

The function f is one-to-one since each y-value corresponds to only one x-value.

2. $g = \{(3, 9), (-4, 2), (-3, 9), (0, 0)\}$

The function g is not one-to-one because the y-value 9 in $(3, 9)$ and $(-3, 9)$ corresponds to two different x-values.

3. $h = \{(1, 1), (2, 2), (10, 10), (-5, -5)\}$

The function h is one-to-one since each y-value corresponds to only one x-value.

4.

Mineral (Input)	Talc	Gypsum	Diamond	Topaz	Stibnite
Hardness on the Mohs Scale (Output)	1	2	10	8	2

This table does not describe a one-to-one function since the output 2 corresponds to two different inputs, gypsum and stibnite.

PRACTICE PROBLEMS 1–5

Determine whether each function described is one-to-one.

1. $f = \{(7, 3), (-1, 1), (5, 0), (4, -2)\}$

2. $g = \{(-3, 2), (6, 3), (2, 14), (-6, 2)\}$

3. $h = \{(0, 0), (1, 2), (3, 4), (5, 6)\}$

4.

State (Input)	Colorado	Mississippi	Nevada	New Mexico	Utah
Number of Colleges and Universities (Output)	16	7	4	10	7

Source: American Educational Guidance Center, 2005.

5.

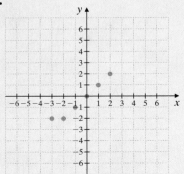

Answers

1. one-to-one, **2.** not one-to-one, **3.** one-to-one, **4.** not one-to-one, **5.** not one-to-one

Objective 🇪 Graphing Inverse Functions

Notice that the graphs of f and f^{-1} in Example 9 are mirror images of each other, and the "mirror" is the dashed line $y = x$. This is true for every function and its inverse. For this reason, we say that the *graphs of f and f^{-1} are symmetric about the line $y = x$.*

To see why this happens, study the graph of a few ordered pairs and their switched coordinates.

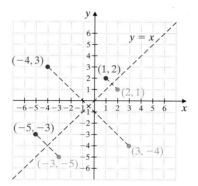

EXAMPLE 10 Graph the inverse of each function.

Solution: The function is graphed in blue and the inverse is graphed in red.

a.

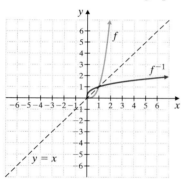

b.

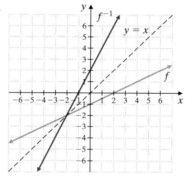

🔲 **Work Practice Problem 10**

PRACTICE PROBLEM 10

Graph the inverse of each function.

a.

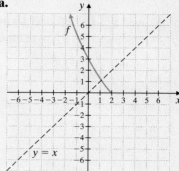

b.

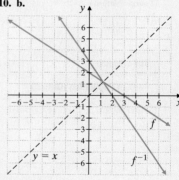

Answers

See page 646 for 10a.

10. b.

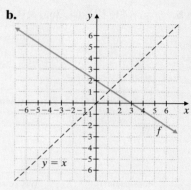

🖩 **CALCULATOR EXPLORATIONS** Graphing

A grapher can be used to visualize functions and their inverses. Recall that the graph of a function f and its inverse f^{-1} are mirror images of each other across the line $y = x$. To see this for the function $f(x) = 3x + 2$, use a square window and graph

the given function: $Y_1 = 3x + 2$

its inverse: $Y_2 = \dfrac{x - 2}{3}$

and the line: $Y_3 = x$

Exercises will follow in Exercise Set 9.2.

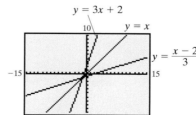

Objectives Ⓐ Ⓒ **Mixed Practice** *Determine whether each function is a one-to-one function. If it is one-to-one, list the inverse function by switching coordinates, or inputs and outputs. See Examples 1 through 5, and 7.*

1. $g = \{(0, 3), (3, 7), (6, 7), (-2, -2)\}$

2. $g = \{(8, 6), (9, 6), (3, 4), (-4, 4)\}$

3. $h = \{(10, 10)\}$

4. $r = \{(1, 2), (3, 4), (5, 6), (6, 7)\}$

 5. $f = \{(11, 12), (4, 3), (3, 4), (6, 6)\}$

6. $f = \{(-1, -1), (1, 1), (0, 2), (2, 0)\}$

7.

Month of 2004 (Input)	July	August	September	October	November	December
Unemployment Rate in Percent (Output)	5.5	5.4	5.4	5.5	5.4	5.4

(*Source:* U.S. Bureau of Labor Statistics)

8.

State (Input)	Wisconsin	Ohio	Georgia	Colorado	California	Arizona
Electoral Votes (Output)	10	20	15	9	55	10

(*Source:* National Archives and Records Administration, based on the 2000 Census)

9.

State (Input)	California	Alaska	Indiana	Louisiana	New Mexico
Rank in Population (Output)	1	47	14	24	36

(*Source:* U.S. Bureau of the Census)

10.

Shape (Input)	Triangle	Pentagon	Quadrilateral	Hexagon	Decagon
Number of Sides (Output)	3	5	4	6	10

Given the one-to-one function $f(x) = x^3 + 2$, find the following. (Hint: You do not need to find the equation for f^{-1}.)

11. a. $f(1)$
 b. $f^{-1}(3)$

12. a. $f(0)$
 b. $f^{-1}(2)$

13. a. $f(-1)$
 b. $f^{-1}(1)$

14. a. $f(-2)$
 b. $f^{-1}(-6)$

Objective **B** *Determine whether the graph of each function is the graph of a one-to-one function. See Example 6.*

15.

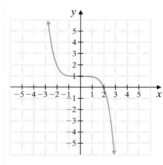

16.

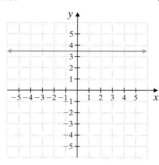

17.

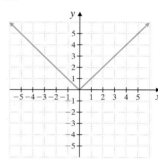

18.

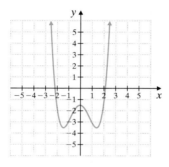

19.

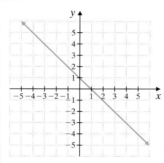

20.

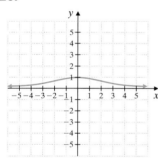

21.

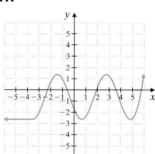

22.

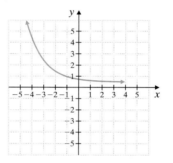

Objectives **D** **E** **Mixed Practice** *Each of the following functions is one-to-one. Find the inverse of each function and graph the function and its inverse on the same set of axes. See Examples 8 and 9.*

23. $f(x) = x + 4$

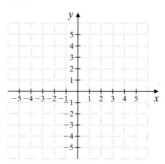

24. $f(x) = x - 5$

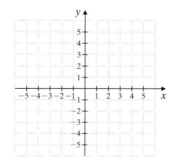

25. $f(x) = 2x - 3$

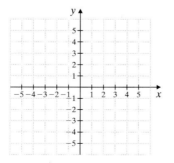

26. $f(x) = 4x + 9$

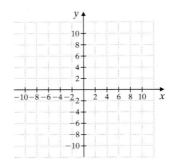

27. $f(x) = \dfrac{1}{2}x - 1$

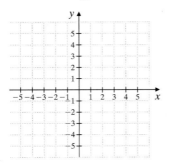

28. $f(x) = -\dfrac{1}{2}x + 2$

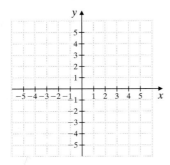

29. $f(x) = x^3$

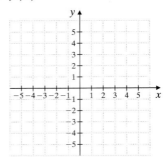

30. $f(x) = x^3 - 1$

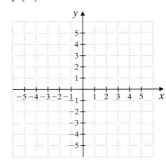

Find the inverse of each one-to-one function. See Examples 8 and 9.

31. $f(x) = \dfrac{x - 2}{5}$

32. $f(x) = \dfrac{4x - 3}{2}$

33. $f(x) = \sqrt[3]{x}$

34. $f(x) = \sqrt[3]{x + 1}$

35. $f(x) = \dfrac{5}{3x + 1}$

36. $f(x) = \dfrac{7}{2x + 4}$

37. $f(x) = (x + 2)^3$

38. $f(x) = (x - 5)^3$

Graph the inverse of each function on the same set of axes. See Example 10.

39.

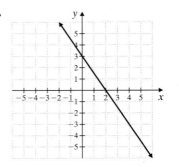

40.

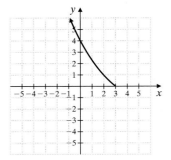

41.

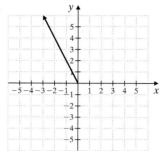

42.

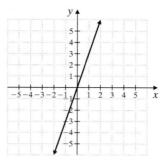

Review

Evaluate each exponential expression. See Section 7.2.

43. $25^{1/2}$

44. $49^{1/2}$

45. $16^{3/4}$

46. $27^{2/3}$

47. $9^{-3/2}$

48. $81^{-3/4}$

If $f(x) = 3^x$, find each value. In Exercises 51 and 52, give an exact answer and a two-decimal-place approximation. See Section 3.6.

49. $f(2)$

50. $f(0)$

51. $f\left(\dfrac{1}{2}\right)$

52. $f\left(\dfrac{2}{3}\right)$

Concept Extensions

Solve. See the Concept Check in this section.

53. Suppose that f is a one-to-one function and that $f(2) = 9$.
   a. Write the corresponding ordered pair.
   b. Name one ordered-pair that we know is a solution of the inverse of f, or f^{-1}.

54. Suppose that F is a one-to-one function and that $F\left(\dfrac{1}{2}\right) = -0.7$.
   a. Write the corresponding ordered pair.
   b. Name one ordered pair that we know is a solution of the inverse of F, or F^{-1}.

For Exercises 55 and 56.

   a. *Write the ordered pairs for f whose points are highlighted. (Include the points whose coordinates are given.)*
   b. *Write the corresponding ordered pairs for the inverse of f, f^{-1}.*
   c. *Graph the ordered pairs for f^{-1} found in part (b).*
   d. *Graph f^{-1} by drawing a smooth curve through the plotted points.*

55. a.

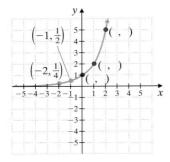

56. a.

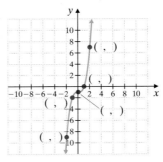

c. d.

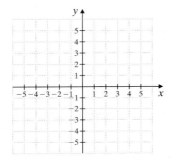

c. d.

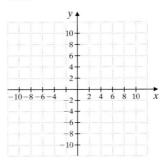

57. If you are given the graph of a function, describe how you can tell from the graph whether a function has an inverse.

58. Describe the appearance of the graphs of a function and its inverse.

Find the inverse of each one-to-one function. Then graph the function and its inverse in a square window.

59. $f(x) = 3x + 1$ **60.** $f(x) = -2x - 6$ **61.** $f(x) = \sqrt[3]{x + 3}$ **62.** $f(x) = x^3 - 3$

9.3 EXPONENTIAL FUNCTIONS

Objectives

A Graph Exponential Functions.

B Solve Equations of the Form $b^x = b^y$.

C Solve Problems Modeled by Exponential Equations.

In earlier chapters, we gave meaning to exponential expressions such as 2^x, where x is a rational number. Recall the following examples.

$2^3 = 2 \cdot 2 \cdot 2$ Three factors; each factor is 2

$2^{3/2} = (2^{1/2})^3 = \sqrt{2} \cdot \sqrt{2} \cdot \sqrt{2}$ Three factors; each factor is $\sqrt{2}$

When x is an irrational number (for example, $\sqrt{3}$), what meaning can we give to $2^{\sqrt{3}}$?

It is beyond the scope of this book to give precise meaning to 2^x if x is irrational. We can confirm your intuition and say that $2^{\sqrt{3}}$ is a real number, and since $1 < \sqrt{3} < 2$, then $2^1 < 2^{\sqrt{3}} < 2^2$. We can also use a calculator and approximate $2^{\sqrt{3}} : 2^{\sqrt{3}} \approx 3.321997$. In fact, as long as the base b is positive, b^x is a real number for all real numbers x. Finally, the rules of exponents apply whether x is rational or irrational, as long as b is positive.

In this section, we are interested in functions of the form $f(x) = b^x$, or $y = b^x$, where $b > 0$. A function of this form is called an *exponential function*.

Exponential Function

A function of the form

$$f(x) = b^x$$

is called an **exponential function**, where $b > 0$, b is not 1, and x is a real number.

Objective **A** Graphing Exponential Functions

Now let's practice graphing exponential functions.

EXAMPLE 1 Graph the exponential functions $f(x) = 2^x$ and $g(x) = 3^x$ on the same set of axes.

Solution: To graph these functions, we find some ordered pair solutions, plot the points, and connect them with a smooth curve. Remember throughout that $y = f(x)$.

$f(x) = 2^x$	x	0	1	2	3	-1	-2
	$f(x)$	1	2	4	8	$\frac{1}{2}$	$\frac{1}{4}$

$g(x) = 3^x$	x	0	1	2	3	-1	-2
	$g(x)$	1	3	9	27	$\frac{1}{3}$	$\frac{1}{9}$

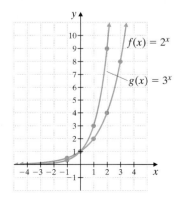

PRACTICE PROBLEM 1

Graph the exponential function $f(x) = 6^x$.

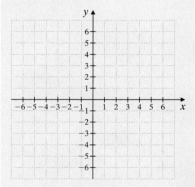

Answer

1.

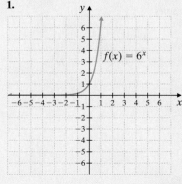

Work Practice Problem 1

653

A number of things should be noted about the two graphs of exponential functions in Example 1. First, the graphs show that $f(x) = 2^x$ and $g(x) = 3^x$ are one-to-one functions since each graph passes the vertical and horizontal line tests. The y-intercept of each graph is $(0, 1)$, but neither graph has an x-intercept. From the graph, we can also see that the domain of each function is all real numbers and that the range is $(0, \infty)$. We can also see that as x-values are increasing, y-values are increasing also.

PRACTICE PROBLEM 2

Graph the exponential function

$$f(x) = \left(\frac{1}{5}\right)^x.$$

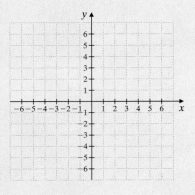

EXAMPLE 2 Graph the exponential functions $y = \left(\frac{1}{2}\right)^x$ and $y = \left(\frac{1}{3}\right)^x$ on the same set of axes.

Solution: As before, we find some ordered pair solutions, plot the points, and connect them with a smooth curve.

$y = \left(\frac{1}{2}\right)^x$	x	0	1	2	3	−1	−2
	y	1	$\frac{1}{2}$	$\frac{1}{4}$	$\frac{1}{8}$	2	4

$y = \left(\frac{1}{3}\right)^x$	x	0	1	2	3	−1	−2
	y	1	$\frac{1}{3}$	$\frac{1}{9}$	$\frac{1}{27}$	3	9

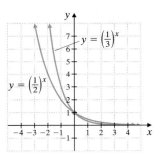

□ **Work Practice Problem 2**

Each function in Example 2 again is a one-to-one function. The y-intercept of both is $(0, 1)$. The domain is the set of all real numbers, and the range is $(0, \infty)$.

Notice the difference between the graphs of Example 1 and the graphs of Example 2. An exponential function is always increasing if the base is greater than 1. When the base is between 0 and 1, the graph is always decreasing. The following figures summarize these characteristics of exponential functions.

Answer

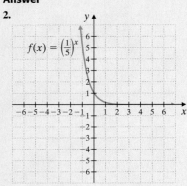

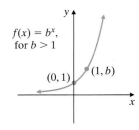

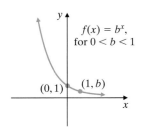

EXAMPLE 3 Graph the exponential function $f(x) = 3^{x+2}$.

Solution: As before, we find and plot a few ordered pair solutions. Then we connect the points with a smooth curve.

$f(x) = 3^{x+2}$	x	0	−1	−2	−3	−4
	y	9	3	1	$\frac{1}{3}$	$\frac{1}{9}$

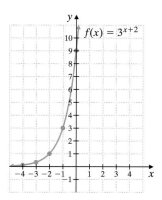

Work Practice Problem 3

✔**Concept Check** Which functions are exponential functions?

a. $f(x) = x^3$ **b.** $g(x) = \left(\dfrac{2}{3}\right)^x$ **c.** $h(x) = 5^{x-2}$ **d.** $w(x) = (2x)^2$

Objective B Solving Equations of the Form $b^x = b^y$

We have seen that an exponential function $y = b^x$ is a one-to-one function. Another way of stating this fact is a property that we can use to solve exponential equations.

> **Uniqueness of b^x**
>
> Let $b > 0$ and $b \neq 1$. Then $b^x = b^y$ is equivalent to $x = y$.

Thus, one way to solve an exponential equation depends on whether it's possible to write each side of the equation with the same base; that is, $b^x = b^y$. We solve by this method first.

EXAMPLE 4 Solve: $2^x = 16$

Solution: We write 16 as a power of 2 so that each side of the equation has the same base. Then we use the uniqueness of b^x to solve.

$$2^x = 16$$
$$2^x = 2^4$$

Since the bases are the same and are nonnegative, by the uniqueness of b^x we then have that the exponents are equal. Thus,

$$x = 4$$

To check, we replace x with 4 in the original equation.
The solution set is $\{4\}$.

Work Practice Problem 4

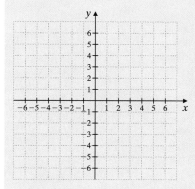
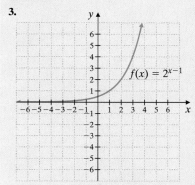

PRACTICE PROBLEM 5

Solve: $4^x = 8$

EXAMPLE 5 Solve: $25^x = 125$

Solution: Since both 25 and 125 are powers of 5, we can use the uniqueness of b^x.

$$25^x = 125$$
$$(5^2)^x = 5^3 \quad \text{Write 25 and 125 as powers of 5.}$$
$$5^{2x} = 5^3$$
$$2x = 3 \quad \text{Use the uniqueness of } b^x.$$
$$x = \frac{3}{2} \quad \text{Divide both sides by 2.}$$

To check, we replace x with $\frac{3}{2}$ in the original equation.

The solution set is $\left\{\frac{3}{2}\right\}$.

🔲 **Work Practice Problem 5**

PRACTICE PROBLEM 6

Solve: $9^{x-1} = 27^x$

EXAMPLE 6 Solve: $4^{x+3} = 8^x$

Solution: We write both 4 and 8 as powers of 2, and then use the uniqueness of b^x.

$$4^{x+3} = 8^x$$
$$(2^2)^{x+3} = (2^3)^x$$
$$2^{2x+6} = 2^{3x}$$
$$2x + 6 = 3x \quad \text{Use the uniqueness of } b^x.$$
$$6 = x \quad \text{Subtract } 2x \text{ from both sides.}$$

Check to see that the solution set is $\{6\}$.

🔲 **Work Practice Problem 6**

There is one major problem with the preceding technique. Often the two sides of an equation, $4 = 3^x$ for example, cannot easily be written as powers of a common base. We explore how to solve such an equation with the help of *logarithms* later.

Objective ⓒ Solving Problems Modeled by Exponential Equations

The bar graph below shows the increase in the number of cellular phone users. Notice that the graph of the exponential function $y = 52.47(1.196)^x$ approximates the heights of the bars. This is just one example of how the world abounds with patterns that can be modeled by exponential functions. To make these applications realistic, we use numbers that warrant a calculator.

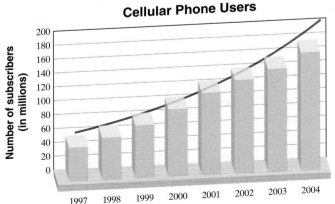

$y = 52.47(1.196)^x$

where $x = 0$ corresponds to 1997, $x = 1$ corresponds to 1998, and so on

Source: Cellular Telecommunications & Internet Association

Answers

5. $\left\{\frac{3}{2}\right\}$,   6. $\{-2\}$

Another application of an exponential function has to do with interest rates on loans. The exponential function defined by $A = P\left(1 + \dfrac{r}{n}\right)^{nt}$ models the pattern relating the dollars A accrued (or owed) after P dollars are invested (or loaned) at an annual rate of interest r compounded n times each year for t years. This function is known as the *compound interest formula*.

EXAMPLE 7 **Using the Compound Interest Formula**

Find the amount owed at the end of 5 years if $1600 is loaned at a rate of 9% compounded monthly.

Solution: Use the formula $A = P\left(1 + \dfrac{r}{n}\right)^{nt}$, with the following values:

$P = \$1600$ (the amount of the loan)
$r = 9\% = 0.09$ (the annual rate of interest)
$n = 12$ (the number of times interest is compounded each year)
$t = 5$ (the duration of the loan, in years)

$A = P\left(1 + \dfrac{r}{n}\right)^{nt}$ Compound interest formula

$= 1600\left(1 + \dfrac{0.09}{12}\right)^{12(5)}$ Substitute known values.

$= 1600(1.0075)^{60}$

To approximate A, use the $\boxed{y^x}$ or $\boxed{\wedge}$ key on your calculator.

$\boxed{2505.0896}$

Thus, the amount A owed is approximately $2505.09.

Work Practice Problem 7

PRACTICE PROBLEM 7

a. As a result of the Chernobyl nuclear accident, radioactive debris was carried through the atmosphere. One immediate concern was the impact that the debris had on the milk supply. The percent y of radioactive material in raw milk t days after the accident is estimated by $y = 100(2.7)^{-0.1t}$. Estimate the expected percent of radioactive material in the milk after 30 days.

b. Find the amount owed at the end of 6 years if $23,000 is loaned at a rate of 12% compounded quarterly. (4 times a year). Round your answer to the nearest cent.

Answers
7. a. approximately 5.08%,
    b. $46,754.26

CALCULATOR EXPLORATIONS Graphing

We can use a graphing calculator and its TRACE feature to solve Practice Problem 7a graphically.

To estimate the percent of radioactive material in the milk after 30 days, enter $Y_1 = 100(2.7)^{-0.1x}$. The graph does not appear on a standard viewing window, so we need to determine an appropriate viewing window. Because it doesn't make sense to look at radioactivity *before* the Chernobyl nuclear accident, we use Xmin = 0. We are interested in finding the percent of radioactive material in the milk when $x = 30$, so we choose Xmax = 35 to leave enough space to see the graph at $x = 30$. Because the values of y are percents, it seems appropriate that $0 \leq y \leq 100$. (We also use Xscl = 1 and Yscl = 10.) Now we graph the function.

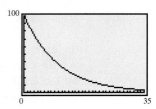

We can use the TRACE feature to obtain an approximation of the expected percent of radioactive material in the milk when $x = 30$. (A TABLE feature may also be used to approximate the percent.) To obtain a better approximation, let's use the ZOOM feature several times to zoom in near $x = 30$.

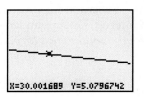

The percent of radioactive material in the milk 30 days after the Chernobyl accident was 5.08%, accurate to two decimal places.

Use a grapher to find each percent. Approximate your solutions so that they are accurate to two decimal places.

1. Estimate the percent of radioactive material in the milk 2 days after the Chernobyl nuclear accident.

2. Estimate the percent of radioactive material in the milk 10 days after the Chernobyl nuclear accident.

3. Estimate the percent of radioactive material in the milk 15 days after the Chernobyl nuclear accident.

4. Estimate the percent of radioactive material in the milk 25 days after the Chernobyl nuclear accident.

9.3 EXERCISE SET

Student Solutions Manual PH Math/Tutor Center CD/Video for Review Math XL MathXL® MyMathLab MyMathLab

Objective Ⓐ *Graph each exponential function. See Examples 1 through 3.*

1. $y = 5^x$

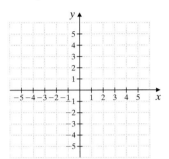

2. $y = 4^x$

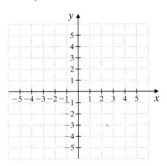

3. $y = 1 + 2^x$

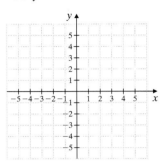

4. $y = 3^x - 1$

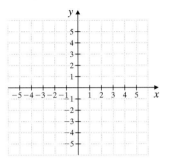

5. $y = \left(\dfrac{1}{4}\right)^x$

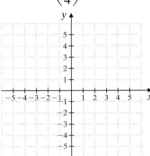

6. $y = \left(\dfrac{1}{5}\right)^x$

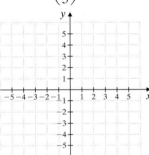

7. $y = \left(\dfrac{1}{2}\right)^x - 2$

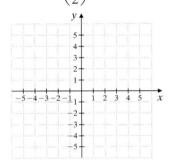

8. $y = \left(\dfrac{1}{3}\right)^x + 2$

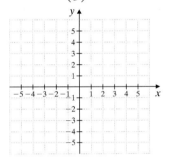

9. $y = -2^x$

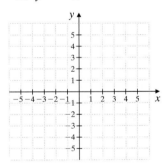

10. $y = -3^x$

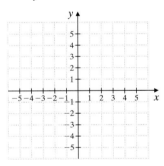

11. $y = 3^x - 2$

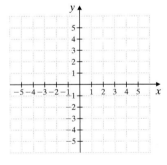

12. $y = 2^x - 3$

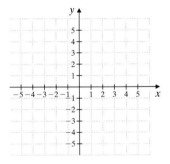

13. $y = -\left(\dfrac{1}{4}\right)^x$

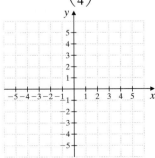

14. $y = -\left(\dfrac{1}{5}\right)^x$

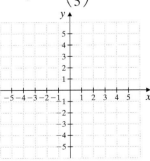

15. $y = \left(\dfrac{1}{3}\right)^x + 1$

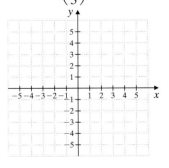

16. $y = \left(\dfrac{1}{2}\right)^x + 2$

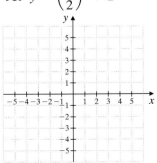

17. $f(x) = 2^{x-2}$

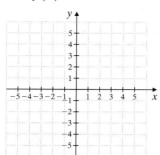

18. $g(x) = 2^{x+1}$

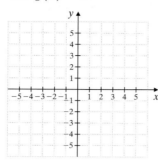

19. $F(x) = 5^{x+1}$

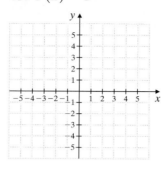

20. $G(x) = 3^{x-2}$

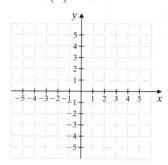

Objective **B** *Solve. See Examples 4 through 6.*

21. $3^x = 27$

22. $6^x = 36$

23. $16^x = 8$

24. $64^x = 16$

25. $32^{2x-3} = 2$

26. $9^{2x+1} = 81$

27. $\dfrac{1}{4} = 2^{3x}$

28. $\dfrac{1}{27} = 3^{2x}$

29. $9^x = 27$

30. $32^x = 4$

31. $27^{x+1} = 9$

32. $125^{x-2} = 25$

33. $81^{x-1} = 27^{2x}$

34. $4^{3x-7} = 32^{2x}$

35. $\left(\dfrac{1}{8}\right)^x = 16^{1-x}$

36. $\left(\dfrac{1}{9}\right)^x = 27^{2-x}$

Objective **C** *Solve. Unless otherwise indicated, round results to one decimal place. See Example 7.*

37. One type of uranium has a daily radioactive decay rate of 0.4%. If 30 pounds of this uranium is available today, how much will still remain after 50 days? Use $y = 30(2.7)^{-0.004t}$, and let t be 50.

38. The nuclear waste from an atomic energy plant decays at a rate of 3% each century. If 150 pounds of nuclear waste is disposed of, how much of it will still remain after 10 centuries? Use $y = 150(2.7)^{-0.03t}$, and let t be 10.

39. This atmospheric pressure p, in millibars, on a weather balloon decreases with increasing height. This pressure is related to the height in kilometers h above sea level and is given by the function $p(h) = 760(2.7)^{-0.145h}$. (*Source:* National Weather Service)

   a. Find the atmospheric pressure on a weather balloon at a height of 2 km.

   b. Find the atmospheric pressure on a weather balloon at its expected maximum altitude of 30.48 km.

   c. What causes the difference in atmospheric pressure at these two heights?

40. The equation $y = 158.97(1.012)^x$ models the population of the United States from 1950 through 2003. In this equation, y is the population in millions and x represents the number of years after 1950. Round answers to the nearest tenth of a million.

(*Source:* Based on data from the U.S. Bureau of the Census)

   a. Estimate the population of the United States in 1970.

   b. Assuming this equation continues to be valid in the future, use the equation to predict the population of the United States in 2020.

41. Retail revenue from shopping on the Internet is currently growing at a rate of 29% per year. In 2002, a total of $44 billion in revenue was collected through Internet retail sales. Answer the following questions using $y = 44(1.29)^t$, where y is Internet revenues in billions of dollars and t is the number of years after 2002. Round answers to the nearest tenth of a billion dollars. (*Source:* U.S. Bureau of the Census)

   a. According to the model, what level of retail revenues from Internet shopping was expected in 2003?

   b. If the given model continues, predict the level of Internet shopping revenues in 2009.

42. Carbon dioxide (CO_2) is a greenhouse gas that contributes to global warming. Partially due to the combustion of fossil fuels, the amount of CO_2 in Earth's atmosphere has been increasing by 0.4% annually over the past century. In 2000, the concentration of CO_2 in the atmosphere was 369.4 parts per million by volume. To make the following predictions, use $y = 369.4(1.004)^t$ where y is the concentration of CO_2 in parts per million and t is the number of years after 2000. (*Sources:* Based on data from the United Nations Environment Programme and the Carbon Dioxide Information Analysis Center)

   a. Predict the concentration of CO_2 in the atmosphere in the year 2006.

   b. Predict the concentration of CO_2 in the atmosphere in the year 2030.

The equation $y = 52.47(1.196)^x$ gives the number of cellular phone users y (in millions) in the United States for the years 1997 through 2004. In this equation, $x = 0$ corresponds to 1997, $x = 1$ corresponds to 1998, and so on. Use this model to solve Exercises 43 and 44. Round answers to the nearest tenth of a million.

43. Predict the number of cellular phone users in the year 2010.

44. Predict the number of cellular phone users in the year 2017.

Solve. Use $A = P\left(1 + \dfrac{r}{n}\right)^{nt}$. Round answers to two decimal places. See Example 7.

45. Find the amount Erica Entada owes at the end of 3 years if $6000 is loaned to her at a rate of 8% compounded monthly.

46. Find the amount owed at the end of 5 years if $3000 is loaned at a rate of 10% compounded quarterly.

Review

Solve each equation. See Section 2.1.

47. $5x - 2 = 18$

48. $3x - 7 = 11$

49. $3x - 4 = 3(x + 1)$

50. $2 - 6x = 6(1 - x)$

Concept Extensions

Which functions are exponential functions? See the Concept Check in this section.

51. $f(x) = 1.5x^2$ **52.** $g(x) = 3^x$ **53.** $h(x) = \left(\dfrac{1}{2}x\right)^2$ **54.** $F(x) = 0.4^{x+1}$

Match each exponential function with its graph.

55. $f(x) = \left(\dfrac{1}{2}\right)^x$ **56.** $f(x) = 2^x$ **57.** $f(x) = \left(\dfrac{1}{4}\right)^x$ **58.** $f(x) = 3^x$

A 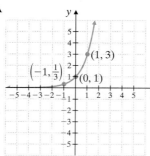   **B** 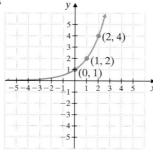   **C** 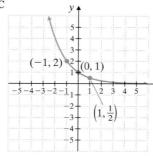   **D**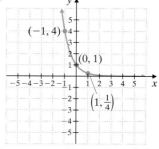

59. Explain why the graph of an exponential function $y = b^x$ contains the point $(1, b)$.

60. Explain why an exponential function $y = b^x$ has a y-intercept of $(0, 1)$.

Use a graphing calculator to solve. Estimate your results to two decimal places.

61. Verify the results of Exercise 37.

62. From Exercise 37, estimate the number of pounds of uranium that will be available after 100 days.

63. From Exercise 37, estimate the number of pounds of uranium that will be available after 120 days.

9.4 LOGARITHMIC FUNCTIONS

Objectives

A Write Exponential Equations with Logarithmic Notation and Write Logarithmic Equations with Exponential Notation.

B Solve Logarithmic Equations by Using Exponential Notation.

C Identify and Graph Logarithmic Functions.

Objective **A** Using Logarithmic Notation

Since the exponential function $f(x) = 2^x$ is a one-to-one function, it has an inverse. We can create a table of values for f^{-1} by switching the coordinates in the accompanying table of values for $f(x) = 2^x$.

x	y = f(x)
−3	$\frac{1}{8}$
−2	$\frac{1}{4}$
−1	$\frac{1}{2}$
0	1
1	2
2	4
3	8

x	y = f⁻¹(x)
$\frac{1}{8}$	−3
$\frac{1}{4}$	−2
$\frac{1}{2}$	−1
1	0
2	1
4	2
8	3

The graphs of f and its inverse are shown in the margin. Notice that the graphs of f and f^{-1} are symmetric about the line $y = x$, as expected.

Now we would like to be able to write an equation for f^{-1}. To do so, we follow the steps for finding the equation of an inverse.

$$f(x) = 2^x$$

Step 1: Replace $f(x)$ by y. $\qquad\qquad y = 2^x$

Step 2: Interchange x and y. $\qquad\quad x = 2^y$

Step 3: Solve for y.

At this point, we are stuck. To solve this equation for y, a new notation, **logarithmic notation,** is needed.

The symbol $\log_b x$ means "the power to which b is raised to produce a result of x." In other words,

$$\log_b x = y \quad \text{means} \quad b^y = x$$

We say that $\log_b x$ is "the logarithm of x to the base b" or "the log of x to the base b."

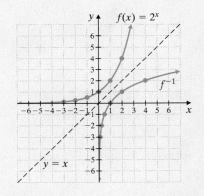

Logarithmic Definition

If $b > 0$, and $b \neq 1$, then

$$y = \log_b x \quad \text{means} \quad x = b^y$$

for every $x > 0$ and every real number y.

Before returning to the function $x = 2^y$ and solving it for y in terms of x, let's practice using the new notation $\log_b x$.

It is important to be able to write exponential equations with logarithmic notation, and vice versa. The following table shows examples of both forms.

Logarithmic Equation	Corresponding Exponential Equation
$\log_3 9 = 2$	$3^2 = 9$
$\log_6 1 = 0$	$6^0 = 1$
$\log_2 8 = 3$	$2^3 = 8$
$\log_4 \dfrac{1}{16} = -2$	$4^{-2} = \dfrac{1}{16}$
$\log_8 2 = \dfrac{1}{3}$	$8^{1/3} = 2$

 Helpful Hint Notice that a *logarithm* is an *exponent*. In other words, $\log_3 9$ is the *power* that we raise 3 to in order to get 9.

PRACTICE PROBLEMS 1–3

Write as an exponential equation.

1. $\log_7 49 = 2$ **2.** $\log_8 \dfrac{1}{8} = -1$

3. $\log_3 \sqrt{3} = \dfrac{1}{2}$

EXAMPLES Write as an exponential equation.

1. $\log_5 25 = 2$ means $5^2 = 25$.

2. $\log_6 \dfrac{1}{6} = -1$ means $6^{-1} = \dfrac{1}{6}$.

3. $\log_2 \sqrt{2} = \dfrac{1}{2}$ means $2^{1/2} = \sqrt{2}$.

🖥 **Work Practice Problems 1–3**

PRACTICE PROBLEMS 4–6

Write as a logarithmic equation.

4. $3^4 = 81$ **5.** $2^{-3} = \dfrac{1}{8}$

6. $7^{1/3} = \sqrt[3]{7}$

EXAMPLES Write as a logarithmic equation.

4. $9^3 = 729$ means $\log_9 729 = 3$.

5. $6^{-2} = \dfrac{1}{36}$ means $\log_6 \dfrac{1}{36} = -2$.

6. $5^{1/3} = \sqrt[3]{5}$ means $\log_5 \sqrt[3]{5} = \dfrac{1}{3}$.

🖥 **Work Practice Problems 4–6**

PRACTICE PROBLEM 7

Find the value of each logarithmic expression.

a. $\log_5 125$ **b.** $\log_7 \dfrac{1}{49}$

c. $\log_{100} 10$

EXAMPLE 7 Find the value of each logarithmic expression.

a. $\log_4 16$ **b.** $\log_{10} \dfrac{1}{10}$ **c.** $\log_9 3$

Solution:

a. $\log_4 16 = 2$ because $4^2 = 16$.

b. $\log_{10} \dfrac{1}{10} = -1$ because $10^{-1} = \dfrac{1}{10}$.

c. $\log_9 3 = \dfrac{1}{2}$ because $9^{1/2} = \sqrt{9} = 3$.

🖥 **Work Practice Problem 7**

Objective B Solving Logarithmic Equations

The ability to interchange the logarithmic and exponential forms of a statement is often the key to solving logarithmic equations.

PRACTICE PROBLEM 8

Solve: $\log_2 x = 4$

EXAMPLE 8 Solve: $\log_5 x = 3$

Solution: $\log_5 x = 3$

$\qquad\quad 5^3 = x$ Write as an exponential equation.

$\qquad 125 = x$

The solution set is $\{125\}$.

🖥 **Work Practice Problem 8**

Answers

1. $7^2 = 49$,   **2.** $8^{-1} = \dfrac{1}{8}$,

3. $3^{1/2} = \sqrt{3}$,   **4.** $\log_3 81 = 4$,

5. $\log_2 \dfrac{1}{8} = -3$,   **6.** $\log_7 \sqrt[3]{7} = \dfrac{1}{3}$,

7. a. 3,   **b.** -2,   **c.** $\dfrac{1}{2}$,   **8.** $\{16\}$

EXAMPLE 9 Solve: $\log_x 25 = 2$

Solution: $\log_x 25 = 2$

$$x^2 = 25 \quad \text{Write as an exponential equation.}$$
$$x = 5$$

Even though $(-5)^2 = 25$, the base b of a logarithm must be positive. The solution set is $\{5\}$.

▣ **Work Practice Problem 9**

EXAMPLE 10 Solve: $\log_3 1 = x$

Solution: $\log_3 1 = x$

$$3^x = 1 \quad \text{Write as an exponential equation.}$$
$$3^x = 3^0 \quad \text{Write 1 as } 3^0.$$
$$x = 0 \quad \text{Use the uniqueness of } b^x.$$

The solution set is $\{0\}$.

▣ **Work Practice Problem 10**

In Example 10, we illustrated an important property of logarithms. That is, $\log_b 1$ is always 0. This property as well as two important others are given below.

Properties of Logarithms

If b is a real number, $b > 0$ and $b \neq 1$, then

1. $\log_b 1 = 0$

2. $\log_b b^x = x$

3. $b^{\log_b x} = x$

To see that $\log_b b^x = x$, we change the logarithmic form to exponential form. Then, $\log_b b^x = x$ means $b^x = b^x$. In exponential form, the statement is true, so in logarithmic form, the statement is also true.

EXAMPLE 11 Simplify.

a. $\log_3 3^2$ **b.** $\log_7 7^{-1}$
c. $5^{\log_5 3}$ **d.** $2^{\log_2 6}$

Solution:

a. From property 2, $\log_3 3^2 = 2$.
b. From property 2, $\log_7 7^{-1} = -1$.
c. From property 3, $5^{\log_5 3} = 3$.
d. From property 3, $2^{\log_2 6} = 6$.

▣ **Work Practice Problem 11**

Objective **C** Graphing Logarithmic Functions

Let us now return to the function $f(x) = 2^x$ and write an equation for its inverse, f^{-1}. Recall our earlier work.

$$f(x) = 2^x$$

Step 1: Replace $f(x)$ by y. $y = 2^x$
Step 2: Interchange x and y. $x = 2^y$

PRACTICE PROBLEM 9
Solve: $\log_x 9 = 2$

PRACTICE PROBLEM 10
Solve: $\log_2 1 = x$

PRACTICE PROBLEM 11
Simplify.
a. $\log_6 6^3$ **b.** $\log_{11} 11^{-4}$
c. $7^{\log_7 13}$ **d.** $3^{\log_3 10}$

PRACTICE PROBLEM 12

Graph the logarithmic function $y = \log_4 x$.

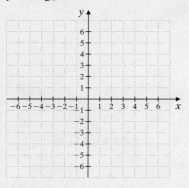

PRACTICE PROBLEM 13

Graph the logarithmic function $f(x) = \log_{1/2} x$.

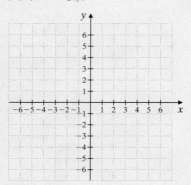

Answers

12.

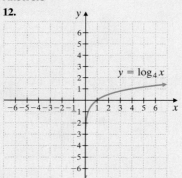

13.

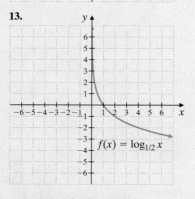

Having gained proficiency with the notation $\log_b x$, we can now complete the steps for writing the inverse equation.

Step 3: Solve for y. $\qquad\qquad\qquad\qquad y = \log_2 x$

Step 4: Replace y with $f^{-1}(x)$. $\quad f^{-1}(x) = \log_2 x$

Thus, $f^{-1}(x) = \log_2 x$ defines a function that is the inverse function of the function $f(x) = 2^x$. The function $f^{-1}(x)$ or $y = \log_2 x$ is called a *logarithmic function*.

Logarithmic Function

If x is a positive real number, b is a constant positive real number, and b is not 1, then a **logarithmic function** is a function that can be defined by

$$f(x) = \log_b x$$

The domain of f is the set of positive real numbers, and the range of f is the set of real numbers.

✔**Concept Check** Let $f(x) = \log_3 x$ and $g(x) = 3^x$. These two functions are inverses of each other. Since $(2, 9)$ is an ordered pair solution of $g(x)$, what ordered pair do we know to be a solution of $f(x)$? Explain why.

We can explore logarithmic functions by graphing them.

EXAMPLE 12 Graph the logarithmic function $y = \log_2 x$.

Solution: First we write the equation with exponential notation as $2^y = x$. Then we find some ordered pair solutions that satisfy this equation. Finally, we plot the points and connect them with a smooth curve. The domain of this function is $(0, \infty)$, and the range is all real numbers.

Since $x = 2^y$ is solved for x, we choose y-values and compute corresponding x-values.

If $y = 0$, $x = 2^0 = 1$.
If $y = 1$, $x = 2^1 = 2$.
If $y = 2$, $x = 2^2 = 4$.
If $y = -1$, $x = 2^{-1} = \dfrac{1}{2}$.

$x = 2^y$	y
1	0
2	1
4	2
$\dfrac{1}{2}$	-1

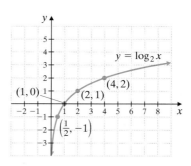

Work Practice Problem 12

EXAMPLE 13 Graph the logarithmic function $f(x) = \log_{1/3} x$.

Solution: We can replace $f(x)$ with y, and write the result with exponential notation.

$$f(x) = \log_{1/3} x$$
$$y = \log_{1/3} x \quad \text{Replace } f(x) \text{ with } y.$$
$$\left(\frac{1}{3}\right)^y = x \qquad \text{Write in exponential form.}$$

Now we can find ordered pair solutions that satisfy $\left(\dfrac{1}{3}\right)^y = x$, plot these points, and connect them with a smooth curve.

If $y = 0$, $x = \left(\dfrac{1}{3}\right)^0 = 1$.

If $y = 1$, $x = \left(\dfrac{1}{3}\right)^1 = \dfrac{1}{3}$.

If $y = -1$, $x = \left(\dfrac{1}{3}\right)^{-1} = 3$.

If $y = -2$, $x = \left(\dfrac{1}{3}\right)^{-2} = 9$.

$x = \left(\dfrac{1}{3}\right)^y$	y
1	0
$\dfrac{1}{3}$	1
3	-1
9	-2

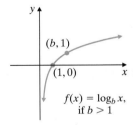

The domain of this function is $(0, \infty)$, and the range is the set of all real numbers.

Work Practice Problem 13

The following figures summarize characteristics of logarithmic functions.

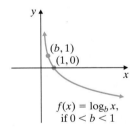

9.4 EXERCISE SET

FOR EXTRA HELP

Student Solutions Manual

PH Math/Tutor Center

CD/Video for Review

Math XL
MathXL®

MyMathLab
MyMathLab

Objective A *Write each as an exponential equation. See Examples 1 through 3.*

1. $\log_6 36 = 2$

2. $\log_2 32 = 5$

3. $\log_3 \dfrac{1}{27} = -3$

4. $\log_5 \dfrac{1}{25} = -2$

5. $\log_{10} 1000 = 3$

6. $\log_{10} 10 = 1$

7. $\log_e x = 4$

8. $\log_e y = 7$

9. $\log_e \dfrac{1}{e^2} = -2$

10. $\log_e \dfrac{1}{e} = -1$

11. $\log_7 \sqrt{7} = \dfrac{1}{2}$

12. $\log_{11} \sqrt[4]{11} = \dfrac{1}{4}$

13. $\log_{0.7} 0.343 = 3$

14. $\log_{1.2} 1.44 = 2$

15. $\log_3 \dfrac{1}{81} = -4$

16. $\log_{1/4} 16 = -2$

Write each as a logarithmic equation. See Examples 4 through 6.

17. $2^4 = 16$

18. $5^3 = 125$

19. $10^2 = 100$

20. $10^4 = 10{,}000$

21. $e^3 = x$

22. $e^5 = y$

23. $10^{-1} = \dfrac{1}{10}$

24. $10^{-2} = \dfrac{1}{100}$

25. $4^{-2} = \dfrac{1}{16}$

26. $3^{-4} = \dfrac{1}{81}$

27. $5^{1/2} = \sqrt{5}$

28. $4^{1/3} = \sqrt[3]{4}$

Find the value of each logarithmic expression. See Example 7.

29. $\log_2 8$

30. $\log_3 9$

31. $\log_3 \dfrac{1}{9}$

32. $\log_2 \dfrac{1}{32}$

33. $\log_{25} 5$

34. $\log_8 \dfrac{1}{2}$

35. $\log_{1/2} 2$

36. $\log_{2/3} \dfrac{4}{9}$

37. $\log_6 1$

38. $\log_9 9$

39. $\log_{10} 100$

40. $\log_{10} \dfrac{1}{10}$ -1

41. $\log_3 81$

42. $\log_2 16$

43. $\log_4 \dfrac{1}{64}$

44. $\log_3 \dfrac{1}{9}$

Objective B *Solve. See Examples 8 through 10.*

45. $\log_3 9 = x$

46. $\log_2 8 = x$

47. $\log_3 x = 4$

48. $\log_2 x = 3$

49. $\log_x 49 = 2$

50. $\log_x 8 = 3$

51. $\log_2 \dfrac{1}{8} = x$

52. $\log_3 \dfrac{1}{81} = x$

53. $\log_3 \dfrac{1}{27} = x$

54. $\log_5 \dfrac{1}{125} = x$

55. $\log_8 x = \dfrac{1}{3}$

56. $\log_9 x = \dfrac{1}{2}$

57. $\log_4 16 = x$

58. $\log_2 16 = x$

59. $\log_{3/4} x = 3$

60. $\log_{2/3} x = 2$

61. $\log_x 100 = 2$

62. $\log_x 27 = 3$

63. $\log_2 2^4 = x$

64. $\log_6 6^{-2} = x$

65. $3^{\log_3 5} = x$

66. $5^{\log_5 7} = x$

67. $\log_x \dfrac{1}{7} = \dfrac{1}{2}$

68. $\log_x 2 = -\dfrac{1}{3}$

Simplify. See Example 11.

69. $\log_5 5^3$

70. $\log_6 6^2$

71. $2^{\log_2 3}$

72. $7^{\log_7 4}$

73. $\log_9 9$

74. $\log_8 (8)^{-1}$

Objective **C** *Graph each logarithmic function. See Examples 12 and 13.*

75. $y = \log_3 x$

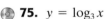
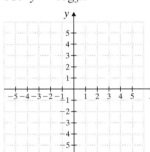

76. $y = \log_2 x$

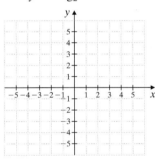

77. $f(x) = \log_{1/4} x$

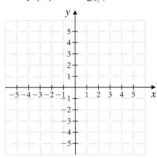

78. $f(x) = \log_{1/2} x$

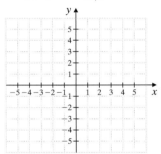

79. $f(x) = \log_5 x$

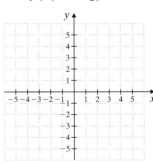

80. $f(x) = \log_6 x$

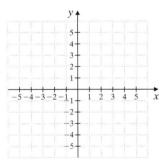

81. $f(x) = \log_{1/6} x$

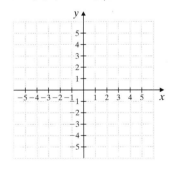

82. $f(x) = \log_{1/5} x$

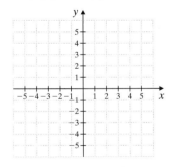

Review

Simplify each rational expression. See Section 6.1.

83. $\dfrac{x + 3}{3 + x}$

84. $\dfrac{x - 5}{5 - x}$

85. $\dfrac{x^2 - 8x + 16}{2x - 8}$

86. $\dfrac{x^2 - 3x - 10}{2 + x}$

Concept Extensions

Solve. See the Concept Check in this section.

87. Let $f(x) = \log_5 x$. Then $g(x) = 5^x$ is the inverse of $f(x)$. The ordered pair $(2, 25)$ is a solution of the function $g(x)$.

   a. Write this solution using function notation.

   b. Write an ordered pair that we know to be a solution of $f(x)$.

   c. Use the answer to part b and write the solution using function notation.

88. Let $f(x) = \log_{0.3} x$. Then $g(x) = 0.3^x$ is the inverse of $f(x)$. The ordered pair $(3, 0.027)$ is a solution of the function $g(x)$.

   a. Write this solution using function notation.

   b. Write an ordered pair that we know to be a solution of $f(x)$.

   c. Use the answer to part b and write the solution using function notation.

89. Explain why negative numbers are not included as logarithmic bases.

90. Explain why 1 is not included as a logarithmic base.

Graph each function and its inverse on the same set of axes.

91. $y = 4^x$; $y = \log_4 x$

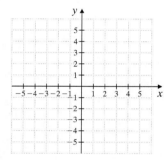

92. $y = 3^x$; $y = \log_3 x$

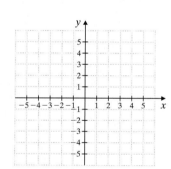

93. $y = \left(\dfrac{1}{3}\right)^x$; $y = \log_{1/3} x$

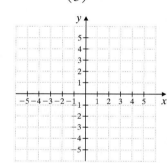

94. $y = \left(\dfrac{1}{2}\right)^x$; $y = \log_{1/2} x$

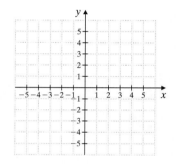

95. Explain why the graph of the function $y = \log_b x$ contains the point $(1, 0)$ no matter what b is.

96. $\log_3 10$ is between which two integers? Explain your answer.

97. The formula $\log_{10}(1 - k) = \dfrac{-0.3}{H}$ models the relationship between the half-life H of a radioactive material and its rate of decay k. Find the rate of decay of the iodine isotope I-131 if its half-life is 8 days. Round to four decimal places.

9.5 PROPERTIES OF LOGARITHMS

In the previous section we explored some basic properties of logarithms. We now introduce and study additional properties. Because a logarithm is an exponent, logarithmic properties are just restatements of exponential properties.

Objective A Using the Product Property

The first of these properties is called the **product property of logarithms** because it deals with the logarithm of a product.

> ### Product Property of Logarithms
>
> If x, y, and b are positive real numbers and $b \neq 1$, then
>
> $$\log_b xy = \log_b x + \log_b y$$

To prove this, we let $\log_b x = M$ and $\log_b y = N$. Now we write each logarithm with exponential notation.

$\log_b x = M$ is equivalent to $b^M = x$

$\log_b y = N$ is equivalent to $b^N = y$

When we multiply the left sides and the right sides of the exponential equations, we have that

$$xy = (b^M)(b^N) = b^{M+N}$$

If we write the equation $xy = b^{M+N}$ in equivalent logarithmic form, we have

$$\log_b xy = M + N$$

But since $M = \log_b x$ and $N = \log_b y$, we can write

$$\log_b xy = \log_b x + \log_b y \quad \text{Let } M = \log_b x \text{ and } N = \log_b y.$$

In other words, the logarithm of a product is the sum of the logarithms of the factors. This property is sometimes used to simplify logarithmic expressions.

EXAMPLE 1 Write as a single logarithm: $\log_{11} 10 + \log_{11} 3$

Solution: $\log_{11} 10 + \log_{11} 3 = \log_{11}(10 \cdot 3)$ Use the product property.

$$= \log_{11} 30$$

▣ **Work Practice Problem 1**

EXAMPLE 2 Write as a single logarithm: $\log_2(x + 2) + \log_2 x$

Solution: $\log_2(x + 2) + \log_2 x = \log_2[(x + 2) \cdot x] = \log_2(x^2 + 2x)$

▣ **Work Practice Problem 2**

Objective B Using the Quotient Property

The second property is the **quotient property of logarithms**.

Objectives

- **A** Use the Product Property of Logarithms.
- **B** Use the Quotient Property of Logarithms.
- **C** Use the Power Property of Logarithms.
- **D** Use the Properties of Logarithms Together.

PRACTICE PROBLEM 1

Write as a single logarithm:
$\log_2 7 + \log_2 5$

PRACTICE PROBLEM 2

Write as a single logarithm:
$\log_3 x + \log_3(x - 9)$

Answers

1. $\log_2 35$,   **2.** $\log_3(x^2 - 9x)$

671

Quotient Property of Logarithms

If x, y, and b are positive real numbers and $b \neq 1$, then

$$\log_b \frac{x}{y} = \log_b x - \log_b y$$

The proof of the quotient property of logarithms is similar to the proof of the product property. Notice that the quotient property says that the logarithm of a quotient is the difference of the logarithms of the dividend and divisor.

✔ **Concept Check** Which of the following is the correct way to rewrite $\log_5 \frac{7}{2}$?

a. $\log_5 7 - \log_5 2$ **b.** $\log_5(7 - 2)$ **c.** $\dfrac{\log_5 7}{\log_5 2}$ **d.** $\log_5 14$

PRACTICE PROBLEM 3

Write as a single logarithm:
$\log_7 40 - \log_7 8$

EXAMPLE 3 Write as a single logarithm: $\log_{10} 27 - \log_{10} 3$

Solution: $\log_{10} 27 - \log_{10} 3 = \log_{10} \dfrac{27}{3}$ Use the quotient property.

$$= \log_{10} 9$$

▣ **Work Practice Problem 3**

PRACTICE PROBLEM 4

Write as a single logarithm:
$\log_3(x^3 + 4) - \log_3(x^2 + 2)$

EXAMPLE 4 Write as a single logarithm: $\log_3(x^2 + 5) - \log_3(x^2 + 1)$

Solution: $\log_3(x^2 + 5) - \log_3(x^2 + 1) = \log_3 \dfrac{x^2 + 5}{x^2 + 1}$ Use the quotient property.

▣ **Work Practice Problem 4**

Objective C Using the Power Property

The third and final property we introduce is the **power property of logarithms.**

Power Property of Logarithms

If x and b are positive real numbers, $b \neq 1$, and r is a real number, then

$$\log_b x^r = r \log_b x$$

PRACTICE PROBLEMS 5–6

Use the power property to
rewrite each expression.
5. $\log_3 x^5$ **6.** $\log_7 \sqrt[3]{4}$

EXAMPLES Use the power property to rewrite each expression.

5. $\log_5 x^3 = 3 \log_5 x$

6. $\log_4 \sqrt{2} = \log_4 2^{1/2} = \dfrac{1}{2} \log_4 2$

▣ **Work Practice Problems 5–6**

Objective D Using More Than One Property

Many times we must use more than one property of logarithms to simplify logarithmic expressions.

EXAMPLES Write as a single logarithm.

7. $2\log_5 3 + 3\log_5 2 = \log_5 3^2 + \log_5 2^3$ Use the power property.

$\qquad = \log_5 9 + \log_5 8$

$\qquad = \log_5(9 \cdot 8)$ Use the product property.

$\qquad = \log_5 72$

8. $3\log_9 x - \log_9(x + 1) = \log_9 x^3 - \log_9(x + 1)$ Use the power property.

$\qquad = \log_9 \dfrac{x^3}{x + 1}$ Use the quotient property.

⬛ **Work Practice Problems 7–8**

PRACTICE PROBLEMS 7–8

Write as a single logarithm.
7. $3\log_4 2 + 2\log_4 5$
8. $5\log_2(2x - 1) - \log_2 x$

EXAMPLES Write each expression as sums or differences of logarithms.

9. $\log_3 \dfrac{5 \cdot 7}{4} = \log_3(5 \cdot 7) - \log_3 4$ Use the quotient property.

$\qquad = \log_3 5 + \log_3 7 - \log_3 4$ Use the product property.

10. $\log_2 \dfrac{x^5}{y^2} = \log_2(x^5) - \log_2(y^2)$ Use the quotient property.

$\qquad = 5\log_2 x - 2\log_2 y$ Use the power property.

⬛ **Work Practice Problems 9–10**

PRACTICE PROBLEMS 9–10

Write each expression as sums or differences of logarithms.
9. $\log_7 \dfrac{6 \cdot 2}{5}$ **10.** $\log_3 \dfrac{x^4}{y^3}$

Helpful Hint

Notice that we are not able to simplify further a logarithmic expression such as $\log_5(2x - 1)$. None of the basic properties gives a way to write the logarithm of a difference (or sum) in some equivalent form.

✔ **Concept Check** What is wrong with the following?

$$\log_{10}(x^2 + 5) = \log_{10} x^2 + \log_{10} 5$$
$$= 2\log_{10} x + \log_{10} 5$$

Use a numerical example to demonstrate that the result is incorrect.

EXAMPLES If $\log_b 2 = 0.43$ and $\log_b 3 = 0.68$, use the properties of logarithms to evaluate each expression.

11. $\log_b 6 = \log_b(2 \cdot 3)$ Write 6 as $2 \cdot 3$.

$\qquad = \log_b 2 + \log_b 3$ Use the product property.

$\qquad = 0.43 + 0.68$ Substitute given values.

$\qquad = 1.11$ Simplify.

12. $\log_b 9 = \log_b 3^2$ Write 9 as 3^2.

$\qquad = 2\log_b 3$ Use the power property.

$\qquad = 2(0.68)$ Substitute the given value.

$\qquad = 1.36$ Simplify.

13. $\log_b \sqrt{2} = \log_b 2^{1/2}$ Write $\sqrt{2}$ as $2^{1/2}$.

$\qquad = \dfrac{1}{2}\log_b 2$ Use the power property.

$\qquad = \dfrac{1}{2}(0.43)$ Substitute the given value.

$\qquad = 0.215$ Simplify.

⬛ **Work Practice Problems 11–13**

PRACTICE PROBLEMS 11–13

If $\log_b 4 = 0.86$ and $\log_b 7 = 1.21$, use the properties of logarithms to evaluate each expression.
11. $\log_b 28$ **12.** $\log_b 49$
13. $\log_b \sqrt[3]{4}$

Answers
7. $\log_4 200$,   **8.** $\log_2 \dfrac{(2x - 1)^5}{x}$,
9. $\log_7 6 + \log_7 2 - \log_7 5$,
10. $4\log_3 x - 3\log_3 y$,   **11.** 2.07,
12. 2.42,   **13.** $0.28\overline{6}$

✔ Concept Check Answer
The properties do not give any way to simplify the logarithm of a sum; answers may vary.

Objective A *Write each sum as a single logarithm. Assume that variables represent positive numbers. See Examples 1 and 2.*

1. $\log_5 2 + \log_5 7$

2. $\log_3 8 + \log_3 4$

3. $\log_4 9 + \log_4 x$

4. $\log_2 x + \log_2 y$

5. $\log_6 x + \log_6 (x + 1)$

6. $\log_5 y^3 + \log_5 (y - 7)$

7. $\log_{10} 5 + \log_{10} 2 + \log_{10} (x^2 + 2)$

8. $\log_6 3 + \log_6 (x + 4) + \log_6 5$

Objective B *Write each difference as a single logarithm. Assume that variables represent positive numbers. See Examples 3 and 4.*

9. $\log_5 12 - \log_5 4$

10. $\log_7 20 - \log_7 4$

11. $\log_3 8 - \log_3 2$

12. $\log_5 12 - \log_5 3$

13. $\log_2 x - \log_2 y$

14. $\log_3 12 - \log_3 z$

15. $\log_2 (x^2 + 6) - \log_2 (x^2 + 1)$

16. $\log_7 (x + 9) - \log_7 (x^2 + 10)$

Objective C *Use the power property to rewrite each expression. See Examples 5 and 6.*

17. $\log_3 x^2$

18. $\log_2 x^5$

19. $\log_4 5^{-1}$

20. $\log_6 7^{-2}$

21. $\log_5 \sqrt{y}$

22. $\log_5 \sqrt[3]{x}$

Objective D **Mixed Practice** *Write each as a single logarithm. Assume that variables represent positive numbers. See Examples 7 and 8.*

23. $\log_2 5 + \log_2 x^3$

24. $\log_5 2 + \log_5 y^2$

25. $3\log_4 2 + \log_4 6$

26. $2\log_3 5 + \log_3 2$

27. $3\log_5 x + 6\log_5 z$

28. $2\log_7 y + 6\log_7 z$

29. $\log_4 2 + \log_4 10 - \log_4 5$

30. $\log_6 18 + \log_6 2 - \log_6 9$

31. $\log_7 6 + \log_7 3 - \log_7 4$

32. $\log_8 5 + \log_8 15 - \log_8 20$

33. $\log_{10} x - \log_{10} (x + 1) + \log_{10} (x^2 - 2)$

34. $\log_9 (4x) - \log_9 (x - 3) + \log_9 (x^3 + 1)$

35. $3\log_2 x + \frac{1}{2}\log_2 x - 2\log_2 (x + 1)$

36. $2\log_5 x + \frac{1}{3}\log_5 x - 3\log_5 (x + 5)$

37. $2\log_8 x - \frac{2}{3}\log_8 x + 4\log_8 x$

38. $5\log_6 x - \frac{3}{4}\log_6 x + 3\log_6 x$

Mixed Practice *Write each expression as a sum or difference of logarithms. Assume that variables represent positive numbers. See Examples 9 and 10.*

39. $\log_3 \dfrac{4y}{5}$

40. $\log_7 \dfrac{5x}{4}$

41. $\log_4 \dfrac{2}{9z}$

42. $\log_9 \dfrac{7}{8y}$

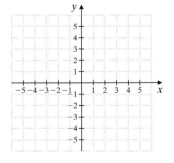 **43.** $\log_2 \dfrac{x^3}{y}$

44. $\log_5 \dfrac{x}{y^4}$

45. $\log_b \sqrt{7x}$

46. $\log_b \sqrt{\dfrac{3}{y}}$

47. $\log_6 x^4 y^5$

48. $\log_2 y^3 z$

49. $\log_5 x^3(x+1)$

50. $\log_3 x^2(x-9)$

51. $\log_6 \dfrac{x^2}{x+3}$

52. $\log_3 \dfrac{(x+5)^2}{x}$

If $\log_b 3 = 0.5$ and $\log_b 5 = 0.7$, evaluate each expression. See Examples 11 through 13.

53. $\log_b \dfrac{5}{3}$

54. $\log_b 25$

55. $\log_b 15$

56. $\log_b \dfrac{3}{5}$

57. $\log_b \sqrt{5}$

58. $\log_b \sqrt[4]{3}$

If $\log_b 2 = 0.43$ and $\log_b 3 = 0.68$, evaluate each expression. See Examples 11 through 13.

59. $\log_b 8$

60. $\log_b 81$

61. $\log_b \dfrac{3}{9}$

62. $\log_b \dfrac{4}{32}$

63. $\log_b \sqrt{\dfrac{2}{3}}$

64. $\log_b \sqrt{\dfrac{3}{2}}$

Review

65. Graph the functions $y = 10^x$ and $y = \log_{10} x$ on the same set of axes. See Section 9.4.

Evaluate each expression. See Section 9.4.

66. $\log_{10} 100$

67. $\log_{10} \dfrac{1}{10}$

68. $\log_7 7^2$

69. $\log_7 \sqrt{7}$

Concept Extensions

Solve. See the Concept Checks in this section.

70. Which of the following is the correct way to rewrite $\log_3 \frac{14}{11}$?

   a. $\dfrac{\log_3 14}{\log_3 11}$

   b. $\log_3 14 - \log_3 11$

   c. $\log_3(14 - 11)$

   d. $\log_3 154$

71. Which of the following is the correct way to rewrite $\log_9 \frac{21}{3}$?

   a. $\log_9 7$

   b. $\log_9(21 - 3)$

   c. $\dfrac{\log_9 21}{\log_9 3}$

   d. $\log_9 21 - \log_9 3$

Determine whether each statement is true or false.

72. $\log_2 x^3 = 3\log_2 x$

73. $\log_3(x + y) = \log_3 x + \log_3 y$

74. $\dfrac{\log_7 10}{\log_7 5} = \log_7 2$

75. $\log_7 \dfrac{14}{8} = \log_7 14 - \log_7 8$

76. $\dfrac{\log_7 x}{\log_7 y} = \log_7 x - \log_7 y$

77. $(\log_3 6) \cdot (\log_3 4) = \log_3 24$

78. It is true that $\log_b 8 = \log_b(8 \cdot 1) = \log_b 8 + \log_b 1$. Explain how $\log_b 8$ can equal $\log_b 8 + \log_b 1$.

Functions and Properties of Logarithms

If $f(x) = x - 6$ and $g(x) = x^2 + 1$, find each value.

1. $(f + g)(x)$ **2.** $(f - g)(x)$ **3.** $(f \cdot g)(x)$ **4.** $\left(\dfrac{f}{g}\right)(x)$

If $f(x) = \sqrt{x}$ and $g(x) = 3x - 1$, find each value.

5. $(f \circ g)(x)$ **6.** $(g \circ f)(x)$

Determine whether each is a one-to-one function. If it is, find its inverse.

7. $f = \{(-2, 6), (4, 8), (2, -6), (3, 3)\}$ **8.** $g = \{(4, 2), (-1, 3), (5, 3), (7, 1)\}$

Determine whether the graph of each function is one-to-one.

9.

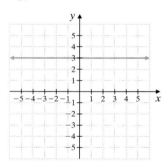

10.

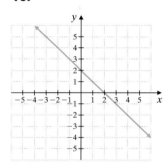

11.

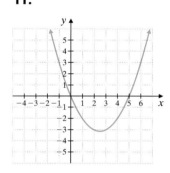

Each function listed is one-to-one. Find the inverse of each function.

12. $f(x) = 3x$ **13.** $f(x) = x + 4$

14. $f(x) = 5x - 1$ **15.** $f(x) = 3x + 2$

Graph each function.

16. $y = \left(\dfrac{1}{2}\right)^x$ **17.** $y = 2^x + 1$

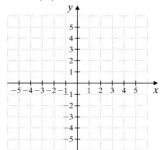   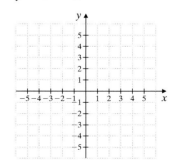

Answers

1. _____

2. _____

3. _____

4. _____

5. _____

6. _____

7. _____

8. _____

9. _____

10. _____

11. _____

12. _____

13. _____

14. _____

15. _____

16. see graph

17. see graph

18. see graph

19. see graph

20. _____

21. _____

22. _____

23. _____

24. _____

25. _____

26. _____

27. _____

28. _____

29. _____

30. _____

31. _____

32. _____

33. _____

34. _____

35. _____

36. _____

37. _____

18. $y = \log_3 x$

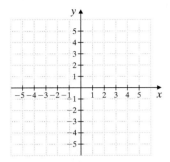

19. $y = \log_{1/3} x$

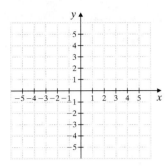

Solve.

20. $2^x = 8$

21. $9 = 3^{x-5}$

22. $4^{x-1} = 8^{x+2}$

23. $25^x = 125^{x-1}$

24. $\log_4 16 = x$

25. $\log_{49} 7 = x$

26. $\log_2 x = 5$

27. $\log_x 64 = 3$

28. $\log_x \dfrac{1}{125} = -3$

29. $\log_3 x = -2$

Write each as a single logarithm.

30. $\log_2 x + \log_2 14$

31. $x \log_2 5 + \log_2 8$

32. $3 \log_5 x - 5 \log_5 y$

33. $9 \log_5 x + 3 \log_5 y$

34. $\log_2 x + \log_2 (x - 3) - \log_2 (x^2 + 4)$

35. $\log_3 y - \log_3 (y + 2) + \log_3 (y^3 + 11)$

Write each expression as a sum or difference of logarithms.

36. $\log_7 \dfrac{9x^2}{y}$

37. $\log_6 \dfrac{5y}{z^2}$

9.6 COMMON LOGARITHMS, NATURAL LOGARITHMS, AND CHANGE OF BASE

Objectives

A Identify Common Logarithms and Approximate Them with a Calculator.

B Evaluate Common Logarithms of Powers of 10.

C Identify Natural Logarithms and Approximate Them with a Calculator.

D Evaluate Natural Logarithms of Powers of e.

E Use the Change of Base Formula.

In this section we look closely at two particular logarithmic bases. These two logarithmic bases are used so frequently that logarithms to their bases are given special names. **Common logarithms** are logarithms to base 10. **Natural logarithms** are logarithms to base e, which we introduce in this section. The work in this section is based on the use of a calculator that has both the "common log" $\boxed{\text{LOG}}$ and the "natural log" $\boxed{\text{LN}}$ keys.

Objective **A** Approximating Common Logarithms

Logarithms to base 10—**common logarithms**—are used frequently because our number system is a base 10 decimal system. The notation $\log x$ means the same as $\log_{10} x$.

> ### Common Logarithm
>
> $$\log x \quad \text{means} \quad \log_{10} x$$

EXAMPLE 1 Use a calculator to approximate $\log 7$ to four decimal places.

Solution: Press the following sequence of keys:

$$\boxed{7}\boxed{\text{LOG}} \quad \text{or} \quad \boxed{\text{LOG}}\boxed{7}\boxed{\text{ENTER}}$$

To four decimal places,

$$\log 7 \approx 0.8451$$

 Work Practice Problem 1

PRACTICE PROBLEM 1

Use a calculator to approximate $\log 21$ to four decimal places.

Objective **B** Evaluating Common Logarithms of Powers of 10

To evaluate the common log of a power of 10, a calculator is not needed. According to the property of logarithms,

$$\log_b b^x = x$$

It follows that if b is replaced with 10, we have

$$\log 10^x = x$$

> **Helpful Hint**
> Remember that the base of this logarithm is understood to be 10.

EXAMPLES Find the exact value of each logarithm.

2. $\log 10 = \log 10^1 = 1$

3. $\log \dfrac{1}{10} = \log 10^{-1} = -1$

4. $\log 100{,}000 = \log 10^5 = 5$

5. $\log \sqrt[4]{10} = \log 10^{1/4} = \dfrac{1}{4}$

 Work Practice Problems 2–5

PRACTICE PROBLEMS 2–5

Find the exact value of each logarithm.

2. $\log 1000$

3. $\log \dfrac{1}{100}$

4. $\log 10{,}000$

5. $\log \sqrt[3]{10}$

Answers

1. 1.3222, **2.** 3, **3.** −2, **4.** 4, **5.** $\dfrac{1}{3}$

PRACTICE PROBLEM 6

Solve: $\log x = 2.9$. Give an exact solution and then approximate the solution to four decimal places.

Helpful Hint

The understood base is 10.

As we will soon see, equations containing common logs are useful models of many natural phenomena.

EXAMPLE 6 Solve: $\log x = 1.2$. Give an exact solution and then approximate the solution to four decimal places.

Solution: Remember that the base of a common log is understood to be 10.

$$\log x = 1.2$$
$$10^{1.2} = x \qquad \text{Write with exponential notation.}$$

The exact solution is $10^{1.2}$ or the solution set is $\{10^{1.2}\}$. To four decimal places, $x \approx 15.8489$.

◻ **Work Practice Problem 6**

Objective ◉ Approximating Natural Logarithms

Natural logarithms are also frequently used, especially to describe natural events; hence the label "natural logarithm." **Natural logarithms** are logarithms to the base e, which is a constant approximately equal to 2.7183. The number e is an irrational number, as is π. The notation $\log_e x$ is usually abbreviated to $\ln x$. (The abbreviation ln is read "el en.")

Natural Logarithm

$$\ln x \qquad \text{means} \qquad \log_e x$$

PRACTICE PROBLEM 7

Use a calculator to approximate ln 11 to four decimal places.

EXAMPLE 7 Use a calculator to approximate ln 8 to four decimal places.

Solution: Press the following sequence of keys:

$\boxed{8}\,\boxed{\ln}$ or $\boxed{\ln}\,\boxed{8}\,\boxed{\text{ENTER}}$

To four decimal places,
$\ln 8 \approx 2.0794$

◻ **Work Practice Problem 7**

Objective ◉ Evaluating Natural Logarithms of Powers of e

As a result of the property $\log_b b^x = x$, we know that $\log_e e^x = x$, or $\ln e^x = x$.

PRACTICE PROBLEMS 8–9

Find the exact value of each natural logarithm.
8. $\ln e^9$
9. $\ln \sqrt[3]{e}$

EXAMPLES Find the exact value of each natural logarithm.

8. $\ln e^3 = 3$

9. $\ln \sqrt[7]{e} = \ln e^{1/7} = \dfrac{1}{7}$

◻ **Work Practice Problems 8–9**

Answers

6. $x = 10^{2.9}$; $x \approx 794.3282$, 7. 2.3979,

8. 9, 9. $\dfrac{1}{3}$

EXAMPLE 10 Solve: $\ln 3x = 5$. Give an exact solution and then approximate the solution to four decimal places.

Solution: Remember that the base of a natural logarithm is understood to be e.

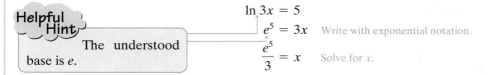

Helpful Hint
The understood base is e.

$$\ln 3x = 5$$
$$e^5 = 3x \quad \text{Write with exponential notation.}$$
$$\frac{e^5}{3} = x \quad \text{Solve for } x.$$

The exact solution is $\dfrac{e^5}{3}$. To four decimal places, $x \approx 49.4711$.

◼ **Work Practice Problem 10**

Recall from Section 9.3 the formula $A = P\left(1 + \dfrac{r}{n}\right)^{nt}$ for compound interest, where n represents the number of compoundings per year. When interest is compounded continuously, we use the formula $A = Pe^{rt}$, where r is the annual interest rate and interest is compounded continuously for t years.

EXAMPLE 11 Finding the Amount Owed on a Loan

Find the amount owed at the end of 5 years if $1600 is loaned at a rate of 9% compounded continuously.

Solution: We use the formula $A = Pe^{rt}$ and the following values of the variables.

$P = \$1600$ (the amount of the loan)
$r = 9\% = 0.09$ (the rate of interest)
$t = 5$ (the 5-year duration of the loan)

$A = Pe^{rt}$
$\quad = 1600e^{0.09(5)} \quad \text{Substitute known values.}$
$\quad = 1600e^{0.45}$

Now we can use a calculator to approximate the solution.

$A \approx 2509.30$

The total amount of money owed is approximately $2509.30.

◼ **Work Practice Problem 11**

Objective E Using the Change of Base Formula

Calculators are handy tools for approximating natural and common logarithms. Unfortunately, some calculators cannot be used to approximate logarithms to bases other than e or 10—at least not directly. In such cases, we use the **change of base formula.**

Change of Base

If a, b, and c are positive real numbers and neither b nor c is 1, then

$$\log_b a = \frac{\log_c a}{\log_c b}$$

PRACTICE PROBLEM 12

Approximate $\log_7 5$ to four decimal places.

EXAMPLE 12 Approximate $\log_5 3$ to four decimal places.

Solution: We use the change of base property to write $\log_5 3$ as a quotient of logarithms to base 10.

$$\log_5 3 = \frac{\log 3}{\log 5} \qquad \text{Use the change of base property.}$$

$$\approx \frac{0.4771213}{0.69897} \qquad \text{Approximate the logarithms by calculator.}$$

$$\approx 0.6826063 \qquad \text{Simplify by calculator.}$$

To four decimal places, $\log_5 3 \approx 0.6826$.

■ **Work Practice Problem 12**

✔**Concept Check** If a graphing calculator cannot directly evaluate logarithms to base 5, describe how you could use the graphing calculator to graph the function $f(x) = \log_5 x$.

Answer

12. 0.8271

✔ **Concept Check Answer**

$$f(x) = \frac{\log x}{\log 5}$$

9.6 EXERCISE SET

Student Solutions Manual   PH Math/Tutor Center   CD/Video for Review   MathXL® MathXL®   MyMathLab MyMathLab

Objectives **A** **C** **Mixed Practice**   *Use a calculator to approximate each logarithm to four decimal places. See Examples 1 and 7.*

1. $\log 8$

2. $\log 6$

3. $\log 2.31$

4. $\log 4.86$

5. $\ln 2$

6. $\ln 3$

7. $\ln 0.0716$

8. $\ln 0.0032$

9. $\log 12.6$

10. $\log 25.9$

11. $\ln 5$

12. $\ln 7$

13. $\log 41.5$

14. $\ln 41.5$

Objectives **B** **D** **Mixed Practice**   *Find the exact value of each logarithm. See Examples 2 through 5, 8, and 9.*

15. $\log 100$

16. $\log 10{,}000$

17. $\log \dfrac{1}{1000}$

18. $\log \dfrac{1}{10}$

19. $\ln e^2$

20. $\ln e^4$

21. $\ln \sqrt[4]{e}$

22. $\ln \sqrt[5]{e}$

23. $\log 10^3$

24. $\ln e^5$

25. $\ln e^{3.1}$

26. $\log 10^7$

27. $\log 0.0001$

28. $\log 0.001$

29. $\ln \sqrt{e}$

30. $\log \sqrt{10}$

Solve each equation. Give an exact solution and a four-decimal-place approximation. See Examples 6 and 10.

31. $\log x = 1.3$

32. $\log x = 2.1$

33. $\ln x = 1.4$

34. $\ln x = 2.1$

35. $\log x = 2.3$

36. $\log x = 3.1$

37. $\ln x = -2.3$

38. $\ln x = -3.7$

39. $\log 2x = 1.1$

40. $\log 3x = 1.3$

41. $\ln 4x = 0.18$

42. $\ln 3x = 0.76$

43. $\ln(3x - 4) = 2.3$

44. $\ln(2x + 5) = 3.4$

45. $\log(2x + 1) = -0.5$

46. $\log(3x - 2) = -0.8$

Use the formula $A = Pe^{rt}$ to solve. See Example 11.

47. How much money does Dana Jones have after 12 years if she invests $1400 at 8% interest compounded continuously?

48. Determine the size of an account in which $3500 earns 6% interest compounded continuously for 1 year.

49. How much money does Barbara Mack owe at the end of 4 years if 6% interest is compounded continuously on her $2000 debt?

50. Find the amount of money for which a $2500 certificate of deposit is redeemable if it has been paying 10% interest compounded continuously for 3 years.

Objective **E** *Approximate each logarithm to four decimal places. See Example 12.*

51. $\log_2 3$ **52.** $\log_3 2$ **53.** $\log_8 6$ **54.** $\log_6 8$ 🔘 **55.** $\log_4 9$

56. $\log_9 4$ **57.** $\log_3 \dfrac{1}{6}$ **58.** $\log_6 \dfrac{2}{3}$ **59.** $\log_{1/2} 5$ **60.** $\log_{1/3} 2$

Review

Solve for x. See Sections 2.1, 2.3, and 5.7.

61. $6x - 3(2 - 5x) = 6$ **62.** $2x + 3 = 5 - 2(3x - 1)$

63. $2x + 3y = 6x$ **64.** $4x - 8y = 10x$

65. $x^2 + 7x = -6$ **66.** $x^2 + 4x = 12$

Concept Extensions

67. Use a calculator to try to approximate $\log 0$. Describe what happens and explain why.

68. Use a calculator to try to approximate $\ln 0$. Describe what happens and explain why.

Graph each function by finding ordered pair solutions, plotting the solutions, and then drawing a smooth curve through the plotted points.

69. $f(x) = e^x$ **70.** $f(x) = e^{2x}$ **71.** $f(x) = \ln x$ **72.** $f(x) = \log x$

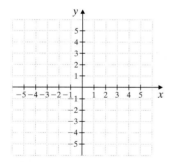   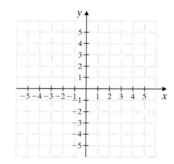   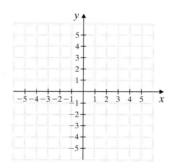   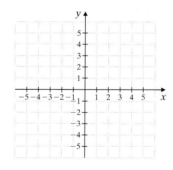

73. Without using a calculator, explain which of $\log 50$ or $\ln 50$ must be larger.

74. Without using a calculator, explain which of $\log 50^{-1}$ or $\ln 50^{-1}$ must be larger.

The Richter scale measures the intensity, or magnitude, of an earthquake. The formula for the magnitude R of an earthquake is $R = \log\left(\dfrac{a}{T}\right) + B$, where a is the amplitude in micrometers of the vertical motion of the ground at the recording station, T is the number of seconds between successive seismic waves, and B is an adjustment factor that takes into account the weakening of the seismic wave as the distance increases from the epicenter of the earthquake.

    Use the Richter scale formula to find the magnitude R of the earthquake that fits the description given. Round answers to one decimal place.

75. Amplitude *a* is 200 micrometers, time *T* between waves is 1.6 seconds, and *B* is 2.1.

76. Amplitude *a* is 150 micrometers, time *T* between waves is 3.6 seconds, and *B* is 1.9.

77. Amplitude *a* is 400 micrometers, time *T* between waves is 2.6 seconds, and *B* is 3.1.

78. Amplitude *a* is 450 micrometers, time *T* between waves is 4.2 seconds, and *B* is 2.7.

 STUDY SKILLS BUILDER

What to Do the Day of an Exam?

On the day of an exam, don't forget to try the following:

- Allow yourself plenty of time to arrive.
- Read the directions on the test carefully.
- Read each problem carefully as you take your test. Make sure that you answer the question asked.
- Watch your time and pace yourself so that you may attempt each problem on your test.
- Check your work and answers.
- ***Do not turn your test in early.*** If you have extra time, spend it double-checking your work.

Good luck!

Answer the following questions based on your most recent mathematics exam, whenever that was.

1. How soon before class did you arrive?
2. Did you read the directions on the test carefully?
3. Did you make sure you answered the question asked for each problem on the exam?
4. Were you able to attempt each problem on your exam?
5. If your answer to Question 4 is no, list reasons why.
6. Did you have extra time on your exam?
7. If your answer to Question 6 is yes, describe how you spent that extra time.

A Solve Exponential Equations.

B Solve Logarithmic Equations.

C Solve Problems That Can Be Modeled by Exponential and Logarithmic Equations.

9.7 EXPONENTIAL AND LOGARITHMIC EQUATIONS AND PROBLEM SOLVING

Objective **A** Solving Exponential Equations

In Section 9.3 we solved exponential equations such as $2^x = 16$ by writing both sides in terms of the same base. Here, we write 16 as a power of 2 and using the uniqueness of b^x.

$$2^x = 16$$
$$2^x = 2^4 \quad \text{Write 16 as } 2^4.$$
$$x = 4 \quad \text{Use the uniqueness of } b^x.$$

How do we solve an exponential equation when the bases cannot easily be written the same? For example, how do we solve an equation such as $3^x = 7$? We use the fact that $f(x) = \log_b x$ is a one-to-one function. Another way of stating this fact is as a property of equality.

Logarithm Property of Equality

Let a, b, and c be real numbers such that $\log_b a$ and $\log_b c$ are real numbers and b is not 1. Then

$$\log_b a = \log_b c \quad \text{is equivalent to} \quad a = c$$

PRACTICE PROBLEM 1

Solve: $2^x = 5$. Give an exact answer and a four-decimal-place approximation.

EXAMPLE 1 Solve: $3^x = 7$. Give an exact answer and a four-decimal-place approximation.

Solution: We use the logarithm property of equality and take the logarithm of both sides. For this example, we use the common logarithm.

$$3^x = 7$$
$$\log 3^x = \log 7 \quad \text{Take the common log of both sides.}$$
$$x \log 3 = \log 7 \quad \text{Use the power property of logarithms.}$$
$$x = \frac{\log 7}{\log 3} \quad \text{Divide both sides by log 3.}$$

The exact solution is $\dfrac{\log 7}{\log 3}$. When we approximate to four decimal places, we have

$$\frac{\log 7}{\log 3} \approx \frac{0.845098}{0.4771213} \approx 1.7712$$

The solution set is $\left\{ \dfrac{\log 7}{\log 3} \right\}$, or approximately $\{1.7712\}$.

▪ **Work Practice Problem 1**

Objective **B** Solving Logarithmic Equations

By applying the appropriate properties of logarithms, we can solve a broad variety of logarithmic equations.

EXAMPLE 2 Solve: $\log_4(x - 2) = 2$

Solution: Notice that $x - 2$ must be positive, so x must be greater than 2. With this in mind, we first write the equation with exponential notation.

$$\log_4(x - 2) = 2$$
$$4^2 = x - 2$$
$$16 = x - 2$$
$$18 = x \qquad \text{Add 2 to both sides.}$$

To check, we replace x with 18 in the original equation.

$$\log_4(x - 2) = 2$$
$$\log_4(18 - 2) \stackrel{?}{=} 2 \qquad \text{Let } x = 18.$$
$$\log_4 16 \stackrel{?}{=} 2$$
$$4^2 = 16 \qquad \text{True}$$

The solution set is $\{18\}$.

■ **Work Practice Problem 2**

PRACTICE PROBLEM 2

Solve: $\log_3(x + 5) = 2$

EXAMPLE 3 Solve: $\log_2 x + \log_2(x - 1) = 1$

Solution: Notice that $x - 1$ must be positive, so x must be greater than 1. We use the product property on the left side of the equation.

$$\log_2 x + \log_2(x - 1) = 1$$
$$\log_2[x(x - 1)] = 1 \qquad \text{Use the product property.}$$
$$\log_2(x^2 - x) = 1$$

Next we write the equation with exponential notation and solve for x.

$$2^1 = x^2 - x$$
$$0 = x^2 - x - 2 \qquad \text{Subtract 2 from both sides.}$$
$$0 = (x - 2)(x + 1) \qquad \text{Factor.}$$
$$0 = x - 2 \quad \text{or} \quad 0 = x + 1 \qquad \text{Set each factor equal to 0.}$$
$$2 = x \qquad\qquad -1 = x$$

Recall that -1 cannot be a solution because x must be greater than 1. If we forgot this, we would still reject -1 after checking. To see this, we replace x with -1 in the original equation.

$$\log_2 x + \log_2(x - 1) = 1$$
$$\log_2(-1) + \log_2(-1 - 1) \stackrel{?}{=} 1 \qquad \text{Let } x = -1.$$

Because the logarithm of a negative number is undefined, -1 is rejected. Check to see that the solution set is $\{2\}$.

■ **Work Practice Problem 3**

PRACTICE PROBLEM 3

Solve: $\log_6 x + \log_6(x + 1) = 1$

Answers

2. $\{4\}$, **3.** $\{2\}$

PRACTICE PROBLEM 4

Solve: $\log(x + 1) - \log x = 1$

EXAMPLE 4 Solve: $\log(x + 2) - \log x = 2$

Solution: We use the quotient property of logarithms on the left side of the equation.

$$\log(x + 2) - \log x = 2$$

$$\log \frac{x + 2}{x} = 2 \qquad \text{Use the quotient property.}$$

$$10^2 = \frac{x + 2}{x} \qquad \text{Write using exponential notation.}$$

$$100 = \frac{x + 2}{x}$$

$$100x = x + 2 \qquad \text{Multiply both sides by } x.$$

$$99x = 2 \qquad \text{Subtract } x \text{ from both sides.}$$

$$x = \frac{2}{99} \qquad \text{Divide both sides by 99.}$$

Check to see that the solution set is $\left\{\dfrac{2}{99}\right\}$.

▣ **Work Practice Problem 4**

Objective C Solving Problems Modeled by Exponential and Logarithmic Equations

Logarithmic and exponential functions are used in a variety of scientific, technical, and business settings. A few examples follow.

PRACTICE PROBLEM 5

Use the equation in Example 5 to estimate the lemming population in 8 months.

EXAMPLE 5 Estimating Population Size

The population size y of a community of lemmings varies according to the relationship $y = y_0 e^{0.15t}$. In this formula, t is time in months, and y_0 is the initial population at time 0. Estimate the population after 6 months if there were originally 5000 lemmings.

Solution: We substitute 5000 for y_0 and 6 for t.

$$y = y_0 e^{0.15t}$$

$$= 5000 e^{0.15(6)} \qquad \text{Let } t = 6 \text{ and } y_0 = 5000.$$

$$= 5000 e^{0.9} \qquad \text{Multiply.}$$

Using a calculator, we find that $y \approx 12{,}298.016$. In 6 months the population will be approximately 12,300 lemmings.

▣ **Work Practice Problem 5**

PRACTICE PROBLEM 6

How long does it take an investment of $1000 to double if it is invested at 6% interest compounded quarterly?

EXAMPLE 6 Doubling an Investment

How long does it take an investment of $2000 to double if it is invested at 5% interest compounded quarterly? The necessary formula is $A = P\left(1 + \dfrac{r}{n}\right)^{nt}$, where A is the accrued amount, P is the principal invested, r is the annual rate of interest, n is the number of compounding periods per year, and t is the number of years.

Answers

4. $\left\{\dfrac{1}{9}\right\}$,

5. approximately 16,600 lemmings,

6. $11\dfrac{3}{4}$ yr

Solution: We are given that $P = \$2000$ and $r = 5\% = 0.05$. Compounding quarterly means 4 times a year, so $n = 4$. The investment is to double, so A must be 4000. We substitute these values and solve for t.

$$A = P\left(1 + \frac{r}{n}\right)^{nt}$$

$$4000 = 2000\left(1 + \frac{0.05}{4}\right)^{4t} \qquad \text{Substitute known values.}$$

$$4000 = 2000(1.0125)^{4t} \qquad \text{Simplify } 1 + \frac{0.05}{4}.$$

$$2 = (1.0125)^{4t} \qquad \text{Divide both sides by 2000.}$$

$$\log 2 = \log 1.0125^{4t} \qquad \text{Take the logarithm of both sides.}$$

$$\log 2 = 4t(\log 1.0125) \qquad \text{Use the power property.}$$

$$\frac{\log 2}{4 \log 1.0125} = t \qquad \text{Divide both sides by 4 log 1.0125.}$$

$$13.949408 \approx t \qquad \text{Approximate by calculator.}$$

It takes approximately 14 years for the money to double in value.

🔲 **Work Practice Problem 6**

🖩 CALCULATOR EXPLORATIONS Graphing

Use a grapher to find how long it takes an investment of $1500 to triple if it is invested at 8% interest compounded monthly. First, let $P = \$1500$, $r = 0.08$, and $n = 12$ (for monthly compounding) in the formula

$$A = P\left(1 + \frac{r}{n}\right)^{nt}$$

Notice that when the investment has tripled, the accrued amount A is $4500. Thus,

$$4500 = 1500\left(1 + \frac{0.08}{12}\right)^{12t}$$

Determine an appropriate viewing window and enter and graph the equations

$$Y_1 = 1500\left(1 + \frac{0.08}{12}\right)^{12x}$$

and

$$Y_2 = 4500$$

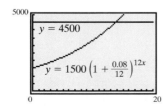

The point of intersection of the two curves is the solution. The x-coordinate tells how long it takes for the investment to triple.

Use a TRACE feature or an INTERSECT feature to approximate the coordinates of the point of intersection of the two curves. It takes approximately 13.78 years, or 13 years and 10 months, for the investment to triple in value to $4500.

Use this graphical solution method to solve each problem. Round each answer to the nearest hundredth.

1. Find how long it takes an investment of $5000 to grow to $6000 if it is invested at 5% interest compounded quarterly.

2. Find how long it takes an investment of $1000 to double if it is invested at 4.5% interest compounded daily. (Use 365 days in a year.)

3. Find how long it takes an investment of $10,000 to quadruple if it is invested at 6% interest compounded monthly.

4. Find how long it takes $500 to grow to $800 if it is invested at 4% interest compounded semiannually.

Objective **A** *Solve each equation. Give an exact solution and a four-decimal-place approximation. See Example 1.*

1. $3^x = 6$ **2.** $4^x = 7$ **3.** $9^x = 5$ **4.** $3^x = 11$

5. $3^{2x} = 3.8$ **6.** $5^{3x} = 5.6$ **7.** $e^{6x} = 5$ **8.** $e^{2x} = 8$

9. $2^{x-3} = 5$ **10.** $8^{x-2} = 12$ **11.** $4^{x+7} = 3$

12. $6^{x+3} = 2$ **13.** $7^{3x-4} = 11$ **14.** $5^{2x-6} = 12$

Objective **B** *Solve each equation. See Examples 2 through 4.*

15. $\log_2(x + 5) = 4$ **16.** $\log_2(x - 5) = 3$ **17.** $\log_4 2 + \log_4 x = 0$ **18.** $\log_3 5 + \log_3 x = 1$

19. $\log_2 6 - \log_2 x = 3$ **20.** $\log_4 10 - \log_4 x = 2$ **21.** $\log_2(x^2 + x) = 1$ **22.** $\log_6(x^2 - x) = 1$

23. $\log_4 x + \log_4(x + 6) = 2$ **24.** $\log_3 x + \log_3(x + 6) = 3$ **25.** $\log_5(x + 3) - \log_5 x = 2$

26. $\log_6(x + 2) - \log_6 x = 2$ **27.** $\log_4(x^2 - 3x) = 1$ **28.** $\log_8(x^2 - 2x) = 1$

29. $\log_2 x + \log_2(3x + 1) = 1$ **30.** $\log_3 x + \log_3(x - 8) = 2$

Objective **C** *Solve. See Example 5.*

31. The size of the wolf population at Isle Royale National Park increases at a rate of 4.3% per year. If the size of the current population is 83 wolves, find how many there should be in 5 years. Use $y = y_0 e^{0.043t}$ and round to the nearest whole number.

32. The number of victims of a flu epidemic is increasing at a rate of 7.5% per week. If 20,000 people are currently infected, in how many days can we expect 45,000 people to have the flu? Use $y = y_0 e^{0.075t}$ and round to the nearest whole number.

33. The population of Paraguay is increasing at a rate of 2.5% per year. The population of Paraguay in 2004 was approximately 6,191,000. Use $y = y_0e^{0.025t}$ to estimate the population of Paraguay in 2010. Round to the nearest whole number. (*Source:* Population Reference Bureau)

34. In 2004, 184.1 million people lived in Brazil. The population of Brazil is growing at a rate of 1.3% per year. Find how long it will take the Brazilian population to reach a size of 200 million people. Use $y = y_0e^{0.013t}$ and round to the nearest tenth. (*Source:* Population Reference Bureau)

35. In 2004, Hungary had a population of about 10,032,000. At that time, Hungary's population was declining at a rate of 0.4% per year. At that rate, how long will it take for Hungary's population to decline to 9,000,000? Use $y = y_0e^{-0.004t}$ and round to the nearest tenth. (*Source:* Population Reference Bureau)

36. The population of Russia has been decreasing at the rate of 0.6% per year. There were about 143,800,000 people living in Russia in 2004. How many inhabitants will there be in 2016? Use $y = y_0e^{-0.006t}$. Round to the nearest whole number. (*Source:* Population Reference Bureau)

Use the formula $A = P\left(1 + \dfrac{r}{n}\right)^{nt}$ to solve these compound interest problems. Round to the nearest tenth. See Example 6.

37. How long does it take for $600 to double if it is invested at 7% interest compounded monthly?

38. How long does it take for $600 to double if it is invested at 12% interest compounded monthly?

39. How long does it take for a $1200 investment to earn $200 interest if it is invested at 9% interest compounded quarterly?

40. How long does it take for a $1500 investment to earn $200 interest if it is invested at 10% interest compounded semiannually?

41. How long does it take for $1000 to double if it is invested at 8% interest compounded semiannually?

42. How long does it take for $1000 to double if it is invested at 8% interest compounded monthly?

The formula $w = 0.00185h^{2.67}$ is used to estimate the normal weight w in pounds of a boy h inches tall. Use this formula to solve Exercises 43 and 44. Round to the nearest tenth.

43. Find the expected height of a boy who weighs 85 pounds.

44. Find the expected height of a boy who weighs 140 pounds.

The formula $P = 14.7e^{-0.21x}$ gives the average atmospheric pressure P, in pounds per square inch, at an altitude x, in miles above sea level. Use this formula to solve Exercises 45 through 58. Round to the nearest tenth.

45. Find the average atmospheric pressure of Denver, which is 1 mile above sea level.

46. Find the average atmospheric pressure of Pikes Peak, which is 2.7 miles above sea level.

47. Find the elevation of a Delta jet if the atmospheric pressure outside the jet is 7.5 pounds per square inch.

48. Find the elevation of a remote Himalayan peak if the atmospheric pressure atop the peak is 6.5 pounds per square inch.

Psychologists call the graph of the formula $t = \dfrac{1}{c}\ln\left(\dfrac{A}{A - N}\right)$ *the learning curve since the formula relates time t passed, in weeks, to a measure N of learning achieved, to a measure A of maximum learning possible, and to a measure c of an individual's learning style. Use this formula to answer Exercises 49 through 52. Round to the nearest whole number.*

49. Norman Weidner is learning to type. If he wants to type at a rate of 50 words per minute ($N = 50$) and his expected maximum rate is 75 words per minute ($A = 75$), how many weeks should it take him to achieve his goal? Assume that c is 0.09.

50. An experiment of teaching chimpanzees sign language shows that a typical chimp can master a maximum of 65 signs. How many weeks should it take a chimpanzee to master 30 signs if c is 0.03?

51. Janine Jenkins is working on her dictation skills. She wants to take dictation at a rate of 150 words per minute and believes that the maximum rate she can hope for is 210 words per minute. How many weeks should it take her to achieve the 150-word level if c is 0.07?

52. A psychologist is measuring human capability to memorize nonsense syllables. How many weeks should it take a subject to learn 15 nonsense syllables if the maximum possible to learn is 24 syllables and c is 0.17?

Review

If $x = -2$, $y = 0$, and $z = 3$, find the value of each expression. See Section 1.5.

53. $\dfrac{x^2 - y + 2z}{3x}$

54. $\dfrac{x^3 - 2y + z}{2z}$

55. $\dfrac{3z - 4x + y}{x + 2z}$

56. $\dfrac{4y - 3x + z}{2x + y}$

Concept Extensions

The formula $y = y_0 e^{kt}$ gives the population size y of a population that experiences an annual rate of population growth k (given as a decimal). In this formula, t is time in years and y_0 is the initial population at time 0. Use this formula to solve Exercises 57 and 58.

57. In 2000, the population of Arizona was 5,130,632. By 2003, the population had grown to 5,880,811. Find the annual rate of population growth over this period. Round your answer to the nearest tenth of a percent. (*Source:* U.S. Census Bureau, State of Arizona)

58. In 2000, the population of Nevada was 1,998,257. By 2003, the population had grown to 2,241,154. Find the annual rate of population growth over this period. Round your answer to the nearest tenth of a percent. (*Source:* U.S. Census Bureau, State of Nevada)

59. When solving a logarithmic equation, explain why you must check possible solutions in the original equation.

60. Solve $5^x = 9$ by taking the common logarithm of both sides of the equation. Next, solve this equation by taking the natural logarithm of both sides. Compare your solutions. Are they the same? Why or why not?

Use a graphing calculator to solve. Round your answers to two decimal places.

61. $e^{0.3x} = 8$

62. $10^{0.5x} = 7$

THE BIGGER PICTURE   Solving Equations and Inequalities

Continue your outline from Sections 2.6, 5.7, 6.4, 7.6, and 8.4. Write how to recognize and how to solve exponential and logarithmic equations in your own words. For example:

Solving Equations and Inequalities

I. Equations

  A. Linear equations (Sec. 2.1)

  B. Absolute value equations (Sec. 2.6)

  C. Quadratic and higher degree equations (Sec. 5.7 and Chapter 8)

  D. Equations with rational expressions (Sec. 6.4)

  E. Equations with radicals (Sec. 7.6)

  F. Exponential Equations—equations with variables in the exponent.

1. If we can write both expressions with the same base, then set the exponents equal to each other and solve

$$9^x = 27^{x+1}$$
$$(3^2)^x = (3^3)^{x+1}$$
$$3^{2x} = 3^{3x+3}$$
$$2x = 3x + 3$$
$$-3 = x$$

2. If we can't write both expressions with the same base, then solve using logarithms

$$5^x = 7$$
$$\log 5^x = \log 7$$
$$x \log 5 = \log 7$$
$$x = \frac{\log 7}{\log 5}$$
$$x \approx 1.2091$$

  G. Logarithmic Equations—equations with logarithms of variable expressions

$\log 7 + \log (x + 3) = 2$ Write equation so that single logarithm on
$\log 7(x + 3) = 2$ one side and constant on the other side.
$10^2 = 7(x + 3)$ Use definition of logarithm.
$100 = 7x + 21$ Multiply.
$79 = 7x$
$\dfrac{79}{7} = x$ Solve.

II. Inequalities

  A. Linear inequalities (Sec. 2.4)

  B. Compound inequalities (Sec. 2.5)

  C. Absolute value inequalities (Sec. 2.6)

  D. Nonlinear inequalities (Sec. 8.4)

   1. Polynomial inequalities

   2. Rational inequalities

Solve. Write solutions to inequalities in interval notation.

1. $8^x = 2^{x-3}$

2. $11^x = 5$

3. $-7x + 3 \le -5x + 13$

4. $-7 \le 3x + 6 \le 0$

5. $|5y + 3| < 3$

6. $(x - 6)(5x + 1) = 0$

7. $\log_{13} 8 + \log_{13}(x - 1) = 1$

8. $\left| \dfrac{3x - 1}{4} \right| = 2$

9. $|7x + 1| > -2$

10. $x^2 = 4$

11. $(x + 5)^2 = 3$

12. $\log_7(4x^2 - 27x) = 1$

CHAPTER 9   Group Activity

Sound Intensity

The decibel (dB) measures sound intensity, or the relative loudness or strength of a sound. One decibel is the smallest difference in sound levels that is detectable by humans. The decibel is a logarithmic unit. This means that for approximately every 3-decibel increase in sound intensity, the relative loudness of the sound is doubled. For example, a 35 dB sound is twice as loud as a 32 dB sound.

In the modern world, noise pollution has increasingly become a concern. Sustained exposure to high sound intensities can lead to hearing loss. Regular exposure to 90 dB sounds can eventually lead to loss of hearing. Sounds of 130 dB and more can cause permanent loss of hearing instantaneously.

The relative loudness of a sound D in decibels is given by the equation

$$D = 10 \log_{10} \frac{I}{10^{-16}}$$

where I is the intensity of a sound given in watts per square centimeter. Some sound intensities of common noises are listed in the table in order of increasing sound intensity.

Group Activity

1. Work together to create a table of the relative loudness (in decibels) of the sounds listed in the table.

2. Research the loudness of other common noises. Add these sounds and their decibel levels to your table. Be sure to list the sounds in order of increasing sound intensity.

Some Sound Intensities of Common Noises	
Noise	Intensity (watts/cm²)
Whispering	10^{-15}
Rustling leaves	$10^{-14.2}$
Normal conversation	10^{-13}
Background noise in a quiet residence	$10^{-12.2}$
Typewriter	10^{-11}
Air conditioning	10^{-10}
Freight train at 50 feet	$10^{-8.5}$
Vacuum cleaner	10^{-8}
Nearby thunder	10^{-7}
Air hammer	$10^{-6.5}$
Jet plane at takeoff	10^{-6}
Threshold of pain	10^{-4}

STUDY SKILLS BUILDER

Are You Prepared for a Test on Chapter 9?

Below I have listed some common trouble areas for students in Chapter 9. After studying for your test—but before taking your test—read these.

- Don't forget how to find the composition of two functions.

  If $f(x) = x^2 + 5$ and $g(x) = 3x$, then
  $$(f \circ g)(x) = f[g(x)] = f(3x)$$
  $$= (3x)^2 + 5 = 9x^2 + 5$$
  $$(g \circ f)(x) = g[f(x)] = g(x^2 + 5)$$
  $$= 3(x^2 + 5) = 3x^2 + 15$$

- Don't forget that f^{-1} is a special notation used to denote the inverse of a function.

  Let's find the inverse of the one-to-one function $f(x) = 3x - 5$.

  $$f(x) = 3x - 5$$
  $$y = 3x - 5 \quad \text{Replace } f(x) \text{ with } y.$$
  $$x = 3y - 5 \quad \text{Interchange } x \text{ and } y.$$
  $$x + 5 = 3y$$
  $$\frac{x + 5}{3} = y \quad \text{Solve for } y.$$
  $$f^{-1}(x) = \frac{x + 5}{3} \quad \text{Replace } y \text{ with } f^{-1}(x).$$

- Don't forget that $y = \log_b x$ means $b^y = x$.

  Thus, $3 = \log_5 125$ means $5^3 = 125$.

- Remember rules for logarithms.

  $$\log_b 3x = \log_b 3 + \log_b x$$
  $\log_b (3 + x)$ cannot be simplified in the same manner.

Remember: This is simply a checklist of common trouble areas. For a review of Chapter 9, see the Highlights and Chapter Review at the end of this chapter.

Chapter 9 Vocabulary Check

Fill in each blank with one of the words or phrases listed below.

inverse   common   composition   symmetric   exponential

vertical   logarithmic   natural   horizontal

1. For a one-to-one function, we can find its _____ function by switching the coordinates of the ordered pairs of the function.
2. The _____ of functions f and g is $(f \circ g)(x) = f(g(x))$.
3. A function of the form $f(x) = b^x$ is called an _____ function if $b > 0$, b is not 1, and x is a real number.
4. The graphs of f and f^{-1} are _____ about the line $y = x$.
5. _____ logarithms are logarithms to base e.
6. _____ logarithms are logarithms to base 10.
7. To see whether a graph is the graph of a one-to-one function, apply the _____ line test to see whether it is a function, and then apply the _____ line test to see whether it is a one-to-one function.
8. A _____ function is a function that can be defined by $f(x) = \log_b x$ where x is a positive real number, b is a constant positive real number, and b is not 1.

Helpful Hint

Are you preparing for your test? Don't forget to take the Chapter 9 Test on page 705. Then check your answers at the back of the text and use the Chapter Test Prep Video CD to see the fully worked-out solutions to any of the exercises you want to review.

Chapter Highlights

DEFINITIONS AND CONCEPTS	EXAMPLES
Section 9.1 The Algebra of Functions	

ALGEBRA OF FUNCTIONS

Let f and g be functions.

Sum	$(f + g)(x) = f(x) + g(x)$
Difference	$(f - g)(x) = f(x) - g(x)$
Product	$(f \cdot g)(x) = f(x) \cdot g(x)$
Quotient	$\left(\dfrac{f}{g}\right)(x) = \dfrac{f(x)}{g(x)}, g(x) \neq 0$

If $f(x) = 7x$ and $g(x) = x^2 + 1$,

$$(f + g)(x) = f(x) + g(x) = 7x + x^2 + 1$$
$$(f - g)(x) = f(x) - g(x) = 7x - (x^2 + 1)$$
$$= 7x - x^2 - 1$$
$$(f \cdot g)(x) = f(x) \cdot g(x) = 7x(x^2 + 1)$$
$$= 7x^3 + 7x^2$$
$$\left(\frac{f}{g}\right)(x) = \frac{f(x)}{g(x)} = \frac{7x}{x^2 + 1}$$

COMPOSITE FUNCTIONS

The notation $(f \circ g)(x)$ means "f composed with g."

$$(f \circ g)(x) = f(g(x))$$
$$(g \circ f)(x) = g(f(x))$$

If $f(x) = x^2 + 1$ and $g(x) = x - 5$, find $(f \circ g)(x)$.

$$(f \circ g)(x) = f(g(x))$$
$$= f(x - 5)$$
$$= (x - 5)^2 + 1$$
$$= x^2 - 10x + 26$$

DEFINITIONS AND CONCEPTS	**EXAMPLES**

Section 9.2   Inverse Functions

ONE-TO-ONE FUNCTION

If f is a function, then f is a **one-to-one function** only if each y-value (output) corresponds to only one x-value (input).

HORIZONTAL LINE TEST

If every horizontal line intersects the graph of a function at most once, then the function is a one-to-one function.

Determine whether each graph is a one-to-one function.

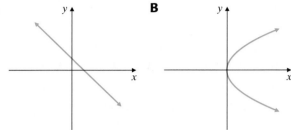

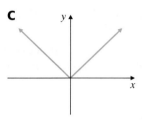

Graphs A and C pass the vertical line test, so only these are graphs of functions. Of graphs A and C, only graph A passes the horizontal line test, so only graph A is the graph of a one-to-one function.

The **inverse** of a one-to-one function f is the one-to-one function f^{-1} that is the set of all ordered pairs (b, a) such that (a, b) belongs to f.

FINDING THE INVERSE OF A ONE-TO-ONE FUNCTION f

Step 1. Replace $f(x)$ with y.
Step 2. Interchange x and y.
Step 3. Solve for the equation for y.
Step 4. Replace y with the notation $f^{-1}(x)$.

Find the inverse of $f(x) = 2x + 7$.

$$y = 2x + 7 \qquad \text{Replace } f(x) \text{ with } y.$$
$$x = 2y + 7 \qquad \text{Interchange } x \text{ and } y.$$
$$2y = x - 7 \qquad \text{Solve for } y.$$
$$y = \frac{x - 7}{2}$$

$$f^{-1}(x) = \frac{x - 7}{2} \qquad \text{Replace } y \text{ with } f^{-1}(x).$$

The inverse of $f(x) = 2x + 7$ is $f^{-1}(x) = \dfrac{x - 7}{2}$.

Section 9.3   Exponential Functions

EXPONENTIAL FUNCTION

A function of the form $f(x) = b^x$ is an **exponential function,** where $b > 0$, $b \neq 1$, and x is a real number.

Graph the exponential function $y = 4^x$.

x	y
-2	$\dfrac{1}{16}$
-1	$\dfrac{1}{4}$
0	1
1	4
2	16

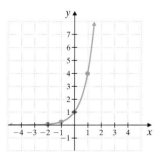

DEFINITIONS AND CONCEPTS	**EXAMPLES**

Section 9.3   Exponential Functions (*continued*)

UNIQUENESS OF b^x

If $b > 0$ and $b \neq 1$, then $b^x = b^y$ is equivalent to $x = y$.

Solve: $2^{x+5} = 8$

$\quad\quad 2^{x+5} = 2^3$ Write 8 as 2^3.

$\quad\quad x + 5 = 3$ Use the uniqueness of b^x.

$\quad\quad\quad\quad x = -2$ Subtract 5 from both sides.

Section 9.4   Logarithmic Functions

LOGARITHMIC DEFINITION

If $b > 0$ and $b \neq 1$, then

$\quad y = \log_b x$ means $x = b^y$

for any positive number x and real number y.

LOGARITHMIC FORM	**CORRESPONDING EXPONENTIAL STATEMENT**
$\log_5 25 = 2$	$5^2 = 25$
$\log_9 3 = \dfrac{1}{2}$	$9^{1/2} = 3$

PROPERTIES OF LOGARITHMS

If b is a real number, $b > 0$ and $b \neq 1$, then

$\quad \log_b 1 = 0, \quad \log_b b^x = x, \quad$ and $\quad b^{\log_b x} = x$

$\log_5 1 = 0, \quad \log_7 7^2 = 2, \quad$ and $\quad 3^{\log_3 6} = 6$

LOGARITHMIC FUNCTION

If $b > 0$ and $b \neq 1$, then a **logarithmic function** is a function that can be defined as

$\quad f(x) = \log_b x$

The domain of f is the set of positive real numbers, and the range of f is the set of real numbers.

Graph: $y = \log_3 x$

Write $y = \log_3 x$ as $3^y = x$. Plot the ordered pair solutions listed in the table, and connect them with a smooth curve.

x	y
3	1
1	0
$\dfrac{1}{3}$	-1
$\dfrac{1}{9}$	-2

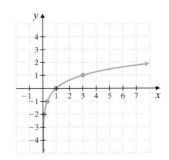

Section 9.5   Properties of Logarithms

Let x, y, and b be positive numbers, $b \neq 1$, and r is a real number.

PRODUCT PROPERTY

$\quad \log_b xy = \log_b x + \log_b y$

QUOTIENT PROPERTY

$\quad \log_b \dfrac{x}{y} = \log_b x - \log_b y$

POWER PROPERTY

$\quad \log_b x^r = r \log_b x$

Write as a single logarithm:

$\quad 2\log_5 6 + \log_5 x - \log_5(y + 2)$

$\quad = \log_5 6^2 + \log_5 x - \log_5(y + 2)$ Power property

$\quad = \log_5 36 \cdot x - \log_5(y + 2)$ Product property

$\quad = \log_5 \dfrac{36x}{y + 2}$ Quotient property

Definitions and Concepts	**Examples**
Section 9.6 Common Logarithms, Natural Logarithms, and Change of Base	

Definitions and Concepts	**Examples**
Common Logarithms $$\log x \quad \text{means} \quad \log_{10} x$$ **Natural Logarithms** $$\ln x \quad \text{means} \quad \log_e x$$ **Continuously Compounded Interest Formula** $$A = Pe^{rt}$$ where r is the annual interest rate for P dollars invested for t years. **Change of Base Formula** If $a, b,$ and c are positive real numbers and neither b nor c is 1, then $$\log_b a = \frac{\log_c a}{\log_c b}$$	$$\log 5 = \log_{10} 5 \approx 0.6990$$ $$\ln 7 = \log_e 7 \approx 1.9459$$ Find the amount in an account at the end of 3 years if $1000 is invested at an interest rate of 4% compounded continuously. Here, $t = 3$ years, $P = \$1000,$ and $r = 0.04.$ $$A = Pe^{rt}$$ $$= 1000e^{0.04(3)}$$ $$\approx \$1127.50$$

Section 9.7 Exponential and Logarithmic Equations and Problem Solving	

Definitions and Concepts	**Examples**
Logarithm Property of Equality Let $\log_b a$ and $\log_b c$ be real numbers and $b \neq 1.$ Then $$\log_b a = \log_b c \quad \text{is equivalent to} \quad a = c$$	Solve: $2^x = 5$ $$\log 2^x = \log 5 \qquad \text{Log property of equality}$$ $$x \log 2 = \log 5 \qquad \text{Power property}$$ $$x = \frac{\log 5}{\log 2} \qquad \text{Divide both sides by log 2.}$$ $$x \approx 2.3219 \qquad \text{Use a calculator.}$$

9 CHAPTER REVIEW

(9.1) *If $f(x) = x - 5$ and $g(x) = 2x + 1$, find the following.*

1. $(f + g)(x)$ **2.** $(f - g)(x)$ **3.** $(f \cdot g)(x)$ **4.** $\left(\dfrac{g}{f}\right)(x)$

If $f(x) = x^2 - 2$, $g(x) = x + 1$, and $h(x) = x^3 - x^2$, find each composition.

5. $(f \circ g)(x)$ **6.** $(g \circ f)(x)$ **7.** $(h \circ g)(2)$

8. $(f \circ f)(x)$ **9.** $(f \circ g)(-1)$ **10.** $(h \circ h)(2)$

(9.2) *Determine whether each function is a one-to-one function. If it is one-to-one, list the elements of its inverse.*

11. $h = \{(-9, 14), (6, 8), (-11, 12), (15, 15)\}$ **12.** $f = \{(-5, 5), (0, 4), (13, 5), (11, -6)\}$

13.

U.S. Region (Input)	West	Midwest	South	Northeast
Rank in Automobile Thefts (Output)	2	4	1	3

14.

Shape (Input)	Square	Triangle	Parallelogram	Rectangle
Number of Sides (Output)	4	3	4	4

Determine whether each function is a one-to-one function.

15.

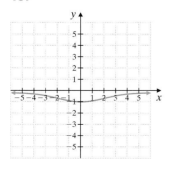

16.

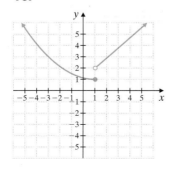

17.

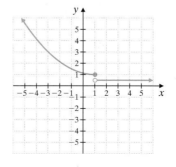

18.

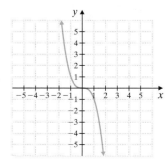

Find an equation defining the inverse function of each one-to-one function.

19. $f(x) = 6x + 11$ **20.** $f(x) = 12x$ **21.** $f(x) = 3x - 5$ **22.** $f(x) = 2x + 1$

Graph each one-to-one function and its inverse on the same set of axes.

23. $f(x) = -2x + 3$

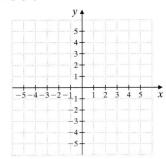

24. $f(x) = 5x - 5$

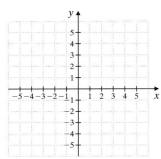

(9.3) *Solve each equation.*

25. $4^x = 64$ **26.** $2^{3x} = \dfrac{1}{16}$ **27.** $9^{x+1} = 243$

Graph each exponential function.

28. $y = 3^x$ **29.** $y = \left(\dfrac{1}{3}\right)^x$ **30.** $y = 2^{x-4}$ **31.** $y = 2^x + 4$

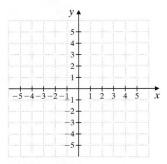   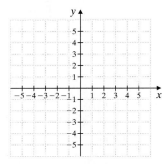   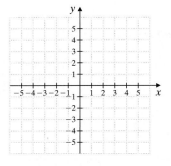   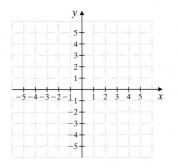

Use the formula $A = P\left(1 + \dfrac{r}{n}\right)^{nt}$ to solve Exercises 32 and 33. In this formula,

    A = amount accrued (or owed)
    P = principal invested (or loaned)
    r = rate of interest
    n = number of compounding periods per year
    t = time in years

32. Find the amount accrued if \$1600 is invested at 9% interest compounded semiannually for 7 years.

33. A total of \$800 is invested in a 7% certificate of deposit for which interest is compounded quarterly. Find the value that this certificate will have at the end of 5 years.

(9.4) *Write each exponential equation with logarithmic notation.*

34. $49 = 7^2$

35. $2^{-4} = \dfrac{1}{16}$

Write each logarithmic equation with exponential notation.

36. $\log_{1/2} 16 = -4$

37. $\log_{0.4} 0.064 = 3$

Solve.

38. $\log_4 x = -3$

39. $\log_3 x = 2$

40. $\log_3 1 = x$

41. $\log_x 64 = 2$

42. $\log_4 4^5 = x$

43. $\log_7 7^{-2} = x$

44. $5^{\log_5 4} = x$

45. $2^{\log_2 9} = x$

46. $\log_3 (2x + 5) = 2$

47. $\log_8 (x^2 + 7x) = 1$

Graph each pair of equations on the same set of axes.

48. $y = 2^x;\ y = \log_2 x$

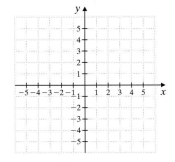

49. $y = \left(\dfrac{1}{2}\right)^x;\ y = \log_{1/2} x$

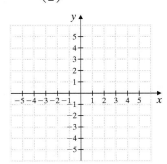

(9.5) *Write each expression as a single logarithm.*

50. $\log_3 8 + \log_3 4$ **51.** $\log_2 6 + \log_2 3$ **52.** $\log_7 15 - \log_7 20$

53. $\log_e 18 - \log_e 12$ **54.** $\log_{11} 8 + \log_{11} 3 - \log_{11} 6$ **55.** $\log_5 14 + \log_5 3 - \log_5 21$

56. $2\log_5 x - 2\log_5 (x + 1) + \log_5 x$ **57.** $4\log_3 x - \log_3 x + \log_3 (x + 2)$

Use properties of logarithms to write each expression as a sum or difference of logarithms.

58. $\log_3 \dfrac{x^3}{x + 2}$ **59.** $\log_4 \dfrac{x + 5}{x^2}$

60. $\log_2 \dfrac{3x^2 y}{z}$ **61.** $\log_7 \dfrac{yz^3}{x}$

If $\log_b 2 = 0.36$ *and* $\log_b 5 = 0.83,$ *evaluate each expression.*

62. $\log_b 50$ **63.** $\log_b \dfrac{4}{5}$

(9.6) *Use a calculator to approximate each logarithm to four decimal places.*

64. $\log 3.6$ **65.** $\log 0.15$ **66.** $\ln 1.25$ **67.** $\ln 4.63$

Find the exact value of each logarithm.

68. $\log 1000$ **69.** $\log \dfrac{1}{10}$ **70.** $\ln \dfrac{1}{e}$ **71.** $\ln e^4$

Solve each equation. Give an exact solution and a four-decimal approximation where necessary.

72. $\log 2x = 2$ **73.** $\ln(3x) = 1.6$ **74.** $\ln(2x - 3) = -1$ **75.** $\ln(3x + 1) = 2$

Approximate each logarithm to four decimal places.

76. $\log_5 1.6$ **77.** $\log_3 4$

Use the formula $A = Pe^{rt}$ to solve Exercises 78 and 79, in which interest is compounded continuously. In this formula,

A = amount accrued (or owed)
P = principal invested (or loaned)
r = rate of interest
t = time in years

78. Bank of New York offers a 5-year 6% continuously compounded investment option. Find the amount accrued if $1450 is invested.

79. Find the amount to which a $940 investment grows if it is invested at 11% interest compounded continuously for 3 years.

(9.7) *Solve each exponential equation. Given an exact solution and a four-decimal-place approximation.*

80. $7^x = 20$

81. $3^{2x} = 7$

82. $3^{2x+1} = 6$

83. $8^{4x-2} = 3$

Solve each equation.

84. $\log_5 2 + \log_5 x = 2$

85. $\log 5x - \log(x + 1) = 4$

86. $\log_2 x + \log_2 2x - 3 = 1$

87. $\log_3(x^2 - 8x) = 2$

Use the formula $y = y_0 e^{kt}$ to solve Exercises 88 through 92. In this formula,

y = size of population
y_0 = initial count of population
k = rate of growth, expressed as a decimal
t = time

Round each answer to the nearest whole number.

88. The population of mallard ducks in Nova Scotia is expected to grow at a rate of 6% per week during the spring migration. If 155,000 ducks are already in Nova Scotia, how many are expected by the end of 4 weeks?

89. The population of Sierra Leone is growing at a rate of 2.7% per year. In 2004, the population of Sierra Leone was about 5,844,000. Find the expected population by 2010. Round to the nearest whole. (*Source:* Population Reference Bureau)

90. France is experiencing an annual growth rate of 0.4%. In 2004, the population of France was 60,424,000. How long will it take for the population to reach 65,000,000? Round to the nearest tenth. (*Source:* Population Reference Bureau)

91. In 2004, Australia had a population of 19,913,000. How long will it take Australia to double in population if its growth rate is 0.6% annually? Round to the nearest tenth. (*Source:* Population Reference Bureau)

92. Mexico's population is increasing at a rate of 2.1% per year. How long will it take for its 2004 population of 104,960,000 to double in size? Round to the nearest tenth. (*Source:* Population Reference Bureau)

Use the compound interest equation $A = P\left(1 + \dfrac{r}{n}\right)^{nt}$ to solve Exercises 93 and 94. (See the directions for Exercises 32 and 33 for an explanation of this formula.) Round answers to the nearest tenth.

93. How long does it take for a $5000 investment to grow to $10,000 if it is invested at 8% interest compounded quarterly?

94. An investment of $6000 has grown to $10,000 while the money was invested at 6% interest compounded monthly. How long was it invested?

Mixed Review

Solve each equation.

95. $3^x = \dfrac{1}{9}$

96. $5^{2x} = 125$

97. $8^{3x-2} = 4$

98. $\log_4 64 = x$

99. $\log_x 81 = 4$

100. $\log_4(x^2 - 3x) = 1$

101. $\log_2(3x - 1) = 4$

102. $\ln x = -1.2$

103. $\log_3 x + \log_3 10 = 2$

104. $\ln 3x - \ln(x - 3) = 2$

105. $\log_6 x - \log_6(4x + 7) = 1$

9 CHAPTER TEST

 Remember to use the Chapter Test Prep Video CD to see the fully worked-out solutions to any of the exercises you want to review.

Answers

If $f(x) = x$ and $g(x) = 2x - 3$, find the following.

1. $(f \cdot g)(x)$

2. $(f - g)(x)$

If $f(x) = x$, $g(x) = x - 7$, and $h(x) = x^2 - 6x + 5$, find each composition.

3. $(f \circ h)(0)$

4. $(g \circ f)(x)$

5. $(g \circ h)(x)$

1. _____

2. _____

Graph the one-to-one function and its inverse on the same set of axes.

6. $f(x) = 7x - 14$

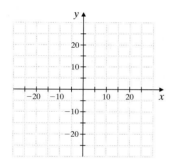

3. _____

4. _____

Determine whether each graph is the graph of a one-to-one function.

7.

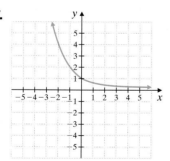

8.

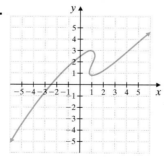

5. _____

6. see graph

7. _____

8. _____

9. _____

Determine whether each function is one-to-one. If it is one-to-one, find an equation or a set of ordered pairs that defines the inverse function of the given function.

9. $y = 6 - 2x$

10. $f = \{(0,0), (2,3), (-1,5)\}$

10. _____

705

11.

Word (Input)	Dog	Cat	House	Desk	Circle
First Letter of Word (Output)	d	c	h	d	c

Use the properties of logarithms to write each expression as a single logarithm.

12. $\log_3 6 + \log_3 4$

13. $\log_5 x + 3\log_5 x - \log_5(x + 1)$

14. Write the expression $\log_6 \dfrac{2x}{y^3}$ as the sum or difference of logarithms.

15. If $\log_b 3 = 0.79$ and $\log_b 5 = 1.16$, find the value of $\log_b \dfrac{3}{25}$.

16. Approximate $\log_7 8$ to four decimal places.

17. Solve $8^{x-1} = \dfrac{1}{64}$ for x. Give an exact solution.

18. Solve $3^{2x+5} = 4$ for x. Give an exact solution and a four-decimal-place approximation.

Solve each logarithmic equation. Give an exact solution.

19. $\log_3 x = -2$

20. $\ln \sqrt{e} = x$

21. $\log_8(3x - 2) = 2$

22. $\log_5 x + \log_5 3 = 2$

23. $\log_4(x + 1) - \log_4(x - 2) = 3$

24. Solve $\ln(3x + 7) = 1.31$ accurate to four decimal places.

25. Graph $f(x) = \left(\dfrac{1}{2}\right)^x + 1$.

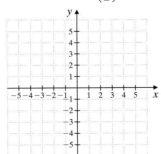

26. Graph the functions $y = 3^x$ and $y = \log_3 x$ on the same set of axes.

Use the formula $A = P\left(1 + \dfrac{r}{n}\right)^{nt}$ to solve Exercises 27 and 28.

27. Find the amount in an account in which $4000 is invested for 3 years at 9% interest compounded monthly.

28. How long will it take $2000 to grow to $3000 if the money is invested at 7% interest compounded semiannually? Round to the nearest whole.

11. _____

12. _____

13. _____

14. _____

15. _____

16. _____

17. _____

18. _____

19. _____

20. _____

21. _____

22. _____

23. _____

24. _____

25. see graph

26. see graph

27. _____

28. _____

Use the population growth formula $y = y_0 e^{kt}$ *to solve Exercises 29 and 30.*

29. The prairie dog population of the Grand Forks area now stands at 57,000 animals. If the population is growing at a rate of 2.6% annually, how many prairie dogs will there be in that area 5 years from now?

30. In an attempt to save an endangered species of wood duck, naturalists would like to increase the wood duck population from 400 to 1000 ducks. If the annual population growth rate is 6.2%, how long will it take the naturalists to reach their goal? Round to the nearest whole year.

The reliability of a new model of CD player can be described by the exponential function $R(t) = 2.7^{-(1/3)t}$, *where the reliability R is the probability (as a decimal) that the CD player is still working t years after it is manufactured. Round answers to the nearest hundredth. Then write your answers as percents.*

31. What is the probability that the CD player will still work half a year after it is manufactured?

32. What is the probability that the CD player will still work 2 years after it is manufactured?

29. _____

30. _____

31. _____

32. _____

Find each root.

1. $\sqrt[3]{27}$

2. $\sqrt[3]{64x^7y^2}$

3. $\sqrt[4]{16}$

4. $\sqrt[4]{162a^4b^5}$

△ **5.** The measure of the largest angle of a triangle is $80°$ more than the measure of the smallest angle, and the measure of the remaining angle is $10°$ more than the measure of the smallest angle. Find the measure of each angle.

6. Line l and line m are parallel lines cut by transversal t. Find the values of x and y.

7. Factor: $7x(x^2 + 5y) - (x^2 + 5y)$

8. Solve: $\dfrac{1}{3}(x - 2) = \dfrac{1}{4}(x + 1)$

9. Subtract: $\dfrac{5k}{k^2 - 4} - \dfrac{2}{k^2 + k - 2}$

10. Perform the indicated operation and simplify if possible.

$$\frac{5}{x - 2} + \frac{3}{x^2 + 4x + 4} - \frac{6}{x + 2}$$

If $f(x) = \sqrt{x - 4}$ and $g(x) = \sqrt[3]{x + 2}$, find each function value.

11. $f(8)$

12. $f(28)$

13. $g(-1)$

14. $g(0)$

Use the properties of exponents to simplify. Write all results using positive exponents.

15. $x^{1/2}x^{1/3}$

16. $(a^{-2}b^3c^{-4})^{-2}$

17. $\dfrac{(2x^{2/5})^5}{x^2}$

18. $\left(\dfrac{3^{-2}}{x}\right)^{-3}$

Use the quotient rule to simplify. Assume that all variables represent positive real numbers.

19. $\sqrt{\dfrac{x}{9}}$

20. $\dfrac{\sqrt{108a^2}}{3\sqrt{3}}$

21. $\sqrt[4]{\dfrac{3}{16y^4}}$

22. $\dfrac{3\sqrt[3]{81a^5b^{10}}}{\sqrt[3]{3b^4}}$

Multiply.

23. $(\sqrt{2x} + 5)(\sqrt{2x} - 5)$ **24.** $\left(\dfrac{1}{2}x + 3\right)\left(\dfrac{1}{2}x - 3\right)$ **25.** $(\sqrt{3} - 1)^2$

26. Multiply.

   a. $a^{1/4}(a^{3/4} - a^8)$

   b. $(x^{1/2} - 3)(x^{1/2} + 5)$

27. Rationalize the numerator of $\dfrac{\sqrt{7}}{\sqrt{45}}$.

28. Rationalize the denominator. $\sqrt[3]{\dfrac{27}{m^4 n^8}}$

29. Solve: $\sqrt{4 - x} = x - 2$

30. Find the equation of a line through $(-2, 6)$ and perpendicular to $f(x) = -3x + 4$. Write the equation using function notation.

31. Add: $(2 + 3i) + (-3 + 2i)$

32. Find the following powers of i.

   a. i^8 **c.** i^{42}

   b. i^{21} **d.** i^{-13}

33. Solve $4x^2 - 24x + 41 = 0$ by completing the square.

34. Solve $4x^2 + 8x - 1 = 0$ by completing the square.

35. Solve: $\dfrac{1}{4}m^2 - m + \dfrac{1}{2} = 0$

36. Solve by using the quadratic formula.

$$\left(x - \dfrac{1}{2}\right)^2 = \dfrac{x}{2}$$

37. Solve: $x^{2/3} - 5x^{1/3} + 6 = 0$

38. Solve:

$$\dfrac{1}{a + 5} = \dfrac{1}{3a + 6} - \dfrac{a + 2}{a^2 + 7a + 10}$$

39. Solve: $\dfrac{5}{x + 1} < -2$

40. Find the length of the unknown side of the triangle.

41. Graph $f(x) = 3x^2 + 3x + 1$. Find the vertex and any intercepts.

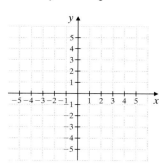

23. _____

24. _____

25. _____

26. a. _____

    b. _____

27. _____

28. _____

29. _____

30. _____

31. _____

32. a. _____ **b.** _____

    c. _____

    d. _____

33. _____

34. _____

35. _____

36. _____

37. _____

38. _____

39. _____

40. _____

41. see graph _____

42. _____

43. a. _____

    b.

    _____

44. a. _____

    b. _____

    c. _____

45. _____

46. _____

47. _____

48. _____

49. _____

50. a. _____

    b. _____

51. _____

52. _____

53. _____

54. _____

42. Divide $x^3 - 8$ by $x - 2$.

43. If $f(x) = |x|$ and $g(x) = x - 2$, find
    a. $(f \circ g)(x)$
    b. $(g \circ f)(x)$

44. Simplify each complex fraction.

    a. $\dfrac{\dfrac{a}{5}}{\dfrac{a-1}{10}}$ **b.** $\dfrac{\dfrac{3}{2+a} + \dfrac{6}{2-a}}{\dfrac{5}{a+2} - \dfrac{1}{a-2}}$ **c.** $\dfrac{x^{-1} + y^{-1}}{xy}$

45. Find an equation of the inverse of $f(x) = 3x - 5$.

46. Use synthetic division to divide:
$(8x^2 - 12x - 7) \div (x - 2)$

47. Solve: $4^{x+3} = 8^x$

48. Suppose that y varies directly as x. If $y = \dfrac{1}{2}$ when $x = 12$, find the constant of variation and the variation equation.

49. Solve: $\log_x 25 = 2$

50. Add or subtract as indicated.

    a. $\dfrac{\sqrt{20}}{3} + \dfrac{\sqrt{5}}{4}$

    b. $\sqrt[3]{\dfrac{24x}{27}} - \dfrac{\sqrt[3]{3x}}{2}$

51. Find the amount owed at the end of 5 years if $1600 is loaned at a rate of 9% compounded continuously.

52. Write as a single logarithm:
$7\log_3 x + 9\log_3 y$

53. Solve: $\log(x + 2) - \log x = 2$

54. Solve: $\log_7(8x - 6) = 2$

10

Conic Sections

In Chapter 8, we analyzed some of the important connections between a parabola and its equation. Parabolas are interesting in their own right but are more interesting still because they are part of a collection of curves known as conic sections. This chapter is devoted to quadratic equations in two variables and their conic section graphs: the parabola, circle, ellipse, and hyperbola.

When the sun rises above the horizon on June 21 or 22, thousands of people gather to witness and celebrate summer solstice at Stonehenge. Stonehenge is a megalithic ruin located on the Salisbury Plain in Wiltshire, England. It is a series of earth, timber and stone structures that were constructed, revised, and reconstructed over a period of 1400 years or so. Although no one can say for certain what its true purpose was, there have been multiple theories. The best-known theory is the one from eighteenth century British antiquarian, William Stukeley. He believed that Stonehenge was a temple, possibly an ancient cult center for the Druids. Despite the fact that we don't know its purpose for certain, Stonehenge acts as a prehistoric timepiece, allowing us to theorize what it would have been like during the Neolithic Period, and who could have built this megalithic wonder. In Exercise 47 on page 723, we will explore the dimensions of the outer stone circle, known as the Sarsen circle, or Stonehenge. (*Source:* The Discovery Channel)

Summer Solstice is Celebrated in Many Locations, Including Stonehenge

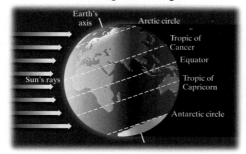

At summer solstice, one hemisphere of the Earth is at a point in its orbit most tilted toward the Sun. It is the day of the year with the longest daylight period, usually occuring on June 21 or 22 in the northern hemisphere and on December 21 or 22 in the southern hemisphere.

A Graph Parabolas of the Forms
$y = a(x - h)^2 + k$ and
$x = a(y - k)^2 + h$.

B Graph Circles of the Form
$(x - h)^2 + (y - k)^2 = r^2$.

C Write the Equation of a Circle,
Given Its Center and Radius.

D Find the Center and the Radius of
a Circle, Given Its Equation.

10.1 THE PARABOLA AND THE CIRCLE

Conic sections are called such because each conic section is the intersection of a right circular cone and a plane. The circle, parabola, ellipse, and hyperbola are the conic sections.

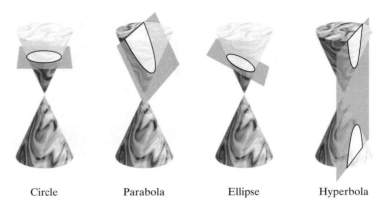

Circle   Parabola   Ellipse   Hyperbola

Objective **A** Graphing Parabolas

Thus far, we have seen that $f(x)$ or $y = a(x - h)^2 + k$ is the equation of a parabola that opens upward if $a > 0$ or downward if $a < 0$. Parabolas can also open left or right, or even on a slant. Equations of these parabolas are not functions of x, of course, since a parabola opening any way other than upward or downward fails the vertical line test. In this section, we introduce parabolas that open to the left and to the right. Parabolas opening on a slant will not be developed in this book.

Just as $y = a(x - h)^2 + k$ is the equation of a parabola that opens upward or downward, $x = a(y - k)^2 + h$ is the equation of a parabola that opens to the right or to the left. The parabola opens to the right if $a > 0$ and to the left if $a < 0$. The parabola has vertex (h, k), and its axis of symmetry is the line $y = k$.

Parabolas

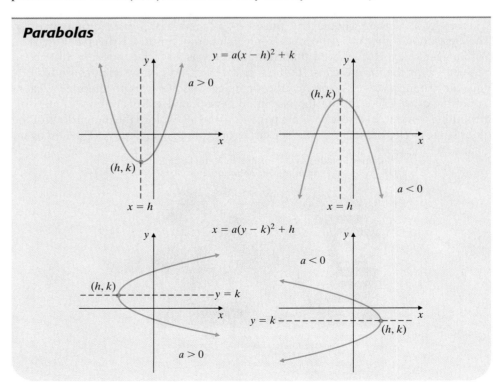

The forms $y = a(x - h)^2 + k$ and $x = a(y - k)^2 + h$ are called **standard forms.**

✔ **Concept Check** Does the graph of the parabola given by the equation $x = -3y^2$ open to the left, to the right, upward, or downward?

EXAMPLE 1 Graph: $x = 2y^2$

Solution: Written in standard form, the equation $x = 2y^2$ is $x = 2(y - 0)^2 + 0$ with $a = 2, h = 0$, and $k = 0$. Its graph is a parabola with vertex $(0, 0)$, and its axis of symmetry is the line $y = 0$. Since $a > 0$, this parabola opens to the right. We use a table to obtain a few more ordered pair solutions to help us graph $x = 2y^2$.

x	y
8	−2
2	−1
0	0
2	1
8	2

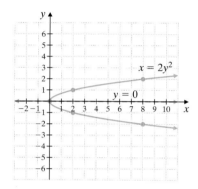

🔲 **Work Practice Problem 1**

EXAMPLE 2 Graph: $x = -3(y - 1)^2 + 2$

Solution: The equation $x = -3(y - 1)^2 + 2$ is in the form $x = a(y - k)^2 + h$ with $a = -3, k = 1$, and $h = 2$. Since $a < 0$, the parabola opens to the left. The vertex (h, k) is $(2, 1)$, and the axis of symmetry is the horizontal line $y = 1$. When $y = 0$, the x-value is -1, so the x-intercept is $(-1, 0)$.

Again, we use a table to obtain a few ordered pair solutions and then graph the parabola.

x	y
2	1
−1	0
−1	2

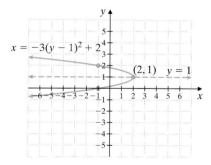

🔲 **Work Practice Problem 2**

PRACTICE PROBLEM 1
Graph: $x = 4y^2$

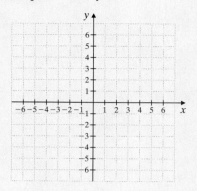

PRACTICE PROBLEM 2
Graph: $x = -2(y - 3)^2 + 1$

Answers
1.

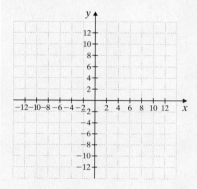

2.

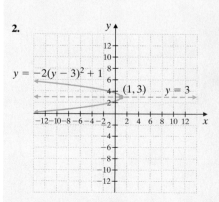

✔ **Concept Check Answer**
to the left

PRACTICE PROBLEM 3

Graph: $y = -x^2 - 4x + 12$

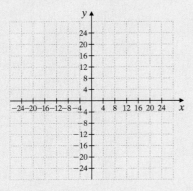

EXAMPLE 3 Graph: $y = -x^2 - 2x + 15$

Solution: Notice that this equation is not written in standard form, $y = a(x - h)^2 + k$. There are two methods that we can use to find the vertex. The first method is completing the square.

$y - 15 = -x^2 - 2x$ Subtract 15 from both sides.

$y - 15 = -1(x^2 + 2x)$ Factor -1 from the terms $-x^2 - 2x$.

The coefficient of x is 2, so we find the square of half of 2.

$$\frac{1}{2}(2) = 1 \quad \text{and} \quad 1^2 = 1$$

$y - 15 - 1(1) = -1(x^2 + 2x + 1)$ Add -1 (1) to both sides.

$y - 16 = -1(x + 1)^2$ Simplify the left side, and factor the right side.

$y = -(x + 1)^2 + 16$ Add 16 to both sides.

The vertex is $(-1, 16)$.

The second method for finding the vertex is by using the expression $\dfrac{-b}{2a}$. Since the equation is quadratic in x, the expression gives us the x-value of the vertex.

$$x = \frac{-(-2)}{2(-1)} = \frac{2}{-2} = -1$$

To find the corresponding y-value of the vertex, replace x with -1 in the original equation.

$$y = -(-1)^2 - 2(-1) + 15 = -1 + 2 + 15 = 16$$

Again, we see that the vertex is $(-1, 16)$, and the axis of symmetry is the vertical line $x = -1$. The y-intercept is $(0, 15)$. Now we can use a few more ordered pair solutions to graph the parabola.

x	y
−5	0
−3	12
−2	15
−1	16
0	15
1	12
3	0

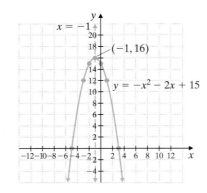

Work Practice Problem 3

Answer

3.

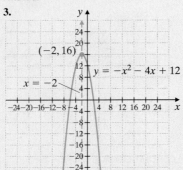

EXAMPLE 4 Graph: $x = 2y^2 + 4y + 5$

Solution: We notice that this equation is quadratic in y so its graph is a parabola that opens to the left or the right. We can complete the square on y or we can use the expression $\dfrac{-b}{2a}$ to find the vertex.

Since the equation is quadratic in y, the expression gives us the y-value of the vertex.

$$y = \frac{-4}{2 \cdot 2} = \frac{-4}{4} = -1$$

$$x = 2(-1)^2 + 4(-1) + 5 = 2 \cdot 1 - 4 + 5 = 3$$

The vertex is $(3, -1)$, and the axis of symmetry is the line $y = -1$. The parabola opens to the right since $a > 0$. The x-intercept is $(5, 0)$.

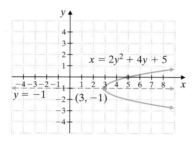

Work Practice Problem 4

Objective B Graphing Circles

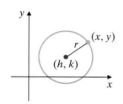

Another conic section is the **circle**. A circle is the set of all points in a plane that are the same distance from a fixed point called the **center**. The distance is called the **radius** of the circle. To find a standard equation for a circle, let (h, k) represent the center of the circle, and let (x, y) represent any point on the circle. The distance between (h, k) and (x, y) is defined to be the radius, r units. We can find this distance r by using the distance formula. (For a review of the distance formula, see Section 7.3.)

$$r = \sqrt{(x - h)^2 + (y - k)^2} \qquad \text{The distance formula.}$$
$$r^2 = (x - h)^2 + (y - k)^2 \qquad \text{Square both sides.}$$

Circle

The graph of $(x - h)^2 + (y - k)^2 = r^2$ is a circle with center (h, k) and radius r.

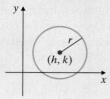

PRACTICE PROBLEM 4

Graph: $x = 3y^2 + 12y + 13$

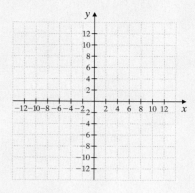

Answer

4.

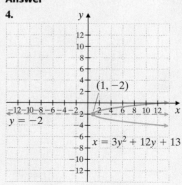

The form $(x - h)^2 + (y - k)^2 = r^2$ is called **standard form.**

If an equation can be written in the standard form

$$(x - h)^2 + (y - k)^2 = r^2$$

then its graph is a circle, which we can draw by graphing the center (h, k) and using the radius r.

> **Helpful Hint**
>
> Notice that the radius is the *distance* from the center of the circle to any point of the circle. Also notice that the *midpoint* of a diameter of a circle is the center of the circle.
>
>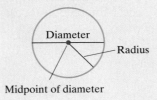
>
> Diameter
>
> Radius
>
> Midpoint of diameter

PRACTICE PROBLEM 5

Graph: $x^2 + y^2 = 36$

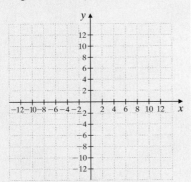

EXAMPLE 5 Graph: $x^2 + y^2 = 4$

Solution: The equation can be written in standard form as

$$(x - 0)^2 + (y - 0)^2 = 2^2$$

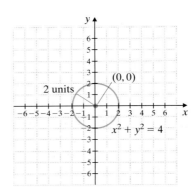

The center of the circle is $(0, 0)$, and the radius is 2. The graph of the circle is shown above.

■ **Work Practice Problem 5**

> **Helpful Hint**
>
> Notice the difference between the equation of a circle and the equation of a parabola. The equation of a circle contains both x^2- and y^2-terms on the same side of the equation with equal coefficients. The equation of a parabola has either an x^2-term or a y^2-term but not both.

Answer

5.

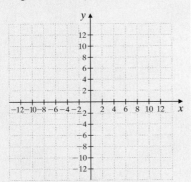

$x^2 + y^2 = 36$

6 units

$(0, 0)$

EXAMPLE 6 Graph: $(x + 1)^2 + y^2 = 8$

Solution: The equation can be written as $(x - (-1))^2 + (y - 0)^2 = 8$ with $h = -1$, $k = 0$, and $r = \sqrt{8}$. The center is $(-1, 0)$, and the radius is $\sqrt{8} = 2\sqrt{2} \approx 2.8$. We use the decimal approximation to approximate the radius when graphing.

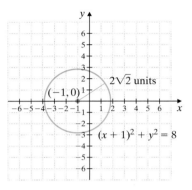

◻ **Work Practice Problem 6**

✔ **Concept Check** In the graph of the equation $(x - 3)^2 + (y - 2)^2 = 5$, what is the distance between the center of the circle and any point on the circle?

Objective C **Writing Equations of Circles**

Since a circle is determined entirely by its center and radius, this information is all we need to write the equation of a circle.

EXAMPLE 7 Write an equation of the circle with center $(-7, 3)$ and radius 10.

Solution: Using the given values $h = -7$, $k = 3$, and $r = 10$, we write the equation

$$(x - h)^2 + (y - k)^2 = r^2$$

or

$$(x - (-7))^2 + (y - 3)^2 = 10^2 \qquad \text{Substitute the given values.}$$

or

$$(x + 7)^2 + (y - 3)^2 = 100$$

◻ **Work Practice Problem 7**

Objective D **Finding the Center and the Radius of a Circle**

To find the center and the radius of a circle from its equation, we write the equation in standard form. To write the equation of a circle in standard form, we complete the square on both x and y.

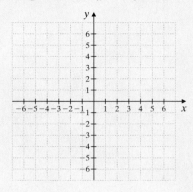
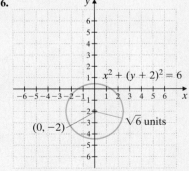

PRACTICE PROBLEM 8

Graph: $x^2 + y^2 - 2x + 6y = 6$

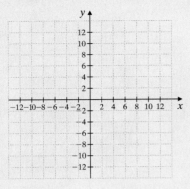

Copyright 2007 Pearson Education, Inc.

Answer

8.

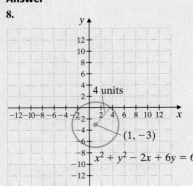

EXAMPLE 8 Graph: $x^2 + y^2 + 4x - 8y = 16$

Solution: Since this equation contains x^2- and y^2-terms on the same side of the equation with equal coefficients, its graph is a circle. To write the equation in standard form, we group the terms involving x and the terms involving y, and then complete the square on each variable.

$$(x^2 + 4x) + (y^2 - 8y) = 16$$

Now, $\frac{1}{2}(4) = 2$ and $2^2 = 4$. Also, $\frac{1}{2}(-8) = -4$ and $(-4)^2 = 16$. We add 4 and then 16 to both sides.

$$(x^2 + 4x + 4) + (y^2 - 8y + 16) = 16 + 4 + 16$$
$$(x + 2)^2 + (y - 4)^2 = 36 \qquad \text{Factor.}$$

This circle has the center $(-2, 4)$ and radius 6, as shown.

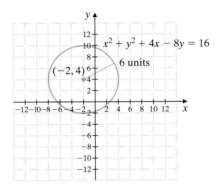

🔲 **Work Practice Problem 8**

CALCULATOR EXPLORATIONS Graphing

To graph an equation such as $x^2 + y^2 = 25$ with a graphing calculator, we first solve the equation for y.

$$x^2 + y^2 = 25$$
$$y^2 = 25 - x^2$$
$$y = \pm\sqrt{25 - x^2}$$

The graph of $y = \sqrt{25 - x^2}$ will be the top half of the circle, and the graph of $y = -\sqrt{25 - x^2}$ will be the bottom half of the circle.

To graph, we press $\boxed{Y=}$ and enter $Y_1 = \sqrt{25 - x^2}$ and $Y_2 = -\sqrt{25 - x^2}$. We insert parentheses about $25 - x^2$ so that $\sqrt{25 - x^2}$ and not $\sqrt{25} - x^2$ is graphed.

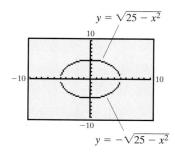

The graph does not appear to be a circle because we are currently using a standard window and the screen is rectangular. This causes the tick marks on the x-axis to be farther apart than the tick marks on the y-axis and thus creates the distorted circle. If we want the graph to appear circular, we define a square window by using a feature of the graphing calculator or redefine the window to show the x-axis from -15 to 15 and the y-axis from -10 to 10. Using a square window, the graph appears as follows:

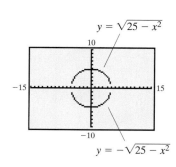

Use a graphing calculator to graph each circle.

1. $x^2 + y^2 = 55$

2. $x^2 + y^2 = 20$

3. $7x^2 + 7y^2 - 89 = 0$

4. $3x^2 + 3y^2 - 35 = 0$

Mental Math

The graph of each equation is a parabola. Determine whether the parabola opens upward, downward, to the left, or to the right.

1. $y = x^2 - 7x + 5$

2. $y = -x^2 + 16$

3. $x = -y^2 - y + 2$

4. $x = 3y^2 + 2y - 5$

5. $y = -x^2 + 2x + 1$

6. $x = -y^2 + 2y - 6$

10.1 EXERCISE SET

FOR EXTRA HELP

Student Solutions Manual PH Math/Tutor Center CD/Video for Review MathXL® MyMathLab

Objective **A** *The graph of each equation is a parabola. Find the vertex of the parabola and then graph it. See Examples 1 through 4.*

1. $x = 3y^2$

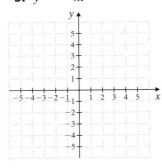

2. $x = 5y^2$

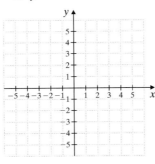

3. $x = -2y^2$

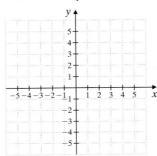

4. $x = -4y^2$

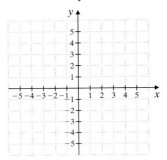

5. $y = -4x^2$

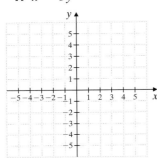

6. $y = -2x^2$

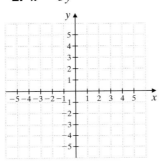

7. $x = (y - 2)^2 + 3$

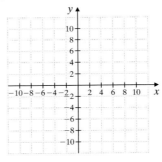

8. $x = (y - 4)^2 - 1$

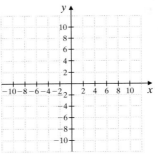

9. $y = -3(x - 1)^2 + 5$

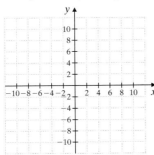

10. $x = -4(y - 2)^2 + 2$

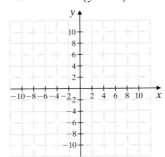

11. $x = y^2 + 6y + 8$

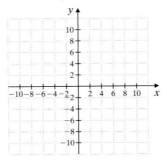

12. $x = y^2 - 6y + 6$

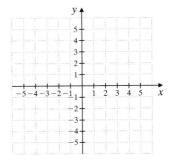

13. $y = x^2 + 10x + 20$

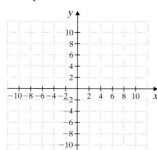

14. $y = x^2 + 4x - 5$

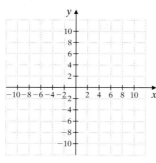

15. $x = -2y^2 + 4y + 6$

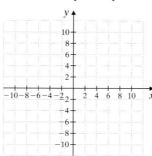

16. $x = 3y^2 + 6y + 7$

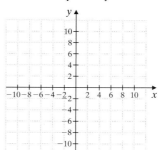

Objectives **B** **D** **Mixed Practice** *The graph of each equation is a circle. Find the center and the radius, and then graph the circle. See Examples 5, 6, and 8.*

17. $x^2 + y^2 = 9$

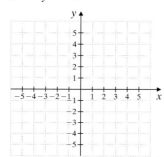

18. $x^2 + y^2 = 25$

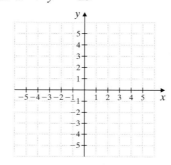

19. $x^2 + (y - 2)^2 = 1$

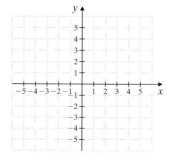

20. $(x - 3)^2 + y^2 = 9$

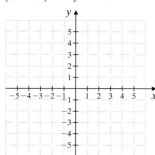

21. $(x - 5)^2 + (y + 2)^2 = 1$

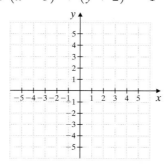

22. $(x + 3)^2 + (y + 3)^2 = 4$

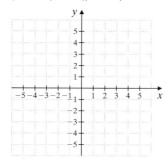

23. $x^2 + y^2 + 6y = 0$

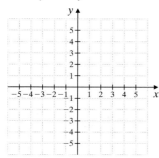

24. $x^2 + 10x + y^2 = 0$

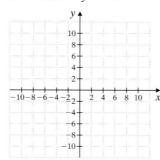

25. $x^2 + y^2 + 2x - 4y = 4$

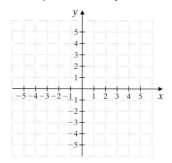

26. $x^2 + y^2 + 6x - 4y = 3$

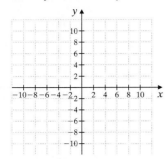

27. $x^2 + y^2 - 4x - 8y - 2 = 0$

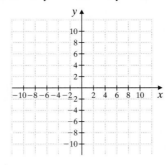

28. $x^2 + y^2 - 2x - 6y - 5 = 0$

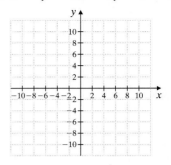

Hint: For Exercises 29 through 32, first divide the equation through by the coefficient of x^2 (or y^2).

29. $3x^2 + 3y^2 = 75$

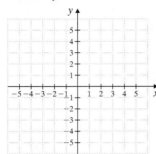

30. $2x^2 + 2y^2 = 18$

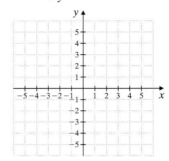

31. $4(x + 1)^2 + 4(y - 3)^2 = 12$

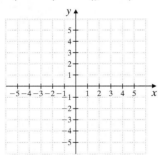

32. $5(x - 2)^2 + 5(y + 1) = 50$

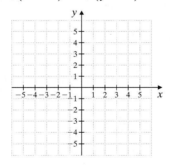

Objective **C** *Write an equation of the circle with the given center and radius. See Example 7.*

33. $(2, 3); 6$

34. $(-7, 6); 2$

35. $(0, 0); \sqrt{3}$

36. $(0, -6); \sqrt{2}$

37. $(-5, 4); 3\sqrt{5}$

38. The origin; $4\sqrt{7}$

Review

Graph each equation. See Section 3.3.

39. $y = 2x + 5$

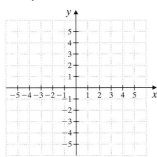

40. $y = -3x + 3$

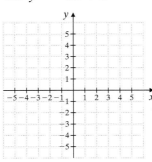

41. $y = 3$

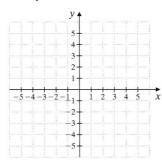

42. $x = -2$

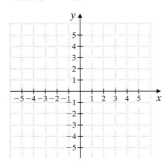

Rationalize each denominator and simplify if possible. See Section 7.5.

43. $\dfrac{1}{\sqrt{3}}$

44. $\dfrac{\sqrt{5}}{\sqrt{8}}$

45. $\dfrac{4\sqrt{7}}{\sqrt{6}}$

46. $\dfrac{10}{\sqrt{5}}$

Concept Extensions

47. The first image that comes to mind when one thinks of Stonehenge is the very large sandstone blocks with sandstone lintels across the top. The Sarsen Circle of Stonehenge is the outer circle of the sandstone blocks, each of which weighs up to 50 tons. There were originally 30 of these monolithic blocks, but only 17 remain upright to this day. The "altar stone" lies at the center of this circle, which has a diameter of 33 meters.

   a. What is the radius of the Sarsen circle?

   b. What is the circumference of the Sarsen circle? Round your result to 2 decimal places.

   c. Since there were originally 30 Sarsen stones located on the circumference, how far apart would they have been? Round to the nearest tenth of a meter.

   d. Using the axes in the drawing, what are the coordinates of the center of the circle?

   e. Use parts **a** and **d** to write the equation of the Sarsen circle.

48. Opened in 2000 to honor the millennium, the British Airways London Eye is the world's biggest observation wheel. Each of the 32 enclosed capsules, which each hold 25 passengers, completes a full rotation every 30 minutes. Its diameter is 135 meters, and it is constructed on London's South Bank, to allow passengers to enter the Eye at ground level.
(*Source: Guinness Book of World Records,* 2005)

   a. What is the radius of the London Eye?

   b. How close is the wheel to the ground?

   c. How high is the center of the wheel from the ground?

   d. Using the axes in the drawing, what are the coordinates of the center of the wheel?

   e. Use parts **a** and **d** to write the equation of the Eye.

49. In 1893, Pittsburgh bridge builder George Ferris designed and built a gigantic revolving steel wheel whose height was 264 feet and diameter was 250 feet. This Ferris wheel opened at the 1893 exposition in Chicago. It had 36 wooden cars, each capable of holding 60 passengers.
(*Source: The Handy Science Answer Book*)

  a. What was the radius of this Ferris wheel?
  b. How close is the wheel to the ground?
  c. How high is the center of the wheel from the ground?
  d. Using the axes in the drawing, what are the coordinates of the center of the wheel?
  e. Use parts **a** and **d** to write the equation of the wheel.

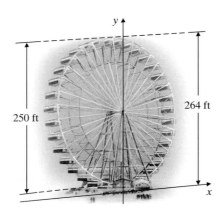

250 ft   264 ft

50. The world's largest-diameter Ferris wheel currently operating is the Cosmoclock 21 at Yokohama City, Japan. It has a 60-armed wheel, its diameter is 100 meters and it has a height of 105 meters.
(*Source: The Handy Science Answer Book*)

  a. What is the radius of this Ferris wheel?
  b. How close is the wheel to the ground?
  c. How high is the center of the wheel from the ground?
  d. Using the axes in the drawing, what are the coordinates of the center of the wheel?
  e. Use parts **a** and **d** to write the equation of the wheel.

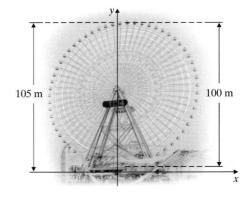

105 m   100 m

51. If you are given a list of equations of circles and parabolas and none are in standard form, explain how you would determine which is an equation of a circle and which is an equation of a parabola. Explain also how you would distinguish the upward or downward parabolas from the left-opening or right-opening parabolas.

52. Determine whether the triangle with vertices $(2, 6)$, $(0, -2)$, and $(5, 1)$ is an isosceles triangle.

Solve.

53. Two surveyors need to find the distance across a lake. They place a reference pole at point *A* in the diagram. Point *B* is 3 meters east and 1 meter north of the reference point *A*. Point *C* is 19 meters east and 13 meters north of point *A*. Find the distance across the lake, from *B* to *C*.

54. Cindy Brown, an architect, is drawing plans on grid paper for a circular pool with a fountain in the middle. The paper is marked off in centimeters, and each centimeter represents 1 foot. On the paper, the diameter of the "pool" is 20 centimeters, and "fountain" is the point $(0, 0)$.

a. Sketch the architect's drawing. Be sure to label the axes.

b. Write an equation that describes the circular pool.

c. Cindy plans to place a circle of lights around the fountain such that each light is 5 feet from the fountain. Write an equation for the circle of lights and sketch the circle on your drawing.

55. A bridge constructed over a bayou has a supporting arch in the shape of a parabola. Find an equation of the parabolic arch if the length of the road over the arch is 100 meters and the maximum height of the arch is 40 meters.

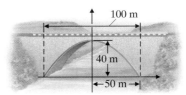

10.2 THE ELLIPSE AND THE HYPERBOLA

Objective **A** Graphing Ellipses

An **ellipse** can be thought of as the set of points in a plane such that the sum of the distances of each of those points from two fixed points is constant. Each of the two fixed points is called a **focus.** The plural of focus is **foci.** The point midway between the foci is called the **center.**

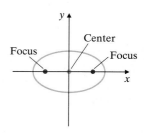

An ellipse may be drawn by hand by using two tacks, a piece of string, and a pencil. Secure the two tacks into a piece of cardboard, for example, and tie each end of the string to a tack. Use your pencil to pull the string tight and draw the ellipse. The two tacks are the foci of the drawn ellipse.

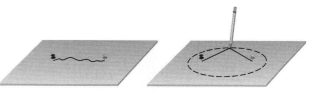

Ellipse with Center (0, 0)

The graph of an equation of the form $\dfrac{x^2}{a^2} + \dfrac{y^2}{b^2} = 1$ is an ellipse with center $(0, 0)$.

The x-intercepts are $(a, 0)$ and $(-a, 0)$ and the y-intercepts are $(0, b)$ and $(0, -b)$.

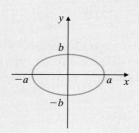

PRACTICE PROBLEM 1

Graph: $\dfrac{x^2}{9} + \dfrac{y^2}{4} = 1$

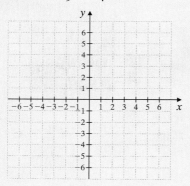

The **standard form** of an ellipse with center $(0, 0)$ is $\dfrac{x^2}{a^2} + \dfrac{y^2}{b^2} = 1$.

EXAMPLE 1 Graph: $\dfrac{x^2}{9} + \dfrac{y^2}{16} = 1$

Solution: The equation is of the form $\dfrac{x^2}{a^2} + \dfrac{y^2}{b^2} = 1$ with $a = 3$ and $b = 4$, so its graph is an ellipse with center $(0, 0)$, x-intercepts $(3, 0)$ and $(-3, 0)$, and y-intercepts $(0, 4)$ and $(0, -4)$.

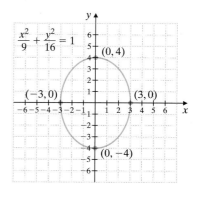

Answer

1.

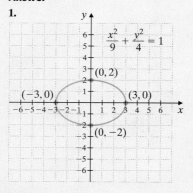

■ **Work Practice Problem 1**

EXAMPLE 2 Graph: $4x^2 + 16y^2 = 64$

Solution: Although this equation contains a sum of squared terms in x and y on the same side of an equation, this is not the equation of a circle since the coefficients of x^2 and y^2 are not the same. When this happens, the graph is an ellipse. Since the standard form of the equation of an ellipse has 1 on one side, we divide both sides of this equation by 64 to get it in standard form.

$$4x^2 + 16y^2 = 64$$

$$\frac{4x^2}{64} + \frac{16y^2}{64} = \frac{64}{64} \quad \text{Divide both sides by 64.}$$

$$\frac{x^2}{16} + \frac{y^2}{4} = 1 \quad \text{Simplify.}$$

We now recognize the equation of an ellipse with center $(0, 0)$, x-intercepts $(4, 0)$ and $(-4, 0)$, and y-intercepts $(0, 2)$ and $(0, -2)$.

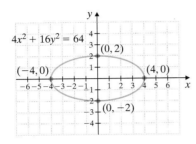

🔲 **Work Practice Problem 2**

The center of an ellipse is not always $(0, 0)$, as shown in the next example. The standard form of an ellipse with center (h, k) is

$$\frac{(x - h)^2}{a^2} + \frac{(y - k)^2}{b^2} = 1$$

EXAMPLE 3 Graph: $\dfrac{(x + 3)^2}{25} + \dfrac{(y - 2)^2}{36} = 1$

Solution: This ellipse has center $(-3, 2)$. Notice that $a = 5$ and $b = 6$. To find four points on the graph of the ellipse, we first graph the center, $(-3, 2)$. Since $a = 5$, we count 5 units right and then 5 units left of the point with coordinates $(-3, 2)$. Next, since $b = 6$, we start at $(-3, 2)$ and count 6 units up and then 6 units down to find two more points on the ellipse.

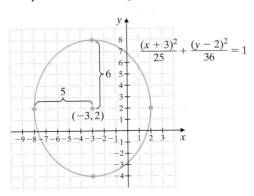

🔲 **Work Practice Problem 3**

PRACTICE PROBLEM 2

Graph: $4x^2 + 36y^2 = 144$

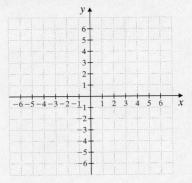

PRACTICE PROBLEM 3

Graph: $\dfrac{(x - 1)^2}{9} + \dfrac{(y - 3)^2}{16} = 1$

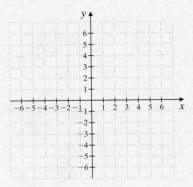

Answers

2.

3.

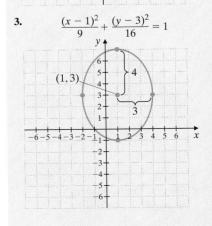

✔**Concept Check** In the graph of the equation $\dfrac{x^2}{64} + \dfrac{y^2}{36} = 1$, which distance is longer: the distance between the x-intercepts or the distance between the y-intercepts? How much longer? Explain.

Objective B Graphing Hyperbolas

The final conic section is the **hyperbola.** A hyperbola is the set of points in a plane such that for each point in the set, the absolute value of the difference of the distances from two fixed points is constant. Each of the two fixed points is called a **focus.** The point midway between the foci is called the **center.**

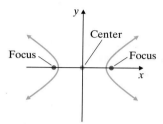

Using the distance formula, we can show that the graph of $\dfrac{x^2}{a^2} - \dfrac{y^2}{b^2} = 1$ is a hyperbola with center $(0, 0)$ and x-intercepts $(a, 0)$ and $(-a, 0)$. Also, the graph of $\dfrac{y^2}{b^2} - \dfrac{x^2}{a^2} = 1$ is a hyperbola with center $(0, 0)$ and y-intercepts $(0, b)$ and $(0, -b)$.

Hyperbola with Center (0, 0)

The graph of an equation of the form $\dfrac{x^2}{a^2} - \dfrac{y^2}{b^2} = 1$ is a hyperbola with center $(0, 0)$ and x-intercepts $(a, 0)$ and $(-a, 0)$.

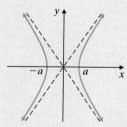

The graph of an equation of the form $\dfrac{y^2}{b^2} - \dfrac{x^2}{a^2} = 1$ is a hyperbola with center $(0, 0)$ and y-intercepts $(0, b)$ and $(0, -b)$.

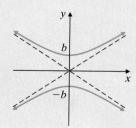

The equations $\dfrac{x^2}{a^2} - \dfrac{y^2}{b^2} = 1$ and $\dfrac{y^2}{b^2} - \dfrac{x^2}{a^2} = 1$ are the **standard forms** for the equation of a hyperbola.

> **Helpful Hint**
>
> Notice the difference between the equation of an ellipse and a hyperbola. The equation of the ellipse contains x^2- and y^2-terms on the same side of the equation with same-sign coefficients. For a hyperbola, the coefficients on the same side of the equation have different signs.

Graphing a hyperbola such as $\dfrac{y^2}{b^2} - \dfrac{x^2}{a^2} = 1$ is made easier by recognizing one of its important characteristics. Examining the figure below, notice how the sides of the branches of the hyperbola extend indefinitely and seem to approach, but not intersect, the dashed lines in the figure. These dashed lines are called the **asymptotes** of the hyperbola.

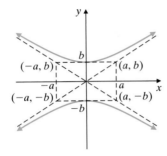

To sketch these lines, or asymptotes, draw a rectangle with vertices (a, b), $(-a, b)$, $(a, -b)$, and $(-a, -b)$. The asymptotes of the hyperbola are the extended diagonals of this rectangle.

EXAMPLE 4 Graph: $\dfrac{x^2}{16} - \dfrac{y^2}{25} = 1$

Solution: This equation has the form $\dfrac{x^2}{a^2} - \dfrac{y^2}{b^2} = 1$, with $a = 4$ and $b = 5$. Thus, its graph is a hyperbola with center $(0, 0)$ and x-intercepts of $(4, 0)$ and $(-4, 0)$. To aid in graphing the hyperbola, we first sketch its asymptotes. The extended diagonals of the rectangle with coordinates $(4, 5)$, $(4, -5)$, $(-4, 5)$, and $(-4, -5)$ are the asymptotes of the hyperbola. Then we use the asymptotes to aid in graphing the hyperbola.

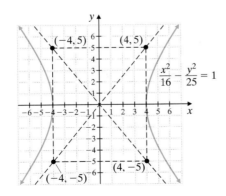

🔲 **Work Practice Problem 4**

PRACTICE PROBLEM 4

Graph: $\dfrac{x^2}{9} - \dfrac{y^2}{4} = 1$

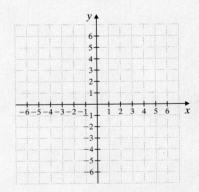

Answer

4.

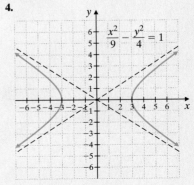

PRACTICE PROBLEM 5

Graph: $9y^2 - 16x^2 = 144$

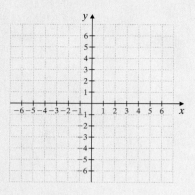

EXAMPLE 5 Graph: $4y^2 - 9x^2 = 36$

Solution: Since this is a difference of squared terms in x and y on the same side of the equation, its graph is a hyperbola, as opposed to an ellipse or a circle. The standard form of the equation of a hyperbola has a 1 on one side, so we divide both sides of the equation by 36 to get it in standard form.

$$4y^2 - 9x^2 = 36$$

$$\frac{4y^2}{36} - \frac{9x^2}{36} = \frac{36}{36} \quad \text{Divide both sides by 36.}$$

$$\frac{y^2}{9} - \frac{x^2}{4} = 1 \quad \text{Simplify.}$$

The equation is of the form $\frac{y^2}{b^2} - \frac{x^2}{a^2} = 1$ with $a = 2$ and $b = 3$, so the hyperbola is centered at $(0, 0)$ with y-intercepts $(0, 3)$ and $(0, -3)$.

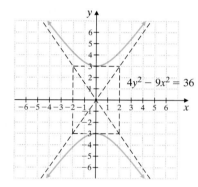

■ **Work Practice Problem 5**

Answer

5.

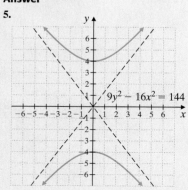

CALCULATOR EXPLORATIONS Graphing

To find the graph of an ellipse by using a graphing calculator, use the same procedure as for graphing a circle. For example, to graph $x^2 + 3y^2 = 22$, first solve for y.

$$3y^2 = 22 - x^2$$

$$y^2 = \frac{22 - x^2}{3}$$

$$y = \pm\sqrt{\frac{22 - x^2}{3}}$$

Next press the $\boxed{Y=}$ key and enter $Y_1 = \sqrt{\frac{22 - x^2}{3}}$ and $Y_2 = -\sqrt{\frac{22 - x^2}{3}}$. (Insert two sets of parentheses in the radicand as in $\sqrt{((22 - x^2)/3)}$ so that the desired graph is obtained.) The graph appears as follows:

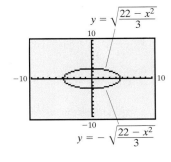

2. $20x^2 + 5y^2 = 100$

3. $7.3x^2 + 15.5y^2 = 95.2$

4. $18.8x^2 + 36.1y^2 = 205.8$

Use a graphing calculator to graph each ellipse.

1. $10x^2 + y^2 = 32$

Mental Math

Identify the graph of each equation as an ellipse or a hyperbola.

1. $\dfrac{x^2}{16} + \dfrac{y^2}{4} = 1$

2. $\dfrac{x^2}{16} - \dfrac{y^2}{4} = 1$

3. $x^2 - 5y^2 = 3$

4. $-x^2 + 5y^2 = 3$

5. $-\dfrac{y^2}{25} + \dfrac{x^2}{36} = 1$

6. $\dfrac{y^2}{25} + \dfrac{x^2}{36} = 1$

10.2 EXERCISE SET

Objective **A** *Graph each ellipse. See Examples 1 and 2.*

1. $\dfrac{x^2}{4} + \dfrac{y^2}{25} = 1$

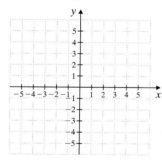

2. $\dfrac{x^2}{16} + \dfrac{y^2}{9} = 1$

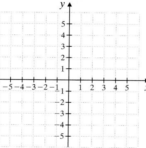

3. $\dfrac{x^2}{9} + y^2 = 1$

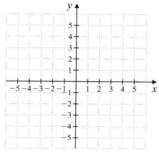

4. $x^2 + \dfrac{y^2}{4} = 1$

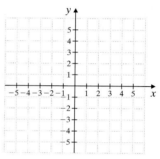

5. $9x^2 + 4y^2 = 36$

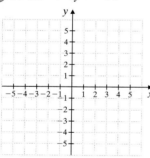

6. $x^2 + 4y^2 = 16$

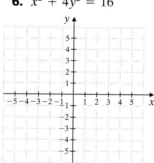

7. $4x^2 + 25y^2 = 100$

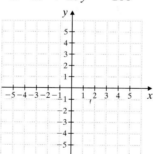

8. $36x^2 + y^2 = 36$

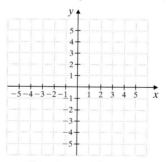

Graph each ellipse. See Example 3.

9. $\dfrac{(x+1)^2}{36} + \dfrac{(y-2)^2}{49} = 1$

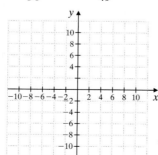

10. $\dfrac{(x-3)^2}{9} + \dfrac{(y+3)^2}{16} = 1$

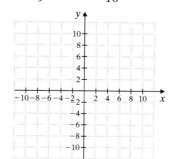

11. $\dfrac{(x-1)^2}{4} + \dfrac{(y-1)^2}{25} = 1$

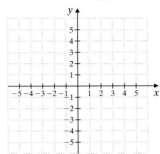

12. $\dfrac{(x+3)^2}{16} + \dfrac{(y+2)^2}{4} = 1$

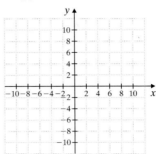

Objective B *Graph each hyperbola. See Examples 4 and 5.*

13. $\dfrac{x^2}{4} - \dfrac{y^2}{9} = 1$

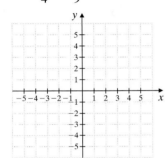

14. $\dfrac{x^2}{36} - \dfrac{y^2}{36} = 1$

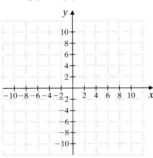

15. $\dfrac{y^2}{25} - \dfrac{x^2}{16} = 1$

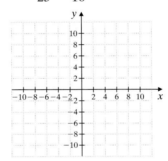

16. $\dfrac{y^2}{25} - \dfrac{x^2}{49} = 1$

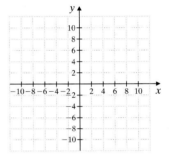

17. $x^2 - 4y^2 = 16$

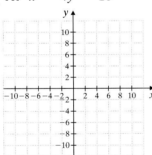

18. $4x^2 - y^2 = 36$

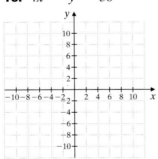

19. $16y^2 - x^2 = 16$

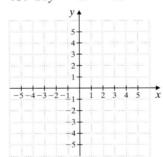

20. $4y^2 - 25x^2 = 100$

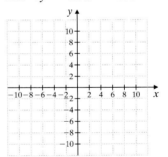

Objectives A B Mixed Practice *Graph each equation. See Examples 1 through 5.*

21. $\dfrac{y^2}{36} = 1 - x^2$

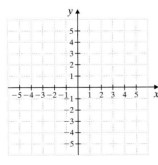

22. $\dfrac{x^2}{36} = 1 - y^2$

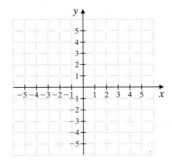

23. $4(x-1)^2 + 9(y+2)^2 = 36$

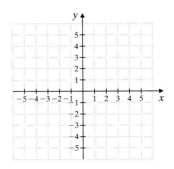

24. $25(x + 3)^2 + 4(y - 3)^2 = 100$

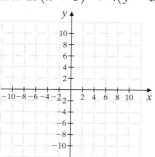

25. $8x^2 + 2y^2 = 32$

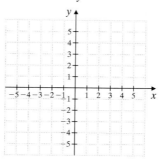

26. $3x^2 + 12y^2 = 48$

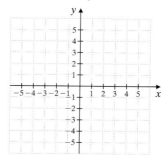

27. $25x^2 - y^2 = 25$

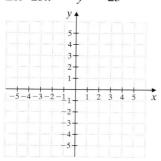

28. $x^2 - 9y^2 = 9$

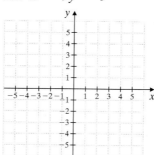

Review

Perform each indicated operation. See Sections 5.1 and 5.2.

29. $(2x^3)(-4x^2)$

30. $2x^3 - 4x^3$

31. $-5x^2 + x^2$

32. $(-5x^2)(x^2)$

Concept Extensions

The graph of each equation is an ellipse. Determine which distance is longer. The distance between the x-intercepts or the distance between the y-intercepts. How much longer? See the Concept Check in this section.

33. $\dfrac{x^2}{16} + \dfrac{y^2}{25} = 1$

34. $\dfrac{x^2}{100} + \dfrac{y^2}{49} = 1$

35. $4x^2 + y^2 = 16$

36. $x^2 + 4y^2 = 36$

37. We know that $x^2 + y^2 = 25$ is the equation of a circle. Rewrite the equation so that the right side is equal to 1. Which type of conic section does this equation form resemble? In fact, the circle is a special case of this type of conic section. Describe the conditions under which this type of conic section is a circle.

The orbits of stars, planets, comets, asteroids, and satellites all have the shape of one of the conic sections. Astronomers use a measure called eccentricity *to describe the shape and elongation of an orbital path. For the circle and ellipse, eccentricity e is calculated with the formula* $e = \dfrac{c}{d}$, *where* $c^2 = |a^2 - b^2|$ *and d is the larger value of a or b. For a hyperbola, eccentricity e is calculated with the formula* $e = \dfrac{c}{d}$, *where* $c^2 = a^2 + b^2$ *and the value of d is equal to a if the hyperbola has x-intercepts or equal to b if the hyperbola has y-intercepts. Use equations A–H to answer Exercises 38–47.*

A. $\dfrac{x^2}{36} - \dfrac{y^2}{13} = 1$

B. $\dfrac{x^2}{4} + \dfrac{y^2}{4} = 1$

C. $\dfrac{x^2}{25} + \dfrac{y^2}{16} = 1$

D. $\dfrac{y^2}{25} - \dfrac{x^2}{39} = 1$

E. $\dfrac{x^2}{17} + \dfrac{y^2}{81} = 1$ **F.** $\dfrac{x^2}{36} + \dfrac{y^2}{36} = 1$ **G.** $\dfrac{x^2}{16} - \dfrac{y^2}{65} = 1$ **H.** $\dfrac{x^2}{144} + \dfrac{y^2}{140} = 1$

38. Identify the type of conic section represented by each of the equations A–H.

39. For each of the equations A–H, identify the values of a^2 and b^2.

40. For each of the equations A–H, calculate the value of c^2 and c.

41. For each of the equations A–H, find the value of d.

42. For each of the equations A–H, calculate the eccentricity e.

43. What do you notice about the values of e for the equations you identified as ellipses?

44. What do you notice about the values of e for the equations you identified as circles?

45. What do you notice about the values of e for the equations you identified as hyperbolas?

46. The eccentricity of a parabola is exactly 1. Use this information and the observations you made in Exercises 43, 44, and 45 to describe a way that could be used to identify the type of conic section based on its eccentricity value.

47. Graph each of the conic sections given in equations A–H. What do you notice about the shape of the ellipses for increasing values of eccentricity? Which is the most elliptical? Which is the least elliptical, that is, the most circular?

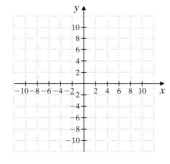
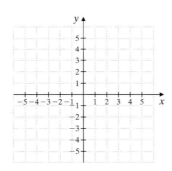
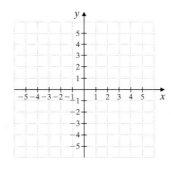
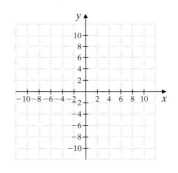

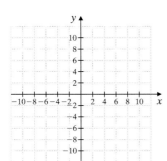
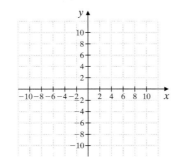
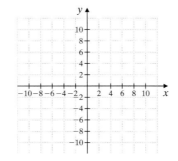
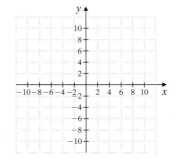

Graphing Conic Sections

Following is a summary of conic sections.

Conic Sections

	Standard Form	Graph
Parabola	$y = a(x - h)^2 + k$	
Parabola	$x = a(y - k)^2 + h$	
Circle	$(x - h)^2 + (y - k)^2 = r^2$	
Ellipse	$\dfrac{x^2}{a^2} + \dfrac{y^2}{b^2} = 1$	
Hyperbola	$\dfrac{x^2}{a^2} - \dfrac{y^2}{b^2} = 1$	
Hyperbola	$\dfrac{y^2}{b^2} - \dfrac{x^2}{a^2} = 1$	

Identify whether each equation, when graphed, will be a parabola, circle, ellipse, or hyperbola. Then graph each equation.

1. $(x - 7)^2 + (y - 2)^2 = 4$

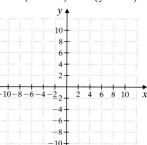

2. $y = x^2 + 4$

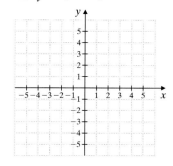

3. $y = x^2 + 12x + 36$

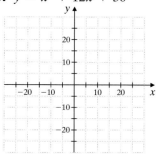

4. $\dfrac{x^2}{4} + \dfrac{y^2}{9} = 1$

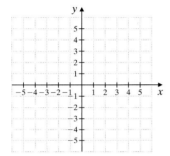

5. $\dfrac{y^2}{9} - \dfrac{x^2}{9} = 1$

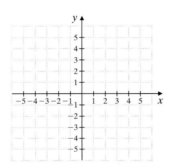

6. $\dfrac{x^2}{16} - \dfrac{y^2}{4} = 1$

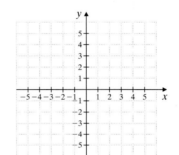

7. $2x^2 + 8y^2 = 32$

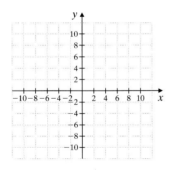

8. $x^2 + y^2 = 16$

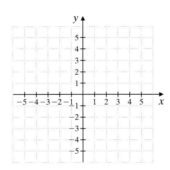

9. $x = y^2 + 4y - 1$

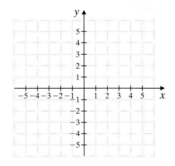

10. $x = -y^2 + 6y$

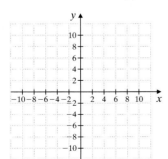

11. $9x^2 - 4y^2 = 36$

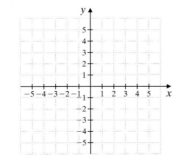

12. $18x^2 + 8y^2 = 72$

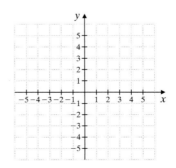

13. $\dfrac{(x-1)^2}{49} + \dfrac{(y+2)^2}{25} = 1$

14. $y^2 = x^2 + 16$

15. $\left(x + \dfrac{1}{2}\right)^2 + \left(y - \dfrac{1}{2}\right)^2 = 1$

4. _____

5. _____

6. _____

7. _____

8. _____

9. _____

10. _____

11. _____

12. _____

13. _____

14. _____

15. _____

10.3 SOLVING NONLINEAR SYSTEMS OF EQUATIONS

In Section 4.1, we used graphing, substitution, and elimination methods to find solutions of systems of linear equations in two variables. We now apply these same methods to nonlinear systems of equations in two variables. A **nonlinear system of equations** is a system of equations at least one of which is not linear. Since we will be graphing the equations in each system, we are interested in real number solutions only.

Objective **A** Solving Nonlinear Systems by Substitution

First we solve nonlinear systems by the substitution method.

EXAMPLE 1 Solve the system:

$$\begin{cases} x^2 - 3y = 1 \\ x - y = 1 \end{cases}$$

Solution: We can solve this system by substitution if we solve one equation for one of the variables. Solving the first equation for x is not the best choice since doing so introduces a radical. Also, solving for y in the first equation introduces a fraction. Thus, we solve the second equation for y.

$x - y = 1$ Second equation
$x - 1 = y$ Solve for y.

Now we replace y with $x - 1$ in the first equation, and then solve for x.

$$x^2 - 3y = 1 \quad \text{First equation}$$
$$x^2 - 3(x - 1) = 1 \quad \text{Replace } y \text{ with } x - 1.$$
$$x^2 - 3x + 3 = 1$$
$$x^2 - 3x + 2 = 0$$
$$(x - 2)(x - 1) = 0$$
$$x = 2 \quad \text{or} \quad x = 1$$

Now we let $x = 2$ and then $x = 1$ in the equation $y = x - 1$ to find corresponding y-values.

Let $x = 2$.   Let $x = 1$.
$y = x - 1$ $y = x - 1$
$y = 2 - 1 = 1$ $y = 1 - 1 = 0$

When we check $(2, 1)$ and $(1, 0)$ in the equations, we find that both ordered pairs satisfy both equations. Thus, the solution set for the system is $\{(2, 1), (1, 0)\}$. The graph of each equation in the system is shown.

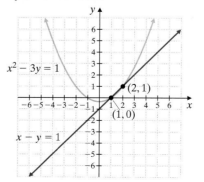

Answer
1. $\{(-3, 2), (1, -2)\}$

■ **Work Practice Problem 1**

EXAMPLE 2 Solve the system:

$$\begin{cases} y = \sqrt{x} \\ x^2 + y^2 = 6 \end{cases}$$

Solution: This system is ideal for the substitution method since y is expressed in terms of x in the first equation. Notice that if $y = \sqrt{x}$, then both x and y must be nonnegative if they are real numbers. Let's substitute $\sqrt{x}$ for y in the second equation, and solve for x.

$$x^2 + y^2 = 6$$
$$x^2 + (\sqrt{x})^2 = 6 \quad \text{Let } y = \sqrt{x}.$$
$$x^2 + x = 6$$
$$x^2 + x - 6 = 0$$
$$(x + 3)(x - 2) = 0$$
$$x = -3 \quad \text{or} \quad x = 2$$

The solution -3 is discarded because we have noted that x must be nonnegative. To see this, we let $x = -3$ and $x = 2$ in the first equation to find the corresponding y-values.

Let $x = -3$.   Let $x = 2$.

$\quad y = \sqrt{x}$ $y = \sqrt{x}$

$\quad y = \sqrt{-3}$ Not a real number $y = \sqrt{2}$

Since we are interested only in real number solutions, the only solution is $(2, \sqrt{2})$. The solution set is $\{(2, \sqrt{2})\}$. Check to see that this solution satisfies both equations. The graph of each equation in this system is shown.

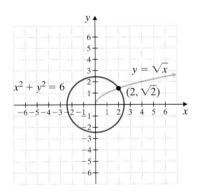

Work Practice Problem 2

EXAMPLE 3 Solve the system:

$$\begin{cases} x^2 + y^2 = 4 \\ x + y = 3 \end{cases}$$

Solution: We use the substitution method and solve the second equation for x

$$x + y = 3 \quad \text{Second equation}$$
$$x = 3 - y$$

Continued on next page

Continued on next page

PRACTICE PROBLEM 2

Solve the system:

$$\begin{cases} y = \sqrt{x} \\ x^2 + y^2 = 30 \end{cases}$$

PRACTICE PROBLEM 3

Solve the system:

$$\begin{cases} x^2 + y^2 = 1 \\ x + y = 4 \end{cases}$$

Answers

2. $\{(5, \sqrt{5})\}$,   **3.** no solutions

Now we let $x = 3 - y$ in the first equation.

$$x^2 + y^2 = 4 \quad \text{First equation}$$

$$(3 - y)^2 + y^2 = 4 \quad \text{Let } x = 3 - y.$$

$$9 - 6y + y^2 + y^2 = 4$$

$$2y^2 - 6y + 5 = 0$$

By the quadratic formula, where $a = 2, b = -6$, and $c = 5$, we have

$$y = \frac{6 \pm \sqrt{(-6)^2 - 4 \cdot 2 \cdot 5}}{2 \cdot 2} = \frac{6 \pm \sqrt{-4}}{4}$$

Since $\sqrt{-4}$ is not a real number, there is no solution. Graphically, the circle and the line do not intersect, as shown.

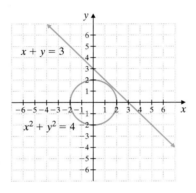

■ **Work Practice Problem 3**

✔ **Concept Check** Without solving, how can you tell that the graphs of $x^2 + y^2 = 9$ and $x^2 + y^2 = 16$ do not have any points of intersection?

Objective **B** Solving Nonlinear Systems by Elimination

Some nonlinear systems may be solved by the elimination method.

EXAMPLE 4 Solve the system:

$$\begin{cases} x^2 + 2y^2 = 10 \\ x^2 - y^2 = 1 \end{cases}$$

Solution: We will use the elimination, or addition, method to solve this system. To eliminate x^2 when we add the two equations, we multiply both sides of the second equation by -1. Then

$$x^2 + 2y^2 = 10 \qquad \text{is}$$

$$(-1)(x^2 - y^2) = -1 \cdot 1 \qquad \begin{matrix} \text{equivalent} \\ \text{to} \end{matrix}$$

$$\begin{cases} x^2 + 2y^2 = 10 \\ -x^2 + y^2 = -1 \end{cases}$$

$$3y^2 = 9 \qquad \text{Add.}$$

$$y^2 = 3$$

$$y = \pm\sqrt{3} \qquad \text{Divide both sides by 3.}$$

PRACTICE PROBLEM 4

Solve the equation:

$$\begin{cases} x^2 + 3y^2 = 21 \\ x^2 - y^2 = 1 \end{cases}$$

Answers

4. $\{(\sqrt{6}, \sqrt{5}), (\sqrt{6}, -\sqrt{5}),$
$(-\sqrt{6}, \sqrt{5}), (-\sqrt{6}, -\sqrt{5})\}$

✔ **Concept Check Answer**

$x^2 + y^2 = 9$ is a circle inside the circle $x^2 + y^2 = 16$, therefore they do not have any points of intersection.

To find the corresponding x-values, we let $y = \sqrt{3}$ and $y = -\sqrt{3}$ in either original equation. We choose the second equation.

Let $y = \sqrt{3}$.

$$x^2 - y^2 = 1$$
$$x^2 - (\sqrt{3})^2 = 1$$
$$x^2 - 3 = 1$$
$$x^2 = 4$$
$$x = \pm\sqrt{4} = \pm 2$$

Let $y = -\sqrt{3}$.

$$x^2 - y^2 = 1$$
$$x^2 - (-\sqrt{3})^2 = 1$$
$$x^2 - 3 = 1$$
$$x^2 = 4$$
$$x = \pm\sqrt{4} = \pm 2$$

The solution set is $\{(2, \sqrt{3}), (-2, \sqrt{3}), (2, -\sqrt{3}), (-2, -\sqrt{3})\}$. Check all four ordered pairs in both equations of the system. The graph of each equation in this system is shown.

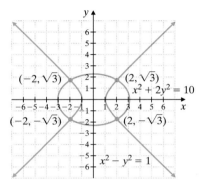

■ **Work Practice Problem 4**

Objectives **A** **B** **Mixed Practice** *Solve each nonlinear system of equations. See Examples 1 through 4.*

1. $\begin{cases} x^2 + y^2 = 25 \\ 4x + 3y = 0 \end{cases}$

2. $\begin{cases} x^2 + y^2 = 25 \\ 3x + 4y = 0 \end{cases}$

3. $\begin{cases} x^2 + 4y^2 = 10 \\ y = x \end{cases}$

4. $\begin{cases} 4x^2 + y^2 = 10 \\ y = x \end{cases}$

5. $\begin{cases} y^2 = 4 - x \\ x - 2y = 4 \end{cases}$

6. $\begin{cases} x^2 + y^2 = 4 \\ x + y = -2 \end{cases}$

7. $\begin{cases} x^2 + y^2 = 9 \\ 16x^2 - 4y^2 = 64 \end{cases}$

8. $\begin{cases} 4x^2 + 3y^2 = 35 \\ 5x^2 + 2y^2 = 42 \end{cases}$

9. $\begin{cases} x^2 + 2y^2 = 2 \\ x - y = 2 \end{cases}$

10. $\begin{cases} x^2 + 2y^2 = 2 \\ x^2 - 2y^2 = 6 \end{cases}$

11. $\begin{cases} y = x^2 - 3 \\ 4x - y = 6 \end{cases}$

12. $\begin{cases} y = x + 1 \\ x^2 - y^2 = 1 \end{cases}$

13. $\begin{cases} y = x^2 \\ 3x + y = 10 \end{cases}$

14. $\begin{cases} 6x - y = 5 \\ xy = 1 \end{cases}$

15. $\begin{cases} y = 2x^2 + 1 \\ x + y = -1 \end{cases}$

16. $\begin{cases} x^2 + y^2 = 9 \\ x + y = 5 \end{cases}$

17. $\begin{cases} y = x^2 - 4 \\ y = x^2 - 4x \end{cases}$

18. $\begin{cases} x = y^2 - 3 \\ x = y^2 - 3y \end{cases}$

19. $\begin{cases} 2x^2 + 3y^2 = 14 \\ -x^2 + y^2 = 3 \end{cases}$

20. $\begin{cases} 4x^2 - 2y^2 = 2 \\ -x^2 + y^2 = 2 \end{cases}$

21. $\begin{cases} x^2 + y^2 = 1 \\ x^2 + (y + 3)^2 = 4 \end{cases}$

22. $\begin{cases} x^2 + 2y^2 = 4 \\ x^2 - y^2 = 4 \end{cases}$

23. $\begin{cases} y = x^2 + 2 \\ y = -x^2 + 4 \end{cases}$

24. $\begin{cases} x = -y^2 - 3 \\ x = y^2 - 5 \end{cases}$

25. $\begin{cases} 3x^2 + y^2 = 9 \\ 3x^2 - y^2 = 9 \end{cases}$

26. $\begin{cases} x^2 + y^2 = 25 \\ x = y^2 - 5 \end{cases}$

27. $\begin{cases} x^2 + 3y^2 = 6 \\ x^2 - 3y^2 = 10 \end{cases}$

28. $\begin{cases} x^2 + y^2 = 1 \\ y = x^2 - 9 \end{cases}$

29. $\begin{cases} x^2 + y^2 = 36 \\ y = \frac{1}{6}x^2 - 6 \end{cases}$

30. $\begin{cases} x^2 + y^2 = 16 \\ y = -\frac{1}{4}x^2 + 4 \end{cases}$

31. $\begin{cases} y = \sqrt{x} \\ x^2 + y^2 = 12 \end{cases}$

32. $\begin{cases} y = \sqrt{x} \\ x^2 + y^2 = 20 \end{cases}$

Review

Graph each inequality in two variables. See Section 3.5.

33. $x > -3$

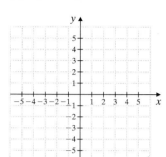

34. $y \leq 1$

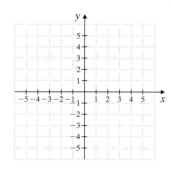

35. $y < 2x - 1$

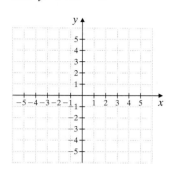

36. $3x - y \leq 4$

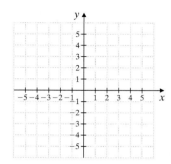

Concept Extensions

For the exercises below, see the Concept Check in this section.

37. Without graphing, how can you tell that the graph of $x^2 + y^2 = 1$ and $x^2 + y^2 = 4$ do not have any points of intersection?

38. Without solving, how can you tell that the graphs of $y = 2x + 3$ and $y = 2x + 7$ do not have any points of intersection?

39. How many real solutions are possible for a system of equations whose graphs are a circle and a parabola?

40. How many real solutions are possible for a system of equations whose graphs are an ellipse and a line?

Solve.

41. The sum of the squares of two numbers is 130. The difference of the squares of the two numbers is 32. Find the two numbers.

42. The sum of the squares of two numbers is 20. Their product is 8. Find the two numbers.

43. During the development stage of a new rectangular keypad for a security system, it was decided that the area of the rectangle should be 285 square centimeters and the perimeter should be 68 centimeters. Find the dimensions of the keypad.

44. A rectangular holding pen for cattle is to be designed so that its perimeter is 92 feet and its area is 525 feet. Find the dimensions of the holding pen.

Recall that in business, a demand function expresses the quantity of a commodity demanded as a function of the commodity's unit price. A supply function expresses the quantity of a commodity supplied as a function of the commodity's unit price. When the quantity produced and supplied is equal to the quantity demanded, then we have what is called **market equilibrium.** *Use this information for Exercises 45–46.*

45. The demand function for a certain compact disc is given by the function $p(x) = -0.01x^2 - 0.2x + 9$ and the corresponding supply function is given by $p(x) = 0.01x^2 - 0.1x + 3$, where $p(x)$ is in dollars and x is in thousands of units. Find the equilibrium quantity and the corresponding price by solving the system consisting of the two given equations.

46. The demand function for a certain style of picture frame is given by the function $p(x) = -2x^2 + 90$ and the corresponding supply function is given by $p(x) = 9x + 34$, where $p(x)$ is in dollars and x is in thousands of units. Find the equilibrium quantity and the corresponding price by solving the system consisting of the two given equations.

Use a grapher to verify the results of each exercise.

 **47.** Exercise 3   **48.** Exercise 4   **49.** Exercise 23   **50.** Exercise 24

 STUDY SKILLS BUILDER

Are You Prepared for Your Final Exam?

To prepare for your final exam, try the following study techniques:

- Review the material that you will be responsible for on your exam. This includes material from your textbook, your notebook, and any handouts from your instructor.

- Review any formulas that you may need to memorize.

- Check to see if your instructor or mathematics department will be conducting a final exam review.

- Check with your instructor to see whether final exams from previous semesters/quarters are available to students for review.

- Use your previously taken exams as a practice final exam. To do so, rewrite the test questions in mixed order on blank sheets of paper. This will help you prepare for exam conditions.

- If you are unsure of a few concepts, see your instructor or visit a learning lab for assistance. Also, view the video segment of any troublesome sections.

- If you need further exercises to work, try the Cumulative Reviews at the end of the chapters.

Once again, good luck! I hope you have enjoyed this textbook and your mathematics course.

10.4 NONLINEAR INEQUALITIES AND SYSTEMS OF INEQUALITIES

Objectives

A Graph a Nonlinear Inequality.

B Graph a System of Nonlinear Inequalities.

Objective **A** Graphing Nonlinear Inequalities

We can graph a nonlinear inequality in two variables such as $\frac{x^2}{9} + \frac{y^2}{16} \leq 1$ in a way similar to the way we graphed a linear inequality in two variables in Section 3.5. First, we graph the related equation $\frac{x^2}{9} + \frac{y^2}{16} = 1$. The graph of the equation is our boundary. Then, using test points, we determine and shade the region whose points satisfy the inequality.

EXAMPLE 1 Graph: $\frac{x^2}{9} + \frac{y^2}{16} \leq 1$

Solution: First we graph the equation $\frac{x^2}{9} + \frac{y^2}{16} = 1$. We sketch a solid curve because of the inequality symbol $\leq$. It means that the graph of $\frac{x^2}{9} + \frac{y^2}{16} \leq 1$ includes the graph of $\frac{x^2}{9} + \frac{y^2}{16} = 1$. The graph is an ellipse, and it divides the plane into two regions, the "inside" and the "outside" of the ellipse. Recall from Section 3.5 that to determine which region contains the solutions, we select a test point in either region and determine whether the coordinates of the point satisfy the inequality. We choose $(0, 0)$ as the test point.

$$\frac{x^2}{9} + \frac{y^2}{16} \leq 1$$

$$\frac{0^2}{9} + \frac{0^2}{16} \leq 1 \quad \text{Let } x = 0 \text{ and } y = 0.$$

$$0 \leq 1 \quad \text{True}$$

Since this statement is true, the solution set is the region containing $(0, 0)$. The graph of the solution set includes the points on and inside the ellipse, as shaded in the figure.

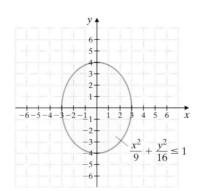

■ **Work Practice Problem 1**

PRACTICE PROBLEM 1

Graph: $\frac{x^2}{25} + \frac{y^2}{4} \leq 1$

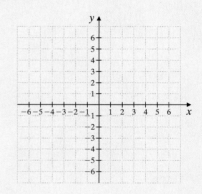

Answer

1.

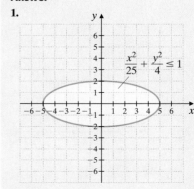

745

PRACTICE PROBLEM 2

Graph: $9x^2 > 4y^2 + 144$

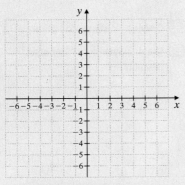

PRACTICE PROBLEM 3

Graph the system:

$$\begin{cases} y \geq x^2 \\ y \leq -4x + 2 \end{cases}$$

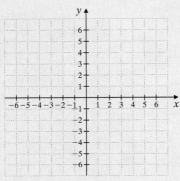

Answers

2.

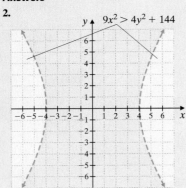

3.

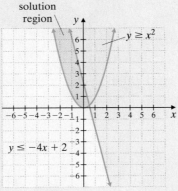

EXAMPLE 2 Graph: $4y^2 > x^2 + 16$

Solution: The related equation is $4y^2 = x^2 + 16$, or $\dfrac{y^2}{4} - \dfrac{x^2}{16} = 1$, which is a hyperbola. We graph the hyperbola as a dashed curve because of the inequality symbol $>$. It means that the graph of $4y^2 > x^2 + 16$ does *not* include the graph of $4y^2 = x^2 + 16$. The hyperbola divides the plane into three regions. We select a test point in each region—not on a boundary line—to determine whether that region contains solutions of the inequality.

Test Region *A* with (0, 4)	Test Region *B* with (0, 0)	Test Region *C* with (0, −4)
$4y^2 > x^2 + 16$	$4y^2 > x^2 + 16$	$4y^2 > x^2 + 16$
$4(4)^2 > 0^2 + 16$	$4(0)^2 > 0^2 + 16$	$4(-4)^2 > 0^2 + 16$
$64 > 16$ True	$0 > 16$ False	$64 > 16$ True

The graph of the solution set includes the shaded regions *A* and *C* only, not the boundary.

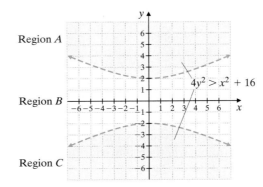

Work Practice Problem 2

Objective **B** Graphing Systems of Nonlinear Inequalities

In Section 4.5 we graphed systems of linear inequalities. Recall that the graph of a system of inequalities is the intersection of the graphs of the inequalities.

EXAMPLE 3 Graph the system:

$$\begin{cases} x \leq 1 - 2y \\ y \leq x^2 \end{cases}$$

Solution: We graph each inequality on the same set of axes. The intersection is the darkest shaded region along with its boundary lines. The coordinates of the points of intersection can be found by solving the related system

$$\begin{cases} x = 1 - 2y \\ y = x^2 \end{cases}$$

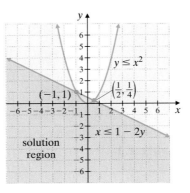

Work Practice Problem 3

EXAMPLE 4 Graph the system:

$$\begin{cases} x^2 + y^2 < 25 \\ \dfrac{x^2}{9} - \dfrac{y^2}{25} < 1 \\ \qquad y < x + 3 \end{cases}$$

Solution: We graph each inequality. The graph of $x^2 + y^2 < 25$ contains points "inside" the circle that has center $(0, 0)$ and radius 5. The graph of $\dfrac{x^2}{9} - \dfrac{y^2}{25} < 1$ is the region between the two branches of the hyperbola with x-intercepts $(-3, 0)$ and $(3, 0)$ and center $(0, 0)$. The graph of $y < x + 3$ is the region "below" the line with slope 1 and y-intercept $(0, 3)$. The graph of the solution set of the system is the intersection of all the graphs. This intersection region is shown as the shaded region on the fourth graph. The boundary of this region is not part of the solution.

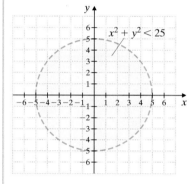

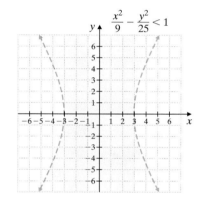

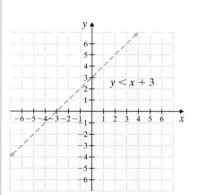

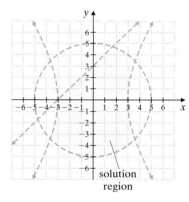

Work Practice Problem 4

PRACTICE PROBLEM 4

Graph the system:

$$\begin{cases} x^2 + y^2 < 9 \\ \dfrac{x^2}{9} - \dfrac{y^2}{4} < 1 \\ \qquad y > x - 2 \end{cases}$$

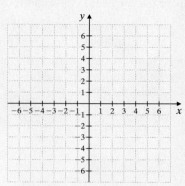

Answer

4.

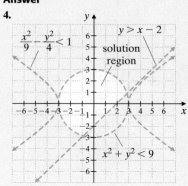

Are You Prepared for a Test on Chapter 10?

Below I have listed some common trouble areas for students in Chapter 10. After studying for your test—but before taking your test—read these.

- Don't forget to review all the standard forms for the conic sections.

- Don't forget that both methods, substitution and elimination, are available for solving nonlinear systems of equations.

$$\begin{cases} x^2 + y^2 = 7 \\ 2x^2 - 3y^2 = 4 \end{cases} \text{ is equivalent to } \begin{cases} 3x^2 + 3y^2 = 21 \\ 2x^2 - 3y^2 = 4 \end{cases}$$
$$\begin{array}{rl} \overline{5x^2} & = 25 \\ x^2 & = 5 \\ x & = \pm\sqrt{5} \end{array}$$

Let $x = \pm\sqrt{5}$ in either original equation, and $y = \pm\sqrt{2}$, the solution set is $\{(\sqrt{5}, \sqrt{2}), (-\sqrt{5}, \sqrt{2}), (\sqrt{5}, -\sqrt{2}), (-\sqrt{5}, -\sqrt{2})\}$.

Remember: This is simply a checklist of common trouble areas. For a review of Chapter 10, see the Highlights and Chapter Review at the end of this chapter.

10.4 EXERCISE SET

Objective A *Graph each inequality. See Examples 1 and 2.*

1. $y < x^2$

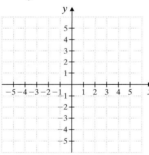

2. $y < -x^2$

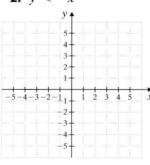

3. $x^2 + y^2 \geq 16$

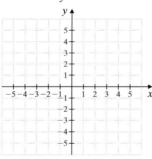

4. $x^2 + y^2 < 36$

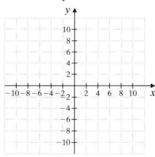

5. $\dfrac{x^2}{4} - y^2 < 1$

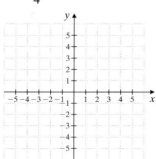

6. $x^2 - \dfrac{y^2}{9} \geq 1$

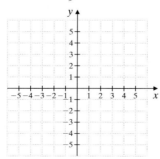

7. $y > (x - 1)^2 - 3$

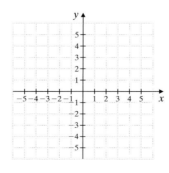

8. $y > (x + 3)^2 + 2$

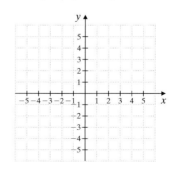

9. $x^2 + y^2 \leq 9$

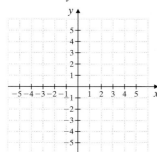

10. $x^2 + y^2 > 4$

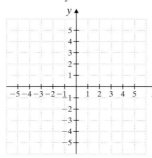

11. $y > -x^2 + 5$

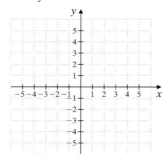

12. $y < -x^2 + 5$

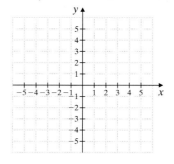

13. $\dfrac{x^2}{4} + \dfrac{y^2}{9} \leq 1$

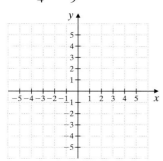

14. $\dfrac{x^2}{25} + \dfrac{y^2}{4} \geq 1$

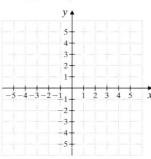

15. $\dfrac{y^2}{4} - x^2 \leq 1$

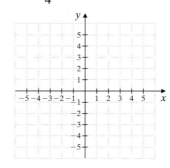

16. $\dfrac{y^2}{16} - \dfrac{x^2}{9} > 1$

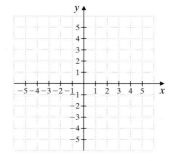

17. $y < (x - 2)^2 + 1$

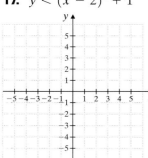

18. $y > (x - 2)^2 + 1$

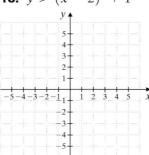

19. $y \leq x^2 + x - 2$

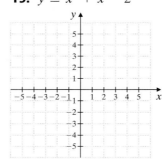

20. $y > x^2 + x - 2$

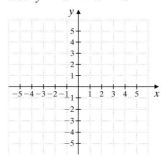

Objective **B** *Graph each system. See Examples 3 and 4.*

21. $\begin{cases} 4x + 3y \geq 12 \\ x^2 + y^2 < 16 \end{cases}$

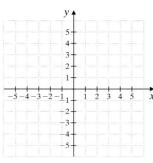

22. $\begin{cases} 3x - 4y \leq 12 \\ x^2 + y^2 < 16 \end{cases}$

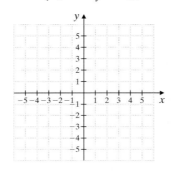

23. $\begin{cases} x^2 + y^2 \leq 9 \\ x^2 + y^2 \geq 1 \end{cases}$

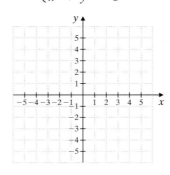

24. $\begin{cases} x^2 + y^2 \geq 9 \\ x^2 + y^2 \geq 16 \end{cases}$

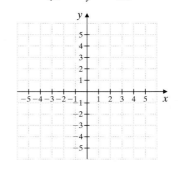

25. $\begin{cases} y > x^2 \\ y \geq 2x + 1 \end{cases}$

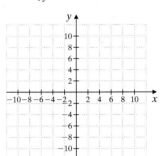

26. $\begin{cases} y \leq -x^2 + 3 \\ y \leq 2x - 1 \end{cases}$

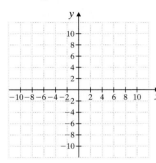

27. $\begin{cases} x^2 + y^2 > 9 \\ y > x^2 \end{cases}$

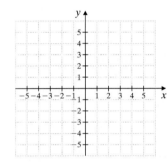

28. $\begin{cases} x^2 + y^2 \leq 9 \\ y < x^2 \end{cases}$

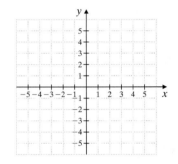

29. $\begin{cases} \dfrac{x^2}{4} + \dfrac{y^2}{9} \geq 1 \\ x^2 + y^2 \geq 4 \end{cases}$

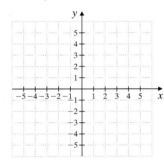

30. $\begin{cases} x^2 + (y - 2)^2 \geq 9 \\ \dfrac{x^2}{4} + \dfrac{y^2}{25} < 1 \end{cases}$

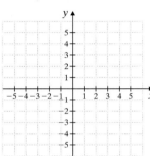

31. $\begin{cases} x^2 - y^2 \geq 1 \\ y \geq 0 \end{cases}$

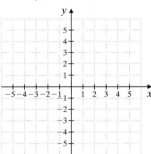

32. $\begin{cases} x^2 - y^2 \geq 1 \\ x \geq 0 \end{cases}$

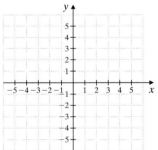

33. $\begin{cases} x + y \geq 1 \\ 2x + 3y < 1 \\ x > -3 \end{cases}$

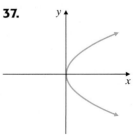

34. $\begin{cases} x - y < -1 \\ 4x - 3y > 0 \\ y > 0 \end{cases}$

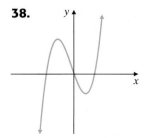

35. $\begin{cases} x^2 - y^2 < 1 \\ \dfrac{x^2}{16} + y^2 \leq 1 \\ x \geq -2 \end{cases}$

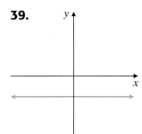

36. $\begin{cases} x^2 - y^2 \geq 1 \\ \dfrac{x^2}{16} + \dfrac{y^2}{4} \leq 1 \\ y \geq 1 \end{cases}$

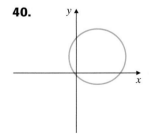

Review

Determine whether each graph is the graph of a function. See Section 3.6.

37.

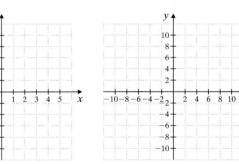

38.

39.

40.

Concept Extensions

41. Discuss how graphing a linear inequality such as $x + y < 9$ is similar to graphing a nonlinear inequality such as $x^2 + y^2 < 9$.

42. Discuss how graphing a linear inequality such as $x + y < 9$ is different from graphing a nonlinear inequality such as $x^2 + y^2 < 9$.

43. Graph the system:
$$\begin{cases} y \le x^2 \\ y \ge x + 2 \\ x \ge 0 \\ y \ge 0 \end{cases}$$

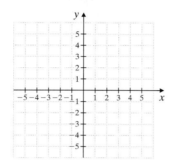

CHAPTER 10   Group Activity

Modeling Conic Sections

Materials

- two thumbtacks (or nails)
- graph paper
- cardboard
- tape
- string
- pencil
- ruler

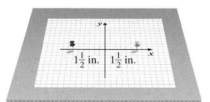

Figure 1

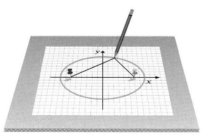

Figure 2

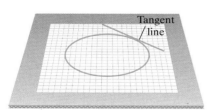

Tangent line

Figure 3

This activity may be completed by working in groups or individually.

1. Draw an x-axis and a y-axis on the graph paper as shown in Figure 1.

2. Place the graph paper on the cardboard and use tape to attach.

3. Locate two points on the x-axis each about $1\frac{1}{2}$ inches from the origin and on opposite sides of the origin (see Figure 1). Insert thumbtacks (or nails) at each of these locations.

4. Fasten a 9-inch piece of string to the thumbtacks as shown in Figure 2. Use your pencil to draw and keep the string taut while you carefully move the pencil in a path all around the thumbtacks.

5. Using the grid of the graph paper as a guide, find an approximate equation of the ellipse you drew.

6. Experiment by moving the tacks closer together or farther apart and drawing new ellipses. What do you observe?

7. Write a paragraph explaining why the figure drawn by the pencil is an ellipse. How might you use the same materials to draw a circle?

8. (Optional) Choose one of the ellipses you drew with the string and pencil. Use a ruler to draw any six tangent lines to the ellipse. (A line is tangent to the ellipse if it intersects, or just touches, the ellipse at only one point. See Figure 3.) Extend the tangent lines to yield six points of intersection among the tangents. Use a straight edge to draw a line connecting each pair of opposite points of intersection. What do you observe? Repeat with a different ellipse. Can you make a conjecture about the relationship among the lines that connect opposite points of intersection?

Chapter 10 Vocabulary Check

Fill in each blank with one of the words or phrases listed below.

circle   ellipse   hyperbola

center   radius   nonlinear system of equations

1. A(n) _____ is the set of all points in a plane that are the same distance from a fixed point, called the _____.
2. A _____ is a system of equations at least one of which is not linear.
3. A(n) _____ is the set of points on a plane such that the sum of the distances of those points from two fixed points is a constant.
4. In a circle, the distance from the center to a point of the circle is called its _____.
5. A(n) _____ is the set of points in a plane such that the absolute value of the difference of the distance from two fixed points is constant.

> **Helpful Hint**
>
> Are you preparing for your test? Don't forget to take the Chapter 10 Test on page 760. Then check your answers at the back of the text and use the Chapter Test Prep Video CD to see the fully worked-out solutions to any of the exercises you want to review.

10 Chapter Highlights

DEFINITIONS AND CONCEPTS	**EXAMPLES**

Section 10.1   The Parabola and the Circle

PARABOLAS

$$y = a(x - h)^2 + k$$

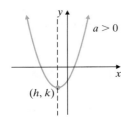

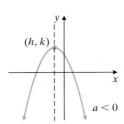

$$x = a(y - k)^2 + h$$

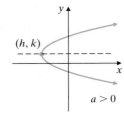

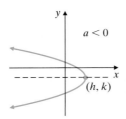

Graph: $x = 3y^2 - 12y + 13$

$$x - 13 = 3(y^2 - 4y)$$
$$x - 13 + 3(4) = 3(y^2 - 4y + 4)$$
$$x = 3(y - 2)^2 + 1$$

Since $a = 3$, this parabola opens to the right with vertex $(1, 2)$. Its axis of symmetry is $y = 2$. The x-intercept is $(13, 0)$.

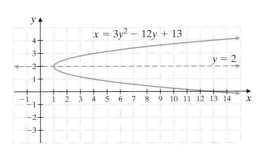

DEFINITIONS AND CONCEPTS	**EXAMPLES**

Section 10.1   The Parabola and the Circle *(continued)*

CIRCLE

The graph $(x - h)^2 + (y - k)^2 = r^2$ is a circle with center (h, k) and radius r.

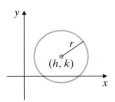

Graph: $x^2 + (y + 3)^2 = 5$

This equation can be written as

$$(x - 0)^2 + (y + 3)^2 = 5$$

with $h = 0, k = -3,$ and $r = \sqrt{5}.$ The center of this circle is $(0, -3)$, and the radius is $\sqrt{5}.$

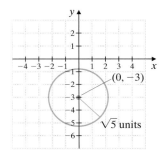

Section 10.2   The Ellipse and the Hyperbola

ELLIPSE WITH CENTER (0, 0)

The graph of an equation of the form $\dfrac{x^2}{a^2} + \dfrac{y^2}{b^2} = 1$ is an ellipse with center $(0, 0)$. The x-intercepts are $(a, 0)$ and $(-a, 0)$, and the y-intercepts are $(0, b)$ and $(0, -b)$.

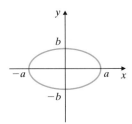

Graph: $4x^2 + 9y^2 = 36$

$$\dfrac{x^2}{9} + \dfrac{y^2}{4} = 1 \qquad \text{Divide both sides by 36.}$$

$$\dfrac{x^2}{3^2} + \dfrac{y^2}{2^2} = 1$$

The ellipse has center $(0, 0)$, x-intercepts $(3, 0)$ and $(-3, 0)$, and y-intercepts $(0, 2)$ and $(0, -2)$.

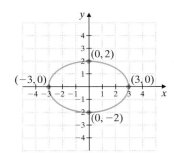

continued

DEFINITIONS AND CONCEPTS	**EXAMPLES**

Section 10.2   The Ellipse and the Hyperbola (continued)

HYPERBOLA WITH CENTER (0, 0)

The graph of an equation of the form $\dfrac{x^2}{a^2} - \dfrac{y^2}{b^2} = 1$ is a hyperbola with center $(0,0)$ and x-intercepts $(a, 0)$ and $(-a, 0)$.

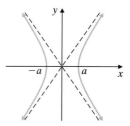

The graph of an equation of the form $\dfrac{y^2}{b^2} - \dfrac{x^2}{a^2} = 1$ is a hyperbola with center $(0,0)$ and y-intercepts $(0, b)$ and $(0, -b)$.

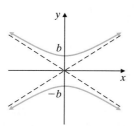

Graph: $\dfrac{x^2}{9} - \dfrac{y^2}{4} = 1$. Here $a = 3$ and $b = 2$.

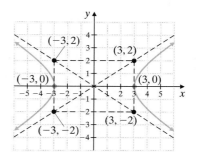

Section 10.3   Solving Nonlinear Systems of Equations

A **nonlinear system of equations** is a system of equations at least one of which is not linear. Both the substitution method and the elimination method may be used to solve a nonlinear system of equations.

Solve the nonlinear system: $\begin{cases} y = x + 2 \\ 2x^2 + y^2 = 3 \end{cases}$

Substitute $x + 2$ for y in the second equation:

$$2x^2 + y^2 = 3$$
$$2x^2 + (x + 2)^2 = 3$$
$$2x^2 + x^2 + 4x + 4 = 3$$
$$3x^2 + 4x + 1 = 0$$
$$(3x + 1)(x + 1) = 0$$
$$x = -\frac{1}{3} \quad \text{or} \quad x = -1$$

If $x = -\dfrac{1}{3}, y = x + 2 = -\dfrac{1}{3} + 2 = \dfrac{5}{3}$.

If $x = -1, y = x + 2 = -1 + 2 = 1$.

The solution set is $\left\{ \left(-\dfrac{1}{3}, \dfrac{5}{3} \right), (-1, 1) \right\}$

DEFINITIONS AND CONCEPTS	**EXAMPLES**
Section 10.4 Nonlinear Inequalities and Systems of Inequalities	

The **graph of a system of inequalities** is the intersection of the graphs of the inequalities.

Graph the system: $\begin{cases} x \geq y^2 \\ x + y \leq 4 \end{cases}$

The graph of the system is the darkest shaded region along with its boundary lines.

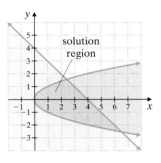

10 CHAPTER REVIEW

(10.1) *Write an equation of the circle with the given center and radius or diameter. For Exercises 3 and 4, begin by writing diameters as radii.*

1. Center $(-4, 4)$, radius 3

2. Center $(-7, -9)$, radius $\sqrt{11}$

3. Center $(5, 0)$, diameter 10

4. Center $(0, 0)$, diameter 7

Graph each equation. If the graph is a circle, find its center and radius. If the graph is a parabola, find its vertex.

5. $x^2 + y^2 = 4$

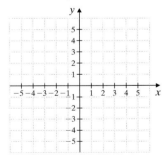

6. $x = 2(y - 5)^2 + 4$

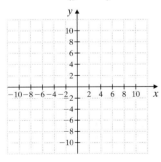

7. $x = -(y + 2)^2 + 3$

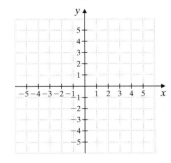

8. $(x - 1)^2 + (y - 2)^2 = 4$

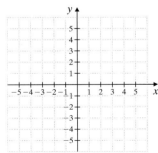

9. $y = -x^2 + 4x + 10$

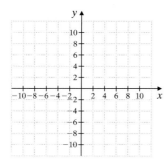

10. $x = -y^2 - 4y + 6$

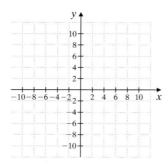

11. $x = \frac{1}{2}y^2 + 2y + 1$

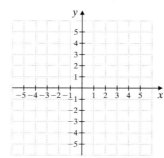

12. $y = -3x^2 + \frac{1}{2}x + 4$

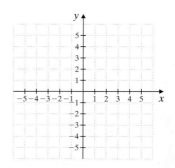

13. $x^2 + y^2 + 2x + y = \frac{3}{4}$

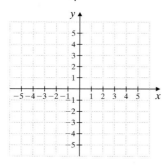

14. $x^2 + y^2 - 3y = \frac{7}{4}$

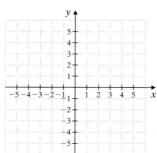

15. $4x^2 + 4y^2 + 16x + 8y = 1$

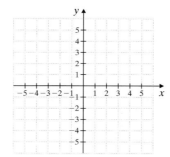

(10.1, 10.2) *Graph each equation.*

16. $x^2 + \frac{y^2}{4} = 1$

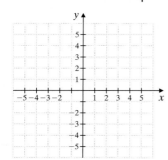

17. $x^2 - \frac{y^2}{4} = 1$

18. $-5x^2 + 25y^2 = 125$

19. $4y^2 + 9x^2 = 36$

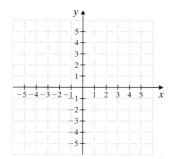

20. $\frac{(x+3)^2}{9} + \frac{(y-4)^2}{25} = 1$

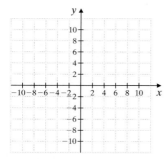

21. $x^2 - y^2 = 1$

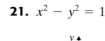
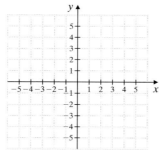

22. $36y^2 - 49x^2 = 1764$

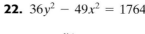
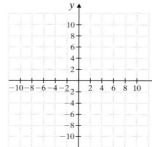

23. $y = x^2 + 9$

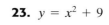
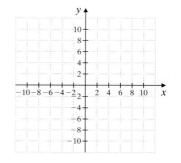

24. $x = 4y^2 - 16$

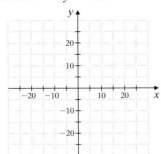

25. $y = x^2 + 4x + 6$

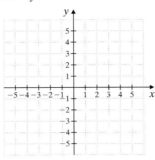

26. $y^2 + 2x^2 = 4x + 6$

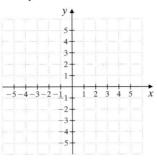

27. $x - 4y = y^2$

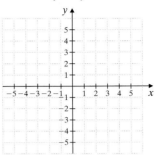

28. $x^2 - 4 = y^2$

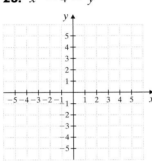

29. $x^2 = 4 - y^2$

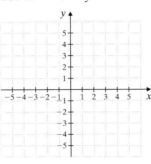

30. $36y^2 = 576 + 16x^2$

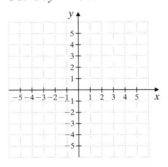

31. $3(x - 7)^2 + 3(y + 4)^2 = 1$

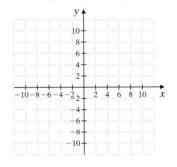

(10.3) *Solve each system of equations.*

32. $\begin{cases} y = 2x - 4 \\ y^2 = 4x \end{cases}$

33. $\begin{cases} x^2 + y^2 = 4 \\ x - y = 4 \end{cases}$

34. $\begin{cases} y = x + 2 \\ y = x^2 \end{cases}$

35. $\begin{cases} 4x - y^2 = 0 \\ 2x^2 + y^2 = 16 \end{cases}$

36. $\begin{cases} x^2 + 4y^2 = 16 \\ x^2 + y^2 = 4 \end{cases}$

37. $\begin{cases} x^2 + 2y = 9 \\ 5x - 2y = 5 \end{cases}$

38. $\begin{cases} y = 3x^2 + 5x - 4 \\ y = 3x^2 - x + 2 \end{cases}$

39. $\begin{cases} x^2 - 3y^2 = 1 \\ 4x^2 + 5y^2 = 21 \end{cases}$

△ **40.** Find the length and the width of a room whose area is 150 square feet and whose perimeter is 50 feet.

41. What is the greatest number of real number solutions possible for a system of two equations whose graphs are an ellipse and a hyperbola?

(10.4) *Graph each inequality or system of inequalities.*

42. $y \leq -x^2 + 3$

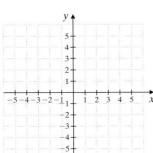

43. $x^2 + y^2 < 9$

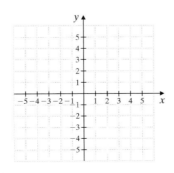

44. $\dfrac{x^2}{4} + \dfrac{y^2}{9} \geq 1$

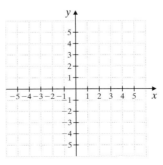

45. $\begin{cases} 2x \leq 4 \\ x + y \geq 1 \end{cases}$

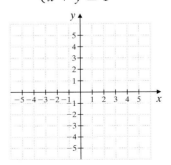

46. $\begin{cases} 3x + 4y \le 12 \\ x - 2y > 6 \end{cases}$

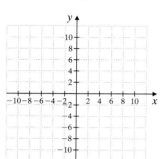

47. $\begin{cases} x^2 + y^2 \le 16 \\ x^2 + y^2 \ge 4 \end{cases}$

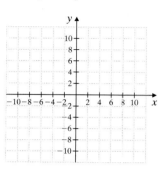

48. $\begin{cases} x^2 + y^2 < 4 \\ x^2 - y^2 \le 1 \end{cases}$

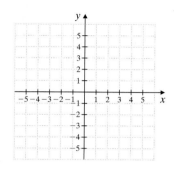

49. $\begin{cases} x^2 + y^2 < 4 \\ y \ge x^2 - 1 \\ x \ge 0 \end{cases}$

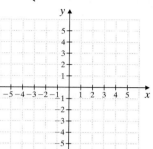

Mixed Review

50. Write an equation of the circle with center $(-7, 8)$ and radius 5.

Graph each equation.

51. $3x^2 + 6x + 3y^2 = 9$

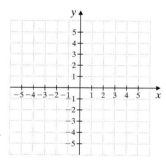

52. $y = x^2 + 6x + 9$

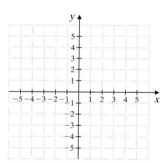

53. $x = y^2 + 6y + 9$

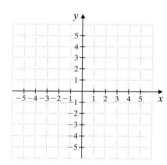

54. $\dfrac{y^2}{4} - \dfrac{x^2}{16} = 1$

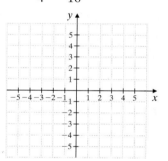

55. $\dfrac{y^2}{4} + \dfrac{x^2}{16} = 1$

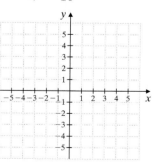

56. $\dfrac{(x-2)^2}{4} + (y-1)^2 = 1$

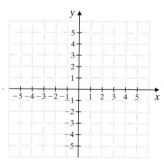

57. $y^2 = x^2 + 6$

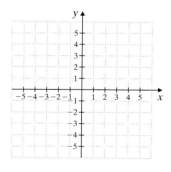

58. $y^2 + x^2 = 4x + 6$

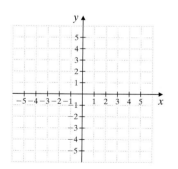

59. $x^2 + y^2 - 8y = 0$

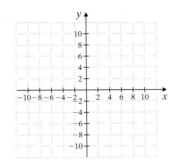

60. $6(x - 2)^2 + 9(y + 5)^2 = 36$

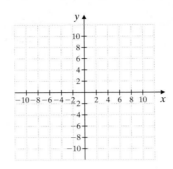

61. $\dfrac{x^2}{16} - \dfrac{y^2}{25} = 1$

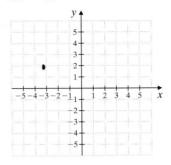

Solve each system of equations.

62. $\begin{cases} y = x^2 - 5x + 1 \\ y = -x + 6 \end{cases}$

63. $\begin{cases} x^2 + y^2 = 10 \\ 9x^2 + y^2 = 18 \end{cases}$

Graph each inequality or system of inequalities.

64. $x^2 - y^2 < 1$

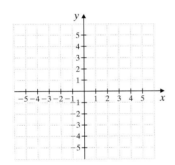

65. $\begin{cases} y > x^2 \\ x + y \geq 3 \end{cases}$

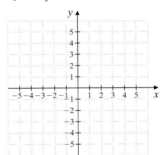

10 CHAPTER TEST

 Remember to use the Chapter Test Prep Video CD to see the fully worked-out solutions to any of the exercises you want to review.

Graph each equation.

1. $x^2 + y^2 = 36$

1. see graph

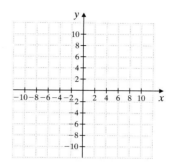

2. $x^2 - y^2 = 36$

2. see graph

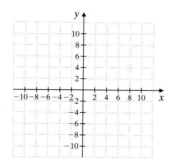

3. $16x^2 + 9y^2 = 144$

3. see graph

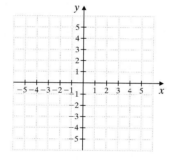

4. $y = x^2 - 8x + 16$

4. see graph

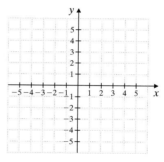

5. $x^2 + y^2 + 6x = 16$

5. see graph

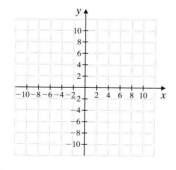

6. $x = y^2 + 8y - 3$

6. see graph

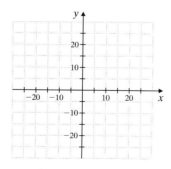

760

7. $\dfrac{(x-4)^2}{16} + \dfrac{(y-3)^2}{9} = 1$

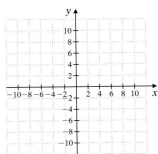

8. $y^2 - x^2 = 1$

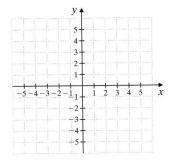

Solve each system.

9. $\begin{cases} x^2 + y^2 = 169 \\ 5x + 12y = 0 \end{cases}$

10. $\begin{cases} x^2 + y^2 = 26 \\ x^2 - 2y^2 = 23 \end{cases}$

11. $\begin{cases} y = x^2 - 5x + 6 \\ y = 2x \end{cases}$

12. $\begin{cases} x^2 + 4y^2 = 5 \\ y = x \end{cases}$

Graph each system.

13. $\begin{cases} 2x + 5y \geq 10 \\ y \geq x^2 + 1 \end{cases}$

14. $\begin{cases} \dfrac{x^2}{4} + y^2 \leq 1 \\ x + y > 1 \end{cases}$

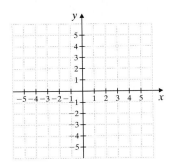

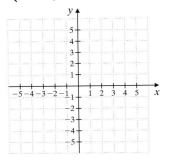

15. $\begin{cases} x^2 + y^2 > 1 \\ \dfrac{x^2}{4} - y^2 \geq 1 \end{cases}$

16. $\begin{cases} x^2 + y^2 \geq 4 \\ x^2 + y^2 < 16 \\ y \geq 0 \end{cases}$

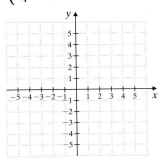

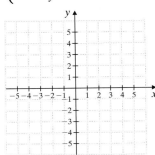

---

7. see graph

8. see graph

9. _____

10. _____

11. _____

12. _____

13. see graph

14. see graph

15. see graph

16. see graph

17. _____

17. Which graph best resembles the graph of $x = a(y - k)^2 + h$ if $a > 0, h < 0,$ and $k > 0$?

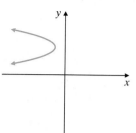

A.

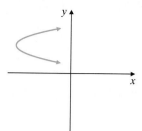

B.

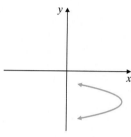

C.

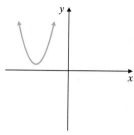

D.

18. A bridge has an arch in the shape of half an ellipse. If the equation of the ellipse, measured in feet, is $100x^2 + 225y^2 = 22{,}500$, find the height of the arch from the road and the width of the arch.

18. _____

1. Add $11x^3 - 12x^2 + x - 3$ and $x^3 - 10x + 5$.

2. Solve $3x + 4 > 1$ and $2x - 5 \le 9$. Write the solution in interval notation.

3. Multiply: $(x + 3)(2x + 5)$

4. Find the slope of the line that goes through $(3, 2)$ and $(1, -4)$.

5. Use synthetic division to divide $x^4 - 2x^3 - 11x^2 + 34$ by $x + 2$.

6. Two planes leave Greensboro, one traveling north and the other south. After 2 hours they are 650 miles apart. If one plane is flying 25 mph faster than the other, what is the speed of each?

Factor.

7. $2(x - 5) + 3a(x - 5)$

8. $3y^2 + 14y + 15$

9. $x^2 - 12x + 35$

10. $20a^5 + 54a^4 + 10a^3$

11. $(x + 3)^2 - 36$

12. $(y - 3)^2 - 2(y - 3) - 8$

13. Solve: $2x^2 + 9x - 5 = 0$

14. Solve $\dfrac{2}{x + 3} = \dfrac{1}{x^2 - 9} - \dfrac{1}{x - 3}$.

Simplify each rational expression.

15. $\dfrac{2 + x}{x + 2}$

16. $\dfrac{8x^3 - 1}{2x - 1}$

17. $\dfrac{2 - x}{x - 2}$

18. $\dfrac{4x^9 - 25x^5y^2}{2x^7 + 5x^5y}$

Perform the indicated operation. Simplify if possible.

19. $\dfrac{x^2}{x + 7} - \dfrac{49}{x + 7}$

20. $\dfrac{2}{3a - 15} - \dfrac{a}{25 - a^2}$

1. _____

2. _____

3. _____

4. _____

5. _____

6. _____

7. _____

8. _____

9. _____

10. _____

11. _____

12. _____

13. _____

14. _____

15. _____

16. _____

17. _____

18. _____

19. _____

20. _____

21. $\dfrac{x}{3y^2} - \dfrac{x+1}{3y^2}$

22. $\dfrac{x}{3x^2 - 13x - 10} + \dfrac{9}{9x^2 - 4}$

23. Simplify: $\dfrac{\dfrac{5x}{x+2}}{\dfrac{10}{x-2}}$

24. Suppose that y varies inversely as x. If $y = 3$ when $x = \dfrac{2}{3}$, find the constant of variation and the variation equation.

25. Solve: $\dfrac{2x}{x-3} + \dfrac{6-2x}{x^2-9} = \dfrac{x}{x+3}$

26. Multiply and simplify if possible.
  a. $\sqrt{5}(2 + \sqrt{15})$
  b. $(\sqrt{3} - \sqrt{5})(\sqrt{7} - 1)$
  c. $(2\sqrt{5} - 1)^2$
  d. $(3\sqrt{2} + 5)(3\sqrt{2} - 5)$

27. If a certain number is subtracted from the numerator and added to the denominator of $\dfrac{9}{19}$, the new fraction is equivalent to $\dfrac{1}{3}$. Find the number.

28. Rationalize the denominator. $\dfrac{-2}{\sqrt{3}+3}$

Find each square root. Assume that all variables represent positive real numbers.

29. $\sqrt{0}$

30. $\sqrt{\dfrac{x^7}{49}}$

31. $\sqrt{0.25}$

32. $\sqrt{90a^5b^2}$

Use rational exponents to simplify. Assume that all variables represent positive real numbers.

33. $\sqrt[8]{x^4}$

34. $\sqrt[4]{x^2y^4}$

35. $\sqrt[6]{r^2s^4}$

36. Solve $\sqrt{2x-3} = x - 3$.

Simplify. Assume that all variables represent positive real numbers.

37. $\sqrt[3]{24}$

38. $\sqrt{144y^2}$

39. $\sqrt[4]{32}$

40. $-\sqrt[3]{-27x^3}$

41. Rationalize the denominator of $\dfrac{2}{\sqrt{5}}$.

42. Use the quotient rule to simplify.
  a. $\dfrac{\sqrt{32}}{\sqrt{4}}$
  c. $\dfrac{\sqrt[5]{64x^9y^2}}{\sqrt[5]{2x^2y^{-8}}}$
  b. $\dfrac{\sqrt[3]{240y^2}}{5\sqrt[3]{3y^{-4}}}$

21. _____

22. _____

23. _____

24. _____

25. _____

26. a. _____

   b. _____

   c. _____

   d. _____

27. _____

28. _____

29. _____

30. _____

31. _____

32. _____

33. _____

34. _____

35. _____

36. _____

37. _____

38. _____

39. _____

40. _____

41. _____

42. a. _____

   b. _____

   c. _____

43. Solve: $\sqrt{-10x - 1} + 3x = 0$

44. Solve $x^3 + 2x^2 - 4x \geq 8$.

45. Multiply: $(2 - 5i)(4 + i)$

46. Find the inverse of $f(x) = \dfrac{x + 1}{2}$.

47. Solve $p^2 + 2p = 4$ by completing the square.

48. Find the maximum value of $f(x) = -x^2 - 6x + 4$.

49. Solve: $2x^2 - 4x = 3$

50. Solve each equation for x.
   a. $64^x = 4$
   b. $125^{x-3} = 25$
   c. $\dfrac{1}{81} = 3^{2x}$

51. Find the distance between $(2, -5)$ and $(1, -4)$. Give an exact distance and a three-decimal-place approximation.

52. Solve $\dfrac{4}{x - 2} - \dfrac{x}{x + 2} = \dfrac{16}{x^2 - 4}$

53. Graph: $\dfrac{x^2}{9} + \dfrac{y^2}{16} = 1$

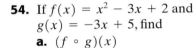

54. If $f(x) = x^2 - 3x + 2$ and $g(x) = -3x + 5$, find
   a. $(f \circ g)(x)$
   b. $(f \circ g)(-2)$
   c. $(g \circ f)(x)$
   d. $(g \circ f)(5)$

Solve each system.

55. $\begin{cases} x^2 + y^2 = 4 \\ x + y = 3 \end{cases}$

56. Solve the system $\begin{cases} x^2 + y^2 = 36 \\ y = x + 6 \end{cases}$

57. $\begin{cases} x^2 + 2y^2 = 10 \\ x^2 - y^2 = 1 \end{cases}$

43. _____

44. _____

45. _____

46. _____

47. _____

48. _____

49. _____

50. a. _____

b. _____

c. _____

51. _____

52. _____

53. see graph _____

54. a. _____

b. _____

c. _____

d. _____

55. _____

56. _____

57. _____

Copyright 2007 Pearson Education, Inc.

Objectives

A Write Equivalent Fractions.

B Write Fractions in Simplest Form.

C Multiply and Divide Fractions.

D Add and Subtract Fractions.

E Perform Operations on Mixed Numbers.

A.1 FRACTIONS

A quotient of two numbers such as $\frac{2}{9}$ is called a **fraction.** The parts of a fraction are:

Fraction bar $\rightarrow \dfrac{2}{9}$ $\begin{array}{l}\leftarrow \text{Numerator}\\ \leftarrow \text{Denominator}\end{array}$

$\frac{2}{9}$ of the circle
is shaded.

A fraction may be used to refer to part of a whole. For example, $\frac{2}{9}$ of the circle in the figure is shaded. The denominator 9 tells us how many equal parts the whole circle is divided into and the numerator 2 tells us how many equal parts are shaded.

In this section, we will use numerators that are **whole numbers** and denominators that are nonzero whole numbers. The whole numbers consist of 0 and the natural numbers.

Whole numbers: 0, 1, 2, 3, 4, 5, and so on

Objective **A** Writing Equivalent Fractions

More than one fraction can be used to name the same part of a whole. Such fractions are called **equivalent fractions.**

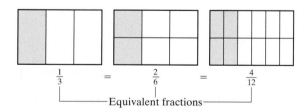

$\frac{1}{3}$ = $\frac{2}{6}$ = $\frac{4}{12}$

Equivalent fractions

Equivalent Fractions

Fractions that represent the same portion of a whole are called **equivalent fractions.**

To write equivalent fractions, we use the **fundamental principle of fractions.** This principle guarantees that, if we multiply both the numerator and the denominator by the same nonzero number, the result is an equivalent fraction. For example, if we multiply the numerator and denominator of $\frac{1}{3}$ by the same number, 2, the result is the equivalent fraction $\frac{2}{6}$.

$$\frac{1 \cdot 2}{3 \cdot 2} = \frac{2}{6}$$

Fundamental Principle of Fractions

If a, b, and c are numbers, then

$$\frac{a}{b} = \frac{a \cdot c}{b \cdot c} \quad \text{or} \quad \frac{a \cdot c}{b \cdot c} = \frac{a}{b}$$

as long as b and c are not 0.

EXAMPLE 1 Write $\frac{2}{5}$ as an equivalent fraction with a denominator of 15.

Solution: Since $5 \cdot 3 = 15$, we use the fundamental principle of fractions and multiply the numerator and denominator of $\frac{2}{5}$ by 3.

$$\frac{2}{5} = \frac{2 \cdot 3}{5 \cdot 3} = \frac{6}{15}$$

Thus $\frac{2}{5}$ is equivalent to $\frac{6}{15}$. They both represent the same part of a whole.

■ **Work Practice Problem 1**

Objective **B** **Simplifying Fractions**

A fraction is said to be **simplified** or in **lowest terms** when the numerator and the denominator have no factors in common other than 1. For example, the fraction $\frac{5}{11}$ is in lowest terms since 5 and 11 have no common factors other than 1.

One way to simplify fractions is to write both the numerator and the denominator as a product of primes and then apply the fundamental principle of fractions.

EXAMPLE 2 Simplify: $\frac{42}{49}$

Solution: We write the numerator and the denominator as products of primes. Then we apply the fundamental principle of fractions to the common factor 7.

$$\frac{42}{49} = \frac{2 \cdot 3 \cdot \boxed{7}}{7 \cdot \boxed{7}} = \frac{2 \cdot 3}{7} = \frac{6}{7}$$

■ **Work Practice Problem 2**

✔ **Concept Check** Explain the error in the following steps.

a. $\frac{15}{55} = \frac{1\boxed{5}}{5\boxed{5}} = \frac{1}{5}$ **b.** $\frac{6}{7} = \frac{\boxed{5} + 1}{\boxed{5} + 2} = \frac{1}{2}$

EXAMPLES Simplify each fraction.

3. $\frac{11}{27} = \frac{11}{3 \cdot 3 \cdot 3}$ There are no common factors other than 1, so $\frac{11}{27}$ is already simplified.

4. $\frac{88}{20} = \frac{\boxed{2} \cdot \boxed{2} \cdot 2 \cdot 11}{\boxed{2} \cdot \boxed{2} \cdot 5} = \frac{22}{5}$

■ **Work Practice Problems 3–4**

PRACTICE PROBLEM 1

Write $\frac{1}{4}$ as an equivalent fraction with a denominator of 20.

PRACTICE PROBLEM 2

Simplify: $\frac{20}{35}$

PRACTICE PROBLEMS 3–4

Simplify each fraction.

3. $\frac{7}{20}$ **4.** $\frac{12}{40}$

Answers

1. $\frac{5}{20}$,   **2.** $\frac{4}{7}$,   **3.** $\frac{7}{20}$,   **4.** $\frac{3}{10}$

✔ **Concept Check Answer**

answers may vary

A **proper fraction** is a fraction whose numerator is less than its denominator. The fraction $\frac{22}{5}$ from Example 4 is called an improper fraction. An **improper fraction** is a fraction whose numerator is greater than or equal to its denominator.

The improper fraction $\frac{22}{5}$ may be written as the mixed number $4\frac{2}{5}$. Notice that a **mixed number** has a whole number part and a fraction part. We review operations on mixed numbers in objective E in this section.

We may simplify some fractions by recalling that the fraction bar means division.

$$\frac{6}{6} = 6 \div 6 = 1 \quad \text{and} \quad \frac{3}{1} = 3 \div 1 = 3$$

PRACTICE PROBLEMS 5–8

Simplify by dividing the numerator by the denominator.

5. $\frac{4}{4}$ **6.** $\frac{9}{3}$

7. $\frac{10}{10}$ **8.** $\frac{5}{1}$

EXAMPLES Simplify by dividing the numerator by the denominator.

5. $\frac{3}{3} = 1$ Since $3 \div 3 = 1$.

6. $\frac{4}{2} = 2$ Since $4 \div 2 = 2$.

7. $\frac{7}{7} = 1$ Since $7 \div 7 = 1$.

8. $\frac{8}{1} = 8$ Since $8 \div 1 = 8$.

Work Practice Problems 5–8

In general, if the numerator and the denominator are the same nonzero number, the fraction is equivalent to 1. Also, if the denominator of a fraction is 1, the fraction is equivalent to the numerator.

If a is any number other than 0, then $\frac{a}{a} = 1$.

Also, if a is any number, $\frac{a}{1} = a$.

Objective C **Multiplying and Dividing Fractions**

To multiply two fractions, we multiply numerator times numerator to obtain the numerator of the product. Then we multiply denominator times denominator to obtain the denominator of the product.

Multiplying Fractions

$$\frac{a}{b} \cdot \frac{c}{d} = \frac{a \cdot c}{b \cdot d} \quad \text{if } b \neq 0 \text{ and } d \neq 0$$

PRACTICE PROBLEM 9

Multiply: $\frac{3}{7} \cdot \frac{3}{5}$. Simplify the product if possible.

EXAMPLE 9 Multiply: $\frac{2}{15} \cdot \frac{5}{13}$. Simplify the product if possible.

Solution: $\frac{2}{15} \cdot \frac{5}{13} = \frac{2 \cdot 5}{15 \cdot 13}$ Multiply numerators.
Multiply denominators.

To simplify the product, we divide the numerator and the denominator by any common factors.

$$\frac{2}{15} \cdot \frac{5}{13} = \frac{2 \cdot 5}{3 \cdot 5 \cdot 13} = \frac{2}{39}$$

Work Practice Problem 9

Answers

5. 1, **6.** 3, **7.** 1, **8.** 5, **9.** $\frac{9}{35}$

Before we divide fractions, we first define **reciprocals.** Two numbers are reciprocals of each other if their product is 1.

The reciprocal of $\frac{2}{3}$ is $\frac{3}{2}$ because $\frac{2}{3} \cdot \frac{3}{2} = \frac{6}{6} = 1$.

The reciprocal of 5 is $\frac{1}{5}$ because $5 \cdot \frac{1}{5} = \frac{5}{1} \cdot \frac{1}{5} = \frac{5}{5} = 1$.

To divide fractions, we multiply the first fraction by the reciprocal of the second fraction. For example,

$$\frac{1}{2} \div \frac{5}{7} = \frac{1}{2} \cdot \frac{7}{5} = \frac{1 \cdot 7}{2 \cdot 5} = \frac{7}{10}$$

Helpful Hint

To divide, multiply by the reciprocal.

Dividing Fractions

$$\frac{a}{b} \div \frac{c}{d} = \frac{a}{b} \cdot \frac{d}{c}, \quad \text{if } b \neq 0, d \neq 0, \text{ and } c \neq 0$$

EXAMPLES Divide and simplify.

10. $\dfrac{4}{5} \div \dfrac{5}{16} = \dfrac{4}{5} \cdot \dfrac{16}{5} = \dfrac{4 \cdot 16}{5 \cdot 5} = \dfrac{64}{25}$

11. $\dfrac{7}{10} \div 14 = \dfrac{7}{10} \div \dfrac{14}{1} = \dfrac{7}{10} \cdot \dfrac{1}{14} = \dfrac{7 \cdot 1}{2 \cdot 5 \cdot 2 \cdot 7} = \dfrac{1}{20}$

12. $\dfrac{3}{8} \div \dfrac{3}{10} = \dfrac{3}{8} \cdot \dfrac{10}{3} = \dfrac{3 \cdot 2 \cdot 5}{2 \cdot 2 \cdot 2 \cdot 3} = \dfrac{5}{4}$

Work Practice Problems 10–12

Objective D Adding and Subtracting Fractions

To add or subtract fractions with the same denominator, we combine numerators and place the sum or difference over the common denominator.

Adding and Subtracting Fractions with the Same Denominator

$$\frac{a}{b} + \frac{c}{b} = \frac{a + c}{b}, \quad \text{if } b \neq 0$$

$$\frac{a}{b} - \frac{c}{b} = \frac{a - c}{b}, \quad \text{if } b \neq 0$$

EXAMPLES Add or subtract as indicated. Then simplify if possible.

13. $\dfrac{2}{7} + \dfrac{4}{7} = \dfrac{2 + 4}{7} = \dfrac{6}{7}$

14. $\dfrac{3}{10} + \dfrac{2}{10} = \dfrac{3 + 2}{10} = \dfrac{5}{10} = \dfrac{5}{2 \cdot 5} = \dfrac{1}{2}$

15. $\dfrac{9}{7} - \dfrac{2}{7} = \dfrac{9 - 2}{7} = \dfrac{7}{7} = 1$

16. $\dfrac{5}{3} - \dfrac{1}{3} = \dfrac{5 - 1}{3} = \dfrac{4}{3}$

Work Practice Problems 13–16

PRACTICE PROBLEMS 10–12

Divide and simplify.

10. $\dfrac{2}{9} \div \dfrac{3}{4}$ **11.** $\dfrac{8}{11} \div 24$

12. $\dfrac{5}{4} \div \dfrac{5}{8}$

PRACTICE PROBLEMS 13–16

Add or subtract as indicated. Then simplify if possible.

13. $\dfrac{2}{11} + \dfrac{5}{11}$ **14.** $\dfrac{1}{8} + \dfrac{3}{8}$

15. $\dfrac{13}{10} - \dfrac{3}{10}$ **16.** $\dfrac{7}{6} - \dfrac{2}{6}$

Answers

10. $\dfrac{8}{27}$, **11.** $\dfrac{1}{33}$, **12.** 2, **13.** $\dfrac{7}{11}$,

14. $\dfrac{1}{2}$, **15.** 1, **16.** $\dfrac{5}{6}$

To add or subtract with different denominators, we first write the fractions as **equivalent fractions** with the same denominator. We use the smallest or **least common denominator,** or **LCD.** The LCD is the same as the least common multiple.

PRACTICE PROBLEM 17

Add: $\dfrac{3}{8} + \dfrac{1}{20}$

EXAMPLE 17 Add: $\dfrac{2}{5} + \dfrac{1}{4}$

Solution: We first must find the least common denominator before the fractions can be added. The least common multiple for the denominators 5 and 4 is 20. This is the LCD we will use.

We write both fractions as equivalent fractions with denominators of 20. Since

$$\frac{2}{5} = \frac{2 \cdot 4}{5 \cdot 4} = \frac{8}{20} \quad \text{and} \quad \frac{1}{4} = \frac{1 \cdot 5}{4 \cdot 5} = \frac{5}{20}$$

then

$$\frac{2}{5} + \frac{1}{4} = \frac{8}{20} + \frac{5}{20} = \frac{13}{20}$$

◼ **Work Practice Problem 17**

PRACTICE PROBLEM 18

Subtract and simplify: $\dfrac{8}{15} - \dfrac{1}{3}$

EXAMPLE 18 Subtract and simplify: $\dfrac{19}{6} - \dfrac{23}{12}$

Solution: The LCD is 12. We write both fractions as equivalent fractions with denominators of 12.

$$\frac{19}{6} - \frac{23}{12} = \frac{19 \cdot 2}{6 \cdot 2} - \frac{23}{12}$$

$$= \frac{38}{12} - \frac{23}{12}$$

$$= \frac{15}{12} = \frac{3 \cdot 5}{2 \cdot 2 \cdot 3} = \frac{5}{4}$$

◼ **Work Practice Problem 18**

Objective E Performing Operations on Mixed Numbers

To perform operations on mixed numbers, first write each mixed number as an improper fraction. To recall how this is done, let's write $3\dfrac{1}{5}$ as an improper fraction.

$$3\frac{1}{5} = 3 + \frac{1}{5} = \frac{15}{5} + \frac{1}{5} = \frac{16}{5}$$

Because of the steps above, notice we can use a shortcut process for writing a mixed number as an improper fraction.

$$3\frac{1}{5} = \frac{5 \cdot 3 + 1}{5} = \frac{16}{5}$$

Answers

17. $\dfrac{17}{40}$, **18.** $\dfrac{1}{5}$

EXAMPLE 19 Divide: $2\dfrac{1}{8} \div 1\dfrac{2}{3}$

Solution: First write each mixed number as an improper fraction.

$$2\dfrac{1}{8} = \dfrac{8 \cdot 2 + 1}{8} = \dfrac{17}{8}; \qquad 1\dfrac{2}{3} = \dfrac{3 \cdot 1 + 2}{3} = \dfrac{5}{3}$$

Now divide as usual.

$$2\dfrac{1}{8} \div 1\dfrac{2}{3} = \dfrac{17}{8} \div \dfrac{5}{3} = \dfrac{17}{8} \cdot \dfrac{3}{5} = \dfrac{51}{40} \quad \text{or} \quad 1\dfrac{11}{40}$$

▣ **Work Practice Problem 19**

As a general rule, if the original exercise contains mixed numbers, write the result as a mixed number, if possible.

EXAMPLE 20 Add: $2\dfrac{1}{8} + 1\dfrac{2}{3}$

Solution:

$$2\dfrac{1}{8} + 1\dfrac{2}{3} = \dfrac{17}{8} + \dfrac{5}{3} = \dfrac{51}{24} + \dfrac{40}{24} = \dfrac{91}{24} \quad \text{or} \quad 3\dfrac{19}{24}$$

▣ **Work Practice Problem 20**

When adding or subtracting larger mixed numbers, you might want to use the following method.

EXAMPLE 21 Subtract: $50\dfrac{1}{6} - 38\dfrac{1}{3}$

Solution:

$$
\begin{array}{rcccl}
50\dfrac{1}{6} & = & 50\dfrac{1}{6} & = & 49\dfrac{7}{6} \\[4pt]
-38\dfrac{1}{3} & = & -38\dfrac{2}{6} & = & -38\dfrac{2}{6} \\[4pt]
& & & & 11\dfrac{5}{6}
\end{array}
$$

$50\dfrac{1}{6} = 49 + 1 + \dfrac{1}{6} = 49\dfrac{7}{6}$

▣ **Work Practice Problem 21**

PRACTICE PROBLEM 19

Multiply: $5\dfrac{1}{6} \cdot 4\dfrac{2}{5}$

PRACTICE PROBLEM 20

Subtract: $7\dfrac{3}{8} - 6\dfrac{1}{4}$

PRACTICE PROBLEM 21

Add: $76\dfrac{1}{9} + 35\dfrac{3}{4}$

Answers

19. $22\dfrac{11}{15}$, **20.** $1\dfrac{1}{8}$, **21.** $111\dfrac{31}{36}$

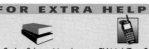

Objective Ⓐ *Write each fraction as an equivalent fraction with the given denominator. See Example 1.*

1. $\dfrac{7}{10}$ with a denominator of 30

2. $\dfrac{2}{3}$ with a denominator of 9

3. $\dfrac{2}{9}$ with a denominator of 18

4. $\dfrac{8}{7}$ with a denominator of 56

5. $\dfrac{4}{5}$ with a denominator of 20

6. $\dfrac{4}{5}$ with a denominator of 25

Objective Ⓑ *Simplify each fraction. See Examples 2 through 8.*

7. $\dfrac{2}{4}$ **8.** $\dfrac{3}{6}$ **9.** $\dfrac{10}{15}$ **10.** $\dfrac{15}{20}$ **11.** $\dfrac{3}{7}$

12. $\dfrac{5}{9}$ **13.** $\dfrac{20}{20}$ **14.** $\dfrac{24}{24}$ **15.** $\dfrac{35}{7}$ **16.** $\dfrac{42}{6}$

17. $\dfrac{18}{30}$ **18.** $\dfrac{42}{45}$ **19.** $\dfrac{16}{20}$ **20.** $\dfrac{8}{40}$ **21.** $\dfrac{66}{48}$

22. $\dfrac{64}{24}$ **23.** $\dfrac{120}{244}$ **24.** $\dfrac{360}{700}$ **25.** $\dfrac{192}{264}$ **26.** $\dfrac{455}{525}$

Objectives Ⓒ Ⓔ **Mixed Practice** *Multiply or divide as indicated. See Examples 9 through 12 and 19.*

27. $\dfrac{1}{2} \cdot \dfrac{3}{4}$ **28.** $\dfrac{10}{6} \cdot \dfrac{3}{5}$ **29.** $\dfrac{2}{3} \cdot \dfrac{3}{4}$ **30.** $\dfrac{7}{8} \cdot \dfrac{3}{21}$ **31.** $5\dfrac{1}{9} \cdot 3\dfrac{2}{3}$

32. $2\dfrac{3}{4} \cdot 1\dfrac{7}{8}$ **33.** $7\dfrac{2}{5} \div \dfrac{1}{5}$ **34.** $9\dfrac{5}{6} \div \dfrac{1}{6}$ **35.** $\dfrac{1}{2} \div \dfrac{7}{12}$ **36.** $\dfrac{7}{12} \div \dfrac{1}{2}$

37. $\dfrac{3}{4} \div \dfrac{1}{20}$ **38.** $\dfrac{3}{5} \div \dfrac{9}{10}$ **39.** $\dfrac{7}{10} \cdot \dfrac{5}{21}$ **40.** $\dfrac{3}{35} \cdot \dfrac{10}{63}$ **41.** $\dfrac{9}{20} \div 12$

42. $\dfrac{25}{36} \div 10$ **43.** $4\dfrac{2}{11} \cdot 2\dfrac{1}{2}$ **44.** $6\dfrac{6}{7} \cdot 3\dfrac{1}{2}$ **45.** $8\dfrac{3}{5} \div 2\dfrac{9}{10}$ **46.** $1\dfrac{7}{8} \div 3\dfrac{8}{9}$

Objectives Ⓓ Ⓔ **Mixed Practice** *Add or subtract as indicated. See Examples 13 through 18, 20 and 21.*

47. $\dfrac{4}{5} + \dfrac{1}{5}$ **48.** $\dfrac{6}{7} + \dfrac{1}{7}$ **49.** $\dfrac{4}{5} - \dfrac{1}{5}$ **50.** $\dfrac{6}{7} - \dfrac{1}{7}$ **51.** $\dfrac{23}{105} + \dfrac{4}{105}$

52. $\dfrac{13}{132} + \dfrac{35}{132}$ **53.** $\dfrac{17}{21} - \dfrac{10}{21}$ **54.** $\dfrac{18}{35} - \dfrac{11}{35}$ **55.** $9\dfrac{7}{8} + 2\dfrac{3}{8}$ **56.** $8\dfrac{1}{8} - 6\dfrac{3}{8}$

772

57. $5\frac{2}{5} - 3\frac{4}{5}$ **58.** $7\frac{3}{4} + 2\frac{1}{4}$ **59.** $\frac{2}{3} + \frac{3}{7}$ **60.** $\frac{3}{4} + \frac{1}{6}$ **61.** $\frac{10}{3} - \frac{5}{21}$

62. $\frac{11}{7} - \frac{3}{35}$ **63.** $\frac{10}{21} + \frac{5}{21}$ **64.** $\frac{11}{35} + \frac{3}{35}$ **65.** $\frac{5}{22} - \frac{5}{33}$ **66.** $\frac{7}{10} - \frac{8}{15}$

67. $8\frac{11}{12} - 1\frac{5}{6}$ **68.** $4\frac{7}{8} - 2\frac{3}{16}$ **69.** $17\frac{2}{5} + 30\frac{2}{3}$ **70.** $26\frac{11}{20} + 40\frac{7}{10}$

71. $\frac{12}{5} - 1$ **72.** $2 - \frac{3}{8}$ **73.** $\frac{2}{3} - \frac{5}{9} + \frac{5}{6}$ **74.** $\frac{8}{11} - \frac{1}{4} + \frac{1}{2}$

75. In your own words, describe how to add or subtract fractions.

76. In your own words, describe how to divide fractions.

Concept Extensions

Each circle below represents a whole, or 1. Determine the unknown part of the circle.

77. **78.** **79.** **80.**

81. During the 2000 Summer Olympic Games, Ellina Zvereva of Belarus took the gold medal in the women's discus throw with a distance of $224\frac{5}{12}$ feet. However, the Olympic record for the women's discus throw was set in 1988 by Martina Hellmann of East Germany with a distance of $237\frac{1}{6}$ feet. How much longer was the Olympic record discus throw than the gold medal throw in 2000? (*Source: World Almanac and Book of Facts,* 2001)

82. Approximately $\frac{41}{50}$ of all American adults agree that the U.S. federal government should support basic scientific research. What fraction of American adults do *not* agree that the U.S. federal government should support such research? (*Source:* National Science Foundation)

B.1 SOLVING SYSTEMS OF EQUATIONS USING DETERMINANTS

We have solved systems of two linear equations in two variables in four different ways: graphically, by substitution, by elimination, and by matrices. Now we analyze another method called **Cramer's rule.**

Objective **A** Evaluating 2 × 2 Determinants

Recall that a matrix is a rectangular array of numbers. If a matrix has the same number of rows and columns, it is called a **square matrix.** Examples of square matrices are

$$\begin{bmatrix} 1 & 6 \\ 5 & 2 \end{bmatrix} \qquad \begin{bmatrix} 2 & 4 & 1 \\ 0 & 5 & 2 \\ 3 & 6 & 9 \end{bmatrix}$$

A **determinant** is a real number associated with a square matrix. The determinant of a square matrix is denoted by placing vertical bars about the array of numbers. Thus,

The determinant of the square matrix $\begin{bmatrix} 1 & 6 \\ 5 & 2 \end{bmatrix}$ is $\begin{vmatrix} 1 & 6 \\ 5 & 2 \end{vmatrix}$.

The determinant of the square matrix $\begin{bmatrix} 2 & 4 & 1 \\ 0 & 5 & 2 \\ 3 & 6 & 9 \end{bmatrix}$ is $\begin{vmatrix} 2 & 4 & 1 \\ 0 & 5 & 2 \\ 3 & 6 & 9 \end{vmatrix}$.

We define the determinant of a 2 × 2 matrix first. (Recall that 2 × 2 is read "two by two." It means that the matrix has 2 rows and 2 columns.)

> ### *Determinant of a 2 × 2 Matrix*
>
> $$\begin{vmatrix} a & b \\ c & d \end{vmatrix} = ad - bc$$

EXAMPLE 1 Evaluate each determinant.

a. $\begin{vmatrix} -1 & 2 \\ 3 & -4 \end{vmatrix}$ **b.** $\begin{vmatrix} 2 & 0 \\ 7 & -5 \end{vmatrix}$

Solution: First we identify the values of a, b, c, and d. Then we perform the evaluation.

a. Here $a = -1, b = 2, c = 3,$ and $d = -4$.

$$\begin{vmatrix} -1 & 2 \\ 3 & -4 \end{vmatrix} = ad - bc = (-1)(-4) - (2)(3) = -2$$

b. In this example, $a = 2, b = 0, c = 7,$ and $d = -5$.

$$\begin{vmatrix} 2 & 0 \\ 7 & -5 \end{vmatrix} = ad - bc = 2(-5) - (0)(7) = -10$$

☐ **Work Practice Problem 1**

PRACTICE PROBLEM 1

Evaluate each determinant.

a. $\begin{vmatrix} -3 & 6 \\ 2 & 1 \end{vmatrix}$

b. $\begin{vmatrix} 4 & 5 \\ 0 & -5 \end{vmatrix}$

Answers

1. a. -15, **b.** -20

Objective **B** Using Cramer's Rule to Solve a System of Two Linear Equations

To develop Cramer's rule, we solve the system $\begin{cases} ax + by = h \\ cx + dy = k \end{cases}$ using elimination. First, we eliminate y by multiplying both sides of the first equation by d and both sides of the second equation by $-b$ so that the coefficients of y are opposites. The result is that

$$\begin{cases} d(ax + by) = d \cdot h \\ -b(cx + dy) = -b \cdot k \end{cases} \text{ simplifies to } \begin{cases} adx + bdy = hd \\ -bcx - bdy = -kb \end{cases}$$

We now add the two equations and solve for x.

$$adx + bdy = hd$$

$$\underline{-bcx - bdy = -kb}$$

$$adx - bcx = hd - kb \quad \text{Add the equations.}$$

$$(ad - bc)x = hd - kb$$

$$x = \frac{hd - kb}{ad - bc} \quad \text{Solve for } x.$$

When we replace x with $\dfrac{hd - kb}{ad - bc}$ in the equation $ax + by = h$ and solve for y, we find that $y = \dfrac{ak - ch}{ad - bc}$.

Notice that the numerator of the value of x is the determinant of

$$\begin{vmatrix} h & b \\ k & d \end{vmatrix} = hd - kb$$

Also, the numerator of the value of y is the determinant of

$$\begin{vmatrix} a & h \\ c & k \end{vmatrix} = ak - hc$$

Finally, the denominators of the values of x and y are the same and are the determinant of

$$\begin{vmatrix} a & b \\ c & d \end{vmatrix} = ad - bc$$

This means that the values of x and y can be written in determinant notation:

$$x = \frac{\begin{vmatrix} h & b \\ k & d \end{vmatrix}}{\begin{vmatrix} a & b \\ c & d \end{vmatrix}} \quad \text{and} \quad y = \frac{\begin{vmatrix} a & h \\ c & k \end{vmatrix}}{\begin{vmatrix} a & b \\ c & d \end{vmatrix}}$$

For convenience, we label the determinants D, D_x, and D_y.

x-coefficients

y-coefficients

$$\begin{vmatrix} a & b \\ c & d \end{vmatrix} = D \qquad \begin{vmatrix} h & b \\ k & d \end{vmatrix} = D_x \qquad \begin{vmatrix} a & h \\ c & k \end{vmatrix} = D_y$$

x-column replaced by constants

y-column replaced by constants

These determinant formulas for the coordinates of the solution of a system are known as **Cramer's rule.**

Cramer's Rule for Two Linear Equations in Two Variables

The solution of the system $\begin{cases} ax + by = h \\ cx + dy = k \end{cases}$ is given by

$$x = \frac{\begin{vmatrix} h & b \\ k & d \end{vmatrix}}{\begin{vmatrix} a & b \\ c & d \end{vmatrix}} = \frac{D_x}{D} \qquad y = \frac{\begin{vmatrix} a & h \\ c & k \end{vmatrix}}{\begin{vmatrix} a & b \\ c & d \end{vmatrix}} = \frac{D_y}{D}$$

as long as $D = ad - bc$ is not 0.

When $D = 0$, the system is either inconsistent or the equations are dependent. When this happens, we need to use another method to see which is the case.

PRACTICE PROBLEM 2

Use Cramer's rule to solve the system.

$$\begin{cases} x - y = -4 \\ 2x + 3y = 2 \end{cases}$$

EXAMPLE 2 Use Cramer's rule to solve the system:

$$\begin{cases} 3x + 4y = -7 \\ x - 2y = -9 \end{cases}$$

Solution: First we find D, D_x, and D_y.

$$\begin{array}{ccc} a & b & h \\ \downarrow & \downarrow & \downarrow \end{array}$$

$$\begin{cases} 3x + 4y = -7 \\ x - 2y = -9 \end{cases}$$

$$\begin{array}{ccc} \uparrow & \uparrow & \uparrow \\ c & d & k \end{array}$$

$$D = \begin{vmatrix} a & b \\ c & d \end{vmatrix} = \begin{vmatrix} 3 & 4 \\ 1 & -2 \end{vmatrix} = 3(-2) - 4(1) = -10$$

$$D_x = \begin{vmatrix} h & b \\ k & d \end{vmatrix} = \begin{vmatrix} -7 & 4 \\ -9 & -2 \end{vmatrix} = (-7)(-2) - 4(-9) = 50$$

$$D_y = \begin{vmatrix} a & h \\ c & d \end{vmatrix} = \begin{vmatrix} 3 & -7 \\ 1 & -9 \end{vmatrix} = 3(-9) - (-7)(1) = -20$$

Then $x = \dfrac{D_x}{D} = \dfrac{50}{-10} = -5$ and $y = \dfrac{D_y}{D} = \dfrac{-20}{-10} = 2$.

The ordered pair solution is $(-5, 2)$.
As always, check the solution in both original equations.

Work Practice Problem 2

PRACTICE PROBLEM 3

Use Cramer's rule to solve the system.

$$\begin{cases} 4x + y = 3 \\ 2x - 3y = -9 \end{cases}$$

EXAMPLE 3 Use Cramer's rule to solve the system:

$$\begin{cases} 5x + y = 5 \\ -7x - 2y = -7 \end{cases}$$

Solution: First we find D, D_x, and D_y.

$$D = \begin{vmatrix} 5 & 1 \\ -7 & -2 \end{vmatrix} = 5(-2) - (-7)(1) = -3$$

$$D_x = \begin{vmatrix} 5 & 1 \\ -7 & -2 \end{vmatrix} = 5(-2) - (-7)(1) = -3$$

$$D_y = \begin{vmatrix} 5 & 5 \\ -7 & -7 \end{vmatrix} = 5(-7) - 5(-7) = 0$$

Answers
2. $(-2, 2)$, **3.** $(0, 3)$

Then

$$x = \frac{D_x}{D} = \frac{-3}{-3} = 1 \qquad y = \frac{D_y}{D} = \frac{0}{-3} = 0$$

The ordered pair solution is $(1, 0)$.

■ **Work Practice Problem 3**

Objective C Evaluating 3 × 3 Determinants

A 3×3 determinant can be used to solve a system of three equations in three variables. The determinant of a 3×3 matrix, however, is considerably more complex than a 2×2 one.

Determinant of a 3 × 3 Matrix

$$\begin{vmatrix} a_1 & b_1 & c_1 \\ a_2 & b_2 & c_2 \\ a_3 & b_3 & c_3 \end{vmatrix} = a_1 \cdot \begin{vmatrix} b_2 & c_2 \\ b_3 & c_3 \end{vmatrix} - a_2 \cdot \begin{vmatrix} b_1 & c_1 \\ b_3 & c_3 \end{vmatrix} + a_3 \cdot \begin{vmatrix} b_1 & c_1 \\ b_2 & c_2 \end{vmatrix}$$

Notice that the determinant of a 3×3 matrix is related to the determinants of three 2×2 matrices. Each determinant of these 2×2 matrices is called a **minor,** and every element of a 3×3 matrix has a minor associated with it. For example, the minor of c_2 is the determinant of the 2×2 matrix found by deleting the row and column containing c_2.

$$\begin{matrix} a_1 & b_1 & c_1 \\ a_2 & b_2 & c_2 \\ a_3 & b_3 & c_3 \end{matrix} \quad \text{The minor of } c_2 \text{ is} \quad \begin{vmatrix} a_1 & b_1 \\ a_3 & b_3 \end{vmatrix}$$

Also, the minor of element a_1 is the determinant of the 2×2 matrix that has no row or column containing a_1.

$$\begin{matrix} a_1 & b_1 & c_1 \\ a_2 & b_2 & c_2 \\ a_3 & b_3 & c_3 \end{matrix} \quad \text{The minor of } a_1 \text{ is} \quad \begin{vmatrix} b_2 & c_2 \\ b_3 & c_3 \end{vmatrix}$$

So the determinant of a 3×3 matrix can be written as

$$a_1 \cdot (\text{minor of } a_1) - a_2 \cdot (\text{minor of } a_2) + a_3 \cdot (\text{minor of } a_3)$$

Finding the determinant by using minors of elements in the first column is called **expanding** by the minors of the first column. *The value of a determinant can be found by expanding by the minors of any row or column.* The following **array of signs** is helpful in determining whether to add or subtract the product of an element and its minor.

$$\begin{matrix} + & - & + \\ - & + & - \\ + & - & + \end{matrix}$$

If an element is in a position marked $+$, we add. If marked $-$, we subtract.

✔ **Concept Check** Suppose you are interested in finding the determinant of a 4×4 matrix. Study the pattern shown in the array of signs for a 3×3 matrix. Use the pattern to expand the array of signs for use with a 4×4 matrix.

EXAMPLE 4 Evaluate by expanding by the minors of the given row or column.

$$\begin{vmatrix} 0 & 5 & 1 \\ 1 & 3 & -1 \\ -2 & 2 & 4 \end{vmatrix}$$

a. First column   **b.** Second row

Continued on next page

PRACTICE PROBLEM 4

Evaluate by expanding by the minors of the given row or column.

a. First column   **b.** Third row

$$\begin{vmatrix} 2 & 0 & 1 \\ -1 & 3 & 2 \\ 5 & 1 & 4 \end{vmatrix}$$

Answers
4. a. 4,   **b.** 4

✔ **Concept Check Answer**

$$\begin{matrix} + & - & + & - \\ - & + & - & + \\ + & - & + & - \\ - & + & - & + \end{matrix}$$

Solution:

a. The elements of the first column are $0, 1$, and -2. The first column of the array of signs is $+, -, +$.

$$\begin{vmatrix} 0 & 5 & 1 \\ 1 & 3 & -1 \\ -2 & 2 & 4 \end{vmatrix} = 0 \cdot \begin{vmatrix} 3 & -1 \\ 2 & 4 \end{vmatrix} - 1 \cdot \begin{vmatrix} 5 & 1 \\ 2 & 4 \end{vmatrix} + (-2) \cdot \begin{vmatrix} 5 & 1 \\ 3 & -1 \end{vmatrix}$$

$$= 0(12 - (-2)) - 1(20 - 2) + (-2)(-5 - 3)$$
$$= 0 - 18 + 16 = -2$$

b. The elements of the second row are $1, 3$, and -1. This time, the signs begin with $-$ and again alternate.

$$\begin{vmatrix} 0 & 5 & 1 \\ 1 & 3 & -1 \\ -2 & 2 & 4 \end{vmatrix} = -1 \cdot \begin{vmatrix} 5 & 1 \\ 2 & 4 \end{vmatrix} + 3 \cdot \begin{vmatrix} 0 & 1 \\ -2 & 4 \end{vmatrix} - (-1) \cdot \begin{vmatrix} 0 & 5 \\ -2 & 2 \end{vmatrix}$$

$$= -1(20 - 2) + 3(0 - (-2)) - (-1)(0 - (-10))$$
$$= -18 + 6 + 10 = -2$$

Notice that the determinant of the 3×3 matrix is the same regardless of the row or column you select to expand by.

Work Practice Problem 4

✔ Concept Check Why would expanding by minors of the second row be a good choice for the determinant $\begin{vmatrix} 3 & 4 & -2 \\ 5 & 0 & 0 \\ 6 & -3 & 7 \end{vmatrix}$?

Objective D Using Cramer's Rule to Solve a System of Three Linear Equations

A system of three equations in three variables may be solved with Cramer's rule also. Using the elimination process to solve a system with unknown constants as coefficients leads to the following.

Cramer's Rule for Three Equations in Three Variables

The solution of the system $\begin{cases} a_1x + b_1y + c_1z = k_1 \\ a_2x + b_2y + c_2z = k_2 \\ a_3x + b_3y + c_3z = k_3 \end{cases}$ is given by

$$x = \frac{D_x}{D} \qquad y = \frac{D_y}{D} \qquad \text{and} \qquad z = \frac{D_z}{D}$$

where

$$D = \begin{vmatrix} a_1 & b_1 & c_1 \\ a_2 & b_2 & c_2 \\ a_3 & b_3 & c_3 \end{vmatrix} \qquad D_x = \begin{vmatrix} k_1 & b_1 & c_1 \\ k_2 & b_2 & c_2 \\ k_3 & b_3 & c_3 \end{vmatrix}$$

$$D_y = \begin{vmatrix} a_1 & k_1 & c_1 \\ a_2 & k_2 & c_2 \\ a_3 & k_3 & c_3 \end{vmatrix} \qquad D_z = \begin{vmatrix} a_1 & b_1 & k_1 \\ a_2 & b_2 & k_2 \\ a_3 & b_3 & k_3 \end{vmatrix}$$

as long as D is not 0.

✔ Concept Check Answer

Two elements of the second row are 0, which makes calculations easier.

EXAMPLE 5 Use Cramer's rule to solve the system:

$$\begin{cases} x - 2y + z = 4 \\ 3x + y - 2z = 3 \\ 5x + 5y + 3z = -8 \end{cases}$$

Solution: First we find $D, D_x, D_y,$ and D_z. Beginning with D, we expand by the minors of the first column.

$$D = \begin{vmatrix} 1 & -2 & 1 \\ 3 & 1 & -2 \\ 5 & 5 & 3 \end{vmatrix} = 1 \cdot \begin{vmatrix} 1 & -2 \\ 5 & 3 \end{vmatrix} - 3 \cdot \begin{vmatrix} -2 & 1 \\ 5 & 3 \end{vmatrix} + 5 \cdot \begin{vmatrix} -2 & 1 \\ 1 & -2 \end{vmatrix}$$

$$= 1(3 - (-10)) - 3(-6 - 5) + 5(4 - 1)$$

$$= 13 + 33 + 15 = 61$$

$$D_x = \begin{vmatrix} 4 & -2 & 1 \\ 3 & 1 & -2 \\ -8 & 5 & 3 \end{vmatrix} = 4 \cdot \begin{vmatrix} 1 & -2 \\ 5 & 3 \end{vmatrix} - 3 \cdot \begin{vmatrix} -2 & 1 \\ 5 & 3 \end{vmatrix} + (-8) \cdot \begin{vmatrix} -2 & 1 \\ 1 & -2 \end{vmatrix}$$

$$= 4(3 - (-10)) - 3(-6 - 5) + (-8)(4 - 1)$$

$$= 52 + 33 - 24 = 61$$

$$D_y = \begin{vmatrix} 1 & 4 & 1 \\ 3 & 3 & -2 \\ 5 & -8 & 3 \end{vmatrix} = 1 \cdot \begin{vmatrix} 3 & -2 \\ -8 & 3 \end{vmatrix} - 3 \cdot \begin{vmatrix} 4 & 1 \\ -8 & 3 \end{vmatrix} + 5 \cdot \begin{vmatrix} 4 & 1 \\ 3 & -2 \end{vmatrix}$$

$$= 1(9 - 16) - 3(12 - (-8)) + 5(-8 - 3)$$

$$= -7 - 60 - 55 = -122$$

$$D_z = \begin{vmatrix} 1 & -2 & 4 \\ 3 & 1 & 3 \\ 5 & 5 & -8 \end{vmatrix} = 1 \cdot \begin{vmatrix} 1 & 3 \\ 5 & -8 \end{vmatrix} - 3 \cdot \begin{vmatrix} -2 & 4 \\ 5 & -8 \end{vmatrix} + 5 \cdot \begin{vmatrix} -2 & 4 \\ 1 & 3 \end{vmatrix}$$

$$= 1(-8 - 15) - 3(16 - 20) + 5(-6 - 4)$$

$$= -23 + 12 - 50 = -61$$

From these determinants, we calculate the solution:

$$x = \frac{D_x}{D} = \frac{61}{61} = 1 \quad y = \frac{D_y}{D} = \frac{-122}{61} = -2 \quad z = \frac{D_z}{D} = \frac{-61}{61} = -1$$

The ordered triple solution is $(1, -2, -1)$. Check this solution by verifying that it satisfies each equation of the system.

🖥 **Work Practice Problem 5**

PRACTICE PROBLEM 5

Use Cramer's rule to solve the system:

$$\begin{cases} x + 2y - z = 3 \\ 2x - 3y + z = -9 \\ -x + y - 2z = 0 \end{cases}$$

Answer

5. $(-1, 3, 2)$

Mental Math

Evaluate each determinant mentally.

1. $\begin{vmatrix} 7 & 2 \\ 0 & 8 \end{vmatrix}$

2. $\begin{vmatrix} 6 & 0 \\ 1 & 2 \end{vmatrix}$

3. $\begin{vmatrix} -4 & 2 \\ 0 & 8 \end{vmatrix}$

4. $\begin{vmatrix} 5 & 0 \\ 3 & -5 \end{vmatrix}$

5. $\begin{vmatrix} -2 & 0 \\ 3 & -10 \end{vmatrix}$

6. $\begin{vmatrix} -1 & 4 \\ 0 & -18 \end{vmatrix}$

B EXERCISE SET

Student Solutions Manual   PH Math/Tutor Center   CD/Video for Review   Math XL MathXL®   MyMathLab MyMathLab

Objective A *Evaluate each determinant. See Example 1.*

 1. $\begin{vmatrix} 3 & 5 \\ -1 & 7 \end{vmatrix}$

2. $\begin{vmatrix} -5 & 1 \\ 1 & -4 \end{vmatrix}$

3. $\begin{vmatrix} 9 & -2 \\ 4 & -3 \end{vmatrix}$

4. $\begin{vmatrix} 4 & -1 \\ 9 & 8 \end{vmatrix}$

5. $\begin{vmatrix} -2 & 9 \\ 4 & -18 \end{vmatrix}$

6. $\begin{vmatrix} -40 & 8 \\ 70 & -14 \end{vmatrix}$

 7. $\begin{vmatrix} \dfrac{3}{4} & \dfrac{5}{2} \\ -\dfrac{1}{6} & \dfrac{7}{3} \end{vmatrix}$

8. $\begin{vmatrix} \dfrac{5}{7} & \dfrac{1}{3} \\ \dfrac{6}{7} & \dfrac{2}{3} \end{vmatrix}$

Objective B *Use Cramer's rule, if possible, to solve each system of linear equations. See Examples 2 and 3.*

9. $\begin{cases} 2y - 4 = 0 \\ x + 2y = 5 \end{cases}$

10. $\begin{cases} 4x - y = 5 \\ 3x - 3 = 0 \end{cases}$

11. $\begin{cases} 3x + y = 1 \\ 2y = 2 - 6x \end{cases}$

12. $\begin{cases} y = 2x - 5 \\ 8x - 4y = 20 \end{cases}$

13. $\begin{cases} 5x - 2y = 27 \\ -3x + 5y = 18 \end{cases}$

14. $\begin{cases} 4x - y = 9 \\ 2x + 3y = -27 \end{cases}$

 15. $\begin{cases} 2x - 5y = 4 \\ x + 2y = -7 \end{cases}$

16. $\begin{cases} 3x - y = 2 \\ -5x + 2y = 0 \end{cases}$

17. $\begin{cases} \dfrac{2}{3}x - \dfrac{3}{4}y = -1 \\ -\dfrac{1}{6}x + \dfrac{3}{4}y = \dfrac{5}{2} \end{cases}$

18. $\begin{cases} \dfrac{1}{2}x - \dfrac{1}{3}y = -3 \\ \dfrac{1}{8}x + \dfrac{1}{6}y = 0 \end{cases}$

Objective C *Evaluate. See Example 4.*

19. $\begin{vmatrix} 2 & 1 & 0 \\ 0 & 5 & -3 \\ 4 & 0 & 2 \end{vmatrix}$

20. $\begin{vmatrix} -6 & 4 & 2 \\ 1 & 0 & 5 \\ 0 & 3 & 1 \end{vmatrix}$

21. $\begin{vmatrix} 4 & -6 & 0 \\ -2 & 3 & 0 \\ 4 & -6 & 1 \end{vmatrix}$

22. $\begin{vmatrix} 5 & 2 & 1 \\ 3 & -6 & 0 \\ -2 & 8 & 0 \end{vmatrix}$

23. $\begin{vmatrix} 1 & 0 & 4 \\ 1 & -1 & 2 \\ 3 & 2 & 1 \end{vmatrix}$

24. $\begin{vmatrix} 0 & 1 & 2 \\ 3 & -1 & 2 \\ 3 & 2 & -2 \end{vmatrix}$

25. $\begin{vmatrix} 3 & 6 & -3 \\ -1 & -2 & 3 \\ 4 & -1 & 6 \end{vmatrix}$

26. $\begin{vmatrix} 2 & -2 & 1 \\ 4 & 1 & 3 \\ 3 & 1 & 2 \end{vmatrix}$

Objective **D** *Use Cramer's rule, if possible, to solve each system of linear equations. See Example 5.*

27. $\begin{cases} 3x \quad\quad + z = -1 \\ -x - 3y + z = \quad 7 \\ \quad\quad 3y + z = \quad 5 \end{cases}$

28. $\begin{cases} \quad 4y - 3z = -2 \\ 8x - 4y \quad\quad = \quad 4 \\ -8x + 4y + \quad z = -2 \end{cases}$

29. $\begin{cases} x + \quad y + \quad z = \quad 8 \\ 2x - \quad y - \quad z = 10 \\ x - 2y + 3z = 22 \end{cases}$

30. $\begin{cases} 5x + y + 3z = \quad 1 \\ x - y - 3z = -7 \\ -x + y \quad\quad = \quad 1 \end{cases}$

31. $\begin{cases} 2x + 2y + \quad z = 1 \\ -x + \quad y + 2z = 3 \\ x + 2y + 4z = 0 \end{cases}$

32. $\begin{cases} 2x - 3y + \quad z = 5 \\ x + \quad y + \quad z = 0 \\ 4x + 2y + 4z = 4 \end{cases}$

33. $\begin{cases} x - 2y + \quad z = -5 \\ \quad\quad 3y + 2z = \quad 4 \\ 3x - \quad y \quad\quad = -2 \end{cases}$

34. $\begin{cases} 4x + 5y \quad\quad = 10 \\ \quad\quad 3y + 2z = -6 \\ x + \quad y + \quad z = \quad 3 \end{cases}$

Concept Extensions

Find the value of x that will make each a true statement.

35. $\begin{vmatrix} 1 & x \\ 2 & 7 \end{vmatrix} = -3$

36. $\begin{vmatrix} 6 & 1 \\ -2 & x \end{vmatrix} = 26$

37. If all the elements in a single row of a determinant are zero, what is the value of the determinant? Explain your answer.

38. If all the elements in a single column of a determinant are 0, what is the value of the determinant? Explain your answer.

Appendix

The Bigger Picture–Solving Equations and Inequalities

I. Equations

A. Linear Equations (Sec. 2.1)

$$5(x - 2) = \frac{4(2x + 1)}{3}$$

$$3 \cdot 5(x - 2) = \cancel{3} \cdot \frac{4(2x + 1)}{\cancel{3}}$$

$$15x - 30 = 8x + 4$$

$$7x = 34$$

$$x = \frac{34}{7}$$

B. Absolute Value Equations (Sec. 2.6)

$$|3x - 1| = 8$$

$3x - 1 = 8$ or $3x - 1 = -8$

$3x = 9$ or $3x = -7$

$x = 3$ or $x = -\dfrac{7}{3}$

$$|x - 5| = |x + 1|$$

$x - 5 = x + 1$ or $x - 5 = -(x + 1)$

$\underbrace{-5 = 1}_{\text{No solution}}$ or $x - 5 = -x - 1$

or $2x = 4$

$x = 2$

C. Quadratic and Higher-Degree Equations (Secs. 5.7, 8.1, 8.2, 8.3)

$$2x^2 - 7x = 9$$

$$2x^2 - 7x - 9 = 0$$

$$(2x - 9)(x + 1) = 0$$

$2x - 9 = 0$ or $x + 1 = 0$

$x = \dfrac{9}{2}$ or $x = -1$

$$2x^2 + x - 2 = 0$$

$$a = 2, \quad b = 1, \quad c = -2$$

$$x = \frac{-1 \pm \sqrt{1^2 - 4(2)(-2)}}{2 \cdot 2}$$

$$x = \frac{-1 \pm \sqrt{17}}{4}$$

D. Equations with Rational Expressions (Sec. 6.4)

$$\frac{7}{x - 1} + \frac{3}{x + 1} = \frac{x + 3}{x^2 - 1}$$

$$\cancel{(x - 1)}(x + 1) \cdot \frac{7}{\cancel{x - 1}} + (x - 1)\cancel{(x + 1)} \cdot \frac{3}{\cancel{x + 1}}$$

$$= \cancel{(x - 1)}\cancel{(x + 1)} \cdot \frac{x + 3}{\cancel{(x - 1)}\cancel{(x + 1)}}$$

$$7(x + 1) + 3(x - 1) = x + 3$$

$$7x + 7 + 3x - 3 = x + 3$$

$$9x = -1$$

$$x = -\frac{1}{9}$$

782

E. Equations with Radicals (Sec. 7.6)

$$\sqrt{5x + 10} - 2 = x$$
$$\sqrt{5x + 10} = x + 2$$
$$(\sqrt{5x + 10})^2 = (x + 2)^2$$
$$5x + 10 = x^2 + 4x + 4$$
$$0 = x^2 - x - 6$$
$$0 = (x - 3)(x + 2)$$
$$x - 3 = 0 \quad \text{or} \quad x + 2 = 0$$
$$x = 3 \quad \text{or} \quad x = -2$$

Both solutions check.

F. Exponential Equations (Secs. 9.3, 9.7)

$$9^x = 27^{x+1}$$
$$(3^2)^x = (3^3)^{x+1}$$
$$3^{2x} = 3^{3x+3}$$
$$2x = 3x + 3$$
$$-3 = x$$

$$5^x = 7$$
$$\log 5^x = \log 7$$
$$x \log 5 = \log 7$$
$$x = \frac{\log 7}{\log 5}$$

G. Logarithmic Equations (Sec. 9.7)

$$\log 7 + \log(x + 3) = \log 5$$
$$\log 7(x + 3) = \log 5$$
$$7(x + 3) = 5$$
$$7x + 21 = 5$$
$$7x = -16$$
$$x = \frac{-16}{7}$$

II. Inequalities

A. Linear Inequalities (Sec. 2.4)

$$-3(x + 2) \geq 6$$
$$-3x - 6 \geq 6$$
$$-3x \geq 12$$
$$\frac{-3x}{-3} \leq \frac{12}{-3}$$
$$x \leq -4 \quad \text{or} \quad (-\infty, -4]$$

B. Compound Inequalities (Sec. 2.5)

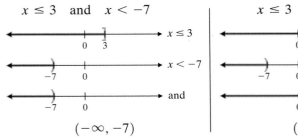

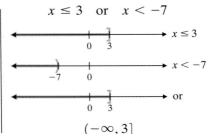

C. Absolute Value Inequalities (Sec. 2.6)

$$|x - 5| - 8 < -2$$
$$|x - 5| < 6$$
$$-6 < x - 5 < 6$$
$$-1 < x < 11$$
$$(-1, 11)$$

$$|2x + 1| \geq 17$$
$$2x + 1 \geq 17 \quad \text{or} \quad 2x + 1 \leq -17$$
$$2x \geq 16 \quad \text{or} \quad 2x \leq -18$$
$$x \geq 8 \quad \text{or} \quad x \leq -9$$
$$(-\infty, -9] \cup [8, \infty)$$

D. Nonlinear Inequalities (Sec. 8.4)

$$x^2 - x < 6$$
$$x^2 - x - 6 < 0$$
$$(x - 3)(x + 2) < 0$$

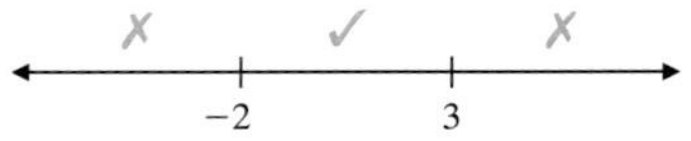

$$(-2, 3)$$

$$\frac{x - 5}{x + 1} \geq 0$$

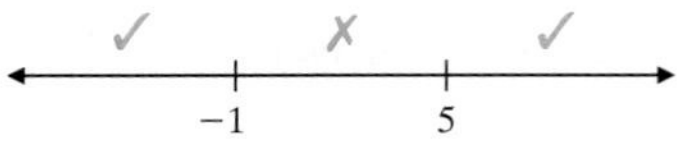

$$(-\infty, -1) \cup [5, \infty)$$

D

D.1 AN INTRODUCTION TO USING A GRAPHING UTILITY

A View Window and Interpret Window Settings.

B Graph Equations and Use Square Viewing Windows.

Objective **A** Viewing Window and Interpreting Window Settings

In this appendix, we will use the term **graphing utility** to mean a graphing calculator or a computer software graphing package. All graphing utilities graph equations by plotting points on a screen. While plotting several points can be slow and sometimes tedious for us, a graphing utility can quickly and accurately plot hundreds of points. How does a graphing utility show plotted points? A computer or calculator screen is made up of a grid of small rectangular areas called **pixels.** If a pixel contains a point to be plotted, the pixel is turned "on"; otherwise, the pixel remains "off." The graph of an equation is then a collection of pixels turned "on." The graph of $y = 3x + 1$ from a graphing calculator is shown in Figure D-1. Notice the irregular shape of the line caused by the rectangular pixels.

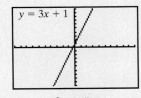

Figure D-1

The portion of the coordinate plane shown on the screen in Figure D-1 is called the **viewing window** or the **viewing rectangle.** Notice the x-axis and the y-axis on the graph. While tick marks are shown on the axes, they are not labeled. This means that from this screen alone, we do not know how many units each tick mark represents. To see what each tick mark represents and the minimum and maximum values on the axes, check the *window setting* of the graphing utility. It defines the viewing window. The window of the graph of $y = 3x + 1$ shown in Figure D-1 has the following setting (Figure D-2):

$Xmin = -10$ The minimum x-value is -10.

$Xmax = 10$ The maximum x-value is 10.

$Xscl = 1$ The x-axis scale is 1 unit per tick mark.

$Ymin = -10$ The minimum y-value is -10.

$Ymax = 10$ The maximum y-value is 10.

$Yscl = 1$ The y-axis scale is 1 unit per tick mark.

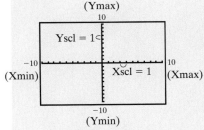

Figure D-2

By knowing the scale, we can find the minimum and the maximum values on the axes simply by counting tick marks. For example, if both the Xscl (x-axis scale) and the Yscl are 1 unit per tick mark on the graph in Figure D-3, we can count the tick marks and find that the minimum x-value is -10 and the maximum x-value is 10. Also, the minimum y-value is -10 and the maximum y-value is 10. If the Xscl (x-axis scale) changes to 2 units per tick mark (shown in Figure D-4), by counting tick marks, we see that the minimum x-value is now -20 and the maximum x-value is now 20.

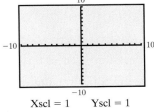

Xscl = 1   Yscl = 1

Figure D-3

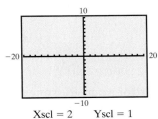

Xscl = 2   Yscl = 1

Figure D-4

785

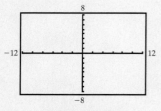

Figure D-5

It is also true that if we know the Xmin and the Xmax values, we can calculate the Xscl by the displayed axes. For example, the Xscl of the graph in Figure D-5 must be 2 units per tick mark for the maximum and minimum x-values to be as shown. Also, the Yscl of that graph must be 1 unit per tick mark for the maximum and minimum y-values to be as shown.

We will call the viewing window in Figure D-3 a *standard* viewing window or rectangle. Although a standard viewing window is sufficient for much of this text, special care must be taken to ensure that all key features of a graph are shown. Figures D-6, D-7, and D-8 show the graph of $y = x^2 + 11x - 1$ on three different viewing windows. Note that certain viewing windows for this equation are misleading.

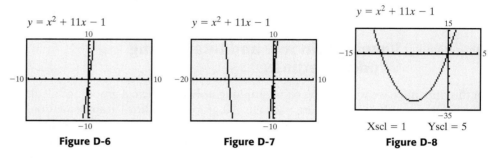

| Figure D-6 | Figure D-7 | Figure D-8 |

How do we ensure that all distinguishing features of the graph of an equation are shown? It helps to know about the equation that is being graphed. For example, the equation $y = x^2 + 11x - 1$ is not a linear equation, and its graph is not a line. This equation is a quadratic equation, and therefore its graph is a parabola. By knowing this information, we know that the graph shown in Figure D-6, although correct, is misleading. Of the three viewing rectangles shown, the graph in Figure D-8 is best because it shows more of the distinguishing features of the parabola. Properties of equations needed for graphing will be studied in this text.

Objective B Graphing Equations and Square Viewing Windows

In general, the following steps may be used to graph an equation on a standard viewing window.

To Graph an Equation in x and y with a Graphing Utility on a Standard Viewing Window

Step 1: Solve the equation for y.

Step 2: Use your graphing utility and enter the equation in the form
$Y = expression\ involving\ x$

Step 3: Activate the graphing utility.

Special care must be taken when entering the *expression involving x* in *Step 2*. You must be sure that the graphing utility you are using interprets the expression as you want it to. For example, let's graph $3y = 4x$. To do so,

Step 1: Solve the equation for y.

$$3y = 4x \qquad \frac{3y}{3} = \frac{4x}{3} \qquad y = \frac{4}{3}x$$

Step 2: Using your graphing utility, enter the expression $\frac{4}{3}x$ after the Y = prompt. In order for your graphing utility to correctly interpret the expression, you may need to enter $(4/3)x$ or $(4 \div 3)x$.

Step 3: Activate the graphing utility. The graph should appear as in Figure D-9.

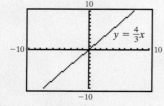

Figure D-9

Distinguishing features of the graph of a line include showing all the intercepts of the line. For example, the window of the graph of the line in Figure D-10 does not show both intercepts of the line, but the window of the graph of the same line in Figure D-11 does show both intercepts.

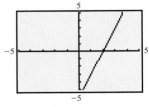

Figure D-10

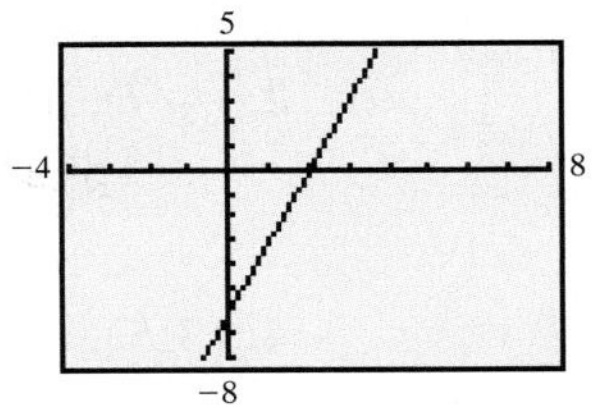

Figure D-11

On a standard viewing window, the tick marks on the y-axis are closer than the tick marks on the x-axis. This happens because the viewing window is a rectangle, and so 10 equally spaced tick marks on the positive y-axis will be closer together than 10 equally spaced tick marks on the positive x-axis. This causes the appearance of graphs to be distorted.

For example, notice the different appearances of the same line graphed using different viewing windows. The line in Figure D-12 is distorted because the tick marks along the x-axis are farther apart than the tick marks along the y-axis. The graph of the same line in Figure D-13 is not distorted because the viewing rectangle has been selected so that there is equal spacing between tick marks on both axes.

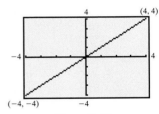

Figure D-12

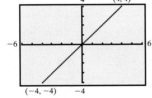

Figure D-13

We say that the line in Figure D-13 is graphed on a *square* setting. Some graphing utilities have a built-in program that, if activated, will automatically provide a square setting. A square setting is especially helpful when we are graphing perpendicular lines, circles, or when a true geometric perspective is desired. Some examples of square screens are shown in Figures D-14 and D-15.

Other features of a graphing utility such as Trace, Zoom, Intersect, and Table are discussed in appropriate Graphing Calculator Explorations in this text.

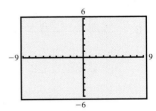

Figure D-14

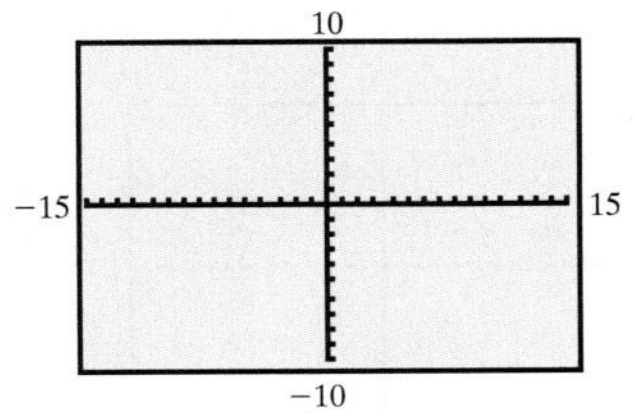

Figure D-15

Objective A *In Exercises 1–4, determine whether all ordered pairs listed will lie within a standard viewing rectangle.*

1. $(-9, 0), (5, 8), (1, -8)$

2. $(4, 7), (0, 0), (-8, 9)$

3. $(-11, 0), (2, 2), (7, -5)$

4. $(3, 5), (-3, -5), (15, 0)$

In Exercises 5–10, choose an Xmin, Xmax, Ymin, and Ymax so that all ordered pairs listed will lie within the viewing rectangle.

5. $(-90, 0), (55, 80), (0, -80)$

6. $(4, 70), (20, 20), (-18, 90)$

7. $(-11, 0), (2, 2), (7, -5)$

8. $(3, 5), (-3, -5), (15, 0)$

9. $(200, 200), (50, -50), (70, -50)$

10. $(40, 800), (-30, 500), (15, 0)$

Write the window setting for each viewing window shown. Use the following format:

Xmin =   Ymin =
Xmax =   Ymax =
Xscl =   Yscl =

11.
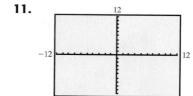

12.
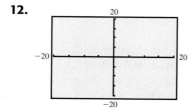

13.
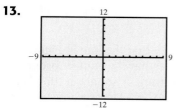

14.
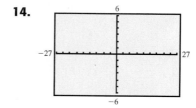

15.
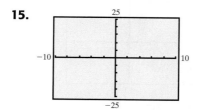

16.
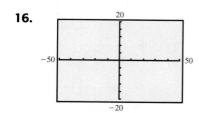

17.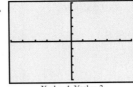
Xscl = 1, Yscl = 3

18.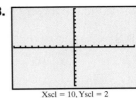
Xscl = 10, Yscl = 2

19.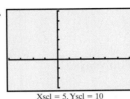
Xscl = 5, Yscl = 10

20.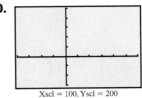
Xscl = 100, Yscl = 200

Objective **B** *Graph each linear equation in two variables, using the two different range settings given. Determine which setting shows all intercepts of a line.*

21. $y = 2x + 12$
    Setting A: $[-10, 10]$ by $[-10, 10]$
    Setting B: $[-10, 10]$ by $[-10, 15]$

22. $y = -3x + 25$
    Setting A: $[-5, 5]$ by $[-30, 10]$
    Setting B: $[-10, 10]$ by $[-10, 30]$

23. $y = -x - 41$
    Setting A: $[-50, 10]$ by $[-10, 10]$
    Setting B: $[-50, 10]$ by $[-50, 15]$

24. $y = 6x - 18$
    Setting A: $[-10, 10]$ by $[-20, 10]$
    Setting B: $[-10, 10]$ by $[-10, 10]$

25. $y = \dfrac{1}{2}x - 15$
    Setting A: $[-10, 10]$ by $[-20, 10]$
    Setting B: $[-10, 35]$ by $[-20, 15]$

26. $y = -\dfrac{2}{3}x - \dfrac{29}{3}$
    Setting A: $[-10, 10]$ by $[-10, 10]$
    Setting B: $[-15, 5]$ by $[-15, 5]$

The graph of each equation is a line. Use a graphing utility and a standard viewing window to graph each equation.

27. $3x = 5y$ **28.** $7y = -3x$ **29.** $9x - 5y = 30$ **30.** $4x + 6y = 20$

31. $y = -7$ **32.** $y = 2$ **33.** $x + 10y = -5$ **34.** $x - 5y = 9$

Graph the following equations using the square setting given. Some keystrokes that may be helpful are given.

35. $y = \sqrt{x}$ $[-12, 12]$ by $[-8, 8]$

    Suggested keystrokes: $\sqrt{x}$

36. $y = \sqrt{2x}$ $[-12, 12]$ by $[-8, 8]$

    Suggested keystrokes: $\sqrt{(2x)}$

37. $y = x^2 + 2x + 1$ $[-15, 15]$ by $[-10, 10]$

    Suggested keystrokes: $x \wedge 2 + 2x + 1$

38. $y = x^2 - 5$ $[-15, 15]$ by $[-10, 10]$

    Suggested keystrokes: $x \wedge 2 - 5$

39. $y = |x|$ $[-9, 9]$ by $[-6, 6]$

    Suggested keystrokes: ABS (x)

40. $y = |x - 2|$ $[-9, 9]$ by $[-6, 6]$

    Suggested keystrokes: ABS $(x - 2)$

Graph the line on a single set of axes. Use a standard viewing window; then, if necessary, change the viewing window so that all intercepts of the line show.

41. $x + 2y = 30$

42. $1.5x - 3.7y = 40.3$

Appendix

E Stretching and Compressing Graphs of Absolute Value Functions

In Section 3.8, we learned to shift and reflect graphs of common functions: $f(x) = x$, $f(x) = x^2$, $f(x) = |x|$ and $f(x) = \sqrt{x}$. Since other common functions are studied throughout this text, in this Appendix we concentrate on the absolute value function.

Recall that the graph of $h(x) = -|x - 1| + 2$, for example, is the same as the graph of $f(x) = |x|$ reflected about the x-axis, moved 1 unit to the right and 2 units upward. In other words,

$$h(x) = -|x - 1| + 2$$

opens downward $(1, 2)$ location of vertex of V-shape

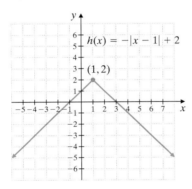

Let's now study the graphs of a few other absolute value functions.

EXAMPLE 1 Graph $h(x) = 2|x|$, and $g(x) = \dfrac{1}{2}|x|$

Solution: Let's find and plot ordered-pair solutions for the functions.

x	h(x)	g(x)
-2	4	1
-1	2	$\dfrac{1}{2}$
0	0	0
1	2	$\dfrac{1}{2}$
2	4	2

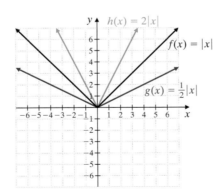

Notice that the graph of $h(x) = 2|x|$ is narrower than the graph of $f(x) = |x|$ and the graph of $g(x) = \dfrac{1}{2}|x|$ is wider than the graph of $f(x) = |x|$.

■ **Work Practice Problem 1**

PRACTICE PROBLEM 1

Graph $h(x) = 4|x|$ and $g(x) = \dfrac{1}{5}|x|$ on the same set of axes.

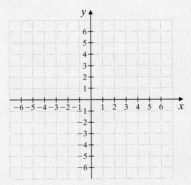

Answer

1.

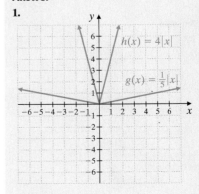

In general, for the absolute function, we have the following

> ### The Graph of the Absolute Value Function
>
> The graph of $f(x) = a|x - h| + k$
>
> - Has vertex (h, k) and is V-shaped.
> - Opens up if $a > 0$ and down if $a < 0$.
> - If $|a| < 1$, the graph is wider than the graph of $y = |x|$.
> - If $|a| > 1$, the graph is narrower than the graph of $y = |x|$.

PRACTICE PROBLEM 2

Graph $f(x) = -\dfrac{1}{2}|x + 1| + 3$

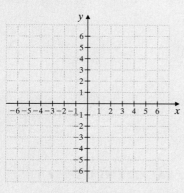

EXAMPLE 2 Graph $f(x) = -\dfrac{1}{3}|x + 2| + 4$

Solution: Let's write this function in the form $f(x) = a|x - h| + k$. For our function, we have $f(x) = -\dfrac{1}{3}|x - (-2)| + 4$. Thus:

- vertex is $(-2, 4)$
- since $a < 0$, V-shape opens down
- since $|a| = \left|-\dfrac{1}{3}\right| = \dfrac{1}{3} < 1$, the graph is wider than $y = |x|$

We will also find and plot ordered-pair solutions.

If $x = -5, f(-5) = -\dfrac{1}{3}|-5 + 2| + 4$, or 3

If $x = 1, f(1) = -\dfrac{1}{3}|1 + 2| + 4$, or 3

If $x = 3, f(3) = -\dfrac{1}{3}|3 + 2| + 4$, or $\dfrac{7}{3}$, or $2\dfrac{1}{3}$

x	f(x)
-5	3
1	3
3	$2\dfrac{1}{3}$

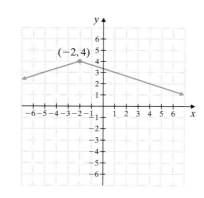

🔲 **Work Practice Problem 2**

Answer

2.

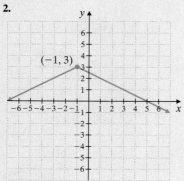

Sketch the graph of each function. Label the vertex of the V-shape.

1. $f(x) = 3|x|$

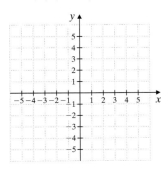

2. $f(x) = 5|x|$

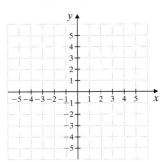

3. $f(x) = \frac{1}{4}|x|$

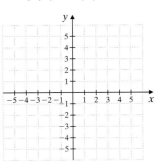

4. $f(x) = \frac{1}{3}|x|$

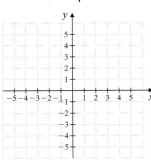

5. $g(x) = 2|x| + 3$

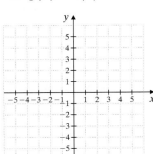

6. $g(x) = 3|x| + 2$

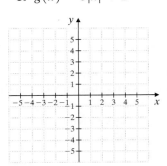

7. $h(x) = -\frac{1}{2}|x|$

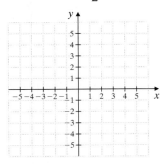

8. $h(x) = -\frac{1}{3}|x|$

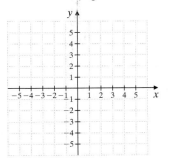

9. $f(x) = 4|x - 1|$

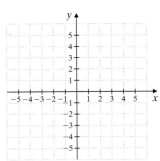

10. $f(x) = 3|x - 2|$

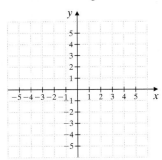

11. $g(x) = -\frac{1}{3}|x| - 2$

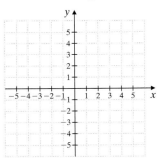

12. $g(x) = -\frac{1}{2}|x| - 3$

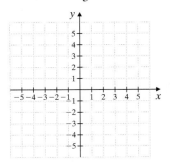

13. $f(x) = -2|x - 3| + 4$ **14.** $f(x) = -3|x - 1| + 5$ **15.** $f(x) = \frac{2}{3}|x + 2| - 5$ **16.** $f(x) = \frac{3}{4}|x + 1| - 4$

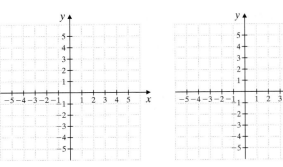

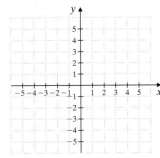

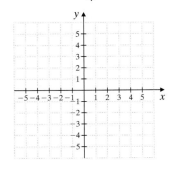

Review of Angles, Lines, and Special Triangles

The word **geometry** is formed from the Greek words, **geo,** meaning earth, and **metron,** meaning measure. Geometry literally means to measure the earth.

This appendix contains a review of some basic geometric ideas. It will be assumed that fundamental ideas of geometry such as point, line, ray, and angle are known. In this appendix, the notation $\angle 1$ is read "angle 1" and the notation $m\angle 1$ is read "the measure of angle 1."

We first review types of angles.

Angles

An angle whose measure is greater than 0° but less than 90° is called an **acute angle.**

A **right angle** is an angle whose measure is 90°. A right angle can be indicated by a square drawn at the vertex of the angle, as shown below.

An angle whose measure is greater than 90° but less than 180° is called an **obtuse angle.**

An angle whose measure is 180° is called a **straight angle.**

Two angles are said to be **complementary** if the sum of their measures is 90°. Each angle is called the **complement** of the other.

Two angles are said to be **supplementary** if the sum of their measures is 180°. Each angle is called the **supplement** of the other.

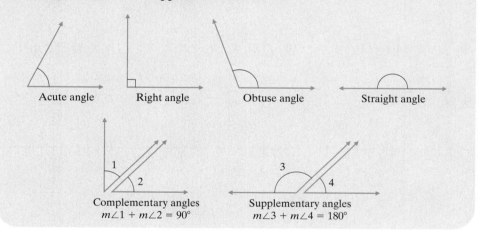

Acute angle   Right angle   Obtuse angle   Straight angle

Complementary angles
$m\angle 1 + m\angle 2 = 90°$

Supplementary angles
$m\angle 3 + m\angle 4 = 180°$

EXAMPLE 1 If an angle measures 28°, find its complement.

Solution: Two angles are complementary if the sum of their measures is 90°. The complement of a 28° angle is an angle whose measure is $90° - 28° = 62°$. To check, notice that $28° + 62° = 90°$.

Plane is an undefined term that we will describe. A plane can be thought of as a flat surface with infinite length and width, but no thickness. A plane is two dimensional. The arrows in the following diagram indicate that a plane extends indefinitely and has no boundaries.

Figures that lie on a plane are called **plane figures.** Lines that lie in the same plane are called **coplanar.**

Lines

Two lines are **parallel** if they lie in the same plane but never meet. **Intersecting lines** meet or cross in one point.

Two lines that form right angles when they intersect are said to be **perpendicular.**

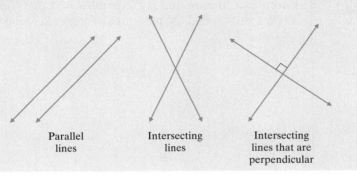

Parallel lines

Intersecting lines

Intersecting lines that are perpendicular

Two intersecting lines form **vertical angles.** Angles 1 and 3 are vertical angles. Also angles 2 and 4 are vertical angles. It can be shown that **vertical angles have equal measures.**

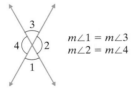

$m\angle 1 = m\angle 3$
$m\angle 2 = m\angle 4$

Adjacent angles have the same vertex and share a side. Angles 1 and 2 are adjacent angles. Other pairs of adjacent angles are angles 2 and 3, angles 3 and 4, and angles 4 and 1.

A **transversal** is a line that intersects two or more lines in the same plane. Line l is a transversal that intersects lines m and n. The eight angles formed are numbered and certain pairs of these angles are given special names.

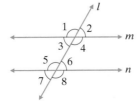

Corresponding angles: $\angle1$ and $\angle5$, $\angle3$ and $\angle7$, $\angle2$ and $\angle6$, and $\angle4$ and $\angle8$.
Exterior angles: $\angle1$, $\angle2$, $\angle7$, and $\angle8$.
Interior angles: $\angle3$, $\angle4$, $\angle5$, and $\angle6$.
Alternate interior angles: $\angle3$ and $\angle6$, $\angle4$ and $\angle5$.

These angles and parallel lines are related in the following manner.

Parallel Lines Cut by a Transversal

1. If two parallel lines are cut by a transversal, then
  a. corresponding angles are equal and
  b. alternate interior angles are equal.

2. If corresponding angles formed by two lines and a transversal are equal, then the lines are parallel.

3. If alternate interior angles formed by two lines and a transversal are equal, then the lines are parallel.

EXAMPLE 2 Given that lines m and n are parallel and that the measure of angle 1 is $100°$, find the measures of angles 2, 3, and 4.

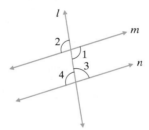

Solution:

$m\angle2 = 100°$ since angles 1 and 2 are vertical angles.

$m\angle4 = 100°$ since angles 1 and 4 are alternate interior angles.

$m\angle3 = 180° - 100° = 80°$ since angles 4 and 3 are supplementary angles.

A **polygon** is the union of three or more coplanar line segments that intersect each other only at each end point, with each end point shared by exactly two segments.

A **triangle** is a polygon with three sides. The sum of the measures of the three angles of a triangle is $180°$. In the following figure, $m\angle1 + m\angle2 + m\angle3 = 180°$.

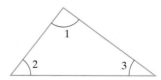

EXAMPLE 3 Find the measure of the third angle of the triangle shown.

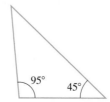

Solution: The sum of the measures of the angles of a triangle is 180°. Since one angle measures 45° and the other angle measures 95°, the third angle measures $180° - 45° - 95° = 40°$.

Two triangles are **congruent** if they have the same size and the same shape. In congruent triangles, the measures of corresponding angles are equal and the lengths of corresponding sides are equal. The following triangles are congruent.

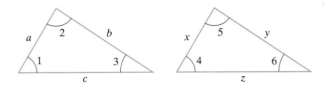

Corresponding angles are equal: $m\angle 1 = m\angle 4, m\angle 2 = m\angle 5$, and $m\angle 3 = m\angle 6$. Also, lengths of corresponding sides are equal: $a = x, b = y$, and $c = z$.

Any one of the following may be used to determine whether two triangles are congruent.

Congruent Triangles

1. If the measures of two angles of a triangle equal the measures of two angles of another triangle and the lengths of the sides between each pair of angles are equal, the triangles are congruent.

$$m\angle 1 = m\angle 3$$
$$m\angle 2 = m\angle 4$$
$$\text{and}$$
$$a = x$$

2. If the lengths of the three sides of a triangle equal the lengths of corresponding sides of another triangle, the triangles are congruent.

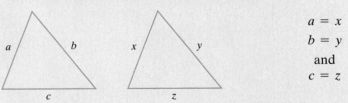

$$a = x$$
$$b = y$$
$$\text{and}$$
$$c = z$$

3. If the lengths of two sides of a triangle equal the lengths of corresponding sides of another triangle, and the measures of the angles between each pair of sides are equal, the triangles are congruent.

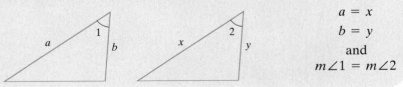

$$a = x$$
$$b = y$$
$$\text{and}$$
$$m\angle 1 = m\angle 2$$

Two triangles are **similar** if they have the same shape but not necessarily the same size. In similar triangles, the measures of corresponding angles are equal and corresponding sides are in proportion. The following triangles are similar. (All similar triangles drawn in this appendix will be oriented the same.)

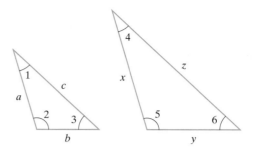

Corresponding angles are equal: $m\angle 1 = m\angle 4$, $m\angle 2 = m\angle 5$, and $m\angle 3 = m\angle 6$. Also, corresponding sides are proportional: $\dfrac{a}{x} = \dfrac{b}{y} = \dfrac{c}{z}$.

Any one of the following may be used to determine whether two triangles are similar.

Similar Triangles

1. If the measures of two angles of a triangle equal the measures of two angles of another triangle, the triangles are similar.

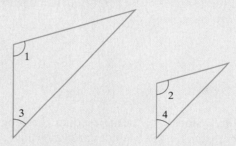

$$m\angle 1 = m\angle 2$$
$$\text{and}$$
$$m\angle 3 = m\angle 4$$

2. If three sides of one triangle are proportional to three sides of another triangle, the triangles are similar.

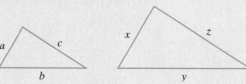

$$\frac{a}{x} = \frac{b}{y} = \frac{c}{z}$$

3. If two sides of a triangle are proportional to two sides of another triangle and the measures of the included angles are equal, the triangles are similar.

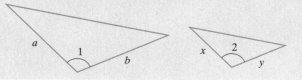

$$m\angle 1 = m\angle 2$$
$$\text{and}$$
$$\frac{a}{x} = \frac{b}{y}$$

EXAMPLE 4   Given that the following triangles are similar, find the missing length x.

Solution: Since the triangles are similar, corresponding sides are in proportion. Thus, $\frac{2}{3} = \frac{10}{x}$. To solve this equation for x, we cross multiply.

$$\frac{2}{3} = \frac{10}{x}$$

$$2x = 30$$

$$x = 15$$

The missing length is 15 units.

A **right triangle** contains a right angle. The side opposite the right angle is called the **hypotenuse,** and the other two sides are called the **legs.** The **Pythagorean theorem** gives a formula that relates the lengths of the three sides of a right triangle.

The Pythagorean Theorem

If a and b are the lengths of the legs of a right triangle, and c is the length of the hypotenuse, then $a^2 + b^2 = c^2$.

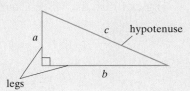

EXAMPLE 5 Find the length of the hypotenuse of a right triangle whose legs have lengths of 3 centimeters and 4 centimeters.

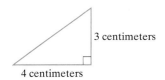

Solution: Because we have a right triangle, we use the Pythagorean theorem. The legs are 3 centimeters and 4 centimeters, so let $a = 3$ and $b = 4$ in the formula.

$$a^2 + b^2 = c^2$$
$$3^2 + 4^2 = c^2$$
$$9 + 16 = c^2$$
$$25 = c^2$$

Since c represents a length, we assume that c is positive. Thus, if c^2 is 25, c must be 5. The hypotenuse has a length of 5 centimeters.

F EXERCISE SET

Find the complement of each angle. See Example 1.

1. $19°$

2. $65°$

3. $70.8°$

4. $45\frac{2}{3}°$

5. $11\frac{1}{4}°$

6. $19.6°$

Find the supplement of each angle.

7. $150°$

8. $90°$

9. $30.2°$

10. $81.9°$

11. $79\frac{1}{2}°$

12. $165\frac{8}{9}°$

13. If lines m and n are parallel, find the measures of angles 1 through 7. See Example 2.

14. If lines m and n are parallel, find the measures of angles 1 through 5. See Example 2.

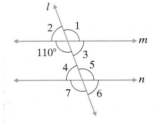

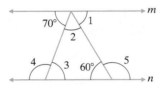

In each of the following, the measures of two angles of a triangle are given. Find the measure of the third angle. See Example 3.

15. $11°, 79°$

16. $8°, 102°$

17. $25°, 65°$

18. $44°, 19°$

19. $30°, 60°$

20. $67°, 23°$

In each of the following, the measure of one angle of a right triangle is given. Find the measures of the other two angles.

21. $45°$

22. $60°$

23. $17°$

24. $30°$

25. $39\frac{3}{4}°$

26. $72.6°$

Given that each of the following pairs of triangles is similar, find the missing length x. See Example 4.

27.

28.

29.

30.
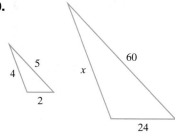

Use the Pythagorean theorem to find the missing lengths in the right triangles. See Example 5.

31.
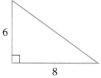

32.
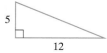

33.
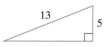

34.
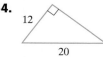

ANSWERS TO SELECTED EXERCISES

CHAPTER 1   Real Numbers and Algebraic Expressions

Exercise Set 1.2   1. $\{1, 2, 3, 4, 5\}$ **3.** $\{11, 12, 13, 14, 15, 16\}$ **5.** $\{0\}$ **7.** $\{0, 2, 4, 6, 8\}$ **9.** $\{3, 0, \sqrt{36}\}$ **11.** $\{3, \sqrt{36}\}$ **13.** $\{\sqrt{7}\}$
15. $\in$ **17.** $\notin$ **19.** $\notin$ **21.** true   **23.** true   **25.** false   **27.** true   **29.** true   **31.** false   **33.** true   **35.** true   **37.** false
39. $2x$ **41.** $x - 10$ **43.** $x + 2$ **45.** $\dfrac{x}{11}$ or $x \div 11$ **47.** $x - 4$ **49.** $x + 2.3$ or $x + 2\dfrac{3}{10}$ **51.** $1\dfrac{1}{3} - x$ **53.** $9x$ **55.** $x + 9$
57. $2x + 5$ **59.** $12 - 3x$ **61.** $1 + 2x$ **63.** $5x - 10$ **65.** $\dfrac{5}{4 - x}$ **67.** $2(x + 3)$ **69.** answers may vary   **71.** answers may vary
73. 0   **75.** 5.4; 2.1; 1.7; 2.6; 1.4; 0.7   **77.** answers may vary

Exercise Set 1.3   1. $10 + x = -12$ **3.** $2x + 5 = -14$ **5.** $\dfrac{n}{5} = 4n$ **7.** $z - \dfrac{1}{2} = \dfrac{1}{2}z$ **9.** $>$ **11.** $>$ **13.** $<$
15. $=$ **17.** $>$ **19.** $<$ **21.** $<$ **23.** $<$ **25.** true   **27.** true   **29.** false   **31.** true   **33.** -9 **35.** 6.2
37. $-\dfrac{4}{7}$ **39.** $\dfrac{5}{11}$ **41.** 0   **43.** $\dfrac{1}{5}$ **45.** $-\dfrac{1}{8}$ **47.** -4 **49.** undefined   **51.** $\dfrac{8}{7}$ **53.** $-25; \dfrac{1}{25}$ **55.** $-10; -\dfrac{1}{10}$ **57.** $\dfrac{1}{7}; -7$
59. 0; undefined   **61.** $\dfrac{16}{19}; -\dfrac{16}{19}$ **63.** $y + 7x$ **65.** $w \cdot z$ **67.** $\dfrac{x}{5} \cdot \dfrac{1}{3}$ **69.** $(5 \cdot 7)x$ **71.** $x + (1.2 + y)$ **73.** $14(z \cdot y)$ **75.** $3x + 15$
77. $4z - 24$ **79.** $16a + 8b$ **81.** $6xy - 24x$ **83.** $0.8x + 2y$ **85.** $2x - \dfrac{9}{2}y$ **87.** $12x + 10y + 4z$ **89.** $6 + 3x$ **91.** 0   **93.** 7
95. $(10 \cdot 2)y$ **97.** $3x + 12$ **99.** $8y + 4$ **101.** zero; answers may vary   **103.** no; answers may vary   **105.** answers may vary

Exercise Set 1.4   1. 2   **3.** 4   **5.** 0   **7.** -3 **9.** $-\dfrac{2}{9}$ **11.** 5   **13.** -24 **15.** -11 **17.** -4 **19.** $\dfrac{4}{3}$ **21.** -2 **23.** -21
25. $-\dfrac{1}{2}$ **27.** -6 **29.** -60 **31.** 80   **33.** 0   **35.** 0   **37.** -3 **39.** 3   **41.** $-\dfrac{1}{6}$ **43.** 0.56   **45.** -7 **47.** -8 **49.** -49
51. 36   **53.** -8 **55.** $-\dfrac{1}{27}$ **57.** 7   **59.** 8   **61.** $\dfrac{1}{3}$ **63.** 4   **65.** $\dfrac{2}{5}$ **67.** 3   **69.** -12 **71.** -18 **73.** -72 **75.** 121
77. -8 **79.** -3 **81.** $\dfrac{1}{21}$ **83.** 7.2   **85.** 15   **87.** -12.2 **89.** 0   **91.** undefined   **93.** $\dfrac{1}{4}$ **95.** $-\dfrac{3}{7}$ **97.** 4   **99.** 5
101. 3   **103.** 2   **105.** $-\dfrac{5}{27}$ **107.** -1 **109.** $-\dfrac{3}{2}$ **111.** -1 **113.** -22 **115.** 35   **117.** 0   **119.** -256 **121.** answers may vary
123. true   **125.** $\dfrac{13}{35}$ **127.** 4205 m   **129.** b   **131.** d   **133.** yes; two players have 6 points each (the third player has 0 points);
or two players have 5 points each (the third player has 2 points)   **135.** 16.5227   **137.** 4.4272   **139.** 17%   **141.** 40%

Integrated Review   1. $\{1, 2, 3\}$ **2.** $\{1, 3, 5\}$ **3.** $\{8, 10, 12, \ldots\}$ **4.** $\{11, 12, 13, 14\}$ **5.** $2(x - 3)$ **6.** $\dfrac{6}{x + 10}$ **7.** $>$ **8.** $=$
9. $<$ **10.** $<$ **11.** $5x = 20$ **12.** $a + 12 = 14$ **13.** $\dfrac{y}{10} = y \cdot 10$ **14.** $x + 1 = x - 1$ **15.** 3   **16.** 9   **17.** -28 **18.** -220
19. 5   **20.** -28 **21.** 5   **22.** -3 **23.** -25 **24.** 25   **25.** 13   **26.** -5 **27.** 0   **28.** undefined   **29.** -24 **30.** 30
31. $-\dfrac{3}{7}$ **32.** $-\dfrac{1}{10}$ **33.** $-\dfrac{1}{2}$ **34.** $-\dfrac{11}{12}$ **35.** -8 **36.** -0.3 **37.** 8   **38.** 4.4   **39.** $-\dfrac{3}{8}$ **40.** $-\dfrac{4}{7}$ **41.** 7   **42.** 3   **43.** -4
44. -8 **45.** $\dfrac{1}{9}$ **46.** $\dfrac{1}{10}$ **47.** $6; -\dfrac{1}{6}$ **48.** $-4; \dfrac{1}{4}$ **49.** $\dfrac{5}{7}; -\dfrac{5}{7}$ **50.** $-\dfrac{7}{30}; -\dfrac{30}{7}$ **51.** $9m + 45$ **52.** $77 + 11r$ **53.** $6y - 9x$
54. $32m - 56n$ **55.** $0.6a + 1.4$ **56.** $1.2n + 3$ **57.** $2x - \dfrac{19}{5}y + 4$ **58.** $5x - \dfrac{19}{2}y + 10$

Exercise Set 1.5   1. 48   **3.** -1 **5.** -9 **7.** 14.4   **9.** 17   **11.** -4 **13.** 40   **15.** -2 **17.** 11   **19.** $-\dfrac{3}{4}$ **21.** 7   **23.** 3
25. -11 **27.** -2.1 **29.** $-\dfrac{1}{3}$ **31.** $-\dfrac{79}{15}$ **33.** $-\dfrac{4}{5}$ **35.** -81 **37.** $-\dfrac{20}{33}$ **39.** -235.5 **41.** 12.25   **43.** 93   **45.** -12
47. $-\dfrac{23}{18}$ **49.** 5   **51.** $-\dfrac{3}{19}$ **53. a.** 18; 22; 28; 208   **b.** increase; answers may vary   **55. a.** 600; 150; 105   **b.** decrease; answers may vary
57. \$3690.88   **59.** $8x$ **61.** $18y$ **63.** $-x - 8$ **65.** $-6x + 9$ **67.** $4a - 13b$ **69.** $2x - 2y$ **71.** $0.8x - 3.6$ **73.** $\dfrac{11}{12}b - \dfrac{7}{6}$
75. $-3ab - 12.8$ **77.** $6x + 14$ **79.** $2k + 10$ **81.** $-3x + 5$ **83.** $4x + 9$ **85.** $4n - 8$ **87.** -24 **89.** $-2x + \dfrac{6}{5}y - 1$
91. $2x + 10$ **93.** $3a + \dfrac{3}{35}$ **95.** $-0.6a + 3.5$ **97.** $-8t + 12$ **99.** $1.91x + 4.32$ **101.** $15.4z + 31.11$ **103.** $(2 + 7) \cdot (1 + 3)$
105. answers may vary   **107.** $(3x + 2) - (5x + 7) = 3x + 2 - 5x - 7 = -2x - 5$ **109.** 20 million   **111.** 70 million
113. increasing; answers may vary   **115.** -0.5876

Calculator Explorations   1. 6×10^{43} **3.** 3.796×10^{28}

Mental Math   1. $\dfrac{5}{xy^2}$ **3.** $\dfrac{a^2}{bc^5}$ **5.** $\dfrac{x^4}{y^2}$

Exercise Set 1.6   **1.** 4^5 **3.** x^8 **5.** m^{14} **7.** $-20x^2y$ **9.** $-16x^6y^3p^2$ **11.** -1 **13.** 1 **15.** -1 **17.** 9 **19.** a^3 **21.** $-13z^4$

23. x **25.** $\dfrac{4}{3}x^3y^2$ **27.** $-6a^4b^4c^6$ **29.** $\dfrac{1}{16}$ **31.** $-\dfrac{1}{27}$ **33.** $\dfrac{1}{x^8}$ **35.** $\dfrac{5}{a^4}$ **37.** $\dfrac{y^2}{x^7}$ **39.** $\dfrac{1}{x^7}$ **41.** $4r^8$ **43.** 1 **45.** $\dfrac{b^7}{9a^7}$

47. $\dfrac{6x^{16}}{5}$ **49.** $-140x^{12}$ **51.** x^{16} **53.** $10x^{10}$ **55.** 6 **57.** $\dfrac{1}{z^3}$ **59.** -2 **61.** y^4 **63.** $\dfrac{13}{36}$ **65.** $\dfrac{3}{x}$ **67.** r^8 **69.** $\dfrac{1}{x^9y^4}$

71. $24x^7y^6$ **73.** $\dfrac{x}{16}$ **75.** 625 **77.** $\dfrac{1}{8}$ **79.** $\dfrac{a^5}{81}$ **81.** $\dfrac{7}{x^3z^5}$ **83.** x^{7a+5} **85.** x^{2t-1} **87.** x^{4a+7} **89.** z^{6x-7} **91.** x^{6t-1}

93. 3.125×10^7 **95.** 1.6×10^{-2} **97.** 6.7413×10^4 **99.** 1.25×10^{-2} **101.** 5.3×10^{-5} **103.** 7.783×10^8 **105.** 6.404×10^9

107. 1.24×10^{11} **109.** 1.0×10^{-3} **111.** 0.0000000036 **113.** $93{,}000{,}000$ **115.** $1{,}278{,}000$ **117.** $7{,}350{,}000{,}000{,}000$ **119.** 0.000000403

121. $300{,}000{,}000$ **123.** $4{,}900{,}000{,}000$ **125.** answers may vary   **127.** answers may vary

Mental Math   **1.** x^{20} **3.** x^9 **5.** y^{42} **7.** z^{36} **9.** z^{18}

Exercise Set 1.7   **1.** $\dfrac{1}{9}$ **3.** $\dfrac{1}{x^{36}}$ **5.** $9x^4y^6$ **7.** $16x^{20}y^{12}$ **9.** $\dfrac{c^{18}}{64a^{12}b^6}$ **11.** $\dfrac{343}{512}$ **13.** $-64y^3$ **15.** $\dfrac{1}{a^2}$ **17.** $\dfrac{36}{p^{12}}$ **19.** $-\dfrac{a^6}{512x^3y^9}$

21. $\dfrac{64}{27}$ **23.** $4a^8b^4$ **25.** $\dfrac{x^{14}y^{14}}{a^{21}}$ **27.** $\dfrac{1}{y^{10}}$ **29.** $\dfrac{1}{125}$ **31.** $\dfrac{1}{x^{63}}$ **33.** $\dfrac{y^{15}}{x^{35}z^{20}}$ **35.** $16x^4$ **37.** $4^8x^2y^6$ **39.** $\dfrac{x^9}{8y^3}$ **41.** $\dfrac{x^4}{4z^2}$

43. $\dfrac{x^4}{16}$ **45.** $\dfrac{1}{y^{15}}$ **47.** $\dfrac{16a^2b^9}{9}$ **49.** $\dfrac{3}{8x^8y^7}$ **51.** $\dfrac{1}{x^{30}b^6c^6}$ **53.** $\dfrac{25}{8x^5y^4}$ **55.** $\dfrac{2}{x^4y^{10}}$ **57.** x^{9a+18} **59.** x^{16a-1} **61.** b^{10x-4} **63.** y^{15a+3}

65. $16x^{4t+4}$ **67.** $5x^{a+2b}y^{a-2b}$ **69.** 1.45×10^9 **71.** 8×10^{15} **73.** 4×10^{-7} **75.** 3×10^{-1} **77.** 2×10^1 **79.** 1×10^1

81. 8×10^{-5} **83.** 1.1×10^7 **85.** 3.5×10^{22} **87.** 2×10^{-3} sec   **89.** 6.232×10^{-11} cu m   **91.** $\dfrac{15y^3}{x^8}$ sq ft   **93.** 1.331928×10^{13} tons

95. no   **97.** 83 people per sq mi   **99.** 3.9

Vocabulary Check   **1.** algebraic expression   **2.** opposite   **3.** distributive   **4.** absolute value   **5.** exponent   **6.** variable
7. inequality   **8.** reciprocals   **9.** commutative   **10.** associative   **11.** whole   **12.** real

Chepter 1 Review   **1.** $\dfrac{x}{7}$ **2.** $7x$ **3.** $4(x+10)$ **4.** $3x-9$ **5.** $\{-1,1,3\}$ **6.** $\{-2,0,2,4,6\}$ **7.** $\{\ \}$ **8.** $\{\ \}$

9. $\{6,7,8,\dots\}$ **10.** $\{\dots,-1,0,1,2\}$ **11.** true   **12.** false   **13.** true   **14.** true   **15.** false   **16.** true   **17.** false   **18.** true

19. $\left\{5,\dfrac{8}{2},\sqrt{9}\right\}$ **20.** $\left\{5,\dfrac{8}{2},\sqrt{9}\right\}$ **21.** $\left\{5,-\dfrac{2}{3},\dfrac{8}{2},\sqrt{9},0.3,1\dfrac{5}{8},-1\right\}$ **22.** $\{\sqrt{7},\pi\}$ **23.** $\left\{5,-\dfrac{2}{3},\dfrac{8}{2},\sqrt{9},0.3,\sqrt{7},1\dfrac{5}{8},-1,\pi\right\}$

24. $\left\{5,\dfrac{8}{2},\sqrt{9},-1\right\}$ **25.** $12=-4x$ **26.** $n+2n=-15$ **27.** $4(y+3)=-1$ **28.** $6(t-5)=4$ **29.** $z-7=6$

30. $9x-10=5$ **31.** $x-5=12$ **32.** $-4=7y$ **33.** $\dfrac{2}{3}=2\left(n+\dfrac{1}{4}\right)$ **34.** $t+6=-12$ **35.** $\dfrac{3}{4}$ **36.** -0.6 **37.** 0 **38.** 1

39. $-\dfrac{4}{3}$,   **40.** 5 **41.** undefined   **42.** -1 **43.** associative property of addition   **44.** distributive property

45. additive inverse property   **46.** commutative property of addition   **47.** associative property of multiplication
48. multiplicative inverse property   **49.** multiplicative identity property   **50.** commutative property of addition
51. additive identity property   **52.** multiplicative identity property   **53.** $5x-15z$ **54.** $(3+x)+(7+y)$ **55.** $2+(-2)$, for example
56. $2\cdot\dfrac{1}{2}$, for example   **57.** $(3.4)[(0.7)5]$ **58.** $7+0$ **59.** $>$ **60.** $>$ **61.** $<$ **62.** $=$ **63.** $<$ **64.** $>$

65. $-6x+21y$ **66.** $-90a-36b$ **67.** $9m-4n+\dfrac{1}{2}$ **68.** $4x-11y+\dfrac{2}{3}$ **69.** -4 **70.** -35 **71.** -2 **72.** 0.31 **73.** 8

74. 13.3 **75.** -4 **76.** -22 **77.** undefined   **78.** 0 **79.** 4 **80.** -5 **81.** $-\dfrac{2}{15}$ **82.** 4 **83.** $\dfrac{5}{12}$

84. 29,852 ft below sea level   **85.** 9 **86.** 13 **87.** 3 **88.** 54 **89.** $-\dfrac{32}{135}$ **90.** $-\dfrac{15}{56}$ **91.** $-\dfrac{5}{4}$ **92.** $-\dfrac{5}{2}$ **93.** $\dfrac{5}{8}$ **94.** $-6\dfrac{1}{2}$

95. -1 **96.** 24 **97.** 1 **98.** 18 **99.** -4 **100.** $\dfrac{7}{3}$ **101.** $\dfrac{5}{7}$ **102.** $-\dfrac{8}{25}$ **103.** $\dfrac{1}{5}$ **104.** 1 **105.** $6.28; 62.8; 628$

106. increase   **107.** $3x-13$ **108.** $80y-1$ **109.** $a+7y-6$ **110.** $-23b+13x-10$ **111.** $n+5$ **112.** $y-4$ **113.** 4

114. 81 **115.** -4 **116.** -81 **117.** 1 **118.** -1 **119.** $-\dfrac{1}{16}$ **120.** $\dfrac{1}{16}$ **121.** $-x^2y^7z$ **122.** $12x^2y^3b$ **123.** $\dfrac{1}{a^9}$ **124.** $\dfrac{1}{a}$

125. $\dfrac{1}{x^{11}}$ **126.** $\dfrac{1}{2a^{17}}$ **127.** $\dfrac{1}{y^5}$ **128.** $9x^{4a+2b}y^{-6b}$ or $\dfrac{9x^{4a+2b}}{y^{6b}}$ **129.** 3.689×10^7 **130.** 3.62×10^{-4} **131.** 0.000001678

132. $410{,}000$ **133.** 8^{15} **134.** $\dfrac{a^2}{16}$ **135.** $27x^3$ **136.** $\dfrac{1}{16x^2}$ **137.** $\dfrac{36x^2}{25}$ **138.** $\dfrac{1}{8^{18}}$ **139.** $\dfrac{9}{16}$ **140.** $-\dfrac{1}{8x^9}$ **141.** $\dfrac{1}{4p^4}$

142. $-\dfrac{27y^6}{x^6}$ **143.** $x^{25}y^{15}z^{15}$ **144.** $\dfrac{xz}{4}$ **145.** $\dfrac{x^2}{625y^4z^4}$ **146.** $\dfrac{2}{27z^3}$ **147.** $27x^{19a}$ **148.** $2y^{x-7}$

149. $\frac{3}{4}; -\frac{4}{3}$ **150.** $5; \frac{1}{5}$ **151.** $-10x + 6.1$ **152.** 9 **153.** $-\frac{6}{11}$ **154.** 15 **155.** $-\frac{4}{15}$ **156.** $\frac{1}{11}$ **157.** $-x + 3y + 1$

158. $-\frac{1}{25}$ **159.** $\frac{8}{x^{30}}$ **160.** $\frac{9b^2c^4}{25a^2}$

Chapter 1 Test **1.** true **2.** false **3.** false **4.** $\{0, 1\}$ **5.** true **6.** false **7.** -3 **8.** -225 **9.** -2 **10.** 1 **11.** $-\frac{3}{2}$

12. 12 **13.** 1 **14.** **a.** 5.75; 17.25; 57.50; 115.00 **b.** increase **15.** $3\left(\frac{n}{5}\right) = -n$ **16.** $20 = 2x - 6$ **17.** $-z = \frac{x}{x + 5}$

18. distributive property **19.** associative property of addition **20.** additive inverse property **21.** reciprocal: $-\frac{11}{7}$; opposite: $\frac{7}{11}$

22. $6x + 12y + 13.8$ **23.** $-x - 11y$ **24.** $\frac{1}{81x^2}$ **25.** $\frac{3a^7}{2b^5}$ **26.** $-\frac{y^{40}}{z^5}$ **27.** $\frac{a^{12}}{4b^{14}}$ **28.** x^{10w} **29.** 6.3×10^8 **30.** 1.2×10^{-2}

31. 0.000005 **32.** 5.76×10^4 **33.** 9×10^{-4} **34.** 8.486×10^8

CHAPTER 2 Equations, Inequalities, and Problem Solving

Mental Math **1.** $\{6\}$ **3.** $\{17\}$ **5.** $\{8\}$ **7.** $\{10\}$

Exercise Set 2.1 **1.** yes **3.** no **5.** yes **7.** no **9.** no **11.** yes **13.** $\{6\}$ **15.** $\{-22\}$ **17.** $\{4.7\}$ **19.** $\{10\}$ **21.** $\{-1.1\}$

23. $\{-5\}$ **25.** $\{-2\}$ **27.** $\{0\}$ **29.** $\{2\}$ **31.** $\{-9\}$ **33.** $\left\{-\frac{10}{7}\right\}$ **35.** $\left\{\frac{9}{10}\right\}$ **37.** $\{4\}$ **39.** $\{1\}$ **41.** $\{5\}$ **43.** $\left\{\frac{40}{3}\right\}$

45. $\{17\}$ **47.** $\{n | n$ is a real number$\}$ **49.** $\varnothing$ **51.** $\{x | x$ is a real number$\}$ **53.** $\varnothing$ **55.** $\left\{\frac{1}{8}\right\}$ **57.** $\{0\}$ **59.** $\{x | x$ is a real number$\}$

61. $\{4\}$ **63.** $\left\{\frac{4}{5}\right\}$ **65.** $\{8\}$ **67.** $\varnothing$ **69.** $\{-8\}$ **71.** $\left\{-\frac{5}{4}\right\}$ **73.** $\{-2\}$ **75.** $\{23\}$ **77.** $\left\{-\frac{2}{9}\right\}$ **79.** $\frac{8}{x}$ **81.** $8x$

83. $2x - 5$ **85.** add 19 instead of subtracting; -3 **87.** $0.4 - 1.6 = -1.2$, not 1.2; -0.24 **89.** **a.** $4x + 5$ **b.** $\{-3\}$ **c.** answers may vary
91. answers may vary **93.** $K = -11$ **95.** $K = -23$ **97.** answers may vary **99.** $\{5.217\}$ **101.** $\{1\}$

Exercise Set 2.2 **1.** $4y$ **3.** $3z + 3$ **5.** $(65x + 30)$ cents **7.** $10x + 3$ **9.** $2x + 14$ **11.** -5 **13.** 45, 145 225
15. approximately 1612.41 million acres **17.** 3145 earthquakes **19.** 1275 shoppers **21.** 23% **23.** 417 employees
25. INVESCO Field: 76,125 seats; Heinz Field: 64,450 seats **27.** 40.5 ft; 202.5 ft; 240 ft **29.** Tokyo: 35.0 million; New York:
18.3 million; Mexico City: 18.7 million **31.** 29°, 35°, 116° **33.** 28 m, 36 m, 38 m **35.** 18 in., 18 in., 27 in., 36 in.
37. 317 thousand; 279 thousand; 184 thousand **39.** medical assistant: 215 thousand; postsecondary teacher jobs: 603 thousand; registered nurses:
623 thousand **41.** B767-300ER: 207 seats; B737-200: 119 seats; F-100: 87 seats **43.** $430.00 **45.** $100.34 **47.** 75, 76, 77
49. 64°, 32°, 84° **51.** square: 18 cm; triangle: 24 cm **53.** 41,741,000 **55.** Fallon's zip code is 89406; Fernley's zip code is 89408; Gardnerville
Ranchos' zip code is 89410 **57.** 40°, 140° **59.** incandescent: 1500 bulb hours; fluorescent: 100,000 bulb hours; halogen: 4000 bulb hours
61. Jeter: 23; Mulder: 25; Jenkins: 27 **63.** 208 **65.** -55 **67.** 3195 **69.** yes; answers may vary **71.** answers may vary
73. **a.** 2029 **b.** 1387.5 **c.** 4; no: this is the average daily number of cigarettes for all American adults—smokers and non smokers
75. no such odd integers exist **77.** 500 boards; $30,000 **79.** company makes a profit

Mental Math **1.** $y = 5 - 2x$ **3.** $a = 5b + 8$ **5.** $k = h - 5j + 6$

Exercise Set 2.3 **1.** $t = \frac{d}{r}$ **3.** $r = \frac{I}{Pt}$ **5.** $c = P - a - b$ **7.** $y = \frac{9x - 16}{4}$ or $y = \frac{9}{4}x - 4$ **9.** $l = \frac{P - 2w}{2}$ or $l = \frac{P}{2} - w$

11. $r = \frac{E}{I} - R$ **13.** $y = \frac{20 - 5x}{4}$ or $y = 5 - \frac{5}{4}x$ **15.** $H = \frac{S - 2LW}{2L + 2W}$ **17.** $r = \frac{C}{2\pi}$ **19.** $F = \frac{9}{5}C + 32$

21.

n	1	2	4	12	365
A	$4703.71	$4713.99	$4719.22	$4722.74	$4724.45

23. **a.** $7313.97 **b.** $7321.14 **c.** $7325.98 **25.** 40°C **27.** 3 hr, 36 min **29.** 171 packages **31.** 9 ft **33.** 2 gal
35. **a.** 1174.86 cu m **b.** 310.34 cu m **c.** 1485.20 cu m **37.** 164,921 mi **39.** 0.42 ft **41.** 41.125π ft; 129.1325 ft
43. $f = \frac{C - 4h - 4p}{9}$ **45.** 178 cal **47.** 1.5 g **49.** $\{-3, -2, -1\}$ **51.** $\{-3, -2, -1, 0, 1\}$ **53.** answers may vary
55. 0.388; 0.723; 1.00; 1.523; 5.202; 9.538; 19.193; 30.065; 39.505 **57.** $6.80 per person **59.** answers may vary
61. $n_e = \frac{N}{R^* \times f_p \times f_l \times f_i \times f_c \times L}$

Mental Math **1.** $\{x | x < 6\}$ **3.** $\{x | x \geq 10\}$ **5.** $\{x | x > 4\}$ **7.** $\{x | x \leq 2\}$

Exercise Set 2.4 **1.** $\longrightarrow$ $(-\infty, -3)$ **3.** $\longrightarrow$ $[0.3, \infty)$ **5.** $\longrightarrow$ $[-7, \infty)$
  -3 0.3 -7

7. $\longrightarrow$ $(-2, 5)$ **9.** $\longrightarrow$ $(-1, 5]$ **11.** $\longrightarrow$ $[-2, \infty)$ **13.** $\longrightarrow$ $(-\infty, 1)$
  -2 5 -1 5 -2 1

15. $\longrightarrow$ $(-\infty, 2]$ **17.** $\longrightarrow$ $[8, \infty)$ **19.** $\longrightarrow$ $(-\infty, -4.7)$ **21.** $\longrightarrow$ $(-\infty, -3]$
  2 8 -4.7 -3

23. $(-\infty, -1]$ **25.** $(-\infty, 11]$ **27.** $(0, \infty)$ **29.** $(-13, \infty)$ **31.** $\left[-\dfrac{79}{3}, \infty\right)$ **33.** $\left(-\infty, -\dfrac{35}{6}\right)$ **35.** $(-\infty, -6)$ **37.** $(4, \infty)$

39. $[-0.5, \infty)$ **41.** $(-\infty, 7]$ **43.** $[0, \infty)$ **45.** $(-\infty, -29]$ **47.** $[3, \infty)$ **49.** $(-\infty, -1]$ **51.** $[-31, \infty)$ **53.** $(-\infty, -2]$

55. $(-\infty, 9)$ **57.** $\left(-\infty, -\dfrac{11}{2}\right]$ **59. a.** $\{x \mid x \geq 81\}$ **b.** A final exam grade of 81 or higher will result in an average of 77 or higher.

61. a. $\{x \mid x \leq 1040\}$ **b.** The luggage and cargo must weight 1040 pounds or less. **63. a.** $\{x \mid x \leq 20\}$ **b.** She can move at most 20 whole boxes at one time. **65. a.** $\{x \mid x > 200\}$ **b.** If you make more than 200 calls, plan 1 is more economical. **67.** $\{F \mid F \geq 932°\}$

69. a. 2000 **b.** answers may vary **71.** decreasing; answers may vary **73.** 43.5 lb **75.** during 2006 **77.** answers may vary

79. a. $\{t \mid t > 32.\overline{54}\}$ **b.** 2029 **81.** $\{0, 1, 2, 3, 4, 5, 6, 7\}$ **83.** $\{\ldots, -9, -8, -7, -6\}$ **85.** $\{5\}$ **87.** $\{0\}$

89. yes **91.** yes **93.** $\{4\}$ **95.** $(4, \infty)$ **97.** $(-\infty, \infty)$ **99.** $\varnothing$ **101.** answers may vary **103.** answers may vary

Integrated Review 1. $\{-5\}$ **2.** $(-5, \infty)$ **3.** $\left[\dfrac{8}{3}, \infty\right)$ **4.** $[-1, \infty)$ **5.** $\{0\}$ **6.** $\left[-\dfrac{1}{10}, \infty\right)$ **7.** $\left(-\infty, -\dfrac{1}{6}\right]$ **8.** $\{0\}$ **9.** $\varnothing$

10. $\left[-\dfrac{3}{5}, \infty\right)$ **11.** $\{4.2\}$ **12.** $\{6\}$ **13.** $\{-8\}$ **14.** $(-\infty, -16)$ **15.** $\left\{\dfrac{20}{11}\right\}$ **16.** $\{1\}$ **17.** $(38, \infty)$ **18.** $\{-5.5\}$ **19.** $\left\{\dfrac{3}{5}\right\}$

20. $(-\infty, \infty)$ **21.** $\{29\}$ **22.** $\{x \mid x \text{ is a real number}\}$ **23.** $(-\infty, 5)$ **24.** $\left\{\dfrac{9}{13}\right\}$ **25.** $(23, \infty)$ **26.** $(-\infty, 6]$ **27.** $\left(-\infty, \dfrac{3}{5}\right]$

28. $\left(-\infty, -\dfrac{19}{32}\right)$

Exercise Set 2.5 1. $\{2, 3, 4, 5, 6, 7\}$ **3.** $\{4, 6\}$ **5.** $\{\ldots, -2, -1, 0, 1, \ldots\}$ **7.** $\{5, 7\}$ **9.** $\{x \mid x \text{ is an odd integer or } x = 2 \text{ or } x = 4\}$

11. $\{2, 4\}$ **13.** **15.** **17.** **19.** $[6, \infty)$ **21.** $(-\infty, -3]$ **23.** $(4, 10)$

25. $(11, 17)$ **27.** $[1, 4]$ **29.** $\left[-3, \dfrac{3}{2}\right]$ **31.** $\left[-\dfrac{7}{3}, 7\right]$ **33.** **35.** **37.**

39. $[2, \infty)$ **41.** $(-\infty, -4) \cup (-2, \infty)$ **43.** $(-\infty, \infty)$ **45.** $\left(-\dfrac{1}{2}, \dfrac{2}{3}\right)$ **47.** $(-\infty, \infty)$ **49.** $\left[\dfrac{3}{2}, 6\right]$ **51.** $\left(\dfrac{5}{4}, \dfrac{11}{4}\right)$ **53.** $\varnothing$

55. $\left(-\infty, -\dfrac{56}{5}\right) \cup \left(\dfrac{5}{3}, \infty\right)$ **57.** $\left(-5, \dfrac{5}{2}\right)$ **59.** $\left(0, \dfrac{14}{3}\right]$ **61.** $(-\infty, -3]$ **63.** $(-\infty, 1] \cup \left(\dfrac{29}{7}, \infty\right)$ **65.** $\varnothing$ **67.** $\left[-\dfrac{1}{2}, \dfrac{3}{2}\right)$

69. $\left(-\dfrac{4}{3}, \dfrac{7}{3}\right)$ **71.** $(6, 12)$ **73.** -12 **75.** -4 **77.** $\{-7, 7\}$ **79.** $\{0\}$ **81.** 1993, 1994, 1995, 1998, 1999, and 2002

83. answers may vary **85.** $(-3, 5)$ **87.** $(2, \infty)$ **89.** $-20.2° \leq F \leq 95°$ **91.** $67 \leq$ final score ≤ 94

Mental Math 1. d **3.** c **5.** a

Exercise Set 2.6 1. $\{7, -7\}$ **3.** $\varnothing$ **5.** $\{4.2, -4.2\}$ **7.** $\{-4, 4\}$ **9.** $\{-9, 9\}$ **11.** $\{-5, 23\}$ **13.** $\{7, -2\}$ **15.** $\{8, 4\}$

17. $\{5, -5\}$ **19.** $\{3, -3\}$ **21.** $\{-3, 6\}$ **23.** $\{0\}$ **25.** $\varnothing$ **27.** $\left\{-\dfrac{1}{3}, \dfrac{7}{3}\right\}$ **29.** $\left\{-\dfrac{1}{2}, 9\right\}$ **31.** $\left\{-\dfrac{5}{2}\right\}$ **33.** $\{3, 2\}$

35. $\{-4, 16\}$ **37.** $\{4\}$ **39.** $\left\{\dfrac{3}{2}\right\}$ **41.** $\left\{\dfrac{32}{21}, \dfrac{38}{9}\right\}$ **43.** $\left\{-8, \dfrac{2}{3}\right\}$ **45.** $[-4, 4]$

47. $(-\infty, -3) \cup (3, \infty)$ **49.** $(-5, -1)$ **51.** $(-\infty, -1] \cup [13, \infty)$

53. $(-5, 1)$ **55.** $[-5, 5]$ **57.** $\left\{-\dfrac{3}{2}\right\}$ **59.** $(-\infty, -4) \cup (4, \infty)$

61. $[-10, 3]$ **63.** $\left(1, \dfrac{13}{3}\right)$ **65.** $(-\infty, -24] \cup [4, \infty)$ **67.** $[-2, 9]$

69. $(-\infty, \infty)$ **71.** $\left[-\dfrac{1}{2}, 1\right]$ **73.** $\left(-\infty, \dfrac{2}{3}\right) \cup (2, \infty)$ **75.** $\varnothing$

77. $(-\infty, -12) \cup (0, \infty)$ **79.** $\{-13, 13\}$ **81.** $(-13, 13)$ **83.** $\varnothing$ **85.** $[-10, 10]$ **87.** $\{5, -2\}$

89. $(-\infty, -7] \cup [17, \infty)$ **91.** $\left\{-\dfrac{9}{4}\right\}$ **93.** $(-2, 1)$ **95.** $(-\infty, -18) \cup (12, \infty)$ **97.** $\left\{2, \dfrac{4}{3}\right\}$ **99.** $\varnothing$ **101.** $\left\{-\dfrac{17}{2}, \dfrac{19}{2}\right\}$

103. $\left(-\infty, -\dfrac{25}{3}\right) \cup \left(\dfrac{35}{3}, \infty\right)$ **105.** $\left\{4, -\dfrac{1}{5}\right\}$ **107.** $\left\{-\dfrac{17}{3}, 5\right\}$ **109.** 32% **111.** 28.8° **113.** $\varnothing$ **115.** $\{x \mid x \text{ is a real number}\}$

117. $|x| = 5$ **119.** $|x| < 7$ **121.** $|x| \leq 5$, answers may vary **123.** $3.45 < x < 3.55$

The Bigger Picture   1. $\{-8\}$ **2.** $\{-13, 21\}$ **3.** $(-\infty, 6]$ **4.** $(-7, 5]$ **5.** $\left\{-\dfrac{13}{2}\right\}$ **6.** $\varnothing$ **7.** $\{-50\}$ **8.** $\{-5\}$
9. $\left[-\dfrac{9}{5}, 1\right]$ **10.** $(\infty, -13) \cup (-9, \infty)$ **11.** $\left\{-3, \dfrac{23}{9}\right\}$ **12.** $\left\{-11, -\dfrac{9}{7}\right\}$

Chapter 2 Vocabulary Check   1. compound inequality   **2.** contradiction   **3.** intersection   **4.** union   **5.** identity   **6.** formula
7. absolute value   **8.** solution   **9.** consecutive integers   **10.** linear inequality in one variable   **11.** linear equation in one variable

Chapter 2 Review   1. $\{3\}$ **2.** $\left\{-\dfrac{23}{3}\right\}$ **3.** $\left\{-\dfrac{45}{14}\right\}$ **4.** $\left\{-\dfrac{7}{11}\right\}$ **5.** $\{0\}$ **6.** $\varnothing$ **7.** $\{6\}$ **8.** $\{7.8\}$ **9.** $\{x \mid x$ is a real number$\}$
10. $\{0\}$ **11.** $\varnothing$ **12.** $\{p \mid p$ is a real number$\}$ **13.** $\{-3\}$ **14.** $\{0\}$ **15.** $\left\{\dfrac{96}{5}\right\}$ **16.** $\{-3\}$ **17.** $\{32\}$ **18.** $\{-8\}$ **19.** $\{8\}$
20. $\{1\}$ **21.** $\varnothing$ **22.** $\{11\}$ **23.** -7 **24.** $140, 145$ **25.** 52 **26.** 0.12 **27.** $\$51,652$ **28.** $10, 11, 12, 13$ **29.** no such integers exist
30. width: 40m; length: 75m   **31.** 258 mi   **32.** 250 calculators   **33.** $W = \dfrac{V}{LH}$ **34.** $r = \dfrac{C}{2\pi}$ **35.** $y = \dfrac{5x + 12}{4}$ **36.** $x = \dfrac{4y - 12}{5}$
37. $m = \dfrac{y - y_1}{x - x_1}$ **38.** $x = \dfrac{y - y_1 + mx_1}{m}$ **39.** $r = \dfrac{E - IR}{I}$ **40.** $g = \dfrac{S - vt}{t^2}$ **41.** $g = \dfrac{T}{r + vt}$ **42.** $P = \dfrac{I}{1 + rt}$ **43. a.** $\$3695.27$
b. $\$3700.81$ **44.** $\left(\dfrac{290}{9}\right)°C \approx 32.2°C$ **45.** length: 10 in.; width: 8 in.   **46.** 16 packages   **47.** $(3, \infty)$ **48.** $(-\infty, -4]$ **49.** $(-4, \infty)$
50. $(-17, \infty)$ **51.** $(-\infty, 7]$ **52.** $\left(\dfrac{1}{2}, \infty\right)$ **53.** $(-\infty, 1)$ **54.** $[-19, \infty)$ **55.** more economical to use housekeeper for more than
35 pounds per week   **56.** $260° \leq C \leq 538°$ **57.** 9.6   **58.** $\$1750$ to $\$3750$ **59.** $\left[2, \dfrac{5}{2}\right]$ **60.** $\left[-2, -\dfrac{9}{5}\right)$ **61.** $\left(\dfrac{1}{8}, 2\right)$ **62.** $\left(-\dfrac{3}{5}, 0\right)$
63. $\left(\dfrac{7}{8}, \dfrac{27}{20}\right]$ **64.** $(-5, 2]$ **65.** $\left(\dfrac{11}{3}, \infty\right)$ **66.** $\{16, -2\}$ **67.** $\{5, 11\}$ **68.** $\{0, -9\}$ **69.** $\left\{-1, \dfrac{11}{3}\right\}$ **70.** $\left\{2, -\dfrac{2}{3}\right\}$ **71.** $\left\{-\dfrac{1}{6}\right\}$
72. $\varnothing$ **73.** $\varnothing$ **74.** $\{1, 5\}$ **75.** $\left\{5, -\dfrac{1}{3}\right\}$ **76.** $\left\{7, -\dfrac{8}{5}\right\}$ **77.** (graph) $\left(-\dfrac{8}{5}, 2\right)$ **78.** (graph) $(-\infty, -4] \cup [1, \infty)$
79. (graph) $(-\infty, -3) \cup (3, \infty)$ **80.** (graph) $(-3, 3)$ **81.** (graph) $\varnothing$
82. (graph) $\left(-\infty, -\dfrac{22}{15}\right] \cup \left[\dfrac{6}{5}, \infty\right)$ **83.** (graph) $(-\infty, -27) \cup (-9, \infty)$ **84.** (graph) $\varnothing$ **85.** $\{2\}$
86. $\left\{\dfrac{37}{8}\right\}$ **87.** China: 137 million; USA: 102 million; France: 93 million   **88.** $B = \dfrac{2A - hb}{h}$ **89.** $h = \dfrac{3V}{\pi r^2}$
90. cylinder holds more ice cream   **91.** 58 mph   **92.** $(-\infty, 4]$ **93.** $(2, \infty)$ **94.** $\left[-\dfrac{4}{3}, \dfrac{7}{6}\right]$ **95.** $(-\infty, \infty)$ **96.** $(5, \infty)$
97. $\{3, -3\}$ **98.** $\varnothing$ **99.** $\left\{-10, -\dfrac{4}{3}\right\}$ **100.** $(-\infty, \infty)$ **101.** $\left(-\dfrac{1}{2}, 2\right)$

Chapter 2 Test   1. $\{10\}$ **2.** $\{-32\}$ **3.** $\varnothing$ **4.** $\{n \mid n$ is a real number$\}$ **5.** $\left\{-\dfrac{80}{29}\right\}$ **6.** $\left\{\dfrac{29}{4}\right\}$ **7.** $\left\{1, \dfrac{2}{3}\right\}$ **8.** $\varnothing$ **9.** $\left\{-4, -\dfrac{1}{3}\right\}$
10. $\left\{\dfrac{3}{2}\right\}$ **11.** $y = \dfrac{3x - 8}{4}$ **12.** $g = \dfrac{S}{t^2 + vt}$ **13.** $C = \dfrac{5}{9}(F - 32)$ **14.** $(5, \infty)$ **15.** $\left(-\infty, -\dfrac{11}{3}\right]$ **16.** $\left(\dfrac{3}{2}, 5\right]$
17. $(-\infty, -2) \cup \left(\dfrac{4}{3}, \infty\right)$ **18.** $(3, 7)$ **19.** $[5, \infty)$ **20.** $[4, \infty)$ **21.** $\left[1, \dfrac{11}{2}\right)$ **22.** $(-\infty, \infty)$ **23.** 9.6   **24.** 7,424,000 vehicles
25. approximately 8 hunting dogs   **26.** 211,468 people   **27.** $\$3542.27$ **28.** Florida: $\$17$ billion; California: $\$13$ billion; New York: $\$9$ billion

Cumulative Review   1. $\{101, 102, 103, \dots\}$; Sec. 1.2, Ex. 2   **2. a.** $\{-2, -1, 0, 1, 2, 3, 4\}$ **b.** $\{4\}$; Sec. 1.2   **3.** $x + 5 = 20$; Sec. 1.3, Ex. 1
4. $x - 19 = 4x$; Sec. 1.3   **5.** $\dfrac{z}{9} = 9 + z$; Sec. 1.3, Ex. 3   **6. a.** $>$ **b.** $=$ **c.** $>$; Sec. 1.3   **7.** -8; Sec. 1.3, Ex. 16   **8.** 0; Sec. 1.3
9. $\dfrac{1}{5}$; Sec. 1.3, Ex. 17   **10.** 7.3; Sec. 1.3   **11.** -14; Sec. 1.4, Ex. 6   **12.** -12.4; Sec. 1.4   **13.** 5; Sec. 1.4, Ex. 8   **14.** 3; Sec. 1.4
15. $-\dfrac{5}{21}$; Sec. 1.4, Ex. 11   **16.** -7.2; Sec. 1.4   **17.** -5; Sec. 1.4, Ex. 25   **18.** undefined; Sec. 1.4   **19.** 0; Sec. 1.4, Ex. 29   **20.** $-\dfrac{1}{3}$; Sec. 1.4
21. $\dfrac{1}{8}$; Sec. 1.4, Ex. 30   **22.** $\dfrac{3}{7}$; Sec. 1.4   **23.** 63; Sec. 1.5, Ex. 1   **24.** $\dfrac{3}{20}$; Sec. 1.5   **25.** $-2x + 4$; Sec. 1.5, Ex. 7   **26.** $-7y - 10$; Sec. 1.5
27. $4y$; Sec. 1.5, Ex. 8   **28.** $14x + 8$; Sec. 1.5   **29.** x^3; Sec. 1.6, Ex. 10   **30.** $\dfrac{1}{a^2 b}$; Sec. 1.6   **31.** $5x$; Sec. 1.6, Ex. 12   **32.** $\dfrac{z^{20}}{9}$; Sec. 1.6
33. $\dfrac{z^2}{9x^4 y^{20}}$; Sec. 1.7, Ex. 14   **34.** $-\dfrac{9}{4a^7 b^2}$; Sec. 1.7   **35.** $\dfrac{27a^4 x^6}{2}$; Sec. 1.7, Ex. 15   **36.** $\{-2\}$; Sec. 2.1   **37.** $\{2\}$; Sec. 2.1, Ex. 3
38. $\{-4.25\}$; Sec. 2.1   **39.** $\{x \mid x$ is a real number$\}$; Sec. 2.1, Ex. 11   **40. a.** $3x + 3$; Sec. 2.2   **b.** $12x + 4$ **41.** $\$2350$; Sec. 2.2, Ex. 2
42. $11, 35$; Sec. 2.2   **43.** $\dfrac{V}{LW} = H$; Sec. 2.3, Ex. 1   **44.** $x = \dfrac{10 + 4y}{7}$; Sec. 2.3   **45.** $(-\infty, 6]$; Sec. 2.4, Ex. 7   **46.** $(3, \infty)$; Sec. 2.4
47. $(-\infty, \infty)$; Sec. 2.5, Ex. 8   **48.** $\varnothing$; Sec. 2.5   **49.** $\{3, -3\}$; Sec. 2.6, Ex. 1   **50.** $\{2\}$; Sec. 2.6

CHAPTER 3   Graphs and Functions

Calculator Explorations   **1.** 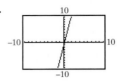   **3.** 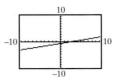   **5.**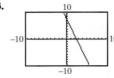

Mental Math   **1.** $(5, 2)$ **3.** $(3, -1)$ **5.** $(-5, -2)$ **7.** $(-1, 0)$

Exercise Set 3.1   **1.** 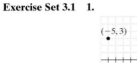 quadrant I; quadrant II; quadrant IV; y-axis; quadrant III   **3.** quadrant IV   **5.** x-axis   **7.** quadrant III   **9.**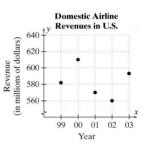

11. no; yes   **13.** yes; yes   **15.** yes; no   **17.** 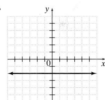   **19.** 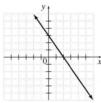   **21.**

23. 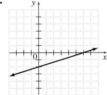   **25.** 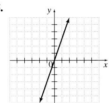   **27.**    **29.**   **31.**

33. 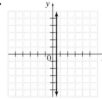   **35.** 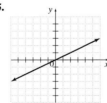   **37.** 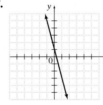   **39.** 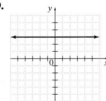   **41.**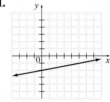

43. 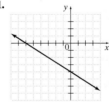   **45.** $\dfrac{3}{2}$ **47.** 6   **49.** $-\dfrac{6}{5}$ **51.** b   **53.** B   **55.** C   **57.** 1991   **59.** answers may vary

61. a.   **b.** 14 in.   **63.** $7000   **65.** $500   **67.** depreciation is the same from year to year

69. a. $(0, 500)$; 0 tables and 500 chairs are produced   **b.** $(750, 0)$; 750 tables and 0 chairs are produced   **c.** 466 chairs   **71.** answers may vary

73. vertical line $x = 0$ has y-intercepts.　**75.** 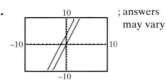　**77.**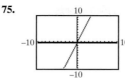

Calculator Explorations　1. 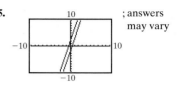 ; answers may vary　**3.** 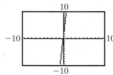 ; answers may vary　**5.** 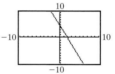 ; answers may vary

Mental Math　1. upward　**3.** horizontally

Exercise Set 3.2　1. $\dfrac{9}{5}$　**3.** $-\dfrac{7}{2}$　**5.** $-\dfrac{5}{6}$　**7.** $\dfrac{1}{3}$　**9.** $-\dfrac{4}{3}$　**11.** 0　**13.** $m = 2$　**15.** $m = -\dfrac{4}{5}$　**17.** $\dfrac{2}{3}$　**19.** $\dfrac{3}{20}$

21. $m = -1; (0, 5)$　**23.** $m = 5; (0, -2)$　**25.** $m = -2; (0, 7)$　**27.** $m = \dfrac{2}{3}; \left(0, -\dfrac{10}{3}\right)$　**29.** m undefined; no y-intercept

31. $m = 0; (0, -2)$　**33.** $m = \dfrac{1}{2}; (0, 0)$　**35.** $m = \dfrac{1}{3}; \left(0, -\dfrac{8}{3}\right)$　**37.** $m = \dfrac{6}{5}; (0, 6)$　**39.** $m = 7; (0, 0)$　**41.** m undefined; no y-intercept

43. A　**45.** B　**47.** l_2　**49.** l_2　**51.** l_2　**53.** neither　**55.** neither　**57.** perpendicular　**59.** parallel　**61.** parallel

63. perpendicular　**65.** perpendicular　**67.** neither　**69.** $\{9, -3\}$　**71.** $(-\infty, -4) \cup (-1, \infty)$　**73.** $\left[\dfrac{2}{3}, 2\right]$

75. incorrect; $m = \dfrac{-14 - 6}{7 - (-2)} = -\dfrac{20}{9}$　**77.** correct　**79.** $-\dfrac{7}{2}$　**81.** $\dfrac{2}{7}$　**83.** $\dfrac{5}{2}$　**85.** $-\dfrac{2}{5}$　**87.** $(10, 13)$　**89.** $\dfrac{3}{2}$ yd per sec

91. answers may vary　**93. a.** $l_1: -2; l_2: -1; l_3: -\dfrac{2}{3}$　**b.** lesser

Calculator Explorations　1. $y = \dfrac{1}{3.5}x$ 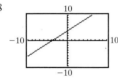　**3.** $y = \dfrac{-5.78}{2.31}x + \dfrac{10.98}{2.31}$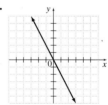

5. $y = x + 3.78$ 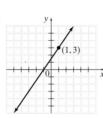　**7.** $y = 13.3x + 1.5$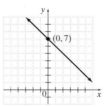

Mental Math　1. $m = -4; (0, 12)$　**3.** $m = 5; (0, 0)$　**5.** $m = \dfrac{1}{2}; (0, 6)$

Exercise Set 3.3　1. 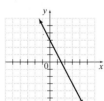　**3.** 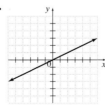　**5.** 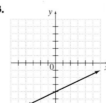　**7.**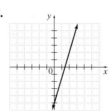

9. 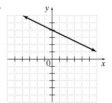　**11.**　**13.**　**15.**　**17.**

19. C　**21.** D　**23.** $y = -x + 1$　**25.** $y = 2x + \dfrac{3}{4}$　**27.** $y = \dfrac{2}{7}x$　**29. a.** $44{,}955$　**b.** $m = 1983$; annual income increases $1983 each year

c. $(0, 42{,}972)$ at year $= 0$, or 1999, annual average income was $42,972　**31. a.** $m = 10.6; (0, 186)$　**b.** number of people employed as network systems and data communications analysts increases 10.6 thousand for every 1 year　**c.** 930 thousand network systems and data communications analysts employed in 2002　**33. a.** 168.8 million　**b.** in 2006　**c.** answers may vary　**35.** $y = 5x + 32$　**37.** $y = 2x - 1$

39. incorrect; $y = 3x + 1.7$　**41.** correct　**43.** $y = 42{,}000x + 2{,}900{,}000$, where y is revenue and x is number of years since 1995

Calculator Explorations   **1.** 18.4   **3.** −1.5   **5.** 8.7; 7.6

Mental Math   **1.** $m = -2; (1, 4)$ **3.** $m = \frac{1}{4}; (2, 0)$ **5.** $m = 5; (3, -2)$

Exercise Set 3.4   **1.** $y = 3x - 1$ **3.** $y = -2x - 1$ **5.** $y = \frac{1}{2}x + 5$ **7.** $y = -\frac{9}{10}x - \frac{27}{10}$ **9.** $3x - y = 6$ **11.** $2x + y = 1$

13. $x + 2y = -10$ **15.** $x - 3y = 21$ **17.** $3x + 8y = 5$ **19.** $x = 2$ **21.** $y = 1$ **23.** $x = 0$ **25.** $y = 4x - 4$ **27.** $y = -3x + 1$

29. $y = 4$ **31.** $y = -\frac{3}{2}x - 6$ **33.** $y = -5$ **35.** $y = -4x + 1$ **37.** $2x - y = -7$ **39.** $y = -x + 7$ **41.** $y = -\frac{1}{2}x + \frac{3}{8}$

43. $2x + 7y = -42$ **45.** $4x + 3y = -20$ **47.** $x = -2$ **49.** $x + 2y = 2$ **51.** $y = 12$ **53.** $8x - y = 47$ **55.** $x = 5$

57. $y = -\frac{3}{8}x - \frac{29}{4}$ **59. a.** $y = 32x$ **b.** 128 ft per sec   **61. a.** $y = 12,000x + 18,000$ **b.** \$102,000   **c.** 9 yr   **63. a.** $y = 9266.7x + 142,200$

b. \$216,334   **65. a.** $y = 0.1x + 70.0$ **b.** 70.7 million subscribers   **67.** 31.   **69.** −8.4   **71.** 4   **73.** $2x + y = 3$ **75.** $2x - 3y = -7$

77. true   **79.** answers may vary   **81.**

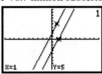

Integrated Review   **1.**

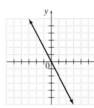

2.

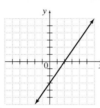

3.

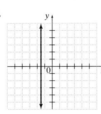

4.

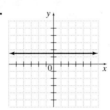

5. 0

6. $-\frac{3}{5}$ **7.** $m = 3; (0, -5)$ **8.** $m = \frac{5}{2}; \left(0, -\frac{7}{2}\right)$ **9.** parallel   **10.** perpendicular   **11.** $y = -x + 7$ **12.** $x = -2$ **13.** $y = 0$

14. $y = -\frac{3}{8}x - \frac{29}{4}$ **15.** $y = -5x - 6$ **16.** $y = -4x + \frac{1}{3}$ **17.** $y = \frac{1}{2}x - 1$ **18.** $y = 3x - \frac{3}{2}$ **19.** $y = 3x - 2$

20. $y = -\frac{5}{4}x + 4$ **21.** $y = \frac{1}{4}x - \frac{7}{2}$ **22.** $y = -\frac{5}{2}x - \frac{5}{2}$ **23.** $x = -1$ **24.** $y = 3$

Exercise Set 3.5   **1.**

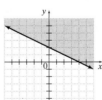

3.

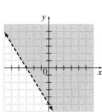

5.

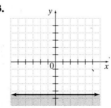

7.

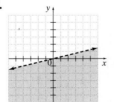

9.

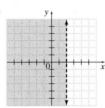

11.

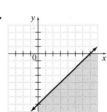

13.

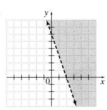

15.

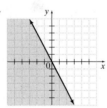

17.

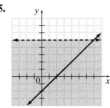 (17)

19.

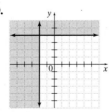

21.

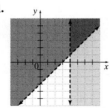

23.

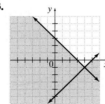

25.

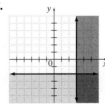 (25)

27.

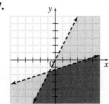

29. D   **31.** A   **33.** yes   **35.** no   **37.** with $<$ or $>$ **39.**

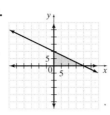

Exercise Set 3.6   1. domain; $\{-1, 0, -2, 5\}$; range: $\{7, 6, 2\}$; function   **3.** domain; $\{-2, 6, -7\}$; range: $\{4, -3, -8\}$; not a function

5. domain: $\{1\}$; range: $\{1, 2, 3, 4\}$; not a function   **7.** domain: $\left\{\frac{3}{2}, 0\right\}$; range: $\left\{\frac{1}{2}, -7, \frac{4}{5}\right\}$; not a function   **9.** domain: $\{-3, 0, 3\}$;

range: $\{-3, 0, 3\}$; function   **11.** domain: $\{-1, 1, 2, 3\}$; range: $\{2, 1\}$; function

13. domain; $\{$Colorado, Alaska, Massachusetts, Delaware, Illinois, Pennsylvania$\}$; range: $\{1, 7, 10, 19\}$; function

15. domain: $\{32°, 104°, 212°, 50°\}$; range: $\{0°, 40°, 10°, 100°\}$; function   **17.** domain: $\{2, -1, 5, 100\}$; range: $\{0\}$; function

19. function   **21.** yes   **23.** no   **25.** yes   **27.** function   **29.** not a function   **31.** function   **33.** not a function

35. not a function   **37.** not a function   **39.** not a function   **41.** not a function   **43.** function   **45.** 15   **47.** 38

49. 7   **51.** 3   **53. a.** 0   **b.** 1   **c.** -1 **55. a.** -5 **b.** -5 **c.** -5 **57.** 25π sq cm   **59.** 2744 cu in.   **61.** 166.38 cm   **63.** 163.2 mg

65. a. 99.42; per capita consumption of poultry was 99.42 lb in 2003.   **b.** 108.54 lb

67.   **69.**   **71.**   **73.** $(-\infty, 14]$ **75.** $\left[\frac{7}{2}, \infty\right)$ **77.** $\left(-\infty, -\frac{1}{4}\right)$

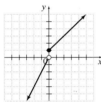   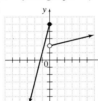   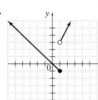

79. no; answers may vary   **81.** yes; answers may vary   **83.** true   **85.** true   **87.** answers may vary   **89.** $f(x) = x + 7$

91. answers may vary   **93. a.** $f(12) = 132$ **b.** $f(a) = a^2 - 12$ **c.** $f(-x) = x^2 - 12$ **d.** $f(x + h) = x^2 + 2xh + h^2 - 12$

Exercise Set 3.7   1. domain; $[0, \infty)$; range: $(-\infty, \infty)$ **3.** domain: $(-\infty, \infty)$; range: $[0, \infty)$ **5.** domain: $(-\infty, \infty)$; range: $(-\infty, -3] \cup [3, \infty)$

7. domain: $[1, 7]$; range: $[1, 7]$ **9.** domain: $\{-2\}$; range: $(-\infty, \infty)$ **11.** domain: $(-\infty, \infty)$; range: $(-\infty, 3]$ **13.** domain: $(-\infty, \infty)$;

range: $(-\infty, 3]$ **15.** domain: $[2, \infty)$; range: $[3, \infty)$

17.   **19.**   **21.**   **23.**

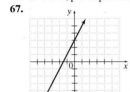   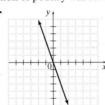   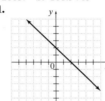   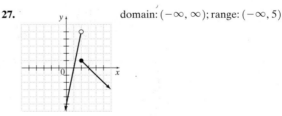

25.   domain: $(-\infty, \infty)$; range: $[0, \infty)$ **27.**   domain: $(-\infty, \infty)$; range: $(-\infty, 5)$

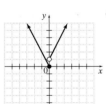   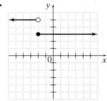

29.   domain: $(-\infty, \infty)$; range: $(-\infty, 6]$ **31.**   domain: $(-\infty, 0] \cup [1, \infty)$; range: $\{-4, -2\}$

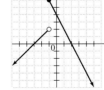   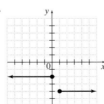

33. A   **35.** D   **37.** answers may vary   **39.**

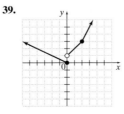

Mental Math   **1.** c   **3.** d

Exercise Set 3.8   **1.** 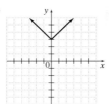   **3.** 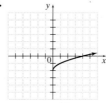   **5.** 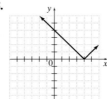   **7.**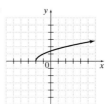

9. 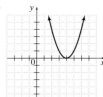   **11.** 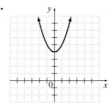   **13.** 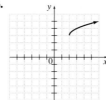   **15.** 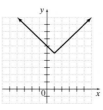   **17.**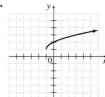

19. 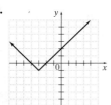   **21.** 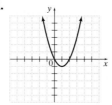   **23.** 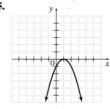   **25.** 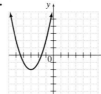   **27.**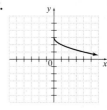

29. 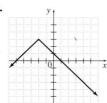   **31.**   **33.** $-15x^8$ **35.** y^{18} **37.** domain: $[2, \infty)$; range: $[3, \infty)$
39. domain: $(-\infty, \infty)$; range: $(-\infty, 3]$ **41.** $[20, \infty)$ **43.** $(-\infty, \infty)$
45. $[-103, \infty)$

47.   domain: $(-\infty, \infty)$; range: $[0, \infty)$ **49.**   domain: $(-\infty, \infty)$; range: $(-\infty, 0] \cup (2, \infty)$

Chapter 3 Vocabulary Check   **1.** relation   **2.** line   **3.** linear inequality   **4.** standard   **5.** range   **6.** parallel   **7.** slope-intercept
8. function   **9.** slope   **10.** perpendicular   **11.** y   **12.** domain   **13.** linear function   **14.** x   **15.** point-slope

Chapter 3 Review   **1.**    A, quadrant IV; B, quadrant II; C, y-axis; D, quadrant III

2. 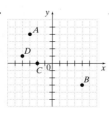   A and D, quadrant II; C, x-axis; B, quadrant IV   **3.**

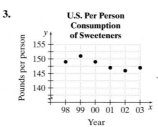

U.S. Per Person Consumption of Sweeteners

4.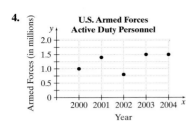

5. no; yes **6.** yes; no **7.** yes; yes **8.** yes; no **9.**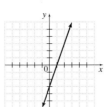

10. 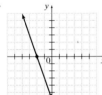 **11.** 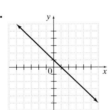 **12.** 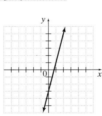 **13.** 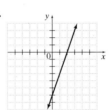 **14.**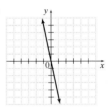

15. -3 **16.** $\dfrac{1}{2}$ **17.** $\dfrac{5}{2}$ **18.** $-\dfrac{3}{4}$ **19.** $\dfrac{3}{2}$ **20.** -3 **21.** $-\dfrac{1}{2}$ **22.** 1 **23.** l_2 **24.** l_2 **25.** l_2 **26.** l_1

27. $m = -3;\left(0, \dfrac{1}{2}\right)$ **28.** $m = 2;(0, 4)$ **29.** $m = \dfrac{2}{5};\left(0, -\dfrac{4}{3}\right)$ **30.** $m = -\dfrac{2}{7};\left(0, \dfrac{3}{2}\right)$ **31.** 0 **32.** undefined **33.** neither

34. parallel **35.** perpendicular **36.** neither **37.** 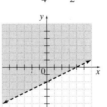 **38.** 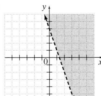 **39.**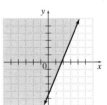

40. 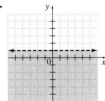 **41.** 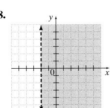 **42.** 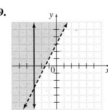 **43.** 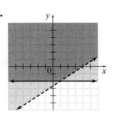 **44.**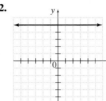

45. a. \$87 **b.** $m = 0.3$; cost increases by \$0.30 for each additional mile driven **c.** $(0, 42)$; cost for 0 miles driven is \$42 **46.** $y = -1$

47. $x = -4$ **48.** $y = 3x + 14$ **49.** $y = -\dfrac{1}{2}x - 4$ **50.** $y = -2x - 2$ **51.** $y = \dfrac{3}{4}x + \dfrac{7}{2}$ **52. a.** $y = -2050x + 18,400$ **b.** \$6100

53. a. $y = 12,000x + 126,000$ **b.** \$342,000 **54.** 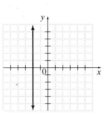 **55.** 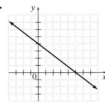 **56.**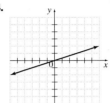

57. 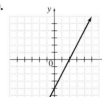 **58.** **59.** **60.**

61. domain: $\left\{-\dfrac{1}{2}, 6, 0, 25\right\}$; range: $\left\{\dfrac{3}{4}, 0.65, -12, 25\right\}$; function **62.** domain: $\left\{\dfrac{3}{4}, 0.65, -12, 25\right\}$; range: $\left\{-\dfrac{1}{2}, 6, 0, 25\right\}$; function

63. domain: $\{2, 4, 6, 8\}$; range: $\{2, 4, 5, 6\}$; not a function **64.** domain: $\{\text{triangle, square, rectangle, parallelogram}\}$; range: $\{3, 4\}$; function

65. domain: $(-\infty, \infty)$; range: $(-\infty - 1] \cup [1, \infty)$; not a function **66.** domain: $\{-3\}$; range: $(-\infty, \infty)$; not a function

67. domain: $(-\infty, \infty)$; range: $\{4\}$; function **68.** domain: $[-1, 1]$; range: $[-1, 1]$; not a function **69.** -3 **70.** 0

71. 18 **72.** 9 **73.** -3 **74.** 0 **75.** 381 lb **76.** 5080 lb

77.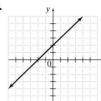

78.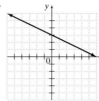

79. domain: $\{2\}$; range: $(-\infty, \infty)$ **80.** domain: $(-\infty, \infty)$; range: $(-\infty, \infty)$

81. domain: $[-4, 4]$; range: $[-1, 5]$ **82.** domain: $(-\infty, \infty)$; range: $\{-5\}$ **83.**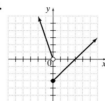

84.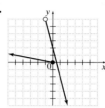

85.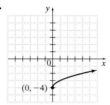

86.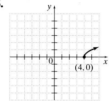

87.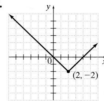

88.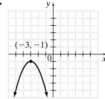

89.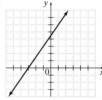

90.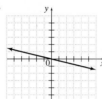

91.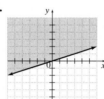

92. $x = -2$ **93.** $y = 5$ **94.** $y = 2x - 12$ **95.** $y = -11x - 52$

96. $y = -\dfrac{3}{2}x - 8$ **97.** $y = -\dfrac{3}{2}x - 1$ **98.** domain: $(-\infty, \infty)$; range: $(-\infty, 0]$ **99.** domain: $(-\infty, \infty)$; range: $(-\infty, \infty)$

100.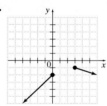

101.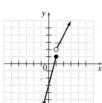

102.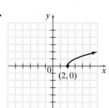

103.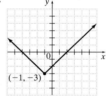

Chapter 3 Test   **1.** 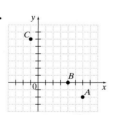 A is in quadrant IV. B is on the x-axis, no quadrant. C is in quadrant II.

2.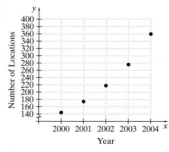

3.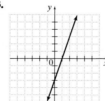

4.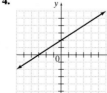

5.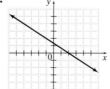

6.
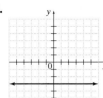
   7. $-\dfrac{3}{2}$ **8.** $m = -\dfrac{1}{4}; \left(0, \dfrac{2}{3}\right)$ **9.** C   **10.** A   **11.** B   **12.** D   **13.** $y = -8$ **14.** $x = -4$

15. $y = -2$ **16.** $3x + y = 11$ **17.** $5x - y = 2$ **18.** $y = -\dfrac{1}{2}x$ **19.** $y = -\dfrac{1}{3}x + \dfrac{5}{3}$ **20.** $y = -\dfrac{1}{2}x - \dfrac{1}{2}$ **21.** neither

22. a. 78 games   **b.** 86 games   **c.** \$134 million   **d.** 0.232; every million dollars spent on payroll increases winnings by 0.232 games.

23.
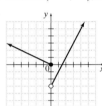
   24.
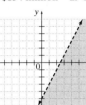
   25.
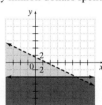
   26. domain: $(-\infty, \infty)$; range: $\{5\}$; function

27. domain: $\{-2\}$; range: $(-\infty, \infty)$; not a function   **28.** domain: $(-\infty, \infty)$; range: $[0, \infty)$; function

29. domain: $(-\infty, \infty)$; range: $(-\infty, \infty)$; function

30.

   domain: $(-\infty, \infty)$; range: $(-3, \infty)$ **31.**
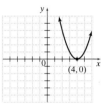
   32.
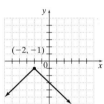
   domain: $(-\infty, \infty)$; range: $(-\infty, -1]$ **33.**
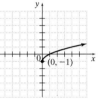

Cumulative Review   1. $\{2, 3, 4, 5\}$; Sec. 1.2, Ex. 1   **2.** $\{-1, 0, 1, 2\}$; Sec. 1.2   **3.** $\{101, 102, 103, \dots\}$; Sec. 1.2, Ex. 5

4. a. 7   **b.** 0   **c.** -10; Sec. 1.4   **5.** $-\dfrac{1}{9}$; Sec. 1.3, Ex. 21   **6.** $\dfrac{1}{10.2}$ or $\dfrac{10}{102} = \dfrac{5}{51}$; Sec. 1.3   **7.** $\dfrac{4}{7}$; Sec. 1.3; Ex. 22   **8. a.** 98   **b.** 98; Sec. 1.5

9. 3; Sec. 1.4, Ex. 1   **10.** $\dfrac{7}{15}$; Sec. 1.4   **11.** -2.7; Sec. 1.4, Ex. 4   **12.** -102.86; Sec. 1.4   **13.** 23; Sec. 1.5, Ex. 5   **14.** 22; Sec. 1.5

15. 2^7; Sec. 1.6, Ex. 1   **16.** $-\dfrac{3}{2}x^{13}$; Sec. 1.6   **17.** y^7; Sec. 1.6, Ex. 3   **18.** $4.86a^8b^{13}$; Sec. 1.6   **19.** $125x^6$; Sec. 1.7, Ex. 5   **20.** $\dfrac{y^8}{81}$; Sec. 1.7

21. $64y^2$; Sec. 1.7, Ex. 8   **22.** $\dfrac{c^{15}}{125a^9b^6}$; Sec. 1.7   **23.** $\{0.4\}$; Sec. 2.1, Ex. 4   **24.** $\{x \,|\, x$ is a real number$\}$; Sec. 2.1   **25.** 23, 49, 92; Sec. 2.2, Ex. 3

26. 69, 71, 73; Sec. 2.2   **27.** \$11,607.55; Sec. 2.3, Ex. 4   **28. a.** -7 **b.** $\dfrac{5}{6}$ **c.** -13; Sec. 1.4

29. $[2, \infty)$; Sec.2.4, Ex.1   **30.** $\left(-\infty, -\dfrac{3}{2}\right)$; Sec. 2.4

31. $(0.5, 3]$; Sec. 2.4, Ex. 3   **32.** $(-25, \infty)$; Sec. 2.4   **33.** $(-\infty, 4)$; Sec. 2.5, Ex. 2

34. $(-\infty, \infty)$; Sec. 2.5   **35.** $\{4\}$; Sec. 2.6, Ex. 8   **36.** $\left\{1, -\dfrac{1}{5}\right\}$; Sec. 2.6   **37.** $(-\infty, -4) \cup (10, \infty)$; Sec. 2.6, Ex. 11   **38.** $[-2, 18]$; Sec. 2.6

39.   Sec. 3.1, Ex. 5   **40.** $m = -\dfrac{7}{2}; (0, 5)$; Sec. 3.2   **41.** $m = 3$; Sec. 3.2, Ex. 3   **42.** $m = 3$; Sec. 3.2

43. $y = \dfrac{2}{3}x - 7$; Sec. 3.3, Ex. 3   **44.** $x + y = 3$; Sec. 3.4   **45.** $y = 3$; Sec. 3.4, Ex. 3   **46.** $x = -2$; Sec. 3.4   **47.** 5; Sec. 3.6, Ex. 16

48. $m = -2.3$; Sec. 3.3   **49.** -2; Sec. 3.6, Ex. 15   **50.** no; Sec. 3.6

CHAPTER 4   Systems of Equations and Inequalities

Calculator Explorations   **1.** $(2.11, 0.17)$ **3.** $(-8.20, -6.30)$

Mental math   **1.** B   **3.** A

Exercise Set 4.1   **1.** yes   **3.** no   **5.** yes   **7.** no   **9.** 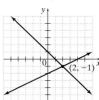   **11.** 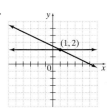   **13.** $\varnothing$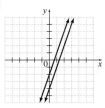

15. 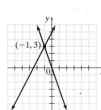   **17.** $(2, 8)$ **19.** $(0, -9)$ **21.** $(1, -1)$ **23.** $\{x \mid x = -3y + 4\}$ **25.** $\left(\dfrac{5}{2}, \dfrac{5}{4}\right)$ **27.** $(1, -2)$

29. $(9, 9)$ **31.** $(7, 2)$ **33.** $\varnothing$ **35.** $\{(x, y) \mid 3x + y = 1\}$ **37.** $\left(\dfrac{3}{2}, 1\right)$ **39.** $(2, -1)$ **41.** $(-5, 3)$

43. $\varnothing$ **45.** $\left(\dfrac{1}{2}, \dfrac{1}{5}\right)$ **47.** $(8, 2)$ **49.** $\{(x, y) \mid x = 3y + 2\}$ **51.** $(-5, 3)$ **53.** $(3, 2)$ **55.** $(7, -3)$

57. $\varnothing$ **59.** $(3, 4)$ **61.** $\left(-\dfrac{1}{4}, \dfrac{1}{2}\right)$ **63.** $(-2, 1)$ **65.** $(1.2, -3.6)$ **67.** true   **69.** false

71. $6y - 4z = 25$ **73.** $x + 10y = 2$ **75.** no solution   **77.** infinite number of solutions   **79.** no   **81.** 5000 DVDs; \$21

83. supply greater than demand   **85.** $(1875, 4687.5)$ **87.** makes money   **89.** for x-values greater than 1875

91. answers may vary; one possibility: $\begin{cases} -2x + y = 1 \\ x - 2y = -8 \end{cases}$ **93. a.** Consumption of red meat is decreasing while consumption of poultry is increasing.

b. $(31, 97)$ **c.** In the year 2028, red meat and poultry consumption will both be about 97 pounds per person.   **95.** $\left(\dfrac{1}{4}, 8\right)$ **97.** $\left(\dfrac{1}{3}, \dfrac{1}{2}\right)$

99. $\left(\dfrac{1}{4}, -\dfrac{1}{3}\right)$ **101.** $\varnothing$

Exercise Set 4.2   **1.** a, b, d   **3.** yes; answers may vary   **5.** $(-1, 5, 2)$ **7.** $(-2, 5, 1)$ **9.** $(-2, 3, -1)$ **11.** $\{(x, y, z) \mid x - 2y + z = -5\}$
13. $\varnothing$ **15.** $(0, 0, 0)$ **17.** $(-3, -35, -7)$ **19.** $(6, 22, -20)$ **21.** $\varnothing$ **23.** $(3, 2, 2)$ **25.** $\{(x, y, z) \mid x + 2y - 3z = 4\}$

27. $(-3, -4, -5)$ **29.** $\left(0, \dfrac{1}{2}, -4\right)$ **31.** $(12, 6, 4)$ **33.** $\{5\}$ **35.** $\left\{-\dfrac{5}{3}\right\}$ **37.** 15 and 30   **39.** answers may vary
41. answers may vary   **43.** $(1, 1, -1)$ **45.** $(1, 1, 0, 2)$ **47.** $(1, -1, 2, 3)$ **49.** answers may vary

Exercise Set 4.3   **1.** 10 and 8   **3. a.** Enterprise class: 1101 ft; Nimitz class: 1092 ft   **b.** 3.67 football fields   **5.** plane: 520 mph; wind: 40 mph

7. 20 qt of 4%; 40 qt of 1%   **9.** United Kingdom: 31,706 students; Italy: 18,936 students   **11.** 9 large frames; 13 small frames

13. -10 and -8 **15.** 2008   **17.** tablets: \$0.80; pens: \$0.20   **19.** speed of plane: 630 mph; speed of wind: 90 mph

21. a. answers may vary   **b.** 1984   **23.** 28 cm; 28 cm; 37 cm   **25.** 600 mi   **27.** $x = 75; y = 105$ **29.** 625 units   **31.** 3000 units

33. 1280 units   **35. a.** $R(x) = 450x$ **b.** $C(x) = 200x + 6000$ **c.** 24 desks   **37.** 2 units of Mix A; 3 units of Mix B; 1 unit of Mix C

39. 5 in.; 7 in.; 10 in.   **41.** 18, 13, and 9   **43.** free throws: 142; two-point field goals: 220; three-point field goals: 52   **45.** $x = 60; y = 55; z = 65$

47. $5x + 5z = 10$ **49.** $-5y + 2z = 2$ **51.** 1993: 897,231; 2003: 1,661,996   **53.** $a = 3, b = 4, c = -1$

55. $a = -\dfrac{7}{60}; b = 32\dfrac{23}{60}; c = 828\dfrac{9}{10}$; or $a = -0.11; b = 32.38; c = 828.9$; 2009—1317 thousand students

Integrated Review   **1.** C   **2.** D   **3.** A   **4.** B   **5.** $(1, 3)$ **6.** $\left(\dfrac{4}{3}, \dfrac{16}{3}\right)$ **7.** $(2, -1)$ **8.** $(5, 2)$ **9.** $\left(\dfrac{3}{2}, 1\right)$ **10.** $\left(-2, \dfrac{3}{4}\right)$

11. $\varnothing$ **12.** $\{(x, y) \mid 2x - 5y = 3\}$ **13.** $\left(1, \dfrac{1}{3}\right)$ **14.** $\left(3, \dfrac{3}{4}\right)$ **15.** $(-1, 3, 2)$ **16.** $(1, -3, 0)$ **17.** $\varnothing$

18. $\{(x, y, z) \mid x - y + 3z = 2\}$ **19.** $\left(2, 5, \dfrac{1}{2}\right)$ **20.** $\left(1, 1, \dfrac{1}{3}\right)$ **21.** 19 and 27   **22.** 70°; 70°; 100°; 120°

Exercise Set 4.4   **1.** $(2, -1)$ **3.** $(-4, 2)$ **5.** $\varnothing$ **7.** $\{(x, y) \mid 3x - 3y = 9\}$ **9.** $(-2, 5, -2)$ **11.** $(1, -2, 3)$ **13.** $(4, -3)$

15. $(2, 1, -1)$ **17.** $(9, 9)$ **19.** $\varnothing$ **21.** $\varnothing$ **23.** $(1, -4, 3)$ **25.** function   **27.** not a function   **29.** c   **31. a.** end of 1984
b. black-and-white sets; microwave ovens; the percent of households owning black-and-white television sets is decreasing and the percent of households owning microwave ovens is increasing; answers may vary   **c.** in 2002   **d.** answers may vary   **33.** answers may vary

Exercise Set 4.5 $\quad$ **1.** 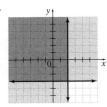 $\quad$ **3.** 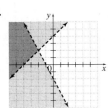 $\quad$ **5.** 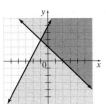 $\quad$ **7.**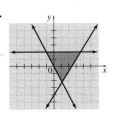

9. 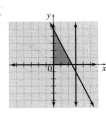 $\quad$ **11.** 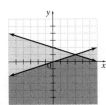 $\quad$ **13.** 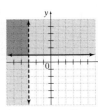 $\quad$ **15.** 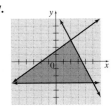 $\quad$ **17.**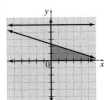

19. 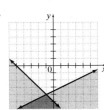 $\quad$ **21.** C $\quad$ **23.** D $\quad$ **25.** 9 $\quad$ **27.** $\dfrac{4}{9}$ $\quad$ **29.** 5 $\quad$ **31.** 59 $\quad$ **33.** the line $y = 3$ $\quad$ **35.** answers may vary

Vocabulary Check $\quad$ **1.** system of equations $\quad$ **2.** solution $\quad$ **3.** consistent $\quad$ **4.** square $\quad$ **5.** inconsistent $\quad$ **6.** matrix

Chapter 4 Review

1. $(-3, 1)$ $\quad$ **2.** $\left(0, \dfrac{2}{3}\right)$ $\quad$ **3.** $\varnothing$ $\quad$ **4.** $\{(x, y) \mid 3x - 6y = 12\}$ $\quad$ **5.** $\left(3, \dfrac{8}{3}\right)$ $\quad$ **6.** 1500 backpacks $\quad$ **7.** $(2, 0, 2)$ $\quad$ **8.** $(2, 0, -3)$

9. $\left(-\dfrac{1}{2}, \dfrac{3}{4}, 1\right)$ $\quad$ **10.** $(-1, 2, 0)$ $\quad$ **11.** $\varnothing$ $\quad$ **12.** $(5, 3, 0)$ $\quad$ **13.** $(1, 1, -2)$ $\quad$ **14.** $(3, 1, 1)$ $\quad$ **15.** 10, 40, and 48 $\quad$ **16.** 63 and 21

17. 58 mph; 65 mph $\quad$ **18.** width: 37 ft; length: 111 ft $\quad$ **19.** 20 L of 10% solution; 30 L of 60% solution $\quad$ **20.** 30 lb of creme-filled; 5 lb of chocolate-covered nuts; 10 lb of chocolate-covered raisins $\quad$ **21.** 17 pennies; 20 nickels; 16 dimes $\quad$ **22.** larger investment: 9.5%; smaller investment: 7.5% $\quad$ **23.** two sides: 22 cm each; third side: 29 cm $\quad$ **24.** 120, 115, and 60 $\quad$ **25.** $(-3, 1)$ $\quad$ **26.** $\{(x, y) \mid x - 2y = 4\}$

27. $\left(-\dfrac{2}{3}, 3\right)$ $\quad$ **28.** $\left(\dfrac{1}{3}, \dfrac{7}{6}\right)$ $\quad$ **29.** $\left(\dfrac{5}{4}, \dfrac{5}{8}\right)$ $\quad$ **30.** $(-7, -15)$ $\quad$ **31.** $(1, 3)$ $\quad$ **32.** $(2, 1)$ $\quad$ **33.** $(1, 2, 3)$ $\quad$ **34.** $(2, 0, -3)$ $\quad$ **35.** $(3, -2, 5)$

36. $(-1, 2, 0)$ $\quad$ **37.** $(1, 1, -2)$ $\quad$ **38.** $\varnothing$ $\quad$ **39.** 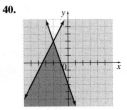 $\quad$ **40.** 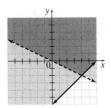 $\quad$ **41.**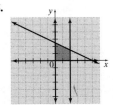

42. 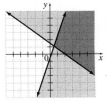 $\quad$ **43.** 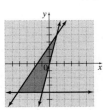 $\quad$ **44.** 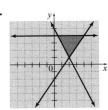 $\quad$ **45.** $\quad$ **46.**

47. $\left(\dfrac{7}{3}, -\dfrac{8}{3}\right)$ $\quad$ **48.** $(10, -4)$ $\quad$ **49.** $\{(x, y) \mid 5x - 2y = 10\}$ $\quad$ **50.** $\varnothing$ $\quad$ **51.** $(-1, 3, 5)$ $\quad$ **52.** 33 and 94 $\quad$ **53.** 28 units, 42 units, 56 units

54. 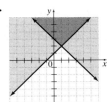 $\quad$ **55.** 2000

Chapter 4 Test   **1.** $(1, 3)$ **2.** $\varnothing$ **3.** $(2, -3)$ **4.** $\{(x, y) \mid 10x + 4y = 10\}$ **5.** $(-1, -2, 4)$ **6.** $\varnothing$ **7.** $\left(\dfrac{7}{2}, -10\right)$

8. $\{(x, y) \mid x - y = -2\}$ **9.** $(5, -3)$ **10.** $(-1, -1, 0)$ **11.** 53 double rooms; 27 single rooms   **12.** 5 gal of 10%; 15 gal of 20%
13. 800 packages   **14.** $23°, 45°, 112°$ **15.**

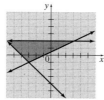

Cumulative Review   **1.** true; Sec. 1.2, Ex. 4   **2.** false; Sec. 1.2   **3.** false; Sec. 1.2, Ex. 7   **4.** true; Sec. 1.2   **5.** $<$; Sec. 1.3, Ex. 4

6. $>$; Sec. 1.3   **7.** $<$; Sec. 1.3, Ex. 7   **8.** $=$; Sec. 1.3   **9.** $<$; Sec. 1.3, Ex. 11   **10.** $<$; Sec. 1.3   **11.** $\dfrac{1}{7}$; Sec. 1.4, Ex. 3   **12.** 7.8; Sec. 1.4

13. -8; Sec. 1.4, Ex. 5   **14.** $-\dfrac{3}{8}$; Sec. 1.4   **15.** -1; Sec. 1.5, Ex. 3   **16. a.** -5 **b.** -24; Sec. 1.4   **17.** $\dfrac{1}{3x}$; Sec. 1.6, Ex. 16   **18.** $\dfrac{1}{81}$; Sec. 1.6

19. $\dfrac{11}{18}$; Sec. 1.6, Ex. 17   **20.** x^{16}; Sec. 1.6   **21.** solution; Sec. 2.1, Ex. 1   **22.** $\{0\}$; Sec. 2.1   **23.** \$2350; Sec. 2.2, Ex. 4

24. $100°, 100°, 110°, 50°$; Sec. 2.2   **25.** $b = \dfrac{2A - Bh}{h}$; Sec. 2.3, Ex. 3   **26.** $\{m \mid m$ is a real number$\}$; Sec. 2.1   **27.** $(-2, \infty)$; Sec. 2.4, Ex. 5

28. $(0, \infty)$; Sec. 2.4   **29.** $\{4, 6\}$; Sec. 2.5, Ex. 1   **30.** $\{2, 3, 4, 5, 6, 8\}$; Sec. 2.5   **31.** $\{0\}$; Sec. 2.5, Ex. 3   **32.** $\{1, 9\}$; Sec. 2.6

33. 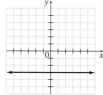 ; Sec. 3.1, Ex. 8   **34.** $(-\infty, -3) \cup (2, \infty)$; Sec. 2.6   **35.** slope: $\dfrac{3}{4}$; y-intercept: $(0, -1)$; Sec. 3.2, Ex. 4

36. a. quadrant III   **b.** quadrant IV   **c.** y-axis; Sec. 3.1   **37.** $y = -3x - 2$; Sec. 3.4, Ex. 1   **38. a.** 75   **b.** 12; Sec. 3.6

39. function; Sec. 3.6, Ex. 7   **40.** $m = \dfrac{3}{2}$; Sec. 3.2   **41.** 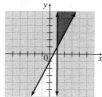 ; Sec. 3.5, Ex. 3   **42.** $y = -2x + 2$; Sec. 3.3

43. $(0, -5)$; Sec. 4.1, Ex. 10   **44.** $(-1, 3)$; Sec. 4.1   **45.** $\varnothing$; Sec. 4.2, Ex. 2   **46.** $(-1, 2, 5)$; Sec. 4.2   **47.** 7 and 11; Sec. 4.3, Ex. 2
48. $\varnothing$; Sec. 4.1

CHAPTER 5   Polynomials and Polynomial Functions

Calculator Explorations   **1.** $x^3 - 4x^2 + 7x - 8$ **3.** $-2.1x^2 - 3.2x - 1.7$ **5.** $7.69x^2 - 1.26x + 5.3$

Mental Math   **1.** $10x$ **3.** $5y$ **5.** $-9z$

Exercise Set 5.1   **1.** 0   **3.** 2   **5.** 3   **7.** 3   **9.** 9   **11.** binomial of degree 1   **13.** trinomial of degree 2   **15.** monomial of degree 3

17. degree 4; none of these   **19.** $6y - 7y^2$ **21.** $-3x^4 + 11x$ **23.** $x^2 y - 5x - \dfrac{1}{2}$ **25.** $18y^2 - 17$ **27.** $3x^2 - 3xy + 6y^2$

29. $x^2 - 4x + 8$ **31.** $12x^3 y + 8x + 8$ **33.** $4.5x^3 + 0.2x^2 - 3.8x + 9.1$ **35.** $y^2 + 3$ **37.** $-2x^2 + 5x$ **39.** $7y^2 - 12y - 3$
41. $7x^3 + 4x^2 + 8x - 10$ **43.** $-20y^2 + 3yx$ **45.** $-3x^2 + 3$ **47.** $2y^4 - 5y^2 + x^2 + 1$ **49.** $5x^2 - 9x - 3$ **51.** $3x^2 + x + 18$
53. $4x - 13$ **55.** $-x^3 + 8a - 12$ **57.** $14ab + 10a^2 b - 18a^2 + 12b^2$ **59.** $5x^2 + 22x + 16$ **61.** 0   **63.** $8xy^2 + 2x^3 + 3x^2 - 3$

65. $3x^2 - 9x + 15$ **67.** $15x^2 + 8x - 6$ **69.** $\dfrac{1}{3}x^2 - x + 1$ **71.** 57   **73.** 499   **75.** 1   **77.** 202 sq in.   **79. a.** 284 ft   **b.** 536 ft

c. 756 ft   **d.** 944 ft   **e.** answers may vary   **f.** 19 sec   **81.** \$80,000   **83.** \$40,000   **85.** $-14z + 42y$ **87.** $-15y^2 - 10y + 35$ **89.** a and c
91. $(12x - 1.7) - (15x + 6.2) = 12x - 1.7 - 15x - 6.2 = -3x - 7.9$ **93.** answers may vary   **95.** answers may vary
97. a. 0.6 million   **b.** 3.1 million   **c.** 9.7 million   **d.** answers may vary   **99.** $4x^2 - 3x + 6$ **101.** $2a - 3; -2x - 3; 2x + 2h - 3$
103. $12z^{5x} + 13z^{2x} - 2z$ **105.** $(2x^2 + 7xy + 10y^2)$ units

Calculator Explorations   **1.** $x^2 - 16$ **3.** $9x^2 - 42x + 49$ **5.** $5x^3 - 14x^2 - 13x - 2$

Exercise Set 5.2   **1.** $-12x^5$ **3.** $86a^5 b^8 c^3$ **5.** $12x^2 + 21x$ **7.** $-24x^2 y - 6xy^2$ **9.** $-4a^3 bx - 4a^3 by + 12ab$ **11.** $2x^2 - 2x - 12$
13. $2x^4 + 3x^3 - 2x^2 + x + 6$ **15.** $15x^2 - 7x - 2$ **17.** $15m^3 + 16m^2 - m - 2$ **19.** $-30a^4 b^4 + 36a^3 b^2 + 36a^2 b^3$ **21.** $x^2 + x - 12$

23. $10x^2 - 21xy - 8y^2$ **25.** $16x^2 - \dfrac{2}{3}x - \dfrac{1}{6}$ **27.** $5x^4 - 17x^2 y^2 + 6y^4$ **29.** $x^2 + 8x + 16$ **31.** $36y^2 - 1$ **33.** $9x^2 - 6xy + y^2$

35. $49a^2b^2 - 9c^2$ **37.** $9x^2 - \dfrac{1}{4}$ **39.** $16b^2 + 32b + 16$ **41.** $4s^2 - 12s + 8$ **43.** $2x^3 + 2x^2y + x^2 + xy - x - y$

45. $x^4 - 8x^3 + 24x^2 - 32x + 16$ **47.** $x^4 - 625$ **49.** $-24a^2b^3 + 40a^2b^2 - 160a^2b$ **51.** $36x^2 + 12x + 1$ **53.** $25x^6 - 4y^2$

55. $10x^5 + 8x^4 + 2x^3 + 25x^2 + 20x + 5$ **57.** $9x^4 + 12x^3 - 2x^2 - 4x + 1$ **59.** $3x^2 + 8x - 3$ **61.** $9x^6 + 15x^4 + 3x^2 + 5$

63. $9x^2 + 6x + 1$ **65.** $9b^2 - 36y^2$ **67.** $49x^2 - 9$ **69.** $9x^3 + 30x^2 + 12x - 24$ **71.** $12x^3 - 2x^2 + 13x + 5$ **73.** $x^2y^2 - 4xy + 4$

75. $22a^3 + 11a^2 + 2a + 1$ **77.** $\dfrac{1}{3}n^2 - 7n + 18$ **79.** $45x^3 + 18x^2y - 5x - 2y$ **81.** $a^2 - 3a$ **83.** $a^2 + 2ah + h^2 - 3a - 3h$

85. $b^2 - 7b + 10$ **87.** $2x^2$ **89.** $\dfrac{10a^2b^3}{9}$ **91.** $\dfrac{2m^3}{3}$ **93.** $7y(3z - 2) + 1 = 21yz - 14y + 1$ **95.** answers may vary

97. $30x^2y^{2n+1} - 10x^2y^n$ **99.** $x^{3a} + 5x^{2a} - 3x^a - 15$ **101. a.** $6x + 12$ **b.** $9x^2 + 36x + 35$ **c.** one operation is addition, the other is multiplication

103. $5x^2 + 25x$ **105.** $a^3 - 2a^2$ **107.** $(9x^2 - 12x + 4)$ sq in. **109.** $\pi(25x^2 - 20x + 4)$ sq km or $(25\pi x^2 - 20\pi x + 4\pi)$ sq km

Exercise Set 5.3 **1.** $2a + 4$ **3.** $3ab + 4$ **5.** $2y + \dfrac{3y}{x} - \dfrac{2y}{x^2}$ **7.** $x + 1$ **9.** $2x - 8$ **11.** $x - \dfrac{1}{2}$ **13.** $2x^2 - \dfrac{1}{2}x + 5$

15. $2x^2 - 6$ **17.** $3x^3 + 5x + 4 - \dfrac{2x}{x^2 - 2}$ **19.** $2x^3 + \dfrac{9}{2}x^2 + 10x + 21 + \dfrac{42}{x - 2}$ **21.** $x + 8$ **23.** $x - 1$ **25.** $x^2 - 5x - 23 - \dfrac{41}{x - 2}$

27. $4x + 8 + \dfrac{7}{x - 2}$ **29.** $x^6y + \dfrac{2}{y} + 1$ **31.** $5x^2 - 6 - \dfrac{5}{2x - 1}$ **33.** $2x^2 + 2x + 8 + \dfrac{28}{x - 4}$ **35.** $2x^3 - 3x^2 + x - 4$

37. $3x^2 + 4x - 8 + \dfrac{20}{x + 1}$ **39.** $3x^2 + 3x - 3$ **41.** $x^2 + x + 1$ **43.** $-\dfrac{5y}{x} - \dfrac{15z}{x} - 25z$ **45.** $3x^4 - 2x$ **47.** $(-9, -1)$

49. $(-\infty, -8] \cup [1, \infty)$ **51.** yes **53.** no **55.** a or d **57.** $(x^4 + 2x^2 - 6)$ m **59.** $(3x - 7)$ in. **61.** 4; answers may vary

63. 372; answers may vary **65.** answers may vary **67.** $x^3 + \dfrac{5}{3}x^2 + \dfrac{5}{3}x + \dfrac{8}{3} + \dfrac{8}{3(x - 1)}$ **69.** $5x - 1 + \dfrac{6}{x}; x \neq 0$

71. $7x^3 + 14x^2 + 25x + 50 + \dfrac{102}{x - 2}; x \neq 2$ **73.** answers may vary **75.** answers may vary

Mental Math **1.** 6 **3.** 5 **5.** x **7.** $7x$

Exercise Set 5.4 **1.** $6(3x - 2)$ **3.** $4y^2(1 - 4xy)$ **5.** $2x^3(3x^2 - 4x + 1)$ **7.** $4ab(2a^2b^2 - ab + 1 + 4b)$ **9.** $(x + 3)(6 + 5a)$

11. $(z + 7)(2x + 1)$ **13.** $(6x^2 + 5)(3x - 2)$ **15.** $(a + 2)(b + 3)$ **17.** $(a - 2)(c + 4)$ **19.** $(x - 2)(2y - 3)$

21. $(4x - 1)(3y - 2)$ **23.** $3(2x^3 + 3)$ **25.** $x^2(x + 3)$ **27.** $4a(2a^2 - 1)$ **29.** $4xy(-5x + 4y^2)$ or $-4xy(5x - 4y^2)$

31. $5ab^2(2ab + 1 - 3b)$ **33.** $3b(3ac^2 + 2a^2c - 2a + c)$ **35.** $(y - 2)(4x - 3)$ **37.** $(2x + 3)(3y + 5)$ **39.** $(x + 3)(y - 5)$

41. $(2a - 3)(3b - 1)$ **43.** $(6x + 1)(2y + 3)$ **45.** $(n - 8)(2m - 1)$ **47.** $3x^2y^2(5x - 6)$ **49.** $(2x + 3y)(x + 2)$

51. $(5x - 3)(x + y)$ **53.** $(x^2 + 4)(x + 3)$ **55.** $(x^2 - 2)(x - 1)$ **57.** $x^2 - 3x - 10$ **59.** $x^2 + 5x + 6$ **61.** $y^2 - 4y + 3$

63. d **65.** $2\pi r(r + h)$ **67.** $A = 5600(1 + rt)$ **69.** answers may vary **71.** none **73.** a **75.** $A = P(1 + RT)$

77. $y^n(3 + 3y^n + 5y^{7n})$ **79.** $3x^{2a}(x^{3a} - 2x^a + 3)$ **81. a.** $h(t) = -16(t^2 - 14)$ **b.** 160 **c.** answers may vary

83. $f(x) = 2(-x^2 + 11x + 14)$ or $f(x) = -2(x^2 - 11x - 14)$

Mental Math **1.** 5 and 2 **3.** 8 and 3

Exercise Set 5.5 **1.** $(x + 3)(x + 6)$ **3.** $(x - 8)(x - 4)$ **5.** $(x + 12)(x - 2)$ **7.** $(x - 6)(x + 4)$ **9.** $3(x - 2)(x - 4)$

11. $4z(x + 2)(x + 5)$ **13.** $2(x^2 - 12x - 32)$ **15.** $(5x + 1)(x + 3)$ **17.** $(2x - 3)(x - 4)$ **19.** prime polynomial **21.** $(2x - 3)^2$

23. $2(3x - 5)(2x + 5)$ **25.** $y^2(3y + 5)(y - 2)$ **27.** $2x(3x^2 + 4x + 12)$ **29.** $(2x + y)(x - 3y)$ **31.** $2(7y + 2)(2y + 1)$

33. $(2x - 3)(x + 9)$ **35.** $(x^2 + 3)(x^2 - 2)$ **37.** $(5x + 8)(5x + 2)$ **39.** $(x^3 - 4)(x^3 - 3)$ **41.** $(a - 3)(a + 8)$ **43.** $(x - 27)(x + 3)$

45. $(x - 18)(x + 3)$ **47.** $3(x - 1)^2$ **49.** $(3x + 1)(x - 2)$ **51.** $(4x - 3)(2x - 5)$ **53.** $3x^2(2x + 1)(3x + 2)$ **55.** $(x + 7z)(x + z)$

57. $(x - 4)(x + 3)$ **59.** $3(a + 2b)^2$ **61.** prime polynomial **63.** $(2x + 13)(x + 3)$ **65.** $(3x - 2)(2x - 15)$ **67.** $(x^2 - 6)(x^2 + 1)$

69. $x(3x + 1)(2x - 1)$ **71.** $(4a - 3b)(3a - 5b)$ **73.** $(3x + 5)^2$ **75.** $y(3x - 8)(x - 1)$ **77.** $2(x + 3)(x - 2)$ **79.** $(x + 2)(x - 7)$

81. $(2x^3 - 3)(x^3 + 3)$ **83.** $2x(6y^2 - z)^2$ **85.** $2xy(x + 3)(x - 2)$ **87.** $(x + 5y)(x + y)$ **89.** $x^2 - 9$ **91.** $4x^2 + 4x + 1$

93. $x^3 - 8$ **95.** $\pm 5, \pm 7$ **97.** $x(x + 4)(x - 2)$ **99. a.** 576 ft; 672 ft; 640 ft; 480 ft **b.** answers may vary **c.** $-16(t + 4)(t - 9)$

101. $(x^n + 2)(x^n + 8)$ **103.** $(x^n - 6)(x^n + 3)$ **105.** $(2x^n + 1)(x^n + 5)$ **107.** $(2x^n - 3)^2$ **109.** $x^2(x + 5)(x + 1)$

111. $3x(5x - 1)(2x + 1)$

Exercise Set 5.6 **1.** $(x + 3)^2$ **3.** $(2x - 3)^2$ **5.** $(2a + 3)^2$ **7.** $3(x - 4)^2$ **9.** $x^2(3y + 2)^2$ **11.** $(4x - 7y)^2$

13. $(x + 5)(x - 5)$ **15.** $\left(\dfrac{1}{3} + 2z\right)\left(\dfrac{1}{3} - 2z\right)$ **17.** $(y + 9)(y - 5)$ **19.** $4(4x + 5)(4x - 5)$ **21.** $(x + 2y + 3)(x + 2y - 3)$

23. $(x + 8 + x^2)(x + 8 - x^2)$ **25.** $(x + 3)(x^2 - 3x + 9)$ **27.** $(z - 1)(z^2 + z + 1)$ **29.** $(m + n)(m^2 - mn + n^2)$

31. $y^2(x - 3)(x^2 + 3x + 9)$ **33.** $q^2(4 - p)(16 + 4p + p^2)$ **35.** $2(5y - 2x)(25y^2 + 10xy + 4x^2)$ **37.** $(x - 6)^2$

39. $2y(3x + 1)(3x - 1)$ **41.** $(3x + 7)(3x - 7)$ **43.** $(x^2 + 1)(x + 1)(x - 1)$ **45.** $(x^2 - y)(x^4 + x^2y + y^2)$

47. $(2x + 3y)(4x^2 - 6xy + 9y^2)$ **49.** $(2x + 1 + z)(2x + 1 - z)$ **51.** $3y^2(x^2 + 3)(x^4 - 3x^2 + 9)$ **53.** $\left(n - \dfrac{1}{3}\right)\left(n^2 + \dfrac{1}{3}n + \dfrac{1}{9}\right)$

55. $-16(y + 2)(y - 2)$ **57.** $(x - 5 + y)(x - 5 - y)$ **59.** $(ab + 5)(a^2b^2 - 5ab + 25)$ **61.** $\left(\dfrac{x}{5} + \dfrac{y}{3}\right)\left(\dfrac{x}{5} - \dfrac{y}{3}\right)$

63. $(x + y + 5)(x^2 + 2xy + y^2 - 5x - 5y + 25)$ **65.** $\{5\}$ **67.** $\left\{-\dfrac{1}{3}\right\}$ **69.** $\{0\}$ **71.** $\{5\}$ **73.** no; $x^2 - 4$ can be factored further

75. yes **77.** $\pi R^2 - \pi r^2 = \pi(R + r)(R - r)$ **79.** $x^3 - y^2x; x(x + y)(x - y)$ **81.** $c = 9$ **83.** $c = 49$

85. a. $(x + 1)(x^2 - x + 1)(x - 1)(x^2 + x + 1)$ **b.** $(x + 1)(x - 1)(x^4 + x^2 + 1)$ **c.** answers may vary **87.** $(x^n + 6)(x^n - 6)$

89. $(5x^n + 9)(5x^n - 9)$ **91.** $(x^{2n} + 25)(x^n + 5)(x^n - 5)$

Integrated Review   1. $2y^2 + 2y - 11$ **2.** $-2z^4 - 6z^2 + 3z$ **3.** $x^2 - 7x + 7$ **4.** $7x^2 - 4x - 5$ **5.** $25x^2 - 30x + 9$ **6.** $x - 3$
7. $2x^3 - 4x^2 + 5x - 5 + \dfrac{8}{x+2}$ **8.** $4x^3 - 13x^2 - 5x + 2$ **9.** $(x - 4 + y)(x - 4 - y)$ **10.** $2(3x + 2)(2x - 5)$
11. $x(x - 1)(x^2 + x + 1)$ **12.** $2x(2x - 1)$ **13.** $2xy(7x - 1)$ **14.** $6ab(4b - 1)$ **15.** $4(x + 2)(x - 2)$ **16.** $9(x + 3)(x - 3)$
17. $(3x - 11)(x + 1)$ **18.** $(5x + 3)(x - 1)$ **19.** $4(x + 3)(x - 1)$ **20.** $6(x + 1)(x - 2)$ **21.** $(2x + 9)^2$ **22.** $(5x + 4)^2$
23. $(2x + 5y)(4x^2 - 10xy + 25y^2)$ **24.** $(3x - 4y)(9x^2 + 12xy + 16y^2)$ **25.** $8x^2(2y - 1)(4y^2 + 2y + 1)$
26. $27x^2y(xy - 2)(x^2y^2 + 2xy + 4)$ **27.** $(x + 5 + y)(x^2 + 10x + 25 - xy - 5y + y^2)$
28. $(y - 1 + 3x)(y^2 - 2y + 1 - 3xy + 3x + 9x^2)$ **29.** $(5a - 6)^2$ **30.** $(4r + 5)^2$ **31.** $7x(x - 9)$ **32.** $(4x + 3)(5x + 2)$
33. $(a + 7)(b - 6)$ **34.** $20(x - 6)(x - 5)$ **35.** $(x^2 + 1)(x + 1)(x - 1)$ **36.** $5x(3x - 4)$ **37.** $(5x - 11)(2x + 3)$
38. $9m^2n^2(5mn - 3)$ **39.** $5a^3b(b^2 - 10)$ **40.** $x(x + 1)(x^2 - x + 1)$ **41.** prime   **42.** $20(x + y)(x^2 - xy + y^2)$
43. $10x(x - 10)(x - 11)$ **44.** $(3y - 7)^2$ **45.** $a^3b(4b - 3)(16b^2 + 12b + 9)$ **46.** $(y^2 + 4)(y + 2)(y - 2)$ **47.** $2(x - 3)(x^2 + 3x + 9)$
48. $(2s - 1)(r + 5)$ **49.** $(y^4 + 2)(3y - 5)$ **50.** prime   **51.** $100(z + 1)(z^2 - z + 1)$ **52.** $2x(5x - 2)(25x^2 + 10x + 4)$
53. $(2b - 9)^2$ **54.** $(a^4 + 3)(2a - 1)$ **55.** $(y - 4)(y - 5)$ **56.** $(c - 3)(c + 1)$ **57.** $A = 9 - 4x^2 = (3 + 2x)(3 - 2x)$

Calculator Explorations   1. $-3.562, 0.562$ **3.** $-0.874, 2.787$ **5.** $-0.465, 1.910$

Mental Math   1. $\{3, -5\}$ **3.** $\{3, -7\}$ **5.** $\{0, 9\}$

Exercise Set 5.7   1. $\left\{-3, \dfrac{4}{3}\right\}$ **3.** $\left\{-\dfrac{3}{4}, \dfrac{5}{2}\right\}$ **5.** $\{-3, -8\}$ **7.** $\left\{\dfrac{1}{4}, -\dfrac{2}{3}\right\}$ **9.** $\{1, 9\}$ **11.** $\left\{\dfrac{3}{5}, -1\right\}$ **13.** $\{0\}$ **15.** $\{6, -3\}$
17. $\left\{\dfrac{2}{5}, -\dfrac{1}{2}\right\}$ **19.** $\left\{\dfrac{3}{4}, -\dfrac{1}{2}\right\}$ **21.** $\left\{-2, 7, \dfrac{8}{3}\right\}$ **23.** $\{0, 3, -3\}$ **25.** $\{-1, 1, 2\}$ **27.** $\left\{-\dfrac{7}{2}, 10\right\}$ **29.** $\{0, 5\}$ **31.** $\{-3, 5\}$
33. $\left\{-\dfrac{1}{2}, \dfrac{1}{3}\right\}$ **35.** $\{-4, 9\}$ **37.** $\left\{\dfrac{4}{5}\right\}$ **39.** $\{-5, 0, 2\}$ **41.** $\left\{-3, 0, \dfrac{4}{5}\right\}$ **43.** $\varnothing$ **45.** $\{-7, 4\}$ **47.** $\{4, 6\}$
49. $\left\{-\dfrac{1}{2}\right\}$ **51.** $\{-4, -3, 3\}$ **53.** $\{-5, 0, 5\}$ **55.** $\{-6, 5\}$ **57.** $\left\{-\dfrac{1}{3}, 0, 1\right\}$ **59.** $\left\{-\dfrac{1}{3}, 0\right\}$ **61.** $\left\{-\dfrac{7}{8}\right\}$ **63.** $\left\{\dfrac{31}{4}\right\}$ **65.** $\{1\}$

67. -11 and -6, or 6 and 11 **69.** 75 ft   **71.** 105 units   **73.** 12 cm and 9 cm   **75.** 2 in.   **77.** 10 sec   **79.** Width: $7\dfrac{1}{2}$ ft; length: 12 ft
81. 10-in. square tier   **83.** E   **85.** F   **87.** B   **89.** $(-3, 0), (0, 2)$ **91.** $(-4, 0), (4, 0), (0, 2), (0, -2)$

93. $x - 5 = 0$ or $x + 2 = 0$ **95.**
$$y(y - 5) = -6$$
$$x = 5 \text{ or } x = -2$$
$$y^2 - 5y + 6 = 0$$
$$(y - 2)(y - 3) = 0$$
$$y - 2 = 0 \text{ or } y - 3 = 0$$
$$y = 2 \text{ or } y = 3$$
97. $\left\{-3, -\dfrac{1}{3}, 2, 5\right\}$ **99.** answers may vary   **101.** no; answers may vary
103. answers may vary

The Bigger Picture   1. $\left\{-\dfrac{1}{2}, 6\right\}$ **2.** $(-7, 3)$ **3.** $\left\{-\dfrac{5}{3}\right\}$ **4.** $\left\{-\dfrac{3}{2}, 6\right\}$ **5.** $(-\infty, \infty)$ **6.** $\{-8, 3\}$ **7.** $\{-3, 10\}$ **8.** $(-\infty, 0]$

Vocabulary Check   1. polynomial   **2.** factoring   **3.** degree of a term   **4.** monomial   **5.** trinomial   **6.** quadratic equation
7. degree of a polynomial   **8.** binomial   **9.** 0   **10.** FOIL   **11.** synthetic division

Chapter 5 Review   1. 5   **2.** 1   **3.** $12x - 6x^2 - 6x^2y$ **4.** $-4xy^3 - 3x^3y$ **5.** $4x^2 + 8y + 6$ **6.** $-4x^2 + 10y^2$ **7.** $8x^2 + 2b - 22$
8. $-4x^3 + 4x^2 + 16xy - 9x + 18$ **9.** $12x^2y - 7xy + 3$ **10.** $x^2 - 6x + 3$ **11.** $x^3 + x - 2xy^2 - y - 7$ **12.** 290   **13.** 58
14. 110   **15.** $x^2 + 4x - 6$ **16.** $-x^2 + 2x + 3$ **17.** $(6x^2y - 12x + 12)$ cm   **18.** $-24x^3 + 36x^2 - 6x$ **19.** $-12a^2b^5 - 28a^2b^3 - 4ab^2$
20. $2x^2 + x - 36$ **21.** $9x^2a^2 - 24xab + 16b^2$ **22.** $36x^3 - 11x^2 - 8x - 3$ **23.** $15x^2 + 18xy - 81y^2$ **24.** $x^2 + \dfrac{1}{3}x - \dfrac{2}{9}$
25. $x^4 + 18x^3 + 83x^2 + 18x + 1$ **26.** $2x^3 + 3x^2 - 12x + 5$ **27.** $9x^2 - 6xy + y^2$ **28.** $16x^2 + 72x + 81$ **29.** $x^2 - 9y^2$
30. $16 - 9a^2 + 6ab - b^2$ **31.** $(9y^2 - 49z^2)$ sq units   **32.** $1 + \dfrac{x}{2y} - \dfrac{9}{4xy}$ **33.** $\dfrac{3}{b} + 4b$ **34.** $3x^3 + 9x^2 + 2x + 6 - \dfrac{2}{x-3}$
35. $2x^3 + 6x^2 + 17x + 56 + \dfrac{156}{x-3}$ **36.** $2x^3 + 2x - 2$ **37.** $x^2 + \dfrac{7}{2}x - \dfrac{1}{4} + \dfrac{15}{8\left(x - \dfrac{1}{2}\right)}$ **38.** $3x^2 + 2x - 1$ **39.** $3x^2 + 6$
40. $3x^2 + 6x + 24 + \dfrac{44}{x-2}$ **41.** $4x^2 - 4x + 2 - \dfrac{5}{x + \dfrac{3}{2}}$ **42.** $x^4 - x^3 + x^2 - x + 1 - \dfrac{2}{x+1}$ **43.** $x^2 + 3x + 9 - \dfrac{54}{x-3}$
44. $3x^3 + 13x^2 + 51x + 204 + \dfrac{814}{x-4}$ **45.** $3x^3 - 6x^2 + 10x - 20 + \dfrac{50}{x+2}$ **46.** $8x^2(2x - 3)$ **47.** $12y(3 - 2y)$
48. $2ab(3b + 4 - 2ab)$ **49.** $7ab(2ab - 3b + 1)$ **50.** $(a + 3b)(6a - 5)$ **51.** $(x - 2y)(4x - 5)$ **52.** $(x - 6)(y + 3)$
53. $(a - 8)(b + 4)$ **54.** $(p - 5)(q - 3)$ **55.** $(x^2 - 2)(x - 1)$ **56.** $x(2y - x)$ **57.** $(x - 18)(x + 4)$ **58.** $(x - 4)(x + 20)$
59. $2(x - 2)(x - 7)$ **60.** $3(x + 2)(x + 9)$ **61.** $x(2x - 9)(x + 1)$ **62.** $(3x + 8)(x - 2)$ **63.** $(6x + 5)(x + 2)$
64. $(15x - 1)(x - 6)$ **65.** $2(2x - 3)(x + 2)$ **66.** $3(x - 2)(3x + 2)$ **67.** $(x + 6)^2(y - 3)(y + 1)$ **68.** $(x + 7)(x + 9)$
69. $(x^2 - 8)(x^2 + 2)$ **70.** $(x^2 - 2)(x^2 + 10)$ **71.** $(x + 10)(x - 10)$ **72.** $(x + 9)(x - 9)$ **73.** $2(x + 4)(x - 4)$
74. $6(x + 3)(x - 3)$ **75.** $(9 + x^2)(3 + x)(3 - x)$ **76.** $(4 + y^2)(2 + y)(2 - y)$ **77.** $(y + 7)(y - 3)$ **78.** $(x - 7)(x + 1)$
79. $(x + 6)(x^2 - 6x + 36)$ **80.** $(y + 8)(y^2 - 8y + 64)$ **81.** $(2 - 3y)(4 + 6y + 9y^2)$ **82.** $(1 - 4y)(1 + 4y + 16y^2)$
83. $6xy(x + 2)(x^2 - 2x + 4)$ **84.** $2x^2(x + 2y)(x^2 - 2xy + 4y^2)$ **85.** $(x - 1 + y)(x - 1 - y)$ **86.** $(x - 3 + 2y)(x - 3 - 2y)$
87. $(2x + 3)^2$ **88.** $(4a - 5b)^2$ **89.** $\pi h(R + r)(R - r)$ cu units   **90.** $\left\{\dfrac{1}{3}, -7\right\}$ **91.** $\left\{-5, \dfrac{3}{8}\right\}$ **92.** $\left\{0, 4, \dfrac{9}{2}\right\}$ **93.** $\left\{-3, -\dfrac{1}{5}, 4\right\}$

94. $\{0, 6\}$ **95.** $\{-3, 0, 3\}$ **96.** $\left\{-\dfrac{1}{3}, 2\right\}$ **97.** $\{2, 10\}$ **98.** $\{-4, 1\}$ **99.** $\left\{\dfrac{7}{2}, -5\right\}$ **100.** $\{0, 6, -3\}$ **101.** $\{-21, 0, 2\}$ **102.** $\{0, -2, 1\}$

103. $\left\{-\dfrac{3}{2}, 0, \dfrac{1}{4}\right\}$ **104.** $-\dfrac{15}{2}$ or 7 **105.** width: 2 m; length: 8 m **106.** 5 sec **107.** $3x^3 + 13x^2 - 9x + 5$ **108.** $-2x^2 + 5x - 6.9$

109. $8x^2 + 3x + 4.5$ **110.** $49a^2b^2 - 7ab + \dfrac{1}{4}$ **111.** -24 **112.** -10 **113.** $6y^4(2y - 1)$ **114.** $(x^2 - 3)(y + 4)$

115. $2(3x + 1)(x - 6)$ **116.** $(4x + 3)^2(y - 20)(y + 1)$ **117.** $z^5(2z + 7)(2z - 7)$ **118.** $(x + 1)(x - 1)(5x^2 + 9)$ **119.** $\{0, 3\}$
120. $\{-2, 13\}$

Chapter 5 Test **1.** $-5x^3 - 11x - 9$ **2.** $-12x^2y - 3xy^2$ **3.** $12x^2 - 5x - 28$ **4.** $25a^2 - 4b^2$ **5.** $36m^2 + 12mn + n^2$

6. $2x^3 - 13x^2 + 14x - 4$ **7.** $\dfrac{4xy}{3z} + \dfrac{3}{z} + \dfrac{1}{3x}$ **8.** $2x^4 + 2x - 2 + \dfrac{1}{2x - 1}$ **9.** $4x^3 - 15x^2 + 47x - 142 + \dfrac{425}{x + 3}$ **10.** $4x^2y(4x - 3y^3)$

11. $(x - 15)(x + 2)$ **12.** $(2y + 5)^2$ **13.** $3(2x + 1)(x - 3)$ **14.** $(2x + 5)(2x - 5)$ **15.** $(x + 4)(x^2 - 4x + 16)$

16. $3y(x + 3y)(x - 3y)$ **17.** $2(2y - 1)(4y^2 + 2y + 1)$ **18.** $(x + 3)(x - 3)(y + 3)$ **19.** $\left\{4, -\dfrac{8}{7}\right\}$ **20.** $\{-3, 8\}$ **21.** $\left\{-\dfrac{5}{2}, -2, 2\right\}$

22. $(x + 2y)(x - 2y)$ **23. a.** 960 ft **b.** 953.44 ft **c.** $-16(t - 11)(t + 5)$ **d.** 11 sec

Cumulative Review **1.** $-2x + 1$; Sec. 1.5, Ex. 13 **2.** $-2x - \dfrac{7}{8}$; Sec. 1.5 **3.** $2x + 23$; Sec. 1.5, Ex. 14 **4.** $16.3x - 0.8$; Sec. 1.5

5. 6×10^{-5}; Sec. 1.7, Ex. 20 **6. a.** $\dfrac{a^3}{64}$ **b.** $\dfrac{1}{a^4}$ **c.** $\dfrac{27}{8}$ **d.** $\dfrac{b^{17}}{9a^6}$; Sec. 1.7 **7.** $\{0\}$; Sec. 2.1, Ex. 6 **8.** $\{8\}$; Sec. 2.1

9. $y = \dfrac{7 + 2x}{3}$ or $y = \dfrac{7}{3} + \dfrac{2}{3}x$; Sec. 2.3, Ex. 2 **10.** 3 gal; Sec. 2.3 **11.** $\left[\dfrac{5}{2}, \infty\right)$; Sec. 2.4, Ex. 9 **12.** $(-\infty, -5)$; Sec. 2.4

13. $\left[-9, -\dfrac{9}{2}\right)$; Sec. 2.5, Ex. 5 **14.** $\left(-1, \dfrac{1}{3}\right]$; Sec. 2.5 **15.** $\left\{\dfrac{3}{4}, 5\right\}$; Sec. 2.6, Ex. 7 **16.** $(-\infty, \infty)$; Sec. 2.6 **17.** undefined; Sec. 3.2, Ex. 5

18. -2; Sec. 3.2 **19.** 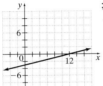 ; Sec. 3.3, Ex. 1 **20.** 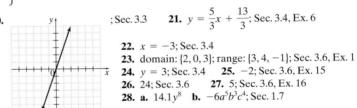 ; Sec. 3.3 **21.** $y = \dfrac{5}{3}x + \dfrac{13}{3}$; Sec. 3.4, Ex. 6

22. $x = -3$; Sec. 3.4
23. domain: $\{2, 0, 3\}$; range: $\{3, 4, -1\}$; Sec. 3.6, Ex. 1
24. $y = 3$; Sec. 3.4 **25.** -2; Sec. 3.6, Ex. 15
26. 24; Sec. 3.6 **27.** 5; Sec. 3.6, Ex. 16
28. a. $14.1y^8$ **b.** $-6a^5b^3c^4$; Sec. 1.7

29. 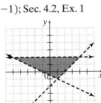 ; Sec. 3.5, Ex. 1 **30.** 37.5 oz. of 20% solution; 12.5 oz. of 60% solution; Sec. 4.3 **31.** $\left(-4, \dfrac{1}{2}\right)$; Sec. 4.1, Ex. 6

32. $(5, 2)$; Sec. 4.1 **33.** $(-4, 2, -1)$; Sec. 4.2, Ex. 1 **34.** $(-3, -2, 5)$; Sec. 4.2 **35.** $(-1, 2)$; Sec. 4.4, Ex. 1
36. $(-3, -4, -5)$; Sec. 4.2 **37.** ; Sec. 4.5, Ex. 2 **38. a.** 8.25×10^6 **b.** 3.46×10^{-5}; Sec. 1.6
39. 4; Sec. 5.1, Ex. 9 **40.** $-2x^2 - 5x$; Sec. 5.1
41. $10x^2 - 8x$; Sec. 5.2, Ex. 3
42. $10x^3 - 17x^2 + 10x - 2$; Sec. 5.2
43. $-7x^3y^2 - 3x^2y^2 + 11xy$; Sec. 5.2, Ex. 5
44. $49x^2 - 7x + \dfrac{1}{4}$; Sec. 5.2 **45.** $2x^2 - x + 4$; Sec. 5.3, Ex. 1

46. $3xy(2x + y)(2x - y)$; Sec. 5.6 **47.** $(x + 2)(x + 8)$; Sec. 5.5, Ex. 1 **48.** $(5a - 1)(a + 3)$; Sec. 5.5 **49.** $\left\{-5, \dfrac{1}{2}\right\}$; Sec. 5.7, Ex. 2

50. $\left\{4, -\dfrac{2}{3}\right\}$; Sec. 5.7

CHAPTER 6 Rational Expressions

Calculator Explorations **1.** $\{x \mid x \text{ is a real number and } x \neq 6\}$ **3.** $\{x \mid x \text{ is a real number and } x \neq -2, x \neq 2\}$

5. $\left\{x \mid x \text{ is a real number and } x \neq -4, x \neq \dfrac{1}{2}\right\}$ **7.** $\{x \mid x \text{ is a real number}\}$

Mental Math **1.** $\dfrac{xy}{10}$ **3.** $\dfrac{2y}{3x}$ **5.** $\dfrac{m^2}{36}$

Exercise Set 6.1 **1.** $\{x \mid x \text{ is a real number}\}$ **3.** $\{t \mid t \text{ is a real number and } t \neq 0\}$ **5.** $\{x \mid x \text{ is a real number and } x \neq 7\}$

7. $\left\{x \mid x \text{ is a real number and } x \neq \dfrac{1}{3}\right\}$ **9.** $\{x \mid x \text{ is a real number and } x \neq -2, x \neq 0, x \neq 1\}$ **11.** $\{x \mid x \text{ is a real number and } x \neq 2, x \neq -2\}$

13. $1 - 2x$ **15.** $3 - x$ **17.** $\dfrac{9}{7}$ **19.** $x - 4$ **21.** -1 **23.** $-(x + 7)$ **25.** $\dfrac{2x + 1}{x - 1}$ **27.** $\dfrac{x^2 + 5x + 25}{2}$ **29.** $\dfrac{x - 2}{2x^2 + 1}$

31. $\dfrac{1}{3x + 5}$ **33.** $-\dfrac{4}{5}$ **35.** $-\dfrac{6a}{2a + 1}$ **37.** $\dfrac{3}{2(x - 1)}$ **39.** $\dfrac{x + 2}{x + 3}$ **41.** $\dfrac{3a}{5(a - b)}$ **43.** $\dfrac{1}{6}$ **45.** $\dfrac{x}{3}$ **47.** $\dfrac{4a^2}{a - b}$ **49.** $\dfrac{4}{(x + 2)(x + 3)}$

51. $\dfrac{1}{2}$ **53.** -1 **55.** $\dfrac{8(a-2)}{3(a+2)}$ **57.** $\dfrac{(x+2)(x+3)}{4}$ **59.** $\dfrac{2(x+3)(x-3)}{5(x^2-8x-15)}$ **61.** $r^2 - rs + s^2$ **63.** $\dfrac{8}{x^2 y}$ **65.** $\dfrac{(y+5)(2x-1)}{(y+2)(5x+1)}$

67. $\dfrac{10}{3}, -8, -\dfrac{7}{3}$ **69.** $-\dfrac{17}{48}, \dfrac{2}{7}, -\dfrac{3}{8}$ **71. a.** \$200 million   **b.** \$500 million   **c.** \$300 million   **d.** $\{x \mid x \text{ is a real number}\}$ **73.** $\dfrac{7}{5}$ **75.** $\dfrac{1}{12}$

77. $\dfrac{11}{16}$ **79.** b and d   **81.** no; answers may vary   **83.** $\dfrac{5}{x-2}$ sq m   **85.** $\dfrac{(x+2)(x-1)^2}{x^5}$ ft   **87.** answers may vary   **89. a.** 1 **b.** -1

c. neither **d.** -1 **e.** -1 **f.** 1   **91.** $(x-5)(2x+7)$ **93.** -1 **95.** $\dfrac{1}{x^n - 4}$

Exercise Set 6.2   1. $-\dfrac{3}{xz^2}$ **3.** $\dfrac{x+2}{x-2}$ **5.** $x-2$ **7.** $\dfrac{-1}{x-2}$ or $\dfrac{1}{2-x}$ **9.** $-\dfrac{5}{x}$ **11.** $35x$ **13.** $x(x+1)$ **15.** $(x+7)(x-7)$

17. $6(x+2)(x-2)$ **19.** $(a+b)(a-b)^2$ **21.** $-4x(x+3)(x-3)$ **23.** $\dfrac{17}{6x}$ **25.** $\dfrac{35-4y}{14y^2}$ **27.** $\dfrac{-13x+4}{(x+4)(x-4)}$ **29.** $\dfrac{3}{x+4}$

31. 0   **33.** $-\dfrac{x}{x-1}$ **35.** $\dfrac{-x+1}{x-2}$ **37.** $\dfrac{y^2+2y+10}{(y+4)(y-4)(y-2)}$ **39.** $\dfrac{5(x^2+x-4)}{(3x+2)(x+3)(2x-5)}$ **41.** $\dfrac{x^2+5x-21}{(x-2)(x+1)(x+3)}$

43. $\dfrac{5(x^2+x-4)}{(3x+2)(x+3)(2x-5)}$ **45.** $\dfrac{5a+1}{(a+1)^2(a-1)}$ **47.** $\dfrac{3}{x^2 y^3}$ **49.** $-\dfrac{5}{x}$ **51.** $\dfrac{25}{6(x+5)}$ **53.** $\dfrac{-2x-1}{x^2(x-3)}$ **55.** $\dfrac{b(2a-b)}{(a+b)(a-b)}$

57. $\dfrac{2(x+8)}{(x+2)^2(x-2)}$ **59.** $\dfrac{3x^2+23x-7}{(2x-1)(x-5)(x+3)}$ **61.** $\dfrac{5-2x}{2(x+1)}$ **63.** $\dfrac{2(x^2+x-21)}{(x+3)^2(x-3)}$ **65.** $\dfrac{6x}{(x+3)(x-3)^2}$ **67.** $\dfrac{4}{3}$ **69.** 10

71. $4+x^2$ **73.** $\dfrac{2x-3}{x^2+1} - \dfrac{x-6}{x^2+1} = \dfrac{2x-3-x+6}{x^2+1} = \dfrac{x+3}{x^2+1}$ **75.** $\dfrac{4x}{x+5}$ ft; $\dfrac{x^2}{(x+5)^2}$ sq ft   **77.** answers may vary

79. answers may vary   **81.** answers may vary   **83.** $\dfrac{(x+6)(2x-3)}{6x^2}$ or $\dfrac{2x^2+9x-18}{6x^2}$ **85.** $\dfrac{4a^2}{9(a-1)}$ **87.** 4   **89.** $-\dfrac{4}{x-1}$

91. $\dfrac{32}{x(x+2)(x-2)}$ **93.** $\dfrac{3}{2x}$ **95.** $\dfrac{4-3x}{x^2}$ **97.**

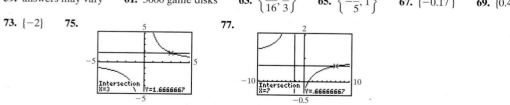

Exercise Set 6.3   1. $\dfrac{7}{13}$ **3.** $\dfrac{4}{x}$ **5.** $\dfrac{9(x-2)}{9x^2+4}$ **7.** 4   **9.** $2x+y$ **11.** $\dfrac{2(x+1)}{2x-1}$ **13.** $\dfrac{2x+3}{4-9x}$ **15.** $\dfrac{1}{x^2-2x+4}$ **17.** $\dfrac{x}{5x-10}$

19. $\dfrac{x-2}{2x-1}$ **21.** $\dfrac{x}{2-3x}$ **23.** $-\dfrac{y}{x+y}$ **25.** $-\dfrac{2x^2}{y(x-y)}$ **27.** $\dfrac{2x+1}{y+x^2}$ **29.** $\dfrac{x-3}{9}$ **31.** $\dfrac{1}{x+2}$ **33.** 2   **35.** $\dfrac{xy^2}{x^2+y^2}$

37. $\dfrac{2b^2+3a}{b(b-a)}$ **39.** $\dfrac{x}{(x+1)(x-1)}$ **41.** $\dfrac{1+a}{1-a}$ **43.** $\dfrac{x(x+6y)}{2y}$ **45.** $\dfrac{5a}{2(a+2)}$ **47.** $5xy^2 + 2x^2 y$ **49.** $\dfrac{xy}{5y+2x}$ **51.** $\left\{-\dfrac{5}{6}\right\}$

53. $\{2\}$ **55.** $\{54\}$ **57.** a and c   **59.** $\dfrac{770a}{770-s}$ **61.** a and b   **63.** $\dfrac{x-3y}{x+3y}$ **65.** $\dfrac{1+x}{2+x}$ **67.** $x(x+1)$ **69. a.** $\dfrac{1}{a+h}$ **b.** $\dfrac{1}{a}$

c. $\dfrac{\frac{1}{a+h} - \frac{1}{a}}{h}$ **d.** $\dfrac{-1}{a(a+h)}$ **71. a.** $\dfrac{3}{a+h+1}$ **b.** $\dfrac{3}{a+1}$ **c.** $\dfrac{\frac{3}{a+h+1} - \frac{3}{a+1}}{h}$ **d.** $\dfrac{-3}{(a+h+1)(a+1)}$

Mental Math   1. equation   **3.** expression   **5.** equation

Exercise Set 6.4   1. $\{72\}$ **3.** $\{2\}$ **5.** $\{6\}$ **7.** $\{2, -2\}$ **9.** $\varnothing$ **11.** $\left\{-\dfrac{28}{3}\right\}$ **13.** $\{3\}$ **15.** $\{-8\}$ **17.** $\{3\}$ **19.** $\varnothing$

21. $\{1\}$ **23.** $\{3\}$ **25.** $\{-1\}$ **27.** $\{6\}$ **29.** $\left\{\dfrac{1}{3}\right\}$ **31.** $\{-5, 5\}$ **33.** $\{3\}$ **35.** $\{7\}$ **37.** $\varnothing$ **39.** $\left\{\dfrac{4}{3}\right\}$ **41.** $\{-12\}$

43. $\left\{1, \dfrac{11}{4}\right\}$ **45.** $\{-5, -1\}$ **47.** $\left\{-\dfrac{7}{5}\right\}$ **49.** 5   **51.** length: 15 in.; width: 10 in.   **53.** 13.5%   **55.** 25–29   **57.** 6370 inmates

59. answers may vary   **61.** 3000 game disks   **63.** $\left\{\dfrac{1}{16}, \dfrac{1}{3}\right\}$ **65.** $\left\{-\dfrac{1}{5}, 1\right\}$ **67.** $\{-0.17\}$ **69.** $\{0.42\}$ **71.** $\{-1, 0\}$

73. $\{-2\}$ **75.**   **77.**

The Bigger Picture   1. $\left(-2, \dfrac{16}{7}\right)$ **2.** $\left\{-2, \dfrac{16}{7}\right\}$ **3.** $\{\pm 11\}$ **4.** $\{5\}$ **5.** $\left\{-\dfrac{8}{5}\right\}$ **6.** $(-\infty, 2]$ **7.** $(-\infty, -5]$

8. $(7, 10]$ **9.** $(-\infty, -17) \cup (18, \infty)$ **10.** $\left\{0, -\dfrac{1}{3}, \dfrac{7}{5}\right\}$

Integrated Review　1. $\left\{\dfrac{1}{2}\right\}$　**2.** $\{10\}$　**3.** $\dfrac{1+2x}{8}$　**4.** $\dfrac{15+x}{10}$　**5.** $\dfrac{2(x-4)}{(x+2)(x-1)}$　**6.** $-\dfrac{5(x-8)}{(x-2)(x+4)}$　**7.** $\{4\}$　**8.** $\{8\}$

9. $\{-5\}$　**10.** $\left\{-\dfrac{2}{3}\right\}$　**11.** $\dfrac{2x+5}{x(x-3)}$　**12.** $\dfrac{5}{2x}$　**13.** $\{-2\}$　**14.** $-\dfrac{y}{x}$　**15.** $\dfrac{(a+3)(a+1)}{a+2}$　**16.** $\dfrac{-a^2+31a+10}{5(a-6)(a+1)}$　**17.** $\left\{-\dfrac{1}{5}\right\}$

18. $\left\{-\dfrac{3}{13}\right\}$　**19.** $\dfrac{4a+1}{(3a+1)(3a-1)}$　**20.** $\dfrac{-a-8}{4a(a-2)}$ or $-\dfrac{a+8}{4a(a-2)}$　**21.** $\left\{-1,\dfrac{3}{2}\right\}$　**22.** $\dfrac{x^2-3x+10}{2(x+3)(x-3)}$　**23.** $\dfrac{3}{x+1}$

24. $\{x\,|\,x$ is a real number and $x\neq 2,\,x\neq -1\}$　**25.** $\{-1\}$　**26.** $\dfrac{22z-45}{3z(z-3)}$　**27. a.** $\dfrac{x}{5}-\dfrac{x}{4}+\dfrac{1}{10}$　**b.** Write each rational expression term so that

the denominator is the LCD, 20.　**c.** $\dfrac{-x+2}{20}$　**28. a.** $\dfrac{x}{5}-\dfrac{x}{4}=\dfrac{1}{10}$　**b.** Clear the equation of fractions by multiplying each term by the LCD, 20.

c. $\{-2\}$　**29.** b　**30.** d　**31.** d　**32.** a　**33.** d

Exercise Set 6.5　1. $C=\dfrac{5}{9}(F-32)$　**3.** $I=A-QL$　**5.** $R=\dfrac{R_1R_2}{R_1+R_2}$　**7.** $n=\dfrac{2s}{a+L}$　**9.** $b=\dfrac{2A-ah}{h}$　**11.** $T_2=\dfrac{P_2V_2T_1}{P_1V_1}$

13. $f_2=\dfrac{f_1f}{f_1-f}$　**15.** $L=\dfrac{n\lambda}{2}$　**17.** $c=\dfrac{2L\omega}{\theta}$　**19.** 1 and 5　**21.** 5　**23.** 4.5 gal　**25.** 4296 women　**27.** 15.6 hr　**29.** 10 min

31. 200 mph　**33.** 15 mph　**35.** -8 and -7　**37.** 36 min　**39.** 45 mph; 60 mph　**41.** 5.9 hr　**43.** 2 hr　**45.** 135 mph　**47.** 12 mi

49. $\dfrac{7}{8}$　**51.** $1\dfrac{1}{2}$ min　**53.** 1 hr　**55.** $2\dfrac{2}{9}$ hr　**57.** 2 hr　**59.** $\{-5\}$　**61.** $\{2\}$　**63.** 164 lb　**65.** higher; F　**67.** 6 ohms

Exercise Set 6.6　1. $k=\dfrac{1}{5};\ y=\dfrac{1}{5}x$　**3.** $k=\dfrac{3}{2};\ y=\dfrac{3}{2}x$　**5.** $k=14;\ y=14x$　**7.** $k=0.25;\ y=0.25x$　**9.** 4.05 lb　**11.** 204,706 tons

13. $k=30;\ y=\dfrac{30}{x}$　**15.** $k=700;\ y=\dfrac{700}{x}$　**17.** $k=2;\ y=\dfrac{2}{x}$　**19.** $k=0.14;\ y=\dfrac{0.14}{x}$　**21.** 54 mph　**23.** 72 amps

25. divided by 4　**27.** $x=kyz$　**29.** $r=kst^3$　**31.** $k=\dfrac{1}{3};\ y=\dfrac{1}{3}x^3$　**33.** $k=0.2;\ y=0.2\sqrt{x}$　**35.** $k=1.3;\ y=\dfrac{1.3}{x^2}$

37. $k=3;\ y=3xz^3$　**39.** 22.5 tons　**41.** 15π cu in.　**43.** 8 ft　**45.** $y=kx$　**47.** $a=\dfrac{k}{b}$　**49.** $y=kxz$　**51.** $y=\dfrac{k}{x^3}$　**53.** $y=\dfrac{kx}{p^2}$

55. $C=12\pi$ cm; $A=36\pi$ sq cm　**57.** $C=14\pi$ m; $A=49\pi$ sq m　**59.** 0　**61.** -1　**63.** a　**65.** c　**67.** multiplied by 2
69. multiplied by 4

Chapter 6 Vocabulary Check　1. complex fraction　**2.** directly　**3.** inversely　**4.** least common denominator　**5.** jointly
6. opposites　**7.** rational expression　**8.** equation; expression

Chapter Review　1. $\{x\,|\,x$ is a real number$\}$　**2.** $\{x\,|\,x$ is a real number$\}$　**3.** $\{x\,|\,x$ is a real number and $x\neq 5\}$　**4.** $\{x\,|\,x$ is a real number and $x\neq 4\}$
5. $\{x\,|\,x$ is a real number and $x\neq 0,\,x\neq -8\}$　**6.** $\{x\,|\,x$ is a real number and $x\neq -4,\,x\neq 4\}$　**7.** -1　**8.** $\dfrac{1}{5}$　**9.** $\dfrac{1}{x-1}$　**10.** $\dfrac{1}{x-7}$

11. $\dfrac{2(x-3)}{x-4}$　**12. a.** \$119　**b.** \$77　**c.** decrease　**13.** $-\dfrac{3}{2}$　**14.** $\dfrac{2}{5}$　**15.** $\dfrac{a-b}{2a}$　**16.** $\dfrac{1}{6}$　**17.** $\dfrac{12}{5}$　**18.** $\dfrac{(x+4)(x+5)}{3}$

19. $\dfrac{a-b}{5a}$　**20.** $\dfrac{3(x+1)}{x-7}$　**21.** $-\dfrac{1}{x}$　**22.** $-\dfrac{x+3}{2(x+2)}$　**23.** $60x^2y^5$　**24.** $2x(x-2)$　**25.** $5x(x-5)$　**26.** $10x^3(x-4)(x+7)(x-3)$

27. $\dfrac{4+x}{x-4}$　**28.** $\dfrac{2}{x^2}$　**29.** $\dfrac{3}{2(x-2)}$　**30.** $\dfrac{x-2}{x-10}$　**31.** $\dfrac{-7x-6}{5(x-3)(x+3)}$　**32.** $\dfrac{2x^2-5x-4}{x-3}$　**33.** $\dfrac{5a-1}{(a-1)^2(a+1)}$

34. $\dfrac{3x^2-7x-4}{(3x-4)(9x^2+12x+16)}$　**35.** $\dfrac{5-2x}{2(x-1)}$　**36.** $\dfrac{11}{x}$　**37.** $\dfrac{4-3x}{8+x}$　**38.** $\dfrac{x^3}{3(x+1)}$　**39.** $\dfrac{5(4x-3)}{2(5x^2-2)}$　**40.** $\dfrac{y}{x-y}$

41. $\dfrac{x(5y+1)}{3y}$　**42.** $\dfrac{x-3}{3}$　**43.** $\dfrac{1+x}{1-x}$　**44.** $-\dfrac{x^2+9}{6x}$　**45.** $\{6\}$　**46.** $\{3\}$　**47.** $\left\{\dfrac{3}{2}\right\}$　**48.** $\{-7,7\}$　**49.** $\dfrac{2x+5}{x(x-7)}$　**50.** $\left\{-\dfrac{1}{3},2\right\}$

51. $\dfrac{-5(x+6)}{2x(x-3)}$　**52.** $a=\dfrac{2A}{h}-b$　**53.** $R_2=\dfrac{RR_1}{R_1-R}$　**54.** $R=\dfrac{E}{I}-r$　**55.** $r=\dfrac{A-P}{Pt}$　**56.** $A=\dfrac{HL}{k(T_1-T_2)}$　**57.** $\{1,2\}$　**58.** 7

59. $1\dfrac{23}{37}$ hr　**60.** 12 hr　**61.** 8 mph　**62.** 4 mph　**63.** 9　**64.** 3.125 cu m　**65.** 2　**66.** $\dfrac{y-3}{x+2}$　**67.** $\dfrac{3}{5x}$　**68.** $\dfrac{7(x-4)}{2(x-2)}$

69. $\dfrac{5(a-2)}{7}$　**70.** $\dfrac{6}{a}$　**71.** $\dfrac{13}{3x}$　**72.** $\dfrac{-x+5}{(x+1)(x-1)}$　**73.** $\dfrac{1}{x-2}$　**74.** $\dfrac{2(7x-20)}{(x+4)^2(x-4)}$　**75.** $\dfrac{2}{15-2x}$　**76.** $\dfrac{2x}{x-2}$　**77.** $\dfrac{2(x-1)}{x+6}$

78. $\dfrac{5x}{2}$　**79.** $\{2\}$　**80.** $\left\{\dfrac{5}{3}\right\}$　**81.** $\dfrac{23}{25}$　**82.** -10 and -8　**83.** 10 hr　**84.** 490 mph　**85.** $63\dfrac{2}{3}$ mph; 45 mph　**86.** 4　**87.** 64π sq in.

Chapter 6 Test　1. $\{x\,|\,x$ is a real number and $x\neq 1\}$　**2.** $\{x\,|\,x$ is a real number and $x\neq -3,\,x\neq -11\}$　**3.** $-\dfrac{7}{8}$　**4.** $\dfrac{x}{x+9}$　**5.** x^2+2x+4

6. $\dfrac{5}{3x}$　**7.** $\dfrac{3}{x^3}$　**8.** $\dfrac{x+2}{2(x+3)}$　**9.** $-\dfrac{4(2x+9)}{5}$　**10.** -1　**11.** $\dfrac{1}{x(x+3)}$　**12.** $\dfrac{5x-2}{(x-3)(x+2)(x-2)}$　**13.** $\dfrac{30-x}{6(x-7)}$　**14.** $\dfrac{3}{2}$

15. $\dfrac{64}{3}$　**16.** $\dfrac{(x-3)^2}{x-2}$　**17.** $\{7\}$　**18.** $\{8\}$　**19.** $\left\{\dfrac{2}{7}\right\}$　**20.** $\{3\}$　**21.** $x=\dfrac{7a^2+b^2}{4a-b}$　**22.** 5　**23.** $\dfrac{6}{7}$ hr　**24.** 16　**25.** 9
26. 256 ft　**27.** 42 earned runs

Cumulative Review **1.** true; Sec. 1.2, Ex. 3 **2. a.** $x - \dfrac{1}{3}$ **b.** $5x - 6$ **c.** $8x + 3$ **d.** $\dfrac{7}{2 - x}$; Sec. 1.2 **3.** -1; Sec. 1.6, Ex. 7 **4.** $x^2 y^2$; Sec. 1.6

5. 1; Sec. 1.6, Ex. 8 **6.** $\dfrac{8a^7}{9b^{11}}$; Sec. 1.7 **7.** $\varnothing$; Sec. 2.5, Ex. 3 **8.** $\varnothing$; Sec. 2.6 **9.** $\{4, -10\}$; Sec. 2.6, Ex. 4 **10.** $\left(-\infty, -\dfrac{5}{3}\right] \cup \left[\dfrac{11}{3}, \infty\right)$; Sec. 2.6

11. $(0, -12)$, solution; $(1, 9)$, not a solution; $(2, -6)$, solution; Sec. 3.1, Ex. 3 **12. a.** -2 **b.** -20 **c.** $-\dfrac{10}{9}$; Sec. 3.6 **13.** parallel; Sec. 3.2, Ex. 8

14. 7; Sec. 2.1 **15.** $-5x + 8y = -20$ or $5x - 8y = 20$; Sec. 3.4, Ex. 2 **16.** $f(x) = \dfrac{1}{2}x + \dfrac{7}{2}$; Sec. 3.6

17. domain: $(-\infty, \infty)$; range: $[0, \infty)$; Sec. 3.7, Ex. 2 **18.** domain: $(-\infty, \infty)$; range: $(-\infty, \infty)$; Sec. 3.7

19. domain: $[-4, 4]$; range: $[-2, 2]$; Sec. 3.7, Ex. 3 **20.** domain: $(-\infty, \infty)$; range: $(-\infty, -1]$; Sec. 3.7 **21.** $\left(-\dfrac{21}{10}, \dfrac{3}{10}\right)$; Sec. 4.1, Ex. 7

22. $(3, 4)$; Sec. 4.1 **23.** $\left(\dfrac{1}{2}, 0, \dfrac{3}{4}\right)$; Sec. 4.2, Ex. 3 **24.** $(2, 1, 1)$; Sec. 4.2 **25.** 52 mph; 47 mph; Sec. 4.3, Ex. 3

26. Paper, \$3.80; folders, \$5.25; Sec. 4.3 **27.** $(1, -1, 3)$; Sec. 4.4, Ex. 3 **28.** $(0, 5, 4)$; Sec. 4.4 **29.** $\dfrac{z^2}{9x^4 y^{20}}$; Sec. 1.7, Ex. 14

30. a. $\dfrac{7}{12}$ **b.** -6 **c.** $\dfrac{1}{x^3}$; Sec. 1.7 **31.** 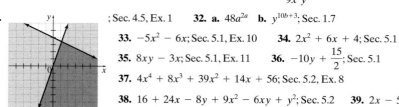 ; Sec. 4.5, Ex. 1 **32. a.** $48a^{2a}$ **b.** y^{10b+3}; Sec. 1.7

33. $-5x^2 - 6x$; Sec. 5.1, Ex. 10 **34.** $2x^2 + 6x + 4$; Sec. 5.1

35. $8xy - 3x$; Sec. 5.1, Ex. 11 **36.** $-10y + \dfrac{15}{2}$; Sec. 5.1

37. $4x^4 + 8x^3 + 39x^2 + 14x + 56$; Sec. 5.2, Ex. 8

38. $16 + 24x - 8y + 9x^2 - 6xy + y^2$; Sec. 5.2 **39.** $2x - 5$; Sec. 5.3, Ex. 3

40. $(y + 2)(x - 5)$; Sec. 5.4 **41.** $xy(-3x^2 + 2x - 5)$ or $-xy(3x^2 - 2x + 5)$; Sec. 5.4, Ex. 5 **42.** $(2x - 5)(3x + 7)$; Sec. 5.5

43. $(x + 2)(x + 8)$; Sec. 5.5, Ex. 1 **44.** $(2x - 1 + 3y)(2x - 1 - 3y)$; Sec. 5.6 **45.** $(p^2 + 4)(p + 2)(p - 2)$; Sec. 5.6, Ex. 9

46. a. domain: $(-\infty, \infty)$; range: $[-4, \infty)$ **b.** x-intercepts: $(-2, 0)$, $(2, 0)$; y-intercept: $(0, -4)$ **c.** There is no such point. **d.** $(0, -4)$ **e.** $-2, 2$

f. between $x = -2$ and $x = 2$ **g.** $\{-2, 2\}$; Sec. 3.6 **47.** $\{-2, 6\}$; Sec. 5.7, Ex. 1 **48.** $\left\{0, -\dfrac{1}{3}, 3\right\}$; Sec. 5.7 **49.** $\dfrac{x(x + 5)}{2(x - 5)}$; Sec. 6.3, Ex. 1

50. $\{-4\}$; Sec. 6.4

CHAPTER 7   Rational Exponents, Radicals, and Complex Numbers

Exercise Set 7.1 **1.** $2, -2$ **3.** no real number square roots **5.** $10, -10$ **7.** 10 **9.** $\dfrac{1}{2}$ **11.** 0.01 **13.** -6 **15.** x^5 **17.** $4y^3$

19. 2.646 **21.** 6.164 **23.** 14.142 **25.** 4 **27.** $\dfrac{1}{2}$ **29.** -1 **31.** x^4 **33.** $-3x^3$ **35.** -2 **37.** not a real number **39.** -2

41. x^4 **43.** $2x^2$ **45.** $9x^2$ **47.** $4x^2$ **49.** 8 **51.** -8 **53.** $2|x|$ **55.** x **57.** $|x - 2|$ **59.** $|x + 2|$ **61.** -11 **63.** $2x$

65. y^6 **67.** $5ab^{10}$ **69.** $-3x^4 y^3$ **71.** $a^4 b$ **73.** $-2x^2 y$ **75.** $\dfrac{5}{7}$ **77.** $\dfrac{x}{2y}$ **79.** $-\dfrac{z^7}{3x}$ **81.** $\dfrac{x}{2}$ **83.** $\sqrt{3}$ **85.** -1 **87.** -3

89. $\sqrt{7}$ **91.** $(-\infty, \infty)$; 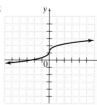 **93.** $(-\infty, \infty)$; 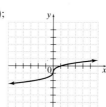 **95.** $-32x^{15} y^{10}$ **97.** $-60x^7 y^{10} z^5$ **99.** $\dfrac{x^9 y^5}{2}$

101. not a real number **103.** not a real number
105. answers may vary **107.** b **109.** b
111. 1.69 sq m **113.** answers may vary

Exercise Set 7.2 **1.** 7 **3.** 3 **5.** $\dfrac{1}{2}$ **7.** 13 **9.** $2\sqrt[3]{m}$ **11.** $3x^2$ **13.** -3 **15.** -2 **17.** 8 **19.** 16 **21.** not a real number

23. $\sqrt[5]{(2x)^3}$ **25.** $\sqrt[3]{(7x + 2)^2}$ or $\left(\sqrt[3]{7x + 2}\right)^2$ **27.** $\dfrac{64}{27}$ **29.** $\dfrac{1}{16}$ **31.** $\dfrac{1}{16}$ **33.** not a real number **35.** $\dfrac{1}{x^{1/4}}$ **37.** $a^{2/3}$

39. $\dfrac{5x^{3/4}}{7}$ **41.** $a^{7/3}$ **43.** x **45.** $3^{5/8}$ **47.** $y^{1/6}$ **49.** $8u^3$ **51.** $-b$ **53.** $\dfrac{1}{x^2}$ **55.** $27x^{2/3}$ **57.** $\dfrac{y}{z^{1/6}}$ **59.** $\dfrac{1}{x^{7/4}}$ **61.** $\sqrt{x}$

63. $\sqrt[3]{2}$ **65.** $2\sqrt{x}$ **67.** $\sqrt{x + 3}$ **69.** $\sqrt{xy}$ **71.** $\sqrt[3]{a^2 b}$ **73.** $\sqrt[15]{y^{11}}$ **75.** $\sqrt[12]{b^5}$ **77.** $\sqrt[24]{x^{23}}$ **79.** $\sqrt{a}$ **81.** $\sqrt[9]{432}$

83. $\sqrt[15]{343y^5}$ **85.** $\sqrt[9]{125r^3 s^2}$ **87.** $25 \cdot 3$ **89.** $16 \cdot 3$ or $4 \cdot 12$ **91.** $8 \cdot 2$ **93.** $27 \cdot 2$ **95.** 1509 calories **97.** 176.1 million

99. answers may vary **101.** $a^{1/3}$ **103.** $x^{1/5}$ **105.** 1.6818 **107.** $\dfrac{t^{1/2}}{u^{1/2}}$

Exercise Set 7.3 **1.** $\sqrt{14}$ **3.** 2 **5.** $\sqrt[3]{36}$ **7.** $\sqrt{6x}$ **9.** $\sqrt{\dfrac{14}{xy}}$ **11.** $\sqrt[4]{20x^3}$ **13.** $\dfrac{\sqrt{6}}{7}$ **15.** $\dfrac{\sqrt{2}}{7}$ **17.** $\dfrac{\sqrt[4]{x^3}}{2}$ **19.** $\dfrac{\sqrt[3]{4}}{3}$

21. $\dfrac{\sqrt[4]{8}}{x^2}$ **23.** $\dfrac{\sqrt[3]{2x}}{3y^4 \sqrt[3]{3}}$ **25.** $\dfrac{x\sqrt{y}}{10}$ **27.** $\dfrac{x\sqrt{5}}{13y}$ **29.** $-\dfrac{z^2 \sqrt[3]{z}}{5x}$ **31.** $4\sqrt{2}$ **33.** $4\sqrt[3]{3}$ **35.** $25\sqrt{3}$ **37.** $2\sqrt{6}$ **39.** $10x^2 \sqrt{x}$

41. $2y^2\sqrt[3]{2y}$ **43.** $a^2b\sqrt[4]{b^3}$ **45.** $y^2\sqrt{y}$ **47.** $5ab\sqrt{b}$ **49.** $-2x^2\sqrt[5]{y}$ **51.** $x^4\sqrt[3]{50x^2}$ **53.** $-4a^4b^3\sqrt{2b}$ **55.** $3x^3y^4\sqrt{xy}$

57. $5r^3s^4$ **59.** $\sqrt{2}$ **61.** 2   **63.** 10   **65.** x^2y **67.** $24m^2$ **69.** $\dfrac{15x\sqrt{2x}}{2}$ or $\dfrac{15x}{2}\sqrt{2x}$ **71.** $2a^2\sqrt[4]{2}$ **73.** 5 units

75. $\sqrt{41}$ units ≈ 6.403 **77.** $\sqrt{5}$ units ≈ 2.236 **79.** $\sqrt{192.58}$ units ≈ 13.877 **81.** $(4, -2)$ **83.** $\left(-5, \dfrac{5}{2}\right)$ **85.** $\left(-\dfrac{1}{2}, \dfrac{1}{2}\right)$

87. $\left(\sqrt{2}, \dfrac{\sqrt{5}}{2}\right)$ **89.** $14x$ **91.** $2x^2 - 7x - 15$ **93.** y^2 **95.** $x^2 - 8x + 16$ **97.** $\dfrac{\sqrt[3]{64}}{\sqrt{64}} = \dfrac{4}{8} = \dfrac{1}{2}$ **99.** $r = 1.6$ meters

101. a. 3.8 times   **b.** 2.9 times   **c.** answers may vary

Mental Math   **1.** $6\sqrt{3}$ **3.** $3\sqrt{x}$ **5.** $12\sqrt[3]{x}$ **7.** 3

Exercise Set 7.4   **1.** $-2\sqrt{2}$ **3.** $10x\sqrt{2x}$ **5.** $17\sqrt{2} - 15\sqrt{5}$ **7.** $-\sqrt[3]{2x}$ **9.** $5b\sqrt{b}$ **11.** $\dfrac{31\sqrt{2}}{15}$ **13.** $\dfrac{\sqrt[3]{11}}{3}$ **15.** $\dfrac{5\sqrt{5x}}{9}$

17. $14 + \sqrt{3}$ **19.** $7 - 3y$ **21.** $6\sqrt{3} - 6\sqrt{2}$ **23.** $-23\sqrt[3]{5}$ **25.** $2b\sqrt{b}$ **27.** $20y\sqrt{2y}$ **29.** $2y\sqrt[3]{2x}$ **31.** $6\sqrt[3]{11} - 4\sqrt{11}$

33. $4x\sqrt[4]{x^3}$ **35.** $\dfrac{2\sqrt{3}}{3}$ **37.** $\dfrac{5x\sqrt[3]{x}}{7}$ **39.** $\dfrac{5\sqrt{7}}{2x}$ **41.** $\dfrac{\sqrt[3]{2}}{6}$ **43.** $\dfrac{14x\sqrt[3]{2x}}{9}$ **45.** $15\sqrt{3}$ in.   **47.** $\sqrt{35} + \sqrt{21}$ **49.** $7 - 2\sqrt{10}$

51. $3\sqrt{x} - x\sqrt{3}$ **53.** $6x - 13\sqrt{x} - 5$ **55.** $\sqrt[3]{a^2} + \sqrt[3]{a} - 20$ **57.** $6\sqrt{2} - 12$ **59.** $2 + 2x\sqrt{3}$ **61.** $-16 - \sqrt{35}$ **63.** $x - y^2$

65. $3 + 2x\sqrt{3} + x^2$ **67.** $5x - 3\sqrt{15x} - 3\sqrt{10x} + 9\sqrt{6}$ **69.** $-\sqrt[3]{4} + 2\sqrt[3]{2}$ **71.** $\sqrt[3]{x^2} - 4\sqrt[6]{x^5} + 8\sqrt[3]{x} - 4\sqrt{x} + 7$

73. $x + 24 + 10\sqrt{x - 1}$ **75.** $2x + 6 - 2\sqrt{2x + 5}$ **77.** $x - 7$ **79.** $\dfrac{7}{x + y}$ **81.** $2a - 3$ **83.** $\dfrac{-2 + \sqrt{3}}{3}$ **85.** $22\sqrt{5}$ ft; 150 sq ft
87. a. $2\sqrt{3}$ **b.** 3   **c.** answers may vary

Mental Math   **1.** $\sqrt{2} - x$ **3.** $5 + \sqrt{a}$ **5.** $7\sqrt{4} - 8\sqrt{x}$

Exercise Set 7.5   **1.** $\dfrac{\sqrt{14}}{7}$ **3.** $\dfrac{\sqrt{5}}{5}$ **5.** $\dfrac{4\sqrt[3]{9}}{3}$ **7.** $\dfrac{3\sqrt{2x}}{4x}$ **9.** $\dfrac{3\sqrt[3]{2x}}{2x}$ **11.** $\dfrac{3\sqrt{3a}}{a}$ **13.** $\dfrac{3\sqrt[3]{4}}{2}$ **15.** $\dfrac{2\sqrt{21}}{7}$ **17.** $\dfrac{\sqrt{10xy}}{5y}$

19. $\dfrac{\sqrt[3]{75}}{5}$ **21.** $\dfrac{\sqrt{6x}}{10}$ **23.** $\dfrac{\sqrt{3z}}{6z}$ **25.** $\dfrac{\sqrt[3]{6xy^2}}{3x}$ **27.** $\dfrac{2\sqrt[4]{9x}}{3x^2}$ **29.** $\dfrac{5\sqrt[5]{4ab^4}}{2ab^3}$ **31.** $-2(2 + \sqrt{7})$ **33.** $\dfrac{7(\sqrt{x} + 3)}{9 - x}$

35. $-5 + 2\sqrt{6}$ **37.** $\dfrac{2a + \sqrt{ab} + 2\sqrt{a} + \sqrt{b}}{4a - b}$ **39.** $-\dfrac{8(1 - \sqrt{10})}{9}$ **41.** $\dfrac{x - \sqrt{xy}}{x - y}$ **43.** $\dfrac{5 + 3\sqrt{2}}{7}$ **45.** $\dfrac{5}{\sqrt{15}}$ **47.** $\dfrac{6}{\sqrt{10}}$

49. $\dfrac{2x}{7\sqrt{x}}$ **51.** $\dfrac{5y}{\sqrt[3]{100xy}}$ **53.** $\dfrac{2}{\sqrt{10}}$ **55.** $\dfrac{2x}{11\sqrt{2x}}$ **57.** $\dfrac{7}{2\sqrt[3]{49}}$ **59.** $\dfrac{3x^2}{10\sqrt[3]{9x}}$ **61.** $\dfrac{6x^2y^3}{\sqrt{6z}}$ **63.** answers may vary

65. $\dfrac{-7}{12 + 6\sqrt{11}}$ **67.** $\dfrac{3}{10 + 5\sqrt{7}}$ **69.** $\dfrac{x - 9}{x - 3\sqrt{x}}$ **71.** $\dfrac{x - 1}{x - 2\sqrt{x} + 1}$ **73.** $\{5\}$ **75.** $\left\{-\dfrac{1}{2}, 6\right\}$ **77.** $\{2, 6\}$ **79.** $r = \dfrac{\sqrt{A\pi}}{2\pi}$

81. answers may vary   **83.** $\sqrt[3]{25}$

Integrated Review   **1.** 9   **2.** -2 **3.** $\dfrac{1}{2}$ **4.** x^3 **5.** y^3 **6.** $2y^5$ **7.** $-2y$ **8.** $3b^3$ **9.** 6   **10.** $\sqrt[4]{3y}$ **11.** $\dfrac{1}{16}$ **12.** $\sqrt[5]{(x + 1)^3}$

13. y **14.** $16x^{1/2}$ **15.** $x^{5/4}$ **16.** $4^{11/15}$ **17.** $2x^2$ **18.** $\sqrt[4]{a^3b^2}$ **19.** $\sqrt[4]{x^3}$ **20.** $\sqrt[6]{500}$ **21.** $2\sqrt{10}$ **22.** $2xy^2\sqrt[4]{x^3y^2}$

23. $3x\sqrt[3]{2x}$ **24.** $-2b^2\sqrt[5]{2}$ **25.** $\sqrt{5x}$ **26.** $4x$ **27.** $7y^2\sqrt{y}$ **28.** $2a^2\sqrt[3]{3}$ **29.** $2\sqrt{5} - 5\sqrt{3} + 5\sqrt{7}$ **30.** $y\sqrt[3]{2y}$

31. $\sqrt{15} - \sqrt{6}$ **32.** $10 + 2\sqrt{21}$ **33.** $4x^2 - 5$ **34.** $x + 2 - 2\sqrt{x + 1}$ **35.** $\dfrac{\sqrt{21}}{3}$ **36.** $\dfrac{5\sqrt[3]{4x}}{2x}$ **37.** $\dfrac{13 - 3\sqrt{21}}{5}$ **38.** $\dfrac{7}{\sqrt{21}}$
39. $\dfrac{3y}{\sqrt[3]{33y^2}}$ **40.** $\dfrac{x - 4}{x + 2\sqrt{x}}$

Graphing Calculator Explorations   **1.** $\{3.19\}$ **3.** $\varnothing$ **5.** $\{3.23\}$

Exercise Set 7.6   **1.** $\{8\}$ **3.** $\{7\}$ **5.** $\varnothing$ **7.** $\{7\}$ **9.** $\{6\}$ **11.** $\left\{-\dfrac{9}{2}\right\}$ **13.** $\{29\}$ **15.** $\{4\}$ **17.** $\{-4\}$ **19.** $\varnothing$

21. $\{7\}$ **23.** $\{9\}$ **25.** $\{50\}$ **27.** $\varnothing$ **29.** $\left\{\dfrac{15}{4}\right\}$ **31.** $\{7\}$ **33.** $\{5\}$ **35.** $\{-12\}$ **37.** $\{9\}$ **39.** $\{-3\}$ **41.** $\{1\}$

43. $\{1\}$ **45.** $\left\{\dfrac{1}{2}\right\}$ **47.** $\{0, 4\}$ **49.** $\left\{\dfrac{37}{4}\right\}$ **51.** $3\sqrt{5}$ ft   **53.** $2\sqrt{10}$ m   **55.** $2\sqrt{131}$ m ≈ 22.9 m   **57.** $\sqrt{100.84}$ mm ≈ 10.0 mm

59. 17 ft   **61.** 13 ft   **63.** 14,657,415 sq mi   **65.** 100 ft   **67.** 100   **69.** $\dfrac{\pi}{2}$ sec ≈ 1.57 sec   **71.** 12.97 ft   **73.** answers may vary

75. $15\sqrt{3}$ sq mi ≈ 25.98 sq mi   **77.** answers may vary   **79.** 0.51 km   **81.** $\dfrac{x}{4x + 3}$ **83.** $-\dfrac{4z + 2}{3z}$ **85.** $\{1\}$

87. $\sqrt{5x - 1} + 4 = 7$ **89.** 2743 deliveries
$$\sqrt{5x - 1} = 3$$
$$\left(\sqrt{5x - 1}\right)^2 = 3^2$$
$$5x - 1 = 9$$
$$5x = 10$$
$$x = 2$$

The Bigger Picture   **1.** $\{-19\}$ **2.** $\left\{-\dfrac{5}{3}, 5\right\}$ **3.** $\left\{-\dfrac{9}{2}, 5\right\}$ **4.** $\left[-\dfrac{11}{5}, 1\right]$ **5.** $\left(-\dfrac{7}{5}, \infty\right)$ **6.** $\{25\}$ **7.** $(-5, \infty)$ **8.** $\varnothing$
9. $(-\infty, -13) \cup (17, \infty)$ **10.** $\left\{\dfrac{17}{25}\right\}$

Mental Math   **1.** $9i$ **3.** $i\sqrt{7}$ **5.** -4 **7.** $8i$

Exercise Set 7.7   **1.** $2i\sqrt{6}$ **3.** $-6i$ **5.** $24i\sqrt{7}$ **7.** $-3\sqrt{6}$ **9.** $-\sqrt{14}$ **11.** $-5\sqrt{2}$ **13.** $4i$ **15.** $i\sqrt{3}$ **17.** $2\sqrt{2}$ **19.** $6 - 4i$
21. $-2 + 6i$ **23.** $-2 - 4i$ **25.** $2 - i$ **27.** $5 - 10i$ **29.** $8 - i$ **31.** -12 **33.** 63 **35.** -40 **37.** $18 + 12i$ **39.** $27 + 3i$
41. $18 + 13i$ **43.** 7 **45.** $12 - 16i$ **47.** 20 **49.** 2 **51.** $17 + 144i$ **53.** $-2i$ **55.** $-4i$ **57.** $\dfrac{28}{25} - \dfrac{21}{25}i$ **59.** $-\dfrac{12}{5} + \dfrac{6}{5}i$
61. $4 + i$ **63.** $-\dfrac{5}{2} - 2i$ **65.** $-5 + \dfrac{16}{3}i$ **67.** $\dfrac{3}{5} - \dfrac{1}{5}i$ **69.** $\dfrac{1}{5} - \dfrac{8}{5}i$ **71.** 1 **73.** i **75.** $-i$ **77.** -1 **79.** -64 **81.** $-243i$
83. 5 people   **85.** 14 people   **87.** 16.7%   **89.** $1 - i$ **91.** 0 **93.** $2 + 3i$ **95.** $2 + i\sqrt{2}$ **97.** $\dfrac{1}{2} - \dfrac{\sqrt{3}}{2}i$ **99.** answers may vary
101. $6 - 6i$ **103.** yes

Chapter 7 Vocabulary Check   **1.** conjugate   **2.** principal square root   **3.** rationalizing   **4.** imaginary unit   **5.** cube root
6. index, radicand   **7.** like radicals   **8.** complex number   **9.** distance   **10.** midpoint

Chapter 7 Review   **1.** 9 **2.** 3 **3.** -2 **4.** not a real number   **5.** $-\dfrac{1}{7}$ **6.** x^{32} **7.** -6 **8.** 4 **9.** $-a^2b^3$ **10.** $4a^2b^6$ **11.** $2ab^2$
12. $-2x^3y^4$ **13.** $\dfrac{x^6}{6y}$ **14.** $\dfrac{3y}{z^4}$ **15.** $|x|$ **16.** $|x^2 - 4|$ **17.** -27 **18.** -5 **19.** $-x$ **20.** $2|2y + z|$ **21.** $5|x - y|$ **22.** y
23. $|x|$ **24.** $3, 6$ **25.** $2, \sqrt[3]{17}$ **26.** $\dfrac{1}{3}$ **27.** $-\dfrac{1}{3}$ **28.** $-\dfrac{1}{3}$ **29.** $-\dfrac{1}{4}$ **30.** -27 **31.** $\dfrac{1}{4}$ **32.** not a real number   **33.** $\dfrac{343}{125}$
34. $\dfrac{9}{4}$ **35.** not a real number   **36.** $x^{2/3}$ **37.** $5^{1/5}x^{2/5}y^{3/5}$ **38.** $\sqrt[5]{y^4}$ **39.** $5\sqrt[3]{xy^2z^5}$ **40.** $\dfrac{1}{\sqrt{x + 2y}}$ **41.** $a^{13/6}$ **42.** $\dfrac{1}{b}$ **43.** $\dfrac{1}{a^{9/2}}$
44. $\dfrac{y^2}{x}$ **45.** a^4b^6 **46.** $\dfrac{1}{x^{11/12}}$ **47.** $\dfrac{b^{5/6}}{49a^{1/4}c^{5/3}}$ **48.** $a - a^2$ **49.** 4.472 **50.** -3.391 **51.** 5.191 **52.** 3.826 **53.** -26.246
54. 0.045 **55.** $\sqrt[6]{1372}$ **56.** $\sqrt[12]{81x^3}$ **57.** $2\sqrt{6}$ **58.** $\sqrt[3]{7x^2yz}$ **59.** $2x$ **60.** ab^3 **61.** $2\sqrt{15}$ **62.** $-5\sqrt{3}$ **63.** $3\sqrt[3]{6}$
64. $-2\sqrt[3]{4}$ **65.** $6x^3\sqrt{x}$ **66.** $2ab^2\sqrt[3]{3a^2b}$ **67.** $\dfrac{p^8\sqrt{p}}{11}$ **68.** $\dfrac{y\sqrt[3]{y^2}}{3x^2}$ **69.** $\dfrac{y\sqrt[4]{xy^2}}{3}$ **70.** $\dfrac{x\sqrt{2x}}{7y^2}$
71. **a.** $\dfrac{5}{\sqrt{\pi}}$ m or $\dfrac{5\sqrt{\pi}}{\pi}$ m   **b.** 5.75 in.   **72.** $\sqrt{197}$ units ≈ 14.036 **73.** $\sqrt{130}$ units ≈ 11.402 **74.** $\sqrt{73}$ units ≈ 8.544
75. $7\sqrt{2}$ units ≈ 9.899 **76.** $2\sqrt{11}$ units ≈ 6.633 **77.** $\sqrt{275.6}$ units ≈ 16.601 **78.** $(-5, 5)$ **79.** $\left(-\dfrac{15}{2}, 1\right)$ **80.** $\left(-\dfrac{11}{2}, -2\right)$
81. $\left(\dfrac{1}{20}, -\dfrac{3}{16}\right)$ **82.** $\left(\dfrac{1}{4}, -\dfrac{2}{7}\right)$ **83.** $(\sqrt{3}, -3\sqrt{6})$ **84.** $-2\sqrt{5}$ **85.** $2x\sqrt{3x}$ **86.** $9\sqrt[3]{2}$ **87.** $3a\sqrt[4]{2a}$ **88.** $\dfrac{15 + 2\sqrt{3}}{6}$
89. $\dfrac{3\sqrt{2}}{4x}$ **90.** $17\sqrt{2} - 15\sqrt{5}$ **91.** $-4ab\sqrt[4]{2b}$ **92.** 6 **93.** $x - 6\sqrt{x} + 9$ **94.** $-8\sqrt{5}$ **95.** $4x - 9y$ **96.** $a - 9$
97. $\sqrt[3]{a^2} + 4\sqrt[3]{a} + 4$ **98.** $\sqrt[3]{25x^2} - 81$ **99.** $a + 64$ **100.** $\dfrac{3\sqrt{7}}{7}$ **101.** $\dfrac{\sqrt{3x}}{6}$ **102.** $\dfrac{5\sqrt[3]{2}}{2}$ **103.** $\dfrac{2x^2\sqrt{2x}}{y}$ **104.** $\dfrac{x^2y^2\sqrt[3]{15yz}}{z}$
105. $\dfrac{3\sqrt[4]{2x^2}}{2x^3}$ **106.** $\dfrac{3\sqrt{y} + 6}{y - 4}$ **107.** $-5 + 2\sqrt{6}$ **108.** $\dfrac{11}{3\sqrt{11}}$ **109.** $\dfrac{6}{\sqrt{2y}}$ **110.** $\dfrac{3}{7\sqrt[3]{3}}$ **111.** $\dfrac{4x^3}{y\sqrt{2x}}$ **112.** $\dfrac{xy}{\sqrt[3]{10x^2yz}}$
113. $\dfrac{x - 25}{-3\sqrt{x} + 15}$ **114.** $\{32\}$ **115.** $\varnothing$ **116.** $\{35\}$ **117.** $\varnothing$ **118.** $\{9\}$ **119.** $\{16\}$ **120.** $3\sqrt{2}$ cm   **121.** $\sqrt{241}$ ft
122. 51.2 ft   **123.** 4.24 ft   **124.** $2i\sqrt{2}$ **125.** $-i\sqrt{6}$ **126.** $6i$ **127.** $-\sqrt{10}$ **128.** $15 - 4i$ **129.** $-13 - 3i$ **130.** -64
131. $-12 - 18i$ **132.** $1 + 5i$ **133.** $-5 - 12i$ **134.** 87 **135.** $\dfrac{3}{2} - i$ **136.** $-\dfrac{1}{3} + \dfrac{1}{3}i$ **137.** x **138.** $|x + 2|$ **139.** -10
140. $-x^4y$ **141.** $\dfrac{y^5}{2x^3}$ **142.** 3 **143.** $\dfrac{1}{8}$ **144.** $\dfrac{16}{9}$ **145.** $\dfrac{1}{x^{13/2}}$ **146.** $10x^4\sqrt{2x}$ **147.** $\dfrac{n\sqrt{3n}}{11m^5}$ **148.** $6\sqrt{5} - 11x\sqrt[3]{5}$
149. $4x - 20\sqrt{x} + 25$ **150.** $\sqrt{41}$ units   **151.** $(4, 16)$ **152.** $\dfrac{7\sqrt{13}}{13}$ **153.** $\dfrac{2\sqrt{x} - 6}{x - 9}$ **154.** $\{4\}$

Chapter 7 Test   **1.** $6\sqrt{6}$ **2.** $-x^{16}$ **3.** $\dfrac{1}{5}$ **4.** 5 **5.** $\dfrac{4x^2}{9}$ **6.** $-a^6b^3$ **7.** $\dfrac{8a^{1/3}c^{2/3}}{b^{5/12}}$ **8.** $a^{7/12} - a^{7/3}$ **9.** $|4xy|$ or $4|xy|$ **10.** -27
11. $\dfrac{3\sqrt{y}}{y}$ **12.** $\dfrac{8 - 6\sqrt{x} + x}{8 - 2x}$ **13.** $\dfrac{2\sqrt[3]{3x^2}}{3x}$ **14.** $\dfrac{6 - x^2}{8(\sqrt{6} - x)}$ **15.** $-x\sqrt{5x}$ **16.** $4\sqrt{3} - \sqrt{6}$ **17.** $x + 2\sqrt{x} + 1$
18. $\sqrt{6} + \sqrt{2} - 4\sqrt{3} - 4$ **19.** -20 **20.** 23.685 **21.** 0.019 **22.** $\{2, 3\}$ **23.** $\varnothing$ **24.** $\{6\}$ **25.** $i\sqrt{2}$ **26.** $-2i\sqrt{2}$
27. $-3i$ **28.** 40 **29.** $7 + 24i$ **30.** $-\dfrac{3}{2} + \dfrac{5}{2}i$ **31.** $x = \dfrac{5\sqrt{2}}{2}$ in.   **32.** $\sqrt{2}, 5$ **33.** $2\sqrt{26}$ units   **34.** $\sqrt{95}$ units   **35.** $\left(-4, \dfrac{7}{2}\right)$
36. $\left(-\dfrac{1}{2}, \dfrac{3}{10}\right)$ **37.** 27 mph   **38.** 360 ft

Cumulative Review   **1.** $4x^b$; Sec. 1.7, Ex. 16   **2.** $\dfrac{2}{15}x^{2a}$; Sec. 1.7   **3.** y^{5a+6}; Sec. 1.7, Ex. 17   **4.** $-1.4y^{9a}$; Sec. 1.7   **5.** $\{-7, 5\}$; Sec. 2.6, Ex. 12

6. $\left\{\dfrac{2}{3}\right\}$; Sec. 2.6   **7.** $y = -2x + 12$; Sec. 3.4, Ex. 5   **8.** $\dfrac{13}{14}$; Sec. 2.1   **9.** $(-2, 2)$; Sec. 4.1, Ex. 8   **10.** 0; Sec. 3.2

11. $\{(x, y, z)|x - 5y - 2z = 6\}$; Sec. 4.2, Ex. 4   **12.** $(6, 0)$; Sec. 4.1   **13.** $\varnothing$; Sec. 4.4, Ex. 2   **14. a.** $7x - 3$ **b.** $-3x + 11$

c. $6x - 14$; Sec. 1.5   **15.** $13x^3y - xy^3 + 7$; Sec. 5.1, Ex. 14   **16.** 5×10^{-2}; Sec. 1.7   **17.** -4; Sec. 5.1, Ex. 18   **18.** -5; Sec. 5.1

19. 35; Sec. 5.1, Ex. 19   **20.** $-\dfrac{85}{16}$; Sec. 5.1   **21.** $x^2 + 10x + 25$; Sec. 5.2, Ex. 12   **22.** $3y^2 - 2y - 8$; Sec. 5.2

23. $16m^4 - 24m^2n + 9n^2$; Sec. 5.2, Ex. 15   **24.** $6y^3 + 7y^2 - 6y + 1$; Sec. 5.2   **25.** $2x^2 + 5x + 2 + \dfrac{7}{x - 3}$; Sec. 5.3, Ex. 8

26. $4y^2 - 1 + \dfrac{9}{y - 3}$; Sec. 5.3   **27.** $(b - 6)(a + 2)$; Sec. 5.4, Ex. 8   **28.** $(x - 1)(x^2 + 4)$; Sec. 5.4   **29.** $2(n^2 - 19n + 40)$; Sec. 5.5, Ex. 4

30. $2(x - 2)(x^2 + 2x + 4)$; Sec. 5.6   **31.** $(4x + 3y)^2$; Sec. 5.5, Ex. 8   **32.** $(x + 1 + y)(x + 1 - y)$; Sec. 5.6

33. $2(5 + 2y)(5 - 2y)$; Sec. 5.6, Ex. 7   **34.** $(x^2 + 1)(x + 1)(x - 1)$; Sec. 5.6   **35.** $\left\{-\dfrac{1}{2}, 4\right\}$; Sec. 5.7, Ex. 3   **36.** $x^2 + 3$; Sec. 5.3

37. $\{-2, 0, 2\}$; Sec. 5.7, Ex. 6   **38. a.** $\dfrac{3(y - 2)}{4(2y + 3)}$ **b.** $\dfrac{x + 4}{16}$; Sec. 6.3   **39.** 1; Sec. 6.1, Ex. 4   **40.** $-a^2 - 2a - 4$; Sec. 6.1

41. $\dfrac{1}{5x - 1}$; Sec. 6.1, Ex. 2   **42.** $\dfrac{3}{a + 5}$; Sec. 6.1   **43.** $\dfrac{5 + x}{7z^2}$; Sec. 6.2, Ex. 1   **44. a.** $\dfrac{9x - 2y}{3x^2y^2}$ **b.** $\dfrac{3x(x - 7)}{(x + 3)(x - 3)}$ **c.** $\dfrac{x + 5}{x - 2}$; Sec. 6.2

45. $\{-2\}$; Sec. 6.4, Ex. 2   **46.** $\{-1, -5\}$; Sec. 6.4   **47.** $\dfrac{1}{8}$; Sec. 7.2, Ex. 12   **48.** $\dfrac{1}{27}$; Sec. 7.2   **49.** $\dfrac{1}{9}$; Sec. 7.2, Ex. 13   **50.** $\dfrac{1}{25}$; Sec. 7.2

CHAPTER 8   Quadratic Equations and Functions

Exercise Set 8.1   **1.** $\{-4, 4\}$ **3.** $\{-\sqrt{7}, \sqrt{7}\}$ **5.** $\{-3\sqrt{2}, 3\sqrt{2}\}$ **7.** $\{-\sqrt{10}, \sqrt{10}\}$ **9.** $\{-8, -2\}$ **11.** $\{6 - 3\sqrt{2}, 6 + 3\sqrt{2}\}$

13. $\left\{\dfrac{3 - 2\sqrt{2}}{2}, \dfrac{3 + 2\sqrt{2}}{2}\right\}$ **15.** $\{-3i, 3i\}$ **17.** $\{-\sqrt{6}, \sqrt{6}\}$ **19.** $\{-2i\sqrt{2}, 2i\sqrt{2}\}$ **21.** $\left\{\dfrac{1 - 4i}{3}, \dfrac{1 + 4i}{3}\right\}$

23. $\{-7 - \sqrt{5}, -7 + \sqrt{5}\}$ **25.** $\{-3 - 2i\sqrt{2}, -3 + 2i\sqrt{2}\}$ **27.** $x^2 + 16x + 64 = (x + 8)^2$ **29.** $z^2 - 12z + 36 = (z - 6)^2$

31. $p^2 + 9p + \dfrac{81}{4} = \left(p + \dfrac{9}{2}\right)^2$ **33.** $r^2 - r + \dfrac{1}{4} = \left(r - \dfrac{1}{2}\right)^2$ **35.** $\{-5, -3\}$ **37.** $\{-3 - \sqrt{7}, -3 + \sqrt{7}\}$

39. $\left\{\dfrac{-1 - \sqrt{5}}{2}, \dfrac{-1 + \sqrt{5}}{2}\right\}$ **41.** $\{-1 - \sqrt{6}, -1 + \sqrt{6}\}$ **43.** $\left\{\dfrac{-1 - \sqrt{29}}{2}, \dfrac{-1 + \sqrt{29}}{2}\right\}$ **45.** $\{-4 - \sqrt{15}, -4 + \sqrt{15}\}$

47. $\left\{\dfrac{6 - \sqrt{30}}{3}, \dfrac{6 + \sqrt{30}}{3}\right\}$ **49.** $\left\{-4, \dfrac{1}{2}\right\}$ **51.** $\left\{\dfrac{-3 - \sqrt{21}}{3}, \dfrac{-3 + \sqrt{21}}{3}\right\}$ **53.** $\{-1 - i, -1 + i\}$ **55.** $\{-2 - i\sqrt{2}, -2 + i\sqrt{2}\}$

57. $\left\{\dfrac{1 - i\sqrt{47}}{4}, \dfrac{1 + i\sqrt{47}}{4}\right\}$ **59.** $\{-5 - i\sqrt{3}, -5 + i\sqrt{3}\}$ **61.** $\{-4, 1\}$ **63.** $\left\{\dfrac{2 - i\sqrt{2}}{2}, \dfrac{2 + i\sqrt{2}}{2}\right\}$ **65.** $\left\{\dfrac{-3 - \sqrt{69}}{6}, \dfrac{-3 + \sqrt{69}}{6}\right\}$

67. 20% **69.** 11% **71.** 9.63 sec   **73.** 8.29 sec   **75.** 15 ft by 15 ft   **77.** $10\sqrt{2}$ cm   **79.** $5 - 10\sqrt{3}$ **81.** $\dfrac{3 - 2\sqrt{7}}{4}$ **83.** $2\sqrt{7}$

85. $\sqrt{13}$ **87.** complex, but not real numbers   **89.** real solutions   **91.** complex, but not real numbers   **93.** compound; answers may vary
95. $-8x, 8x$ **97.** 6 thousand scissors

Calculator Explorations   **1.** $\{-1.27, 6.27\}$ **3.** $\{-1.10, 0.90\}$ **5.** $\varnothing$

Mental Math   **1.** $a = 1, b = 3, c = 1$ **3.** $a = 7, b = 0, c = -4$ **5.** $a = 6, b = -1, c = 0$

Exercise Set 8.2   **1.** $\{-6, 1\}$ **3.** $\left\{-\dfrac{3}{5}, 1\right\}$ **5.** $\{3\}$ **7.** $\left\{\dfrac{-7 - \sqrt{33}}{2}, \dfrac{-7 + \sqrt{33}}{2}\right\}$ **9.** $\left\{\dfrac{1 - \sqrt{57}}{8}, \dfrac{1 + \sqrt{57}}{8}\right\}$

11. $\left\{\dfrac{7 - \sqrt{85}}{6}, \dfrac{7 + \sqrt{85}}{6}\right\}$ **13.** $\{1 - \sqrt{3}, 1 + \sqrt{3}\}$ **15.** $\left\{-\dfrac{3}{2}, 1\right\}$ **17.** $\left\{\dfrac{3 - \sqrt{11}}{2}, \dfrac{3 + \sqrt{11}}{2}\right\}$ **19.** $\left\{\dfrac{-5 - \sqrt{17}}{2}, \dfrac{-5 + \sqrt{17}}{2}\right\}$

21. $\left\{\dfrac{5}{2}, 1\right\}$ **23.** $\{-3 - 2i, -3 + 2i\}$ **25.** $\{-2 - \sqrt{11}, -2 + \sqrt{11}\}$ **27.** $\left\{\dfrac{3 - i\sqrt{87}}{8}, \dfrac{3 + i\sqrt{87}}{8}\right\}$ **29.** $\left\{\dfrac{3 - \sqrt{29}}{2}, \dfrac{3 + \sqrt{29}}{2}\right\}$

31. $\left\{\dfrac{-5 - i\sqrt{5}}{10}, \dfrac{-5 + i\sqrt{5}}{10}\right\}$ **33.** $\left\{\dfrac{-1 - \sqrt{19}}{6}, \dfrac{-1 + \sqrt{19}}{6}\right\}$ **35.** $\left\{\dfrac{-1 - i\sqrt{23}}{4}, \dfrac{-1 + i\sqrt{23}}{4}\right\}$ **37.** $\{1\}$ **39.** $\{3 + \sqrt{5}, 3 - \sqrt{5}\}$

41. two real solutions   **43.** one real solutions   **45.** two real solutions   **47.** two complex but not real solutions   **49.** two real solutions

51. 14 ft   **53.** $(2 + 2\sqrt{2})$ cm, $(2 + 2\sqrt{2})$ cm, $(4 + 2\sqrt{2})$ cm   **55.** width: $(-5 + 5\sqrt{17})$ ft; length: $(5 + 5\sqrt{17})$ ft

57. a. $50\sqrt{2}$ m   **b.** 5000 sq m   **59.** 37.4 ft by 38.5 ft   **61.** base: $(2 + 2\sqrt{43})$ cm; height: $(-1 + \sqrt{43})$ cm   **63.** 8.9 sec   **65.** 2.8 sec

67. $\left\{\dfrac{11}{5}\right\}$ **69.** $\{15\}$ **71.** $(x^2 + 5)(x + 2)(x - 2)$ **73.** $(z + 3)(z - 3)(z + 2)(z - 2)$ **75.** b   **77.** answers may vary

79. $\{0.6, 2.4\}$　　**81.** Sunday to Monday　　**83.** Wednesday　　**85.** $f(4) = 33$; answers may vary　　**87. a.** $6406.8 million　**b.** 2006
89. a. 8684 thousand barrels per day　**b.** 2006 or 2011　　**91.** $\{0.6, 2.4\}$

Exercise Set 8.3　**1.** $\{2\}$　　**3.** $\{16\}$　　**5.** $\{1, 4\}$　　**7.** $\{3 - \sqrt{7}, 3 + \sqrt{7}\}$　　**9.** $\left\{\dfrac{3 - \sqrt{57}}{4}, \dfrac{3 + \sqrt{57}}{4}\right\}$　　**11.** $\left\{\dfrac{1 - \sqrt{29}}{2}, \dfrac{1 + \sqrt{29}}{2}\right\}$

13. $\{-2, 2, -2i, 2i\}$　　**15.** $\{-3, 3, -2, 2\}$　　**17.** $\left\{-\dfrac{1}{2}, \dfrac{1}{2}, -i\sqrt{3}, i\sqrt{3}\right\}$　　**19.** $\{125, -8\}$　　**21.** $\left\{-\dfrac{4}{5}, 0\right\}$　　**23.** $\left\{-\dfrac{1}{8}, 27\right\}$　　**25.** $\left\{-\dfrac{2}{3}, \dfrac{4}{3}\right\}$

27. $\left\{-\dfrac{1}{125}, \dfrac{1}{8}\right\}$　　**29.** $\{-\sqrt{2}, \sqrt{2}, -\sqrt{3}, \sqrt{3}\}$　　**31.** $\left\{\dfrac{-9 - \sqrt{201}}{6}, \dfrac{-9 + \sqrt{201}}{6}\right\}$　　**33.** $\{2, 3\}$　　**35.** $\{3\}$　　**37.** $\{27, 125\}$　　**39.** $\{5\}$

41. $\left\{\dfrac{1}{8}, -8\right\}$　　**43.** $\{-5, 1\}$　　**45.** $\{4\}$　　**47.** $\{-3\}$　　**49.** $\{-\sqrt{5}, \sqrt{5}, -2i, 2i\}$　　**51.** $\{6, 12\}$　　**53.** $\left\{-\dfrac{1}{3}, \dfrac{1}{3}, -\dfrac{i\sqrt{6}}{3}, \dfrac{i\sqrt{6}}{3}\right\}$

55. 5 mph, then 4 mph　　**57.** inlet pipe: 15.5 hr; hose: 16.5 hr　　**59.** 55 mph; 66 mph　　**61.** 8.5 hr　　**63.** 12 or -8
65. a. $(x - 6)$ in.　**b.** $300 = (x - 6) \cdot (x - 6) \cdot 3$　**c.** 16 in. by 16 in.　　**67.** 22 feet　　**69.** $(-\infty, 3]$　　**71.** $(-5, \infty)$　　**73.** $\{1, -3i, 3i\}$

75. $\left\{-\dfrac{1}{2}, \dfrac{1}{3}\right\}$　　**77.** $\left\{-3, \dfrac{3 - 3i\sqrt{3}}{2}, \dfrac{3 + 3i\sqrt{3}}{2}\right\}$　　**79.** answers may vary　　**81. a.** 189.265 feet/sec　**b.** 189.149 feet/sec　**c.** 128.965 mph

Integrated Review　**1.** $\{-\sqrt{10}, \sqrt{10}\}$　　**2.** $\{-2i\sqrt{2}, 2i\sqrt{2}\}$　　**3.** $\{1 - 2\sqrt{2}, 1 + 2\sqrt{2}\}$　　**4.** $\left\{\dfrac{-5 - 2\sqrt{3}}{2}, \dfrac{-5 + 2\sqrt{3}}{2}\right\}$

5. $\{-1 - \sqrt{13}, -1 + \sqrt{13}\}$　　**6.** $\{1, 11\}$　　**7.** $\left\{\dfrac{-3 - \sqrt{17}}{2}, \dfrac{-3 + \sqrt{17}}{2}\right\}$　　**8.** $\left\{\dfrac{-2 - \sqrt{5}}{4}, \dfrac{-2 + \sqrt{5}}{4}\right\}$　　**9.** $\left\{\dfrac{2 - \sqrt{2}}{2}, \dfrac{2 + \sqrt{2}}{2}\right\}$

10. $\{-3 - \sqrt{5}, -3 + \sqrt{5}\}$　　**11.** $\{-2 + i\sqrt{3}, -2 - i\sqrt{3}\}$　　**12.** $\left\{\dfrac{-3 - i\sqrt{6}}{5}, \dfrac{-3 + i\sqrt{6}}{5}\right\}$　　**13.** $\left\{\dfrac{-3 + i\sqrt{15}}{2}, \dfrac{-3 - i\sqrt{15}}{2}\right\}$

14. $\{3i, -3i\}$　　**15.** $\{0, -17\}$　　**16.** $\left\{\dfrac{1 + \sqrt{13}}{4}, \dfrac{1 - \sqrt{13}}{4}\right\}$　　**17.** $\{2 + 3\sqrt{3}, 2 - 3\sqrt{3}\}$　　**18.** $\{2 + \sqrt{3}, 2 - \sqrt{3}\}$　　**19.** $\left\{-2, \dfrac{4}{3}\right\}$

20. $\left\{\dfrac{-5 + \sqrt{17}}{4}, \dfrac{-5 - \sqrt{17}}{4}\right\}$　　**21.** $\{1 - \sqrt{6}, 1 + \sqrt{6}\}$　　**22.** $\{-\sqrt{31}, \sqrt{31}\}$　　**23.** $\{-2\sqrt{3}, 2\sqrt{3}\}$　　**24.** $\{-i\sqrt{11}, i\sqrt{11}\}$

25. $\{-11, 6\}$　　**26.** $\left\{\dfrac{-3 + \sqrt{19}}{5}, \dfrac{-3 - \sqrt{19}}{5}\right\}$　　**27.** $\left\{\dfrac{-3 + \sqrt{17}}{4}, \dfrac{-3 - \sqrt{17}}{4}\right\}$　　**28.** $\{4\}$　　**29.** $\left\{\dfrac{-1 + \sqrt{17}}{8}, \dfrac{-1 - \sqrt{17}}{8}\right\}$

30. $10\sqrt{2}$ ft $\approx$ 14.1 ft　　**31.** Diane: 9.1 hr; Lucy: 7.1 hr　　**32.** 5 mph during the first part, then 6 mph

Exercise Set 8.4　**1.** $(-\infty, -5) \cup (-1, \infty)$　　**3.** $[-4, 3]$　　**5.** $(-\infty, -5] \cup [-3, \infty)$　　**7.** $\left(-5, -\dfrac{1}{3}\right)$　　**9.** $(2, 4) \cup (6, \infty)$

11. $(-\infty, -4] \cup [0, 1]$　　**13.** $(-\infty, -3) \cup (-2, 2) \cup (3, \infty)$　　**15.** $(-7, 2)$　　**17.** $(-1, \infty)$　　**19.** $(-\infty, -1] \cup (4, \infty)$

21. $(-\infty, 2) \cup \left(\dfrac{11}{4}, \infty\right)$　　**23.** $(0, 2] \cup [3, \infty)$　　**25.** $(-\infty, 3)$　　**27.** $\left[-\dfrac{5}{4}, \dfrac{3}{2}\right]$　　**29.** $(-\infty, 0) \cup (1, \infty)$　　**31.** $(0, 10)$

33. $(-\infty, -4] \cup [4, 6]$　　**35.** $\left(-\infty, -\dfrac{2}{3}\right] \cup \left[\dfrac{3}{2}, \infty\right)$　　**37.** $(-\infty, -4) \cup [5, \infty)$　　**39.** $(-\infty, 1) \cup (2, \infty)$　　**41.** $\left(-4, -\dfrac{3}{2}\right) \cup \left(\dfrac{3}{2}, \infty\right)$

43. $(-\infty, -5] \cup [-1, 1] \cup [5, \infty)$　　**45.** $(-\infty, -6] \cup (-1, 0] \cup (7, \infty)$　　**47.** $(-\infty, -8] \cup (-4, \infty)$　　**49.** $\left(-\infty, -\dfrac{5}{3}\right) \cup \left(\dfrac{7}{2}, \infty\right)$

51. $(-\infty, 0] \cup \left(5, \dfrac{11}{2}\right]$　　**53.** $(0, \infty)$　　**55.** 0; 1; 1; 4; 4　　**57.** 0; -1; -1; -4; -4　　**59.** answers may vary

61. $(-\infty, -1) \cup (0, 1)$　　**63.** when x is between 2 and 11　　**65.** $(-\infty, -7) \cup (8, \infty)$

The Bigger Picture　**1.** $\left\{-9, \dfrac{7}{3}\right\}$　　**2.** $(4, 7)$　　**3.** $\{4, 5\}$　　**4.** $\left\{\dfrac{-1 - \sqrt{13}}{6}, \dfrac{-1 + \sqrt{13}}{6}\right\}$　　**5.** $[-2, 7)$　　**6.** $\left\{\dfrac{5}{4}\right\}$　　**7.** $\left\{\dfrac{1}{5}, 7\right\}$

8. $\left(-\infty, -\dfrac{1}{2}\right] \cup [4, \infty)$　　**9.** $(-\infty, -8) \cup (22, \infty)$　　**10.** $(7, \infty)$

Calculator Explorations　**1.**　　　　　　　　**3.**　　　　　　**5.**

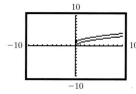　　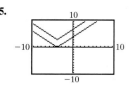

Mental Math　**1.** $(0, 0)$　　**3.** $(2, 0)$　　**5.** $(0, 3)$　　**7.** $(-1, 5)$

Exercise Set 8.5　**1.**　　　　**3.**　　　　**5.**　　　　**7.**

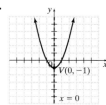　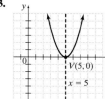　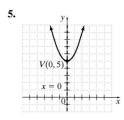　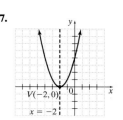

9.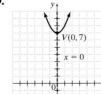

11.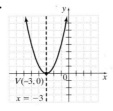

13.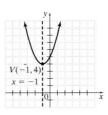

15.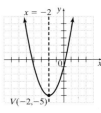

17.

19.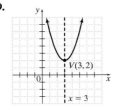

21.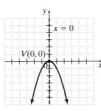

23.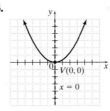

25.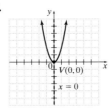

27.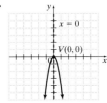

29.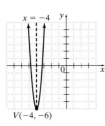

31.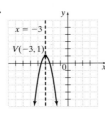

33.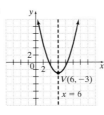

35.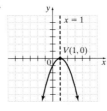

37.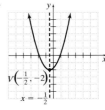

39.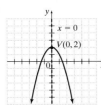

41. $x^2 + 8x + 16$ **43.** $z^2 - 16z + 64$ **45.** $y^2 + y + \dfrac{1}{4}$

47. $g(x) = 5(x - 2)^2 + 3$ **49.** $g(x) = 5(x + 3)^2 + 6$

51.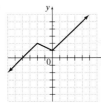

53.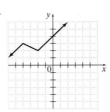

55.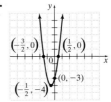

57. c

Exercise Set 8.6   **1.** $(-4, -9)$ **3.** $(5, 30)$ **5.** $(1, -2)$ **7.** $\left(\dfrac{1}{2}, \dfrac{5}{4}\right)$ **9.** D   **11.** B

13.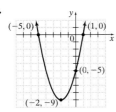

vertex: $(-2, -9)$; opens upward;
x-intercepts: $(-5, 0)$, $(1, 0)$;
y-intercept: $(0, -5)$

15.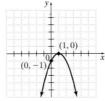

vertex: $(1, 0)$; opens downward;
x-intercept: $(1, 0)$;
y-intercept: $(0, -1)$

17.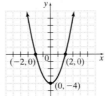

vertex: $(0, -4)$; opens upward;
x-intercepts: $(-2, 0)$, $(2, 0)$;
y-intercept: $(0, -4)$

19.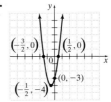

vertex: $\left(-\dfrac{1}{2}, -4\right)$; opens upward;

x-intercepts: $\left(-\dfrac{3}{2}, 0\right)$, $\left(\dfrac{1}{2}, 0\right)$;

y-intercept: $(0, -3)$

21.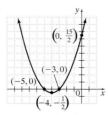

23.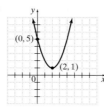

25.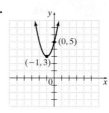

27.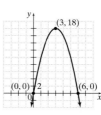

vertex: $\left(-4, -\dfrac{1}{2}\right)$; opens upward;
x-intercepts: $(-5, 0), (-3, 0)$;
y-intercept: $\left(0, \dfrac{15}{2}\right)$

vertex: $(2, 1)$; opens upward;
y-intercept: $(0, 5)$

vertex: $(-1, 3)$; opens upward;
y-intercept: $(0, 5)$

vertex: $(3, 18)$; opens downward;
x-intercepts: $(0, 0), (6, 0)$;
y-intercept: $(0, 0)$

29. 144 ft **31. a.** 200 bicycles **b.** $12,000 **33.** 30, 30 **35.** 5, -5 **37.** length: 20 units; width: 20 units **39.** $(0, 2)$
41. $(-2, 0)$ **43.** $(-5, 2)$ **45.** $(4, 1)$ **47.** minimum value **49.** maximum value
51. vertex: $(-5, -10)$; opens upward; y-intercept: $(0, 15)$; x-intercepts: $(-1.8, 0), (-8.2, 0)$ **55.** -0.84
57. a. maximum, answers may vary **b.** 2003 **c.** 2,220,000 or 2220 thousand

Chapter 8 Vocabulary Check **1.** discriminant **2.** $\pm\sqrt{b}$ **3.** $\dfrac{-b}{2a}$ **4.** quadratic inequality **5.** completing the square **6.** $(0, k)$
7. $(h, 0)$ **8.** (h, k) **9.** quadratic formula **10.** quadratic

Chapter 8 Review **1.** $\{14, 1\}$ **2.** $\left\{-\dfrac{6}{7}, 5\right\}$ **3.** $\{-7, 7\}$ **4.** $\left\{\dfrac{2 - \sqrt{2}}{5}, \dfrac{2 + \sqrt{2}}{5}\right\}$ **5.** $\left\{\dfrac{-3 - \sqrt{5}}{2}, \dfrac{-3 + \sqrt{5}}{2}\right\}$
6. $\left\{\dfrac{-3 - i\sqrt{7}}{8}, \dfrac{-3 + i\sqrt{7}}{8}\right\}$ **7.** 4.25% **8.** $75\sqrt{2}$ mi; 106.1 mi **9.** two complex but not real solutions **10.** two real solutions
11. two real solutions **12.** one real solution **13.** $\{8\}$ **14.** $\{-5, 0\}$ **15.** $\left\{-\dfrac{5}{2}, 1\right\}$ **16.** $\left\{\dfrac{5 - i\sqrt{143}}{12}, \dfrac{5 + i\sqrt{143}}{12}\right\}$
17. $\left\{\dfrac{1 - i\sqrt{35}}{9}, \dfrac{1 + i\sqrt{35}}{9}\right\}$ **18.** $\left\{1, \dfrac{9}{4}\right\}$ **19. a.** 20 ft **b.** $\dfrac{15 + \sqrt{321}}{16}$ sec; 2.1 sec **20.** $(6 + 6\sqrt{2})$ cm
21. $\left\{3, \dfrac{-3 + 3i\sqrt{3}}{2}, \dfrac{-3 - 3i\sqrt{3}}{2}\right\}$ **22.** $\{-4, 2 - 2i\sqrt{3}, 2 + 2i\sqrt{3}\}$ **23.** $\left\{\dfrac{2}{3}, 5\right\}$ **24.** $\{-5, 5, -2i, 2i\}$ **25.** $\left\{-\dfrac{16}{5}, 1\right\}$
26. $\{1, 125\}$ **27.** $\{-1, 1, -i, i\}$ **28.** $\left\{-\dfrac{1}{5}, \dfrac{1}{4}\right\}$ **29.** Jerome: 10.5 hr; Tim: 9.5 hr **30.** -5 **31.** $[-5, 5]$ **32.** $\left(-\dfrac{1}{2}, \dfrac{1}{2}\right)$
33. $(-\infty, -4) \cup (-1, 1) \cup (4, \infty)$ **34.** $(5, 6)$ **35.** $(-\infty, -6) \cup \left(-\dfrac{3}{4}, 0\right) \cup (5, \infty)$ **36.** $(-\infty, -5] \cup [-2, 6]$
37. $(-5, -3) \cup (5, \infty)$ **38.** $(-\infty, 0)$ **39.** $\left(-\dfrac{6}{5}, 0\right) \cup \left(\dfrac{5}{6}, 3\right)$

40.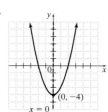

41.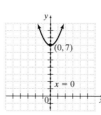

42.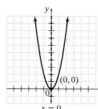

43.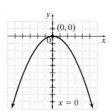

44.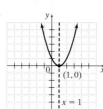

45.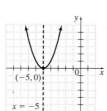

46.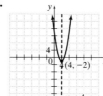

47.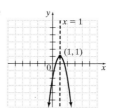

48. vertex: $(-5, 0)$;
x-intercept: $(-5, 0)$;
y-intercept: $(0, 25)$

49. vertex: $(3, 0)$;
x-intercept: $(3, 0)$;
y-intercept: $(0, -9)$

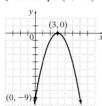

50. vertex: $(0, -1)$;
x-intercepts: $\left(-\frac{1}{2}, 0\right), \left(\frac{1}{2}, 0\right)$;
y-intercept: $(0, -1)$

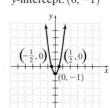

51. vertex: $(0, 5)$;
x-intercepts: $(-1, 0), (1, 0)$;
y-intercept: $(0, 5)$

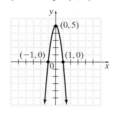

52. vertex: $\left(-\frac{5}{6}, \frac{73}{12}\right)$; opens downward; x-intercepts: $(-2.3, 0), (0.6, 0)$; y-intercept: $(0, 4)$

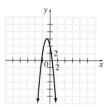

53. a. 0.4 sec and 7.1 sec   **b.** answers may vary   **54.** The numbers are both 210.   **55.** $\{-5, 6\}$ **56.** $\left\{\frac{4}{5}, -\frac{1}{2}\right\}$ **57.** $\{-2, 2\}$

58. $\left\{-\frac{4}{9}, \frac{2}{9}\right\}$ **59.** $\left\{\frac{-1 - 3i\sqrt{3}}{2}, \frac{-1 + 3i\sqrt{3}}{2}\right\}$ **60.** $\left\{\frac{17 - \sqrt{145}}{9}, \frac{17 + \sqrt{145}}{9}\right\}$ **61.** $\{-i\sqrt{11}, i\sqrt{11}\}$

62. $\left\{\frac{21 - \sqrt{41}}{50}, \frac{21 + \sqrt{41}}{50}\right\}$ **63.** $\left\{-\frac{8\sqrt{7}}{7}, \frac{8\sqrt{7}}{7}\right\}$ **64.** $\{8, 64\}$ **65.** $\left(-\infty, -\frac{5}{4}\right] \cup \left[\frac{3}{2}, \infty\right)$ **66.** $[-5, 0] \cup \left(\frac{3}{4}, \infty\right)$

67. $\left(2, \frac{7}{2}\right)$ **68. a.** 37,077,000   **b.** 2010

Chapter 8 Test

1. $\left\{\frac{7}{5}, -1\right\}$ **2.** $\{-1 - \sqrt{10}, -1 + \sqrt{10}\}$ **3.** $\left\{\frac{1 + i\sqrt{31}}{2}, \frac{1 - i\sqrt{31}}{2}\right\}$ **4.** $\{3 - \sqrt{7}, 3 + \sqrt{7}\}$ **5.** $\left\{-\frac{1}{7}, -1\right\}$

6. $\left\{\frac{3 + \sqrt{29}}{2}, \frac{3 - \sqrt{29}}{2}\right\}$ **7.** $\{-2 - \sqrt{11}, -2 + \sqrt{11}\}$ **8.** $\{-3, 3, -i, i\}$ **9.** $\{-1, 1, -i, i\}$ **10.** $\{6, 7\}$ **11.** $\{3 - \sqrt{7}, 3 + \sqrt{7}\}$

12. $\left\{\frac{2 - i\sqrt{6}}{2}, \frac{2 + i\sqrt{6}}{2}\right\}$ **13.** $\left(-\infty, -\frac{3}{2}\right) \cup (5, \infty)$ **14.** $(-\infty, -5) \cup (-4, 4) \cup (5, \infty)$ **15.** $(-\infty, -3) \cup (2, \infty)$

16. $(-\infty, -3) \cup [2, 3)$

17.

18.

19.

20.

21. 7 ft

22. $(5 + \sqrt{17})$ hr ≈ 9.12 hr   **23. a.** 272 ft   **b.** 5.12 sec

Cumulative Review   **1.** 7.3×10^5; Sec. 1.6, Ex. 28   **2.** 0.002068; Sec. 1.6   **3.** $\{0, 1\}$; Sec. 2.6, Ex. 6   **4.** $\varnothing$; Sec. 2.6

5. not a function; Sec. 3.6, Ex. 8   **6.** vertex: $\left(-\frac{1}{2}, -\frac{49}{4}\right)$; y-intercept: $(0, -12)$; x-intercepts: $(3, 0), (-4, 0)$; Sec. 8.6   **7.** $\varnothing$; Sec. 4.1, Ex. 9

8. $(-2, -7)$; Sec. 4.1   **9.** $6x^3 - 4x^2 - 2$; Sec. 5.1, Ex. 16   **10.** $2 - x$; Sec. 6.1   **11.** $6x^2 - 29x + 28$; Sec. 5.2, Ex. 10

12. $28a^2 - 29a + 6$; Sec. 5.2   **13.** $15x^4 - x^2y - 2y^2$; Sec. 5.2, Ex. 11   **14.** $6a^2 - 7ab - 5b^2$; Sec. 5.2, Ex. 10

15. $3x^2 + 2x + 3 + \frac{-6x + 9}{x^2 - 1}$; Sec. 5.3, Ex. 6   **16. a.** $\frac{a^6}{b^3c^9}$ **b.** $\frac{a^8c^6}{b^4}$ **c.** $\frac{16b^6}{a^6}$; Sec. 1.7   **17.** $3x^2(2 + x)$; Sec. 5.4, Ex. 2

18. $3x(3x^2 + 9x - 5)$; Sec. 5.4   **19.** $(2a - 1)(a + 4)$; Sec. 5.5, Ex. 7   **20.** $(3y - 2)(2x - 5)$; Sec. 5.5   **21.** $3x(a - 2b)^2$; Sec. 5.6, Ex. 4

22. $(2x - 1)(y + 3)$; Sec. 5.4   **23.** $\left\{-\frac{1}{6}, 3\right\}$; Sec. 5.7, Ex. 5   **24.** $\left\{\frac{1}{2}\right\}$; Sec. 5.7   **25.** $\frac{n - 2}{n(n - 1)}$; Sec. 6.1, Ex. 9   **26.** $-2 - 6\sqrt{3}$; Sec. 7.4

27. $-5x^2$; Sec. 6.1, Ex. 10   **28.** $\frac{1}{x - 2}$; Sec. 6.2   **29.** $\frac{6x + 5}{3x^2}$; Sec. 6.2, Ex. 6   **30.** $\frac{a^2 + 8a - 2}{(a - 2)(a - 4)(a + 4)}$; Sec. 6.2   **31.** $\frac{x^2}{y^2}$; Sec. 6.3, Ex. 4

32. $\frac{2a + b}{a + 2b}$; Sec. 6.3   **33.** $\{-6, -1\}$; Sec. 6.4, Ex. 5   **34.** $\{-1\}$; Sec. 6.4   **35.** $x = \frac{yz}{y - z}$; Sec. 6.5, Ex. 1   **36.** $x^2 - 6x + 8$; Sec. 5.3

37. $S = krh$; Sec. 6.6, Ex. 5 **38.** 192; $y = \dfrac{192}{x}$; Sec. 6.6 **39.** -4; Sec. 7.1, Ex. 14 **40.** $2a^2b^3\sqrt[3]{3a}$; Sec. 7.1 **41.** $\dfrac{2}{5}$; Sec. 7.1, Ex. 15

42. $\dfrac{8\sqrt{x}}{3x}$; Sec. 7.5 **43.** 2; Sec. 7.3, Ex. 3 **44.** $5 - 2x\sqrt{5} + x^2$; Sec. 7.4 **45.** $\sqrt{\dfrac{2b}{3a}}$; Sec. 7.3, Ex. 5 **46.** $a - b^2$; Sec. 7.4

47. $-6\sqrt[3]{2}$; Sec. 7.4, Ex. 5 **48.** $\{2, 6\}$; Sec. 7.6 **49.** $\dfrac{5\sqrt[3]{7x}}{2}$; Sec. 7.4, Ex. 10 **50.** $2x^2\sqrt[3]{3x}$; Sec. 7.3

51. $\{-1 + \sqrt{5}, -1 - \sqrt{5}\}$; Sec. 8.1, Ex. 7 **52.** $\{1 + 2\sqrt{6}, 1 - 2\sqrt{6}\}$; Sec. 8.1 **53.** $\left\{\dfrac{2 + \sqrt{10}}{2}, \dfrac{2 - \sqrt{10}}{2}\right\}$; Sec. 8.2, Ex. 2

54. $\{2 - 2\sqrt{3}, 2 + 2\sqrt{3}\}$; Sec. 8.2 **55.** 3.9 sec; Sec. 8.2, Ex. 9 **56.** 15 hours; Sec. 6.5

CHAPTER 9 Exponential and Logarithmic Functions

Exercise Set 9.1 **1. a.** $3x - 6$ **b.** $-x - 8$ **c.** $2x^2 - 13x - 7$ **d.** $\dfrac{x - 7}{2x + 1}$ where $x \neq -\dfrac{1}{2}$ **3. a.** $x^2 + 5x + 1$ **b.** $x^2 - 5x + 1$ **c.** $5x^3 + 5x$

d. $\dfrac{x^2 + 1}{5x}$ where $x \neq 0$ **5. a.** $\sqrt{x} + x + 5$ **b.** $\sqrt{x} - x - 5$ **c.** $x\sqrt{x} + 5\sqrt{x}$ **d.** $\dfrac{\sqrt{x}}{x + 5}$ where $x \neq -5$ **7. a.** $5x^2 - 3x$ **b.** $-5x^2 - 3x$

c. $-15x^3$ **d.** $-\dfrac{3}{5x}$ where $x \neq 0$ **9.** 42 **11.** -18 **13.** 0 **15.** $(f \circ g)(x) = 25x^2 + 1$; $(g \circ f)(x) = 5x^2 + 5$

17. $(f \circ g)(x) = 2x + 11$; $(g \circ f)(x) = 2x + 4$ **19.** $(f \circ g)(x) = -8x^3 - 2x - 2$; $(g \circ f)(x) = -2x^3 - 2x + 4$

21. $(f \circ g)(x) = |10x - 3|$; $(g \circ f)(x) = 10|x| - 3$ **23.** $(f \circ g)(x) = \sqrt{-5x + 2}$; $(g \circ f)(x) = -5\sqrt{x} + 2$ **25.** $H(x) = (g \circ h)(x)$

27. $F(x) = (h \circ f)(x)$ **29.** $G(x) = (f \circ g)(x)$ **31.** answers may vary **33.** answers may vary **35.** answers may vary

37. $y = x - 2$ **39.** $y = \dfrac{x}{3}$ **41.** $y = -\dfrac{x + 7}{2}$ **43.** $P(x) = R(x) - C(x)$ **45.** answers may vary

Exercise Set 9.2 **1.** not one-to-one **3.** one-to-one; $h^{-1} = \{(10, 10)\}$ **5.** one-to-one; $f^{-1} = \{(12, 11), (3, 4), (4, 3), (6, 6)\}$

7. not one-to-one **9.** one-to-one;

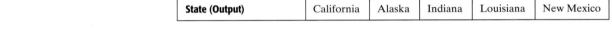

Rank in Population (Input)	1	47	14	24	36
State (Output)	California	Alaska	Indiana	Louisiana	New Mexico

11. a. 3 **b.** 1 **13. a.** 1 **b.** -1 **15.** one-to-one **17.** not one-to-one **19.** one-to-one **21.** not one-to-one

23. 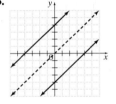 **25.** 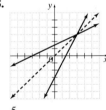 **27.** 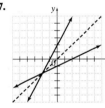 **29.** 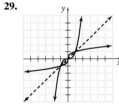 **31.** $f^{-1}(x) = 5x + 2$

33. $f^{-1}(x) = x^3$ **35.** $f^{-1}(x) = \dfrac{\dfrac{5}{x} - 1}{3}$ **37.** $f^{-1}(x) = \sqrt[3]{x} - 2$ **39.** 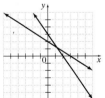 **41.** 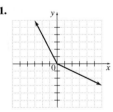 **43.** 5 **45.** 8

47. $\dfrac{1}{27}$ **49.** 9 **51.** $3^{1/2} \approx 1.73$ **53. a.** $(2, 9)$ **b.** $(9, 2)$ **55. a.** $\left(-2, \dfrac{1}{4}\right), \left(-1, \dfrac{1}{2}\right), (0, 1), (1, 2), (2, 5)$

b. $\left(\dfrac{1}{4}, -2\right), \left(\dfrac{1}{2}, -1\right), (1, 0), (2, 1)(5, 2)$ **c.** 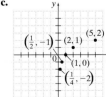 **d.** 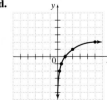 **57.** answers may vary

59. 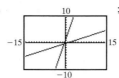 ; $f^{-1}(x) = \dfrac{x-1}{3}$ **61.** 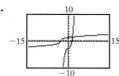 ; $f^{-1}(x) = x^3 - 3$

Calculator Explorations **1.** 81.98% **3.** 22.54%

Exercise Set 9.3 **1.** 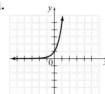 **3.** 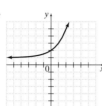 **5.** 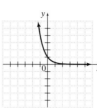 **7.**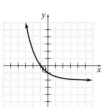

9. 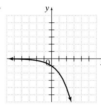 **11.** 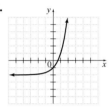 **13.** 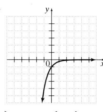 **15.** 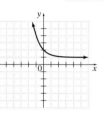 **17.**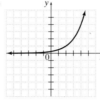

19. 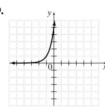 **21.** $\{3\}$ **23.** $\left\{\dfrac{3}{4}\right\}$ **25.** $\left\{\dfrac{8}{5}\right\}$ **27.** $\left\{-\dfrac{2}{3}\right\}$ **29.** $\left\{\dfrac{3}{2}\right\}$ **31.** $\left\{-\dfrac{1}{3}\right\}$ **33.** $\{-2\}$ **35.** $\{4\}$

37. 24.6 lb **39. a.** 568.7 millibars **b.** 9.1 millibars **c.** answers may vary **41. a.** $56.8 billion **b.** $261.6 billion
43. 537.6 million cellular phone users **45.** $7621.42 **47.** $\{4\}$ **49.** $\varnothing$ **51.** no **53.** no **55.** C **57.** D **59.** answers may vary
61. 24.6 lb **63.** 18.62 lb

Exercise Set 9.4 **1.** $6^2 = 36$ **3.** $3^{-3} = \dfrac{1}{27}$ **5.** $10^3 = 1000$ **7.** $e^4 = x$ **9.** $e^{-2} = \dfrac{1}{e^2}$ **11.** $7^{1/2} = \sqrt{7}$ **13.** $0.7^3 = 0.343$

15. $3^{-4} = \dfrac{1}{81}$ **17.** $\log_2 16 = 4$ **19.** $\log_{10} 100 = 2$ **21.** $\log_e x = 3$ **23.** $\log_{10} \dfrac{1}{10} = -1$ **25.** $\log_4 \dfrac{1}{16} = -2$ **27.** $\log_5 \sqrt{5} = \dfrac{1}{2}$

29. 3 **31.** -2 **33.** $\dfrac{1}{2}$ **35.** -1 **37.** 0 **39.** 2 **41.** 4 **43.** -3 **45.** $\{2\}$ **47.** $\{81\}$ **49.** $\{7\}$ **51.** $\{-3\}$

53. $\{-3\}$ **55.** $\{2\}$ **57.** $\{2\}$ **59.** $\left\{\dfrac{27}{64}\right\}$ **61.** $\{10\}$ **63.** $\{4\}$ **65.** $\{5\}$ **67.** $\left\{\dfrac{1}{49}\right\}$ **69.** 3 **71.** 3 **73.** 1

75. 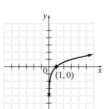 **77.** 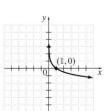 **79.** 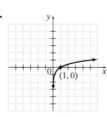 **81.** 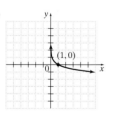 **83.** 1 **85.** $\dfrac{x-4}{2}$

87. a. $g(2) = 25$ **b.** $(25, 2)$ **c.** $f(25) = 2$ **89.** answers may vary **91.** 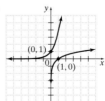 **93.**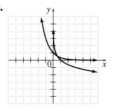

95. answers may vary **97.** 0.0827

Exercise Set 9.5 **1.** $\log_5 14$ **3.** $\log_4 9x$ **5.** $\log_6(x^2 + x)$ **7.** $\log_{10}(10x^2 + 20)$ **9.** $\log_5 3$ **11.** $\log_3 4$ **13.** $\log_2\dfrac{x}{y}$ **15.** $\log_2\dfrac{x^2 + 6}{x^2 + 1}$

17. $2\log_3 x$ **19.** $-1\log_4 5 = -\log_4 5$ **21.** $\dfrac{1}{2}\log_5 y$ **23.** $\log_2 5x^3$ **25.** $\log_4 48$ **27.** $\log_5 x^3 z^6$ **29.** $\log_4 4$, or 1 **31.** $\log_7\dfrac{9}{2}$

33. $\log_{10}\dfrac{x^3 - 2x}{x + 1}$ **35.** $\log_2\dfrac{x^{7/2}}{(x + 1)^2}$ **37.** $\log_8 x^{16/3}$ **39.** $\log_3 4 + \log_3 y - \log_3 5$ **41.** $\log_4 2 - \log_4 9 - \log_4 z$ **43.** $3\log_2 x - \log_2 y$

45. $\dfrac{1}{2}\log_b 7 + \dfrac{1}{2}\log_b x$ **47.** $4\log_6 x + 5\log_6 y$ **49.** $3\log_5 x + \log_5(x + 1)$ **51.** $2\log_6 x - \log_6(x + 3)$ **53.** 0.2 **55.** 1.2 **57.** 0.35

59. 1.29 **61.** -0.68 **63.** -0.125 **65.** **67.** -1 **69.** $\dfrac{1}{2}$ **71.** a and d **73.** false **75.** true **77.** false

Integrated Review **1.** $x^2 + x - 5$ **2.** $-x^2 + x - 7$ **3.** $x^3 - 6x^2 + x - 6$ **4.** $\dfrac{x - 6}{x^2 + 1}$ **5.** $\sqrt{3x - 1}$ **6.** $3\sqrt{x} - 1$

7. one-to-one; $\{(6, -2), (8, 4), (-6, 2), (3, 3)\}$, **8.** not one-to-one **9.** not one-to-one **10.** one-to-one **11.** not one-to-one

12. $f^{-1}(x) = \dfrac{x}{3}$ **13.** $f^{-1}(x) = x - 4$ **14.** $f^{-1}(x) = \dfrac{x + 1}{5}$ **15.** $f^{-1}(x) = \dfrac{x - 2}{3}$ **16.**

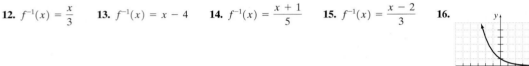

17. **18.** **19.** **20.** $\{3\}$ **21.** $\{7\}$ **22.** $\{-8\}$ **23.** $\{3\}$

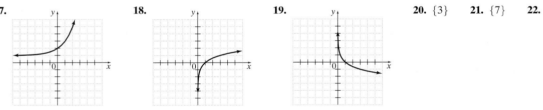

24. $\{2\}$ **25.** $\left\{\dfrac{1}{2}\right\}$ **26.** $\{32\}$ **27.** $\{4\}$ **28.** $\{5\}$ **29.** $\left\{\dfrac{1}{9}\right\}$ **30.** $\log_2 14x$ **31.** $\log_2 5^x \cdot 8$ **32.** $\log_5\dfrac{x^3}{y^5}$ **33.** $\log_5 x^9 y^3$

34. $\log_2\dfrac{x^2 - 3x}{x^2 + 4}$ **35.** $\log_3\dfrac{y^4 + 11y}{y + 2}$ **36.** $\log_7 9 + 2\log_7 x - \log_7 y$ **37.** $\log_6 5 + \log_6 y - 2\log_6 z$

Exercise Set 9.6 **1.** 0.9031 **3.** 0.3636 **5.** 0.6931 **7.** -2.6367 **9.** 1.1004 **11.** 1.6094 **13.** 1.6180 **15.** 2 **17.** -3 **19.** 2

21. $\dfrac{1}{4}$ **23.** 3 **25.** 3.1 **27.** -4 **29.** $\dfrac{1}{2}$ **31.** $\{10^{1.3}\}$; $\{19.9526\}$ **33.** $\{e^{1.4}\}$; $\{4.0552\}$ **35.** $\{10^{2.3}\}$; $\{199.5262\}$

37. $\{e^{-2.3}\}$; $\{0.1003\}$ **39.** $\left\{\dfrac{10^{1.1}}{2}\right\}$; $\{6.2946\}$ **41.** $\left\{\dfrac{e^{0.18}}{4}\right\}$; $\{0.2993\}$ **43.** $\left\{\dfrac{4 + e^{2.3}}{3}\right\}$; $\{4.6581\}$ **45.** $\left\{\dfrac{10^{-0.5} - 1}{2}\right\}$; $\{-0.3419\}$

47. $\$3656.38$ **49.** $\$2542.50$ **51.** 1.5850 **53.** 0.8617 **55.** 1.5850 **57.** -1.6309 **59.** -2.3219 **61.** $\left\{\dfrac{4}{7}\right\}$ **63.** $x = \dfrac{3y}{4}$

65. $\{-6, -1\}$ **67.** answers may vary **69.** **71.** **73.** answers may vary **75.** 4.2 **77.** 5.3

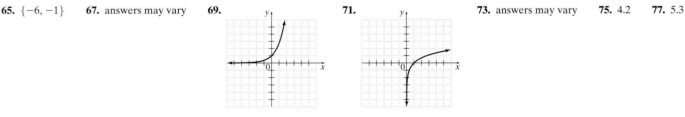

Calculator Explorations **1.** 3.67 yr, or 3 yr and 8 mo **3.** 23.16 yr, or 23 yr and 2 mo

Exercise Set 9.7 **1.** $\left\{\dfrac{\log 6}{\log 3}\right\}$; $\{1.6309\}$ **3.** $\left\{\dfrac{\log 5}{\log 9}\right\}$; $\{0.7325\}$ **5.** $\left\{\dfrac{\log 3.8}{2\log 3}\right\}$; $\{0.6076\}$ **7.** $\left\{\dfrac{\ln 5}{6}\right\}$; $\{0.2682\}$ **9.** $\left\{3 + \dfrac{\log 5}{\log 2}\right\}$; $\{5.3219\}$

11. $\left\{\dfrac{\log 3}{\log 4} - 7\right\}$; $\{-6.2075\}$ **13.** $\left\{\dfrac{1}{3}\left(4 + \dfrac{\log 11}{\log 7}\right)\right\}$; $\{1.7441\}$ **15.** $\{11\}$ **17.** $\left\{\dfrac{1}{2}\right\}$ **19.** $\left\{\dfrac{3}{4}\right\}$ **21.** $\{-2, 1\}$ **23.** $\{2\}$ **25.** $\left\{\dfrac{1}{8}\right\}$

27. $\{4, -1\}$ **29.** $\left\{\dfrac{2}{3}\right\}$ **31.** 103 wolves **33.** 7,192,916 people **35.** 27.1 years **37.** 9.9 yr **39.** 1.7 yr **41.** 8.8 yr **43.** 55.7 in.

45. 11.9 lb per sq in. **47.** 3.2 mi **49.** 12 weeks **51.** 18 weeks **53.** $-\dfrac{5}{3}$ **55.** $\dfrac{17}{4}$ **57.** 4.5% **59.** answers may vary

61. 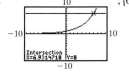 ; {6.93}

The Bigger Picture **1.** $\left\{-\dfrac{3}{2}\right\}$ **2.** $\left\{\dfrac{\log 5}{\log 11} \approx 0.6712\right\}$ **3.** $[-5, \infty)$ **4.** $\left[-\dfrac{13}{3}, -2\right]$ **5.** $\left(-\dfrac{6}{5}, 0\right)$ **6.** $\left\{6, -\dfrac{1}{5}\right\}$

7. $\left\{\dfrac{21}{8}\right\}$ **8.** $\left\{-\dfrac{7}{3}, 3\right\}$ **9.** $(-\infty, \infty)$ **10.** $\{-2, 2\}$ **11.** $\left\{-5 + \sqrt{3}, -5 - \sqrt{3}\right\}$ **12.** $\left\{-\dfrac{1}{4}, 7\right\}$

Chapter 9 Vocabulary Check **1.** inverse **2.** composition **3.** exponential **4.** symmetric **5.** natural **6.** common
7. vertical; horizontal **8.** logarithmic

Chapter 9 Review **1.** $3x - 4$ **2.** $-x - 6$ **3.** $2x^2 - 9x - 5$ **4.** $\dfrac{2x + 1}{x - 5}$ where $x \neq 5$ **5.** $x^2 + 2x - 1$ **6.** $x^2 - 1$ **7.** 18

8. $x^4 - 4x^2 + 2$ **9.** -2 **10.** 48 **11.** one-to-one; $h^{-1} = \{(14, -9), (8, 6), (12, -11), (15, 15)\}$ **12.** not one-to-one

13. one-to-one;

Rank in Automobile Thefts (Input)	2	4	1	3
U.S. Region (Output)	West	Midwest	South	Northeast

14. not one-to-one **15.** not one-to-one **16.** not one-to-one **17.** not one-to-one **18.** one-to-one **19.** $f^{-1}(x) = \dfrac{x - 11}{6}$

20. $f^{-1}(x) = \dfrac{x}{12}$ **21.** $f^{-1}(x) = \dfrac{x + 5}{3}$ **22.** $f^{-1}(x) = \dfrac{x - 1}{2}$ **23.** **24.**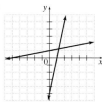

25. $\{3\}$ **26.** $\left\{-\dfrac{4}{3}\right\}$ **27.** $\left\{\dfrac{3}{2}\right\}$ **28.** 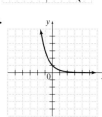 **29.** 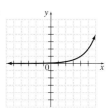 **30.**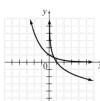

31. 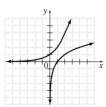 **32.** $2963.11 **33.** $1131.82 **34.** $\log_7 49 = 2$ **35.** $\log_2 \dfrac{1}{16} = -4$ **36.** $\left(\dfrac{1}{2}\right)^{-4} = 16$

37. $0.4^3 = 0.064$ **38.** $\left\{\dfrac{1}{64}\right\}$ **39.** $\{9\}$ **40.** $\{0\}$ **41.** $\{8\}$ **42.** $\{5\}$ **43.** $\{-2\}$ **44.** $\{4\}$ **45.** $\{9\}$ **46.** $\{2\}$ **47.** $\{-8, 1\}$

48. 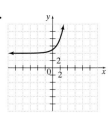 **49.** 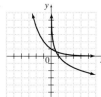 **50.** $\log_3 32$ **51.** $\log_2 18$ **52.** $\log_7 \dfrac{3}{4}$ **53.** $\log_e \dfrac{3}{2}$ **54.** $\log_{11} 4$ **55.** $\log_5 2$

56. $\log_5 \dfrac{x^3}{(x + 1)^2}$ **57.** $\log_3 (x^4 + 2x^3)$ **58.** $3\log_3 x - \log_3 (x + 2)$ **59.** $\log_4 (x + 5) - 2\log_4 x$ **60.** $\log_2 3 + 2\log_2 x + \log_2 y - \log_2 z$

61. $\log_7 y + 3\log_7 z - \log_7 x$ **62.** 2.02   **63.** -0.11 **64.** 0.5563   **65.** -0.8239 **66.** 0.2231   **67.** 1.5326   **68.** 3   **69.** -1

70. -1 **71.** 4   **72.** $\{50\}$ **73.** $\left\{\dfrac{e^{1.6}}{3}\right\}$; $\{1.6510\}$ **74.** $\left\{\dfrac{e^{-1}+3}{2}\right\}$; $\{1.6839\}$ **75.** $\left\{\dfrac{e^2-1}{3}\right\}$; $\{2.1297\}$ **76.** 0.2920   **77.** 1.2619

78. $1957.30   **79.** $1307.51   **80.** $\left\{\dfrac{\log 20}{\log 7}\right\}$; $\{1.5395\}$ **81.** $\left\{\dfrac{\log 7}{2\log 3}\right\}$; $\{0.8856\}$ **82.** $\left\{\dfrac{1}{2}\left(\dfrac{\log 6}{\log 3}-1\right)\right\}$; $\{0.3155\}$

83. $\left\{\dfrac{1}{4}\left(\dfrac{\log 3}{\log 8}+2\right)\right\}$; $\{0.6321\}$ **84.** $\left\{\dfrac{25}{2}\right\}$ **85.** $\varnothing$ **86.** $\{2\sqrt{2}\}$ **87.** $\{9, -1\}$ **88.** 197,044 ducks   **89.** 6,871,727   **90.** 18.3 years

91. 115.5 years   **92.** 33.0 yr   **93.** 8.8 yr   **94.** 8.5 yr   **95.** $\{-2\}$ **96.** $\left\{\dfrac{3}{2}\right\}$ **97.** $\left\{\dfrac{8}{9}\right\}$ **98.** $\{3\}$ **99.** $\{3\}$ **100.** $\{-1, 4\}$

101. $\left\{\dfrac{17}{3}\right\}$ **102.** $\{e^{-1.2}\}$ **103.** $\left\{\dfrac{9}{10}\right\}$ **104.** $\left\{\dfrac{3e^2}{e^2-3}\right\}$ **105.** $\varnothing$

Chapter 9 Test   **1.** $2x^2 - 3x$ **2.** $3 - x$ **3.** 5   **4.** $x - 7$ **5.** $x^2 - 6x - 2$ **6.**   **7.** one-to-one

8. not a function   **9.** one-to-one; $f^{-1}(x) = \dfrac{-x + 6}{2}$ **10.** one-to-one; $f^{-1} = \{(0, 0), (3, 2),(5, -1)\}$ **11.** not one-to-one   **12.** $\log_3 24$

13. $\log_5 \dfrac{x^4}{x + 1}$ **14.** $\log_6 2 + \log_6 x - 3\log_6 y$ **15.** -1.53 **16.** 1.0686   **17.** $\{-1\}$ **18.** $\left\{\dfrac{1}{2}\left(\dfrac{\log 4}{\log 3} - 5\right)\right\}$; $\{-1.8691\}$

19. $\left\{\dfrac{1}{9}\right\}$ **20.** $\left\{\dfrac{1}{2}\right\}$ **21.** $\{22\}$ **22.** $\left\{\dfrac{25}{3}\right\}$ **23.** $\left\{\dfrac{43}{21}\right\}$ **24.** $\{-1.0979\}$ **25.** 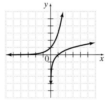   **26.**

27. $5234.58   **28.** 6 yr   **29.** 64,913 prairie dogs   **30.** 15 yr   **31.** 85%   **32.** 52%

Cumulative Review   **1.** 3; Sec. 1.4, Ex. 42   **2.** $4x^2\sqrt[3]{xy^2}$; Sec. 7.2   **3.** 2; Sec. 1.4, Ex. 44   **4.** $3ab\sqrt[4]{2b}$; Sec. 7.2

5. $30°, 40°$, and $110°$; Sec. 4.3, Ex. 6   **6.** $x = 110, y = 70$; Sec. 4.3   **7.** $(x^2 + 5y)(7x - 1)$; Sec. 5.4, Ex. 7   **8.** $\{11\}$; Sec. 2.1

9. $\dfrac{5k^2 - 7k + 4}{(k + 2)(k - 2)(k - 1)}$; Sec. 6.2, Ex. 9   **10.** $\dfrac{-x^2 + 23x + 38}{(x - 2)(x + 2)^2}$; Sec. 6.2   **11.** 2; Sec. 7.1, Ex. 30   **12.** $2\sqrt{6}$; Sec. 7.2   **13.** 1; Sec. 7.1, Ex. 32

14. $\sqrt[3]{2}$; Sec. 7.1   **15.** $x^{5/6}$; Sec. 7.2, Ex. 14   **16.** $\dfrac{a^4c^8}{b^6}$; Sec. 1.7   **17.** 32; Sec. 7.2, Ex. 18   **18.** $729x^3$; Sec. 1.7   **19.** $\dfrac{\sqrt{x}}{3}$; Sec. 7.3, Ex. 7

20. $2a$; Sec. 7.3   **21.** $\dfrac{\sqrt[4]{3}}{2y}$; Sec. 7.3, Ex. 9   **22.** $9ab^2\sqrt[3]{a^2}$; Sec. 7.3   **23.** $2x - 25$; Sec. 7.4, Ex. 13   **24.** $\dfrac{1}{4}x^2 - 9$; Sec. 5.4

25. $4 - 2\sqrt{3}$; Sec. 7.4, Ex. 14   **26.** **a.** $a - a^2$ **b.** $x + 2x^{1/2} - 15$; Sec. 7.2   **27.** $\dfrac{7}{3\sqrt{35}}$; Sec. 7.5, Ex. 9   **28.** $\dfrac{3\sqrt[3]{m^2n}}{m^2n^3}$; Sec. 7.5

29. $\{3\}$; Sec. 7.6, Ex. 4   **30.** $f(x) = \dfrac{x + 20}{3}$; Sec. 3.4   **31.** $-1 + 5i$; Sec. 7.7, Ex. 8   **32.** **a.** 1 **b.** i **c.** -1 **d.** $-i$; Sec. 7.7

33. $\left\{\dfrac{6 + i\sqrt{5}}{2}, \dfrac{6 - i\sqrt{5}}{2}\right\}$; Sec. 8.1, Ex. 9   **34.** $\left\{\dfrac{-2 + \sqrt{5}}{2}, \dfrac{-2 - \sqrt{5}}{2}\right\}$; Sec. 8.1   **35.** $\{2 + \sqrt{2}, 2 - \sqrt{2}\}$; Sec. 8.2, Ex. 3

36. $\left\{\dfrac{3 - \sqrt{5}}{4}, \dfrac{3 + \sqrt{5}}{4}\right\}$; Sec. 8.2   **37.** $\{8, 27\}$; Sec. 8.3, Ex. 5   **38.** $\left\{-\dfrac{7}{5}\right\}$; Sec. 6.4   **39.** $\left(-\dfrac{7}{2}, -1\right)$; Sec. 8.4, Ex. 5   **40.** $4\sqrt{3}$ in.; Sec. 7.6

41. ; Sec. 8.6, Ex. 2   **42.** $x^2 + 2x + 4$; Sec. 5.3

43. **a.** $(f \circ g)(x) = |x - 2|$ **b.** $(g \circ f)(x) = |x| - 2$; Sec. 9.1, Ex. 3   **44.** **a.** $\dfrac{2a}{a - 1}$ **b.** $\dfrac{-3(a + 6)}{4(a - 3)}$ **c.** $\dfrac{y + x}{x^2y^2}$; Sec. 6.3

45. $f^{-1}(x) = \dfrac{x+5}{3}$; Sec. 9.2, Ex. 9   **46.** $8x + 4 + \dfrac{1}{x-2}$; Sec. 5.3   **47.** $\{6\}$; Sec. 9.3, Ex. 6   **48.** $k = \dfrac{1}{24}$; $y = \dfrac{1}{24}x$; Sec. 6.6

49. $\{5\}$; Sec. 9.4, Ex. 9   **50. a.** $\dfrac{11\sqrt{5}}{12}$ **b.** $\dfrac{\sqrt[3]{3x}}{6}$; Sec. 7.4   **51.** \$2509.30; Sec. 9.6, Ex. 11   **52.** $\log_3 x^7 y^9$; Sec. 9.5

53. $\left\{\dfrac{2}{99}\right\}$; Sec. 9.7, Ex. 4   **54.** $\left\{\dfrac{55}{8}\right\}$; Sec. 9.7

CHAPTER 10   Conic Sections

Calculator Explorations 1. 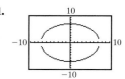 **3.**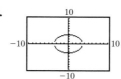

Mental Math 1. upward   **3.** to the left   **5.** downward

Exercise Set 10.1 1. 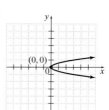 **3.** 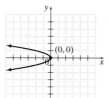 **5.** 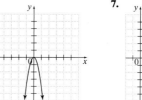 **7.**

9. 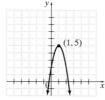 **11.** 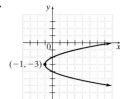 **13.** 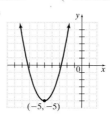 **15.**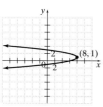

17. 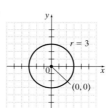 **19.** 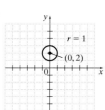 **21.** 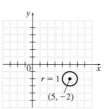 **23.**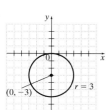

25. 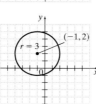 **27.** 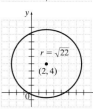 **29.** 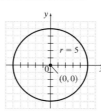 **31.**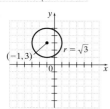

33. $(x-2)^2 + (y-3)^2 = 36$ **35.** $x^2 + y^2 = 3$ **37.** $(x+5)^2 + (y-4)^2 = 45$

39. 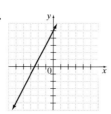 **41.** 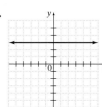   **43.** $\dfrac{\sqrt{3}}{3}$ **45.** $\dfrac{2\sqrt{42}}{3}$

47. a. 16.5 meters **b.** 103.67 meters **c.** 3.5 meters apart **d.** $(0, 16.5)$ **e.** $x^2 + (y - 16.5)^2 = (16.5)^2$

49. a. 125 ft **b.** 14 ft **c.** 139 ft **d.** $(0, 139)$ **e.** $x^2 + (y - 139)^2 = 125^2$ **51.** answers may vary   **53.** 20 m   **55.** $y = -\dfrac{2}{125}x^2 + 40$

Calculator Explorations   1. 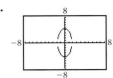   **3.**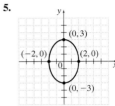

Mental Math   1. ellipse   **3.** hyperbola   **5.** hyperbola

Exercise Set 10.2   1. 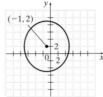   **3.** 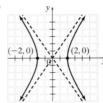   **5.** 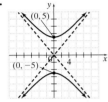   **7.**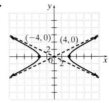

9. 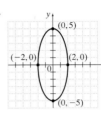   **11.** 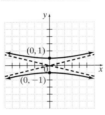   **13.** 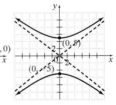   **15.** 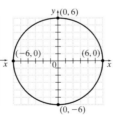   **17.**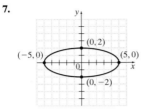

19. 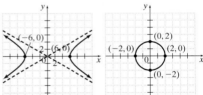   **21.** 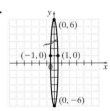   **23.** 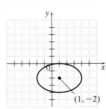   **25.** 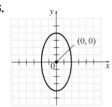   **27.**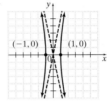

29. $-8x^5$ **31.** $-4x^2$ **33.** y-intercepts; 2 units   **35.** y-intercepts; 4 units   **37.** $\dfrac{x^2}{25} + \dfrac{y^2}{25} = 1$; ellipse; when $a = b$

39. A: 36, 13; B: 4, 4; C: 25, 16; D: 39, 25; E: 17, 81; F: 36, 36; G: 16, 65; H: 144, 140   **41.** A: 6; B: 2; C: 5; D: 5; E: 9; F: 6; G: 4; H: 12

43. greater than 0 and less than 1   **45.** greater than 1

47.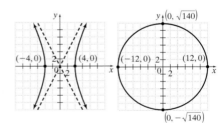

Integrated Review   1. circle 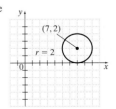   **2.** parabola 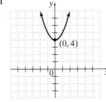   **3.** parabola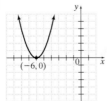

4. ellipse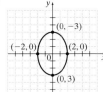

5. hyperbola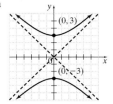

6. hyperbola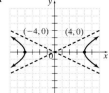

7. ellipse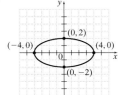

8. circle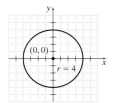

9. parabola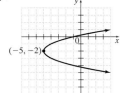

10. parabola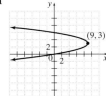

11. hyperbola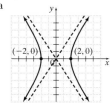

12. ellipse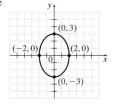

13. ellipse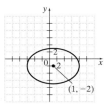

14. hyperbola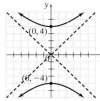

15. circle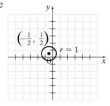

Exercise Set 10.3 **1.** $\{(3,-4),(-3,4)\}$ **3.** $\{(\sqrt{2},\sqrt{2}),(-\sqrt{2},-\sqrt{2})\}$ **5.** $\{(4,0),(0,-2)\}$

7. $\{(-\sqrt{5},-2),(-\sqrt{5},2),(\sqrt{5},-2),(\sqrt{5},2)\}$ **9.** $\varnothing$ **11.** $\{(1,-2),(3,6)\}$ **13.** $\{(2,4),(-5,25)\}$ **15.** $\varnothing$ **17.** $\{(1,-3)\}$
19. $\{(-1,-2),(-1,2),(1,-2),(1,2)\}$ **21.** $\{(0,-1)\}$ **23.** $\{(-1,3),(1,3)\}$ **25.** $\{(\sqrt{3},0),(-\sqrt{3},0)\}$ **27.** $\varnothing$
29. $\{(-6,0),(6,0),(0,-6)\}$ **31.** $\{(3\sqrt{3})\}$ **33.** 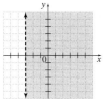 **35.** 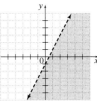 **37.** answers may vary

39. $0, 1, 2, 3,$ or 4 **41.** -9 and 7; 9 and -7; 9 and 7; -9 and -7 **43.** 15 cm by 19 cm **45.** 15 thousand compact discs; price: $3.75

Exercise Set 10.4 **1.** 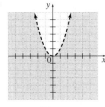 **3.** **5.** 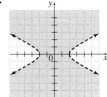 **7.**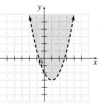

9. 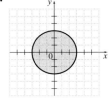 **11.** 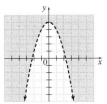 **13.** 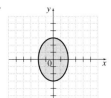 **15.**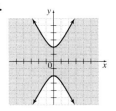

17.
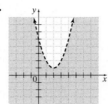

19.
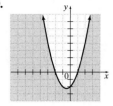

21.
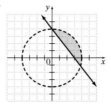

23.
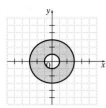

25.
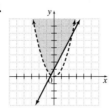

27.
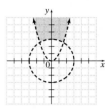

29.
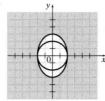

31.
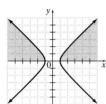

33.
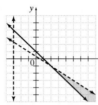

35.
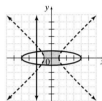

37. not a function   **39.** function   **41.** answers may vary

43.

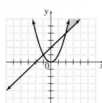

Chapter 10 Vocabulary Check   **1.** circle; center   **2.** nonlinear system of equations   **3.** ellipse   **4.** radius   **5.** hyperbola

Chapter 10 Review   **1.** $(x + 4)^2 + (y - 4)^2 = 9$ **2.** $(x + 7)^2 + (y + 9)^2 = 11$ **3.** $(x - 5)^2 + y^2 = 25$ **4.** $x^2 + y^2 = \dfrac{49}{4}$

5.
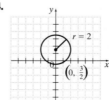

6.
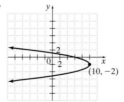

7.
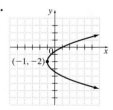

8.
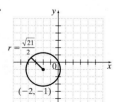

9.
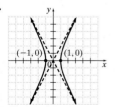

10.
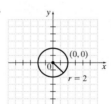

11.
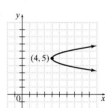

12.
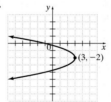

13.
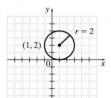

14.

15.

16.
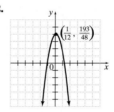

17.
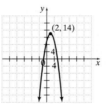

18.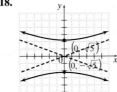

19.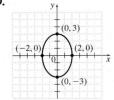

20.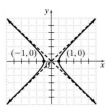

21.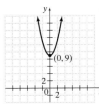

22.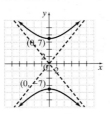

23.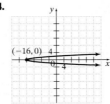

24.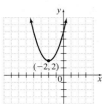

25.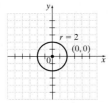

26.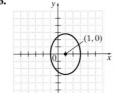

27.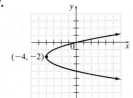

28.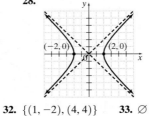

29.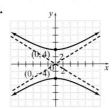

30.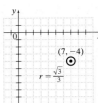

31.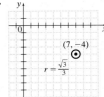

32. $\{(1, -2), (4, 4)\}$ **33.** $\varnothing$ **34.** $\{(-1, 1), (2, 4)\}$

35. $\{(2, 2\sqrt{2}), (2, -2\sqrt{2})\}$ **36.** $\{(0, 2), (0, -2)\}$ **37.** $\left\{\left(2, \dfrac{5}{2}\right), (-7, -20)\right\}$ **38.** $\{(1, 4)\}$ **39.** $\{(-2, -1), (-2, 1), (2, -1), (2, 1)\}$

40. length: 15 ft; width: 10 ft **41.** 4 **42.**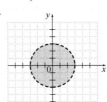

43.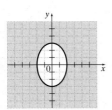

44.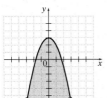

45.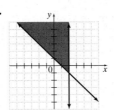

46.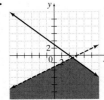

47.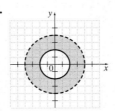

48.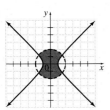

49.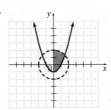

50. $(x + 7)^2 + (y - 8)^2 = 25$ **51.**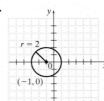

52.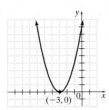

53.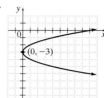

54.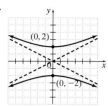

55.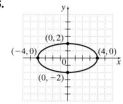

56.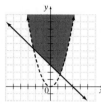

57.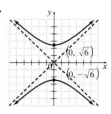

58.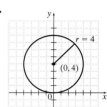

59.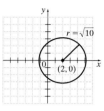

60.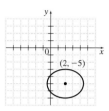

61.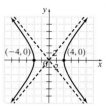

62. $\{(5, 1), (-1, 7)\}$ **63.** $\{(-1, 3), (-1, -3), (1, 3), (1, -3)\}$ **64.**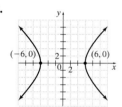

65.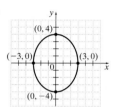

Chapter 10 Test   1.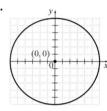

2.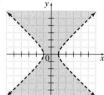

3.

4.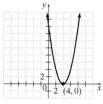

5.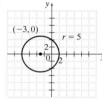

6.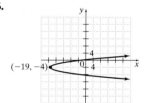

7.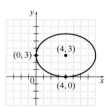

8.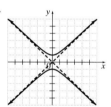

9. $\{(-12, 5), (12, -5)\}$ **10.** $\{(-5, -1), (-5, 1), (5, -1), (5, 1)\}$ **11.** $\{(6, 12), (1, 2)\}$ **12.** $\{(1, 1), (-1, -1)\}$

13.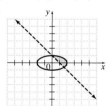

14.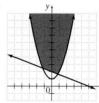

15.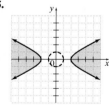

16.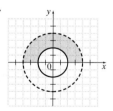

17. B

18. height: 10 ft; width: 30 ft

Cumulative Review   1. $12x^3 - 12x^2 - 9x + 2$; Sec. 5.1, Ex. 12   **2.** $(-1, 7]$; Sec. 2.5   **3.** $2x^2 + 11x + 15$; Sec. 5.2, Ex. 6

4. 3; Sec. 3.2   **5.** $x^3 - 4x^2 - 3x + 6 + \dfrac{22}{x + 2}$; Sec. 5.3, Ex. 9   **6.** 150 mph, 175 mph; Sec. 4.3   **7.** $(x - 5)(2 + 3a)$; Sec. 5.4, Ex. 6

8. $(3y + 5)(y + 3)$; Sec. 5.5   **9.** $(x - 5)(x - 7)$; Sec. 5.5, Ex. 2   **10.** $2a^3(2a + 5)(5a + 1)$; Sec. 5.5

11. $(x + 9)(x - 3)$; Sec. 5.6, Ex. 10   **12.** $(y - 1)(y - 7)$; Sec. 5.5   **13.** $\left\{-5, \dfrac{1}{2}\right\}$; Sec. 5.7, Ex. 2   **14.** $\left\{\dfrac{4}{3}\right\}$; Sec 6.4

15. 1; Sec. 6.1, Ex. 4   **16.** $4x^2 + 2x + 1$; Sec. 6.1   **17.** -1; Sec. 6.1, Ex. 5   **18.** $2x^2 - 5y$; Sec. 6.1   **19.** $x - 7$; Sec. 6.2, Ex. 3

20. $\dfrac{5(a + 2)}{3(a + 5)(a - 5)}$; Sec. 6.2   **21.** $-\dfrac{1}{3y^2}$; Sec. 6.2, Ex. 4   **22.** $\dfrac{3x^2 + 7x - 45}{(3x + 2)(x - 5)(3x - 2)}$; Sec. 6.2   **23.** $\dfrac{x(x - 2)}{2(x + 2)}$; Sec. 6.3, Ex. 1

24. $k = 2; y = \dfrac{2}{x}$; Sec. 6.6   **25.** $\{-6, -1\}$; Sec. 6.4, Ex. 5

26. a. $2\sqrt{5} + 5\sqrt{3}$ **b.** $\sqrt{21} - \sqrt{3} - \sqrt{35} + \sqrt{5}$ **c.** $21 - 4\sqrt{5}$ **d.** -7; Sec. 7.4

27. 2; Sec. 6.5, Ex. 2   **28.** $\dfrac{\sqrt{3} - 3}{3}$; Sec. 7.5   **29.** 0; Sec. 7.1, Ex. 5   **30.** $\dfrac{x^3\sqrt{x}}{7}$; Sec. 7.1

31. 0.5; Sec. 7.1, Ex. 7   **32.** $3a^2b\sqrt{10a}$; Sec. 7.1   **33.** $\sqrt{x}$; Sec. 7.2, Ex. 19   **34.** $y\sqrt{x}$; Sec 7.2   **35.** $\sqrt[3]{rs^2}$; Sec. 7.2, Ex. 21

36. 6; Sec. 7.6   **37.** $2\sqrt[3]{3}$; Sec. 7.3, Ex. 11   **38.** $12y$; Sec. 7.3   **39.** $2\sqrt[4]{2}$; Sec. 7.3, Ex. 13   **40.** $3x$; Sec. 7.3   **41.** $\dfrac{2\sqrt{5}}{5}$; Sec. 7.5, Ex. 1

42. a. $2\sqrt{2}$ **b.** $\dfrac{2y^2\sqrt[3]{10}}{5}$ **c.** $2xy^2\sqrt[5]{x^2}$; Sec. 7.3   **43.** $\left\{-\dfrac{1}{9}, -1\right\}$; Sec. 7.6, Ex. 2   **44.** $[2, \infty)$; Sec. 8.4

45. $13 - 8i$; Sec. 7.7, Ex. 13   **46.** $f^{-1}(x) = 2x - 1$; Sec. 9.2   **47.** $\{-1 + \sqrt{5}, -1 - \sqrt{5}\}$; Sec. 8.1, Ex. 7   **48.** 13; Sec. 8.6

49. $\left\{\dfrac{2 + \sqrt{10}}{2}, \dfrac{2 - \sqrt{10}}{2}\right\}$; Sec. 8.2, Ex. 2   **50. a.** $\left\{\dfrac{1}{3}\right\}$ **b.** $\left\{\dfrac{11}{3}\right\}$ **c.** $\{-2\}$; Sec. 9.3   **51.** $\sqrt{2} \approx 1.414$; Sec. 7.3, Ex. 21   **52.** $\{4\}$; Sec. 6.4

53. 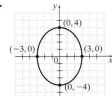 ; Sec. 10.2, Ex. 1   **54. a.** $9x^2 - 21x + 12$ **b.** 90   **c.** $-3x^2 + 9x - 1$ **d.** -31; Sec. 9.1

55. $\varnothing$; Sec. 10.3, Ex. 3   **56.** $\{(-6, 0), (0, 6)\}$; Sec. 10.3   **57.** $\{(2, \sqrt{3}), (-2, \sqrt{3}), (2, -\sqrt{3}), (-2, -\sqrt{3})\}$; Sec. 10.3, Ex. 4

APPENDIX

Exercise Set Appendix A   1. $\dfrac{21}{30}$ **3.** $\dfrac{4}{18}$ **5.** $\dfrac{16}{20}$ **7.** $\dfrac{1}{2}$ **9.** $\dfrac{2}{3}$ **11.** $\dfrac{3}{7}$ **13.** 1   **15.** 5   **17.** $\dfrac{3}{5}$ **19.** $\dfrac{4}{5}$ **21.** $\dfrac{11}{8}$ **23.** $\dfrac{30}{61}$

25. $\dfrac{8}{11}$ **27.** $\dfrac{3}{8}$ **29.** $\dfrac{1}{2}$ **31.** $18\dfrac{20}{27}$ **33.** 37   **35.** $\dfrac{6}{7}$ **37.** 15   **39.** $\dfrac{1}{6}$ **41.** $\dfrac{3}{80}$ **43.** $10\dfrac{5}{11}$ **45.** $2\dfrac{28}{29}$ **47.** 1   **49.** $\dfrac{3}{5}$

51. $\dfrac{9}{35}$ **53.** $\dfrac{1}{3}$ **55.** $12\dfrac{1}{4}$ **57.** $1\dfrac{3}{5}$ **59.** $\dfrac{23}{21}$ **61.** $\dfrac{65}{21}$ **63.** $\dfrac{5}{7}$ **65.** $\dfrac{5}{66}$ **67.** $7\dfrac{1}{12}$ **69.** $48\dfrac{1}{15}$ **71.** $\dfrac{7}{5}$ **73.** $\dfrac{17}{18}$

75. answers may vary   **77.** $\dfrac{1}{5}$ **79.** $\dfrac{3}{8}$ **81.** $12\dfrac{3}{4}$ ft

Mental Math   1. 56   **3.** -32 **5.** 20

Exercise Set Appendix B   1. 26   **3.** -19 **5.** 0   **7.** $\dfrac{13}{6}$ **9.** $(1, 2)$ **11.** $\{(x, y) \mid 3x + y = 1\}$ **13.** $(9, 9)$ **15.** $(-3, -2)$

17. $(3, 4)$ **19.** 8   **21.** 0   **23.** 15   **25.** 54   **27.** $(-2, 0, 5)$ **29.** $(6, -2, 4)$ **31.** $(-2, 3, -1)$ **33.** $(0, 2, -1)$ **35.** 5

37. 0; answers may vary

Exercise Set Appendix D   1. yes   **3.** no   **5.** answers may vary   **7.** answers may vary   **9.** answers may vary

11. Xmin $= -12$ Ymin $= -12$ **13.** Xmin $= -9$ Ymin $= -12$ **15.** Xmin $= -10$ Ymin $= -25$ **17.** Xmin $= -5$ Ymin $= -15$
      Xmax $= 12$ Ymax $= 12$ Xmax $= 9$ Ymax $= 12$ Xmax $= 10$ Ymax $= 25$ Xmax $= 5$ Ymax $= 15$
      Xscl $= \dfrac{6}{5}$ Yscl $= \dfrac{6}{5}$ Xscl $= 1$ Yscl $= 2$ Xscl $= 2$ Yscl $= 5$ Xscl $= 1$ Yscl $= 3$

19. Xmin $= -20$ Ymin $= -30$ **21.** Setting B   **23.** Setting B   **25.** Setting B
      Xmax $= 30$ Ymax $= 50$
      Xscl $= 5$ Yscl $= 10$

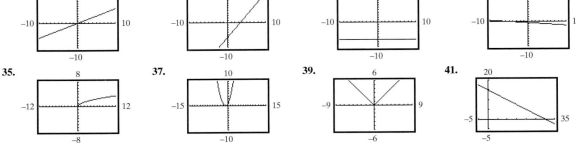

Exercise Set Appendix E

1.
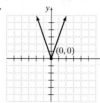

3.
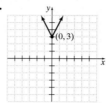

5.
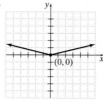

7.
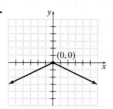

9.
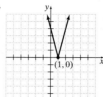

11.
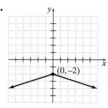

13.
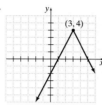

15.
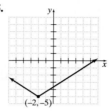

Exercise Set Appendix F   **1.** $71°$ **3.** $19.2°$ **5.** $78\frac{3}{4}°$ **7.** $30°$ **9.** $149.8°$ **11.** $100\frac{1}{2}°$

13. $m\angle1 = m\angle5 = m\angle7 = 110°$, $m\angle2 = m\angle3 = m\angle4 = m\angle6 = 70°$ **15.** $90°$ **17.** $90°$ **19.** $90°$ **21.** $45°, 90°$ **23.** $73°, 90°$

25. $50\frac{1}{4}°, 90°$ **27.** $x = 6$ **29.** $x = 4.5$ **31.** 10 **33.** 12

SOLUTIONS TO SELECTED EXERCISES

CHAPTER 1

Exercise Set 1.2

1. $\{x|x \text{ is a natural number less than } 6\} = \{1, 2, 3, 4, 5\}$
5. $\{x|x \text{ is a whole number that is not a natural number}\} = \{0\}$
9. The whole numbers in the set are $\{3, 0, \sqrt{36}\}$.
13. The irrational number in the set is $\{\sqrt{7}\}$.
17. $0 \notin \{x|x \text{ is a positive integer}\}$
21. Since the set of whole numbers is contained in the set of real numbers, the statement is true.
25. Since the number 0 is a whole number but not a natural number, the statement is false.
29. Since 0 is an integer, the statement is true.
33. Since there are no days of the week which start with the letter B, the statement is true.
37. Since irrational numbers are numbers which cannot be written as rational numbers, the statement is false.
41. Ten less than a number is written as $x - 10$
45. A number divided by 11 is written as $\dfrac{x}{11}$ or $x \div 11$.
49. A number plus two and three-tenths is written as $x + 2.3$ or $x + 2\dfrac{3}{10}$.
53. Nine times a number is written as $9x$.
57. Five more than twice a number is written as $2x + 5$.
61. One plus twice a number is written as $1 + 2x$.
65. The quotient of five and the difference of four and a number is written as $\dfrac{5}{4 - x}$.
69. answers may vary
73. The only whole number that is not a natural number is the number 0.
77. answers may vary

Exercise Set 1.3

1. The sum of 10 and x is -12 is written as $10 + x = -12$.
5. The quotient of n and 5 is 4 times n is written as $\dfrac{n}{5} = 4n$.
9. 0 is to the right of -2, so $0 > -2$.
13. $\dfrac{12}{3} = 4, \dfrac{12}{2} = 6$, and 4 is the left of 6, so $\dfrac{12}{3} < \dfrac{12}{2}$.
17. 8.6 is to the right of -3.5, so $8.6 > -3.5$.
21. By dividing, we see that $\dfrac{1}{2} = 0.5$ and $\dfrac{5}{8} = 0.625$. Thus, $\dfrac{1}{2} < \dfrac{5}{8}$, since $0.5 < 0.625$.
25. True, since $-6 < 0$ is true.
29. False, since neither $-14 > -1$ nor $-14 = -1$ is true.
33. The opposite of 9 is -9.
37. The opposite of $\dfrac{4}{7}$ is $-\dfrac{4}{7}$.
41. The opposite of 0 is 0.
45. The reciprocal of -8 is $-\dfrac{1}{8}$.

49. The reciprocal of 0 is undefined, since there is no number that when multiplied by 0 gives a product of 1.

	Number	Opposite	Reciprocal
53.	25	-25	$\dfrac{1}{25}$
57.	$-\dfrac{1}{7}$	$\dfrac{1}{7}$	-7
61.	$\dfrac{16}{19}$	$-\dfrac{16}{19}$	$\dfrac{19}{16}$

65. $z \cdot w = w \cdot z$ by the commutative property of multiplication.
69. $5 \cdot (7x) = (5 \cdot 7)x$ by the associative property of multiplication.
73. $(14z) \cdot y = 14(z \cdot y)$ by the associative property of multiplication.
77. $4(z - 6) = 4 \cdot z - 4 \cdot 6 = 4z - 24$
81. $6x(y - 4) = 6x \cdot y - 6x \cdot 4 = 6xy - 24x$
85. $\dfrac{1}{2}(4x - 9y) = \dfrac{1}{2} \cdot 4x - \dfrac{1}{2} \cdot 9y = 2x - \dfrac{9}{2}y$
89. $3x + 6 = 6 + 3x$ by the commutative property of addition.
93. $7 \cdot 1 = 7$ by the multiplicative identity property.
97. $3(x + 4) = 3x + 12$
101. zero; answers may vary
105. answers may vary

Exercise Set 1.4

1. $|2| = 2$
5. $|0| = 0$
9. $-\left|-\dfrac{2}{9}\right| = -\dfrac{2}{9}$
13. $-14 + (-10) = -24$
17. $13 - 17 = 13 + (-17) = -4$
21. $19 - 10 - 11 = 19 + (-10) + (-11)$
$\qquad = 9 + (-11)$
$\qquad = -2$
25. $-\dfrac{4}{5} - \left(-\dfrac{3}{10}\right) = -\dfrac{4}{5} \cdot \dfrac{2}{2} + \dfrac{3}{10}$
$\qquad = -\dfrac{8}{10} + \dfrac{3}{10}$
$\qquad = -\dfrac{5}{10}$
$\qquad = -\dfrac{1}{2}$
29. $-5 \cdot 12 = -60$
33. $-17 \cdot 0 = 0$
37. $\dfrac{-9}{3} = -3$
41. $3\left(-\dfrac{1}{18}\right) = -\dfrac{1}{6}$
45. $9.1 \div (-1.3) = -7$
49. $-7^2 = -(7 \cdot 7) = -49$
53. $(-2)^3 = (-2)(-2)(-2) = -8$
57. $\sqrt{49} = 7$, since $7^2 = 49$.
61. $\sqrt{\dfrac{1}{9}} = \dfrac{1}{3}$, since $\left(\dfrac{1}{3}\right)^2 = \dfrac{1}{9}$.

65. $\sqrt{\dfrac{4}{25}} = \dfrac{2}{5}$, since $\left(\dfrac{2}{5}\right)^2 = \dfrac{4}{25}$.

69. $-9 + (-3) = -12$

73. $-9 \cdot 8 = -72$

77. $\dfrac{16}{-2} = -8$

81. $-\dfrac{2}{7} \cdot \left(-\dfrac{1}{6}\right) = \dfrac{2}{42} = \dfrac{1}{21}$

85. $-4 - (-19) = -4 + 19 = 15$

89. $\dfrac{0}{-5} = 0$

93. $\sqrt{\dfrac{1}{16}} = \dfrac{1}{4}$

97. $\dfrac{-5.2}{-1.3} = 4$

101. $\sqrt[4]{81} = 3$, since $3^4 = 81$.

105. $-\dfrac{1}{6} \div \left(\dfrac{9}{10}\right) = -\dfrac{1}{6} \cdot \left(\dfrac{10}{9}\right) = -\dfrac{10}{54} = -\dfrac{5}{27}$

109. $\dfrac{3}{5} \div \left(-\dfrac{2}{5}\right) = \dfrac{3}{5} \cdot \left(-\dfrac{5}{2}\right) = -\dfrac{15}{10} = -\dfrac{3}{2}$

113. $-5 + (-7) - 10 = -12 + (-10) = -22$

117. $-6(-5)(0) = 30(0) = 0$

121. answers may vary

125. $1 - \dfrac{1}{5} - \dfrac{3}{7} = \dfrac{35}{35} - \dfrac{7}{35} - \dfrac{15}{35} = \dfrac{28}{35} - \dfrac{15}{35} = \dfrac{13}{35}$

129. Since each color has an equal sector of the spinner in b, spinner b would lead to a fair game.

133. Yes, a two-way tie could occur if two players have 6 points each (the third player has 0 points); or two players have 5 points each (the third player has 2 points).

137. $\sqrt{19.6} \approx 4.4272$

141. After 5 years, the highest average annual return is 23% and the lowest average annual return is -17%.

$23 - (-17) = 23 + 17 = 40$

The difference is 40%.

Exercise Set 1.5

1. $3(5 - 7)^4 = 3(-2)^4 = 3 \cdot 16 = 48$

5. $\dfrac{3.1 - (-1.4)}{-0.5} = \dfrac{4.5}{-0.5} = -9$

9. $(-3)^2 + 2^3 = 9 + 8 = 17$

13. $4[8 - (2 - 4)] = 4[8 - (-2)] = 4[10] = 40$

17. $2 - [(7 - 6) + (9 - 19)] = 2 - [1 + (-10)]$
$= 2 - [-9]$
$= 11$

21. $(\sqrt[3]{8})(-4) - (\sqrt{9})(-5) = (2)(-4) - (3)(-5)$
$= -8 - (-15)$
$= 7$

25. $25 - [(3 - 5) + (14 - 18)]^2 = 25 - [(-2) + (-4)]^2$
$= 25 - [-6]^2$
$= 25 - 36$
$= -11$

29. $\dfrac{|3 - 9| - |-5|}{-3} = \dfrac{|-6| - |-5|}{-3}$
$= \dfrac{6 - 5}{-3}$
$= \dfrac{1}{-3}$
$= -\dfrac{1}{3}$

33. $\dfrac{\frac{1}{3} \cdot 9 - 7}{3 + \frac{1}{2} \cdot 4} = \dfrac{3 - 7}{3 + 2} = \dfrac{-4}{5} = -\dfrac{4}{5}$

37. $\dfrac{-4\sqrt{80 + 1} + (-4)^2}{3^3 + |-2(3)|} = \dfrac{-4\sqrt{81} + (-4)^2}{3^3 + |-6|}$
$= \dfrac{-4 \cdot 9 + 16}{27 + 6}$
$= \dfrac{-36 + 16}{33}$
$= \dfrac{-20}{33}$
$= -\dfrac{20}{33}$

41. $\left(\dfrac{5.6 - 8.4}{1.9 - 2.7}\right)^2 = \left(\dfrac{-2.8}{-0.8}\right)^2 = (3.5)^2 = 12.25$

45. Replace y with -2.
$-3y^2 = -3(-2)^2 = -3 \cdot 4 = -12$

49. Replace x with 9 and y with -2.
$\dfrac{3 + 2|x - y|}{x + 2y} = \dfrac{3 + 2|9 - (-2)|}{9 + 2(-2)}$
$= \dfrac{3 + 2|11|}{9 + (-4)}$
$= \dfrac{3 + 22}{5}$
$= \dfrac{25}{5}$
$= 5$

53. a.

Length, y	Perimeter, $8 + 2y$
5	$8 + 2(5) = 8 + 10 = 18$
7	$8 + 2(7) = 8 + 14 = 22$
10	$8 + 2(10) = 8 + 20 = 28$
100	$8 + 2(100) = 8 + 200 = 208$

b. As the length increases, the perimeter increases; answers may vary.

57. Replace z with 8.4.
$439.39z = 439.39 \cdot 8.4 = 3690.876$
The average cost of 8.4 ounces of gold at that time was $3690.88.

61. $19y - y = 19y - 1y = (19 - 1)y = 18y$

65. $-9 + 4x + 18 - 10x = 4x - 10x - 9 + 18$
$= -6x + 9$

69. $x - y + x - y = x + x - y - y = 2x - 2y$

73. $\dfrac{3}{4}b - \dfrac{1}{2} + \dfrac{1}{6}b - \dfrac{2}{3} = \dfrac{3}{4}b + \dfrac{1}{6}b - \dfrac{1}{2} - \dfrac{2}{3}$
$= \dfrac{9}{12}b + \dfrac{2}{12}b - \dfrac{3}{6} - \dfrac{4}{6}$
$= \dfrac{11}{12}b - \dfrac{7}{6}$

77. $2(3x + 7) = 2 \cdot 3x + 2 \cdot 7 = 6x + 14$

81. $(3x + 4) - (6x - 1) = 3x + 4 - 6x + 1$
$= 3x - 6x + 4 + 1$
$= -3x + 5$

85. $-(n + 5) + (5n - 3) = -n - 5 + 5n - 3$
$= -n + 5n - 5 - 3$
$= 4n - 8$

89. $\dfrac{1}{4}(8x - 4) - \dfrac{1}{5}(20x - 6y) = 2x - 1 - 4x + \dfrac{6}{5}y$

$$= 2x - 4x + \dfrac{6}{5}y - 1$$

$$= -2x + \dfrac{6}{5}y - 1$$

93. $\dfrac{1}{6}(24a - 18b) - \dfrac{1}{7}(7a - 21b - 2) - \dfrac{1}{5}$

$$= 4a - 3b - a + 3b + \dfrac{2}{7} - \dfrac{1}{5}$$

$$= 4a - a - 3b + 3b + \dfrac{10}{35} - \dfrac{7}{35}$$

$$= 3a + \dfrac{3}{35}$$

97. $-4[6(2t + 1) - (9 + 10t)] = -4[12t + 6 - 9 - 10t]$

$$= -4[12t - 10t + 6 - 9]$$

$$= -4[2t - 3]$$

$$= -8t + 12$$

101. $8.1z + 7.3(z + 5.2) - 6.85$

$$= 8.1z + 7.3z + 37.96 - 6.85$$

$$= 15.4z + 31.11$$

105. answers may vary

109. From the graph the population that was over 65 in 1970 was about 20 million.

113. The population over 65 is increasing; answers may vary.

Exercise Set 1.6

1. $4^2 \cdot 4^3 = 4^{2+3} = 4^5$

5. $m \cdot m^7 \cdot m^6 = m^1 \cdot m^7 \cdot m^6 = m^{1+7+6} = m^{14}$

9. $(-4x^3p^2)(4y^3x^3) = -4(4)x^3x^3y^3p^2$

$$= -16x^{3+3}y^3p^2$$

$$= -16x^6y^3p^2$$

13. $(4x + 5)^0 = 1$

17. $4x^0 + 5 = 4 \cdot 1 + 5 = 4 + 5 = 9$

21. $\dfrac{-26z^{11}}{2z^7} = -\dfrac{26}{2}z^{11-7} = -13z^4$

25. $\dfrac{12x^4y^7}{9xy^5} = \dfrac{12}{9}x^{4-1}y^{7-5} = \dfrac{4}{3}x^3y^2$

29. $4^{-2} = \dfrac{1}{4^2} = \dfrac{1}{16}$

33. $\dfrac{x^7}{x^{15}} = x^{7-15} = x^{-8} = \dfrac{1}{x^8}$

37. $\dfrac{x^{-7}}{y^{-2}} = x^{-7} \cdot \dfrac{1}{y^{-2}} = \dfrac{1}{x^7} \cdot y^2 = \dfrac{y^2}{x^7}$

41. $\dfrac{8r^4}{2r^{-4}} = \dfrac{8}{2}r^{4-(-4)} = 4r^8$

45. $\dfrac{2a^{-6}b^2}{18ab^{-5}} = \dfrac{2}{18}a^{-6-1}b^{2-(-5)} = \dfrac{1}{9}a^{-7}b^7 = \dfrac{b^7}{9a^7}$

49. $-7x^3 \cdot 20x^9 = -7 \cdot 20x^{3+9} = -140x^{12}$

53. $2x^3 \cdot 5x^7 = 2 \cdot 5x^{3+7} = 10x^{10}$

57. $\dfrac{z^{12}}{z^{15}} = z^{12-15} = z^{-3} = \dfrac{1}{z^3}$

61. $\dfrac{y^{-3}}{y^{-7}} = y^{-3-(-7)} = y^4$

65. $3x^{-1} = 3 \cdot \dfrac{1}{x} = \dfrac{3}{x}$

69. $\dfrac{x^{-7}y^{-2}}{x^2y^2} = x^{-7-2}y^{-2-2} = x^{-9}y^{-4} = \dfrac{1}{x^9y^4}$

73. $2^{-4}x = \dfrac{1}{2^4} \cdot x = \dfrac{1}{16}x = \dfrac{x}{16}$

77. $\dfrac{8^{-7}}{8^{-6}} = 8^{-7-(-6)} = 8^{-1} = \dfrac{1}{8}$

81. $\dfrac{14x^{-2}yz^{-4}}{2xyz} = \dfrac{14}{2}x^{-2-1}y^{1-1}z^{-4-1} = 7x^{-3}y^0z^{-5} = \dfrac{7}{x^3z^5}$

85. $\dfrac{x^{3t-1}}{x^t} = x^{3t-1-t} = x^{2t-1}$

89. $\dfrac{z^{6x}}{z^7} = z^{6x-7}$

93. $31{,}250{,}000 = 3.125 \times 10^7$

97. $67{,}413 = 6.7413 \times 10^4$

101. $0.000053 = 5.3 \times 10^{-5}$

105. $6{,}404{,}000{,}000 = 6.404 \times 10^9$

109. $0.001 = 1.0 \times 10^{-3}$

113. $9.3 \times 10^7 = 93{,}000{,}000$

117. $7.35 \times 10^{12} = 7{,}350{,}000{,}000{,}000$

121. $3.0 \times 10^8 = 300{,}000{,}000$

125. answers may vary

Exercise Set 1.7

1. $(3^{-1})^2 = 3^{-2} = \dfrac{1}{3^2} = \dfrac{1}{9}$

5. $(3x^2y^3)^2 = 3^2(x^2)^2(y^3)^2 = 9x^4y^6$

9. $(2a^2bc^{-3})^{-6} = 2^{-6}(a^2)^{-6}b^{-6}(c^{-3})^{-6}$

$$= 2^{-6}a^{-12}b^{-6}c^{18}$$

$$= \dfrac{c^{18}}{2^6a^{12}b^6}$$

$$= \dfrac{c^{18}}{64a^{12}b^6}$$

13. $(-2^{-2}y^{-1})^{-3} = [-1 \cdot (2)^{-2}y^{-1}]^{-3}$

$$= (-1)^{-3}(2^{-2})^{-3}(y^{-1})^{-3}$$

$$= (-1) \cdot 2^6y^3$$

$$= -64y^3$$

17. $\left(\dfrac{6p^6}{p^{12}}\right)^2 = (6p^{6-12})^2 = (6p^{-6})^2 = 6^2p^{-12} = \dfrac{36}{p^{12}}$

21. $\left(\dfrac{3}{4}\right)^{-3} = \dfrac{3^{-3}}{4^{-3}} = \dfrac{4^3}{3^3} = \dfrac{64}{27}$

25. $\left(\dfrac{x^{-2}y^{-2}}{a^{-3}}\right)^{-7} = \dfrac{(x^{-2})^{-7}(y^{-2})^{-7}}{(a^{-3})^{-7}} = \dfrac{x^{14}y^{14}}{a^{21}}$

29. $(5^{-1})^3 = 5^{-3} = \dfrac{1}{5^3} = \dfrac{1}{125}$

33. $\left(\dfrac{x^7y^{-3}}{z^{-4}}\right)^{-5} = \dfrac{(x^7)^{-5}(y^{-3})^{-5}}{(z^{-4})^{-5}} = \dfrac{x^{-35}y^{15}}{z^{20}} = \dfrac{y^{15}}{x^{35}z^{20}}$

37. $\left(\dfrac{4^{-4}}{y^3x}\right)^{-2} = \dfrac{(4^{-4})^{-2}}{(y^3)^{-2}x^{-2}} = \dfrac{4^8}{y^{-6}x^{-2}} = 4^8x^2y^6$

41. $\dfrac{4^{-1}x^2yz}{x^{-2}yz^3} = 4^{-1}x^{2-(-2)}y^{1-1}z^{1-3} = \dfrac{1}{4}x^4y^0z^{-2} = \dfrac{x^4}{4z^2}$

45. $\dfrac{(y^3)^{-4}}{y^3} = \dfrac{y^{-12}}{y^3} = y^{-12-3} = y^{-15} = \dfrac{1}{y^{15}}$

49. $(4x^6y^5)^{-2}(6x^4y^3) = 4^{-2}(x^6)^{-2}(y^5)^{-2} \cdot 6x^4y^3$

$$= \dfrac{1}{4^2}x^{-12}y^{-10} \cdot 6x^4y^3$$

$$= \dfrac{6}{16}x^{-12+4}y^{-10+3}$$

$$= \dfrac{3}{8}x^{-8}y^{-7}$$

$$= \dfrac{3}{8x^8y^7}$$

53. $\dfrac{2^{-3}x^2y^{-5}}{5^{-2}x^7y^{-1}} = \dfrac{5^2}{2^3}x^{2-7}y^{-5-(-1)} = \dfrac{25}{8}x^{-5}y^{-4} = \dfrac{25}{8x^5y^4}$

57. $(x^{3a+6})^3 = x^{(3a+6)\cdot 3} = x^{9a+18}$

61. $(b^{5x-2})^2 = b^{(5x-2)\cdot 2} = b^{10x-4}$

65. $\left(\dfrac{2x^{3t}}{x^{2t-1}}\right)^4 = (2x^{3t-(2t-1)})^4$

$= (2x^{t+1})^4$

$= 2^4x^{(t+1)\cdot 4}$

$= 16x^{4t+4}$

69. $(5 \times 10^{11})(2.9 \times 10^{-3}) = 5 \times 2.9 \times 10^{11} \times 10^{-3}$

$= 14.5 \times 10^8$

$= (1.45 \times 10^1) \times 10^8$

$= 1.45 \times 10^9$

73. $\dfrac{3.6 \times 10^{-4}}{9 \times 10^2} = \dfrac{3.6}{9} \cdot \dfrac{10^{-4}}{10^2}$

$= 0.4 \times 10^{-6}$

$= (4 \times 10^{-1}) \times 10^{-6}$

$= 4 \times 10^{-7}$

77. $\dfrac{18,200 \times 100}{91,000} = \dfrac{1.82 \times 10^4 \times 10^2}{9.1 \times 10^4}$

$= \dfrac{1.82}{9.1} \cdot \dfrac{10^6}{10^4}$

$= 0.2 \times 10^2$

$= (2 \times 10^{-1}) \times 10^2$

$= 2 \times 10^1$

81. $\dfrac{0.00064 \times 2000}{16,000} = \dfrac{6.4 \times 10^{-4} \times 2 \times 10^3}{1.6 \times 10^4}$

$= \dfrac{12.8 \times 10^{-1}}{1.6 \times 10^4}$

$= \dfrac{12.8}{1.6} \cdot \dfrac{10^{-1}}{10^4}$

$= 8 \times 10^{-5}$

85. $\dfrac{9.24 \times 10^{15}}{(2.2 \times 10^{-2})(1.2 \times 10^{-5})} = \dfrac{9.24}{(2.2)(1.2)} \cdot \dfrac{10^{15}}{10^{-2} \cdot 10^{-5}}$

$= \dfrac{9.24}{2.64} \times 10^{22}$

$= 3.5 \times 10^{22}$

89. $(3.8 \times 10^{-6})(1.64 \times 10^{-5}) = 3.8 \times 1.64 \times 10^{-6} \times 10^{-5}$

$= 6.232 \times 10^{-11}$

The average size of a grain of salt is 6.232×10^{-11} cubic meter.

93. Since $D = \dfrac{M}{V}$, then $D \cdot V = M$.

$(3.12 \times 10^{-2})(4.269 \times 10^{14}) = 3.12 \times 4.269 \times 10^{-2} \times 10^{14}$

$= 13.31928 \times 10^{12}$

$= (1.331928 \times 10^1) \times 10^{12}$

$= 1.331928 \times 10^{13}$

The mass of water in Lake Superior is 1.331928×10^{13} tons.

97. $\dfrac{2.95 \times 10^8}{3.536 \times 10^6} = \dfrac{2.95}{3.536} \cdot \dfrac{10^8}{10^6} \approx 0.834276 \times 10^2 \approx 83$

The population density of the United States in 2004 was approximately 83 people per square mile.

Chapter 1 Test

1. Since -2.3 is to the right of -2.33, $-2.3 > -2.33$, so the statement is true.

5. Since the set of natural numbers is contained in the set of integers, the statement is true.

9. $[3|4 - 5|^5 - (-9)] \div (-6) = [3|-1|^5 - (-9)] \div (-6)$

$= [3(1)^5 - (-9)] \div (-6)$

$= [3 \cdot 1 - (-9)] \div (-6)$

$= [3 - (-9)] \div (-6)$

$= [3 + 9] \div (-6)$

$= [12] \div (-6)$

$= -2$

13. Replace q with 4, r with -2, and t with 1.

$\dfrac{5t - 3q}{3r - 1} = \dfrac{5(1) - 3(4)}{3(-2) - 1} = \dfrac{5 - 12}{-6 - 1} = \dfrac{-7}{-7} = 1$

17. Negative two is equal to x divided by the sum of x and 5 is written as $-2 = \dfrac{x}{x + 5}$.

21. The reciprocal of $-\dfrac{7}{11}$ is $-\dfrac{11}{7}$; the opposite of $-\dfrac{7}{11}$ is $\dfrac{7}{11}$.

25. $\dfrac{6^{-1}a^2b^{-3}}{3^{-2}a^{-5}b^2} = \dfrac{3^2}{6^1}a^{2-(-5)}b^{-3-2} = \dfrac{9}{6}a^7b^{-5} = \dfrac{3a^7}{2b^5}$

29. $630,000,000 = 6.3 \times 10^8$

33. $\dfrac{(0.0024)(0.00012)}{0.00032} = \dfrac{(2.4 \times 10^{-3})(1.2 \times 10^{-4})}{3.2 \times 10^{-4}}$

$= \dfrac{(2.4)(1.2)}{3.2} \cdot \dfrac{(10^{-3})(10^{-4})}{10^{-4}}$

$= 0.9 \times 10^{-3}$

$= (9 \times 10^{-1}) \times 10^{-3}$

$= 9 \times 10^{-4}$

CHAPTER 2

Exercise Set 2.1

1. $\dfrac{x}{-6} = 4$

$\dfrac{-24}{-6} \overset{?}{=} 4$

$4 = 4$ True

Yes, -24 is a solution.

5. $5 + 3x = -1$

$5 + 3(-2) \overset{?}{=} -1$

$-1 = -1$ True

Yes, -2 is a solution.

9. $4(x - 3) = 12$

$4(5 - 3) \overset{?}{=} 12$

$8 = 12$ False

No, 5 is not a solution.

13. $-5x = -30$

$\dfrac{-5x}{-5} = \dfrac{-30}{-5}$

$x = 6$

Check: $-5x = -30$

$-5(6) \overset{?}{=} -30$

$-30 = -30$

The solution set is $\{6\}$.

17. $x - 2.8 = 1.9$

$x - 2.8 + 2.8 = 1.9 + 2.8$

$x = 4.7$

Check: $x - 2.8 = 1.9$

$4.7 - 2.8 \overset{?}{=} 1.9$

$1.9 = 1.9$ True

The solution set is $\{4.7\}$.

21.
$$-4.1 - 7z = 3.6$$
$$-4.1 - 7z + 4.1 = 3.6 + 4.1$$
$$-7z = 7.7$$
$$\frac{-7z}{-7} = \frac{7.7}{-7}$$
$$z = -1.1$$

Check:
$$-4.1 - 7z = 3.6$$
$$-4.1 - 7(-1.1) \stackrel{?}{=} 3.6$$
$$-4.1 + 7.7 \stackrel{?}{=} 3.6$$
$$3.6 = 3.6 \quad \text{True}$$

The solution set is $\{-1.1\}$.

25.
$$3x - 4 - 5x = x + 4 + x$$
$$-2x - 4 = 2x + 4$$
$$-2x - 4 + 4 = 2x + 4 + 4$$
$$-2x = 2x + 8$$
$$-2x - 2x = 2x + 8 - 2x$$
$$-4x = 8$$
$$\frac{-4x}{-4} = \frac{8}{-4}$$
$$x = -2$$

Check:
$$3x - 4 - 5x = x + 4 + x$$
$$3(-2) - 4 - 5(-2) \stackrel{?}{=} -2 + 4 + (-2)$$
$$-6 - 4 + 10 \stackrel{?}{=} 0$$
$$0 = 0 \quad \text{True}$$

The solution set is $\{-2\}$.

29.
$$5x + 12 = 2(2x + 7)$$
$$5x + 12 = 4x + 14$$
$$5x + 12 - 4x = 4x + 14 - 4x$$
$$x + 12 = 14$$
$$x + 12 - 12 = 14 - 12$$
$$x = 2$$

Check:
$$5x + 12 = 2(2x + 7)$$
$$5(2) + 12 \stackrel{?}{=} 2(2 \cdot 2 + 7)$$
$$10 + 2 \stackrel{?}{=} 2(11)$$
$$22 = 22 \quad \text{True}$$

The solution set is $\{2\}$.

33.
$$-2(5y - 1) - y = -4(y - 3)$$
$$-10y + 2 - y = -4y + 12$$
$$-11y + 2 = -4y + 12$$
$$-11y + 2 + 11y = -4y + 12 + 11y$$
$$2 = 7y + 12$$
$$2 - 12 = 7y + 12 - 12$$
$$-10 = 7y$$
$$\frac{-10}{7} = \frac{7y}{7}$$
$$-\frac{10}{7} = y$$

Check:
$$-2(5y - 1) - y = -4(y - 3)$$
$$-2\left[5\left(-\frac{10}{7}\right) - 1\right] - \left(-\frac{10}{7}\right) \stackrel{?}{=} -4\left(-\frac{10}{7} - 3\right)$$
$$-2\left(-\frac{57}{7}\right) + \frac{10}{7} \stackrel{?}{=} -4\left(-\frac{31}{7}\right)$$
$$\frac{114}{7} + \frac{10}{7} \stackrel{?}{=} \frac{124}{7}$$
$$\frac{124}{7} = \frac{124}{7} \quad \text{True}$$

The solution set is $\left\{-\dfrac{10}{7}\right\}$.

37.
$$\frac{3t}{4} - \frac{t}{2} = 1$$
$$4\left(\frac{3t}{4} - \frac{t}{2}\right) = 4(1)$$
$$4\left(\frac{3t}{4}\right) - 4\left(\frac{t}{2}\right) = 4$$
$$3t - 2t = 4$$
$$t = 4$$

Check:
$$\frac{3t}{4} - \frac{t}{2} = 1$$
$$\frac{3(4)}{4} - \frac{4}{2} \stackrel{?}{=} 1$$
$$3 - 2 \stackrel{?}{=} 1$$
$$1 = 1 \quad \text{True}$$

The solution set is $\{4\}$.

41.
$$0.6x - 10 = 1.4x - 14$$
$$10(0.6x - 10) = 10(1.4x - 14)$$
$$10(0.6x) - 10(10) = 10(1.4x) - 10(14)$$
$$6x - 100 = 14x - 140$$
$$6x - 100 - 6x = 14x - 140 - 6x$$
$$-100 = 8x - 140$$
$$-100 + 140 = 8x - 140 + 140$$
$$40 = 8x$$
$$\frac{40}{8} = \frac{8x}{8}$$
$$5 = x$$

Check: $\quad 0.6x - 10 = 1.4x - 14$
$$0.6(5) - 10 \stackrel{?}{=} 1.4(5) - 14$$
$$3 - 10 \stackrel{?}{=} 7 - 14$$
$$-7 = -7 \quad \text{True}$$

The solution set is $\{5\}$.

45.
$$1.5(4 - x) = 1.3(2 - x)$$
$$10[1.5(4 - x)] = 10[1.3(2 - x)]$$
$$15(4 - x) = 13(2 - x)$$
$$60 - 15x = 26 - 13x$$
$$60 - 15x + 15x = 26 - 13x + 15x$$
$$60 = 26 + 2x$$
$$60 - 26 = 26 + 2x - 26$$
$$34 = 2x$$
$$\frac{34}{2} = \frac{2x}{2}$$
$$17 = x$$

Check: $\quad 1.5(4 - x) = 1.3(2 - x)$
$$1.5(4 - 17) \stackrel{?}{=} 1.3(2 - 17)$$
$$1.5(-13) \stackrel{?}{=} 1.3(-15)$$
$$-19.5 = -19.5 \quad \text{True}$$

The solution set is $\{17\}$.

49.
$$3(x + 1) + 5 = 3x + 2$$
$$3x + 3 + 5 = 3x + 2$$
$$3x + 8 = 3x + 2$$
$$3x + 8 - 3x = 3x + 2 - 3x$$
$$8 = 2 \quad \text{False}$$

The equation $8 = 2$ is a false statement, so the original equation has no solution. The solution set is $\varnothing$.

53.
$$4(x + 5) = 3(x - 4) + x$$
$$4x + 20 = 3x - 12 + x$$
$$4x + 20 = 4x - 12$$
$$4x + 20 - 4x = 4x - 12 - 4x$$
$$20 = -12 \quad \text{False}$$

The equation $20 = -12$ is a false statement, so the original equation has no solution. The solution set is $\varnothing$.

57.
$$x - 10 = -6x - 10$$
$$x - 10 + 6x = -6x - 10 + 6x$$
$$7x - 10 = -10$$
$$7x - 10 + 10 = -10 + 10$$
$$7x = 0$$
$$\frac{7x}{7} = \frac{0}{7}$$
$$x = 0$$
The solution set is $\{0\}$.

61.
$$y + 0.2 = 0.6(y + 3)$$
$$10(y + 0.2) = 10[0.6(y + 3)]$$
$$10y + 2 = 6(y + 3)$$
$$10y + 2 = 6y + 18$$
$$10y + 2 - 6y = 6y + 18 - 6y$$
$$4y + 2 = 18$$
$$4y + 2 - 2 = 18 - 2$$
$$4y = 16$$
$$\frac{4y}{4} = \frac{16}{4}$$
$$y = 4$$
The solution set is $\{4\}$.

65.
$$2y + 5(y - 4) = 4y - 2(y - 10)$$
$$2y + 5y - 20 = 4y - 2y + 20$$
$$7y - 20 = 2y + 20$$
$$7y - 20 - 2y = 2y + 20 - 2y$$
$$5y - 20 = 20$$
$$5y - 20 + 20 = 20 + 20$$
$$5y = 40$$
$$\frac{5y}{5} = \frac{40}{5}$$
$$y = 8$$
The solution set is $\{8\}$.

69.
$$\frac{m - 4}{3} - \frac{3m - 1}{5} = 1$$
$$15\left(\frac{m - 4}{3} - \frac{3m - 1}{5}\right) = 15(1)$$
$$15\left(\frac{m - 4}{3}\right) - 15\left(\frac{3m - 1}{5}\right) = 15$$
$$5(m - 4) - 3(3m - 1) = 15$$
$$5m - 20 - 9m + 3 = 15$$
$$-4m - 17 = 15$$
$$-4m - 17 + 17 = 15 + 17$$
$$-4m = 32$$
$$\frac{-4m}{-4} = \frac{32}{-4}$$
$$m = -8$$
The solution set is $\{-8\}$.

73.
$$-(3x - 5) - (2x - 6) + 1 = -5(x - 1) - (3x + 2) + 3$$
$$-3x + 5 - 2x + 6 + 1 = -5x + 5 - 3x - 2 + 3$$
$$-5x + 12 = -8x + 6$$
$$-5x + 12 + 8x = -8x + 6 + 8x$$
$$3x + 12 = 6$$
$$3x + 12 - 12 = 6 - 12$$
$$3x = -6$$
$$\frac{3x}{3} = \frac{-6}{3}$$
$$x = -2$$
The solution set is $\{-2\}$.

77.
$$2[7 - 5(1 - n)] + 8n = -16 + 3[6(n + 1) - 3n]$$
$$2[7 - 5 + 5n] + 8n = -16 + 3[6n + 6 - 3n]$$
$$2[2 + 5n] + 8n = -16 + 3[3n + 6]$$
$$4 + 10n + 8n = -16 + 9n + 18$$
$$4 + 18n = 9n + 2$$

$$4 + 18n - 9n = 9n + 2 - 9n$$
$$4 + 9n = 2$$
$$4 + 9n - 4 = 2 - 4$$
$$9n = -2$$
$$\frac{9n}{9} = \frac{-2}{9}$$
$$n = -\frac{2}{9}$$
The solution set is $\left\{-\dfrac{2}{9}\right\}$.

81. The product of 8 and a number is written as $8x$.

85.
$$2x + 19 = 13$$
$$2x + 19 - 19 = 13 - 19$$
$$2x = -6$$
$$\frac{2x}{2} = \frac{-6}{2}$$
$$x = -3$$

89. a. $4(x + 1) + 1 = 4x + 4 + 1 = 4x + 5$

b.
$$4(x + 1) + 1 = -7$$
$$4x + 4 + 1 = -7$$
$$4x + 5 = -7$$
$$4x = -7 - 5$$
$$4x = -12$$
$$x = \frac{-12}{4}$$
$$x = -3$$
The solution set is $\{-3\}$.

c. answers may vary

93.
$$3.2x + 4 = 5.4x - 7$$
$$3.2x + 4 - 4 = 5.4x - 7 - 4$$
$$3.2x = 5.4x - 11$$
The equations are equivalent when $K = -11$.

97. answers may vary

101.
$$x(x - 6) + 7 = x(x + 1)$$
$$x^2 - 6x + 7 = x^2 + x$$
$$x^2 - 6x + 7 - x^2 = x^2 + x - x^2$$
$$-6x + 7 = x$$
$$6x + 7 + 6x = x + 6x$$
$$7 = 7x$$
$$\frac{7}{7} = \frac{7x}{7}$$
$$1 = x$$
Check: $\qquad x(x - 6) + 7 = x(x + 1)$
$$1(1 - 6) + 7 \stackrel{?}{=} 1(1 + 1)$$
$$-5 + 7 \stackrel{?}{=} 2$$
$$2 = 2 \quad \text{True}$$
The solution set is $\{1\}$.

Exercise Set 2.2

1. The perimeter is the sum of the lengths of the sides.
$$y + y + y + y = 4y$$

5. x nickels have a value of $5x$ cents. $(x + 3)$ dimes have a value of $10(x + 3)$ cents, and $2x$ quarters have a value of $25(2x)$ cents.
$$5x + 10(x + 3) + 25(2x) = 5x + 10x + 30 + 50x$$
$$= 65x + 30$$
The total amount of money is $(65x + 30)$ cents.

9. The length of the unknown vertical side is $10 - 2 = 8$. The length of the unknown horizontal side is
$$x - 3 - (x - 10) = x - 3 - x + 10 = 7.$$
$$10 + (x - 10) + 2 + 7 + 8 + (x - 3) = 2x + 14$$
The perimeter is $2x + 14$.

13. Let x be the first number. The other two numbers are then $5x$ and $100 + x$.

$$x + 5x + 100 + x = 415$$
$$7x + 100 = 415$$
$$7x = 315$$
$$x = 45$$

$5x = 5(45) = 225$

$x + 100 = 45 + 100 = 145$

The numbers are 45, 225, and 145.

17. Let x be the number of minor earthquakes in 2004. Then x is 91% of 3456.

$$x = 0.91(3456) = 3144.96$$

There were 3145 minor earthquakes in the United States in 2004.

21. Let x be the percent of e-mail users that spend less than 15 minutes on e-mail at work per day.

$$x + 50 + 8 + 10 + 9 = 100$$
$$x + 77 = 100$$
$$x = 23$$

23% of e-mail users at work spend less than 15 minutes on e-mail each day.

25. Let x be the number of seats at Heinz Field. Then INVESCO Field has $x + 11{,}675$ seats.

$$x + (x + 11{,}675) = 140{,}575$$
$$2x + 11{,}675 = 140{,}575$$
$$2x = 128{,}900$$
$$x = 64{,}450$$

$x + 11{,}675 = 64{,}450 + 11{,}675 = 76{,}125$

INVESCO Field at Mile High has 76,125 seats, while Heinz Field has 64,450 seats.

29. Let x be the population of the New York metropolitan region. Then Mexico City's population is $x + 0.4$, and Tokyo's population is $2x - 1.6$.

$$x + (x + 0.4) + 2x - 1.6 = 72$$
$$4x - 1.2 = 72$$
$$4x = 73.2$$
$$x = 18.3$$

$x + 0.4 = 18.3 + 0.4 = 18.7$

$2x - 1.6 = 2(18.3) - 1.6 = 36.6 - 1.6 = 35.0$

The population of New York is 18.3 million, the population of Mexico City is 18.7 million, and the population of Tokyo is 35.0 million.

33.
$$4x + (5x + 1) + (5x + 3) = 102$$
$$14x + 4 = 102$$
$$14x = 98$$
$$x = 7$$

$4x = 4(7) = 28$

$5x + 1 = 5(7) + 1 = 35 + 1 = 36$

$5x + 3 = 5(7) + 3 = 35 + 3 = 38$

The sides have lengths 28 meters, 36 meters, and 38 meters.

37.
$$(2x - 51) + \left(\frac{3}{2}x + 3\right) + x = 780$$

$$2(2x - 51) + 2\left(\frac{3}{2}x + 3\right) + 2(x) = 2(780)$$

$$4x - 102 + 3x + 6 + 2x = 1560$$
$$9x - 96 = 1560$$
$$9x = 1656$$
$$x = 184$$

$2x - 51 = 2(184) - 51 = 368 - 51 = 317$

$\frac{3}{2}x + 3 = \frac{3}{2}(184) + 3 = 276 + 3 = 279$

Occupation	Increase in Number of Jobs
Security guards	317 thousand
Home Health aides	279 thousand
Computer systems analysts	184 thousand
Total	780 thousand

41. Let x be the number of seats in a B737-200 aircraft. Then the number of seats in a B767-300ER is $x + 88$, and the number of seats in a F-100 is $x - 32$.

$$x + (x + 88) + (x - 32) = 413$$
$$3x + 56 = 413$$
$$3x = 357$$
$$x = 119$$

$x + 88 = 119 + 88 = 207$

$x - 32 = 119 - 32 = 87$

The B737-200 has 119 seats, the B767-300ER has 207 seats, and the F-100 has 87 seats.

45. Let x be the closing price on August 19, 2004. Then the closing price on December 28, 2004, is x increased by 92.1% of x.

$$x + 0.921x = 192.76$$
$$1.921x = 192.76$$
$$x \approx 100.34$$

The closing price on August 19, 2004 was $100.34.

49. Let $x°$ be the measure of the second angle. Then the first angle measures $(2x)°$, and the third angle measures $(3x - 12)°$.

$$x + 2x + (3x - 12) = 180$$
$$6x - 12 = 180$$
$$6x = 192$$
$$x = 32$$

$2x = 2(32) = 64$

$3x - 12 = 3(32) - 12 = 96 - 12 = 84$

The angles measure 64°, 32°, and 84°.

53. Let x be the expected population of South Africa in 2050. Then x is 46,900,000 less the decrease of 11% of 46,900,000.

$$x = 46{,}900{,}000 - 0.11(46{,}900{,}000)$$
$$= 46{,}900{,}000 - 5{,}159{,}000$$
$$= 41{,}741{,}000$$

The population of South Africa is expected to be 41,741,000 in 2050.

57. Let $x°$ be the measure of the first angle. Then its supplement measures $(3x + 20)°$.

$$x + (3x + 20) = 180$$
$$4x + 20 = 180$$
$$4x = 160$$
$$x = 40$$

$3x + 20 = 3(40) + 20 = 120 + 20 = 140$

The angles measure 40° and 140°.

61. Let x be the number of home runs that Derek Jeter hit. Then Mark Mulder hit $(x + 2)$ home runs, and Geoff Jenkins hit $(x + 4)$ home runs.

$$x + (x + 2) + (x + 4) = 75$$
$$3x + 6 = 75$$
$$3x = 69$$
$$x = 23$$

$x + 2 = 23 + 2 = 25$

$x + 4 = 23 + 4 = 27$

Jeter hit 23 home runs, Mulder hit 25 home runs, and Jenkins hit 27 home runs in 2004.

65. $n^2 - m^2 = (-3)^2 - (-8)^2 = 9 - 64 = -55$

69. yes; answers may vary

73. a. $\quad y = -72.8x + 2843.5$

$\quad\quad 0 = -72.8x + 2843.5$

$\quad 72.8x = 2843.5$

$\quad\quad x \approx 39.06$

The average number of cigarettes smoked will be 0 39 years after 1990, or in 2029.

b. $y = -72.8x + 2843.5$

$\quad y = -72.8(20) + 2843.5$

$\quad\quad = -1456 + 2843.5$

$\quad\quad = 1387.5$

The average number of cigarettes smoked by an American adult is predicted to be 1387.5 in 2010.

c. The average number of cigarettes smoked daily in 2010 is predicted to be $\dfrac{1387.5}{365} \approx 4$. This does not represent the average number of cigarettes smoked by an American smoker, because it is the average for *all* Americans, both smokers and non-smokers.

77. $\quad R = C$

$\quad 60x = 50x + 5000$

$\quad 10x = 5000$

$\quad\quad x = 500$

$\quad 50x + 5000 = 50(500) + 5000$

$\quad\quad\quad\quad\quad = 25,000 + 5000$

$\quad\quad\quad\quad\quad = 30,000$

500 computer boards must be sold to break even. It costs $30,000 to produce the 500 boards.

Exercise Set 2.3

1. $d = rt$

$\dfrac{d}{r} = \dfrac{rt}{r}$

$\dfrac{d}{r} = t$

5. $\quad\quad P = a + b + c$

$P - a - b = c$

9. $\quad\quad P = 2l + 2w$

$\quad P - 2w = 2l$

$\dfrac{P - 2w}{2} = l \text{ or } l = \dfrac{P}{2} - w$

13. $5x + 4y = 20$

$\quad\quad 4y = 20 - 5x$

$\quad\quad y = \dfrac{20 - 5x}{4} \text{ or } y = 5 - \dfrac{5}{4}x$

17. $\quad C = 2\pi r$

$\dfrac{C}{2\pi} = \dfrac{2\pi r}{2\pi}$

$\dfrac{C}{2\pi} = r$

21. Use $A = P\left(1 + \dfrac{r}{n}\right)^{nt}$ with $P = 3500$,

$r = 3\% = 0.03$, and $t = 10$.

$n = 1: \quad A = 3500\left(1 + \dfrac{0.03}{1}\right)^{1 \cdot 10}$

$\quad\quad\quad A = 3500(1.03)^{10}$

$\quad\quad\quad A \approx \$4703.71$

$n = 2: \quad A = 3500\left(1 + \dfrac{0.03}{2}\right)^{2 \cdot 10}$

$\quad\quad\quad A = 3500(1.015)^{20}$

$\quad\quad\quad A \approx \$4713.99$

$n = 4: \quad A = 3500\left(1 + \dfrac{0.03}{4}\right)^{4 \cdot 10}$

$\quad\quad\quad A = 3500(1.0075)^{40}$

$\quad\quad\quad A \approx \$4719.22$

$n = 12: \quad A = 3500\left(1 + \dfrac{0.03}{12}\right)^{12 \cdot 10}$

$\quad\quad\quad A = 3500(1.0025)^{120}$

$\quad\quad\quad A \approx \$4722.74$

$n = 365: \quad A = 3500\left(1 + \dfrac{0.03}{365}\right)^{365 \cdot 10}$

$\quad\quad\quad A \approx 4724.45$

25. $C = \dfrac{5}{9}(F - 32)$

$C = \dfrac{5}{9}(104 - 32)$

$C = \dfrac{5}{9}(72)$

$C = 40$

$104°F$ is $40°C$.

29. $A = l \cdot w$

$A = 64 \cdot 64$

$A = 4096 \text{ sq ft}$

$\text{Packages} = \dfrac{4096}{24} \approx 170.6$

Buy 171 packages of tiles.

33. $A = 2(l \cdot h) + 2(w \cdot h)$

$A = 2(14 \cdot 8) + 2(16 \cdot 8)$

$A = 224 + 256 = 480 \text{ sq ft}$

$\text{Two coats} = 2 \cdot 480 = 960 \text{ sq ft}$

$\text{Gallons} = \dfrac{960}{500} = 1.92$

Buy 2 gallons of paint.

37. Radius of satellite orbit $= 22,248 + 4000$

$\quad\quad\quad\quad\quad\quad\quad\quad = 26,248 \text{ mi}$

$C = 2\pi r = 2 \cdot \pi \cdot 26,248 \approx 164,921 \text{ mi}$

41. $C = \pi d = \pi \cdot 41,125 = 41.125\pi \text{ ft}$

$\quad\quad\quad\quad\quad\quad \approx 41.125 \cdot 3.14$

$\quad\quad\quad\quad\quad\quad = 129.1325 \text{ ft}$

45. $C = 4h + 9f + 4p$

$C = 4 \cdot 7 + 9 \cdot 14 + 4 \cdot 6$

$C = 28 + 126 + 24 = 178$

A serving of cashews contains 178 calories.

49. $\{-3, -2, -1\}$

53. answers may vary

57. $C = \dfrac{1,700,000,000}{250,000,000} = 6.8$

Cost is $6.80 per person.

61. $N = R^* \times f_p \times n_e \times f_l \times f_i \times f_c \times L$

$\dfrac{N}{R^* \times f_p \times f_l \times f_i \times f_c \times L} = \dfrac{R^* \times f_p \times n_e \times f_l \times f_i \times f_c \times L}{R^* \times f_p \times f_l \times f_i \times f_c \times L}$

$\dfrac{N}{R^* \times f_p \times f_l \times f_i \times f_c \times L} = n_e$

Exercise Set 2.4

1. $\{x \mid x < -3\}$ is $(-\infty, -3)$.

5. $\{x \mid -7 \le x\}$ is $[-7, \infty)$.

9. $\{x \mid 5 \ge x > -1\}$ is $(-1, 5]$.

13.
$$7x < 6x + 1$$
$$7x - 6x < 6x + 1 - 6x$$
$$x < 1$$
The solution set is $(-\infty, 1)$. ⟵——————→
$\qquad\qquad\qquad\qquad\qquad\qquad\quad 1$

17.
$$\frac{3}{4}x \geq 6$$
$$\frac{4}{3}\left(\frac{3}{4}x\right) \geq \frac{4}{3}(6)$$
$$x \geq 8$$
The solution set is $[8, \infty)$. ⟵——[——————→
$\qquad\qquad\qquad\qquad\qquad\qquad\quad 8$

21. $-3x \geq 9$
$$\frac{-3x}{-3} \leq \frac{9}{-3}$$
$$x \leq -3$$
The solution set is $(-\infty, -3]$. ⟵—————]——→
$\qquad\qquad\qquad\qquad\qquad\qquad\quad -3$

25.
$$15 + 2x \geq 4x - 7$$
$$15 + 2x - 4x \geq 4x - 7 - 4x$$
$$15 - 2x \geq -7$$
$$15 - 2x - 15 \geq -7 - 15$$
$$-2x \geq -22$$
$$\frac{-2x}{-2} \leq \frac{-22}{-2}$$
$$x \leq 11$$
The solution set is $(-\infty, 11]$.

29.
$$3(x - 5) < 2(2x - 1)$$
$$3x - 15 < 4x - 2$$
$$3x - 15 - 3x < 4x - 2 - 3x$$
$$-15 < x - 2$$
$$-15 + 2 < x - 2 + 2$$
$$-13 < x$$
The solution set is $(-13, \infty)$.

33.
$$-3(2x - 1) < -4[2 + 3(x + 2)]$$
$$-6x + 3 < -4[2 + 3x + 6]$$
$$-6x + 3 < -4[3x + 8]$$
$$-6x + 3 < -12x - 32$$
$$-6x + 3 + 12x < -12x - 32 + 12x$$
$$6x + 3 < -32$$
$$6x + 3 - 3 < -32 - 3$$
$$6x < -35$$
$$\frac{6x}{6} < \frac{-35}{6}$$
$$x < -\frac{35}{6}$$
The solution set is $\left(-\infty, -\frac{35}{6}\right)$.

37. $-x < -4$
$$\frac{-x}{-1} > \frac{-4}{-1}$$
$$x > 4$$
The solution set is $(4, \infty)$.

41.
$$\frac{1}{2} + \frac{2}{3} \geq \frac{x}{6}$$
$$6\left(\frac{1}{2} + \frac{2}{3}\right) \geq 6\left(\frac{x}{6}\right)$$
$$6\left(\frac{1}{2}\right) + 6\left(\frac{2}{3}\right) \geq x$$
$$3 + 4 \geq x$$
$$7 \geq x$$
The solution set is $(-\infty, 7]$.

45.
$$\frac{3}{4}(x - 7) \geq x + 2$$
$$4\left[\frac{3}{4}(x - 7)\right] \geq 4(x + 2)$$
$$3(x - 7) \geq 4(x + 2)$$
$$3x - 21 \geq 4x + 8$$
$$3x - 21 - 3x \geq 4x + 8 - 3x$$
$$-21 \geq x + 8$$
$$-21 - 8 \geq x + 8 - 8$$
$$-29 \geq x$$
The solution set is $(-\infty, -29]$.

49.
$$4(x - 6) + 2x - 4 \geq 3(x - 7) + 10x$$
$$4x - 24 + 2x - 4 \geq 3x - 21 + 10x$$
$$6x - 28 \geq 13x - 21$$
$$6x - 28 - 6x \geq 13x - 21 - 6x$$
$$-28 \geq 7x - 21$$
$$-28 + 21 \geq 7x - 21 + 21$$
$$-7 \geq 7x$$
$$\frac{-7}{7} \geq \frac{7x}{7}$$
$$-1 \geq x$$
The solution set is $(-\infty, -1]$.

53.
$$\frac{1}{2}(3x - 4) \leq \frac{3}{4}(x - 6) + 1$$
$$4\left[\frac{1}{2}(3x - 4)\right] \leq 4\left[\frac{3}{4}(x - 6) + 1\right]$$
$$2(3x - 4) \leq 3(x - 6) + 4$$
$$6x - 8 \leq 3x - 18 + 4$$
$$6x - 8 \leq 3x - 14$$
$$6x - 8 - 3x \leq 3x - 14 - 3x$$
$$3x - 8 \leq -14$$
$$3x - 8 + 8 \leq -14 + 8$$
$$3x \leq -6$$
$$\frac{3x}{3} \leq \frac{-6}{3}$$
$$x \leq -2$$
The solution set is $(-\infty, -2]$.

57.
$$\frac{2}{5}x - \frac{1}{4} \leq \frac{3}{10}x - \frac{4}{5}$$
$$20\left(\frac{2}{5}x - \frac{1}{4}\right) \leq 20\left(\frac{3}{10}x - \frac{4}{5}\right)$$
$$20\left(\frac{2}{5}x\right) - 20\left(\frac{1}{4}\right) \leq 20\left(\frac{3}{10}x\right) - 20\left(\frac{4}{5}\right)$$
$$8x - 5 \leq 6x - 16$$
$$8x - 5 - 6x \leq 6x - 16 - 6x$$
$$2x - 5 \leq -16$$
$$2x - 5 + 5 \leq -16 + 5$$
$$2x \leq -11$$
$$\frac{2x}{2} \leq \frac{-11}{2}$$
$$x \leq -\frac{11}{2}$$
The solution set is $\left(-\infty, -\frac{11}{2}\right]$.

61. a. Let x be the weight of luggage and cargo. The six passengers have a total weight of $6(160)$ pounds.
$$x + 6(160) \leq 2000$$
$$x + 960 \leq 2000$$
$$x \leq 1040$$
The solution is $\{x | x \leq 1040\}$.

b. The luggage and cargo must weigh no more than 1040 pounds.

65. a. Let x be the number of monthly calls. The monthly charge for plan 2 is $\$(13 + 0.06x)$, and plan 1 is more economical when the monthly charge for plan 2 is more than $25.

$$13 + 0.06x > 25$$
$$0.06x > 12$$
$$x > 200$$

The solution is $\{x \mid x > 200\}$.

b. Plan 1 is more economical for more than 200 monthly calls.

69. a.
$$s > 35{,}000$$
$$651.2t + 27{,}821 > 35{,}000$$
$$651.2t > 7179$$
$$t > \frac{7179}{651.2} \approx 11$$

Beginning salaries will be greater than $35,000 11 years after 1989, which is in the year 2000.

b. answers may vary

73. 2010 is 14 years after 1996, so 2010 corresponds to $t = 14$.

$$w = -1.9t + 70.1$$
$$w = -1.9(14) + 70.1$$
$$w = -26.6 + 70.1$$
$$w = 43.5$$

The average consumption of whole milk is expected to be 43.5 pounds per person per year in 2010.

77. answers may vary

81. The integers that are both greater than or equal to 0 and less than or equal to 7 are $\{0, 1, 2, 3, 4, 5, 6, 7\}$.

85. $3x - 12 = 3$
$$3x = 15$$
$$x = 5$$
The solution set is $\{5\}$.

89. To solve $-3x \le 14$, both sides must be divided by -3, so the inequality symbol must be reversed.

93. $2x - 3 = 5$
$$2x = 8$$
$$x = 4$$
The solution set is $\{4\}$.

97.
$$4(x - 1) \ge 4x - 8$$
$$4x - 4 \ge 4x - 8$$
$$4x - 4 - 4x \ge 4x - 8 - 4x$$
$$-4 \ge -8$$

Since the statement $-4 \ge -8$ is true regardless of the value of x, every real number is a solution. The solution set is $(-\infty, \infty)$.

101. answers may vary

Exercise Set 2.5

1. $C \cup D = \{2, 3, 4, 5\} \cup \{4, 5, 6, 7\}$
$$= \{2, 3, 4, 5, 6, 7\}$$

5. $A \cup B = \{x \mid x \text{ is an even number}\} \cup \{x \mid x \text{ is an odd integer}\}$
$$= \{x \mid x \text{ is an integer}\}$$
$$= \{\ldots, -2, -1, 0, 1, 2, \ldots\}$$

9. $B \cup C = \{x \mid x \text{ is an odd integer}\} \cup \{2, 3, 4, 5\}$
$$= \{x \mid x \text{ is an odd integer or } x = 2 \text{ or } x = 4\}$$

13.

$x < 1$:

$x > -3$:

$x < 1 \text{ and } x > -3$:

17.

$x < -1$:

$x < 1$:

$x < -1 \text{ and } x < 1$:

21. $4x + 2 \le -10$ and $2x \le 0$
$$4x \le -12 \quad \text{and} \quad x \le 0$$
$$x \le -3 \quad \text{and} \quad x \le 0$$
The solution set is $(-\infty, -3] \cap (-\infty, 0] = (-\infty, -3]$.

25. $5 < x - 6 < 11$
$$5 + 6 < x - 6 + 6 < 11 + 6$$
$$11 < x < 17$$
The solution set is $(11, 17)$.

29. $1 \le \dfrac{2}{3}x + 3 \le 4$
$$1 - 3 \le \frac{2}{3}x + 3 - 3 \le 4 - 3$$
$$-2 \le \frac{2}{3}x \le 1$$
$$\frac{3}{2}(-2) \le \frac{3}{2}\left(\frac{2}{3}x\right) \le \frac{3}{2}(1)$$
$$-3 \le x \le \frac{3}{2}$$
The solution set is $\left[-3, \dfrac{3}{2}\right]$.

33.

$x < 4$:

$x < 5$:

$x < 4 \text{ or } x < 5$:

37.

$x > 0$:

$x < 3$:

$x > 0 \text{ or } x < 3$:

41. $x + 4 < 0$ or $6x > -12$
$$x < -4 \quad \text{or} \quad x > -12$$
The solution set is $(-\infty, -4) \cup (-2, \infty)$.

45. $x < \dfrac{2}{3}$ and $x > -\dfrac{1}{2}$

The solution set is $\left(-\infty, \dfrac{2}{3}\right) \cap \left(-\dfrac{1}{2}, \infty\right) = \left(-\dfrac{1}{2}, \dfrac{2}{3}\right)$.

49. $0 \le 2x - 3 \le 9$
$$0 + 3 \le 2x - 3 + 3 \le 9 + 3$$
$$3 \le 2x \le 12$$
$$\frac{3}{2} \le \frac{2x}{2} \le \frac{12}{2}$$
$$\frac{3}{2} \le x \le 6$$
The solution set is $\left[\dfrac{3}{2}, 6\right]$.

53. $x + 3 \ge 3$ and $x + 3 \le 2$
$$x \ge 0 \quad \text{and} \quad x \le -1$$
The solution set is $[0, \infty) \cap (-\infty, -1] = \varnothing$.

57. $0 < \dfrac{5 - 2x}{3} < 5$

$$3(0) < 3\left(\dfrac{5 - 2x}{3}\right) < 3(5)$$

$$0 < 5 - 2x < 15$$

$$0 - 5 < 5 - 2x - 5 < 15 - 5$$

$$-5 < -2x < 10$$

$$\dfrac{-5}{-2} > \dfrac{-2x}{-2} > \dfrac{10}{-2}$$

$$\dfrac{5}{2} > x > -5$$

$$-5 < x < \dfrac{5}{2}$$

The solutions set is $\left(-5, \dfrac{5}{2}\right)$.

61. $-x + 5 > 6 \quad$ and $\quad 1 + 2x \le -5$

$\qquad -x > 1 \quad$ and $\qquad 2x \le -6$

$\qquad x < -1 \quad$ and $\qquad x \le -3$

The solution set is $(-\infty, -1) \cap (-\infty, -3] = (-\infty, -3]$.

65. $5 - x > 7 \quad$ and $\quad 2x + 3 \ge 13$

$\qquad -x > 2 \quad$ and $\qquad 2x \ge 10$

$\qquad x < -2 \quad$ and $\qquad x \ge 5$

The solution set is $(-\infty, -2) \cap [5, \infty) = \varnothing$.

69. $\dfrac{1}{15} < \dfrac{8 - 3x}{15} < \dfrac{4}{5}$

$$15\left(\dfrac{1}{15}\right) < 15\left(\dfrac{8 - 3x}{15}\right) < 15\left(\dfrac{4}{5}\right)$$

$$1 < 8 - 3x < 12$$

$$1 - 8 < 8 - 3x - 8 < 12 - 8$$

$$-7 < -3x < 4$$

$$\dfrac{-7}{-3} > \dfrac{-3x}{-3} > \dfrac{4}{-3}$$

$$\dfrac{7}{3} > x > -\dfrac{4}{3}$$

$$-\dfrac{4}{3} < x < \dfrac{7}{3}$$

The solution set is $\left(-\dfrac{4}{3}, \dfrac{7}{3}\right)$.

73. $|-7| - |19| = 7 - 19 = -12$

77. $|x| = 7$ when $x = 7$ or $x = -7$. The solution set is $\{-7, 7\}$.

81. Both lines are above the level representing 48 pounds per person for the years 1993, 1994, 1995, 1998, 1999, and 2002.

85. $x - 6 < 3x < 2x + 5$

$\quad x - 6 < 3x \quad$ and $\quad 3x < 2x + 5$

$\qquad -6 < 2x \quad$ and $\qquad x < 5$

$\qquad -3 < x \quad$ and $\qquad x < 5$

The solution set is $(-3, \infty) \cap (-\infty, 5) = (-3, 5)$.

89. $-29 \le C \le 35$

$$-29 \le \dfrac{5}{9}(F - 32) \le 35$$

$$\dfrac{9}{5}(-29) \le \dfrac{9}{5}\left[\dfrac{5}{9}(F - 32)\right] \le \dfrac{9}{5}(35)$$

$$-52.2 \le F - 32 \le 63$$

$$-52.2 + 32 \le F - 32 + 32 \le 63 + 32$$

$$-20.2 \le F \le 95$$

The temperatures ranged from $-20.2°$F to $95°$F.

Exercise Set 2.6

1. $|x| = 7$

$\quad x = 7 \quad$ or $\quad x = -7$

The solution set is $\{7, -7\}$.

5. $|3x| = 12.6$

$\quad 3x = 12.6 \quad$ or $\quad 3x = -12.6$

$\qquad x = 4.2 \quad$ or $\qquad x = -4.2$

The solution set is $\{4.2, -4.2\}$.

9. $-6|x| + 44 = -10$

$\qquad -6|x| = -54$

$\qquad\quad |x| = 9$

$\quad x = 9$ or $x = -9$

The solution set is $\{-9, 9\}$.

13. $|2x - 5| = 9$

$\quad 2x - 5 = 9 \quad$ or $\quad 2x - 5 = -9$

$\qquad 2x = 14 \quad$ or $\qquad 2x = -4$

$\qquad\quad x = 7 \quad$ or $\qquad\quad x = -2$

The solution set is $\{7, -2\}$.

17. $|z| + 4 = 9$

$\quad |z| = 5$

$\quad z = 5 \quad$ or $\quad z = -5$

The solution set is $\{5, -5\}$.

21. $\left|\dfrac{4x - 6}{3}\right| = 6$

$\quad \dfrac{4x - 6}{3} = 6 \quad$ or $\quad \dfrac{4x - 6}{3} = -6$

$\quad 4x - 6 = 18 \quad$ or $\quad 4x - 6 = -18$

$\qquad 4x = 24 \quad$ or $\qquad 4x = -12$

$\qquad\quad x = 6 \quad$ or $\qquad\quad x = -3$

The solution set is $\{-3, 6\}$.

25. $|4n + 1| + 10 = 4$

$\qquad |4n + 1| = -6$

Since $|4n + 1| \ge 0$, there are no values satisfying $|4n + 1| = -6$. The solution set is $\varnothing$.

29. $|5x - 7| = |3x + 11|$

$\quad 5x - 7 = 3x + 11 \quad$ or $\quad 5x - 7 = -(3x + 11)$

$\quad 2x - 7 = 11 \qquad\quad$ or $\quad 5x - 7 = -3x - 11$

$\qquad 2x = 18 \qquad\quad$ or $\qquad 8x - 7 = -11$

$\qquad\quad x = 9 \qquad\quad$ or $\qquad\quad 8x = -4$

$\qquad\qquad\qquad\qquad\qquad\qquad x = -\dfrac{4}{8} = -\dfrac{1}{2}$

The solution set is $\left\{-\dfrac{1}{2}, 9\right\}$.

33. $|2y - 3| = |9 - 4y|$

$\quad 2y - 3 = 9 - 4y \quad$ or $\quad 2y - 3 = -(9 - 4y)$

$\qquad 6y = 12 \qquad\quad$ or $\quad 2y - 3 = -9 + 4y$

$\qquad\quad y = 2 \qquad\quad$ or $\qquad\quad 6 = 2y$

$\qquad\quad y = 2 \qquad\quad$ or $\qquad\quad 3 = y$

The solution set is $\{2, 3\}$.

37. $|2x - 6| = |10 - 2x|$

$\quad 2x - 6 = 10 - 2x \quad$ or $\quad 2x - 6 = -(10 - 2x)$

$\qquad 4x - 6 = 10 \qquad$ or $\quad 2x - 6 = -10 + 2x$

$\qquad\quad 4x = 16 \qquad\quad$ or $\qquad -6 = -10 \quad$ False

$\qquad\qquad x = 4$

The solution set is $\{4\}$.

41. $\left|\dfrac{2x+1}{5}\right| = \left|\dfrac{3x-7}{3}\right|$

$\dfrac{2x+1}{5} = \dfrac{3x-7}{3}$ or $\dfrac{2x+1}{5} = -\left(\dfrac{3x-7}{3}\right)$

$15 \cdot \dfrac{2x+1}{5} = 15 \cdot \dfrac{3x-7}{3}$ or $\dfrac{2x+1}{5} = \dfrac{-3x+7}{3}$

$3(2x+1) = 5(3x-7)$ or $15 \cdot \dfrac{2x+1}{5} = 15 \cdot \dfrac{-3x+7}{3}$

$6x+3 = 15x-35$ or $3(2x+1) = 5(-3x+7)$

$-9x+3 = -35$ or $6x+3 = -15x+35$

$-9x = -38$ or $21x+3 = 35$

$x = \dfrac{-38}{-9} = \dfrac{38}{9}$ or $21x = 32$

$x = \dfrac{32}{21}$

The solution set is $\left\{\dfrac{32}{21}, \dfrac{38}{9}\right\}$.

45. $|x| \le 4$
$-4 \le x \le 4$
The solution set is $[-4, 4]$.

49. $|x+3| < 2$
$-2 < x+3 < 2$
$-5 < x < -1$
The solution set is $(-5, -1)$.

53. $\left|\dfrac{x+2}{3}\right| < 1$

$-1 < \dfrac{x+2}{3} < 1$

$-3 < x+2 < 3$

$-5 < x < 1$

The solution set is $(-5, 1)$.

57. $|2x+3| \le 0$
Since absolute value can never be negative, the only solution is when $|2x+3| = 0$.
$|2x+3| = 0$
$2x+3 = 0$
$2x = -3$
$x = -\dfrac{3}{2}$
The solution set is $\left\{-\dfrac{3}{2}\right\}$.

61. $|2x+7| \le 13$
$-13 \le 2x+7 \le 13$
$-20 \le 2x \le 6$
$-10 \le x \le 3$
The solution set is $[-10, 3]$.

65. $|x+10| \ge 14$
$x+10 \le -14$ or $x+10 \ge 14$
$x \le -24$ or $x \ge 4$
The solution set is $(-\infty, -24] \cup [4, \infty)$.

69. $|x| > -4$
The absolute value of a number is always nonnegative. Thus, it will always be greater than -4. The solution set is all real numbers or $(-\infty, \infty)$.

73. $|6x-8| - 7 > -3$
$|6x-8| > 4$
$6x-8 < -4$ or $6x-8 > 4$
$6x < 4$ or $6x > 12$
$x < \dfrac{2}{3}$ or $x > 2$
The solution set is $\left(-\infty, \dfrac{2}{3}\right) \cup (2, \infty)$.

77. $\left|\dfrac{x+6}{3}\right| > 2$

$\dfrac{x+6}{3} < -2$ or $\dfrac{x+6}{3} > 2$

$x+6 < -6$ or $x+6 > 6$

$x < -12$ or $x > 0$

The solution set is $(-\infty, -12) \cup (0, \infty)$.

81. $|x| < 13$
$-13 < x < 13$
The solution set is $(-13, 13)$.

85. $2|x| - 9 \le 11$
$2|x| \le 20$
$|x| \le 10$
$-10 \le x \le 10$
The solution set is $[-10, 10]$.

89. $|x-5| \ge 12$
$x-5 \le -12$ or $x-5 \ge 12$
$x \le -7$ or $x \ge 17$
The solution set is $(-\infty, -7] \cup [17, \infty)$.

93. $|2x+1| - 7 < -4$
$|2x+1| < 3$
$-3 < 2x+1 < 3$
$-4 < 2x < 2$
$-2 < x < 1$
The solution set is $(-2, 1)$.

97. $|3x-5| + 4 = 5$
$|3x-5| = 1$
$3x-5 = -1$ or $3x-5 = 1$
$3x = 4$ or $3x = 6$
$x = \dfrac{4}{3}$ or $x = 2$
The solution set is $\left\{\dfrac{4}{3}, 2\right\}$.

101. $\left|\dfrac{1-2x}{3}\right| = 6$

$\dfrac{1-2x}{3} = -6$ or $\dfrac{1-2x}{3} = 6$

$1-2x = -18$ or $1-2x = 18$

$-2x = -19$ or $-2x = 17$

$x = \dfrac{19}{2}$ or $x = -\dfrac{17}{2}$

The solution set is $\left\{-\dfrac{17}{2}, \dfrac{19}{2}\right\}$.

105. $|6x-3| = |4x+5|$
$6x-3 = 4x+5$ or $6x-3 = -(4x+5)$
$6x-3 = 4x+5$ or $6x-3 = -4x-5$
$2x-3 = 5$ or $10x-3 = -5$
$2x = 8$ or $10x = -2$
$x = 4$ or $x = -\dfrac{1}{5}$
The solution set is $\left\{-\dfrac{1}{5}, 4\right\}$.

109. From the circle graph, cheddar cheese made up 32% of U.S. cheese production in 2003.

113. Since absolute value is never negative, the solution set is $\varnothing$.

117. All numbers whose distance from 0 is 5 units is written as $|x| = 5$; answers may vary.

121. $-5 \le x \le 5$ is $|x| \le 5$.

Chapter 2 Test

1.
$$8x + 14 = 5x + 44$$
$$8x + 14 - 5x = 5x + 44 - 5x$$
$$3x + 14 = 44$$
$$3x + 14 - 14 = 44 - 14$$
$$3x = 30$$
$$\frac{3x}{3} = \frac{30}{3}$$
$$x = 10$$
The solution set is $\{10\}$.

5.
$$\frac{7w}{4} + 5 = \frac{3w}{10} + 1$$
$$20\left(\frac{7w}{4} + 5\right) = 20\left(\frac{3w}{10} + 1\right)$$
$$20\left(\frac{7w}{4}\right) + 20(5) = 20\left(\frac{3w}{10}\right) + 20(1)$$
$$35w + 100 = 6w + 20$$
$$35w + 100 - 6w = 6w + 20 - 6w$$
$$29w + 100 = 20$$
$$29w + 100 - 100 = 20 - 100$$
$$29w = -80$$
$$\frac{29w}{29} = \frac{-80}{29}$$
$$w = -\frac{80}{29}$$
The solution set is $\left\{-\frac{80}{29}\right\}$.

9. $|2x - 3| = |4x + 5|$
$$2x - 3 = 4x + 5 \quad \text{or} \quad 2x - 3 = -(4x + 5)$$
$$2x - 3 = 4x + 5 \quad \text{or} \quad 2x - 3 = -4x - 5$$
$$-2x - 3 = 5 \qquad \text{or} \quad 6x - 3 = -5$$
$$-2x = 8 \qquad \text{or} \qquad 6x = -2$$
$$x = -4 \qquad \text{or} \qquad x = -\frac{1}{3}$$
The solution set is $\left\{-4, -\frac{1}{3}\right\}$.

13.
$$F = \frac{9}{5}C + 32$$
$$F - 32 = \frac{9}{5}C$$
$$\frac{5}{9}(F - 32) = C$$

17. $|3x + 1| > 5$
$$3x + 1 < -5 \quad \text{or} \quad 3x + 1 > 5$$
$$3x < -6 \quad \text{or} \qquad 3x > 4$$
$$x < -2 \quad \text{or} \qquad x > \frac{4}{3}$$
The solution set is $(-\infty, -2) \cup \left(\frac{4}{3}, \infty\right)$.

21.
$$-1 \le \frac{2x - 5}{3} < 2$$
$$3(-1) \le 3\left(\frac{2x - 5}{3}\right) < 3(2)$$
$$-3 \le 2x - 5 < 6$$
$$-3 + 5 \le 2x < 6 + 5$$
$$2 \le 2x < 11$$
$$\frac{2}{2} \le \frac{2x}{2} < \frac{11}{2}$$
$$1 \le x < \frac{11}{2}$$
The solution set is $\left[1, \frac{11}{2}\right)$.

25. Recall that $C = 2\pi r$. Here $C = 78.5$.
$$78.5 = 2\pi r$$
$$r = \frac{78.5}{2\pi} = \frac{39.25}{\pi}$$
Also recall that $A = \pi r^2$.
$$A = \pi\left(\frac{39.25}{\pi}\right)^2 = \frac{39.25^2}{\pi} \approx \frac{39.25^2}{3.14} \approx 491$$
The area of the pen is about 491 square feet. Each dog requires at least 60 square feet of space, and $\frac{491}{60} \approx 8.18$. At most 8 dogs could be kept in the pen.

CHAPTER 3

Exercise Set 3.1

1. $(3, 2)$ is in quadrant I.

$(-5, 3)$ is in quadrant II.

$\left(5\frac{1}{2}, -4\right)$ is in quadrant IV.

$(0, 3.5)$ is on the y-axis.

$(-2, -4)$ is in quadrant III.

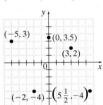

5. $(x, 0)$ is on the x-axis.

9.

Domestic Airline Revenues in U.S.

13. $(1, 0)$: $\quad -6 + 5y = -6$
$$6(1) + 5(0) \overset{?}{=} -6$$
$$-6 + 0 \overset{?}{=} -6$$
$$-6 = -6 \quad \text{True}$$
$(1, 0)$ is a solution.

$\left(2, \frac{6}{5}\right)$: $\quad -6x + 5y = -6$
$$-6(2) + 5\left(\frac{6}{5}\right) \overset{?}{=} -6$$
$$-12 + 6 \overset{?}{=} -6$$
$$-6 = -6 \quad \text{True}$$
$\left(2, \frac{6}{5}\right)$ is a solution.

17. $x - 2y = 4$

Find three ordered pair solutions.

x	y
4	0
0	−2
−2	−3

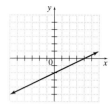

21. $x = 4$

Any ordered pair with an x-coordinate of 4 is a solution to $x = 4$.

x	y
4	−1
4	0
4	1

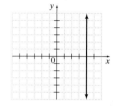

25. $y = 3x$

Find three ordered pair solutions.

x	y
0	0
1	3
−1	−3

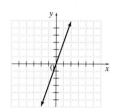

29. $4x + 5y = 15$

Find three ordered pair solutions.

x	y
0	3
5	−1
−5	7

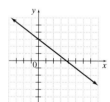

33. $x = \dfrac{1}{2}$

This is a vertical line with an x-intercept at $\left(\dfrac{1}{2}, 0\right)$ and no y-intercept.

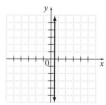

37. $y = -4x + 1$

Find three ordered pair solutions.

x	y
0	1
1	−3
−1	5

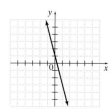

41. $y = -\dfrac{2}{3}x - 4$

Find three ordered pair solutions.

x	y
−3	−2
0	−4
3	−6

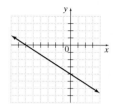

45. $\dfrac{-6 - 3}{2 - 8} = \dfrac{-9}{-6} = \dfrac{3}{2}$

49. $\dfrac{0 - 6}{5 - 0} = \dfrac{-6}{5} = -\dfrac{6}{5}$

53. The description matches graph B.

57. From the graph, the minimum hourly wage went from $3.80 to $4.25 in 1991.

61. a. $y = 2x + 6$

Find three ordered pair solutions.

x	y
0	6
1	8
2	10

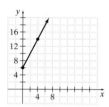

b. The line goes through the point $(4, 14)$, so the perimeter is 14 inches when the length is 4 inches.

65. The value was originally $7000. After one year, the value was $6500, so the loss in value during the first year was $500.

69. a. $2x + 3y = 1500$

Let $x = 0$.

$2(0) + 3y = 1500$

$3y = 1500$

$y = 500$

The ordered pair is $(0, 500)$. This corresponds to manufacturing 0 tables and 500 chairs.

b. $2x + 3y = 1500$

Let $y = 0$.

$2x + 3(0) = 1500$

$2x = 1500$

$x = 750$

The ordered pair is $(750, 0)$. This corresponds to manufacturing 750 tables and 0 chairs.

c. Let $x = 50$.

$2x + 3y = 1500$

$2(50) + 3y = 1500$

$100 + 3y = 1500$

$3y = 1400$

$y = \dfrac{1400}{3} = 466\dfrac{2}{3}$

At most 466 chairs can be produced.

73. The vertical line $x = 0$ has y-intercepts.

77.

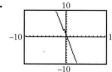

Exercise Set 3.2

1. $m = \dfrac{11 - 2}{8 - 3} = \dfrac{9}{5}$

5. $m = \dfrac{3 - 8}{4 - (-2)} = -\dfrac{5}{6}$

9. $m = \dfrac{11 - (-1)}{-12 - (-3)} = \dfrac{12}{-9} = -\dfrac{4}{3}$

13. The two points are $(0, -2)$ and $(2, 2)$.

$m = \dfrac{2 - (-2)}{2 - 0} = \dfrac{4}{2} = 2$

17. $m = \dfrac{8}{12} = \dfrac{2}{3}$

The pitch of the roof is $\dfrac{2}{3}$.

21. $y = -x + 5$
$m = -1$; y-intercept $= (0, 5)$

25. $2x + y = 7$
$y = -2x + 7$
$m = -2$; y-intercept $= (0, 7)$

29. $x = 4$

This is a vertical line, so the slope is undefined. Since the line is not the y-axis, there is no y-intercept.

33. $y = \dfrac{1}{2}x$

$y = \dfrac{1}{2}x + 0$

$m = \dfrac{1}{2}$; y-intercept $= (0, 0)$

37. $-6x + 5y = 30$
$5y = 6x + 30$
$y = \dfrac{6}{5}x + 6$

$m = \dfrac{6}{5}$; y-intercept $= (0, 6)$

41. $x + 2 = 0$
$x = -2$

This is a vertical line, so the slope is undefined. Since the line is not the y-axis, there is no y-intercept.

45. $y = -2x + 3$
$m = -2$; y-intercept $= (0, 3)$
The graph that matches these conditions is graph B.

49. l_2 has the greater slope because the slope of l_2 is 0 and the slope of l_1 is negative.

53. $y = -3x + 6$ $\qquad$ $y = 3x + 5$
$m = -3$ $\qquad\qquad$ $m = 3$
y-int $= (0, 6)$ $\qquad$ y-int $= (0, 5)$

Neither, since their slopes are not equal, nor does their product equal -1.

57. $2x - y = -10$ $\qquad$ $2x + 4y = 2$
$\quad -y = -2x - 10$ $\qquad$ $4y = -2x + 2$
$\quad\;\; y = 2x + 10$ $\qquad$ $y = -\dfrac{1}{2}x + \dfrac{1}{2}$

$m = 2$ $\qquad\qquad\qquad$ $m = -\dfrac{1}{2}$

y-int $= (0, 10)$ $\qquad$ y-int $= \left(0, \dfrac{1}{2}\right)$

The lines are perpendicular since the product of their slopes is -1.

61. $-4x + 2y = 5$ $\qquad$ $2x - y = 7$
$\quad\quad 2y = 4x + 5$ $\qquad$ $-y = -2x + 7$
$\quad\quad\; y = 2x + \dfrac{5}{2}$ $\qquad$ $y = 2x - 7$

$m = 2$ $\qquad\qquad\qquad$ $m = 2$

y-int $= \left(0, \dfrac{5}{2}\right)$ $\qquad$ y-int $= (0, -7)$

The lines are parallel, since they have the same slope, but different y-intercepts.

65. Line 1: $m = \dfrac{0 - 5}{10 - 0} = \dfrac{-5}{10} = -\dfrac{1}{2}$

$\quad\;$ Line 2: $m = \dfrac{0 - (-10)}{5 - 0} = \dfrac{10}{5} = 2$

The lines are perpendicular since the product of their slopes is -1.

69. $|x - 3| = 6$
$\quad x - 3 = -6$ or $x - 3 = 6$
$\qquad\quad x = -3$ or $\qquad x = 9$

The solution set is $\{-3, 9\}$.

73. $|3x - 4| \leq 2$
$-2 \leq 3x - 4 \leq 2$
$\;\; 2 \leq 3x \leq 6$
$\;\; \dfrac{2}{3} \leq x \leq 2$

The solution set is $\left\{ x \middle| \dfrac{2}{3} \leq x \leq 2 \right\}$ or $\left[\dfrac{2}{3}, 2\right]$.

77. The calculation is correct.

81. $y = -\dfrac{7}{2}x - 6$

$m = -\dfrac{7}{2}$

The slope of a perpendicular line is

$-\left(-\dfrac{2}{7}\right) = \dfrac{2}{7}$.

85. $5x - 2y = 6$
$\quad -2y = -5x + 6$
$\qquad\; y = \dfrac{5}{2}x - 3$

The slope of a perpendicular line is

$-\left(\dfrac{2}{5}\right) = -\dfrac{2}{5}$.

89. $F(22, 2)$ and $G(26, 8)$

$m = \dfrac{8 - 2}{26 - 22} = \dfrac{6}{4} = \dfrac{3}{2}$

The rate of change in altitude is $\dfrac{3}{2}$ yards per second.

93. a. l_1: $\;\; m = \dfrac{-2 - 4}{2 - (-1)} = \dfrac{-6}{3} = -2$

$\quad\;$ l_2: $\;\; m = \dfrac{6 - 2}{-8 - (-4)} = \dfrac{4}{-4} = -1$

$\quad\;$ l_3: $\;\; m = \dfrac{-4 - 0}{0 - (-6)} = \dfrac{-4}{6} = -\dfrac{2}{3}$

b. Since l_1 is the steepest line, for lines with negative slopes, the steeper line has the lesser slope.

Exercise Set 3.3

1.

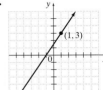

5.

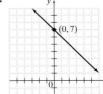

9.

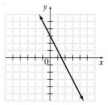

13.

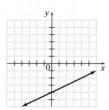

17.

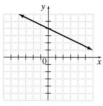

21. $y = -2x + 3$

$m = -2$; y-intercept $= (0, 3)$

This is graph D.

25. $y = mx + b$

$y = 2x + \dfrac{3}{4}$

29. a. Use $y = 1983x + 42{,}972$ with $x = 1$, since 2000 is one year after 1999.

$y = 1983(1) + 42{,}972 = 44{,}955$

The average annual income is \$44,955.

b. From $y = 1983x + 42{,}972$, $m = 1983$. The average annual income increases by \$1983 each year.

c. From $y = 1983x + 42{,}972$, the y-intercept is $(0, 42{,}972)$. At year 0, which is 1999, the average annual income was \$42,972.

33. a. Use $y = 17.7x + 44.9$ with $x = 7$, since 2004 is 7 years after 1997.

$y = 17.7(7) + 44.9 = 168.8$

In 2004, there were approximately 168.8 million cellular subscribers in the U.S.

b. Let $y = 220$ and solve for x.

$220 = 17.7x + 44.9$

$175.1 = 17.7x$

$9.89 \approx x$

The number of cellular subscribers will exceed 220 million about 9.9 years after 1997, or in 2006.

c. answers may vary

37. $y - (-1) = 2(x - 0)$

$y + 1 = 2x$

$\qquad y = 2x - 1$

41. The equation is correct.

Exercise Set 3.4

1. $y - y_1 = m(x - x_1)$

$y - 2 = 3(x - 1)$

$y - 2 = 3x - 3$

$\qquad y = 3x - 1$

5. $y - y_1 = m(x - x_1)$

$y - 2 = \dfrac{1}{2}[x - (-6)]$

$y - 2 = \dfrac{1}{2}(x + 6)$

$y - 2 = \dfrac{1}{2}x + 3$

$\qquad y = \dfrac{1}{2}x + 5$

9. $m = \dfrac{6 - 0}{4 - 2} = \dfrac{6}{2} = 3$

$y - y_1 = m(x - x_1)$

$y - 0 = 3(x - 2)$

$\qquad y = 3x - 6$

$\qquad 6 = 3x - y$

$3x - y = 6$

13. $m = \dfrac{-3 - (-4)}{-4 - (-2)} = \dfrac{1}{-2} = -\dfrac{1}{2}$

$y - y_1 = m(x - x_1)$

$y - (-4) = -\dfrac{1}{2}[x - (-2)]$

$\quad y + 4 = -\dfrac{1}{2}(x + 2)$

$-2y - 8 = x + 2$

$\quad -10 = x + 2y$

$x + 2y = -10$

17. $m = \dfrac{\dfrac{7}{10} - \dfrac{4}{10}}{-\dfrac{1}{5} - \dfrac{3}{5}} = \dfrac{\dfrac{3}{10}}{-\dfrac{4}{5}} = \dfrac{3}{10}\left(-\dfrac{5}{4}\right) = -\dfrac{3}{8}$

$y - y_1 = m(x - x_1)$

$y - \dfrac{4}{10} = -\dfrac{3}{8}\left(x - \dfrac{3}{5}\right)$

$y - \dfrac{4}{10} = -\dfrac{3}{8}x + \dfrac{9}{40}$

$\qquad y = -\dfrac{3}{8}x + \dfrac{5}{8}$

$\qquad 8y = -3x + 5$

$3x + 8y = 5$

21. A horizontal line has an equation of the form $y = b$. Since the line passes through the point $(-3, 1)$, its equation is $y = 1$.

25. $y = 4x - 2$; $m = 4$

A parallel line will have slope 4.

$y - y_1 = m(x - x_1)$

$y - 8 = 4(x - 3)$

$y - 8 = 4x - 12$

$\qquad y = 4x - 4$

29. The line $y = 7$ is horizontal, so a parallel line will also be horizontal, with equation $y = 4$.

33. The line $x = 3$ is vertical, so a perpendicular line will be horizontal, with equation $y = -5$.

37. $y - y_1 = m(x - x_1)$

$y - 3 = 2[x - (-2)]$

$y - 3 = 2(x + 2)$

$y - 3 = 2x + 4$

$\quad -7 = 2x - y$

$2x - y = -7$

41. $y = mx + b$

$y = -\dfrac{1}{2}x + \dfrac{3}{8}$

45. $y - y_1 = m(x - x_1)$

$y - 0 = -\dfrac{4}{3}[x - (-5)]$

$y = -\dfrac{4}{3}(x + 5)$

$3y = -4(x + 5)$

$3y = -4x - 20$

$4x + 3y = -20$

49. $2x + 4y = 9$

$4y = -2x + 9$

$y = -\dfrac{1}{2}x + 9$

$m = -\dfrac{1}{2}$

A parallel line will have slope $-\dfrac{1}{2}$.

$y - y_1 = m(x - x_1)$

$y - (-2) = -\dfrac{1}{2}(x - 6)$

$y + 2 = -\dfrac{1}{2}x + 3$

$y = -\dfrac{1}{2}x + 1$

$2y = -x + 2$

$x + 2y = 2$

53. $8x - y = 9$

$-y = -8x + 9$

$y = 8x - 9$

$m = 8$

A parallel line will have slope 8.

$y - y_1 = m(x - x_1)$

$y - 1 = 8(x - 6)$

$y - 1 = 8x - 48$

$y = 8x - 47$

$47 = 8x - y$

$8x - y = 47$

57. $m = \dfrac{-5 - (-8)}{-6 - 2} = \dfrac{3}{-8} = -\dfrac{3}{8}$

$y - y_1 = m(x - x_1)$

$y - (-8) = -\dfrac{3}{8}(x - 2)$

$y + 8 = -\dfrac{3}{8}x + \dfrac{3}{4}$

$y = -\dfrac{3}{8}x - \dfrac{29}{4}$

61. a. Use the points $(1, 30{,}000)$ and $(4, 66{,}000)$.

$m = \dfrac{66{,}000 - 30{,}000}{4 - 1} = \dfrac{36{,}000}{3} = 12{,}000$

$y - y_1 = m(x - x_1)$

$y - 30{,}000 = 12{,}000(x - 1)$

$y - 30{,}000 = 12{,}000x - 12{,}000$

$y = 12{,}000x + 18{,}000$

b. Let $x = 7$.

$y = 12{,}000(7) + 18{,}000$

$= 84{,}000 + 18{,}000$

$= 102{,}000$

The profit at the end of the seventh year is expected to be $102,000.

c. Find x when $y = 126{,}000$.

$126{,}000 = 12{,}000x + 18{,}000$

$108{,}000 = 12{,}000x$

$9 = x$

The profit should reach $126,000 at the end of the ninth year.

65. a. Use the points $(0, 70)$ and $(3, 70.3)$.

$m = \dfrac{70.3 - 70}{3 - 0} = \dfrac{0.3}{3} = 0.1$

$y - y_1 = m(x - x_1)$

$y - 70 = 0.1(x - 0)$

$y = 0.1x + 70$

b. Let $x = 7$, since 2008 is 7 years after 2001.

$y = 0.1(7) + 70 = 0.7 + 70 = 70.7$

There will be approximately 70.7 million basic cable subscribers in 2008.

69. $y = 4.2x$

$y = 4.2(-2)$

$y = -8.4$

$(-2, -8.4)$

73. $m = \dfrac{1 - 3}{1 - 0} = \dfrac{-2}{1} = -2$

$y - y_1 = m(x - x_1)$

$y - 3 = -2(x - 0)$

$y - 3 = -2x$

$2x + y = 3$

77. True; a vertical line is always perpendicular to a horizontal line.

81.

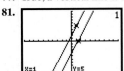

Exercise Set 3.5

1. $x < 2$

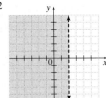

5. $3x + y > 6$

Test $(0, 0)$

$3(0) + 0 > 6$

$0 > 6$ False

Shade the half-plane that does not contain $(0, 0)$.

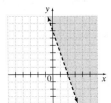

9. $2x + 4y \geq 8$

Test $(0, 0)$

$2(0) + 4(0) \geq 8$

$0 \geq 8$ False

Shade the half-plane that does not contain the point $(0, 0)$.

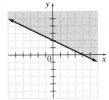

13. $y \leq -4.5$

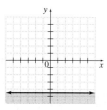

17. $x \geq 3$ and $y \leq -2$

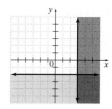

21. $x - y < 3$ and $x > 4$

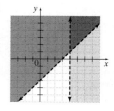

25. $x - y \geq 2$ or $y < 5$

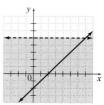

29. $y \leq 2x + 3$

The graph will have a solid line. Testing $(0, 0)$ results in a true inequality, so the graph is graph D.

33.
$$x - y = 4$$
$$3 - (-1) \stackrel{?}{=} 4$$
$$4 = 4 \quad \text{True}$$
$$x + 2y = 1$$
$$3 + 2(-1) \stackrel{?}{=} 1$$
$$3 - 2 \stackrel{?}{=} 1$$
$$1 = 1 \quad \text{True}$$

Yes, $(3, -1)$ is a solution of both equations.

37. A dashed boundary line should be used when the inequality involves the $>$ symbol or the $<$ symbol.

Exercise Set 3.6

1. Domain $= \{-1, 0, -2, 5\}$
Range $= \{7, 6, 2\}$
The relation is a function.

5. Domain $= \{1\}$
Range $= \{1, 2, 3, 4\}$
The relation is not a function since 1 is paired with 1, 2, 3 and 4.

9. Domain $= \{-3, 0, 3\}$
Range $= \{-3, 0, 3\}$
The relation is a function.

13. Domain $= \{$Colorado, Alaska, Massachusetts, Delaware, Illinois, Pennsylvania$\}$
Range $= \{1, 7, 10, 19\}$
The relation is a function.

17. Domain $= \{-1, 2, 5, 100\}$
Range $= \{0\}$
The relation is a function.

21. Since each x-value corresponds to exactly one y-value, the relation is a function.

25. $y - x = 7$
$$y = x + 7$$
Since each x-value corresponds to exactly one y-value, the relation is a function.

29. A vertical line can be drawn that intersects the graph more than once, so the graph is not the graph of a function.

33. A vertical line can be drawn that intersects the graph more than once, so the graph is not the graph of a function.

37. A vertical line can be drawn that intersects the graph more than once, so the graph is not the graph of a function.

41. A vertical line can be drawn that intersects the graph more than once, so the graph is not the graph of a function.

45. $f(x) = 3x + 3$
$$f(4) = 3(4) + 3 = 12 + 3 = 15$$

49. $g(x) = 4x^2 - 6x + 3$
$$g(2) = 4(2)^2 - 6(2) + 3$$
$$= 4(4) - 12 + 3$$
$$= 16 - 12 + 3$$
$$= 7$$

53. $f(x) = \dfrac{1}{2}x$

  a. $f(0) = \dfrac{1}{2}(0) = 0$

  b. $f(2) = \dfrac{1}{2}(2) = 1$

  c. $f(-2) = \dfrac{1}{2}(-2) = -1$

57. $A(r) = \pi r^2$
$$A(5) = \pi(5)^2 = 25\pi$$
The area is 25π square centimeters.

61. Use $H(f) = 2.59f + 47.24$ with $f = 46$.
$$H(46) = 2.59(46) + 47.24$$
$$= 119.14 + 47.24$$
$$= 166.38$$
The height of the woman is 166.38 cm.

65. a. $C(x) = 2.28x + 94.86$
$$C(2) = 2.28(2) + 94.86$$
$$= 4.56 + 94.86$$
$$= 99.42 \text{ cm}$$
The per capita consumption of poultry was 99.42 pounds in 2003.

  b. Use $x = 6$ since 2007 is 6 years after 2001.
$$C(6) = 2.28(6) + 94.86$$
$$= 13.68 + 94.86$$
$$= 108.54$$
The per capita consumption of poultry is expected to be 108.54 pounds in 2007.

69. $f(x) = -3x$ or $y = -3x = -3x + 0$
$m = -3$; y-intercept $= (0, 0)$

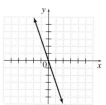

73. $2x - 7 \leq 21$
$$2x \leq 28$$
$$x \leq 14$$
$$(-\infty, 14]$$

77. $\dfrac{x}{2} + \dfrac{1}{4} < \dfrac{1}{8}$

$\dfrac{x}{2} < \dfrac{1}{8} - \dfrac{1}{4}$

$\dfrac{x}{2} < -\dfrac{1}{8}$

$x < -\dfrac{1}{4}$

$\left(-\infty, -\dfrac{1}{4}\right)$

81. The graph of $y = 2x$ is a line which is not vertical, so it passes the vertical line test. $y = 2x$ represents a function.

85. $f(x) = x^2 + 1$

$f(7) = (7)^2 + 1 = 49 + 1 = 50$

Since $f(7) = 50$ when $f(x) = x^2 + 1$, a possible function is $f(x) = x^2 + 1$; true.

89. $y = x + 7$ can be written as $f(x) = x + 7$.

93. $f(x) = x^2 - 12$

   a. $f(12) = (12)^2 - 12 = 144 - 12 = 132$

   b. $f(a) = (a)^2 - 12 = a^2 - 12$

   c. $f(-x) = (-x)^2 - 12 = x^2 - 12$

   d. $f(x + h) = (x + h)^2 - 12$
$$= x^2 + 2xh + h^2 - 12$$

Exercise Set 3.7

1. Domain $= [0, \infty)$
Range $= (-\infty, \infty)$

5. Domain $= (-\infty, \infty)$
Range $= (-\infty, -3] \cup [3, \infty)$

9. Domain $= \{-2\}$
Range $= (-\infty, \infty)$

13. Domain $= (-\infty, \infty)$
Range $= (-\infty, 3]$

17. $f(x) = \begin{cases} 2x & \text{if} \quad x < 0 \\ x + 1 & \text{if} \quad x \geq 0 \end{cases}$

Graph $y = 2x$ for $x < 0$, with an open circle at $x = 0$.
Graph $y = x + 1$ for $x \geq 0$, with a closed circle at $x = 0$.

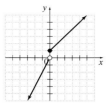

21. $g(x) = \begin{cases} -x & \text{if} \quad x \leq -1 \\ 2x + 1 & \text{if} \quad x > 1 \end{cases}$

Graph $y = -x$ for $x \leq -1$, with a closed circle at $x = -1$.
Graph $y = 2x + 1$ for $x > 1$, with an open circle at $x = 1$.

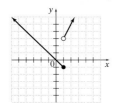

25. $f(x) = \begin{cases} -2x & \text{if} \quad x \leq 0 \\ 2x + 1 & \text{if} \quad x > 0 \end{cases}$

Graph $y = -2x$ for $x \leq 0$, with a closed circle at $x = 0$.
Graph $y = 2x + 1$ for $x > 0$, with an open circle at $x = 0$.
Domain $= (-\infty, \infty)$
Range $= [0, \infty)$

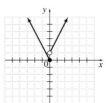

29. $f(x) = \begin{cases} x + 3 & \text{if} \quad x < -1 \\ -2x + 4 & \text{if} \quad x \geq -1 \end{cases}$

Graph $y = x + 3$ for $x < -1$, with an open circle at $x = -1$.
Graph $y = -2x + 4$ for $x \geq -1$, with a closed circle at $x = -1$.
Domain $= (-\infty, \infty)$
Range $= (-\infty, 6]$

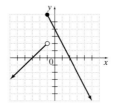

33. $y = -1$ is a horizontal line that crosses the y-axis at $(0, -1)$. This is a graph A.

37. answers may vary

Exercise Set 3.8

1. The graph of $f(x) = |x| + 3$ is the graph of $y = |x|$ shifted 3 units up.

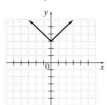

5. The graph of $f(x) = |x - 4|$ is the graph of $y = |x|$ shifted 4 units to the right.

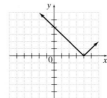

9. The graph of $y = (x - 4)^2$ is the graph of $y = x^2$ shifted 4 units to the right.

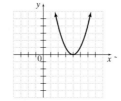

13. The graph of $f(x) = \sqrt{x - 2} + 3$ is the graph of $y = \sqrt{x}$ shifted 2 units to the right and 3 units up.

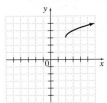

17. The graph of $f(x) = \sqrt{x + 1} + 1$ is the graph of $y = \sqrt{x}$ shifted 1 unit to the left and 1 unit up.

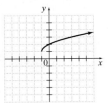

21. The graph of $g(x) = (x - 1)^2 - 1$ is the graph of $y = x^2$ shifted 1 unit to the right and 1 unit down.

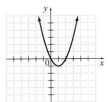

25. The graph of $f(x) = -(x - 1)^2$ is the graph of $y = x^2$ reflected about the x-axis and shifted 1 unit to the right.

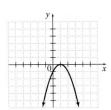

29. The graph of $h(x) = -|x + 2| + 3$ is the graph of $y = |x|$ reflected about the x-axis then shifted 2 units to the left and 3 units up.

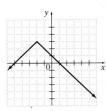

33. $-3x^4 \cdot 5x^4 = -3 \cdot 5x^{4+4} = -15x^8$

37. Domain $= [2, \infty)$
Range $= [3, \infty)$

41. Since the square root of a negative number is not a real number, $x - 20 \geq 0$ or $x \geq 20$. The domain is $[20, \infty)$.

45. Since the square root of a negative number is not a real number, $x + 103 \geq 0$ or $x \geq -103$. The domain is $[-103, \infty)$.

49. $g(x) = \begin{cases} |x - 2| & \text{if } x < 0 \\ -x^2 & \text{if } x \geq 0 \end{cases}$

Graph $y = |x - 2|$ for $x < 0$, with an open circle at $x = 0$.
Graph $y = -x^2$ for $x \geq 0$, with a closed circle at $x = 0$.

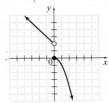

Chapter 3 Test

1.

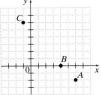

A is in quadrant IV.
B is on the x-axis, no quadrant.
C is in quadrant II.

5.

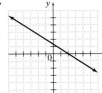

9. $f(x) = 3x + 1$ or $y = 3x + 1$
$m = 3$; y-intercept $= (0, 1)$
The equation matches graph C.

13. A horizontal line has equation $y = b$. Since the line passes through $(2, -8)$, its equation is $y = -8$.

17. $y = mx + b$
            $y = 5x + (-2)$
            $y = 5x - 2$
            $2 = 5x - y$
$5x - y = 2$

21. L_1: $2x - 5y = 8$
                $-5y = -2x + 8$
                $y = \dfrac{2}{5}x - \dfrac{8}{5}$

$m = \dfrac{2}{5}$

$y\text{-int} = \left(0, -\dfrac{8}{5}\right)$

L_2: $m = \dfrac{-1 - 4}{-1 - 1} = \dfrac{-5}{-2} = \dfrac{5}{2}$

Since the slopes of the lines are different, and the product of the slopes is not -1, the lines are neither parallel nor perpendicular.

25. $2x + 4y < 6$ and $y \leq -4$

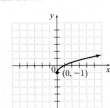

29. Domain $= (-\infty, \infty)$
Range $= (-\infty, \infty)$
Since no vertical line can intersect the graph more than once, the graph is the graph of a function.

33. The graph of $h(x) = \sqrt{x} - 1$ is the graph of $y = \sqrt{x}$ shifted 1 unit down.

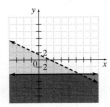

CHAPTER 4

Exercise Set 4.1

1. $\begin{cases} x - y = 3 \\ 2x - 4y = 8 \end{cases}$

$$
\begin{array}{ll}
x - y = 3 & 2x - 4y = 8 \\
2 - (-1) \overset{?}{=} 3 & 2(2) - 4(-1) \overset{?}{=} 8 \\
2 + 1 \overset{?}{=} 3 & 4 + 4 \overset{?}{=} 8 \\
\quad 3 = 3 \quad \text{True} & \quad 8 = 8 \quad \text{True}
\end{array}
$$

Yes, $(2, -1)$ is a solution of the system.

5. $\begin{cases} y = -5x \\ x = -2 \end{cases}$

$$
\begin{array}{ll}
y = -5x & \\
10 \overset{?}{=} -5(-2) & x = -2 \\
10 = 10 \quad \text{True} & -2 = -2 \quad \text{True}
\end{array}
$$

Yes, $(-2, 10)$ is a solution of the system.

9. $\begin{cases} x + y = 1 \\ x - 2y = 4 \end{cases}$

The solution of the system is $(2, -1)$.

13. $\begin{cases} 3x - y = 4 \\ 6x - 2y = 4 \end{cases}$

The lines are parallel. The system has no solution.

17. $\begin{cases} x + y = 10 \\ \quad\quad y = 4x \end{cases}$

Replace y by $4x$ in the first equation.

$$
\begin{aligned}
x + y &= 10 \\
x + 4x &= 10 \\
5x &= 10 \\
x &= 2
\end{aligned}
$$

Replace x with 2 in the second equation.

$y = 4x = 4(2) = 8$

The solution of the system is $(2, 8)$.

21. $\begin{cases} \dfrac{1}{2}x + \dfrac{3}{4}y = -\dfrac{1}{4} \\ \dfrac{3}{4}x - \dfrac{1}{4}y = 1 \end{cases}$

Multiply both equations by 4 to clear the fractions.

$\begin{cases} 2x + 3y = -1 \\ 3x - \ y = 4 \end{cases}$

Solve the second equation for y.

$$
\begin{aligned}
3x - y &= 4 \\
-y &= -3x + 4 \\
y &= 3x - 4
\end{aligned}
$$

Replace y with $3x - 4$ in the (simplified) first equation.

$$
\begin{aligned}
2x + 3y &= -1 \\
2x + 3(3x - 4) &= -1 \\
2x + 9x - 12 &= -1 \\
11x &= 11 \\
x &= 1
\end{aligned}
$$

Replace x by 1 in the equation $y = 3x - 4$.

$y = 3x - 4 = 3(1) - 4 = 3 - 4 = -1$

The solution of the system is $(1, -1)$.

25. $\begin{cases} 2x - 4y = 0 \\ \ x + 2y = 5 \end{cases}$

Multiply the second equation by 2.

$\begin{cases} 2x - 4y = 0 \\ 2x + 4y = 10 \end{cases}$

Add the equations.

$$
\begin{array}{r}
2x - 4y = 0 \\
2x + 4y = 10 \\
\hline
4x \quad\quad = 10 \\
x \quad\quad = \dfrac{5}{2}
\end{array}
$$

Replace x with $\dfrac{5}{2}$ in the first equation.

$$
\begin{aligned}
2x - 4y &= 0 \\
2\left(\frac{5}{2}\right) - 4y &= 0 \\
5 - 4y &= 0 \\
5 &= 4y \\
\frac{5}{4} &= y
\end{aligned}
$$

The solution of the system is $\left(\dfrac{5}{2}, \dfrac{5}{4}\right)$.

29. $\begin{cases} \ 5x - 2y = 27 \\ -3x + 5y = 18 \end{cases}$

Multiply the first equation by 3 and the second equation by 5.

$\begin{cases} \ 15x - \ 6y = 81 \\ -15x + 25y = 90 \end{cases}$

Add the equations.

$$
\begin{array}{r}
15x - \ 6y = 81 \\
-15x + 25y = 90 \\
\hline
19y = 171 \\
y = 9
\end{array}
$$

Replace y with 9 in the first equation.

$$
\begin{aligned}
5x - 2y &= 27 \\
5x - 2(9) &= 27 \\
5x - 18 &= 27 \\
5x &= 45 \\
x &= 9
\end{aligned}
$$

The solution of the system is $(9, 9)$.

33. $\begin{cases} \ x - 2y = 4 \\ 2x - 4y = 4 \end{cases}$

Multiply the first equation by -2.

$\begin{cases} -2x + 4y = -8 \\ \ 2x - 4y = 4 \end{cases}$

Add the equations.

$$
\begin{array}{r}
-2x + 4y = -8 \\
2x - 4y = 4 \\
\hline
0 = -4
\end{array}
$$

The statement $0 = -4$ is false, so the system has no solution.

37. $\begin{cases} 2x + 5y = 8 \\ 6x + y = 10 \end{cases}$

Multiply the first equation by -3, then add the resulting equations.

$-6x - 15y = -24$
$\underline{6x + y = 10}$
$-14y = -14$
$y = 1$

Replace y with 1 in the second equation.

$6x + y = 10$
$6x + 1 = 10$
$6x = 9$

$x = \dfrac{3}{2}$

The solution of the system is $\left(\dfrac{3}{2}, 1\right)$.

41. $\begin{cases} \dfrac{1}{3}x + y = \dfrac{4}{3} \\ -\dfrac{1}{4}x - \dfrac{1}{2}y = -\dfrac{1}{4} \end{cases}$

Multiply the first equation by 3 and the second equation by 4 to clear fractions, then add the resulting equations.

$x + 3y = 4$
$\underline{-x - 2y = -1}$
$y = 3$

Replace y with 3 in the equation $x + 3y = 4$.

$x + 3y = 4$
$x + 3(3) = 4$
$x + 9 = 4$
$x = -5$

The solution of the system is $(-5, 3)$.

45. $\begin{cases} 10y - 2x = 1 \\ 5y = 4 - 6x \end{cases}$

$\begin{cases} 10y - 2x = 1 \\ 5y + 6x = 4 \end{cases}$

Multiply the first equation by 3, then add the resulting equations.

$30y - 6x = 3$
$\underline{5y + 6x = 4}$
$35y = 7$

$y = \dfrac{1}{5}$

Replace y with $\dfrac{1}{5}$ in the first equation.

$10y - 2x = 1$
$10\left(\dfrac{1}{5}\right) - 2x = 1$
$2 - 2x = 1$
$-2x = -1$
$x = \dfrac{1}{2}$

The solution of the system is $\left(\dfrac{1}{2}, \dfrac{1}{5}\right)$.

49. $\begin{cases} x = 3y + 2 \\ 5x - 15y = 10 \end{cases}$

Replace x by $3y + 2$ in the second equation.

$5x - 15y = 10$
$5(3y + 2) - 15y = 10$
$15y + 10 - 15y = 10$
$10 = 10$

The statement $10 = 10$ is true, so the system has infinitely many solutions. The solution set is $\{x \mid x = 3y + 2\}$.

53. $\begin{cases} 2x = 6 \\ y = 5 - x \end{cases}$

Solve the first equation for x.

$2x = 6$
$x = 3$

Replace x with 3 in the second equation.

$y = 5 - x = 5 - 3 = 2$

The solution of the system is $(3, 2)$.

57. $\begin{cases} 4x - 7y = 7 \\ 12x - 21y = 24 \end{cases}$

Multiply the first equation by -3, then add the resulting equations.

$-12x + 21y = -21$
$\underline{12x - 21y = 24}$
$0 = 3$

The statement $0 = 3$ is false, so the system has no solutions.

61. $\begin{cases} 2x - y = -1 \\ y = -2x \end{cases}$

Replace y by $-2x$ in the first equation:

$2x - y = -1$
$2x - (-2x) = -1$
$4x = -1$
$x = -\dfrac{1}{4}$

Replace x with $-\dfrac{1}{4}$ in the second equation.

$y = -2x = -2\left(-\dfrac{1}{4}\right) = \dfrac{1}{2}$

The solution of the system is $\left(-\dfrac{1}{4}, \dfrac{1}{2}\right)$.

65. $\begin{cases} 4x - 1.5y = 10.2 \\ 2x + 7.8y = -25.68 \end{cases}$

Multiply the second equation by -2 and add the resulting equations.

$4x - 1.5y = 10.2$
$\underline{-4x - 15.6y = 51.36}$
$-17.1y = 61.56$
$y = -3.6$

Replace y with -3.6 in the first equation.

$4x - 1.5y = 10.2$
$4x - 1.5(-3.6) = 10.2$
$4x + 5.4 = 10.2$
$4x = 4.8$
$x = 1.2$

The solution of the system is $(1.2, -3.6)$.

69. Replace x with 0, y with -1, and z with 5.

$-x - 5y + 3z = 15$
$-0 - 5(-1) + 3(5) \stackrel{?}{=} 15$
$5 + 15 \stackrel{?}{=} 15$
$20 = 15 \quad$ False

The given replacement values make the equation false.

73. $10x + 5y + 6z = 14$
$\underline{-9x + 5y - 6z = -12}$
$x + 10y + 0z = 2$
$x + 10y = 2$

77. $\begin{cases} x + y = 3 \\ 5x + 5y = 15 \end{cases}$

The result of multiplying the first equation by 5 is $5x + 5y = 15$, which is the second equation. The system is dependent, so it has an infinite number of solutions.

81. The lines intersect at the point $(5, 21)$, which corresponds to 5000 DVDs at a price of \$21 per DVD.

85. $\begin{cases} y = 2.5x \\ y = 0.9x + 3000 \end{cases}$

Replace y with $2.5x$ in the second equation.

$y = 0.9x + 3000$

$2.5x = 0.9x + 3000$

$1.6x = 3000$

$x = 1875$

Replace x with 1875 in the first equation.

$y = 2.5x = 2.5(1875) = 4687.5$

The break-even point is $(1875, 4687.5)$.

89. The revenue graph is above the cost graph to the right of the intersection point. Thus, for x-values greater than 1875, the company will make a profit.

93. a. The slope of the red meat equation is negative, so the consumption of red meat is decreasing. The slope of the poultry line is positive, so the consumption of poultry is increasing.

b. $\begin{cases} y = -0.56x + 114 \\ y = 0.98x + 67 \end{cases}$

Replace y with $-0.56x + 114$ in the second equation.

$y = 0.98x + 67$

$-0.56x + 114 = 0.98x + 67$

$-0.56x + 47 = 0.98x$

$47 = 1.54x$

$30.5 \approx x$

Replace x with 30.5 in the first equation.

$y = -0.56x + 114 = -0.56(30.5) + 114 \approx 97.$

The solution of the system is approximately $(31, 97)$.

c. The meaning of the intersection point $(31, 97)$ is that 31 years after 1997, or in 2028, the red meat and poultry consumption will both be about 97 pounds per person per year.

97. $\begin{cases} \dfrac{1}{x} + \dfrac{1}{y} = 5 \\ \dfrac{1}{x} - \dfrac{1}{y} = 1 \end{cases}$

Let $a = \dfrac{1}{x}$ and $b = \dfrac{1}{y}$ to get the system:

$\begin{cases} a + b = 5 \\ a - b = 1 \end{cases}$

Add the equations.

$a + b = 5$

$\underline{a - b = 1}$

$2a = 6$

$a = 3$

Replace a with 3 in $a + b = 5$.

$a + b = 5$

$3 + b = 5$

$b = 2$

Since $a = \dfrac{1}{x}$, then $\dfrac{1}{x} = 3$, and $x = \dfrac{1}{3}$.

Since $b = \dfrac{1}{y}$, then $\dfrac{1}{y} = 2$, and $y = \dfrac{1}{2}$.

The solution of the system is $\left(\dfrac{1}{3}, \dfrac{1}{2} \right)$.

101. $\begin{cases} \dfrac{2}{x} - \dfrac{4}{y} = 5 \\ \dfrac{1}{x} - \dfrac{2}{y} = \dfrac{3}{2} \end{cases}$

Let $a = \dfrac{1}{x}$ and $b = \dfrac{1}{y}$ to get the system:

$\begin{cases} 2a - 4b = 5 \\ a - 2b = \dfrac{3}{2} \end{cases}$

Multiply the second equation by -2 and add the resulting equations.

$2a - 4b = 5$

$\underline{-2a + 4b = -3}$

$0 = 2$

The statement $0 = 2$ is false, so the system has no solution.

Exercise Set 4.2

1. a. $x + y + z = 3$

$-1 + 3 + 1 \overset{?}{=} 3$

$3 = 3$ True

b. $-x + y + z = 5$

$-(-1) + 3 + 1 \overset{?}{=} 5$

$1 + 3 + 1 \overset{?}{=} 5$

$5 = 5$ True

c. $-x + y + 2z = 0$

$-(-1) + 3 + 2(1) \overset{?}{=} 0$

$1 + 3 + 2 \overset{?}{=} 0$

$6 = 0$ False

d. $x + 2y - 3z = 2$

$-1 + 2(3) - 3(1) \overset{?}{=} 2$

$-1 + 6 - 3 \overset{?}{=} 2$

$2 = 2$ True

$(-1, 3, 1)$ is a solution to the equations a, b, and d.

5. $\begin{cases} x - y + z = -4 & (1) \\ 3x + 2y - z = 5 & (2) \\ -2x + 3y - z = 15 & (3) \end{cases}$

Add the equations (1) and (2) to eliminate z.

$x - y + z = -4$

$\underline{3x + 2y - z = 5}$

$4x + y = 1$ (4)

Add equations (1) and (3) to eliminate z.

$x - y + z = -4$

$\underline{-2x + 3y - z = 15}$

$-x + 2y = 11$ (5)

Multiply equation (5) by 4 and add it to equation (4).

$4x + y = 1$

$\underline{-4x + 8y = 44}$

$9y = 45$

$y = 5$

Replace y with 5 in equation (4).

$4x + y = 1$

$4x + 5 = 1$

$4x = -4$

$x = -1$

Replace x with -1 and y with 5 in equation (1).

$x - y + z = -4$

$-1 - 5 + z = -4$

$-6 + z = -4$

$z = 2$

The solution of the system is $(-1, 5, 2)$.

9. $\begin{cases} 2x + 2y + z = 1 & (1) \\ -x + y + 2z = 3 & (2) \\ x + 2y + 4z = 0 & (3) \end{cases}$

Add equations (2) and (3) to eliminate x.

$\begin{array}{r} -x + y + 2z = 3 \\ x + 2y + 4z = 0 \\ \hline 3y + 6z = 3 \\ y + 2z = 1 \quad (4) \end{array}$

Multiply equation (2) by 2 and add it to equation (1) to eliminate x.

$\begin{array}{r} 2x + 2y + z = 1 \\ -2x + 2y + 4z = 6 \\ \hline 4y + 5z = 7 \quad (5) \end{array}$

Multiply equation (4) by -4 and add it to equation (5).

$\begin{array}{r} -4y - 8z = -4 \\ 4y + 5z = 7 \\ \hline -3z = 3 \\ z = -1 \end{array}$

Replace z with -1 in equation (4).

$y + 2z = 1$
$y + 2(-1) = 1$
$y - 2 = 1$
$y = 3$

Replace y with 3 and z with -1 in equation (3).

$x + 2y + 4z = 0$
$x + 2(3) + 4(-1) = 0$
$x + 6 - 4 = 0$
$x + 2 = 0$
$x = -2$

The solution of the system is $(-2, 3, -1)$.

13. $\begin{cases} 4x - y + 2z = 5 & (1) \\ 2y + z = 4 & (2) \\ 4x + y + 3z = 10 & (3) \end{cases}$

Multiply equation (1) by -1 and add it to equation (3) to eliminate x.

$\begin{array}{r} -4x + y - 2z = -5 \\ 4x + y + 3z = 10 \\ \hline 2y + z = 5 \quad (4) \end{array}$

Multiply equation (2) by -1 and add it to equation (4).

$\begin{array}{r} -2y - z = -4 \\ 2y + z = 5 \\ \hline 0 = 1 \end{array}$

The statement $0 = 1$ is false, so the system has no solution.

17. $\begin{cases} 6x - 5z = 17 & (1) \\ 5x - y + 3z = -1 & (2) \\ 2x + y = -41 & (3) \end{cases}$

Add equations (2) and (3) to eliminate y.

$\begin{array}{r} 5x - y + 3z = -1 \\ 2x + y = -41 \\ \hline 7x + 3z = -42 \quad (4) \end{array}$

Multiply equation (1) by 3 and equation (4) by 5, then add the resulting equations.

$\begin{array}{r} 18x - 15z = 51 \\ 35x + 15z = -210 \\ \hline 53x = -159 \\ x = -3 \end{array}$

Replace x with -3 in equation (1).

$6x - 5z = 17$
$6(-3) - 5z = 17$
$-18 - 5z = 17$
$-5z = 35$
$z = -7$

Replace x with -3 in equation (3).

$2x + y = -41$
$2(-3) + y = -41$
$-6 + y = -41$
$y = -35$

The solution of the system is $(-3, -35, -7)$.

21. $\begin{cases} x + 2y - z = 5 & (1) \\ 6x + y + z = 7 & (2) \\ 2x + 4y - 2z = 5 & (3) \end{cases}$

Add equations (1) and (2) to eliminate z.

$\begin{array}{r} x + 2y - z = 5 \\ 6x + y + z = 7 \\ \hline 7x + 3y = 12 \quad (4) \end{array}$

Multiply equation (2) by 2 and add to equation (3) to eliminate z.

$\begin{array}{r} 12x + 2y + 2z = 14 \\ 2x + 4y - 2z = 5 \\ \hline 14x + 6y = 19 \quad (5) \end{array}$

Multiply equation (4) by -2 and add to equation (5).

$\begin{array}{r} -14x - 6y = -24 \\ 14x + 6y = 19 \\ \hline 0 = -5 \end{array}$

The statement $0 = -5$ is false, so the system has no solution.

25. $\begin{cases} -2x - 4y + 6z = -8 & (1) \\ x + 2y - 3z = 4 & (2) \\ 4x + 8y - 12z = 16 & (3) \end{cases}$

Multiply equation (2) by -2.

$-2x - 4y + 6z = -8$

This is equation (1).

Multiply equation (2) by 4.

$4x + 8y - 12z = 16$

This is equation (3).

The system is dependent and the solution set is
$\{(x, y, z) \mid x + 2y - 3z = 4\}$.

29. $\begin{cases} x + 2y - z = 5 & (1) \\ -3x - 2y - 3z = 11 & (2) \\ 4x + 4y + 5z = -18 & (3) \end{cases}$

Add equations (1) and (2) to eliminate y.

$\begin{array}{r} x + 2y - z = 5 \\ -3x - 2y - 3z = 11 \\ \hline -2x - 4z = 16 \\ x + 2z = -8 \quad (4) \end{array}$

Multiply equation (2) by 2 and add to equation (3) to eliminate y.

$\begin{array}{r} -6x - 4y - 6z = 22 \\ 4x + 4y + 5z = -18 \\ \hline -2x - z = 4 \quad (5) \end{array}$

Multiply equation (5) by 2 and add to equation (4).

$\begin{array}{r} x + 2z = -8 \\ -4x - 2z = 8 \\ \hline -3x = 0 \\ x = 0 \end{array}$

Replace x with 0 in equation (4).

$$x + 2z = -8$$
$$0 + 2z = -8$$
$$2z = -8$$
$$z = -4$$

Replace x with 0 and z with -4 in equation (1).

$$x + 2y - z = 5$$
$$0 + 2y - (-4) = 5$$
$$2y + 4 = 5$$
$$2y = 1$$
$$y = \frac{1}{2}$$

The solution of the system is $\left(0, \frac{1}{2}, -4\right)$.

33. $2(x - 1) - 3x = x - 12$
$$2x - 2 - 3x = x - 12$$
$$-x - 2 = x - 12$$
$$-2 = 2x - 12$$
$$10 = 2x$$
$$5 = x$$

The solution set is $\{5\}$.

37. Let x be one number. Then $2x$ is the other number.

$$x + 2x = 45$$
$$3x = 45$$
$$x = 15$$
$$2x = 2(15) = 30$$

The numbers are 15 and 30.

41. answers may vary

45. $\begin{cases} x + y \quad\ - w = 0 & (1) \\ \quad\ y + 2z + w = 3 & (2) \\ x \quad\ - z \quad\ = 1 & (3) \\ 2x - y \quad\ - w = -1 & (4) \end{cases}$

Add equations (1) and (2) to eliminate w.

$$x + y \qquad - w = 0$$
$$\underline{\qquad y + 2z + w = 3}$$
$$x + 2y + 2z \qquad = 3 \quad (5)$$

Add equations (2) and (4) to eliminate w.

$$y + 2z + w = 3$$
$$\underline{2x - y \qquad - w = -1}$$
$$2x \qquad + 2z \qquad = 2$$
$$x + z = 1 \qquad (6)$$

Add equations (3) and (6).

$$x - z = 1$$
$$\underline{x + z = 1}$$
$$2x \quad = 2$$
$$x = 1$$

Replace x with 1 in equation (3).

$$x - z = 1$$
$$1 - z = 1$$
$$-z = 0$$
$$z = 0$$

Replace x with 1 and z with 0 in equation (5).

$$x + 2y + 2z = 3$$
$$1 + 2y + 2(0) = 3$$
$$1 + 2y = 3$$
$$2y = 2$$
$$y = 1$$

Replace x with 1 and y with 1 in equation (1).

$$x + y - w = 0$$
$$1 + 1 - w = 0$$
$$2 - w = 0$$
$$-w = -2$$
$$w = 2$$

The solution of the system is $(1, 1, 0, 2)$.

49. answers may vary

Exercise Set 4.3

1. Let $m =$ the first number, $n =$ the second number.

$$\begin{cases} m = n + 2 \\ 2m = 3n - 4 \end{cases}$$

Substitute $m = n + 2$ in the second equation.

$$2(n + 2) = 3n - 4$$
$$2n + 4 = 3n - 4$$
$$n = 8$$

Replace n with 8 in the first equation.

$$m = 8 + 2 = 10$$

The numbers are 10 and 8.

5. Let $p =$ speed of the plane in still air, $w =$ speed of the wind.

$$\begin{cases} p + w = 560 \\ p - w = 480 \end{cases}$$

Add the equations.

$$2p = 1040$$
$$p = 520$$

Replace p with 520 in the first equation.

$$520 + w = 560$$
$$w = 40$$

The speed of the plane is 520 mph and the speed of the wind is 40 mph.

9. Let x be the number of students studying in the United Kingdom, and y be the number studying in Italy.

$$\begin{cases} x + y = 50{,}642 \\ x = y + 12{,}770 \end{cases}$$

Replace x with $y + 12{,}770$ in the first equation.

$$x + y = 50{,}642$$
$$y + 12{,}770 + y = 50{,}642$$
$$2y + 12{,}770 = 50{,}642$$
$$2y = 37{,}872$$
$$y = 18{,}936$$

Replace y with 18,936 in the second equation.

$$x = y + 12{,}770 = 18{,}936 + 12{,}770 = 31{,}706$$

31,706 students studied in the United Kingdom, and 18,936 studied in Italy.

13. Let $m =$ the first number, $n =$ the second number.

$$\begin{cases} m = n - 2 \\ 2m = 3n + 4 \end{cases}$$

Substitute $m = n - 2$ in the second equation.

$$2(n - 2) = 3n + 4$$
$$2n - 4 = 3n + 4$$
$$-8 = n$$

Replace n with -8 in the first equation.

$$m = -8 - 2 = -10$$

The numbers are -10 and -8.

17. Let p be the price of a pen and w be the price of a writing tablet.
$$\begin{cases} 7w + 4p = 6.40 \\ 2w + 19p = 5.40 \end{cases}$$
Multiply the first equation by -2 and the second equation by 7 and add the resulting equations.

$$\begin{array}{r} -14w - 8p = -12.8 \\ \underline{14w + 133p = 37.8} \\ 125p = 25 \\ p = 0.2 \end{array}$$

Replace p with 0.2 in the first equation.
$$\begin{aligned} 7w + 4p &= 6.40 \\ 7w + 4(0.2) &= 6.40 \\ 7w + 0.8 &= 6.4 \\ 7w &= 5.6 \\ w &= 0.8 \end{aligned}$$
The price of a writing tablet is \$0.80 and the price of a pen is \$0.20.

21. a. answers may vary

b. $\begin{cases} y = -0.513x + 33.5 \\ y = 0.56x + 18.4 \end{cases}$

Replace y with $0.56x + 18.4$ in the first equation.
$$\begin{aligned} y &= -0.513x + 33.5 \\ 0.56x + 18.4 &= -0.513x + 33.5 \\ 1.073x + 18.4 &= 33.5 \\ 1.073x &= 15.1 \\ x &\approx 14.07 \end{aligned}$$
The percents of jobs in the two sectors were the same 14 years after 1970, or in 1984.

25. Let m be the number of miles.
Hertz $= 25 + 0.10m$
Budget $= 20 + 0.25m$
Using Budget $= 2 \cdot$ Hertz gives
$$\begin{aligned} 20 + 0.25m &= 2(25 + 0.10m) \\ 20 + 0.25m &= 50 + 0.20m \\ 0.25m &= 30 + 0.20m \\ 0.05m &= 30 \\ m &= \frac{30}{0.05} = 600 \end{aligned}$$
The Budget charge is twice the Hertz charge for a daily mileage of 600 miles.

29. The break-even point is where $C(x) = R(x)$.
$$\begin{aligned} 30x + 10{,}000 &= 46x \\ 10{,}000 &= 16x \\ 625 &= x \end{aligned}$$
625 units must be sold to break even.

33. The break-even point is where $C(x) = R(x)$.
$$\begin{aligned} 75x + 160{,}000 &= 200x \\ 160{,}000 &= 125x \\ 1280 &= x \end{aligned}$$
1280 units must be sold to break even.

37. Let $x =$ number of units of Mix A,
$y =$ number of units of Mix B,
$z =$ number of units of Mix C.
$$\begin{cases} 4x + 6y + 4z = 30 & (1) \\ 6x + y + z = 16 & (2) \\ 3x + 2y + 12z = 24 & (3) \end{cases}$$
Multiply equation (2) by -6 and add to equation (1).
$-32x - 2z = -66$ or $16x + z = 33$ (4)
Multiply equation (2) by -2 and add to equation (3).
$-9x + 10z = -8$ (5)

Multiply equation (4) by -10 and add to equation (5).
$$\begin{aligned} -169x &= -338 \\ x &= 2 \end{aligned}$$
Replace x with 2 in equation (4).
$$\begin{aligned} 16(2) + z &= 33 \\ 32 + z &= 33 \\ z &= 1 \end{aligned}$$
Replace x with 2 and z with 1 in equation (2).
$$\begin{aligned} 6(2) + y + 1 &= 16 \\ 12 + y + 1 &= 16 \\ y + 13 &= 16 \\ y &= 3 \end{aligned}$$
Combine 2 units of Mix A, 3 units of Mix B, and 1 unit of Mix C.

41. Let $x =$ the first number, $y =$ the second number, and $z =$ the third number.
$$\begin{cases} x + y + z = 40 \\ x = y + 5 \\ x = 2z \end{cases}$$
$$\begin{cases} x + y + z = 40 \\ y = x - 5 \\ z = \dfrac{1}{2}x \end{cases}$$
Substitute $y = x - 5$ and $z = \dfrac{1}{2}x$ in the first equation.
$$x + x - 5 + \frac{1}{2}x = 40$$
$$\frac{5}{2}x - 5 = 40$$
$$\frac{5}{2}x = 45$$
$$x = \frac{2}{5}(45) = 18$$
$$y = x - 5 = 18 - 5 = 13$$
$$z = \frac{1}{2}x = \frac{1}{2}(18) = 9$$
The three numbers are 18, 13, and 9.

45. $x + y + z = 180$
$2x + 5 + y = 180$
$2x - 5 + z = 180$
$z = 185 - 2x$
$y = 175 - 2x$
Replace y with $175 - 2x$ and z with $185 - 2x$ in the first equation.
$$\begin{aligned} x + (175 - 2x) + (185 - 2x) &= 180 \\ 360 - 3x &= 180 \\ -3x &= -180 \\ x &= 60 \end{aligned}$$
$z = 185 - 2(60) = 185 - 120 = 65$
$y = 175 - 2(60) = 175 - 120 = 55$
The values are $x = 60$, $y = 55$, and $z = 65$.

49. $-3x - 6y + 3z = 0$
$\underline{3x + y - z = 2}$
$-5y + 2z = 2$

53. $y = ax^2 + bx + c$
For $(1, 6)$, use $x = 1$ and $y = 6$.
$6 = a + b + c$ (1)
For $(-1, -2)$, use $x = -1$ and $y = -2$.
$-2 = a - b + c$ (2)

For $(0, -1)$, use $x = 0$ and $y = -1$.

$-1 = a \cdot 0 + b \cdot 0 + c$

$-1 = c$ (3)

The system is

$6 = a + b + c$ (1)

$-2 = a - b + c$ (2)

$-1 = c$ (3)

From equation (3), we see that $c = -1$. Multiply equation (2) by -1 and add to equation (1).

$8 = 2b$

$4 = b$

Replace b with 4 and c with -1 in equation (1).

$6 = a + 4 - 1$

$6 = a + 3$

$3 = a$

The solution is $a = 3, b = 4,$ and $c = -1$.

Exercise Set 4.4

1. $\begin{cases} x + y = 1 \\ x - 2y = 4 \end{cases}$

$\begin{bmatrix} 1 & 1 & \vdots & 1 \\ 1 & -2 & \vdots & 4 \end{bmatrix}$

Multiply row 1 by -1 and add to row 2.

$\begin{bmatrix} 1 & 1 & \vdots & 1 \\ 0 & -3 & \vdots & 3 \end{bmatrix}$

Divide row 2 by -3.

$\begin{bmatrix} 1 & 1 & \vdots & 1 \\ 0 & 1 & \vdots & -1 \end{bmatrix}$

This corresponds to $\begin{cases} x + y = 1 \\ y = -1 \end{cases}$.

$x + (-1) = 1$

$x - 1 = 1$

$x = 2$

The solution is $(2, -1)$.

5. $\begin{cases} x - 2y = 4 \\ 2x - 4y = 4 \end{cases}$

$\begin{bmatrix} 1 & -2 & \vdots & 4 \\ 2 & -4 & \vdots & 4 \end{bmatrix}$

Multiply row 1 by -2 and add to row 2.

$\begin{bmatrix} 1 & -2 & \vdots & 4 \\ 0 & 0 & \vdots & -4 \end{bmatrix}$

This is an inconsistent system. The solution is $\varnothing$.

9. $\begin{cases} x + y = 3 \\ 2y = 10 \\ 3x + 2y - 4z = 12 \end{cases}$

$\begin{bmatrix} 1 & 1 & 0 & \vdots & 3 \\ 0 & 2 & 0 & \vdots & 10 \\ 3 & 2 & -4 & \vdots & 12 \end{bmatrix}$

Multiply row 1 by -3 and add to row 3.

$\begin{bmatrix} 1 & 1 & 0 & \vdots & 3 \\ 0 & 2 & 0 & \vdots & 10 \\ 0 & -1 & -4 & \vdots & 3 \end{bmatrix}$

Divide row 2 by 2.

$\begin{bmatrix} 1 & 1 & 0 & \vdots & 3 \\ 0 & 1 & 0 & \vdots & 5 \\ 0 & -1 & -4 & \vdots & 3 \end{bmatrix}$

Add row 2 to row 3.

$\begin{bmatrix} 1 & 1 & 0 & \vdots & 3 \\ 0 & 1 & 0 & \vdots & 5 \\ 0 & 0 & -4 & \vdots & 8 \end{bmatrix}$

Multiply row 3 by $-\dfrac{1}{4}$.

$\begin{bmatrix} 1 & 1 & 0 & \vdots & 3 \\ 0 & 1 & 0 & \vdots & 5 \\ 0 & 0 & 1 & \vdots & -2 \end{bmatrix}$

This corresponds to $\begin{cases} x + y = 3 \\ y = 5 \\ z = -2 \end{cases}$.

$x + 5 = 3$

$x = -2$

The solution is $(-2, 5, -2)$.

13. $\begin{cases} x - 4 = 0 \\ x + y = 1 \end{cases}$ or $\begin{cases} x = 4 \\ x + y = 1 \end{cases}$

$\begin{bmatrix} 1 & 0 & \vdots & 4 \\ 1 & 1 & \vdots & 1 \end{bmatrix}$

Multiply row 1 by -1 and add to row 2.

$\begin{bmatrix} 1 & 0 & \vdots & 4 \\ 0 & 1 & \vdots & -3 \end{bmatrix}$

This corresponds to $\begin{cases} x = 4 \\ y = -3 \end{cases}$.

The solution is $(4, -3)$.

17. $\begin{cases} 5x - 2y = 27 \\ -3x + 5y = 18 \end{cases}$

$\begin{bmatrix} 5 & -2 & \vdots & 27 \\ -3 & 5 & \vdots & 18 \end{bmatrix}$

Multiply row 1 by $\dfrac{1}{5}$.

$\begin{bmatrix} 1 & -\dfrac{2}{5} & \vdots & \dfrac{27}{5} \\ -3 & 5 & \vdots & 18 \end{bmatrix}$

Multiply row 1 by 3 and add to row 2.

$\begin{bmatrix} 1 & -\dfrac{2}{5} & \vdots & \dfrac{27}{5} \\ 0 & \dfrac{19}{5} & \vdots & \dfrac{171}{5} \end{bmatrix}$

Multiply row 2 by $\dfrac{5}{19}$.

$\begin{bmatrix} 1 & -\dfrac{2}{5} & \vdots & \dfrac{27}{5} \\ 0 & 0 & \vdots & 9 \end{bmatrix}$

This corresponds to $\begin{cases} x - \dfrac{2}{5}y = \dfrac{27}{5} \\ \phantom{x-\frac{2}{5}} y = 9 \end{cases}$

$x - \dfrac{2}{5}(9) = \dfrac{27}{5}$

$x - \dfrac{18}{5} = \dfrac{27}{5}$

$x = 9$

The solution is $(9, 9)$.

21. $\begin{cases} 4x - y + 2z = 5 \\ 2y + z = 4 \\ 4x + y + 3z = 10 \end{cases}$

$\begin{bmatrix} 4 & -1 & 2 & \vdots & 5 \\ 0 & 2 & 1 & \vdots & 4 \\ 4 & 1 & 3 & \vdots & 10 \end{bmatrix}$

Multiply row 1 by $\frac{1}{4}$.

$$\begin{bmatrix} 1 & -\frac{1}{4} & \frac{1}{2} & \vdots & \frac{5}{4} \\ 0 & 2 & 1 & \vdots & 4 \\ 4 & 1 & 3 & \vdots & 10 \end{bmatrix}$$

Multiply row 1 by -4 and add to row 3.

$$\begin{bmatrix} 1 & -\frac{1}{4} & \frac{1}{2} & \vdots & \frac{5}{4} \\ 0 & 2 & 1 & \vdots & 4 \\ 0 & 2 & 1 & \vdots & 5 \end{bmatrix}$$

Multiply row 2 by $\frac{1}{2}$.

$$\begin{bmatrix} 1 & -\frac{1}{4} & \frac{1}{2} & \vdots & \frac{5}{4} \\ 0 & 1 & \frac{1}{2} & \vdots & 2 \\ 0 & 2 & 1 & \vdots & 5 \end{bmatrix}$$

Multiply row 2 by -2 and add to row 3.

$$\begin{bmatrix} 1 & -\frac{1}{4} & \frac{1}{2} & \vdots & \frac{5}{4} \\ 0 & 1 & \frac{1}{2} & \vdots & 2 \\ 0 & 0 & 0 & \vdots & 1 \end{bmatrix}$$

Row 3 corresponds to the equation $0 = 1$, which is false, so the system has no solution.

25. No vertical line intersects the graph more than once, so the graph is the graph of a function.

29. $\begin{cases} x + + z = 7 \\ + y + 2z = -6 \\ 3x - y = 0 \end{cases}$ can be written as

$\begin{cases} x + 0y + z = 7 \\ 0x + y + 2z = -6 \\ 3x - y + 0z = 0 \end{cases}$ which corresponds to the

matrix $\begin{bmatrix} 1 & 0 & 1 & \vdots & 7 \\ 0 & 1 & 2 & \vdots & -6 \\ 3 & -1 & 0 & \vdots & 0 \end{bmatrix}$; choice c.

33. answers may vary

Exercise Set 4.5

1. $\begin{cases} y \geq x + 1 \\ y \geq 3 - x \end{cases}$

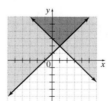

5. $\begin{cases} y < -2x - 2 \\ y > x + 4 \end{cases}$

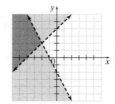

9. $\begin{cases} x \geq 3y \\ x + 3y \leq 6 \end{cases}$

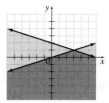

13. $\begin{cases} y \geq 1 \\ x < -3 \end{cases}$

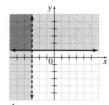

17. $\begin{cases} 3x - 4y \geq -6 \\ 2x + y \leq 7 \\ y \geq -3 \end{cases}$

21. The graph of the system $\begin{cases} y < 5 \\ x > 3 \end{cases}$ is graph C.

25. $(-3)^2 = (-3)(-3) = 9$

29. $(-2)^2 - (-3) + 2(-1) = 4 - (-3) + 2(-1)$
$= 4 - (-3) + (-2)$
$= 4 + 3 + (-2)$
$= 7 + (-2)$
$= 5$

33. The solution of the system $\begin{cases} y \leq 3 \\ y \geq 3 \end{cases}$ is the line $y = 3$.

Chapter 4 Test

1. $\begin{cases} 2x - y = -1 \\ 5x + 4y = 17 \end{cases}$

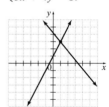

Solve the first equation for y.
$2x - y = -1$
$2x + 1 = y$

Replace y by $2x + 1$ in the second equation.
$5x + 4y = 17$
$5x + 4(2x + 1) = 17$
$5x + 8x + 4 = 17$
$13x = 13$
$x = 1$

Replace x with 1 in the first equation.

$2x - y = -1$

$2(1) - y = -1$

$2 - y = -1$

$-y = -3$

$y = 3$

The solution is $(1, 3)$.

5. $\begin{cases} 2x - 3y = 4 \quad (1) \\ 3y + 2z = 2 \quad (2) \\ x - z = -5 \quad (3) \end{cases}$

Add equation (1) to equation (2).

$2x + 2z = 6 \quad \text{or} \quad x + z = 3 \quad (5)$

Add equation (3) to equation (5).

$2x = -2$

$x = -1$

Replace x with -1 in equation (3).

$-1 - z = -5$

$-z = -4$

$z = 4$

Replace x with -1 in equation (1).

$2(-1) - 3y = 4$

$-2 - 3y = 4$

$-3y = 6$

$y = -2$

The solution is $(-1, -2, 4)$.

9. $\begin{cases} x + 2y = -1 \\ 2x + 5y = -5 \end{cases}$

$\begin{bmatrix} 1 & 2 & \vdots & -1 \\ 2 & 5 & \vdots & -5 \end{bmatrix}$

Multiply row 1 by -2 and add to row 2.

$\begin{bmatrix} 1 & 2 & \vdots & -1 \\ 0 & 1 & \vdots & -3 \end{bmatrix}$

This corresponds to $\begin{cases} x + 2y = -1 \\ y = -3 \end{cases}$.

$x + 2(-3) = -1$

$x - 6 = -1$

$x = 5$

The solution is $(5, -3)$.

13. Let x be the number of packages of gift boxes sold. Then the cost of producing x boxes is $C(x) = 2000 + 1.50x$ and the revenue from selling x boxes is $R(x) = 4x$. The break-even point is where $C(x) = R(x)$.

$2000 + 1.5x = 4x$

$2000 = 2.5x$

$800 = x$

800 packages of gift boxes must be sold for the company to break even.

CHAPTER 5

Exercise Set 5.1

1. 4 can be written as $4x^0$. The exponent on x is 0, so the degree of 4 is 0.

5. The degree is the sum of the exponents on the variables, so the degree of $-3xy^2$ is $1 + 2 = 3$.

9. The degree is the sum of the exponents on the variables, so the degree of $3.78ab^3c^5$ is $1 + 3 + 5 = 9$.

13. $3x^2 - 2x + 5$ has three terms, so it is a trinomial. The term with the greatest degree is $3x^2$, which has degree 2, so the degree of the trinomial is 2.

17. $x^2y - 4xy^2 + 5x + y^4$ has 4 terms so it is not a monomial, binomial, or trinomial. The term with the greatest degree is y^4, which has degree 4, so the degree of the polynomial is 4.

21. $4x + 7x - 3x^4 = (4 + 7)x - 3x^4$

$= 11x - 3x^4 \text{ or } -3x^4 + 11x$

25. $(9y^2 + y - 8) + (9y^2 - y - 9)$

$= 9y^2 + y - 8 + 9y^2 - y - 9$

$= 9y^2 + 9y^2 + y - y - 8 - 9$

$= 18y^2 - 17$

29. $\begin{array}{r} x^2 - 6x + 3 \\ + \quad (2x + 5) \\ \hline x^2 - 4x + 8 \end{array}$

33. $(0.6x^3 + 1.2x^2 - 4.5x + 9.1) + (3.9x^3 - x^2 + 0.7x)$

$= 0.6x^3 + 1.2x^2 - 4.5x + 9.1 + 3.9x^3 - x^2 + 0.7x$

$= 0.6x^3 + 3.9x^3 + 1.2x^2 - x^2 - 4.5x + 0.7x + 9.1$

$= 4.5x^3 + 0.2x^2 - 3.8x + 9.1$

37. $(4x^2 + 2x) - (6x^2 - 3x) = 4x^2 + 2x - 6x^2 + 3x$

$= 4x^2 - 6x^2 + 2x + 3x$

$= -2x^2 + 5x$

41. $(9x^3 - 2x^2 + 4x - 7) - (2x^3 - 6x^2 - 4x + 3)$

$= 9x^3 - 2x^2 + 4x - 7 - 2x^3 + 6x^2 + 4x - 3$

$= 9x^3 - 2x^3 - 2x^2 + 6x^2 + 4x + 4x - 7 - 3$

$= 7x^3 + 4x^2 + 8x - 10$

45. $(-3x + 8) + (-3x^2 + 3x - 5)$

$= -3x + 8 - 3x^2 + 3x - 5$

$= -3x^2 - 3x + 3x + 8 - 5$

$= -3x^2 + 3$

49. $(4x^2 - 6x + 2) - (-x^2 + 3x + 5)$

$= 4x^2 - 6x + 2 + x^2 - 3x - 5$

$= 4x^2 + x^2 - 6x - 3x + 2 - 5$

$= 5x^2 - 9x - 3$

53. $(5x - 11) + (-x - 2)$

$= 5x - 11 - x - 2$

$= 5x - x - 11 - 2$

$= 4x - 13$

57. $(14ab - 10a^2b + 6b^2) - (18a^2 - 20a^2b - 6b^2)$

$= 14ab - 10a^2b + 6b^2 - 18a^2 + 20a^2b + 6b^2$

$= 14ab - 10a^2b + 20a^2b - 18a^2 + 6b^2 + 6b^2$

$= 14ab + 10a^2b - 18a^2 + 12b^2$

61. $(7x^2 - 5) + (-3x^2 - 2) - (4x^2 - 7)$

$= 7x^2 - 5 - 3x^2 - 2 - 4x^2 + 7$

$= 7x^2 - 3x^2 - 4x^2 - 5 - 2 + 7$

$= 0$

65. $\begin{array}{r} 3x^2 - 4x + 8 \\ - \quad (5x - 7) \end{array}$

is equivalent to $\begin{array}{r} 3x^2 - 4x + 8 \\ + \quad (-5x + 7) \\ \hline 3x^2 - 9x + 15 \end{array}$

69. $\left(\frac{2}{3}x^2 - \frac{1}{6}x + \frac{5}{6} \right) - \left(\frac{1}{3}x^2 + \frac{5}{6}x - \frac{1}{6} \right)$

$= \frac{2}{3}x^2 - \frac{1}{6}x + \frac{5}{6} - \frac{1}{3}x^2 - \frac{5}{6}x + \frac{1}{6}$

$= \frac{2}{3}x^2 - \frac{1}{3}x^2 - \frac{1}{6}x - \frac{5}{6}x + \frac{5}{6} + \frac{1}{6}$

$= \frac{1}{3}x^2 - x + 1$

73. $Q(x) = 5x^2 - 1$

$Q(-10) = 5(-10)^2 - 1$

$\quad = 5(100)-1$

$\quad = 500 - 1$

$\quad = 499$

77. Use $2HL + 2LW + 2HW$ with $L = 5, W = 4,$ and $H = 9.$

$2HL + 2LW + 2HW = 2(9)(5) + 2(5)(4) + 2(9)(4)$

$\quad = 90 + 40 + 72$

$\quad = 202$

The surface area is 202 square inches.

81. $P(x) = 45x - 100,000$

$P(4000) = 45(4000)-100,000$

$\quad = 180,000 - 100,000$

$\quad = 80,000$

This profit from selling 4000 computer briefcases is \$80,000.

85. $-7(2z - 6y) = -7\cdot 2z - (-7)(6y) = -14z + 42y$

89. The opposite of $8x - 6$ is $-(8x - 6) = -8x + 6,$ which are choices a and c.

93. answers may vary

97. $f(x) = 0.014x^2 + 0.043x + 0.584$

   a. June 2003 is 0 months after June 2003, so $x = 0.$

$\quad f(0) = 0.014(0)^2 + 0.043(0) + 0.584$

$\quad\quad = 0.584$

$\quad\quad \approx 0.6$

There were approximately 0.6 million Web logs on the Internet in June 2003.

   b. June 2004 is 12 months after June 2003, so $x = 12.$

$\quad f(12) = 0.014(12)^2 + 0.043(12) + 0.584$

$\quad\quad = 2.016 + 0.516 + 0.584$

$\quad\quad = 3.116$

$\quad\quad \approx 3.1$

There were approximately 3.1 million Web logs on the Internet in June 2004.

   c. June 2005 is 24 months after June 2003, so $x = 24.$

$\quad f(24) = 0.014(24)^2 + 0.043(24) + 0.584$

$\quad\quad = 8.064 + 1.032 + 0.584$

$\quad\quad = 9.68$

$\quad\quad \approx 9.7$

There were approximately 9.7 million Web logs on the Internet in June 2005.

   d. answers may vary

101. $P(x) = 2x - 3$

$P(a) = 2a - 3$

$P(-x) = 2(-x) - 3 = -2x - 3$

$P(x + h) = 2(x + h) - 3 = 2x + 2h - 3$

105. $(2x^2 - xy + 7y^2) + (5xy - y^2) + (3xy + 4y^2)$

$\quad = 2x^2 - xy + 7y^2 + 5xy - y^2 + 3xy + 4y^2$

$\quad = 2x^2 - xy + 5xy + 3xy + 7y^2 - y^2 + 4y^2$

$\quad = 2x^2 + 7xy + 10y^2$

The perimeter is $(2x^2 + 7xy + 10y^2)$ units.

Exercise Set 5.2

1. $(-4x^3)(3x^2) = -4(3)(x^3)(x^2)$

$\quad = -12x^{3+2}$

$\quad = -12x^5$

5. $3x(4x + 7) = 3x(4x) + 3x(7) = 12x^2 + 21x$

9. $-4ab(xa^2 + ya^2 - 3)$

$\quad = -4ab(xa^2) + (-4ab)(ya^2) + (-4ab)(-3)$

$\quad = -4a^3bx - 4a^3by + 12ab$

13. $(2x + 3)(x^3 - x + 2)$

$\quad = 2x(x^3 - x + 2) + 3(x^3 - x + 2)$

$\quad = 2x^4 - 2x^2 + 4x + 3x^3 - 3x + 6$

$\quad = 2x^4 + 3x^3 - 2x^2 + x + 6$

17.

$$3m^2 + 2m - 1$$
$$\times \quad\quad\quad\quad 5m + 2$$
$$\overline{6m^2 + 4m - 2}$$
$$15m^3 + 10m^2 - 5m$$
$$\overline{15m^3 + 16m^2 - \ m - 2}$$

21. $(x - 3)(x + 4) = x^2 + 4x - 3x - 12 = x^2 + x - 12$

25. $\left(4x + \dfrac{1}{3}\right)\left(4x - \dfrac{1}{2}\right) = 16x^2 - 2x + \dfrac{4}{3}x - \dfrac{1}{6}$

$\quad\quad\quad\quad\quad\quad\quad = 16x^2 - \dfrac{2}{3}x - \dfrac{1}{6}$

29. $(x + 4)^2 = x^2 + 2\cdot x\cdot 4 + 4^2 = x^2 + 8x + 16$

33. $(3x - y)^2 = (3x)^2 - 2\cdot 3x\cdot y + y^2$

$\quad\quad\quad\quad = 9x^2 - 6xy + y^2$

37. $\left(3x + \dfrac{1}{2}\right)\left(3x - \dfrac{1}{2}\right) = (3x)^2 - \left(\dfrac{1}{2}\right)^2 = 9x^2 - \dfrac{1}{4}$

41. $[(2s - 3)-1][(2s - 3) + 1] = (2s - 3)^2 - 1^2$

$\quad\quad\quad\quad\quad\quad\quad = (2s)^2 - 2\cdot 2s\cdot 3 + 3^2 - 1$

$\quad\quad\quad\quad\quad\quad\quad = 4s^2 - 12s + 9 - 1$

$\quad\quad\quad\quad\quad\quad\quad = 4s^2 - 12s + 8$

45. $(x - 2)^4 = (x - 2)(x - 2)(x - 2)(x - 2)$

$\quad\quad\quad = [(x - 2)^2][(x - 2)^2]$

$\quad\quad\quad = [x^2 - 2\cdot x\cdot 2 + 2^2][x^2 - 2\cdot x\cdot 2 + 2^2]$

$\quad\quad\quad = (x^2 - 4x + 4)(x^2 - 4x + 4)$

$\quad\quad\quad = x^2(x^2 - 4x + 4) - 4x(x^2 - 4x + 4) + 4(x^2 - 4x + 4)$

$\quad\quad\quad = x^4 - 4x^3 + 4x^2 - 4x^3 + 16x^2 - 16x + 4x^2 - 16x + 16$

$\quad\quad\quad = x^4 - 8x^3 + 24x^2 - 32x + 16$

49. $-8a^2b(3b^2 - 5b + 20) = -24a^2b^3 + 40a^2b^2 - 160a^2b$

53. $(5x^3 + 2y)(5x^3 - 2y) = (5x^3)^2 - (2y)^2$

$\quad\quad\quad\quad\quad\quad\quad = 25x^6 - 4y^2$

57. $(3x^2 + 2x - 1)^2 = (3x^2 + 2x - 1)(3x^2 + 2x - 1)$

$\quad\quad\quad\quad = 3x^2(3x^2 + 2x - 1) + 2x(3x^2 + 2x - 1)$

$\quad\quad\quad\quad\quad - 1(3x^2 + 2x - 1)$

$\quad\quad\quad\quad = 9x^4 + 6x^3 - 3x^2 + 6x^3 + 4x^2 - 2x$

$\quad\quad\quad\quad\quad - 3x^2 - 2x + 1$

$\quad\quad\quad\quad = 9x^4 + 12x^3 - 2x^2 - 4x + 1$

61. $(3x^4 + 1)(3x^2 + 5) = 9x^6 + 15x^4 + 3x^2 + 5$

65. $(3b - 6y)(3b + 6y) = (3b)^2 - (6y)^2 = 9b^2 - 36y^2$

69.

$$3x^2 + 4x - 4$$
$$\times \quad\quad\quad\quad 3x + 6$$
$$\overline{18x^2 + 24x - 24}$$
$$9x^3 + 12x^2 - 12x$$
$$\overline{9x^3 + 30x^2 + 12x - 24}$$

73. $[(xy + 4) - 6]^2 = [xy + 4 - 6]^2$

$\quad\quad\quad\quad = (xy - 2)^2$

$\quad\quad\quad\quad = (xy)^2 - 2\cdot xy\cdot 2 + 2^2$

$\quad\quad\quad\quad = x^2y^2 - 4xy + 4$

77. $\left(\dfrac{2}{3}n - 2\right)\left(\dfrac{1}{2}n - 9\right) = \dfrac{1}{3}n^2 - 6n - n + 18$

$\qquad\qquad\qquad\qquad = \dfrac{1}{3}n^2 - 7n + 18$

81. $f(x) = x^2 - 3x$

$\quad f(a) = a^2 - 3a$

85. $f(x) = x^2 - 3x$

$\quad f(b - 2) = (b - 2)^2 - 3(b - 2)$

$\qquad\qquad = b^2 - 4b + 4 - 3b + 6$

$\qquad\qquad = b^2 - 7b + 10$

89. $\dfrac{20a^3b^5}{18ab^2} = \dfrac{20}{18} \cdot \dfrac{a^3}{a} \cdot \dfrac{b^5}{b^2}$

$\qquad\quad = \dfrac{10}{9} \cdot a^{3-1} \cdot b^{5-2}$

$\qquad\quad = \dfrac{10}{9}a^2b^3$

$\qquad\quad = \dfrac{10a^2b^3}{9}$

93. $7y(3z - 2) + 1 = 7y \cdot 3z + 7y(-2) + 1$

$\qquad\qquad\qquad = 21yz - 14y + 1$

97. $5x^2y^n(6y^{n+1} - 2) = 5x^2y^n \cdot 6y^{n+1} + 5x^2y^n(-2)$

$\qquad\qquad\qquad = 30x^2y^{n+n+1} - 10x^2y^n$

$\qquad\qquad\qquad = 30x^2y^{2n+1} - 10x^2y^n$

101. a. $(3x + 5) + (3x + 7) = 3x + 5 + 3x + 7 = 6x + 12$

b. $(3x + 5)(3x + 7) = 9x^2 + 21x + 15x + 35$

$\qquad\qquad\qquad\qquad = 9x^2 + 36x + 35$

The first operation is addition, the second is addition.

105. $f(x) = x^3 - 2x^2$

$\quad f(a) = a^3 - 2a^2$

109. $A = \pi r^2$

$\quad = \pi(5x - 2)^2$

$\quad = \pi[(5x)^2 - 2 \cdot 5x \cdot 2 + 2^2]$

$\quad = \pi(25x^2 - 20x + 4)$

The area is $\pi(25x^2 - 20x + 4)$ or $(25\pi x^2 - 20\pi x + 4\pi)$ square kilometers.

Exercise Set 5.3

1. $\dfrac{4a^2 + 8a}{2a} = \dfrac{4a^2}{2a} + \dfrac{8a}{2a} = 2a + 4$

5. $\dfrac{4x^2y^2 + 6xy^2 - 4y^2}{2x^2y} = \dfrac{4x^2y^2}{2x^2y} + \dfrac{6xy^2}{2x^2y} - \dfrac{4y^2}{2x^2y}$

$\qquad\qquad\qquad\qquad = 2y + \dfrac{3y}{x} - \dfrac{2y}{x^2}$

9. $x + 1\overline{)2x^2 - 6x - 8}$ with quotient $2x - 8$

$\quad\underline{2x^2 + 2x}$

$\qquad -8x - 8$

$\qquad \underline{-8x - 8}$

$\qquad\qquad 0$

$\dfrac{2x^2 - 6x - 8}{x + 1} = 2x - 8$

13. $2x + 4\overline{)4x^3 + 7x^2 + 8x + 20}$

$\quad\underline{4x^3 + 8x^2}$

$\qquad -x^2 + 8x$

$\qquad \underline{-x^2 - 2x}$

$\qquad\qquad 10x + 20$

$\qquad\qquad \underline{10x + 20}$

$\qquad\qquad\qquad 0$

$\dfrac{4x^3 + 7x^2 + 8x + 20}{2x + 4} = 2x^2 - \dfrac{1}{2}x + 5$

17. $x^2 - 2\overline{)3x^5 - x^3 + 4x^2 - 12x - 8}$ with quotient $3x^3 + 5x + 4$

$\quad\underline{3x^5 - 6x^3}$

$\qquad 5x^3 + 4x^2 - 12x$

$\qquad \underline{5x^3 \qquad - 10x}$

$\qquad\qquad 4x^2 - 2x - 8$

$\qquad\qquad \underline{4x^2 \qquad - 8}$

$\qquad\qquad\qquad -2x$

$\dfrac{3x^5 - x^3 + 4x^2 - 12x - 8}{x^2 - 2} = 3x^3 + 5x + 4 - \dfrac{2x}{x^2 - 2}$

21. $x - 5 = x - c$ where $c = 5$.

$\underline{5\rfloor\ \ 1\quad 3\quad -40}$

$\qquad\qquad 5\quad 40$

$\qquad \overline{\ 1\quad 8}$

$\dfrac{x^2 + 3x - 40}{x - 5} = x + 8$

25. $x - 2 = x - c$ where $c = 2$.

$\underline{2\rfloor\ \ 1\quad -7\quad -13\quad 5}$

$\qquad\qquad 2\quad -10\quad -46$

$\qquad \overline{\ 1\quad -5\quad -23\quad -41}$

$\dfrac{x^3 - 7x^2 - 13x + 5}{x - 2} = x^2 - 5x - 23 - \dfrac{41}{x - 2}$

29. $\dfrac{4x^7y^4 + 8xy^2 + 4xy^3}{4xy^3} = \dfrac{4x^7y^4}{4xy^3} + \dfrac{8xy^2}{4xy^3} + \dfrac{4xy^3}{4xy^3}$

$\qquad\qquad\qquad\qquad = x^6y + \dfrac{2}{y} + 1$

33. $x - 4$ has the form $x - c$, where $c = 4$, so synthetic division can be used.

$2x^3 - 6x^2 - 4 = 2x^3 - 6x^2 + 0x - 4$

$\underline{4\rfloor\ \ 2\quad -6\quad 0\quad -4}$

$\qquad\qquad 8\quad 8\quad 32$

$\qquad \overline{\ 2\quad 2\quad 8\quad 28}$

$\dfrac{2x^3 - 6x^2 - 4}{x - 4} = 2x^2 + 2x + 8 + \dfrac{28}{x - 4}$

37. $x + 1$ has the form $x - c$, where $c = -1$, so synthetic division can be used.

$7x^2 - 4x + 12 + 3x^3 = 3x^3 + 7x^2 - 4x + 12$

$\underline{-1\rfloor\ \ 3\quad 7\quad -4\quad 12}$

$\qquad\qquad -3\quad -4\quad 8$

$\qquad \overline{\ 3\quad 4\quad -8\quad 20}$

$\dfrac{7x^2 - 4x + 12 + 3x^3}{x + 1} = 3x^2 + 4x - 8 + \dfrac{20}{x + 1}$

41. $x - 1$ has the form $x - c$, where $c = 1$, so synthetic division can be used.

$x^3 - 1 = x^3 + 0x^2 + 0x - 1$

$\underline{1\rfloor\ \ 1\quad 0\quad 0\quad -1}$

$\qquad\qquad 1\quad 1\quad 1$

$\qquad \overline{\ 1\quad 1\quad 1\quad 0}$

$\dfrac{x^3 - 1}{x - 1} = x^2 + x + 1$

45. $9x^5 + 6x^4 - 6x^2 - 4x$

$= 9x^5 + 6x^4 + 0x^3 - 6x^2 - 4x + 0$

$$
\begin{array}{r}
3x^4 - 2x \\
3x + 2\overline{)9x^5 + 6x^4 + 0x^3 - 6x^2 - 4x + 0} \\
\underline{9x^5 + 6x^4} \\
0 + 0x^3 - 6x^2 - 4x \\
\underline{-6x^2 - 4x} \\
0
\end{array}
$$

$(9x^5 + 6x^4 - 6x^2 - 4x) \div (3x + 2) = 3x^2 - 2x$

49. $|2x + 7| \geq 9$

$2x + 7 \leq -9 \quad$ or $\quad 2x + 7 \geq 9$

$2x \leq -16 \qquad\qquad 2x \geq 2$

$x \leq -8 \qquad\qquad x \geq 1$

The solution set is $(-\infty, -8] \cup [1, \infty)$.

53. $(x^7 - 2) \div (x^5 + 1)$ is not a candidate for synthetic division since $x^5 + 1$ does not have the form $x - c$.

57. $\dfrac{3x^4 + 6x^2 - 18}{3} = \dfrac{3x^4}{3} + \dfrac{6x^2}{3} - \dfrac{18}{3} = x^4 + 2x^2 - 6$

The length of each piece is $(x^4 + 2x^2 - 6)$ meters.

61. $P(x) = 3x^3 + 2x^2 - 4x + 3$

$P(1) = 3(1)^3 + 2(1)^2 - 4(1) + 3$

$= 3 + 2 - 4 + 3$

$= 4$

Use synthetic division to divide $3x^3 + 2x^2 - 4x + 3$ by $x - 1$.

$$
\begin{array}{r|rrrr}
1 & 3 & 2 & -4 & 3 \\
  &   & 3 & 5 & 1 \\
\hline
  & 3 & 5 & 1 & 4
\end{array}
$$

The remainder is 4; answers may vary.

65. answers may vary

69. $\dfrac{f(x)}{g(x)} = \dfrac{25x^2 - 5x + 30}{5x}$

$= \dfrac{25x^2}{5x} - \dfrac{5x}{5x} + \dfrac{30}{5x}$

$= 5x - 1 + \dfrac{6}{x}$

Since $5x$ cannot be 0, x cannot be 0.

73. answers may vary

Exercise Set 5.4

1. $18x - 12 = 6 \cdot 3x - 6 \cdot 2 = 6(3x - 2)$

5. $6x^5 - 8x^4 + 2x^3 = 2x^3 \cdot 3x^2 - 2x^3 \cdot 4x + 2x^3 \cdot 1$

$= 2x^3(3x^2 - 4x + 1)$

9. $6(x + 3) + 5a(x + 3) = (x + 3)(6 + 5a)$

13. $3x(6x^2 + 5) - 2(6x^2 + 5) = (6x^2 + 5)(3x - 2)$

17. $ac + 4a - 2c - 8 = (ac + 4a) + (-2c - 8)$

$= a(c + 4) + (-2)(c + 4)$

$= (c + 4)(a - 2)$

21. $12xy - 8x - 3y + 2 = (12xy - 8x) + (-3y + 2)$

$= 4x(3y - 2) + (-1)(3y - 2)$

$= (3y - 2)(4x - 1)$

25. $x^3 + 3x^2 = x^2 \cdot x + x^2 \cdot 3 = x^2(x + 3)$

29. $-20x^2y + 16xy^3 = 4xy \cdot (-5x) + 4xy \cdot 4y^2$

$= 4xy(-5x + 4y^2)$

or

$-20x^2y + 16xy^3 = -4xy \cdot 5x + (-4xy)(-4y^2)$

$= -4xy(5x - 4y^2)$

33. $9abc^2 + 6a^2bc - 6ab + 3bc$

$= 3b \cdot 3ac^2 + 3b \cdot 2a^2c + 3b(-2a) + 3b \cdot c$

$= 3b(3ac^2 + 2a^2c - 2a + c)$

37. $6xy + 10x + 9y + 15 = (6xy + 10x) + (9y + 15)$

$= 2x(3y + 5) + 3(3y + 5)$

$= (3y + 5)(2x + 3)$

41. $6ab - 2a - 9b + 3 = (6ab - 2a) + (-9b + 3)$

$= 2a(3b - 1) + (-3)(3b - 1)$

$= (3b - 1)(2a - 3)$

45. $2m(n - 8) - (n - 8) = 2m(n - 8) - 1(n - 8)$

$= (n - 8)(2m - 1)$

49. $2x^2 + 3xy + 4x + 6y = (2x^2 + 3xy) + (4x + 6y)$

$= x(2x + 3y) + 2(2x + 3y)$

$= (2x + 3y)(x + 2)$

53. $x^3 + 3x^2 + 4x + 12 = (x^3 + 3x^2) + (4x + 12)$

$= x^2(x + 3) + 4(x + 3)$

$= (x + 3)(x^2 + 4)$

57. $(x + 2)(x - 5) = x^2 - 5x + 2x - 10 = x^2 - 3x - 10$

61. $(y - 3)(y - 1) = y^2 - y - 3y + 3 = y^2 - 4y + 3$

65. $2\pi r^2 + 2\pi rh = 2\pi r \cdot r + 2\pi r \cdot h = 2\pi r(r + h)$

69. answers may vary

73. $12x^2 + 9x + 3 = 3 \cdot 4x^2 + 3 \cdot 3x + 3 \cdot 1$

$= 3(4x^2 + 3x + 1)$

Choice a is correct.

77. $3y^n + 3y^{2n} + 5y^{8n} = y^n \cdot 3 + y^n \cdot 3y^n + y^n \cdot 5y^{7n}$

$= y^n(3 + 3y^n + 5y^{7n})$

81. a. $-16t^2 + 224 = -16 \cdot t^2 + (-16)(-14)$

$= -16(t^2 - 14)$

b. $h(t) = -16t^2 + 224$

$h(2) = -16(2)^2 + 224 = -64 + 224 = 160$

$h(t) = -16(t^2 - 14)$

$h(2) = -16(2^2 - 14)$

$= -16(4 - 14)$

$= -16(-10)$

$= 160$

c. answers may vary

Exercise Set 5.5

1. $x^2 + 9x + 18$

$3 \cdot 6 = 18$ and $3 + 6 = 9$

$x^2 + 9x + 18 = (x + 3)(x + 6)$

5. $x^2 + 10x - 24$

$12 \cdot (-2) = -24$ and $12 + (-2) = 10$

$x^2 + 10x - 24 = (x + 12)(x - 2)$

9. $3x^2 - 18x + 24 = 3(x^2 - 6x + 8)$

$-2(-4) = 8$ and $-2 + (-4) = -6$

$3x^2 - 18x + 24 = 3(x - 2)(x - 4)$

13. $2x^2 - 24x - 64 = 2(x^2 - 12x - 32)$

There is no pair of numbers whose product is -32 and whose sum is -12, so $x^2 - 12x - 32$ is prime.

$2x^2 - 24x - 64 = 2(x^2 - 12x - 32)$

17. $2x^2 - 11x + 12$

Factors of $2x^2$: $2x \cdot x$

Negative factors of 12: $-12(-1), -6(-2), -4(-3)$

$2x^2 - 11x + 12 = (2x - 3)(x - 4)$

21. $4x^2 - 12x + 9$

Factors of $4x^2$: $4x \cdot x, 2x \cdot 2x$

Negative factors of 9: $-9(-1), -3(-3)$

$4x^2 - 12x + 9 = (2x - 3)(2x - 3) = (2x - 3)^2$

25. $3y^4 - y^3 - 10y^2 = y^2(3y^2 - y - 10)$

Factors of $3y^2$: $3y \cdot y$

Factors of -10: $10(-1), -10 \cdot 1, 5(-2), -5 \cdot 2$

$3y^4 - y^3 - 10y^2 = y^2(3y + 5)(y - 2)$

29. $2x^2 - 5xy - 3y^2$

Factors of $2x^2$: $2x \cdot x$

Factors of $-3y^2$: $3y(-y), -3y(y)$

$2x^2 - 5xy - 3y^2 = (2x + y)(x - 3y)$

33. $2x^2 + 15x - 27$

Factors of $2x^2$: $2x \cdot x$

Factors of -27: $27(-1), -27 \cdot 1, 9(-3), -9 \cdot 3$

$2x^2 + 15x - 27 = (2x - 3)(x + 9)$

37. Substitute y for $5x + 1$.

$(5x + 1)^2 + 8(5x + 1) + 7 = y^2 + 8y + 7$

$1 \cdot 7 = 7$ and $1 + 7 = 8$

$y^2 + 8y + 7 = (y + 7)(y + 1)$

Now replace y with $5x + 1$.

$(y + 7)(y + 1) = [(5x + 1) + 7][(5x + 1) + 1]$

$\qquad\qquad = (5x + 8)(5x + 2)$

$(5x + 1)^2 + 8(5x + 1) + 7 = (5x + 8)(5x + 2)$

41. Substitute x for $a + 5$.

$(a + 5)^2 - 5(a + 5) - 24 = x^2 - 5x - 24$

$-8 \cdot 3 = -24$ and $-8 + 3 = -5$

$x^2 - 5x - 24 = (x - 8)(x + 3)$

Now replace x with $a + 5$.

$(x - 8)(x - 3) = [(a + 5) - 8][(a + 5) - 3]$

$\qquad\qquad = (a - 3)(a + 2)$

$(a + 5)^2 - 5(a + 5) - 24 = (a - 3)(a + 2)$

45. $x^2 - 15x - 54$

$-18 \cdot 3 = -54$ and $-18 + 3 = -15$

$x^2 - 15x - 54 = (x - 18)(x + 3)$

49. $3x^2 - 5x - 2$

Factors of $3x^2$: $3x \cdot x$

Factors of -2: $2(-1)$, $-2 \cdot 1$

$3x^2 - 5x - 2 = (3x + 1)(x - 2)$

53. $18x^4 + 21x^3 + 6x^2 = 3x^2(6x^2 + 7x + 2)$

Factors of $6x^2$: $6x \cdot x$, $3x \cdot 2x$

Factors of 2: $2 \cdot 1$

$18x^4 + 21x^3 + 16x^2 = 3x^2(2x + 1)(3x + 2)$

57. $x^2 - x - 12$

$-4 \cdot 3 = -12$ and $-4 + 3 = -1$

$x^2 - x - 12 = (x - 4)(x + 3)$

61. $x^2 + 4x + 5$

There is no pair of numbers whose product is 5 and whose sum is 4. $x^2 + 4x + 5$ is a prime polynomial.

65. $6x^2 - 49x + 30$

Factors of $6x^2$: $6x \cdot x$, $3x \cdot 2x$

Negative factors of 30: $-30(-1)$, $-15(-2)$, $-10(-3)$, $-6(-5)$

$6x^2 - 49x + 30 = (3x - 2)(2x - 15)$

69. $6x^3 - x^2 - x = x(6x^2 - x - 1)$

Factors of $6x^2$: $6x \cdot x$, $3x \cdot 2x$

Factors of -1: $-1 \cdot 1$

$6x^3 - x^2 - x = x(3x + 1)(2x - 1)$

73. $9x^2 + 30x + 25$

Factors of $9x^2$: $9x \cdot x$, $3x \cdot 3x$

Factors of 25: $25 \cdot 1$, $5 \cdot 5$

$9x^2 + 30x + 25 = (3x + 5)(3x + 5) = (3x + 5)^2$

77. $2x^2 + 2x - 12 = 2(x^2 + x - 6)$

$3(-2) = -6$ and $3 + (-2) = 1$

$2x^2 + 2x - 12 = 2(x + 3)(x - 2)$

81. $2x^6 + 3x^3 - 9 = 2(x^3)^2 + 3x^3 - 9$

Substitute y for x^3.

$2x^6 + 3x^3 - 9 = 2y^2 + 3y - 9$

Factors of $2y^2$: $2y \cdot y$

Factors of -9: $9(-1)$, $-9 \cdot 1$, $-3 \cdot 3$

$2y^2 + 3y - 9 = (2y - 3)(y + 3)$

Replace y with x^3.

$(2y - 3)(y + 3) = (2x^3 - 3)(x^3 + 3)$

$2x^6 + 3x^3 - 9 = (2x^3 - 3)(x^3 + 3)$

85. $2x^3y + 2x^2y - 12xy = 2xy(x^2 + x - 6)$

$3(-2) = -6$ and $3 + (-2) = 1$

$2x^3y + 2x^2y - 12xy = 2xy(x + 3)(x - 2)$

89. $(x - 3)(x + 3) = x^2 - 3^2 = x^2 - 9$

93. $(x - 2)(x^2 + 2x + 4)$

$\qquad = x(x^2 + 2x + 4) - 2(x^2 + 2x + 4)$

$\qquad = x^3 + 2x^2 + 4x - 2x^2 - 4x - 8$

$\qquad = x^3 - 8$

97. $x^3 + 2x^2 - 8x = x(x^2 + 2x - 8)$

$4(-2) = -8$ and $4 + (-2) = 2$

$x^3 + 2x^2 - 8x = x(x + 4)(x - 2)$

101. $x^{2n} + 10x^n + 16 = (x^n)^2 + 10x^n + 16$

$2 \cdot 8 = 16$ and $2 + 8 = 10$

$x^{2n} + 10x^n + 16 = (x^n + 2)(x^n + 8)$

105. $2x^{2n} + 11x^n + 5 = 2(x^n)^2 + 11x^n + 5$

Factors of $2x^{2n}$: $2x^n \cdot x^n$

Factors of 5: $5 \cdot 1$

$2x^{2n} + 11x^n + 5 = (2x^n + 1)(x^n + 5)$

109. $x^4 + 6x^3 + 5x^2 = x^2(x^2 + 6x + 5)$

$\qquad\qquad = x^2(x + 5)(x + 1)$

Exercise Set 5.6

1. $x^2 + 6x + 9 = (x)^2 + 2 \cdot x \cdot 3 + (3)^2 = (x + 3)^2$

5. $4a^2 + 12a + 9 = (2a)^2 + 2 \cdot 2a \cdot 3 + (3)^2$

$\qquad\qquad = (2a + 3)^2$

9. $9y^2x^2 + 12yx^2 + 4x^2$

$\qquad = x^2(9y^2 + 12y + 4)$

$\qquad = x^2[(3y)^2 + 2 \cdot 3y \cdot 2 + (2)^2]$

$\qquad = x^2(3y + 2)^2$

13. $x^2 - 25 = (x)^2 - (5)^2 = (x + 5)(x - 5)$

17. $(y + 2)^2 - 49 = (y + 2)^2 - (7)^2$

$\qquad\qquad = [(y + 2) + 7][(y + 2) - 7]$

$\qquad\qquad = (y + 9)(y - 5)$

21. $(x + 2y)^2 - 9 = (x + 2y)^2 - (3)^2$

$\qquad\qquad = [(x + 2y) + 3][(x + 2y) - 3]$

$\qquad\qquad = (x + 2y + 3)(x + 2y - 3)$

25. $x^3 + 27 = x^3 + (3)^3 = (x + 3)(x^2 - 3x + 9)$

29. $m^3 + n^3 = (m + n)(m^2 - mn + n^2)$

33. $64q^2 - q^2p^3 = q^2(64 - p^3)$

$\qquad\qquad = q^2[(4)^3 - p^3]$

$\qquad\qquad = q^2(4 - p)(4^2 + 4p + p^2)$

$\qquad\qquad = q^2(4 - p)(16 + 4p + p^2)$

37. $x^2 - 12x + 36 = x^2 - 2 \cdot x \cdot 6 + 6^2 = (x - 6)^2$

41. $9x^2 - 49 = (3x)^2 - 7^2 = (3x + 7)(3x - 7)$

45. $x^6 - y^3 = (x^2)^3 - y^3$

$\qquad = (x^2 - y)[(x^2)^2 + x^2y + y^2]$

$\qquad = (x^2 - y)(x^4 + x^2y + y^2)$

49. $4x^2 + 4x + 1 - z^2 = (4x^2 + 4x + 1) - z^2$

$\qquad\qquad = [(2x)^2 + 2 \cdot 2x \cdot 1 + 1^2] - z^2$

$\qquad\qquad = (2x + 1)^2 - z^2$

$\qquad\qquad = (2x + 1 + z)(2x + 1 - z)$

53. $n^3 - \dfrac{1}{27} = n^3 - \left(\dfrac{1}{3}\right)^3$

$\qquad = \left(n - \dfrac{1}{3}\right)\left[n^2 + n \cdot \dfrac{1}{3} + \left(\dfrac{1}{3}\right)^2\right]$

$\qquad = \left(n - \dfrac{1}{3}\right)\left(n^2 + \dfrac{1}{3}n + \dfrac{1}{9}\right)$

57. $x^2 - 10x + 25 - y^2 = (x^2 - 10x + 25) - y^2$

$\qquad\qquad = (x^2 - 2 \cdot x \cdot 5 + 5^2) - y^2$

$\qquad\qquad = (x - 5)^2 - y^2$

$\qquad\qquad = (x - 5 + y)(x - 5 - y)$

61. $\dfrac{x^2}{25} - \dfrac{y^2}{9} = \left(\dfrac{x}{5}\right)^2 - \left(\dfrac{y}{3}\right)^2 = \left(\dfrac{x}{5} + \dfrac{y}{3}\right)\left(\dfrac{x}{5} - \dfrac{y}{3}\right)$

65. $x - 5 = 0$
$$x = 5$$
The solution set is {5}.

69. $-2x = 0$
$$x = \frac{0}{-2}$$
$$x = 0$$
The solution set is {0}.

73. Since $x^2 - 4 = x^2 - 2^2 = (x + 2)(x - 2)$, the polynomial $5x(x^2 - 4)$ is not factored completely.

77. If the washer didn't have a hole, the area would be πR^2. Subtracting the hole, the area is $\pi R^2 - \pi r^2$.
$$\pi R^2 - \pi r^2 = \pi(R^2 - r^2) = \pi(R + r)(R - r)$$

81. $6x = 2 \cdot x \cdot 3$
$$(x + 3)^2 = x^2 + 6x + 9$$
The value of c is 9.

85. a. $x^6 - 1 = (x^3)^2 - 1^2$
$$= (x^3 + 1)(x^3 - 1)$$
$$= (x^3 + 1^3)(x^3 - 1^3)$$
$$= (x + 1)(x^2 - x + 1)(x - 1)(x^2 + x + 1)$$
b. $x^6 - 1 = (x^2)^3 - 1$
$$= (x^2 - 1)[(x^2)^2 + x^2 + 1]$$
$$= (x^2 - 1^2)(x^4 + x^2 + 1)$$
$$= (x + 1)(x - 1)(x^4 + x^2 + 1)$$
c. answers may vary

89. $25x^{2n} - 81 = (5x^n)^2 - 9^2 = (5x^n + 9)(5x^n - 9)$

Exercise Set 5.7

1. $(x + 3)(3x - 4) = 0$
$$x + 3 = 0 \quad \text{or} \quad 3x - 4 = 0$$
$$3x = 4$$
$$x = -3 \quad \text{or} \quad x = \frac{4}{3}$$
The solution set is $\left\{-3, \frac{4}{3}\right\}$.

5. $x^2 + 11x + 24 = 0$
$$(x + 3)(x + 8) = 0$$
$$x + 3 = 0 \quad \text{or} \quad x + 8 = 0$$
$$x = -3 \quad \text{or} \quad x = -8$$
The solution set is {-3, -8}.

9. $z^2 + 9 = 10z$
$$z^2 - 10z + 9 = 0$$
$$(z - 1)(z - 9) = 0$$
$$z - 1 = 0 \quad \text{or} \quad z - 9 = 0$$
$$z = 1 \quad \text{or} \quad z = 9$$
The solution set is {1, 9}.

13. $x^2 - 6x = x(8 + x)$
$$x^2 - 6x = 8x + x^2$$
$$0 = 14x$$
$$x = 0$$
The solution set is {0}.

17.
$$\frac{x^2}{2} + \frac{x}{20} = \frac{1}{10}$$
$$10x^2 + x = 2$$
$$10x^2 + x - 2 = 0$$
$$(5x - 2)(2x + 1) = 0$$
$$5x - 2 = 0 \quad \text{or} \quad 2x + 1 = 0$$
$$5x = 2 \quad \text{or} \quad 2x = -1$$
$$x = \frac{2}{5} \quad \text{or} \quad x = -\frac{1}{2}$$
The solution set is $\left\{\frac{2}{5}, -\frac{1}{2}\right\}$.

21. $(x + 2)(x - 7)(3x - 8) = 0$
$$x + 2 = 0 \quad \text{or} \quad x - 7 = 0 \quad \text{or} \quad 3x - 8 = 0$$
$$x = -2 \quad \text{or} \quad x = 7 \quad \text{or} \quad x = \frac{8}{3}$$
The solution set is $\left\{-2, 7, \frac{8}{3}\right\}$.

25.
$$x^3 - x = 2x^2 - 2$$
$$x^3 - 2x^2 - x + 2 = 0$$
$$x^2(x - 2) - (x - 2) = 0$$
$$(x^2 - 1)(x - 2) = 0$$
$$(x + 1)(x - 1)(x - 2) = 0$$
$$x + 1 = 0 \quad \text{or} \quad x - 1 = 0 \quad \text{or} \quad x - 2 = 0$$
$$x = -1 \quad \text{or} \quad x = 1 \quad \text{or} \quad x = 2$$
The solution set is {-1, 1, 2}.

29. $3x(x - 5) = 0$
$$3x = 0 \quad \text{or} \quad x - 5 = 0$$
$$x = 0 \quad \text{or} \quad x = 5$$
The solution set is {0, 5}.

33.
$$12x^2 + 2x - 2 = 0$$
$$2(6x^2 + x - 1) = 0$$
$$2(2x + 1)(3x - 1) = 0$$
$$2x + 1 = 0 \quad \text{or} \quad 3x - 1 = 0$$
$$2x = -1 \quad \text{or} \quad 3x = 1$$
$$x = -\frac{1}{2} \quad \text{or} \quad x = \frac{1}{3}$$
The solution set is $\left\{-\frac{1}{2}, \frac{1}{3}\right\}$.

37. $25x^2 - 40x + 16 = 0$
$$(5x - 4)(5x - 4) = 0$$
$$5x - 4 = 0 \quad \text{or} \quad 5x - 4 = 0$$
$$5x = 4 \quad \quad 5x = 4$$
$$x = \frac{4}{5} \quad \text{or} \quad x = \frac{4}{5}$$
The solution set is $\left\{\frac{4}{5}\right\}$.

41. $z(5z - 4)(z + 3) = 0$
$$z = 0 \quad \text{or} \quad 5z - 4 = 0 \quad \text{or} \quad z + 3 = 0$$
$$5z = 4$$
$$z = 0 \quad \text{or} \quad z = \frac{4}{5} \quad \text{or} \quad z = -3$$
The solution set is $\left\{-3, 0, \frac{4}{5}\right\}$.

45. $(x - 1)(x + 4) = 24$
$$x^2 + 3x - 4 = 24$$
$$x^2 + 3x - 28 = 0$$
$$(x + 7)(x - 4) = 0$$
$$x + 7 = 0 \quad \text{or} \quad x - 4 = 0$$
$$x = -7 \quad \text{or} \quad x = 4$$
The solution set is {-7, 4}.

49.
$$y^2 + \frac{1}{4} = -y$$
$$4y^2 + 1 = -4y$$
$$4y^2 + 4y + 1 = 0$$
$$(2y + 1)(2y + 1) = 0$$
$$2y + 1 = 0 \quad \text{or} \quad 2y + 1 = 0$$
$$2y = -1 \quad \text{or} \quad 2y = -1$$
$$y = -\frac{1}{2} \quad \text{or} \quad y = -\frac{1}{2}$$
The solution set is $\left\{-\frac{1}{2}\right\}$.

53.
$$2x^3 = 50x$$
$$2x^3 - 50x = 0$$
$$2x(x^2 - 25) = 0$$
$$2x(x + 5)(x - 5) = 0$$
$$2x = 0 \quad \text{or} \quad x + 5 = 0 \quad \text{or} \quad x - 5 = 0$$
$$x = 0 \quad \text{or} \quad x = -5 \quad \text{or} \quad x = 5$$
The solution set is $\{-5, 0, 5\}$.

57.
$$m^2(3m - 2) = m$$
$$3m^3 - 2m^2 = m$$
$$3m^3 - 2m^2 - m = 0$$
$$m(3m^2 - 2m - 1) = 0$$
$$m(3m + 1)(m - 1) = 0$$
$$m = 0 \quad \text{or} \quad 3m + 1 = 0 \quad \text{or} \quad m - 1 = 0$$
$$3m = -1$$
$$m = 0 \quad \text{or} \quad m = -\frac{1}{3} \quad \text{or} \quad m = 1$$
The solution set is $\left\{-\frac{1}{3}, 0, 1\right\}$.

61.
$$x(x - 3) = x^2 + 5x + 7$$
$$x^2 - 3x = x^2 + 5x + 7$$
$$0 = 8x + 7$$
$$-7 = 8x$$
$$-\frac{7}{8} = x$$
The solution set is $\left\{-\frac{7}{8}\right\}$.

65.
$$-3(x - 4) + x = 5(3 - x)$$
$$-3x + 12 + x = 15 - 5x$$
$$12 - 2x = 15 - 5x$$
$$12 + 3x = 15$$
$$3x = 3$$
$$x = 1$$
The solution set is $\{1\}$.

69. Let x be the length of cable needed. By the Pythagorean theorem, $45^2 + 60^2 = x^2$.
$$x^2 = 45^2 + 60^2$$
$$x^2 = 2025 + 3600$$
$$x^2 = 5625$$
$$x^2 - 5625 = 0$$
$$x^2 - 75^2 = 0$$
$$(x + 75)(x - 75) = 0$$
$$x + 75 = 0 \quad \text{or} \quad x - 75 = 0$$
$$x = -75 \quad \text{or} \quad x = 75$$
Since the length of a cable cannot be negative, discard $x = -75$. The cable should be 75 feet long.

73. Let the length of the longer leg be x. Then the shorter leg has length $x - 3$.
$$x^2 + (x - 3)^2 = 15^2$$
$$x^2 + x^2 - 6x + 9 = 225$$
$$2x^2 - 6x - 216 = 0$$
$$2(x^2 - 3x - 108) = 0$$
$$2(x - 12)(x + 9) = 0$$
$$x - 12 = 0 \quad \text{or} \quad x + 9 = 0$$
$$x = 12 \quad \text{or} \quad x = -9$$
Since a length cannot be negative, discard $x = -9$.
$$x - 3 = 12 - 3 = 9$$
The legs of the triangle have length 9 centimeters and 12 centimeters.

77. Find t when $h(t) = 0$.
$$-16t^2 + 1600 = 0$$
$$-16(t^2 - 100) = 0$$
$$-16(t + 10)(t - 10) = 0$$
$$t + 10 = 0 \quad \text{or} \quad t - 10 = 0$$
$$t = -10 \quad \text{or} \quad t = 10$$
Since the time cannot be negative, discard $t = -10$. The sunglasses hit the ground after 10 seconds.

81. Find x when $W(x) = 50$.
$$0.5x^2 = 50$$
$$2(0.5x^2) = 2(50)$$
$$x^2 = 100$$
$$x^2 - 100 = 0$$
$$(x + 10)(x - 10) = 0$$
$$x + 10 = 0 \quad \text{or} \quad x - 10 = 0$$
$$x = -10 \quad \text{or} \quad x = 10$$
Since a length cannot be negative, discard $x = -10$. The tiers of the cake are 10-inch squares.

85. The graph of $h(x)$ has three x-intercepts, $(-3, 0)$, $(0, 0)$, and $(3, 0)$ because the equation $0 = x(x + 3)(x - 3)$ has three solutions, -3, 0, and 3. The graph of $h(x)$ is graph F.

89. x-intercept: $(-3, 0)$
y-intercept: $(0, 2)$

93. The error is in the last line.
$$(x - 5)(x + 2) = 0$$
$$x - 5 = 0 \quad \text{or} \quad x + 2 = 0$$
$$x = 5 \quad \text{or} \quad x = -2$$

97.
$$(x^2 + x - 6)(3x^2 - 14x - 5) = 0$$
$$[(x + 3)(x - 2)][(3x + 1)(x - 5)] = 0$$
$$(x + 3)(x - 2)(3x + 1)(x - 5) = 0$$
$$x + 3 = 0 \quad \text{or} \quad x - 2 = 0 \quad \text{or} \quad 3x + 1 = 0 \quad \text{or} \quad x - 5 = 0$$
$$3x = -1$$
$$x = -3 \quad \text{or} \quad x = 2 \quad \text{or} \quad x = -\frac{1}{3} \quad \text{or} \quad x = 5$$

101. no; answers may vary

Chapter 5 Test

1.
$$(4x^3 - 3x - 4) - (9x^3 + 8x + 5)$$
$$= 4x^3 - 3x - 4 - 9x^3 - 8x - 5$$
$$= 4x^3 - 9x^3 - 3x - 8x - 4 - 5$$
$$= -5x^3 - 11x - 9$$

5.
$$(6m + n)^2 = (6m + n)(6m + n)$$
$$= 36m^2 + 6mn + 6mn + n^2$$
$$= 36m^2 + 12mn + n^2$$

9. $x + 3$ is $x - c$ where $c = -3$.

$$\begin{array}{r|rrrrr} -3 & 4 & -3 & 2 & -1 & -1 \\ & & -12 & 45 & -141 & 426 \\ \hline & 4 & -15 & 47 & -142 & 425 \end{array}$$

$$\frac{4x^4 - 3x^3 + 2x^2 - x - 1}{x + 3} = 4x^3 - 15x^2 + 47x - 142 + \frac{425}{x + 3}$$

13. $6x^2 - 15x - 9 = 3(2x^2 - 5x - 3) = 3(2x + 1)(x - 3)$

17.
$$16y^3 - 2 = 2(8y^3 - 1)$$
$$= 2[(2y)^3 - 1^3]$$
$$= 2(2y - 1)[(2y)^2 + 2y(1) + 1^2]$$
$$= 2(2y - 1)(4y^2 + 2y + 1)$$

21.
$$2x^3 + 5x^2 = 8x + 20$$
$$2x^3 + 5x^2 - 8x - 20 = 0$$
$$(2x^3 + 5x^2) + (-8x - 20) = 0$$
$$x^2(2x + 5) + (-4)(2x + 5) = 0$$
$$(2x + 5)(x^2 - 4) = 0$$

$$(2x + 5)(x^2 - 2^2) = 0$$
$$(2x + 5)(x + 2)(x - 2) = 0$$
$$2x + 5 = 0 \quad \text{or} \quad x + 2 = 0 \quad \text{or} \quad x - 2 = 0$$
$$2x = -5$$
$$x = -\frac{5}{2} \quad \text{or} \quad x = -2 \quad \text{or} \quad x = 2$$

The solution set is $\left\{ -\frac{5}{2}, -2, 2 \right\}$.

CHAPTER 6

Exercise Set 6.1

1. 4 is never 0, so the domain of $f(x) = \dfrac{5x - 7}{4}$ is

$\{x|x \text{ is a real number}\}$.

5. $7 - x = 0$
$\quad\quad 7 = x$

The domain of $f(x) = \dfrac{3x}{7 - x}$ is $\{x|x \text{ is a real number and } x \neq 7\}$.

9. $x^3 + x^2 - 2x = 0$

$x(x^2 + x - 2) = 0$

$x(x + 2)(x - 1) = 0$

$x = 0 \quad \text{or} \quad x + 2 = 0 \quad \text{or} \quad x - 1 = 0$

$x = 0 \quad \text{or} \quad\quad x = -2 \quad \text{or} \quad\quad x = 1$

The domain of $R(x) = \dfrac{3 + 2x}{x^3 + x^2 - 2x}$ is
$\{x|x \text{ is a real number and } x \neq -2, x \neq 0, x \neq 1\}$.

13. $\dfrac{8x - 16x^2}{8x} = \dfrac{8x(1 - 2x)}{8x} = 1 - 2x$

17. $\dfrac{9y - 18}{7y - 14} = \dfrac{9(y - 2)}{7(y - 2)} = \dfrac{9}{7}$

21. $\dfrac{x - 9}{9 - x} = \dfrac{-1(9 - x)}{9 - x} = -1$

25. $\dfrac{2x^2 - 7x - 4}{x^2 - 5x + 4} = \dfrac{(2x + 1)(x - 4)}{(x - 1)(x - 4)} = \dfrac{2x + 1}{x - 1}$

29. $\dfrac{3x^2 - 5x - 2}{6x^3 + 2x^2 + 3x + 1} = \dfrac{(3x + 1)(x - 2)}{2x^2(3x + 1) + 1(3x + 1)}$

$\quad\quad = \dfrac{(3x + 1)(x - 2)}{(3x + 1)(2x^2 + 1)}$

$\quad\quad = \dfrac{x - 2}{2x^2 + 1}$

33. $\dfrac{2x - 4}{15} \cdot \dfrac{6}{2 - x} = \dfrac{2(x - 2)}{3 \cdot 5} \cdot \dfrac{2 \cdot 3}{-(x - 2)}$

$\quad\quad = \dfrac{2 \cdot 2}{5(-1)}$

$\quad\quad = -\dfrac{4}{5}$

37. $\dfrac{9x + 9}{4x + 8} \cdot \dfrac{2x + 4}{3x^2 - 3} = \dfrac{9(x + 1)}{4(x + 2)} \cdot \dfrac{2(x + 2)}{3(x^2 - 1)}$

$\quad\quad = \dfrac{3 \cdot 3(x + 1)}{2 \cdot 2(x + 2)} \cdot \dfrac{2(x + 2)}{3(x + 1)(x - 1)}$

$\quad\quad = \dfrac{3}{2(x - 1)}$

41. $\dfrac{a^3 + a^2b + a + b}{5a^3 + 5a} \cdot \dfrac{6a^2}{2a^2 - 2b^2}$

$\quad = \dfrac{a^2(a + b) + 1(a + b)}{5a(a^2 + 1)} \cdot \dfrac{6a^2}{2(a^2 - b^2)}$

$\quad = \dfrac{(a + b)(a^2 + 1)}{5a(a^2 + 1)} \cdot \dfrac{2 \cdot 3 \cdot a \cdot a}{2(a + b)(a - b)}$

$\quad = \dfrac{3a}{5(a - b)}$

45. $\dfrac{2x}{5} \div \dfrac{6x + 12}{5x + 10} = \dfrac{2x}{5} \cdot \dfrac{5x + 10}{6x + 12}$

$\quad\quad = \dfrac{2x}{5} \cdot \dfrac{5(x + 2)}{6(x + 2)}$

$\quad\quad = \dfrac{2x}{5} \cdot \dfrac{5(x + 2)}{2 \cdot 3(x + 2)}$

$\quad\quad = \dfrac{x}{3}$

49. $\dfrac{x^2 - 6x + 9}{x^2 - x - 6} \div \dfrac{x^2 - 9}{4} = \dfrac{x^2 - 6x + 9}{x^2 - x - 6} \cdot \dfrac{4}{x^2 - 9}$

$\quad\quad = \dfrac{(x - 3)^2}{(x - 3)(x + 2)} \cdot \dfrac{4}{(x + 3)(x - 3)}$

$\quad\quad = \dfrac{4}{(x + 2)(x + 3)}$

53. $\dfrac{3x - x^2}{x^3 - 27} \div \dfrac{x}{x^2 + 3x + 9} = \dfrac{3x - x^2}{x^3 - 27} \cdot \dfrac{x^2 + 3x + 9}{x}$

$\quad\quad = \dfrac{-x(x - 3)}{(x - 3)(x^2 + 3x + 9)} \cdot \dfrac{x^2 + 3x + 9}{x}$

$\quad\quad = -1$

57. $\dfrac{x^2 - 9}{4} \cdot \dfrac{x^2 - x - 6}{x^2 - 6x + 9} = \dfrac{(x + 3)(x - 3)}{4} \cdot \dfrac{(x + 2)(x - 3)}{(x - 3)^2}$

$\quad\quad = \dfrac{(x + 3)(x + 2)}{4}$

61. $\dfrac{r^3 + s^3}{r + s} = \dfrac{(r + s)(r^2 - rs + s^2)}{r + s} = r^2 - rs + s^2$

65. $\dfrac{3x^2 - 5x - 2}{y^2 + y - 2} \cdot \dfrac{y^2 + 4y - 5}{12x^2 + 7x + 1} \div \dfrac{5x^2 - 9x - 2}{8x^2 - 2x - 1}$

$\quad = \dfrac{3x^2 - 5x - 2}{y^2 + y - 2} \cdot \dfrac{y^2 + 4y - 5}{12x^2 + 7x + 1} \cdot \dfrac{8x^2 - 2x - 1}{5x^2 - 9x - 2}$

$\quad = \dfrac{(3x + 1)(x - 2)(y + 5)(y - 1)(4x + 1)(2x - 1)}{(y + 2)(y - 1)(4x + 1)(3x + 1)(5x + 1)(x - 2)}$

$\quad = \dfrac{(y + 5)(2x - 1)}{(y + 2)(5x + 1)}$

69. $g(x) = \dfrac{x^2 + 8}{x^3 - 25x}$

$\quad g(3) = \dfrac{3^2 + 8}{3^3 - 25(3)} = \dfrac{9 + 8}{27 - 75} = \dfrac{17}{-48} = -\dfrac{17}{48}$

$\quad g(-2) = \dfrac{(-2)^2 + 8}{(-2)^3 - 25(-2)} = \dfrac{4 + 8}{-8 + 50} = \dfrac{12}{42} = \dfrac{2}{7}$

$\quad g(1) = \dfrac{1^2 + 8}{1^3 - 25(1)} = \dfrac{1 + 8}{1 - 25} = \dfrac{9}{-24} = -\dfrac{3}{8}$

73. $\dfrac{4}{3} + \dfrac{3}{5} = \dfrac{7}{5}$

77. $\dfrac{3}{8} + \dfrac{1}{2} - \dfrac{3}{16} = \dfrac{6}{16} + \dfrac{8}{16} - \dfrac{3}{16} = \dfrac{11}{16}$

81. no; answers may vary

85. Since $A = b \cdot h$, $b = \dfrac{A}{h}$.

$$b = \frac{\dfrac{x^2 + x - 2}{x^3}}{\dfrac{x^2}{x - 1}}$$

$$b = \frac{(x + 2)(x - 1)}{x^3} \cdot \frac{(x - 1)}{x^2}$$

$$b = \frac{(x + 2)(x - 1)^2}{x^5} \text{ feet}$$

89. a. $\dfrac{x + 5}{5 + x} = \dfrac{x + 5}{x + 5} = 1$

b. $\dfrac{x - 5}{5 - x} = \dfrac{x - 5}{-(x - 5)} = -1$

c. $\dfrac{x + 5}{x - 5}$ neither

d. $\dfrac{-x - 5}{x + 5} = \dfrac{-(x + 5)}{x + 5} = -1$

e. $\dfrac{x - 5}{-x + 5} = \dfrac{x - 5}{-(x - 5)} = -1$

f. $\dfrac{-5 + x}{x - 5} = \dfrac{x - 5}{x - 5} = 1$

93. $\dfrac{p^x - 4}{4 - p^x} = \dfrac{p^x - 4}{-(p^x - 4)} = -1$

Exercise Set 6.2

1. $\dfrac{2}{xz^2} - \dfrac{5}{xz^2} = \dfrac{2 - 5}{xz^2} = \dfrac{-3}{xz^2} = -\dfrac{3}{xz^2}$

5. $\dfrac{x^2}{x + 2} - \dfrac{4}{x + 2} = \dfrac{x^2 - 4}{x + 2} = \dfrac{(x + 2)(x - 2)}{x + 2} = x - 2$

9. $\dfrac{x - 5}{2x} - \dfrac{x + 5}{2x} = \dfrac{x - 5 - x - 5}{2x} = \dfrac{-10}{2x} = -\dfrac{5}{x}$

13. $x = x$

$x + 1 = x + 1$

LCD $= x(x + 1)$

17. $3x + 6 = 3(x + 2)$

$2x - 4 = 2(x - 2)$

LCD $= 3 \cdot 2(x + 2)(x - 2) = 6(x + 2)(x - 2)$

21. $x^2 - 9 = (x + 3)(x - 3)$

$x = x$

$12 - 4x = -4(x - 3)$

LCD $= -4x(x + 3)(x - 3)$

25. $\dfrac{5}{2y^2} - \dfrac{2}{7y} = \dfrac{5 \cdot 7}{2y^2 \cdot 7} - \dfrac{2 \cdot 2y}{7y \cdot 2y}$

$= \dfrac{35}{14y^2} - \dfrac{4y}{14y^2}$

$= \dfrac{35 - 4y}{14y^2}$

29. $\dfrac{1}{x - 5} + \dfrac{2x - 19}{(x - 5)(x + 4)}$

$= \dfrac{1 \cdot (x + 4)}{(x - 5) \cdot (x - 4)} + \dfrac{2x - 19}{(x - 5)(x + 4)}$

$= \dfrac{(x + 4) + 2x - 19}{(x - 5)(x + 4)}$

$= \dfrac{3x - 15}{(x - 5)(x + 4)}$

$= \dfrac{3(x - 5)}{(x - 5)(x + 4)}$

$= \dfrac{3}{x + 4}$

33. $\dfrac{x + 1}{1 - x} + \dfrac{1}{x - 1} = \dfrac{x + 1}{-(x - 1)} + \dfrac{1}{x - 1}$

$= \dfrac{-(x + 1)}{x - 1} + \dfrac{1}{x - 1}$

$= \dfrac{-x - 1 + 1}{x - 1}$

$= \dfrac{-x}{x - 1}$

$= -\dfrac{x}{x - 1}$

37. $\dfrac{y + 1}{y^2 - 6y + 8} - \dfrac{3}{y^2 - 16}$

$= \dfrac{y + 1}{(y - 2)(y - 4)} - \dfrac{3}{(y + 4)(y - 4)}$

$= \dfrac{(y + 1)(y + 4)}{(y - 2)(y - 4)(y + 4)} - \dfrac{3(y - 2)}{(y - 2)(y + 4)(y - 4)}$

$= \dfrac{(y + 1)(y + 4) - 3(y - 2)}{(y - 2)(y - 4)(y + 4)}$

$= \dfrac{y^2 + 5y + 4 - 3y + 6}{(y - 2)(y - 4)(y + 4)}$

$= \dfrac{y^2 + 2y + 10}{(y - 2)(y - 4)(y + 4)}$

41. $\dfrac{7}{x^2 - x - 2} + \dfrac{x}{x^2 + 4x + 3}$

$= \dfrac{7}{(x - 2)(x + 1)} + \dfrac{x}{(x + 3)(x + 1)}$

$= \dfrac{7(x + 3)}{(x - 2)(x + 1)(x + 3)} + \dfrac{x(x - 2)}{(x + 3)(x + 1)(x - 2)}$

$= \dfrac{7(x + 3) + x(x - 2)}{(x - 2)(x + 1)(x + 3)}$

$= \dfrac{7x + 21 + x^2 - 2x}{(x - 2)(x + 1)(x + 3)}$

$= \dfrac{x^2 + 5x + 21}{(x - 2)(x + 1)(x + 3)}$

45. $\dfrac{2}{a^2 + 2a + 1} + \dfrac{3}{a^2 - 1}$

$= \dfrac{2}{(a + 1)^2} + \dfrac{3}{(a + 1)(a - 1)}$

$= \dfrac{2(a - 1)}{(a + 1)^2(a - 1)} + \dfrac{3(a + 1)}{(a + 1)(a - 1)(a + 1)}$

$= \dfrac{2(a - 1) + 3(a + 1)}{(a + 1)^2(a - 1)}$

$= \dfrac{2a - 2 + 3a + 3}{(a + 1)^2(a - 1)}$

$= \dfrac{5a + 1}{(a + 1)^2(a - 1)}$

49. $\dfrac{13x - 5}{2x} - \dfrac{13x + 5}{2x} = \dfrac{13x - 5 - 13x - 5}{2x} = \dfrac{-10}{2x} = -\dfrac{5}{x}$

53. $\dfrac{-2}{x^2 - 3x} - \dfrac{1}{x^3 - 3x^2} = \dfrac{-2}{x(x - 3)} - \dfrac{1}{x^2(x - 3)}$

$= \dfrac{-2(x)}{x^2(x - 3)} - \dfrac{1}{x^2(x - 3)}$

$= \dfrac{-2x - 1}{x^2(x - 3)}$

57. $\dfrac{5}{x^2 - 4} - \dfrac{3}{x^2 + 4x + 4}$

$= \dfrac{5}{(x + 2)(x - 2)} - \dfrac{3}{(x + 2)^2}$

$= \dfrac{5(x + 2)}{(x + 2)(x - 2)(x + 2)} - \dfrac{3(x - 2)}{(x + 2)^2(x - 2)}$

$= \dfrac{5(x + 2) - 3(x - 2)}{(x + 2)^2(x - 2)}$

$= \dfrac{5x + 10 - 3x + 6}{(x + 2)^2(x - 2)}$

$= \dfrac{2x + 16}{(x + 2)^2(x - 2)}$

$= \dfrac{2(x + 8)}{(x + 2)^2(x - 2)}$

61. $\dfrac{2}{x + 1} - \dfrac{3x}{3x + 3} + \dfrac{1}{2x + 2}$

$= \dfrac{2}{x + 1} - \dfrac{3x}{3(x + 1)} + \dfrac{1}{2(x + 1)}$

$= \dfrac{2}{x + 1} - \dfrac{x}{x + 1} + \dfrac{1}{2(x + 1)}$

$= \dfrac{2 \cdot 2}{2(x + 1)} - \dfrac{2 \cdot x}{2(x + 1)} + \dfrac{1}{2(x + 1)}$

$= \dfrac{4 - 2x + 1}{2(x + 1)}$

$= \dfrac{5 - 2x}{2(x + 1)}$

65. $\dfrac{x}{x^2 - 9} + \dfrac{3}{x^2 - 6x + 9} - \dfrac{1}{x + 3}$

$= \dfrac{x}{(x + 3)(x - 3)} + \dfrac{3}{(x - 3)^2} - \dfrac{1}{x + 3}$

$= \dfrac{x(x - 3)}{(x + 3)(x - 3)^2} + \dfrac{3(x + 3)}{(x - 3)^2(x + 3)} - \dfrac{1(x - 3)^2}{(x + 3)(x - 3)^2}$

$= \dfrac{x(x - 3) + 3(x + 3) - (x - 3)^2}{(x + 3)(x - 3)^2}$

$= \dfrac{x^2 - 3x + 3x + 9 - (x^2 - 6x + 9)}{(x + 3)(x - 3)^2}$

$= \dfrac{x^2 + 9 - x^2 + 6x - 9}{(x + 3)(x - 3)^2}$

$= \dfrac{6x}{(x + 3)(x - 3)^2}$

69. $12\left(\dfrac{2}{3} + \dfrac{1}{6}\right) = 12\left(\dfrac{2}{3}\right) + 12\left(\dfrac{1}{6}\right)$

$= 4(2) + 2(1)$

$= 8 + 2$

$= 10$

73. $\dfrac{2x - 3}{x^2 + 1} - \dfrac{x - 6}{x^2 + 1} = \dfrac{2x - 3 - (x - 6)}{x^2 + 1}$

$= \dfrac{2x - 3 - x + 6}{x^2 + 1}$

$= \dfrac{x + 3}{x^2 + 1}$

77. answers may vary

81. answers may vary

85. $\left(\dfrac{2a}{3}\right)^2 \div \left(\dfrac{a^2}{a + 1} - \dfrac{1}{a + 1}\right) = \dfrac{4a^2}{9} \div \dfrac{a^2 - 1}{a + 1}$

$= \dfrac{4a^2}{9} \div \dfrac{(a + 1)(a - 1)}{a + 1}$

$= \dfrac{4a^2}{9} \div (a - 1)$

$= \dfrac{4a^2}{9} \cdot \dfrac{1}{a - 1}$

$= \dfrac{4a^2}{9(a - 1)}$

89. $\left(\dfrac{x}{x + 1} - \dfrac{x}{x - 1}\right) \div \dfrac{x}{2x + 2}$

$= \left(\dfrac{x \cdot (x - 1)}{(x + 1)(x - 1)} - \dfrac{x \cdot (x + 1)}{(x - 1)(x + 1)}\right) \div \dfrac{x}{2(x + 1)}$

$= \dfrac{x(x - 1) - x(x + 1)}{(x + 1)(x - 1)} \div \dfrac{x}{2(x + 1)}$

$= \dfrac{x^2 - x - x^2 - x}{(x + 1)(x - 1)} \div \dfrac{x}{2(x + 1)}$

$= \dfrac{-2x}{(x + 1)(x - 1)} \cdot \dfrac{2(x + 1)}{x}$

$= -\dfrac{4}{x - 1}$

93. $x^{-1} + (2x)^{-1} = \dfrac{1}{x} + \dfrac{1}{2x} = \dfrac{1 \cdot 2}{x \cdot 2} + \dfrac{1}{2x} = \dfrac{2 + 1}{2x} = \dfrac{3}{2x}$

97. answers may vary

Exercise Set 6.3

1. $\dfrac{1 + \dfrac{2}{5}}{2 + \dfrac{3}{5}} = \dfrac{5\left(1 + \dfrac{2}{5}\right)}{5\left(2 + \dfrac{3}{5}\right)} = \dfrac{5 + 2}{10 + 3} = \dfrac{7}{13}$

5. $\dfrac{1 - \dfrac{2}{x}}{x + \dfrac{4}{9x}} = \dfrac{9x\left(1 - \dfrac{2}{x}\right)}{9x\left(x + \dfrac{4}{9x}\right)} = \dfrac{9x - 18}{9x^2 + 4} = \dfrac{9(x - 2)}{9x^2 + 4}$

9. $\dfrac{\dfrac{4x^2 - y^2}{xy}}{\dfrac{2}{y} - \dfrac{1}{x}} = \dfrac{\left(\dfrac{4x^2 - y^2}{xy}\right) \cdot xy}{\left(\dfrac{2}{y} - \dfrac{1}{x}\right) \cdot xy}$

$= \dfrac{4x^2 - y^2}{2x - y}$

$= \dfrac{(2x - y)(2x + y)}{2x - y}$

$= 2x + y$

13. $\dfrac{\dfrac{2}{x} + \dfrac{3}{x^2}}{\dfrac{4}{x^2} - \dfrac{9}{x}} = \dfrac{\left(\dfrac{2}{x} + \dfrac{3}{x^2}\right)x^2}{\left(\dfrac{4}{x^2} - \dfrac{9}{x}\right)x^2} = \dfrac{2x + 3}{4 - 9x}$

17. $\dfrac{\dfrac{4}{5 - x} + \dfrac{5}{x - 5}}{\dfrac{2}{x} + \dfrac{3}{x - 5}} = \dfrac{-\dfrac{4}{x - 5} + \dfrac{5}{x - 5}}{\dfrac{2(x - 5) + 3x}{x(x - 5)}}$

$= \dfrac{\dfrac{1}{x - 5}}{\dfrac{2x - 10 + 3x}{x(x - 5)}}$

$= \dfrac{1}{x - 5} \cdot \dfrac{x(x - 5)}{5x - 10}$

$= \dfrac{x}{5x - 10}$

21. $\dfrac{\dfrac{2}{x}+3}{\dfrac{4}{x^2}-9} = \dfrac{\left(\dfrac{2}{x}+3\right)\cdot x^2}{\left(\dfrac{4}{x^2}-9\right)\cdot x^2}$

$= \dfrac{2x+3x^2}{4-9x^2}$

$= \dfrac{x(2+3x)}{(2+3x)(2-3x)}$

$= \dfrac{x}{2-3x}$

25. $\dfrac{\dfrac{-2x}{x^2-xy}}{\dfrac{y}{x^2}} = \dfrac{-2x}{x^2-xy} \div \dfrac{y}{x^2}$

$= \dfrac{-2x}{x^2-xy}\cdot\dfrac{x^2}{y}$

$= \dfrac{-2x}{x(x-y)}\cdot\dfrac{x^2}{y}$

$= \dfrac{-2x\cdot x}{(x-y)y}$

$= -\dfrac{2x^2}{y(x-y)}$

29. $\dfrac{\dfrac{x}{9}-\dfrac{1}{x}}{1+\dfrac{3}{x}} = \dfrac{\left(\dfrac{x}{9}-\dfrac{1}{x}\right)\cdot 9x}{\left(1+\dfrac{3}{x}\right)\cdot 9x}$

$= \dfrac{x^2-9}{9x+27}$

$= \dfrac{(x+3)(x-3)}{9(x+3)}$

$= \dfrac{x-3}{9}$

33. $\dfrac{\dfrac{2}{x+5}+\dfrac{4}{x+3}}{\dfrac{3x+13}{x^2+8x+15}} = \dfrac{\dfrac{2}{x+5}+\dfrac{4}{x+3}}{\dfrac{3x+13}{(x+5)(x+3)}}$

$= \dfrac{\left(\dfrac{2}{x+5}+\dfrac{4}{x+3}\right)(x+5)(x+3)}{\left(\dfrac{3x+13}{(x+5)(x+3)}\right)(x+5)(x+3)}$

$= \dfrac{2(x+3)+4(x+5)}{3x+13}$

$= \dfrac{2x+6+4x+20}{3x+13}$

$= \dfrac{6x+26}{3x+13}$

$= \dfrac{2(3x+13)}{3x+13}$

$= 2$

37. $\dfrac{2a^{-1}+3b^{-2}}{a^{-1}-b^{-1}} = \dfrac{\dfrac{2}{a}+\dfrac{3}{b^2}}{\dfrac{1}{a}-\dfrac{1}{b}}$

$= \dfrac{ab^2\left(\dfrac{2}{a}+\dfrac{3}{b^2}\right)}{ab^2\left(\dfrac{1}{a}-\dfrac{1}{b}\right)}$

$= \dfrac{2b^2+3a}{b^2-ab}$

$= \dfrac{2b^2+3a}{b(b-a)}$

41. $\dfrac{a^{-1}+1}{a^{-1}-1} = \dfrac{\dfrac{1}{a}+1}{\dfrac{1}{a}-1} = \dfrac{a\left(\dfrac{1}{a}+1\right)}{a\left(\dfrac{1}{a}-1\right)} = \dfrac{1+a}{1-a}$

45. $\dfrac{2a^{-1}+(2a)^{-1}}{a^{-1}+2a^{-2}} = \dfrac{\dfrac{2}{a}+\dfrac{1}{2a}}{\dfrac{1}{a}+\dfrac{2}{a^2}}$

$= \dfrac{2a^2\left(\dfrac{2}{a}+\dfrac{1}{2a}\right)}{2a^2\left(\dfrac{1}{a}+\dfrac{2}{a^2}\right)}$

$= \dfrac{4a+a}{2a+4}$

$= \dfrac{5a}{2(a+2)}$

49. $\dfrac{5x^{-1}-2y^{-1}}{25x^{-2}-4y^{-2}} = \dfrac{\dfrac{5}{x}-\dfrac{2}{y}}{\dfrac{25}{x^2}-\dfrac{4}{y^2}}$

$= \dfrac{x^2y^2\left(\dfrac{5}{x}-\dfrac{2}{y}\right)}{x^2y^2\left(\dfrac{25}{x^2}-\dfrac{4}{y^2}\right)}$

$= \dfrac{5xy^2-2x^2y}{25y^2-4x^2}$

$= \dfrac{xy(5y-2x)}{(5y+2x)(5y-2x)}$

$= \dfrac{xy}{5y+2x}$

53. $\qquad x^2 = 4x-4$

$\qquad x^2-4x+4 = 0$

$\qquad (x-2)(x-2) = 0$

$\quad x-2=0 \quad$ or $\quad x-2=0$

$\qquad x=2 \quad$ or $\qquad x=2$

The solution set is $\{2\}$.

57. $\dfrac{\dfrac{x+1}{9}}{\dfrac{y-2}{5}} = \dfrac{x+1}{9} \div \dfrac{y-2}{5} = \dfrac{x+1}{9}\cdot\dfrac{5}{y-2}$

Both a and c are equivalent to the original expression.

61. $\dfrac{\dfrac{1}{x}}{\dfrac{3}{y}} = \dfrac{1}{x} \div \dfrac{3}{y} = \dfrac{1}{x}\cdot\dfrac{y}{3}$

Both a and b are equivalent to the original expression.

65. $\dfrac{1}{1+(1+x)^{-1}} = \dfrac{1}{1+\dfrac{1}{1+x}}$

$= \dfrac{(1+x)\cdot 1}{(1+x)\left(1+\dfrac{1}{1+x}\right)}$

$= \dfrac{1+x}{1+x+1}$

$= \dfrac{1+x}{2+x}$

69. $f(x) = \dfrac{1}{x}$

a. $f(a+h) = \dfrac{1}{a+h}$

b. $f(a) = \dfrac{1}{a}$

c. $\dfrac{f(a + h) - f(a)}{h} = \dfrac{\dfrac{1}{a + h} - \dfrac{1}{a}}{h}$

d. $\dfrac{\dfrac{1}{a + h} - \dfrac{1}{a}}{h} = \dfrac{a(a + h)\left(\dfrac{1}{a + h} - \dfrac{1}{a}\right)}{a(a + h)h}$

$\qquad\qquad = \dfrac{a - (a + h)}{ah(a + h)}$

$\qquad\qquad = \dfrac{-h}{ah(a + h)}$

$\qquad\qquad = \dfrac{-1}{a(a + h)}$

Exercise Set 6.4

1. The LCD of 2 and 3 is 6.

$\dfrac{x}{2} - \dfrac{x}{3} = 12$

$6\left(\dfrac{x}{2} - \dfrac{x}{3}\right) = 6(12)$

$6 \cdot \dfrac{x}{2} - 6 \cdot \dfrac{x}{3} = 72$

$3x - 2x = 72$

$x = 72$

The solution set is $\{72\}$.

5. The LCD of x and 2 is $2x$.

$\dfrac{2}{x} + \dfrac{1}{2} = \dfrac{5}{x}$

$2x\left(\dfrac{2}{x} + \dfrac{1}{2}\right) = 2x\left(\dfrac{5}{x}\right)$

$2x \cdot \dfrac{2}{x} + 2x \cdot \dfrac{1}{2} = 10$

$4 + x = 10$

$x = 6$

The solution set is $\{6\}$.

9. The LCD is $x + 3$.

$\dfrac{x + 5}{x + 3} = \dfrac{2}{x + 3}$

$(x + 3)\left(\dfrac{x + 5}{x + 3}\right) = (x + 3)\left(\dfrac{2}{x + 3}\right)$

$x + 5 = 2$

$x = -3$

The number -3 makes the denominator $x + 3$ equal 0, so it is not a solution. The solution set is $\varnothing$.

13. The LCD is $(x - 1)(x + 1)$.

$\dfrac{1}{x - 1} = \dfrac{2}{x + 1}$

$(x - 1)(x + 1)\left(\dfrac{1}{x - 1}\right) = (x - 1)(x + 1)\left(\dfrac{2}{x + 1}\right)$

$x + 1 = 2(x - 1)$

$x + 1 = 2x - 2$

$1 = x - 2$

$3 = x$

The solution set is $\{3\}$.

17. $x^2 - 16 = (x + 4)(x - 4)$

The LCD is $x^2 - 16 = (x + 4)(x - 4)$.

$\dfrac{1}{x - 4} - \dfrac{3x}{x^2 - 16} = \dfrac{2}{x + 4}$

$(x + 4)(x - 4)\left(\dfrac{1}{x - 4}\right) - (x^2 - 16)\left(\dfrac{3x}{x^2 - 16}\right)$

$\qquad\qquad = (x + 4)(x - 4)\left(\dfrac{2}{x + 4}\right)$

$x + 4 - 3x = 2(x - 4)$

$4 - 2x = 2x - 8$

$4 = 4x - 8$

$12 = 4x$

$3 = x$

The solution set is $\{3\}$.

21. $x^2 - 2x = x(x - 2)$

The LCD is $x^2 - 2x = x(x - 2)$.

$\dfrac{1}{x - 2} - \dfrac{2}{x^2 - 2x} = 1$

$x(x - 2)\left(\dfrac{1}{x - 2}\right) - (x^2 - 2x)\left(\dfrac{2}{x^2 - 2x}\right) = (x^2 - 2x)1$

$x - 2 = x^2 - 2x$

$0 = x^2 - 3x + 2$

$0 = (x - 2)(x - 1)$

$x - 2 = 0 \quad \text{or} \quad x - 1 = 0$

$x = 2 \quad \text{or} \qquad x = 1$

The number 2 makes the denominators $x - 2$ and $x^2 - 2x$ equal 0, so it is not a solution. The solution set is $\{1\}$.

25. The LCD is a.

$1 - \dfrac{4}{a} = 5$

$a \cdot 1 - a \cdot \dfrac{4}{a} = a \cdot 5$

$a - 4 = 5a$

$-4 = 4a$

$-1 = a$

The solution set is $\{-1\}$.

29. $3x^2 + 3x = 3x(x + 1)$

The LCD is $2 \cdot 3x(x + 1) = 6x(x + 1)$.

$\dfrac{1}{2x} - \dfrac{1}{x + 1} = \dfrac{1}{3x^2 + 3x}$

$2 \cdot 3x(x + 1) \cdot \dfrac{1}{2x} - 6x(x + 1) \cdot \dfrac{1}{x + 1} = 2 \cdot 3x(x + 1) \cdot \dfrac{1}{3x(x + 1)}$

$3(x + 1) - 6x = 2$

$3x + 3 - 6x = 2$

$3 - 3x = 2$

$-3x = -1$

$x = \dfrac{1}{3}$

The solution set is $\left\{\dfrac{1}{3}\right\}$.

33. The LCD is $2y - 5$.

$5 - \dfrac{2}{2y - 5} = \dfrac{3}{2y - 5}$

$(2y - 5) \cdot 5 - (2y - 5)\left(\dfrac{2}{2y - 5}\right) = (2y - 5)\left(\dfrac{3}{2y - 5}\right)$

$5(2y - 5) - 2 = 3$

$10y - 25 - 2 = 3$

$10y - 27 = 3$

$10y = 30$

$y = 3$

The solution set is $\{3\}$.

37. The LCD is $x + 2$.

$\dfrac{x + 3}{x + 2} = \dfrac{1}{x + 2}$

$(x + 2)\left(\dfrac{x + 3}{x + 2}\right) = (x + 2)\left(\dfrac{1}{x + 2}\right)$

$x + 3 = 1$

$x = -2$

The number -2 makes the denominator $x + 2$ equal 0, so it is not a solution. The solution set is $\varnothing$.

41. $x^2 - 16 = (x + 4)(x - 4)$

The LCD is $x^2 - 16 = (x + 4)(x - 4)$.

$$\frac{64}{x^2 - 16} + 1 = \frac{2x}{x - 4}$$

$$(x^2 - 16)\left(\frac{64}{x^2 - 16}\right) + (x^2 - 16)(1) = (x + 4)(x - 4)\left(\frac{2x}{x - 4}\right)$$

$$64 + x^2 - 16 = 2x(x + 4)$$

$$x^2 + 48 = 2x^2 + 8x$$

$$0 = x^2 + 8x - 48$$

$$0 = (x - 4)(x + 12)$$

$x - 4 = 0$ or $x + 12 = 0$

$x = 4$ $x = -12$

The number 4 makes the denominators $x^2 - 16$ and $x - 4$ equal 0, so it is not a solution. The solution set is $\{-12\}$.

45. $x^2 - 9 = (x + 3)(x - 3)$

The LCD is $x^2 - 9 = (x + 3)(x - 3)$.

$$\frac{28}{x^2 - 9} + \frac{2x}{x - 3} + \frac{6}{x + 3} = 0$$

$$(x^2 - 9)\left(\frac{28}{x^2 - 9}\right) + (x + 3)(x - 3)\left(\frac{2x}{x - 3}\right)$$

$$+ (x + 3)(x - 3)\left(\frac{6}{x + 3}\right) = (x^2 - 9)(0)$$

$$28 + 2x(x + 3) + 6(x - 3) = 0$$

$$28 + 2x^2 + 6x + 6x - 18 = 0$$

$$2x^2 + 12x + 10 = 0$$

$$2(x^2 + 6x + 5) = 0$$

$$2(x + 5)(x + 1) = 0$$

$x + 5 = 0$ or $x + 1 = 0$

$x = -5$ $x = -1$

The solution set is $\{-5, -1\}$.

49. Let x be the unknown number.

$$4 + 3x = 19$$

$$3x = 15$$

$$x = 5$$

The number is 5.

53. From the graph, 13.5% of state and federal prison inmates are age 45 to 54.

57. 17.4% of 36,612 $= 0.174(36,612) = 6370.488$

Approximately 6370 25- to 29-year-old inmates would be expected in Louisiana at the end of 2003.

61. Find x when $C(x) = 5.1$.

$$5.1 = 3.3 + \frac{5400}{x}$$

$$x(5.1) = x(3.3) + x\left(\frac{5400}{x}\right)$$

$$5.1x = 3.3x + 5400$$

$$1.8x = 5400$$

$$x = 3000$$

3000 games disks must be produced for the average cost to be $5.10.

65.
$$p^{-2} + 4p^{-1} - 5 = 0$$

$$\frac{1}{p^2} + \frac{4}{p} - 5 = 0$$

$$p^2\left(\frac{1}{p^2}\right) + p^2\left(\frac{4}{p}\right) - p^2(5) = p^2(0)$$

$$1 + 4p - 5p^2 = 0$$

$$(1 + 5p)(1 - p) = 0$$

$1 + 5p = 0$ or $1 - p = 0$

$$p = -\frac{1}{5} \qquad 1 = p$$

The solution set is $\left\{-\dfrac{1}{5}, 1\right\}$.

69.
$$\frac{10.6}{y} - 14.7 = \frac{9.92}{3.2} + 7.6$$

$$\frac{10.6}{y} - 14.7 = 3.1 + 7.6$$

$$\frac{10.6}{y} - 14.7 = 10.7$$

$$\frac{10.6}{y} = 25.4$$

$$y\left(\frac{10.6}{y}\right) = y(25.4)$$

$$10.6 = 25.4y$$

$$\frac{10.6}{25.4} = y$$

$$0.42 \approx y$$

The solution set is $\{0.42\}$.

73. Let $u = \dfrac{3}{x - 1}$.

$$\left(\frac{3}{x - 1}\right)^2 + 2\left(\frac{3}{x - 1}\right) + 1 = 0$$

$$u^2 + 2u + 1 = 0$$

$$(u + 1)^2 = 0$$

$$u + 1 = 0$$

$$u = -1$$

Now solve for x.

$$-1 = \frac{3}{x - 1}$$

$$(x - 1)(-1) = (x - 1)\left(\frac{3}{x - 1}\right)$$

$$-x + 1 = 3$$

$$-x = 2$$

$$x = -2$$

The solution set is 0.42.

77. answers may vary

Exercise Set 6.5

1.
$$F = \frac{9}{5}C + 32$$

$$F - 32 = \frac{9}{5}C$$

$$C = \frac{5}{9}(F - 32)$$

5.
$$\frac{1}{R} = \frac{1}{R_1} + \frac{1}{R_2}$$

$$RR_1R_2\left(\frac{1}{R}\right) = RR_1R_2\left(\frac{1}{R_1} + \frac{1}{R_2}\right)$$

$$R_1R_2 = RR_2 + RR_1$$

$$R_1R_2 = R(R_2 + R_1)$$

$$R = \frac{R_1R_2}{R_1 + R_2}$$

9.
$$A = \frac{h(a+b)}{2}$$
$$2A = ah + bh$$
$$2A - ah = bh$$
$$\frac{2A - ah}{h} = b$$

13.
$$f = \frac{f_1 f_2}{f_1 + f_2}$$
$$(f_1 + f_2)f = \left(\frac{f_1 f_2}{f_1 + f_2}\right)(f_1 + f_2)$$
$$f_1 f + f_2 f = f_1 f_2$$
$$f_1 f = f_1 f_2 - f f_2$$
$$f_1 f = f_2(f_1 - f)$$
$$\frac{f_1 f}{f_1 - f} = f_2$$

17.
$$\frac{\theta}{\omega} = \frac{2L}{c}$$
$$(c\omega)\left(\frac{\theta}{\omega}\right) = \frac{2L}{c}(c\omega)$$
$$c\theta = 2L\omega$$
$$c = \frac{2L\omega}{\theta}$$

21. Let $x =$ the number.
$$\frac{12 + x}{41 + 2x} = \frac{1}{3}$$
$$3(12 + x) = (41 + 2x) \cdot 1$$
$$36 + 3x = 41 + 2x$$
$$x = 5$$
The number is 5.

25. Let x be the number of women. Solve the proportion $\frac{x}{40,151} = \frac{10.7}{100}$.
$$\frac{x}{40,151} = \frac{10.7}{100}$$
$$100x = 10.7 \cdot 40,151$$
$$x = \frac{429,615.7}{100}$$
$$x = 4296.157 \approx 4269$$
There were approximately 4296 women in the Coast Guard in 2004.

29. Convert each time to a rate.
$$\frac{1 \text{ stack}}{20 \text{ minutes}} = \frac{1}{20} \frac{\text{stack}}{\text{minute}}$$
$$\frac{1 \text{ stack}}{30 \text{ minutes}} = \frac{1}{30} \frac{\text{stack}}{\text{minute}}$$
$$\frac{1 \text{ stack}}{60 \text{ minutes}} = \frac{1}{60} \frac{\text{stack}}{\text{minute}}$$
Let $x =$ amount of time required for all three workers to sort the stack.

So $\frac{1 \text{ stack}}{x \text{ minute}} = \frac{1}{x} \frac{\text{stack}}{\text{minute}}$. Adding we get,
$$\frac{1}{20} + \frac{1}{30} + \frac{1}{60} = \frac{1}{x}$$
$$60x\left(\frac{1}{20}\right) + 60x\left(\frac{1}{30}\right) + 60x\left(\frac{1}{60}\right) = 60x\left(\frac{1}{x}\right)$$
$$3x + 2x + x = 60$$
$$6x = 60$$
$$x = 10$$
It takes 10 minutes for the three workers to sort the stack.

33. Let $r =$ the speed of the boat in still water.

Recall that $d = rt$ or $t = \frac{d}{r}$. Using the latter equation we get $\frac{20}{r+5} = \frac{10}{r-5}$, where $r + 5$ is the rate of the boat traveling downstream and $r - 5$ is the rate of the boat traveling upstream.
$$\frac{20}{r+5} = \frac{10}{r-5}$$
$$20(r - 5) = 10(r + 5)$$
$$20r - 100 = 10r + 50$$
$$10r = 150$$
$$r = 15$$
The speed of the boat in still water is 15 mph.

37. Convert times to rates.
$$\frac{1 \text{ pond}}{45 \text{ minutes}} = \frac{1}{45} \frac{\text{pond}}{\text{minute}}$$
$$\frac{1 \text{ pond}}{20 \text{ minutes}} = \frac{1}{20} \frac{\text{pond}}{\text{minute}}$$
Let $x =$ the number of minutes required for the second hose to fill the pond. Then,
$$\frac{1 \text{ pond}}{x \text{ minute}} = \frac{1}{x} \frac{\text{pond}}{\text{minute}}.$$
$$\frac{1}{20} = \frac{1}{45} + \frac{1}{x}$$
$$\frac{1}{20} = \frac{x + 45}{45x}$$
$$45x = 20(x + 45)$$
$$45x = 20x + 900$$
$$25x = 900$$
$$x = \frac{900}{25} = 36$$
Thus, the second hose alone can fill the pond in 36 minutes.

41. Let $x =$ time.
$$\frac{x \text{ hour}}{1 \text{ mile}} = \frac{1 \text{ hour}}{0.17 \text{ mile}} \quad \text{or}$$
$$\frac{x}{1} = \frac{1}{0.17}$$
$$0.17x = 1$$
$$x = \frac{1}{0.17} \approx 5.9$$
It takes approximately 5.9 hours to travel 1 mile.

45.

	distance	=	rate	·	time
With Wind	465		$x + 20$		$\frac{465}{x + 20}$
Against Wind	345		$x - 20$		$\frac{345}{x - 20}$

$$\frac{465}{x + 20} = \frac{345}{x - 20}$$
$$465(x - 20) = 345(x + 20)$$
$$465x - 9300 = 345x + 6900$$
$$120x = 16,200$$
$$x = 135 \text{ mph}$$
The speed of the plane in still air is 135 miles per hour.

49. Let x be the numerator of the fraction. Then the denominator is $x + 1$.

$$\frac{x - 3}{(x + 1) - 3} = \frac{4}{5}$$

$$\frac{x - 3}{x - 2} = \frac{4}{5}$$

$$5(x - 3) = 4(x - 2)$$

$$5x - 15 = 4x - 8$$

$$x - 15 = -8$$

$$x = 7$$

$$x + 1 = 7 + 1 = 8$$

The original fraction is $\frac{7}{8}$.

53.

	distance	=	rate	·	time
Upstream	3		$x - 6$		$\dfrac{3}{x - 6}$
Downstream	9		$x + 6$		$\dfrac{9}{x + 6}$

$$\frac{3}{x - 6} = \frac{9}{x + 6}$$

$$3(x + 6) = 9(x - 6)$$

$$3x + 18 = 9x - 54$$

$$3x + 72 = 9x$$

$$72 = 6x$$

$$12 = x$$

Time upstream $= \dfrac{3}{12 - 6} = \dfrac{3}{6} = \dfrac{1}{2}$ hour.

Total time is $2\left(\dfrac{1}{2}\right) = 1$ hour.

57.

	Time	Part done in one hour
Experienced	3	$\dfrac{1}{3}$
Apprentice	6	$\dfrac{1}{6}$
Together	x	$\dfrac{1}{x}$

$$\frac{1}{3} + \frac{1}{6} = \frac{1}{x}$$

$$6x\left(\frac{1}{3}\right) + 6x\left(\frac{1}{6}\right) = 6x\left(\frac{1}{x}\right)$$

$$2x + x = 6$$

$$3x = 6$$

$$x = 2$$

It takes them 2 hours if they work together.

61. The LCD is 6.

$$\frac{x - 3}{2} = \frac{x - 5}{6}$$

$$6\left(\frac{x - 3}{2}\right) = 6\left(\frac{x - 5}{6}\right)$$

$$3(x - 3) = x - 5$$

$$3x - 9 = x - 5$$

$$2x - 9 = -5$$

$$2x = 4$$

$$x = 2$$

The solution set is $\{2\}$.

65. Use $h = \dfrac{a}{1 - \dfrac{s}{770}}$ with $a = 329.63$ and $s = 50$.

$$h = \frac{a}{1 - \dfrac{s}{770}} = \frac{329.63}{1 - \dfrac{50}{770}} \approx 352.52$$

A pitch of 352.52 is closest to the pitch of the musical note F. The observer hears a higher pitch than the actual siren.

Exercise Set 6.6

1. $y = kx$

$4 = k(20)$

$k = \dfrac{1}{5}$

$y = \dfrac{1}{5}x$

5. $y = kx$

$7 = k\left(\dfrac{1}{2}\right)$

$k = 14$

$y = 14x$

9. $W = kr^3$

$1.2 = k \cdot 2^3$

$k = \dfrac{1.2}{8} = 0.15$

$W = 0.15r^3 = 0.15(3)^3 = 0.15(27) = 4.05$ pounds

13. $y = \dfrac{k}{x}$

$6 = \dfrac{k}{5}$

$k = 30$

$y = \dfrac{30}{x}$

17. $y = \dfrac{k}{x}$

$\dfrac{1}{8} = \dfrac{k}{16}$

$k = 2$

$y = \dfrac{2}{x}$

21. $R = \dfrac{k}{T}$

$45 = \dfrac{k}{6}$

$k = 270$

$R = \dfrac{270}{5} = 54$ mph

25. $I_1 = \dfrac{k}{d^2}$

Replace d by $2d$.

$$I_2 = \frac{k}{(2d)^2} = \frac{k}{4d^2} = \frac{1}{4}I_1$$

Thus, the intensity is divided by 4.

29. $r = kst^3$

33. $y = k\sqrt{x}$

$0.4 = k\sqrt{4}$

$0.4 = 2k$

$\dfrac{0.4}{2} = k$

$0.2 = k$

$y = 0.2\sqrt{x}$

37. $y = kxz^3$

$120 = k(5)(2^3)$

$120 = k(5)(8)$

$120 = 40k$

$3 = k$

$y = 3xz^3$

41. $V = kr^2h$

$32\pi = k(4)^2(6)$

$32\pi = k(16)(6)$

$32\pi = 96k$

$\dfrac{32\pi}{96} = k$

$\dfrac{\pi}{3} = k$

$V = \dfrac{\pi}{3}r^2h$

$V = \dfrac{\pi}{3}(3)^2(5)$

$V = 15\pi$ cu in.

45. y varies directly as x is written as $y = kx$.

49. y varies jointly as x and z is written as $y = kxz$.

53. y varies directly as x and inversely as p^2 is written as $y = \dfrac{kx}{p^2}$.

57. Circumference: $C = 2\pi \cdot$ radius $= 2\pi(7$ meters$) = 14\pi$ meters

Area:

$A = \pi(\text{radius})^2$

$= \pi(7 \text{ meters})^2$

$= 49\pi$ square meters

61. $m = \dfrac{y_2 - y_1}{x_2 - x_1} = \dfrac{-2 - (-1)}{5 - 4} = \dfrac{-2 + 1}{1} = -1$

65. $y = 9ab$ is an example of joint variation; c.

69. Since y varies directly as x^2, $y = kx^2$.

Replace x by $2x$.

$y = k(2x)^2$

$y = 4kx^2$

If x is doubled, y is multiplied by 4.

Chapter 6 Test

1. $1 - x = 0$

$1 = x$

The domain of $f(x) = \dfrac{5x^2}{1 - x}$ is $\{x | x$ is a real number and $x \neq 1\}$.

5. $\dfrac{x^3 - 8}{x - 2} = \dfrac{x^3 - 2^3}{x - 2}$

$= \dfrac{(x - 2)(x^2 + 2x + 4)}{x - 2}$

$= x^2 + 2x + 4$

9. $\dfrac{4x - 12}{2x - 9} \div \dfrac{3 - x}{4x^2 - 81} \cdot \dfrac{x + 3}{5x + 15}$

$= \dfrac{4x - 12}{2x - 9} \cdot \dfrac{4x^2 - 81}{3 - x} \cdot \dfrac{x + 3}{5x + 15}$

$= \dfrac{4(x - 3)}{2x - 9} \cdot \dfrac{(2x + 9)(2x - 9)}{-(x - 3)} \cdot \dfrac{x + 3}{5(x + 3)}$

$= \dfrac{4(2x + 9)}{-5}$

$= -\dfrac{4(2x + 9)}{5}$

13. $3x - 21 = 3(x - 7)$

$2x - 14 = 2(x - 7)$

The LCD is $3 \cdot 2(x - 7) = 6(x - 7)$.

$\dfrac{5}{x - 7} - \dfrac{2x}{3x - 21} + \dfrac{x}{2x - 14}$

$= \dfrac{6(5)}{6(x - 7)} - \dfrac{2(2x)}{2(3x - 21)} + \dfrac{3(x)}{3(2x - 14)}$

$= \dfrac{30 - 4x + 3x}{6(x - 7)}$

$= \dfrac{30 - x}{6(x - 7)}$

17. The LCD is $7(3x - 7)$.

$\dfrac{5x + 3}{3x - 7} = \dfrac{19}{7}$

$7(3x - 7)\left(\dfrac{5x + 3}{3x - 7}\right) = 7(3x - 7)\left(\dfrac{19}{7}\right)$

$7(5x + 3) = 19(3x - 7)$

$35x + 21 = 57x - 133$

$21 = 22x - 133$

$154 = 22x$

$7 = x$

The solution set is $\{7\}$.

21. The LCD is ab.

$\dfrac{x + b}{a} = \dfrac{4x - 7a}{b}$

$ab\left(\dfrac{x + b}{a}\right) = ab\left(\dfrac{4x - 7a}{b}\right)$

$b(x + b) = a(4x - 7a)$

$bx + b^2 = 4ax - 7a^2$

$b^2 = 4ax - 7a^2 - bx$

$7a^2 + b^2 = 4ax - bx$

$7a^2 + b^2 = x(4a - b)$

$\dfrac{7a^2 + b^2}{4a - b} = x$

25. $Q = kRS^2$

Use $Q = 24$, $R = 3$, and $S = 4$.

$24 = k \cdot 3 \cdot 4^2$

$24 = 48k$

$\dfrac{1}{2} = k$

$Q = \dfrac{1}{2}RS^2$

$R = 2$ and $S = 3$

$Q = \dfrac{1}{2} \cdot 2 \cdot 3^2 = 9$

CHAPTER 7

Exercise Set 7.1

1. Since $2^2 = 4$ and $(-2)^2 = 4$, the square roots of 4 are 2 and -2.

5. Since $10^2 = 100$ and $(-10)^2 = 100$, the square roots of 100 are 10 and -10.

9. $\sqrt{\dfrac{1}{4}} = \dfrac{1}{2}$ because $\left(\dfrac{1}{2}\right)^2 = \dfrac{1}{4}$.

13. $-\sqrt{36} = -6$ because $(6)^2 = 36$.

17. $\sqrt{16y^6} = \sqrt{16}\sqrt{y^6} = 4y^3$ because $(4y^3)^2 = 16y^6$.

21. $\sqrt{38} \approx 6.164$

Since $36 < 38 < 49$, then $\sqrt{36} < \sqrt{38} < \sqrt{49}$ or $6 < \sqrt{38} < 7$. The approximation is between 6 and 7 and is thus reasonable.

25. $\sqrt[3]{64} = 4$ because $(4)^3 = 64$.

29. $\sqrt[3]{-1} = -1$ because $(-1)^3 = -1$.

33. $\sqrt[3]{-27x^9} = -3x^3$ because $(-3x^3)^3 = -27x^9$.

37. $\sqrt[4]{-16}$ is not a real number.

41. $\sqrt[5]{x^{20}} = x^4$ because $(x^4)^5 = x^{20}$.

45. $\sqrt{81x^4} = 9x^2$ because $(9x^2)^2 = 81x^4$

49. $\sqrt{(-8)^2} = |-8| = 8$

53. $\sqrt{4x^2} = |2x| = 2|x|$

57. $\sqrt[4]{(x-2)^4} = |x-2|$

61. $-\sqrt{121} = -11$ because -11 is the opposite of $\sqrt{121}$.

65. $\sqrt{y^{12}} = y^6$ because $(y^6)^2 = y^{12}$.

69. $\sqrt[3]{-27x^{12}y^9} = -3x^4y^3$ because $(-3x^4y^3)^3 = -27x^{12}y^9$.

73. $\sqrt[5]{-32x^{10}y^5} = -2x^2y$ because $(-2x^2y)^5 = -32x^{10}y^5$.

77. $\sqrt{\dfrac{x^2}{4y^2}} = \dfrac{x}{2y}$ because $\left(\dfrac{x}{2y}\right)^2 = \dfrac{x^2}{4y^2}$.

81. $\sqrt[4]{\dfrac{x^4}{16}} = \dfrac{x}{2}$ because $\left(\dfrac{x}{2}\right)^4 = \dfrac{x^4}{16}$.

85. $g(x) = \sqrt[3]{x-8}$

$g(7) = \sqrt[3]{7-8} = \sqrt[3]{-1} = -1$

89. $f(x) = \sqrt{2x+3}$

$f(2) = \sqrt{2(2)+3} = \sqrt{7}$

93. The domain of $g(x) = \sqrt[3]{x-1}$ is $(-\infty, \infty)$.

x	$g(x) = \sqrt[3]{x-1}$
0	$\sqrt[3]{0-1} = \sqrt[3]{-1} = -1$
1	$\sqrt[3]{1-1} = \sqrt[3]{0} = 0$
2	$\sqrt[3]{2-1} = \sqrt[3]{1} = 1$
7	$\sqrt[3]{7-1} = \sqrt[3]{6} \approx 1.8$
-5	$\sqrt[3]{-5-1} = \sqrt[3]{-6} \approx -1.8$
9	$\sqrt[3]{9-1} = \sqrt[3]{8} = 2$
-7	$\sqrt[3]{-7-1} = \sqrt[3]{-8} = -2$

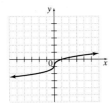

97. $(-3x^2y^3z^5)(20x^5y^7) = -3 \cdot 20x^2x^5y^3y^7z^5$

$\qquad = -60x^7y^{10}z^5$

101. $\sqrt{-17}$ is not a real number.

105. answers may vary

109. $25 < 30 < 36$ so $5 = \sqrt{25} < \sqrt{30} < \sqrt{36} = 6$

$9 < 10 < 16$ so $3 = \sqrt{9} < \sqrt{10} < \sqrt{16} = 4$

$81 < 90 < 100$ so $9 = \sqrt{81} < \sqrt{90} < \sqrt{100} = 10$

The perimeter is $\sqrt{30} + \sqrt{10} + \sqrt{90}$ which is between $5 + 3 + 9 = 17$ and $6 + 4 + 10 = 20$. The perimeter is closest to 18; b.

113. answers may vary

Exercise Set 7.2

1. $49^{1/2} = \sqrt{49} = 7$

5. $\left(\dfrac{1}{16}\right)^{1/4} = \sqrt[4]{\dfrac{1}{16}} = \dfrac{1}{2}$

9. $2m^{1/3} = 2\sqrt[3]{m}$

13. $(-27)^{1/3} = \sqrt[3]{-27} = -3$

17. $16^{3/4} = \left(\sqrt[4]{16}\right)^3 = 2^3 = 8$

21. $(-16)^{3/4} = \left(\sqrt[4]{-16}\right)^3$ is not a real number.

25. $(7x + 2)^{2/3} = \sqrt[3]{(7x+2)^2}$ or $\left(\sqrt[3]{7x+2}\right)^2$

29. $8^{-4/3} = \dfrac{1}{8^{4/3}} = \dfrac{1}{(8^{1/3})^4} = \dfrac{1}{2^4} = \dfrac{1}{16}$

33. $(-4)^{-3/2} = \dfrac{1}{(-4)^{3/2}} = \dfrac{1}{\left(\sqrt{-4}\right)^3}$ is not a real number.

37. $\dfrac{1}{a^{-2/3}} = a^{2/3}$

41. $a^{2/3}a^{5/3} = a^{2/3+5/3} = a^{7/3}$

45. $3^{1/4} \cdot 3^{3/8} = 3^{1/4+3/8} = 3^{2/8+3/8} = 3^{5/8}$

49. $(4u^2)^{3/2} = 4^{3/2}(u^2)^{3/2}$

$\qquad = (4^{1/2})^3 u^{2 \cdot 3/2}$

$\qquad = 2^3 \cdot u^3$

$\qquad = 8u^3$

53. $\dfrac{(x^3)^{1/2}}{x^{7/2}} = \dfrac{x^{3 \cdot 1/2}}{x^{7/2}} = \dfrac{x^{3/2}}{x^{7/2}} = x^{3/2-7/2} = x^{-2} = \dfrac{1}{x^2}$

57. $\dfrac{(y^3z)^{1/6}}{y^{-1/2}z^{1/3}} = \dfrac{y^{3/6}z^{1/6}}{y^{-1/2}z^{1/3}}$

$\qquad = y^{3/6-(-1/2)}z^{1/6-1/3}$

$\qquad = y^{1/2+1/2}z^{1/6-2/6}$

$\qquad = y^1z^{-1/6}$

$\qquad = \dfrac{y}{z^{1/6}}$

61. $\sqrt[6]{x^3} = x^{3/6} = x^{1/2} = \sqrt{x}$

65. $\sqrt[4]{16x^2} = 16^{1/4}x^{2/4} = 2x^{1/2} = 2\sqrt{x}$

69. $\sqrt[8]{x^4y^4} = x^{4/8}y^{4/8} = x^{1/2}y^{1/2} = \sqrt{xy}$

73. $\sqrt[3]{y} \cdot \sqrt[5]{y^2} = y^{1/3} \cdot y^{2/5}$

$\qquad = y^{5/15} \cdot y^{6/15}$

$\qquad = y^{11/15}$

$\qquad = \sqrt[15]{y^{11}}$

77. $\sqrt[3]{x} \cdot \sqrt[4]{x} \cdot \sqrt[8]{x^3} = x^{1/3} \cdot x^{1/4} \cdot x^{3/8}$

$\qquad = x^{8/24} \cdot x^{6/24} \cdot x^{9/24}$

$\qquad = x^{23/24}$

$\qquad = \sqrt[24]{x^{23}}$

81. $\sqrt{3} \cdot \sqrt[3]{4} = 3^{1/2} \cdot 4^{1/3}$

$\qquad = 3^{3/6} \cdot 4^{2/6}$

$\qquad = (3^3 \cdot 4^2)^{1/6}$

$\qquad = (432)^{1/6}$

$\qquad = \sqrt[6]{432}$

85. $\sqrt{5r} \cdot \sqrt[3]{s} = (5r)^{1/2} \cdot s^{1/3}$

$\qquad = (5r)^{3/6} \cdot s^{2/6}$

$\qquad = [(5r)^3 \cdot s^2]^{1/6}$

$\qquad = (125r^3s^2)^{1/6}$

$\qquad = \sqrt[6]{125r^3s^2}$

89. $48 = 16 \cdot 3$ or $4 \cdot 12$

93. $54 = 27 \cdot 2$

97. $f(x) = 12x^{7/6}$

2004 is 10 years after 1994, so $x = 10$.

$f(10) = 12(10)^{7/6} \approx 176.1$

There were approximately 176.1 million cellular telephone subscriptions in 2004.

101. Since $\dfrac{1}{3} + \dfrac{2}{3} = \dfrac{3}{3}$, then

$a^{1/3} \cdot a^{2/3} = a^{1/3+2/3} = a^{3/3} = a^1 = a$.

105. $8^{1/4} \approx 1.6818$

Exercise Set 7.3

1. $\sqrt{7} \cdot \sqrt{2} = \sqrt{7 \cdot 2} = \sqrt{14}$

5. $\sqrt[3]{4} \cdot \sqrt[3]{9} = \sqrt[3]{4 \cdot 9} = \sqrt[3]{26}$

9. $\sqrt{\dfrac{7}{x}} \cdot \sqrt{\dfrac{2}{y}} = \sqrt{\dfrac{7 \cdot 2}{x \cdot y}} = \sqrt{\dfrac{14}{xy}}$

13. $\sqrt{\dfrac{6}{49}} = \dfrac{\sqrt{6}}{\sqrt{49}} = \dfrac{\sqrt{6}}{7}$

17. $\sqrt[4]{\dfrac{x^3}{16}} = \dfrac{\sqrt[4]{x^3}}{\sqrt[4]{16}} = \dfrac{\sqrt[4]{x^3}}{2}$

21. $\sqrt[4]{\dfrac{8}{x^8}} = \dfrac{\sqrt[4]{8}}{\sqrt[4]{x^8}} = \dfrac{\sqrt[4]{8}}{x^2}$

25. $\sqrt{\dfrac{x^2 y}{100}} = \dfrac{\sqrt{x^2} \cdot \sqrt{y}}{\sqrt{100}} = \dfrac{x\sqrt{y}}{10}$

29. $-\sqrt[3]{\dfrac{z^7}{125x^3}} = \dfrac{-\sqrt[3]{z^7}}{\sqrt[3]{125x^3}}$

$= \dfrac{-\sqrt[3]{z^6 z}}{\sqrt[3]{125} \cdot \sqrt[3]{x^3}}$

$= \dfrac{-\sqrt[3]{z^6} \cdot \sqrt[3]{z^3}}{5x}$

$= -\dfrac{z^2 \sqrt[3]{z}}{5x}$

33. $\sqrt[3]{192} = \sqrt[3]{64(3)} = \sqrt[3]{64} \cdot \sqrt[3]{3} = 4\sqrt[3]{3}$

37. $\sqrt{24} = \sqrt{4 \cdot 6} = \sqrt{4} \cdot \sqrt{6} = 2\sqrt{6}$

41. $\sqrt[3]{16y^7} = \sqrt[3]{(8y^6)(2y)}$

$= \sqrt[3]{8} \cdot \sqrt[3]{y^6} \cdot \sqrt[3]{2y}$

$= 2y^2 \sqrt[3]{2y}$

45. $\sqrt{y^5} = \sqrt{y^4 y} = \sqrt{y^4} \cdot \sqrt{y} = y^2 \sqrt{y}$

49. $\sqrt[5]{-32x^{10}y} = \sqrt[5]{-32} \cdot \sqrt[5]{x^{10}} \cdot \sqrt[5]{y} = -2x^2 \sqrt[5]{y}$

53. $-\sqrt{32a^8 b^7} = -\sqrt{16a^8 b^6(2b)}$

$= -\sqrt{16} \cdot \sqrt{a^8} \cdot \sqrt{b^6} \cdot \sqrt{2b}$

$= -4a^4 b^3 \sqrt{2b}$

57. $\sqrt[3]{125r^9 s^{12}} = 5r^3 s^4$

61. $\dfrac{\sqrt[3]{24}}{\sqrt[3]{3}} = \sqrt[3]{\dfrac{24}{3}} = \sqrt[3]{8} = 2$

65. $\dfrac{\sqrt{x^5 y^3}}{\sqrt{xy}} = \sqrt{\dfrac{x^5 y^3}{xy}} = \sqrt{x^4 y^2} = x^2 y$

69. $\dfrac{3\sqrt{100x^2}}{2\sqrt{2x^{-1}}} = \dfrac{3}{2} \sqrt{\dfrac{100x^2}{2x^{-1}}}$

$= \dfrac{3}{2} \sqrt{50x^3}$

$= \dfrac{3}{2} \sqrt{25x^2 \cdot 2x}$

$= \dfrac{3}{2} \cdot 5x\sqrt{2x}$

$= \dfrac{15x}{2} \sqrt{2x}$

73. $(x_1, y_1) = (5, 1), (x_2, y_2) = (8, 5)$

$d = \sqrt{(x_2 - x_1)^2 + (y_2 - y_1)^2}$

$= \sqrt{(5 - 1)^2 + (8 - 5)^2}$

$= \sqrt{4^2 + 3^2}$

$= \sqrt{16 + 9}$

$= \sqrt{25}$

$= 5$

The distance is 5 units.

77. $(x_1, y_1) = (0, -\sqrt{2}), (x_2, y_2) = (\sqrt{3}, 0)$

$d = \sqrt{(x_2 - x_1)^2 + (y_2 - y_1)^2}$

$= \sqrt{(\sqrt{3} - 0)^2 + [0 - (-\sqrt{2})]^2}$

$= \sqrt{(\sqrt{3})^2 + (\sqrt{2})^2}$

$= \sqrt{3 + 2}$

$= \sqrt{5}$

The distance is $\sqrt{5}$ units, approximately 2.236 units.

81. $(x_1, y_1) = (6, -8), (x_2, y_2) = (2, 4)$

$\text{midpoint} = \left(\dfrac{x_1 + x_2}{2}, \dfrac{y_1 + y_2}{2} \right)$

$= \left(\dfrac{6 + 2}{2}, \dfrac{-8 + 4}{2} \right)$

$= \left(\dfrac{8}{2}, \dfrac{-4}{2} \right)$

$= (4, -2)$

85. $(x_1, y_1) = \left(\dfrac{1}{2}, \dfrac{3}{8} \right), (x_2, y_2) = \left(-\dfrac{3}{2}, \dfrac{5}{8} \right)$

$\text{midpoint} = \left(\dfrac{x_1 + x_2}{2}, \dfrac{y_1 + y_2}{2} \right)$

$= \left(\dfrac{\dfrac{1}{2} + \left(-\dfrac{3}{2} \right)}{2}, \dfrac{\dfrac{3}{8} + \dfrac{5}{8}}{2} \right)$

$= \left(\dfrac{-1}{2}, \dfrac{1}{2} \right)$

$= \left(-\dfrac{1}{2}, \dfrac{1}{2} \right)$

89. $6x + 8x = 14x$

93. $9y^2 - 8y^2 = y^2$

97. $\dfrac{\sqrt[3]{64}}{\sqrt{64}} = \dfrac{4}{8} = \dfrac{1}{2}$

101. $F(x) = 0.6\sqrt{49 - x^2}$

   a. $F(3) = 0.6\sqrt{49 - 3^2}$

     $= 0.6\sqrt{49 - 9}$

     $= 0.6\sqrt{40}$

     ≈ 3.8

   b. $F(5) = 0.6\sqrt{49 - 5^2}$

     $= 0.6\sqrt{49 - 25}$

     $= 0.6\sqrt{24}$

     ≈ 2.9

   c. answers may vary

Exercise Set 7.4

1. $\sqrt{8} - \sqrt{32} = \sqrt{4(2)} - \sqrt{16(2)}$

$= \sqrt{4}\sqrt{2} - \sqrt{16}\sqrt{2}$

$= 2\sqrt{2} - 4\sqrt{2}$

$= -2\sqrt{2}$

5. $2\sqrt{50} - 3\sqrt{125} + \sqrt{98}$

$= 2\sqrt{25}\sqrt{2} - 3\sqrt{25}\sqrt{5} + \sqrt{49}\sqrt{2}$

$= 2(5)\sqrt{2} - 3(5)\sqrt{5} + 7\sqrt{2}$

$= 10\sqrt{2} - 15\sqrt{5} + 7\sqrt{2}$

$= 17\sqrt{2} - 15\sqrt{5}$

9. $\sqrt{9b^3} - \sqrt{25b^3} + \sqrt{49b^3}$

$= \sqrt{9b^2}\sqrt{b} - \sqrt{25b^2}\sqrt{b} + \sqrt{49b^2}\sqrt{b}$

$= 3b\sqrt{b} - 5b\sqrt{b} + 7b\sqrt{b}$

$= 5b\sqrt{b}$

13. $\sqrt[3]{\dfrac{11}{8}} - \dfrac{\sqrt[3]{11}}{6} = \dfrac{\sqrt[3]{11}}{\sqrt[3]{8}} - \dfrac{\sqrt[3]{11}}{6}$

$\qquad\qquad = \dfrac{\sqrt[3]{11}}{2} - \dfrac{\sqrt[3]{11}}{6}$

$\qquad\qquad = \dfrac{3\sqrt[3]{11} - \sqrt[3]{11}}{6}$

$\qquad\qquad = \dfrac{2\sqrt[3]{11}}{6}$

$\qquad\qquad = \dfrac{\sqrt[3]{11}}{3}$

17. $7\sqrt{9} - 7 + \sqrt{3} = 7(3) - 7 + \sqrt{3}$

$\qquad\qquad\qquad = 21 - 7 + \sqrt{3}$

$\qquad\qquad\qquad = 14 + \sqrt{3}$

21. $3\sqrt{108} - 2\sqrt{18} - 3\sqrt{48} = 3\sqrt{36}\sqrt{3} - 2\sqrt{9}\sqrt{2} - 3\sqrt{16}\sqrt{3}$

$\qquad\qquad\qquad\qquad = 3(6)\sqrt{3} - 2(3)\sqrt{2} - 3(4)\sqrt{3}$

$\qquad\qquad\qquad\qquad = 18\sqrt{3} - 6\sqrt{2} - 12\sqrt{3}$

$\qquad\qquad\qquad\qquad = 6\sqrt{3} - 6\sqrt{2}$

25. $\sqrt{9b^3} - \sqrt{25b^3} + \sqrt{16b^3}$

$\qquad = \sqrt{9b^2}\sqrt{b} - \sqrt{25b^2}\sqrt{b} + \sqrt{16b^2}\sqrt{b}$

$\qquad = 3b\sqrt{b} - 5b\sqrt{b} + 4b\sqrt{b}$

$\qquad = 2b\sqrt{b}$

29. $\sqrt[3]{54xy^3} - 5\sqrt[3]{2xy^3} + y\sqrt[3]{128x}$

$\qquad = \sqrt[3]{27y^3}\sqrt[3]{2x} - 5\sqrt[3]{y^3}\sqrt[3]{2x} + y\sqrt[3]{64}\sqrt[3]{2x}$

$\qquad = 3y\sqrt[3]{2x} - 5y\sqrt[3]{2x} + y(4)\sqrt[3]{2x}$

$\qquad = -2y\sqrt[3]{2x} + 4y\sqrt[3]{2x}$

$\qquad = 2y\sqrt[3]{2x}$

33. $-2\sqrt[4]{x^7} + 3\sqrt[4]{16x^7} = -2\sqrt[4]{x^4}\sqrt[4]{x^3} + 3\sqrt[4]{16x^4}\sqrt[4]{x^3}$

$\qquad\qquad\qquad = -2x\sqrt[4]{x^3} + 3(2x)\sqrt[4]{x^3}$

$\qquad\qquad\qquad = -2x\sqrt[4]{x^3} + 6x\sqrt[4]{x^3}$

$\qquad\qquad\qquad = 4x\sqrt[4]{x^3}$

37. $\dfrac{\sqrt[3]{8x^4}}{7} + \dfrac{3x\sqrt[3]{x}}{7} = \dfrac{\sqrt[3]{8x^3}\sqrt[3]{x} + 3x\sqrt[3]{x}}{7}$

$\qquad\qquad\qquad = \dfrac{2x\sqrt[3]{x} + 3x\sqrt[3]{x}}{7}$

$\qquad\qquad\qquad = \dfrac{5x\sqrt[3]{x}}{7}$

41. $\sqrt[3]{\dfrac{16}{27}} - \dfrac{\sqrt[3]{54}}{6} = \dfrac{\sqrt[3]{16}}{\sqrt[3]{27}} - \dfrac{\sqrt[3]{27}\sqrt[3]{2}}{6}$

$\qquad\qquad\qquad = \dfrac{\sqrt[3]{8}\sqrt[3]{2}}{3} - \dfrac{3\sqrt[3]{2}}{6}$

$\qquad\qquad\qquad = \dfrac{2(2)\sqrt[3]{2}}{6} - \dfrac{3\sqrt[3]{2}}{6}$

$\qquad\qquad\qquad = \dfrac{4\sqrt[3]{2} - 3\sqrt[3]{2}}{6}$

$\qquad\qquad\qquad = \dfrac{\sqrt[3]{2}}{6}$

45. $P = 2\sqrt{12} + \sqrt{12} + 2\sqrt{27} + 3\sqrt{3}$

$\qquad = 2\sqrt{4}\sqrt{3} + \sqrt{4}\sqrt{3} + 2\sqrt{9}\sqrt{3} + 3\sqrt{3}$

$\qquad = 2\cdot 2\sqrt{3} + 2\sqrt{3} + 2\cdot 3\sqrt{3} + 3\sqrt{3}$

$\qquad = (4 + 2 + 6 + 3)\sqrt{3}$

$\qquad = 15\sqrt{3}$

The perimeter of the trapezoid is $15\sqrt{3}$ inches.

49. $\left(\sqrt{5} - \sqrt{2}\right)^2 = \left(\sqrt{5}\right)^2 - 2\sqrt{5}\sqrt{2} + \left(\sqrt{2}\right)^2$

$\qquad\qquad\qquad = 5 - 2\sqrt{10} + 2$

$\qquad\qquad\qquad = 7 - 2\sqrt{10}$

53. $\left(2\sqrt{x} - 5\right)\left(3\sqrt{x} + 1\right)$

$\qquad = \left(2\sqrt{x}\right)\left(3\sqrt{x}\right) + \left(2\sqrt{x}\right)1 - 5\left(3\sqrt{x}\right) - 5\cdot 1$

$\qquad = 6x + 2\sqrt{x} - 15\sqrt{x} - 5$

$\qquad = 6x - 13\sqrt{x} - 5$

57. $6\left(\sqrt{2} - 2\right) = 6\sqrt{2} - 6\cdot 2 = 6\sqrt{2} - 12$

61. $\left(2\sqrt{7} + 3\sqrt{5}\right)\left(\sqrt{7} - 2\sqrt{5}\right)$

$\qquad = 2\left(\sqrt{7}\right)^2 - \left(2\sqrt{7}\right)\left(2\sqrt{5}\right) + \left(3\sqrt{5}\right)\sqrt{7} - 3\cdot 2\left(\sqrt{5}\right)^2$

$\qquad = 2\cdot 7 - 4\sqrt{35} + 3\sqrt{35} - 6\cdot 5$

$\qquad = 14 - \sqrt{35} - 30$

$\qquad = -16 - \sqrt{35}$

65. $\left(\sqrt{3} + x\right)^2 = \left(\sqrt{3}\right)^2 + 2\sqrt{3}x + x^2$

$\qquad\qquad\qquad = 3 + 2x\sqrt{3} + x^2$

69. $\left(\sqrt[3]{4} + 2\right)\left(\sqrt[3]{2} - 1\right) = \sqrt[3]{4}\sqrt[3]{2} - \sqrt[3]{4}\cdot 1 + 2\sqrt[3]{2} - 2\cdot 1$

$\qquad\qquad\qquad\qquad = \sqrt[3]{8} - \sqrt[3]{4} + 2\sqrt[3]{2} - 2$

$\qquad\qquad\qquad\qquad = 2 - \sqrt[3]{4} + 2\sqrt[3]{2} - 2$

$\qquad\qquad\qquad\qquad = -\sqrt[3]{4} + 2\sqrt[3]{2}$

73. $\left(\sqrt{x - 1} + 5\right)^2 = \left(\sqrt{x - 1}\right)^2 + 2\sqrt{x - 1}(5) + 5^2$

$\qquad\qquad\qquad\qquad = x - 1 + 10\sqrt{x - 1} + 25$

$\qquad\qquad\qquad\qquad = x + 24 + 10\sqrt{x - 1}$

77. $\dfrac{2x - 14}{2} = \dfrac{2(x - 7)}{2} = x - 7$

81. $\dfrac{6a^2b - 9ab}{3ab} = \dfrac{3ab(2a - 3)}{3ab} = 2a - 3$

85. $P = 2\left(3\sqrt{20}\right) + 2\sqrt{125}$

$\qquad = 6\sqrt{4}\sqrt{5} + 2\sqrt{25}\sqrt{5}$

$\qquad = 6(2)\sqrt{5} + 2(5)\sqrt{5}$

$\qquad = 12\sqrt{5} + 10\sqrt{5}$

$\qquad = 22\sqrt{5}$ feet

$A = 3\sqrt{20}\cdot\sqrt{125}$

$\qquad = 3\cdot 2\sqrt{5}\cdot 5\sqrt{5}$

$\qquad = 30\left(\sqrt{5}\right)^2$

$\qquad = 30\cdot 5$

$\qquad = 150$ square feet

Exercise Set 7.5

1. $\dfrac{\sqrt{2}}{\sqrt{7}} = \dfrac{\sqrt{2}\cdot\sqrt{7}}{\sqrt{7}\cdot\sqrt{7}} = \dfrac{\sqrt{14}}{7}$

5. $\dfrac{4}{\sqrt[3]{3}}\cdot\dfrac{\sqrt[3]{9}}{\sqrt[3]{9}} = \dfrac{4\sqrt[3]{9}}{\sqrt[3]{27}} = \dfrac{4\sqrt[3]{9}}{3}$

9. $\dfrac{3}{\sqrt[3]{4x^2}} = \dfrac{3}{\sqrt[3]{4x^2}}\cdot\dfrac{\sqrt[3]{2x}}{\sqrt[3]{2x}} = \dfrac{3\sqrt[3]{2x}}{\sqrt[3]{8x^3}} = \dfrac{3\sqrt[3]{2x}}{2x}$

13. $\dfrac{3}{\sqrt[3]{2}} = \dfrac{3}{\sqrt[3]{2}}\cdot\dfrac{\sqrt[3]{4}}{\sqrt[3]{4}} = \dfrac{3\sqrt[3]{4}}{\sqrt[3]{8}} = \dfrac{3\sqrt[3]{4}}{2}$

17. $\sqrt{\dfrac{2x}{5y}} = \dfrac{\sqrt{2x}}{\sqrt{5y}} = \dfrac{\sqrt{2x}\cdot\sqrt{5y}}{\sqrt{5y}\cdot\sqrt{5y}} = \dfrac{\sqrt{10xy}}{5y}$

21. $\sqrt{\dfrac{3x}{50}} = \dfrac{\sqrt{3x}}{\sqrt{50}} = \dfrac{\sqrt{3x}}{5\sqrt{2}} = \dfrac{\sqrt{3x}\cdot\sqrt{2}}{5\sqrt{2}\cdot\sqrt{2}} = \dfrac{\sqrt{6x}}{5\cdot 2} = \dfrac{\sqrt{6x}}{10}$

25. $\dfrac{\sqrt[3]{2y^2}}{\sqrt[3]{9x^2}} = \dfrac{\sqrt[3]{2y^2}\cdot\sqrt[3]{3x}}{\sqrt[3]{9x^2}\cdot\sqrt[3]{3x}} = \dfrac{\sqrt[3]{6xy^2}}{3x}$

29. $\dfrac{5a}{\sqrt[5]{8a^9b^{11}}} = \dfrac{5a}{ab^2\sqrt[5]{8a^4b}}$

$\qquad\qquad = \dfrac{5a\sqrt[5]{4ab^4}}{ab^2\sqrt[5]{8a^4b}\cdot\sqrt[5]{4ab^4}}$

$\qquad\qquad = \dfrac{5a\sqrt[5]{4ab^4}}{2a^2b^3}$

$\qquad\qquad = \dfrac{5\sqrt[5]{4ab^4}}{2ab^3}$

33. $\dfrac{-7}{\sqrt{x}-3} = \dfrac{(-7)(\sqrt{x}+3)}{(\sqrt{x}+3)(\sqrt{x}+3)}$

$= \dfrac{-7(\sqrt{x}+3)}{(\sqrt{x})^2 - 3^2}$

$= \dfrac{-7(\sqrt{x}+3)}{x-9}$

$= \dfrac{7(\sqrt{x}+3)}{9-x}$

37. $\dfrac{\sqrt{a}+1}{2\sqrt{a}-\sqrt{b}} = \dfrac{(\sqrt{a}+1)(2\sqrt{a}+\sqrt{b})}{(2\sqrt{a}-\sqrt{b})(2\sqrt{a}+\sqrt{b})}$

$= \dfrac{\sqrt{a}(2\sqrt{a}) + \sqrt{a}\sqrt{b} + 1(2\sqrt{a}) + 1\sqrt{b}}{(2\sqrt{a})^2 - (\sqrt{b})^2}$

$= \dfrac{2a + \sqrt{ab} + 2\sqrt{a} + \sqrt{b}}{4a - b}$

41. $\dfrac{\sqrt{x}}{\sqrt{x}+\sqrt{y}} = \dfrac{(\sqrt{x})(\sqrt{x}-\sqrt{y})}{(\sqrt{x}+\sqrt{y})(\sqrt{x}-\sqrt{y})}$

$= \dfrac{\sqrt{x}\sqrt{x} - \sqrt{x}\sqrt{y}}{(\sqrt{x})^2 - (\sqrt{y})^2}$

$= \dfrac{x - \sqrt{xy}}{x - y}$

45. $\sqrt{\dfrac{5}{3}} = \dfrac{\sqrt{5}}{\sqrt{3}} = \dfrac{\sqrt{5}\cdot\sqrt{5}}{\sqrt{3}\cdot\sqrt{5}} = \dfrac{5}{\sqrt{15}}$

49. $\dfrac{\sqrt{4x}}{7} = \dfrac{2\sqrt{x}}{7} = \dfrac{2\sqrt{x}\cdot\sqrt{x}}{7\cdot\sqrt{x}} = \dfrac{2x}{7\sqrt{x}}$

53. $\sqrt{\dfrac{2}{5}} = \dfrac{\sqrt{2}}{\sqrt{5}} = \dfrac{\sqrt{2}\cdot\sqrt{2}}{\sqrt{5}\cdot\sqrt{2}} = \dfrac{2}{\sqrt{10}}$

57. $\sqrt[3]{\dfrac{7}{8}} = \dfrac{\sqrt[3]{7}}{\sqrt[3]{8}} = \dfrac{\sqrt[3]{7}}{2} = \dfrac{\sqrt[3]{7}\cdot\sqrt[3]{7^2}}{2\cdot\sqrt[3]{7^2}} = \dfrac{\sqrt[3]{7^3}}{2\sqrt[3]{7^2}} = \dfrac{7}{2\sqrt[3]{49}}$

61. $\sqrt{\dfrac{18x^4y^6}{3z}} = \dfrac{\sqrt{18x^4y^6}}{\sqrt{3z}}$

$= \dfrac{3x^2y^3\sqrt{2}}{\sqrt{3z}}$

$= \dfrac{3x^2y^3\sqrt{2}\cdot\sqrt{2}}{\sqrt{3z}\cdot\sqrt{2}}$

$= \dfrac{6x^2y^3}{\sqrt{6z}}$

65. $\dfrac{2-\sqrt{11}}{6} = \dfrac{(2-\sqrt{11})(2+\sqrt{11})}{6(2+\sqrt{11})}$

$= \dfrac{4-11}{12+6\sqrt{11}}$

$= \dfrac{-7}{12+6\sqrt{11}}$

69. $\dfrac{\sqrt{x}+3}{\sqrt{x}} = \dfrac{(\sqrt{x}+3)(\sqrt{x}-3)}{\sqrt{x}(\sqrt{x}-3)}$

$= \dfrac{x-9}{(\sqrt{x})^2 - 3\sqrt{x}}$

$= \dfrac{x-9}{x - 3\sqrt{x}}$

73. $2x - 7 = 3(x-4)$

$2x - 7 = 3x - 12$

$2x + 5 = 3x$

$5 = x$

The solution set is {5}.

77. $x^2 - 8x = -12$

$x^2 - 8x + 12 = 0$

$(x-2)(x-6) = 0$

$x - 2 = 0 \text{ or } x - 6 = 0$

$x = 2 \qquad\quad x = 6$

The solution set is {2, 6}.

81. answers may vary

Exercise Set 7.6

1. $\sqrt{2x} = 4$

$2x = 4^2$

$2x = 16$

$x = 8$

The solution set is {8}.

5. $\sqrt{2x} = -4$

There is no solution since a principle square root does not yield a negative number. The solution set is $\varnothing$.

9. $\sqrt{2x-3} - 2 = 1$

$\sqrt{2x-3} = 3$

$2x - 3 = 3^2$

$2x - 3 = 9$

$2x = 12$

$x = 6$

The solution set is {6}.

13. $\sqrt[3]{x-2} - 3 = 0$

$\sqrt[3]{x-2} = 3$

$x - 2 = 3^3$

$x - 2 = 27$

$x = 29$

The solution set is {29}.

17. $x - \sqrt{4-3x} = -8$

$x + 8 = \sqrt{4-3x}$

$(x+8)^2 = 4 - 3x$

$x^2 + 16x + 64 = 4 - 3x$

$x^2 + 19x + 60 = 0$

$(x+4)(x+15) = 0$

$x + 4 = 0 \quad\text{or}\quad x + 15 = 0$

$x = -4 \qquad\qquad x = -15$

We discard $x = -15$ as extraneous, leaving $x = -4$ as the only solution. The solution set is $\{-4\}$.

21. $\sqrt{x-3} + \sqrt{x+2} = 5$

$\sqrt{x-3} = 5 - \sqrt{x+2}$

$x - 3 = (5 - \sqrt{x+2})^2$

$x - 3 = 25 - 10\sqrt{x+2} + x + 2$

$-3 = 27 - 10\sqrt{x+2}$

$-30 = -10\sqrt{x+2}$

$3 = \sqrt{x+2}$

$3^2 = x + 2$

$9 = x + 2$

$7 = x$

The solution set is {7}.

25. $-\sqrt{2x} + 4 = -6$

$10 = \sqrt{2x}$

$10^2 = 2x$

$100 = 2x$

$x = 50$

The solution set is {50}.

29. $\sqrt[4]{4x+1} - 2 = 0$

$\sqrt[4]{4x+1} = 2$

$4x + 1 = 2^4$

$4x + 1 = 16$

$4x = 15$

$x = \dfrac{15}{4}$

The solution set is $\left\{\dfrac{15}{4}\right\}$.

33. $\sqrt[3]{6x - 3} - 3 = 0$
$\sqrt[3]{6x - 3} = 3$
$6x - 3 = 3^3$
$6x - 3 = 27$
$6x = 30$
$x = 5$
The solution set is $\{5\}$.

37. $\sqrt{x + 4} = \sqrt{2x - 5}$
$x + 4 = 2x - 5$
$9 = x$
$x = 9$
The solution set is $\{9\}$.

41. $\sqrt[3]{-6x - 1} = \sqrt[3]{-2x - 5}$
$-6x - 1 = -2x - 5$
$4 = 4x$
$x = 1$
The solution set is $\{1\}$.

45. $\sqrt{2x - 1} = \sqrt{1 - 2x}$
$2x - 1 = 1 - 2x$
$4x = 2$
$x = \dfrac{1}{2}$
The solution set is $\left\{\dfrac{1}{2}\right\}$.

49. $\sqrt{y + 3} - \sqrt{y - 3} = 1$
$\sqrt{y + 3} = 1 + \sqrt{y - 3}$
$(\sqrt{y + 3})^2 = (1 + \sqrt{y - 3})^2$
$y + 3 = 1 + 2\sqrt{y - 3} + y - 3$
$5 = 2\sqrt{y - 3}$
$25 = 4(y - 3)$
$\dfrac{25}{4} = y - 3$
$\dfrac{25}{4} + \dfrac{12}{4} = y$
$\dfrac{37}{4} = y$
The solution set is $\left\{\dfrac{37}{4}\right\}$.

53. Let $b =$ the length of the unknown leg of the right triangle. Use the Pythagorean theorem.
$7^2 = 3^2 + b^2$
$49 = 9 + b^2$
$b^2 = 40$
$b = \sqrt{40}$
$= \sqrt{4}\sqrt{10}$
$= 2\sqrt{10}$ meters

57. Let $c =$ the length of the hypotenuse of the right triangle. Use the Pythagorean theorem.
$c^2 = 7^2 + (7.2)^2$
$c^2 = 49 + 51.84$
$c^2 = 100.84$
$c = \sqrt{100.84} \approx 10.0$ mm

61. $x^2 = 5^2 + 12^2$
$x^2 = 25 + 144$
$x^2 = 169$
$x = \sqrt{169}$
$x = 13$
A 13 foot ladder is needed.

65. $v = \sqrt{2gh}$
$80 = \sqrt{2(32)h}$
$(80)^2 = \left(\sqrt{2(32)h}\right)^2$
$6400 = 2(32) \cdot h$
$100 = h$
The object fell 100 feet.

69. $P = 2\pi\sqrt{\dfrac{l}{32}}$
$= 2\pi\sqrt{\dfrac{2}{32}}$
$= 2\pi\sqrt{\dfrac{1}{16}}$
$= 2\pi\left(\dfrac{1}{\sqrt{16}}\right)$
$= 2\pi\left(\dfrac{1}{4}\right)$
$= \dfrac{2\pi}{4}$
$= \dfrac{\pi}{2}$ sec ≈ 1.57 sec

73. answers may vary

77. answers may vary

81. $\dfrac{\dfrac{x}{6}}{\dfrac{2x}{3} + \dfrac{1}{2}} = \dfrac{\left(\dfrac{x}{6}\right)6}{\left(\dfrac{2x}{3} + \dfrac{1}{2}\right)6} = \dfrac{x}{\left(\dfrac{2x}{3}\right)6 + \left(\dfrac{1}{2}\right)6} = \dfrac{x}{4x + 3}$

85. $\sqrt{\sqrt{x + 3} + \sqrt{x}} = \sqrt{3}$
$\left(\sqrt{\sqrt{x + 3} + \sqrt{x}}\right)^2 = \left(\sqrt{3}\right)^2$
$\sqrt{x + 3} + \sqrt{x} = 3$
$\sqrt{x + 3} = 3 - \sqrt{x}$
$\left(\sqrt{x + 3}\right)^2 = \left(3 - \sqrt{x}\right)^2$
$x + 3 = 9 - 6\sqrt{x} + x$
$-6 = -6\sqrt{x}$
$1 = \sqrt{x}$
$1^2 = \left(\sqrt{x}\right)^2$
$1 = x$
The solution set is $\{1\}$.

89. $80\sqrt[3]{x} + 500 < 1620$
$80\sqrt[3]{x} < 1120$
$\sqrt[3]{x} < 14$
$\left(\sqrt[3]{x}\right)^3 < 14^3$
$x < 2744$
The company can make at most 2743 deliveries.

Exercise Set 7.7

1. $\sqrt{-24} = \sqrt{4}\sqrt{6}\sqrt{-1} = 2i\sqrt{6}$

5. $8\sqrt{-63} = 8\sqrt{9}\sqrt{7}\sqrt{-1} = 8 \cdot 3i\sqrt{7} = 24i\sqrt{7}$

9. $\sqrt{-2} \cdot \sqrt{-7} = \left(i\sqrt{2}\right)\left(i\sqrt{7}\right)$
$= i^2\sqrt{14}$
$= (-1)\sqrt{14}$
$= -\sqrt{14}$

13. $\sqrt{16} \cdot \sqrt{-1} = 4i$

17. $\dfrac{\sqrt{-80}}{\sqrt{-10}} = \dfrac{i\sqrt{80}}{i\sqrt{10}} = \sqrt{\dfrac{80}{10}} = \sqrt{8} = \sqrt{4}\sqrt{2} = 2\sqrt{2}$

21. $(6 + 5i) - (8 - i) = (6 - 8) + [5 - (-1)]i = -2 + 6i$

25. $(6 - 3i) - (4 - 2i) = 6 - 3i - 4 + 2i$
$= (6 - 4) + (-3 + 2)i$
$= 2 - i$

29. $(2 + 4i) + (6 - 5i) = (2 + 6) + (4 - 5)i = 8 - i$

33. $-9i \cdot 7i = -63i^2 = -63(-1) = 63$

37. $6i(2 - 3i) = 6i(2) - 6i(3i)$
$$= 12i - 18i^2$$
$$= 12i - 18(-1)$$
$$= 18 + 12i$$

41. $(4 + i)(5 + 2i) = 20 + 8i + 5i + 2i^2$
$$= 20 + 13i + 2(-1)$$
$$= 20 + 13i - 2$$
$$= 18 + 13i$$

45. $(4 - 2i)^2 = 16 - 2 \cdot 4 \cdot 2i + 4i^2$
$$= 16 - 16i + 4(-1)$$
$$= 16 - 4 - 16i$$
$$= 12 - 16i$$

49. $(1 - i)(1 + i) = 1^2 - i^2 = 1^2 - (-1) = 1 + 1 = 2$

53. $(1 - i)^2 = 1^2 - 2(1)(i) + i^2 = 1 - 2i - 1 = -2i$

57. $\dfrac{7}{4 + 3i} = \dfrac{7}{4 + 3i} \cdot \dfrac{4 - 3i}{4 - 3i}$
$$= \dfrac{28 - 21i}{4^2 + 3^2}$$
$$= \dfrac{28 - 21i}{16 + 9}$$
$$= \dfrac{28 - 21i}{25}$$
$$= \dfrac{28}{25} - \dfrac{21}{25}i$$

61. $\dfrac{3 + 5i}{1 + i} = \dfrac{3 + 5i}{1 + i} \cdot \dfrac{1 - i}{1 - i}$
$$= \dfrac{3 - 3i + 5i - 5i^2}{1^2 + 1^2}$$
$$= \dfrac{3 + 2i - 5(-1)}{1 + 1}$$
$$= \dfrac{3 + 5 + 2i}{2}$$
$$= \dfrac{8 + 2i}{2}$$
$$= 4 + i$$

65. $\dfrac{16 + 15i}{-3i} = \dfrac{(16 + 15i)i}{-3i^2}$
$$= \dfrac{16i + 15i^2}{-3(-1)}$$
$$= \dfrac{16i + 15(-1)}{3}$$
$$= \dfrac{-15}{3} + \dfrac{16}{3}i$$
$$= -5 + \dfrac{16}{3}i$$

69. $\dfrac{2 - 3i}{2 + i} = \dfrac{(2 - 3i)(2 - i)}{(2 + i)(2 - i)}$
$$= \dfrac{4 - 2i - 6i + 3i^2}{2^2 + 1^2}$$
$$= \dfrac{4 - 8i + 3(-1)}{4 + 1}$$
$$= \dfrac{4 - 3 - 8i}{5}$$
$$= \dfrac{1}{5} - \dfrac{8}{5}i$$

73. $i^{21} = i^{20}i = (i^4)^5i = 1^5i = 1i = i$

77. $i^{-6} = (i^2)^{-3} = (-1)^{-3} = \dfrac{1}{(-1)^3} = \dfrac{1}{-1} = -1$

81. $(-3i)^5 = (-3)^5i^5$
$$= -243 \cdot i^4 \cdot i$$
$$= -243(1)(i)$$
$$= -243i$$

85. $5 + 9 = 14$

14 people reported an average balance of \$200 or less

89. $i^3 + i^4 = -i + 1 = 1 - i$

93. $2 + \sqrt{-9} = 2 + 3i$

97. $\dfrac{5 - \sqrt{-75}}{10} = \dfrac{5 - 5i\sqrt{3}}{10} = \dfrac{5}{10} - \dfrac{5i\sqrt{3}}{10} = \dfrac{1}{2} - \dfrac{\sqrt{3}}{2}i$

101. $\left(8 - \sqrt{-4}\right) - \left(2 + \sqrt{-16}\right) = (8 - 2i) - (2 + 4i)$
$$= (8 - 2) + (-2i - 4i)$$
$$= 6 - 6i$$

Chapter 7 Test

1. $\sqrt{216} = \sqrt{36 \cdot 6} = 6\sqrt{6}$

5. $\left(\dfrac{8x^3}{27}\right)^{2/3} = \dfrac{8^{2/3}(x^3)^{2/3}}{27^{2/3}}$
$$= \dfrac{(8^{1/3})^2 x^2}{(27^{1/3})^2}$$
$$= \dfrac{2^2 x^2}{3^2}$$
$$= \dfrac{4x^2}{9}$$

9. $\sqrt[4]{(4xy)^4} = |4xy|$ or $4|xy|$

13. $\sqrt[3]{\dfrac{8}{9x}} = \dfrac{\sqrt[3]{8}}{\sqrt[3]{9x}} = \dfrac{2 \cdot \sqrt[3]{3x^2}}{\sqrt[3]{9x} \cdot \sqrt[3]{3x^2}} = \dfrac{2\sqrt[3]{3x^2}}{3x}$

17. $\left(\sqrt{x} + 1\right)^2 = \left(\sqrt{x}\right)^2 + 2\sqrt{x} + 1 = x + 2\sqrt{x} + 1$

21. $386^{-2/3} \approx 0.019$

25. $\sqrt{-2} = i\sqrt{2}$

29. $(4 + 3i)^2 = 16 + 24i + 9i^2$
$$= 16 + 24i + 9(-1)$$
$$= (16 - 9) + 24i$$
$$= 7 + 24i$$

33. $d = \sqrt{(-8 - (-6))^2 + (-7 - 3)^2}$
$$= \sqrt{(-2)^2 + (-10)^2}$$
$$= \sqrt{4 + 100}$$
$$= \sqrt{104}$$
$$= \sqrt{4 \cdot 26}$$
$$= 2\sqrt{26} \text{ units}$$

37. $V = \sqrt{2.5(300)} \approx 27$ mph

CHAPTER 8

Exercise Set 8.1

1. $x^2 = 16$
$$x = \pm\sqrt{16}$$
$$x = \pm 4$$
The solution set is $\{-4, 4\}$.

5. $x^2 = 18$
$$x = \pm\sqrt{18}$$
$$x = \pm\sqrt{9}\sqrt{2}$$
$$x = \pm 3\sqrt{2}$$
The solution set is $\left\{-3\sqrt{2}, 3\sqrt{2}\right\}$.

9. $(x + 5)^2 = 9$

$x + 5 = \pm\sqrt{9}$

$x + 5 = \pm 3$

$x = -5 \pm 3$

$x = -8 \quad \text{or} \quad x = -2$

The solution set is $\{-8, -2\}$.

13. $(2x - 3)^2 = 8$

$2x - 3 = \pm\sqrt{8}$

$2x - 3 = \pm\sqrt{4}\sqrt{2}$

$2x - 3 = \pm 2\sqrt{2}$

$2x = 3 \pm 2\sqrt{2}$

$x = \dfrac{3 \pm 2\sqrt{2}}{2}$

The solution set is $\left\{\dfrac{3 - 2\sqrt{2}}{2}, \dfrac{3 + 2\sqrt{2}}{2}\right\}$.

17. $x^2 - 6 = 0$

$x^2 = 6$

$x = \pm\sqrt{6}$

The solution set is $\left\{-\sqrt{6}, \sqrt{6}\right\}$.

21. $(3x - 1)^2 = -16$

$3x - 1 = \pm\sqrt{-16}$

$3x - 1 = \pm 4i$

$3x = 1 \pm 4i$

$x = \dfrac{1 \pm 4i}{3}$

The solution set is $\left\{\dfrac{1 - 4i}{3}, \dfrac{1 + 4i}{3}\right\}$.

25. $(x + 3)^2 + 8 = 0$

$(x + 3)^2 = -8$

$x + 3 = \pm\sqrt{-8}$

$x + 3 = \pm\sqrt{4}\sqrt{2}\sqrt{-1}$

$x + 3 = \pm 2i\sqrt{2}$

$x = -3 \pm 2i\sqrt{2}$

The solution set is $\left\{-3 - 2i\sqrt{2}, -3 + 2i\sqrt{2}\right\}$.

29. $z^2 - 12z + \left(-\dfrac{12}{2}\right)^2 = z^2 - 12z + 36 = (z - 6)^2$

33. $r^2 - r + \left(-\dfrac{1}{2}\right)^2 = r^2 - r + \dfrac{1}{4} = \left(r - \dfrac{1}{2}\right)^2$

37. $x^2 + 6x + 2 = 0$

$x^2 + 6x = -2$

$x^2 + 6x + \left(\dfrac{6}{2}\right)^2 = -2 + 9$

$(x + 3)^2 = 7$

$x + 3 = \pm\sqrt{7}$

$x = -3 \pm \sqrt{7}$

The solution set is $\left\{-3 - \sqrt{7}, -3 + \sqrt{7}\right\}$.

41. $x^2 + 2x - 5 = 0$

$x^2 + 2x = 5$

$x^2 + 2x + \left(\dfrac{2}{2}\right)^2 = 5 + 1$

$(x + 1)^2 = 6$

$x + 1 = \pm\sqrt{6}$

$x = -1 \pm \sqrt{6}$

The solution set is $\left\{-1 - \sqrt{6}, -1 + \sqrt{6}\right\}$.

45. $x^2 + 8x + 1 = 0$

$x^2 + 8x = -1$

$x^2 + 8x + \left(\dfrac{8}{2}\right)^2 = -1 + 16$

$(x + 4)^2 = 15$

$x + 4 = \pm\sqrt{15}$

$x = -4 \pm \sqrt{15}$

The solution set is $\left\{-4 - \sqrt{15}, -4 + \sqrt{15}\right\}$.

49. $2x^2 + 7x = 4$

$x^2 + \dfrac{7}{2}x = 2$

$x^2 + \dfrac{7}{2}x + \left(\dfrac{\frac{7}{2}}{2}\right)^2 = 2 + \dfrac{49}{16}$

$\left(x + \dfrac{7}{4}\right)^2 = \dfrac{81}{16}$

$x + \dfrac{7}{4} = \pm\sqrt{\dfrac{81}{16}}$

$x = -\dfrac{7}{4} \pm \dfrac{9}{4}$

$x = -4 \quad \text{or} \quad x = \dfrac{1}{2}$

The solution set is $\left\{-4, \dfrac{1}{2}\right\}$.

53. $y^2 + 2y + 2 = 0$

$y^2 + 2y = -2$

$y^2 + 2y + \left(\dfrac{2}{2}\right)^2 = -2 + 1$

$(y + 1)^2 = -1$

$y + 1 = \pm\sqrt{-1}$

$y + 1 = \pm i$

$y = -1 \pm i$

The solution set is $\{-1 - i, -1 + i\}$.

57. $2x^2 - x + 6 = 0$

$2x^2 - x = -6$

$x^2 - \dfrac{1}{2}x = -3$

$x^2 - \dfrac{1}{2}x + \left(\dfrac{-\frac{1}{2}}{2}\right)^2 = -3 + \dfrac{1}{16}$

$\left(x - \dfrac{1}{4}\right)^2 = -\dfrac{47}{16}$

$x - \dfrac{1}{4} = \pm\sqrt{-\dfrac{47}{16}}$

$x - \dfrac{1}{4} = \pm\dfrac{\sqrt{47}\sqrt{-1}}{\sqrt{16}}$

$x - \dfrac{1}{4} = \pm\dfrac{i\sqrt{47}}{4}$

$x = \dfrac{1 \pm i\sqrt{47}}{4}$

The solution set is $\left\{\dfrac{1 - i\sqrt{47}}{4}, \dfrac{1 + i\sqrt{47}}{4}\right\}$.

61. $z^2 + 3z - 4 = 0$

$z^2 + 3z = 4$

$z^2 + 3z + \left(\dfrac{3}{2}\right)^2 = 4 + \dfrac{9}{4}$

$\left(z + \dfrac{3}{2}\right)^2 = \dfrac{25}{4}$

$z + \dfrac{3}{2} = \pm\sqrt{\dfrac{25}{4}}$

$z + \dfrac{3}{2} = \pm\dfrac{5}{2}$

$z = -\dfrac{3}{2} \pm \dfrac{5}{2}$

$z = -4 \quad \text{or} \quad z = 1$

The solution set is $\{-4, 1\}$.

65.
$$3x^2 + 3x = 5$$
$$x^2 + x = \frac{5}{3}$$
$$x^2 + x + \left(\frac{1}{2}\right)^2 = \frac{5}{3} + \frac{1}{4}$$
$$\left(x + \frac{1}{2}\right)^2 = \frac{23}{12}$$
$$x + \frac{1}{2} = \pm\sqrt{\frac{23}{12}}$$
$$x + \frac{1}{2} = \pm\frac{\sqrt{23}}{2\sqrt{3}} \cdot \frac{\sqrt{3}}{\sqrt{3}}$$
$$x + \frac{1}{2} = \pm\frac{\sqrt{69}}{6}$$
$$x = -\frac{1}{2} \pm \frac{\sqrt{69}}{6}$$
$$x = -\frac{3}{6} \pm \frac{\sqrt{69}}{6}$$
$$x = \frac{-3 \pm \sqrt{69}}{6}$$

The solution set is $\left\{\dfrac{-3 - \sqrt{69}}{6}, \dfrac{-3 + \sqrt{69}}{6}\right\}$.

69. Use $A = P(1 + r)^t$ with $A = 1000$, $P = 810$, and $t = 2$.
$$A = P(1 + r)^t$$
$$1000 = 810(1 + r)^2$$
$$\frac{1000}{810} = (1 + r)^2$$
$$\frac{100}{81} = (1 + r)^2$$
$$\pm\sqrt{\frac{100}{81}} = 1 + r$$
$$\pm\frac{10}{9} = 1 + r$$
$$-1 \pm \frac{10}{9} = r$$
$$-\frac{9}{9} \pm \frac{10}{9} = r$$
$$r = \frac{1}{9} \quad \text{or} \quad r = -\frac{19}{9}$$

The rate cannot be negative, so $-\dfrac{19}{9}$ is rejected. $\dfrac{1}{9} = 0.1\overline{1} \approx 11\%$, so \$810 grows to approximately \$1000 in 2 years with an interest rate of 11%.

73.
$$s(t) = 16t^2$$
$$1100 = 16t^2$$
$$\frac{1100}{16} = t^2$$
$$\frac{275}{4} = t^2$$
$$\pm\sqrt{\frac{275}{4}} = t^2$$
$$\pm\frac{5\sqrt{11}}{2} = t$$

The time cannot be negative, so $-\dfrac{5\sqrt{11}}{2}$ is rejected.

It takes the object $\dfrac{5\sqrt{11}}{2} \approx 8.29$ seconds to fall from the top to the base of the dam.

77.
$$a^2 + b^2 = c^2$$
$$x^2 + x^2 = 20^2$$
$$2x^2 = 400$$
$$x^2 = 200$$
$$x = \pm\sqrt{200}$$
$$x = \pm10\sqrt{2}$$

The length cannot be negative, so $-10\sqrt{2}$ is rejected. The length of each leg is $10\sqrt{2}$ centimeters.

81. $\dfrac{12 - 8\sqrt{7}}{16} = \dfrac{4\left(3 - 2\sqrt{7}\right)}{4 \cdot 4} = \dfrac{3 - 2\sqrt{7}}{4}$

85. $\sqrt{b^2 - 4ac} = \sqrt{(-3)^2 - 4(1)(-1)}$
$$= \sqrt{9 + 4}$$
$$= \sqrt{13}$$

89.
$$4x^2 = 17$$
$$x^2 = \frac{17}{4}$$

Since the value that the squared term equals is positive, the equation has real solutions.

93. Compound interest is preferable; answers may vary.

97.
$$p = -x^2 + 47$$
$$11 = -x^2 + 47$$
$$-36 = -x^2$$
$$36 = x^2$$
$$\pm\sqrt{36} = x$$
$$\pm6 = x$$

Since the quantity demanded cannot be negative, $x = 6$ and the demand is 6 thousand scissors.

Exercise Set 8.2

1. $m^2 + 5m - 6 = 0$
$$a = 1, b = 5, c = -6$$
$$m = \frac{-b \pm \sqrt{b^2 - 4ac}}{2a}$$
$$= \frac{-5 \pm \sqrt{5^2 - 4(1)(-6)}}{2(1)}$$
$$= \frac{-5 \pm \sqrt{25 + 24}}{2}$$
$$= \frac{-5 \pm \sqrt{49}}{2}$$
$$= \frac{-5 \pm 7}{2}$$
$$m = \frac{-5 - 7}{2} = -6 \quad \text{or} \quad m = \frac{-5 + 7}{2} = 1$$

The solution set is $\{-6, 1\}$.

5. $x^2 - 6x + 9 = 0$
$$a = 1, b = -6, c = 9$$
$$x = \frac{-b \pm \sqrt{b^2 - 4ac}}{2a}$$
$$= \frac{-(-6) \pm \sqrt{(-6)^2 - 4(1)(9)}}{2(1)}$$
$$= \frac{6 \pm \sqrt{36 - 36}}{2}$$
$$= \frac{6 \pm \sqrt{0}}{2}$$
$$= 3$$

The solution set is $\{3\}$.

9.
$$8m^2 - 2m = 7$$
$$8m^2 - 2m - 7 = 0$$
$$a = 8, b = -2, c = -7$$
$$m = \frac{-b \pm \sqrt{b^2 - 4ac}}{2a}$$
$$= \frac{-(-2) \pm \sqrt{(-2)^2 - 4(8)(-7)}}{2(8)}$$
$$= \frac{2 \pm \sqrt{4 + 224}}{16}$$
$$= \frac{2 \pm \sqrt{228}}{16}$$
$$= \frac{2 \pm 2\sqrt{57}}{16}$$
$$= \frac{2(1 \pm \sqrt{57})}{16}$$
$$= \frac{1 \pm \sqrt{57}}{8}$$

The solution set is $\left\{ \dfrac{1 - \sqrt{57}}{8}, \dfrac{1 + \sqrt{57}}{8} \right\}$.

13. $\dfrac{1}{2}x^2 - x - 1 = 0$
$$x^2 - 2x - 2 = 0$$
$$a = 1, b = -2, c = -2$$
$$x = \frac{-b \pm \sqrt{b^2 - 4ac}}{2a}$$
$$= \frac{-(-2) \pm \sqrt{(-2)^2 - 4(1)(-2)}}{2(1)}$$
$$= \frac{2 \pm \sqrt{4 + 8}}{2}$$
$$= \frac{2 \pm \sqrt{12}}{2}$$
$$= \frac{2 \pm 2\sqrt{3}}{2}$$
$$= \frac{2(1 \pm \sqrt{3})}{2}$$
$$= 1 \pm \sqrt{3}$$

The solution set is $\left\{ 1 - \sqrt{3}, 1 + \sqrt{3} \right\}$.

17. $\dfrac{1}{3}y^2 - y - \dfrac{1}{6} = 0$
$$2y^2 - 6y - 1 = 0$$
$$a = 2, b = -6, c = -1$$
$$y = \frac{-b \pm \sqrt{b^2 - 4ac}}{2a}$$
$$= \frac{-(-6) \pm \sqrt{(-6)^2 - 4(2)(-1)}}{2(2)}$$
$$= \frac{6 \pm \sqrt{36 + 8}}{4}$$
$$= \frac{6 \pm \sqrt{44}}{4}$$
$$= \frac{6 \pm 2\sqrt{11}}{4}$$
$$= \frac{2(3 + \sqrt{11})}{4}$$
$$= \frac{3 \pm \sqrt{11}}{2}$$

The solution set is $\left\{ \dfrac{3 - \sqrt{11}}{2}, \dfrac{3 + \sqrt{11}}{2} \right\}$.

21. $(m + 2)(2m - 6) = 5(m - 1) - 12$
$$2m^2 - 2m - 12 = 5m - 5 - 12$$
$$2m^2 - 7m + 5 = 0$$
$$a = 2, b = -7, c = 5$$
$$m = \frac{-b \pm \sqrt{b^2 - 4ac}}{2a}$$
$$= \frac{-(-7) \pm \sqrt{(-7)^2 - 4(2)(5)}}{2(2)}$$
$$= \frac{7 \pm \sqrt{49 - 40}}{4}$$
$$= \frac{7 \pm \sqrt{9}}{4}$$
$$= \frac{7 \pm 3}{4}$$
$$m = \frac{7 - 3}{4} = 1 \quad \text{or} \quad m = \frac{7 + 3}{4} = \frac{10}{4} = \frac{5}{2}$$

The solution set is $\left\{ 1, \dfrac{5}{2} \right\}$.

25. $(x + 5)(x - 1) = 2$
$$x^2 + 4x - 5 = 2$$
$$x^2 + 4x - 7 = 2$$
$$a = 1, b = 4, c = -7$$
$$x = \frac{-b \pm \sqrt{b^2 - 4ac}}{2a}$$
$$= \frac{-4 \pm \sqrt{4^2 - 4(1)(-7)}}{2(1)}$$
$$= \frac{-4 \pm \sqrt{16 + 28}}{2}$$
$$= \frac{-4 \pm \sqrt{44}}{2}$$
$$= \frac{-4 \pm 2\sqrt{11}}{2}$$
$$= \frac{2(-2 \pm \sqrt{11})}{2}$$
$$= -2 \pm \sqrt{11}$$

The solution set is $\left\{ -2 - \sqrt{11}, -2 + \sqrt{11} \right\}$.

29. $\dfrac{x^2}{3} - x = \dfrac{5}{3}$
$$\frac{x^2}{3} - x - \frac{5}{3} = 0$$
$$x^2 - 3x - 5 = 0$$
$$a = 1, b = -3, c = -5$$
$$x = \frac{-b \pm \sqrt{b^2 - 4ac}}{2a}$$
$$= \frac{-(-3) \pm \sqrt{(-3)^2 - 4(1)(-5)}}{2(1)}$$
$$= \frac{3 \pm \sqrt{9 + 20}}{2}$$
$$= \frac{3 \pm \sqrt{29}}{2}$$

The solution set is $\left\{ \dfrac{3 - \sqrt{29}}{2}, \dfrac{3 + \sqrt{29}}{2} \right\}$.

33.
$$x(6x + 2) = 3$$
$$6x^2 + 2x = 3$$
$$6x^2 + 2x - 3 = 0$$
$$a = 6, b = 2, c = -3$$
$$x = \frac{-b \pm \sqrt{b^2 - 4ac}}{2a}$$
$$= \frac{-2 \pm \sqrt{2^2 - 4(6)(-3)}}{2(6)}$$
$$= \frac{-2 \pm \sqrt{4 + 72}}{12}$$
$$= \frac{-2 \pm \sqrt{76}}{12}$$
$$= \frac{-2 \pm 2\sqrt{19}}{12}$$
$$= \frac{2(-1 \pm \sqrt{19})}{12}$$
$$= \frac{-1 \pm \sqrt{19}}{6}$$

The solution set is $\left\{ \dfrac{-1 - \sqrt{19}}{6}, \dfrac{-1 + \sqrt{19}}{6} \right\}$.

37.
$$\frac{1}{2}y^2 = y - \frac{1}{2}$$
$$\frac{1}{2}y^2 - y + \frac{1}{2} = 0$$
$$y^2 - 2y + 1 = 0$$
$$a = 1, b = -2, c = 1$$
$$y = \frac{-b \pm \sqrt{b^2 - 4ac}}{2a}$$
$$= \frac{-(-2) \pm \sqrt{(-2)^2 - 4(1)(1)}}{2(1)}$$
$$= \frac{2 \pm \sqrt{4 - 4}}{2}$$
$$= \frac{2 \pm \sqrt{0}}{2}$$
$$= 1$$

The solution set is $\{1\}$.

41.
$$x^2 - 5 = 0$$
$$x^2 + 0x - 5 = 0$$
$$a = 1, b = 0, c = -5$$
$$b^2 - 4ac = 0^2 - 4(1)(-5) = 0 + 20 = 20$$
Since $b^2 - 4ac$ is positive, the equation has two real solutions.

45. $3x = -2x^2 + 7$
$$0 = -2x^2 - 3x + 7$$
$$a = -2, b = -3, c = 7$$
$$b^2 - 4ac = (-3)^2 - 4(-2)(7) = 9 + 56 = 65$$
Since $b^2 - 4ac$ is positive, the equation has two real solutions.

49.
$$9x - 2x^2 + 5 = 0$$
$$-2x^2 + 9x + 5 = 0$$
$$a = -2, b = 9, c = 5$$
$$b^2 - 4ac = 9^2 - 4(-2)(5) = 81 + 40 = 121$$
Since $b^2 - 4ac$ is positive, the equation has two real solutions.

53. Let x be the length of the legs, then the length of the hypotenuse is $x + 2$.
$$x^2 + x^2 = (x + 2)^2$$
$$2x^2 = x^2 + 4x + 4$$
$$x^2 - 4x - 4 = 0$$
$$a = 1, b = -4, c = -4$$
$$x = \frac{-b \pm \sqrt{b^2 - 4ac}}{2a}$$
$$= \frac{-(-4) \pm \sqrt{(-4)^2 - 4(1)(-4)}}{2(1)}$$
$$= \frac{4 \pm \sqrt{16 + 16}}{2}$$
$$= \frac{4 \pm \sqrt{32}}{2}$$
$$= \frac{4 \pm 4\sqrt{2}}{2}$$
$$= \frac{2(2 \pm 2\sqrt{2})}{2}$$
$$= 2 \pm 2\sqrt{2}$$

Since the lengths cannot be negative, reject $2 - 2\sqrt{2}$.
$$x + 2 = (2 + 2\sqrt{2}) + 2 = 4 + 2\sqrt{2}$$

The legs both have length $(2 + 2\sqrt{2})$ centimeters, and the hypotenuse has length $(4 + 2\sqrt{2})$ centimeters.

57. a. Let x be the length of the sides.
$$x^2 + x^2 = 100^2$$
$$2x^2 = 10,000$$
$$x^2 = 5000$$
$$x^2 - 5000 = 0$$
$$a = 1, b = 0, c = -5000$$
$$x = \frac{-b \pm \sqrt{b^2 - 4ac}}{2a}$$
$$= \frac{-0 \pm \sqrt{0^2 - 4(1)(-5000)}}{2(1)}$$
$$= \frac{\pm\sqrt{20,000}}{2}$$
$$= \frac{\pm 100\sqrt{2}}{2}$$
$$= \pm 50\sqrt{2}$$

Since the length cannot be negative, reject $-50\sqrt{2}$. The sides of the pen are $50\sqrt{2}$ meters long.

b. $A = s^2$
$$= (50\sqrt{2})^2$$
$$= 50^2(\sqrt{2})^2$$
$$= 2500(2)$$
$$= 5000$$
The area of the pen is 5000 square meters.

61. Let h be the height of the triangle. Then the base is $2h + 4$.

$$A = \frac{1}{2}bh$$

$$42 = \frac{1}{2}(2h + 4) \cdot h$$

$$42 = h^2 + 2h$$

$$0 = h^2 + 2h - 42$$

$$a = 1, b = 2, c = -42$$

$$h = \frac{-b \pm \sqrt{b^2 - 4ac}}{2a}$$

$$= \frac{-2 \pm \sqrt{2^2 - 4(1)(-42)}}{2(1)}$$

$$= \frac{-2 \pm \sqrt{4 + 168}}{2}$$

$$= \frac{-2 \pm \sqrt{172}}{2}$$

$$= \frac{-2 \pm 2\sqrt{43}}{2}$$

$$= \frac{2(-1 \pm \sqrt{43})}{2}$$

$$= -1 \pm \sqrt{43}$$

Since the height cannot be negative, reject $-1 - \sqrt{43}$.

$$2h + 4 = 2(-1 + \sqrt{43}) + 4$$

$$= -2 + 2\sqrt{43} + 4$$

$$= 2 + 2\sqrt{43}$$

The height of the triangle is $(-1 + \sqrt{43})$ centimeters and the base is $(2 + 2\sqrt{43})$ centimeters.

65. Find t when $h = 0$.

$$h = -16t^2 - 20t + 180$$

$$0 = -16t^2 - 20t + 180$$

$$0 = -4t^2 - 5t + 45$$

$$a = -4, b = -5, c = 45$$

$$t = \frac{-b \pm \sqrt{b^2 - 4ac}}{2a}$$

$$= \frac{-(-5) \pm \sqrt{(-5)^2 - 4(-4)(45)}}{2(-4)}$$

$$= \frac{5 \pm \sqrt{25 + 720}}{-8}$$

$$= \frac{5 \pm \sqrt{745}}{-8}$$

$$= \frac{-5 \pm \sqrt{745}}{8}$$

Since the time cannot be negative, reject $\frac{-5 - \sqrt{745}}{8}$. The ball strikes the ground $\frac{-5 + \sqrt{745}}{8} \approx 2.8$ seconds after it is thrown.

69. The LCD is $5x$.

$$\frac{1}{x} + \frac{2}{5} = \frac{7}{x}$$

$$5x\left(\frac{1}{x}\right) + 5x\left(\frac{2}{5}\right) = 5x\left(\frac{7}{x}\right)$$

$$5 + 2x = 35$$

$$2x = 30$$

$$x = 15$$

The solution set is $\{15\}$.

73. $z^4 - 13z^2 + 36 = (z^2 - 9)(z^2 - 4)$
$$= (z^2 - 3^2)(z^2 - 2^2)$$
$$= (z + 3)(z - 3)(z + 2)(z - 2)$$

77. answers may vary

81. The greatest decrease occurs from Sunday to Monday.

85. $f(x) = 3x^2 - 18x + 57$
Thursday is 4 days from Sunday.

$$f(4) = 3(4)^2 - 18(4) + 57$$
$$= 48 - 72 + 57$$
$$= 33$$

The temperature was approximately 33°F on Thursday; answers may vary.

89. a. 2001 is 1 year after 2000, so $x = 1$.

$$y = -13x^2 + 221x + 8476$$
$$y = -13(1)^2 + 221(1) + 8476 = 8684$$

The average total daily supply of motor gasoline in 2001 was 8684 thousand barrels per day.

b. Find x when $y = 9334$.

$$y = -13x^2 + 221x + 8476$$
$$9334 = -13x^2 + 221x + 8476$$
$$0 = -13x^2 + 221x - 858$$
$$0 = x^2 - 17x + 66$$
$$a = 1, b = -17, c = 66$$
$$x = \frac{-b \pm \sqrt{b^2 - 4ac}}{2a}$$
$$= \frac{-(-17) \pm \sqrt{(-17)^2 - 4(1)(66)}}{2(1)}$$
$$= \frac{17 \pm \sqrt{289 - 264}}{2}$$
$$= \frac{17 \pm \sqrt{25}}{2}$$
$$= \frac{17 \pm 5}{2}$$
$$x = \frac{17 - 5}{2} = 6 \quad \text{or} \quad x = \frac{17 + 5}{6} = 11$$

The average total daily supply of motor gasoline will be 9334 thousand barrels per day 6 or 11 years after 2000, so in 2006 or 2011.

Exercise Set 8.3

1.
$$2x = \sqrt{10 + 3x}$$
$$(2x)^2 = \left(\sqrt{10 + 3x}\right)^2$$
$$4x^2 = 10 + 3x$$
$$4x^2 - 3x - 10 = 0$$
$$(4x + 5)(x - 2) = 0$$
$$4x + 5 = 0 \quad \text{or} \quad x - 2 = 0$$
$$4x = -5$$
$$x = -\frac{5}{4} \qquad\qquad x = 2$$

The value $-\frac{5}{4}$ does not check, so the solution set is $\{2\}$.

5. $\sqrt{9x} = x + 2$

$(\sqrt{9x})^2 = (x + 2)^2$

$9x = x^2 + 4x + 4$

$0 = x^2 - 5x + 4$

$0 = (x - 1)(x - 4)$

$x - 1 = 0 \quad \text{or} \quad x - 4 = 0$

$x = 1 \qquad\qquad x = 4$

The solution set is $\{1, 4\}$.

9. The LCD is $x(x + 2)$.

$$\frac{3}{x} + \frac{4}{x + 2} = 2$$

$$x(x + 2)\left(\frac{3}{x}\right) + x(x + 2)\left(\frac{4}{x + 2}\right) = x(x + 2)(2)$$

$$3(x + 2) + 4x = 2x(x + 2)$$

$$3x + 6 + 4x = 2x^2 + 4x$$

$$0 = 2x^2 - 3x - 6$$

$a = 2, \ b = -3, \ c = -6$

$x = \dfrac{-b \pm \sqrt{b^2 - 4ac}}{2a}$

$= \dfrac{-(-3) \pm \sqrt{(-3)^2 - 4(2)(-6)}}{2(2)}$

$= \dfrac{3 \pm \sqrt{9 + 48}}{4}$

$= \dfrac{3 \pm \sqrt{57}}{4}$

Neither value will make a denominator 0, so the solution set is

$\left\{\dfrac{3 - \sqrt{57}}{4}, \dfrac{3 + \sqrt{57}}{4}\right\}$.

13. $p^4 - 16 = 0$

$(p^2 + 4)(p^2 - 4) = 0$

$(p^2 + 4)(p + 2)(p - 2) = 0$

$p^2 + 4 = 0 \qquad\qquad \text{or} \quad p + 2 = 0 \quad \text{or} \quad p - 2 = 0$

$p^2 = 4$

$p = \pm\sqrt{-4} = \pm 2i \qquad\qquad p = -2 \qquad\qquad p = 2$

The solution set is $\{-2, 2, -2i, 2i\}$.

17. $4x^4 + 11x^2 = 3$

$4x^4 + 11x^2 - 3 = 0$

$(4x^2 - 1)(x^2 + 3) = 0$

$(2x + 1)(2x - 1)(x^2 + 3) = 0$

$2x + 1 = 0 \quad \text{or} \quad 2x - 1 = 0 \quad \text{or} \quad x^2 + 3 = 0$

$2x = -1 \qquad\qquad 2x = 1 \qquad\qquad x^2 = -3$

$x = -\dfrac{1}{2} \qquad\qquad x = \dfrac{1}{2} \qquad\qquad x = \pm\sqrt{-3} = \pm 3i$

The solution set is $\left\{-\dfrac{1}{2}, \dfrac{1}{2}, -3i, 3i\right\}$.

21. Let $y = 5n + 1$.

$(5n + 1)^2 + 2(5n + 1) - 3 = 0$

$y^2 + 2y - 3 = 0$

$(y + 3)(y - 1) = 0$

$y + 3 = 0 \quad \text{or} \quad y - 1 = 0$

$y = -3 \qquad\qquad y = 1$

$5n + 1 = -3 \qquad 5n + 1 = 1$

$5n = -4 \qquad\qquad 5n = 0$

$n = -\dfrac{4}{5} \qquad\qquad n = 0$

The solution set is $\left\{-\dfrac{4}{5}, 0\right\}$.

25. Let $x = \dfrac{1}{3t - 2}$.

$$1 + \frac{2}{3t - 2} = \frac{8}{(3t - 2)^2}$$

$$1 + 2x = 8x^2$$

$$0 = 8x^2 - 2x - 1$$

$$0 = (2x - 1)(4x + 1)$$

$2x - 1 = 0 \qquad \text{or} \qquad 4x + 1 = 0$

$2x = 1 \qquad\qquad 4x = -1$

$x = \dfrac{1}{2} \qquad\qquad x = -\dfrac{1}{4}$

$\dfrac{1}{3t - 2} = \dfrac{1}{2} \qquad \dfrac{1}{3t - 2} = -\dfrac{1}{4}$

$2 = 3t - 2 \qquad\qquad -4 = 3t - 2$

$4 = 3t \qquad\qquad\qquad -2 = 3t$

$\dfrac{4}{3} = t \qquad\qquad\qquad -\dfrac{2}{3} = t$

The solution set is $\left\{-\dfrac{2}{3}, \dfrac{4}{3}\right\}$.

29. $a^4 - 5a^2 + 6 = 0$

$(a^2 - 2)(a^2 - 3) = 0$

$a^2 - 2 = 0 \qquad \text{or} \qquad a^2 - 3 = 0$

$a^2 = 2 \qquad\qquad a^2 = 3$

$a = \pm\sqrt{2} \qquad\qquad a = \pm\sqrt{3}$

The solution set is $\{-\sqrt{2}, \sqrt{2}, -\sqrt{3}, \sqrt{3}\}$.

33. $(p + 2)^2 = 9(p + 2) - 20$

$(p + 2)^2 - 9(p + 2) + 20 = 0$

Let $x = p + 2$.

$x^2 - 9x + 20 = 0$

$(x - 4)(x - 5) = 0$

$x - 4 = 0 \qquad \text{or} \qquad x - 5 = 0$

$x = 4 \qquad\qquad x = 5$

$p + 2 = 4 \qquad\qquad p + 2 = 5$

$p = 2 \qquad\qquad\qquad p = 3$

The solution set is $\{2, 3\}$.

37. Let $y = x^{1/3}$.

$x^{2/3} - 8x^{1/3} + 15 = 0$

$(x^{1/3})^2 - 8x^{1/3} + 15 = 0$

$y^2 - 8y + 15 = 0$

$(y - 3)(y - 5) = 0$

$y - 3 = 0 \qquad \text{or} \qquad y - 5 = 0$

$y = 3 \qquad\qquad y = 5$

$x^{1/3} = 3 \qquad\qquad x^{1/3} = 5$

$(x^{1/3})^3 = 3^3 \qquad (x^{1/3})^3 = 5^3$

$x = 27 \qquad\qquad x = 125$

The solution set is $\{27, 125\}$.

41. Let $y = x^{1/3}$.

$$2x^{2/3} + 3x^{1/3} - 2 = 0$$
$$2(x^{1/3})^2 + 3x^{1/3} - 2 = 0$$
$$2y^2 + 3y - 2 = 0$$
$$(y + 2)(2y - 1) = 0$$

$y + 2 = 0$ or $2y - 1 = 0$

$y = -2$ $2y = 1$

$$y = \frac{1}{2}$$

$x^{1/3} = -2$ $x^{1/3} = \dfrac{1}{2}$

$(x^{1/3})^3 = (-2)^3$ $(x^{1/3})^3 = \left(\dfrac{1}{2}\right)^3$

$x = -8$ $x = \dfrac{1}{8}$

The solution set is $\left\{-8, \dfrac{1}{8}\right\}$.

45. Let $y = \sqrt{x}$.

$$x - \sqrt{x} = 2$$
$$x - 2 = \sqrt{x}$$
$$(x - 2)^2 = \left(\sqrt{x}\right)^2$$
$$x^2 - 4x + 4 = x$$
$$x^2 - 5x + 4 = 0$$
$$(x - 1)(x - 4) = 0$$

$x - 1 = 0$ or $x - 4 = 0$

$x = 1$ or $x = 4$

The value $x = 1$ does not check, so the solution set is $\{4\}$.

49.
$$p^4 - p^2 - 20 = 0$$
$$(p^2 - 5)(p^2 + 4) = 0$$

$p^2 - 5 = 0$ or $p^2 + 4 = 0$

$p^2 = 5$ $p^2 = -4$

$p = \pm\sqrt{5}$ $p = \pm\sqrt{-4} = \pm 2i$

The solution set is $\left\{-\sqrt{5}, \sqrt{5}, -2i, 2i\right\}$.

53.
$$27y^4 + 15y^2 = 2$$
$$27y^4 + 15y^2 - 2 = 0$$
$$(3y^2 + 2)(9y^2 - 1) = 0$$

$3y^2 + 2 = 0$ or $9y^2 - 1 = 0$

$3y^2 = -2$ $9y^2 = 1$

$y^2 = -\dfrac{2}{3}$ $y^2 = \dfrac{1}{9}$

$y = \pm\sqrt{-\dfrac{2}{3}}$ $y = \pm\sqrt{\dfrac{1}{9}}$

$y = \pm\dfrac{i\sqrt{2}}{\sqrt{3}} \cdot \dfrac{\sqrt{3}}{\sqrt{3}} = \pm\dfrac{i\sqrt{6}}{3}$ $y = \pm\dfrac{1}{3}$

The solution set is $\left\{-\dfrac{1}{3}, \dfrac{1}{3}, -\dfrac{i\sqrt{6}}{3}, \dfrac{i\sqrt{6}}{3}\right\}$.

57. Let x be the amount of time it takes the hose alone to fill the pond. Then the inlet pipe can fill the pond in $(x - 1)$ hours.

	Time to Fill	Part Filled in 1 Hour
Hose	x hours	$\dfrac{1}{x}$
Inlet Pipe	$(x - 1)$ hours	$\dfrac{1}{x - 1}$
Together	8 hours	$\dfrac{1}{8}$

The LCD is $8x(x - 1)$.

$$\frac{1}{x} + \frac{1}{x - 1} = \frac{1}{8}$$
$$8x(x - 1)\left(\frac{1}{x}\right) + 8x(x - 1)\left(\frac{1}{x - 1}\right) = 8x(x - 1)\left(\frac{1}{8}\right)$$
$$8(x - 1) + 8x = x(x - 1)$$
$$8x - 8 + 8x = x^2 - x$$
$$0 = x^2 - 17x + 8$$

$a = 1, b = -17, c = 8$

$$x = \frac{-b \pm \sqrt{b^2 - 4ac}}{2a}$$
$$= \frac{-(-17) \pm \sqrt{(-17)^2 - 4(1)(8)}}{2(1)}$$
$$= \frac{17 \pm \sqrt{289 - 32}}{2}$$
$$= \frac{17 \pm \sqrt{257}}{2}$$

Since the time cannot be negative, reject $\dfrac{17 - \sqrt{257}}{2}$.

The hose fills the pond in $\dfrac{17 + \sqrt{257}}{2} \approx 16.5$ hours. The inlet pipe takes about 15.5 hours.

61. Let x be the amount of time it takes Billy to clean the house. Then Billy's dad takes $(x - 1)$ hours.

	Time	Part in 1 Hour
Dad	$x - 1$	$\dfrac{1}{x - 1}$
Billy	x	$\dfrac{1}{x}$
Together	4	$\dfrac{1}{4}$

The LCD is $4x(x - 1)$.

$$\frac{1}{x} + \frac{1}{x - 1} = \frac{1}{4}$$
$$4x(x - 1)\left(\frac{1}{x}\right) + 4x(x - 1)\left(\frac{1}{x - 1}\right) = 4x(x - 1)\left(\frac{1}{4}\right)$$
$$4(x - 1) + 4x = x(x - 1)$$
$$4x - 4 + 4x = x^2 - x$$
$$0 = x^2 - 9x + 4$$

$a = 1, b = -9, c = 4$

$$x = \frac{-b \pm \sqrt{b^2 - 4ac}}{2a}$$

$$= \frac{-(-9) \pm \sqrt{(-9)^2 - 4(1)(4)}}{2(1)}$$

$$= \frac{9 \pm \sqrt{81 - 16}}{2}$$

$$= \frac{9 \pm \sqrt{65}}{2}$$

$$x = \frac{9 - \sqrt{65}}{2} \approx 0.5 \text{ or } x = \frac{9 + \sqrt{65}}{2} \approx 8.5$$

Since Billy's dad takes one hour less to clean the house, reject 0.5. It takes Billy 8.5 hours to clean the house.

65. a. The ? in the drawing is the original length, x, less the two 3-inch squares, or $x - 6$.

b. The height of the box is 3 inches, while both the length and the width are $(x - 6)$ inches, and the volume is 300 cubic inches.

$$V = l \cdot w \cdot h$$

$$300 = (x - 6) \cdot (x - 6) \cdot 3$$

c. $300 = 3(x^2 - 12x + 36)$

$$100 = x^2 - 12x + 36$$

$$0 = x^2 - 12x - 64$$

$$0 = (x + 4)(x - 16)$$

$$x + 4 = 0 \quad \text{or} \quad x - 16 = 0$$

$$x = -4 \qquad\qquad x = 16$$

Since the length cannot be negative, reject -4.

The piece of cardboard is 16 inches by 16 inches. The sides of the box have length $16 - 6 = 10$ inches each and the volume is $V = lwh = 10 \cdot 10 \cdot 3 = 300$ cubic inches.

69. $\dfrac{5x}{3} + 2 \leq 7$

$$\frac{5x}{3} \leq 5$$

$$5x \leq 15$$

$$x \leq 3$$

The solution set is $(-\infty, 3]$.

73. $y^3 + 9y - y^2 - 9 = 0$

$$y(y^2 + 9) - (y^2 + 9) = 0$$

$$(y^2 + 9)(y - 1) = 0$$

$$y^2 + 9 = 0 \qquad \text{or} \quad y - 1 = 0$$

$$y^2 = -9 \qquad\qquad\qquad y = 1$$

$$y = \pm\sqrt{-9} = \pm 3i$$

The solution set is $\{1, -3i, 3i\}$.

77. $2x^3 = -54$

$$2x^3 + 54 = 0$$

$$x^3 + 27 = 0$$

$$x^3 + 3^3 = 0$$

$$(x + 3)(x^2 - 3x + 9) = 0$$

$$x + 3 = 0 \quad \text{or} \quad x^2 - 3x + 9 = 0$$

$$x = -3$$

$$x^2 - 3x + 9 = 0$$

$$a = 1, b = -3, c = 9$$

$$x = \frac{-b \pm \sqrt{b^2 - 4ac}}{2a}$$

$$= \frac{-(-3) \pm \sqrt{(-3)^2 - 4(1)(9)}}{2(1)}$$

$$= \frac{3 \pm \sqrt{9 - 36}}{2}$$

$$= \frac{3 \pm \sqrt{-27}}{2}$$

$$= \frac{3 \pm 3i\sqrt{3}}{2}$$

The solution set is $\left\{ -3, \dfrac{3 - 3i\sqrt{3}}{2}, \dfrac{3 + 3i\sqrt{3}}{2} \right\}$.

81. a. Let x be Junqueira's fastest speed. Using $d = r \cdot t$ or $t = \dfrac{d}{r}$, Junqueira's time for that lap was $\dfrac{11{,}119.68}{x}$ seconds. Bourdais' fastest speed is then $(x - 0.1159)$ feet per second and the time for Bourdais' fastest lap was $\dfrac{11{,}119.68}{x - 0.1159}$ seconds, which is also $\dfrac{11{,}119.68}{x} + 0.036$.

The LCD is $x(x - 0.1159)$.

$$\frac{11{,}119.68}{x} + 0.036 = \frac{11{,}119.68}{x - 0.1159}$$

$$x(x - 0.1159)\left(\frac{11{,}119.68}{x} + 0.036\right) = x(x - 0.1159)\left(\frac{11{,}119.68}{x - 0.1159}\right)$$

$$(11{,}119.68 + 0.036x)(x - 0.1159) = 11{,}119.68x$$

$$11{,}119.68x - 1288.770912 + 0.036x^2 - 0.0041724x = 11{,}119.68x$$

$$0 = -0.036x^2 + 0.0041724x + 1288.770912$$

$$a = -0.036, b = 0.0041724, c = 1288.770912$$

$$x = \frac{-b \pm \sqrt{b^2 - 4ac}}{2a}$$

$$= \frac{-0.0041724 \pm \sqrt{(0.0041724)^2 - 4(-0.036)(1288.770912)}}{2(-0.036)}$$

$$x \approx -189.1488 \quad \text{or} \quad x \approx 189.2647$$

Since the speed cannot be negative, reject -189.1488.

Junqueira's fastest speed was approximately 189.265 feet per second.

b. Bourdais' fastest speed was $189.2647 - 0.1159 \approx 189.149$ feet per second.

c. 189.149 feet/sec $= \dfrac{189.149 \text{ feet}}{1 \text{ sec}} \cdot \dfrac{3600 \text{ sec}}{1 \text{ hour}} \cdot \dfrac{1 \text{ mile}}{5280 \text{ feet}}$

$$\approx 128.965 \text{ miles/hour}$$

Bourdais' fastest speed was about 128.965 miles per hour.

Exercise Set 8.4

1. $(x + 1)(x + 5) > 0$

$(x + 1)(x + 5) = 0$

$x + 1 = 0$ or $x + 5 = 0$

$x = -1$ $\qquad$ $x = -5$

Region	Test Point Value	$(x + 1)(x + 5) > 0$	Result
$(-\infty, -5)$	-6	$(-5)(-1) > 0$	True
$(-5, -1)$	-2	$(-1)(3) > 0$	False
$(-1, \infty)$	0	$(1)(5) > 0$	True

The solution set is $(-\infty, -5) \cup (-1, \infty)$.

5. $x^2 + 8x + 15 \geq 0$

$(x + 5)(x + 3) \geq 0$

$(x + 5)(x + 3) = 0$

$x + 5 = 0$ or $x + 3 = 0$

$x = -5$ $\qquad$ $x = -3$

Region	Test Point Value	$(x + 5)(x + 3) \geq 0$	Result
$(-\infty, -5)$	-6	$(-1)(-3) \geq 0$	True
$(-5, -3)$	-4	$(1)(-1) \geq 0$	False
$(-3, \infty)$	0	$(5)(3) \geq 0$	True

The solution set is $(-\infty, -5] \cup [-3, \infty)$.

9. $(x - 6)(x - 4)(x - 2) > 0$

$(x - 6)(x - 4)(x - 2) = 0$

$x - 6 = 0$ or $x - 4 = 0$ or $x - 2 = 0$

$x = 6$ $\qquad$ $x = 4$ $\qquad$ $x = 2$

Region	Test Point Value	$(x - 6)(x - 4)(x - 2) > 0$	Result
$(-\infty, 2)$	0	$(-6)(-4)(-2) > 0$	False
$(2, 4)$	3	$(-3)(-1)(1) > 0$	True
$(4, 6)$	5	$(-1)(1)(3) > 0$	False
$(6, \infty)$	7	$(1)(3)(5) > 0$	True

The solution set is $(2, 4) \cup (6, \infty)$.

13. $(x^2 - 9)(x^2 - 4) > 0$

$(x + 3)(x - 3)(x + 2)(x - 2) > 0$

$(x + 3)(x - 3)(x + 2)(x - 2) = 0$

$x + 3 = 0$ or $x - 3 = 0$ or $x + 2 = 0$ or $x - 2 = 0$

$x = -3$ $\qquad$ $x = 3$ $\qquad$ $x = -2$ $\qquad$ $x = 2$

Region	Test Point Value	$(x + 3)(x - 3)(x + 2)(x - 2) > 0$	Result
$(-\infty, -3)$	-4	$(-1)(-7)(-2)(-6) > 0$	True
$(-3, -2)$	$-\dfrac{5}{2}$	$\left(\dfrac{1}{2}\right)\left(-\dfrac{11}{2}\right)\left(-\dfrac{1}{2}\right)\left(-\dfrac{9}{2}\right) > 0$	False
$(-2, 2)$	0	$(3)(-3)(2)(-2) > 0$	True
$(2, 3)$	$\dfrac{5}{2}$	$\left(\dfrac{11}{2}\right)\left(-\dfrac{1}{2}\right)\left(\dfrac{9}{2}\right)\left(\dfrac{1}{2}\right) > 0$	False
$(3, \infty)$	4	$(7)(1)(6)(2) > 0$	True

The solution set is $(-\infty, -3) \cup (-2, 2) \cup (3, \infty)$.

17. $\dfrac{5}{x + 1} > 0$

$x + 1 = 0$

$x = -1$ makes the denominator 0.

Region	Test Point Value	$\dfrac{5}{x + 1} > 0$	Result
$(-\infty, -1)$	-2	$\dfrac{5}{-1} > 0$	False
$(-1, \infty)$	0	$\dfrac{5}{1} > 0$	True

The solution set is $(-1, \infty)$.

21. $\dfrac{3}{x - 2} < 4$

$x - 2 = 0$

$x = 2$ makes the denominator 0.

$$\dfrac{3}{x - 2} = 4$$

$$(x - 2)\left(\dfrac{3}{x - 2}\right) = (x - 2)(4)$$

$$3 = 4x - 8$$

$$11 = 4x$$

$$\dfrac{11}{4} = x$$

Region	Test Point Value	$\dfrac{3}{x - 2} < 4$	Result
$(-\infty, 2)$	0	$\dfrac{3}{-2} < 4$	True
$\left(2, \dfrac{11}{4}\right)$	$\dfrac{5}{2}$	$\dfrac{3}{\frac{1}{2}} < 4$	False
$\left(\dfrac{11}{4}, \infty\right)$	3	$\dfrac{3}{1} < 4$	True

The solution set is $(-\infty, 2) \cup \left(\dfrac{11}{4}, \infty\right)$.

25. $\dfrac{x + 2}{x - 3} < 1$

$x - 3 = 0$

$x = 3$ makes the denominator 0.

$$\dfrac{x + 2}{x - 3} = 1$$

$$(x - 3)\left(\dfrac{x + 2}{x - 3}\right) = (x - 3)(1)$$

$$x + 2 = x - 3$$

$$2 = -3 \quad \text{False}$$

The related equation has no solution.

Region	Test Point Value	$\dfrac{x + 2}{x - 3} < 1$	Result
$(-\infty, 3)$	0	$\dfrac{2}{-3} < 1$	True
$(3, \infty)$	4	$\dfrac{6}{1} < 1$	False

The solution set is $(-\infty, 3)$.

29.
$$x^2 > x$$
$$x^2 = x$$
$$x^2 - x = 0$$
$$x(x - 1) = 0$$
$$x = 0 \quad \text{or} \quad x - 1 = 0$$
$$x = 1$$

Region	Test Point Value	$x^2 > x$	Result
$(-\infty, 0)$	-1	$(1)^2 > -1$	True
$(0, 1)$	$\dfrac{1}{2}$	$\left(\dfrac{1}{2}\right)^2 > \dfrac{1}{2}$	False
$(1, \infty)$	2	$(2)^2 > 2$	True

The solution set is $(-\infty, 0) \cup (1, \infty)$.

33. $(2x - 8)(x + 4)(x - 6) \leq 0$
$$(2x - 8)(x + 4)(x - 6) = 0$$
$$2x - 8 = 0 \quad \text{or} \quad x + 4 = 0 \quad \text{or} \quad x - 6 = 0$$
$$2x = 8$$
$$x = 4 \qquad\qquad x = -4 \qquad\qquad x = 6$$

Region	Test Point Value	$(2x - 8)(x + 4)(x - 6) \leq 0$	Result
$(-\infty, -4)$	-5	$(-18)(-1)(-11) \leq 0$	True
$(-4, 4)$	0	$(-8)(4)(-6) \leq 0$	False
$(4, 6)$	5	$(2)(9)(-1) \leq 0$	True
$(6, \infty)$	7	$(6)(11)(1) \leq 0$	False

The solution set is $(-\infty, -4] \cup [4, 6]$.

37. $\dfrac{x - 5}{x + 4} \geq 0$
$$x + 4 = 0$$
$$x = -4 \text{ makes the denominator } 0.$$
$$\frac{x - 5}{x + 4} = 0$$
$$(x + 4)\left(\frac{x - 5}{x + 4}\right) = (x + 4)(0)$$
$$x - 5 = 0$$
$$x = 5$$

Region	Test Point Value	$\dfrac{x - 5}{x + 4} \geq 0$	Result
$(-\infty, -4)$	-5	$\dfrac{-10}{-1} \geq 0$	True
$(-4, 5)$	0	$\dfrac{-5}{4} \geq 0$	False
$(5, \infty)$	6	$\dfrac{1}{10} \geq 0$	True

The solution set is $(-\infty, -4) \cup [5, \infty)$.

41.
$$4x^3 + 16x^2 - 9x - 36 > 0$$
$$4x^2(x + 4) - 9(x + 4) > 0$$
$$(x + 4)(4x^2 - 9) > 0$$
$$(x + 4)(2x + 3)(2x - 3) > 0$$
$$(x + 4)(2x + 3)(2x - 3) = 0$$
$$x + 4 = 0 \quad \text{or} \quad 2x + 3 = 0 \quad \text{or} \quad 2x - 3 = 0$$
$$2x = -3 \qquad\qquad 2x = 3$$
$$x = -4 \qquad\qquad x = -\frac{3}{2} \qquad\qquad x = \frac{3}{2}$$

Region	Test Point Value	$(x + 4)(2x + 3)(2x - 3) > 0$	Result
$(-\infty, -4)$	-5	$(-1)(-7)(-13) > 0$	False
$\left(-4, -\dfrac{3}{2}\right)$	-2	$(2)(-1)(-7) > 0$	True
$\left(-\dfrac{3}{2}, \dfrac{3}{2}\right)$	0	$(4)(3)(-3) > 0$	False
$\left(\dfrac{3}{2}, \infty\right)$	2	$(6)(7)(1) > 0$	True

The solution set is $\left(-4, -\dfrac{3}{2}\right) \cup \left(\dfrac{3}{2}, \infty\right)$.

45. $\dfrac{x(x + 6)}{(x - 7)(x + 1)} \geq 0$
$$(x - 7)(x + 1) = 0$$
$$x - 7 = 0 \quad \text{or} \quad x + 1 = 0$$
$$x = 7 \text{ and } x = -1 \text{ make the denominator } 0.$$
$$\frac{x(x + 6)}{(x - 7)(x + 1)} = 0$$
$$(x - 7)(x + 1)\left(\frac{x(x + 6)}{(x - 7)(x + 1)}\right) = (x - 7)(x + 1)(0)$$
$$x(x + 6) = 0$$
$$x = 0 \quad \text{or} \quad x + 6 = 0$$
$$x = -6$$

Region	Test Point Value	$\dfrac{x(x + 6)}{(x - 7)(x + 1)} \geq 0$	Result
$(-\infty, -6)$	-7	$\dfrac{(-7)(-1)}{(-14)(-6)} \geq 0$	True
$(-6, -1)$	-2	$\dfrac{(-2)(4)}{(-9)(-1)} \geq 0$	False
$(-1, 0)$	$-\dfrac{1}{2}$	$\dfrac{\left(-\dfrac{1}{2}\right)\left(\dfrac{11}{2}\right)}{\left(-\dfrac{15}{2}\right)\left(\dfrac{1}{2}\right)} \geq 0$	True
$(0, 7)$	1	$\dfrac{(1)(7)}{(-6)(2)} \geq 0$	False
$(7, \infty)$	8	$\dfrac{8(14)}{(1)(9)} \geq 0$	True

The solution set is $(-\infty, -6] \cup (-1, 0] \cup (7, \infty)$.

49. $(2x - 7)(3x + 5) > 0$
$(2x - 7)(3x + 5) = 0$
$2x - 7 = 0$ or $3x + 5 = 0$
$2x = 7$ $\qquad$ $3x = -5$
$x = \dfrac{7}{2}$ $\qquad$ $x = -\dfrac{5}{3}$

Region	Test Point Value	$(2x - 7)(3x + 5) > 0$	Result
$\left(-\infty, -\dfrac{5}{3}\right)$	-2	$(-11)(-1) > 0$	True
$\left(-\dfrac{5}{3}, \dfrac{7}{2}\right)$	0	$(-7)(5) > 0$	False
$\left(\dfrac{7}{2}, \infty\right)$	4	$(1)(17) > 0$	True

The solution set is $\left(-\infty, -\dfrac{5}{3}\right) \cup \left(\dfrac{7}{2}, \infty\right)$.

53. $\dfrac{(x + 1)^2}{5x} > 0$
$5x = 0$
$x = 0$ makes the denominator 0.
$\dfrac{(x + 1)^2}{5x} = 0$
$(5x)\left(\dfrac{(x + 1)^2}{5x}\right) = (5x)(0)$
$(x + 1)^2 = 0$
$x + 1 = 0$
$x = -1$

Region	Test Point Value	$\dfrac{(x + 1)^2}{5x} > 0$	Result
$(-\infty, -1)$	-2	$\dfrac{(-1)^2}{-10} > 0$	False
$(-1, 0)$	$-\dfrac{1}{2}$	$\dfrac{\left(-\dfrac{1}{2}\right)^2}{-\dfrac{5}{2}} > 0$	False
$(0, \infty)$	1	$\dfrac{2^2}{5} > 0$	True

The solution set is $(0, \infty)$.

57.

x	$y = f(x) = -x^2$
0	$-0^2 = 0$
1	$-1^2 = -1$
-1	$-(-1)^2 = -1$
2	$-2^2 = -4$
-2	$-(-2)^2 = -4$

61. A number minus its reciprocal is written as $x - \dfrac{1}{x}$.

$x - \dfrac{1}{x} < 0$
$x = 0$ makes the denominator 0.
$x - \dfrac{1}{x} = 0$
$x(x) - x\left(\dfrac{1}{x}\right) = x(0)$
$x^2 - 1 = 0$
$(x + 1)(x - 1) = 0$
$x + 1 = 0$ or $x - 1 = 0$
$x = -1$ $\qquad$ $x = 1$

Region	Test Point Value	$x - \dfrac{1}{x} < 0$	Result
$(-\infty, -1)$	-2	$-2 - \dfrac{1}{-2} = -2 + \dfrac{1}{2} < 0$	True
$(-1, 0)$	$-\dfrac{1}{2}$	$-\dfrac{1}{2} - \dfrac{1}{-\dfrac{1}{2}} = -\dfrac{1}{2} + 2 < 0$	False
$(0, 1)$	$\dfrac{1}{2}$	$\dfrac{1}{2} - \dfrac{1}{\dfrac{1}{2}} = \dfrac{1}{2} - 2 < 0$	True
$(1, \infty)$	2	$2 - \dfrac{1}{2} < 0$	False

The solution set is $(-\infty, -1) \cup (0, 1)$.

65. $x^2 - x - 56 > 0$
$(x + 7)(x - 8) > 0$
$(x + 7)(x - 8) = 0$
$x + 7 = 0$ or $x - 8 = 0$
$x = -7$ $\qquad$ $x = 8$

Region	Test Point Value	$(x + 7)(x - 8) > 0$	Result
$(-\infty, -7)$	-8	$(-1)(-16) > 0$	True
$(-7, 8)$	0	$(7)(-8) > 0$	False
$(8, \infty)$	9	$(16)(1) > 0$	True

The solution set is $(-\infty, -7) \cup (8, \infty)$.

Exercise Set 8.5

1. $f(x) = x^2 - 1$

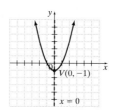

5. $h(x) = x^2 + 5$

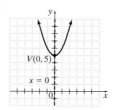

9. $g(x) = x^2 + 7$

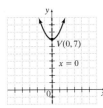

13. $f(x) = (x - 2)^2 + 5$

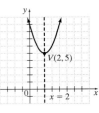

17. $g(x) = (x + 2)^2 - 5$

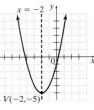

21. $g(x) = -x^2$

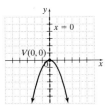

25. $H(x) = 2x^2$

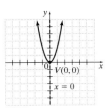

29. $f(x) = 10(x + 4)^2 - 6$

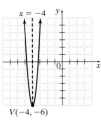

33. $H(x) = \dfrac{1}{2}(x - 6)^2 - 3$

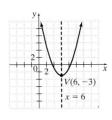

37. $F(x) = \left(x + \dfrac{1}{2}\right)^2 - 2$

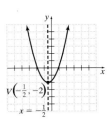

41. $x^2 + 8x$

$$\left[\frac{1}{2}(8)\right]^2 = 4^2 = 16$$

$x^2 + 8x + 16$

45. $y^2 + y$

$$\left[\frac{1}{2}(1)\right]^2 = \left[\frac{1}{2}\right]^2 = \frac{1}{4}$$

$y^2 + y + \dfrac{1}{4}$

49. $g(x) = 5[(x - (-3))]^2 + 6$
$g(x) = 5(x + 3)^2 + 6$

53. $y = f(x - 3)$
Shift the graph 3 units to the right.

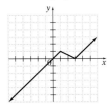

57. Since $a < 0$, the graph of $f(x)$ opens downward, with vertex $(0.1, 3.6)$, which is choice c.

Exercise Set 8.6

1. $f(x) = x^2 + 8x + 7$

$$\frac{-b}{2a} = \frac{-8}{2(1)} = -4 \text{ and}$$

$f(-4) = (-4)^2 + 8(-4) + 7 = 16 - 32 + 7 = -9$

Thus, the vertex is $(-4, -9)$.

5. $f(x) = 5x^2 - 10x + 3$

$$\frac{-b}{2a} = \frac{-(-10)}{2(5)} = 1 \text{ and}$$

$f(1) = 5(1)^2 - 10(1) + 3 = 5 - 10 + 3 = -2$

Thus, the vertex is $(1, -2)$.

9. $f(x) = x^2 - 4x + 3$

$$\frac{-b}{2a} = \frac{-(-4)}{2(1)} = 2$$

$f(2) = 2^2 - 4(2) + 3 = -1$

The vertex is $(2, -1)$, which is graph D.

33. answers may vary

37. $x = y + 2$
$x - 2 = y$
$y = x - 2$

41. $x = -2y - 7$
$x + 7 = -2y$
$\dfrac{x + 7}{-2} = y$
$y = -\dfrac{x + 7}{2}$

45. answers may vary

Exercise Set 9.2

1. Since the y-value 7 corresponds to two x-values, 3 and 6, the function is not one-to-one.

5. $f = \{(11, 12), (4, 3), (3, 4), (6, 6)\}$ is a one-to-one function.
$f^{-1} = \{(12, 11), (3, 4), (4, 3), (6, 6)\}$

9. The given function is one-to-one.

Rank in Population (Input)	1	47	14	24	36
State (Output)	California	Alaska	Indiana	Louisiana	New Mexico

13. $f(x) = x^3 + 2$
a. $f(-1) = (-1)^3 + 2 = 1$
b. $f^{-1}(1) = -1$

17. The graph does not represent a one-to-one function because it does not pass the horizontal line test.

21. The graph does not represent a one-to-one function because it does not pass the horizontal line test.

25. $f(x) = 2x - 3$
$y = 2x - 3$
$x = 2y - 3$
$2y = x + 3$
$y = \dfrac{x + 3}{2}$
$f^{-1}(x) = \dfrac{x + 3}{2}$

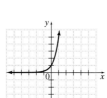

29. $f(x) = x^3$
$y = x^3$
$x = y^3$
$y = \sqrt[3]{x}$
$f^{-1}(x) = \sqrt[3]{x}$

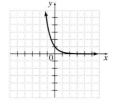

33. $f(x) = \sqrt[3]{x}$
$y = \sqrt[3]{x}$
$x = \sqrt[3]{y}$
$x^3 = y$
$f^{-1}(x) = x^3$

37. $f(x) = (x + 2)^3$
$y = (x + 2)^3$
$x = (y + 2)^3$
$\sqrt[3]{x} = y + 2$
$\sqrt[3]{x} - 2 = y$
$f^{-1}(x) = \sqrt[3]{x} - 2$

41.

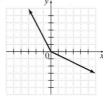

45. $16^{3/4} = (16^{1/4})^3 = (2)^3 = 8$

49. $f(x) = 3^x$
$f(2) = 3^2 = 9$

53. a. The ordered pair corresponding to $f(2)$ is $(2, 9)$.
b. One ordered pair that we know is a solution of f^{-1} is $(9, 2)$.

57. answers may vary

61. $f(x) = \sqrt[3]{x + 3}$
$y = \sqrt[3]{x + 3}$
$x = \sqrt[3]{y + 3}$
$x^3 = (\sqrt[3]{y + 3})^3$
$x^3 = y + 3$
$x^3 - 3 = y$
$f^{-1}(x) = x^3 - 3$

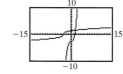

Exercise Set 9.3

1.

x	$y = 5^x$
-2	$5^{-2} = \dfrac{1}{5^2} = \dfrac{1}{25}$
-1	$5^{-1} = \dfrac{1}{5}$
0	$5^0 = 1$
1	$5^1 = 5$
2	$5^2 = 25$

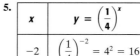

5.

x	$y = \left(\dfrac{1}{4}\right)^x$
-2	$\left(\dfrac{1}{4}\right)^{-2} = 4^2 = 16$
-1	$\left(\dfrac{1}{4}\right)^{-1} = 4^1 = 4$
0	$\left(\dfrac{1}{4}\right)^0 = 1$
1	$\left(\dfrac{1}{4}\right)^1 = \dfrac{1}{4}$
2	$\left(\dfrac{1}{4}\right)^2 = \dfrac{1}{16}$

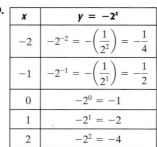

9.

x	$y = -2^x$
-2	$-2^{-2} = -\left(\dfrac{1}{2^2}\right) = -\dfrac{1}{4}$
-1	$-2^{-1} = -\left(\dfrac{1}{2^1}\right) = -\dfrac{1}{2}$
0	$-2^0 = -1$
1	$-2^1 = -2$
2	$-2^2 = -4$

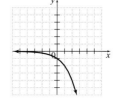

13.

x	$y = -\left(\dfrac{1}{4}\right)^x$
-2	$-\left(\dfrac{1}{4}\right)^{-2} = -(4)^2 = -16$
-1	$-\left(\dfrac{1}{4}\right)^{-1} = -(4)^1 = -4$
0	$-\left(\dfrac{1}{4}\right)^0 = -1$
1	$-\left(\dfrac{1}{4}\right)^1 = -\dfrac{1}{4}$
2	$-\left(\dfrac{1}{4}\right)^2 = -\dfrac{1}{16}$

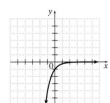

17.

x	$f(x) = 2^{x-2}$
-1	$2^{-1-2} = 2^{-3} = \dfrac{1}{2^3} = \dfrac{1}{8}$
0	$2^{0-2} = 2^{-2} = \dfrac{1}{2^2} = \dfrac{1}{4}$
1	$2^{1-2} = 2^{-1} = \dfrac{1}{2}$
2	$2^{2-2} = 2^0 = 1$
3	$2^{3-2} = 2^1 = 2$
4	$2^{4-2} = 2^2 = 4$

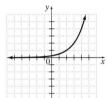

21. $3^x = 27$
$3^x = 3^3$
$x = 3$
The solution set is $\{3\}$.

25. $32^{2x-3} = 2$
$(2^5)^{2x-3} = 2^1$
$10x - 15 = 1$
$10x = 16$
$x = \dfrac{8}{5}$
The solution set is $\left\{\dfrac{8}{5}\right\}$.

29. $9^x = 27$
$(3^2)^x = 3^3$
$3^{2x} = 3^3$
$2x = 3$
$x = \dfrac{3}{2}$
The solution set is $\left\{\dfrac{3}{2}\right\}$.

33. $81^{x-1} = 27^{2x}$
$(3^4)^{x-1} = (3^3)^{2x}$
$3^{4x-4} = 3^{6x}$
$4x - 4 = 6x$
$-4 = 2x$
$-2 = x$
The solution set is $\{-2\}$.

37. $y = 30(2.7)^{-0.004t}$
$y = 30(2.7)^{-0.004(50)} = 30(2.7)^{-0.2} \approx 24.6$
There will be about 24.6 pounds left after 50 days.

41. a. 2003 is 1 year after 2002, so $t = 1$.
$y = 44(1.29)^t$
$y = 44(1.29)^1 = 44(1.29) = 56.76$
The retail revenues from Internet shopping were predicted to be $56.8 billion in 2003.

b. 2009 is 7 years after 2002, so $t = 7$.
$y = 44(1.29)^t$
$y = 44(1.29)^7 \approx 261.57$
The retail revenues from Internet shopping are predicted to be $261.6 billion in 2009.

45. Use $P = 6000, t = 3, n = 12$, and $r = 8\% = 0.08$.
$$A = P\left(1 + \frac{r}{n}\right)^{nt}$$
$$A = 6000\left(1 + \frac{0.08}{12}\right)^{(12)(3)}$$
$$= 6000\left(1 + \frac{0.08}{12}\right)^{36}$$
$$\approx 7621.42$$
Erica owes $7621.42.

49. $3x - 4 = 3(x + 1)$
$3x - 4 = 3x + 3$
$-4 = 3$ False
Since the equation $-4 = 3$ is false, the equation has no solution. The solution set is $\varnothing$.

53. Since there are no variables in the exponent in $h(x) = \left(\dfrac{1}{2}x\right)^2$, it is not an exponential function.

57. $f(x) = \left(\dfrac{1}{4}\right)^x$
$f(1) = \left(\dfrac{1}{4}\right)^1 = \dfrac{1}{4}$
This is graph D.

Exercise Set 9.4

1. $\log_6 36 = 2$
$6^2 = 36$

5. $\log_{10} 1000 = 3$
$10^3 = 1000$

9. $\log_e \dfrac{1}{e^2} = -2$
$e^{-2} = \dfrac{1}{e^2}$

13. $\log_{0.7} 0.343 = 3$
$0.7^3 = 0.343$

17. $2^4 = 16$
$\log_2 16 = 4$

21. $e^3 = x$
$\log_e x = 3$

25. $4^{-2} = \dfrac{1}{16}$
$\log_4 \dfrac{1}{16} = -2$

29. $\log_2 8 = 3$ since $2^3 = 8$

33. $\log_{25} 5 = \dfrac{1}{2}$ since $25^{1/2} = 5$

37. $\log_6 1 = 0$ since $6^0 = 1$

41. $\log_3 81 = 4$ since $3^4 = 81$

45. $\log_3 9 = x$
$3^x = 9$
$3^x = 3^2$
$x = 2$
The solution set is $\{2\}$.

49. $\log_x 49 = 2$
$x^2 = 49$
$x^2 = 7^2$
$x = 7$
The solution set is $\{7\}$.

53. $\log_3 \dfrac{1}{27} = x$

$3^x = \dfrac{1}{27}$

$3^x = \dfrac{1}{3^3}$

$3^x = 3^{-3}$

$x = -3$

The solution set is $\{-3\}$.

57. $\log_4 16 = x$

$4^x = 16$

$4^x = 4^2$

$x = 2$

The solution set is $\{2\}$.

61. $\log_x 100 = 2$

$x^2 = 100$

$x^2 = 10^2$

$x = 10$

The solution set is $\{10\}$.

65. $3^{\log_3 5} = x$

$5 = x$

The solution set is $\{5\}$

69. $\log_5 5^3 = 3$

73. $\log_9 9 = 1$

77. $f(x) = \log_{1/4} x$ or $y = \log_{1/4} x$

$\left(\dfrac{1}{4}\right)^y = x$

$x = \left(\dfrac{1}{4}\right)^y$	y
$\left(\dfrac{1}{4}\right)^{-1} = 4^1 = 4$	-1
$\left(\dfrac{1}{4}\right)^0 = 1$	0
$\left(\dfrac{1}{4}\right)^1 = \dfrac{1}{4}$	1

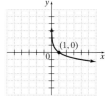

81. $f(x) = \log_{1/6} x$ or $y = \log_{1/6} x$

$\left(\dfrac{1}{6}\right)^y = x$

$x = \left(\dfrac{1}{6}\right)^y$	y
$\left(\dfrac{1}{6}\right)^{-1} = 6^1 = 6$	-1
$\left(\dfrac{1}{6}\right)^0 = 1$	0
$\left(\dfrac{1}{6}\right)^1 = \dfrac{1}{6}$	1

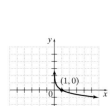

85. $\dfrac{x^2 - 8x + 16}{2x - 8} = \dfrac{(x-4)(x-4)}{2(x-4)} = \dfrac{x-4}{2}$

89. answers may vary

93. $y = \left(\dfrac{1}{3}\right)^x$; $y = \log_{1/3} x$

x	$y = \left(\dfrac{1}{3}\right)^x$
-2	9
-1	3
0	1
1	$\dfrac{1}{3}$
2	$\dfrac{1}{9}$

x	$y = \log_{1/3} x$
9	-2
3	-1
1	0
$\dfrac{1}{3}$	1
$\dfrac{1}{9}$	2

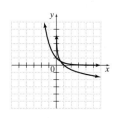

97. Use $H = 8$.

$\log_{10}(1 - k) = \dfrac{-0.3}{H}$

$\log_{10}(1 - k) = \dfrac{-0.3}{8}$

$\log_{10}(1 - k) = -0.0375$

$1 - k = 10^{-0.0375}$

$1 - 10^{-0.0375} = k$

$0.0827 \approx k$

The decay rate is 0.0827.

Exercise Set 9.5

1. $\log_5 2 + \log_5 7 = \log_5(2 \cdot 7) = \log_5 14$

5. $\log_6 x + \log_6(x + 1) = \log_6[x \cdot (x + 1)]$
$= \log_6(x^2 + x)$

9. $\log_5 12 - \log_5 4 = \log_5 \dfrac{12}{4} = \log_5 3$

13. $\log_2 x - \log_2 y = \log_2 \dfrac{x}{y}$

17. $\log_3 x^2 = 2\log_3 x$

21. $\log_5 \sqrt{y} = \log_5 y^{1/2} = \dfrac{1}{2}\log_5 y$

25. $3\log_4 2 + \log_4 6 = \log_4 2^3 + \log_4 6$
$= \log_4 8 + \log_4 6$
$= \log_4(8 \cdot 6)$
$= \log_4 48$

29. $\log_4 2 + \log_4 10 - \log_4 5 = \log_4(2 \cdot 10) - \log_4 5$
$= \log_4 20 - \log_4 5$
$= \log_4 \dfrac{20}{5}$
$= \log_4 4$
$= 1$

33. $\log_{10} x - \log_{10}(x + 1) + \log_{10}(x^2 - 2)$
$= \log_{10} \dfrac{x}{x + 1} + \log_{10}(x^2 - 2)$
$= \log_{10} \dfrac{x(x^2 - 2)}{x + 1}$
$= \log_{10} \dfrac{x^3 - 2x}{x + 1}$

37. $2\log_8 x - \dfrac{2}{3}\log_8 x + 4\log_8 x$

$\quad = \log_8 x^2 - \log_8 x^{2/3} + \log_8 x^4$

$\quad = \log_8 \dfrac{x^2}{x^{2/3}} + \log_8 x^4$

$\quad = \log_8 \dfrac{x^2 \cdot x^4}{x^{2/3}}$

$\quad = \log_8 x^{2+4-2/3}$

$\quad = \log_8 x^{16/3}$

41. $\log_4 \dfrac{2}{9z} = \log_4 2 - \log_4 9z$

$\quad = \log_4 2 - (\log_4 9 + \log_4 z)$

$\quad = \log_4 2 - \log_4 9 - \log_4 z$

45. $\log_b \sqrt{7x} = \log_b (7x)^{1/2}$

$\quad = \dfrac{1}{2}\log_b 7x$

$\quad = \dfrac{1}{2}(\log_b 7 + \log_b x)$

$\quad = \dfrac{1}{2}\log_b 7 + \dfrac{1}{2}\log_b x$

49. $\log_5 x^3(x + 1) = \log_5 x^3 + \log_5 (x + 1)$

$\quad = 3\log_5 x + \log_5 (x + 1)$

53. $\log_b \dfrac{5}{3} = \log_b 5 - \log_b 3 = 0.7 - 0.5 = 0.2$

57. $\log_b \sqrt{5} = \log_b 5^{1/2} = \dfrac{1}{2}\log_b 5 = \dfrac{1}{2}(0.7) = 0.35$

61. $\log_b \dfrac{3}{9} = \log_b \dfrac{1}{3}$

$\quad = \log_b 3^{-1}$

$\quad = -1 \cdot \log_b 3$

$\quad = -1 \cdot (0.68)$

$\quad = -0.68$

65.

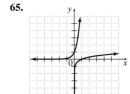

69. $\log_7 \sqrt{7} = \log_7 7^{1/2} = \dfrac{1}{2}\log_7 7 = \dfrac{1}{2}(1) = \dfrac{1}{2}$

73. $\log_3 x + \log_3 y = \log_3 xy \neq \log_3 (x + y)$

The statement is false.

77. $\log_3 24 = \log_3 (6 \cdot 4)$

$\quad = \log_3 6 + \log_3 4$

$\quad \neq (\log_3 6) \cdot (\log_3 4)$

The statement is false.

Exercise Set 9.6

1. $\log 8 \approx 0.9031$

5. $\log 2 \approx 0.6931$

9. $\log 12.6 \approx 1.1004$

13. $\log 41.5 \approx 1.6180$

17. $\log \dfrac{1}{1000} = \log 10^{-3} = -3$

21. $\ln \sqrt[4]{e} = \ln e^{1/4} = \dfrac{1}{4}$

25. $\ln e^{3.1} = 3.1$

29. $\ln \sqrt{e} = \ln e^{1/2} = \dfrac{1}{2}$

33. $\ln x = 1.4$

$\quad x = e^{1.4} \approx 4.0552$

The solution set is $\{e^{1.4}\}$ or approximately $\{4.0552\}$.

37. $\ln x = -2.3$

$\quad x = e^{-2.3} \approx 0.1003$

The solution set is $\{e^{-2.3}\}$ or approximately $\{0.1003\}$.

41. $\ln 4x = 0.18$

$\quad 4x = e^{0.18}$

$\quad x = \dfrac{1}{4}e^{0.18} \approx 0.2993$

The solution set is $\left\{\dfrac{1}{4}e^{0.18}\right\}$ or approximately $\{0.2993\}$.

45. $\log (2x + 1) = -0.5$

$\quad 2x + 1 = 10^{-0.5}$

$\quad 2x = 10^{-0.5} - 1$

$\quad x = \dfrac{10^{-0.5} - 1}{2} \approx -0.3419$

The solution set is $\left\{\dfrac{10^{-0.5} - 1}{2}\right\}$ or approximately $\{-0.3419\}$.

49. Use $A = Pe^{rt}$ with $P = 2000, r = 6\% = 0.06$, and $t = 4$.

$\quad A = Pe^{rt}$

$\quad = 2000e^{0.06(4)}$

$\quad = 2000e^{0.24}$

$\quad \approx 2542.50$

She owes \$2542.50.

53. $\log_8 6 = \dfrac{\log 6}{\log 8}$

$\quad \approx \dfrac{0.7782}{0.9031} \approx 0.8617$

57. $\log_3 \dfrac{1}{6} = \log_3 6^{-1}$

$\quad = -\log_3 6$

$\quad = -\dfrac{\ln 6}{\ln 3}$

$\quad \approx -\dfrac{1.79176}{1.09861}$

$\quad \approx -1.6309$

61. $6x - 3(2 - 5x) = 6$

$\quad 6x - 6 + 15x = 6$

$\quad 21x = 12$

$\quad x = \dfrac{12}{21}$

$\quad x = \dfrac{4}{7}$

The solution set is $\left\{\dfrac{4}{7}\right\}$.

65. $\quad x^2 + 7x = -6$

$\quad x^2 + 7x + 6 = 0$

$\quad (x + 6)(x + 1) = 0$

$\quad x + 6 = 0 \quad$ or $\quad x + 1 = 0$

$\quad\quad x = -6 \quad\quad\quad\quad x = -1$

The solution set is $\{-6, -1\}$.

5. $\begin{cases} y^2 = 4 - x \\ x - 2y = 4 \end{cases}$

Rewrite the second equation as $-2y = 4 - x$, then substitute.

$$y^2 = -2y$$
$$y^2 + 2y = 0$$
$$y(y + 2) = 0$$
$$y = 0 \quad \text{or} \quad y + 2 = 0$$
$$y = -2$$
$$y = 0: x - 2(0) = 4$$
$$x = 4$$
$$y = -2: x - 2(-2) = 4$$
$$x + 4 = 4$$
$$x = 0$$

The solution set is $\{(4, 0), (0, -2)\}$.

9. $\begin{cases} x^2 + 2y^2 = 2 \\ x - y = 2 \end{cases}$

Rewrite the second equation as $x = y + 2$, then substitute.

$$(y + 2)^2 + 2y^2 = 2$$
$$y^2 + 4y + 4 + 2y^2 = 2$$
$$3y^2 + 4y + 2 = 0$$
$$b^2 - 4ac = 4^2 - 4(3)(2) = 16 - 24 = -8 < 0$$

No real solutions exist. The solution set is $\varnothing$.

13. $\begin{cases} y = x^2 \\ 3x + y = 10 \end{cases}$

Substitute.

$$3x + x^2 = 10$$
$$x^2 + 3x - 10 = 0$$
$$(x + 5)(x - 2) = 0$$
$$x + 5 = 0 \quad \text{or} \quad x - 2 = 0$$
$$x = -5 \quad \text{or} \quad x = 2$$
$$x = -5: y = (-5)^2 = 25$$
$$x = 2: y = 2^2 = 4$$

The solution set is $\{(-5, 25), (2, 4)\}$.

17. $\begin{cases} y = x^2 - 4 \\ y = x^2 - 4x \end{cases}$

Substitute.

$$x^2 - 4 = x^2 - 4x$$
$$-4 = -4x$$
$$x = 1$$
$$y = 1^2 - 4 = -3$$

The solution set is $\{(1, -3)\}$.

21. $\begin{cases} x^2 + y^2 = 1 \\ x^2 + (y + 3)^2 = 4 \end{cases}$

Subtract the first equation from the second.

$$(y + 3)^2 - y^2 = 3$$
$$y^2 + 6y + 9 - y^2 = 3$$
$$6y + 9 = 3$$
$$6y = -6$$
$$y = -1$$

Substitute back.

$$x^2 + (-1)^2 = 1$$
$$x^2 + 1 = 1$$
$$x^2 = 0$$
$$x = 0$$

The solution set is $\{(0, -1)\}$.

25. $\begin{cases} 3x^2 + y^2 = 9 \\ 3x^2 - y^2 = 9 \end{cases}$

Subtract the second equation from the first.

$$2y^2 = 0$$
$$y^2 = 0$$
$$y = 0$$

Substitute back.

$$3x^2 + 0 = 9$$
$$3x^2 = 9$$
$$x^2 = 3$$
$$x = \pm\sqrt{3}$$

The solution set is $\{(\sqrt{3}, 0), (-\sqrt{3}, 0)\}$.

29. $\begin{cases} x^2 + y^2 = 36 \\ y = \dfrac{1}{6}x^2 - 6 \end{cases}$

$$y + 6 = \frac{1}{6}x^2$$
$$x^2 = 6(y + 6)$$

Substitute.

$$6(y + 6) + y^2 = 36$$
$$6y + 36 + y^2 = 36$$
$$6y + y^2 = 0$$
$$y(6 + y) = 0$$
$$y = 0 \quad \text{or} \quad 6 + y = 0$$
$$y = -6$$
$$y = 0: x^2 + 0^2 = 36$$
$$x^2 = 36$$
$$x = \pm 6$$
$$y = -6: x^2 + (-6)^2 = 36$$
$$x^2 + 36 = 36$$
$$x^2 = 0$$
$$x = 0$$

The solution set is $\{(6, 0), (-6, 0), (0, -6)\}$.

33. $x > -3$

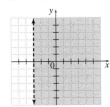

37. answers may vary

41. $\begin{cases} x^2 + y^2 = 130 \\ x^2 - y^2 = 32 \end{cases}$

Add.

$$2x^2 = 162$$
$$x^2 = 81$$
$$x = \pm 9$$

Substitute back.

$$9^2 + y^2 = 130 \qquad\qquad (-9)^2 + y^2 = 130$$
$$y^2 = 49 \qquad\qquad\qquad\qquad y^2 = 49$$
$$y = \pm 7 \qquad\qquad\qquad\qquad y = \pm 7$$

The numbers are 9 and 7, 9 and -7, -9 and 7, -9 and -7.

45. $\begin{cases} p = -0.01x^2 - 0.2x + 9 \\ p = 0.01x^2 - 0.1x + 3 \end{cases}$

Substitute.

$-0.01x^2 - 0.2x + 9 = 0.01x^2 - 0.1x + 3$

$\qquad 0 = 0.02x^2 + 0.1x - 6$

$\qquad 0 = x^2 + 5x - 300$

$\qquad 0 = (x + 20)(x - 15)$

$x + 20 = 0 \quad$ or $\quad x - 15 = 0$

$\qquad x = -20 \quad$ or $\qquad x = 15$

Disregard the negative.

$p = -0.01(15)^2 - 0.2(15) + 9$

$p = 3.75$

The equilibrium quantity is 15 thousand compact discs and the equilibrium price is $3.75.

Exercise Set 10.4

1. $y < x^2$

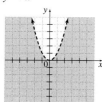

5. $\dfrac{x^2}{4} - y^2 < 1$

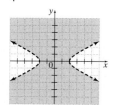

9. $x^2 + y^2 \le 9$

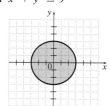

13. $\dfrac{x^2}{4} + \dfrac{y^2}{9} \le 1$

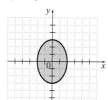

17. $y < (x - 2)^2 + 1$

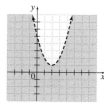

21. $\begin{cases} 4x + 3y \ge 12 \\ x^2 + y^2 < 16 \end{cases}$

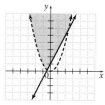

25. $\begin{cases} y > x^2 \\ y \ge 2x + 1 \end{cases}$

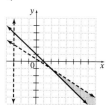

29. $\begin{cases} \dfrac{x^2}{4} + \dfrac{y^2}{9} \ge 1 \\ x^2 + y^2 \ge 4 \end{cases}$

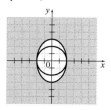

33. $\begin{cases} x + y \ge 1 \\ 2x + 3y < 1 \\ x > -3 \end{cases}$

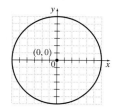

37. This is not a function because a vertical line can cross the graph in two places.

41. answers may vary

Chapter 10 Test

1. $x^2 + y^2 = 36 \quad$ or $\quad (x - 0)^2 + (y - 0)^2 = 6^2$

This is a circle with center $(0, 0)$ and radius 6.

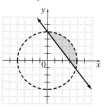

5.
$$x^2 + y^2 + 6x = 16$$
$$x^2 + 6x + y^2 = 16$$
$$(x^2 + 6x + 9) + y^2 = 16 + 9$$
$$(x + 3)^2 + y^2 = 5^2$$

This is a circle with center $(-3, 0)$ and radius 5.

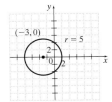

9. $\begin{cases} x^2 + y^2 = 169 \\ 5x + 12y = 0 \end{cases}$

$$12y = -5x$$
$$y = -\frac{5x}{12}$$

Substitute.

$$x^2 + \left(-\frac{5x}{12}\right)^2 = 169$$
$$x^2 + \frac{25x^2}{144} = 169$$
$$\frac{169x^2}{144} = 169$$
$$x^2 = 144$$
$$x = \pm 12$$

Substitute back.

$$x = 12: y = -\frac{5}{12}(12) = -5$$
$$x = -12: y = -\frac{5}{12}(-12) = 5$$

The solution set is $\{(12, -5), (-12, 5)\}$.

13. $\begin{cases} 2x + 5y \geq 10 \\ y \geq x^2 + 1 \end{cases}$

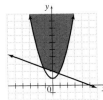

17. Since the equation is quadratic in y and $a > 0$, the parabola opens to the right. Since $h < 0$ and $k > 0$, the vertex is in quadrant II, which is graph B.

APPENDIX

Exercise Set Appendix A

1. $\frac{7}{10} = \frac{7 \cdot 3}{10 \cdot 3} = \frac{21}{30}$

5. $\frac{4}{5} = \frac{4 \cdot 4}{5 \cdot 4} = \frac{16}{20}$

9. $\frac{10}{15} = \frac{2 \cdot 5}{3 \cdot 5} = \frac{2}{3}$

13. $\frac{20}{20} = 1$

17. $\frac{18}{30} = \frac{2 \cdot 3 \cdot 3}{2 \cdot 3 \cdot 5} = \frac{3}{5}$

21. $\frac{66}{48} = \frac{2 \cdot 3 \cdot 11}{2 \cdot 2 \cdot 2 \cdot 2 \cdot 3} = \frac{11}{2 \cdot 2 \cdot 2} = \frac{11}{8}$

25. $\frac{192}{264} = \frac{2 \cdot 2 \cdot 2 \cdot 2 \cdot 2 \cdot 2 \cdot 3}{2 \cdot 2 \cdot 2 \cdot 3 \cdot 11} = \frac{2 \cdot 2 \cdot 2}{11} = \frac{8}{11}$

29. $\frac{2}{3} \cdot \frac{3}{4} = \frac{2 \cdot 3}{3 \cdot 2 \cdot 2} = \frac{1}{2}$

33. $7\frac{2}{5} \div \frac{1}{5} = \frac{37}{5} \div \frac{1}{5} = \frac{37}{5} \cdot \frac{5}{1} = \frac{37}{1} = 37$

37. $\frac{3}{4} \div \frac{1}{20} = \frac{3}{4} \cdot \frac{20}{1} = \frac{3 \cdot 4 \cdot 5}{4 \cdot 1} = \frac{3 \cdot 5}{1} = \frac{15}{1} = 15$

41. $\frac{9}{20} \div 12 = \frac{9}{20} \div \frac{12}{1}$
$$= \frac{9}{20} \cdot \frac{1}{12}$$
$$= \frac{3 \cdot 3 \cdot 1}{20 \cdot 3 \cdot 4}$$
$$= \frac{3}{20 \cdot 4}$$
$$= \frac{3}{80}$$

45. $8\frac{3}{5} \div 2\frac{9}{10} = \frac{43}{5} \div \frac{29}{10}$
$$= \frac{43}{5} \cdot \frac{10}{29}$$
$$= \frac{43 \cdot 2 \cdot 5}{5 \cdot 29}$$
$$= \frac{43 \cdot 2}{29}$$
$$= \frac{86}{29}$$
$$= 2\frac{28}{29}$$

49. $\frac{4}{5} - \frac{1}{5} = \frac{4 - 1}{5} = \frac{3}{5}$

53. $\frac{17}{21} - \frac{10}{21} = \frac{17 - 10}{21} = \frac{7}{21} = \frac{7}{3 \cdot 7} = \frac{1}{3}$

57. $5\frac{2}{5} - 3\frac{4}{5} = \frac{27}{5} - \frac{19}{5} = \frac{8}{5} = 1\frac{3}{5}$

61. $\frac{10}{3} - \frac{5}{21} = \frac{10 \cdot 7}{3 \cdot 7} - \frac{5}{21} = \frac{70}{21} - \frac{5}{21} = \frac{65}{21}$

65. $\frac{5}{22} - \frac{5}{33} = \frac{5 \cdot 3}{22 \cdot 3} - \frac{5 \cdot 2}{33 \cdot 2} = \frac{15}{66} - \frac{10}{66} = \frac{5}{66}$

69.
$$\begin{array}{r} 17\frac{2}{5} \qquad 17\frac{6}{15} \\ +30\frac{2}{3} \quad +30\frac{10}{15} \\ \hline 47\frac{16}{15} = 47 + 1\frac{1}{15} = 48\frac{1}{15} \end{array}$$

73. $\frac{2}{3} - \frac{5}{9} + \frac{5}{6} = \frac{2 \cdot 6}{3 \cdot 6} - \frac{5 \cdot 2}{9 \cdot 2} + \frac{5 \cdot 3}{6 \cdot 3}$
$$= \frac{12}{18} - \frac{10}{18} + \frac{15}{18}$$
$$= \frac{12 - 10 + 15}{18}$$
$$= \frac{17}{18}$$

77. $1 - \frac{3}{10} - \frac{5}{10} = \frac{10}{10} - \frac{3}{10} - \frac{5}{10}$
$$= \frac{10 - 3 - 5}{10}$$
$$= \frac{2}{10}$$
$$= \frac{2}{2 \cdot 5}$$
$$= \frac{1}{5}$$

81.

$$237\frac{1}{6} \qquad 237\frac{2}{12} \qquad 236\frac{14}{12}$$

$$\underline{-224\frac{5}{12}} \qquad \underline{-224\frac{5}{12}} \qquad \underline{-224\frac{5}{12}}$$

$$12\frac{9}{12} = 12\frac{3}{4}$$

The Olympic record throw is $12\frac{3}{4}$ feet longer.

Exercise Set Appendix B

1. $\begin{vmatrix} 3 & 5 \\ -1 & 7 \end{vmatrix} = ad - bc$

$$= 3(7) - (5)(-1)$$
$$= 21 - (-5)$$
$$= 21 + 5$$
$$= 26$$

5. $\begin{vmatrix} -2 & 9 \\ 4 & -18 \end{vmatrix} = ad - bc$

$$= (-2)(-18) - 9(4)$$
$$= 36 - 36$$
$$= 0$$

9. $\begin{cases} 2y - 4 = 0 \\ x + 2y = 5 \end{cases} \rightarrow \begin{cases} 0x + 2y = 4 \\ x + 2y = 5 \end{cases}$

$$D = \begin{vmatrix} 0 & 2 \\ 1 & 2 \end{vmatrix} = 0(2) - (2)(1) = 0 - 2 = -2$$

$$D_x = \begin{vmatrix} 4 & 2 \\ 5 & 2 \end{vmatrix} = 4(2) - (2)(5) = 8 - 10 = -2$$

$$D_y = \begin{vmatrix} 0 & 4 \\ 1 & 5 \end{vmatrix} = 0(5) - 4(1) = 0 - 4 = -4$$

$$x = \frac{D_x}{D} = \frac{-2}{-2} = 1$$

$$y = \frac{D_y}{D} = \frac{-4}{-2} = 2$$

The solution is $(1, 2)$.

13. $\begin{cases} 5x - 2y = 27 \\ -3x + 5y = 18 \end{cases}$

$$D = \begin{vmatrix} 5 & -2 \\ -3 & 5 \end{vmatrix} = 5(5) - (-2)(-3) = 25 - 6 = 19$$

$$D_x = \begin{vmatrix} 27 & -2 \\ 18 & 5 \end{vmatrix} = 27(5) - (-2)(18)$$
$$= 135 - (-36)$$
$$= 135 + 36$$
$$= 171$$

$$D_y = \begin{vmatrix} 5 & 27 \\ -3 & 18 \end{vmatrix} = 5(18) - (27)(-3)$$
$$= 90 - (-81)$$
$$= 90 + 81$$
$$= 171$$

$$x = \frac{D_x}{D} = \frac{171}{19} = 9$$

$$y = \frac{D_y}{D} = \frac{171}{19} = 9$$

The solution is $(9, 9)$.

17. $\begin{cases} \dfrac{2}{3}x - \dfrac{3}{4}x = -1 \\ -\dfrac{1}{6}x + \dfrac{3}{4}y = \dfrac{5}{2} \end{cases}$

$$D = \begin{vmatrix} \dfrac{2}{3} & -\dfrac{3}{4} \\ -\dfrac{1}{6} & \dfrac{3}{4} \end{vmatrix} = \frac{2}{3}\left(\frac{3}{4}\right) - \left(-\frac{3}{4}\right)\left(-\frac{1}{6}\right)$$

$$= \frac{1}{2} - \frac{3}{24}$$
$$= \frac{4}{8} - \frac{1}{8}$$
$$= \frac{3}{8}$$

$$D_x = \begin{vmatrix} -1 & -\dfrac{3}{4} \\ \dfrac{5}{2} & \dfrac{3}{4} \end{vmatrix} = (-1)\left(\frac{3}{4}\right) - \left(-\frac{3}{4}\right)\left(\frac{5}{2}\right)$$

$$= -\frac{3}{4} - \left(-\frac{15}{8}\right)$$
$$= -\frac{6}{8} + \frac{15}{8}$$
$$= \frac{9}{8}$$

$$D_y = \begin{vmatrix} \dfrac{2}{3} & -1 \\ -\dfrac{1}{6} & \dfrac{5}{2} \end{vmatrix} = \frac{2}{3}\left(\frac{5}{2}\right) - (-1)\left(-\frac{1}{6}\right)$$

$$= \frac{5}{3} - \frac{1}{6}$$
$$= \frac{10}{6} - \frac{1}{6}$$
$$= \frac{9}{6}$$
$$= \frac{3}{2}$$

$$x = \frac{D_x}{D} = \frac{\frac{9}{8}}{\frac{3}{8}} = \frac{9}{8} \div \frac{3}{8} = \frac{9}{8} \cdot \frac{8}{3} = \frac{9}{3} = 3$$

$$y = \frac{D_y}{D} = \frac{\frac{3}{2}}{\frac{3}{8}} = \frac{3}{2} \div \frac{3}{8} = \frac{3}{2} \cdot \frac{8}{3} = \frac{8}{2} = 4$$

The solution is $(3, 4)$.

21. Expand by the minors of the third column.

$$\begin{vmatrix} 4 & -6 & 0 \\ -2 & 3 & 0 \\ 4 & -6 & 1 \end{vmatrix} = 0\begin{vmatrix} -2 & 3 \\ 4 & -6 \end{vmatrix} - 0\begin{vmatrix} 4 & -6 \\ 4 & -6 \end{vmatrix} + 1\begin{vmatrix} 4 & -6 \\ -2 & 3 \end{vmatrix}$$

$$= 0 - 0 + 1(12 - 12)$$
$$= 0 - 0 + 1(0)$$
$$= 0$$

25. Expand by the minors of the first row.

$$\begin{vmatrix} 3 & 6 & -3 \\ -1 & -2 & 3 \\ 4 & -1 & 6 \end{vmatrix}$$

$$= 3\begin{vmatrix} -2 & 3 \\ -1 & 6 \end{vmatrix} - 6\begin{vmatrix} -1 & 3 \\ 4 & 6 \end{vmatrix} + (-3)\begin{vmatrix} -1 & -2 \\ 4 & -1 \end{vmatrix}$$

$$= 3(-12 - (-3)) - 6(-6 - 12) - 3(1 - (-8))$$
$$= -27 + 108 - 27$$
$$= 54$$

29. $\begin{cases} x + y + z = 8 \\ 2x - y - z = 10 \\ x - 2y + 3z = 22 \end{cases}$

$$D = \begin{vmatrix} 1 & 1 & 1 \\ 2 & -1 & -1 \\ 1 & -2 & 3 \end{vmatrix}$$

$$= 1\begin{vmatrix} -1 & -1 \\ -2 & 3 \end{vmatrix} - 1\begin{vmatrix} 2 & -1 \\ 1 & 3 \end{vmatrix} + 1\begin{vmatrix} 2 & -1 \\ 1 & -2 \end{vmatrix}$$

$$= 1(-3 - 2) - 1(6 - (-1)) + 1(-4 - (-1))$$
$$= -5 - 7 - 3$$
$$= -15$$

$$D_x = \begin{vmatrix} 8 & 1 & 1 \\ 10 & -1 & -1 \\ 22 & -2 & 3 \end{vmatrix}$$

$$= 8\begin{vmatrix} -1 & -1 \\ -2 & 3 \end{vmatrix} - 10\begin{vmatrix} 1 & 1 \\ -2 & 3 \end{vmatrix} + 22\begin{vmatrix} 1 & 1 \\ -1 & -1 \end{vmatrix}$$

$$= 8(-3 - 2) - 10(3 - (-2)) + 22(-1 - (-1))$$
$$= -40 - 50 + 0$$
$$= -90$$

$$D_y = \begin{vmatrix} 1 & 8 & 1 \\ 2 & 10 & -1 \\ 1 & 22 & 3 \end{vmatrix}$$

$$= -8\begin{vmatrix} 2 & -1 \\ 1 & 3 \end{vmatrix} + 10\begin{vmatrix} 1 & 1 \\ 1 & 3 \end{vmatrix} - 22\begin{vmatrix} 1 & 1 \\ 2 & -1 \end{vmatrix}$$

$$= -8(6 - (-1)) + 10(3 - 1) - 22(-1 - 2)$$
$$= -56 + 20 + 66$$
$$= 30$$

$$D_z = \begin{vmatrix} 1 & 1 & 8 \\ 2 & -1 & 10 \\ 1 & -2 & 22 \end{vmatrix}$$

$$= 8\begin{vmatrix} 2 & -1 \\ 1 & -2 \end{vmatrix} - 10\begin{vmatrix} 1 & 1 \\ 1 & -2 \end{vmatrix} + 22\begin{vmatrix} 1 & 1 \\ 2 & -1 \end{vmatrix}$$

$$= 8(-4 - (-1)) - 10(-2 - 1) + 22(-1 - 2)$$
$$= -24 + 30 - 66$$
$$= -60$$

$$x = \frac{D_x}{D} = \frac{-90}{-15} = 6$$

$$y = \frac{D_y}{D} = \frac{30}{-15} = -2$$

$$z = \frac{D_z}{D} = \frac{-60}{-15} = 4$$

The solution is $(6, -2, 4)$.

33. $\begin{cases} x - 2y + z = -5 \\ 3y + 2z = 4 \\ 3x - y = -2 \end{cases}$

$$D = \begin{vmatrix} 1 & -2 & 1 \\ 0 & 3 & 2 \\ 3 & -1 & 0 \end{vmatrix}$$

$$= 1\begin{vmatrix} 3 & 2 \\ -1 & 0 \end{vmatrix} - (-2)\begin{vmatrix} 0 & 2 \\ 3 & 0 \end{vmatrix} + 1\begin{vmatrix} 0 & 3 \\ 3 & -1 \end{vmatrix}$$

$$= 1(0 - (-2)) + 2(0 - 6) + 1(0 - 9)$$
$$= 2 - 12 - 9$$
$$= -19$$

$$D_x = \begin{vmatrix} -5 & -2 & 1 \\ 4 & 3 & 2 \\ -2 & -1 & 0 \end{vmatrix}$$

$$= -5\begin{vmatrix} 3 & 2 \\ -1 & 0 \end{vmatrix} - 4\begin{vmatrix} -2 & 1 \\ -1 & 0 \end{vmatrix} + (-2)\begin{vmatrix} -2 & 1 \\ 3 & 2 \end{vmatrix}$$

$$= -5(0 - (-2)) - 4(0 - (-1)) - 2(-4 - 3)$$
$$= -10 - 4 + 14$$
$$= 0$$

$$D_y = \begin{vmatrix} 1 & -5 & 1 \\ 0 & 4 & 2 \\ 3 & -2 & 0 \end{vmatrix}$$

$$= -(-5)\begin{vmatrix} 0 & 2 \\ 3 & 0 \end{vmatrix} + 4\begin{vmatrix} 1 & 1 \\ 3 & 0 \end{vmatrix} - (-2)\begin{vmatrix} 1 & 1 \\ 0 & 2 \end{vmatrix}$$

$$= 5(0 - 6) + 4(0 - 3) + 2(2 - 0)$$
$$= -30 - 12 + 4$$
$$= -38$$

$$D_z = \begin{vmatrix} 1 & -2 & -5 \\ 0 & 3 & 4 \\ 3 & -1 & -2 \end{vmatrix}$$

$$= -5\begin{vmatrix} 0 & 3 \\ 3 & -1 \end{vmatrix} - 4\begin{vmatrix} 1 & -2 \\ 3 & -1 \end{vmatrix} + (-2)\begin{vmatrix} 1 & -2 \\ 0 & 3 \end{vmatrix}$$

$$= -5(0 - 9) - 4(-1 - (-6)) - 2(3 - 0)$$
$$= 45 - 20 - 6$$
$$= 19$$

$$x = \frac{D_x}{D} = \frac{0}{-19} = 0$$

$$y = \frac{D_y}{D} = \frac{-38}{-19} = 2$$

$$z = \frac{D_z}{D} = \frac{19}{-19} = -1$$

The solution is $(0, 2, -1)$.

37. 0; answers may vary

Exercise Set Appendix D

1. Since the x- and y-values are between -10 and 10, the ordered pairs will lie within a standard viewing rectangle.

5. answers may vary

9. answers may vary

13. Xmin $= -9$ Ymin $= -12$
Xmax $= 9$ Ymax $= 12$
Xscl $= 1$ Yscl $= 2$

17. Xmin $= -5$ Ymin $= -15$
Xmax $= 5$ Ymax $= 15$
Xscl $= 1$ Yscl $= 3$

21. Setting A:

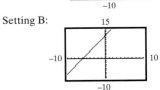

Setting B:

Setting B shows all the intercepts.

25. Setting A:

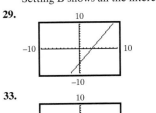

Setting B:

Setting B shows all the intercepts.

29.

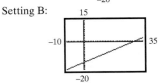

33.

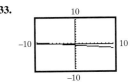

37.

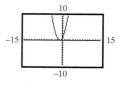

41.

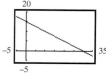

Exercise Set Appendix E

1. $f(x) = 3|x| = 3|x - 0| + 0$

The vertex is $(0, 0)$, the graph opens up, and the graph is narrower than $y = |x|$.

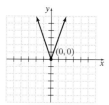

5. $g(x) = 2|x| + 3 = 2|x - 0| + 3$

The vertex is $(0, 3)$, the graph opens up, and the graph is narrower than $y = |x|$.

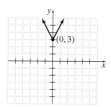

9. $f(x) = 4|x - 1| = 4|x - 1| + 0$

The vertex is $(1, 0)$, the graph opens up, and the graph is narrower than $y = |x|$.

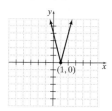

13. $f(x) = -2|x - 3| + 4$

The vertex is $(3, 4)$, the graph opens down, and the graph is narrower than $y = |x|$.

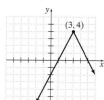

Exercise Set Appendix F

1. $90° - 19° = 71°$

The complement of a 19° angle is a 71° angle.

5. $90° - 11\frac{1}{4}° = 78\frac{3}{4}°$

The complement of an $11\frac{1}{4}°$ angle is a $78\frac{3}{4}°$ angle.

9. $180° - 30.2° = 149.8°$

The supplement of a 30.2° angle is a 149.8° angle.

13. $\angle 1$ and the angle marked 110° are vertical angles, so $m\angle 1 = 110°$.

$\angle 2$ and the angle marked 110° are supplementary angles, so $m\angle 2 = 180° - 110° = 70°$.

$\angle 3$ and the angle marked 110° are supplementary angles, so $m\angle 3 = 180° - 110° = 70°$.

$\angle 4$ and $\angle 3$ are alternate interior angles, so $m\angle 4 = m\angle 3 = 70°$.

$\angle 5$ and the angle marked 110° are alternate interior angles, so $m\angle 5 = 110°$.

$\angle 6$ and $\angle 5$ are supplementary angles, so $m\angle 6 = 180° - m\angle 5 = 180° - 110° = 70°$.

$\angle 7$ and the angle marked 110° are corresponding angles, so $m\angle 7 = 110°$.

17. $180° - 25° - 65° = 90°$

The third angle measures 90°.

21. Since the triangle is a right triangle, one angle measures 90°.
$180° - 45° - 90° = 45°$

The other two angles of the triangle measure 45° and 90°.

25. Since the triangle is a right triangle, one angle measures 90°.

$180° - 39\frac{3}{4}° - 90° = 50\frac{1}{4}°$

The other two angles of the triangle measure $50\frac{1}{4}°$ and 90°.

29. $\frac{6}{9} = \frac{3}{x}$

$6x = 27$

$x = \frac{27}{6}$

$x = 4.5$

33. $a^2 + b^2 = c^2$

$a^2 + 5^2 = 13^2$

$a^2 + 25 = 169$

$a^2 = 144$

Since a represents a length, we assume that a is positive. Since $a^2 = 144$, $a = 12$. The other leg of the right triangle has length 12.

INDEX

Photo Credits